Frontiers of Life

Frontiers of Life

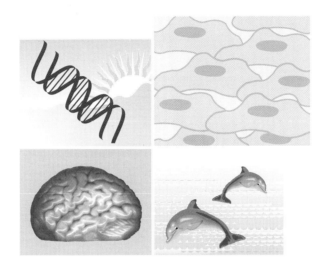

VOLUME II
Cells and Organisms

A Harcourt Science and Technology Company

San Diego   San Francisco   New York   Boston   London   Sydney   Tokyo

ISTITUTO DELLA ENCICLOPEDIA ITALIANA
ROMA

Academic Press
A Harcourt Science and Technology Company
525 B Street, Suite 1900, San Diego, California 92101-4495, USA
http://www.academicpress.com

Academic Press
Harcourt Place, 32 Jamestown Road, London NW1 7BY, UK
http://www.academicpress.com

Library of Congress Catalog Card Number: 2001091214

International Standard Book Number: 0-12-077340-6 (set)
International Standard Book Number: 0-12-077341-4 (volume 1)
International Standard Book Number: 0-12-077342-2 (volume 2)
International Standard Book Number: 0-12-077343-0 (volume 3)
International Standard Book Number: 0-12-077344-9 (volume 4)

PRINTED IN THE UNITED STATES OF AMERICA
01   02   03   04   05   06   WP   9   8   7   6   5   4   3   2   1

FRONTIERS OF LIFE

EDITORS-IN-CHIEF:
David Baltimore, Renato Dulbecco, François Jacob, Rita Levi-Montalcini

CONTENTS OF VOLUME II

SECTION IV / **FORMATION OF ORGANS AND TISSUES**

PART TWO

THE IMMUNOLOGICAL SYSTEMS

SECTION I / **IMMUNOLOGICAL THOUGHT**

SECTION II / MOLECULES OF THE IMMUNE SYSTEM

**SECTION III / ANTIGEN PRESENTATION
AND LYMPHOCYTE ACTIVATION**

INTRODUCTION TO VOLUME II

Dᴀᴠɪᴅ Bᴀʟᴛɪᴍᴏʀᴇ

When we think of the biology of the 20th century, usually what comes to mind is the great advances in the fields of genetics and molecular biology. However, an equally deep revolution has occurred in the understanding of the cell. One of the most important steps forward made by 19th-century science is represented by the cell theory of biological organization, according to which all living systems consist of one or several cells and the cell is the unit of independent biological existence. This theory has been amply confirmed and has now become a truism; however, until the 1960s, there was still much uncertainty about certain fundamental issues, such as the organization of the membrane that surrounds the cell, the way the various components find their correct locations in cell architecture, and the way in which cell division is controlled.

In this volume, we will see the solutions to the problems related to the cell structure of living organisms. The cells are always highly specialized, but there is no system in which specialization is better exemplified than in the immune system. Since immune cells spend part of their life as individual cells in the blood and spend another part, the longest, adhering only loosely to tissues, they can be isolated and described more easily than most cells and are consequently better understood. For this reason, our volume moves from the examination of cells in general to a more in-depth study of the cells of the immune system.

The first part of the volume examines individual cells and their interactions. The technological key that opened up the path to our knowledge of the cell was the invention of the electron microscope immediately after World War II. This instrument has opened up the cell to analysis at a level previously unimaginable, proposing extraordinary challenges. The existence of organelles inside the cell was already known, but it was discovered that they were finely structured and that "protoplasm" was not a homogeneous whole, but contained highly organized regions. The existence of such a complex reality is the origin of numerous problems that became some of the most challenging for biology at the end of the 20th century: how the structure develops; what the functions of each of its parts are; how the information contained in DNA manifests itself in visible entities; how cells control the position of substructural elements on their surfaces to facilitate intercellular communication. Today we know many answers to these questions. Cells grow by synthesis of components such as lipids, carbohydrates, and, in particular, proteins. All proteins are produced by the same ribosomal device that translates the messenger RNA, but the proteins that form the visible structures must be targeted to specific sites. Such proteins are normally subject to systematic modifications obtained by cutting segments and adding chemical residues. Proteins move through compartments inside the cells, often enclosed in vesicles that bud from one compartment and merge their contents with another.

In a multicellular organism such as a human being, cells must communicate with each other to coordinate activities and maintain functional synchrony. This happens mostly through the release of chemical substances, which may act locally or at a great distance. These substances — proteins or smaller molecules — allow communication because their release is regulated by the emitting cell, and the receiving cell is equipped with receptors that enable it to interpret the message. Decoding these signals and their preparation in cells is the subject of ongoing research, still far from being completed. Only in recent years have we obtained a rich knowledge of the transduction of these signals. Another aspect of signaling is the transmission of signals from the nervous system, which implies the release of chemical signals to the connection sites between neurons, while long-distance transmission — whose source and target must be extremely precise — is entrusted to electrical signals sent during elaborate nervous processes.

The most feared pathology of the cell is cancer, caused by cell growth at inappropriate times and places. The most stupefying thing, however, is that cells are equipped with mechanisms that control their growth and that they have a number of devices for protection against error, so they fail quite rarely. Our bodies are equipped with approximately 10^{17} cells during the course of our lives, and practically none of these cells, or perhaps one or two of them, modifies its own growth to become a clone of cancerous cells. How our body controls cell growth and how cells organize their division cycle — with multiple control points to guarantee intercellular fidelity and synchrony — are among the great discoveries made by modern biology. A key element in the maintenance of the body in a functional state is the fact that cells die and are replaced by new cells. For this purpose, each cell contains a suicide mechanism. The harmonious process of cell birth and death serves not only to remove those that have exhausted their function, but also to eliminate those that have developed irregularities likely to make them become cancerous.

All this complex cell regulation and architecture exists for the sole purpose of allowing cells to produce tissues, organs, and organisms that operate as a unit within an ecological system. This implies that cells are different from each other during embryonic life and the development process, as well as that they die by the activation of their suicide program (known as apoptosis).

In the second part of the volume, we will see how the immune system develops and functions to protect us from external aggressors. The world literally contains millions of potentially pathogenic microorganisms capable of invading the bodies of multicellular organisms. These potential invaders may easily destroy the animal or plant infected by them, and defense mechanisms are crucial for survival in this dangerous world. The immune system must fight extremely powerful adversaries, because microorganisms may multiply and evolve extremely quickly, while complex organisms live a long time and evolve only slowly. For this reason, the immune system has had to develop the capacity to prepare defenses with a speed equal to that of the evolution of the pathogenic agents that attack it. In short, the body needs a subsystem with its own evolutionary capacity, which is exactly the case of the immune system. This system is especially significant because it cannot transmit what it learns to the next generation and therefore, in any new animal that is born, the system is implemented *ex novo* to start evolving. However, although it is deprived of any memory of the specific protective mechanisms of its parents — except for a temporary protection around the moment of birth — the immune system normally succeeds in giving us the necessary superiority over the pathogenic agents in the surrounding environment. In the less developed regions of the planet, microorganisms often defeat the immune response and many children die, but when it is sustained by correct modern hygienic rules and antibiotics, the system proves to be extraordinarily efficacious.

The immune system consists of many parts, described from time to time in individual specific essays and therefore presented as elements of an organic whole. There are two fundamentally different types of protection: the secreted antibodies active in the blood and other bodily fluids, and the mediated immunity of the cell, thanks to which an immune cell immediately attacks a cell that has been infected by a pathogenic agent. There are mechanisms through which the pathogenic agents are recognized as external invaders and therefore guided toward the immune system.

Humans have learned to mobilize the immune system in advance, to prepare it for the attack of potential pathogenic agents, through vaccination. This technique consists of presenting the crucial parts of certain pathogenic agents to the immune system, allowing it to register their characteristics so that, when it is called to review them as part of true and proper pathogenic agents, it may react promptly and effectively. Vaccination is counted among the highest successes ever achieved by public health.

There is a pathogenic agent that has completely eluded our counter-attack strength, the human immunodeficiency virus, HIV. Perversely, it infects precisely the cells of the immune system and therefore, by vocation, it is virtually impervious to antibodies. This is an infectious disease that modern medicine has not been successful in disarming and is therefore rightly the subject of a specific contribution to the volume.

PART ONE

Cells and Cell Communities

Edited by Nica Borgese, Francesco Clementi, and Pietro De Camill

INTRODUCTION

Nica Borgese

Francesco Clementi

Pietro De Camilli

Genetic language is linear, whereas the three-dimensional spatial organization is a universal feature of all organisms so that it is not even possible to conceive of life without thinking in the three dimensions. Proteins represent the class of macromolecules responsible for the transition from a single-dimensional organization to the three-dimensional one. Every protein assumes a characteristic spatial conformation determined by its linear amino acid sequence, in its turn translated by the genetic information. Recognition by different proteins and between proteins and other biological molecules leads to the constitution of the cell's complex spatial organization and to that of multicellular organisms. "Cells and Cell Communities" addresses the problem of the organization and the regulation of three-dimensional supramolecular structures, focusing on the recent discoveries in the biology of the animal eukaryotic cell.

This part of the volume begins with the chapter by Peter J. T. Dekker, Wolfgang Voos, Nikolaus Pfanner, and Joachim Rassow that is dedicated to the problem of the spatial conformation of proteins. What are the rules that link the amino acid sequence to the spatial conformation? How is the great efficiency of the folding of the proteins neosynthetized *in vivo* ensured, avoiding the intracellular accumulation of aggregates of proteins with mistaken conformations? The second chapter, by Mark Hochstrasser, is dedicated to a problem specular to the one treated in the first chapter: regulation of protein degradation, a complex phenomenon that makes a decisive contribution toward regulating the intracellular abundance of every single protein.

In the next three chapters, the internal organization of the eukaryotic cell is analyzed. Kai Simons and Marino Zerial examine the way in which the cell constructs and maintains its organelles and the regulation of the multiple interactions between the various compartments, which takes place without the loss of their individuality. The cytoskeleton, constituted by the regulated polymerization of globular proteins, is analyzed both in its role in cell motility, in the chapter by Chris M. Coppin, Daniel W. Pierce, and Ronald D. Vale, and as the factor determining the acquisition of specific spatial forms characteristic of the various differentiated cells — forms essential for the performance of complex functions, characteristic of the higher organisms — in the chapter by Eric Karsenti.

Having laid the foundations of the cell's spatial organization, the way in which, in response to extracellular stimuli, it regulates relations among its macromolecules and, consequently, its behavior is discussed. Thus, we begin to consider the cell no longer in isolation but as forming part of a community of cells. First, Flavia Valtorta examines the molecular mechanisms responsible for the secretion of chemical mediators by neuronal and endocrine cells. Next is a description of recent progress in our knowledge of the transduction of the signal, i.e., of the way in which cells respond to the presence of a chemical agent (mediator) in the extracellular environment: Dario Di Francesco describes the production of electrical signals, whereas in the chapters by Lucia Vallar and Atanasio Pandiella the intracellular molecules (second messengers) that bring about the appropriate cellular response are treated. The impressive progress in the field of signal transduction has been made in part to the recent development of innovative technologies which enable the second messengers inside extracellular compartments to be measured. Some of these developments are reported by Rosario Rizzuto.

Next, there are three chapters that address a central theme for understanding the relations of cells in the multicellular organism: cell division and death. In the chapter by Giulio F. Draetta, the molecular protagonists of the cell reproductive cycle are described, plausible models to explain the result of their actions are proposed, and conservation of the essential characteristics of the cell cycle through evolution is examined, from the simplest eukaryotic cells (yeasts) to those of the higher organisms. Thomas Cotter deals with the recently discovered, fascinating question of active cell death (apoptosis) which occurs by means of the execution of a precise genetic program put into effect by the single cell, which under certain conditions will sacrifice itself for the good of the whole organism. Last, in the chapter by Paolo M. Comoglio and Carla Boccaccio, the particular requisites of cell division in the higher animals are considered as well as the way in which the loss of this regulation leads to the occurrence of tumors.

The final chapters in "Cells and Cell Communities" analyze the way in which eukaryotic cells interact, in accordance with precise spatial relations, to form tissues and organs in the multicellular organism. Guido Tarone describes recent progress in the study of surface molecules involved in cell–cell recognition and in the adhesion of cells to the extracellular matrix, whereas Roberto Bruzzone and Paolo Meda analyze the construction of the epithelia that delimit the animal organism with respect to the outside world. The last two chapters, by Edoardo Boncinelli and Mark E. Fortini, are dedicated to the mechanisms of construction of the architectural plan of multicellular organisms during development, and they complete this part of the volume — intended to provide a vision of life starting from the spatial organization of the molecules.

Peter J. T. Dekker
Wolfgang Voos
Nikolaus Pfanner
Joachim Rassow

Institute of Biochemistry and Molecular Biology
University of Freiburg
Hermann-Herder-Str. 7
D-79104 Freiburg, Germany

The Folding of Proteins inside the Cell

Although proteins are synthesized as unstructured, extended polypeptides, they have to obtain a stable three-dimensional shape in order to perform their biological function. It has been manifest for decades that all information that is required to fold a denatured polypeptide into a structured functional conformation is contained within the amino acid sequence of the polypeptide. Theoretically, the biologically active form of a protein should therefore be predictable when the primary structure is known. Although numerous proteins have been shown to fold spontaneously in the test tube, in vivo protein folding involves an intricate interplay between different enzymes and chaperones that assist the polypeptide in obtaining its final native conformation. In this article, we will discuss the possible functions of the different helpers of protein folding.

Introduction

In recent years, our understanding of the mechanisms employed in the folding of polypeptides has been greatly aided by plentiful studies on the physical chemistry and cell biology of the protein folding process. Although knowledge on this subject is far from complete, the general idea is that folding proceeds through several distinct stages. Numerous interactions within the folding chain, and *in vivo* with specialized folding enzymes and "helper" proteins (molecular chaperones), must be created and broken during this highly complex process. The attention that the protein folding problem has received in the past and the challenge it will undoubtedly give scientists in the future are justified by the basic nature of the subject. In addition to the elucidation of

the genetic code several decades ago, the understanding of how proteins fold will aid our understanding of how nature manages to translate one-dimensional genetic information into biologically relevant structures. Success in the discovery of a "folding code" would greatly enhance the information content of the huge genome projects which are currently producing DNA sequences at an ever-increasing pace. This increasing demand for reliable prediction of protein structure cannot be met by protein crystallography and nuclear magnetic resonance (NMR) spectroscopy because of practical limitations. Thus, the comprehension of folding pathways might be an invaluable tool in the translation of amino acid sequences into three-dimensional structures.

In addition to being of fundamental interest to scientists,

defective protein folding has recently been implicated in the occurrence of several human diseases. In principle any error that leads to the loss of a functional protein might lead to a defective phenotype. Defects may include desta- bilization of the native or an intermediate state of a pro- tein, prolonged association with molecular chaperones or folding enzymes, preferential formation of off-pathway or toxic conformations, or folding in an incorrect compart- ment. Diseases correlated to defective protein folding can be divided into three classes: proteins that are unable to fold, proteins that acquire toxic folds, or proteins that are mislocalized owing to misfolding (Thomas *et al.,* 1995).

One example of these diseases is the occurrence of cys- tic fibrosis (CF), which is caused by mutations in the gene encoding an ATP-dependent chloride channel (CFTR). The most common CF-causing mutation is a deletion of a single amino acid in the nucleotide-binding domain of CFTR, which leads to defective protein folding. The muta- tion entails temperature sensitivity of the maturation pro- cess not by destabilization of the native CFTR protein but probably by destabilizing a folding intermediate, indicating that the defect may be in the folding pathway. As a conse- quence, at the normal body temperature, the partly folded protein is unable to reach its final location at the cell sur- face but rather accumulates in the endoplasmatic reticulum in association with chaperones and is degraded. Signifi- cantly, mutant CFTR folds correctly at a lower tempera- ture, and once folded it is then functional even at 37°C.

In the case of scrapie (an encephalopathy of sheep and goat) and BSE (the "mad cow disease") the infectious agent is an incorrectly folded protein [prion protein (PrP)]. En- dogenous PrP, a protein present in neurons, contains many α-helices. The infectious scrapie prion protein is mainly in a β-sheet conformation. Infectious PrP is thought to induce the refolding of endogenous PrP into the conformation of infectious PrP. Subsequent increase of the β-sheet content of the protein leads to infectivity of PrP and disease. Defec- tive folding leads in this case to an extremely stable struc- ture that is toxic to the cell.

Since the aim of this article is to give a general descrip- tion of the protein folding process, the reader is referred to more specialized literature concerning the occurrence of diseases related to protein structure. A more detailed un- derstanding of protein folding pathways will undoubtedly have an influence on the design of therapies for treatment of these diseases. In addition to this medical aspect, protein folding is also of importance in biotechnology due to the in- creasing demand for the refolding of recombinant proteins.

The *in Vitro* Protein Folding Problem

Although the three-dimensional structure of numerous proteins has been solved in great detail, the pathways that linear polypeptides follow to attain their native, biologi- cally active conformations are less well defined. From the early experiments of Anfinsen on the *in vitro* refolding of denatured ribonuclease it has become clear that all essen- tial information required to determine the final conforma- tion of a protein resides in the polypeptide chain (Anfin- sen, 1973). Since then, similar results have been obtained with several other small, single-domain polypeptides and also with larger, more complex proteins. If the correct re- folding conditions are chosen a denatured enzyme will re- fold into its native conformation in the absence of any other proteins. However, how exactly the information con- tained within the polypeptide chain is translated into a unique three-dimensional structure is unclear.

Recent comparisons of known protein structures have identified numerous polypeptides that have an overall simi- lar tertiary structure or similar domains, although their pri- mary amino acid sequence shows no obvious similarity. Furthermore, mutations in the amino acid sequence often do not influence the final folded structure of a protein. Therefore, not only are a few key amino acids required for correct folding but also the general nature of the amino acid sequence seems to be important in attaining the native conformation.

On first sight the possible structures that can be folded from a single polypeptide chain are innumerable. It has been calculated that the time required for a large polypep- tide to scan all possible conformations on its way to the na- tive state would be on the order of the age of the universe. Refolding of polypeptides when diluted from denaturant into an aqueous solution, however, occurs spontaneously on a physiological time scale. Therefore, the studies on pro- tein folding have been focused on the examination of fold- ing intermediates and the characterization of pathways that are followed during the refolding of denatured proteins. Biophysicists have made considerable progress in defining the pathways a polypeptide takes during the folding into a biologically active conformation.

The Stable Conformational States of Proteins

During the folding of proteins by dilution from denatu- rant, a polypeptide chain will theoretically adopt numer- ous different conformational states. Many of these folding states are insufficiently stable to be detected by physical means and therefore remain unidentified. However, some folding states are apparently more stable and can be iso- lated in adequate amounts to allow examination. The oc- currence of these so-called "folding intermediates" has led to the assumption that folding occurs via defined pathways with cooperativity of multiple interactions that stabilize the native state of a protein.

The Unfolded State. The ideal unfolded state is the random coil in which there are very many conformations possible with even a small protein. The polypeptide is in a very extended conformation in this state in which noncovalent interactions that stabilize the native state are disrupted. However, dependent on the nature of the denaturant, very weak, local structures might occur in this state. Although unfolding in chaotropic agents (e.g., urea or guanidine hydrochloride) almost leads to a random coil, in polypeptides unfolded by extremes of pH or temperature localized structures could be detected. This indicates that harsh physical or chemical treatments, which should hardly be called physiological, have to be used to keep a protein in the unfolded state. *In vivo*, proteins that are synthesized as linear polypeptides on ribosomes will therefore start folding during their synthesis. Proteins that are transported to distinct cellular compartments such as chloroplasts, mitochondria, and the endoplasmic reticulum apparently pass membranes in an extended or loosely folded conformation, even when transport occurs posttranslationally. As discussed later, molecular chaperones are thought to keep the newly synthesized proteins in an unfolded, translocation-competent state.

The Molten Globule. A variety of proteins have been observed under certain conditions to exist in stable conformations that are neither fully folded nor fully unfolded (Fig. 1). The existence of such a structure was first proposed on theoretical grounds (Ptitsyn, 1995). Today, the occurrence of a "molten globule" state is well documented. The following are the most common properties of the molten globule: (i) The molten globule is much more compact than the random coil and only marginally larger than the fully folded protein; (ii) the content of secondary structure is similar to that of the folded protein; (iii) molten globules exposure hydrophobic surfaces and are therefore susceptible to aggregation; (iv) its enthalpy is nearly the same as that of the fully unfolded state; and (v) conversions to the fully unfolded state are rapid and noncooperative, but they are slow and cooperative with the fully folded protein.

These observations suggest that the molten globule is a collapsed molecule with native-like secondary structure but lacking stable tertiary structure. Although the number of possible conformations of the molten globule is innumerable, it is obviously lower than the number of conformations that an unfolded protein can adopt. The biological relevance of the molten globule state is a matter of debate. Although it was reported that proteins fold at the surface of the chaperonin GroEL through a molten globule-like intermediate, it is not clear if molten globules are essential intermediates during protein folding *in vivo*.

The Native State. The native state of numerous different proteins has been examined by X-ray crystallography and NMR spectroscopy. From these studies some general aspects of fully folded globular proteins can be deduced. The most notable similar aspects of globular protein structures are the nonpolar character of the side chains that comprise the folded interior and the general prevalence of

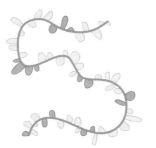

unfolded polypeptide

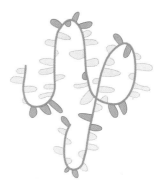

molten globule structure

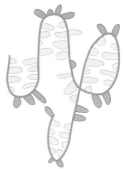

native conformation

FIGURE I Schematic representation of three different folding states that a protein can adopt. (Top) The unfolded polypeptide is a relaxed, extended molecule which passes randomly between multiple conformations. (Middle) The molten globule is much more compact but still lacks the rigidity of the fully folded protein. It is also characterized by the exposure of hydrophobic side chains (yellow), in addition to hydrophilic residues (blue), at its surface. (Bottom) The native fold is even more compact and mainly exposes hydrophilic residues.

hydrophilic side chains at the surface. The hydrophobic core seems to be the most critical aspect for stability of the normal folded state, whereas flexibility is most common at the protein surface. Most folded proteins are not known to adopt alternative fully folded conformations. The only substantial conformational differences that occur within known protein domains result from proteolysis or, as determined recently, in the infectious prion proteins. The interactions that stabilize the native state are intrinsically weak, but many stabilizing interactions are simultaneously present and they may cooperate to produce a stable structure (Creighton, 1995).

The *in Vitro* Folding Pathway

Refolding *in vitro* is generally thought to occur via diverse routes involving one or more relatively stable folding intermediates. The process may be initiated by (i) collapse of hydrophobic regions into the interior of the molecule, (ii) formation of stable secondary structures that provide a framework for subsequent folding, and (iii) formation of covalent interactions, such as disulfide bonds, that stabilize the polypeptide in specific conformations. These mechanisms may operate in conjunction during the early stages of refolding. As a consequence, a folding intermediate is formed that resembles the molten globule state described previously. These intermediates seem to be in rapid equilibrium with the fully denatured state (within milliseconds) and are converted only slowly to the native state. Thus, the rate-limiting step in the refolding process frequently occurs at a late stage, just before the protein adopts its final native conformation. The conversion of the folding intermediate into the native state is thought to be cooperative, mean-

ing that the occurrence of one interaction will stabilize the next, which will in turn stabilize the first interaction (Creighton, 1995). The combined stability will therefore be greater than the sum of the two interactions. This model implies a stepwise acquisition of protein structure, in which a cooperative interaction is thermodynamically favorable to a noncooperative interaction.

This simple linear folding pathway might not be applicable to the folding of most polypeptides. Recently, evidence for the existence of multiple independent folding pathways has been accumulating (Fig. 2; Weissman, 1995). In this scheme, folding of a denatured protein might occur via distinct parallel pathways involving different intermediates but resulting in one defined native state. During the folding of bovine pancreatic trypsin inhibitor (BPTI), which forms three disulfide bridges, nonnative disulfide bonds have been detected. Obviously, these bonds have to be broken again in the further folding process, indicating that folding does not necessarily follow one route, or even the most direct route.

The question of whether the observed intermediates represent true folding states or are products of off-pathway reactions has been raised recently as a result of the finding that the folding of some small proteins occurs extremely fast

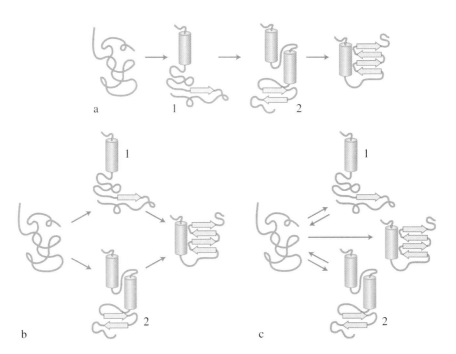

FIGURE 2 Representation of three different folding pathways. The unfolded protein can adopt two different intermediate states while folding into the native conformation. (a) The protein always first adopts intermediate 1, then intermediate 2, before folding into the native conformation. (b) Both intermediates can directly fold into the same native form, leading to parallel folding pathways in which the formation of a certain intermediate is not absolutely required for the completion of the folding reaction. (c) The intermediates are not productive and act as kinetic traps. The dead-end species must be partly unfolded before folding can be completed (Weismann, 1995). Cylinders represent α-helices, and arrows represent β-sheets.

(~10–50 ms) in an all-or-none manner that does not involve detectable folding intermediates. Since normal folding intermediates persist for hundreds of milliseconds, it is possible that some of them act as kinetic traps, having structures that are irrelevant or even detrimental to the formation of the final native state (Fig. 2). Furthermore, the discussed intermediates in the folding of BPTI were shown to fold faster to the native state when they were first unfolded, suggesting that they represent a kinetically trapped species (Weissman, 1995). The question of the kinetic importance of observed intermediates is particularly critical when a folding reaction proceeds through multiple phases since (if unproductive intermediates exist) precisely those intermediates that persist at longer times are most likely to be unproductive.

The *in Vivo* Protein Folding Problem

Although *in vitro* experiments have significant value in defining the types of intramolecular interactions that drive protein folding, they do not accurately reflect the process of folding of nascent proteins in the interior of the cell. As discussed previously, unfolded proteins and folding intermediates tend to expose hydrophobic surfaces to the aqueous surrounding and are therefore especially prone to aggregation. Because aggregation is a major problem in any protein folding assay, *in vitro* experiments are often performed in extremely diluted solutions at low temperature. Since aggregation involves intermolecular contacts, it is obviously a second-order process, in contrast to folding. Therefore, the amount of aggregation compared to the amount of folding will exponentially increase at higher protein concentration and elevated temperature. *In vivo*, the physiological temperature and high concentration of both total protein (100–150 mg/ml) and unfolded polypeptides would strongly favor unproductive interactions over the correct folding pathway. A further difference between folding of proteins *in vitro* and *in vivo* is the complexity of many proteins in cells. Highly hydrophobic membrane proteins, proteins that assemble into enzyme complexes or into microfilaments, and proteins that are co- or posttranslationally modified by attachment of lipids or carbohydrates are hardly expected to follow the folding pathways deduced for small globular proteins. An additional consideration is that in the cell proteins are gradually synthesized on ribosomes or in an extended state translocated over membranes and should, according to the cooperative folding theory, stay in the unfolded form until synthesis of at least the first domain is complete. Although the concentration of nascent chains in the cell is high (30–50 μM in the *Escherichia coli* cytosol), they would be especially prone to aggregation. In other words, inside cell aggregation of nascent chains has to be prevented. Moreover, it was found that some rate-limiting steps in protein folding are catalyzed by specific enzymes. In recent years, it has become increasingly clear how cells solve these folding problems.

Enzymes Involved in Protein Folding

In vitro, two folding rate-determining steps involving isomerization of covalent bonds can be catalyzed by purified cellular enzymes. Protein disulfide isomerase (PDI) catalyzes thiol/disulfide interchange and promotes protein disulfide formation, isomerization, or reduction. Peptidyl prolyl *cis–trans* isomerases (PPIases) catalyze the otherwise slow isomerization of peptide bonds preceding prolines. Both enzymes do not determine the polypeptides folding pathway but rather accelerate the folding of disulfide bond-forming or proline-containing polypeptides.

Protein Disulfide Isomerases

Formation of disulfide bonds in proteins is a rate-limiting step in the correct folding of several proteins (Fig. 3). Typically, only secreted proteins contain disulfide bonds; therefore, PDI is localized in cell compartments that are part of the secretory path. In fact, PDI amounts to about 2% of the

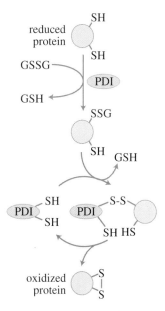

FIGURE 3 Mode of action of the eukaryotic protein PDI, which acts as a catalyst of oxidation of the protein substrate by means of the dimeric oxidized glutathione (GSSG); this is reduced (GSH), forming an intermediate oxidized protein characterized by mixed disulfide bonds between glutathione and the protein. In the second stage, PDI catalyzes the formation of internal disulfide bonds in the protein with liberation of the molecules of bound GSH (modified from Freedman, 1994).

entire protein content of rat liver endoplasmatic reticulum. *In vivo*, the formation of the correct disulfide bonds can be very rapid. In some cases, such as with the immunoglobulin chains, it is even a cotranslational event: As soon as an entire domain is translocated into the lumen of the endoplasmatic reticulum, the single intradomain disulfide bond is formed. A specific association between PDI and newly synthesized immunoglobulin chains in the endoplasmatic reticulum has been demonstrated.

The role of PDI appears to be clear. It facilitates the formation of the correct set of disulfide bonds during *de novo* folding of secreted proteins. It does not determine the folding pathway but rather catalyzes slow steps, presumably by rapid reshuffling of incorrect disulfide bonds in the presence of a low-molecular-weight thiol compound, which may be reduced and oxidized glutathione. The direction of folding and the end product are determined by the protein, i.e., by the stable, native set of disulfide bonds and suitable solvent and redox conditions.

Mammalian PDI is a dimer of identical subunits of relative molecular mass 57,000, each of which contains duplications of domains showing strong homology to thioredoxin, a small protein catalyzing many redox reactions that is present in all classes of organisms from bacteria to higher eukaryotes. Computer modeling studies based on the known three-dimensional structure of *E. coli* thioredoxin indicate that a functional PDI dimer contains four thioredoxin-like domains, each having a dithiol active site located on a prominent loop at the surface of the molecule.

In bacteria, the formation of native disulfide bonds requires DsbA, a protein that resembles thioredoxin in three-dimensional structure but contains an additional helical domain (Freedman, 1994). DsbA seems to be much less effective than PDI as an isomerase on both unfolded and folded protein substrates, and it may act as an oxidant directly transferring its disulfide to substrate proteins (Fig. 4).

PDI effectively catalyzes the conversion of the reduced peptide to its disulfide form in the presence of a glutathione redox buffer, but analysis of intermediates showed that the process occurred predominantly via the formation of mixed disulfides between the polypeptide and glutathione, which isomerize to release reduced glutathione and yield the polypeptide with an intramolecular disulfide bond. Both the initial formation of the mixed disulfide and its rearrangement are catalyzed by PDI so that PDI acts in oxidation mainly by catalyzing successive thiol–disulfide interchanges with an exogenous oxidant, in contrast to DsbA.

Peptidyl Prolyl *cis–trans* Isomerases

Peptide bonds can exist in the *cis* and *trans* isomers (Fig. 5). Typically, the *trans* form is energetically favored about 1000-fold over the *cis* form. The situation is different if the next residue is proline. In this case, the *trans* form is favored only 4-fold. However, in crystallographic models of protein structures, only approximately 6.5% of the proline residues are in the *cis* isomeric form, in contrast to the expected 20% (Creighton, 1990). Of course, any individual proline in a folded protein will be in the same isomeric form in every molecule. Spontaneous *cis–trans* isomerization of proline peptide bonds is slow, with a half-time at 0°C of 20 min. The isomerization of incorrect X-Pro peptide bonds is therefore one of the slow, rate-determining steps in *in vitro* refolding experiments. This rate of isomerization is slow relative to the rate of protein folding in the cell. The reaction can be catalyzed by enzymes with peptidyl prolyl *cis–trans* isomerase (PPIase) activity. PPIases are highly abundant and widely distributed, being found in virtually all tissues and organisms from bacteria to mammals and in all cellular compartments. *In vitro* studies on PPIases have shown that they accelerate refolding of a wide range of proteins but with different efficiencies. The best catalysis of protein refolding by these enzymes is a 100-fold increase in the rate constant for ribonuclease T1; a similar degree of catalysis is seen with immunoglobulin light chains. However, proline bonds which are located in the interior of folding intermediates are not accessible for PPIases.

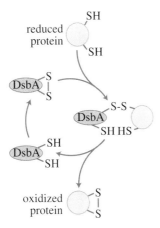

FIGURE 4 Mode of action of the bacterial protein DsbA, which acts as a direct oxidant in the formation of disulfide bonds in a protein substrate (modified from Freedman, 1994).

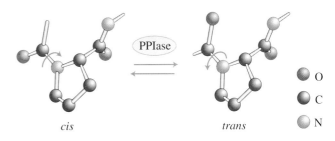

FIGURE 5 *Cis* and *trans* isomers of prolyl peptide bonds. The majority of peptide bonds favor the *trans* isomer to avoid steric hindrance. However, a significant minority of prolyl peptide bonds are in the *cis* isomer. PPIases catalyze the transition of *cis* into *trans* isomers and vice versa.

The PPIases characterized to date can be classified into three structurally unrelated families, which are named after the clinically important immunosuppresive agents that inhibit their isomerase activity. Thus, the cyclophilins bind cyclosporin A, whereas the FK506-binding proteins (FKBPs) bind the structurally distinct compounds FK506 and rapamycin and the parvulins do not bind cyclosporin A or FK506. The function of PPIases in the immune system is not related to their function in proline folding but to their partial involvement in signal transduction.

The protein folding reaction that these enzymes catalyze is a 180° rotation about the C–N linkage of the peptide bond preceding proline, which involves neither net cleavage nor net formation of covalent bonds. Thus, PPIases can be classified as "conformases" with a very high efficiency (Fischer and Schmid, 1990).

Molecular Chaperones

A set of universally conserved, structurally unrelated families of proteins help the newly synthesized proteins to obtain their native conformations. These proteins, collectively referred to as "molecular chaperones," bind to unfolded or partly folded polypeptides and thereby prevent their aggregation. Molecular chaperones are very abundant and the expression of several of them is highly upregulated under a variety of cellular stress conditions (Fig. 6). In fact, the accumulation of abnormally folded proteins in the cell leads to an increased expression of these chaperones. For historic reasons, many of the molecular chaperones are classified as "heat shock proteins" (Hsps). Indeed, in the case of *E. coli* at 46°C, a temperature at which bacterial growth almost ceases, more than 20% of all cellular polypeptides belong to this class of proteins. Such a tremendous accumulation most likely reflects the extra need for molecular chaperones to deal with the increased protein misfolding and aggregation that occurs at this relatively high temperature. It should be emphasized that the term "heat shock" or "stress proteins" is somewhat misleading since most of these proteins are important for cell growth at all temperature ranges. Although the binding and release from some of the Hsps is regulated by ATP binding and hydrolysis, the molecular chaperones cannot be regarded as catalysts of protein folding. More likely, as discussed later, they prevent aggregation and unproductive interactions by providing a regulated and protected environment for folding and thereby increase folding efficiency.

The majority of the currently identified chaperones belong to highly conserved protein families. Although more classes of chaperones have been identified, some of which are also Hsps (e.g., Hsp110, Hsp90, and Hsp20), we focus our attention on the Hsp60 (GroEL) and Hsp70 (DnaK) families of chaperones (numbers indicate the approximate relative molecular weight in thousands of the family mem-

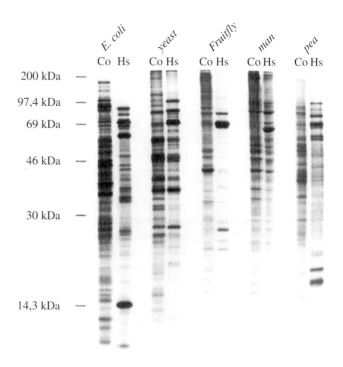

FIGURE 6 Gel electrophoresis analysis of total protein extracts, metabolically marked with a radioactive amino acid, from cells growing at normal temperatures (Co) or after shift to elevated temperatures (Hs). The polypeptides are separated according to molecular mass; each band corresponds to a single polypeptide or several polypeptides of the same molecular mass. The major classes of heat shock proteins are clearly visible in all organisms in the right-hand columns. Reproduced with permission from Parsell and Lindquist (1994).

bers), which are studied most intensively. In the past few years, it has become clear that these major intracellular chaperone proteins do not function in isolation. Rather, to ensure their proper and efficient function, many so-called cochaperones have evolved. In some cases, these cochaperones are bona fide chaperones, such as the DnaJ (Hsp40) protein of *E. coli* that works synergistically with DnaK. In other cases, the sole role of the cochaperones appears to be to ensure the efficient function and recycling of the chaperone protein, e.g., the *E. coli* GroES (Hsp10) and GrpE proteins, working with the Hsp60 and Hsp70 chaperones, respectively. As will be discussed later, in addition to the individual function of these "chaperone machines," Hsp60 and Hsp70 might functionally cooperate in protein folding in mitochondria, chloroplasts, and bacteria.

How Molecular Chaperones Might Function

Before we discuss the properties of the different chaperone machines, we have to define the way we think about the molecular mechanism of chaperone function. First, we should make the distinction between the facilitation of a process and catalysis.

During folding, polypeptides proceed through different stages which are usually separated by only very low energy barriers. Small, single-domain proteins often fold very fast while no kinetic intermediates can be trapped. Larger proteins start their final folding step from an intermediate which is separated from the native state by a high energy barrier. This final step is therefore rate limiting and highly cooperative. In this case the final folding intermediate can be trapped and analyzed. The rate of appearance of the native product will be dependent on the concentration of this intermediate and the rate constant. The rate constant is related to the energy barrier between the intermediate and the final state (the activation energy). Enzymatic catalysis, like the enzymes discussed previously, increases the rate constant by lowering the barrier, often by stabilizing the transition state relative to the substrate and products. Molecular chaperones, however, are not known for their ability to stimulate the folding rate. The rate of formation of the native product is in many cases even slowed. However, chaperones do increase the amount of correctly folded protein. Some clues to how chaperones might be able to do so derive from the work on SecB (Randall and Hardy, 1995). SecB is a tetrameric chaperone in *E. coli* that specifically associates with polypeptides that have to be secreted. Essential for its function is that the polypeptide stays unfolded until it is delivered to SecA, which regulates the transport across the plasma membrane. Therefore, SecB has to compete with both aggregation and folding of proteins. To do so effectively, the chaperone has to be abundant, have a high binding rate, and have a high specificity for its substrates.

Abundance. Since aggregation involves contacts between two or more molecules, it is obviously a second- or higher order reaction. As a consequence, the aggregation rate is very sensitive to the concentration of folding intermediates. Folding, however, is a first-order process, much less influenced by concentration. An increase in free folding intermediates will therefore lead to increased aggregation and a decrease in intermediates to preferential folding. A chaperone might therefore act by decreasing the concentration of free intermediates. Interaction of polypeptides with a chaperone, however, is also a second-order process, meaning that the concentration of the chaperone in the cell should be equal to or higher than the concentration of unfolded polypeptide. The concentration of SecB (in the micromolar range) present in a normally growing cell appears to be sufficiently high to favor binding of its substrate over aggregation. The concentration of other molecular chaperones in the cell is also constitutively high. Under stress conditions, when the concentration of malfolded polypeptides increases, expression of chaperones of the Hsp60 and Hsp70 class is even induced to compensate for this effect.

Binding Rate. If chaperone binding has to compete effectively with aggregation, the rate of binding to the chaperone should approach the encounter limit between chaperone and polypeptide. It has been calculated from results obtained with SecB that this is indeed the case (Randall and Hardy, 1995). Macromolecules of the size of an average protein collide with a rate constant of approximately 10^9 $M^{-1}s^{-1}$. It is calculated that the rate constant for association of SecB with an unfolded polypeptide might be in the range of 10^8 $M^{-1}s^{-1}$. The chaperone might therefore effectively compete for aggregation.

In addition to competing for aggregation, SecB also has to compete for folding of polypeptides. Collapse of a polypeptide to a compact state and acquisition of secondary structure can occur on a millisecond timescale. However, since the intermediates along the folding pathway have very little energy of stabilization and are in rapid equilibrium, conclusions concerning the "structure" a chaperone might prefer cannot be drawn. It is clear, however, that polypeptides are no longer ligands for chaperones as soon as they have passed the rate-limiting step in folding, which is for many proteins a late step involving formation of native structure. For the precursor proteins that interact with SecB, the rate constant for this step is decreased by the presence of the leader peptide. In the case of SecB this would strongly favor binding over folding (Randall and Hardy, 1995). Other chaperones assist the folding of unfolded polypeptides, and therefore folding is not the competing but rather the preferred process.

Specificity. All chaperones have in common that they specifically recognize a polypeptide only in its nonnative state. This recognition must result in tight and rapid binding of the ligand and sequestration of its hydrophobic surfaces so that it is prevented from aggregation. Direct recognition of proteins as nonnative by virtue of their hydrophobic character has the inherent problem that a chaperone with a permanently exposed hydrophobic area would aggregate. For SecB it is therefore proposed that binding is not due to the hydrophobic character but rather recognition of flexible regions. After initial binding of several flexible regions of an unfolded protein by SecB, a conformational change in the chaperone occurs. Only hereafter is a hydrophobic site on the chaperone exposed, thereby identifying the target protein as nonnative. This guarantees the proximity of the substrate and thereby avoids the problem of aggregation of the uncomplexed chaperone. The concerted interactions at several sites within one ligand result in binding of high affinity. Because each of the initial contacts is of low specificity, the rate of binding approaches the encounter limit, allowing a kinetic partitioning of polypeptides into the export pathway rather than into the alternative fates of aggregation or folding. How other chaperones solve the problem of aggregation is less clear. As will be discussed later, chaperonins of the Hsp60 class are ring shaped and unfolded polypeptides bind at the inside or entrance of the ring. The chaperonins, therefore, also have a shielded hydrophobic binding site which might prevent the aggregation of the chaperonin.

Although the molecular mechanism of chaperone activity is far from clear, the case of SecB provides an example of how a chaperone might solve the problem of the specificity, velocity, and affinity of binding to nonnative polypeptides.

The Hsp70 Family

Members of the Hsp70 family are found throughout prokaryotic and eukaryotic cells in the cytoplasm, nucleus, endoplasmatic reticulum, mitochondria, and chloroplasts. Some members of this extensive family are essential for cell growth, some are highly expressed upon heat shock (Hsp), whereas others are expressed constitutively [heat shock cognate (Hsc)]. The high degree of sequence conservation in the Hsp70 family has resulted in a consensus model for Hsp70 structure and function. All members of the family carry a highly conserved amino-terminal ATPase domain followed by a less conserved carboxy-terminal portion containing the peptide binding site. Some members of the family also contain an additional amino-terminal targeting signal or a carboxy-terminal retention signal, dependent on the final location of the Hsp70 protein. The crystal structure of the amino-terminal ATPase domain of bovine Hsc70 has been determined to 2.2Å resolution (Flaherty et al., 1990). It has four structural domains that form two lobes with a cleft between them (Fig. 7). The binding pocket for the nucleotide and a magnesium ion is located at the base of the cleft. Interestingly, the ATPase domain has an overall tertiary structure that is strikingly similar to that of G-actin, despite negligible similarity between these proteins at the amino acid sequence level. Additionally, the nucleotide-binding fold resembles that of hexokinase; therefore; these proteins can be regarded as members of a common structural superfamily. The amino acid side chains of the residues that are observed to interact with the Mg^{2+}-nucleotide complex in the active site are rigorously conserved throughout the heat shock protein family and can be candidates that may influence the ATPase activity of the protein.

An 18-kDa fragment located immediately after the ATPase domain is sufficient for high-affinity substrate binding. The substrate specificity of Hsp70 is still a matter of debate. Analysis of peptides that were eluted from BiP, the mammalian Hsp70 of the endoplasmatic reticulum, showed a preference for hydrophobic residues. The crystal structure of the peptide binding domain of bacterial DnaK showed a β-sandwich that carries the binding site for peptides and an α-helical lid that can open and close the peptide entry site, probably regulated by the ATPase domain (Zhu et al., 1996) (Fig. 7).

The Hsp70 Reaction Cycle

The model protein for Hsp70 action has been the DnaK protein of E. coli, which acts in concert with two other Hsps, DnaJ and GrpE (Georgopoulos, 1992). As might be expected, similar phenotypes result from mutations in any of the three components. Both DnaK and DnaJ can bind polypeptides, but their affinities for substrates differ. Although DnaK seems to interact preferentially with unfolded proteins, it has been suggested that DnaJ prefers binding to protein substrates exhibiting secondary and tertiary structure (Cyr et al., 1994). These differences in substrate specificity apparently allow DnaJ-like proteins to facilitate the interaction of nascent proteins with Hsp70, which would not normally be bound. Escherichia coli DnaJ can recognize a structural feature of compact protein folding intermediates and stabilize them in a conformation in which binding sites for Hsp70 are exposed and thus substrate binding to DnaK is facilitated. DnaJ also acts on DnaK, stimulating the rate-limiting chemical step in the DnaK ATPase reaction cycle. The amino-terminal region of DnaJ, containing the conserved J domain and an adjacent glycine/phenylalanine-rich region, is sufficient for the interaction with DnaK because this segment stimulates the ATP hydrolysis of DnaK. Unlike DnaJ, GrpE promotes the exchange of nucleotides, acting as a nucleotide release factor. GrpE interacts with a conserved loop next to the ATP binding site of DnaK, and it is thereby thought to influence the nucleotide binding conditions of the chaperone.

Recent studies have elucidated some of the important aspects of the DnaK ATPase cycle. From these results a model for the reaction cycle of Hsp70s can be deduced (Fig. 8). ATP binding to Hsp70 accelerates both the rate of binding and the release of peptide substrates. Hsp70 in the ATP-bound form can therefore be regarded as the fast-binding/fast-release form (McCarty et al., 1995). Hsp70 in the ADP form releases polypeptides only slowly, and it was therefore previously termed the "high-affinity" state but also binds very slow to polypeptides. The ADP-bound form of Hsp70 can therefore be regarded as the slow-binding/slow-release form. Since the ATP content of growing cells is relatively high, it is expected that a substantial fraction of the Hsp70 molecules are in the ATP-bound form. Since the binding rate for unfolded substrate proteins is higher in the ATP-bound form than in the ADP-bound form (McCarty et al., 1995), Hsp70 will probably mainly interact with unfolded polypeptides in the ATP-bound form. Binding of polypeptides stimulates the hydrolysis of ATP, a reaction that is also stimulated by DnaJ. The resultant ADP-bound form of Hsp70 will have a stable interaction with the polypeptide, and it is therefore the principal polypeptide-bound form found in in vitro experiments. The presence of GrpE will result in the release of ADP, resulting in a nucleotide-free form. The biological significance of the nucleotide-free form is unclear, but in several aspects it seems to resemble

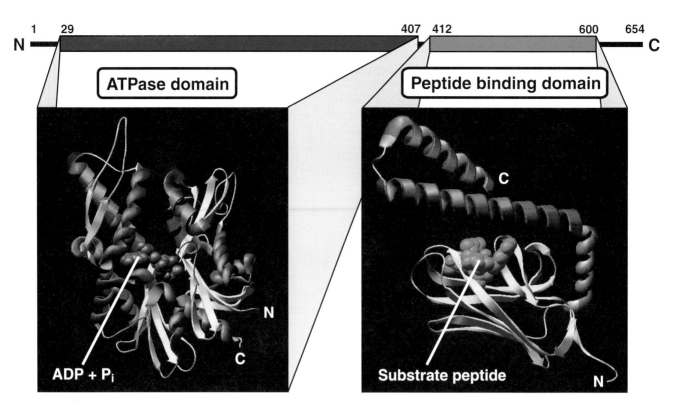

FIGURE 7 Structure of Ssc1, the mitochondrial member of the Hsp70 from the yeast *Saccharomyces cerevisiae.* (Top) A diagram of the linear amino acid chain representing the two major domains: the amino-terminal ATPase domain (blue) consisting of about 400 amino acid residues and the carboxy-terminal peptide-binding domain (green) of about 200 amino acid residues. In addition, Ssc1 contains an amino-terminal presequence of 23 amino acid residues carrying the mitochondrial targeting information that is cleaved off in the mitochondrial matrix. Indicated are the numbers of the amino acid residues corresponding to the borders of the structural models below. (Bottom left) Structure of the ATPase domain of Ssc1 modeled after the structure of bovine Hsc70 (Flaherty *et al.,* 1990). A molecule of ADP (blue) binds in the deep cleft between the two lobes of the ATPase domain. (Bottom right) Structure of the peptide-binding domain of Ssc1 modeled after the related DnaK molecule from *E. coli* (Zhu *et al.,* 1996). The peptide substrate (green) is bound in an extended conformation by a β-sandwich subdomain. An α-helical domain covers the substrate binding channel and stabilizes the complex. The red helices represent α-helices, and the yellow arrows represent β-sheets.

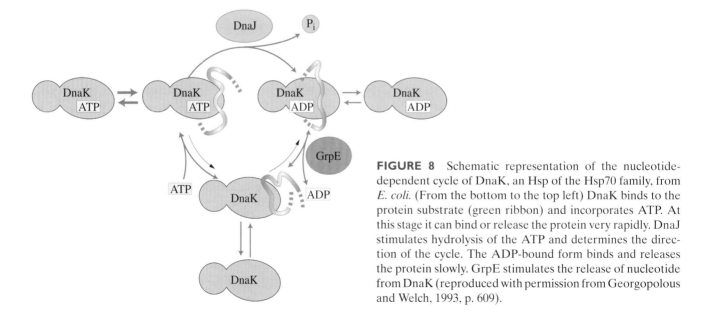

FIGURE 8 Schematic representation of the nucleotide-dependent cycle of DnaK, an Hsp of the Hsp70 family, from *E. coli.* (From the bottom to the top left) DnaK binds to the protein substrate (green ribbon) and incorporates ATP. At this stage it can bind or release the protein very rapidly. DnaJ stimulates hydrolysis of the ATP and determines the direction of the cycle. The ADP-bound form binds and releases the protein slowly. GrpE stimulates the release of nucleotide from DnaK (reproduced with permission from Georgopolous and Welch, 1993, p. 609).

the ADP-bound form. Free phosphate is probably released before ADP, but the relevance of this reaction is not known. Binding of ATP will complete the cycle and leads to the release of the polypeptide chain. The resultant ATP-bound Hsp70 is then again available for another cycle of polypeptide binding and release.

This reaction cycle predicts that there must be some kind of communication between the ATPase domain and the peptide-binding domain of Hsp70. ATP binding leads to a more "open" conformation, whereas ADP binding leads to a "closed" conformation of the peptide-binding domain. Indeed, a different protease digestion pattern and tryptophane fluorescence in the ATP- and ADP-bound forms has been detected, indicating that there is a conformational change in the molecule upon binding to ATP. It must be emphasized that this model does not necessarily apply to all Hsp70s per se since it is mainly based on studies on the *E. coli* Hsp70 DnaK. For instance, Ssa1p, a yeast cytoplasmic Hsp70, seems to have a stable interaction with unfolded proteins in the presence of ATP, and hydrolysis leads to release.

An apparent dilemma in this model is the seemingly nonproductiveness of the ATP cycling. The ATP-bound state of Hsp70 is the form that both binds and releases polypeptide; therefore, it should be sufficient for chaperone protein folding. A cycling by hydrolysis of ATP would only delay this process. However, this delay in folding might not be detrimental to the folding process. By the production of a stably bound folding intermediate, the effective concentration of aggregation-prone unfolded protein decreases, leading to a more efficient, albeit slower, folding process. Due to the ATP cycle Hsp70 has a high affinity for unfolded polypeptides (in the ATP form) but also binds with high stability (in the ADP form) and is still able to release the protein quickly (in the ATP form). This regulated binding–release cycle gives the chaperone the opportunity to perform multiple tasks in the biogenesis of proteins. Chaperones of the Hsp70 class are involved in numerous different processes inside the cell. Since for the purpose of this review a discussion of all these functions is too indepth, we will concentrate on one example in which different Hsp70s have different functions.

Hsp70s in the Biogenesis of Mitochondrial Proteins

An elegant example of the different functions that Hsp70s can perform during the synthesis and maturation of proteins is the biogenesis of mitochondrial proteins in their pathway from the cytosol to their functional location. Mitochondrial preproteins contain an amino-terminal presequence and are produced on cytosolic ribosomes, transported to mitochondria, translocated across the two mitochondrial membranes in an extended conformation, and finally folded into their native state inside the mito-

chondrial matrix. Hsp70s influence every step of this process and are even essential in driving the translocation of polypeptides across the mitochondrial membranes (Fig. 9).

Protein Synthesis. The genome of the yeast *Saccharomyces cerevisiae* codes for 14 different Hsp70s. Like in other eukaryotes, these Hsp70s are localized in the cytosol, mitochondria, and the endoplasmic reticulum (ER). There are three cytosolic subfamilies — SSA (four members), SSB (two members), and SSE (two members) — and several mitochondrial and ER-localized Hsp70s. The cytosolic subfamilies have distinct functions in the cell. SSA members have been implicated in the transport of proteins and function in the regulation of the heat response and protein degradation. SSB members are thought to have a role during protein synthesis. The Ssb proteins associate with translating ribosomes and are released with puromycin, an antibiotic that leads to the release of the nascent chain. Furthermore, mutants in the SSB genes are hypersensitive to antibiotics that inhibit polypeptide chain elongation. One could imagine that an Ssb protein binds to a nascent chain as it emerges from the exit tunnel of the ribosomal 60S subunit, preventing intramolecular interactions of the nascent chain with itself or with the surface of the ribosome and thus facilitating the movement of the remainder of the polypeptide through the tunnel as it is synthesized. Notably, expression of the SSB genes is not induced upon heat shock. Instead, they are most highly expressed under optimal growth conditions, when the protein synthesis rate is high.

Protein Targeting. Proteins must attain a loosely folded conformation to traverse biological membranes. Studies of mitochondrial protein import provided direct experimental evidence for the requirement of a "translocation-competent" state of precursor proteins during the translocation process. Tight folding into a stable tertiary structure, e.g., induced by the binding of substrate analogs or cofactors, prevents the import of precursor proteins into mitochondria. About 50 amino acid residues are sufficient to span both mitochondrial membranes. This excludes that precursor proteins traverse membranes in their native conformation, meaning that these molecules should remain in an unfolded or loosely folded structure prior to translocation.

Molecular chaperones of the Hsp70 family help to maintain a translocation-competent conformation of proteins targeted to mitochondria, the ER, chloroplasts, and the nucleus. In yeast, this function is performed by the Ssa proteins. Genetic depletion of Ssa1p results in the accumulation of precursor forms of the mitochondrial inner membrane protein $F_1\beta$ (the β-subunit of the ATP-synthase) and of the secreted protein α factor in the cytosol, suggesting a common step in the posttranslational protein transport across mitochondrial and ER membranes. Essential in this step is that the Hsp70s stay bound to the precursor proteins

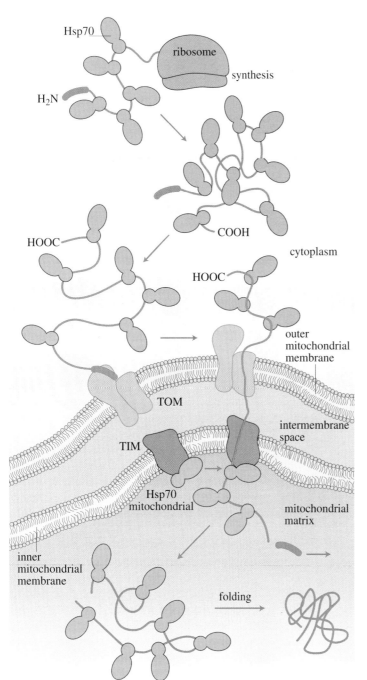

FIGURE 9 Hsp70s are involved in different steps of the biogenesis of mitochondrial proteins. After synthesis of preproteins on cytoplasmic ribosomes, Hsp70s bind to the polypeptide chains and prevent premature folding and aggregation. Preproteins are targeted to receptor molecules which are part of the translocation complex of the mitochondrial outer membrane (TOM), after which they are translocated in an extended state over the two mitochondrial membranes by the combined action of the outer membrane translocase and that of the inner membrane (TIM). Mitochondrial Hsp70 in the mitochondrial matrix binds to the emerging polypeptide and facilitates further translocation. After completion of the process, mitochondrial Hsp70 remains bound to the translocated protein and may also assist its folding.

to prevent premature folding into their native conformation. The significance for cytoplasmic Hsp70s to acquire a "slow-release" form is in this case evident. For transport to mitochondria, the preproteins are transferred to specific "receptors" located on the outside of the mitochondrial outer membrane, which in turn present the preprotein to the import channel.

The activity of the yeast cytoplasmic Hsp70s is regulated by the DnaJ homologs Ydj1p and Sis1p. Ydj1p interacts with Ssa proteins, whereas Sis1p interacts with Ssb proteins. Since Sis1p is associated with small polysomes and 40S subunits, it is possible that the DnaJ homolog is involved in targeting of Ssb proteins to polysomes. *ydj1* mutants are defective in translocation of proteins into the ER and mitochondria. Ydj1p stimulates the ATPase activity of Ssa1p and is attached to membranes via a carboxy-terminal farnesyl group, thus present at a site where it could interact with Ssa proteins during the translocation process. Surprisingly, no GrpE homolog has until now been detected in the eukaryotic cytoplasm, although numerous Hsp70 and DnaJ homologs have been identified.

Protein Translocation. Translocation of precursor proteins across the two mitochondrial membranes requires both ATP and a membrane potential over the inner mitochondrial membrane. The membrane potential (negative inside) is thought to drive the translocation of the presequence part of mitochondrial preproteins, which is positively charged. The need for ATP can be largely attributed to the requirement for the function of mitochondrial Hsp70 (Ssc1p in yeast). Experiments on different *ssc1* mutants, and by matrix ATP depletion, have established the requirement for binding of Ssc1p to incoming precursor proteins in order to complete the import reaction. Currently, two models for the action of Ssc1p are popular (Fig. 10; Pfanner and Meijer, 1995).

In the first model, termed the "Brownian ratchet," Ssc1p has a more passive function during the translocation process. By Brownian motion, the unfolded precursor protein slides back and forth through the import channel and is trapped at the matrix side by the binding of Ssc1p. This activity is termed the "trapping" function of Ssc1p. Eventually, the entire polypeptide chain can be translocated across the two mitochondrial membranes by further oscillation of the polypeptide chain and binding of additional Ssc1p molecules. ATP is then required to recycle bound Ssc1p proteins. This model predicts an unfolded, or at least loosely folded, state of the preproteins prior to translocation and can therefore account for the import of unfolded precursor proteins.

This Brownian ratchet model, however, cannot readily explain the finding that precursors such as cytochrome b_2 are actively unfolded by Ssc1p during import into mitochondria. Therefore, a second model was proposed in which Ssc1p has a more active role in the translocation of mitochondrial precursor proteins. Incoming precursor is bound to Ssc1p on the matrix site and is then pulled inside mitochondria by a conformational change in Ssc1p that remains bound to both the precursor protein and the membrane. Such a conformational change would generate a force that could drive the unfolding and inward translocation of a precursor protein. Indeed, a membrane localized protein (Tim44) that interacts with Ssc1p and could serve as mem-

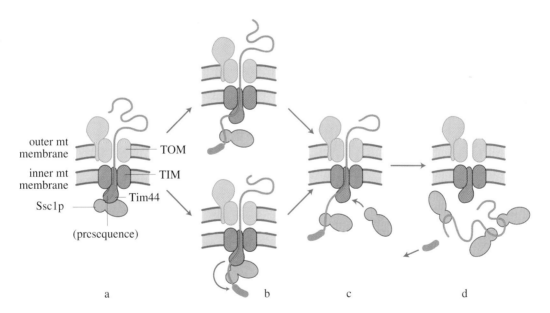

FIGURE 10 Two alternative models for the translocation of preproteins across the mitochondrial inner membrane. (a) The membrane potential across the mitochondrial inner membrane drives the translocation of the presequence. The presequence interacts with Tim44 located in the TIM complex. (b) (Top) The preprotein is handed over to mitochondrial Hsp70 (Ssc1p), which prevents retrograde translocation acting as a ratchet (ratchet model). Alternatively (bottom), mitochondrial Hsp70 could interact with the preprotein while it is still bound to Tim44 and change conformation generating the active force necessary for translocation (pulling model). (c) Mitochondrial Hsp70 releases from Tim44, thereby enabling the binding of additional mitochondrial Hsp70 molecules to Tim44 and the preprotein. (d) Further translocation occurs by Brownian motion and trapping of the translocated preprotein by mitochondrial Hsp70 or by repeated mitochondrial Hsp70 pulling cycles, leading to complete translocation and removal of the presequence.

brane anchor has been identified. Interestingly, the interaction between Ssc1p and Tim44 is disrupted by binding of ATP to the chaperone. Ssc1p might therefore function in a dynamic equilibrium between membrane-bound and -soluble states. The action of Ssc1p in this second model has been described as the "pulling" function of Ssc1p. This model predicts that Ssc1p can obtain different conformational states during the translocation reaction. Indeed, recently Ssc1p was found to attain a different conformation upon binding to ATP, which might represent the change required for the "pulling" action.

Homologs of the *E. coli* DnaJ and GrpE cochaperones have also been identified in yeast mitochondria. Mitochondrial GrpE (Mge1p) is essential for cell growth and involved in the import process, suggesting a similar function in the ATPase cycle of Hsp70 as the *E. coli* homolog. Surprisingly, the DnaJ homolog (Mdj1p) is not essential and a role in the translocation of mitochondrial precursor proteins could not be established. However, Mdj1p might function in the subsequent folding of the Ssc1p-bound proteins.

Protein Folding. After import and removal of the presequence, the Ssc1p-bound proteins have to be folded into their native conformation. Therefore, while Hsp70 molecules on the cytoplasmic site have to prevent the folding of precursor proteins, on the matrix side closely related Hsp70 must facilitate this process. Many imported mitochondrial proteins have to be assembled into the large complexes in the mitochondrial inner membrane. Many gene products might function as specialized chaperones to help the assembly of these inner membrane complexes. A more general role in the folding of mitochondrial proteins has been described for mitochondrial Hsp60. How mitochondrial Hsp70 and Hsp60 cooperate in the folding of newly imported proteins will be described later. First, we will discuss how Hsp60 is believed to function.

Chaperonins

Traditionally, the chaperonins are a family of chaperones with subunits of about 60 kDa which assemble in 14-subunit oligomers of characteristic double-ring shape (Fig. 11). Chaperonins are highly conserved and can be found in eubacteria (GroEL), mitochondria (Hsp60), and chloroplasts (Rubisco-binding protein). Chaperonins are thus a subgroup of the molecular chaperone proteins (e.g., Hsp70s are not chaperonins). The cytosol contains TCP ring complexes (TRiC), which also have a double-ring-like structure and are proposed to be functionally similar to GroEL/Hsp60 but are built from only weakly related subunits. In contrast, the TCP proteins show strong homology to archaebacterial chaperonins.

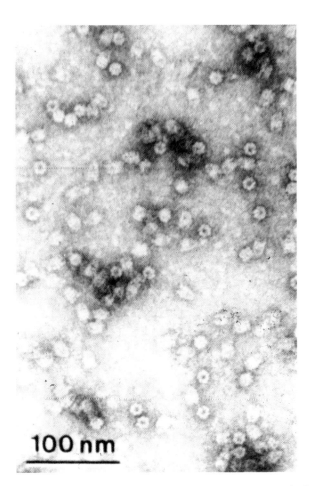

FIGURE 11 Electron microscograph of the negatively stained GroEL–GroES complex (reproduced with permission from Georgopoulos and Welch, 1993, p. 609).

The crystal structure of the most thoroughly investigated chaperonin, GroEL of *E. coli,* has been elucidated at 2.8 Å (Braig *et al.,* 1994). GroEL is composed of two heptameric rings of 57-kDa subunits stacked back to back. Structural and mutational studies have identified a large portion of the functional surfaces concentrated on the surface of the inner channel and its invaginations. The function of GroEL is dependent on the cochaperonin GroES, of which the crystal structure has been elucidated only recently. GroES is an essential protein composed of a single heptameric ring of 10 kDa subunits and forms a 1:1 asymmetric complex with GroEL by binding to one end of the GroEL cylinder. GroES has a small hole in the middle, which might allow diffusion of metabolites in and out of the GroEL cylinder. Unfolded substrate polypeptides bind to the same domain as GroES, whereas the ATP-binding pocket is located on the inner surface of the equatorial domain, which is close to the center of the cylinder in the crystal structure of GroEL. An allosteric adjustment could therefore couple the presence of the appropriate nucleotide to the process of GroES-assisted polypeptide binding and release.

Mechanism of GroEL Action

GroEL can form a tight complex with a variety of unfolded proteins, but it has only slight affinity for native proteins. Electron microscopic and mutagenesis studies indicate that nonnative polypeptides are held within the central cavity formed by the GroEL rings. At physiological ATP and ADP concentrations, GroEL and GroES form a stable but dynamic complex. The asymmetric GroES–GroEL complex containing bound ADP is likely to act as the polypeptide acceptor state since it is a relatively long-lived species and has high affinity for polypeptides (Fig. 12; Weissman *et al.,* 1995). Since one end of the cylinder is capped by bound GroES (the *cis* side), which is expected to prevent a protein from accessing the peptide-binding regions within the central cavity at that end of the cylinder, polypeptides might bind initially to the GroEL ring not occupied by GroES (the *trans* side). During the folding cycle, release and rebinding of GroES allows the formation of a *cis* complex in which GroES and polypeptide bind to the same ring of GroEL. GroES acts as a cap in this *cis* complex as well, but here, instead of excluding polypeptide, it sequesters polypeptide underneath it within the central cavity of GroEL. The transition of a *trans* complex to a *cis* complex with a single GroES cap (the bullet model) might involve a complex without any GroES bound or, alternatively, a complex with a GroES cap on both sides of the cylinder (the football intermediate). Binding and hydrolysis of ATP in the GroEL ring opposite that occupied by GroES and the polypeptide induces release of GroES bound to the *cis* side and allows release of the polypeptide into the cytosol.

The model in which polypeptides are productively released from a sequestered position under the GroES co-chaperonin has as a consequence that there is a physical limit to the size of a polypeptide whose folding can be assisted by chaperonins. Indeed, in the absence of the co-

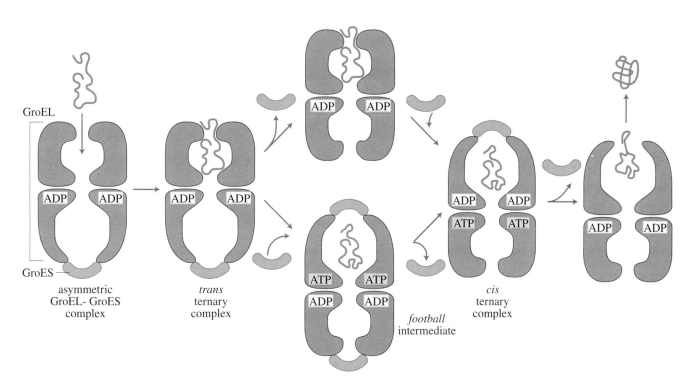

FIGURE 12 Schematic model for a GroEL-mediated folding reaction. The unfolded polypeptide (green) most likely binds to a GroEL–GroES asymmetric complex when this is bound to ADP, leading to the formation of the *trans* ternary complex. Release and rebinding of GroES to the other end of the cylinder leads to the formation of the *cis* ternary complex. Two possible pathways are depicted. In the upper pathway GroES dissociates before rebinding occurs in *cis*. In the lower pathway binding of a second GroES cap occurs before release of the first, leading to the formation of the *football* intermediate. Binding of GroES in *cis* leads to a conformational change in the complex, which might result in the release of the polypeptide in the interior of the GroEL ring complex. Hydrolysis of ATP in the opposite ring to that occupied by GroES leads to dissociation of GroES and release of the polypeptide. Folding of the polypeptide might occur inside the GroEL ring or after release of the protein into the surrounding medium (reproduced with permission from Weissman *et al.,* 1995).

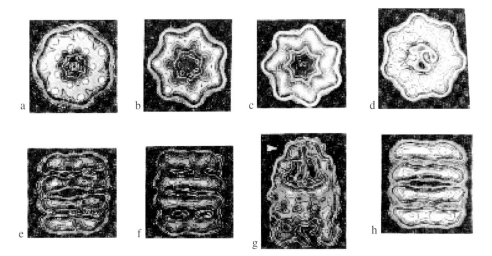

FIGURE 13 Electron micrographs of the GroEL protein, seen from above (top row) and from the side (bottom row). (a) The protein GroEL, (b) GroEL in the presence of Mg–ATP, (c) GroEL–GroES complex, and (d) GroEL bound to rhodanese, a protein substrate (reproduced with permission from Georgopoulos and Welch, 1993, p. 609).

chaperonin, polypeptides up to ~35 kDa can be accommodated within a single ring of GroEL. Binding of the cochaperonin, however, induces a large conformational change in GroEL, leading to an approximate doubling of volume in the central cavity of the chaperonin ring to which it is bound (Fig. 13). The known substrates of chaperonins are generally small enough to be comfortably accommodated within the expanded interior of the complex. However, the folding of many proteins that are near or beyond the physical limit predicted to be accommodated in the chaperonin is not assisted by the GroEL–GroES complex, even though they can form stable complexes with GroEL. Chaperonins might therefore only act in the folding of cellular proteins which are sufficiently small, whereas the mechanism of folding of larger proteins has until recently been unclear. Moreover, under normal growth conditions, many or perhaps most of the smaller proteins reach their native folding state without involvement of GroEL. The exact percentage of GroEL-dependent and GroEL-independent folding pathways has not been determined.

How Do Chaperonins Stimulate Folding?

A Passive Function. Although it is clear from this model that chaperonins act by binding and release of unfolded polypeptides in a regulated manner, it is not obvious how this might lead to an increased folding efficiency. Currently, two contradictory opinions regarding the folding of polypeptides on GroEL exist and are a major topic of discussion. One possibility is that the polypeptide is extruded from the chaperonin in an unfolded state, and refolds outside the chaperonin, in the cytosol (Todd *et al.*, 1994; Weissman *et al.*, 1995). An alternative view is that the polypep-

tide is released and folds into the native conformation or a conformation committed to fold inside the interior of the GroEL–GroES complex and is only extruded from the chaperonin when no rebinding occurs (Mayhem *et al.*, 1996).

In the first model chaperonins might act, as discussed for Hsp70 and SecB, by sequestration of folding intermediates and thereby reduce the concentration of these aggregation-prone polypeptides. Competition with aggregation in this case leads to an increased folding yield. The extraordinary structural properties of chaperonins, however, makes a different mechanism feasible. Since unfolded polypeptides are thought to bind inside the chaperonin cylinder, they are protected from all other nonnative structures with which they might aggregate. Chaperonins might therefore provide a shielded environment in which polypeptides fold without a chance for aggregation. The multisubunit character of chaperonins, with several polypeptide and ATP binding sites, might lead to a regulated binding and hydrolysis of multiple ATP molecules, meaning that the folding of the polypeptide may be sequential. In this alternative model, the polypeptide might not be released from the chaperone completely at time but rather different domains might be released separately. Polypeptides may therefore start folding while still bound to the chaperone.

It has been calculated that the full folding of a single polypeptide molecule on GroEL requires the hydrolysis of approximately 100 ATP. This result implies that several binding and release cycles are required for a polypeptide to adopt its final native state. The chaperonin might therefore rebind domains that do not immediately acquire a native conformation and still expose hydrophobic surfaces. Domains that are folded correctly do not have affinity for the chaperonin anymore and remain unbound. This rebinding does not necessarily have to occur to the same chaperonin molecule but might instead, after release of a nonnative polypeptide, occur to a different GroEL molecule (Weissman *et al.,* 1996). Release of nonnative forms of a protein might also have a physiologic role. Many substrate proteins of GroEL are subunits of oligomeric complexes, of which the subunits must occupy a partially nonnative form to oligomerize efficiently. Furthermore, release of nonnative forms gives substrate proteins the opportunity to interact with other chaperone systems (e.g., the Hsp70 chaperone machinery), which may be essential for efficient folding of some proteins.

An Active Function? As discussed previously, the most plausible model for the action of chaperonins is by sequestration of polypeptides in a protected environment, which thereby increases the yield of folding. However, several observations argue against a solely passive mechanism for chaperonin action and suggest that, in some circumstances, chaperonins can actively lower the free energy barrier of folding. First, the folding of malate dehydrogenase was reported to be assisted by GroEL at stoichiometries as low as 10:1. Thus, 90% of the substrate must be unbound in solution so that the effect on the rate of aggregation by sequestration of unfolded polypeptides is negligible. Second, in some cases GroEL rescues stable off-pathway folding intermediates, which neither readily fold nor form irreversible aggregates by themselves. By this process, the chaperonin substantially increases the rate of folding (Todd *et al.,* 1994). How chaperonins accelerate folding without having any specific knowledge of the structure of the substrate is unclear. One possibility is that chaperonins might actively unfold aggregation-prone or kinetically trapped intermediates.

Cooperation between Different Chaperones during Protein Folding

The different classes of chaperones are ubiquitous proteins in a single compartment, are apparently unrelated, and have a broad substrate specificity. They might therefore compete with each other for the same substrate and assist the folding of similar proteins. Indeed, several chaperones are not essential for the cell; for example, mutations in DnaK are not lethal at normal temperatures. Apparently, the lack of function of this chaperone can be overcome by expression of a different, unrelated, chaperone. At elevated temperatures, DnaK becomes essential for life, suggesting that under these harsh conditions the prevention of protein aggregation requires the full potential of all chaperones in the cell. Other chaperones, however, are essential at all temperatures (e.g., GroEL, Hsp60, and Hsp70 of the mitochondrial matrix). This suggests that they perform a unique function that cannot be mastered by another chaperone.

For the mitochondrial Hsp70 this unique function is obvious since it drives the import of all mitochondrial matrix proteins, including the essential ones. The chaperonins, however, are involved in the prevention of aggregation or assistance of folding of unfolded proteins. Apparently, they perform an essential step in this process that cannot be fulfilled by another chaperone.

The substrate specificity of different classes of chaperones should therefore be partly different. Indeed, whereas the Hsp70 class of chaperones primarily interact with proteins in their fully denatured state, Hsp60 chaperonins have been reported to interact with secondary structure elements and intermediates on the protein folding pathway, notably the molten globule state. In contrast to Hsp70s, chaperonins do not recognize short peptides and polypeptides in extended conformations with high affinity. Both chaperones might therefore act sequentially in the folding of newly synthesized or translocated proteins.

Folding of Mitochondrial Preproteins. During the biogenesis of mitochondrial matrix proteins, precursor proteins are translocated across the mitochondrial membranes in an unfolded state. On the matrix side the polypeptide chain interacts directly with mitochondrial Hsp70 (Ssc1p in yeast). This interaction was shown to be only transient (Manning-Krieg *et al.,* 1991); folding to a protease-resistant conformation occurred after a lag and was much slower than the release from Hsp70. Binding of an imported protein to mitochondrial Hsp60 was delayed compared to binding to mtHsp70. A single preprotein was shown to bind first to mtHsp70 and then to mtHsp60 and folded into its native state only after release from the chaperonin. Release of polypeptides from mtHsp70 and binding to mtHsp60 both require ATP.

The following model for the folding of mitochondrial proteins can therefore be envisioned. During translocation the extended polypeptide first interacts with mtHsp70 in an ATP-dependent manner. After release from Hsp70, some proteins do not fold immediately into their native state but instead are delivered to Hsp60, which mediates the further folding. Hsp70 and Hsp60 therefore have a partly nonoverlapping, sequential function in the folding of mitochondrial proteins. The influence of the different cochaperones in the mitochondrial matrix on this folding process until recently has been less clear. It must be emphasized that not all proteins necessarily require the presence of Hsp60 for folding since several fold into their native state after release from Hsp70.

Folding of E. coli Proteins. A similar situation might exist for the folding of proteins in the *E. coli* cytoplasm. Protein aggregation in a mutant lacking heat shock proteins is prevented by overexpression of either the GroEL–GroES or DnaK–DnaJ chaperone machines. However, at physiological concentrations the expression of all four heat shock proteins is required, indicating that *in vivo* both chaperone machines are required at elevated temperatures. This also suggests that in *E. coli* the two chaperone machines in some instances are nonoverlapping in their specificity or function.

The role of protein folding of both chaperone machines has been reconstituted *in vitro* with isolated components from *E. coli*. Refolding of denatured rhodanese was shown to depend on both the DnaK–DnaJ–GrpE and GroEL–GroES machines (Langer *et al.*, 1992). DnaK prevents the aggregation of refolding rhodanese in the presence of both DnaJ and ATP. Only after release of ADP from DnaK by GrpE does ATP binding to DnaK release the unfolded protein permitting its transfer to GroEL for folding to the native state. The basis for this sequential action of Hsp70 and Hsp60 appears to be the differential specificity of these chaperones for structural features of a polypeptide chain. Hsp70s preferentially recognize extended peptide segments enriched in hydrophobic residues, whereas the chaperonins were reported to stabilize proteins in a conformation that represents the molten globule state. As in the case of the folding of mitochondrial preproteins, in the case of the *E. coli* chaperone machines this route from DnaK to GroEL is not absolute; some proteins can fold in the absence of GroEL.

It has been speculated that this sequence of events also represents the folding of new polypeptide chains as they emerge from the ribosome during synthesis. However, the situation with nascent chains seems to be more complicated. The first proteins with which a nascent chain interacts are NAC (nascent chain-associated complex) in eukaryotes and trigger factor in *E. coli*, which are ribosome associated and do not show any polypeptide specificity (Rassow and Pfanner, 1996). Next, targeting sequence-containing polypeptides are transferred to SecB in *E. coli* or, for example, SRP (signal recognition particle for targeting to the ER membrane) in eukaryotes. Polypeptides lacking targeting sequences might then be assisted in folding by the DnaK–DnaJ and GroEL–GroES machines discussed previously. However, a clear picture of the cooperation of different chaperones, the general and the specific, in the folding or targeting of newly synthesized proteins is lacking.

Since the chaperone machines are conserved in the compartments of most organisms, the folding pathway described for mitochondria and *E. coli* might represent a general mechanism for the prevention of aggregation and the stimulation of folding of proteins. Even in the eukaryotic cytosol, although lacking direct homologs of Hsp60 and GrpE, the sequence of folding events might be conserved. In this case, however, the function of Hsp60 might be performed by TRiC and that of GrpE by Hip. The specific function of other chaperones and heat shock proteins is currently less well understood. Although some of them might perform specialized functions, such as SecB in the prevention of folding of preproteins in *E. coli*, others might partly substitute for the action of the Hsp70 or Hsp60 chaperone machines. Some proteins fail to fold even though they bind and are released from GroEL, indicating that the Hsp70–Hsp60 pathway is not applicable to the folding of all polypeptides. Future research will be aimed at the elucidation of the mechanism of chaperone action and will hopefully resolve the question of whether chaperones are really involved in active protein folding or merely prevent aggregation and thereby increase the yield of correctly folded proteins.

Acknowledgments

The work in the authors' laboratory is supported by funds from the Deutsche Forschungs-gemeinschaft (Schwerpunktprogramm "Molekulare Zellbiologie der Hitzestreßantwort") and the Fonds der Chemischen Industrie (N.P.). P.J.T.D. is supported by a fellowship from the Human Frontiers Science Program.

References Cited

ANFINSEN, C. B. (1973). Principles that govern the folding of protein chains. *Science* **181**, 223–230.

BRAIG, K., OTWINOWSKI, Z., HEGDE, R., BOISVERT, D. C., JOACHIMIAK, A., HORWICH, A. L., and SIGLER, P. B. (1994). The crystal structure of the bacterial chaperonin GroEL at 2.8 Å. *Nature* **371**, 578–586.

CREIGHTON, T. E. (1990). Protein folding. *Biochem. J.* **270**, 1–16.

CREIGHTON, T. E. (1995). Protein folding. An unfolding story. *Curr. Biol.* **5**, 353–356.

CYR, D. M., LANGER, T., and DOUGLAS, M. G. (1994). DnaJ-like proteins: Molecular chaperones and specific regulators of Hsp70. *Trends Biochem. Sci.* **19**, 176–181.

FISCHER, G., AND SCHMID, F. X. (1990). The mechanism of protein folding. Implications of *in vitro* refolding models for *de novo* protein folding and translocation in the cell. *Biochemistry* **29**, 2205–2212.

FLAHERTY, K. M., DeLUCA-FLAHERTY, C., and McKAY, D. B. (1990). Three-dimensional structure of the ATPase fragment of a 70 k heat-shock cognate protein. *Nature* **346**, 623–628.

FREEDMAN, R. B. (1994). Protein folding: Folding helpers and unhelpful folders. *Curr. Biol.* **4**, 933–935.

GEORGOPOULOS, C. (1992). The emergence of the chaperone machine. *Trends Biochem. Sci.* **17**, 295–299.

GEORGOPOULOS, C., and WELCH, W. J. (1993). Role of the major heat shock proteins as molecular chaperones. *Annu. Rev. Cell Biol.* **9**, 601–634.

LANGER, T., LU, C., ECHOLS, H., FLANAGAN, J., HAYER, M. K., and HARTL, F. U. (1992). Successive action of DnaK, DnaJ and GroEL along the pathway of chaperone-mediated protein folding. *Nature* **356**, 683–689.

MANNING-KRIEG, U. C., SCHERER, P. E., and SCHATZ, G. (1991). Sequential action of mitochondrial chaperones in protein import into the matrix. *EMBO J.* **10**, 3273–3280.

MAYHEM, M., DA SILVA, A. C. R., MARTIN, J., ERDJUMENT-BROMAGE, H., TEMPST, P., and HARTL, F.-U. (1996). Protein folding in the central cavity of the GroEL–GroES chaperonin complex. *Nature* **379**, 420–426.

McCARTY, J. S., BUCHBERGER, A., REINSTEIN, J., and BUKAU, B.

(1995). The role of ATP in the functional cycle of the DnaK chaperone system. *J. Mol. Biol.* **249**, 126–137.

PFANNER, N., and MEIJER, M. (1995). Protein sorting. Pulling in the proteins. *Curr. Biol.* **5**, 132–135.

PTITSYN, O. B. (1995). How the molten globule became. *Trends Biochem. Sci.* 20, 376–379.

RANDALL, L. L. and HARDY, S. J. SS. J. (1995) *Trends Biochem. Sci.* **20**, 65–69.

RASSOW, J., and PFANNER, N. (1996). Protein biogenesis: Chaperones for nascent polypeptides. *Curr. Biol.* **6**, 115–118.

THOMAS, P. J., QU, B.-H., and PEDERSEN, P. L. (1995). Defective protein folding as a basis of human disease. *Trends Biochem. Sci.* **20**, 456–459.

TODD, M. J., VIITANEN, P. V., and LORIMER, G. H. (1994). Dynamics of the chaperonin ATPase cycle: Implications for facilitated protein folding. *Science* **265**, 659–666.

WEISSMAN, J. S. (1995). All roads lead to Rome? The multiple pathways of protein folding. *Chem. Biol.* **2**, 255–260.

WEISSMAN, J. S., HOHL, C. M., KOVALENKO, O., KASHI, Y., CHEN, S., BRAIG, K., SAIBIL, H. R., FENTON, W. A., and HORWICH, A. L. (1995). Mechanism of GroEL action: Productive release of polypeptide from a sequestered position under GroES. *Cell* **83**, 577–587.

WEISSMAN, J. S., RYE, H. S., FENTON, W. A., BEECHEM, J. M., and HORWICH, A. L. (1996). Characterization of the active intermediate of a GroEL–GroES-mediated protein folding reaction. *Cell* **84**, 481–490.

ZHU, X., ZHAO, X., BURKHOLDER, W. F., GRAGEROV, A., OGATA, C. M., GOTTESMAN, M. E., and HENDRICKSON, W. A. (1996). Structural analysis of substrate binding by the molecular chaperone DnaK. *Science* **272**, 1606–1614.

General References

CREIGHTON, T. E. (1993). *Proteins: Structures and Molecular Properties.* Freeman, New York.

ELLIS, R. J., and VAN DER VIES, S. M. (1991). Molecular chaperones. *Annu. Rev. Biochem.* **60**, 321–347.

GETHING, M.-J., and SAMBROOK, J. (1992). Protein folding in the cell. *Nature* **355**, 33–45.

HENDRICK, J. P., and HARTL, F. U. (1993). Molecular chaperone functions of heat-shock proteins. *Annu. Rev. Biochem.* **62**, 349–384.

MORIMOTO, R. I., TISSIÈRES, A., and GEORGOPOULOS, C. (Eds.) (1994). *The Biology of Heat Shock Proteins and Molecular Chaperones.* CSHL Press, New York.

PARSELL, D. A., and LINDQUIST, S. (1994). In *The Biology of Heat Shock Proteins and Molecular Chaperones* (R. I. Morimoto, A. Tissières, and C. Georgopoulos, Eds.), pp. 457–494. CSHL Press, New York.

MARK HOCHSTRASSER

Yale University
Department of Molecular Biophysics
and Biochemistry
266 Whitney Avenue
New Haven, Connecticut 06520-8114, USA

Intracellular Protein Degradation*

Different proteins are degraded at vastly different rates in both eukaryotic and prokaryotic cells, with half-lives ranging from seconds to years. Rapid intracellular turnover is an important characteristic of many regulatory proteins, and the rate of degradation is often modulated by signaling pathways, growth conditions, or cell cycle stage. Among the short-lived regulatory proteins characterized to date are those controlling cell cycle progression, a variety of transcription factors, and growth regulators such as the p53 tumor suppressor and various protooncogene products. In almost all of these examples, it is the ubiquitin–proteasome pathway that is responsible for proteolysis. Understanding of this pathway has increased tremendously during the past decade and will be the focus of this article.

Introduction

Cells often switch from one cellular state to another, either in response to environmental cues or as part of regulated developmental pathways. Such switches generally require rapid dismantlement of an existing regulatory network, a process frequently dependent on protein degradation (Hochstrasser, 1996). The specificity of intracellular protein degradation must be extremely high to avoid mistargeting of essential proteins or degradation of proteins at inappropriate times. Important phenotypic alterations may result even from small changes in the rates of protein turnover. Many protooncogene products, for instance, are very

*This review is an expanded version of an article written for *Ann. Rev. Genet.* **30**, 405–439 (1996) by M. Hochstrasser.

short-lived *in vivo,* and relatively small increases in their intracellular concentrations can be tumorigenic.

Selective protein turnover offers several advantages over other kinds of genetic regulatory mechanisms (and frequently functions in concert with other controls). One advantage is speed. The shorter a protein's half-life, the shorter the time it takes to reach a new steady state level following a change in its rate of synthesis. A second advantage is irreversibility. Elimination of a protein removes any chance of its being reactivated inappropriately. These features help explain why selective protein degradation is almost always a component of regulatory mechanisms that involve timing controls. Examples range from cell cycle progression, circadian rhythms, and various signal transduction pathways to cell lineage specification, metabolic control, and embryogenesis (Hochstrasser, 1996; Gottesman and Maurizi, 1992;

Deshaies, 1995). The price paid for such a sensitive and rapid mechanism of regulating protein levels is the relatively large amount of energy consumed, principally for resynthesizing the destroyed proteins (Gottesman and Maurizi, 1992). However, only a small fraction of cell proteins — almost all of which are key regulatory proteins — normally undergo rapid and continuous turnover in the cell.

Work during the past two decades has demonstrated that a single highly complex proteolytic pathway is responsible for much of the selective degradation of soluble cellular proteins under most conditions. This pathway is the ubiquitin–proteasome pathway. Development in the 1970s of a cell-free lysate from rabbit reticulocytes capable of ATP-dependent degradation of specific protein substrates allowed the biochemical dissection of this pathway to be initiated (Hershko and Ciechanover, 1992). This work identified covalent modification of substrate by ubiquitin as a critical step in degradation and established the basic enzymological framework of protein ubiquitination and breakdown. The widespread impact of the ubiquitin-dependent proteolytic system on various physiological processes was made clear first in mammalian cells and then in more extensive studies in yeast (Hochstrasser, 1996).

Although lysosomes/vacuoles are important degradative organelles, particularly under stress conditions, most of the proteolysis of cytosolic proteins that occurs in lysosomes is relatively nonspecific. Such proteins generally gain access to the hydrolytic enzymes within the lysosome by autophagic processes that engulf random portions of cytoplasm. There is evidence for more specific uptake mechanisms as well (Chiang and Dice, 1988), but it is likely that these will account for only a small percentage of the specific proteolysis that occurs *in vivo* (on the other hand, many membrane proteins may be targeted to the vacuolar/lysosomal compartment by ubiquitin-dependent endocytic routes). Of course, there are many other proteolytic enzymes within cells, and much attention is being trained on some of them (e.g., the calpains and the ICE-like proteases crucial for apoptosis), but here again the existing evidence suggests a more circumscribed role, being limited either in the range of substrates attacked or in the range of conditions under which they operate at significant rates.

There has been an explosion of information on the ubiquitin system during the past several years (Hochstrasser, 1996). In this article, I will emphasize the mechanistic and physiological logic of the ubiquitin–proteasome pathway using some of the more well-developed model systems for illustration. The theme will be the intricate, multistep nature of ubiquitin-dependent proteolysis. It appears that the root of this complexity is the need both for extremely high substrate specificity and for the ability to alter quickly the rates of proteolysis in response to or to engender rapid changes in cell state, development, or the cellular environment.

Overview of the Ubiquitin–Proteasome Pathway

As stated previously, for many short-lived eukaryotic proteins conjugation to the polypeptide ubiquitin is an obligatory step in their degradation. Ubiquitin is joined reversibly to proteins by covalent (isopeptide) linkage between the carboxyl terminus of ubiquitin and lysine ε amino groups of the acceptor proteins. A simplified view of the ubiquitin pathway, which is highly conserved among diverse eukaryotes, is depicted in Fig. 1. The C terminus of ubiquitin must be activated before it can form isopeptide bonds with other proteins (Pickart, 1988). Initially, ubiquitin is adenylated by the ubiquitin-activating enzyme (E1). The ubiquitin–AMP intermediate of the ternary complex is then attacked by a sulfhydryl group of the enzyme, yielding an E1–ubiquitin thiolester. The activated ubiquitin is then passed to one of a large number of distinct ubiquitin-conjugating (Ubc or E2) enzymes by transthiolation. The E2 proteins catalyze substrate ubiquitination either alone or in conjunction with a ubiquitin–protein ligase (E3). As will be described later, recent data suggest E3 proteins may have a more direct mechanistic role in ubiquitin transfer to substrate. For proteolytic substrates, assembly of a ubiquitin chain(s) on the protein is generally observed, which appears to accelerate degradation relative to monoubiquitination. Ubiquitinated proteins are in a dynamic state, subject to further rounds of ubiquitin addition, ubiquitin removal by deubiquitinating enzymes, or degradation by a complex multicatalytic proteinase called the 26S proteasome (Fig. 1). The proteasome breaks down targeted substrates to short peptides but recycles the ubiquitin molecules.

Protein Ubiquitination

The initial and often most important factor that determines the probability of a particular protein within the cell becoming degraded by the proteasome is its modification by ubiquitin. As outlined in the preceding section, protein ubiquitination depends on the action of a series of distinct enzymes. Despite the wealth of enzymological studies, however, there is still only a dim understanding of the mechanics of ubiquitin transfer among these enzymes, ubiquitin chain assembly, and ubiquitin ligation to protein substrates.

Ubiquitin-Activating Enzyme (E1)

Activation of the α-carboxyl group of ubiquitin by E1 is an obligatory step in all processes dependent on ubiquitin ligation. Insofar as ubiquitin is essential for cell viability (Finley *et al.,* 1994), it was not surprising to learn that E1

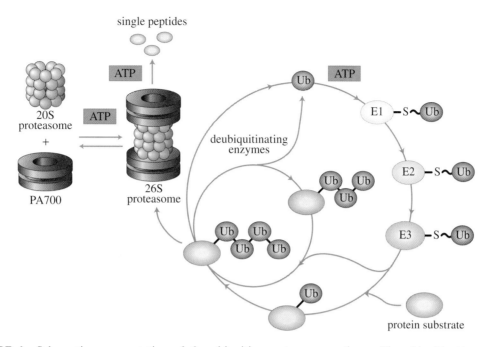

FIGURE 1 Schematic representation of the ubiquitin–proteasome pathway. The ubiquitinating enzymes E1–E3 transfer ubiquitin (red) to the protein substrate. In the presence of ATP, E1 forms a thioester with ubiquitin, which is then transferred to E2 and E3 by transthiolation. Additional ubiquitin molecules are added to the monoubiquitinated protein by repetition of the ubiquitination process represented by the external circle. The polyubiquitinated chains are dynamic structures since they are rapidly modified (internal circle) by ubiquitinating (E1–E3) and deubiquitinating enzymes. Once polyubiquitinated, the protein is degraded by the 26S proteasome, formed from the 20S proteasome and PA700 subunits.

is essential as well. Many conditionally lethal alleles of E1-encoding genes have been isolated in mammalian cell lines, and a gene encoding a yeast E1, *UBA1*, was also found to be essential for growth. However, there are two additional genes found in the yeast genome, which has been completely sequenced, that encode fairly divergent E1-like polypeptides. These proteins share a nucleotide-binding consensus sequence and a conserved Cys residue thought to form a thiolester with ubiquitin. One of these genes, *UBA2*, was shown to be required for viability (Dohmen *et al.,* 1995). Mutation of the conserved Cys residue in Uba2 destroyed the essential function(s) of the protein, although attempts to demonstrate thiolester formation between wild-type Uba2 and ubiquitin were unsuccessful.

A growing family of ubiquitin-related proteins is being ncovered in yeast and other organisms. Many do not appear to have the canonical sequence elements thought to be required for activation by E1 and/or for cleavage from larger proteins by deubiquitinating enzymes. However, a number may be posttranslationally conjugated to other proteins. The best studied of these is the mammalian ubiquitin cross-reactive protein, UCRP (Narasimhan *et al.,* 1996). Pre-UCRP comprises a head-to-tail pair of ubiquitin-like

sequences followed by a short C-terminal peptide tail. This tail is removed from UCRP, allowing UCRP to be activated and covalently ligated to proteins. Thiolester formation with purified mammalian E1 is very inefficient, however; this and other data suggest that a UCRP-specific activating enzyme must exist (Narasimhan *et al.,* 1996). Different members of the E1 enzyme family may have specialized functions, perhaps preferentially transferring ubiquitin or ubiquitin-like proteins to particular E2 enzymes or directly to certain substrate proteins. Distinct E1s may also reside in separate cellular compartments.

Ubiquitin-Conjugating Enzyme (E2)

In the presence of ubiquitin and ATP, E1 enzyme will form a thiolester with ubiquitin. In contrast, the E2 proteins will not forge thiolester linkages with ubiquitin in the absence of E1. Hershko and colleagues (1983) showed that ubiquitin is transferred from E1 to E2 by transthiolation. Fractionation of either rabbit reticulocyte or yeast extracts revealed the presence of multiple E2 isoforms (Hershko *et al.,* 1983). Subsequent genetic analyses confirmed that these are encoded by a family of related genes. In yeast, 13 E2-related enzymes can be recognized in the genome. Mutations in many of them lead to distinct phenotypes, indicating that E2 proteins have different functions and presumably different substrate specificities. Degradation of several short-lived proteins has now been shown to depend on particular E2 enzymes, and for some of them multiple E2 isozymes are necessary for wild-type rates of degradation *in vivo*. This is clearest in the case of the MATα2 transcriptional repressor. Four enzymes participate in two distinct α2 ubiquitination pathways, one involving a complex between the Ubc6 and Ubc7 E2 proteins (Hochstrasser, 1996). Interestingly, mutants lacking Ubc6 or Ubc7 have disparate phenotypes, indicating that these enzymes work independently in some cellular processes. These observations led to the idea that the ubiquitin conjugation system could expand its repertoire of substrate specificities by association of a limited set of E2 (and E3) proteins into multiple heterooligomeric complexes.

Whereas the genetic data in yeast indicate multiple E2 enzymes may be necessary for maximal degradation of a substrate, analysis of protein ubiquitination *in vitro* has suggested that different E2 enzymes may sometimes serve redundant functions in the ubiquitination of a particular substrate (Hochstrasser, 1996). Any one of at least three different E2 enzymes, which do not all appear to be part of the same E2 sequence subfamily, is able to ubiquitinate the p53 tumor suppressor protein *in vitro*. For mitotic cyclin ubiquitination *in vitro*, either Ubc4 or UBCx/E2-C is sufficient. Ubiquitination by either of these enzymes is kinetically similar and is dependent on the substrate having an intact cyclin destruction box, a sequence element necessary for cell cycle-regulated degradation; in both cases, ubiquitination also requires a large protein complex, called the anaphase promoting complex (APC) or cyclosome. Interestingly, the pattern of ubiquitin–cyclin conjugates generated with the different E2s differs, with UBCx generating conjugates with fewer ubiquitin molecules than Ubc4. Although no synergistic effects on ubiquitination were observed from mixing the E2s, it remains possible that the two enzymes play distinct roles in cyclin ubiquitination.

Whether the E2s implicated in cyclin B ubiquitination *in vitro* function *in vivo* in these overlapping ways is uncertain. Perhaps both are necessary or possibly neither is necessary. *In vivo* degradation of two yeast B cyclins, Clb2 and Clb5, was found to be inhibited by conditional mutations in an E2 enzyme, Ubc9, that is distinct from both Ubc4 and E2-C. Ubc9 could not function in place of Ubc4 or UBCx in the *Xenopus in vitro* system. Conversely, there are no compelling data implicating Ubc4 (or the closely related Ubc5 enzyme) in yeast cyclin degradation. Yeast *ubc4 ubc5* strains grow very poorly and have a pleiotropic phenotype which may obscure the cell division cycle arrest expected from a failure to degrade cyclin.

Ubiquitin–Protein Ligase (E3)

The component of the ubiquitin conjugation system that is generally thought to be the most directly involved in substrate recognition is E3. At the same time, the E3 enzymes are the least well-understood factors involved in ubiquitin–protein ligation. First, the E3 ligases are likely to fall into distinct groups, at least in terms of sequence class and possibly with regard to mechanism as well. Indeed, exactly what constitutes an E3 is a matter of debate. The first characterized E3, E3α from rabbit reticulocytes, was defined (Hershko *et al.*, 1983) as a factor that (i) stimulates substrate ubiquitination when combined with E1 and the appropriate E2, (ii) binds to E2, and (iii) binds to substrate (Hershko and Ciechanover, 1992). Experiments with yeast Ubr1, the likely counterpart to E3α, suggest the yeast enzyme has these properties as well. On the other hand, studies on the stimulation of p53 degradation *in vitro* by the oncogenic human papillomavirus protein E6 have led to the identification of a novel protein, E6-associated protein (E6-AP), that, together with E6, binds to p53 and triggers p53 ubiquitination (Scheffner *et al.*, 1995). E6-AP has not been shown to bind any E2 enzymes, at least not when measured by coimmunoprecipitation assays. As such, E6-AP has not met all the formal criteria of an E3 as used to define E3α. Unexpectedly, E6-AP was found to form a thiolester with ubiquitin, and this thiolester appears to serve as an intermediate in the transfer of ubiquitin from E2 to substrate (Scheffner *et al.*, 1995). E6-AP must therefore be able to interact with E2 at least transiently. One can imagine that for some ubiquitination pathways, E3–substrate interactions may be correspondingly transient or weak so that again it might be difficult to meet the formal definition of an E3 experimentally. Another interesting example of an enzyme that does not conform simply to the previous definition of an E3 is a large protein identified in rabbit reticulocytes that actually has properties that suggest it may combine features of both E2 and E3 enzymes (Berleth and Pickart, 1996).

The family of known or suspected E3s has begun to expand significantly during the past several years. The C-terminal, ~350-residue region of E6-AP shows similarity to at least a dozen proteins in organisms ranging from yeast to mammals and was named the hect domain (*h*omologous to *E*6-AP *c*arboxyl *t*erminus) (Huibregtse *et al.*, 1995). The hect domain includes an absolutely conserved cysteine, which in the case of E6-AP is essential for ubiquitin–thiolester formation and substrate ubiquitination. Several other proteins of this class have been shown to form thiolesters with ubiquitin (Huibregtse *et al.*, 1995), as has yeast Ubr1 (Hochstrasser, 1996), and have been implicated in the ubiquitination of specific proteins.

Other, quite different proteins may also turn out to facilitate protein ubiquitination, and as such may be regarded as E3-like factors, at least until a clearer mechanistic basis for defining E3s can be found (Hochstrasser, 1996). The cyclosome/APC has properties of an E3, but none of the three identified subunits of the complex, which has at least eight different subunits, bears similarity to known E3s (King *et al.,* 1995). Although the yeast Rad6 E2 enzyme, which is important for DNA repair, combines with the Ubr1 protein to ubiquitinate substrates of the N-end rule pathway, Ubr1 plays no known role in DNA repair. For its function in repair, Rad6 forms a distinct complex with Rad18, a DNA-binding protein. It is therefore possible that Rad18 operates by an E3-like mechanism, bringing Rad6 to chromatin sites bearing DNA lesions, perhaps to help clear out histones or other chromatin proteins and thereby provide access to repair enzymes. Rad18 bears no similarity to Ubr1 and lacks a hect domain. It would not be surprising if other novel E3-like factors were to be identified during the next few years as well. Such factors may include, for example, certain heat shock proteins, which have long been suspected to facilitate degradation of denatured or misfolded proteins.

Mechanism and Regulation of Protein Ubiquitination

Although it is clear that several distinct enzymes are required for ubiquitinating protein substrates, precisely how the substrate is recognized and how it becomes multiubiquitinated remain uncertain. Early *in vitro* experiments suggested the E3 ubiquitin–protein ligase directly recognized the substrate and functioned as an adaptor-like molecule, juxtaposing the thiolester-linked ubiquitin–E2 complex and the substrate (Hershko and Ciechanover, 1992). However, the ability of the E3 to form a thiolester with ubiquitin had not been detected in these studies. Moreover, it appears that some E3s (e.g., E6-AP/E6) do not form tight complexes with their cognate E2s. These results suggest that E3s do not function simply as adaptor molecules. The E2 enzymes may also contribute to substrate binding since at least some of them can ubiquitinate substrate in the absence of E3. Hence, it may be more appropriate to regard the entire ubiquitination complex, which includes E2- and E3-like proteins (and perhaps even E1), as the recognition factor, rather than just the E3 protein (Fig. 2).

The ubiquitination complex may either assemble a ubiquitin chain(s) directly on the substrate by repeated additions of ubiquitin units or transfer a preassembled chain onto the targeted protein. Both modes of multiubiquitin–protein conjugate formation have been observed *in vitro* (Chen and Pickart, 1990). However, it has been argued that, because the intracellular concentration of free ubiquitin is much greater than that of unanchored ubiquitin chains, transfer of preformed ubiquitin chain to substrate *in vivo* may be negligible relative to the transfer of single ubiquitin moieties (Chen and Pickart, 1990). There may be ubiquitination complexes that prefer ubiquitin chains over free ubiquitin in such transfer reactions, but this has not been observed.

What distinguishes an efficiently ubiquitinated protein from proteins that are rarely if ever so modified? This remains a fundamental but largely unanswered question. It is becoming clear that proteolytic substrates of the ubiquitin system bear sequence elements or "signals" that target them for rapid turnover and that such signals differ from substrate to substrate. Analysis of artificial model substrates (Varshavsky, 1992) revealed that at least some degradation signals contain separable determinants — one that functions as a binding site for a substrate recognition factor and another that serves as a ubiquitin ligation site. This was shown most dramatically by placing these distinct determinants on different subunits in a heteromeric protein (Johnson *et al.,* 1990). For many natural substrate proteins, the position of ubiquitin ligation sites (generally, ubiquitinatable lysine side chains) has proven to be surprisingly plastic. For example, the ζ subunit of the transmembrane T cell antigen receptor (TCR) can be ubiquitinated on any of multiple intracellular lysines in response to a variety of stimuli; the introduction of Lys residues at positions not normally occupied by this residue can also restore ubiquitination of an otherwise Lys-free TCR ζ chain (Hou *et al.,* 1994).

Putative recognition sites for ubiquitinating enzymes are just beginning to be defined. The first example was the N-terminal region of certain artificial protein substrates (Varshavsky, 1992). The half-lives of these proteins show a remarkable dependence on the nature of the amino-terminal residue, an effect called the "N-end rule" (Varshavsky,

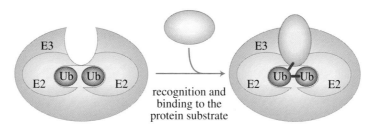

FIGURE 2 Substrate recognition by the ubiquitin conjugation system. The figure depicts the possible binding between the substrate (in green) and both the E2 and E3 enzymes.

1992). Peptide inhibition studies suggested that the E3 factor directly recognizes the N-terminal residue. Endogenous substrates recognized in this way have not been reported.

A genetic screen in yeast for random peptides that yield relatively short-lived β-galactosidase fusion proteins suggested there are several kinds of proteolytic recognition motifs, including hydrophobic and amphipathic sequence elements (Sadis *et al.*, 1995). Interestingly, the amphipathic degradation signal described in this study bears some resemblance to a ubiquitin-dependent degradation signal called *Deg1* in the yeast MATα2 repressor (Chen *et al.*, 1993), and substrate degradation depends on the same E2 enzymes. Random mutagenesis of the *Deg1* signal also indicates the importance of an amphipathic sequence element (L. Rakhilina and M. H., unpublished data); moreover, α2 degradation was recently found to be strongly reduced in **a**/α cells, apparently due to binding of the **a**1 protein to α2, which is predicted to block access of *Deg1* to the ubiquitination complex (P. Johnson and M. H., unpublished data) (Table 1).

Another example of a conditionally active degradation signal comes from studies of mitotic cyclin turnover. Many of the B cyclins are degraded only in mitosis and (at least in yeast) in the early G_1 phase of the cell cycle. A moderately conserved nine-residue sequence within the N-terminal cyclin degradation signal, the destruction box, is required for cyclin ubiquitination. How cyclin stability is regulated as a function of cell cycle progression is currently a subject of intense study. Biochemical fractionation studies from two different groups indicate it is the cyclosome/APC, an E3-like factor, whose activity is cell cycle regulated (King *et al.*, 1995; Sudakin *et al.*, 1995). Cyclosome activity in clam egg extracts is activated by Cdc2 kinase-dependent phosphorylation and can be inactivated by an okadaic acid-sensitive phosphatase. Because Cdc2 kinase activity requires a Cdc2–cyclin heterodimer, activation of the cyclosome, which leads to the elimination of cyclin, will eventually also turn off the Cdc2 kinase (and then the cyclosome). Which component of the multisubunit cyclosome is cell cycle regulated is not known. Additional controls are likely to exist at the level of the substrate. This is suggested both by the finding that degradation of some B cyclins depends on their ability to inter-

TABLE I			

Examples of Proteolytic Substrates of the Ubiquitin System

SUBSTRATE	**FUNCTION**	**UBIQUI-TINATED?**	**REGULATION OF TURNOVER**
Phytochrome	plant light-responsive regulator	Yes	Red light stimulates turnover
MATα2	transcription factor	Yes	Regulated by cell-type
G2 cyclins	cell cycle regulation	Yes	Turnover is cell-cycle dependent
p53	tumor suppressor	Yes	HPV E6 protein stimulates turnover
Mos	oocyte Ser/Thr protein kinase	Yes	Phosphorylation inhibits turnover
Ornithine decarboxylase	synthesis of polyamines	No	Polyamines stimulate turnover
IgE,T cell receptors	immune system surface receptors	Yes	Ubiquitination regulated by ligand
Ste2	pheromone receptor	Yes	Ubiquitination important for endocytosis pheromone binding stimulates endocytosis
Gα	signal transduction	Yes	Unknown
NFκB, IκB	transcription regulators	Yes	Modulated by a variety of factors
c-jun	transcription factor	Yes	Turnover depends on δ element
Fructose bisphophatase	gluconeogenesis (yeast)	Yes	Turnover induced by glucose
Sic1	inhibitor of Cln-CDC28 (yeast)	ND	Turnover is cell-cycle dependent
Itr1	inositol permease (yeast)	ND	Inositol stimulates turnover§
Gcn4	transcription factor (yeast)	Yes	Stabilized by amino acid starvation
Cln2,3	G1 cyclins (yeast)	Yes	Turnover depends on phosphorylation
CFTR	cystic fibrosis transmembrane conductance regulator	Yes	Common mutation in cystic fibrosis patients causes increased ubiquitination & degradation

ND, not determined; §, Pat McGraw, personal communication.

act with Cdc2 and by the fact that degradation of A cyclins is destruction box dependent but shows a cell cycle dependency distinct from that of B cyclins (Deshaies, 1995). Furthermore, a destruction box-like sequence has recently been shown to be critical for ubiquitination of uracil permease in yeast. Ubiquitination of the permease is stimulated under stress conditions but does not appear to be cell cycle regulated (Hein *et al.,* 1995). Interestingly, ubiquitination depends on Npi1/Rsp5, an essential E3-like hect protein. Whether this E3 protein is a component of the yeast cyclosome, which has similarities to the vertebrate cyclosome, is not known. Based on available evidence, its seems likely that different destruction box-containing proteins are targets for distinct ubiquitination factors, which may or may not contain common components.

The meiotic cell cycle regulatory protein Mos has been shown to be degraded by the ubiquitin system in meiosis I during oogenesis and, later, upon egg activation or fertilization (Nishizawa *et al.,* 1993). Later in oogenesis, ubiquitination of the Mos kinase is inhibited by autophosphorylation of a conserved Ser residue (Ser3). The Pro residue immediately preceding this Ser was shown to be critical for Mos ubiquitination, particularly during egg activation, whereas the predominant site of ubiquitin ligation is Lys34. In this case, there is evidently much less flexibility in the position of the ubiquitin acceptor site than was observed, for example, with the ζ subunit of TCR. For some substrates, a more rigid framework for recognition or modification by the ubiquitination complex may exist. This may be advantageous when ubiquitination must be modulated by posttranslational modification, as is true for Mos.

Support for this last idea comes from the analysis of IκBα ubiquitination. IκBα is phosphorylated in response to various signals at two specific serines near its N terminus, Ser32 and Ser36; this leads to its ubiquitination and subsequent degradation (Thanos and Maniatis, 1995). Two adjacent lysines, residues 21 and 22, are the major targets for IκBα ubiquitination. Thus, as with Mos, IκB ubiquitination is largely limited to a single preferred site and is controlled by phosphorylation, although the control is positive rather than negative. Phosphorylation-dependent degradation of the yeast Cln3 protein, a G_1 cyclin, has also been suggested (Deshaies, 1995). In this case, rapid degradation requires the Cdc34 E2 enzyme and depends on sequences rich in Pro, Glu, Ser, and Thr (PEST elements), which were previously postulated to function in protein degradation. The function of the PEST elements in degradation may be indirect; they may serve principally as preferred target sites for certain Ser/Thr kinases. Degradation of Cln3–β-galactosidase derivatives, for instance, was shown to depend on a Cdc28 kinase site within one of the Cln3 PEST elements.

It is interesting that in several of the examples cited previously, a polypeptide substrate of the ubiquitin system is part of a heteromeric complex that includes subunits which are not short-lived. This is best documented for the IκB–NF-κB complex. The IκB subunit is phosphorylated and multiubiquitinated while still bound to the NF-κB subunits and is then selectively degraded by the proteasome (Thanos and Maniatis, 1995); the NF-κB subunits escape degradation and translocate into the cell nucleus. The ability of the ubiquitin–proteasome pathway to direct subunit-specific degradation had been documented previously with engineered N-end rule substrates and MATα2 derivatives (Hochstrasser, 1996); recent data indicate this remarkable property is important for the functional remodeling of natural heteromeric proteins.

Ubiquitin Chains

Addition of multiple ubiquitins to a proteolytic substrate, usually as a ubiquitin oligomer(s), appears to stimulate proteolysis, but monoubiquitination is clearly sufficient, at least *in vitro,* for appreciable rates of proteasomal degradation of some proteins, e.g., α-globin in reticulocyte extracts (Shaeffer and Kania, 1995). On the other hand, some proteins clearly depend on multiubiquitination for rapid turnover *in vitro* (Chau *et al.,* 1989), and this is likely to be true for the majority of substrates *in vivo.* For example, some proteins may be more susceptible to deubiquitination; therefore, assembly of ubiquitin chains may be needed for them to stand a reasonable chance of binding the proteasome prior to complete deubiquitination (Fig. 3).

It also has been suggested, although there are little data on this point, that ubiquitin could trigger substrate unfolding rather than simply serving as a passive tag for proteasome targeting. A requirement for ubiquitin-assisted unfolding would also be expected to vary between substrates.

FIGURE 3 Stereo view of the structure of a tetraubiquitin chain, highlighting the β-sheets (yellow) and the α-helices (red). The translation symmetry of the structure allows extension by addition of other ubiquitin molecules. The ubiquitin oligomers formed on proteolytic substrates sometimes include several dozen ubiquitin molecules.

Ubiquitin could also facilitate substrate denaturation indirectly. The 19S regulatory complexes within the 26S proteasome are thought to unfold substrates, possibly through the action of the half-dozen or more ATPase subunits found in the complex (Rubin and Finley, 1995), in order to funnel the protein into the 20S proteasome. Substrate-attached ubiquitin chains could serve as proteasome tethers, with proteins that are difficult to denature perhaps requiring more extensive ubiquitination to keep them bound to the proteasome long enough to be fully denatured and thereby "committed" to complete breakdown.

Recent genetic and molecular studies strongly suggest that ubiquitin chains with different ubiquitin–ubiquitin linkages can be formed both *in vitro* and *in vivo* (Hochstrasser, 1996). The most common isopeptide bond between ubiquitins involves Lys48; this is the only ubiquitin–ubiquitin linkage directly verified by physical mapping (Chau et al., 1989). The most intriguing feature of the apparent "alternative" ubiquitin chains that have been seen in yeast is that mutations expected to specifically block formation of certain chain types are associated with distinct phenotypes. For instance, mutation of Lys63 to Arg in ubiquitin results in sensitivity to stress in cells overexpressing this variant and to a defect in DNA repair in cells for which this variant is the only source of ubiquitin. Thus, it is conceivable that attachment of a protein to alternative ubiquitin chains signals alternative metabolic fates. A simple idea is that only Lys48-linked chains target proteins efficiently to the 26S proteasome. Although this idea is attractive, at least three distinct kinds of ubiquitin chains can bind equally well to the S5a ubiquitin chain-binding subunit of the 26S proteasome based on far-Western assays. However, genetic analysis in yeast indicates that S5a is not the only and perhaps not even the major ubiquitin–protein binding site in the 26S proteasome; therefore, the significance of the S5a *in vitro* binding assays could be questioned.

Another interesting aspect of the studies on alternative ubiquitin chains is the finding that the same E2 enzyme can catalyze different ubiquitin–ubiquitin linkages in different circumstances. The yeast Ubc4 and Ubc5 enzymes are necessary *in vivo* for the Lys48-linked ubiquitin chains that form on certain ubiquitin–protein fusions and for the formation of unanchored Lys63-linked ubiquitin chains. *In vitro*, the type of ubiquitin chain assembled by Rad6 E2 enzyme depends on whether or not it is associated with E3.

Protein Deubiquitination

Most work on the regulation of intracellular protein turnover has focused on the enzymes that attach ubiquitin to proteins, but recent results strongly suggest that regulatory events will also center on rates of substrate deubiquitination. Ubiquitin chains that assemble on various proteins are highly dynamic, with rapid addition and removal of ubiquitin from the proteins. Ubiquitin is cleaved from substrates by enzymes known as deubiquitinating enzymes, ubiquitin carboxyl-terminal hydrolases, or ubiquitin isopeptidases. These enzymes are classified into two distinct families: a set of relatively small proteins that may preferentially cleave ubiquitin from peptides and small adducts and a group of larger proteins that can generally cleave ubiquitin from a range of protein substrates, at least *in vitro*. This latter family of enzymes, the so-called Ubp (*ubiquitin-specific processing protease*) class, is extremely divergent, but all members contain several short consensus sequences, called the Cys and His boxes, that are likely to help form the enzyme active site (Papa and Hochstrasser, 1993). Although the Cys and His boxes are the most conserved elements common to all Ubps, additional short sequences show some conservation as well (Papa and Hochstrasser, 1993; Wilkinson *et al.*, 1995).

Analysis of sequences from the various protein sequence databases has revealed that the Ubp enzyme family is remarkably large. In yeast there are 16 Ubp enzymes (Hochstrasser, 1996), which exceeds the number of E2 enzymes in this organism. Changing the rate of ubiquitin removal from a substrate will alter the probability of the multiubiquitinated intermediate being recognized by the 26S proteasome. The unexpectedly high number of deubiquitinating enzymes being uncovered therefore raises the possibility that specific protein turnover rates can be differentially regulated by these enzymes. Such regulation would imply that Ubps possess a considerable degree of substrate specificity and/or are differentially compartmentalized within the cell. Many yeast *ubp* mutants do not display striking phenotypic abnormalities, which means either many Ubps function in a very restricted set of metabolic processes (that were not tested by the standard phenotypic assays) or there may be considerable overlap in Ubp functions.

Genetic analysis of a Ubp-type deubiquitinating enzyme in *Drosophila*, the product of the *fat facets* (*faf*) gene, provides evidence in favor of the notion of substrate-specific modulation of degradation rates by Ubps (Huang et al., 1995). The *faf* gene is required for normal eye development, and *faf* mutations also have a maternal effect on embryogenesis, with embryos from homozygous mutant mothers dying at an early stage in development. Mutations in Faf of the conserved Cys and His boxes behave as null mutations in transgenic flies, suggesting that it is the deubiquitinating activity of Faf that is critical to its biological function. Most interestingly, Huang *et al.* (1995) demonstrated that several different mutant alleles of a 20S proteasome subunit gene could dramatically suppress the defect in eye development seen in *faf* flies. These data suggest that Faf functions to reverse the ubiquitination of a specific cellular protein(s), thereby preventing or slowing its degradation by the proteasome. Ubiquitin-dependent degradation of specific proteins may be regulated by changes in activity of Ubps such as Faf and/or by concomitant changes in the activity of ubiquitinating enzymes, which would shift the dynamic balance between ubiquitination and deubiquitination.

It seems reasonable to suppose that most deubiquitinating enzymes will work in the way just described, namely, as negative regulators of the ubiquitin system. However, there are several ways in which deubiquitinating enzymes may function to enhance protein ubiquitination and/or degradation. First, ubiquitin is always synthesized in precursor forms, which require the removal of C-terminal peptides or amino acids. Deubiquitinating enzymes are therefore needed to generate ubiquitin monomers. Second, activated ubiquitin can readily form adducts with abundant intracellular nucleophiles, e.g., amines or glutathione. It has been estimated that such adducts could consume the pool of free ubiquitin in less than 1 min in mammalian cells were they not rapidly broken down again. Third, long ubiquitin chains are often assembled onto intracellular proteins, in some cases perhaps even on inappropriately targeted proteins. A failure to regenerate ubiquitin from these conjugates may also deplete the intracellular pool of free ubiquitin. Finally, proteasomes and certain enzymes of the ubiquitin conjugation system (e.g., E3α) appear to bind ubiquitin chains avidly (Hershko and Ciechanover, 1992; Deveraux *et al.*, 1994). Such chains, either generated *de novo* or in the course of substrate proteolysis, may need to be rapidly disassembled to prevent extensive binding to and therefore inhibition of proteasomes or other ubiquitin system enzymes.

Evidence is accumulating for several different deubiquitinating enzymes that indicates they function *in vivo* to facilitate ubiquitin-dependent degradation by the previously described mechanism, i.e., to prevent the excessive accumulation of inhibitory ubiquitin oligomers. Mammalian isopeptidase T (isoT) is a Ubp-type enzyme that acts largely, if not exclusively, on unanchored ubiquitin chains, i.e., ubiquitin oligomers with a free ubiquitin C terminus (Wilkinson *et al.*, 1995). *In vitro* studies suggested isoT could facilitate proteasome-mediated degradation, possibly by preventing the accumulation of ubiquitin chains generated as intermediates in substrate degradation. Many E2 enzymes can also synthesize unanchored ubiquitin chains from free ubiquitin. Genetic experiments in yeast have recently provided results consistent with the *in vitro* data on isoT (A. Amerik, S. Swaminathan, K. D. Wilkinson, and M. Hochstrasser, 1997). Human isoT is the functional homolog of yeast Ubp14. The two proteins are approximately 31% identical and share similar enzymological properties. A yeast mutant lacking Ubp14 has defects in the ubiquitin-dependent degradation of many distinct proteins and shows a striking accumulation of unanchored ubiquitin oligomers. The defects of the yeast *ubp14* mutant can be reversed by expression of human isoT. That it is the excess unanchored ubiq-

uitin chains which cause the proteolytic defects is suggested by experiments in which an overabundance of ubiquitin chains were generated in otherwise wild-type yeast cells; these cells show defects in proteolysis as well.

Another yeast deubiquitinating enzyme, Doa4, has been shown to have a very broad role in ubiquitin-dependent degradation *in vivo* (Papa and Hochstrasser, 1993). Molecular and genetic analyses strongly suggest that Doa4 works in conjunction with the 26S proteasome, and purified preparations of the yeast 26S proteasome contain Doa4 (F. R. Papa and M. H. Hochstrasser, 1999). In *doa4* mutants, small ubiquitinated species accumulate; these species are all slightly larger than unanchored ubiquitin chains, suggesting that they may be the ubiquitinated proteolytic remnants of 26S proteasome action (Papa and Hochstrasser, 1993). Such remnants may collect on the protease in *doa4* cells, preventing its recycling for further rounds of protein degradation.

Several points concerning isoT/Ubp14 and Doa4 function *in vivo* are worthy of elaboration. First, given the abundance of different deubiquitinating enzymes in the cell, it is somewhat surprising that any ubiquitinated species should accumulate in *ubp14* or *doa4* mutants. This may be due to very high substrate specificity among the Ubps, but *in vitro* cleavage experiments (as well as the minimal phenotypic abnormalities of many *ubp* mutants noted previously) suggest there will be significant overlap in specificities among Ubps. An alternative explanation is based on a compartmentalization model. Specifically, ubiquitinated species may accumulate in compartments in which they are inaccessible to the bulk of the deubiquitinating enzymes in the cell. As mentioned previously, the ubiquitinated species that build up in *doa4* cells were proposed to bind the 26S proteasome in which they might be sequestered from cellular isopeptidases other than Doa4. An analogous explanation may account for the accumulation of unanchored ubiquitin chains in *ubp14* cells. This does not require that Ubp14 be an integral component of the proteasome but simply that it, unlike most other Ubps (including Doa4), can access specific sites of ubiquitin chain binding on the proteasome.

A second interesting point about Doa4 and Ubp14/isoT is that both may have as their primary substrates late intermediates of proteolysis. These products may provide a natural brake on proteasome activity in the cell, which could have several important consequences. First, by restraining the action of the 26S proteasome complex, it could confer on the protease a degree of substrate discrimination that it would not have if rates of proteolysis greatly exceeded rates of substrate ubiquitination. In fact, elimination of Doa4 or Ubp14 from yeast cells stabilizes various proteins to very different degrees, as might be predicted if ubiquitinated substrates were competing for limiting amounts of active protease. A second consequence of this feedback loop on the 26S proteasome is that changes in relative protein degradation rates could be engendered by varying the level of Doa4 or Ubp14. For instance, Doa4 levels are partially rate limiting for $\alpha 2$ degradation but not for N-end rule substrates (Papa and Hochstrasser, 1993). Hence, an increase in Doa4 levels, such as occurs when cells enter stationary phase, could change the relative degradation rates of different proteins.

Further evidence that deubiquitinating enzymes can have regulatory functions comes from recent work in mammalian cells. Several proteins implicated in tumorigenesis or growth control have recently been shown to be deubiquitinating enzymes. Interestingly, for the human *tre-2* oncogene, it appears that it is an inactive form of the deubiquitinating enzyme that is tumorigenic (Papa and Hochstrasser, 1993). The inactive protein may act in a dominant-negative fashion, interfering with *tre-2* enzyme-mediated degradation of one or more positive regulators of cell proliferation, e.g., the G_1 cyclins. Alternatively, the *tre-2* enzyme may normally limit the degradation of a negative regulator(s) of growth, such as p53, by rapidly disassembling ubiquitinated intermediates. Several additional mammalian growth regulators have also been found to be deubiquitinating enzymes. One such protein is a mouse enzyme called DUB-1 (Zhu *et al.*, 1996). *DUB-1* is an erythroid cell-specific, immediate early gene induced by the cytokine interleukin-3. *DUB-1* rapidly disappears after a short burst of expression. Interestingly, if the turnoff of *DUB-1* expression is prevented, cells arrest in the G_1 phase of the cell cycle, suggesting a growth regulatory role for this deubiquitinating enzyme.

Collectively, these findings indicate that deubiquitinating enzymes play central but as yet largely unexplored regulatory roles in the growth and development of eukaryotic organisms.

....•

The 20S and 26S Proteasomes

As shown in Fig. 4, the 26S proteasome is a large multisubunit complex composed of a core proteinase known as the 20S proteasome and a pair of symmetrically disposed 19S regulatory particles (Peters, 1994; Rubin and Finley, 1995). The latter particles are closely related to a protein known as PA700 (*p*roteasome *a*ctivator of 700 kDa), although their exact composition and functions and their precise relationship to PA700 have yet to be fully defined. The structure of the 19S particles has only been glimpsed from averaged electron microscopic projection images of negatively stained 26S proteasomes (Peters, 1994). The 19S particles are severely flattened under these conditions, making 3-D reconstructions unfeasible.

Four seven-subunit rings stack into a cylinder to form the 20S proteasome (Fig. 5). These subunits are encoded by a family of related but distinct genes. In the archaeon *Ther-*

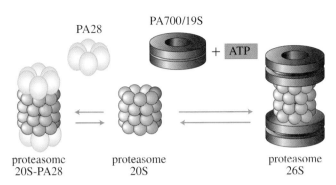

FIGURE 4 Interaction of the 20S proteasome (gray) with some of the complexes observed *in vitro*. (Right) A complex of approximately 700 kDa, called PA700 (also known as the 19S or particle), can bind to both ends of the 20S proteasome, forming the ATP-dependent 26S proteasome which interacts with ubiquitin. (Left) The activator PA28, consisting of a hexamer or heptamer of 28-kDa subunits, which stimulates the proteasome peptidase (but not protease) activity (modified from Lupas *et al.*, 1996).

moplasma acidophilum, a proteasome-like enzyme has also been described; this protein has two subunit types, α and β, with α subunits occupying the outer two heptameric rings and β subunits the inner rings.

All the eukaryotic 20S proteasome subunits are related in sequence to either the α or β subunit of the prokaryotic protease, and based on 3-D electron microscopic reconstructions the eukaryotic and prokaryotic particles are also extremely similar in overall structure. In yeast there are 14 genes that encode subunits of the 20S enzyme (Hochstrasser, 1996). The proteins predicted to be encoded by these genes, all but one of which is essential for growth, can be grouped into seven α-type and seven β-type subunits.

A key recent development in our understanding of the proteasome was the solution of the archaeon *T. acidophilum* 20S proteasome crystal structure at 3.4 Å (Löwe *et al.*, 1995) (Fig. 5), which has provided insight into the mechanism of proteasome action and will permit much more rigorous structure–function studies on the enzyme in the future. This remarkable structure has been the subject of numerous reviews during the past few year (Rubin and Finley, 1995), so only a few key features will be reiterated here. From a cocrystal structure of the proteasome with a peptide aldehyde inhibitor, the active sites could be located within the β subunits on the inner surface of a central chamber in the proteasome particle. The walls of the cylinder appear to have no large openings, limiting entry, and egress of substrates and products to a set of very narrow channels leading to the two ends of the particle; the outer set of channels formed by the α subunits are only 13 Å in diameter. Thus, the proteasome, like the GroEL chaperonin with which it shares a gross architectural similarity, shields its

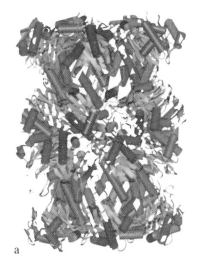

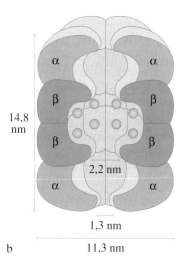

FIGURE 5 The 20S proteasome from the archaeon *Thermoplasma acidophilum.* (a) Model of the three-dimensional structure of the proteasome showing the β-sheets (yellow ribbons) and α-helices (red cylinders). (b) Schematic section of the proteasome made up of four rings, each composed of seven subunits (α in the outer rings and β in the inner rings). The three interior chambers and the narrow channels linking the chambers to each other and to the proteasome exterior are visible. The small circles indicate the positions of the active sites.

catalytic groups from the rest of the cell, a feature likely to be of advantage for an intracellular protease.

Localization of these catalytic groups in the crystal structure yielded another surprise: The proteasome is a threonine protease, the first such protease to be described. This inference received strong support from mutagenesis experiments (Seemüller *et al.,* 1995). The catalytic Thr residue is at the β subunit N terminus, and the free α amino group appears to serve as the general base in the catalytic mechanism. Interestingly, an irreversible inhibitor of the eukaryotic proteasome, lactacystin, was later shown to covalently

modify two sites in one of the β subunits, with one site being the N-terminal Thr residue (Rubin and Finley, 1995). What is puzzling from these new structural data is the fact that only three of the seven eukaryotic β-type subunits would be predicted to have appropriate N-terminal catalytic residues (for all three subunits, a propeptide preceding the catalytic Thr at the mature N terminus is removed posttranslationally); the remaining four subunits would therefore be "inactive" in the simple model for active site formation suggested by the prokaryotic structure (Seemüller *et al.,* 1995).

A significant distinction between the *Thermoplasma* and eukaryotic 20S proteasomes is that the prokaryotic enzyme has a single kind of active site, whereas the more complex eukaryotic particle has at least three (Orlowski, 1990). This inference is based largely on experiments with model peptide substrates. It is not immediately obvious why the eukaryotic 20S proteasome has so many distinct subunits, in contrast to the simple pair of subunits in *Thermoplasma,* and what their relationship is to the different kinds of catalytic sites. A recently discovered proteasome from another archaeon also has just one α and one β subunit type per particle but has two distinct kinds of peptidase activities. Conversely, genetic studies in yeast suggest that multiple nonidentical β subunits are required for particular active sites (Hochstrasser, 1996).

Because the eukaryotic 20S proteasome must associate with several different kinds of regulatory factors at the two ends of the proteasome cylinder (Fig. 4), one might suppose that additional variability in the α subunits could provide a greater range of regulator binding capabilities. The presence of multiple β subunits, on the other hand, might allow for subtle regulation of protein cleavage specificities or for more efficient degradation of a wider array of substrates. Adding further to the puzzle of 20S proteasome subunit heterogeneity is the finding that in mammals proteasomal populations with distinct subunit compositions can exist in the same cell. Three variant β subunits — LMP2, LMP7, and Z — are inducible by γ-interferon and can specifically replace certain "constitutive" proteasome subunits during incorporation into newly assembled proteasome particles (Hochstrasser, 1996).

Experiments aimed at testing the mechanistic consequences of these subunit replacements remain controversial (Groettrup *et al.,* 1995). Both mutant and inhibitor studies suggest that the 20S or 26S proteasome helps generate the peptides presented by class I major histocompatibility complex (MHC) molecules. Evidence has also been presented which favors the notion that the range of possible protein cleavages can be modulated by altering the composition of these enzymes, perhaps tailoring cleavage specificity to match the peptide-binding specificity of MHC molecules. However, the reported interferon-induced changes in peptide cleavage preferences were discrepant among various laboratories. Unfortunately, cleavage preferences identified in model peptides also do not reflect the preferences seen in proteins. Despite the mechanistic uncertainties, a deficit in presentation of at least certain antigens has been shown with cells from mice carrying a deletion of either *LMP2* or *LMP7*, suggesting that these subunits do modulate proteasome activity in a way that is significant immunologically (Groettrup *et al.*, 1995).

Recent developments suggest, however, that the major mechanism by which mammalian cells alter proteasome activity to facilitate class I antigen presentation is to increase 20S proteasome association with a specific regulatory factor, PA28 (Groettrup *et al.*, 1996). The 28-kDa PA28 subunits associate into hexameric or heptameric rings which, like PA700, bind to the ends of the proteasome cylinder. Two closely related genes encode PA28 and are inducible by γ-interferon. Mammalian cells appear to have separate populations of proteasomes containing either PA28 or PA700 caps. Exactly how PA28 stimulates antigen presentation is not known. PA28 stimulates the peptidase activities of the 20S proteasome but apparently not its ability to break down full-size proteins. Conceivably, PA28–20S proteasome complexes could preferentially bind and process peptides suitable for presentation by cells of a certain haplotype. It might be more appropriate, however, to view PA28 as an allosteric regulator of the 20S proteasome, an idea which recalls the analogy to the GroEL chaperonin and its allosteric regulator, GroES. Cooperative structural changes in the 20S proteasome induced through PA28 binding to the outer α subunit rings may be transmitted to the active sites in the β subunit rings, allowing more efficient cleavage of certain substrates and/or more rapid release of peptide products. Congruent with this possibility, it has been inferred from biochemical analyses of mutant yeast 20S proteasomes that single point mutations can engender significant cooperative changes in the particle (Chen and Hochstrasser, 1995). Perhaps proteasomes bearing the alternative β subunits respond differently to PA28 binding, accounting for the coinduction of PA28 and these subunits by γ-interferon and their importance in antigen presentation (Groettrup *et al.*, 1995).

These data would seem to discount the importance in antigen presentation of the 26S proteasome, which contains the PA700-like complexes rather than PA28. Because PA700 complexes confer ubiquitin dependence on protein degradation by the 20S proteasome, ubiquitin–protein conjugation may also be less important for this process. There are sharply conflicting results on the involvement of ubiquitination in class I antigen presentation. Inasmuch as the antigens for this pathway derive from proteins and protein breakdown by the proteasome is not stimulated by PA28, there must be other cellular proteases involved in the initial breakdown of the proteins. These may well include the 26S proteasome.

Unlike the 26S proteasome, the 20S proteasome is unable to degrade stably folded proteins, but it can break these proteins down completely (in the absence of ATP) if they are first denatured. This result is readily rationalized by the structure of the archaeal proteasome which, as noted previously, interposes a sieve with a 13-Å pore size between its central hydrolytic chamber and the rest of the cell's contents. Therefore, one potential function of the 19S particle–PA700, which is positioned at the 20S proteasome opening, is as a protein unfoldase. The requirement for an unfoldase activity brings to mind the action of certain molecular chaperones. Recent experiments with the ClpAP and ClpXP proteases of *Escherichia coli* support this analogy. ClpA and ClpX are ATP-dependent activators of the ClpP protease, a property with some similarity to the effect of PA700 on the 20S proteasome. Surprisingly, in the absence of clpP, clpA and clpX function as chaperones that help dissociate multisubunit proteins (Wickner *et al.*, 1994; Levchenko *et al.*, 1995). Hence, an ATP-dependent chaperone may conformationally alter a substrate protein in such a way that it either changes the protein's activity or, if the chaperone is coupled to a proteolytic enzyme, causes the protein to be destroyed. Also interesting in this regard is the presumptive mechanism by which GroEL/ES facilitates protein folding. It is thought that GroEL/ES actually causes transient protein unfolding, providing a protein trapped in a misfolded state new opportunities to find a productive folding pathway. When juxtaposed to a protease, however, such a (partially) unfolded protein may be rendered susceptible to proteolysis rather than refolding.

At least six PA700 subunits are members of a closely related family of ATPases, and PA700 is known to have ATPase activity. All the studied ATPases are essential for viability in yeast, and point mutations in several of them have been shown to cause defects in ubiquitin-dependent proteolysis *in vivo* (Rubin and Finley, 1995). The number of different ATPases in PA700 is surprising. They may provide alternative binding sites that favor binding (unfolding) of particular proteins or they may have functions beyond protein denaturation, e.g., substrate translocation from binding sites in PA700 to the proteasome interior.

Another anticipated function of PA700, the binding of multiubiquitinated proteins, was traced to a specific subunit in the 26S proteasome called S5a (Deveraux *et al.*, 1994). The gene encoding this subunit has been cloned from many organisms, but its sequence is unremarkable (Haracska and Udvardy, 1995; van Nocker *et al.*, 1996). S5a is only loosely associated with PA700 based on fractionation studies in *Drosophila* embryonic extracts (Haracska and Udvardy, 1995). Surprisingly, deletion of the gene encoding the yeast homolog of S5a, *SUN1*, is without major consequence for cell growth or stress resistance (M. Hochstrasser, unpublished data); no other sequences obviously related to *SUN1* are present in the yeast genome. Because there are strong reasons to believe ubiquitin-conjugate binding to the proteasome will be essential for cell viability, other means of conjugate binding to the proteasome must exist, at least in yeast.

Ubiquitin is usually recycled after targeting of the ubiquitinated protein to the proteasome, indicating the proteasome contains or can recruit deubiquitinating enzymes. An important question is when in the degradative cycle is ubiquitin released from the substrate. A mechanism that removes ubiquitin only after the substrate is largely broken down into peptides has the advantage of keeping incompletely digested substrate fragments from dissociating prematurely. A seeming disadvantage is the topological complication of having part(s) of the protein pinned to the 19S particle while the protein is being unfolded and fed into the proteasome core. This may not be such a serious problem if the ubiquitin chain tether is flexible or undergoes repeated cycles of partial release and rebinding. Such a tether may even help to corral the unfolding polypeptide chain near the outer 20S proteasome pore such that the probability of entry of an end (or extended loop) of the polypeptide into the channel will be increased. In this highly speculative scenario, deubiquitination would somehow be triggered only late in the degradation of substrate. The point at which a substrate is severed from its ubiquitin tether(s) may also be substrate dependent, e.g., early release may allow a partly degraded protein to be released, as appears to occur in NF-κB p105 processing (Palombella et al., 1994).

Currently, there is little evidence for these hypotheses. One intriguing result, however, is the finding that yeast cells lacking the Doa4 deubiquitinating enzyme accumulate what appear to be peptides attached to either ubiquitin or short ubiquitin oligomers (proposed to be proteolytic remnants generated by proteasomes) (Papa and Hochstrasser, 1993). These data suggest that most proteins are substantially degraded prior to complete deubiquitination (by Doa4). The small size of the putative peptides further suggests that if ubiquitin is indeed tethered to the proteasome during degradation, the tethering site(s) must be very close to the opening of the 20S particle.

An obvious exception to the general importance of ubiquitin to proteasomal degradation is the class of proteins whose degradation by the 26S proteasome is ubiquitin independent. To date, however, the only known example of such a substrate is mammalian ornithine decarboxylase (ODC) (Murakami et al., 1992). In order to be degraded, ODC must associate with another protein, antizyme, which targets ODC to the 26S proteasome for ATP-dependent proteolysis; antizyme is not degraded in the process. The 20S proteasome cannot replace the 26S particle in this reaction. How far the analogy between antizyme and ubiquitin action can be taken is not clear. Both appear to facilitate interaction of the proteins to which they are bound with the 26S proteasome, either by providing proteasome binding

sites or by altering the associated protein in a way that promotes proteasome binding (or both). It is hoped that the relative simplicity of the antizyme–ODC model will provide some important mechanistic clues as to how ubiquitin functions in proteolysis as well.

New Cellular Functions for Ubiquitin

Recently, there have been reports of several novel cellular regulatory mechanisms involving the ubiquitin system. In several cases, ubiquitination was shown to function independently of proteasome action, confirming a long suspected but previously unproven role for ubiquitin in the modification of protein function. Another remarkable study has implicated the proteasome in the degradation of a protein initially targeted to the endoplasmic reticulum (ER), apparently following the ejection of the protein from the ER. These exciting discoveries, which are briefly summarized here, suggest our knowledge of cellular processes regulated by the ubiquitin system will continue to expand in unanticipated directions.

Ubiquitin-Dependent Endocytosis of Cell Surface Proteins

Many cell surface proteins are rapidly endocytosed and, if not recycled to the plasma membrane, are routed to the lysosome/vacuole, where they are destroyed by vacuolar proteases (Fig. 6.) Endocytosis of the S. cerevisiae α factor pheromone receptor, Ste2, takes place constitutively at a slow rate but is stimulated 5- to 10-fold by α factor binding (Hicke and Riezman, 1996). Pheromone binding to the G protein-coupled Ste2 receptor activates a MAP kinase signal transduction pathway, resulting ultimately in alterations in transcription needed by the cell to mate with α factor-secreting cells. Internalized Ste2 follows the endocytic pathway to the vacuole, where it is degraded. For endocytosis, the cytoplasmic C-terminal tail of the Ste2 membrane protein is critical; within this tail, a nine-residue sequence, SINNDAKSS, is sufficient for receptor internalization (Hicke and Riezman, 1996). Degradation of Ste2 is strongly inhibited in the yeast end4 mutant, which is defective for receptor endocytosis. Surprisingly, much of the Ste2 in this mutant accumulates in a ubiquitinated form. These species of Ste2 are also detectable as transient intermediates in pheromone-stimulated wild-type cells. Internalization of ligand-bound receptors and Ste2 ubiquitination are both markedly inhibited in yeast cells lacking specific E2 enzymes, but degradation is not perturbed by proteasome mutations. Importantly, mutation of the lysine in the SINNDAKSS internalization signal blocks both Ste2 endocytosis and ubiquitination (Hicke and Riezman, 1996). In addition, mutation to alanine of all three serines in the endocytosis signal eliminates receptor hyperphosphoryla-

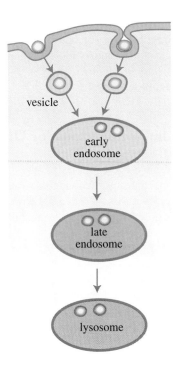

FIGURE 6 The endocytic pathway from the plasma membrane to the lysosomes, responsible for the internalization of proteins from the cell surface and their routing to the different endosomes and to the lysosomes, where they are degraded by lysosomal proteases. Ubiquitination of membrane proteins appears to play an important role at a very early step in the endocytic pathway (modified from Alberts *et al.*, 1995, p. 717).

tion in response to α factor and inhibits receptor internalization. Correspondingly, ubiquitination of Ste2 is much reduced. The simplest interpretation of these data is that endocytosis of Ste2 specifically requires ubiquitination of the lysine in the SINNDAKSS element and that this modification depends on phosphorylation of the flanking serines.

It appears very likely that ubiquitination is used as a signal for endocytosis and vacuolar targeting for many membrane proteins. For example, the yeast general amino acid transporter Gap1 appears to follow a degradation pathway very similar to that which works on Ste2 (Hein *et al.*, 1995). For references to additional examples, see Hochstrasser (1996).

The previously discussed results raise some interesting questions. First, how does ubiquitination of Ste2 (or other membrane proteins) lead to its endocytosis? The ubiquitin (chain?) may provide a binding site for a component of the endocytic machinery or may alter Ste2 structure in a way that facilitates such binding. Alternatively, ubiquitin may promote movement of Ste2 into regions of membrane that actively endocytose. Another question is why ubiquitinated Ste2 is not targeted to the proteasome. What determines

the alternative metabolic fates of ubiquitinated cellular proteins is clearly a question of considerable importance that will need to be addressed in future studies.

The NF-κB Signal Transduction Pathway

In response to a wide array of extracellular signals, the transcription factor NF-κB or related members of the Rel protein family are activated, resulting in a variety of transcriptional responses (Thanos and Maniatis, 1995). Under noninducing conditions, NF-κB and its kin reside in the cytoplasm in inactive heteromeric complexes (Fig. 7). NF-κB is rendered inactive either because the NF-κB subunits are bound to a member of the IκB protein family or because one of its subunits is in a precursor form (p105) that prevents translocation of the complex to the nucleus. The IκB protein is related to the C-terminal prodomain of p105, and both function by masking the NF-κB nuclear localization signal. Work by several groups has shown that NF-κB is activated either by processing p105 by the ubiquitin–proteasome pathway to remove the inhibitory C-terminal domain or by destruction of IκB by the same proteolytic pathway.

Degradation of IκBα *in vivo* requires signal-induced phosphorylation of both of two serines near its amino terminus, which in turn triggers multiubiquitination of nearby lysine residues and subsequent IκB degradation. A protein kinase with the appropriate specificity toward these critical N-terminal serines has been described (Chen *et al.*, 1996). The partially purified kinase is ~700 kDa in size and comprises multiple heterologous subunits. Remarkably, phosphorylation of IκB was found to depend on the ubiquitination of the kinase (or a copurifying regulatory factor). In other words, protein ubiquitination is needed not only for the phosphorylation-dependent ubiquitination of IκB but also for phosphorylation. Phosphorylation of IκB requires E1 and a specific E2 but does not depend on proteolytic activity in the partially purified system. The precise target of ubiquitination in the kinase fraction still needs to be identified.

Degradation of Proteins Ejected from the Endoplasmic Reticulum

Viruses have evolved diverse strategies to elude immunological detection of virus-infected cells. Published data suggest that human cytomegalovirus (CMV) can thwart the immune system by reversing the translocation into the ER of the MHC class I heavy chain, a transmembrane protein that is required for antigen presentation to T cells (Wiertz *et al.*, 1996). Synthesis of a single CMV protein, US11, is sufficient to cause the dislocation; US11 is also a transmembrane glycoprotein that localizes to the ER. Unlike in control cells, the MHC class I molecules are extremely short-lived in US11-transfected cells ($t_{1/2} < 1$ min). Use of protease inhibitors known to inhibit the proteasome strongly inhibited turnover of a deglycosylated form of the

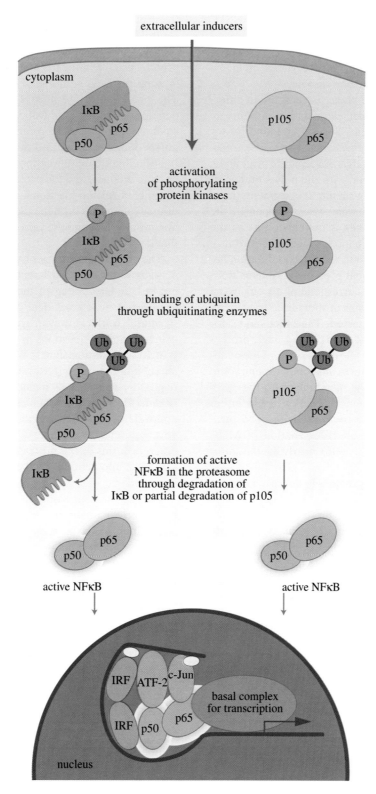

FIGURE 7 The NF-κB transcription factor and its regulation. Different extracellular inducers (virus and TNF) are able to activate different signal transducer pathways leading to the activation of protein kinases which phosphorylate IκB (left) or p105 (right). The phosphorylated IκB and p105 are recognized by the ubiquitinating enzymes. Ubiquitinated IκB is degraded, whereas p105 is proteolytically matured. In both cases, active NF-κB is obtained, which can be translocated to the nucleus in which, together with other factors (IRF, ATF-2, and c-Jun), it activates gene transcription (modified from Thanos and Maniatis, 1995).

class I heavy chain. Wiertz *et al.* (1996) argue that US11, by an unknown mechanism, causes the ejection of class I molecules (but not other ER proteins) from the ER back into the cytosol where they are deglycosylated by an N-glycanase and degraded by the proteasome.

Because parts of transmembrane proteins are in different cellular compartments, their degradation presents an interesting topological problem. Two general kinds of solution to this difficulty can be envisioned. In the first, transmembrane proteins could be targets of separate proteolytic pathways in each compartment, with the polypeptide getting shaved down to the transmembrane segments from each side of the membrane (what would happen to the short transmembrane fragments is unknown, just as the general fate of cleaved signal peptides remains uncertain). Such a mechanism would seem to require coordination between the cytosolic and lumenal proteolytic systems since no large cleavage fragments corresponding to either the cytosolic or the lumenal domains of short-lived transmembrane proteins have been detected. The other model for transmembrane protein degradation posits a single proteolytic pathway working on one side or the other of the membrane. The protein segments not in this compartment would somehow get pulled across the membrane into the "proteolytic compartment." Protein dislocation could either be coupled to degradation or could occur prior to turnover. The CMV US11-stimulated degradation of MHC class I molecules (Wiertz *et al.*, 1996) clearly fits this latter model best.

It remains to be seen how commonly this mechanism is used and whether it can work on proteins stably inserted in the membrane rather than just those in the process of translocation. *In vitro* data suggest a potentially analogous ER ejection mechanism precedes the degradation of an unglycosylated derivative of a yeast protein that is normally secreted from cells (McCracken and Brodsky, 1996). The same unglycosylated protein is subject to rapid degradation *in vivo*. Interestingly, preliminary *in vitro* and *in vivo* results (cited in McCracken and Brodsky, 1996) suggest that the proteasome is critical to this pathway as well.

Conclusions and Future Directions

An increasing number of cellular regulatory mechanisms are being linked to protein modification by the phylogenetically conserved polypeptide ubiquitin. These include key transitions in the cell cycle, processing of foreign proteins in class I antigen presentation, signal transduction pathways, receptor-mediated endocytosis, and various growth control mechanisms. In most, but not all, of these examples, ubiquitination of a protein leads to its rapid degradation by the 26S proteasome. The central portion of the proteasome is a cylindrical particle whose innermost cavity is lined with the protease active sites. Access of cellular proteins to this central proteasomal chamber is gated at multiple levels. Initial steps involve attachment of ubiquitin to the substrate protein, usually in the form of a ubiquitin oligomer(s), and binding of the ubiquitinated protein to the proteasome. The bound substrate must be unfolded (and eventually deubiquitinated) and translocated through a narrow set of channels that lead to the proteasome interior, in which the polypeptide is cleaved into short peptides. Protein ubiquitination and deubiquitination are both mediated by large enzyme families, and the proteasome appears to comprise a family of related but functionally distinct particles. This diversity underlies both the high substrate specificity of the ubiquitin system and the variety of regulatory mechanisms that it serves.

The initial description of ubiquitin-dependent protein degradation was drawn largely from biochemical studies, but our understanding of the ubiquitin system has been fleshed out by a great deal of genetic, biochemical, and structural work. However, we still lack clear answers to many of the key questions in the field: What constitutes a protein degradation signal? By what mechanisms are protein ubiquitination and degradation regulated? Which cell regulatory mechanisms include elements of the ubiquitin system? How are the dynamics of ubiquitin addition to and removal from substrates regulated? What mechanistic roles do ubiquitin chains play? What are the functions of the different forms of the proteasome? How is a protein broken down into peptides by the proteasome and how is this regulated?

Exciting and surprising findings continue to punctuate studies on the ubiquitin system. Because of its central biological importance, a dramatic increase in medical and biotechnological research on this system has also occurred. It is safe to say that there is much left to discover before we can claim a respectable understanding of how this complex mechanism of protein modification and metabolism operates and how it functions in the regulatory circuitry of eukaryotic cells.

References Cited

ALBERTS, B., *et al.* (1995). *Biologia Molecolare della Cellula*, 3rd ed. Zanichelli, Milan, Italy.

AMERIK, A. YU., SWAMINATHAN, S., KRANTZ, B. A., WILKINSON, K. D., and HOCHSTRASSER, M. (1997). *In vivo* disassembly of free polyubiquitin chains by yeast Ubp14 modulates rates of protein degradation by the proteasome. *EMBO J.* **16,** 4826–4838.

BERLETH, E. S., and PICKART, C. M. (1996). Mechanism of ubiquitin conjugating enzyme E2-230K: Catalysis involving a thiol relay? *Biochemistry* **35,** 1664–1671.

CHAU, V., TOBIAS, J. W., BACHMAIR, A., MARRIOTT, D., ECKER, D. J., *et al.* (1989). A multiubiquitin chain is confined to specific lysine in a targeted short-lived protein. *Science* **243,** 1576–1583.

CHEN, P., and HOCHSTRASSER, M. (1995). Biogenesis, structure, and function of the yeast 20S proteasome. *EMBO J.* **14**, 2620–2630.

CHEN, P., JOHNSON, P., SOMMER, T., JENTSCH, S., and HOCHSTRASSER, M. (1993). Multiple ubiquitin-conjugating enzymes participate in the *in vivo* degradation of the yeast MATα2 repressor. *Cell* **74**, 357–369.

CHEN, Z., and PICKART, C. M. (1990). A 25-kilodalton ubiquitin carrier protein (E2) catalyzes multiubiquitin chain synthesis via lysine 48 of ubiquitin. *J. Biol. Chem.* **265**, 21835–21842.

CHEN, Z. I., PARENT, L., and MANIATIS, T. (1996). Site-specific phosphorylation of IkBα by a novel ubiquitination-dependent protein kinase activity. *Cell* **84**, 853–862.

CHIANG, H.-L., and DICE, J. (1988). Peptide sequences that target proteins to lysosomes for enhanced degradation during serum withdrawal. *J. Biol. Chem.* **263**, 6797–6805.

DESHAIES, R. J. (1995). The self-destructive personality of a cell cycle in transition. *Curr. Opin. Cell Biol.* **7**, 781–789.

DEVERAUX, Q., USTRELL, V., PICKART, C., and RECHSTEINER, M. (1994). A 26 S protease subunit that binds ubiquitin conjugates. *J. Biol. Chem.* **269**, 7059–7061.

DOHMEN, R. J., STAPPEN, R., MCGRATH, J. P., FORROVA, H., KOLAROV, J., *et al.* (1995). An essential yeast gene encoding a homolog of ubiquitin-activating enzyme. *J. Biol. Chem.* **270**, 18099–18109.

FINLEY, D., SADIS, S., MONIA, B. P., BOUCHER, P., ECKER, D. J., *et al.* (1994). Inhibition of proteolysis and cell cycle progression in a multiubiquitination-deficient yeast mutant. *Mol. Cell Biol.* **14**, 5501–5509.

GOTTESMAN, S., and MAURIZI, M. (1992). Regulation by proteolysis: Energy-dependent proteases and their targets. *Microbiol. Rev.* **56**, 592–621.

GROETTRUP, M., RUPPERT, T., KUEHN, L., SEEGER, M., STANDERA, S., *et al.* (1995). The interferon-γ-inducible 11S regulator (PA28) and the LMP2/LMP7 subunits govern the peptide production by the 20S proteasome *in vitro. J. Biol. Chem.* **270**, 23808–23815.

GROETTRUP, M., SOZA, A., EGGERS, M., KUEHN, L., DICK, T. P., *et al.* (1996). A role for the proteasome regulator PA28α in antigen presentation. *Nature* **381**, 166–168.

HARACSKA, L., and UDVARDY, A. (1995). Cloning and sequencing a non-ATPase subunit of the regulatory complex of the *Drosophila* 26S protease. *Eur. J. Biochem.* **231**, 720–725.

HEIN, C., SPRINGAEL, J.-Y., VOLLAND, C., HAGUENAUER-TSAPIS, R., and ANDRE, B. (1995). NPI1, an essential yeast gene involved in induced degradation of Gap1 and Fur1 permeases, encodes the Rsp5 ubiquitin–protein ligase. *Mol. Microbiol.* **18**, 77–87.

HERSHKO, A., and CIECHANOVER, A. (1992). The ubiquitin system for protein degradation. *Annu. Rev. Biochem.* **61**, 761–807.

HERSHKO, A., HELLER, H., ELIAS, S., and CIECHANOVER, A. (1983). Components of ubiquitin–protein ligase system: Resolution, affinity purification, and role in protein breakdown. *J. Biol. Chem.* **258**, 8206–8214.

HICKE, L., and RIEZMAN, H. (1996). Ubiquitination of a yeast plasma membrane receptor signals its ligand-stimulated endocytosis. *Cell* **84**, 277–287.

HOCHSTRASSER, M. (1996). Ubiquitin-dependent protein degradation. *Annu. Rev. Genet.* **30**, 405–439.

HOU, D., CENCIARELLI, C., JENSEN, J. P., NGUYGEN, H. B., and

WEISSMAN, A. M. (1994). Activation-dependent ubiquitination of a T cell antigen receptor subunit on multiple intracellular lysines. *J. Biol. Chem.* **269**, 14244–14247.

HUANG, Y., BAKER, R. T., and FISCHER-VIZE, J. A. (1995). Control of cell fate by a deubiquitinating enzyme encoded by the fat facets gene. *Science* **270**, 1828–1831.

HUIBREGTSE, J. M., SCHEFFNER, M., BEAUDENON, S., and HOWLEY, P. M. (1995). A family of proteins structurally and functionally related to the E6–AP ubiquitin–protein ligase. *Proc. Natl. Acad. Sci. USA* **92**, 2563–2567.

JOHNSON, E. S., GONDA, D. K., and VARSHAVSKY, A. (1990). *cis*–*trans* recognition and subunit-specific degradation of short-lived proteins. *Nature* **346**, 287–291.

KING, R. W., PETERS, J. M., TUGENDREICH, S., ROLFE, M., HIETER, P., and KIRSCHNER, M. W. (1995). A 20S complex containing CDC27 and CDC16 catalyzes the mitosis-specific conjugation of ubiquitin to cyclin B. *Cell* **81**, 279–288.

LEVCHENKO, I., LUO, L., and BAKER, T. A. (1995). Disassembly of the Mu transposase tetramer by the ClpX chaperone. *Genes Dev.* **9**, 2399–2408.

LÖWE, J., STOCK, D., JAP, B., ZWICKL, P., BAUMEISTER, W., and HUBER, R. (1995). Crystal structure of the 20S proteasome from the archaeon *T. acidophilum* at 3.4 Å resolution. *Science* **268**, 533–539.

LUPAS, *et al.* (1996). *Cold Spring Harbor Symp. Quant. Biol.* **60**, 515–524.

MCCRACKEN, A. A., and BRODSKY, J. L. (1996). Assembly of ER-associated protein degradation *in vitro:* Dependence on cytosol, calnexin, and ATP. *J. Cell Biol.* **132**, 291–298.

MURAKAMI, Y., MATSUFUJI, S., KAMEJI, T., HAYASHI, S., IGARASHI, K., *et al.* (1992). Ornithine decarboxylase is degraded by the 26S proteasome without ubiquitination. *Nature* **360**, 597–599.

NARASIMHAN, J., RASMUSSEN, J. L., and HAAS, A. L. (1996). Conjugation of the 15-kDa interferon-induced ubiquitin homolog is distinct from that of ubiquitin. *J. Biol. Chem.* **271**, 324–330.

NISHIZAWA, M., FURUNO, N., OKAZAKI, K., TANAKA, H., OGAWA, Y., and SAGATA, N. (1993). Degradation of Mos by the N-terminal proline (Pro^2)-dependent ubiquitin pathway on fertilization of *Xenopus* eggs: Possible significance of natural selection for Pro^2 in Mos. *EMBO J.* **12**, 4021–4027.

ORLOWSKI, M. (1990). The multicatalytic proteinase complex, a major extralysosomal proteolytic system. *Biochemistry* **29**, 10289–10297.

PALOMBELLA, V. J., RANDO, O. J., GOLDBERG, A. L., and MANIATIS, T. (1994). The ubiquitin–proteasome pathway is required for processing the NF-κB1 precursor protein and the activation of NF-κB. *Cell* **78**, 773–785.

PAPA, F., and HOCHSTRASSER, M. (1993). The yeast *DOA4* gene encodes a deubiquitinating enzyme related to a product of the human *tre-2* oncogene. *Nature* **366**, 313–319.

PAPA, F. R., AMERIK, A. Y., and HOCHSTRASSER, M. (1999). Interaction of the Doa4 deubiquitinating enzyme with the yeast 26S proteasome. *Mol. Biol. Cell* **10**, 741–756.

PETERS, J. M. (1994). Proteasomes: Protein degradation machines of the cell. *Trends Biochem. Sci.* **19**, 377–382.

PICKART, C. M. (1988). Ubiquitin activation and ligation. In *Ubiquitin* (M. Rechsteiner, Ed.), pp. 77–100. Plenum, New York.

RUBIN, D. M., and FINLEY, D. (1995). Proteolysis. The proteasome: A protein-degrading organelle? *Curr. Biol.* **5,** 854–858.

SADIS, S., ATIENZA, C., and FINLEY, D. (1995). Synthetic signals for ubiquitin-dependent proteolysis. *Mol. Cell. Biol.* **15,** 1265–1273.

SCHEFFNER, M., NUBER, U., and HUIBREGTSE, J. M. (1995). Protein ubiquitination involving an E1–E2–E3 enzyme ubiquitin thioester cascade. *Nature* **373,** 81–83.

SEEMÜLLER, E., LUPAS, A., STOCK, D., LÖWE, J., HUBER, R., and BAUMEISTER, W. (1995). Proteasome from *Thermoplasma acidophilum:* A threonine protease. *Science* **268,** 579–582.

SHAEFFER, J. R., and KANIA, M. A. (1995). Degradation of monoubiquitinated alpha-globin by 26S proteasomes. *Biochemistry* **34,** 4015–4021.

SUDAKIN, V., GANOTH, D., DAHAN, A., HELLER, H., HERSHKO, J., *et al.* (1995). The cyclosome, a large complex containing cyclin-selective ubiquitin ligase activity, targets cyclins for destruction at the end of mitosis. *Mol. Biol. Cell* **6,** 185–197.

THANOS, D., and MANIATIS, T. (1995). NF-κB: A lesson in family values. *Cell* **80,** 529–532.

VAN NOCKER, S., DEVERAUX, Q., RECHSTEINER, M., and VIERSTRA, R. D. (1996). Arabidopsis MBP1 gene encodes a conserved ubiquitin recognition component of the 26S proteasome. *Proc. Natl. Acad. Sci. USA* **93,** 856–860.

VARSHAVSKY, A. (1992). The N-end rule. *Cell* **69,** 725–735.

WICKNER, S., GOTTESMAN, S., SKOWYRA, D., HOSKINS, J., MCKENNEY, K., and MAURIZI, M. R. (1994). A molecular chaperone, ClpA, functions like DnaK and DnaJ. *Proc. Natl. Acad. Sci. USA* **91,** 12218–12222.

WIERTZ, E. J. H. J., JONES, T. R., SUN, L., BOGYO, M., GEUZE, H. J., and PLOEGH, H. L. (1996). The human cytomegalovirus US11 gene product dislocates MHC class I heavy chains from the endoplasmic reticulum to the cytosol. *Cell* **84,** 769–780.

WILKINSON, K. D., TASHAYEV, V. L., O'CONNOR, L. B., LARSEN, C. N., KASPEREK, E., and PICKART, C. M. (1995). Metabolism of the polyubiquitin degradation signal: Structure, mechanism, and role of isopeptidase T. *Biochemistry* **34,** 14535–14546.

ZHU, Y., CARROLL, M., PAPA, F. R., HOCHSTRASSER, M., and D'ANDREA, A. D. (1996). DUB-1, a novel deubiquitinating enzyme with growth-suppressing activity. *Proc. Natl. Acad. Sci. USA* **93,** 3275–3279.

General References

ALBERTS, B., BRAY, D., LEWIS, J., RAFF, M., ROBERTS, K., and WATSON, J. D. (1994). *Molecular Biology of the Cell,* 3rd ed. Garland, London.

FINLEY, D., and CHAU, V. (1991). Ubiquitination. *Annu. Rev. Cell Biol.* **7,** 25–69.

Kai Simons
Marino Zerial

*Max-Planck Institute of Molecular Cell Biology
and Genetics
01307 Dresden, Germany*

Traffic of Proteins and Compart- mentalization of Eukaryotic Cells

Every eukaryotic cell is compartmentalized into various small organisms, kept in dynamic communication through the use of vesicles, bordered by a membrane, that transport proteins, lipids, and fluids from one cell site to another, in a strictly controlled manner. In order to be incorporated in the correct vesicle, the proteins must exhibit specific signs for their destination. In turn, the vesicles must contain molecular information that sends them to the precisely corresponding target compartment. There are also special molecular interrupters that give specificity, directionality, and time control to the formation of vesicles, at the correct address and fusion. The movement of the vesicles does not occur in a random manner; different types of intracellular filaments are used as "wheels" to send the vesicles towards their acceptor sites and to increase the effectiveness of the vesicular transport. Viruses may also use these traffic paths for their cellular cycle. Under controlled experimental conditions, the virus is used to study intracellular traffic paths.

The Organism Is a Life Cycle

A multicellular living being begins as a single cell. It grows in size by cell division and divides the labor among the newly formed cells by differentiation. An adult human being has about 10^{14} cells. These are organized into different organs, such as the heart, the brain, and the digestive tract, each made of different subpopulations of cells forming tissues. The different cells in each tissue have the same complement of genes as the fertilized egg from which they derive. However, only a part of the genetic program is expressed in each tissue cell. The set of proteins that is produced in the cells of the stomach is different from those of the neurons in the brain. These proteins give each cell its functional properties. For instance, the cells lining the stomach secrete into the gastric juice both hydrochloric acid and the digestive enzyme pepsin, a protein. At the same time these cells form a tight layer lining the stomach wall which is impermeable to this acid juice of pepsin, which, if it could penetrate between the cells into the "milieu interieur," would have catastrophic consequences. The perforation of a gastric ulcer is an example of what happens when this lining fails. The nerve cells in the brain, on the other hand, form long processes called axons and dendrites which connect with other nerve cells. The axon of one neuron contacts the dendrite of another neuron and these contact sites, which are responsible for the transmission of the information, are called synapses. To assemble the synaptic networks in the brain each neuron has to deliver one set of proteins to the axon and another to the dendrites.

Currently, very little is known about the mechanisms responsible for execution of the developmental program in any organism. To investigate this problem we must define the complex sequence of events which lead to the commitment of each cell type to a specific genetic program. However, differentiation is not understood simply by defining how genes are turned on and off during embryonic development. We also have to elucidate the mechanisms by which cells execute the genetic program and organize themselves to fulfill their functions.

Nerve cells and gastric cells look completely different under the microscope. The cells have a characteristic size, form, and architecture. Moreover, cells interact in a precise way with their neighboring cells to form a tissue which has a specific size and form. We have already mentioned the case of neurons that form synapses, but many other cell types in the body communicate by different modes, leading to a remarkable array of intercellular interactions. These are three-dimensional problems with the additional fourth dimension of time included to effect change. We are starting to get glimpses of the mechanisms that cells employ to organize themselves during embryonic development, but we are still far from understanding how cells accomplish their complex tasks.

....●

The Cell Is Compartmentalized

An animal cell is like a city suspended under water and surrounded by a flexible shell. More than 90% of every vertebrate is water. Life on Earth in all likelihood has its origins in the primeval sea. Every living organism on this planet is probably the descendant of a single cell dividing in the primeval soup more than 3 billion years ago as the earth was cooling. Without water, life in its present form would be inconceivable. The outer shell of the cell is called the plasma membrane. This membrane is not totally impermeable but regulates import and export activities of the cell. Inside the cell there are several compartments with different functions. The most obvious, because of its size, is the nucleus (Fig. 1). The nucleus is a repository of the cell's genetic library, the information being encoded in a four-letter code in the nucleic acid molecules organized into chromosomes. In the nucleus the decisions are made as to which set of genes is to be transcribed and then translated into specific proteins. These decisions are not made without external influence. Gene expression is regulated by chemical signals which reach the nucleus from the plasma membrane, which is in continuous contact with the surrounding environment and with neighboring cells. Hormones and growth factors circulating in the body fluids bind to specific proteins in the plasma membrane. The binding stimulates the release of chemical messengers inside the cell which relay signals across the cytosol to the nucleus.

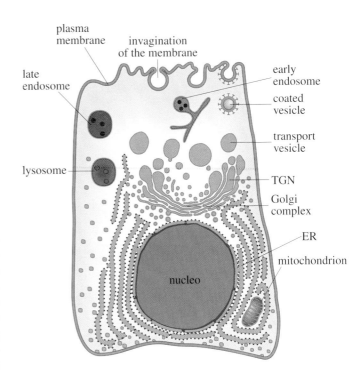

FIGURE I The principal compartments of the cell involved in the transport pathway. The nucleus is surrounded by the nuclear envelope which is continuous with the endoplasmic reticulum (ER), the site of synthesis of secretory proteins. These proteins and lipids pass through the Golgi complex, successively reaching the trans-Golgi network (TGN), from which transport vesicles target them to the plasma membrane. The caveolae and the coated vesicles are involved in the opposite process from the membrane to the cell interior. Early and late endosomes and lysosomes play different roles in the sorting and degradation of proteins, lipids, and solutes internalized in the cell or coming from the biosynthetic pathway.

Another conspicuous compartment is the mitochondrion. The mitochondria are the power plants of the cell which use oxygen to burn foodstuffs providing the cell with chemical energy. The endoplasmic reticulum (ER) forms a large network of membranes that produce proteins for export from the cell as well as for other intracellular destinations. These proteins have to pass the Golgi complex, which is a postal office for the distribution of cellular components to their addresses. Among these are the lysosomes, which are garbage dumps with built-in recycling capacity.

All these cellular compartments have specific sizes, shapes, and locations within the cell depending on the cell type in which they are harbored. They are all bounded by membranes which are similar in structure to that of the plasma membrane. The membranes consist of two closely apposed layers of lipids and they form a framework as flexible as soap bubbles but stable in structure. Most of the membrane lipids, such as phospholipids and cholesterol, are synthesized in the ER. From there, these lipids are, like

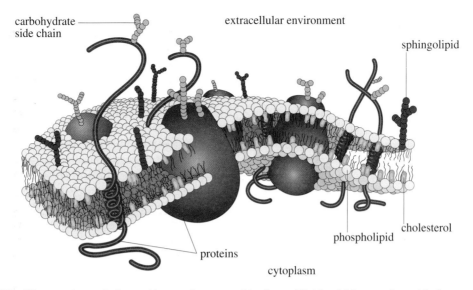

carbohydrate side chain

extracellular environment

sphingolipid

proteins

phospholipid

cholesterol

cytoplasm

FIGURE 2   The membrane is formed by two juxtaposed leaflets of lipids which, together with the proteins that float there, contribute to the asymmetry of this fluid but stable structure. Phospholipids and cholesterol are present on both sides of the lipid bilayer, whereas sphingolipids are located only on the outer leaflet. Some proteins also have distinct topological orientations: For example, the carbohydrate side chains of the glycoproteins face the outside. The membranes of the different cell compartments have a similar basic organization, but they vary in their lipid and protein composition.

proteins, transported to other membrane-bound compartments. Another lipid class is the sphingolipids. These are produced in the Golgi complex. The lipid bilayer in each membrane behaves like a two-dimensional liquid in which the membrane proteins swim around (Fig. 2). As discussed later, the bilayer is not a homogeneous mixture of its lipid constituents. They are asymmetrically disposed over the outer and inner leaflets of the bilayer. Also, the proteins have a precise topology which is generated during synthesis. Parts of the proteins are facing the lumen of the compartment and others are exposed to the interior of the cell. It is important to note that each cellular compartment has its own characteristic set of proteins which carry out functions specific for each compartment. The internal fluid inside the cell into which these cellular compartments are suspended is not water but rather a formless jelly called the cytosol. It is also bolstered by numerous cables crossing the cytosol in different and often carefully specified directions. These cables form the cytoskeleton, and some of these bridge different compartments to each other. The cytoskeleton is made of three different types of filaments that have different diameters and play different roles: Actin filaments and microtubules are the best characterized and are the thinnest and largest, respectively. In addition, there are intermediate filaments whose function is less clear. Unlike what is generally assumed, these cables are not static but extremely dynamic. Actin filaments and microtubules undergo continuous remodeling to serve the needs of the cell in helping to generate the exterior and architecture of

cells. Microtubules play a pivotal role in cell division when they disappear from the cell periphery and polymerize to form the mitotic spindle, which is the cellular apparatus that controls the segregation of the chromosomes to the two daughter cells. Microtubules also constitute tracks which are used to transport cargo across long distances, for example, from the basal to the apical pole in epithelial cells or along the axon in neurons. Actin filaments are well-known to function in muscle contraction. They are also important for cell motility, cell shape, and vesicular transport.

Proteins Must Find Their Compartment in the Cell

Cells are continuously renewing their constituents, like the servicing of modern aircraft: The parts are checked and replaced at specified intervals. This is the case with all the proteins which a cell contains. The cellular proteins are synthesized in the cytosol. The only exceptions are proteins belonging to the mitochondrion. Mitochondria are special because they are probably derived from primitive bacteria which swam into the ancestral precursors of animal cells and remained there to specialize in power plants. They retained some of their genes and also the machinery for transcribing genes and synthesizing proteins. This means that some mitochondrial proteins are specified by nuclear genes and some by the mitochondrial genes, and these latter ones are transcribed and translated inside the mitochondrion into proteins.

Newly synthesized proteins in the cytosol have to find the site in the cell where they belong. This poses no problem for the proteins which perform their function in the cytosol but concerns all the others destined to other locations in the cell. This sorting puzzle which comprises thousands of different proteins in each cell is being studied in laboratories throughout the world. The solutions to this problem lie at the heart of cellular organization. The first clue came from studies by Cesar Milstein in Cambridge and the problem was elucidated in New York by Günther Blobel and Bernhard Dobberstein (Blobel and Dobberstein, 1975; Walter and Lingappa, 1986). They found that proteins secreted by cells have a small extra segment which functions as a postal address for the ER into which these proteins are delivered from the cytosol. The same postal code is used not only by proteins belonging to the ER but also by proteins destined for the Golgi complex, the lysosomes, the plasma membrane, and export from the cells. Other studies have revealed additional categories of postal addresses which are used for transport of mitochondrial and of nuclear proteins from the cytosol to their respective destinations. These three different classes of postal tags are usually not only necessary but also sufficient. When they are tagged artificially onto a bona fide cytosolic protein, the hybrid protein is routed to the ER, the mitochondria, or the nucleus, depending on the tag that was used. Such experiments are routinely performed by splicing the gene segment coding for the postal tag to a gene coding for the cytosolic protein and then expressing this hybrid gene in the test tube or in a suitable animal cell. Usually, the postal addresses can be read correctly by cells from many different species.

The cellular mechanisms for reading the postal codes and routing the proteins to different addresses in the cell have only partially been elucidated. In the case of proteins addressed to the ER, it is known that there is a an escort system to guarantee that proteins that have to be secreted cannot accumulate in the cytosol. This is achieved by a multiprotein complex assembled around a RNA backbone called the signal recognition particle in the cytosol which binds to the postal tag on the protein while it is still being synthesized (Fig. 3). This complex docks to another protein protruding from the ER called the signal recognition par-

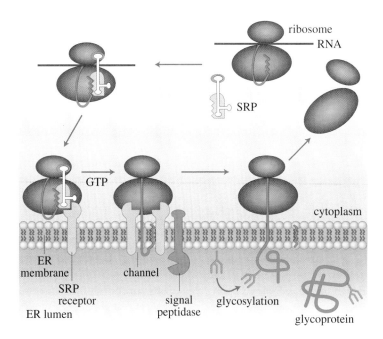

FIGURE 3 Schematic representation of translocation of a protein across the membrane of the endoplasmic reticulum. A secretory protein (green chain) is translated from the corresponding mRNA by a ribosome; as synthesis proceeds a hydrophobic fragment (signal peptide, wavy line) of the protein protrudes from the ribosome. Successively, a ribonucleoprotein complex SRP binds to this signal peptide in the cytosol. At this point protein translation is arrested to allow the ribosome–protein–SRP complex to dock to the SRP receptor present on the ER membrane. Protein translation is resumed in a reaction that requires GTP and the opening of a protein channel through which translocation across the membrane of the ER takes place. A specific protease, signal peptidase, cleaves the signal peptide off the protein. Specialized enzymes then catalyze the addition of carbohydrate side chains to the protein (glycosylation) which then becomes a glycoprotein. After translocation is complete the ribosome is released into the cytosol ready to bind another mRNA molecule to initiate a new cycle of protein synthesis.

ticle receptor. This docking event initiates the passage of the protein being synthesized into the ER. This translocation across the membrane takes place through a protein pore in the bilayer and is often accompanied by the addition of carbohydrate side chains to the protein, which then becomes a glycoprotein. Similar recognition devices and docking proteins exist for mitochondria and nuclei.

The sorting puzzle is further complicated by the fact that proteins routed to one membrane-bound compartment have destinations beyond the first membrane they encounter. The mitochondrion consists of two membranes, each with a specific set of proteins. Also, the nucleus has two membranes. The problem is even more complicated for those proteins that are routed from the ER to the Golgi complex, the lysosomes, or the plasma membrane. These proteins make use of membrane-bound containers called vesicles which bud off from one cellular compartment like soap bubbles to be fused with the next. In fact, the cytosol is full of these membrane bubbles which move about inside the cell.

Transport vesicles allow proteins that follow the biosynthetic pathway (Fig. 4) to be transported from the ER to the Golgi complex and from its exit site, the trans-Golgi network, to the plasma membrane (Rothman and Orci, 1992). From the plasma membrane vesicles containing fluid lipids and proteins are continuously internalized (Gruenberg and

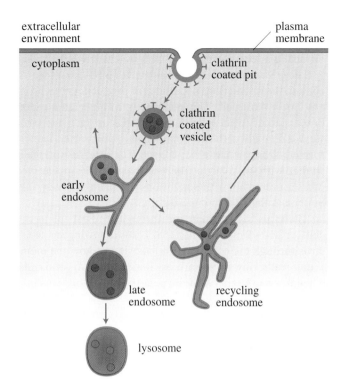

FIGURE 5   The endocytic pathway via transport vesicles. The cytosolic protein clathrin and a set of adaptors assemble on the inner surface of the plasma membrane causing the invagination of the lipid bilayer (clathrin-coated pit). Subsequently, this area of the plasma membrane pinches off in the form of a sealed membrane container (clathrin-coated vesicle) which carries membrane proteins, lipids, and solutes from the surface. These vesicles deliver their contents to the early endosomes. The internalized molecules can return to the plasma membrane either directly or by passing through the recycling endosome. Other molecules are transported into the late endosomes or into the lysosomes where they are degraded.

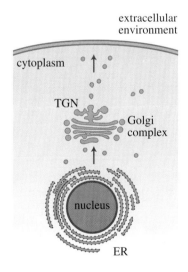

FIGURE 4   The biosynthetic pathway. Proteins destined to the biosynthetic pathway are first translocated across the endoplasmic reticulum (ER) membrane, packed into transport vesicles, and delivered to the Golgi apparatus, in which they undergo various posttranslational modifications. The exit site from the Golgi complex is through the trans-Golgi network (TGN). Here, other vesicles are formed that deliver their cargo of proteins to the plasma membrane as well as vesicles carrying lysosomal enzymes to the endosomes.

Maxfield, 1995). Clathrin-coated vesicles depart and transport their cargo to the first station of the endocytic pathway (Fig. 5), the early endosome. Depending on their function, some molecules can be returned from here to the plasma membrane either directly or by passing through a recycling endosome. Molecules which are destined for degradation continue their journey into compartments which are progressively more acidic and richer in hydrolases, late endosomes, and lysosomes.

Viruses as Blood Bounds

Making sense out of this hubble of bubbles has taken many years. The groundwork was laid by George Palade (1975), who studied the export of proteins from cells. Further insight into how these traffic routes connect the ER with the plasma membrane has been obtained by employing membrane viruses as guides inside the cells. They act as bloodhounds to follow the trail of membrane bubbles connecting intracellular compartments.

These viruses consist of a package of RNA genes which contain the instructions necessary for producing virus progeny in the host cell. This package is wrapped in a membrane similar to the plasma membrane. In fact, the virus steals a segment of the host plasma membrane when a newly formed virus particle leaves the cell. The external surface is studied with spikes giving the virus particle the appearance of a sea urchin but with dimensions so small that it can only be seen in the electron microscope.

The viruses start their tour through the host cell at the cell surface (Simons *et al.*, 1982). By following an RNA virus through the cell, we get a snapshot of how the membrane trafficking routes are organized in eukaryotic cells. The life cycle begins with the penetration of the virus into

the cell. A prerequisite for this interaction to occur is that the virus recognizes chemical structures on the plasma membrane to which the virus particle can attach. Only cell types with the appropriate receptors for the virus can be infected by the virus. Soon after this attachment has occurred, the virus particle is engulfed into an invagination which forms a membrane bubble with a virus inside. The virus is now on its way into the cell interior (Fig. 6). The bubble has a coating of protein organized in a beautifully ordered lattice which plays an important role in its formation. The protein component of this clustering lattice was first discovered by Barbara Pearse (1980). The clustering coat is short-lived and is thrown off before the bubble containing the virus bumps into an intracellular compartment called the endosome and fuses with its membrane. The fusion event is like a small soap bubble fusing with a large one and melting together to form a larger bubble. After the

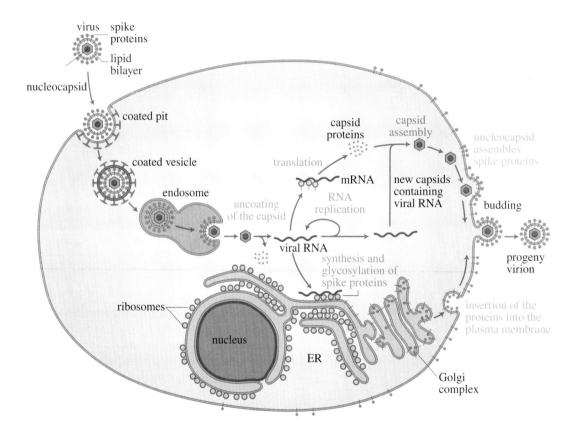

FIGURE 6 Life cycle of a virus. The virus enters the cell through coated pits and coated vesicles and reaches the endosomes, where the viral membrane fuses with that of the endosome. In this way the proteins that constitute the viral envelope are lost but the nucleocapsid, containing the viral nucleic acid, is released into the cytosol. With the uncoating of the capsid, the viral RNA is exposed and the cell unwittingly becomes a factory for virus production. Large amounts of the proteins that constitute the viral spikes are synthesized and glycosylated in the endoplasmic reticulum and transported successively to the Golgi complex (where their glycosylation is completed) and hence to the plasma membrane. In the cytosol the viral RNA is replicated into thousands of copies and is packaged again into new capsids. All viral components are now ready to be assembled into new virions. The virus gene packet can recruit the spike proteins on the plasma membrane and a newly formed virion can bud outside the cell ready to spread the infection (modified from Alberts *et al.*, 1989).

two cellular bubbles have fused with each other the virus is dumped into the endosomal interior. This causes a shock to the virus because the inside fluid of the endosome is acidic. This acid bath triggers a second fusion reaction. This time the virus is the small soap bubble which melts from the inside with a large endosome bubble. The first fusion event delivers the virus into the endosome and the second releases the virus gene package from the virus membrane into the cytosol.

This process is amazingly clever, but somehow it does not make sense. Why would the cell have such a handy mechanism to facilitate virus entry? It seems suicidal. Unfortunately for the cell, the virus relies on a Trojan horse type of deception. The virus makes use of a mechanism of engulfment which every animal cell employs normally to remove hormones and growth factors from the cell surface after they have initiated their action on cellular metabolism and growth. Also, nutrients such as iron and cholesterol are taken up into the cell by this route. There is a continuous stream of protein-coated membrane vesicles from the plasma membrane into the endosomes. The extent of this traffic is so great that the entire plasma membrane surface would be consumed in the formation of these bubbles in an hour or two depending on the cell if it were not for a recycling mechanism. On the other hand, the plasma membrane would grow too much if portions of membrane were not constantly removed. In contrast, the plasma membrane surface area is kept constant by a compensatory traffic of membrane vesicles from the endosomes. This continuous stream of vesicles in the opposite directions can be likened to a pair of escalators. The virus deceives the cell by sneaking onto the escalator into the cell which is normally reserved for other passengers. Most of the normal passengers are routed further than the endosomes; they continue on another escalator to the lysosomes, where they are destroyed as waste. For instance, the cholesterol in the blood circulation is carried by a lipoprotein that binds to receptor proteins on the cell surface which are subsequently internalized using the escalator to the endosomes. In the endosomal acid bath the lipoprotein is released from its receptor, which returns on the escalator to the cell surface to function in another round. The lipoprotein is routed to lysosomes where it is degraded and the cholesterol is released for use in the cell. Michael Brown and Joseph Goldstein found patients with genetic defects in the internalization route (Brown *et al.*, 1983). The lipoprotein in these patients can bind to its receptor on the cell surface but the complex cannot enter the cells because the receptor is defective and therefore cannot board on the escalator to the endosomes. This entry defect leads to the accumulation of cholesterol in the circulation and premature death through atherosclerosis.

After entering the endosomes, the viruses escape the stairway to death in the lysosomes by rapidly fusing with the endosome and releasing their contents into the cytosol.

They leave their "coats" in the endosomal membrane, but the gene package is now in the cytosol. If an efficient way to stop the virus membrane from fusing in the acid bath was known, it would constitute a route to chemotherapy of membrane virus infection. However, this has to be done without impeding the traffic of normal passengers to the endosomes and the lysosomes, and to date this has not been possible.

The virus gene packet in the cytosol causes a grim change in cellular metabolism. The cell is reprogrammed by the virus genes to produce virus progeny. These genes are truly ruthlessly selfish. New gene packets are synthesized by the thousands, as are the spikes of the virus membrane. These spikes are made of proteins which belong to the class routed to the ER. They are translated with a normal postal tag for this destination. From the ER the spike proteins are further transported over the Golgi complex to the plasma membrane. These virus proteins follow exactly the same route to the cell surface as normal plasma membrane proteins. For instance, transplantation antigens and hormone receptors have been shown to follow the trail tracked down by the virus spike proteins. However, the virus spike proteins are produced in large numbers and they are easier for the investigator to trace in the cell than the normal multitude of different plasma membrane proteins. They have therefore been used extensively to study the pathway and the mechanism of transport to the cell surface.

The final stage in the life cycle of a membrane virus takes place at the plasma membrane of the infected cell. Here, the assembly of the virus particle is completed. Once again, the formation of a membrane bubble occurs, only this time it forms not in the cytosol but in the external world outside the cell. A virus-infected cell looks like an enormous soap bubble from which a continuous stream of small bubbles are pinching off, each with one packet of virus gene inside. This is a beautiful spectacle from a mechanistic point of view, but as a result the cell dies. However, the virus-infected cell has taught cell biologists many important lessons.

Moving Proteins from One Compartment to Another

Little is known of the mechanisms by which proteins are selectively transported from one site to another in the pathway of intracellular traffic which starts from the ER. The major sorting function is attributed to the Golgi complex. Here, proteins destined for the lysosomes, for export from the cell, and for the plasma membrane are sorted into separate membrane containers which deliver them to their correct addresses. One postal code in this pathway that has been deciphered is that involved in lysosomal routing (Kornfeld and Mellman, 1989). Enzymes active in waste disposal and recycling in the lysosomes receive a second address tag in the Golgi complex as was discovered by Kurt von Figura and by William Sly and Stuart Kornfeld. This stamp is on the carbohydrate side chains linked to lysosomal enzymes and consists of a phosphorylation reaction of mannose residues attached to the glycoprotein. The lysosomal enzymes are subsequently bound to receptor proteins in the Golgi complex before they are packed into the membrane container which will take them to the lysosomes.

The mechanism by which a bubble, a membrane transport vesicle, forms in one compartment and is moved to the next compartment is under intensive investigation. The process can be divided into five stages: (i) the inclusion of cargo proteins into the membrane patch that forms the vesicle, (ii) the formation and the release of the vesicle from the donor compartment, (iii) the movement of the vesicle to the target compartment, (iv) the docking of the vesicle, and (v) the fusion of the vesicle with the membrane of the target compartment followed by the release of cargo (cf. the process of virus entry). The inclusion of cargo consisting of membrane proteins is often guided by signals present in the domains of the proteins exposed to the cytosol. These bind to adaptor proteins, which in turn assemble a protein scaffold, such as the clathrin coat, that bends the membrane to form a vesicle. Release of the vesicle is dependent on specific proteins that form rings around the neck of the budding vesicle and squeeze the vesicle out from the donor compartment. The lipid bilayers in the vesicle and in the donor membrane have to reseal immediately. Also, this resealing is probably facilitated by specific unidentified proteins.

After the transport vesicle has been released, it must find the correct acceptor compartment. In many cases, this movement occurs along cables formed by the cytoskeleton, usually microtubules. The transport vesicles cling like cable cars to the microtubule cables and carry "motors" which propel them along the cable to the correct destination. After arrival at the target compartment specific docking of the vesicle has to be ensured. Proteins called SNAREs, identified by James Rothman, form specific couples: one in the vesicle called a v-SNARE and another on the target membrane called a t-SNARE (Fig. 7; Rothman, 1994). After docking has occurred through the binding of a v-SNARE to a t-SNARE, fusion of the membrane of the transport vesicle with the target membrane is triggered. It is assumed that the SNAREs mediate this process as well; in addition, however, a switch mechanism must exist to control the formation of the v–tSNARE pair (Fig. 7) and its subsequent disassembly after the vesicle delivery process has been completed. Indeed, in vesicular transport, as in many other biological processes, there are chemical switches, called Rab proteins, which hydrolyze the trinucleotide guanosine triphosphate to diphosphate. This hydrolysis event causes a conformational change on the molecule which can be

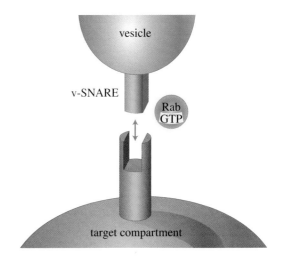

FIGURE 7 Model based on the SNARE hypothesis. Each vesicle contains a protein called vesicle SNARE (v-SNARE) which is specifically recognized by another protein located on the target compartment, t-SNARE. Only if the correct v-SNARE and t-SNARE match like a key and its lock can the vesicle dock and fuse with its acceptor compartment. This interaction, however, is regulated by a molecular switch, called Rab protein, which has to be in the active form, bound to GTP, for SNARE pairing to occur.

viewed as an on/off switch that signals to the transport machinery where and when to release the cargo proteins from the fused vesicular container (Zerial and Stenmark, 1993). Although the molecular principle is the same, every vesicular route utilizes its specific switch. Many Rab proteins (about 30) are expressed in a typical mammalian cells. These proteins are localized very specifically to distinct stations along the biosynthetic and endocytic pathways where they control trafficking from one site to another. What happens if one interferes with their switch function? This can be done by introducing mutations that alter the nucleotide binding or the hydrolysis so that the switch is blocked in the "on" or "off" position. In the case of Rab5, a Rab protein that regulates the transport from the plasma membrane to the early endosomes (Fig. 8) when the switch is blocked in the off position, endocytosis is reduced and the endosomes break down into small fragments. When the switch is blocked in the on position, then endocytosis is stimulated and the endosomes become large vacuoles. In addition to the switch function, Rab proteins also function as timers because they can remain active only for a certain period of time. This mechanism ensures that the proteins stay active for a period of time long enough to allow vesicle transport but not too long to have exaggerated organelle expansion, as in the case of the "on" mutant.

The fascinating conclusion from studies on the complex regulation of the traffic system for intracellular transport from the ER over the Golgi complex to the lysosomes and

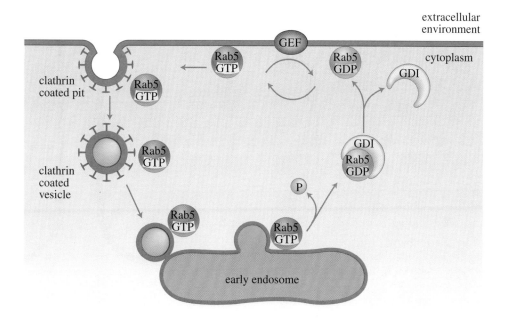

FIGURE 8 Vesicular transport regulated by the protein Rab5, one of a family of molecular switches called Rab GTPases which regulate transport from the plasma membrane to the early endosome. This regulation is controlled by the passage between the active form of Rab5, bound to GTP, and the inactive form bound to GDP. Clathrin-coated pits become clathrin-coated vesicles, which after uncoating fuse with the early endosome. For this to occur, Rab5 must be on the vesicle in the active form. Due to the GTPase activity of Rab5, GTP is hydrolyzed to GDP and Rab5 binds to GDI, an inhibitor of GDP dissociation. GDI transports the protein to the membrane, where a guanine nucleotide exchange factor, GEF (in red) converts Rab 5 to the active state, ready to initiate another cycle of vesicular transport.

the cell surface or beyond is that similar machinery is used at each vesicular transport step. Specificity is ensured by each vesicle and each target membrane having specific Rab proteins and SNAREs.

Important insights into the functioning of the Rab/SNARE machinery have come from studies of yeast cells, which are popular to molecular cell biologists because these simple unicellular organisms provide powerful genetical means to manipulate cell function. One can use yeast cells and obtain results which apply to mammalian cells because all eukaryotic cells are built on the same principles. Once a molecular solution to a central mechanism of life was "invented" during the course of evolution, it was fixed, improved, and reiterated.

Not All Cells Are the Same

It is obvious that yeast cells do not contain the solution to all problems of cellular organization. One question that has received little attention is how membrane trafficking is modulated to serve the requirements of cellular organization in different cell types. All cells have to transport newly synthesized proteins and lipids from the ER over the Golgi complex to the cell surface and to the endosome–lysosome compartments. They also have to endocytose cell surface components into endosomes and lysosomes. However, when one considers the needs of cells such as those lining the digestive tract or of neurons in the brain, radical changes in the traffic routes have to be introduced, especially in the pathways leading from the Golgi complex to the plasma membrane. Epithelial cells line the digestive tract. This lining consists of a single layer of cells connected to each other by cellular junctions. Each epithelial cell has a cell surface divided into two territories (Fig. 9): the exterior apical surface facing the lumen of the gut and the basolateral domain which receives nutrients from the blood supply. The specific traffic problem that the epithelial cells have to solve is how to provide both the apical and the basolateral plasma membrane domains with different proteins and lipids and how to organize the endocytosis routes from the two sides of the cell layer (Simons and Fuller, 1985).

The solution to this problem is novel because only the basolateral route from the Golgi complex employs the SNARE/Rab mechanism, whereas the apical pathway is of a new type. The details of this latter delivery mechanism have not been worked out. However, a central element is the involvement of lipids in the apical sorting process (Simons

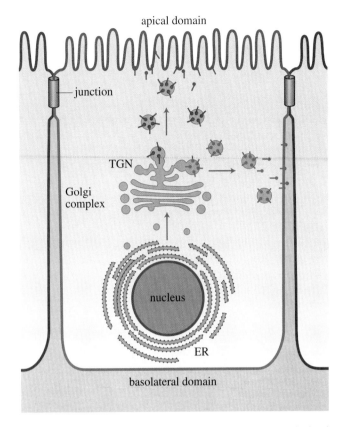

FIGURE 9 The biosynthetic pathway in a typical polarized epithelial cell. The cell surface is divided into two domains: the apical surface faces the lumen of the gut and the basolateral surface faces the bloodstream. The two plasma membrane domains have different protein and lipid contents and intermixing is prevented by the junctions which connect the cells. In the biosynthetic pathway, proteins destined for the apical or basolateral surface share a common route of transport from the ER to the Golgi complex until they reach the TGN. Here, the two sets of proteins, which have different destinations (red and blue), are sorted into distinct vesicles to be targeted to the apical (red) or basolateral (blue) plasma membrane domain.

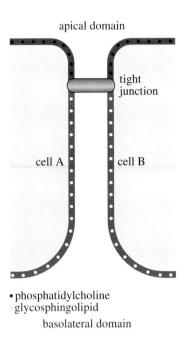

FIGURE 10 Asymmetry of the plasma membrane of polarized epithelial cells. The first level of asymmetry is determined by the fact that the apical membrane (in red) is highly enriched in sphingolipids, especially glycosphingolipids, whereas the basolateral membrane (in blue) has a higher content of phospholipids, mainly phosphatidylcholine. These cannot mix due to the presence of the tight junction. The second level of asymmetry (not shown) is determined by the fact that the glycosphingolipids and phosphatidylcholine are localized exclusively in the outer leaflet of the plasma membrane.

and Ikonen, 1997). We must again consider the organization of the lipid bilayer to understand the principles of the apical mode of membrane transport. The apical membrane (Fig. 10) is highly enriched in sphingolipids, mainly glycosphingolipids (lipids with oligosaccharides as polar headgroups), whereas the basolateral membrane contains much more phospholipids than the apical membrane, mainly phosphatidylcholine. These lipid species are localized to the outer leaflet of the plasma membrane. The tight junctions linking the epithelial cells to each other in the cell layer also form a fence that prevents mixing of the lipids and the protein, but only in the external leaflets of the apical and the basolateral lipid bilayers. The inner leaflet phospholipids (facing the cytosol) freely communicate by lateral diffusion. This segregation of sphingolipids and phospha-

tidylcholine takes place in the membrane of the trans-Golgi network followed by segregated delivery to the correct surface domain. Mechanistically, segregation in the external leaflet of the membrane of the Golgi complex is achieved by a clustering of the glycosphingolipids with each other. The association is facilitated by cholesterol molecules intercalating between the hydrocarbon chains of the sphingolipid molecules. These sphingolipid–cholesterol clusters behave like rafts in the fluid bilayer (Fig. 11). The most interesting property of these lipid rafts is that they specifically associate with proteins that are routed to the apical membrane in epithelial cells. Apical proteins have specific affinity for raft lipids. Thus, the sphingolipid–cholesterol rafts function as platforms for delivery of specific protein cargo to the apical membrane.

This novel mode of membrane trafficking is not restricted to epithelial cells; neurons also employ this mode of delivery for proteins directed to the axon (Fig. 12; Dotti and Simons, 1990). Surprisingly, recent studies have demonstrated that nonpolarized cells such as fibroblasts, which are present in connected tissue, have apical and basolateral cognate routes from the Golgi complex to the cell surface

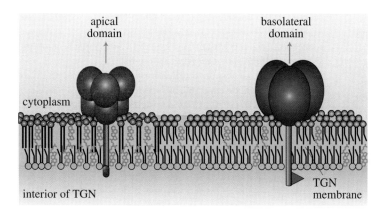

apical
domain

basolateral
domain

cytoplasm

interior of TGN

TGN
membrane

FIGURE 11 The raft model and the microdomain organization of the lipid bilayer. In the TGN, apically directed proteins (in red) are sorted from proteins destined to the basolateral surface (in blue). Concomitantly, sphingolipids and phosphatidylcholine become segregated. This model proposes that the two events are coupled. Glycosphingolipids (in light red) cluster together in a process facilitated by cholesterol molecules (in green) intercalating between the hydrocarbon chains of the sphingolipid molecules. These sphingolipid–cholesterol clusters behave like rafts (the red molecules) in the fluid bilayer. The apical proteins have specific affinity for these lipids and are incorporated in these rafts. The basolateral proteins instead are segregated in the regions of the fluid bilayer rich in phosphatidylcholine (in blue).

(Yoshimori *et al.*, 1996). The difference between the plasma membranes of epithelial cells and fibroblasts is that in the former cells the sphingolipid–cholesterol rafts are preferentially segregated to apical domain, whereas in the latter cells the rafts float about in a continuous fluid bilayer.

Evidence indicates that there are two trafficking circuits connecting the Golgi complex with the plasma membrane — one employing the SNARE/Rab mechanism and another using sphingolipid–cholesterol rafting. This raises new possibilities for subcompartmentalization of protein action in the plasma membrane endosome and the Golgi complex. Previously, the lipid bilayer was considered to be a uniform solvent for membrane proteins. The sphingolipid–cholesterol rafts concept imposes organization into the two-dimensional liquid. Proteins could preferentially associate with the rafts or be excluded into the phosphatidylcholine-dominated fluid regions of the bilayer. Not only membrane trafficking but also processes such as cell surface signaling through hormones and growth factors have been demonstrated to make use of these lipid microdomains for increasing specificity and efficiency. If cholesterol is removed artificially from the membrane, the sphingolipid rafts with their associated proteins disintegrate and cannot perform their functions. The raft-stabilizing function of cholesterol is probably its most important cellular function.

Evidence also indicates that disorders leading to deranged cholesterol metabolism and atherosclerosis as well

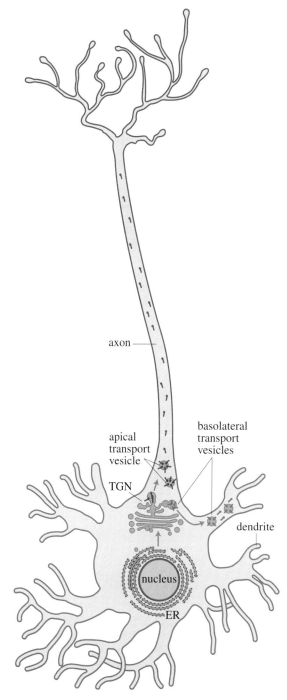

axon

apical
transport
vesicle

basolateral
transport
vesicles

TGN

dendrite

nucleus

ER

FIGURE 12 The biosynthetic pathway in a neuron. Like epithelial cells, neurons have two domains, the dendrites and the axon — structures which are specialized to ensure the transmission of information from one cell to another. The contact site between the axon of one neuron and the dendrite of another neuron is called the synapse. This specific function requires the delivery of distinct sets of proteins to the axon (in red) and to the dendrites (in blue). The segregation of the trafficking routes in the biosynthetic pathway occurs in the TGN.

as to brain dysfunction seen in Alzheimer's disease and in prion-related diseases (scrapies) involve the functioning of sphingolipid–cholesterol rafts. The rafts ensure fidelity and proper processing of key molecules involved in regulation of processes in the brain that are defective in these disease states. It is hoped that further research will provide new insight into effective remedies for these devastating diseases.

References Cited

ALBERTS, B., BRAY, D., LEWIS, J., RAFF, M., ROBERT, K., and WATSON, J. D. (1989). *Molecular Biology of the Cell,* 2nd ed.. Garland, New York.

BLOBEL, G., and DOBBERSTEIN, B. (1975). Presence of proteolytically processed and unprocessed nascent immunoglobulin light chains on membrane-bound ribosomes of murine myeloma. *J. Cell Biol.* **67,** 852–862.

BROWN, M. S., ANDERSON, R. G. W., and GOLDSTEIN, J. L. (1983). Recycling receptors: The round trip itinerary of migrant membrane proteins. *Cell* **32,** 663–667.

DOTTI, C. G., and SIMONS, K. (1990). Polarized sorting of viral glycoproteins to the axon and dendrites of hippocampal neurons in culture. *Cell* **62,** 63–72.

GRUENBERG, J., and MAXFIELD, F. R. (1995). Membrane transport in the endocytic pathway. *Curr. Opin. Cell Biol.* **7,** 552–563.

KORNFELD, S., and MELLMAN, I. (1989). The biogenesis of lysosomes. *Annu. Rev. Cell Biol.* **5,** 483–525.

PALADE, G. E. (1975). Intracellular aspects of the process of protein secretion. *Science* **189,** 347–358.

PEARSE, B. M. F. (1980). Coated vesicles. *Trends Biochem. Sci.* **5,** 131–134.

ROTHMAN, J. E. (1994). Mechanisms of intracellular protein transport. *Nature* **372,** 55–63.

ROTHMAN, J. E., and ORCI, L. (1992). Molecular dissection of the secretory pathway. *Nature* **355,** 409–415.

SIMONS, K., and FULLER, S. D. (1985). Cell surface polarity in epithelia. *Annu. Rev. Cell Biol.* **1,** 243–288.

SIMONS, K., and IKONEN, E. (1997). Functional rafts in cell membranes. *Nature* **387,** 569–572.

SIMONS, K., GAROFF, H., and HELENIUS, A. (1982). How the virus comes in and out from the host cell. *Sci. Am.* **246,** 57–66.

WALTER, P., and LINGAPPA, V. R. (1986). Mechanisms of protein translocation across the endoplasmic reticulum. *Annu. Rev. Cell Biol.* **2,** 499–516.

YOSHIMORI, T., KELLER, P., ROTH, M. G., and SIMONS, K. (1996). Different biosynthetic transport routes to the plasma membrane in BHK and CHO cells. *J. Cell Biol.* **133,** 247–256.

ZERIAL, M., and STENMARK, H. (1993). Rab GTPases in vesicular transport. *Curr. Opin. Cell Biol.* **5,** 613–620.

CHRIS M. COPPIN
DANIEL W. PIERCE

Department of Pharmacology
University of California
San Francisco, California 94143, USA

RONALD D. VALE

Department of Pharmacology and Howard Hughes
Medical Institute
University of California
San Francisco, California 94143, USA

The Molecular Basis of Biological Movements

Movement and reproduction are the two elementary properties that are most commonly associated with life. The mechanism of biological motility that distinguishes life forms from their inanimate environment has long captivated the imagination of scientists. Theories on the propulsion of animals date back in history to ancient Greece, and therefore the scientific investigation of motility is one of the oldest endeavors in the biological sciences. The development of microscopes several centuries ago resulted in an appreciation of new forms of motility at the cellular and subcellular level. The molecular machines that generate movements of a ballet dancer or the equally graceful and purposeful meandering of a bacterium are now known. However, the properties of these protein motors have been found to be remarkably diverse. Some motors make use of an electrochemical gradient as an energy source to spin a propeller, whereas other motors use high-energy chemical compounds [adenosine triphosphate (ATP)] to move linearly along polymeric tracks (cytoskeletal filaments or nucleic acid). The proteins that constitute these machines have been identified, their amino acid sequences are known, and in some cases their detailed three-dimensional structure has been solved. Remarkably, the output (force and displacement) of a single protein motor can now be studied with advanced high-resolution microscopy. This article focuses on how these fascinating molecular machines work and the experimental approaches that are being applied to dissect their mechanism.

Introduction

In contrast to inanimate objects, most living organisms move with purpose, seeking out hospitable environments, nutrients, and mates and moving away from environments that endanger their existence. Naturally, they could rely on diffusion or convection to move around like inanimate objects, and a few organisms do so. However, once the most rudimentary forms of motility arose, those that lacked it must have found themselves at a considerable disadvantage in the competitive world of natural selection. Today, the ability to move purposefully and efficiently is one of the most ubiquitous characteristics of life.

All living cells also need to control their shape. At the very least, they need to be able to cinch down around an annulus in order to divide; therefore, movements of the cell surface are indispensable. Furthermore, immediately before cell division the genetic material needs to be partitioned into the two incipient daughter cells, a process that requires active movement of chromosomes. Eukaryotic (nonbacterial) cells are considerably larger than their bacterial cousins and cannot rely on diffusion to move their intracellular organelles on a reasonable timescale. Therefore, purposeful movements of objects inside the cell are also essential.

How do living organisms produce movement? Nature has evolved a class of highly specialized proteins designed to convert chemical energy into mechanical work. Some work in groups and others work alone, and they all undergo a repeating thermodynamic cycle analogous to that of an internal combustion engine. The demands placed on a molecular motor differ significantly from those placed on a macroscopic engine, however. In the microscopic world, overheating is of no concern because heat dissipation is almost instantaneous. Also, the masses involved are minuscule, rendering inertia and gravity completely irrelevant. However, to a motor trying to propel a bacterium or an intracellular organelle, water is so viscous as to be almost impenetrable and thermal agitation (Brownian movement) is a significant impediment. Nonetheless, natural molecular motors have evolved to such perfection as to vastly outperform the most sophisticated man-made engines. For example, kinesin, a typical molecular motor, functions with 50% efficiency, travels 100 times its own length per second (comparable to a fighter jet), and produces more power per unit mass than a formula one race car engine. This article is an introduction to the remarkable proteins that generate movement, with emphasis on current research into the mechanism of movement generation. The rotary flagellar motor that propels many types of swimming bacteria is presented first. Then the cytoskeleton-based motors myosin, kinesin, and dynein are discussed, with particular attention to novel technologies that permit the manipulation and investigation of a single isolated functioning molecule. Next, we discuss the helicases and polymerases that move along strands of DNA. Finally, the article concludes with examples of cellular movement generated by protein polymerization.

The Bacterial Flagellar Motor

Many bacteria swim toward chemoattractants and away from repellents by making use of a propulsion system reminiscent of the propeller on a submarine (Fig. 1). The machinery involved has been the subject of extensive biochem-

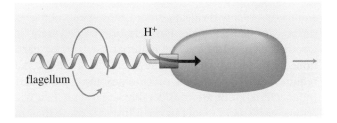

FIGURE I Schematic illustration of a bacterium making use of its helical flagellum to propel itself through an aqueous environment. The spinning of the flagellum is driven by a rotary motor (red) powered by protons (H$^+$) flowing into the cell down an electrochemical energy gradient. This gradient is maintained by ion pumps (not shown) embedded elsewhere in the membrane. Although only a single motor/flagellum assembly is shown, such assemblies are often found in clusters.

ical, biophysical, and genetic investigation, and a general understanding of its structure is emerging although the detailed mechanism of force generation remains unresolved. The motor is composed of an assembly of numerous different protein molecules, lodged in the cell wall, which function together to exert a torque on a rigid screw-shaped flagellum that plays the role of a propeller. Unlike other molecular motors, the flagellar motor does not derive its energy directly from a chemical reaction such as the hydrolysis of ATP. Instead, it functions as an electrical motor, relying on a flux of hydrogen ions (protons) that funnel through the motor down an electrochemical energy potential (Larsen *et al.,* 1974). This potential is maintained by ATP-hydrolyzing enzymes residing elsewhere on the cell membrane that pump protons back out of the cell.

The flagellum usually rotates counterclockwise, as viewed from its tip, and the rate of spin is proportional to the electrochemical potential because the motor transports a fixed number of protons (~1000) per rotation (Meister *et al.,* 1987). However, abrupt reversals in the direction of spin are a frequent occurrence, even though the rotation rate remains unchanged. Such events disrupt the normal forward progress of the swimming bacterium and cause it to tumble aimlessly (Macnab, 1987). The tumbling serves, in part, to allow the bacterium to reset its directional course. Reversal of the rotation sense can also be achieved experimentally by reversing either the proton concentration gradient or the electrical potential (Eisenbach and Adler, 1981). However, attempts at forcing a sense reversal by means of an external torque, without changing the direction of the electrochemical gradient, are opposed by significant resistance and usually result in irreversible damage to the motor (Berg and Turner, 1993). In the absence of a proton gradient no spontaneous rotation is seen. Under these conditions, the motor can be made to act as a proton pump if an external torque is applied to the flagellum, which indicates that the thermodynamic process driving the motor is reversible (Caplan and Kara-Ivanov, 1993). This characteristic sets the bacterial flagellar motor apart from all others.

The overall structure of the motor, as it is currently understood, is shown in Fig. 2 (Caplan and Kara-Ivanov, 1993; Schuster and Khan, 1994). It consists of two main components: an outer ring-like stator, rigidly fixed to the cell wall, and an enclosed rotor that spins with the flagellum. The torque arises from an interaction between the stator and the rotor. Some bacteria (gram-negative) are surrounded by an additional outer membrane, in which case the motor also contains a bushing through which the base of the flagellum (the rod) can spin; this bushing, however, is not thought to take part in torque generation. The rod is connected to the corkscrew-like filament via a universal joint called the hook. It is remarkable that the bacterium, the simplest of all organisms, has developed the biological motor with the most complex and sophisticated architecture.

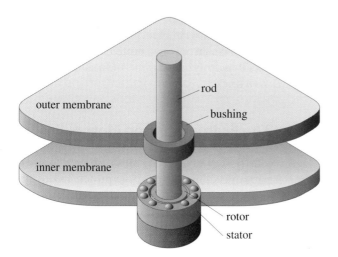

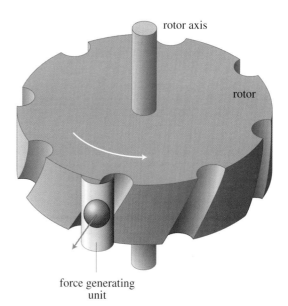

FIGURE 2 Schematic representation of the bacterial flagellar motor illustrating the outer membrane and the bushing (green) system found only in gram-negative bacteria, the rod, the inner membrane, the stator including 12 force-generating units (yellow), the rotor, and the base of the complex which functions as a switch (red).

FIGURE 3 The Läuger model relative to one of the 12 force-generating units of the stator. Each force-generating unit forms a half-channel parallel to the axis of rotation, whereas the outer surface of the rotor is composed of many complementary half-channels that are tilted with respect to this axis. As the proton (blue) moves down the channel, the rotor is forced to turn because of a "scissors" effect. Reversal of the rotational direction depends on a reversal of the angle of tilt of the rotor's half-channels.

Although more than 40 gene products are known to take part in the assembly of the motor and propeller, only 5 are known to be directly essential for rotation. Two of these, motA and motB, pair up into 12 heterodimers which are distributed symmetrically like beads on a ring within the stator (Fig. 2) (Stallmeyer *et al.,* 1989). Each motA/motB complex is thought to constitute an individual force-generating unit. The other three essential gene products, FliG, FliM, and FliN, form the so-called switch complex in the rotor and at the cytoplasmic interface and are presumably responsible both for determining the direction of rotation and for providing a "handle" which can be grasped by the motA/motB force generators. However, the precise geometrical distribution and individual roles of these polypeptides remain to be elucidated. An intricate signal transduction pathway, starting with cell surface receptors for chemoattractants or repellents, acts upon the switch complex to determine the direction of rotation.

How does the flagellar motor couple the proton flow with torque generation? This question has received considerable attention during the past 20 years, and numerous models have been proposed, but the issue remains unresolved. While remaining consistent with the general structural information available, a successful model would have to (i) account for the measured relationship between rotation speed and torque, (ii) account for the high degree of coupling between the transport of a proton and the incremental rotation of the rod, and (iii) provide a mechanism for rapid directional switching.

Since the torque is generally thought to arise from specific interactions between the rotor and individual force-producing units in the stator, the geometrical distribution of the putative stator binding sites on the surface of the rotor has received careful consideration (Läuger, 1977, 1988; Oosawa and Hayashi, 1986; Oosawa and Masai, 1982). A force-generating unit in the stator would bind consecutively to two or more sites (or to a continuous gutter) on the rotor while channeling a proton across the membrane. If the line joining these binding sites is tilted relative to the axis of the rotor, a rotation of the latter would be necessary to complete the transport of the proton across the membrane (Fig. 3). Switching of the rotational sense would require a conformational change on the surface of the rotor such that the relative position of the binding sites is tilted the other way.

Alternatively, the force-producing unit could function as a proton gate, associating simultaneously with a pair of adjacent proton binding sites which would be distributed like beads on a ring around the rotor (Fig. 4) (Berg and Khan, 1983; Khan and Berg, 1983). A proton would reach a binding site on the rotor by entering through the extracellular half of a channel, and it would remain confined there until a stochastic lateral movement of this site relative to the force-generating unit brought it into register with the cytoplasmic half of the channel, allowing the proton to escape into the cell. Directional switching of the sense of rotation would

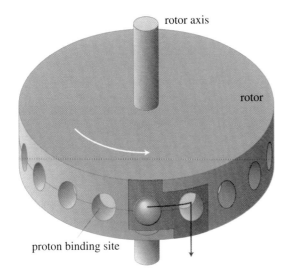

rotor axis

rotor

proton binding site

FIGURE 4 The Berg–Khan type of model. The stator is assumed to contain sigmoidally shaped channels (one of which is shown in green) and the rotor is thought to contain discrete proton binding sites along its outer surface. A proton (blue sphere) can enter through the top of the green channel and lodge in one of these binding sites. At this point, the rotational freedom of the rotor is constrained by the confinement of the proton to the channel. The rotor can undergo a discrete clockwise rotation to bring the proton into register with the downward exit portion of the channel, thereby permitting the trapped proton to enter the cell and making a new empty site available within the channel for a new proton to access from outside the cell. Reversal of the rotational direction depends on an inversion of the sigmoidal channel.

require a conformational change in the stator components, resulting in a reversed gating configuration. Numerous other models, many of them invoking a variety of electrostatic effects, are thoroughly discussed in comprehensive reviews by Caplan and Kara-Ivanov (1993) and Schuster and Khan (1994).

Although considerable progress has been achieved in dissecting the components of the bacterial flagellar motor, much additional work will be needed before a complete understanding of how this remarkable device functions is achieved. The road ahead will be challenging. The resolution of the three-dimensional structure of the stator and rotor proteins by X-ray scattering would be invaluable in elucidating the proton transfer pathway and the interactions between the rotor and the stator. However, this goal will require the crystallization of MotA and MotB, a daunting task because membrane proteins are notoriously difficult to crystallize. Although the elegance of the bacterial flagellar motor is without question, the number of proteins that constitute this motor has presented a significant difficulty in dissecting its mechanism. Ideally, in the future, it would be desirable to measure discrete rotations, torque, and proton flow using modified motors containing only a single active

MotA/MotB force-generating complex. Such simple systems are currently being studied in the context of cytoskeletal motors, in which only two proteins are required for motility: an ATP-hydrolyzing enzyme and a simple polymer that serves as a track.

The Filament-Based Motors

All eukaryotic (i.e., nonbacterial) cells contain a dynamic internal framework composed of various proteinaceous filaments collectively known as the cytoskeleton. These filaments serve, in part, as highways along which motor proteins transport subcellular components such as vesicles or chromosomes. Motor proteins also use these filaments as cables to propagate tension in order to modulate rapidly the shape of the cell. All the movements occurring in the human body depend on the cumulative action of a vast number of motor proteins, each generating only a few piconewtons of force (comparable to the gravitational attraction between two slices of bread) and traveling more than 10 billion miles of filaments. The current understanding of filament-based motors follows in the wake of decades of research on the molecular basis of muscle contraction. The mechanism of force production that has emerged from these studies serves as a general paradigm for the functioning of this class of motors.

Myosin

The mechanism of movement in humans and animals has been a subject of fascination since antiquity. Although the central role of muscles in movement was appreciated early on, it was not until the mid-twentieth century that a rudimentary mechanistic understanding began to emerge. Much of this understanding derives from advances in light and electron microscopy that revealed the fine structure of the muscle's basic contractile units.

The muscle is an organ composed of unusually large, multinucleated syncytial cells, called fibers, that can grow to several centimeters in length and 100 μm in diameter (Fig. 5). The inside of the fiber is packed with parallel longitudinal thread-like structures called myofibrils. Under the light microscope, each myofibril can be seen to contain a highly organized repeating pattern of bright and dark bands, and the bright bands were found to shorten upon contraction of the fiber. It was proposed that each set of bands represents a basic contractile unit about 2.5 μm in length, which was named the sarcomere. It was suggested that the simultaneous shortening of numerous sarcomeres linked in series in a myofibril would lead to length changes on a macroscopic scale.

Electron microscopy revealed the remarkable internal organization of the sarcomere, represented schematically in Fig. 6 (Huxley and Niedergerke, 1954; Huxley and Han-

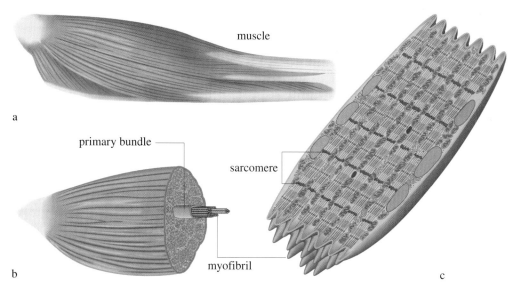

muscle

primary bundle

sarcomere

myofibril

a

b

c

FIGURE 5 Schematic representation of muscle structure, at different magnifications, illustrating the successive levels of structural organization. A muscle (a) consists of a bundle of fibers (b), each of which is composed of a bundle of myofibrils. Each myofibril consists of numerous sarcomeres (c) lined up in series. The sarcomeres are visible in the optical microscope as a regular series of bright and dark bands.

son, 1954). It is composed of a quasi-crystalline array of very fine filaments running parallel to the myofibril's longitudinal axis and distributed symmetrically about the center of the sarcomere. Two kinds of filaments were identified: thin ones, anchored at either end of the sarcomere and extending about half-way toward the center, and thicker ones straddling the center of the sarcomere. Both ends of the

thick filaments were seen to interdigitate with thin filaments in a zone of overlap about 0.5 μm wide.

Upon contraction, the filaments retained their normal lengths, whereas the zone of thick and thin filament overlap grew significantly, which led to the "sliding filament theory" of muscle contraction (Huxley and Niedergerke, 1954; Huxley and Hanson, 1954). In cross-section, the interdigi-

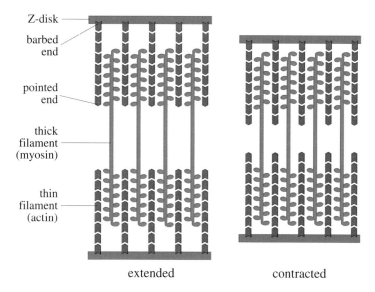

Z-disk

barbed end

pointed end

thick filament (myosin)

thin filament (actin)

extended

contracted

FIGURE 6 Schematic representation of a sarcomere. The sarcomere is bounded at both ends by a proteinaceous anchoring plate called the Z disk. Thin filaments, composed mainly of polymerized actin, are anchored to the Z disk via their barbed end. According to the widely accepted sliding filament theory, bipolar myosin filaments draw the thin filaments toward the center of the sarcomere during muscle contraction.

tated filaments were found to be distributed in a hexagonal array, with six thin filaments surrounding each thick one. Higher magnification electron micrographs of longitudinal sections revealed the existence of periodic cross-bridges between thick and thin filaments in the region of overlap (Huxley and Niedergerke, 1954; Huxley and Hanson, 1954), and their potential importance in promoting the sliding of thick and thin filaments past each other was immediately recognized.

Of the many proteins found in muscle, two are extremely abundant: a 460-kDa) protein called myosin and a 42-kDa protein called actin. Purified actin and myosin can both self-assemble into filaments that take on the appearance of the thin and thick filaments of the sarcomere, respectively. Purified myosin filaments spontaneously bind to purified actin filaments, and the filaments slide past each other in the presence of the nucleotide ATP. It is now known that myosin is an enzyme that converts the chemical energy of ATP's gamma-phosphate bond into mechanical work by exerting a parallel force on an actin filament.

The polarity of the filaments is important. The myosin (thick) filament is a bipolar polymer (symmetrical about its midpoint) containing approximately 300 molecules. In contrast, each actin filament is unipolar, presenting asymmetrical binding sites to myosin and having distinct ends called "barbed" and "pointed." The thin filament in the sarcomere contains approximately 380 actin molecules aligned in a head-to-tail manner. Actin filaments on opposite sides of the sarcomere are oriented in opposite directions, with their barbed end anchored at the sarcomere's boundary (Fig. 6). Each half of the bipolar myosin filament travels toward the barbed end of its respective adjacent actin filaments, thereby drawing antiparallel actin filaments toward each other and shortening the sarcomere.

An isolated myosin molecule contains two globular heads joined at the neck, which are responsible for its actin-binding and mechanoenzymatic activity, and a long tail capable of oligomerizing in a staggered side-by-side manner with other myosin molecules to form the backbone of the myosin filament (Fig. 7). This molecule is a dimer containing two identical heavy chains whose long α-helical domains coil together to form the tail (Cooke, 1993). The heads, each associated with a pair of small light chains involved in enzymatic regulation, are the cross-bridges that were seen connecting the overlapping actin and myosin filaments in early electron micrographs of the sarcomere.

Although muscle myosin has been the subject of intensive study for half a century, it is only one member of a large family of homologous proteins known collectively as the myosin "superfamily" which, like actin, are ubiquitous in eukaryotic cells. To date, at least 11 major classes of myosins have been identified in a wide variety of cell types, and this number is increasing steadily as a result of the discovery of novel myosin-related genes (Mooseker and Cheney,

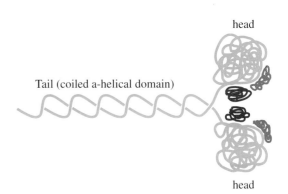

FIGURE 7 Schematic representation of the overall structure of muscle myosin. The two heavy chains (light blue) dimerize to form a single tail formed from two α-helices twisted together (coiled coil) from which emerges a pair of globular heads endowed with mechanoenzymatic activity. The heads are each associated with a pair of much smaller light chains called essential and regulatory (purple and dark blue).

1995). The motor domains (heads) are highly conserved throughout the superfamily, whereas there is significant amino acid sequence variability between the tails, probably reflecting variations in functional specificity (e.g., some tail regions bind to membrane surfaces). Interestingly, some myosins do not dimerize, which suggests that a single head is sufficient to fulfill the mechanical function.

Although the specific function of most nonmuscle myosins is unclear, immunolocalization studies and phenotypic characterization of myosin mutants have suggested probable roles for some of them (Fig. 8). For example, in the important process of cell division known as cytokinesis, myosin is responsible for the contraction of a peripheral belt of actin filaments resulting in the separation of the two daughter cells. Myosin is also involved in the process of phagocytosis, when a cell undergoes a local surface deformation to surround and ingest a small particle or bacterium. Some cells crawl efficiently over solid surfaces by extending and anchoring temporary "limbs," known as pseudopods, which contain high concentrations of myosin at their tips. Other static cells feature permanent specialized protrusions, such as microvilli on intestinal cells (for maximizing absorptive surface area) and stereocilia in auditory hair cells (for resonating with acoustic vibrations), that contain highly organized bundles of actin filaments which are linked to the plasma membrane by myosin. Some myosins have also been shown to transport membrane-bound intracellular vesicles along actin filament.

Kinesin

In addition to actin filaments, all eukaryotic cells contain another kind of cytoskeletal filament, the microtubule, which is a polymer of the protein tubulin. These filaments are four times wider and considerably stiffer than actin fila-

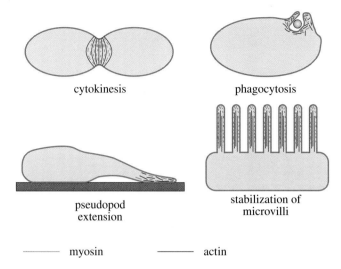

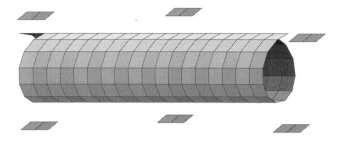

FIGURE 9 Structure of a microtubule. Individual 8-nm-long tubulin dimers (green and blue) are shown dispersed in solution. Dimers polymerize into a hollow helix consisting of 13 protofilaments running parallel to the helical axis.

myosin ——— actin ———

FIGURE 8 Four examples of the roles of nonmuscle myosins. In cytokinesis myosin serves to tighten a belt of actin filaments, leading to the separation of two daughter cells during cell division. Myosin also plays an important role in the formation of the leading foot (pseudopod) when a cell is crawling. In phagocytosis actin and myosin promote the formation of cellular extensions that embrace a particle destined to be engulfed by the cell. Lastly, myosin links bundles of parallel actin filaments to the cell membrane to stabilize microvilli, which are finger-like cellular extensions that are essential for efficient intestinal absorption of nutrients.

ments, and they usually radiate out from a central location in the cell called the centrosome. Like actin filaments, microtubules are polarized structures and have distinct ends (designated "+" and "−") and present asymmetrical binding sites to microtubule-binding proteins. Microtubules play an important role in the spatial organization of the intracellular milieu, and they serve as roadways for shuttling intracellular components such as membrane-bound organelles or chromosomes to the correct intracellular destination at the correct time. This movement depends on the activity of direction-specific motor proteins, two of which (kinesin and dynein) have been the subject of intensive study.

Kinesin was first identified as the enzyme responsible for fast anterograde (toward the microtubule's + end) axonal transport, which is the transport of vesicles laden with neurotransmitter molecules from the body of a nerve cell down its long axon to the point of contact with another nerve cell (Fig. 9) (Brady, 1985; Vale *et al.,* 1985). This motor binds readily to a variety of intracellular organelles and carries them along microtubules to specific locations, although the regulatory mechanism that determines which organelle goes where, and at what time, remains unresolved. The molecular geometry of kinesin resembles that of myosin but on a slightly smaller scale (360 kDa). It is a

dimer composed of two globular heads possessing mechanoenzymatic activity: a long helical tail capped by another pair of small globular domains and a pair of light chains (Vale, 1993). In contrast to myosin, the light chains are associated with the globular end of the tail and are thought to play a role in interacting with the cargo organelle. Like myosin, kinesin is considered the prototype for a family of homologous motor proteins known as the kinesin superfamily, some of which travel exclusively toward the microtubule's + end while others move toward the − end (Goldstein, 1993). The majority of the kinesin-related proteins appear to be involved in chromosome movement and segregation. Interestingly, although the kinesins and the myosins share the same overall geometry, there is no apparent amino acid sequence homology between these families. It is very likely that they diverged a long time ago in evolution from a common ancestral "protomotor."

Dynein

Whereas kinesin carries out + end-directed transport of vesicles along microtubules, cytoplasmic dynein is the most ubiquitous − end-directed transporter (Holzbaur and Vallee, 1994; Vallee, 1993). For example, this motor is responsible for retrograde axonal transport, an important process designed to recycle membranous structures (packaging material) that are deposited at the tip of the axon by kinesin. Members of the dynein family of motor proteins also play a central role in remarkable propulsion devices called cilia and flagella (which bear no relation to the bacterial flagella discussed previously). Both of these structures are long, thin (~200 nm in diameter) extensions of the cell surface that propagate a bending wave, like a whip, in a repeating motion called beating. Their function is to propel the cell relative to the medium or vice versa.

Cilia are relatively short (~10 μm) and are found crowded together on the cell surface. They beat in unison, standing erect when swinging in one direction and bending flaccidly during the return stroke, in order to move a layer of fluid or microscopic particles in a particular direction

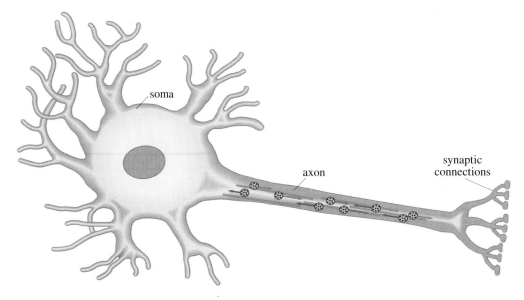

FIGURE 10 Schematic representation of axonal transport along microtubules (in blue) which are all oriented with their + end distal to the cell body. Kinesin transports outbound vesicles from the cell toward the synaptic connections (red arrows), whereas cytoplasmic dynein transports inbound vesicles in the opposite direction (blue arrows).

past the cell (Fig. 10). Cilia play diverse roles, such as the swimming of the unicellular organism *Paramecium,* the outward transport of contaminants trapped in the mucous of respiratory airways, and transport of the ovum along the fallopian tube.

In contrast to cilia, flagella are usually solitary, extremely long (up to 500 μm), and specialized for high-velocity propulsion of a single cell. Their most celebrated role is in the propulsion of sperm cells swimming toward an ovum. This form of propulsion depends on the propagation of a sinusoidal wave, like a snake (Fig. 10).

Cilia and flagella contain the same internal structure, the axoneme (Satir, 1992), which is a highly organized bundle of microtubules linked together by several kinds of proteins including ciliary dynein (which is closely related to cytoplasmic dynein). A normal microtubule is a hollow, cylindrical polymer arising from the linear self-assembly of $\alpha\beta$ tubulin dimers (Fig. 11). These dimers are arranged in 13 parallel rows ("protofilaments") which form the wall of the cylinder; each protofilament is composed of a sequence of alternating α and β tubulin molecules. The center of the axoneme contains two such microtubules surrounded by nine axonemal doublets, each of which consists of a fused pair of microtubules (Fig. 12). In each doublet, one microtubule (designated "A") is complete, whereas the other one ("B") contains only 10 protofilaments. Along the length of each A tubule are found two rows of dynein motors that are permanently attached via their tail in an orderly manner with a constant spacing. The globular heads can bind reversibly to the B tubule of the adjacent doublet, and when they do so they exert a longitudinal force which promotes

the sliding of the two doublets relative to each other. Since the doublets are firmly anchored at their base, this sliding results in the bending of the axoneme (Fig. 13). However, it is important that not all the dyneins in the axoneme exert a force simultaneously; otherwise the axoneme would simply twist into a helix. Despite decades of intensive re-

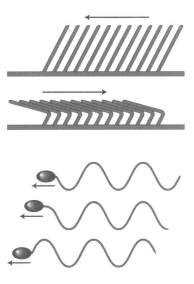

FIGURE 11 Schematic representation of the movement of cilia and flagella. (Top) Cilia are hair-like extensions of the cell surface that work in groups. They all beat in unison to move fluid or particles past the cell surface, using a high profile in one direction (top) and a low profile during the return stroke (bottom. (Bottom) Propulsion of a sperm cell determined by a flagellum which propagates an undulating wave.

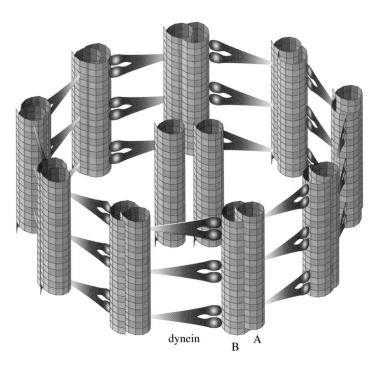

dynein   A
    B

FIGURE 12 Schematic representation of a section through an axoneme. A central pair of microtubules is surrounded by nine doublets of microtubules fused together in pairs. Dynein motors (in red) are firmly anchored by their tails to the A tubules in each outer doublet, whereas their force-producing heads interact with the incomplete B tubules of the adjacent doublets. For simplicity, only one of the two rows of dynein molecules is shown between adjacent microtubular outer doublets.

search on the movement of cilia and flagella, the exquisite spatial and temporal regulation of the dynein activity necessary for effective wave propagation remains obscure. Currently, much attention is focused on the potential regulatory role of an array of proteins radiat-

ing out of the central pair to connect with the outer doublets, forming structures called radial spokes that are visible in electron micrographs (Fig. 12) (Holzbaur and Vallee, 1994).

All dyneins are enormous molecular assemblies compared to myosin and kinesin. They contain two (in some cases three)

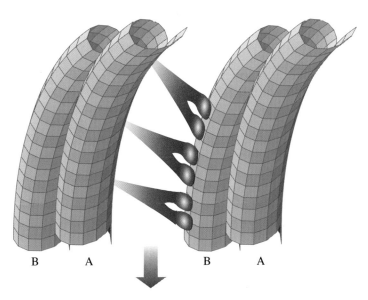

B   A   B   A

FIGURE 13 Schematic representation of an isolated pair of outer doublets which bend as the dynein heads move downward toward the − end of their B tubule. Since both doublets are firmly anchored in the same base (called the basal body), the force exerted by dynein produces a bending torque.

sizable heavy chains (400–550 kDa), which consist of a globular head possessing mechanoenzymatic activity connected to a putative elongated stalk domain, and a highly variable assortment of at least seven intermediate and light chains (8–70 kDa) that are thought to associate with this stalk (Holzbaur and Vallee, 1994; Vallee, 1993). Nonetheless, the overall geometry of dynein resembles that of myosin and kinesin because it has multiple globular enzymatic heads that emerge from a common elongated body.

Mechanism of Force Production

The foremost role of molecular motors is to produce work, a task that involves both movement and force generation. Naturally, a source of energy is needed. Myosin, kinesin, and dynein obtain this energy by catalyzing the hydrolysis of the gamma-phosphate bond of ATP and then converting it into mechanical work, which is why they are sometimes called "mechanoenzymes." The work produced by a single molecule is achieved incrementally through a repeating mechanochemical cycle in which at least one ATP is hydrolyzed and a discrete microscopic movement occurs along the filament. The mechanism of the mechanochemical cycle has been a subject of widespread fascination for decades (Huxley, 1981; Jiang and Sheetz, 1994; Vale, 1994). For all three prototypical motors the broad mechanistic principles have been elucidated; however, the precise molecular event responsible for force production remains elusive.

The general principles of the chemomechanical cycle are as follows. Each cycle is composed of many different states that the motor adopts in an obligatory sequence. At the end of the cycle, the motor is in its original state, an ATP has been hydrolyzed, and a discrete movement has occurred. The "rotating cross-bridge" hypothesis, developed originally to explain the mechanism of muscle contraction (Huxley, 1969), has gained widespread acceptance as the preeminent mechanochemical paradigm for myosin, kinesin, and dynein. The central tenet of this hypothesis is that the motor head binds to a specific site on the filament, undergoes a reorientation (either en bloc or through a conformational change) called the power stroke which pulls the tail forward, and then dissociates from the filament. The cycle then resumes with a rebinding to a different site on the filament. During this sequence, an ATP molecule binds to the motor and is hydrolyzed, and the products (ADP and Pi) are released, but the timing of these events relative to the formation and dissociation of the motor–filament complex varies among different motors. Figure 14 illustrates the sequence of events thought to occur in the mechanical cycles of the myosin and the dynein families (left) and the kinesin family (right). The major difference is that myosin and dynein hydrolyze ATP after dissociating from the filament, whereas kinesin does so while bound to the filament.

The fraction of the cycle time spent bound to the filament, called the duty ratio, is well adapted to the particular function of each motor. For example, a myosin molecule from skeletal muscle is designed to work in concert with numerous identical partners in the thick filament to bring about rapid movement relative to the thin filament. In order to avoid impeding the efforts of its partners through drag, each myosin head must spend as little time as possible bound to the thin filament (Uyeda *et al.*, 1990). In contrast to the high concentration of motors found in sliding filament configurations, transported vesicles are thought to have a relatively sparse distribution of motors on their surface. Therefore, it is advantageous for organelle-transport motors to maximize the time spent attached to the filament; otherwise the vesicle would run a high risk of diffusing away from the filament. The duty ratio has a profound effect on the motor's processivity, i.e., the ability of a single molecule to travel long distances along a filament in an *in vitro* motility assay. For example, myosin usually diffuses away after a single cycle (nonprocessive) (Finer *et al.*, 1994) and axonemal dynein is thought to also do so, whereas kinesin typically undergoes approximately 100 cycles before dissociating (highly processive) (Hackney, 1995; Vale *et al.*, 1995). The processivity of dimeric motors might be further enhanced through a putative coordination between the two heads, which would ensure that one head is bound to the filament when the other one dissociates. Because its high processivity permits the tracking of a single molecule through many consecutive chemomechanical cycles, kinesin is an excellent motor for study by *in vitro* biophysical techniques.

The following are some of the most intriguing questions about molecular motors: What determines their directionality along a polarized filament? Why do all known myosins proceed toward the barbed end of the actin filament? Why does kinesin travel toward the + end of the microtubule while dynein moves toward the − end? Astonishingly, some motors in the kinesin superfamily, such as the fruit fly protein NCD, travel in the opposite direction of kinesin even though the amino acid sequences of their motor domains are 40% identical (Goldstein, 1993). Although there is no shortage of ideas on this subject, very little experimental evidence has been obtained to support any of them. The answer may depend on the elucidation of the three-dimensional structure of these proteins in each of the states of their respective cycles. Examination of such structures may not only offer insight into the basis of directionality but also confirm or refute the hypothesis of a significant conformational change during the power stroke. Atomic-level resolution three-dimensional structures have been determined for myosin, kinesin, and NCD by X-ray crystallography (Kull *et al.*, 1996; Rayment and Holden, 1994; Sablin *et al.*, 1996). Although only a few chemomechanical cycle states are represented, these structures are providing invaluable insight into a unified mechanism of motor function, and

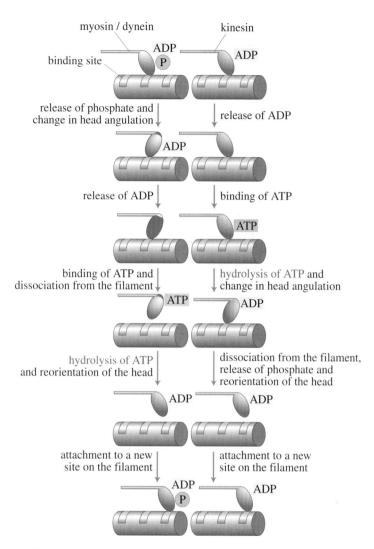

FIGURE 14   Diagram illustrating the sequence of events in the chemomechanical cycles of myosin and dynein (left, in blue) and of kinesin (right, in green). The main difference is that myosin and dynein hydrolyze ATP after the dissociation from the filament, whereas kinesin hydrolyzes ATP while it is attached to the filament. For simplicity, some intermediate states are not shown.

they reveal some remarkable similarities with the mechanism of an entirely different class of extremely important proteins specialized for signal transduction.

What should one expect to find in the three-dimensional structure of a filament-based motor? A priori, four elements are required: a filament binding site, a nucleotide (ATP) binding site, a pivot point, and a lever arm to amplify small conformational changes. The heart of the motor is the nucleotide binding site, where ATP is hydrolyzed and its energy passed on to the other structural elements of the molecule. Not surprisingly, the amino acid sequences and the crystal structures reveal that this vital "pocket," in which the nucleotide sits, is very similar among motors, and its general layout even resembles that found in signal-transducing G proteins, which use nucleotide hydrolysis to

toggle a molecular signaling switch (Sablin *et al.*, 1996). There are also striking similarities among these classes of proteins in the secondary structure elements (loops and helices) that serve to sense the state of the nucleotide (whether it is hydrolyzed) and relay that information to the part of the protein that mediates surface interactions with other proteins, such as those in a filament. Based on these observations, it appears that a universal nucleotide-sensing switch mechanism may exist.

With regard to the general mechanism of rotation or bending about an intramolecular hinge, it is helpful to compare the structure of the myosin family of motors with that of the kinesins. Although these families exhibit little amino acid sequence homology, the crystal structures of the core of the motor domains were unexpectedly found to be nearly

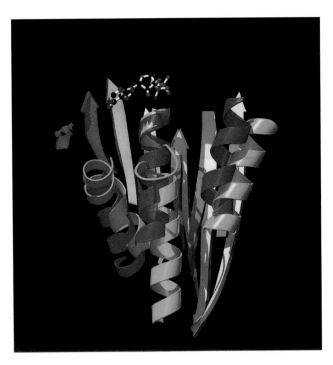

FIGURE 15 Partial crystal structures of kinesin and myosin showing the high degree of secondary and tertiary structural homology in the region adjacent to the nucleotide binding site. This (ADP) is represented as the red, white, and blue stick figure at the top. Seven β-sheets (arrows) and three α-helices are shown for each protein. Kinesin is colored yellow and cyan, whereas myosin is colored green and magenta.

identical (Fig. 15), which suggests the existence of a general strategy for movement generation (Kull *et al.*, 1996).

For myosin, it has been proposed that a modest relative movement between two subdomains introduces a rotation at the base of an 8-nm-long α-helix (Fisher *et al.*, 1995), the other end of which would describe an arc corresponding approximately to the measured unitary displacement step size of ~4 nm (Molloy *et al.*, 1995). Kinesin was found to contain an analogous hinge point at the end of its "neck," a putative α-helical segment that extends out of the globular head to converge with the neck of the partner head onto the dimerized tail (Kull *et al.*, 1996). Thus, the evidence could be consistent with a conformational bending mechanism for force production, in which one part of the motor molecule remains immobilized on the filament while the rest of the molecule rotates forward about a hinge point. It must be noted, however, that one cannot rule out the possibility that the hinge point resides at the interface between the motor and the filament, and that the entire motor rotates as a rigid body following a small conformational change at the filament binding site. Therein lies one of the fundamental questions regarding the generalized mechanism of motor function: Does the power stroke correspond

to a large conformational change within the molecule or does it arise from a rotation at the binding site with the filament? Furthermore, it remains to be determined whether the filament plays a passive role, like a paved roadway, or whether it takes an active part in the movement of the motor. Making use of the full armament of mutagenesis, crystallography, and single-molecule *in vitro* motility assays, it will be possible to answer this question in the near future. Finally, it appears that the determinants of directionality involve subtle structural properties of the motor because no dramatic difference was found between the crystal structures of kinesin and ncd (Kull *et al.*, 1996; Sablin *et al.*, 1996). It seems plausible that the events in and around the nucleotide pocket are identical in these two motors, but that they trigger oppositely directed rotations about the putative hinge point.

Single-Molecule Motility Assays

During the past decade, methods have been developed for examining the motile behavior of purified molecular motors *in vitro*. In one of the first "*in vitro* motility assays," purified muscle myosin molecules were adsorbed onto microscopic synthetic beads which were then observed to move unidirectionally along actin cables extracted from an algae called *Nitella* (Sheetz and Spudich, 1983). In a subsequent refinement of the assay, myosin molecules were adsorbed onto the surface of a glass microscope slide. Fluorescently labeled actin filaments diffusing close to the surface were readily captured by myosin, which then propelled them in a rapid repetition along this surface (Harada *et al.*, 1987; Kron and Spudich, 1986; Toyoshima *et al.*, 1987). Using a fluorescence microscope equipped with a video camera, the mean velocity of actin movement could easily be determined, and it was found to increase as a function of the concentration of myosin on the glass slide (Uyeda *et al.*, 1991). This observation is consistent with myosin's small duty ratio. Each myosin/actin interaction can be thought of as a sharp impulse: The speed increases as a function of the impulse frequency, which increases as a function of the number of myosin molecules in contact with the actin. However, because of myosin's small duty ratio, this kind of assay did not lend itself to the investigation of the movement produced by a single myosin molecule. The time interval between successive myosin/actin interactions proved to be sufficiently long to allow the filament to diffuse away from the surface and out of reach of the motor. In contrast to myosin, kinesin is endowed with a high duty ratio, which makes it ideally suited for single-motor motility assays. A gliding microtubule, propelled by a single kinesin molecule, could easily be tracked as it traveled many micrometers before escaping from the motor's grip and diffusing away from the surface. As expected, increasing the kinesin concentration proved to be detrimental to the velocity because of the drag caused by the excessive number of motors bound to the microtubule at the same time (Howard *et al.*, 1989).

A biophysical characterization of the mechanical behavior or a single motor molecule calls for measurement of such quantities as its step size (magnitude of the unitary displacement), the maximum load under which it can work, the force that it can generate in a single power stroke, and the load dependence of its mechanical cycling rate. In pursuit of these measurements, investigators have been confronted with significant technological challenges because the movements are so small (a few nanometers), the forces so feeble (a few piconewtons), and the events so sudden (a few milliseconds). The two main problems were that the motor molecule cannot be observed directly while it is operating and the inherent noise in the measurement arising from thermal agitation of the molecules in the assay (Brownian movement).

The first problem was solved by observing a larger object that served as a positional reporter of the movements of the motor. For example, a single motor traveling along a fixed surface-bound filament could be attached to a micrometer-size bead (Block *et al.*, 1990; Gelles *et al.*, 1988; Sheetz and Spudich, 1983); conversely, the bead could be attached to a filament being propelled by a single surface-bound motor molecule (Kuo and Sheetz, 1993; Malik *et al.*, 1994). The noise due to Brownian movement can be attributed in large part to the slack in the motor's tail. It was found that the application of a modest external load to stretch the motor molecule could sufficiently attenuate the Brownian movement and reveal the underlying discrete movements along the filament (Finer *et al.*, 1994; Svoboda *et al.*, 1993). An external load has been applied in two different ways. In one approach, a filament being translocated by the motor was attached at its end to the tip of an ultrafine glass needle (Ishijima *et al.*, 1996; Meyhöfer and Howard, 1995). As the needle was slightly flexed by tension in the filament, it exerted an elastic restoring force which was transmitted through the filament to the motor (Fig. 16).

In a second approach, a bead attached either to the motor or to the tip of a filament is placed in the weak force field of an optical trap (laser tweezers) (Block *et al.*, 1990; Finer *et al.*, 1994; Svoboda *et al.*, 1993). As the motor pulls the bead away from the center of the trap, it experiences an elastic restoring force analogous to the one exerted by the flexing of a microneedle (Fig. 17).

One might wonder how discrete movements on a scale of a few nanometers can be measured with an optical microscope whose resolution is subject to a diffraction limit of several hundred nanometers. Although it is true that an optical microscope cannot resolve two objects spaced a few nanometers apart (they would appear as one), it is nonetheless possible to detect movements of a single object with

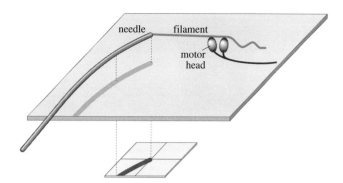

FIGURE 16 Schematic representation of microneedle-based high-resolution single-motor motility assay. A filament is pulled by a motor bound to the surface of the microscope slide. The other end of the filament is attached to the tip of a fine glass needle (gray) oriented perpendicular to the filament. The tension in the filament flexes the needle which exerts an elastic restoring force, thereby removing the slack and attenuating the unwanted Brownian movement. The tip of the needle is imaged with a microscope's objective lens (not shown) onto a photodiode detector which measures its movements with nanometer accuracy. The detector can have four quadrants, as shown, but two quadrants are sufficient for measuring one-dimensional movements.

nanometer accuracy by using an appropriate detector such as a quadrant photodiode. The image of the bead or needle tip is formed on the center of a square pattern of four independent photodiodes, each of which (called a quadrant) generates a current proportional to the intensity of light illuminating it. Movements as small as 1 nm can be detected in two dimensions because they produce measurable differences in current between the four quadrants (Finer *et al.*, 1994; Malik *et al.*, 1994; Meyhöfer and Howard, 1995).

High-resolution optical microscopy has been used successfully in conjunction with optical traps to demonstrate that both myosin and kinesin move along their respective filaments in abrupt discrete steps. Myosin steps are in the range of 3–11 nm, which is geometrically consistent with the swinging lever arm hypothesis proposed on the basis of its crystal structure (Finer *et al.*, 1994; Molloy *et al.*, 1995). Kinesin pauses at regular 8-nm intervals along the microtubule, which neatly corresponds to the intertubulin dimer spacing along a protofilament (Svoboda *et al.*, 1993). Kinesin was also found to occasionally undertake backward steps that become more frequent as the load increases, which indicates that overall forward directionality of movement arises from a modulatable forward bias in the mechanical cycle. Attempts are under way to measure the "isometric" force produced by these motors, i.e., the load sufficient to arrest a motor in mid-power stroke. One approach implements a rapid feedback circuit that displaces the trap away from the motor as soon as the motor starts

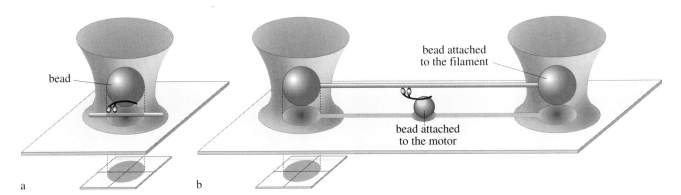

FIGURE 17 Schematic representation of two laser trap-based high-resolution single-motor motility assays. (a) A protein motor (in red) is attached to a microscopic latex or silica bead that is trapped in the force field of a highly focused laser beam. The trap is manipulated to bring the bead in contact with a filament (in green) bound to the surface of the microscope slide. As the bead is pulled by the motor along the filament, its position is tracked by a quadrant photodiode detector. The trap is held fixed and exerts an elastic restoring force that pulls back on the motor. (b) An alternative configuration designed to minimize unwanted interactions with the glass slide, where the motor is attached to a bead that is immobilized on the surface (gray). A bead is also attached to each end of the filament. These end-linked beads are pulled apart by two laser traps in such a way as to stretch the filament like a guitar string. By manipulating the traps, the middle of the filament is brought into contact with the bead serving as the motor's pedestal. The displacements of the filament produced by the motor are measured by tracking the position of one of the end-linked beads with a quadrant photodiode detector.

pulling on the bead in such a way as to immobilize the bead (Coppin *et al.*, 1995; Finer *et al.*, 1994). The movement of the trap is proportional to the force exerted by the motor. However, in practice this kind of measurement is problematic because it requires an extraordinarily fast feedback and very stiff molecular linkages.

Although originally motor proteins were purified solely from organisms and tissues that expressed them in nature, these proteins are now being routinely cloned and expressed in bacteria in the laboratory. This approach offers two major advantages: Certain proteins that are very scarce in nature can be obtained in abundance for experimentation, and genetic engineering techniques can be used to produce specific mutations to test mechanistic models. For example, the mechanical activity of genetically engineered single-headed kinesin motors is being investigated to determine whether processive movement requires coordination between two heads (Berliner *et al.*, 1995).

Another technology that holds great promise for the study of motor proteins is evanescent wave microscopy, also known at total internal reflection (TIR) microscopy (Axelrod, 1989; Funatsu *et al.*, 1995). This is a type of video-enhanced fluorescence microscopy which permits the visualization of an individual fluorescently labeled motor molecule as it travels along a filament. Although it lacks the spatial and temporal resolution of a quadrant photodiode detector, this approach has proven useful in demonstrating that kinesin can move processively without carrying any cargo (Vale *et al.*, 1995). Equally impressive was the observation of individual fluorescently labeled nucleotides binding to a single myosin molecule and their subsequent dissociation as hydrolysis products after the expected dwell time (Funatsu *et al.*, 1995). Future experiments will combine high-resolution optical trap microscopy technology with the TIR approach to permit the long-awaited direct investigation of the correlation between individual events of nucleotide hydrolysis and movement. Specifically, it will be possible to resolve the controversial issue of a motor's "coupling", i.e., whether or not each motor–nucleotide interaction produces a movement.

Polymerases and Helicases

DNA and RNA polymerases travel along a DNA template strand to replicate it into new DNA or transcribe it into RNA. Helicases also travel along DNA, separating the DNA double helix into single strands as they travel. Although generation of movement and force is not the biological purpose of these enzymes, polymerases and helicases move vectorially along DNA tracks in a manner similar to the movement of kinesins, myosins, and dyneins along protein polymers. Like motors, polymerases and helicases must move quickly in order to perform their roles. A significant fraction of the cell cycle of eukaryotes is devoted to replication of the genome, and in bacteria such as *Escherichia coli* it takes almost three times the minimum duration of the cell cycle to replicate the chromosome. The bacterium resolves this paradox by starting the next two rounds of replication before the last one has finished

(Lewin, 1987). In addition to the requirement for speed, polymerases must produce force sufficient to overcome an opposing force (due to DNA supercoiling) estimated at approximately 6 pN (Cook *et al.,* 1992). Additional force may be required to displace DNA-binding proteins bound to the template in front of the polymerase (Polyakov *et al.,* 1995). Nonetheless, the *E. coli* replication apparatus succeeds in moving along DNA at approximately 50,000 base pairs (17 μm) per minute *in vivo* (Lewin, 1987), comparable to kinesin-driven motility along microtubules.

Helicases

Helicases are a family of enzymes that travel along double-stranded DNA in advance of an RNA or DNA polymerase, thereby giving them access to an isolated single strand to use as a template. They undergo rapid (500–1000 base pairs per second), processive motion along DNA and use the energy of ATP hydrolysis to destabilize the hydrogen bonds that hold the DNA double helix together. Most known helicases are believed to be either dimeric or hexameric in their active forms (Lohman and Bjornson, 1996). The best characterized helicase, the *E. coli* Rep homodimer, is a 76-kDa enzyme that dimerizes only in the presence of DNA. There is one DNA binding site per monomer, and this site can bind either double-stranded or single-stranded DNA. The relative affinity of the site for the two forms of DNA was found to depend on whether ATP or ADP is present in the helicase active site (Wong and Lohman, 1992). These observations are consistent with the "rolling dimer" model for the function of this enzyme (Fig. 18) (Moore and Lohman, 1995). Essentially, the dimer "rolls" along the polyphosphate backbone of the DNA duplex, allowing each monomer to interact with the DNA in an alternating fashion. The energy of ATP hydrolysis allows the helicase to cyclically modulate its relative affinities for single-stranded versus double-stranded polyphosphate backbone, much as ATP modulates the affinity of myosin, kinesin, and dynein for their respective filaments. In addition, ATP energy allows the helicase to break the Watson–Crick base pair hydrogen bonds as it moves.

At least one hexameric helicase, the T7 helicase-primase, has been shown by electron microscopy to form a ring structure encircling a single strand of DNA (Egelman *et al.,* 1995). The manner in which this helicase moves directionally along DNA is not clear; one possibility is that it rotates about the long axis of the DNA while moving forward like a nut moving on a bolt (Patel and Hingorani, 1995).

Polymerases

Crystal structures are currently available for two DNA polymerases (the Klenow fragment of *E. coli* DNA polymerase I and HIV reverse transcriptase, a polymerase that transcribes the RNA genome of the virus into DNA) and two accessory proteins (the β subunit of *E. coli* Pol III and

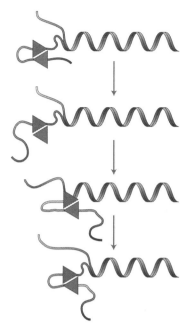

FIGURE 18 Schematic representation of the "rolling dimer" model of *E. coli* Rep helicase-catalyzed unwinding of the DNA double helix. The dimer (blue and green triangles) "rolls" along the polyphosphate backbone of the DNA duplex, allowing each monomer to interact with the DNA in an alternating fashion. The energy of ATP hydrolysis allows the helicase to cyclically modulate its relative affinities for single-stranded versus double-stranded polyphosphate backbone. The net effect of the cycle shown is the unwinding of approximately 14–16 base pairs, approximately equal to the dimensions of a Rep monomer (5 nm) (modified from Moore and Lohman, 1995).

the proliferating cell nuclear antigen from yeast) (Johnson, 1993; Wyman and Botchan, 1995). The structures of the polymerases are broadly similar to each other and contain a deep groove in which DNA binds. Both accessory proteins enhance the processivity of their partner polymerase, and both form a ring around DNA (Kong *et al.,* 1992). This has led to the suggestion that they function as "ring clamps" that tether the polymerase to the DNA substrate.

For RNA polymerases, low-resolution electron crystallography of two-dimensional crystals of *E. coli* RNA polymerase and yeast RNA polymerase II has revealed the overall shapes of these molecules (Darst *et al.,* 1988, 1991). RNA polymerases contain a deep cleft, similar to that seen in DNA polymerases, that almost certainly binds DNA. However, after initiation of transcription the polymerase may change shape so as to convert this groove into a ring surrounding the DNA template (Polyakov *et al.,* 1995). This change in shape may enhance processivity in the same way as the accessory protein ring clamp described previously. However, processive movement by RNA polymerases is a complex process and may depend on a variety of other protein cofactors (Aso *et al.,* 1995).

Polymerases and Helicases as Motor Proteins

Studying polymerases and helicases with the biophysical tools developed for the investigation of myosin and kinesin is a promising route to greater mechanistic understanding of these enzymes. The force produced by a polymerase has been measured (Yin *et al.,* 1995). In this experiment, *E. coli* RNA polymerase adsorbed to a glass surface moved along a DNA template attached to a polystyrene bead in an optical trap assay similar to that used for a single myosin molecule (Fig. 17, top). The RNA polymerase could exert a force of up to 14 pN, at least twice the maximal force value obtained for kinesin (Coppin *et al.,* 1995; Meyhöfer and Howard, 1995; Svoboda and Block, 1994) and myosin (Finer *et al.,* 1994). Since this force is exerted over a smaller, 1-base pair distance (0.34 nm), the overall thermodynamic efficiency as calculated from the work done per cycle (force times distance) is slightly lower than that of kinesin and myosin (~15 versus ~50% for kinesin and myosin).

Single-molecule evanescent wave fluorescence microscopy is a novel technique that also shows considerable promise in the study of polymerases and helicases. For instance, a direct demonstration that helicases are capable of vectorial movement along single-stranded DNA may be achieved by this method (Moore and Lohman, 1995). In addition, the single-molecule processivity of transcription and replication complexes and the effects of various accessory proteins on processivity could be readily evaluated.

Polymerization-Powered Processes

Microtubules and actin filaments are not merely tracks along which motor proteins run. On the contrary, both types of filaments undergo dynamic rearrangements in space, orientation, and concentration (Bray, 1992). The forces necessary to bring about these rearrangements are in many cases provided by motor proteins, but an interesting question is whether the filaments function as force generators. Like motor proteins, actin monomers bind ATP, and after polymerization the ATP is hydrolyzed to ADP. Exchange of bound ADP for ATP follows depolymerization. Tubulin undergoes the same cycle except that GTP is the preferred nucleoside triphosphate. Any polymerization occurring as part of an overall exergonic (thermodynamically favorable) process can do work. Since their cycle is exergonic and results in the net hydrolysis of ATP or GTP, actin and tubulin can in principle generate force and movement. An *in vitro* demonstration of polymerization-induced force generation consists of the deformation of actin- or tubulin-filled membranous vesicles (liposomes) upon polymerization of the protein (Miyata and Hotani, 1992).

The question of whether polymerization-powered processes occur in the cell is still unanswered. If one assumes (as the field generally does) that motor protein activity is the only plausible alternative power source, the problem is that of proving that a movement is motor independent, which requires that all necessary components be identified and shown to lack motor activity. In the following sections, four likely but unproven examples of polymerization-powered processes are briefly described: the crawling of nematode sperm, acrosomal process elongation, the movement of the pathogenic bacterium *Listeria monocytogenes* within its host cell, and the much more general process of membrane surface extension in eukaryotic cells.

Nematode Sperm

Currently, the best candidate for a polymerization-powered movement is a biological oddity and comes from the sperm of the nematode *Ascaris lumbricoides* (Italiano *et al.,* 1996; Theriot, 1996). The sperm of this major human parasite crawl rather than swim, and they do so by means of a unique motile apparatus. Like other crawling cells, the sperm cells extend membrane-bound protrusions (pseudopods) forward from the cell body, and these pseudopods exert a traction force that drags the rest of the cell along. The pseudopods are filled with filaments composed of major sperm protein (MSP), a 14-kDa protein unique to nematode sperm cells and without evident homology to actin or tubulin. In other crawling eukaryotes, the analogous process is lamellipod extension, which is actin dependent (Bray, 1992). In nematode sperm, the filaments are assembled from MSP monomers at the leading edge of the pseudopods and disassembled near the cell body. Once assembled, the filaments remain stationary relative to the substrate on which the cell is crawling, so the rate of movement of the leading edge of the pseudopod is the same as the rate of polymerization (approximately 1 μm/s).

The filament assembly process at the leading edge of the pseudopod has been reconstituted *in vitro* (Italiano *et al.,* 1996). In a solution containing MSP, ATP, and plasma membrane fragments, membrane fragments deriving from the leading edge of the pseudopod were observed to be pushed through the solution by growing filaments of MSP. Since the MSP does not bind ATP, the ATP-requiring component must be associated with the membrane and may act by transiently phosphorylating MSP and thereby inducing polymerization (Theriot, 1996). Other mechanisms are possible; what is important for present purposes is that ATP hydrolysis is coupled to the polymerization–depolymerization cycle of MSP and renders the process exergonic. Since the membrane-associated components have not been characterized, it is still possible that an unidentified filament-walking motor is involved in pushing the membrane away from the newly formed filament tip, but a simpler mechanism in which addition of MSP monomers does the pushing is most consistent with the data. In a Brownian ratchet model of membrane movement (Peskin *et al.,* 1993), thermal fluctuations in the position of the membrane could

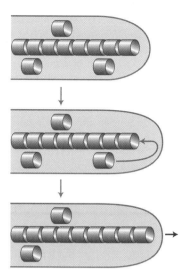

FIGURE 19 The Brownian ratchet model of membrane extension. Brownian fluctuations in the position of the plasma membrane create enough space for new monomers to add on to the tip of a growing filament. The membrane is then unable to return to its former position, so it has effectively been pushed forward.

create enough space to allow monomer addition at the end of a growing filament (Fig. 19).

Acrosomal Process Elongation

The spermatozoa of many marine invertebrates rapidly extend a bundle of membrane-bound actin fibers, called an acrosomal process, on contact with the extracellular matrix surrounding an egg (Bray, 1992). The role of this actin bundle is to penetrate the extracellular matrix and make first contact with the plasma membrane of the egg, and it does so with remarkable speed. In the sea cucumber *Thyone*, the actin is initially unpolymerized but the acrosomal process can polymerize to a length of 90 μm in 10 s (Tilney and Inoue, 1982). Candidate mechanisms for generating the force that drives the front of the (membrane-enclosed) acrosome forward include the Brownian ratchet mentioned previously, osmotic swelling of the membrane, and the action of a myosin motor.

Motility of *Listeria monocytogenes*

Listeria monocytogenes is a pathogenic bacterium that invades the cytoplasm of the host cell (Theriot, 1995). To move about within the host cytoplasm and push their way into neighboring cells, the bacteria induce the polymerization and cross-linking of host cell actin at one end of the bacterium. As a result, the bacterium is pushed forward by a "comet tail" structure made of actin filaments and filament cross-linking proteins in a manner similar to that described for membrane fragments of nematode sperm. Once polymerized, the actin filament comet tail remains station-

ary within the host cell cytosol; therefore, the bacterium moves forward at a speed equal to the rate of actin polymerization. This observation is predicted by a Brownian ratchet model for motility. It appears that the only component necessary for movement that is contributed by *L. monocytogenes* is the actA protein since expression of actA in unrelated bacteria causes them to exhibit indistinguishable motility (Kocks *et al.*, 1995). The function of this protein is not clear, but it is believed to be involved (at least indirectly) in actin filament nucleation and growth. No myosins are known to associate with the bacterium, but the involvement of a myosin motor cannot be ruled out.

Membrane Surface Extensions

The motility of nematode sperm and that of *L. monocytogenes* have been studied in part because they are convenient model systems for a much more widespread process, the extension of membrane-bound, actin-rich protrusions by many types of eukaryotic cells (Alberts *et al.*, 1983). Examples of such protrusions are lamellipodia, which extend the leading edge of a crawling cell forward, and neuronal growth cones, which extend processes out of neurons in order to make contact with other neurons. In these structures, actin polymerization occurs at or near the membrane at the leading edge. Unlike the MSP filaments in the pseudopods of nematode sperm, however, the actin filaments are not stationary with respect to the substrate but flux backwards toward the cell body (Wang, 1985). Recent experiments in neural growth cones have shown that, at least in this system, the rate of advance of the leading edge of the membrane is simply the difference between the rate of filament lengthening and the rate of filament rearward flux (Lin and Forscher, 1995). The rearward flux can be arrested by addition of myosin inhibitors, resulting in a burst of forward movement of the leading edge. This suggests that the rearward flux is driven by myosin motors but leading edge extension is not. These observations are consistent with actin polymerization being the energy source for membrane extension, but again the involvement of a motor is difficult to rule out.

Coupling Depolymerization to Movement

Although the previous examples highlighted pushing forces associated with the polymerization of filaments, pulling forces can also be generating by filament disassembly (Desai and Mitchison, 1995). This idea has often been evoked as a mechanism for moving chromosomes during mitosis (cell division). The chromosomes contain a specialized region of DNA (the centromere) that attracts a group of proteins forming a specialized structure, the kinetochore, which can be recognized by electron microscopy as a series of plate-like objects. One role of the kinetochore is to capture the + ends of microtubules, which constitute the major filamentous component of the mitotic spindle (fusiform framework of tracks for chromosome transport). The − end remains anchored in a large proteinaceous structure called the spindle pole. This process has been reconstituted and studied with purified chromosomes and microtubules *in vitro.* It was found that, under conditions that promote microtubule depolymerization, the tubulin subunits dissociate rapidly from the filament's + end. Interestingly, the chromosome could maintain attachment to the depolymerizing end, and consequently it was drawn toward the opposite end of the shrinking microtubule (Desai and Mitchison, 1995). However, whether depolymerization is the motive force for anaphase chromosome movement in living cells remains controversial.

Conclusion

Few properties of living organisms are as fascinating to observe and study as spontaneous, purposeful movement. The realization that nature has devised a multitude of specialized mechanical machines on a molecular scale that outperform the most sophisticated man-made motors has mesmerized scientists and engineers alike. Although the biochemical characterization of the proteins involved in movement is currently well under way, an understanding of their physics, which determines their performance in producing work, is still lacking. There is considerable hope that arriving at such an understanding could revolutionize the manner in which man-made motors are conceived and designed. Achieving this kind of understanding will depend heavily on the elucidation of the intricate mechanism by which subtle but precise changes in the protein's three-dimensional structure carefully orchestrate a sequence of state transitions. It will also be necessary to understand exactly how the energy of ATP hydrolysis or of proton flow is "spent" as the motor passes through the various obligatory states in its mechanical cycle. Research on the mechanism of motility has always relied heavily on advances in instrumentation technology. Since the pace of such advances has been increasing during the past few decades, there is reason to be optimistic about the prospects for elucidating nature's most closely held secrets about its remarkable molecular machines.

References Cited

ALBERTS, B., BRAY, D., LEWIS, J., RAFF, M., ROBERTS, K., and WATSON, J. D. (1983). *Molecular Biology of the Cell.* Garland, New York.

ASO, T., CONAWAY, J. W., and CONAWAY, R. C. (1995). The RNA polymerase II elongation complex. *FASEB J.* **9,** 1419–1428.

AXELROD, D. (1989). Total internal reflection fluorescence microscopy. *Methods Cell Biol.* **30,** 245–270.

BERG, H. C., and KHAN, S. (1983). A model for the flagellar rotary motor. In *Mobility and Recognition in Cell Biology* (H. Sund and C. Veeger, Eds.), pp. 485–497. deGruyter, Berlin.

BERG, H. C., and TURNER, L. (1993). Torque generated by the flagellar motor of *Escherichia coli. Biophys. J.* **65,** 2201–2216.

BERLINER, E., YOUNG, E. C., ANDERSON, K., MAHTANI, H. K., and GELLES, J. (1995). Failure of a single-headed kinesin to track parallel to the microtubule protofilaments. *Nature (London)* **373,** 718–721.

BLOCK, S. M., GOLDSTEIN, L. S., and SCHNAPP, B. J. (1990). Bead movement by single kinesin molecules studied with optical tweezers. *Nature (London)* **348,** 348–352.

BRADY, S. T. (1985). A novel brain ATPase with properties expected for the fast axonal transport motor. *Nature (London)* **317,** 73–75.

BRAY, D. (1992). *Cell Movements.* Garland, New York.

CAPLAN, S. R., and KARA-IVANOV, M. (1993). The bacterial flagellar motor. *Int. Rev. Cytol.* **147,** 97–164.

COOK, D. N., MA, D., PON, N. G., and HEARST, J. E. (1992). Dynamics of DNA supercoiling by transcription in *Escherichia coli. Proc. Natl. Acad. Sci. USA* **89,** 10603–10607.

COOKE, R. (1993). Sarcomeric myosins. In *Guidebook to the Cytoskeletal and Motor Proteins* (T. Kreis and R. Vale, Eds.), pp. 207–209. Oxford Univ. Press, Oxford.

COPPIN, C. M., FINER, J. T., SPUDICH, J. A., and VALE, R. D. (1995). Measurement of the isometric force exerted by a single kinesin molecule. *Biophys. J.* **68,** 242s–244s.

DARST, S. A., RIBI, H. O., PIERCE, D. W., and KORNBERG, R. D. (1988). Two-dimensional crystals of *Escherichia coli* RNA polymerase holoenzyme on positively charged lipid layers. *J. Mol. Biol.* **203,** 269–273.

DARST, S. A., EDWARDS, A. M., KUBALEK, E. W., and KORNBERG, R. D. (1991). Three-dimensional structure of yeast RNA polymerase II at 16 Å resolution. *Cell* **66,** 121–128

DESAI, A., and MITCHISON, T. J. (1995). A new role for motor proteins as couplers to depolymerizing microtubules. *J. Cell Biol.* **128,** 1–4.

EGELMAN, H. H., YU, X., WILD, R., HINGORANI, M. M., and PATEL, S. (1995). Bacteriophage T7 helicase/primase proteins form rings around single-stranded DNA that suggest a general structure for hexameric helicases. *Proc. Nat. Acad. Sci. USA* **92,** 3869–3873.

EISENBACH, M., and ADLER, J. (1981). Bacterial cell envelopes with functional flagella. *J. Biol. Chem.* **256,** 8807–8814.

FINER, J. T., SIMMONS, R. M., and SPUDICH, J. A. (1994). Single myosin molecule mechanics: Piconewton forces and nanometre steps. *Nature (London)* **368,** 113–118.

FISHER, A. J., SMITH, C. A., THODEN, J., SMITH, R. K. K. S., HOLDEN, H. M., and RAYMENT, I. (1995). Structural studies of myosin: Nucleotide complexes: A revised model for the molecular basis of muscle contraction. *Biophys. J.* **68,** 19S–26S.

FUNATSU, T., HARADA, Y., TOKUNAGA, M., SAITO, K., and YANAGIDA, T. (1995). Imaging of single fluorescent molecules and individual ATP turnovers by single myosin molecules in aqueous solution. *Nature (London)* **374,** 555–559.

GELLES, J., SCHNAPP, B. J., and SHEETZ, M. P. (1988). Tracking kinesin-driven movements with nanometre scale precision. *Nature* **331,** 450–453.

GOLDSTEIN, L. S. B. (1993). With apologies to Scheherazade: Tails of 1001 kinesin motors. *Annu. Rev. Genet.* **27,** 319–351.

HACKNEY, D. D. (1995). Highly processive microtubule-stimulated ATP hydrolysis by dimeric kinesin head domains. *Nature* **377,** 448–450.

HARADA, Y., NOGUCHI, A., KISHINO, A., and YANAGIDA, T. (1987). Sliding movement of single actin filaments on one-headed myosin filaments. *Nature* **326**, 805–808.

HOLZBAUR, E. L. F., and VALLEE, R. B. (1994). Dyneins: Molecular structure and cellular function. *Annu. Rev. Cell Biol.* **10**, 339–372.

HOWARD, J., HUDSPETH, A. J., and VALE, R. D. (1989). Movement of microtubules by single kinesin molecules. *Nature (London)* **342**, 154–158.

HUXLEY, A. F. (1981). *Reflections on Muscle.* Liverpool Univ. Press, Liverpool, UK.

HUXLEY, A. F., and NIEDERGERKE, R. (1954). Interference microscopy of living muscle fiber. *Nature* **173**, 971–973.

HUXLEY, H. E. (1969). The mechanism of muscular contraction. *Science* **164**, 1356–1366.

HUXLEY, H. E., and HANSON, J. (1954). Changes in the cross-striations of muscle during contraction and stretch and their structural interpretation. *Nature* **173**, 973–976.

ISHIJIMA, A., KOJIMA, H., HIGUCHI, H., HARADA, Y., FUNATSU, T., and YANAGIDA, T. (1996). Multiple- and single-molecule analysis of the actomyosin motor by nanometer-piconewton manipulation with a microneedle: Unitary steps and forces. *Biophys. J.* **70**, 383–400.

ITALIANO, J. E., ROBERTS, T. M., STEWART, M., and FONTANA, C. A. (1996). Reconstitution *in vitro* of the motile apparatus from the amoeboid sperm of ascaris shows that filament assembly and bundling move membranes. *Cell* **84**, 105–114.

JIANG, M., and SHEETZ, M. (1994). Mechanics of myosin motor: Force and step size. *Bioessays* **16**, 531–532.

JOHNSON, K. A. (1993). Conformational coupling in DNA polymerase fidelity. *Annu. Rev. Biochem.* **62**, 685–713.

KHAN, S., and BERG, H. C. (1983). Isotope and thermal effects in chemiosmotic coupling to the flagellar motor of *Streptococcus*. *Cell* **32**, 913–919.

KOCKS, C., MARCHAND, J. B., GOUIN, E., and DHAUTEVILLE, H. (1995). The unrelated surface proteins actA of *Listeria monocytogenes* and icsA of *Shigella flexneri* and sufficient to confer actin-based motility on *Listeria innocua* and *Escherichia coli* respectively. *Mol. Microbiol.* **18**, 413–423.

KONG, X. P., ONRUST, R., O'DONNELL, M., and KURIYAN, J. (1992). Three-dimensional structure of the beta subunit of *E. coli* DNA polymerase III holoenzyme: A sliding DNA clamp. *Cell* **69**, 425–437.

KRON, S. J., and SPUDICH, J. A. (1986). Fluorescent actin filaments move on myosin fixed to a glass surface. *Proc. Natl. Acad. Sci. USA* **83**, 6272–6276.

KULL, F. J., SABLIN, E. P., LAU, R., FLETTERICK, R. J., and VALE, R. D. (1996). Crystal structure of the kinesin motor domain reveals a structural similarity to myosin. *Nature* **380**, 550–555.

KUO, S. C., and SHEETZ, M. P. (1993). Force of single kinesin molecules measured with optical tweezers. *Science* **260**, 232–234.

LARSEN, S. H., ADLER, J., GARGUS, J. J., and HOGG, R. W. (1974). Chemomechanical coupling without ATP: The source of energy for motility and chemotaxis in bacteria. *Proc. Natl. Acad Sci. USA* **71**, 1239–1243.

LÄUGER, P. (1977). The proton pump is a molecular engine for motile bacteria. *Nature* **268**, 360–362.

LÄUGER, P. (1988). Torque and rotation rate of the bacterial flagellar motor. *Biophys. J.* **53**, 53–66.

LEWIN, B. (1987). *Genes,* 3rd ed. Wiley, New York.

LIN, C. H., and FORSCHER, P. (1995). Growth cone advance is inversely proportional to retrograde F-actin flow. *Neuron* **14**, 763–771.

LOHMAN, T. M., and BJORNSON, K. P. (1996). Mechanisms of helicase-catalyzed DNA unwinding. *Annu. Rev. Biochem.* **65**, 169–214.

MACNAB, R. M. (1987). In *Escherichia coli and Salmonella typhimurium: Cellular and Molecular Biology* (F. C. Neidhardt, J. Ingraham, K. B. Low, B. Magasanik, M. Schaechter, and H. E. Umbareger, Eds.), pp. 70–83, 732–759. ASM Press, Washington, DC.

MALIK, F., BRILLINGER, D., and VALE, R. D. (1994). High-resolution tracking of microtubule motility driven by a single kinesin motor. *Proc. Natl. Acad. Sci. USA* **91**, 4584–4588.

MEISTER, M., LOWE, G., and BERG, H. C. (1987). The proton flux through the bacterial flagellar motor. *Cell* **49**, 643–650.

MEYHÖFER, E., and HOWARD, J. (1995). The force generated by a single kinesin molecule against an elastic load. *Proc. Natl. Acad. Sci. USA* **92**, 574–578.

MIYATA, H., and HOTANI, H. (1992). Morphological changes in liposomes caused by polymerization of encapsulated actin and spontaneous formation of actin bundles. *Proc. Natl. Acad. Sci. USA* **89**, 11547–11551.

MOLLOY, J. E., BURNS, J. E., KENDRICK-JONES, J., TREGEAR, R. T., and WHITE, D. C. (1995). Movement and force produced by a single myosin head. *Nature* **378**, 209–212.

MOORE, J., and LOHMAN, T. M. (1995). Helicase-catalyzed DNA unwinding: Energy coupling by DNA motor proteins. *Biophys. J.* **68**, 180s–184s.

MOOSEKER, M. S., and CHENEY, R. E. (1995). Unconventional myosins. *Annu. Rev. Cell Dev. Biol.* **11**, 633–675.

OOSAWA, F., and HAYASHI, S. (1986). Coupling between flagellar motor rotation and proton flux in bacteria. *J. Phys. Soc. Jpn.* **52**, 4019–4028.

OOSAWA, F., and MASAI, J. (1982). Mechanism of flagellar motor rotation in bacteria. *J. Phys. Soc. Jpn.* **51**, 631–641.

PATEL, S. S., and HINGORANI., M. M. (1995). Nucleotide binding studies of bacteriophage T7 DNA helicase-primase protein. *Biophys. J.* **68**, 186s–190s.

PESKIN, C. S., ODELL, G. M., and GF, G. F. O. (1993). Cellular motions and thermal fluctuations: The Brownian ratchet. *Biophys. J.* **65**, 316–324.

POLYAKOV, A., SEVERINOVA, E., and DARST, S. A. (1995). Three-dimensional structure of *E. coli* core RNA polymerase: Promoter binding and elongation conformations of the enzyme. *Cell* **83**, 365–373.

RAYMENT, I., and HOLDEN, H. M. (1994). The three-dimensional structure of a molecular motor. *Trends Biochem. Sci.* **19**, 129–134.

SABLIN, E. P., KULL, F. J., COOKE, R., VALE, R. D., and FLETTERICK, R. J. (1996). Crystal structure of the motor domain of the kinesin-related motor ncd. *Nature* **380**, 555–559 .

SATIR, P. (1992). Mechanisms ciliary movement: Contributions from electron microscopy. *Scanning Microscopy* **6**, 573–579.

SCHUSTER, S. C., and KHAN, S. (1994). The bacterial flagellar motor. *Annu. Rev. Biophys. Biomol. Struct.* **23**, 509–539.

SHEETZ, M. P., and SPUDICH, J. A. (1983). Movement of myosin-coated fluorescent beads on actin cables *in vitro*. *Nature* **303**, 31–35.

STALLMEYER, M. J., AIZAWA, S., MACNAB, M., and DeROSIER, D. J. (1989). Image reconstruction of the flagellar lasal body of *Salmonella typhimurium. J. Mol. Biol.* **205,** 519–528.

SVOBODA, K., and BLOCK, S. M. (1994). Force and velocity measured for single kinesin molecules. *Cell* **77,** 773–784.

SVOBODA, K., SCHMIDT, C. F., SCHNAPP, B. J., and BLOCK, S. M. (1993). Direct observation of kinesin stepping by optical trapping interferometry. *Nature (London)* **365,** 721–727.

THERIOT, J. A. (1995). The cell biology of infection by intracellular bacterial pathogens. *Annu. Rev. Cell Dev. Biol.* **11,** 213–239.

THERIOT, J. A. (1996). Worm sperm and advances in cell locomotion. *Cell* **84,** 1–4.

TILNEY, L. G., and INOUE, S. (1982). Acrosomal reaction of thyone sperm. II. The kinetics and possible mechanism of acrosomal process elongation. *J. Cell Biol.* **93,** 820–827.

TOYOSHIMA, Y. Y., KRON, S. J., MCNALLY, E. M., NIEBLING, K. R., TOYOSHIMA, C., and SPUDICH, J. A. (1987). Myosin subfragment-1 is sufficient to move actin filaments *in vitro. Nature* **328,** 536–539.

UYEDA, T. Q., KRON, S. J., and SPUDICH, J. A. (1990). Myosin step size. Estimation from slow sliding movement of actin over low densities of heavy meromyosin. *J. Mol. Biol.* **214,** 699–710.

UYEDA, T. Q. P., WARRICK, H. M., KRON, S. J., and SPUDICH, J. A. (1991). Quantized velocities at low myosin densities in an *in vitro* motility assay. *Nature* **352,** 307–311.

VALE, R. D. (1993). Kinesin. In *Guidebook to the Cytoskeletal and Motor Proteins* (T. Kreis and R. Vale, Eds.), pp. 199–201. Oxford Univ. Press, Oxford.

VALE, R. D. (1994). Getting a grip on myosin. *Cell* **78,** 733–737.

VALE, R. D., REESE, T. S., and SHEETZ, M. P. (1985). Identification of a novel force-generating protein, kinesin, involved in microtubule-based motility. *Cell* **42,** 39–50.

VALE, R. D., FUNATSU, T., PIERCE, D. W., ROMBERG, L., HARADA, Y., and YANAGIDA, T. (1995). Direct observation of single kinesin molecules moving along microtubules. *Nature* **380,** 451–453.

VALLEE, R. B. (1993). Cytoplasmic dynein. In *Guidebook to the Cytoskeletal and Motor Proteins* (T. Kreis and R. Vale, Eds.), pp. 191–193. Oxford Univ. Press, Oxford.

WANG, Y. L. (1985). Exchange of actin subunits at the leading edge of living fibroblasts: Possible role of treadmilling. *J. Cell Biol.* **101,** 597–602.

WONG, I., and LOHMAN, T. M. (1992). Allosteric effects of nucleotide cofactors on *Esherichia coli* Rep helicase-DNA binding. *Science* **256,** 350–355.

WYMAN, C., and BOTCHAN, M. (1995). DNA replication. A familiar ring to DNA polymerase processivity. *Curr. Biol.* **5,** 334–337.

YIN, H., WANG, M. D., SVOBODA, K., LANDICK, R., BLOCK, S. M., and GELLES, J. (1995). Transcription against an applied force. *Science* **270,** 1653–1657.

General References

BERG, H. C. (1983). *Random Walks in Biology.* Princeton Univ. Press, Princeton, NJ.

KREIS, T., and VALE, R. D. (1993). *Guidebook to the Extracellular Matrix and Adhesion Proteins.* Oxford Univ. Press, Oxford.

MCMAHON, T. A. (1984). *Muscles, Reflexes, and Locomotion.* Princeton Univ. Press, Princeton, NJ.

SVOBODA, K., and BLOCK, S. M. (1994). Biological applications of optical forces. *Annu. Rev. Biophys. Biomol. Struct.* **23,** 247–285.

Eric Karsenti

European Molecular Biology Laboratory
Meyerhofstrasse 1
D-69117 Heidelberg, Germany

Cell Form as an Essential Determinant of Cell Function

The main function of simple organisms such as bacteria is restricted to the transformation of energy absorbed from simple sources into living matter. Accordingly, their shapes are fairly simple and uniform. The shapes of unicellular eukaryotes, which are much larger than bacteria, can be extremely complex and they can display complex behaviors. In such cells, organs such as sensory cilia, eye spots, and mouthparts begin to appear,. These organs are shaped at the single cell level from cytoskeletal components. In multicellular organisms, the various functions essential to survival are fulfilled by specific organs in which cells have differentiated in sophisticated shapes. These shapes are largely determined by variations in the three-dimensional organization of the cytoskeleton. The two main components of the cytoskeleton that are important in morphogenetic processes are the microtubules and the microfilaments. Microtubules are mostly involved in the organization of the overall cell shape, directionality of movements, and positioning of organelles in the cytoplasm as well as in the formation of specific structures such as beating and sensory cilia or the mitotic spindle. Microfilaments are more involved in all aspects of cell motility and in transducing information from the outside world to the interior of the cell. Both microfilaments and microtubules are dynamic and oriented polymers. These characteristics are responsible for their morphogenetic properties. Their asymmetry (or polarity) is read by molecular motors that can move cellular components to specific cellular domains. Therefore, to understand the diversity of eukaryotic cell shapes and functions one must , at least in part, understand the principles that underlie the three-dimensional organization of cytoskeletal fibers. These systems have self-organization properties that partly explain how cell shapes and functions are determined.

Shape and Function

Although it is common to relate structure and function in biology, it should not be forgotten that this concept is mostly restricted to living matter. The reason for this is that the structure of a protein determines in some way its activity. For example, a kinase will transfer phosphate groups on specific proteins because the chemistry of the active site has been shaped during evolution for this purpose. When one thinks about the shape of a cell and how it is related to its function, one is immediately confronted with the following question: What is the function of a cell? The basic function of any living system is to transform energy into ordered complexity. In multicellular organisms, many different types of cells have appeared during evolution which have refined functions essential to the survival of the whole organism. However, all cells must in some way absorb energy and transform it into a specific shape that is either compatible with or essential to a certain function. Therefore, the three-dimensional organization of a cell should allow the accomplishment of at least two essential functions. One deals with the absorption of energy and the other deals with the transformation of this energy into specific behaviors.

There are three kinds of questions concerning the relation that exists between shape and function. One has to do with evolution: How do forms evolve and how does selection pressure act on the evolution of forms? The second kind of question deals with the transformation of the linear information stored in the genome into three- or even four-dimensional information, and the third question concerns how shape and function are related at the cellular level.

The evolution of cell forms must in some way be determined by changes in the combination of gene products that are involved in the determination of shapes. Some proteins have a direct morphogenetic role because they can self-assemble into large supramolecular structures. These proteins constitute the cell cytoskeleton. Others also have a morphogenetic function but this is not obvious because they are enzymes that govern the organization of the cytoskeleton in three dimensions. Therefore, the evolution of cell shapes relies directly on the coevolution of a large number of genes coding both for structural proteins and for enzymes. It is the evolution of a network.

The transformation of the linear information stored in the genome into forms and associated functions is more amenable to examination with the current conceptual knowledge and technology. We can begin to understand how the physical properties of the cytoskeleton endow it with such a versatility that it can generate the cell shapes we know. The role of membrane organization in morphogenesis, although very important, is less well worked out, although exciting progress has recently been made in this field.

In this article, I will show how the molecules that participate in the establishment of cell shape could at the same time participate in specific cell functions. Therefore, it will become clear that the dogma one gene, one protein, one function is not valid at the cellular level. One protein can participate in many different cellular functions. However, it will also be shown how the physical properties of the cytoskeleton and their regulation lead to the generation of forms that have a specific function.

The Diversity of Cell Forms and Functions

Prokaryotic cells have very simple forms ranging from spherical to rod-shaped, commonly several micrometers in linear dimensions. They usually possess a protective coat (a cell wall), beneath which a plasma membrane encloses a single cytoplasmic compartment containing DNA, RNA, proteins, and small molecules (Fig. 1a). Prokaryotic cells can absorb their energy from very simple sources such as glucose, transform it into ATP, and use the carbon atoms to make all the enzymes and other organic molecules they need.

Unicellular eukaryotes have much more complex shapes (Fig. 1b). One major difference between prokaryotes and eukaryotes is the source of energy from which they feed. Whereas prokaryotes feed from very simple sources, eukaryotes feed either from other living organisms or directly from light and molecules present in the soil. Consequently, their size and shape vary greatly and they are much larger than prokaryotes (typically more than 10 μm). Eukaryotes have several subcellular compartments. The DNA is packaged into a nucleus and the cytoplasm contains an elaborate network of membranes involved in the secretion and absorption of matter into and from the surrounding environment. Energy is generated by mitochondria or by chloroplasts in the case of photosynthetic organisms. These organelles contain DNA and are thought to have originated as symbiotic bacteria. Interestingly, the cytoplasm of eukaryotic cells is organized by the cytoskeleton composed of actin filaments, intermediate filaments, and microtubules. Intermediate filaments act as a structural support for the cytoplasm, microtubules organize the cytoplasm and act as tracks to transport various compounds in the cell, and actin filaments are usually involved in cell motility and cell shape in conjunction with microtubules. The eukaryotes shown in Fig. 1 are called protists. Although they are unicellular, their anatomy is complex and includes such structures as sensory bristles, flagella, photoreceptors, mouthparts, and muscle-like contractile bundles. The various protists shown in Fig. 1 behave in different ways. Some are carnivorous, whereas others are photosynthetic — they can be motile or sedentary. Their shape is adapted accordingly. Those that are motile have flagella or cilia, and sedentary forms have a foot to anchor on the ground and a mouth surrounded with cilia to capture prey. Therefore, in this case, each cell shape determines a certain behavior that encompasses several functions. There are different ways of moving about or being sedentary and the shapes change accordingly. However, each shape could be slightly different without affecting the life of the organism. This suggests that during evolution shapes arise, driving cells toward specific behaviors.

Multicellularity permits the assembly of large organisms that, at least for the animal kingdom, must feed on other forms of life. As a consequence, feeding, food processing, reproduction, movement, and complex behaviors rely on the generation of specific organs that are built of different cell types with specific shapes and functions (Fig. 2) (Cross and Mercer, 1993).

Epithelial cells line body cavities and constitute the skin, forming the interface between a multicellular organism and the outside world. Digestive track cells, kidney, secretory gland, and sensory organ cells are of the epithelial

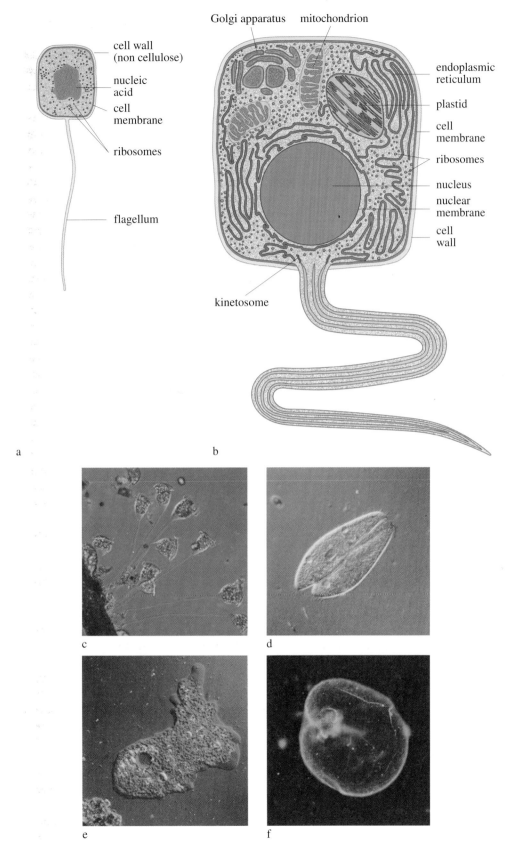

cell wall
(non cellulose)

nucleic
acid

cell
membrane

ribosomes

flagellum

Golgi apparatus   mitochondrion

endoplasmic
reticulum

plastid

cell
membrane

ribosomes

nucleus

nuclear
membrane

cell
wall

kinetosome

a

b

c

d

e

f

FIGURE I   The prokaryotic and eukaryotic cells, with examples of eukaryotes. (a) Structure of a prokaryotic cell. (b) Structure of a plant eukaryotic cell. (c) *Vorticella* (ciliate) (courtesy of A. Margiocco). (d) Paramecium (ciliate)(courtesy of D. J. Patterson/OSF). (e) Amoeba (sarcodina)(courtesy of P. Parks/OSF). (f) *Noctiluca* (dinoflagellate) (courtesy of P. Parks/OSF).

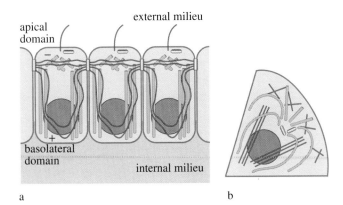

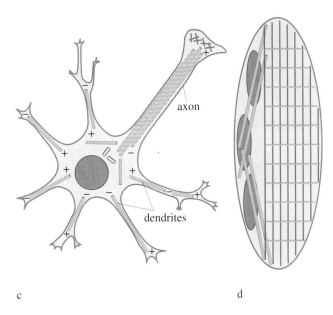

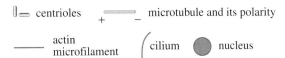

FIGURE 2 A few basic cell types in multicellular organisms: (a) polarized epithelial cell, (b) a fibroblast, (c) a nerve cell, and (d) a multinucleated muscle cell. The organization of the microtubules (green) and the actin filaments (red) is schematized in each cell type. The same basic components have different spatial organizations in each cell type.

type. All these cells are polarized with an apical domain facing the external milieu and a basolateral domain facing the internal milieu (Fig. 2). This asymmetry is essential for their function: the exchange of matter or information with the surrounding environment. They are tightly bound to each other, which is also essential for their fence function (Mays *et al.,* 1994). Sensory cells have developed special structures such as cilia or complex membrane networks that are adapted to the kind of physical or chemical signal they must detect.

Nerve cells are also of epithelial origin, but they display several processes which are classified under two categories of neurons: the axon and dendrites. It has been demonstrated that the axon corresponds to the apical, whereas the dendrites are more akin to the basolateral domain of epithelial cells (Lafont and Simons, 1996). It is clear that the organization of these cells fits their function, which is to generate interconnected networks storing and processing information.

Lymphocytes, macrophages, and fibroblasts are all either free circulating or actively crawling cells moving on other tissues. Lymphocytes are spherical, a shape that fits well their function, which is to secrete antibodies or recognize foreign bodies to be killed by secretion of toxic compounds. Macrophages and fibroblasts have elongated shapes and are equipped to spread over various substrata. Accordingly, both cell types have to move around either to find foreign objects and destroy them by phagocytosis or to fill gaps in a tissue.

Muscle cells are a striking example of cell differentiation and morphogenesis. Cells called myoblasts fuse and form multinucleated myotubes. Myofibrils then form in which actin and myosin get organized into an apparatus, the sarcomere, that can contract in response to nerve impulse. Here, the interplay between shape and function is particularly obvious and this illustrates well how the physical properties of a particular set of proteins can explain the generation of cell shape and cell function (Franzini-Armstrong and Fischman, 1994).

Finally, both male and female gametes seem to have the same function: to fuse and form a new organism. However, they have entirely different shapes. Again, the notion of function is ambiguous. For example, in frogs the egg is a very large spherical cell that contains all the materials required to make approximately 20,000 cells. This cell therefore has the potential for the development of a new organism while simultaneously having a storage function. The sperm, however, with its tiny cytoplasmic compartment and large motile tail, is tailored to swim and bring the male genome into the egg.

It is interesting to note that most of the functions carried out by specific cells in multicellular organisms, such as endocytosis, sensory functions, muscle-dependent motility, and decision making, do exist in unicellular eukaryotes. In fact, each of these functions involves specific proteins or subcellular structures. For example, the same kind of proteins participate in the motility of unicellular organisms and in muscle cells of multicellular eukaryotes: actin and myosin. The same supramolecular assembly based on tubulin participates in the formation of the cilia of protozoa and in the formation of the flagellum of the sperm. It therefore seems that during evolution a restricted set of proteins have been used to fulfill specific cellular functions and that in combination with other regulatory proteins they determine both the shape and function of the various cell types one can observe in nature today.

Diversification of Cell Forms and Functions during Development

During embryogenesis, the first cells to differentiate are those present at the surface of the embryo at the blastula stage. In all embryos, from *Drosophila* to man, these become polarized epithelial cells (Fig. 3). During gastrulation,

this epithelium invaginates to form the future endoderm. Some cells from the endoderm migrate in the cavity formed between the two cell layers and will give rise to the future mesoderm (Fig. 3).

The ectoderm is at the origin of the epidermis, central nervous system, and sensory organs. The endoderm is at the origin of the digestive track. Lungs and liver develop from the wall of the digestive track which, at the beginning

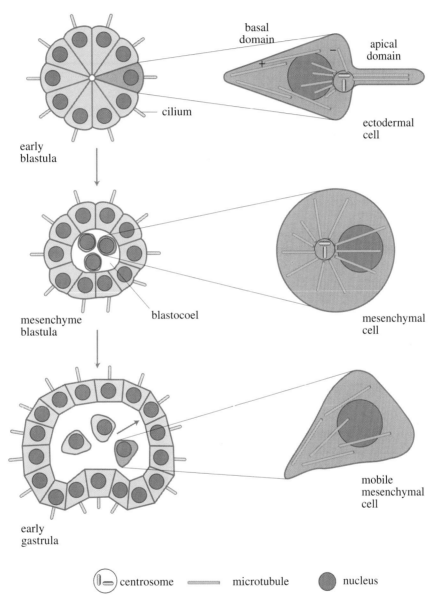

FIGURE 3 Generation of primary mesenchymal cells during gastrulation in echinoderms (e.g., sea urchin). (Left) Three successive stages of development are shown: an early blastula, a mesenchymal blastula, and an early gastrula, the stage during which the ectoderm invaginates into the blastocoel. At the mesenchymal blastula stage, cells begin to migrate into the blastocoel, where they lose their apicobasal asymmetry. (Right) Enlargements of the cells are shown at the three stages, highlighting their forms and the organization of the microtubules. In ectodermal cells the microtubules are oriented as in epithelial cells with the negative pole outwards. In mesenchymal cells they are oriented as in fibroblasts in a radial distribution, at the center of which, near the nucleus, a centrosome is positioned with its relative centrioles. In motile cells the microtubules appear organized according to the direction of migration of the cell (arrows).

of development, is a fairly simple tube joining the future mouth and anus. All these cells, including neurons, have the same organization, an apicobasal asymmetry (Fig. 3), and they all originate from the surface of the embryo, that is, the interface between the external world and the internal medium. Therefore, it is likely that the apicobasal asymmetry has its origin in the physical asymmetry existing at the surface of the blastula. This asymmetry is obviously maintained for epidermal and endodermal cells that always face two different media and form the surface of the exterior and interior of the embryo. A good comparison of the gastrulation process in various organisms is provided by Gilbert (1988).

In contrast to ectodermal and endodermal cells, the cells that will constitute the mesoderm lose their apicobasal asymmetry while migrating into the " internal milieu" between the endoderm and ectoderm. These cells give rise to mesenchyme, bones, muscles, heart, and blood cells as well as most cells of the urogenital track. Interestingly, although most of these cells lose the typical apicobasal organization of epithelial cells, some will acquire it again during kidney morphogenesis. This happens through interactions between mesenchymal and epithelial cells.

All adult organs, probably in all organisms, are formed through interactions between cells of ectoderm/endoderm and mesoderm origin as shown in Fig. 4. This indicates that morphogenesis occurs through a progressive amplification of cell shape diversity during embryogenesis. In other words, the initial asymmetry of the spherical embryo is essential for the generation of all cell shapes in the organism by generating the first type of differentiated cells which are of the epithelial type. These cells, by pumping ions into the blastocoel, create a homogeneous internal milieu in which new cells can differentiate through interactions between each other while being sheltered from the "external milieu." When these cells acquire an asymmetry, it is generated by other cues rather than a difference between external and internal milieu.

The mechanism by which cells acquire their various shapes and functions during organogenesis (cell differentiation) is far from understood. It is clear, though, that cells become different not because their genomes become different but because the combination of expressed genes changes. During development, differentiation is due to a regional control of gene expression. This is achieved in part by the heterogeneous localization of cytoplasmic determinants in the egg before gastrulation and by induction processes involving cell–cell interactions as described previously. The relative contribution of each mechanism varies from species to species. In mammals, the cytoplasm of all blastomeres appears to be identical at the beginning of development and cell differentiation appears to rely mostly on induction processes. In contrast, in ascidians the cytoplasm of each blastomere is different and determines which type of cells to which it will give rise. In frogs, there is an intermediate situation in which determinants are localized through cytoplasmic movements during the first cell cycle following fertilization at the one-cell stage, leading to the generation of two types of blastomeres: those that will give rise to ectodermal cells (in the animal part of the embryo)

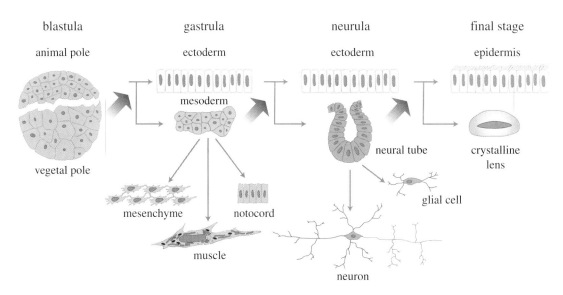

FIGURE 4 Sequential induction, regulated by interactions between cells of different layers, in amphibian embryonal development. The different types of tissue (some of which are represented under the layers from which they derive) are formed successively. In the frog, the formation of the mesoderm is induced by the interaction between the vegetal tissue and the animal tissue, that of the nervous tissue by the action of the mesoderm on the ectoderm, whereas the ectoderm, controlled by the neural tube, gives rise to the epidermis and to the cells of the crystalline lens.

and those that will give rise to endodermal cells (in the vegetal part). Mesodermal cells are then produced by induction of ectodermal cells interacting with the endoderm. Following gastrulation, new interactions lead to new inductions and to the generation of all cell types as shown in Fig. 4 (Gurdon, 1992).

In addition to these apparently different pathways, it is necessary to determine whether there are basic principles responsible for the generation of cell diversity in all phyla. It was previously mentioned that an initial asymmetry is essential to generate diversity. The question is therefore the following: When is such an asymmetry initially generated and how? In mosaic embryos, the asymmetry is generated in the oocyte through the localization of determinants and this is thought to be mediated by the cytoskeleton. In *Drosophila*, experiments have shown that in the oocyte determinants in the form of mRNA coding for specific proteins and possibly specific proteins are localized by the cytoskeleton (Lehman, 1995; Theurkauf, 1994). The heterogeneity of the cytoplasm generated in this way results in the generation of cells that are heterogeneous in cytoplasmic content during segmentation. It remains to be determined how the cytoskeleton of the oocyte is organized in such a way as to position determinants for future developmental pathways. Elegant experiments have shown that this is achieved through the localization of signaling molecules positioned by interaction between the cytoplasm of the oocyte and the follicular cells that surround it (Lehman, 1995). Therefore, even in mosaic embryos, it is the end cellular interactions that determine the localization of determinants governing the fate of the embryonic cells in the future adult. In *Xenopus*, a similar situation exists: Some determinants are positioned in the animal and others in the vegetal pole of the oocyte (Hyatt *et al.*, 1996). This depends on the organization of microtubules which are laid down during oogenesis. In this case, there is another asymmetry

which is imposed by the sperm entrance point and, probably, the orientation of microtubules growing from the centrosome associated with the sperm. This second asymmetry is at the origin of the dorsoventral axis of the embryo (Gilbert, 1988). In mammals, all the cells of the early embryo are equivalent until the eight-cell stage. At this point, each cell becomes polarized through its interaction with neighbors. This results in an asymmetric division at the next cell cycle with each cell producing one daughter inside the embryo, which is completely sheltered from the outside environment (Fig. 5). These cells will give rise to the inner cell mass (and later to the adult), whereas the cells at the surface will give rise to the trophectoderm (and later the placenta). The cells from the inner cell mass then differentiate entirely through successive induction processes following gastrulation.

The general scheme of this very broad and coarse overview of the potential origin of cell diversity during embryogenesis is the following: In all organisms, cell differentiation derives from an initial asymmetry. Since all embryos begin as a sphere (or an ovoid), the most obvious source of initial asymmetry is the surface of the embryo. In mosaic embryos such as *Drosophila*, it is the interactions between the surface of the oocyte and the follicle cells that provide the initial asymmetry required for further cell differentiation, whereas in mammals this occurs later; however, in essence the principles are not very different. Then, three types of cells are generated: the ectoderm, endoderm, and mesoderm. From interactions between these three layers and by successive inductions, the diverse cell types are generated (Fig. 4; Gurdon, 1992). There are three important problems to address in order to understand how cell diversity is generated during development. One concerns the mechanisms by which the initial cell asymmetry leads to cell differentiation. The second concerns the molecular mechanisms of cell–cell interactions during induction processes and how

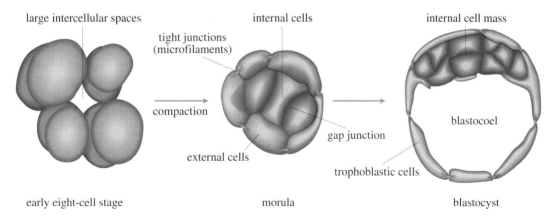

FIGURE 5 Compaction at the eight-cell stage in the mouse embryo leads to asymmetric cell divisions and cell determination. The formation of tight junctions leads to a change in cell division axis, which in turn leads to the generation of two different types of cells: cells in contact with the external medium which become trophoblast cells and the internal cells which will form the inner cell mass.

these affect gene expression in an irreversible way. The third concerns the mechanism by which a given combination of genes can lead to the generation of a specific cell shape and function. We are far from understanding any of these events in detail. However, there are some indications. It seems that cell asymmetry is generated by cell surface events that lead to cytoskeletal rearrangements which result in the asymmetric localization of molecules involved in the regulation of gene expression. How a change in gene expression becomes irreversible so that future generations of cells remain differentiated remains to be determined. The cell surface events that lead to cell differentiation could be physical, such as during compaction in the early mammalian embryos or during the formation of the first epithelial cells in amphibians, or mediated by factors that bind to receptors such as growth factors (Gurdon, 1992). This latter mechanism occurs mostly during induction processes between cells originating from one of the three germ layers. Cell differentiation results from the expression of a specific combination of genes. Particular cytoskeletal protein isoforms and regulatory proteins lead, in conjunction with cell–cell interactions, to the generation of organs. We know very little about this level of multicellular morphogenesis, but we are beginning to understand how specific cell shapes are generated and how this relates to their function. Cytoskeletal proteins have amazing physical properties that endow them with morphogenetic potential. In fact, the cytoskeleton is the major architect in the generation of specific cell shapes.

The Role of Cytoskeletal Components and External Cues in Generating Cell Form

In this section, I will review the main properties of microfilaments and microtubules and describe how these polymers participate in cell morphogenesis. I will first describe how microtubules form a mitotic spindle because this process follows a pathway which is based on general morphogenetic principles. Then I will describe the morphogenesis of epithelial cells to illustrate the generality of some of the morphogenetic principles involved in mitotic spindle assembly. Neuronal and muscle cells are also very interesting because they represent two extreme cases. In neurons, the microtubule system is largely hypertrophied and plays a fundamental role in their organization and function. Also, there are some similarities between the way the growth cone of neurons develops and the way the apical domain of epithelial cells is stabilized. In muscle cells, the actomyosin system is dominant, although microtubules seem to play an essential role in myofibril morphogenesis.

Morphogenetic Properties of the Cytoskeleton

There are three well-described cytoskeletal systems in eukaryotic cells: microfilaments, intermediate filaments, and microtubules. The function of intermediate filaments in the generation and maintenance of cell shape is not well understood. In contrast, microfilaments and microtubules have physical properties that endow them with tremendous morphogenetic potential. Both are highly dynamic fibrous systems, both are oriented polymers, and both interact with motors that can either organize the filaments relative to each other in space or move cargoes along the filaments. Therefore, I will only consider these two filamentous systems in this article.

Microfilaments. Microfilaments are composed of actin monomer subunits. This molecule has an asymmetric structure that results in the assembly of an asymmetric polymer (Fig. 6). Each microfilament is composed of two protofilaments of actin molecules generating a helix. Actin contains one ATP binding site in which ATP is hydrolyzed into ADP during assembly. This is not required for assembly but used to build some instability into the polymer. The asymmetry of the polymer has two consequences: Assembly occurs preferentially at one end of the filaments, and polymers can be oriented in the cell. There is only one type of motor that can move along microfilaments: the myosin molecule (Fig. 6), of which several sorts exist. Myosin moves along microfilaments toward their plus ends. Muscle contraction occurs because the plus end of thin filaments is localized at the Z lines and myosin thick filaments are bipolar (Fig. 6). In addition to skeletal myosin, there are many different types of myosin which are targeted to different localization in the cells by localization domains attached to the motor domain (Fig. 7; Mooseker and Cheney, 1995).

Organization of Microfilaments by Associated Proteins. Many proteins can interact with actin and affect the length of individual filaments and their organization in space (Fig. 8). Some, such as tropomyosin, bind along the length of actin filaments, whereas others such as α-actinin bind only at the plus ends. These molecules are involved in the organization of actin filaments into contractile structures. Filamin cross-links microfilaments in a loose meshwork such as that found in the cell cortex, in which actin filaments are tethered to the plasma membrane by molecules such as spectrin. It is clear from these observations that the 3-D organization and texture of microfilament networks is determined by the expression of a specific combination of microfilament-associated proteins. The combination will be different in muscle cells, fibroblasts, and epithelial and neuronal cells. Actin can also exist in different 3-D organizations in a single cell. Therefore, the local environment plays an important role in determining which kind of molecules will interact with the microfilament system and affect its assembly state. It is also noteworthy that the interaction between myosin molecules and actin filaments in nonmuscle cells can contribute to the ordering of actin microfilaments into specific orientations in addition to producing forces. This is due to the fact that myosin binds in a specific orientation to the filaments and slides along

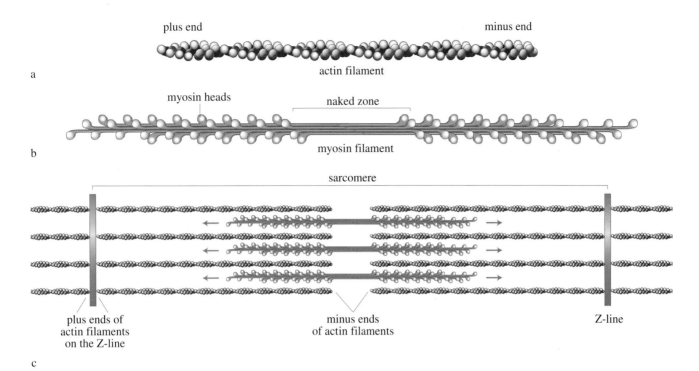

FIGURE 6 The actomyosin system. (a) Structure of actin microfilaments, consisting of actin subunits folded into a spiral. (b) Structure of myosin thick filaments, formed from many myosin molecules arranged head-to-tail in a bipolar filament. (c) Organization of actomyosin in striated muscles. The actin filaments are anchored at the Z-line so that the organization is antiparallel in the sarcomere. The bipolar myosin filaments are placed so that, when they move along the antiparallel actin filaments, the muscle fiber is shortened, producing a contractile force.

them in a given direction (Fig. 8). It will be shown that a similar mechanism participates in the organization of microtubules in three dimensions.

The Cell Cortex: An Example of Multiple Arrangements of Microfilaments. The cell cortex is a layer of cy-

toplasm that is just below the plasma membrane. This area is obviously extremely important for the determination of cell shape because it is where forces are produced on the membrane and where external interactions with the plasma membrane are transduced to the cytoplasm. Actin micro-

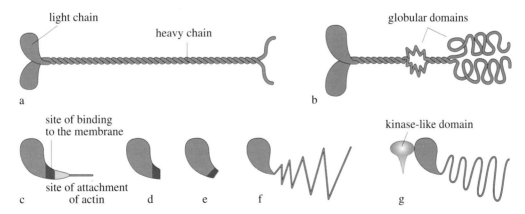

FIGURE 7 Examples of different myosins with their motor domains and targeting tails. (a) Conventional myosin, forming dimers and bipolar filaments, permitting muscle contraction and cell division. (b) Yeast myosin (MYO2), which forms dimers, organizes the actin filaments, and is implicated in exocytosis. (c–e) Unconventional forms of monomeric myosin I, found in amoebae and in vertebrates, which bind to the membrane through their tails. (f) Another type of unconventional myosin, whose function is unknown, found in the acanthamoeba. (g) NinaC, a hybrid, found in *Drosophila* photoreceptors between a myosin and the monomeric domain of a kinase which may have a role in phototransduction.

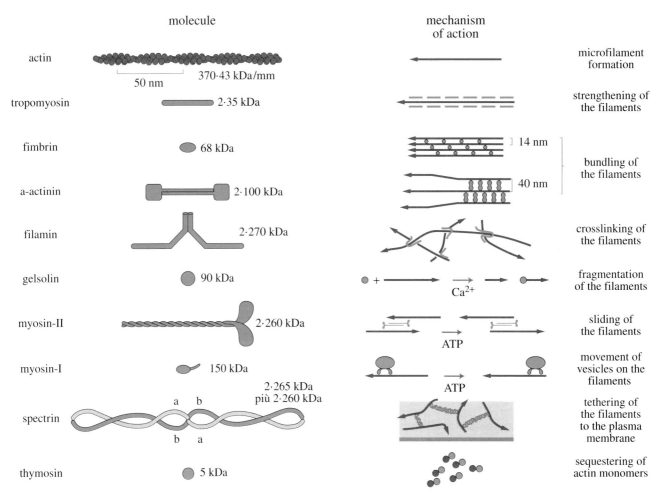

molecule

mechanism
of action

actin   370·43 kDa/mm   50 nm   microfilament formation

tropomyosin   2·35 kDa   strengthening of the filaments

fimbrin   68 kDa   14 nm   bundling of the filaments

a-actinin   2·100 kDa   40 nm

filamin   2·270 kDa   crosslinking of the filaments

gelsolin   90 kDa Ca^{2+} fragmentation of the filaments

myosin-II   2·260 kDa   ATP   sliding of the filaments

myosin-I   150 kDa   ATP   movement of vesicles on the filaments

spectrin   a   b   2·265 kDa più 2·260 kDa   b   a   tethering of the filaments to the plasma membrane

thymosin   5 kDa   sequestering of actin monomers

FIGURE 8   Some proteins which bind to actin (left) and their effects (right) on the 3-D organization of the microfilaments. Actin is shown in red, some actin-binding proteins are shown in green, and myosin is shown in blue.

filaments play an essential role in the shaping of this area. In most cells, there is a dense network of microfilaments cross-linked by spectrin molecules connected to the plasma membrane through ankyrine molecules. Ankyrine binds directly to integral membrane proteins. The apical membrane of epithelial cells forms numerous microvilli which contain a rigid array of microfilaments. These microfilaments have a very precise organization, determined by specific cross-linking proteins and the lateral attachment of the filaments to the plasma membrane (Mays *et al.,* 1994; Louvard *et al.,* 1992). This attachment involves a certain form of minimyosin which has an unknown function. These examples illustrate well the diversity of 3-D arrays generated by the interaction of actin filaments with a combination of associated proteins. However, it is not understood how these complex interactions generate specific shapes.

Microtubules. Microtubules are built of tubulin dimers. This molecule, like actin, has an asymmetric structure that results in the assembly of an asymmetric polymer

(Fig. 9). However, each microtubule is composed of a variable number of protofilaments (usually 13 *in vivo*) that form a hollow tube. Tubulin molecules are dimers, each of which has a GTP binding site. One of the two sites can exchange GTP, and upon assembly the GTP of this site is hydrolyzed into GDP. Again, this is not required for assembly but used to build some instability into the polymer. The GTP-liganded form of free tubulin dimers has a much higher affinity for GTP-liganded subunits than for GDP-liganded dimers at the extremity of the polymer. Whereas in the absence of associated proteins actin polymers are highly stable, microtubules are fairly unstable and display a peculiar behavior, important for their morphogenetic properties and termed dynamic instability (Kirschner and Mitchison, 1986). This means that under certain conditions, microtubules can rapidly alternate between growing and shrinking phases (Figs. 9 and 10). This behavior is important because it can be modulated by local changes in the cytoplasmic state. It is believed that microtubules become unstable

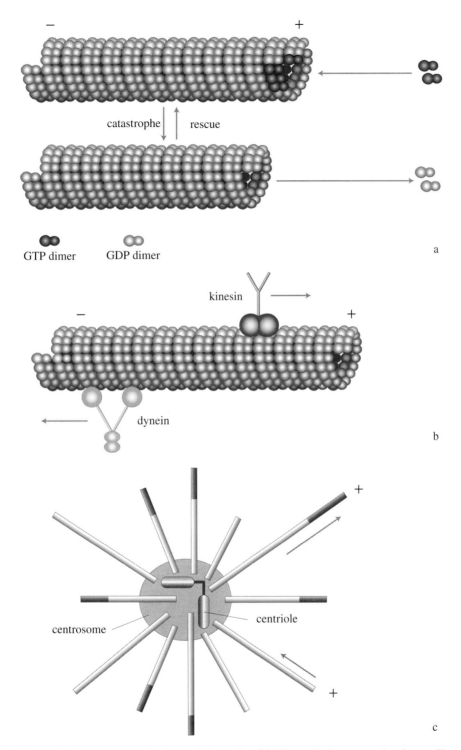

FIGURE 9 Microtubule structure, polarity, and dynamics. (a) Microtubules are made of protofilaments, each built of tubulin dimers. They grow by addition of GTP tubulin (dark green) at their plus end which is hydrolyzed following binding so that the microtubule wall is made of GDP tubulin (light green). Microtubules can alternate between growing phases (rescue) and shrinking phases (catastrophe), during which they lose GDP tubulin from the plus end. (b) Two types of motor move along the surface of the microtubules: Proteins such as dynein move toward the minus end, and proteins such as kinesin move toward the plus end. (c) Microtubule have organization centers composed of the centrosomes, which are composed of a pair of centrioles surrounded by nucleating material containing γ-tubulin complexes. The centrioles always grow away from the plus end.

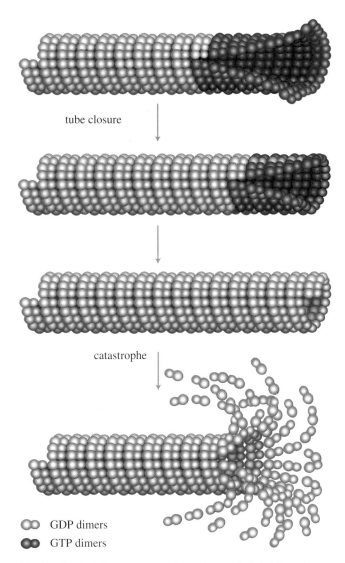

tube closure

catastrophe

⊙⊙ GDP dimers

⊙⊙ GTP dimers

FIGURE 10 The structure of growing and shrinking microtubule ends. Microtubules grow as sheets of protofilaments and shrink by explosive disassembly. The tubulin subunits in growing sheets are believed to be GTP liganded. In the curled oligomers at the tip of shrinking microtubules they are GDP liganded. The tubulin subunits undergo a conformational change during GTP hydrolysis. It is possible that GTP hydrolysis is forced by tube closure so that a catastrophe occurs when a tube closes all the way to the tip of the sheet (modified from Hyman and Karsenti, 1996).

when all GTP is hydrolyzed in the terminal subunits of the polymer and this seems to be related to structural changes that occur in the growing microtubule wall (Fig. 10; Hyman and Karsenti, 1996). The dynamic end of microtubules is also called the plus end (Fig. 9). Microtubule dynamic instability is defined by four parameters that can be measured by video microscopy: the growth rate (V_g), the shrinking rate (V_s), the catastrophe rate (transition from growth to shrinkage or f_{cat}), and the rescue rate (transition from shrinkage to growth or f_{res}). Each parameter can be modulated independently and thereby the behavior of microtubules affected.

In contrast to actin microfilaments, there are two types of motors that move along microtubules: dynein-like proteins that move toward the minus end and kinesin-like proteins (KLPs) that move toward the plus end with a few exceptions (Fig. 9). Both motors are involved in many aspects of cell organization, from the transport of vesicles to the plasma membrane or chromosomes during mitosis to the organization of microtubule arrays in both interphase and mitosis (Hyams and Llyod, 1994; Moore and Endow, 1996). This diversity of tasks is made possible by the adjunction to the motor domain, which is fairly standard, of targeting domains that localize the motor to different cargoes (Fig. 11). There are different families of motors that are defined by specific sequences in the motor domain outside the highly conserved regions forming the ATP and microtubule-binding domains. Usually, these motors are made of a globular domain attached to a coiled coil stalk allowing dimerization. In turn, the stalk is connected to a globular tail that of-

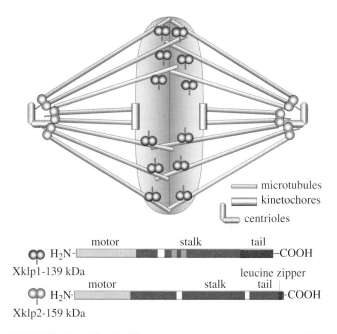

microtubules
kinetochores
centrioles

motor   stalk   tail
⊙⊙ H₂N-[   ]–COOH
Xklp1-139 kDa   leucine zipper
motor   stalk   tail
⊙⊙ H₂N-[   ]–COOH
Xklp2-159 kDa

FIGURE 11 Kinesin-like proteins are targeted to specific cellular domains by signals present in their tails.(Top) A simplified mitotic spindle is shown with a single chromosome indicated in blue. (Bottom) The schematic structures of the two proteins are shown. Xklp1 (magenta) is targeted to chromosome arms by an unknown specific sequence. Xklp2 is targeted to spindle poles and centrosomes by a sequence in its tail that contains a leucine zipper. These zippers are protein sequences containing many leucines in a particular configuration. These sequences are involved in homodimerization of proteins (i.e., formation of dimers of the same molecule) or in the interactions between proteins of different molecular type.

ten contains the targeting sequence. Figure 11 shows Xklp1 and Xklp2, two kinesin-like proteins targeted to the chromosomes and centrosomes, respectively. Xklp1 is required for chromosome positioning and spindle assembly, whereas Xklp2 is required for the separation of spindle poles during prophase. In both cases, the targeting sequences are localized in the tail (Boleti *et al.,* 1996; Vernos and Karsenti, 1995).

Organization of Microtubules in Three Dimensions. Compared to actin microfilaments, microtubules are fairly rigid polymers. Whereas actin is used to modulate the consistency of the cytoplasm, in contractile processes and in the local transport of vesicles microtubules are involved in the long-range organization, shape, and polarity of cells. Various structures and organelles are transported along microtubules. Therefore, the ways by which microtubules are organized in three dimensions and oriented in the cytoplasm are key events in cellular morphogenesis. One important aspect of microtubule organization in all cells deals with the mechanism by which they are oriented in a collective manner. There are two ways in which microtubules can be oriented into arrays of uniform polarity. One way is by being nucleated from specific nucleation centers, and the other is by being arranged by motors that sort and cross-link microtubules in oriented arrays (Hyman and Karsenti, 1996).

There are different kinds of nucleation centers in cells, but the most common in higher eukaryotes are made of a centriole surrounded by a cloud of material capable of nucleating microtubules (Fig. 9). Centrioles are cylinders which are built of microtubule triplets. Their actual function remains a mystery apart from the fact that they are used as seeds to form cilia and flagella. The nature of the pericentriolar material that nucleates microtubules has remained mysterious, but we are beginning to determine some of the molecules present in this material and involved in microtubule nucleation. One form of tubulin that cannot form microtubules (called γ-tubulin) is present in a complex with other proteins in the pericentriolar material. Such γ-turcs form rings which have approximately the diameter of microtubules. Functional experiments have shown that these structures can nucleate microtubule assembly (Raff, 1996). Because microtubule growth is oriented with the plus end leading, such nucleation centers form asters of microtubules with a homogeneous polarity — minus end in the center, and plus end at the periphery. Because microtubules can undergo dynamic instability, the length of microtubules nucleated by such nucleating centers can vary from infinity (limited only by the supply of free tubulin subunit) to a steady-state value defined by the dynamic state of the population. Indeed, as already mentioned, the parameters of microtubule dynamic instability can be modulated in the cell by regulatory factors such as microtubule-associated proteins (MAPs) or destabilizing factors. Phosphorylation

of such factors can also regulate their activity. Therefore, conditions can be created in the cell (e.g., during mitosis) in which short asters of microtubules having well-defined average steady-state length are formed (Verde *et al.,* 1992). This is important because, in principle, this allows for local modulation of microtubule dynamics and therefore modulation of the shape of the asters.

In the previous mechanism, microtubules are organized into an aster of uniform polarity by nucleation centers and the regulation of microtubule dynamics. Indeed, in order to generate only centrosome-nucleated microtubules, the conditions should be such that there is no spontaneous polymer assembly in addition to that nucleated by the centrosomes. There is another way to produce microtubule asters of uniform polarity in the absence of focused nucleation. This can be achieved by motors (Fig. 12). Such a self-organization of microtubules into polar arrays in the absence of preformed centrosomes has been clearly seen in several systems in interphase. Experiments carried out in mitotic *Xenopus* egg extracts also showed that microtubules stabilized by taxol progressively reorganize from an initial random distribution into small asters containing centrosomal material in their centers. Since inactivation of the minus end-directed motor dynein blocked aster assembly, it is clear that motors can organize microtubules into asters. Figure 12 shows how this could happen. If a multisubunit microtubule-based motor binds to two microtubules, it will necessarily move to the end of the two microtubules. If more microtubules are caught in the process, an aster composed of microtubules with uniform polarity will be created. Although minus ends of microtubules are always clustered together *in vivo,* the general property that unidirectional motors form asters of uniform polarity has been demonstrated in pure systems composed of microtubules and the plus end-directed motor kinesin, in which asters assemble with microtubule plus ends at the center (Fig. 12). This mechanism requires two properties of the motors involved. First, the motors must have multiple motor heads. Second, the motor should not run off the minus end of microtubules, or at least there should be an accumulation of motor or cross-linking factor at the minus end so that the steady-state pattern observed is radial and not a parallel bundle.

The Mitotic Spindle: An Example of Morphogenesis by Microtubule Self-Organization

The mitotic spindle is the apparatus by which chromosomes are segregated during mitosis. This apparatus functions because the spindle is bipolar, with two sets of microtubules with opposite polarity being assembled around the chromosomes. Because chromosomes are attached to microtubules through kinetochores, each chromosome can be transported to the pole it faces, which will become the future new cell.

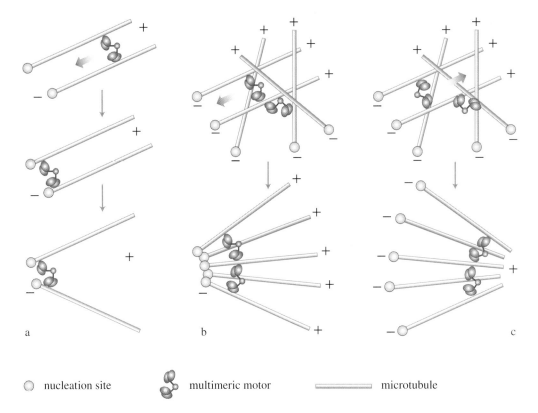

nucleation site   multimeric motor   microtubule

FIGURE 12 The generation of polar microtubule arrays by motors. (a) Multimeric minus end-directed motors can cross-link two microtubules and accumulate at one end. (b) If more than two microtubules are caught in the process, an aster of microtubules is generated with minus ends in the center. (c) A plus end-directed motor could also produce asters, but in this case with plus ends at the center. This behavior is important since it is a morphogenetic event: Starting from randomly oriented polymers, the formation of particular orientations is possible using energy derived by the motors from hydrolysis of ATP as they move along the microtubules (modified from Hyman and Karsenti, 1996).

In animal mitotic cells, uniform polarity in each half-spindle derives from the uniform polarity of growth from the centrosome. In this pathway, the centrosome-nucleated microtubules become highly dynamic at the onset of mitosis, mostly as a result of an increase in the catastrophe frequency. Concomitantly, the duplicated centrosomes migrate around the nucleus through the activity of motors (Fig. 13a). The chromosomes exert a local effect on the cytoplasm that results in the stabilization of microtubules. This local effect may be exerted through a gradient of protein phosphorylation that regulates microtubule dynamics (MAPs and destabilizing factors). The result is that microtubules elongate preferentially toward the chromosomes (Fig. 13a) and are captured by kinetochores and motors present on chromosome arms. As the microtubules from the two half-spindles mingle, they bind together through antiparallel interactions, further stabilizing the spindle in a bipolar configuration with the poles 180° apart.

In plants and meiotic cells, microtubules are nucleated randomly around chromosomes, probably from dispersed nucleation sites not organized in a centrosome (Fig. 13b).

They elongate only around chromosomes due to the local stabilization effect, and they become organized into poles by the minus end-directed motor dynein according to the mechanism described in the previous section. Microtubules are then pushed apart by plus end-directed motors present on chromosome arms as well as by plus end-directed motors interacting with antiparallel microtubules. These mechanisms sort microtubules into two antiparallel populations of uniform polarity and a bipolar spindle with chromosomes located at the equator is generated (Hyman and Karsenti, 1996; Vernos and Karsenti, 1995).

Both meiotic and mitotic spindles assemble according to similar principles: local modulation of microtubule dynamics by chromosomes and sorting of microtubules into a bipolar array by motors that read their polarity. The presence of organized centrosomes is not essential for bipolarity. The bipolarity is centered and stabilized around chromosomes that create a local asymmetry in the cytoplasm. In essence, the mitotic spindle assembles inside a "morphogenetic field" defined around the chromosomes by the localization of enzymes (kinases and phosphatases) that

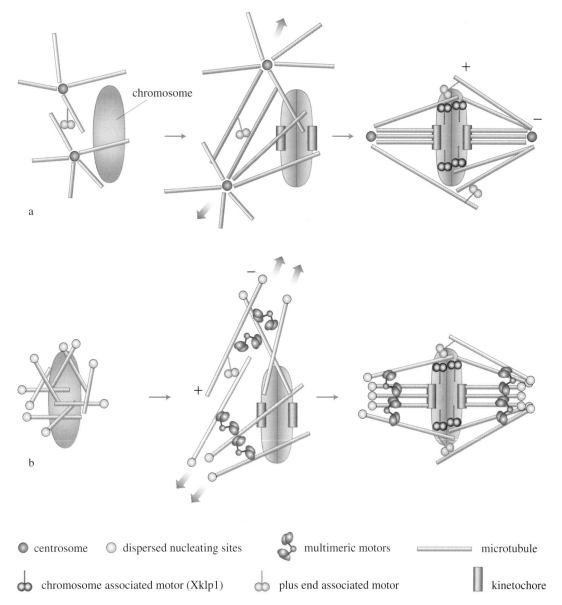

centrosome ○ dispersed nucleating sites multimeric motors ▭▭▭ microtubule

chromosome associated motor (Xklp1) plus end associated motor ▌ kinetochore

FIGURE 13 Pathway of assembly of mitotic and meiotic spindles. (a) In the mitotic spindle of animal cells, centrosomes are separated by motors in prophase when microtubules start to shrink in response to the activation of cdc2 kinase. After the nuclear envelope breaks down, microtubules start to grow preferentially toward chromosomes, where they become captured by chromosome arms and kinetochores. Finally, bipolarity is stabilized by motors that move apart and cross-link antiparallel microtubules originating from each centrosome. (b) In meiotic spindles, there are no centrosomes. Microtubules originate from dispersed nucleating sites around chromosomes and become captured by kinetochores and Xklp1 motors on chromosome arms. The minus ends are pushed away from chromosomes by plus end-directed motors localized on chromosome arms and motors that move antiparallel microtubules apart. The poles are then focused by minus end-directed motors as shown (modified from Hyman and Karsenti, 1996).

govern the activity of proteins regulating microtubule dynamics and motors that move microtubules in a coordinated fashion. It is likely that similar principles guide the organization of the cytoskeleton in interphase during cell diversification and differentiation. The difference is that other kinases and phosphatases localized in specific cell domains

will regulate the activity of other motors and regulators of microtubule dynamics.

Microtubules and the Shapes of Protists

The shapes of the protists shown in Fig. 1b are determined mostly by the arrangement of microtubules that participate in the organization of sensory cilia, beating cilia, and flagella used for movement. The organization of the mouth is also shaped by microtubules (Fig. 1b). Using the example of mitotic spindle assembly, one could think that these various 3-D microtubule arrangements are determined by the coexpression of a combination of molecules that affect their dynamics, position, and orientation in the cell and that a particular morphogenetic field involving localized enzymes is responsible. However, what is the origin of the asymmetry around which such a field could be generated? Most interestingly, in protists the arrangement of microtubules in space is not just a self-assembly process relying on the expression of a combination of molecules in proper stoichiometry. It has been shown that there is a structural memory governing the orientation of ciliary rows in the cortex of ciliates. Therefore, the asymmetry is transmitted from generation to generation (Frankel, 1984). Also, the determination of embryonic axes (anteroposterior and dorsoventral) relies heavily on microtubule orientation in many embryos and this is partly determined by cues that are transmitted from the mother to the oocyte. Therefore, just as mitochondria and chloroplasts are propagated in eukaryotes after a symbiotic event, the shape of cells may be determined in part by the self-organization properties of molecules such as tubulin and in part by the transmission from generation to generation of certain patterns that were once formed spontaneously and then stabilized during evolution. These patterns provide a physical asymmetry around which the cytoskeleton can be organized in specific ways.

Microfilaments, Microtubules, External Cues, and Epithelial Cell Polarity

As discussed previously, epithelial cells are the first to differentiate at the boundary between the embryo and the external medium. From these initial cells, many different kinds of epithelial cells arise and we can develop some ideas about the molecular mechanisms involved in the morphogenesis of these cell types. In order to confront two very different environments, epithelial cells have the capacity to achieve a regulated and vectorial transport of fluids and solutes between these environments. They accomplish these functions by partitioning their plasma membranes into two domains with dramatically different protein and lipid compositions. These cells express proteins that allow the formation of junctions between neighboring cells. Tight junctions establish a boundary between the apical surface, which faces the external milieu, and the basolateral surface, which rests on a basement membrane and is bathed by interstitial fluid. The establishment of epithelial cell polarity requires that the cells orient themselves relative to the asymmetric features of their environment. This happens through interaction with the substratum and neighboring cells, and this information must be transmitted to the interior of the cell so that the proper lipids and proteins become transported toward the apical and basolateral domains of the cell. The

barrier established by tight junctions in the membrane is important in allowing the progressive accumulation of specific proteins in both domains. Newly synthesized apical and basolateral proteins are sorted into different vesicles at the level of the trans-Golgi network before being sent to their respective cell surface domains. This sorting involves specific signals present on the proteins destined to be inserted in each domain. Since vesicle transport requires microtubules, the orientation of microtubules in epithelial cells is extremely important for the generation and maintenance of epithelial cell polarity (Lafont and Simons, 1996). In epithelial cells, microtubules have a fairly uniform orientation, with their plus end facing the basolateral domain and some randomly oriented microtubules in the apical domain (Mays et al., 1994) (Figs. 2a and 14c). How could microtubules become oriented in such a way? When the junctions are broken, the cells lose their polarity. Under such conditions, microtubules take on a very different organization: They become organized in a radial array originating from an ill-defined point localized close to the nucleus. If one allows the cells to reestablish junctions and polarize again, microtubules reorganize and recover their original orientation (Fig. 14a). This process requires the establishment of junctions and actin microfilaments. From what is known of the molecular mechanisms involved in the organization of microtubules in three dimensions, one could envision two mechanisms by which they become oriented in epithelial cells. The nucleating centers could be moved by actin toward the apical domain of the cells, thereby orienting microtubule growth. Another mechanism would be to position minus end-directed motors such as dynein in the cortex of the cells (Fig. 14). In this way, microtubule plus ends would be pushed down toward the basolateral domain of the cells. Although there is no evidence for such a mechanism, it is quite likely to happen because cortical dynein seems to interact with microtubules to orient the mitotic spindle and to transport microtubules in neurons (Baas and Wenqian, 1996). Again, in this hypothesis dynein should be localized to the cortex of the cell and it may be bound to cortical actin filaments.

The way cell–cell interactions and junctions act on the interior organization and polarity of the cytoplasm has begun to be understood. In many epithelia, cell–cell adhesion is regulated by a family of calcium-dependent cell adhesion molecules termed cadherins. For example, in kidney epithelial cells in culture, ankyrin, fodrin, and the Na^+/K^+-ATPase become localized to lateral membranes after induction of calcium-dependent cell–cell adhesion. Impressively, the transfection of nonepithelial cells (fibroblasts) with E-cadherin is sufficient to localize Na^+/K^+-ATPase, ankyrin, and fodrin to sites of cell–cell contacts. In epithelial cells, there is a dense actin cortex just below the apical membrane (Figs. 2a and 14), and in some epithelial cell types (such as in the intestine epithelium) microvilli develop. Both become organized as a result of the establish-

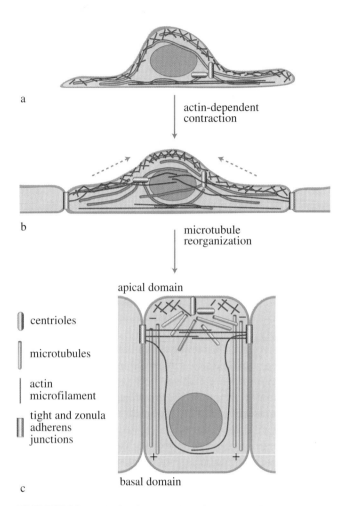

centrioles

microtubules

actin
microfilament

tight and zonula
adherens
junctions

apical domain

basal domain

a

b

c

FIGURE 14 Polarization of epithelial cells and cytoskeleton rearrangements. (a) Epithelial cells in culture are initially dissociated and allowed to reform an epithelium. (b) Tight and zonula adherens junctions reassemble and actin microfilaments begin to grow from these junctions. The centrioles split and microtubules become organized parallel to the apicobasal axis, with their plus end facing the basal domain. (c) Successively, the centrioles split apart and move to the apical domain while the cells start to rise in height, due in part to microtubule growth and in part to a contraction mediated by the apical actin. It is hypothesized that microtubule motors are anchored in the cortical and apical actin network, leading to the organization of microtubules along the apicobasal axis with their plus end facing the basal domain.

ment of cell–cell contacts (Louvard *et al.*, 1992; Mays *et al.*, 1994). Therefore, one could propose a sequential model for the generation of epithelial cell organization (Fig. 14) in which cell adhesion molecules induce the assembly of a cortical actin network. Minus end-directed microtubule motors (dynein) are targeted and anchored into this network. At the same time as the actin filaments begin to contract under the action of myosin, microtubules are pushed downwards and aligned through the action of the cortical dynein moving toward the minus end of microtubules which are

stabilized and made fairly rigid by MAPs that cross-link them. How all this is coordinated in a total wonder.

Microfilaments and Microtubules in the Behavior of Neurons

Development of the nervous system involves the migration and interconnection of billions of neurons. Neurons produce neurites, processes that grow out from the cell body to establish all the interconnections necessary to the function of the nervous system. There are axons and dendrites. During outgrowth, a specific subcellular domain, the growth cone, forms at the tip of the neurite. Growth cones may be viewed as navigation devices that probe the environment to direct the growth of the neurite. Extracellular signals (chemical or physical, such as interaction with other neurites) must be sensed by the growth cone and transmitted to the cytoskeleton to direct its growth. As in epithelial cells, actin seems to react to these signals and act on the microtubules to direct their organization. As shown in Fig. 15, in a typical growth cone, F-actin is distributed predominantly in the peripheral lamellar region, whereas microtubules are mostly located in the more central region (Lin *et al.*, 1994). The peripheral lamella is made of two major subdomains: filopodia, containing bundled unidirectional actin filaments (plus end at the extremity), and lamellipodia, in which microfilaments are randomly oriented. All these microfilaments are in a highly dynamic state while the growth cone is exploring the environment. F-actin assembles at the leading edge of the growth cone and disassembles in the zone between the growth cone and the body of the neurite where microtubules seem to end (Fig. 15). In addition, there is a constant flow of filamentous actin from the leading edge toward this same area of the neurite. These two systems of microfilaments appear to have different functions. The long bundled filament population may be more linked to transmembrane receptors and more involved in the transmission of tension and maybe other signals between peripheral and central domains of the growth cone. The disorganized microfilaments may be involved more in the progression of the growth cone and neurite elongation. In other words, bundled filaments may act as exploratory devices capable of transmitting information to the interior of the growth cone, whereas other microfilaments may respond by producing tension in order to promote elongation of the growth cone.

Parallel microtubules grow in the body of neurites. Most microtubules are probably assembled in the cell body and transported by motors into the neurite, a process that may also orient the microtubules in the neurites (Baas and Wenqian, 1996). It seems that dynein anchored into the actin cortex is involved in microtubule transport and orientation. Microtubules end near the growth cone where they must interact in some way with the actin cytoskeleton (Fig. 15). Microtubule dynamics occurs throughout the neurite but dynamic instability seems to be particularly active close to the

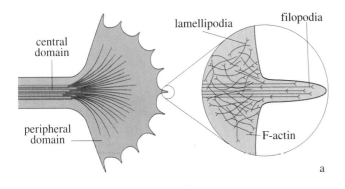

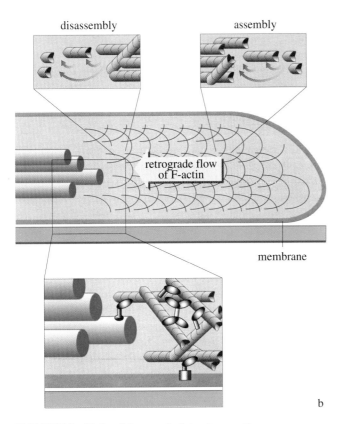

FIGURE 15 Role of the cytoskeleton in growth cone progression. (a) Top view of the structural domains of the growth cone showing the central domain, rich in microtubules (green), and the F-actin-rich (red) peripheral domain. The latter contains both bundled actin filaments (filopodia) and cross-linked networks of unpolarized actin (lamellipodia). (b) Retrograde actin flow functions by filament assembly at the forward domain and disassembly in the rearward domain. Myosin motors (blue) are thought to be bound to the membrane and to the microtubules by protein complexes.

growth cone. In this area, in rapidly advancing growth cones microtubules tend to be straight and bundled, whereas in slowly moving growth cones they splay out and loop. These observations are in agreement with the finding that microtubules are translocated from the cell body into the neu-rite by motors and may be anchored onto actin or on the plasma membrane. Indeed, if microtubules are translocated at a constant speed that depends on the motor involved, when progression of the growth cone slows down microtubules will tend to keep advancing in it and therefore curl up while hitting the actin meshwork of the growth cone.

Growth Cone Steering. The progression of neurites has been observed *in vivo* in limb bud (Fig. 16). They grow on a substrate of epithelial cells and their path is partly determined through interactions with "guidepost cells" (Bentley and O'Connor, 1994). When filopodia contact one such cell, F-actin accumulates at the tip and migrates toward the base of filopodia. Then, microtubule plus ends tend to orient preferentially in this area. As a consequence, a new direction of growth of the neurite is taken toward the guidepost cell. Similar reactions have been observed for neurites contacting other neurons. It is likely that the interaction between specific receptors, or the effect of chemical signals on cell surface receptors, leads to a local increase in actin assembly. In turn, the increased F-actin concentration leads to a stronger interaction with microtubules, resulting in a stabilization of neurite outgrowth in the new direction. Since microtubules are involved in the transport of vesicles and other materials toward the growth cone, this phenomenon is reminiscent of what happens in epithelial cells: External cues act first on the actin cytoskeleton, which in turn acts on the orientation of microtubules which function as tracks on which materials destined to specific areas of the cell can be transported selectively. This again raises the problem of cross-talk between the microtubule system and microfilaments. It is clear that some motor(s) capable of interacting with both microtubules and microfilaments must exist. Such a motor(s) should be bound to the microfilaments and interact with microtubules through the motor domain in order to capture and reel microtubules in toward the growth cone. A minus end-directed motor such as dynein seems indeed to function in this way. Myosin bound to microtubules could also have the same result.

Growth Cone Elongation. Here, interactions between the cytoskeleton and the external substrate through transmembrane proteins are a key issue. This is a good example of morphogenesis involving the regulation of polymer dynamics by membrane proteins which interact with cues on the substrate on which the neurite is growing. As already mentioned, part of the F-actin dynamics is due to a bulk flow of polymer toward the rear end of the growth cone. It seems that this flow is part of the mechanism by which the growth cone progresses. Progression depends on a regulated interaction between microfilaments and transmembrane receptors (Fig. 15). When growth cones are not progressing, there is a robust flow of polymer toward the rear of the growth cone. In other words, actin polymer assembles at the tip of the growth cone and disassembles at the rear without producing force, like an engine uncoupled from wheels (disengaged clutch). When growth cones are

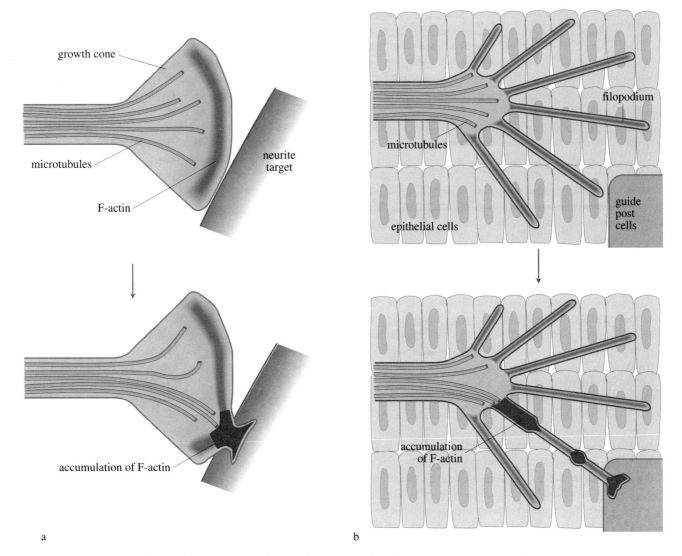

growth cone

microtubules

F-actin

neurite target

accumulation of F-actin

microtubules

epithelial cells

filopodium

guide post cells

accumulation of F-actin

a

b

FIGURE 16 Role of the cytoskeleton in growth cone steering. (a) A growth cone progressing toward a target (another neurite). Following interaction with the neurite through specific receptors, F-actin accumulates at the site of contact in the growth cone. This results in strong binding of microtubules to the accumulated actin and change in direction of growth of the cone. (b) In a growth cone advancing on a carpet of epithelial cells, neurons move by interacting with guidepost cells. When one of the filopodia meets a guidepost cell, F-actin accumulates at the site of contact and then migrates to the base of the filopodium where it binds microtubules and stabilizes them, leading to a turn in the direction of growth of the neurite.

progressing, F-actin becomes linked to transmembrane proteins that interact with the substratum and the rearwards flow stops. Since there is continuous polymer assembly at the tip of the growth cone, this results in net forward progression of the polymer relative to the substratum. Interaction between the polymer and transmembrane proteins is supposed to be governed by regulators such as signaling molecules on the substrate or diffusible chemicals (Lin *et al.,* 1994).

At least four major classes of proteins may be involved in the transduction of rearward F-actin flow into forward growth: extracellular substrates, receptors such as adhesion molecules, proteins mediating interaction between F-actin and transmembrane proteins (the clutch), and F-actin interacting proteins. Three major classes of adhesion proteins have been characterized in neurons: calcium-independent cell adhesion molecules (CAMs), cadherins (calcium-dependent cell adhesion molecules), and integrins (heterodimeric receptors for the extracellular matrix). Members of all three classes seem to associate with the actin cytoskeleton to some extent and to participate in axonal elongation. Spectrin and ankyrin, which are actin-binding proteins of the cortical membrane cytoskeleton, are potential NCAM-binding proteins. Proteins of the catenin family are also good candidates to mediate interaction between F-actin and cadherins. A general picture emerges in which homologous classes of proteins appear to be involved in regulating the organization of cytoskeleton assembly in response to external cues in epithelial and neuronal cells. Also, it seems that, in general, actin participates in the transduction of external signals to the interior of the cell and to microtubule organization, both in epithelial and in neuronal cells, although in a different manner.

Morphogenesis of Skeletal Muscle Fibers

Epithelial cells and neurons originate from the ectoderm and they both display the same kind of asymmetry, probably because of the common origin. Muscle cells originate from the mesoderm and their differentiation behavior is radically different from that of the other two kinds of cells (Franzini-Armstrong and Fischman, 1994). The morphogenesis of skeletal muscle fibers is of particular interest because it involves the alignment and fusion of individual cells (myoblasts). Myoblasts differentiate in the embryo at sites destined to form striated muscles (somites, limb rudiments, etc.). One of the initial signs of myogenesis is the assumption of a bipolar spindle shape by the myoblasts in contrast to the more flattened, multiform adjacent mesenchymal population. This seems to reflect a reorganization of microtubules which is related to substrate interactions and cell migration. Immediately before and during fusion *in vivo,* the myogenic cells cease DNA synthesis and initiate the transcription required for full differentiation.

Fusion. Successive fusion events result in the formation of long multinucleated "myotubes" (Fig. 17). There is a short time window during which both myoblasts and myotubes are fusion competent. Fusion requires calcium ions because N-cadherin, a calcium-dependent cell–cell adhesion molecule, is involved in myoblasts interactions and fusion. Before fusion, microtubule nucleation occurs from centers, containing centrioles surrounded by microtubule nucleating material, localized close to the nucleus of the myoblast. Soon after the fusion event, centrioles are destroyed and microtubules seem to be nucleated at least in part from the nuclear envelope. There are now two kinds of microtubules: those that radiate from the nuclei and a dense population of elongated microtubules that run parallel to the long axis of the myotubes (Tassin *et al.,* 1985). It is not clear how these microtubules are organized in terms of polarity and length. It is possible that they grow from the nuclear envelope and become organized by motors along the length of the myotubes. Following fusion, the nuclei are progressively relocalized just below the plasma membrane. At this point, there is no actomyosin organization into typical myofibrils, although most of the necessary components are being synthesized. The major forms of F-actin present are cytoskeletal forms of microfilaments arranged into stress fibers running just below the plasma membrane. Although microtubules are essential to maintain the shape of myotubes and for further myogenesis, it is not known how they participate in this process.

Myofibrillogenesis. The adult muscle fiber contains a highly ordered arrangement of three major components: myofibrils, membranes, and a basic cytoskeletal network anchored at the plasma membrane and closely associated with sarcoplasmic reticulum and myofibrils. Therefore, an understanding of myogenesis will require an analysis of the mechanism of assembly of each of these components and how they are organized in space relative to each other. However, I restrict this discussion to a description of what is known about myofibril morphogenesis. Myofibrils are composed of precisely aligned sarcomeres which are composed of antiparallel actin filaments of precise lengths (thin filaments) anchored into the Z-line (mostly composed of α-actinin) and of bipolar myosin filaments (thick filaments) cross-linked at the level of the M-line. Two proteins of the sarcomeric scaffolding, titin and nebulin, are associated with thick and thin filaments, respectively. These are among the largest polypetides known. Titin is closely associated with the thick filaments and spans the length of a half sarcomere, starting at the M-line or in close proximity to it and reaching the Z-line, to which it is anchored. Nebulin is associated with thin filaments through interactions with actin. The long nebulin molecules span the length of the entire thin filaments and bind to the Z-line α-actinin molecule at one end. The sarcomere is therefore held together by a scaffold made of titin, nebulin, and the M- and Z-line proteins which respectively cross-link thick and thin filaments to one another. There are few microtubules in adult muscles, but they are abundant in developing muscles, as already mentioned.

Assembly of myofibrils apparently occurs as soon as sufficient concentrations of the necessary proteins accumulate in the cytoplasm. The problem is to understand how thin and thick filaments, together with their associated proteins and the scaffolding proteins, become ordered. We know that both actin and myosin have self-assembly properties. In fact, myosin molecules tend to aggregate *in vitro* in thick filaments that have approximately the length of the thick filaments observed in muscle cells. Actin also has self-assembly properties. The specific length of the actin filaments found in muscle cells is probably determined in part by the relative concentration of actin and other molecules controlling actin assembly present in muscle cells at the time of myofibillogenesis. In the presence of α-actinin and other proteins involved in Z-line formation, the Z-line probably starts to assemble and to nucleate actin microfilaments which reach a determined steady-state length. Actin-capping proteins may be involved in this process. At the same time, myosin thick filaments self-assemble. Titin, nebulin, and M-line proteins must then in some way interact with the other self-assembled components. Structural observations made by both electron microscopy and immunofluorescence suggest that there is first formation of pre-

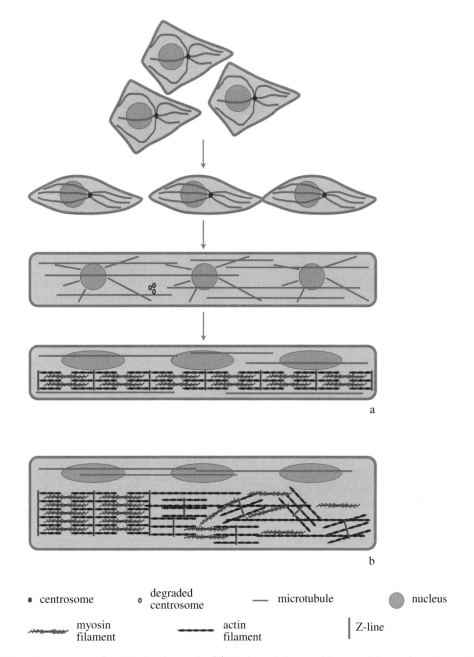

FIGURE 17 Morphogenesis of skeletal muscle. (a) Fusion of the myoblasts and formation of myofibrils. Before fusion (top), myoblasts have a fibroblast–mesenchymal form: the microtubules (in green) are organized in a radial distribution with the centrosome (in brown) at the center. Immediately prior to fusion, the cells assume a bipolar spindle shape with the microtubules aligned along the longitudinal axis. After fusion, the centrosomes are destroyed and the microtubules are concentrated by the nuclei and aligned according to the major axis of the myotube. Muscle proteins (actin, myosin, etc.) are not visible in this phase because, although present in large quantities and actively being synthesized, they are not yet organized. In the days after fusion, the muscle cell proteins organize into myofibrils. (b) Hypothetical mechanism of the formation of myofibrils. (Left) A completely formed myofibril in which the thick filaments of myosin (in blue) are positioned between the actin filaments (in red) anchored at the Z-line (in violet). The microtubules may serve as a scaffold for the autoassembled thick filaments, for the Z-line complexes, and for the actin filaments of the myofibrils.

myofibrils in which the spacing between sarcomeres is very irregular and that these premyofibrils are then quickly transformed into myofibrils. It is unknown how all these self-assembled components become properly aligned with

each other. One possibility is that the abundant microtubules present during early myogenesis carry out this function. In this case, some of the components, such as prefabricated Z-line/thin filament complexes and thick filaments, could interact with microtubules, even move along them before being properly positioned. This would require that the microtubules are properly oriented. How this is achieved is unknown, but it is clear that microtubules play an essential role in shaping myotubes following fusion. This pathway is entirely speculative, but it could easily be studied with current techniques. A detailed description of myogenesis is reported by Franzini-Armstrong and Fischman (1994).

Again, it is clear that an understanding of cellular morphogenesis relies heavily on our knowledge of the self-assembly properties of cytoskeletal proteins and how these are regulated in the cell by environmental cues.

From Form to Function: The Duality of the Cytoskeleton

When one examines how shapes are generated at the cellular level, it becomes clear that the link between shape and function is largely provided by the cytoskeleton. Also, one realizes that most cellular functions that are related to their shape are also carried out by a specific organization of the cytoskeleton which at the same time affects the shape of the cell accordingly. This is why I discussed a few examples of cells with typical shapes such as epithelial cells, neurons, and muscle cells to discuss how cell shape and function are related at the level of the autoorganization of the cytoskeleton. Clearly, the different 3-D assemblies of cytoskeleton proteins in different cells are generated by the expression of different sets of regulatory proteins capable of modulating the self-organization properties of microfilaments and microtubules. In turn, microtubules and microfilaments fulfill specific cellular functions which depend on their 3-D organization. Therefore, one can start to understand how a specific function is tightly linked to the shape of the cellular structure that fulfills it. Because biological shapes can be quite complex, one has to choose some aspect of a shape to relate it to a given function. As noted at the beginning of this article, asymmetry is a key determinant in the generation of cell shape and functions. In cells, asymmetry is largely determined by the polarity of microtubules and microfilaments. This is key to the generation of specific cellular shapes and to the organization and distribution of organelles in space.

Microtubule Orientation and Polarized Transport

Microtubule Orientation and Chromosome Distribution during Mitosis. It has been shown how, during mitosis, microtubules become organized into a bipolar array. This has a clear function: to distribute chromosomes to the two opposite domains of the dividing cell which will become the future daughter cells. This is possible because motors

associated with the kinetochores read the polarity of the microtubules connected to them and pull the chromosomes toward the poles of the spindle. It is interesting to note that spindle bipolarity is generated by the chromosomes, which determine microtubule orientation by a local modulation of microtubule dynamics and through the action of motors. In turn, the apparatus built in this way becomes operational to segregate the chromosomes. There are different kinds of spindles and one could try to classify them according to their morphology. However, they all have a bipolar organization with microtubule plus ends at the chromosomes and minus ends at the poles. Because we now understand the mechanism of mitotic spindle assembly, it is clear in this example how the intrinsic physical properties of microtubules can be responsible for the generation of a very precise shape which can fulfill a dedicated function: the segregation of chromosomes.

Microtubule Orientation and Vesicular Transport. Polarized epithelial cells aid vectorial absorption and secretion of ions and solutes between different biological compartments. This requires the development and maintenance of polarized distribution of proteins and lipids in the apical and basolateral plasma membrane domains of epithelial cells. These distributions appear to be regulated by targeted delivery of transport vesicles from the Golgi complex to specific plasma membrane domains and by protein sorting in both Golgi and cell surface membranes. Protein sorting seems to depend largely on intrinsic protein signals, whereas targeted delivery of transport vesicles relies on the organization of the cytoskeleton. We have seen that in epithelial cells microtubules are oriented with their minus end facing the apical plasma membrane and the plus end facing the basolateral membrane. Integrity of the Golgi apparatus and its localization depend on microtubules and the activity of the minus end-directed motor dynein. In fibroblasts or mesenchymal cells, the Golgi apparatus is localized around the centrosome, from which most microtubules grow in a radial array. In epithelial cells, the Golgi apparatus has a different localization, with the stacks extending along the minus end of the microtubules in the apical domain of the cells. Microtubules are also required for efficient transport of apical secretory and membrane proteins and for the transport of vesicles between the basolateral and apical domains (transcytosis), more specifically in the basolateral to apical direction. Kinesin (or kinesin-like proteins) is required for the transport of vesicles from the Golgi apparatus to both the basolateral and apical domains, whereas dynein is required only for transport to the apical domain. Therefore, the orientation of microtubules along the apicobasal axis is part of the sorting process that allows the distribution of the correct vesicles to their appropriate domain. The transport of components along microtubules is necessary not only because viscosity is high in the cytoplasm but also to the proper function of differentiated cells — in this case, exocytosis and transcytosis, which are essential functions of epithelial cells (Lafont and Simons, 1996).

There is another important function fulfilled by microtubules in epithelial cells during the development of mammalian embryos: the asymmetric distribution of cytoplasmic determinants that will distinguish the inner cell mass from the trophectoderm.

Epithelial cells are an interesting case because the generation of cell polarity requires directed transport of membrane proteins along microtubules, and this directed transport is also the main function of the formed epithelium. There is a sort of positive feedback loop: Extracellular cues lead to the organization of microtubules into an apicobasal orientation that results in the specific transport of membrane proteins required to establish membrane polarity and this system continues to be used afterwards for steady-state cell function.

Microtubule Orientation and Neuronal Cell Functions. The regulated flow of materials between the cell body and the axon and to and from synaptic terminals is essential for neuronal function. Since proteins are synthesized in the cell body, which is separated from the synaptic terminal by large distances, materials must be actively transported (and sorted) between the cell body and the nerve terminals. Moreover, materials should also be transported between axons and dendrites. It was previously discussed how the cytoskeleton is involved in governing the elongation of neurites in the growth cone and how microtubules may be oriented in the axons and dendrites. The orientation of microtubules, together with the activity of motors that read microtubule polarity, is key to the function of neurons. Both kinesin [as well as kinesin-related proteins (KRPs)] and dynein are present in neurons and associated with moving organelles in axons. Some organelles move from the cell body to the extremity of neurites (anterograde movement), whereas others move toward the cell body (retrograde movement). In axons, all microtubules have the same polarity — plus end away from the cell body. This suggests that most anterograde movements are due to KRPs (plus end motors), whereas retrograde transport is due to dynein-related motors. There might be exceptions to this rule because some KRPs move toward microtubule minus ends. This introduces some level of logical understanding of the mechanism of distribution of organelles in neurons: Those that should move toward the cell body need to be attached to dynein and the others to KRPs. However, it is more complicated. Since there are many different KRPs with tails that interact with specific organelles, one could envisage that the motors play a role in the sorting of components to different domains of the neuron. For example, in rat optic ganglion cells, two kinesin heavy chain variants associate with different components traveling at different rates. One kinesin is associated with the transport of vesicle precursors, whereas the other form is associated with mi-

tochondria transport. In another example, genetic experiments in *Caenorhabditis elegans* have shown that, in the absence of conventional kinesin, worms are paralyzed and secretory vesicles accumulate at the cell periphery, although synaptic vesicles appear to be transported normally. In the same worm, in the absence of a KRP different from conventional kinesin, the transport of synaptic vesicles to synaptic terminal regions is impaired (Coy and Howard, 1994).

Therefore, in neurons as well as in epithelial cells, the orientation and organization of microtubules in three dimensions generates a specific shape compatible with specific transport functions. Morphogenesis and function are tightly linked.

Microfilament Orientation and Cellular Functions

Vesicular Transport. Organelles are not only transported along microtubules. They can also move along actin filaments and myosin can bind to membranes. In fact, both in neurons and in epithelial cells, it seems likely that organelles move sequentially along microtubules and microfilaments. In epithelial cells as in neurons, actin is concentrated under the entire cell membrane. In epithelial cells, there is a dense mesh of microfilaments under the apical plasma membrane, and in neurons the growth cone or synaptic terminals are also extremely rich in actin but microtubules are scarce. There is evidence indicating that vesicles are first transported along microtubules until they reach the cortical actin network that functions as a local distribution system to transport vesicles to and from the plasma membrane. Microtubules are tracks for high-speed trains with few stops, whereas actin microfilaments are the local tracks for regional trains. This functional specialization is related to the structure of the two polymers: Actin filaments are thin and flexible, making it possible to generate complex networks over short distances, whereas microtubules are fairly rigid polymers perfectly adapted to form long tracks. Vesicular transport along actin filaments is achieved by specific forms of myosin (Mooseker and Cheney, 1995). Therefore, this is another potential regulatory mechanism for the targeting of membrane proteins to specific cellular domains. In epithelial cells, for example, vesicles lacking a docking site for the myosin required to travel through the apical actin network would rather be routed toward the basolateral domain. Here again, it can be seen that the spatial distribution of actin in the cytoplasm results in a specialized function.

Neuronal Growth. The control of actin assembly and microtubule dynamics by external cues acting on the membrane of the growth cone results in steering of neurite growth and is responsible for the generation of the appropriate neuronal network. The orientation of microfilaments in filopodia at the extremity of growth cones is important because this determines the directionality of growth through polymer assembly at the tip of the cone. This also allows for transport of membrane vesicles toward the membrane since microfilaments are oriented with their plus end at the plasma membrane and myosins move toward microfilaments plus ends.

Muscle Contraction. Muscle contraction is certainly the most striking example of the considerable morphogenetic power of polymer polarity. The antiparallel organization of microfilaments into each sarcomere, in conjunction with the bipolar organization of the myosin molecule, results in the most efficient machine to produce work from chemical energy.

Conclusion and General Principles

In this article, I have examined how cell shape and function are related and how the shape of cells can determine their function. By examining this problem at the molecular level, it can be realized how the 3-D organization of the cytoskeleton determines at the same time a given shape and its associated function. In the mitotic spindle, chromosomes are segregated on a bipolar spindle; in epithelia, the vectorial transport of solutes depends on the orientation of microtubules. In neurons, growth cone navigation and transport of essential components in neurites rely on the activity and organization of microfilaments and microtubules. Finally, in muscles, the production of force at the macroscopic level depends on the specific organization of actomyosin into sarcomeres. The question is how these shapes are generated and how cell differentiation is coupled to cellular morphogenesis. In other words, what is the strict contribution of the genetic information and of physical influence of the surrounding environment on the organization of proteins and lipids in three dimensions? The function is derived from this interplay. The contribution of the information built into proteins to 3-D organization of cells is obvious: It is the folding of proteins in a particular conformation that generates molecules with particular physical properties: enzymes and asymmetric structural proteins such as tubulin and actin. The contribution of physical parameters is less obvious. For example, the organization of actin and tubulin polymers depends on their elongated shape and their dynamic properties which endow them with specific potentials, such as dynamic instability and self-organization through the action of motors.

It appears that there are two essential principles that govern the generation of shapes by the cytoskeleton: symmetry breaking and self-organization. Both principles appear to act in "morphogenetic fields."

The self-organization of polymers such as microfilaments and microtubules into 3-D arrays of various shapes relies first on the dynamic properties of these polymers that are not in thermodynamic equilibrium and second on their polarity. As long as they assemble in a homogeneous medium, no shape can be generated. If an asymmetry is introduced affecting their dynamic properties or if motors bound in an asymmetric manner in the cell capture and pull on the polymer, shapes can be created (Figs. 12–16). The shape will obviously depend on the nature and combination of the factors present in the cell at a given time. This is how microtubules can be organized in different patterns during mitosis in epithelial cells or in myotubes during myogenesis (Fig. 2). It is this combination of factors and the physical constraints imposed on the cells during differentiation, such as their localization at the surface of the embryo, that I call a morphogenetic field. Such a field is not an abstraction. It can, in principle, be defined for each cell type by the naming of the factors and the forces that are applied onto the cytoskeleton and to which it reacts.

In conclusion, one can start to envision how the linear information stored in the genome is used to generate 3-D structures in space and time. The final structures have clear and precise functions (chromosome segregation, muscle contraction, thinking, etc.). All this occurs due to the use of a very limited number of basic components, the structural proteins of the cytoskeleton, in interaction with a multitude of other factors. However, the morphogenetic principles that underlie cellular morphogenesis seem to be few: There is a sort of morphogenetic code that will soon emerge. One can also realize that the dogma one gene, one function is extremely naive. Indeed, since cell shapes play an essential role in cell function and since the same genes (i.e., the actin and tubulin genes) participate in the generation of hundreds of different shapes, the correlation between the gene and the function is lost.

References Cited

BAAS, P. W., and WENQIAN, Y. (1996). A composite model for establishing the microtubule arrays of the neuron. *Mol. Neurobiol.* **12,** 145–161.

BENTLEY, D., and O'CONNOR, T. P. (1994). Cytoskeletal events in growth cone steering. *Curr. Opin. Neurobiol.* **4,** 43–48.

BOLETI, H., KARSENTI, E., and VERNOS, I. (1996). Xklp2, a novel *Xenopus* centrosomal kinesin-like protein required for centrosome separation during mitosis. *Cell* **12,** 49–59.

COY, D. L., and HOWARD, J. (1994). Organelle transport and sorting in axons. *Curr. Opin. Neurobiol.* **4,** 662–667.

CROSS, P. C., and MERCER, K. L. (1993). *Cell and Tissue Ultrastructure.* Freeman, New York.

FRANKEL, J. (1984). Pattern formation in ciliated protozoa. In *Pattern Formation* (G. M. Malacinsky and S. V. Bryant, Eds.), pp. 163–196. McMillan, New York.

FRANZINI-ARMSTRONG, C., and FISCHMAN, D. A. (1994). Morphogenesis of skeletal muscle fibers. In *Myology* (A. Engel and C. Franzini-Armstrong, Eds.), pp. 74–96. McGraw-Hill, New York.

GILBERT, S. F. (1988). *Developmental Biology*. Sinauer, New York.

GURDON, J. B. (1992). The generation of diversity and pattern in animal development. *Cell* **68**, 185–199.

HYAMS, J. S., and LLOYD, C. W. (1994). *Microtubules*. Wiley–Liss, New York.

HYATT, B. A., LOHR, J. L., and YOST, H. J. (1996). Initiation of vertebrate left–right axis formation by maternal Vg1. *Nature* **384**, 63–65.

HYMAN, A. A., and KARSENTI, E. (1996). Morphogenetic properties of microtubules and mitotic spindle assembly. *Cell* **84**, 401–410.

KIRSCHNER, M., and MITCHISON, T. (1986). Beyond self-assembly: From microtubules to morphogenesis. *Cell* **45**, 329–342.

LAFONT, F., and SIMONS, K. (1996). The role of microtubule based motors in the exocytic transport of polarized cells. *Sem. Cell Dev. Biol.* **7**, 343–355.

LEHMAN, R. (1995). Establishment of embryonic polarity during *Drosophila* oogenesis. *Sem. Dev. Biol.* **6**, 25–38.

LIN, C.-H., THOMPSON, C. A., and FORSCHER, P. (1994). Cytoskeletal reorganization underlying growth cone motility. *Curr. Opin. Neurobiol.* **4**, 640–647.

LOUVARD, D., KEDINGER, M., and HAURI, H. P. (1992). The differentiating intestinal epithelial cell: Establishment and maintenance of functions through interactions between cellular structures. *Annu. Rev. Cell Biol.* **8**, 157–195.

MAYS, R. W., BECK, K. A., and NELSON, W. J. (1994). Organization and function of the cytoskeleton in polarized epithelial cells: A component of the protein sorting machinery. *Curr. Opin. Cell Biol.* **6**, 16–24.

MOORE, J. D., and ENDOW, S. A. (1996). kinesin-related proteins: A phylum of motors for microtubule based motility. *Bioessays* **18**, 207–219.

MOOSEKER, M. S., and CHENEY, R. E. (1995). Unconventional myosins. *Annu. Rev. Cell Dev. Biol.* **11**, 633–675.

RAFF, J. W. (1996). Centrosomes and microtubules: Wedded with a ring. *TICB* **6**, 248–251.

TASSIN, A. M., MARO, B., and BORNENS, M. (1985). Fate of microtubule organizing centers during *in vitro* myogenesis. *J Cell Biol.* **100**, 35–46.

THEURKAUF, W. E. (1994). Microtubules and cytoplasm organization during *Drosophila* oogenesis. *Dev. Biol.* **165**, 352–360.

VERDE, F., DOGTEROM, M., STELZER, E., KARSENTI, E., and LEIBLER, S. (1992). Control of microtubule dynamics and length by cyclin A and cyclin B dependent kinases in *Xenopus* egg extracts. *J. Cell Biol.* **118**, 1097–1108.

VERNOS, I., and KARSENTI, E. (1995). Chromosomes take the lead in spindle assembly. *TICB* **5**, 297–301.

General References

ALBERTS, B., BRAY, D., LEWIS, J., RAFF, M., ROBERTS, K., and WATSON, J. (1994). In *Molecular Biology of the Cell* (M. Robertson and R. Adams, Eds.), 3rd ed. Garland, New York.

FOWLER, V. M., and VALE, R. (1996). Cytoskeleton. *Curr. Opin. Cell Biol.* **8**, 1–3.

FLAVIA VALTORTA

Department of Neuroscience
San Raffaele Vita-Salute
Milan, Italy

Intercellular Communication through the Secretion of Hormones and Neurotransmitters

Neuronal cells are specialized for communication and information processing. The principal route of communication involves exocytotic release of substances for which specific receptors are located on nearby neurons. These substances are either small hydrophilic molecules (classical neurotransmitters) that are released extremely rapidly and mediate the point-to-point communication between neurons or proteinaceous substances (neuropeptides) that are released more slowly and can also act on distant neurons. Although the latter mode of release closely resembles the release of hormones from endocrine cells, the former is typical of and exclusive to neurons. The molecular basis that underlies the release of neurotransmitter and its regulation is beginning to be clarified. What is emerging is a portrait of a complex phenomenon that represents an evolution of the secretory processes operating in other cell types from which it differs not only by the immediacy of coupling between stimulus and secretion but also by the high level of plasticity and the capacity to recall previous stimuli.

Characteristics of the Exocytotic Pathway

The release to the external environment of substances synthesized by the cell constitutes an important means of communication with its surroundings that is essential for unicellular and multicellular organisms, which depend on the existence of a well-coordinated system of cell–cell communication to develop and survive. Therefore, it is not surprising that eukaryotic cells have developed sophisticated mechanisms for the control of this process. The substances released by a cell have widely varied chemical structures and often are capable of acting as signals. In this case such substances are called chemical messengers and serve to interact with specific receptor structures located on target cells. Such interactions trigger appropriate responses in the form of altered ionic currents, metabolic changes, and variations in the spectrum of gene expression.

Depending on the location of the cells that release chemical messengers with respect to the corresponding receptors, found on target cells, several types of intercellular communication can be distinguished. In the case of chemical messengers released from neurons and recognized as neurotransmitters, the substance released must diffuse only a few tens of nanometers before encountering the appropriate receptors. An extremely focused type of signaling results called synaptic transmission, which is characterized by its extreme rapidity. In other cases, the released messengers pass into the blood and spread throughout the organism encountering the corresponding receptors in multiple tissues. This is the case of endocrine signaling in which the messengers, hormones, are produced and released by dedicated cells, the endocrine cells, that are generally organized to form structures called endocrine glands. Hormones play an important role in the coordination of the activity of cells in the entire organism, and their synthesis and release are strictly regulated according to physiological needs. Intermediate situations also exist. For example, in paracrine signaling messengers spread across an area immediately surrounding the point of release, thereby influencing only the activity of cells found in that area. This is the case, for example, for substances that are released in response to injury or infection of a tissue and that are responsible for the activation of a complex, local response known as inflammation. The various types of intercellular communication are schematically represented in Fig. 1.

Lipid-soluble molecules (e.g., steroid hormones) are able to freely cross the lipid bilayer that constitutes the plasma membrane. The release of these substances is regulated by modulating their synthesis since once a molecule has been synthesized it may immediately escape from the cell; the intracellular concentration tends to equilibrate with the external concentration but is generally slightly less. On the contrary, water-soluble molecules that cannot traverse the plasma membrane must be transported extracellularly. For this purpose, eukaryotic cells have developed a complex system of internal membranes called the exocytotic pathway. This system derives its name from its final step, exocytosis. Exocytosis consists of the fusion of a membrane-delimited organelle (a vesicle or secretory granule) with the plasma membrane. Following fusion of membranes, the content of the vesicle is released to the exterior of the cell.

The exocytotic pathway is constituted by a series of specialized cellular compartments (Rothman and Orci, 1992). The various compartments communicate by means of transport vesicles that carry proteins from membrane to membrane and from lumen to lumen. The transport of molecules through the various compartments is vectorial and highly organized; transport originates in the endoplasmic reticulum and proceeds through the different sections of the Golgi apparatus to secretory vesicles and finally the plasma membrane. Furthermore, transport is selective with respect to molecular composition in that both the membrane components and the contents of the different compartments are specified. In other words, a transport vesicle that originates in a specific compartment must have an appropriate membrane composition and must be able to accurately select molecules to transport into the subsequent compartment. In addition, the transport vesicle must be able to recognize the destination compartment. Only in this manner is it possible to avoid creating molecular mixtures that could eventually lead to the loss of the identities of the different compartments. The molecular mechanisms that underlie the specificity of transport are not completely clear. In other

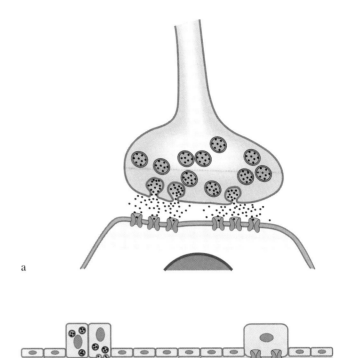

a

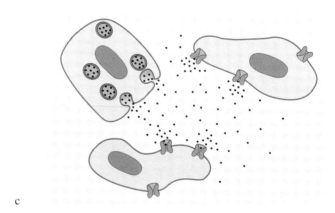

b

c

FIGURE I Synaptic, endocrine, and paracrine secretion. (a) The secretion of neurotransmitters that occurs at the synaptic level represents an extremely specialized form of point-to-point secretion. The molecules of neurotransmitter (black dots) that are secreted must diffuse only a few tens of nanometers before encountering receptors (in green) to which they bind. (b) In hormonal secretion, the hormones are instead released into the blood and spread throughout the organism, binding to all those cells, even very distant and of different types, that possess the appropriate receptors. (c) In paracrine secretion, the secreted substances spread within the tissue in which they were released without achieving a broad distribution throughout the organism.

articles in this encyclopedia, the most substantiated hypotheses are discussed.

Proteins destined for secretion are synthesized in the endoplasmic reticulum. This occurs on ribosomes that adhere to the cytoplasmic surface of the reticulum, and the nascent proteins are transported to the lumen of the reticulum as they are synthesized. In the lumen of the reticulum, the protein molecules undergo important modifications often essential for their biological activities. In particular, the newly synthesized proteins are correctly refolded with the help of "chaperone" proteins, become associated with other proteins to form oligomers, and form intra- or intermolecular disulfide bonds, and oligosaccharides are added to asparagine residues. Inside the reticulum a quality control system is also present to ensure that only those protein molecules that have been correctly refolded and modified can proceed along the exocytotic pathway, whereas the others are destined for degradation.

Subsequent to the endoplasmic reticulum, stages of the exocytotic pathway are constituted by the various compartments of the Golgi apparatus, whose main functions are the modification of oligosaccharide residues bound to proteins and the concentration and sorting of proteins and polysaccharides among the different secretory organelles, the lysosomes, or other destinations. In addition, precisely at the exit from the Golgi apparatus secretory vesicles are formed. These vesicles then undergo a maturation process before reaching their final destination, the plasma membrane, with which they fuse (Fig. 2).

Small water-soluble molecules destined for secretion (e.g., the neurotransmitters) follow a different pathway with respect to proteinaceous secretory molecules. In fact, they are synthesized in the cytoplasm and subsequently loaded by active transport mechanisms into fully formed secretory vesicles. Inside the vesicles they generally complex with macromolecules, which allows them to be stored at high concentrations without causing excessive osmotic pressure.

There is also evidence of retrograde transport, for example, from the Golgi apparatus to the endoplasmic reticulum, that maintains the sizes of the various compartments in addition to allowing repeated use of the membrane components of the transport vesicles.

Constitutive and Regulated Secretion

There exist two separate secretory pathways (Fig. 2) that diverge at the exit from the Golgi apparatus (Burgess and Kelly, 1987). The so-called constitutive secretion pathway functions in all eukaryotic cells. In this case, there is a constant flow, not subject to regulation, of newly formed vesicles from the Golgi apparatus toward the plasma membrane. The constitutive secretory pathway is also called the default pathway since it seems to be the pathway followed

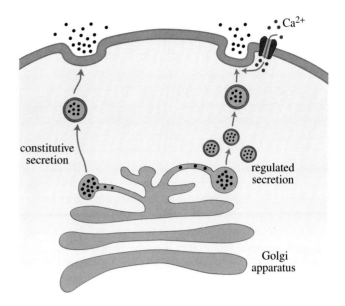

FIGURE 2 Constitutive and regulated secretion. The constitutive secretory pathway is present in all eukaryotic cells. Vesicles, originally formed from the Golgi apparatus, fuse with the plasma membrane. The contents of these vesicles are released outside the cell while proteins and lipids of the vesicular membrane are utilized to replenish components of the plasma membrane. Regulated secretion is characteristic of exocrine, endocrine, and neuronal cells. In this case, secretory vesicles (or granules) tend to accumulate in the cytoplasm until appropriately stimulated, for example, by an influx of Ca^{2+}. Following stimulation, rapid fusion of many vesicles occurs, followed by fusion with the plasma membrane and release of the secretory product to the exterior.

in the absence of a specific signal to direct a vesicle or its contents toward another destination. The incorporation of constitutive exocytotic vesicles into the plasma membrane plays an important role both in normal maintenance and especially in cellular growth and differentiation as a means to add lipidic and proteinaceous components to the plasma membrane. Furthermore, this type of exocytosis is used to release outside the cell diverse molecules that may follow different destinies. Some may become incorporated in the extracellular matrix, whereas others may spread throughout the interstitial fluid and enter the circulation to become components of the blood. Examples of secreted molecules that follow the constitutive pathway are serum albumins, antibodies, and blood-clotting factors.

The regulated pathway is unique to those cell types that are specialized for secretion — neurons and endo- and exocrine cells. If one considers, for example, the secretion of digestive enzymes or neurotransmitters, it obviously would not be very useful to have continuous extracellular release. For this reason, cells specialized for secretion are capable of producing vesicles that have low probabilities of spontaneous fusion with the plasma membrane. These vesicles therefore accumulate in the cytoplasm, undergoing fusion only in response to the occurrence of an appropriate stimulus, such as the activation of receptors or the depolarization of the plasma membrane. In this manner, it is possible to quickly release a large quantity of a given substance. The substances thus released may constitute extracellular signals and interact with receptors present on other cells or with receptors present on the same cell from which the substances were released (autoreceptors). These substances are therefore considered to be mediators of intercellular communication.

The molecular mechanism of the process of membrane fusion is not completely clear. The ability of the Ca^{2+} ion to induce fusion of artificial membranes and the activity of fusogenic proteins such as viral hemagglutinin demonstrate that fusion between vesicles and membranes can be induced by close contact between the two membranes with the exposure of the hydrophobic core and subsequent fusion of the two lipid bilayers. However, the selectivity and vectorial nature of transport processes along the exocytotic pathway suggest that fusion between membranes is regulated and is not a spontaneous process. In addition, from a thermodynamic point of view fusion of biological membranes appears to be energetically unfavorable, such that it cannot occur at a physiologically useful speed in the absence of a catalyzing mechanism (White, 1992). In fact, recent research has identified a "fusion machine" composed of membrane proteins, which differ for the separate organelles of the exocytotic pathway, and soluble factors present in excess quantities in the cytosol which are able to interact with them (Söllner and Rothman, 1994). The fusion machine seems to be able to operate in the case of both constitutive and regulated exocytosis. In the latter case, however, one must assume the existence of an inhibitory control that prevents fusion when the cell is quiescent. Stimulation would therefore act more to remove an inhibitory control than to stimulate directly fusion of secretory vesicles with the plasma membrane. The molecular composition and function of the fusion machine will be described in greater detail for the case of exocytosis of synaptic vesicles.

Characteristics of Neurosecretion

In the nervous system, the evolution of secretory mechanisms has reached the highest level of complexity. Neurons, for example, in addition to the constitutive secretory pathway, have at least two systems for regulated secretion. The system dedicated to the secretion of neuropeptides is characterized by secretory vesicles, granules, very similar to those in endocrine cells that secrete peptide hormones. These vesicles are called large, dense-core vesicles due to

their appearance in electron microscopy, and they have a variable diameter between 100 and 300 nm. The cellular cycle of the granules will be described in conjunction with that of secretory granules which contain peptide hormones.

Within the nervous system, the release of neuropeptides has a prevalently modulatory significance. Rapid transmission of information is achieved by secretion of other types of chemical mediators, the so-called "classical" neurotransmitters. They are constituted by small hydrophilic molecules, such as glutamic acid, glycine, GABA, acetylcholine, and biogenic amines, that are synthesized in the cytoplasm and subsequently loaded into preformed secretory vesicles, which are then referred to as synaptic vesicles or small synaptic vesicles. Synaptic vesicles are translucent organelles 40–50 nm in diameter that accumulate in large quantities in nerve terminals.

The unique features of the synthesis of neurotransmitters and of the cycle followed by synaptic vesicles enable neurons to perform with high efficiency and rapidity. Neurons are in fact highly polarized cells characterized by the presence of very long processes. As a consequence, synaptic terminals are found at great distances from the cell body (up to 1 m for motor neurons of the sciatic nerve in man) and therefore the replenishment of secretory products at the terminals would present great difficulties if they had to be synthesized in the cell body. These problems are overcome by using as neurotransmitters small molecules that can be synthesized locally by cytoplasmic enzymes present in the terminals and by developing a storage and release organelle for neurotransmitters, the synaptic vesicle, that may be repeatedly utilized in an economic and efficient cycle of local exo- and endocytosis.

The Spatial and Temporal Accuracy of Neurotransmitter Release

With respect to other forms of secretion, the release of neurotransmitter is characterized by extreme spatial and temporal accuracy. In a mature neuron, synaptic vesicles are concentrated in nerve terminals (Fig. 3). Inside the terminal, vesicles cluster at sites on the presynaptic membrane called "active zones" (Fig. 4). In physiological conditions, exocytosis occurs exclusively at active zones and involves these clustered vesicles (Ceccarelli and Hurlbut, 1980). It is not clear if the organization of the vesicles in clusters is due to direct interactions between their membranes or to bonds with structures of the cellular matrix. In either case, the protein synapsin I is likely to be involved in mediating these effects. In experiments *in vitro*, this protein has in fact been demonstrated to be able to mediate both the aggregation of vesicles and their linkage to cytoskeletal structures (Valtorta *et al.*, 1992).

The molecular basis for the preferential fusion of synap-

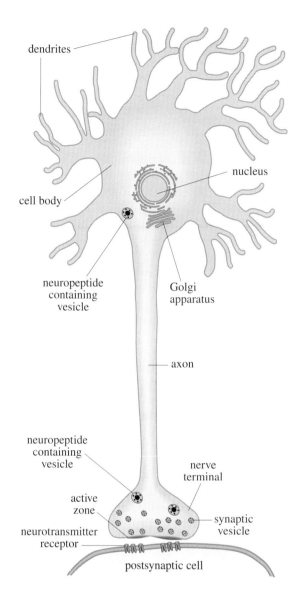

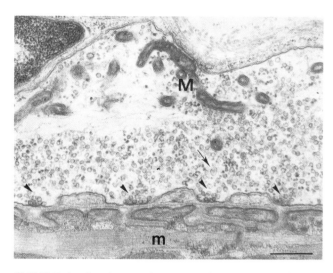

FIGURE 4 A micrograph representing a synapse (in this case a neuromuscular junction) visualized with an electron microscope. The nerve terminal contains numerous synaptic vesicles, some of which (arrowheads) are clustered near a specialized part of the plasma membrane (active zone) where fusion preferentially occurs. The arrow indicates a dense-core, secretory granule. M, mitochondria; m, muscle.

FIGURE 3 Schematic representation of a neuron. Neurons are highly polarized cells characterized by the presence of a cell body from which long processes (dendrites and axons) extend. Axons terminate as cytoplasmic enlargements (nerve terminals) where small synaptic vesicles accumulate. The neurotransmitter contained in such vesicles is released externally when they fuse with the plasma membrane at specialized structures (active zones). The receptors for neurotransmitter are found on the membrane of the postsynaptic neuron and are concentrated in the area of membrane facing the active zones. Neuropeptides are contained in large, dense-core secretory granules, which may also undergo exocytosis in areas of the neuron apart from the nerve terminals.

tic vesicles with the plasma membrane at active zones is not completely clear. It has been hypothesized that the molecular composition of the zones differs from that of other areas of the presynaptic membrane, although a preferential local-

ization has not been conclusively demonstrated for any of the membrane proteins involved in the regulation of fusion.

Exocytosis of synaptic vesicles is triggered by the influx of Ca^{2+} through voltage-dependent channels that occurs following invasion of the nerve terminal by an action potential. Exocytosis is an extremely rapid event in that the delay between the influx of calcium and the release of neurotransmitter is only 200 μs (Zucker, 1993). The retrieval of the vesicular membrane also occurs in a rapid and precise manner, without the mixing of its components with those of the presynaptic membrane. The vesicles recovered from the plasma membrane can then be reloaded *in loco* with neurotransmitter and utilized for successive secretion. This efficient exo-endocytotic recycling of synaptic vesicles, associated with a noteworthy enrichment of vesicles in the terminals, enables prolonged periods of intense activity without exhaustion or interruption of synaptic transmission. From an energetic point of view this system of secretion is very efficient, with the ability to recover both the secretory substance and the organelle dedicated to its storage and release and therefore minimizes the transport necessary between the cell body and the terminals.

Although an action potential is characterized by a constant amplitude, the release of neurotransmitter is a highly modulated process. In other words, neurons can secrete greater or lesser quantities of neurotransmitter in response to a single action potential depending on the recent activity of the cell and on the chemical environment of the presynaptic terminal. This plastic behavior allows the integration of the information that reaches a neuron and is the basis of phenomena such as learning and memory.

Biogenesis of Synaptic Vesicles

The membrane proteins of synaptic vesicles are synthesized and modified in the cell body of the neuron and from there they are transported to the nerve terminals by fast axonal transport. It is not completely clear if the vesicles are transported in a mature form or as precursors that then undergo a process of maturation in the terminals (McPherson and De Camilli, 1994). Results from studies conducted in PC12 cells, a pheochromocytoma cell line from rat often used as a model of a neuroendocrine cell or a facsimile of a neuron, tend to favor the late maturation of synaptic vesicles. In this system, the newly synthesized synaptic vesicle protein synaptophysin seems to exit from the Golgi apparatus with vesicles destined for constitutive secretion. Only after the vesicles of constitutive secretion have undergone fusion with the plasma membrane is synaptophysin retrieved and incorporated into synaptic vesicle membranes. These data suggest that the maturation of vesicles requires a cycle of endocytosis. However, these observations have not been extended to either other vesicle proteins or bona fide neurons.

The idea that the final state of maturation of synaptic vesicles is reached in the nerve terminals and not in the cell body, despite their formation in the Golgi apparatus, is also supported by studies of the kinetics of axonal transport of vesicular proteins. Although integral membrane proteins of the vesicular membrane travel exclusively by fast axonal transport, which implies a membrane-bound form, the synapsin I protein, which associates peripherally with the membrane, travels mostly by slow transport typical of soluble proteins. Therefore, it seems that vesicles acquire their full complement of proteins only after arriving in the terminals.

In some developing neurons maintained in culture it has been observed that synaptic vesicles (or their precursors) continuously undergo cycles of exo-endocytosis. However, before synaptic maturation this process is uniformly distributed over the entire surface of the neuron and not limited to the nerve terminals. Mature synaptic vesicles spend most of their lifetime in the terminals, where they repeatedly cycle exo- and endocytotically. Retrograde transport of vesicular proteins also occurs, constituting a mechanism for replacing vesicles which are transported to the cell body to be degraded. It is not known if degradation occurs on a random basis or if the vesicles can traverse only a limited number of exo-endocytotic cycles. Accordingly, it is not clear if a signal exists that indicates when a particular vesicle has exhausted its life cycle. However, it is interesting to observe that synapsin I and the small GTP-binding protein

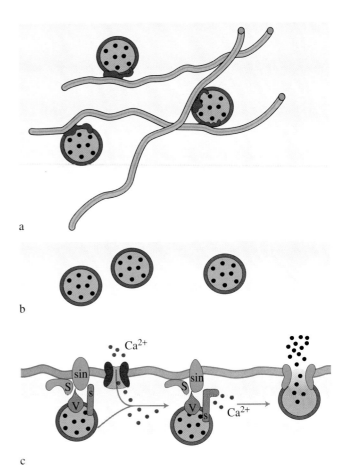

FIGURE 5 Schematic view of functional groups of synaptic vesicles and the probable mechanism of the function of the fusion machine. Inside the nerve terminal, synaptic vesicles can be found in different functional states. (a) Some vesicles, at times the majority of vesicles, are anchored to the cytoskeleton by the protein synapsin I (in blue) and constitute the so-called reserve group. (b and c) Other vesicles, having lost the coat of synapsin I, are no longer anchored to the cytoskeleton and are hypothetically available for exocytosis. Nevertheless, the Ca^{2+} influx that accompanies the arrival of a stimulus will induce exocytosis of only those vesicles that are anchored at the active zone. The anchorage is established through interactions between proteins of the vesicles [V (VAMP) and s (synaptotagmin), in pink] and proteins of the plasma membrane [syn (syntaxin) and S (SNAP-25), in green]. The opening of a vesicle to the extracellular space requires the removal of a fusion blockade, effected by Ca^{2+} ions and probably involving a conformational change of the protein synaptotagmin.

Rab-3, which are associated with the membranes of synaptic vesicles (Fig. 5), are not transported in a retrograde direction and presumably are degraded in the terminal. It is therefore possible, although not yet demonstrated, that the loss of one or both of these proteins from the vesicular membrane constitutes a signal that permits the vesicle to leave the terminal and return to the cell body.

Loading Neurotransmitter into Synaptic Vesicles

Synaptic vesicles are organelles characterized by their surprisingly uniform size. This property dictates that the number of neurotransmitter molecules that can be loaded into each vesicle is constant. Borrowing a concept from physics, this amount is called a "quantum" because it represents the smallest quantity of neurotransmitter that can be released (by exocytosis of a single synaptic vesicle). The release of one quantum evokes an electrophysiologically measurable effect in the postsynaptic membrane, denoted a miniature postsynaptic potential. The presence of one quantum of neurotransmitter in each synaptic vesicle constitutes an important constraint for the processing of signal in the brain since it restricts the variability of a signal that reaches a postsynaptic neuron by the frequency of impulses and the number of synaptic vesicles released per impulse (del Castillo and Katz, 1954).

The number of neurotransmitter molecules that comprise a quantum has been estimated to be approximately 10,000, which corresponds to an intravesicular concentration of 0.1 M. Obviously, such a concentration can only be achieved by employing an active system of loading neurotransmitter into the vesicles (Kelly, 1993). The energy required for this system is provided by a vacuolar type of adenosine triphosphatase (V-ATPase). This enzyme functions as a pump that transfers protons to the inside of the vesicle, creating an electrochemical gradient. The simultaneous activity of a chloride channel attenuates the potential difference as it further increases the proton gradient. These mechanisms are common to all synaptic vesicles. The specificity of the process regards the type of neurotransmitter loaded into a vesicle and is determined both by the type of neurotransmitter synthesized in the cytoplasm and by the presence of specific transporters on the vesicle membrane that exploit the electrochemical gradient generated from the activity of the pump for secondary, active transport of neurotransmitter.

Exocytosis of Synaptic Vesicles: Preparatory Stages

The arrival of an action potential determines the exocytosis of only a small fraction of the synaptic vesicles present in a terminal. The fraction of vesicles exocytosed differs in various types of synapses and is altered at a single terminal depending on its recent activity and the frequency and duration of preceding stimuli. In the case of the synapses of the central nervous system, some researchers hypothesize that the probability of fusion for each vesicle is so low that normally a stimulus causes the fusion of only one vesicle or even no vesicles. Thus, with a certain frequency one observes failures in synaptic transmission, i.e., stimuli without subsequent fusion of a vesicle. In this case, a preceding stimulus might be able to influence the frequency with which a given terminal responds to an action potential with a secretory event. This mode of signal transmission, if verified, would be very interesting, because it would imply the existence of a binary type of language (0, 1) in the communication of signals.

The fact that only a small fraction of the synaptic vesicles present in a terminal undergo fusion with the arrival of a stimulus can be explained assuming the existence of two functional groups of vesicles — one immediately available for exocytosis and another "reserve" group composed of vesicles that are not readily available but that can be recruited to the "ready" group according to physiological needs. Such recruitment of vesicles allows the proportion between available and reserve synaptic vesicles to be variable, forming the basis of synaptic plasticity phenomena. The most accredited hypothesis to explain the existence of the two functional groups of vesicles maintains that the reserve vesicles are held in place by interactions with the matrix of cytoskeletal proteins that occupy a large part of the terminal cytoplasm (Fig. 5a). Liberation from the cytoskeleton would constitute a prerequisite to permit a vesicle to execute the successive phases that lead to exocytosis. The passage of each vesicle from one to the other group may occur at random since the two groups do not seem to be physically separated within the terminal and all vesicles appear to be identically competent for fusion.

The attachment to the cytoskeleton appears to be mediated by vesicular proteins such as synapsin I, which interacts with actin filaments and dissociates from binding sites both on the vesicles and on actin when it becomes phosphorylated by a protein kinase, Ca^{2+}/calmodulin-dependent protein kinase II. Therefore, the availability of the vesicles for exocytosis depends on the phosphorylation state of synapsin I, which in turn is regulated by intraterminal levels of Ca^{2+} and thus by the level of activity of the synapse (Figs. 5b and 5c). This process can contribute to variation in the probability that a vesicle is found in the reserve group and thereby bring about the modulation of subsequent secretory activity (Greengard et al., 1993).

The extremely rapid stimulus-secretion coupling that characterizes the release of neurotransmitter (approximately 200-μs elapse between the influx of Ca^{2+} and exocytosis) suggests that only a few molecular steps occur in the process. It has been postulated that on the arrival of a stimulus, only those vesicles that are already anchored at active zones can undergo exocytosis. It must be emphasized that the number of vesicles anchored at active zones probably includes only a fraction of the group of vesicles available for exocytosis. It is not known if the number of vesicles anchored is limited by the availability of anchorage sites or the kinetics of the process.

It is probable that anchorage at the active zones is facilitated by the existence of a phenomenon that guides vesicles to them. The identification of a small GTP-binding protein, Rab-3, in association with the synaptic vesicle membrane has led to the hypothesis that this protein is responsible for such an "addressing" phenomenon. Rab-3 reversibly associates with the membrane of vesicles in a GTP-bound form and remains associated until the vesicle is correctly positioned in correspondence with fusion sites on the presynaptic membrane. The docking of a vesicle stimulates the intrinsic GTPase activity common to G proteins, and the subsequent hydrolysis of GTP to GDP induces the dissociation of Rab-3 that liberates it for a new cycle of addressing through substitution of GDP by GTP (Simons and Zerial, 1993). Although this hypothesis maintains some validity, experiments conducted by microinjection of peptides corresponding to sequences of Rab-3, in addition to observations of mutant mice that do not express the protein, suggest that Rab-3 is involved in steps which occur after addressing to the active zones such as docking and/or fusion. In particular, Rab-3 may facilitate the assembly of SNARE complexes.

Exocytosis of Synaptic Vesicles: Final Stages

Anchorage at active zones by definition predicts the existence of close molecular interactions between vesicular components and components of the presynaptic membrane. Nerve terminals contain proteins that are highly correlated with components of the multimolecular complex called the "fusion machine" that regulates the fusion of intracellular membranes in all eukaryotic cells (Söllner and Rothman, 1994). This complex is composed of soluble proteins, *N*-ethylmaleimide-sensitive factors (NSF) and soluble NSF attachment proteins (SNAPs), and of their corresponding receptors (called SNAP receptors or SNAREs) on the vesicular and plasma membranes (Fig. 5c). In the case of synaptic vesicle exocytosis, the SNAREs are constituted by a small protein, synaptobrevin, and by two proteins of the presynaptic membrane, syntaxin and SNAP-25 (Table 1). The interaction between these proteins may determine docking of synaptic vesicles at the active zone, creating, in such a way, a binding site for soluble fusogenic factors, NSF and SNAPs. It seems probable that the interaction between SNAREs, and therefore the anchorage of the vesicles, is not a spontaneous process but is regulated instead by the activity of accessory proteins, some of which have already been identified.

The observation that the blockage of neurotransmitter secretion that is observed after tetanus or botulism poisoning is due to the selective proteolysis of a SNARE protein by the bacterial toxins emphasizes the importance of the fusion machine in synaptic vesicle exocytosis (Schiavo *et al.*, 1995). Nonetheless, it is not clear if the role of the fusion machine is limited to anchoring synaptic vesicles or if it also participates in successive steps that lead to the release of the vesicular content into the extracellular space. The NSF protein is an enzyme with adenosine triphosphatase activity and therefore it is possible that the energy produced from ATP hydrolysis contributes to membrane fusion by weakening repulsive forces between phospholipids or perhaps by activating some other protein involved in fusion. Of note is that synaptobrevin, a synaptic vesicle SNARE, is able to interact with another protein of the synaptic vesicle

TABLE I

Proteins Involved in Exocytosis

PHASE OF EXOCYTOSIS	PROTEIN	SUBCELLULAR LOCATION	DISTRIBUTION
Storage of neurotransmitter	Proton pump	Vesicle	Ubiquitous
Release from the cytoskeleton	Synapsins	Vesicle	Neurospecific
Guidance to active zones	Rab-3	Vesicle	Neurospecific
Docking	Synaptobrevin	Vesicle	Ubiquitous isoforms
	Syntaxin	Plasma membrane	Ubiquitous isoforms
	SNAP-25	Plasma membrane	Ubiquitous isoforms
Fusion pore	Synaptophysin	Vesicle	Ubiquitous isoforms
Fusion with the plasma membrane	SNAPs + NSF	Soluble	Ubiquitous
Inhibition of exocytosis	Synaptotagmin	Vesicle	Nerve and endocrine cells

membrane, synaptophysin, which is the most likely to form a fusion pore. In both cases, however, the existence of a mechanism to arrest fusion must be assumed to ensure that fusion only occurs following an appropriate stimulus, when an action potential invades the nerve terminal. At that point, the voltage-dependent Ca^{2+} channels open and Ca^{2+}, driven down an enormous concentration gradient, enters the terminal.

A widely accepted hypothesis is that the vesicular protein synaptotagmin plays a key role in the prevention of vesicle fusion in the absence of appropriate stimulation (Popov and Poo, 1993). This protein is in fact able to interact with the SNAREs as a competitor of the soluble factors, SNAPs and NSF, in addition to binding Ca^{2+} with an affinity commensurate with the concentration reached in the vicinity of active zones after stimulation. According to this hypothesis, the Ca^{2+} that enters through voltage-dependent channels binds to synaptotagmin and induces a conformational change in the protein. This exposes a binding site for fusion factors that attach to the anchored complex and cause ATP hydrolysis. The energy released is used to induce exocytosis, stimulating the opening of a proteinaceous pore, called a fusion pore, and/or inducing the fusion of the phospholipid layers of the two tightly juxtaposed membranes. In retrospect, it is likely that the system of inhibitory control is more complex than previously imagined and that it involves various sequential steps of protein–protein interactions regulated by Ca^{2+} (Fesce et al., 1996). In any case, the presence of a preformed complex between synaptic vesicles and active zones and of a constitutive fusion machine under inhibitory control accounts for the extreme rapidity of the process of neurotransmitter release in contrast to other processes involving regulated exocytosis.

Recent data suggest that the liberation of neurotransmitter from anchored vesicles can occur in the absence of a true fusion between the vesicular and plasma membranes. This may involve a connection between the inside of a vesicle and the extracellular environment through a "fusion pore" with a conductance similar to that of an ion channel and formed of oligomeric membrane proteins (Neher, 1993). Furthermore, it is possible that after an initial phase of unstable pore opening, there is a stable phase followed by partial dissociation of the channel component subunits with mixing of phospholipids and finally fusion between the two contacting membranes. This possibility is very interesting in that not only does it reconcile the two preceding hypotheses but also it suggests that the release of secreted products into the extracellular space does not necessarily imply complete fusion between vesicular and plasma membranes. This consideration is important from a functional point of view, especially when the secreted substance consists of small, rapidly diffusing molecules such as classical neurotransmitters, which can easily pass through a proteinaceous channel.

Endocytosis of Synaptic Vesicles

Following exocytosis, the vesicular membrane becomes incorporated in the plasma membrane. The quantity of membrane that is added to the plasma membrane can be very high, particularly after prolonged and intense stimulation, and may actually exceed several-fold the area of the plasma membrane. In normal conditions, however, the incorporation of vesicular membrane is only a transient event since a mechanism for the recovery of this membrane, endocytosis, exists (Ceccarelli and Hurlbut, 1980). The capability for endocytosis permits repetitive utilization of lipids and proteins of the vesicular membrane, when reincorporated in new vesicles, with a considerable energy saving by the cell. In addition, such endocytosis appears to be selective for components of the vesicular membrane and may prevent mixing of molecular components that would alter the properties of the plasma membrane.

In the case of exocytosis of synaptic vesicles, direct recovery of the vesicular membrane may be possible. In other words, the membrane of the vesicle, after releasing its neurotransmitter content, perhaps through a fusion pore and thereby without true fusion of the phospholipid layers, may be recovered and immediately reloaded (Fig. 6). This cycle of rapid exo-endocytosis has been dubbed "kiss and run" and it has the advantage of being highly selective since the vesicular membrane is recovered before there is molecular mixing with elements of the plasma membrane (Fesce et al., 1994).

However, there also seems to be some amount of an alternative mechanism of endocytosis operating in nerve terminals, i.e., the "classical" endocytosis utilized by all cell types. In this case, endocytosis requires the participation of accessory proteins that assist the recovery of vesicular membrane components from the plasma membrane, selecting them specifically by the action of a "molecular filter" (De Camilli, 1995). The principal accessory proteins are those belonging to the clathrin complex consisting of three polypeptide chains of high molecular weight and three polypeptide chains of low molecular weight assembled to form a three-legged structure called a "triskelion" (from the Greek word σκέλος). In appropriate conditions, several triskelion complexes can associate to form basket- or cage-like polyhedral structures. In the cell, the assembly of clathrin complexes in baskets occurs at the plasma membrane and requires the presence of adapter molecules, the adaptins, that link clathrin and the portion of membrane to be recovered, which is recognized by the presence of a signal sequence on the membrane proteins called a "consensus sequence," for endocytosis. In this way, the adaptins ensure the specificity of the recovery process. The assembly of clathrin complexes on portions of the plasma membrane is followed by the formation of invaginations or pits (Fig. 7).

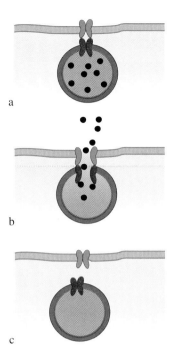

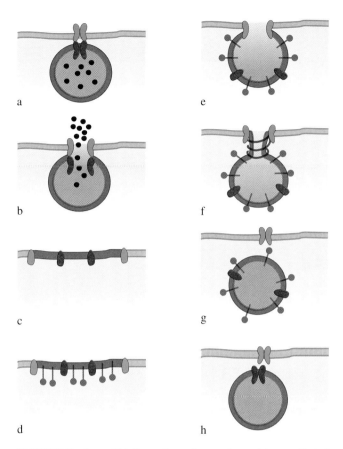

FIGURE 6 Rapid exo-endocytosis of synaptic vesicles: the kiss and run model. In the case of synaptic vesicles, a rapid cycle of exo-endocytosis may be possible in three steps: (a) attachment of the vesicle to the plasma membrane, (b) release of neurotransmitter through a fusion pore constituted by vesicular proteins interacting with proteins of the plasma membrane, and (c) recovery of the vesicle by dissociation of these proteins.

FIGURE 7 (a and b) Recycling of synaptic vesicles mediated by coated vesicles. In analogy with what has been described for other systems, in the case of synaptic vesicles it is possible that exocytosis is followed by the collapse of the vesicular membrane into the plasma membrane (c). Triskelions of clathrin (in blue) assemble on the collapsed vesicular membrane (d) with the subsequent formation of a coated pit (e). Around the collar of the coated pit, rings of dynamin (purple spirals) assemble (f) that promote fission of the membranes to produce a coated vesicle (g) which then sheds the clathrin coating (h).

Such invaginations are referred to as "coated" pits due to the presence of clathrin complexes which are well visualized in an electron microscope. The coated pits then detach from the plasma membrane to form coated vesicles. Coated vesicles have a brief half-life since the clathrin coating is rapidly disassembled by the action of an "uncoating" adenosine triphosphatase. After shedding the clathrin mantle, the vesicles become competent to fuse with other organelles, called endosomes, which are intermediate structures in the endocytotic pathway. From this stage, the components can be recovered to form new vesicles or may proceed to lysosomes for degradation.

Recent studies of the protein dynamin, which has guanine triphosphatase activity, suggest that this protein can play a pivotal role in the formation of coated vesicles. Dynamin is in fact able to self-assemble in tubules around the collars of coated pits, causing them to constrict. The hydrolysis of GTP may then bring about closure of the rings with consequent detachment of the coated vesicles from the plasma membrane (Fig. 7). In effect, fruit flies that carry a mutation in the gene that codes for dynamin, thermosensitive *shibire* mutants, suffer a motor paralysis caused by the interruption of synaptic vesicle recycling when they are kept at the nonpermissive temperature.

The mechanisms that activate and modulate the recovery of the vesicle membrane from the plasma membrane are not clear. The observation that at submaximal levels of stimulation of exocytosis appreciable increases of the plasma membrane do not occur suggests the existence of an equilibrium between the insertion and recovery of vesicular membrane. It is possible that exocytosis and endocytosis are tightly coupled such that the occurrence of the former phenomenon activates the latter. Nonetheless, at maximal levels of stimulation of exocytosis the two phenomena uncouple, suggesting that the maximum velocity of endocytosis is lower than that of exocytosis. This observation, together with the demonstration of selective pharmacological interference with one of the two phenomena, suggests that exocytosis and endocytosis proceed at least in part through different molecular mechanisms.

Secretion of Neuropeptides and Hormones

As previously mentioned, neuropeptides are contained in large, dense-core synaptic vesicles that differ from small synaptic vesicles in many aspects of their life cycle, including the mechanism of release. These large vesicles are analogous to secretory granules of endocrine cells, and for this reason these two types of organelles will be described in parallel.

In most neurons, large, dense-core synaptic vesicles are relatively few in number and their purification in sufficient quantities for a molecular analysis is difficult. Also, the life cycle of large, dense-core synaptic vesicles has not been analyzed in detail and the bulk of our knowledge is an extrapolation from the analysis of secretory granules from so-called neuroendocrine cells. Among these, the most often utilized is the previously noted PC12 cell line. PC12 cells undergo neuronal differentiation when exposed to neurotrophic growth factor. The fact remains, however, that PC12 cells never mature synaptically and therefore represent a somewhat spurious neuronal model.

The budding of secretory granules occurs at the *trans* side of the Golgi complex by a mechanism that has been the object of study of numerous laboratories but is not completely clear (Bauerfeind and Huttner, 1993). In the lumen of the *trans* reticulum of the Golgi apparatus, secretory substances are subject to an aggregation phenomenon favored by low pH. Aggregation involves such strong binding that often the content of a granule requires a certain time to dissolve even after being released into the surroundings of the cell. Such binding permits the storage of the material in high concentrations while avoiding the generation of excessive osmotic pressure. It is also probable that aggregation constitutes an essential signal for recognition of the secretory material by the nascent granule membrane. Another point of view, which is not necessarily exclusive of the preceding one, maintains that recognition requires the presence of a signal in the form of a specific amino acid sequence. The membrane of secretory granules has a characteristic molecular composition, and it is likely that one or more of the proteins of this membrane function as "receptors" for the aggregated material.

The budding of a granule from the Golgi apparatus is an energetically unfavorable process that is catalyzed by the assembly of clathrin around the nascent granule. Once the granule has detached, the clathrin coat is shed and the granule rapidly condenses in a process that is probably correlated with the concomitant acidification of the lumen by the activation of the adenosine triphosphatase proton pump.

The material that is loaded into secretory granules consists not only of neuropeptides and hormones but also of proteins, chromogranins, or secretogranins, which help the condensation process, and proteolytic enzymes necessary for posttranslational processing of the neuropeptides and hormones. In fact, in analogy to what is observed in the case of hydrolytic enzymes secreted by exocrine cells, many neuropeptides and hormones are synthesized as inactive precursors, prohormones, that are later activated by proteolytic cleavage. Sometimes, the cleavage involves only an amino-terminal sequence of the precursor, a "prosequence." In other cases, the situation is more complex; for example, a prohormone may consist of a prosequence followed by several copies of the hormonal sequence. This applies especially to the case of a hormone/neuropeptide that is a short peptide such that if it were synthesized as a single molecule it might not possess the necessary information for aggregation and storage in the granule. It may also be that the polypeptide contains a series of sequences of different neuropeptides/hormones. In this case, depending on the enzymatic capabilities of the granule, proteolytic cleavage can occur at different points, thereby giving rise to a spectrum of active molecules (Fig. 8). The maturation of hormones through proteolytic cleavage begins in the Golgi apparatus but continues inside the secretory granule, which evidently contains the appropriate proteolytic enzymes in addition to prohormones. Upon exocytosis all the peptides that have been cleaved from prosequences will be released outside the cell independently of whether they are biologically active or not.

In contrast to the active zones described for small synaptic vesicles, the large, dense-core vesicles fuse with the plasma membrane at sites which at least morphologically do not appear to be specialized. In fact, in neurons, large, dense-core vesicles undergoing exocytosis can also be found in cellular compartments other than the synaptic terminal, for example, in dendrites (Fig. 3).

Due to the proteinaceous nature of their contents, these vesicles cannot be immediately reloaded after exo-endocytosis, and we must therefore assume that they return to the cell body to be reassembled and refilled in the Golgi apparatus.

The exocytosis of small synaptic vesicles and large, dense-core vesicles appears to be differentially regulated within a single nerve terminal. For example, α-latrotoxin, a toxin from the venom of black widow spiders, is able to stimulate a massive release of neurotransmitter exclusively from small vesicles. When secretion is instead stimulated by the application of electrical impulses to the nerve, the exocytosis of large, dense-core vesicles can be obtained only

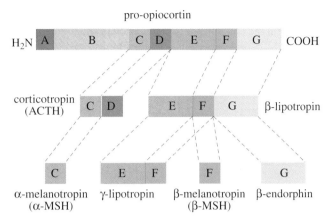

FIGURE 8 Proteolytic maturation of a hormone/neuropeptide. Proopiocortin is synthesized as a long polypeptide within which differing domains may be distinguished (represented here as rectangles labeled with letters). Sequence A is a signal sequence necessary for synthesis in the rough endoplasmic reticulum. Sequences C–G contain sequences of various neuropeptides/hormones with different although partially redundant biological activities. According to the enzymatic pathway of the cell in which proopiocortin is synthesized, cleavage of the molecule occurs at different points. Thus, each cell type will produce a characteristic mixture of active peptides.

if the stimulation is applied at high frequency. This latter observation is probably correlated with the different sensitivities of the two types of exocytosis to the intraterminal concentration of Ca^{2+} (Zucker, 1993). In fact, the exocytosis of small synaptic vesicles requires a very high concentration of Ca^{2+} (on the order of 10^{-4} M). Such concentrations are easily reached, although only for brief periods of time, even with a single stimulus but are limited to the cytoplasmic volume at the openings of the voltage-dependent Ca^{2+} channels. On the other hand, this is the space in which vesicles anchored to the plasma membrane are situated. Exocytosis of large, dense-core vesicles seems to also occur with more modest Ca^{2+} concentrations. Furthermore, since the majority of these vesicles are situated in areas far from active zones and consequently far from Ca^{2+} channels, the Ca^{2+} concentration must rise throughout the cytoplasm of the terminal and not only near the Ca^{2+} channels. For this to occur it is necessary to stimulate the nerve repeatedly.

It is likely that also in the case of secretory granules/large, dense-core vesicles there exist groups of fusion-ready and reserve vesicles. However, their exocytosis is notably slower with respect to that of synaptic vesicles (about 50 ms vs 100–200 μs). This difference could be due in part to the slow rate at which the average concentration of intraterminal Ca^{2+} increases in response to electrical activity and in part to systems of anchorage and exocytosis different from those described for synaptic vesicles.

Plasticity of Secretory Mechanisms

On the basis of the previous discussion, it is clear that regulated secretion in neurons is characterized by a high level of structural and functional complexity and by the presence of a series of interconnected steps that must occur in a well-defined order. Although such complexity is rather costly from an energetic viewpoint, it confers the advantage of creating multiple levels for potential regulation by a cascade of intra- and extracellular signals in addition to permitting the integration of various impinging signals to generate a final response (Jessell and Kandel, 1993).

With respect to electrical signaling mechanisms, intercellular communication mediated by the release of chemical substances is undoubtedly advantageous for its great modulatory potential. Although action potentials which stimulate nerve terminals are typically characterized by a constant amplitude, the release of neurotransmitter evoked by action potentials varies in time depending on factors such as the duration and magnitude of preceding stimuli, the activity of the postsynaptic cell, and the presence of modulatory substances in the external medium. In turn, the postsynaptic response to neurotransmitter is modulated by the amount and frequency of neurotransmitter release.

Modulation of the secretory response and of the evoked postsynaptic effects represents a means of adaptation to physiological needs and protects the central nervous system against overstimulation. In addition, it constitutes a formidable means for the conservation of information in neuronal circuits, forming the basis for learning and memory phenomena. According to the period of time for which information is conserved, modulation phenomena may be distinguished during short periods, in which the variations in the response are maintained for a few seconds, or long periods, in which modifications last for minutes, hours, or even days.

The molecular mechanisms that underlie synaptic plasticity phenomena appear to be extremely complex and are probably distinct for each form of synaptic plasticity. Ca^{2+} ions play a central role in the activation of exocytosis as already discussed and also appear to be involved in the creation or maintenance of most phenomena of synaptic plasticity. The simplest hypothesis among those proposed to explain the modulatory role of Ca^{2+} is the so-called "theory of residual Ca^{2+}." In this model, when a terminal is repetitively stimulated at very short time intervals, a new stimulus may arrive before the Ca^{2+} increase below the plasma membrane caused by the preceding stimulus has completely subsided. Assuming that the new stimulus causes an influx of Ca^{2+} similar to that of the preceding stimulus, the amplitude of the secretory response will be larger since some of the Ca^{2+} binding sites on the receptor are still occupied.

Due to the nonlinear relationship between the concentration of Ca^{2+} and secretion, this effect can be particularly prominent.

Perhaps this elementary model can explain some of the simplest forms of short-term plasticity, but it is unlikely to play an important role in the more complex forms that seem to depend linearly on prolonged increases of the average Ca^{2+} concentration in the terminal. In these cases, it is likely that the effects of Ca^{2+} are mediated by its binding to other receptors. One of these receptors is the cytosolic protein calmodulin, which in its Ca^{2+}-bound state binds to a protein kinase, duly called Ca^{2+}/calmodulin-dependent protein kinase II, to cause its activation. The activated protein kinase phosphorylates one of the synaptic vesicle proteins (synapsin I) and thereby promotes the detachment of a vesicle from the cytoskeleton to increase the number of synaptic vesicles available for exocytosis. Consistent with this model are alterations in some forms of synaptic plasticity in mutant mice that do not express synapsin I.

In the more complex forms of synaptic plasticity, such as long-term potentiation, the response of the postsynaptic neuron seems to be important. The postsynaptic cell, when activated in a prolonged manner, is capable of modulating its response to free neurotransmitter, in addition to liberating lipid-soluble substances, so-called "retrograde messengers," that penetrate the nerve terminal and regulate the exocytosis of synaptic vesicles.

Conclusions

It is clear from the previous discussion that the process of neurotransmitter release constitutes a very highly regulated form of secretion in which it is possible to observe the integration of electrical signals, biochemical messages, previous activity, and information arising from nearby neuronal networks to produce optimal responses that can be conserved and recalled. In other words, although secretory cells modified the basic mechanisms of constitutive secretion to obtain one form of regulated secretion, neurons have utilized the general mechanisms of regulated secretion and, by adding various levels of complexity, have learned to modulate the release of neurotransmitter with great efficiency.

References Cited

BAUERFEIND, R., and HUTTNER, W. B. (1993). Biogenesis of constitutive secretory vesicles, secretory granules and synaptic vesicles. *Curr. Opin. Cell Biol.* **5,** 628–635.

BURGESS, T., and KELLY, R. B. (1987). Constitutive and regulated secretion of proteins. *Annu. Rev. Cell Biol.* **3,** 243–293.

CECCARELLI, B., and HURLBUT, W. P. (1980). Vesicle hypothesis of the release of quanta of acetylcholine. *Physiol. Rev.* **60,** 396–441.

DE CAMILLI, P. (1995). The eighth Datta lecture. Molecular mechanisms in synaptic vesicle recycling. *FEBS Lett.* **369,** 3–12.

DEL CASTILLO, J., and KATZ, B. (1954). Quantal components of the end-plate potential. *J. Physiol. London* **124,** 560–573.

FESCE, R., GROHOVAZ, F., VALTORTA, F., and MELDOLESI, J. (1994). Neurotransmitter release: Fusion or "kiss-and-run"? *Trends Cell Biol.* **4,** 1–4.

FESCE, R., VALTORTA, F., and MELDOLESI, J. (1996). The membrane fusion machine and neurotransmitter release. *Neurochem. Intl.* **28,** 15–21.

GREENGARD, P., VALTORTA, F., CZERNIK, A. J., and BENFENATI, F. (1993). Synaptic vesicle phosphoproteins and regulation of synaptic function. *Science* **259,** 780–785.

JESSELL, T. M., and KANDEL, E. R. (1993). Synaptic transmission: A bidirectional self-modifiable form of cell–cell communication. *Cell* **72**/*Neuron* **10,** 1–30.

KELLY, R. B. (1993). Loading synaptic vesicles with neurotransmitter. *Curr. Biol.* **3,** 59–61.

McPHERSON, P., and DE CAMILLI, P. (1994). *Sem. Neurosci.* **6,** 137–147.

POPOV, S. V., and POO, M.-M. (1993). Synaptotagmin: A calcium-sensitive inhibitor of exocytosis? *Cell* **73,** 1247–1249.

ROTHMAN, J. E., and ORCI, L. (1992). Molecular dissection of the secretory pathway. *Nature* **355,** 409–415.

SCHIAVO, G., ROSSETTO, O., TONELLO, F., and MONTECUCCO, C. (1995). Intracellular targets and metalloprotease activity of tetanus and botulism neurotoxins. *Curr. Topics Microbiol. Immunol.* **195,** 257–275.

SIMONS, K., and ZERIAL, M. (1993). rab proteins and the road maps for intracellular transport. *Neuron* **11,** 789–799.

SÖLLNER, T., and ROTHMAN, J. E. (1994). Neurotransmission: Harnessing fusion machinery at the synapse. *Trends Neurosci.* **17,** 344–348.

VALTORTA, F., BENFENATI, F., and GREENGARD, P. (1992). Structure and function of the synapsins. *J. Biol. Chem.* **267,** 7195–7198.

WHITE, J. M. (1992). Membrane fusion. *Science* **258,** 917–924.

ZUCKER, R. S. (1993). Calcium and transmitter release. *J. Physiol. Paris* **887,** 25–36.

General References

KANDEL, E. R., SCHWARTZ, J. H., and JESSELL, T. M. (1991). *Principles of Neural Sciences,* 3rd ed. Elsevier, New York.

ROTHMAN, J. E., and WARREN, G. (1994). Implications of the SNARE hypothesis for intracellular membrane topology and dynamics. *Curr. Biol.* **4,** 220–233.

Special issue on synaptic transmission (1993). *Cell* **72.**

VALTORTA, F., and BENFENATI, F. (1995). Membrane trafficking in nerve terminals. *Adv. Pharmacol.* **32,** 505–557.

DARIO DI FRANCESCO

University of Milan
Department of Physiology and
General Biochemistry, Electrophysiology
via Celoria 26
20133 Milan, Italy

How Cells Respond to Chemical Mediators: Electrical Signals

Electrical phenomena control the activity of excitable cells such as neurons, in which they control the acquisition, processing, and transfer of information, and muscle cells, in which they control contraction. Furthermore, electrical phenomena affect many physiological processes in nonexcitable tissues. The resting membrane electrical potential and action potential are generated by the flow of ionic electrical charges across the membrane: This occurs due to the presence of special protein structures, the ionic channels, which form aqueous pores across the lipid bilayer of cell membranes. The transfer of electrical signals across neighboring cells occurs directly through electrical synapses or indirectly through chemical synapses. In the latter case, the chemical transmitters (mediators) released by presynaptic membranes activate specific receptors on postsynaptic membranes and by various membrane and/or intracellular mechanisms modify the activity of certain ionic channels. There are several types of ionic channels with diverse kinetic and permeability characteristics and with different regulatory properties. Recently developed dedicated molecular biological techniques have allowed the sequencing and cloning of ionic channels, receptors, and other protein components of the membrane and intracellular compartments that have a role in the transfer of information and substances among cells. Finally, anomalies in the functioning of ion channels have been shown to be responsible for some known pathologies.

Introduction

Electrical signals are the basis of the activity of an important class of cells, the excitable cells, which include neurons and muscle cells of striated and smooth muscle. The functioning of the entire central and peripheral nervous systems, the cardiovascular system, the motor system, and the gastrointestinal system, for example, depends on the cellular electrical activity in these systems.

However, the biological relevance of the electrical signals extends well beyond the activity of excitable cells. Indeed, electrically based signaling plays a key role in the regulation of several physiological processes in nonexcitable tissues: Egg fertilization, the release of some hormones in endocrine glands and the secretion of fluids and substances in exocrine glands, and the transfer of material across epithelial membranes are a few examples.

Underlying these and other related phenomena are elementary mechanisms for the transfer of substances and information across biological membranes; here, a basic role is played by membrane receptors, by various mechanisms interfacing receptors to intracellular targets such as G proteins and second messengers, and by membrane channels. In this article, the basic processes supporting the generation of electrical signals and the cellular components regulating the exchange of information and substances among cells and between cell and environment are discussed. The relevance of these phenomena to the regulation of fundamental physiological processes is also discussed.

....●

The Electrical Potentials

The Resting Potential

When an excitable cell is in the resting state, an electrical gradient exists across its membrane: The cell interior is at a voltage of approximately -70 to -80 mV (1 mV $= 10^{-3}$ V) relative to the cell exterior. This difference is not accidental but rather represents an essential condition for the cell to perform its primary function in excitability. The resting potential of excitable cells is the result of the action of specialized transport systems, called pumps, which move ions across the membrane and are able to transform the energy carried by adenosine triphosphate (ATP) into ionic gradients for different types of ions, such as the cations Na^+, K^+, and Ca^{2+} and the anion Cl^-.

The best characterized ionic "pump" is the ATP-dependent Na^+/K^+ pump, whose activity consists of exchanging three Na^+ ions (carried outside) for two K^+ ions (carried inside) at each work "cycle." The extracellular milieu is normally rich in Na^+ (e.g., in the human plasma its concentration is about 140 mM; 1 m$M = 10^{-3}$ mol/liter) and poor in K^+ ions (about 4 mM), whereas the opposite occurs inside excitable cells, which are poor in Na^+ and rich in K^+ ions. The function of the Na^+/K^+ pump in excitable cells is to create and maintain an intracellular milieu poor in Na^+ (about 10 mM) and rich in K^+ (about 130 mM). This determines a chemical gradient which favors the flow of Na^+ ions from the outside to the inside and of K^+ ions from the inside to the outside of the cell; as described later, this flow occurs through ionic channels, transmembrane proteins with highly specialized structures. An important consequence of the diffusion of ions across membranes is the generation of a voltage decrease between inside and outside of the cell.

Cell membranes are composed of two layers of polar lipid molecules: the lipid bilayer, which acts as a barrier to the passage of most water-soluble molecules, spanned by integral proteins such as the ionic channels, whose structure is characterized by an internal groove (the pore) which, on the contrary, allows ion flow. The ionic channel structure can be schematically represented by an equivalent circuit composed of resistive branches, simulating the flow of different ionic species, and by a capacitive branch simulating the accumulation of charges on the surfaces of the lipid bilayer (Fig. 1a).

The ionic currents (in pA $= 10^{-12}$ A) can thus be described by Ohm's law relations for the resistive branches (I_K, I_{Na}, and I_L) and by a capacitive component relation (I_C) according to the following equations:

$$I_K = g_K(E - E_K)$$
$$I_{Na} = g_{Na}(E - E_{Na})$$
$$I_L = g_L(E - E_L)$$
$$I_C = C\frac{dE}{dt}, \qquad [1]$$

where E is the voltage decrease across the membrane (in mV), C is the membrane capacity (in nF $= 10^{-9}$ F), g_K, g_{Na}, and g_L are the conductances (nS $= 10^{-9}$ S) of the three ionic species involved, and $-E_K$, $-E_{Na}$, and $-E_L$ (mV) are their chemical potentials.

The chemical (or osmotic) potentials are functions of the concentration gradients and represent the tendency of ions to move from compartments in which their concentration is high to compartments in which their concentration is low. In addition to concentration gradients, electrically charged ions are sensitive to the electrical gradient such that, as shown in Eq. (1), the flux of one ion is proportional to the sum of the electrical potential (E) and the chemical potential for that ion ($-E_K$ for K^+, $-E_{Na}$ for Na^+, and so on). For example, when the electrical potential equals E_K (about -100 mV), the flux of K^+ ions is abolished since electrical and osmotic potentials are exactly opposite each other. This explains why E_K, E_{Na}, and E_L are termed equilibrium potentials and are represented by batteries in the resistive branches. Notice that the conductance g_L represents a "leakage" flux, that is, a flux of an unidentified ion species whose equilibrium potential is negative, although less negative than the K^+ equilibrium potential (about -50 mV). A component of this current is normally carried by Cl^- ions.

Under normal conditions no charge accumulation or depletion occurs in either intracellular or extracellular compartments, and the net current transferred (I_t) is thus nil. From Eq. (1), the following equation thus follows:

$$-C\frac{dE}{dt} = g_K(E - E_K) + g_{Na}(E - E_{Na}) + g_L(E - E_L). \quad (2)$$

In the resting state, when the derivative dE/dt vanishes, the membrane potential has the following value:

$$E = E_K\frac{g_K}{g_K + g_{Na} + g_L} + E_{Na}\frac{g_{Na}}{g_K + g_{Na} + g_L} + E_L\frac{g_L}{g_K + g_{Na} + g_L}.$$

This equation expresses the important fact that the resting membrane potential approximates the equilibrium potential of the ion of larger conductance. In other words, the larger the permeability of a given ionic species, the closer the membrane potential to the equilibrium potential for that ionic species. Although Eq. (3) has been developed to describe the equilibrium states of excitable membranes, it can give important information regarding the kinetics of the membrane potential when this changes as a function of membrane conductance changes. Indeed, whereas at rest the K^+ conductance is larger than the Na^+ conductance, and therefore the membrane potential approximates the K^+ equilibrium potential, a rapid increase of the Na^+ conductance and the consequent rapid depolarization toward voltages close to the Na^+ equilibrium potential are triggers for the onset of the action potential.

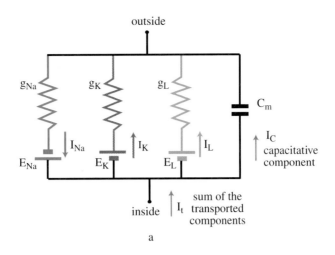

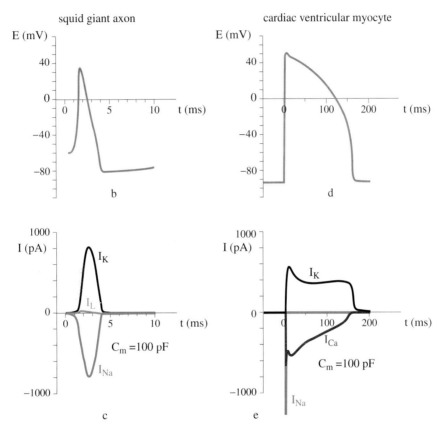

FIGURE 1 Ion flow through a membrane and action potentials. (a) Electrical circuit equivalent to the cell membrane, originally proposed by Hodgkin and Huxley (1952) for the membrane of the squid giant axon. Conductances g_{Na}, g_K, and g_L represent the flow of Na^+ ions, K^+ ions, and a generic "leakage" conductance, respectively. The batteries (or "equilibrium potentials") E_{Na}, E_K, and E_L represent the concentration, or chemical, gradients which exist across the membrane for the respective ionic species, whereas the cell membrane capacity (C_m) is concentrated into a capacitor in parallel with the resistive branches. Since the external Na^+ concentration is higher than the internal one, the chemical gradient for this ion tends to generate a flux of inwardly directed ions, and its equilibrium potential (E_{Na}) is therefore positive (approximately 60 mV). In contrast, for K^+, whose concentration is higher inside than outside, the equilibrium potential is negative (approximately −100 mV). A negative value (−50 mV) is used as the equilibrium potential for the "leakage" conductance; in fact, this component is carried by ions (including Cl^- ions) whose overall chemical gradient tends to generate an outward current flow (at zero membrane potential). I_t is the sum of ionic and capacitive currents. (b, c) Action potential E and ionic currents I during the action potential in the squid giant axon. The currents plotted here correspond to those indicated in a. Action potential and currents result from a numerical computation based on experimental data and reproduce satisfactorily the time courses recorded experimentally. The computation was made on the basis of the Hodgkin and Huxley (1952) equation system (including Eqs. 1 and 2 of the text) for a membrane with a capacity of 100 pF (1 pF = 10^{-12} F). (d, e) Action potential E and currents I during the action potential in a cardiac ventricular myocyte. The computation was made using the OXSOFT HEART software (Di Francesco and Noble, 1985) for a ventricular cell with a capacity of 100 pF. The currents I_K and I_{Ca} represent the net contribution of all K^+ and Ca^{2+} currents.

The Action Potential

The electrical activity of an excitable cell is revealed by the action potential. This is a rapid modification of the membrane potential which propagates along the axon in neurons and causes contraction in muscle cells. Two typical action potentials from a neuron and a cardiac ventricular myocyte are exemplified in Figs. 1b and 1d. Their different duration is obvious and indicates a different function: In nerve, the action potential represents an elementary all-or-nothing type of message which is carried along the axon and is most frequently translated into the release of a neuro-mediator at the synaptic terminal; in a ventricular myocyte, on the other hand, the action potential determines the duration and intensity of the mechanical contraction of the same myocyte.

The action potential results from a set of rapid changes of membrane ionic permeabilities. Following an appropriate depolarizing stimulus, a sudden entry of cations such as Na^+ and Ca^{2+} rapidly brings the membrane potential toward more positive values; this rapid depolarization is then followed by a repolarizing process which brings the membrane voltage back to its resting level and which is normally due to an increase in hyperpolarizing currents (K^+) and to inhibition of the depolarizing currents responsible for the initiation of the action potential. The details of this phenomenon obviously vary in different cell types.

In Figs. 1c and 1e the time courses of ionic currents important in the generation of action potentials in the two cell types under consideration are shown. Equation (2) describes the temporal development of the action potential when differential equations are introduced that govern the dependence of the different ionic conductances upon membrane voltage. In other words, the action potential results from the voltage dependence of the ionic conductances; for example, in nerve (Fig. 1c) a depolarizing stimulus induces an increase in the fast Na^+ conductance, which due to the entry of positive ionic charges generates further depolarization and thus an extra Na^+ entry, and so on. This self-generating process, a sort of "chain reaction," is terminated only when the membrane potential approaches the Na^+ equilibrium potential, thus zeroing the electrochemical gradient for this ion and consequently the Na^+ current. When positive voltages are reached, the opposite process, membrane repolarization, begins; this is caused by the "delayed" activation of a K^+ conductance and by the simultaneous inactivation of the Na^+ conductance, which rapidly brings the action potential to an end. The action potential thus generated propagates along the axonal length at a speed reaching up to 100 m/s due to a self-excitatory membrane mechanism. Indeed, the membrane depolarization caused at a given location by the action potential propagates electrotonically to the contiguous membrane areas via local electrical circuits which, much like a proper electrical cable, are able to "excite" the Na^+ conductances of proximal regions and give rise to the propagation of the signal as a depolarizing wave. These processes have been described in detail since the 1950s (Hodgkin and Huxley, 1952) and are the result of the activity of voltage-dependent ionic channels.

In heart, as it appears in Fig. 1d for a ventricular myocyte, the action potential is different. The most evident difference between cardiac and nerve action potentials is the duration, which in heart is approximately 40-fold longer than in nerve (160 ms as opposed to 4 ms). Furthermore, the cardiac action potential has a different shape, characterized by a long depolarizing phase called "plateau," during which the membrane remains positive. These different features are due to the different ionic currents present in the cardiac membrane. Specifically, an important Ca^{2+} conductance is present (red line in Fig. 1e) whose effect is to allow the entry of Ca^{2+} ions into the ventricular cell during the action potential; this is indeed the process responsible for the initiation of the cascade of events leading to cardiac contraction. This process requires relatively long times, which explains the long duration of the action potential, and it is modulated by the autonomic nervous system, which can therefore modify through the Ca^{2+} current the duration of the action potential and/or the amplitude of the membrane depolarization during the plateau phase.

In contrast to the nerve action potential, the cardiac action potential is not an all-or-nothing process but rather a process able to vary, at least partly, in a continuous way so as to allow through the control of ionic conductances the phenomenon of cardiac contraction.

Electrical Signals
and Intracellular Communication

The Flow of Electrical Signaling across Two Cells

Excitable cells can receive stimuli from contiguous cells through two types of specific junctions, the electrical and chemical synapses. The communication across specialized junctions is the main mechanism for the transfer of information across nerve cells and plays a key role in the integrative function of several physiological processes, such as muscle contraction, hormonal signal transmission, and gland secretion. Synapses are specialized structures at the junction between contiguous membranes, where cells are in contact and can thus communicate by way of special mechanisms involving either the direct transfer of an electrical signal or its transformation into a chemical signal.

Electrical Synapses. Electrical synapses are particularly important in the striated cardiac and smooth muscle, where they provide the structural support to allow propagation of the action potential across different regions of the organ, but they are present in several different cell types such as glial cells, which surround neurons so as to provide protection against shocks, mechanical support, and nutrition. In other words, due to the presence of electrical synapses, a multicellular organ behaves like a syncitium, an aggregate of cells with homogeneous electrical properties. This is possible due to the speed with which voltage changes are transferred from one cell to another. In the mammalian heart, for example, the action potential is generated in the region called cardiac pacemaker and quickly propagates across the two atria before passing with a short delay to the ventricles, across which it again propagates with a fast rate. This temporal succession allows the ventricular and atrial contractions to be out of phase with each other for a time sufficiently long to let the ventricles fill up with blood before ejection occurs. It also allows an almost instantaneous contraction of the whole atrial tissue to be followed by an equally rapid contraction of the whole ventricular muscle, according to the fixed time sequence which is necessary for an efficient pump action.

The regions in which the membranes of two adjacent cells make contact in electrical synapses are called gap junctions and are composed of gap junction pores which connect the two cells by allowing the passage of electrical current and small molecules.

Chemical Synapses. Chemical synapses are described in detail elsewhere in this encyclopedia and here only the essentials are given. In chemical synapses the membranes of two communicating cells do not directly make contact but rather are separated by an intercellular region (Fig. 2). The transfer of electrical signals occurs by a chemical messenger, a substance called transmitter which is released by the synaptic terminal and induces a voltage change in the postsynaptic membrane by way of chemical binding to specialized membrane proteins, the receptors. As will be discussed later, the binding of transmitters to receptors initiates a set of events involving other membrane and cytoplasmic components which in excitable cells eventually leads to a modification of the membrane potential.

Chemical synapses are the choice system for the exchange of information among neurons, even if the mechanism by which a substance is released by specialized cells with the aim of spreading a message is common to other important biological processes, such as those controlled by hormones.

The signal triggering the transfer of information from one neuron to another is the invasion of a presynaptic terminal by one or more action potentials. The main features of a chemical synapse are shown in Fig. 2. Following the entry of Ca^{2+} into the presynaptic membrane, which occurs through Ca^{2+}-permeable membrane channels upon depolarization of the membrane potential, neurotransmitter-containing vesicles fuse with the membrane and release the substance in the intersynaptic space (synaptic cleft). Here, neurotransmitter molecules bind to specific receptors lined up on the postsynaptic membrane and activate them, thus

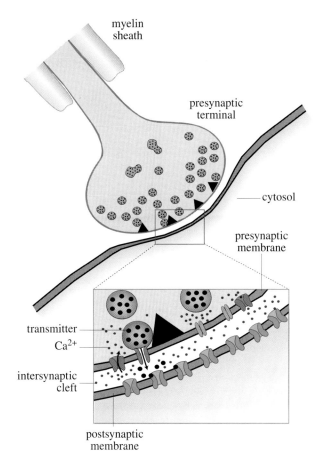

FIGURE 2 Properties of chemical synapses. The chemical transmitter is synthesized and stored in vesicles within the presynaptic terminal. When an action potential invades the synaptic terminal, Ca^{2+} ions enter the presynaptic membrane through Ca^{2+} channels (in blue) in the terminal and cause presynaptic vesicles to fuse with membrane particular proteins (in pink), which eventually leads to release of transmitter molecules (in black) into the intersynaptic cleft. Special structures (dark triangles) favor the "docking" of vesicles to the membrane. The released molecules of the neurotransmitter bind to postsynaptic receptors (in green) which become activated and initiate a cascade of processes in the postsynaptic neuron.

initiating a set of events which terminate with the opening or closing of special ionic channels and the generation of a postsynaptic potential. The postsynaptic potential can be excitatory (EPSP), in which case the membrane will experience a depolarization, or inhibitory, in which case it will experience a hyperpolarization.

The nature of the postsynaptic response is associated with the type of ionic channel, which is either activated or inhibited by the neurotransmitter. In man, each neuron of the central nervous system normally receives many stimuli through thousands of synaptic contacts with other neurons, and the electrical activity of each neuron depends on the integration of all signals simultaneously converging on it.

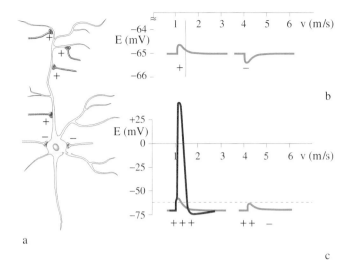

FIGURE 3 Excitatory and inhibitory synapses. (a) Excitatory synapses (+, in blue) are normally concentrated on the dendritic tree of a neuron, whereas the inhibitory ones (−, in yellow) are concentrated on the soma. (b) The changes of postsynaptic membrane potential induced by activation of a single synapse are of small amplitude, usually less than 1 mV; the excitatory synapses are depolarizing (left), whereas the inhibitory ones are hyperpolarizing (right). (c) The net sum of depolarizing inputs from various excitatory synapses (+ + +) can overcome a reference voltage level (threshold, dashed line) and thus trigger an action potential (red trace); the firing threshold may not be reached in the presence of even a few hyperpolarizing inputs from inhibitory synapses (+ + −).

When the excitatory stimuli prevail, the neuron will fire one or more action potentials, whereas no activity will be present whenever the inhibitory stimuli succeed in avoiding the threshold of action potential activation reached by excitatory inputs.

Figure 3 schematically represents one neuron and some of its synapses (Fig. 3a). The excitatory synapses (+) are mostly localized on the dendritic tree, and although their activation gives rise to individually small signals (Fig. 3b), their summation can, if large enough, generate a postsynaptic action potential (red trace in Fig. 3c). Inhibitory synapses (−) are concentrated on the cell body (or soma) and their activation, if strong enough, is able to fully inhibit the postsynaptic activity (Fig. 3c).

The main neurotransmitters are low-weight molecules: acetylcholine, the biogenic amines (such as dopamine, noradrenaline, adrenaline, serotonin, and histamine), and the amino acids (such as γ-aminobutyric acid, glycine, and glutamic acid). There are also several peptides synthesized by neurons which can act as hormones when they exert their action far from the site of release or as true neurotransmitters, such as the endorphins, substance P, the vasoactive intestinal peptide, insulin, and vasopressin.

A Special Chemical Synapse: The Neuromuscular Junction (Motor End Plate). Chemical transmitters play an important role not only in the signal transmission across nerve cells but also in the transmission between neurons and other cell types, such as skeletal muscle cells. In this case, the presynaptic signal has the function of inducing a postsynaptic action potential and hence to trigger muscle contraction.

The neuromuscular junction (or end plate) is the specialized structure responsible for this function in vertebrates and has several presynaptic features in common with interneuronal synapses. The neurotransmitter used here, acetylcholine (ACh), is stored in densely packed vesicles localized proximally to the presynaptic membrane; upon invasion of the terminal by an action potential, the vesicles fuse with the membrane and release the transmitter into the synaptic cleft by a process called exocytosis. Binding of ACh to specific postsynaptic receptors gives rise to an end-plate potential (epp), which is the equivalent of the excitatory postsynaptic potential of interneuronal synapses.

However, there are important differences between chemical synapses and the motor end plate. For example, in the vertebrate end plate only excitatory signals are present (whereas inhibitory signals are found in invertebrate muscle); furthermore, stimulation of the presynaptic neuron (motoneuron) induces a postsynaptic potential of approximately 70 mV (end-plate potential), an amplitude much larger than that of an interneuronal monosynaptic EPSP which therefore normally ensures the generation of a postsynaptic action potential and the associated contraction of the innervated muscle fiber. The buildup of the force of contraction in skeletal muscle does not rely on the gradual control of end-plate potentials but rather on the number of muscle fibers that are simultaneously activated. Induction of the end-plate potential is based on activation by ACh of a receptor channel (see Fig. 6) unselectively permeable to both mono- and divalent cations, which therefore will tend to depolarize the membrane by shifting its resting potential toward 0 mV.

Electrical Signals and the Control of Visceral and Gland Activity by the Autonomous Nervous System. The part of the nervous system called autonomous nervous system uses chemical synapses to control the function of viscera (cardiovascular system, respiratory system, gastrointestinal system, and urogenital system). The two main branches of the autonomous nervous system, sympathetic and parasympathetic, usually exert opposite effects onto target organs by use of the neurotransmitters noradrenaline (NA) and ACh, respectively. For example, the cardiac rate is accelerated by NA and slowed by ACh; the motility of the intestinal smooth muscle, a mechanism essential for the phenomena of digestion and absorption of nutrients, is stimulated by ACh and reduced by NA; and the light adaptation mechanism of the pupil (pupil reflex) regulates the diameter of the iris muscle as a function of the light intensity so as to allow a dilation (in dim light) or a contraction (in bright light) of the pupil upon release of NA or ACh, respectively.

In these examples, the modifications induced by the autonomic transmitters are due to alteration of ionic currents occurring in the postsynaptic membranes and the consequent changes of postsynaptic electrical activity. As discussed later, these changes are brought about by switching on or off ion channels, integral protein structures whose activity can be regulated by the action of chemical transmitters.

Electrical Signals and the Control of Endocrine Gland Secretions by Hormones. The activity of endocrine glands secreting hormones can in some cases be controlled by mechanisms utilizing electrical signals. Hormones are chemical transmitters important in maintaining the stability of the internal environment (homeostasis) in the face of changes due to the assumption of nutritive substances and liquids or changes in temperature, and they control several important phenomena throughout growth and development.

An example is the secretion of insulin by the pancreatic β cells. Insulin is a hormone important for maintaining a constant plasma glucose level (glycemia). Insulin is rapidly released when plasmatic glucose concentration increases. Its action is exerted onto liver cells, skeletal muscle cells, and fat tissue and essentially consists of the activation of mechanisms for the absorption of glucose into the cell, the inhibition of its release, and the accumulation of glucose as glycogen in cellular stores. In the presence of extracellular glucose, insulin is released as a result of Ca^{2+} entry due to depolarization of the β-pancreatic cell; the depolarization is in turn caused by a decreased permeability of the membrane to K^+. It has been shown (Ashcroft *et al.*, 1984) that the decreased K^+ permeability results from the block of K^+ channels generated by intracellular ATP, which is produced by glycolysis of glucose after it has been carried into the cell.

Electrical Signals and the Communication to the Outside World

The Transduction of Chemical Signals into Electrical Signals. The electrical signals are important not only in the information trafficking among cells but also in sensory perception, i.e., the acquisition of information from the outside world. Sensory stimuli of various nature (mechanical, chemical, and optical) are perceived by specialized cells by means of highly refined molecular structures which are capable of translating these external stimuli into an electrical signal.

In the perception of sound, for example, mechanical vibrations of the stereocilia of sensory cells (hair cells) are translated into membrane potential fluctuations by the ac-

tion of specific ion channels, the so-called stretch-activated channels. These channels are located on the cell membrane in close proximity with the cilia and open or close according to the direction and extent of membrane stretching, thus reporting a signal that depends on the tilt of the stereocilia.

Transduction of light stimuli by the retina is another sophisticated mechanism based on the transformation of a sensory stimulus into a membrane potential change. Through an enzymatic cascade beginning with the activation of opsins, photosensitive enzymes able to modify their isomeric structure upon illumination, light exposure generates in vertebrates a hyperpolarization of the photoreceptor membrane. This is caused by the light-induced closing of unspecific cation channels [permeable to both monovalent (i.e., Na^+ and K^+) and divalent (i.e., Ca^{2+}) cations] which are open in the dark.

The presence of chemicals is revealed by the sensory organs mediating taste and smell by way of specific receptors located on the surface of sensitive cells. In smell, sensitive cells are neurons which have developed chemosensitive terminals in the shape of cilia, whereas in taste, chemoreceptors are located on specialized sensory cells, which are in turn innervated by neurons which ultimately receive the taste stimuli and relay them to the brain. Electrical signals therefore mediate both the peripheral perception and the transfer up to sensory cortical centers of information concerning chemical substances.

Receptors, Second Messengers, and Ionic Channels

Membrane Receptors

As described previously, the extracellular chemical mediators induce modifications in the target cells through a set of mechanisms initiating with the activation of specific membrane receptors. A membrane receptor is a glycoprotein binding with high specificity a given mediator at an extracellular site. Binding of the agonist induces a conformational change in the structure of the receptor protein, the first step toward the transformation of the chemical signal into a membrane potential signal. Some receptors are ionic channels (Fig. 4a); in this case, the structural modification caused by the association with the chemical mediator causes a kinetic modification (opening or closing) of the channel, yielding a permeability change which varies the cell membrane potential.

Other receptors bind with high affinity to multimeric membrane complexes called G proteins (Fig. 4b) once they are activated by their specific agonists. G proteins are composed of a catalytic and a regulatory unit (in fact, they have three subunits — α, β, and γ — the last two of which are tightly coupled), and in their inactive form bind guanosine diphosphate (GDP). Following the activation by stimulated

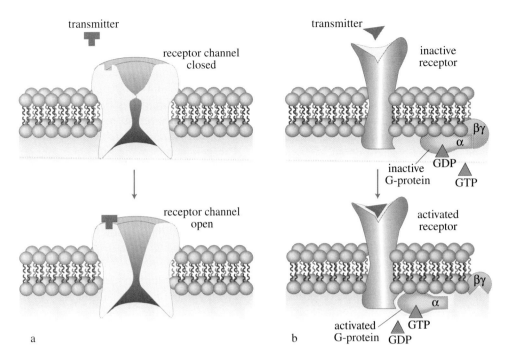

FIGURE 4 Receptor channels and receptors coupled via G proteins. (a) Schematic representation of a receptor channel which is closed in the absence of transmitter molecules (top) and which opens when a transmitter molecule binds to its binding site (bottom). (b) The membrane receptor depicted here is inactive in the absence of binding of a transmitter molecule (top); when binding occurs (bottom), the receptor becomes activated and interacts with a G protein composed of the subunits α and $\beta\gamma$. This causes the substitution of GTP for the GDP molecule normally bound to subunit (α-GDP) of the inactive G protein and the separation of the subunits α-GTP and $\beta\gamma$. Both these complexes are able to activate other processes, including opening or closing of ion channels.

receptors, GDP is exchanged with guanosine triphosphate (GTP) and the catalytic and regulatory subunits split. Activated G proteins modify the membrane potential acting on membrane channels by way of two mechanisms: direct activation (or inhibition) and activation (or inhibition) mediated by second messengers, which are the intracellular chemical mediators whose concentrations are controlled by the action of G proteins on enzymes responsible for their synthesis or hydrolysis. As discussed in the following section, several second messengers acting on ionic channels have been identified.

Second Messengers

As previously discussed, stimulation of a certain type of membrane receptor by extracellular chemical mediators leads to activation of G proteins. The aim of G protein activation is to trigger a set of biochemical reactions able to control the intracellular concentration of second messengers — molecules which in excitable cells have, among others, a controlling function on certain ion channels. Second messengers can be simple ions (such as Ca^{2+} ions), small molecules (such as the cyclic nucleotides cAMP and cGMP), or more complex molecules [such as inositol triphosphate (IP_3) or diacylglycerol (DG)].

In Fig. 5, two G protein-dependent biochemical pathways are shown which are functional in a variety of cell types and which control the second messengers cAMP (Fig. 5a) and IP_3 and DG (Fig. 5b). Intracellular cAMP is synthesized starting from ATP by the enzyme adenylate cyclase, which is activated by stimulatory G proteins (G_s) and inhibited by inhibitory G proteins (G_i). Several types of receptors can be coupled to the G proteins G_s and G_i, and specific receptors exist in every cell type. Therefore, many of the external chemical substances are able to induce an increase and/ or a decrease in the intracellular cAMP levels. cAMP is an ubiquitous second messenger, used in a variety of cellular processes, which often acts via a cytosolic enzyme able to induce phosphorylation of several substrates (protein kinase A). Some cases indicating how protein kinases, and in particular protein kinase A, can phosphorylate ionic channels and modify their properties are discussed later. Furthermore, cAMP and other cyclic nucleotides are able to modulate the activity of certain ionic channels by binding to the channel proteins.

A further biochemical pathway is composed, as shown in Fig. 5b, of a G protein (different from the G proteins G_s and G_i of adenylate cyclase) coupled to the membrane enzyme phospholipase C, which when activated is able to hydrolyze an important component of the membrane bilayer, the phosphatidylinositol biphosphate (PIP_2). Hydrolysis

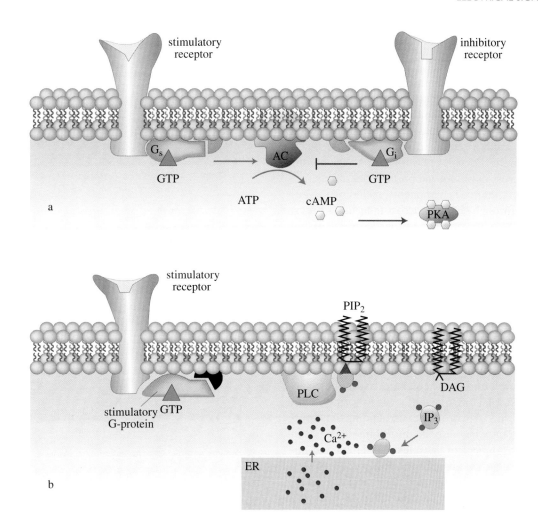

FIGURE 5 Second messenger systems. (a) Adenylate cyclase (AC, in red) pathway. This enzyme synthesizes cAMP from ATP and can be both activated by stimulatory G proteins (G_s) and inhibited by inhibitory G proteins (G_i), each paired in turn with different types of membrane receptors (in orange). cAMP is a second messenger able to modify many cellular processes; for example, it activates a specific protein kinase (PKA). (b) Phospholipase C (PLC, in yellow) pathway. In the scheme, this enzyme hydrolyzes PIP_2 with production of two second messengers, IP_3 and DAG. Activation of the enzyme PLC takes place following activation of a cascade of specific membrane receptors and G proteins coupled to PLC. By binding to specific receptors, IP_3 is able to induce the release of Ca^{2+} ions from the endoplasmic reticulum (ER).

of PIP_2 leads to the formation of two second messengers, the cytoplasmic IP_3 and the membrane-bound DG. Diacylglycerol is an activator of protein kinases, including protein kinase C (PKC), and can thus induce phosphorylation of several substrates. IP_3 has in turn an important role in stimulating the release of Ca^{2+} from the cell sarcoplasmic reticulum. The actions of the two intracellular second messengers can converge, as occurs for Ca^{2+}-dependent PKC isoforms: In this case, the PKC-dependent phosphorylation processes are strongly potentiated by the IP_3-induced Ca^{2+} release.

Ionic Channels

As discussed previously, chemical mediators control the cellular electrical activity generally through a complicated set of processes beginning with the stimulation of a membrane receptor and ending with a modification of one or more ionic currents flowing through the cell membrane. Ionic channels are complex integral membrane proteins, often composed of thousands of amino acid residues, which extend fully across the cell membrane. The characteristic feature distinguishing ionic channels from other membrane proteins, such as membrane receptors, is their ability to assume structural configurations which allow the flow of electrically charged ions through them and hence across the membrane. Cell membranes are composed of a lipid bilayer (e.g., a double layer of phospholipids oriented such that their charged, hydrophilic heads face either the cytoplasm or the outer solution), whereas the hydrocarbon, lipophilic tails occupy the inner core of the layer. The arrangement of phospholipids in a membrane is schematically illustrated in Figs. 4 and 5. The lipophilic composition of the internal part of the lipid bilayer allows the diffusion of lipid-soluble substances such as uncharged amino acids but prevents the flow of charged molecules. The ionic channels provide the specialized "trails" along which ions such as K^+, Na^+, Ca^{2+}, and Cl^- cross the otherwise tightly sealed plasmatic membrane. Specific paths for the passage of ions exist not only across plasmatic membranes but also across the membrane of other cell components, including the nucleus. Ions flow through channels across their inner pores, which are troughs extending across the two sides of the membrane and connecting internal and external cell compartments. The way pores work is very sophisticated, and is based on the functioning principle of the so-called amphipatic substances. These are substances with two distinct surfaces — one polar and hydrophilic and the other apolar and hydrophobic — which can, if assembled properly, physically separate lipophilic regions from hydrophilic ones. For example, the bile acids produced by the liver adhere with their lipophilic surfaces onto the surface of fat droplets resulting from the digestion of lipids, whereas the hydrophilic surfaces face the external environment so as to form the micelles. By conferring an hydrophilic surrounding to fat particles, micelles greatly increase the diffusion rate of fats in the aqueous environment of the intestinal lumen and improve the efficiency of fat absorption.

Similarly, although with a functionally opposite arrangement, the membrane fraction of ionic channels is mostly composed of neutral amino acids but also of a few charged amino acids (such as aspartate and glutamate residues, which are negatively charged). The charged residues are arranged within the membrane such that their charges face the internal part of the channel so as to form a "freeway" for the flow of ions — the pore. An example of such an assembly is shown in Fig. 6 for the ACh-activated nicotinic channel of the neuromuscular junction.

This receptor-operated channel (ROC), so called since the same protein fulfills simultaneously the functions of receptor for the agonist ACh and of ionic channel, is composed of five protein subunits (two α units and one each of β, γ, and δ units), each composed of four hydrophobic segments (M1–M4) which span the membrane in parallel arrays, and of hydrophilic segments which connect consecutive hydrophobic segments and which include the amino- and carboxyl-terminal ends, as shown in Fig. 6a (Changeux et al., 1984). The subunits are spatially organized such that their M2 segments are aligned along the central part of the pore, a necessary condition to allow the flow of charged ions within the pore. The M2 segments have a special property, i.e., they are partly made of charged amino acid residues and form, as shown in Fig. 6c, two charged circular structures along the external and internal mouths of the channel. It is interesting to note that the charged residues belonging

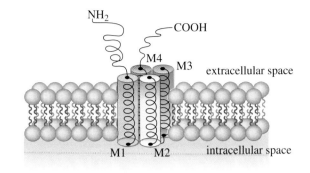

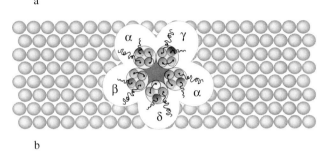

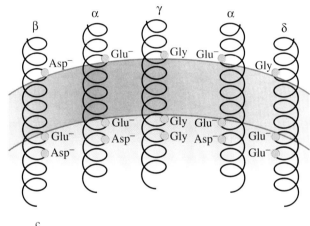

FIGURE 6 Nicotinic receptor channel of the neuromuscular junction. (a) Side view of one of the five channel subunits. The subunit is composed of four hydrophobic segments extending through the membrane (M1–M4) and the hydrophilic carboxyl- and amino-terminals which extend into the extracellular milieu. (b) Top view of the channel, which results from the assembly of five subunits (two α's, one β, one γ, and one δ) surrounding the aqueous pore in the central part (dark gray). Note that the five subunits have the M2 segments (blue) aligned along the pore. (c) View of the inner part of the pore opened to show the alignment of the M2 segments of the various subunits. The M2 segments of subunits α, β, and δ (the last only on the intracellular side) are enclosed by charged amino acid residues (glutamate and aspartate), whereas neutral residues (glycine) are found at corresponding positions in the γ subunits. Negatively charged amino acid residues form two charged rings at the entrance and at the exit of the aqueous pore and allow the flow of cations across the membrane. Furthermore, the amino acid sequences of the M2 segments are important in determining the properties of the ionic "filter," that is, for the selection of ionic species that can pass through the pore.

to segment M2 of subunits α, β, and δ (the latter only at the inner pore side) are electrically negative (glutamate and aspartate); this may explain the selectivity of the channel to cations (positively charged ions) such as Na^+ and Ca^{2+}. Accordingly, the substitution of some of these residues with positively charged residues (Galzi *et al.,* 1992) is able to induce a permeability to anions, namely, to Cl^-.

Various Types of Ionic Channels

Ionic channels are protein structures which allow a controlled passage of charged ions across membranes. The idea of a specialized path for the traffic of ions originated from early measurements of the electrical properties of membranes (Cole and Curtis, 1939) and was developed by the fundamental work of Hodgkin and Huxley (1952), who showed how the nerve action potential could be accounted for by time- and voltage-dependent kinetic changes of specific ionic currents (in particular, Na^+ and K^+ currents). Conceptually, the idea of ion channels therefore arose in relation to the electrical phenomena of excitable cells. A first classification of channels can then be based on criteria concerning their electrical properties. One of these is permeability, and we can thus distinguish among Na^+, K^+, Ca^{2+}, and Cl^- channels, channels with a mixed permeability to Na^+ and K^+, or unspecific cation channels (i.e., permeable to both monovalent and divalent cations). Other important properties of channels concern their "kinetics," that is, how they open and close (activation and deactivation; inactivation and removal of inactivation). In this case channels are regulated by changes in membrane voltage (voltage-operated channels), which may be activated either by depolarization or by hyperpolarization; the previously mentioned ROCs, which are directly activated by molecules (agonists) binding to the extracellular aspect of the protein; channels activated by G proteins; channels activated directly by intracellular second messengers; and channels regulated, through block and/or activation, by certain intracellular molecules (such as ions and ATP). As discussed later, much progress has been made recently in the knowledge of the structure of many ion channels and of the relation between structure and function due to the development of techniques of recombinant DNA. In the following sections, some of the main types of ionic channels and their contributions to cell functions are discussed.

Control of the Activity of Voltage-Dependent Channels by the Membrane Potential

Among the different types of ionic channels, those dependent on voltage play a main role in controlling the electrical activity of excitable cells. As described previously, voltage-dependent channels are responsible for the generation of nerve action potentials. Since the early 1950s, the method used to study the properties of action potentials, the voltage clamp, has allowed (by means of intracellular microelectrodes inserted into cells) the membrane poten-

tial to be controlled by the experimenter and the transmembrane ionic currents to be directly measured. A sophisticated version of the voltage clamp, the patch clamp, was developed in the 1970s by Neher and Sackmann (Hamill *et al.,* 1981). Due to a much greater resolution, this technique allows the recording of current across tiny portions of the cell membrane (patches). If the membrane "patch" contains only one or just a few functional channel proteins, recording of single-channel currents is thus possible. This technique has led to a great advance in the knowledge of the molecular properties of ionic channels.

The Sodium and Potassium Channels of the Squid Giant Axon

The two main current components identified in the squid giant axon by Hodgkin and Huxley (1952) were suggestive of the existence of specific membrane spanning structures (the ionic channels) permeable to Na^+ and K^+ ions, respectively, as schematically illustrated in Fig. 1. The analysis of these authors, as later confirmed by a wealth of evidence (Hille, 1992), allowed the dissection of the Na^+ and K^+ components and the numerical reconstruction of the action potential. In Fig. 7 the time courses of the two components during step potentials applied from a reference (holding) potential of -70 to -50, -30, and -10 mV are shown.

A few observations can be made by comparing the time course of the Na^+ component with that of the K^+ component: (i) Both currents are activated by depolarization of the membrane potential; (ii) at the voltages indicated, the

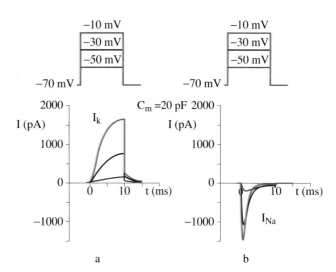

FIGURE 7 Ionic current in the squid giant axon during voltage clamp steps from the reference level (-70 mV) to -50, -30, and -10 mV (top). The currents (bottom) have been calculated according to the Hodgkin and Huxley equations for a membrane capacity (C_m) = 20 pF. Conventionally, outward currents are positive and inward currents are negative. (a) Potassium current (I_K). (b) Sodium current (I_{Na}) (reproduced with permission from Hille, 1992).

Na$^+$ current is inward, whereas the K$^+$ current is outward; (iii) following activation, the Na$^+$ current rapidly declines (inactivation process), whereas no inactivation is observed for the K$^+$ current; and (iv) activation of the Na$^+$ current is faster than that of the K$^+$ current. The axonal action potential (Fig. 1) is a direct result of the previously mentioned properties. Whenever a depolarizing stimulus is sufficiently large (see Fig. 3c), some Na$^+$ channels open, giving rise to a further depolarization; this in turn causes more channels to open, according to a self-amplifying process which results in a fast increase of the Na$^+$ conductance and hence a fast membrane depolarization toward the Na$^+$ equilibrium potential (see Eq. 3). After some delay, the same depolarization determines the activation of K$^+$ channels and the inactivation of Na$^+$ channels and the ensuing repolarization toward the resting membrane potential. A detailed reconstruction of the action potential can be made based on Eqs. (1) and (2).

Although the concept of ionic channels had not been fully developed in the 1950s — indeed, the fluid mosaic model of membrane predicting the existence of integral proteins inserted in the membrane and spanning it side to side was developed only in the 1970s (Singer and Nicolson, 1972) — the Hodgkin and Huxley (1952) description of the squid giant axon action potential represents a fully coherent description of what would only later be identified as the Na$^+$ and K$^+$ ion channels.

The K$^+$ Inwardly Rectifying Channels

As noted in the discussion of Eq. (3), the resting membrane potential of excitable cells is close to the K$^+$ equilibrium potential since the K$^+$ permeability is the prevailing one at rest. By studying the permeation of K$^+$ across the membrane of skeletal muscle fibers, Katz (1949) found that the K$^+$ conductance (g_K) is high at voltages more negative than the resting potential (E in Eq. 3) but becomes increasingly lower at more depolarized voltages. This property was termed inward rectification since it was reminiscent of the behavior of electronic diodes, which rectify alternate current by allowing electrical flow in one direction only. Inward rectification is important in muscle cells in which it inhibits excessive loss of intracellular K$^+$ ions during the action potential, which is particularly long in cardiac myocytes (see Fig. 1d). This implies a substantial energy saving since the K$^+$ ions lost during electrical activity must be recovered by the Na$^+$/K$^+$ pump using ATP as energy source.

The identification of the K$^+$ rectifying channels made by recording single-channel activity by the patch clamp technique has allowed the investigation of the properties of rectification at the single-channel level (Sackmann et al., 1983). This approach led to the observation that the K$^+$ inward rectifier channels close rapidly when the membrane voltage is more positive than the resting membrane potential, not just because of an intrinsic gating mechanism (such as the one operating in Na$^+$ channels) but mostly because of a

voltage-dependent block of the channels by nonpermeable cations different from K$^+$. In other words, at depolarized voltages, charged molecules such as Mg^{2+}, Ca^{2+}, and polyamines enter the channels and block the flow of K$^+$ ions (Matsuda et al., 1987; Mazzanti and Di Francesco, 1989; Ficker et al., 1994).

Voltage-Dependent Ca^{2+} Channels

A most important aspect of the Ca^{2+} channel function relates to the fact that, contrary to other ionic species carried through membranes, Ca^{2+} ions act as intracellular second messengers in a variety of cellular processes. Therefore, the entry of Ca^{2+} ions into cells does not only interfere with the cell's electrical activity but also promotes complex biochemical events. For example, as previously mentioned, Ca^{2+} ions are essential for the activation of processes leading to transmitter release at synaptic terminals, which explains why Ca^{2+} channels are densely expressed in presynaptic terminals.

There are several types of Ca^{2+} channels with similar properties in many excitable and nonexcitable cells (Reuter, 1983; Tsien, 1983; Trautwein and Hescheler, 1990; Eckert and Chad, 1984). Ca^{2+} channels were initially described in mammalian heart (Reuter, 1968), in which they have a fundamental role in contraction since the mechanical contraction of cardiac myocytes, and thus of the whole heart, depends essentially on the entry of Ca^{2+} through these channels during the action potential. The kinetics of the Ca^{2+} current in cardiac cells are qualitatively similar to those of the Na$^+$ current in nerve, with activation/inactivation processes which, although slower, are also induced by depolarization. A large fraction of the Ca^{2+} current in cardiac cells flows through channels termed L-type or high-threshold Ca^{2+} channel. The activation threshold for these channels is more depolarized than that of Na$^+$ channels (about −40 compared to −80 mV), and the inactivation kinetics are fairly slow, which allows a particularly long duration of the cardiac action potential (see Figs. 1d and 1e). The L-type Ca^{2+} channel can also be modulated by chemical mediators, such as the neurotransmitters of the autonomic nervous system noradrenaline and acetylcholine which, in cardiac myocytes, increase and decrease, respectively, the probability of channel opening upon depolarization.

Given the dependence of cardiac contraction by the entry of Ca^{2+} ions through the membrane, the modulation, at least partly mediated by cAMP-dependent phosphorylation of L-type Ca^{2+} channels, is the basis for the autonomic control of cardiac inotropism (strength of contraction).

In addition to the L-type, a T-type (transient or low-threshold) Ca^{2+} current has been described in the myocardium; this is activated at more negative voltages (about −80 mV) and is characterized by rapid inactivation kinetics. It is therefore thought to contribute, together with the Na$^+$ channel, to the rapid depolarization of the action potential.

L-type and T-type Ca^{2+} channels are expressed in a variety of tissues other than the cardiac tissue, including the skeletal muscle, the brain, the smooth muscle, and the endocrine tissue, in which they play different roles together with other types of Ca^{2+} channel more recently described (such as type N, P, Q, and R Ca^{2+} channels). N-, P-, and Q-type Ca^{2+} channels appear to be specifically important in presynaptic terminals of neurons, in which they mediate the Ca^{2+} entry required to initiate the release of neurotransmitters.

The Glutamatergic Receptor Channel NMDA

As discussed previously, the nicotinic ACh receptor channels of the neuromuscular junction belong to the family of ROCs, characterized by the coexistence of the agonist binding site, whose activation leads to channel opening, and of the channel structure itself, in one protein only (Fig. 6). In addition to the nicotinic ACh receptor, several other receptor channels exist. Particularly important are those which in the nervous system mediate the rapid synaptic transmission of excitatory stimuli and inhibitory stimuli, as illustrated in Fig. 3. I now consider the NMDA channel, so called after its agonist N-methyl-D-aspartate.

The investigation of this channel has generated much interest since the channel was shown to be involved in the phenomenon of long-term potentiation (LTP). In LTP, a synapse undergoing a period of intense activity is later able to "remember" this event by an augmented efficiency in signal transmission. LTP is thought to be the cellular basis of the surely much more complex set of events underlying learning and memory (Malgaroli and Tsien, 1992; Malinow *et al.*, 1988; Silva *et al.*, 1992; Bliss and Collingridge, 1993). The phenomenon of LTP has apparently two components: a presynaptic one, associated with an increase in transmitter release, and a postsynaptic one, associated with an increase in the sensitivity of excitatory glutamate receptors (both NMDA and AMPA; Fig. 8). The properties of the mechanisms involved in LTP are such that, following bursts of intense synaptic activity, synaptic transmission is strengthened for up to several hours to days.

The physiological neurotransmitter mediating the opening of NMDA and other channels responsible for the EPSP in the excitatory synapses is glutamate. When activated, all excitatory receptors induce a depolarization in the postsynaptic membrane by inducing an inward current carried mainly by Na^+ and K^+. The reversal potential of this current is indeed close to zero since it is intermediate between the Na^+ and K^+ reversal potentials, and its activation thus tends to move the membrane potential from the (negative) resting level toward zero level. In contrast to other glutamatergic channels, however, the NMDA channel is permeable to Ca^{2+}, which makes the activation of NMDA channels particularly relevant to the mechanism of LTP. Another special feature of NMDA channels which determines its anomalous behavior is related to the channel block by extracellular Mg^{2+} ions. Under normal conditions, during moderate synaptic activity and hence in the presence of a

moderate concentration of glutamate, the EPSP is of limited amplitude; under these conditions, NMDA channels are blocked by Mg^{2+} and do not contribute to the EPSP. When the synaptic activity is intense, such as during a high-frequency burst of action potentials (tetanus), the EPSP size increases to a level sufficient to overcome the Mg^{2+}-induced block of NMDA channels. This further contributes to depolarizing the membrane and eventually results in a massive entry of Ca^{2+} ions into the cell. Opening of the NMDA channels is therefore controlled under these conditions by the removal of the voltage-dependent channel block by Mg^{2+}; in other words, NMDA channels only pass current when the postsynaptic depolarization has reached a threshold level. The main features of NMDA channels are shown in Fig. 8.

Entry of Ca^{2+} ions into the postsynaptic neuron is responsible for initiating the cascade of events leading to LTP. The LTP phenomenon is a complex, not entirely understood process which includes both pre- and postsynaptic events, characterized by an increased activity of several enzymes (e.g., PKC in the postsynaptic cell) and by the presence of one or more retrograde messengers (e.g., nitric oxide, which are able to "brief" the presynaptic membrane on postsynaptic events and to consequently determine an increased release of neurotransmitter.

The ACh-Activated K^+ Channel

In addition to mediating the transduction of the electrical stimulus of the motoneuron into skeletal muscle contraction through the action of nicotinic receptors, the neurotransmitter ACh is used in several other instances in the central nervous system through the stimulation of muscarinic receptors, which are normally localized onto target organs. In the heart, for example, stimulation of the vagus nerve induces release of ACh and activation of muscarinic receptors present in the atria and in other specialized cardiac regions, such as the sinoatrial node (SAN), the natural "pacemaker" region of the heart. In cardiac myocytes, muscarinic receptors are linked to different transduction systems, including a G protein termed G_K which is able to activate specific K^+ channels (current $I_{K,ACh}$; Logothetis *et al.*, 1987; Brown and Birnbaumer, 1990). A peculiar property of the mechanism of the G_K-mediated activation of $I_{K,ACh}$ channels is that it involves a direct interaction between G proteins and channels, without activation of intracellular second messengers. By directly opening K^+ channels, ACh causes a rapid hyperpolarization of cardiac cells which contributes to the slowing of cardiac rhythm associated with vagal activity.

The K^+ Channel Involved in the Release of Pancreatic Insulin

Insulin, a pancreatic hormone essential to the control of plasma glucose content, is released by pancreatic β cells. The main stimulus for insulin release is plasma glucose: The

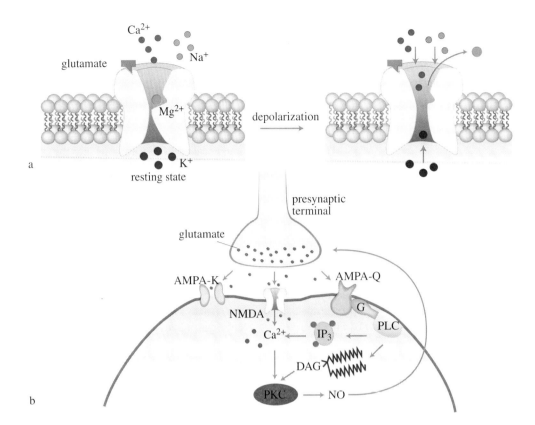

FIGURE 8 NMDA channel and long-term potentiation (LTP). (a) The NMDA receptor channel is normally closed at rest, even if the excitatory neurotransmitter glutamate is bound to the stimulatory site, since the channel is blocked by Mg^{2+} ions. On membrane depolarization, Mg^{2+} ions are expelled from the channel, which is thus unblocked. This allows the flow through the channel of permeable ions, including Ca^{2+}, which can enter massively into the neuron. (b) Simplified scheme of the factors playing a role in LTP; glutamate released by the synapse activates NMDA receptor channels and other receptors (AMPA receptors). AMPA-K (in green), whose selective agonist is kainate, is involved in the depolarization of the postsynaptic membrane, and AMPA-Q's (in blue) agonist is quisqualate. AMPA-Q (type B) receptors, in contrast to the other receptors, are not channels and are coupled through G proteins to a phospholipase C (PLC); their stimulation produces IP_3 and DAG (see Fig. 5), thus increasing the intracellular Ca^{2+} concentration and the activity of protein kinase C (PKC). PKC and Ca^{2+} are in turn able to stimulate the synthesis of, among others, nitric oxide (NO), which is believed to act as a retrograde messenger, that is, able to enter into the presynaptic membrane and potentiate its capacity to release the neurotransmitter.

amount of insulin released into the bloodstream is directly proportional to the concentration of glucose in the plasma. The mechanism of hormone release is based on the sensitivity to glucose of a specific K^+ channel expressed on the membrane of pancreatic β cells ($I_{K,ATP}$ current). An important role is played in this process by intracellular ATP, which behaves as a proper second messenger in the control of K^+ channels. ATP is a molecule of high energy content utilized in biological processes as cellular "fuel," which is synthesized through different catabolic processes starting from glucose, including glycolysis. The essential steps of insulin release are therefore as follows: Plasmatic glucose crosses the membrane of β cells (probably via a specific carrier) and produces ATP; through block of ATP-sensitive K^+ channels, ATP reduces the K^+ current $I_{K,ATP}$ and consequently

causes the membrane to depolarize; and membrane depolarization is associated with opening of Ca^{2+} channels and entry of Ca^{2+} into the cell, which is in turn responsible for the exocytosis of insulin vesicles.

The Cation Pacemaker Channel Involved in the Control of Cardiac Rhythm

It is amazing that a human heart beats as many as one hundred thousand times every 24 h, at a rate which is finely regulated by the autonomic nervous system to allow a constant tuning to the variable needs of the body. Indeed, the heart rate varies during the sleep–wake cycle, during physical activity, in the presence of special emotional or emergency loads, and so on.

How does such a refined modulation take place? At the

basis of the control of cardiac rate is a cation current (I_f current; Di Francesco, 1993) which is strongly expressed in only some specialized regions of the cardiac muscle, particularly in the so-called pacemaker region located in the SAN, approximately at the junction where the superior vena cava (carrying the venous blood back from the body organs to the heart) meets the right atrium. Myocytes from this area have the peculiar property of beating spontaneously; the stimulus autonomously generated in the SAN is then electrically propagated to the atrial myocytes and ultimately, after a short delay which allows the full atrial muscle contraction to occur earlier than the ventricular one, to the ventricles through structures specialized for the fast conduction of action potentials. The spontaneous frequency of pacemaker cells is therefore that of the whole heart.

The SAN region is densely innervated by the two branches of the autonomic nervous system. Sympathetic stimuli accelerate heart rate by the release of NA, whereas parasympathetic stimuli lead to the slowing action by the release of ACh. This control occurs via modulation of the spontaneous activity rate of pacemaker cells by the autonomic neurotransmitters.

In Fig. 9 the spontaneous activity of a mammalian (rabbit) SAN cell is shown in normal conditions and in the presence of ACh or isoprenaline, a β-adrenergic agonist simulating the action of noradrenaline on these cells. The action potential of pacemaker cells differs substantially from that recorded in other cardiac myocytes (e.g., compare with Fig. 1d) because of the presence of a typical phase of the action potential, the slow diastolic (or pacemaker) depolarization, which is responsible for the autorhythmicity of these cells. Following termination of an action potential, the membrane potential does not settle to a constant (resting) value as it does, for example, in ventricular myocytes (Fig. 1d); rather, it undergoes a slow depolarization that leads to threshold for initiation of another action potential.

The diastolic depolarization is due to the slow activation of an inward current, called I_f (also referred to as pacemaker current), which takes place when the membrane potential becomes more negative than approximately $-40\,mV$. This current has several unusual properties ("f" stands for "funny"): For example, it is carried by both Na^+ and K^+ ions, activated on hyperpolarization, and has very slow activation/deactivation kinetics. The I_f current is not expressed functionally in either atrial or ventricular cells, in accordance with the fact that neither diastolic depolarization nor spontaneous activity are present in these cells.

Another important property of the I_f current is its sensitivity to autonomic neurotransmitters (Di Francesco et al., 1986, 1989; Di Francesco and Tromba, 1988). SAN cells richly express β-adrenergic and muscarinic receptors. In the presence of β-adrenergic stimulation, I_f increases as shown in Fig. 9c. This increase is mediated by an augmented intracellular concentration of the second messenger cAMP, which has an activating action on the I_f channels, resulting from an increased probability of channel opening upon hyperpolarization (Di Francesco, 1986). cAMP molecules do not act, as in the case of other channels (e.g., the L-type Ca^{2+} channels), by a phosphorylation process mediated by PKA but rather by a direct binding to the channel protein, thus adopting a more direct and quicker controlling mechanism (Di Francesco and Tortora, 1991).

In the presence of colinergic stimulation, a symmetrical set of events occurs (Fig. 9d). The I_f current is inhibited by ACh as a consequence of a decrease of the intracellular concentration of cAMP, leading to a lower probability of channel opening during hyperpolarization. β-Adrenergic and colinergic effects on the I_f current underlie acceleration and slowing of cardiac rhythm by sympathetic and parasympathetic stimuli, respectively (Fig. 9a). This happens simply because the diastolic depolarization phase of the pacemaker cell action potential becomes faster when I_f (being an inward current) increases and slower when it decreases.

The autonomic control of the I_f current and of cardiac rhythm occurs as illustrated in the simplified scheme of Fig. 9e. The direct target of β-adrenergic and muscarinic colinergic receptors is the membrane adenylate cyclase, to which the two types of receptors are linked through stimulatory and inhibitory G proteins, respectively (see Fig. 5). cAMP production increases and decreases under the action of sympathetic and parasympathetic stimuli, respectively, and as a consequence so does the degree of activation of the I_f current. In this way, the rate of spontaneous activity of pacemaker cells can be controlled by the neurotransmitters of the autonomic nervous system directly, taking advantage of the constitutive properties of an ionic channel.

Electrical Signals from Molecules to Therapy: Future Directions

Several Pharmacological Substances Interact with Ionic Channels

The ionic flow across ion channels, and therefore the alterations induced by opening or closing of channels upon the electrical properties of cells, can be modified by molecules interacting with the channel. Therapeutical or toxic actions of several plants have been known for thousands of years, and the poisonous action of many molecules of animal origin has also long been known. Many of these effects are due to interaction with ion channels. For example, scorpion venoms can contain up to 20 different molecules that act by blocking inactivation of Na^+ channels (β-toxins) by reinforcing their activation (β-toxins), or by blocking K^+ channels (charybdotoxin). Similar substances are present

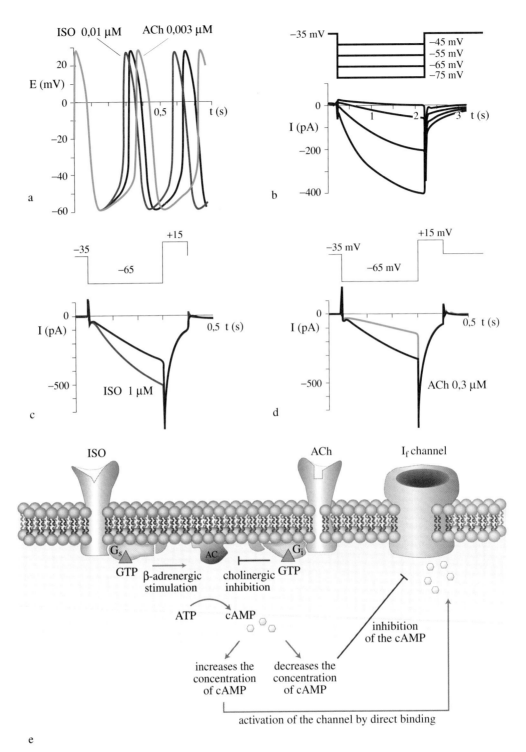

FIGURE 9 Control of cardiac rhythm in mammals. (a) Action potentials recorded in the sino-atrial (SA) node of the rabbit heart (pacemaker region) in control conditions (blue) and in the presence of isoprenaline (ISO; red) or acetylcholine (ACh; green). The rhythm is accelerated by ISO and slowed by ACh. (b) Recordings of the pacemaker current (I_f current) during hyperpolarizing voltage steps to the voltages indicated near each trace (from −45 to −75 mV, starting from a reference level of −35 mV) in an isolated rabbit SA node cell. The I_f current is inward and activates slowly upon hyperpolarization, and it is relevant to the generation and control of the "slow depolarization" (or pacemaker) phase of the action potential in the SA node. For this reason it is called the pacemaker current. (c, d) Effects of ISO and ACh on the I_f current in an SA node cell during hyperpolarizations from −35 to −65 mV; the I_f current is increased by ISO (c) and decreased by ACh (d). Depolarization to 15 mV activates a Ca^{2+} current important for triggering the action potential. (e) Scheme illustrating the modulation of the I_f channel. The channel is directly controlled by the second messenger cAMP and is therefore activated by stimulation of the β-adrenergic receptor (ISO) and inhibited by the muscarinic receptor (ACh) through the membrane adenylate cyclase (see Fig. 5a).

in sea anemones. Ca^{2+} channels can also be targets of toxins contained in the venom of spiders or sea mollusks (e.g., the ν-conotoxin, which blocks preferentially type N Ca^{2+} channels). In other cases, the toxic action of a venom results from the insertion of new ionic channels into the membrane. One such example is the venom of the black widow, whose devastating effect involves a massive release of neurotransmitters from central and peripheral synapses. The effect depends on the presence of a component of the venom (α-latrotoxin) which binds to special membrane proteins and is then inserted into the presynaptic membrane. It appears that the protein is composed of aggregates of unspecific channels, whose insertion into the membrane causes a large entry of Ca^{2+} into the neuron and consequently the uncontrolled release of neurotransmitter.

These substances are often utilized for defense to scare away a possible predator or in an assault to paralyze a possible prey. Given their action on ionic channels, various natural molecules have found application in the study and characterization of the properties and, recently, the molecular structure of the ionic channels they act on. The existence of molecules able to specifically target a given ion channel is indeed essential to the accurate identification of the channel. The best know example is tetrodotoxin (TTX), a molecule extracted from, among other sources, the puffer fish which blocks with great efficacy the neuronal Na^+ channels. The puffer fish of the order Tetraodontiforms is a true delicacy in Japan, and the TTX block is so efficient that improper removal of the TTX-containing organ can be fatal to the unfortunate diner.

Substances which interact with the Na^+ channel by blocking ionic flow through the channel when this is open are the local anesthetics (procaine, quinidine, tetracaine, lidocaine, and so on), which are liposoluble amino compounds typically used in anesthetic therapy because of their ability to stop action potential propagation, in which case they are applied in high concentrations. When they are injected intravenously at lower concentrations, some of these compounds have an important clinical application in the cure of cardiac tachyarrhythmias, particularly when they are associated with anomalous action potential propagation in damaged areas of the cardiac muscle. In heart, a therapeutical action for the treatment of supraventricular arrhythmias, angina pectoris, and hypertension is exerted by another class of compounds — the Ca^{2+} antagonists such as verapamil, diltiazem, and nifedipine. These substances exert their therapeutical action by partial block of Ca^{2+} channels and particularly of type L Ca^{2+} channels which, as mentioned previously, are important in controlling the duration of the cardiac action potential. Molecules able to prolong the action potential duration by partial blockade of the delayed K^+ current, such as sotalol or amiodarone, are, again in cardiac muscle, agents able to efficiently control the onset of potentially lethal ventricular arrhythmias.

Cloning of Ion Channels

A classification of ion channels which has recently replaced those dependent on permeability or kinetic properties is based on the amino acidic composition of channel proteins. Molecular biological techniques (Watson *et al.*, 1992) have led to fast progress in the understanding of molecular properties of channels and have allowed the sequencing and cloning of several ion channels. In Fig. 10 some of the most typical classes of ionic channels, as characterized by their sequence properties, are schematically shown.

In these sequences, hydrophobic regions are represented by cylinders fully spanning the cell membrane, whereas hydrophilic tracts are represented by solid lines within either the cytoplasm or the extracellular space. Hydrophobic segments within the chain normally have an α-helix structure and are about 20 amino acids long.

Voltage-dependent, depolarization-activated channels carrying an inward current, such as Na^+ and Ca^{2+} channels, are composed of several subunits. The best described subunits are those able to form "pores": α subunits for Na^+ channels and $\alpha 1$ for L-type Ca^{2+} channels (Fig. 10a). Pore-forming subunits are long polypeptide chains [about 2400 residues for the Na^+ channel (Catterall, 1988) and about 2200 for the rabbit cardiac Ca^{2+} channel (Mikami *et al.*, 1989)], each composed of four replicas of six membrane-spanning segments (S1–S6). The four replicas are highly homologous to each other and, as shown in the cross section of the channel in Fig. 10a, flank each other around a central region where the pore is localized.

The tract connecting segments S5 and S6 (P tract) is composed partly of charged and partly of hydrophobic residues, which allows its partial insertion as a β-sheet "hairpin" in the double lipid bilayer of the membrane. The presence of charged residues is important; these residues are believed to line the channel pore, thus providing the most peculiar structural element of ion channels — the membrane-spanning trough which allows charged ions to cross from one side of the membrane to the other according to a mechanism similar to the one described for the ACh nicotinic receptor (see Fig. 6).

Similar considerations apply to the voltage-dependent K^+ channels (Fig. 10b), the main difference being that K^+ channels are homotetramers made of four identical subunits aligned according to a circular structure around the pore (Jan and Jan, 1994). Here, too, the P tracts are believed to line the pore (Yellen *et al.*, 1991).

The inward rectifier K^+ channels belong to a different family. As illustrated in Fig. 10c, the subunits of these channels are composed of two transmembrane segments connected by a P region representing the pore. Experimentation on native channels suggests that inward rectifier K^+ channels are also composed of four identical subunits (Mazzanti *et al.*, 1996).

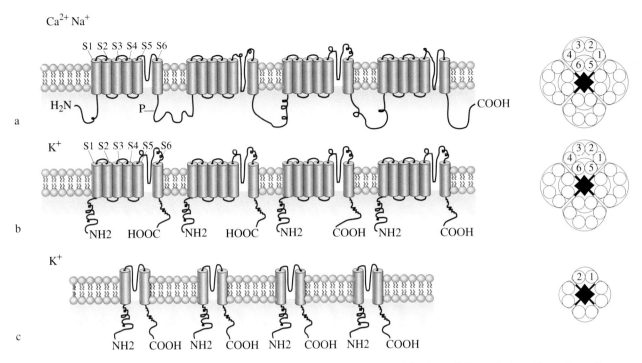

FIGURE 10 Families of ion channel clones. The schemes represent side views (left) and views from the cell exterior (right). (a) Family of voltage-dependent, depolarization-activated channels, such as some Ca^{2+} and Na^+ channels; the structure shown corresponds to one of the several subunits of which channels are made. Each subunit is a monomer composed of four homologous repeats. Each repeat is in turn composed of six transmembrane segments and of regions partly embedded in the membrane (P tracts) which are thought to line the pore. (b) Voltage-dependent K^+ channels; these are formed from four identical, distinct subunits (homotetramers), again composed of six membrane segments, arranged in a similar way to the monomers in a. (c) Inwardly rectifying K^+ channels; here, the monomers are shorter, composed of only two transmembrane segments. Still present, however, are the P tracts, which are thought to form the pore.

Cloning of ion channels has allowed, by means of structure–function studies, the identification of the segments of the polypeptide chain which are responsible for the different channel properties. For example (Catterall, 1988), the kinetic behavior of voltage-dependent channels depends on the sequence of tract S4, which contains charged residues at regular intervals (one charged residue every three residues). Modification of these charged residues in mutant clones expressed via heterologous expression experiments has shown that the voltage dependence of Na^+ (Stühmer *et al.*, 1989) and K^+ (Papazian *et al.*, 1991) channel kinetics is related to the presence of specific charged residues.

Channelopathies

The importance of a correct functioning of processes based on electrical signals which underlie the transfer of information is evident if one considers the variety of biological phenomena dependent on these processes. The activity of the entire central nervous system in man, for example, is based on electrical signals. Joint efforts of electrophysiological and molecular biological techniques have led to the identification of specific ion channel abnormalities as the main causes of some pathologies.

For example, cystic fibrosis is a genetic disease affecting secretory epithelia of various organs, such as lungs, pancreas, and sweat glands. It is common among Caucasians (1 in every 2000 affected) and has a normally malignant outcome, often due to the pulmonary complications arising after a long phase of recurrent infections and progressive degeneration of lung tissue accompanied by persisting accumulation of dense mucus in the airways. It has been shown that this pathology is associated with a defective function of Cl^- transport mechanisms in the epithelium (Quinton, 1990; Welsh, 1990). The Cl^- channel mediating Cl^- transport in these epithelia is activated by cAMP-dependent protein kinase; in individuals affected by cystic fibrosis, the activation of Cl^- channel by PKA does not occur. Molecular biological techniques have allowed the identification, cloning, and sequencing of a gene coding for the CFTR protein (cystic fibrosis transmembrane conductance regulator), which represents the Cl^- channel responsible, if defective, for cystic fibrosis (Riordan *et al.*, 1989). Indeed, from tissue belonging to cystic fibrosis individuals a mutated CFTR protein can be extracted which in most cases differs from the normal one in a single amino acid; a single mutation is therefore able to inhibit the normal PKA-dependent phosphorylation of the CFTR channel.

Other examples of "channelopathies" are some types of nondistrophic myotonias such as the periodic iperkalemic paralysis, which gives rise to paralysis of the skeletal muscle following ingestion of food with high potassium content (hyperkalemia), and the congenital paramyotonia, which leads to involuntary contraction of muscles following a decrease in body temperature. Both diseases are caused by a congenitally defective muscle Na^+ channel carrying one or several mutations in the amino acid sequence which lead to loss of the ability of the channels to inactivate once they are open (Hoffman and Spier, 1993). This leads to a depolarization of the muscle fibers whenever a triggering cause is active (such as high external K^+ concentration or a cold shock) and consequent paralysis.

The long QT syndrome (LQTS) is a congenital cardiac disease characterized by a delayed repolarization of the cardiac ventricular action potential and a consequent pathological prolongation of the action potential duration which results in a prolongation of the QT phase of the electrocardiogram (this is composed of a succession of waves, whose peaks are termed Q, R, S, T, and P). Using mapping strategies, it has been possible to identify in chromosomes 3 and 7 of individuals affected by LQTS some mutant genes which are responsible for the disease (Jiang *et al.*, 1994). One of these, termed HERG (human ether-à-go-go related gene), has been mapped in chromosome 7 and found to code for a K^+ channel with rapid activation kinetics (I_{Kr} current; Curran *et al.*, 1995) known to mediate the action potential repolarization process. The prolongation of the action potential in this case is attributable to a reduced functionality of mutant K^+ channels (Sanguinetti *et al.*, 1995).

Another genetic mutation responsible for a second type of LQTS is localized in chromosome 3 and is associated with a mutant Na^+ channel (Wang *et al.*, 1995); here, the pathological action is attributable to the imperfect inactivation of the mutant channels, which remain partially open during the action potential and thus prolong it. The clinical relevance of these results is emphasized by the fact that specific therapeutic interventions can be developed for the different types of LQT syndromes (Rosen, 1995).

References Cited

ASHCROFT, F. M., HARRISON, D. E., and ASHCROFT, S. J. H. (1984). Glucose induces closure of single potassium channels in isolated rat pancreatic β-cells. *Nature,* **312,** 446–448.

BLISS, T. V., and COLLINGRIDGE, G. L. (1993). A synaptic model of memory: Long-term potentiation in the hippocampus. *Nature,* **361,** 31–39.

BROWN, A. M., and BIRNBAUMER, L. (1990). Ionic channels and their regulation by G protein subunits. *Annu. Rev. Physiol.,* **52,** 197–213.

CATTERALL, W. A. (1988). Structure and function of voltage-sensitive ion channels. *Science,* **242,** 50–61.

CHANGEUX, J-P., DEVILLERS-THIERY, A., and CHEMOUILLI, P. (1984). Acetylcholine receptor: an allosteric protein. *Science,* **225,** 1335–1345.

COLE, K. S., and CURTIS, H. J. (1939). Electrical impedence of the squid giant axon during activity. *J. Gen. Physiol.,* **22,** 649–670.

CURRAN, M. E., SPEAWSKI, I., TIMOTHY, K. W., VINCENT, G. M., GREEN, E. D., and KEATING, M. T. (1995). A molecular basis for cardiac arrhythmia: HERG mutations cause long QT syndrome. *Cell,* **80,** 795–803.

DI FRANCESCO, D. (1986). Characterization of single pacemaker channels in cardiac sino-atrial node cells. *Nature,* **324,** 470–473.

DI FRANCESCO, D. (1993). Pacemaker mechanisms in cardiac tissue. *Annu. Rev. Physiol.,* **55,** 455–472.

DI FRANCESCO, D., and TORTORA, P. (1991). Direct activation of cardiac pacemaker channels by intracellular cyclic AMP. *Nature,* **351,** 145–147.

DI FRANCESCO, D., and TROMBA, C. (1988). Inhibition of the hyperpolarizing-activated current, if, induced by acetylcholine in rabbit sino-atrial node myocytes. *J. Physiol.,* **405,** 477–491.

DI FRANCESCO, D., DUCOURET, P., and ROBINSON, R. B. (1989). Muscarinic modulation of cardiac rate at low acetylcholine concentrations. *Science,* **243,** 669–671.

DI FRANCESCO, D., FERRONI, A., MAZZANTI, M., and TROMBA, C. (1986). Properties of the hyperpolarization-activated current (i_f) in cells isolated from the rabbit sino-atrial node. *J. Physiol.,* **377,** 61–88.

ECKERT, R., and CHAD, J. E. (1984). Inactivation of Ca channels. *Prog. Biophys. molec. Biol.,* **44,** 215–267.

FICKER, E., TAGLIALATELA, M., WIBLE, B. A., HENLEY, C. M., and BROWN, A. M. (1994). Spermine and spermidine as gating molecules for inward rectifier K^+ channels. *Science,* **266,** 1068–1072.

GALZI, J.-L., DEVILLERS-THIÉRY, A., HUSSY, N., BERTRAND, S., CHANGEUX, J.-P., and BERTRAND, D. (1992). Mutations in the channel domain of a neuronal nicotinic receptor convert ion selectivity from cationic to anionic. *Nature,* **359,** 500–505.

HAMILL, O. P., MARTY, A., NEHER, E., SAKMANN, B., and SIGWORTH, F. J. (1981). Improved patch-clamp techniques for high-resolution current recording from cells and cell-free membrane patches. *Pflügers Arch.,* **391,** 85–100.

HILLE, B. (1992). *Ionic channels of excitable membranes.* 2ª ed., Sinauer, Sunderland Associates.

HODGKIN, A. L., and HUXLEY, A. F. (1952). A quantitative description of membrane current and its application to conduction and excitation in nerve. *J. Physiol.,* **117,** 500–544.

HOFFMAN, E. P., and SPIER, S. J. (1993). Sodium channelopathies: dramatic diseases caused by subtle genetic changes. *NIPS,* **8,** 38–41.

JAN. L. Y., and JAN, Y. N. (1994). Potassium channels and their evolving gates. *Nature,* **371,** 119–122.

JIANG, C. *et al.* (1994). Two Long QT Syndrome loci map to chromosomes 3 and 7 with evidence for further heterogeneity. *Nat. Genet.,* **8,** 141–147.

KATZ, B. (1949). Les constantes electriques de la membrane du muscle. *Archives des Sciences Physiologiques,* **3,** 285–299.

LOGOTHETIS, D. E., KURACHI, Y., GALPER, J., NEER, E. J., and CLAPHAM, D. E. (1987). The $\beta\gamma$ subunits of GTP-binding proteins activate the muscarinic K$^+$ channel in heart. *Nature*, **325**, 321–326.

MALGAROLI, A., and TSIEN, R. W. (1992). Glutamate-induced long-term potentiation of the frequency of miniature synaptic currents in cultured hippocampal neurons. *Nature*, **357**, 134–139.

MALINOW, R., MADISON, D. V., and TSIEN, R. W. (1988). Persistent protein kinase activity underlying long-term potentiation. *Nature*, **335**, 820–824.

MATSUDA, H., SAIGUSA, A., and IRISAWA, H. (1987). Ohmic conductance through the inwardly rectifying K channel and blocking by internal Mg^{2+}. *Nature*, **325**, 156–159.

MAZZANTI, M., ASSANDRI, R., FERRONI, A., and DI FRANCESCO, D. (1996). Cytoskeletal control of rectification and expression of four substates in cardiac inward rectifier K$^+$ channels. *FASEB J.*, **10**, 357–361.

MAZZANTI, M., and DI FRANCESCO, D. (1989). Intracellular Ca modulates K-inward rectification in cardiac myocytes. *Pflügers Arch.*, **413**, 322–324.

MIKAMI, A., IMOTO, K., TANABE, T., NIIDOME, T., MORI, Y., TAKESHIMA, H., NARUMIYA, S., and NUMA, S. (1989). Primary structure and functional expression of the cardiac dihydropyridine-sensitive calcium channel. *Nature*, **340**, 230–233.

PAPAZIAN, D. M., TIMPE, L. C., JAN, Y. N., and JAN, L. Y. (1991). Alteration of voltage-dependence of Shaker potassium channel by mutations in the S4 sequence. *Nature*, **349**, 305–310.

QUINTON, P. M. (1990). Cystic fibrosis: a desease in electrolyte transport. *FASEB J.*, **4**, 2709–2717.

REUTER, H. (1968). Slow inactivation of currents in cardiac Purkinje fibres. *J. Physiol.*, **197**, 233–253.

REUTER, H. (1983). Calcium channel modulation by neurotransmitters, enzymes and drugs. *Nature*, **301**, 569–574.

RIORDAN, J. R., ROMMENS, J. M., KEREM, B., ALON, N., ROZMAHEL, R., GRZELCZAK, Z., ZIELENSKI, J., LÜK, S., PLAVSIC, N., and CHÜU, J. L. (1989). Identification of the cystic fibrosis gene: cloning and characterization of complementary DNA. *Science*, **245**, 1066–1073.

ROSEN, M. R. (1995). Long QT syndrome patients with gene mutations. *Circulation*, **92**, 3373–3375.

SACKMANN, B., NOMA, A., and TRAUTWEIN, W. (1983). Acetylcholine activation of single muscarinic K$^+$ channels in isolated pacemaker cells of the mammalian heart. *Nature*, **303**, 250–253.

SANGUINETTI, M. C., JIANG, C., CURRAN, M. E., and KEATING, M. T. (1995). A mechanistic link between an inherited and an acquired cardiac arrhythmia: HERG encodes the I$_{Kr}$ potassium channel. *Cell*, **81**, 299–307.

SILVA, A. J., STEVENS, C. F., TONEGAWA, S., and WANG, Y. (1992). Deficient hippocampal long-term potentiation in α-calcium-calmodulin kinase II mutant mice. *Science*, **257**, 201–206.

SINGER, S. J., and NICOLSON, G. L. (1972). The fluid mosaic model of the structure of cell membranes. *Science*, **175**, 720–731.

STÜHMER, W., CONTI, F., SUZUKI, H., WANG, X. D., NODA, M., YAHAGI, N., KUBO, H., and NUMA, S. (1989). Structural parts involved in activation and inactivation of the sodium channel. *Nature*, **339**, 597–603.

TRAUTWEIN, W., and HESCHELER, J. (1990). Regulation of cardiac L-type calcium current by phosphorylation and G proteins. *Annu. Rev. Physiol.*, **52**, 257–274.

TSIEN, R. W. (1983). Calcium channels in excitable cell membranes. *Annu. Rev. Physiol.*, **45**, 341–358.

WANG, Q., SHEN, J., SPLAWSKI, I., ATKINSON, D., LI, Z., ROBINSON, J. L., MOSS, A. J., TOWBIN, J. A., and KEATING, M. T. (1995). SCN5A mutations associated with an inherited cardiac arrhythmia, long QT syndrome. *Cell*, **80**, 805–811.

WATSON, J. D., GILMAN, M., WITKOWSKI, J., and ZOLLER, M. *Recombinant DNA*. 2^a ed., New York, Scientific American Books, 1992.

WELSH, M. J. (1990). Abnormal regulation of ion channels in cystic fibrosis epithelia. *FASEB J.*, **4**, 2718–2725.

YELLEN, G., JURMAN, M. E., ABRAMSON, T., and MACKINNON, R. (1991). Mutations affecting internal TEA blockade identify the probable pore-forming region of a K$^+$ channel. *Science*, **251**, 939–942.

General References

ANDERSON, M. P., GREGORY, R. J., THOMPSON, S., SOUZA, D. W., PAUL, S., MULLIGAN, R. C., SMITH, A. E., and WELSH, M. J. (1991). Demonstration that CFTR is a chloride channel by alteration of its anion selectivity. *Science*, **253**, 202–205.

DRUMM, M. L., POPE, H. A., CLIFF, W. H., ROMMENS, J. M., MARVIN, S. A., TSUI, L. C., COLLINS, F. S., FRIZZEL, R. A., and WILSON, J. M. (1990). Correction of the cystic fibrosis defect *in vitro* by retrovirus-mediated gene transfer. *Cell*, **62**, 1227–1233.

KANDEL, E. R., SCHWARTZ, J. H., and JESSEL, T. M. *Principi di neuro-scienze*. 2^a ed., Milano, Casa Editrice Ambrosiana, 1994.

KATZ, B. *Nervi, muscoli e sinapsi*. Bologna, Zanichelli, 1971.

RICH, D. P., ANDERSON, M. P., GREGORY, R. J., CHENG, S. H., PAUL, S., JEFFERSON, D. M., MCCANN, J. D., KLINGER, K. W., SMITH, A. E., and WELSH, M. J. (1990). Expression of cystic fibrosis trans-membrane conductance regulator corrects defective chloride channel regulation in cystic fibrosis airway epithelial cells. *Nature*, **347**, 358–363.

Lucia Vallar

Department of Pharmacology
University of Milan
Milan, Italy

Department for Biological
and Technological Research
San Raffaele Scientific Institute
Milan, Italy

How Cells Respond to Chemical Mediators: Second Messengers

Many chemical substances used for communication between cells (neurotransmitters, hormones, and growth factors) induce a cellular response by activating membrane receptors that associate with certain proteins, G proteins, which in turn are able to regulate enzymes or ion channels. In most cases, the activation of a receptor associated with G proteins causes a change in the intracellular concentration of a small molecule or ion which is able to spread the signal within the cell. In transmission of the message originally carried by the chemical mediator or first messenger, such a small molecule or ion plays the role of an "intracellular messenger" also known as a "second messenger." Often, second messengers operate through a cascade of phosphorylations that changes the activity of specific cellular proteins. Scientific research has allowed the characterization of the actions of multiple intracellular messengers, interconnected in a complex network that guarantees the coordination of cellular responses.

General Response Mechanisms to Chemical Mediators

The many cells that constitute an organism operate with continuous communication between them; in each instant, certain cells send messages and others receive them, interpret them, and then respond with appropriate modifications to their behavior. One of the principal systems that cells adopt to transmit messages is the release into the extracellular space of molecules, such as hormones, neurotransmitters, and growth factors that act as chemical "mediators" or "messengers." In some cases, the chemical mediator enters cells to directly induce processes that lead to a cellular response. The vast majority of chemical mediators are limited to contact with proteins on the external surface of the cell

membrane which act as receptors. Receptors bind with high affinity to hormones, neurotransmitters, or growth factors and then transmit the message from the outside to the inside of the cell in a process defined as signal transduction. Thus, the interaction between a chemical mediator and a receptor triggers one or more intracellular events. Modifications of specific functions follow, as a result of a cascade that generally involves numerous intermediates, to produce a behavioral change in the cell in response to the message received.

Each chemical mediator is recognized exclusively by one receptor or by a small group of similar receptors, programmed in a specific way for this purpose. Given that the molecules that act as chemical mediators are numerous and varied, the same is true of the types of receptors present in an organism. However, in the past decade molecular biological and biochemical studies have revealed that the hundreds of known receptors employ a very limited number of signal transduction mechanisms. According to these mechanisms it is possible to categorize the majority of receptors in three principal classes. In the first class are receptors for neurotransmitters which act at the level of the synapse in electrically excitable cells. These proteins are ion channels which open following the binding of neurotransmitters, producing rapid changes in intracellular ion concentrations and the electric potential of the membrane. The second class includes receptors that, when bound to a chemical mediator, actuate an intrinsic enzymatic function or interact with intracellular enzymes. Often, in this type of signal transduction, which is adopted by the majority of growth factor receptors, the receptor or the enzyme associated with it is a protein kinase that phosphorylates tyrosine residues in a series of cellular proteins. To the third class belong a large group of receptors that do not transmit the signal directly but interact with G proteins, which in turn are able to modify the activity of molecules, the effectors, which generate intracellular events. In this case, the transmission of the message to the interior of the cell requires three distinct components: the receptor, the effector, and the G protein that acts as an intermediary between the two. The effectors regulated by G proteins are both enzymes which catalyze specific biochemical reactions and channels in the plasma membrane that control the concentrations of intracellular ions.

In most cases, the activation of a receptor coupled to G proteins changes the intracellular concentration of a small molecule or an ion that diffuses and conveys inside the cell the message originally carried by a chemical mediator and is therefore called a "second messenger," "intracellular messenger," or "intracellular mediator." Each second messenger acts by triggering a specific cascade of molecular reactions that change the activities of particular proteins to bring about the cellular response. The best characterized second messengers are the cyclic nucleotides of adenine (cAMP) and guanine (cGMP), inositol 1,4,5-trisphosphate (IP_3), diacylglycerol, and the calcium ion (Ca^{2+}). All of these second messengers can be regulated by receptors coupled to G proteins. Many receptors affect the activity of the enzyme that produces cAMP, adenylate cyclase, thereby modulating the intracellular concentration of this second messenger. A large group of receptors associated with G proteins are instead able to activate the phospholipase C enzyme. Phospholipase C simultaneously generates two second messengers: IP_3, which in turn produces an increase in the cytoplasmic concentration of Ca^{2+}, and diacylglycerol. Various G protein-coupled receptors can also induce variations in the cytosolic levels of Ca^{2+} by altering the activity of ion channels in the plasma membrane. Finally, in certain systems G protein-coupled receptors alter intracellular concentrations of cGMP by activating cGMP phosphodiesterase, which hydrolyzes the nucleotide.

It is important to remember that the use of second messengers to induce a cellular response is a typical, but not exclusive, property of G protein-coupled receptors. Intracellular levels of some second messengers are also changed by other types of receptors. In electrically excitable cells, channel proteins that also act as receptors are able to regulate the flux of Ca^{2+} across the plasma membrane and therefore the cytosolic concentration of the ion. Growth factor receptors with tyrosine kinase activity induce their effects through phosphorylation of cellular proteins. However, one such protein is a particular form of phospholipase C. Thus, part of the response to growth factors that bind to receptors with tyrosine kinase activity depends on an increase in intracellular IP_3, Ca^{2+}, and diacylglycerol. Currently, it also seems that G protein-associated receptors may induce intracellular events that do not involve second messengers. In fact, recent studies indicate that some G proteins can activate tyrosine kinases to phosphorylate specific cellular substrates. In this case, a signal induced by a G protein-coupled receptor is similar to those employed by receptors with tyrosine kinase activity.

This article focuses on the roles played by second messengers in the response to chemical mediators that activate G protein-coupled receptors. Consequently, what follows is a description of how the activation of G protein-coupled receptors regulates enzymes and ion channels, how such enzymes alter the concentrations of the principal second messengers, and through what mechanisms these messengers cause important behavioral changes in cells.

G Protein-Coupled Receptors

G protein-coupled receptors are by far the largest family of membrane receptors (Baldwin, 1994; Dohlman *et al.*, 1987). In fact, more than 1000 of these receptors have been identified to date. G protein-coupled receptors are employed by numerous neurotransmitters such as catecholamines, acetylcholine, and glutamate; hormones such as thyroid-stimulating hormone (TSH) and luteinizing hormone (LH); and other chemical mediators. The use of these molecules in the transduction of a signal across the membrane also extends to processes apart from intercellular communication through chemical mediators. For instance, in the rods and cones of the retina, rhodopsins and opsins, which are G protein-coupled receptors, actuate the response to light and color vision. Multiple G protein-coupled receptors located in olfactory neurons are instead involved in the perception of odors. Furthermore, it was discovered that G protein-coupled receptors exposed on the cell surface to detect chemical mediators or sensory stimuli can in some cases be utilized by viruses. In fact, the entrance into a cell of the virus responsible for AIDS requires the presence of particular G protein-coupled receptors that normally recognize chemical mediators known as chemokines. A spe-

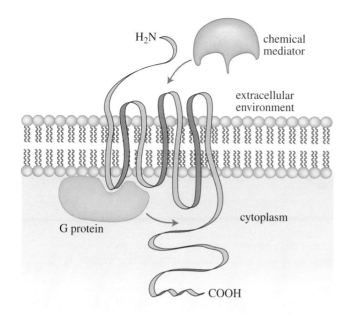

FIGURE 1 Schematic representation of a typical receptor (in purple) associated with a G protein that is activated by binding of a chemical mediator. The receptor is a single polypeptide chain that spans the cell membrane seven times.

cific viral protein was shown to bind to these receptors. It remains to be seen, however, whether the virus uses the receptor molecule only as a docking point on the cell membrane or if it is also able to activate the receptor to transmit a signal like a true mediator.

Receptors that bind to G proteins have a characteristic serpentine shape (Fig. 1). Each receptor is a single polypeptide chain with the amino terminal in the extracellular space, a central part that traverses the plasma membrane seven times, and the carboxy terminus in the cytoplasm (Baldwin, 1994; Dohlman *et al.*, 1987). However, the various receptors differ greatly in their amino acid sequences, especially in the region that contacts the chemical mediator, and in their strategies of interactions with mediators. For instance, in catecholamine receptors the transmembrane regions unite to form a binding pocket for the chemical mediator. In other cases, the extreme amino terminus and other extracellular portions are essential for interaction with the ligand. Binding of the chemical mediator in some way produces a conformational change of the intracellular portion of the receptor that may then activate G proteins. The final two cytoplasmic loops between transmembrane segments and a part of the carboxy-terminal tail of the molecule are thought to be important for this process. Each receptor can react efficiently with only particular types of G proteins. Often, however, a certain chemical mediator "recognizes" several receptors capable of activating G proteins, which in turn alter the activity of diverse effectors. Catecholamines, for instance, can bind to adrenergic receptors type β, which stimulate adenylate cyclase, type α_1, which stimulate phos-

pholipase C, and type α_2, which induce other effects. Therefore, the cellular response to a chemical mediator depends strictly on the types of receptors present on the cell surface.

The G Protein Family

The existence of an intermediate between receptors and intracellular effectors was initially suspected in the 1970s when some experiments on cell membranes revealed that the action of a hormone, glucagon, required the presence of the guanine nucleotide GTP. In the following years it was shown that GTP was utilized by proteins able to bind and subsequently hydrolyze it. Such proteins were called GTP-binding proteins or simply G proteins. For the discovery of the role played by these proteins in coordinating the various activities of the cell through signal transduction, M. Rodbell and A. Gilman received the 1994 Nobel prize in physiology and medicine. G proteins are bound to the internal face of the plasma membrane and are formed of three subunits: the α chain, which binds and hydrolyzes GTP, and the β and γ chains, which closely associate to form a complex (Gilman, 1987; Neer, 1995; Simon *et al.*, 1991). To date, approximately 20 α chains, with different structural and functional properties, and various associations of at least 5 β chains and approximately 10 γ chains have been identified. However, it is not clear if the various α subunits interact with only certain $\beta\gamma$ complexes, giving rise to heterotrimers of a precise composition, or form G proteins indiscriminately with various combinations of β and γ subunits. The different G proteins are therefore distinguished on the basis of the type of α chain present in the heterotrimer. G proteins transmit information from receptors to effector molecules through an elaborate cycle of activation and deactivation (Gilman, 1987; Neer, 1995). When the receptor is not bound to a chemical mediator, the G protein is inactive (Fig. 2). In this situation the binding site for the guanine nucleotides is occupied by GDP and the α subunit remains tightly associated with the $\beta\gamma$ complex. Following hormone or neurotransmitter binding, the receptor interacts with the G protein to induce the dissociation of GDP from the α subunit. GDP is replaced by GTP, which is much more abundant in the cell, and this bond provokes a conformational change in the α subunit which then separates from the $\beta\gamma$ complex. At this point both α-GTP and the $\beta\gamma$ chains may interact with enzymes or ion channels to modify their activity. Stimulation or inhibition of the effector ends when the intrinsic GTPase function of the α subunit transforms GTP to GDP and the α-GDP chain reassociates with the $\beta\gamma$ complex to form an inactive heterotrimer.

The Multiple Effects of G Protein Subunits

The role of the α subunit in the action of G proteins is the best characterized since it was defined first. The various α subunits currently known have a similar structure. In par-

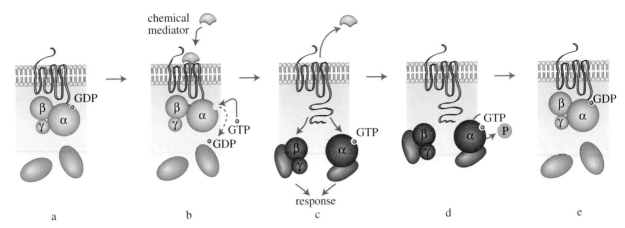

FIGURE 2 G proteins, constituted of α, β, and γ subunits, transfer the signal from receptors to effectors through a complex cycle. (a) In resting conditions, the G protein is an inactive heterotrimer in which the α subunit is bound to GDP. (b) Following the interaction between the receptor (in purple) and the chemical mediator (in blue), the receptor interacts with the G protein and GDP is replaced by GTP. (c) The α and $\beta\gamma$ subunits dissociate and regulate the activity of effectors (in brown). (d) The signal ceases when the α chain hydrolyzes GTP. (e) The α chain reassociates with the $\beta\gamma$ complex to reform the inactive heterotrimer. For simplicity, the diagram does not accurately depict the position of G proteins and effectors in the cell. In reality, these molecules are found in or associated with the cell membrane.

ticular, they possess short amino acid sequences that have been highly conserved during evolution and unite in the three-dimensional structures of the protein to create a binding site for the guanine nucleotides. It is interesting that these sequences are present in many other cellular proteins that act as GTPases to perform various functions, such as the control of proliferation and mechanisms of intracellular transport. In addition, all α subunits of G proteins possess particular regions that permit interaction with receptors and effectors. These regions differ in the various types

of α subunits such that each α chain couples with one specific group of receptors and equally specific effectors. Of the various zones through which the α subunit contacts the intracellular part of the receptor, one of the most important is the carboxy terminus. In fact, it has been observed that the modification of a few of these residues suffices to completely switch the preferred type of receptor. The various α subunits have been classified into four principal groups (Table 1) on the basis of amino acid similarities which correspond to functional similarities (Neer, 1995; Simon *et al.*,

TABLE I			
Characteristics of the G Protein Subunits[a]			
SUBUNIT	**SOME MEMBERS**	**SOME EFFECTS**	**SENSITIVITY TO BACTERIAL TOXINS**
α			
α_s	α_s, α_{olf}	Stimulation of adenylate cyclase	Cholera toxin
α_i	α_{i1}, α_{i2}, α_{i3}, α_o, α_t	Inhibition of adenylate cyclase	Pertussis toxin
		Activation of K^+ channels	
		Stimulation of cGMP phosphodiesterase	
α_q	α_q, α_{11}, α_{14}, α_{16}	Stimulation of phospholipase C	
α_{12}	α_{12}, α_{13}		
$\beta\gamma$		Regulation of adenylate cyclase	
		Stimulation of phospholipase C	
		Activation of K^+ channels	
		Inhibition of Ca^{+2} channels	

[a]Only in some cases are the names of the various subunits particularly meaningful. The α_s and α_i subunits were so named due to the "stimulatory" or "inhibitory" effect on adenylate cyclase. The other subunits identified thereafter were named using arbitrary letters or numbers in a series.

1991). The first group includes the α_s and α_{olf} subunits. The former is expressed ubiquitously and couples the stimulation of adenylate cyclase to specific receptors including the receptor for TSH and the β-adrenergic receptor for catecholamines. Instead, the α_{olf} protein mediates adenylate cyclase activation by olfactory stimuli in specialized neurons. The second group of α chains includes α_i and α_o proteins. These subunits interact with diverse receptors for neurotransmitters and peptides, such as α_2-adrenergic receptors, and probably have multiple effects, including inhibition of adenylate cyclase and activation of K^+ channels. α_i chains are abundant in many tissues, whereas the α_o subunit is mainly expressed in neurons. Transducin, α_t, belongs to the same class of α subunits and couples rhodopsin to phosphodiesterases that degrade cGMP in the rod cells of the retina. The third group of α subunits that has α_q as a prototype is involved in the stimulation of phospholipase C by specific receptors, such as the α_1-adrenergic receptors for catecholamines. Finally, α_{12} and α_{13} constitute another class of α chains, the effects of which have not been completely classified.

An interesting characteristic of some α subunits is the sensitivity to the actions of bacterial toxins. For example, cholera toxin modifies a residue of the α_s subunit that is critical for the hydrolysis of GTP. The functional consequence of this effect is the inhibition of the GTPase activity of the protein which remains in the active form and persistently stimulates adenylate cyclase. In contrast, the α_i and α_o subunits are modified by pertussis toxin at a specific amino acid in the carboxy terminus. This interferes with the activation of the G proteins by receptors and thereby interrupts signal transduction.

The scheme in which both α-GTP and the $\beta\gamma$ complex become active components in the transduction of a signal following receptor-induced dissociation of the heterotrimers is relatively recent. It was long thought that only the α subunits could regulate effector molecules in G protein-mediated processes, whereas the $\beta\gamma$ complex assumed a regulatory role with respect to the α subunit. However, results from various laboratories have clearly shown that the $\beta\gamma$ complex can interact directly with specific enzyme and ion channels (Neer, 1995). $\beta\gamma$ subunits control, for example, the activity of K^+ and Ca^{2+} channels, regulate adenylate cyclase, and stimulate phospholipase C. Many effectors are therefore regulated by both α and $\beta\gamma$ subunits of the G protein hetertrimers. Of considerable interest is the finding that the $\beta\gamma$ complex can also activate cellular tyrosine kinases (Bourne, 1995). By this mechanism chemical mediators that activate G protein-coupled receptors can play an important role in the control of cellular growth. In fact, stimulation of tyrosine kinases by $\beta\gamma$ complexes has been observed to initiate a cascade of cellular reactions similar to that induced by tyrosine kinases receptors for classical growth factors. The demonstration of signal transduction by $\beta\gamma$ complexes is no longer disputed, but there are asso-

ciated questions for which there are not answers. For example, the selectivity of the interaction between various $\beta\gamma$ complexes and α subunits or receptors is uncertain. On the other hand, substantial differences in the capability of various $\beta\gamma$ complexes to regulate effectors have not been demonstrated. How might the $\beta\gamma$ subunits guarantee specificity of the receptor signal? It seems reasonable to hypothesize that functional differences between the different $\beta\gamma$ complexes will be discovered soon. In addition, other factors may be important. Various types of experiments have demonstrated that the concentrations of $\beta\gamma$ subunits required to regulate an enzyme or an ion channel are very much higher than those necessary in the case of α-GTP. Therefore, it is possible that only the activation of particularly abundant heterotrimers in the cell, such as those containing α_i chains, may generate a concentration of $\beta\gamma$ complexes sufficient to induce a signal. In effect, the majority of receptor signals attributed to $\beta\gamma$ complexes identified thus far can be blocked by pertussis toxin and therefore involve G proteins of the G_i and G_o class.

Characteristics of G Protein-Mediated Responses

One may ask why cells developed a system as complicated as that of G proteins in order to respond to extracellular signals. Would it not be simpler for a cell to use as receptors the enzymes and ion channels? Actually, the use of G proteins as intermediates in cellular responses confers numerous advantages. First, the transmission of information from a receptor to G proteins ensures a noteworthy amplification of the signal reaching the interior of the cell. A single receptor molecule can activate many G proteins. For example, it has been calculated that a molecule of rhodopsin in the retina generates, in less than 1 s, hundreds of α_t-GTP molecules capable of stimulating effector enzymes. Furthermore, the use of G proteins enables a single receptor to initiate numerous, distinct intracellular signals (Neer, 1995). In many cases, a receptor activates not just a single G protein but also various types of heterotrimers. This generates different active α chains, which often are able to regulate more than one effector, and $\beta\gamma$ complexes that may stimulate or inhibit other enzymes or ion channels (Fig. 3). For example, some receptors for mediators that act as growth factors in fibroblasts, such as those for thrombin, can activate both G_q and G_i proteins. The activation of these receptors induces stimulation of phospholipase C by α_q and other effects mediated by the subunits liberated from the G_i proteins. The α_i chains inhibit adenylate cyclase, whereas the $\beta\gamma$ subunits are able to stimulate phospholipase C and activate cellular tyrosine kinases. This produces simultaneously a reduction of the intracellular level of cAMP concentrations, increases of the concentrations of Ca^{2+} and diacylglycerol, and the activation of tyrosine kinases. All these events contribute in a coordinated way to the stimulation of cell growth. The system of G proteins is also ex-

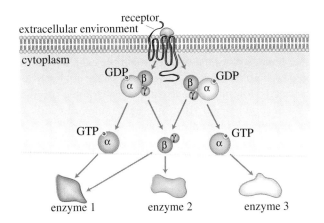

FIGURE 3 A single receptor (in purple) associated with G proteins is able to activate multiple intracellular signals. The receptor can interact with various types of G proteins; the different α subunits (in red and yellow) and the βγ complexes can regulate diverse effector enzymes.

tremely versatile and enables the adjustment of the response to the specific needs of a particular cell type. Not all cells possess the same ensemble of G proteins, enzymes, and ion channels. In addition, many variants of a certain enzyme or ion channel often exist. Thus, identical receptors or G proteins may induce different responses in different cells. For example, in neurons the principal effectors of G_i proteins, besides adenylate cyclase, are ion channels that effect a reduction of the intracellular Ca^{2+} concentration. Thus, in these cells the effects produced by receptors coupled to G_i are profoundly different from those observed in fibroblasts.

The extracellular stimuli that reach a cell change constantly. In order for a cell to adapt to changes, the intracellular signals produced by certain receptors must be of limited duration and not permanent. The mechanisms that guarantee the transient nature of the receptor signal act at many stages. In the case of G proteins, the duration of the stimulation of the effector was initially thought to depend exclusively on the characteristics of the intrinsic GTPase activity of the α subunit. The α chain was believed to function as a timer that automatically interrupted the signal by hydrolyzing GTP after a few seconds of interaction with the effector. However, biochemical studies have shown that in some cases the cellular response decayed faster than the rate at which the involved α subunit degraded GTP. One of these cases was the response of retinal cells to light. Subsequently, some effectors, such as the cGMP phosphodiesterase activated by transducin in the retina, were discovered to act as GTPase-activating proteins (GAPs) which increase the rate of hydrolysis of GTP by the α subunit. Recently, an entire family of proteins called regulators of G protein signaling (RGS) has been identified (Dohlman and Thorner, 1997). RGS proteins act as GAPs on various α subunits and therefore may play important roles in con-

trolling the duration of the cellular response. The fact that in certain conditions the cell may increase the content of RGS and by doing so greatly reduce the effects of receptor activation suggests that these proteins are involved in the modulation of responses to extracellular stimuli.

The Role of Receptors and G Proteins in Human Disease

As a result of an increased understanding of receptors and the associated G proteins, many researchers suspected a role for these molecules in the development of some human diseases. This hypothesis has been proven correct in many cases. Today we know that defects in the normal function of "serpentine" receptors and specific G protein subunits are causes of some pathologies (Spiegel, 1996). Mutations of G protein-coupled receptors are present in various congenital maladies and have oncogenic roles in certain tumors (Table 2). In some cases the mutations impede synthesis, localization in the plasma membrane, or the normal function of the molecule to give a phenotype characterized by a loss of receptor function. Other mutations instead cause increased activity of the receptor. In such cases the receptor molecule maintains an active conformation even in the absence of the chemical mediator and, through interaction with G proteins, induces a persistent intracellular signal. Mutations identified in human diseases involve various types of receptors. For example, in nephrogenic diabetes insipidus diverse mutations of the V_2-type receptor for vasopressin have been identified. In this case the loss of function of the receptor, that through the G_s protein and stimulation of adenylate cyclase increases reabsorption of water in the kidney, leads to the excessive dilution of urine that characterizes the disease. Many cases of retinitis pigmentosa, a disease characterized by degeneration of retinal photoreceptors and progressive blindness, are instead due to mutations in rhodopsin in the rod cells of the retina. Often, the mutation produces a rhodopsin that has an anomalous structure and accumulates intracellularly, provoking the degenerative process. Clear examples of pathological conditions caused by mutations that produce constitutive activation of a receptor are endocrine diseases stemming from abnormal function of receptors for TSH and LH. Somatic mutations that activate the TSH receptor are present in numerous hyperactive thyroid tumors. In the thyroid, the receptor for TSH stimulates both cell growth and hormone production through activation of G_s and adenylate cyclase. The constitutive activity of the receptor induced by the mutation therefore provokes uncontrolled proliferation of cells that release an excessive quantity of thyroid hormone. The receptor for LH, which is also linked to G_s and adenylate cyclase, induces the production of testosterone by the testicular Leydig cells. Mutations that cause an LH-independent activation of the receptor are at the basis of familial forms of male precocious puberty.

TABLE 2

Human Diseases Associated with Mutations in Receptors or G Proteins

RECEPTOR	EFFECT OF THE MUTATION	DISEASE
Rhodopsin	Loss of function	Retinitis pigmentosa
Opsins	Loss of function	Color blindness
Vasopressin receptor V_2	Loss of function	Nephrogenic diabetes insipidus
Thyroid-stimulating hormone receptor	Loss of function	Familial hypothyroidism
	Constitutive activation	Hyperactive thyroid adenomas
Luteinizing hormone receptor	Constitutive activation	Familial male precocious puberty
G protein α Subunits		
α_s	Loss of function	Albright's hereditary osteodystrophy
	Constitutive activation	Pituitary adenomas
		Hyperactive thyroid adenomas
		McCume–Albright syndrome
α_{i2}	Constitutive activation	Ovarian and adrenal tumors

In some pathologies, despite normal receptor function, signal transduction is altered by defective or abnormal function of G proteins (Table 2). It has long been known that activation of α_s by the toxin produced by the bacteria responsible for cholera is linked to the characteristic symptoms of this infection. The cholera toxin penetrates intestinal cells in which persistent stimulation of the α_s-adenylate cyclase signal causes excessive loss of electrolytes and water. The presence of mutations of the α subunits of G proteins has recently been discovered in some human diseases. Albright's hereditary osteodystrophy or type 1a pseudohypoparathyroidism, a disease characterized by the resistance to parathyroid hormone and other hormones that act through receptors coupled to G_s and adenylate cyclase, is due to mutations that cause a defect in the synthesis or function of α_s and impede normal cellular responses. In other cases, mutations that inhibit the GTPase activity of the α chains provoke constitutive activation of G_s or G_{i2}. Mutations that activate α_s were shown in growth hormone-secreting pituitary tumors, some thyroid tumors, and in McCune–Albright's syndrome, a disease characterized by hyperplasia or tumors of the pituitary, thyroid, and adrenal glands. Analogous mutations of α_{i2} seem to be present, although rarely, in ovarian or adrenal tumors. As in mutations that affect the TSH receptor, constitutive activation of α_s causes uncontrolled growth and hyperactivity of pituitary and thyroid cells through persistent activation of adenylate cyclase and the consequent accumulation of cAMP. However, the precise mechanism involved in the oncogenic action of α_{i2} remains to be clarified.

The Adenylate Cyclase Enzyme and the Second Messenger cAMP

The Regulation of Adenylate Cyclase

Numerous hormones and neurotransmitters that act through G protein-coupled receptors regulate the activity of adenylate cyclase, the enzyme which produces cAMP from ATP (Sunahara *et al.*, 1996). Many receptors cause a stimulation of adenylate cyclase by association with G_s and release of α_s-GTP which can activate the enzyme. Instead, receptors associated with G_i and G_o proteins inhibit adenylate cyclase through the α and in some cases the $\beta\gamma$ subunits of these G proteins (Fig. 4). More complex regulatory mechanisms also exist. In certain cells the activation of the G_i and G_o proteins does not inhibit adenylate cyclase; rather, it enhances the stimulatory effect of receptors coupled to G_s. In this situation $\beta\gamma$ complexes freed from G_i or G_o stimulate the enzyme in concert with activated α_s. The potential for such multiple and apparently contradictory mechanisms for the regulation of adenylate cyclase by G protein-coupled receptors is due to the existence of a large family of variants of the enzyme.

To date, nine different forms of adenylate cyclase have been identified. All are integral membrane proteins with similar overall structures (Sunahara *et al.*, 1996). These include a short, cytoplasmic amino terminus followed by six regions that traverse the plasma membrane, a large cytoplasmic portion, six more transmembrane regions, and a long cytoplasmic carboxy terminus (Fig. 5). Among the

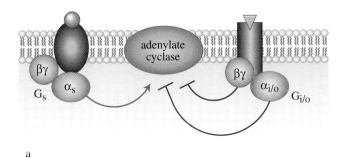

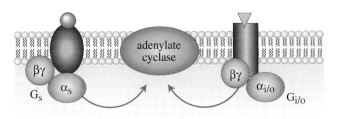

FIGURE 4 Regulation of adenylate cyclase, an integral membrane protein. (a) In many systems the receptors (in purple) coupled to G_s stimulate the enzyme (green arrow) through the α_s subunit and receptors coupled to G_i and G_o have an inhibitory effect (red arrow) via the $\alpha_i, \alpha_o,$ or $\beta\gamma$ subunits. (b) In other cases, receptors associated with G_i and G_o are able to enhance the stimulatory effect of α_s.

different forms of the enzyme, the regions that present the greatest homology are cytoplasmic and are fundamental for the catalytic activity of the protein. The nine adenylate cyclases are distributed differently in the tissues of an organism. Some enzymatic forms are expressed in many cells,

extracellular environment

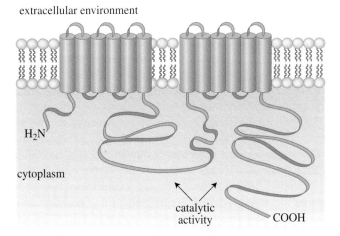

FIGURE 5 Schematic representation of the complex structure of adenylate cyclase. The two large cytoplasmic regions of the protein are fundamental for its enzymatic activity.

whereas others are synthesized exclusively or preferentially in neurons. The diverse forms of adenylate cyclase also differ with respect to regulation by G protein subunits. The α_s chain is able to stimulate all types of adenylate cyclase, but the sensitivity of the various forms to the α_i and α_o subunits and $\beta\gamma$ complexes is extremely variable. For example, α_i subunits inhibit adenylate cyclases type I, V, and VI but have no effect on the type II enzyme, whereas α_o reduces the activity of type I adenylate cyclase but not that of types II, V, and VI. On the other hand, the $\beta\gamma$ complexes can inhibit some types of enzymes, such as type I, and stimulate others, such as type II, together with α_s subunits. This apparent confusion actually guarantees a coherent and exclusive role for the various enzymes in the signal transduction process. For example, one may imagine the consequences of activation of a receptor coupled to G_i and G_o in cells that express either type I or type II adenylate cyclase. In the former, all the subunits freed from G_i and G_o proteins inhibit the enzyme, and so the effect of the receptor would be a reduction of cAMP production. In the latter, the α_i and α_o subunits have no effect and $\beta\gamma$ complexes stimulate the enzyme if simultaneously activated by α_s. The receptor therefore would not have an inhibitory action but would potentiate the stimulation of adenylate cyclase by G_s-coupled receptors.

cAMP and Protein Kinase A

The regulation of the activity of adenylate cyclase effects marked and rapid changes of the intracellular concentration of cAMP. In the late 1950s, cAMP was identified as a second messenger. Since then, many extremely diverse cellular processes have been discovered to be regulated by intracellular levels of cAMP: the increase in the contractile rate and force of the heart induced in conditions of stress by adrenaline through β-adrenergic receptors; metabolic reactions such as the degradation of glycogen in the liver and triglycerides in adipose cells; control of the activity of neurons; specialized functions of the kidney; synthesis and secretion of thyroid hormones and the production of steroid hormones by the adrenal glands, the ovaries, and the testicles; stimulation of the proliferation of particular cell types, as previously described, and the arrest of the growth of others; finally, the capacity to perceive odors.

cAMP induces the majority of these responses through activation of a specific protein kinase, protein kinase A, that in turn causes the phosphorylation of serine and threonine residues in particular proteins (Taylor *et al.*, 1990). Most cells express two forms of protein kinase A that have similar structures and functions but localize differently within cells. Whereas type I protein kinase A is cytosolic, type II associates with the cell membrane. Both enzymes are composed of two regulatory subunits and two catalytic subunits, denoted R and C, respectively. When cAMP occupies the two binding sites present on each R subunit, the R dimer undergoes a conformational change and releases active C subunits (Fig. 6). The C subunits diffuse freely throughout

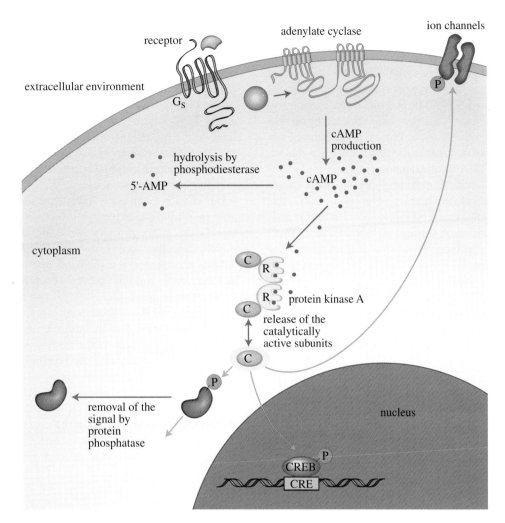

FIGURE 6 Schematic diagram of cAMP action. Adenylate cyclase activated by the G_s protein stimulates the production of cAMP (in green). cAMP binds to the regulatory subunits (R) of protein kinase A causing the detachment of the active catalytic subunits (C), which spread throughout the cytoplasm and the nucleus. The C subunits can phosphorylate (blue arrows) enzymes, ion channels, and the CREB protein that activates the transcription of specific genes by binding to the regulatory sequence CRE. The signal is removed by various phosphodiesterases (which hydrolyze cAMP to 5' AMP) and by protein phosphatases (that dephosphorylate the substrates).

the cytoplasm and the nucleus, in which they phosphorylate specific cellular substrates. Through the phosphorylation mechanism, protein kinase A can modify the activity of enzymes, ion channels, and other cellular proteins. In the liver, for example, it activates a kinase, phosphorylase kinase, which in turn phosphorylates and activates the enzyme that induces the degradation of glycogen. By phosphorylating transcription factors, protein kinase A is also able to alter the expression of some proteins. The regulatory regions of certain genes contain a specific sequence called the cAMP response element (CRE). The regulatory cAMP response element-binding protein (CREB) binds to this sequence and, when phosphorylated by the C subunit of protein kinase A, activates the transcription of the gene. This relatively simple protein kinase A system can mediate the wide

variety of responses induced by cAMP. For example, the phosphorylase kinase responsible for the degradation of glycogen is abundant in the liver, whereas the lipase that, when phosphorylated by protein kinase A, degrades triglycerides is abundant in adipocytes. Responses to cAMP that are independent of protein kinase A also exist. The best characterized is that of receptors present in olfactory neurons, in which cAMP produced by a stimulus directly opens cationic channels in the membrane to trigger depolarization of the cell and initiate a nervous impulse to the brain.

The cAMP system, like those of other second messengers, guarantees amplification of the signal originating from the receptor–G protein complex. Each molecule of adenylate cyclase activated by G proteins in fact produces many molecules of cAMP, and each molecule of protein kinase A recruited by cAMP can phosphorylate numerous copies of a substrate protein. Once the cellular response is induced, the cAMP–protein kinase A signal is removed by a series of specific enzymes. Numerous phosphodiesterases hydrolyze cAMP to produce 5′-AMP, and substrates phosphorylated by protein kinase A are dephosphorylated by diverse groups of protein phosphatases. Protein phosphatase I has a particularly important role in the control of the signal induced by cAMP. This enzyme dephosphorylates many substrates of protein kinase A, including phosphorylase kinase and CREB. However, the activity of protein phosphorylase I is counteracted by an inhibitory protein that is activated by protein kinase A through phosphorylation. Thus, protein kinase A protects its substrates from premature dephosphorylation.

The Signaling System of Phospholipase C

The role of the reactions induced by phospholipase C in signal transduction was first suggested in 1953 and well characterized in recent years. Phospholipase C (Fig. 7) catalyzes the hydrolysis of particular phospholipids, phosphoinositides (Lee and Rhee, 1995). Phosphatidylinositol (PI) is produced in the cell membrane and transformed into phosphatidylinositol-4-phosphate (PIP) and then into phosphatidylinositol-4,5-bisphosphate (PIP_2) by kinases that spe-

cifically act on inositol rings. PIP_2 is the least abundant of these phospholipids and is the preferred substrate of phospholipase C. The enzyme cleaves PIP_2 to produce two intracellular messengers, IP_3 and diacylglycerol. IP_3, a hydrophilic molecule which is free to diffuse in the cytoplasm, can in turn provoke the release of Ca^{2+} from intracellular stores. Diacylglycerol, which is lipidic, instead remains confined to the plasma membrane where it activates a specific protein kinase, protein kinase C. Phospholipase C activity can be stimulated by both G protein-coupled receptors and receptors with tyrosine kinase activity. Other receptors associated with G proteins have been identified which are able to inhibit the enzyme. However, it is not clear if this effect is one of the primary events induced by the receptor nor what mechanism is involved. The enzymatic cascade involving phospholipase C mediates many cellular responses induced by G protein-coupled receptors. By the activation of this enzyme, the α_1-adrenergic receptors for catecholamines and the angiotensin receptors trigger vascular smooth muscle contractions and consequent increases in arterial pressure. Receptors for acetylcholine work through phospholipase C-mediated mechanisms to increase secretion of pancreatic enzymes. Through a similar route receptors for chemotactic peptides activate leukocytes in inflammatory processes and neuropeptide receptors induce hormonal secretion. Finally, in the case of some G protein-associated receptors and some growth factor receptors with tyrosine kinase activity, the generation of IP_3 and diacyl-

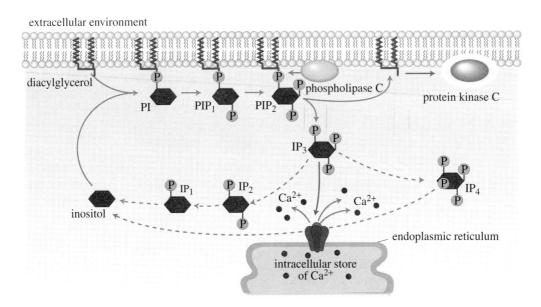

FIGURE 7 Schematic diagram of the phosphoinositide system. In the cell membrane two successive phosphorylations form PIP_2 from PI. Phospholipase C specifically cleaves PIP_2 to generate IP_3 and diacylglycerol. IP_3 binds to a receptor that is also an ion channel (in brown) to induce the release of Ca^{2+} from intracellular stores, whereas diacylglycerol activates protein kinase C. The diagram also illustrates (dashed arrows) how IP_3 can produce another potential intracellular messenger, IP_4, or be transformed into inositol, which the cell recycles for the synthesis of PI.

glycerol acts in conjunction with other intracellular signals to stimulate proliferation.

That which until recently has been called phospholipase C is actually a numerous family of enzymes that have common characteristics (Lee and Rhee, 1995). All the phospholipase C enzymes have a partially or predominantly cytoplasmic localization, but under the appropriate conditions they may associate with the plasma membrane. All are characterized in their amino acid sequences by two highly homologous, well-separated regions referred to as X and Y in which the catalytic activity of the protein resides. Finally, all have more or less highly Ca^{2+}-dependent activity due to a Ca^{2+}-specific binding site. Aside from these similarities, the various enzymes differ notably in size, structure, and characteristics of regulation and therefore have been divided into three principal classes: β, γ, and δ. The role of phospholipase C-δ in intracellular signaling has not been completely clarified. Instead it is known that phospholipases C-γ are the enzymes that become activated through phosphorylation by receptors for growth factors (Fig. 8). Phospholipases C-γ are in fact characterized by a particular sequence, located in the X and Y regions, that enables the interaction with the cytoplasmic portion with tyrosine kinase activity of this type of receptor. Thus, the enzyme becomes activated by phosphorylation of specific tyrosine residues. Finally, the set of type β phospholipase C enzymes constitute the family that is regulated by G protein-associated receptors. Phospholipases C-β possess a long carboxy-terminal portion that is believed to interact with the α subunit of G proteins. In some cases, particular sequences of the amino terminus of the molecule allow the enzyme to "communicate" with the $\beta\gamma$ complex. Four forms of phospholipase C-β have been identified which are differentially expressed in tissues and regulated in different ways by G proteins.

G protein-coupled receptors can stimulate the activity of phospholipase C-β by two mechanisms (Lee and Rhee, 1995). The first, utilized for example by α_1-adrenegic receptors and angiotensin receptors in smooth muscle, involves the activation of G_q proteins. The α-GTP chains released by these G proteins activate, with variable efficiency, all the type β enzymes and may therefore act in various types of cells. The second route of stimulation of phospholipase C-β is mediated by $\beta\gamma$ subunits, which in the majority of cases originate from G_i proteins. However, the $\beta\gamma$ complex is able to activate the β_2 and β_3 but not the β_1 and β_4 enzymes. This route of signal transduction is therefore only operable in cells that express the appropriate forms of the enzyme. For example, G_i-coupled receptors for chemotactic peptides stimulate the activity of phospholipase C in leukocytes which are extremely rich in phospholipase C-β_2.

The Action of IP$_3$

IP$_3$ liberated in the cytoplasm by the action of phospholipase C induces the release of Ca^{2+} from specialized por-

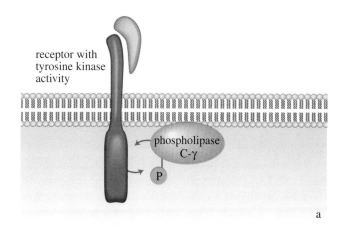

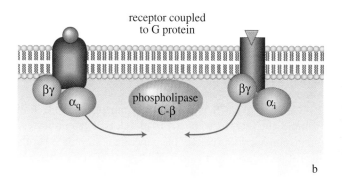

FIGURE 8 Schematic representation of the activity of phospholipase C. (a) Type γ phospholipases C interact with receptors with tyrosine kinase activity and become activated by phosphorylation of specific tyrosine residues. (b) The activity of type β phospholipases C is instead stimulated by G protein-coupled receptors. These enzymes are activated by α_q subunits and, in some cases, by $\beta\gamma$ complexes that originate from G_i proteins.

tions of the endoplasmic reticulum that function as reservoirs of the ion (Berridge, 1993; Fig. 7). IP$_3$ binds specifically to a receptor channel on these intracellular membranes which is constructed of four identical subunits. Each subunit has a large cytoplasmic portion that contains the binding site for IP$_3$ and several transmembrane segments which form a channel that allows the exit of Ca^{2+}. The interaction with IP$_3$ triggers the opening of the channel, thereby causing a rapid increase in the concentration of Ca^{2+} in the cytosol. The effect of IP$_3$ on the ion channel is strongly influenced by the calcium ion. It is believed that in the initial phase of the response, low levels of Ca^{2+} in the cytoplasm potentiate the effect of IP$_3$. Next, the high cytoplasmic concentration of Ca^{2+} that is created in the cell inhibits the further release of Ca^{2+} through the channel. In this way, Ca^{2+} release from the internal reservoirs is disactivated even in the presence of IP$_3$. The IP$_3$ produced by the stimulation of the receptor is in turn removed from the cytoplasm by a series of enzymatic steps. Through progressive elimination of phosphates on the inositol ring, various phosphatases transform IP_3 first into IP_2, then into IP_1, and finally into inositol that may be used for the synthesis of the phospholipid PI. IP_3 may also be transformed into IP_4 if a kinase adds a phophate to the inositol ring of the molecule. Some experiments provide evidence that this derivative of IP_3 is also an intracellular messenger with specific effects.

Diacylglycerol

The activation of phospholipase C generates, simultaneously with IP_3, the intracellular messenger diacylglycerol. The most important mechanism through which diacylglycerol exerts its effects is the activation of a protein kinase called protein kinase C (Jaken, 1996). The effects of protein kinase C have been extensively studied by employing vegetable compounds, phorbol esters, that directly bind to and stimulate the enzyme. Thus, it has been possible to determine that this enzyme is involved in the control of multiple cellular processes, such as neuronal activity, secretion, and cellular growth and differentiation. In many of these processes protein kinase C collaborates efficiently with the IP_3–Ca^{2+} signal which is activated by phospholipase C. For example, the secretory activity of platelets, T lymphocytes, and insulin-secreting cells of the pancreas is stimulated both by protein kinase C activators and by Ca^{2+} but is enhanced greatly by the combination of the two signals. Some longterm responses mediated by protein kinase C, such as cellular growth and differentiation, require the persistent activation of the enzyme. After diacylglycerol generated by phospholipase C becomes rapidly metabolized by specific enzymes, the cell may produce a new wave of second messenger to follow the initial response. This is achieved by the hydrolysis of an abundant membrane phospholipid, phosphatidylcholine, by the phospholipase D enzyme which is activated by mechanisms that have not been completely clarified (Fig. 9). Certainly, it is not surprising that the protein kinase C enzyme is also actually a group of variants with diverse tissue distributions and modes of operation (Jaken, 1996). In general, protein kinases C are cytoplasmic, singlepolypeptide chains that associate with the plasma membrane under the appropriate conditions. The activation of the various types of enzymes requires, in addition to diacylglycerol, the specific membrane phospholipid phosphatidylserine and, at least in some cases, high Ca^{2+} concentrations. Once activated, protein kinases C cause the phosphorylation of serine and threonine of specific protein substrates present in various types of cells, and by this mechanism they may regulate the activity of enzymes or ion channels or control the expression of specific genes.

In addition to diacylglycerol, cells utilize other intracellular messengers of a lipidic nature (Liscovitch and Cantley, 1994). One of these is arachidonic acid (Fig. 9), which is able to generate various metabolites. Arachidonic acid is produced by the action of an enzyme, phospholipase A_2,

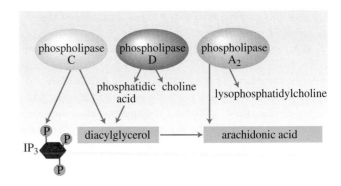

FIGURE 9 Some phospholipases involved in the generation of second messengers of a lipidic nature. Both phospholipase C and phospholipase D can cause an increase of the production of diacylglycerol. The reaction catalyzed by phospholipase D generates choline and phosphatidic acid. Dephosphorylation of the latter gives diacylglycerol. Arachidonic acid, together with lysophosphatidyl choline, is produced instead by the hydrolysis of membrane phospholipids by phospholipase A_2 but may also be derived from diacylglycerol by the action of a lipase.

which hydrolyzes specific membrane phospholipids. G protein-coupled receptors can activate phospholipase A_2 indirectly through an increase of the intracellular Ca^{2+} concentration or by undefined direct mechanisms. Through the action of a lipase, the cell can even generate arachidonic acid from diacylglycerol produced as a result of the activation of phospholipases C and D. Often, the compounds derived from arachidonic acid are used by cells as chemical mediators that permit communication with neighboring cells. However, arachidonic acid and its metabolites can also act as intracellular messengers involved, for instance, in the control of neuronal activity. Recent studies indicate that the products of the hydrolysis of membrane sphingolipids, such as ceramide and some derivatives of sphingosine, and PIP_3, originating from the phosphorylation of PIP_2, also can or could play roles as second messengers.

The Role of Ca^{2+} as an Intracellular Messenger

Mechanisms of Regulation of the Cytosolic Concentration of Ca^{2+}

The intracellular role of Ca^{2+} was first demonstrated in the 1940s through experiments that revealed that small quantities of the ion were able to induce contractions of muscle fibers. Today, it is well-known that Ca^{2+} is the most widely used second messenger among the various types of cells in an organism. Modification of the intracellular Ca^{2+} concentrations, in addition to inducing contractions

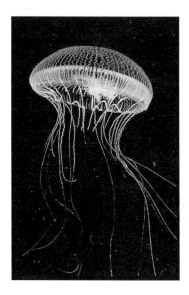
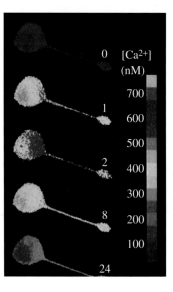

FIGURE 10 Natural proteins, such as aequorin, and synthetic compounds, such as Fura-2, emit characteristic fluorescent signals that depend on the binding of Ca^{2+}. After introduction into cells, they permit the measurement of the concentration of free Ca^{2+} in the cytoplasm or in intracellular organelles. (a) The jellyfish *Aequorea victoria*, which produces aequorin. (b) An example of the rapid increase of the cytoplasmic concentration of Ca^{2+} that one may observe following stimulation with bradykinin in a pheochromocytoma PC12 cell that contains Fura-2. The concentration of Ca^{2+} is color encoded. The numbers beside the cells indicate the time elapsed (seconds) following the application of the stimulus (modified from Pozzan, 1992, and from Grohovaz *et al.*, 1991).

in skeletal muscles and the heart, in fact regulates many cellular processes, such as neuronal activity, the secretion of hormones and other chemical mediators, and proliferation. The total amount of Ca^{2+} inside a cell is very high. The ion is found primarily bound to cytoplasmic proteins or sequestered inside intracellular organelles, such as the endoplasmic reticulum and mitochondria, which function as intracellular reservoirs. Therefore, only a small fraction of the total Ca^{2+} is free in the cytosol, and it is to this small fraction that one refers when speaking of Ca^{2+} as an intracellular messenger. The work of several researchers has led to the development of experimental techniques to permit the detection of free Ca^{2+} inside a cell (Fig. 10). These techniques, which utilize proteins or compounds that emit particular fluorescent signals when bound to Ca^{2+}, have contributed greatly to the clarification of the mechanisms that control the cytosolic concentration of the ion. In each instant of the life of a cell, the cytosolic concentrations of free Ca^{2+} result from a complex equilibrium among various factors (Pietrobon *et al.*, 1990; Pozzan *et al.*, 1994; Tsien and Tsien, 1990). On the plasma membrane there exist various types of ion channels that conduct Ca^{2+} (Fig. 11). Since the concentration of the ion in the extracellular environment ($\sim 10^{-3}$ M) is much higher than that of the cytosol, Ca^{2+} flows into the cell through such channels. Channels through which the ion may reach the cytosol are also found in the membrane of cellular organelles that accumulate noteworthy quantities of Ca^{2+}. The most important intracellular channels for Ca^{2+} that have been characterized are the receptors for IP_3 in the endoplasmic reticulum and channels that bind ryanodine, a vegetable compound (Berridge, 1993; Pozzan *et al.*, 1994). The latter operate in the sarcoplasmic reticulum of muscle tissue and in the endoplasmic reticulum of various cell types.

The action of systems that tend to increase the cytosolic concentration of free Ca^{2+}, however, is counteracted by other mechanisms. In fact, the plasma membrane has pumps and exchangers that transport Ca^{2+} from the cytosol to the extracellular fluid. Other specific pumps and transporters in intracellular membranes instead withdraw Ca^{2+} from the cytosol into storage organelles. Finally, there exist cytoplasmic proteins that bind Ca^{2+} and reduce the amount of the free ion. In the absence of extracellular stimulation, these latter mechanisms enable the cell to maintain very low concentrations of free Ca^{2+} ($\sim 10^{-7}$ M). Generally, the activation of a receptor alters this situation by increasing the activity of Ca^{2+} channels in the plasma membrane, activating the release of Ca^{2+} from intracellular stores, or inducing both effects. In this way the cytoplasmic concentrations of Ca^{2+} increase greatly; mean values may reach 10^{-6} M, and higher concentrations are likely in some parts of the cell. In rare cases, the activation of a receptor causes the inhibition of the activity of Ca^{2+} channels in the plasma membrane and consequently a decrease of the cytoplasmic concentration of the ion. In all cases in which the receptor signal induces an accumulation of free Ca^{2+} in the cytosol, the ion is rapidly transported to the exterior of a cell or into intracellular reservoirs by pumps and transporters or becomes sequestered in an inactive form by cytoplasmic proteins, since it is essential for a cell to have perfect control of the concentration of free Ca^{2+} in the cytosol. Marked and prolonged increases in the cytosolic concentration of Ca^{2+} may damage membranes, organelles, and other structures and even cause the death of a cell.

As mentioned previously, all the principal classes of membrane receptors are able to induce alterations of the cytoplasmic Ca^{2+} concentration. The following is a generalized description of the regulation of this second messenger

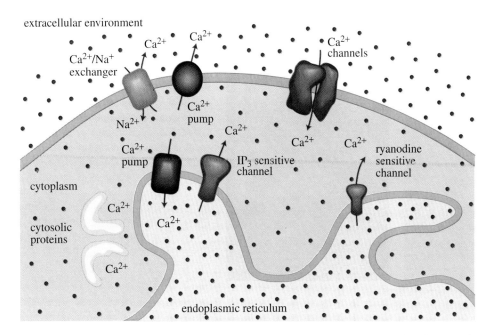

FIGURE 11 A schematic representation of the principal mechanisms for the control of the cytoplasmic concentration of Ca^{2+}. Depicted are the Ca^{2+} channels, pumps, and Ca^{2+}/Na^+ exchangers that operate across the plasma membrane; the IP_3- and ryanodine-sensitive Ca^{2+} channels and the Ca^{2+} pumps of the endoplasmic reticulum; and the cytosolic proteins that sequester Ca^{2+}. The speckling indicates the concentration of free Ca^{2+} in the extracellular space, the endoplasmic reticulum, and the cytosol.

by receptors that act through G proteins. Many G protein-coupled receptors activate the phospholipase C–IP_3 system. The obvious consequence of this type of receptor signal is a rapid increase of the cytosolic concentration of Ca^{2+}. Due to mechanisms that control the IP_3 signal and efficient systems for the removal of Ca^{2+} from the cytosol, one would expect a rapid return to the normal situation. In reality, in the majority of cases the effects of the receptor on the cytoplasmic concentration of Ca^{2+} are more complex (Berridge, 1993; Pozzan *et al.*, 1994; Tsien and Tsien, 1990). The recent capability to detect calcium in single cells has made it possible to observe that the initial increase in Ca^{2+} is often localized in a particular region of the cell and then spreads as a wave through the cytoplasm. Furthermore, the initial transient increase of the Ca^{2+} concentration is often followed by periodic oscillations that persist for long periods of time (Fig. 12). Multiple and controversial mechanisms form the basis of these fascinating phenomena. A further source of complexity in the signal is the fact that very often the receptors that stimulate phospholipase C not only induce the mobilization of Ca^{2+} from intracellular stores but also cause the entrance of Ca^{2+} through plasma membrane channels. The molecular identity and the regulatory mechanisms of the channels involved are not well-known. It is believed that the activation of these channels by receptors is mediated in some cases by G protein subunits or intracellular mediators and in other cases by a retrograde signal

arising from the intracellular Ca^{2+} stores emptied by the action of IP_3 (Berridge, 1993; Pozzan *et al.*, 1994). In addition to receptors that activate the "Ca^{2+} signal," the family of receptors associated with G proteins also includes members that, through the control of ion channels in the membrane, cause a reduction of the cytosolic concentration of the ion. Receptors associated with G_i or G_o have such an effect in neurons and other electrically excitable cells. Through actions that seem to involve both the α_i and α_o subunits and the $\beta\gamma$ complex, these receptors can directly inhibit specific channels for Ca^{2+} and activate K^+ channels, which in turn provoke a hyperpolarization of the membrane and the

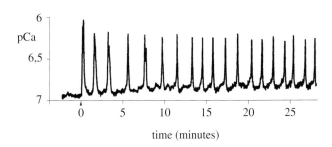

FIGURE 12 Graph of the oscillations of the cytosolic concentration of Ca^{2+} (pCa^{2+}) induced by histamine in a human endothelial cell. The arrow indicates the moment of addition of histamine (reproduced with permission from Jacob *et al.*, 1988).

consequent interruption of the entrance of Ca^{2+} through voltage-regulated channels. By these mechanisms, acetylcholine receptors retard the cardiac rhythm and dopamine receptors inhibit neuronal activity and the secretion of pituitary hormones.

Intracellular Effects of Ca^{2+}

Variations of the cytosolic Ca^{2+} concentration can regulate many proteins which modify their activity after binding to the ion. One of the most important cellular proteins regulated by Ca^{2+} is calmodulin. This protein is present in all cells of an organism and can be considered the intracellular receptor that mediates the majority of Ca^{2+}-induced processes. The calmodulin molecule has four binding sites for Ca^{2+} (Fig. 13). When these sites are occupied by the ion, calmodulin undergoes a conformational change that permits interactions with other proteins. Binding to the Ca^{2+}–calmodulin complex directly regulates the function of en-

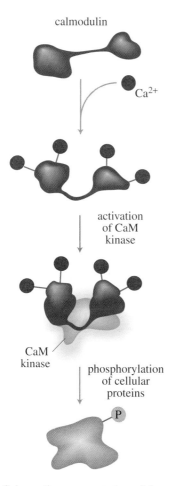

calmodulin

Ca^{2+}

activation
of CaM
kinase

CaM
kinase

phosphorylation
of cellular
proteins

P

FIGURE 13 Schematic representation of the conformational modifications of calmodulin following the binding of four Ca^{2+} ions. The protein can activate CaM kinases, a group of Ca^{2+}–calmodulin-dependent protein kinases, that in turn phosphorylate specific cellular proteins.

zymes and other proteins and activates a group of protein kinases, the Ca^{2+}–calmodulin-dependent protein kinases (CaM kinases). Ca^{2+} exerts many of its effects as a second messenger through the phosphorylation of serine and threonine residues in specific proteins by CaM kinases. The CaM kinase family of enzymes is numerous, with varied structural and functional characteristics. Some CaM kinases act only on particular substrates and therefore are involved in very specific cellular responses. Example of such enzymes are the kinases that act on the light chain of myosin to cause muscle contraction and the phosphorylase kinases that stimulate the degradation of glycogen. The CaM kinase II enzyme, or multifunctional CaM kinase, has instead much less specific activity and can phosphorylate numerous cellular proteins (Braun and Schulman, 1995). CaM kinase II is expressed in various cell types and is particularly abundant in cells of the nervous system, in which it is involved in numerous processes including the synthesis and release of neurotransmitters. As in the case of other protein kinases that mediate receptor signals, the responses induced by CaM kinase also depend on the specific substrates present in each cell type and are inactivated by the action of numerous cellular protein phosphatases. The CaM kinase system also constitutes a possible form of cellular memory after the removal of a stimulus; once activated by the Ca^{2+}–calmodulin complex, CaM kinase II is in fact able to phosphorylate itself and remain autonomously active even after the cytosolic concentrations of Ca^{2+} have returned to normal values. Through the action of calmodulin, the various CaM kinases, and other Ca^{2+}-binding proteins, the signaling system that depends on this ion can regulate a wide variety of cellular functions, including the transcription of specific genes. Therefore, it is not by chance that almost all cells have included this mechanism in their repertoire of responses to external stimuli.

cGMP as a Second Messenger in the Retinal Response to Light

Another second messenger involved in various types of cellular responses is cGMP. There are diverse mechanisms through which membrane receptors can induce variations of intracellular cGMP, and there exist specific protein kinases that mediate many of the effects of the nucleotide. However, only the interesting role played by cGMP in the rod cells of the retina is described here (Stryer, 1986). In these cells, the response to light is due to the activation of rhodopsin that causes the dissociation of the α and $\beta\gamma$ subunits of transducin. The α_t-GTP chain activates a phosphodiesterase that causes a transient reduction of the intracellular concentration of cGMP. In rods exposed to darkness, the concentration of cGMP is very high and the nucleotide binds to sodium channels in the plasma membrane and keeps them open. The cell is therefore constantly depolarized. Destruction of cGMP induced by light causes the Na^+ channels to close and hyperpolarizes the membrane. This signal in turn generates an electrical impulse that carries the visual stimulus to the brain. Complex mechanisms, including the inactivation of transducin, guarantee that the rods return to their resting conditions. Thus, both in the case of olfaction, in which cAMP intervenes, and in that of sight, an organism uses G proteins and nucleotides that act rapidly upon appropriate targets to convert sensory stimuli into nervous impulses.

Integration of Intracellular Signals

In any moment a cell must respond to various chemical mediators that exert their effects through intracellular messengers. In order for a cell to respond in a coherent manner to these multiple signals, it is necessary to have a fine network of reciprocal interactions between the diverse second messengers. One well-characterized example of such interactions are those occurring between cAMP and Ca^{2+}. These two signal systems make contact on multiple levels in a cell. For example, through phosphorylation mediated by protein kinase A, cAMP can activate Ca^{2+} channels in the plasma membrane, inhibit some forms of phospholipase C, and modulate IP_3-sensitive channels. The Ca^{2+}–calmodulin complex is able to stimulate particular types of adenylate cyclases and some of the phosphodiesterases responsible for the degradation of cAMP. In addition, the activity of specific cellular proteins can be regulated by both second messengers. For example, the phosphorylase kinase responsible for the degradation of glycogen is a CaM kinase that is activated through phosphorylation by protein kinase A. Through one or more of these mechanisms cAMP and Ca^{2+} can act to produce complementary or opposing cellular responses. In each cell the action of various second messengers is also strictly connected to those of other intracellular signals. In some cell types, for example, an increase of cAMP blocks the action of growth factors that activate receptors with tyrosine kinase activity through the phosphorylation of an essential protein in this signal transduction pathway. Thus, in every instant, the behavior of a cell in an organism is the result of the complex integration of multiple signals emanating from the surrounding environment.

References Cited

BALDWIN, J. M. (1994). Structure and function of receptors coupled to G proteins. *Curr. Opin. Cell Biol.,* **6,** 180–190.

BERRIDGE, M. J. (1993). Inositol trisphosphate and calcium signalling. *Nature,* **361,** 315–325.

BOURNE, H. R. (1995). Signal transduction. Team blue sees red. *Nature,* **376,** 727–729.

BRAUN, A. P., and SCHULMAN, H. (1995). The multifunctional calcium/calmodulin-dependent protein kinase: from form to function. *Annu. Rev. Physiol.,* **57,** 417–445.

DOHLMAN, H. G., and THORNER, J. (1997). RGS proteins and signaling by heterotrimeric G proteins. *J. Biol. Chem.,* **272,** 3871–3874.

DOHLMAN, H. G., CARON, M. G., and LEFKOWITZ, R. J. (1987). A family of receptors coupled to guanine nucleotide regulatory proteins. *Biochemistry,* **26,** 2657–2664.

GILMAN, A. G. (1987). G proteins: transducers of receptor-generated signals. *Am. Rev. Biochem.,* **56,** 615–649.

GROHOVAZ, F., *et al.* (1991). *J. Cell Biol.* **113,** 1341–1350.

JACOB, R., *et al.* (1988). *Nature* **355,** 35–40.

JAKEN, S. (1996). Protein kinase C isozymes and substrates. *Curr. Opin. Cell Biol.,* **8,** 168–173.

LEE, S. B., and RHEE, S. G. (1995). Significance of PIP_2 hydrolysis and regulation of phospholipase C isozymes. *Curr. Opin. Cell Biol.,* **7,** 183–189.

LISCOVITCH, M., and CANTLEY, L. C. (1994). Lipid second messengers. *Cell,* **77,** 329–334.

NEER, E. J. (1995). Heterotrimeric G proteins: organizers of trans-membrane signals. *Cell,* **80,** 249–257.

PIETROBON, D., DI VIRGILIO, F., and POZZAN, T. (1990). Structural and functional aspects of calcium homeostasis in eukaryotic cells. *Eur. J. Biochem.,* **193,** 599–622.

POZZAN, T. (1992). *Nature,* **358,** 325–327.

POZZAN, T., RIZZUTO, R., VOLPE, P., and MELDOLESI, J. (1994). Molecular and cellular physiology of intracellular calcium stores. *Physiol. Rev.,* **74,** 595–636.

SIMON, M. I., STRATHMANN, M. P., and GAUTAM, N. (1991). Diversity of G proteins in signal transduction. *Science,* **252,** 802–808.

SPIEGEL, A. M. (1996). Defects in G protein-coupled signal transduction in human disease. *Annu. Rev. Physiol.,* **58,** 143–170.

STRYER, L. (1986). Cyclic GMP cascade of vision. *Annu. Rev. Neurosci.,* **9,** 87–119.

SUNAHARA, R. K., DESSAUER, C. W., and GILMAN, A.G. (1996). Complexity and diversity of mammalian adenylyl cyclases. *Annu. Rev. Pharmacol. Toxicol.,* **36,** 461–480.

TAYLOR, S. S., BUECHLER, J. A., and YONEMOTO, W. (1990). cAMP-dependent protein kinase: framework for a diverse family of regulatory enzymes. *Ann. Rev. Biochem.,* **59,** 971–1005.

TSIEN, R. W., and TSIEN, R. Y. (1990). Calcium channels, stores and oscillations. *Annu. Rev. Cell Biol.,* **6,** 715–760.

General References

ALBERTS, B., BRAY, D., LEWIS, J., RAFF, M., ROBERTS, K., and WATSON, J. D. *Molecular biology of the cell.* 3ª ed., New York, Garland Publishing Inc., 1994.

Molecular biology of signal transduction. *Cold Spring Harbor Symposia on Quantitative Biology,* vol. 53, Cold Spring Harbor, Cold Spring Harbor Laboratory Press, 1988.

Signal transduction review issue, *Cell,* **80,** 1995.

WIRTZ, K. W. A., a c. di, *New developments in lipid-protein interactions and receptor function.* NATO ASI Series A, vol. 246, New York, Plenum Press, 1993.

Atanasio Pandiella

Institute of Microbiology and Biochemistry
Avda. del Campo Charro s/n
37007 Salamanca, Spain

How Cells Respond to Chemical Mediators: The Response to Trophic Factors

The fate of cells in the tissues of multicellular organisms is controlled by external molecules that direct genetic programs of cell growth, duplication, differentiation, growth arrest, or even cell death. Factors that promote cell division stimulate the cell machinery to duplicate its contents and to enter a complete round of the cell division cycle. Differentiation and growth arrest factors drive cells out of the cell cycle, thus making them enter a latent but reversible state of arrested cell growth (growth inhibitors), or modify the expression of certain genes that irreversibly drive cells into a differentiated status (differentiation factors). Alterations in the control of these cellular events may lead to disease. Exaggerated and uncontrolled cell growth and duplication, which is the hallmark of neoplastic transformation, represents such a situation in which cell growth controls are unbalanced. In this article, I focus on factors that contribute to these trophic effects and that have generically been termed growth factors. Since the list of these factors has increased in the past few years sufficiently to fill several books, my purpose is to illustrate the general concepts that have emanated from studies using the most prototypical factors as models.

The Response to Trophic Factors Depends on the Target Tissue

In general, polypeptide growth factors exert their action on target cells by using several types of intercellular communication (Fig. 1). One type uses soluble forms of the factor that interact with, and activate, their specific receptors located at the surface of responsive cells. Depending on the distance that the factor must travel, and the type of target cell that contains the receptor, this type of intercellular communication has been categorized as follows: autocrine (in which the factor and the receptor are located in the same cell), paracrine (in which the factor and the cell containing the receptor are located in proximity to each other, usually within the same tissue), and endocrine (in which the target cell is distantly located from the growth factor-producing cell, and the factor must be transported by the bloodstream or other extracellular fluids). Another type of intercellular communication uses forms of the growth factors that are anchored to the plasma membrane of the cell that biosynthesizes the factor. In this type, termed juxtacrine, the cell that produces the factor and the target cell must be in physical contact. This type of intercellular communication is typical of membrane-anchored growth factors.

The response of a cell to a trophic factor depends on several aspects, such as cellular context, the presence of the growth factor receptor, interaction with multiple receptors, the effect of a growth factor on the expression of other growth factors or their receptors, and the intracellular signaling cascades activated (Cross and Dexter, 1991). In fact,

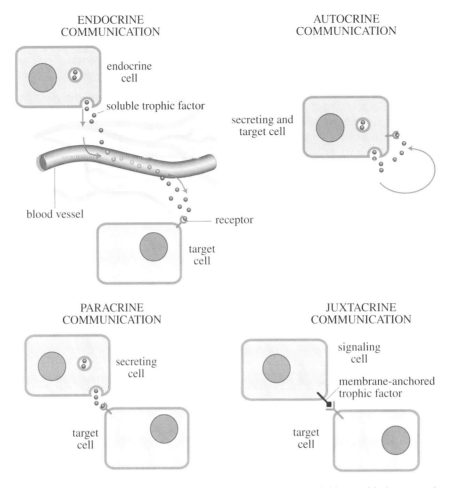

ENDOCRINE
COMMUNICATION

endocrine
cell

soluble trophic factor

blood vessel

receptor

target
cell

AUTOCRINE
COMMUNICATION

secreting and
target cell

PARACRINE
COMMUNICATION

secreting
cell

target
cell

JUXTACRINE
COMMUNICATION

signaling
cell

membrane-anchored
trophic factor

target
cell

FIGURE I   Different types of intercellular communication mediated by soluble trophic factors or by membrane-anchored factors.

several growth factors, even by apparently triggering the same intracellular cascades, may lead to different cellular responses, as occurs with type β transforming growth factor (TGF-β), which is an antimitogen by excellence but triggers growth in some specific situations, or the neuronal survival and differentiation factor nerve growth factor (NGF), which may induce mitogenesis in fibroblasts. The reasons that may explain this multifunctional aspect of growth factors are various. The factors may encounter a cell type whose fate has previously been dictated, at least partially, by other factors that prepare the cell to respond in one direction or another upon activation of a particular growth factor receptor. Furthermore, different domains found in the cytosolic domains of a receptor may transduce a growth or a differentiation signal, as occurs with the cytokine receptor for the granulocyte-colony stimulation factor (G-CSF).

Another important property of some trophic factors/ receptors is their ability to interact with several receptors/ ligands, respectively. Most factors of the epidermal growth factor (EGF) family are able to interact with, and activate,

the EGF receptor (EGF-R). On the other hand, some neurotrophins can interact with their four receptor proteins, although with different affinities. This sharing of ligands suggests a predominant role of receptors in normal physiology. Since the receptor is the final transducer of the signal, the absence of a ligand in a tissue could be irrelevant, provided that another ligand could substitute for it. In fact, ablations of the receptor genes carried out in mice give animals with abnormal phenotypes that are at least identical, and often more severe, than those obtained by the disruption of genes that code for the trophic factor that binds to the receptor.

Many Growth Factors Share Structural Modules

Although their primary structure may be divergent, several growth factors usually share structural characteristics responsible for common properties, such as chemical resistance, rigidity, or the ability to interact with other proteins.

Such common structural motifs may be present in factors whose biological actions are diverse or even opposite. For example, the platelet-derived growth factor (PDGF), NGF, and the TGF-β that direct processes of cell duplication, differentiation, and growth, respectively, all share a structure called the "cystine knot." The sequence identity between these soluble growth factors is less than 10%; but the three-dimensional structure of the cores of the monomers of these factors is highly similar (McDonald and Chao, 1995). The cystine knot structure is a substantially planar molecule that consists of two pairs of antiparallel β-sheets connected at one end by a relatively flexible conformation. On the other side, six clustered cysteines create three disulfide bonds that give a threaded-ring configuration. Two of the disulfide bonds form a ring structure through which the third disulfide bridge passes, creating a tightly packed cystine knot. In the TGF-β superfamily of growth factors, the most conserved region between the different members corresponds to the C-terminal domain, which contains seven cysteine residues, six of which contribute to the cystine knot and the remaining one forms a disulfide bond with the corresponding cysteine of a sister molecule to generate the homodimeric or heterodimeric TGF-β molecule. This cysteine, however, is missing in GDF-3 and GDF-9, two other TGF-β-related factors.

Another structural motif present in several growth factors is the EGF unit, named after its identification in the EGF. Several growth factors, mainly membrane anchored, contain this unit and have been included in the EGF family of growth factors (Fig. 2). This family includes TGF-α, human amphiregulin (hAR), schwannoma-derived growth factor (SDGF), the heparin-binding EGF-like factor (HB-EGF), the neu differentiation factors (NDF)/heregulins (HRG), the vaccinia virus growth factor (VGF), and several proteins isolated from worms and flies that play crucial roles in the developmental processes of *Caenorhabditis elegans* and *Drosophila*. The different members of this family are characterized by the presence, in the extracellular domain, of one or several EGF-like structural units. This EGF unit is defined by a motif containing a set of six cysteine residues characteristically spaced ($CX_7CX_{2-3}GXCX_{10-13}CX-CX_5GXRC$) over a sequence of 35–40 amino acids. The six cysteines pair to form three disulfide bonds whose positions are known in EGF and TGF-α.

Growth Factors Containing an EGF Unit Are Synthesized with a Transmembrane Anchor

EGF, one of the first growth factors isolated, was initially identified and purified from the mouse male submaxillary gland as a soluble hormonal compound able to stimulate

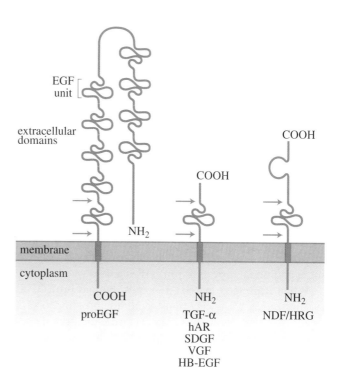

FIGURE 2 Schematic representation of the structure of the EGF precursor (proEGF) and some growth factors of the EGF family. The extracellular domain of these mature factors (center) includes one EGF unit with trophic factor activity (the arrows indicate the cleavage sites for release of the EGF). This unit is linked to the cell by a transmembrane hydrophobic domain, and this is followed by a relatively short cytosolic domain. (Left) The EGF precursor, proEGF, with multiple EGF units; (right) the extracellular domain of the NDF/HRG contains other structural modules in addition to the EGF unit.

eyelid opening in newborn animals. Although this gland is likely to be a major source of the factor, its ablation does not decrease serum levels of EGF, suggesting that other tissues can produce enough quantities to substitute for it. A 4.9-kb hEGF mRNA has also been found in kidney, pancreas, small intestine, brain, and other tissues. The human homolog of EGF, urogastrone, was later isolated from human urine by its gastric antisecretory potential. Its action is mediated by a transmembrane tyrosine kinase, the EGF-R, whose activation leads to an increase in the kinase activity of the intracellular domain. The wide distribution of the EGF-R is responsible for the ample spectrum of biological actions of EGF, which include stimulation of mitogenesis, nutrient and electrolyte transport, phosphoinositide turnover, and glycolysis and morphological changes. These cellular responses result in a variety of different physiological changes depending on the tissue involved. The mitogenic properties of EGF both *in vivo* and *in vitro*, together with the increased coexpression of the factor and its receptor in a number of solid tumors and tumor-derived cell lines, suggest participation of this growth factor in tumor progression.

Mouse and human EGFs are synthesized as precursor proteins (proEGF) with nine EGF units in the extracellular domain (Fig. 2). cDNA clones for both species have been identified, and they encode proteins of 1207 (human) and 1217 (mouse) amino acids, respectively. The ninth EGF unit of the precursor corresponds to the sequence of mature hEGF and is the most similar to the EGF units of other members of this family. This sequence is followed by a 10-amino acid linker region that connects the C-terminal cleavage site of EGF with the transmembrane and cytosolic domains. The large extracellular domain contains eight additional EGF units, N-terminal to EGF, organized in two large regions separated by 260 amino acids. The function of these additional repeats in proEGF is unknown.

When expressed in NIH-3T3 cells, human proEGF is synthesized as a 170-kDa transmembrane protein, 36 kDa larger than that predicted (134 kDa) from the primary sequence. This is probably due to the addition of sugar chains to the precursor since glycosidase treatment increases the mobility of proEGF in electrophoresis gels. The expressed precursor is initially cell associated, but it has a relatively short half-life (<4 h) in a membrane-anchored form. Proteolytic processing occurs preferentially at the C terminus, generating a soluble fragment similar in size to the cell-associated precursor. Processing at the N terminus in cultured cells is very inefficient, as suggested by the difficulty in detecting 6-kDa mature EGF in media supernatants. Interestingly, the intact cell-associated precursor inhibits binding of [^{125}I]EGF to the receptor.

Analogous structural properties are shared by other EGF family members (Massagué and Pandiella, 1993). Molecular cloning of these factors, together with careful biochemical analysis, demonstrated that EGF family growth factors are biosynthesized as membrane-anchored precursors containing one EGF unit instead of several, as occurs with EGF.

......................................●

The Release of Transmembrane Growth Factors Is a Regulated Process

Although molecular cloning of membrane-anchored growth factors demonstrated that these factors are bound to the cell surface by a membrane anchor, the factors have been classically identified as soluble polypeptides in extracellular fluids (Massagué and Pandiella, 1993). Therefore, a mechanism of solubilization of the factor must exist. Extensive studies carried out mainly on TGF-α and its precursor, proTGF-α, have provided information on this process.

TGF-α was initially isolated from the culture supernatants of oncogenically transformed cells as a factor that promoted reversible transformation of mammalian cells in culture. The soluble factor purified from this source is a 50-amino acids single-chain protein that shares 33 and 44% identity with mouse and human EGF, respectively. Human TGF-α is synthesized as a larger precursor protein of 160 amino acids. The N-terminal region of proTGF-α contains a hydrophobic signal sequence that is cotranslationally cleaved by the endoplasmic reticulum signal peptidase at a site between residues 19 and 25 of proTGF-α. The N-terminal region also contains sites for both N- and O-linked glycosylation (Fig. 3). In Chinese hamster ovary cells, N- and O-glycosylation are dispensable for proper targeting of the protein to the cell surface and cleavage into the soluble form. However, inhibition of glycosylation in retrovirally transformed fibroblasts markedly inhibits TGF-α accumulation in media supernatants.

The sequence of mature TGF-α starts at residue 40 and extends to residue 89 within the precursor. Ten amino acids separate the C terminus of TGF-α from the transmembrane domain. The 71 C-terminal amino acids of the precursor, which include the extracellular linker and the transmembrane and cytosolic domains, are nearly identical between the rat and human sequences, with only one conservative amino acid substitution at residue 137. This high degree of homology with respect to the mature sequence suggests an important biological function of this region in the overall physiology of proTGF-α. Another structural feature of proTGF-α is palmitoylation of cysteine at the endodomain, a modification whose structural relevance in proTGF-α physiology is unknown.

The release of membrane-anchored growth factors occurs by a proteolytic attack of the ectodomain mediated by a cell surface endoprotease (Massagué and Pandiella, 1993). Proteolytic processing of membrane-anchored growth factors is complex and includes cleavage of specific peptide bonds at both the N and C terminus of the mature sequence of the factor within its precursor (Fig. 3). In the case of proTGF-α, the initial step consists of cleavage at the N terminus of the mature sequence with the release of the heterogeneously glycosylated N-terminal segment of the precursor. This process is relatively rapid under basal conditions ($t_{1/2} \sim 15$ min) and generates an unglycosylated cell-associated precursor. In the second step, cleavage of proTGF-α occurs at the C terminus of the TGF-α sequence, releasing this factor into the medium and leaving a cell-associated fragment that includes the transmembrane and cytoplasmic domains of the precursor. In several different cell lines this is the rate-limiting step, and in nonstimulated cells it occurs at a slow rate ($t_{1/2} \sim 4$ h). This leads to the accumulation of a steady-state level of proTGF-α at the cell surface, much of which turns over without being converted to soluble factor.

Conversion of membrane proTGF-α to soluble TGF-α occurs rapidly when cells are exposed to agents that increase the activation state of cells, such as stimulators of

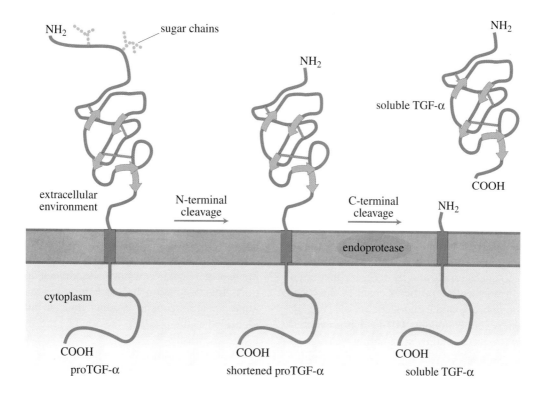

FIGURE 3 Cleavage of membrane-anchored growth factors which usually occurs in several steps. In proTGF-α, cleavage occurs by first eliminating an amino-terminal extension that contains the glycosylation sites; this cleavage is rapid and constitutive and produces accumulation of a shorter membrane-anchored form of proTGF-α at the cell surface. Cleavage at the carboxy terminus occurs very slowly, and the holoprotein may even undergo cycles of internalization without being cleaved. However, in circumstances of cell activation, a membrane endoprotease cleaves the membrane proTGF-α, converting it into a soluble factor (soluble TGF-α).

protein phosphorylation. These treatments activate cleavage through various independent mechanisms, including activation of protein kinase C (PKC) or increasing intracellular calcium levels. The stimulation of the cleavage event occurs within a few seconds of stimulation and almost completely cleaves all cell surface proTGF-α into soluble TGF-α.

Therefore, cells contain mechanisms for switching between two active forms of growth factors: a membrane-anchored one, which works only in juxtacrine stimulation, and a soluble form that can carry out other forms of intercellular communication (endocrine/autocrine/paracrine), classical of soluble factors.

Transmembrane Growth Factors Are Biologically Active and May Act as Receptors

The presence of the membrane anchor in the factors of the EGF family prompted the interesting question of whether the factor in this conformation could be biologically active and function as a juxtacrine factor. Membrane-anchored proTGF-α can activate autophosphorylation of EGF-R in adjacent cells and various other early responses mediated by activation of this receptor. In addition, proTGF-α can stimulate mitogenesis of EGF-R-containing cells.

The existence of transmembrane forms of growth factors and their cognate receptors suggested that their interaction could also result in cell–cell adhesion. Evidence for such stable cell–cell contacts mediated by proTGF-α and the EGF-R is derived from studies carried out on a binary cell system of hemopoietic and stromal cells. Transfection of the EGF-R into the hemopoietic progenitor cells allowed them to adhere and proliferate on the surface of stromal cells transfected with proTGF-α. Analogous experimental designs carried out with the precursors for the EGF, the stem cell factor (SCF) and the colony stimulating factor-1 (CSF-1), corroborated the finding that the membrane-bound form of these growth factors is biologically active.

The structure of membrane-anchored growth factors

very much resembles that of a transmembrane receptor. Careful analysis of the intracellular region of the different growth factors does not show any domain that could be recognized as a signaling module. On the other hand, interspecies comparison of the proTGF-α sequences has reveled that the most conserved region corresponds to the transmembrane and intracellular domains, suggesting a role of this region in the physiology of proTGF-α. Initial clues about the possible role of membrane-anchored growth factors as receptors came from studies with a different member of the EGF family, the HB-EGF. The HB-EGF has been demonstrated to be one of the binding proteins for diphtheria toxin (Naglich *et al.,* 1992). This bacterial toxin consists of two subunits, A and B, that derive from a single-chain peptide by limited proteolysis. Entry of this protein into cells is sequential, requiring first the binding of the B subunit to the cell surface followed by receptor-mediated endocytosis and translocation of the A chain to the cytosol through the membrane of the endocytic vesicle (Fig. 4). Entry of the toxin in cells containing HB-EGF is prevented by antibodies to the factor and by stimulating cleavage of HB-EGF by the endoproteolytic system that participates in the solubilization of growth factors.

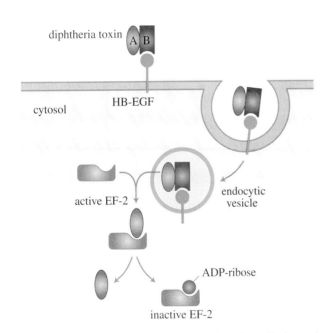

FIGURE 4 Entry of diphtheria toxin into the cells depends on the interaction of the B subunit with the cellular receptor (HB-EGF). This allows internalization of the toxin/HB-EGF complex into endocytic vesicles. A progressive decrease in the pH of these vesicles promotes the uncoupling of the A subunit from the complex and facilitates its translocation to the cell cytosol. The released A subunit transfers an ADP-ribose group onto the EF-2 elongation factor, thus preventing protein synthesis.

Soluble Stem Cell Factor Does Not Replace the Membrane-Bound Factor in Mouse Development

An example of the importance of membrane-anchored growth factors in mammalian development has been documented by studies carried out with the SCF (Massagué and Pandiella, 1993). The SCF (also called Kit ligand) has been identified and isolated by genetic and biochemical analysis of two loci that are essential for normal mouse development: the dominant *white spotting* (*W*) and the *steel* (*Sl*) loci. The close overlapping of phenotypic alterations in the two classes of noncomplementing mutations suggested that the two loci encoded proteins whose interaction was necessary for the activity of a pathway involved in the development of particular cell lineages. Molecular evidence supports this prediction because the c-kit receptor tyrosine kinase protooncogene product and its ligand SCF were identified as the products encoded by the *W* and *Sl* genetic loci, respectively.

Mice with mutations at either loci carry abnormalities of three independent cell lineages: neural crest-derived melanocytes, hematopoietic stem cells, and germ cells. As a consequence, the phenotype of these mice includes white spotting coat, severe macrocytic anemia, and sterility. Homozygous *Sl/Sl* mice die *in utero* with severe macrocytic anemia. Heterozygous animals (*Sl/Sl*^{D}) with mutations in a second allele, the Steel–Dickie (*Sl*^{D}), are able to survive, although they are also anemic. Although the phenotype of mice carrying either of these mutations is similar, several lines of evidence suggest that the *W* locus acts in an autonomous manner and, thus, is required in the affected cells, whereas the *Sl* locus might act nonautonomously and is present in stromal cells, whose apparent phenotype is normal. This has been confirmed by *in situ* hybridization studies that showed localization of the ligand in stromal cells but not germ or hematopoietic precursor cells. In contrast, c-kit is only located in the parenchymal cells.

Interaction of membrane-associated forms of SCF with the c-kit receptor has been suggested by several different results. Coculture of mast cells (that express the c-kit receptor) with fibroblasts supports the growth of the former, whereas culture media from the latter does not. In addition, alkaline phosphatase-tagged c-kit can specifically bind to monolayers of fibroblasts, and the SCF expressed on the surface of COS-7 cells can mediate adhesion to mast cells.

The *Sl*^{D} allele encodes a mutant SCF protein in which a stop codon has produced a large deletion of the C-terminal portion of the protein, including the transmembrane and cytosolic domains. Interestingly, the biological activity of the SCF-Sl^D mutant protein is indistinguishable from that of wild-type SCF, as measured *in vitro* using a mast cell proliferation assay. The expression of the mRNAs for SCF in

Sl/Sl^D mice is similar to that of +/+ animals, suggesting that their altered phenotype is due to the anomalous molecular structure rather than an abnormal expression of the protein. The phenotype of *Sl/Sl^D* mice includes pigment abnormalities, macrocytic anemia, and sterility, strongly suggesting that the membrane forms of the factor are essential for normal development of these cell lineages.

Tumor Necrosis Factor Is a Membrane-Anchored Growth Factor with Killing Properties

Tumor necrosis factor-α (TNF-α; also called cachectin) is a cytokine initially described by two different properties: tumor cell killing and shock. As occurs with some membrane-anchored growth factors, the wide tissue distribution of the TNF-α receptors leads to a broad range of biological and physiological actions of TNF-α, depending on the organ system. In neutrophils, the factor stimulates granule secretion and adhesive properties that, together with the morphological effects of the factor on endothelial cells, might favor extravasation of blood cells to perivascular tissues. Other targets for TNF-α action include adipose and muscular tissue, gastrointestinal tract and liver, central nervous system, adrenal gland, bone, and skin. When recombinant material is administered to animals it induces the classical symptomatology of shock, including hypotension, tachypnea (accelerated breathing), acidosis, and end organ failure.

The cDNA of the precursor for human TNF-α encodes a protein of 233 amino acids. A particular characteristic of the molecular structure of TNF-α is the presence of only one hydrophobic segment. This segment is located N-terminal to the mature sequence of the factor within what was initially suspected to be an unusually long signal peptide but was later demonstrated to be a transmembrane domain. Since the N-terminal sequence acts as a membrane anchor, TNF-α belongs to the type II transmembrane proteins class. Proteolytic cleavage of this precursor generates several soluble forms, depending on the cleavage site and the glycosylation pattern.

Mechanistically related to juxtacrine stimulation of cell growth, but with distinct biological consequences, is the killing by cell–cell contact mediated by the interaction of TNF-α with its receptors (Perez *et al.*, 1990). Experimental models created to investigate this possibility included mutated versions of the human proTNF-α precursor that prevented cleavage of membrane-anchored TNF-α precursor. The cleavage site in the TNF-α precursor occurs between an Ala–Val bond, identical to the sequence cleaved by the proTGF-α processing endoprotease. Cells transfected with these mutated forms express the 26-kDa protein precursor at the cell surface, are unable to generate soluble TNF-α, and can kill sensitive cells by cell–cell contact *in vitro.*

A Growth Factor with Seven Membrane-Spanning Domains Participates in *Drosophila* Eye Development

The bride of Sevenless (Boss) protein, a potential ligand for the Sevenless receptor tyrosine kinase, is an interesting membrane-anchored factor involved in *Drosophila* eye development (Rubin, 1991). The compound eye in *Drosophila* is formed by approximately 800 units, or ommatidia. Each of these units consists of eight parenchymal photoreceptor cells, R1–R8. The ommatidia in *sevenless* and *boss* mutant flies is deficient in R7 cells, which become lens-secreting cone cells instead. Genetic mosaic analyses have demonstrated that Sevenless is required in the R7 cell, whereas Boss is present in the adjacent R8 cell. A unique property of this factor with respect to the other growth factors is the membrane anchor. Structural analysis of the sequence has identified seven hydrophobic segments large enough to be membrane-spanning regions. The C-terminal tail of the protein comprises the last 115 amino acids and is located intracellularly. This structure makes this factor a strong candidate for being both a ligand and a receptor. The seven membrane-spanning domains are characteristic of several receptors coupled to GTP-binding proteins, whose properties are discussed elsewhere in this volume.

What is the fate of membrane-anchored factors upon interaction with the receptor? For some, cleavage of the factor would allow receptor-mediated endocytosis, with termination of juxtacrine stimulation. For proteins in which ectodomain cleavage is not apparent, termination of ligand–receptor interaction might involve a different mechanism that includes receptor-mediated internalization of the whole protein (Fig. 5). Boss colocalizes with Sevenless in intracellular compartments of the R7 cells as analyzed by immunocytochemistry with antibodies directed to both the C and N terminus, supporting the fact that Boss is internalized as a whole molecule, including the membrane segments in which Boss is embedded (Cagan *et al.*, 1992). Boss-containing multivesicular bodies cannot be detected in *sev^{x1}* mutant flies that do not express the Sevenless protein but are present in *sev^{Met2242}*, a kinase-inactive form of Sevenless, suggesting that the protein but not its kinase activity is required for receptor-mediated endocytosis of Boss.

The Receptors for Trophic Factors Have Kinase Activity

The response of a cell to a trophic factor depends on the presence, at the surface of the cell, of its specific receptors. Upon ligand binding, the receptor oligomerizes and the signal is transduced (Heldin, 1995). Most of the receptors for the trophic factors are endowed with enzymatic properties,

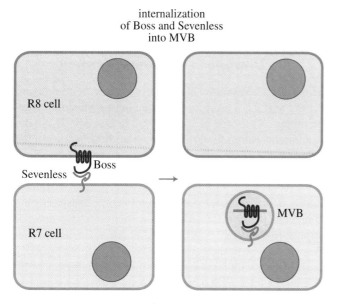

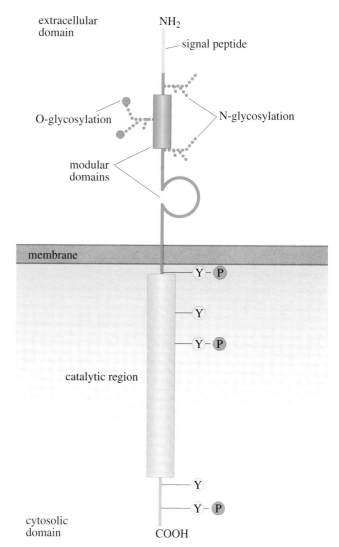

FIGURE 5 The Boss ligand, present on the R8 cell, interacts in a juxtacrine fashion with the Sevenless receptor present on the R7 cell. This interaction provokes the internalization of Boss and Sevenless into multivesicular bodies (MVB) of the R7 cell. Since both the amino and the carboxy termini of Boss are present in these MVBs, it is likely that this occurs by a special type of phagocytosis that has been termed cellular cannibalism.

FIGURE 6 The tyrosine kinase receptor RTK, a type I transmembrane protein with a glycosylated extracellular domain. This domain contains several structural modules important for receptor properties such as chemical resistance or ligand binding. The cytosolic domain contains a tyrosine kinase catalytic region and several tyrosine residues (Y). Some of these become phosphorylated (P) upon receptor activation and act as docking sites for substrates.

mainly phosphotransferase (kinase). Two major types of transmembrane kinases have been associated with the action of trophic factors: tyrosine kinases, which stimulate proliferation and/or differentiation (Ullrich and Schlessinger, 1990; van der Geer *et al.,* 1994), and serine/threonine kinases, which are directly implicated in growth arrest (Massagué, 1996).

Some receptors do not have an intrinsic kinase but can associate or otherwise activate cytosolic itinerant kinases that contribute to propagate the signal inside the cell. Other receptors for the trophic factors do not have a cytosolic domain that has been associated with signaling; however, they can influence the overall response of the cell to the factor. This occurs with the type III receptor for TGF-β or with the p75 neurotrophin receptor, which may increase target cell responsiveness by mechanisms that include the establishment of high-affinity ligand binding or ligand presentation to transducing receptors (Chao, 1994; Massagué, 1996).

The general structure of receptor tyrosine kinases (RTKs) is that of type I transmembrane proteins (Fig. 6). The ectodomain usually contains one or several structural motifs and is usually glycosylated by N- and O-linked addition of sugar chains. The extracellular domain is responsible for ligand binding and is the most divergent domain between the different RTKs. Several receptors have one or several cysteine-based structural motifs that probably offer

structural integrity and resistance but that are not directly involved in ligand binding. Rather, as occurs with the EGF-R, ligand binding occurs in the cleft between the two large cysteine clusters of the ectodomain. Other domains that appear in RTKs are the immunoglobulin-, fibronectin-, cadherin-, and discoidin-like domains, so named because of their similarity to domains previously identified in these molecules. The extracellular domain is followed by a single transmembrane region of approximately 25 amino acids.

The cytosolic region contains the tyrosine kinase domain as well as other biologically essential regions, such as those containing a tyrosine-phosphorylated residue. The latter may act as the target for interaction with several substrates. Using mutated versions of the RTKs, other motifs that do not contain a tyrosine-phosphorylated residue but that are important for signal transduction have been identified.

Signaling Molecules Contain Structural Modules That Allow Interaction with Receptor Tyrosine Kinases and Other Proteins

Interaction of growth factor receptors with substrates occurs by specific protein–protein interactions mediated by particular domains within the receptor and the target signaling molecule (Pawson, 1995). In the growth factor receptors, these sequences often surround (and include) a tyrosine phosphorylated residue whose phosphorylation is triggered upon receptor activation.

Many growth factor receptor-interacting proteins contain related sequences of 50–100 amino acids referred to as src homology regions SH2 or SH3 and pleckstrin homology (PH) domains. SH2 domains interact with short peptide sequences containing a phosphotyrosine residue. This interaction is of high affinity, whereas the affinity for the un-phosphorylated peptide sequence is very low.

SH3 domains have been found not only in many proteins involved in growth factor receptor signaling but also in some cytoskeletal proteins that play important roles in maintaining cell architecture. SH3 domains bind to short peptides of about 10 amino acids containing proline residues. These sequences often adopt a polyproline helix, with three residues per turn. Examples of SH3 interacting proteins that may be of relevance in RTK signaling are the Ras exchange factors sos1 and sos2, which bind to the SH3 domain of the Grb2 adaptor protein.

The PH domains are present in several proteins, including serine/threonine kinases and tyrosine kinases, phospholipase isoforms, dynamin, small regulators of GTPases, and some cytoskeletal proteins. The structure of these domains includes two antiparallel β-sheets followed by an amphipathic α-helix. Some data suggest that this domain may participate in targeting proteins to membranes: Several PH-containing proteins are cytosolic but usually associate to membranes. The best evidence, however, for the role of PH domains in cell physiology comes from a genetic source. A missense mutation of the *btk* gene (a cytoplasmic tyrosine kinase) that affects the Btk PH domain impairs murine B cell development.

Another binding domain, the phosphotyrosine-binding domain (PTB), that interacts with tyrosine-phosphorylated residues has recently been identified in the adaptor protein Shc and in the insulin receptor substrate-1 (IRS-1). This domain in Shc is located at the N terminus and binds several RTKs at a sequence of Asn-X-X-Tyr. Interestingly, this sequence has also been involved in targeting of certain membrane proteins to degradative pathways.

Several Pathways Participate in Receptor Tyrosine Kinase Signaling

Activation of RTKs triggers several intracellular pathways that result in signals that affect cell growth, differentiation, or metabolism. Activation of these pathways usually starts by association and/or phosphorylation of proteins/substrates with the receptor (van der Geer *et al.*, 1994). In fact, some of these intracellular signaling proteins coprecipitate with the receptor, and many are tyrosine phosphorylated upon receptor activation. The phosphorylation of these proteins is expected to affect their biological properties. Tyrosine phosphorylation of those endowed with enzymatic properties may increase or decrease their activity, whereas tyrosine phosphorylation of other proteins may affect signaling by receptor-induced translocation of the protein from one compartment to another.

Proteins That Affect Phospholipid Metabolism

Two direct targets of RTKs are phospholipase Cγ (PLCγ) and phosphatidylinositol-3-kinase (PI3-K). These enzymes act on phosphatidylinositol, a low-abundant membrane phospholipid. PI3-K phosphorylates phosphatidylinositol, phosphatidylinositol-4-phosphate, and phosphatidylinositol-4,5-bisphosphate at position 3 in the inositol ring. The enzyme is a complex of two subunits of 85 and 110 kDa. The former contains an amino-terminal SH3 domain, followed by two SH2 domains, with the latter sequences being involved in the interaction with the RTK. In addition, the region contained between the two SH2 domains contains the site for interaction with the p110 subunit, which is responsible for the catalytic activity. Stimulation of several RTKs leads to a rapid accumulation of the 3-phosphorylated phosphoinositides. However, the interaction with RTKs, and the phosphorylation of p85, is not always clearly detected, and other mechanism (including translocation to the plasma membrane) may be implicated in the activation of the enzymatic complex.

A role for PI3-K in cell growth and morphogenesis is supported by studies using mutated versions of receptors, such as PDGF-R and CSF-1R, lacking the intracellular binding site for PI3-K. In addition, it has been reported that the enzyme is essential for NGF-induced survival of neurons since the PI3-K inhibitor wortmannin induces apoptosis when neuronal cells are cultured in the presence of NGF as the only trophic factor. Interestingly, the survival effect does not appear to depend on Ras activity since mutant versions of Ras, which act as dominant-negative forms, do not prevent the survival-supporting effect of NGF and other growth factors.

The second phosphoinositide-modifying enzyme is PLCγ. The PLCs act on phosphatidylinositol-4,5-bisphosphate to generate two second messengers. One of them, 1,2-*sn*-diacylglycerol (DAG) is a physiological activator of the protein kinase C (PKC) family of enzymes. These enzymes have been implicated in cell growth because they are cellular receptors for tumor promoters of the phorbol ester family, and because overexpression of PKC is followed by the appearance of a transformed phenotype. The other second messenger generated, inositol-1,4,5-trisphosphate, mediates increases in the cytosolic concentration of free calcium. Some, but not all, activated RTKs interact with PLCγ through the SH2 domains of the latter. Receptor activation is followed by tyrosine phosphorylation of PLCγ and its phosphorylation on Tyr771, -783, and -1254. Phosphorylation of Tyr783 is essential for PLCγ activation, and phosphorylation at Tyr1254 has a positive effect on activation. This phosphorylation results in an increased activity of the enzyme *in vitro* and is likely to have the same consequence *in vivo*. Together with the possible activation step that follows tyrosine phosphorylation of PLCγ, an additional mechanism for activation could be positional: that is, by bringing the enzyme to the membrane, it is closer to its substrates.

The treatment of cells with growth and differentiation factors may also result in the stimulation of phospholipase A_2 (PLA$_2$) and phospholipase D (PLD) activities. PLA$_2$ releases fatty acids, preferentially arachidonic acid, whereas PLD hydrolyzes phosphatidylcholine, generating choline and phosphatidic acid. Phosphatidic acid may be transformed to DAG by the action of a hydrolase, creating an alternative output of PKC-activating DAG.

Adaptor Proteins

The adaptor proteins are signaling proteins that lack a catalytic domain (Kazlauskas, 1994). They usually contain one or several SH2 or PTB domains and, commonly, another protein motif such as SH3 that interacts with other proteins.

An especially relevant adaptor protein is Grb2. This protein, present in all tissues, is a relatively small intracellular protein (24 kDa) and contains an SH2 domain flanked at both ends by SH3 domains (Fig. 7). Grb2 associates with several RTKs, including EGF-R, PDGF-R, and CSF-1R, and this interaction can be direct or mediated by intermediate proteins. The SH2 domain in Grb2 is responsible for binding RTKs, with the sequence P-Tyr-Val/Leu-Asn-X being the preferred interaction site in EGF-R and CSF-1R. Genetic evidence in *C. elegans* and *Drosophila* has shown Grb2 to be located in an intermediate position between the

receptor and the Ras pathway, the latter involved in RTK signaling. The Grb2 *C. elegans* analog, sem-5, links the Let-23 RTK to the Ras analog protein Let-60. Several loss of function mutants of the sem-5 protein are characterized by point mutations in highly conserved regions of the SH2 or SH3 domains.

Biochemical evidence for a role of Grb2 in Ras activation has been obtained with cells overexpressing Grb2. Grb2 enhances activation of Ras induced by the EGF-R, and overexpression of Grb2 increases MAP kinase activation upon insulin receptor stimulation. Microinjection of Grb2 and Ras stimulates mitogenesis in fibroblasts, and when anti-Grb2 antibodies are injected the mitogenic response to EGF or PDGF is blocked. In addition, mutation of the Grb2 binding site in CSF-1R diminishes growth stimulation upon CSF-1 treatment. Therefore, Grb2 may act as an intermediate adaptor protein that couples RTK to the Ras pathway.

Another adaptor protein is Shc. Biologically, the overexpression of this protein leads to different phenotypical responses depending on the cellular background. In fibroblasts, overexpression causes transformation, and in PC12 pheochromocytoma cells induces cessation of cell duplication and differentiation toward a neuronal phenotype. This protein is present as three forms of 46, 52, and 66 kDa. Structurally, Shc contains an N-terminal PTB domain, a C-terminal SH2 domain, and a central Pro/Gly-rich region. In different cell lines, Shc is tyrosine phosphorylated under resting conditions, and its phosphorylation is increased upon stimulation with several RTK agonists. This also results in the tight association of Shc to the cytoplasmic domain of the activated receptors. In addition to being regulated by the activity of RTK, Shc is also phosphorylated by soluble cytosolic tyrosine kinases. Mutational analyses performed on different RTKs have identified the residues of the receptor important for the binding and function of Shc in signal transduction. Tyr490 is the binding site for Shc in the TrkA neurotrophin receptor. Point mutation of this site prevents Shc association with TrkA but does not prevent neurite outgrowth in response to NGF, suggesting that Shc is not essential for this biological response. On the other hand, combined mutation of Shc and PLCγ sites completely prevents neurite outgrowth, suggesting that these two receptor sites may act in concert to stimulate neuronal differentiation.

The Ras Pathway

The Ras protein is a small GTP-binding protein. The first indication for the involvement of this protein in mitogenic signal transduction came from studies in tumors in which this protein appeared constitutively activated. Definitive proof for the participation of this protein in mitogenesis came from microinjection studies and genetic analyses.

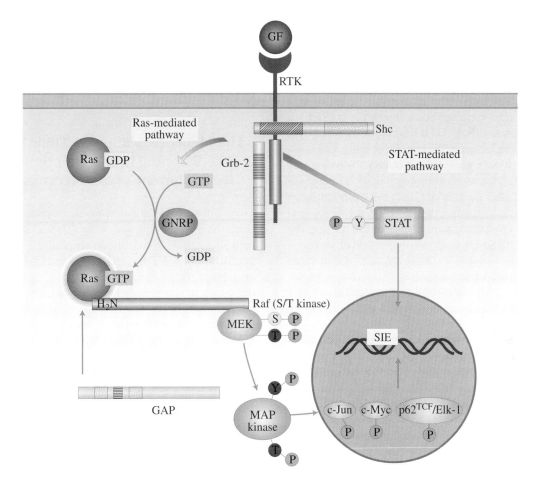

FIGURE 7 Signal transduction pathways mediated by Ras and STAT. Binding of GF to RTK causes association of the receptors with several molecules, such as the adapter protein Shc [via SH2 domains (dotted areas) or PTB domains (diagonally lined areas)]. In the STAT-mediated pathway, this protein, after tyrosine (Y) phosphorylation, translocates to the nucleus in which it interacts with specific DNA sequences and activates, with other protein factors such as SIE, the transcription of genes involved in the cellular response. In the Ras-mediated pathway, activation depends on the proteins Grb2 and GNRP, on the presence of GTP, and on the interaction with the protein GAP, characterized by two SH2 domains (dotted areas) and one SH3 domain (horizontally lined area). The Ras molecule in turn determines, via Raf, the activation of MEK by phosphorylation of serine (S) and threonine (T); MEK then phosphorylates the tyrosine and the threonine of the MAP kinase, which, now activated, translocates to the nucleus in which it phosphorylates various substrates, including the transcription factors c-Jun, c-Myc, and p62^{TCF}/Elk-1, which regulate gene expression.

Ras cycles between an inactive GDP-bound state and an active GTP-bound state (Fig. 7). Ras has high affinity for GDP; therefore, most of the cellular Ras in the cell appears as GDP bound. Activation of several RTKs results in a transient increase in the GTP:GDP ratio of guanine nucleotides bound to Ras, a measure of Ras activation. The levels of Ras–GTP are regulated by different types of enzymes that affect guanine nucleotide metabolism, i.e., exchange factors and GTPases. The GTPase activity of Ras is upregulated by the Ras GTPase-activating protein (GAP). GAP is tyrosine phosphorylated in response to PDGF-R activa-

tion and associates with the receptor through the SH2 domains of GAP. In response to EGF, GAP and the GAP-associated proteins p62 and p190 are phosphorylated on tyrosine, although the kinetics of the phosphorylation are different for GAP and p62 (5 min) than for p190 (1 or 2 h after stimulation). The interaction of this protein with other RTKs has been reported, although CSF-1 receptor and kit do not appear to bind and phosphorylate GAP. However, p62 and p190 are tyrosine phosphorylated in response to CSF-1. p62 is an RNA-binding protein that binds to several SH2 proteins but its precise role in mitogenic signal transduction remains to be elucidated. Overexpression of full-length GAP or just the catalytic region suppresses focus formation and the morphological-transformed phenotype of fibroblasts transformed by *ras, src,* and *v-fms* and the expression of several CSF-1 response genes.

Due to the high stability of Ras coupled with GDP, a guanine nucleotide exchange/releasing factor (GNRF) is necessary to generate Ras–GTP. Several of these GNRFs have been reported, isolated by genetic or molecular studies. One of them, Sos, generates active Ras by promoting the release of GDP and allowing binding of GTP. Sos interacts with the SH3 domain of Grb2, and overexpression of Grb2 increases the fraction of Ras–GTP, suggesting that Grb2 may have a role in regulating Ras function, perhaps by its interaction with Sos. Indeed, overexpression of Sos transforms fibroblasts. Sos activity, however, does not increase following EGF treatment, suggesting that a change in Sos localization (i.e., from the cytosol to the membrane) may be involved in EGF-induced, Sos-mediated Ras activation.

The mechanism of signal transduction that follows Ras activation includes a kinase cascade that starts with Ras–GTP binding to the serine threonine kinase Raf, a process that may be essential for Raf activation. Raf activation can also be triggered by phosphorylation through other pathways, such as the PKC pathway, that are independent of Ras function. Raf, in turn, activates another kinase, the MAP kinase kinase (MEK), that phosphorylates and activates another cytosolic kinase, the MAP kinase. The latter, once phosphorylated, is translocated to the nucleus, where it phosphorylates several substrates, including the transcription factors c-Jun, c-Myc, and p62^{TCF}/Elk-1 that play a role in growth regulation.

Tyrosine Phosphorylation of Transcription Factor Subunits

Recent analysis of gene activation in cells treated with growth factors and cytokines has uncovered a direct signaling pathway from the cell surface receptors to the nucleus that involves tyrosine phosphorylation of subunits of transcription factors that are located in the cell cytosol [the signal transducers and activators of transcription (STATs)]. Activation of cells with factors that act through RTKs has been shown to induce a transcription factor, SIF, to bind to the regulatory element, SIE, present in the c-*fos* promoter. This event has been reconstituted *in vitro* and can be blocked by exogenous addition of soluble SH2 domains or antiphosphotyrosine antibodies. STAT[91], an SH2-containing protein originally identified as a component of the interferon-α-inducible transcription complex, ISGF-3, appears to mediate this event. EGF stimulates tyrosine phosphorylation of STAT[91], and this results in association with the EGF-R. Later, STAT[91] is expected to dissociate from the receptor and be transported to the nucleus, where it accumulates and binds to the SIE element stimulating tran-

scription. Other factors such as PDGF and CSF-1 stimulate STAT[91] binding to SIE. The activation of the transcription occurs in conditions in which Ras function is neutralized, suggesting that Ras activation is not required for this response to growth factors.

The Neurotrophins and Their Receptors Are Essential in the Maintenance and Development of the Nervous System

The neurotrophins are a family of trophic factors that include NGF, brain-derived neurotrophic factor (BDNF), neurotrophin 3 (NT-3), neurotrophin 4/5 (NT-4/5), and neurotrophin 6 (NT-6) (McDonald and Chao, 1995). These proteins participate in nervous system development by acting on particular sets of neurons and neuronal precursor cells (Snider, 1994). Their actions are carried out by binding to the neurotrophin receptors (Fig. 8), of which two types have been identified: p75 and the Trk family of RTKs. p75 acts as a low-affinity receptor for all the neurotrophins and belongs to a family of receptors, such as the p55 TNF-R and the Fas receptor, directly related to signaling cell death. Trk receptors, on the other hand, are RTKs that bind neurotrophins with different degrees of affinity, and this affinity may be increased in collaboration with p75.

A unique feature of NGF physiology is that this factor, in some instances, is synthesized at a distance from the cell body of the responsive neuron. This distance can be considerable, and therefore the action of NGF on the neuron depends on axonal transport of NGF from the site of synthesis (target tissue) to the soma of the neuron, a process called "retrograde transport." In some tissues, however, the pattern of gene expression of the neurotrophins and their receptors is coincident, suggesting that neurotrophins may also act in a paracrine or autocrine mode of intercellular communication. In addition, the response may be more complex since a neurotrophin receptor may interact with several ligands (Fig. 8). Thus, the TrkA tyrosine kinase receptor can be activated by NGF and NT-3; the TrkB receptor can be activated by BDNF, NT-4/5, and NT-3; and the TrkC receptor binds only NT-3. The affinity of NT-3 for TrkA and TrkB receptors, however, is lower than that for TrkC, and NT-3 is expected to substitute other ligands only when its concentration surpasses local concentrations of NGF/BDNF and when TrkC receptors are saturated or absent from the tissue.

Although the effect of neurotrophins on adult neuronal cells is mainly to sustain the viability, in precursors of differentiated neurons the response may be multifunctional, involving survival, differentiation, and even proliferation. A major advance in the knowledge of the physiological importance of neurotrophins and their receptors in nor-

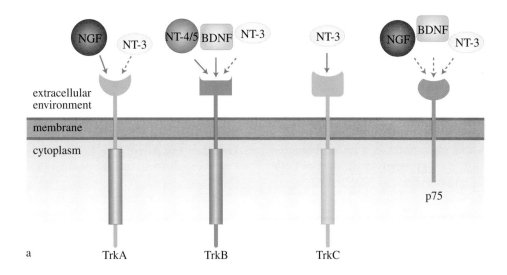

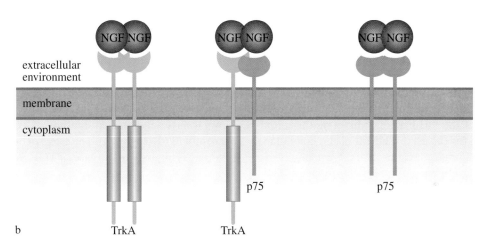

FIGURE 8 The neurotrophin receptors interact with several ligands. (a) Tyrosine kinase receptors of the Trk family, the low-affinity neurotrophin receptor p75, and their respective ligands (NGF, NT-3, NT-4/5, and BDNF). The solid arrows indicate high affinity in the ligand–receptor interaction, whereas dashed arrows represent low-affinity interactions. (b) The interactions between Trk receptors, between TrkA and p75, or between p75 receptors can determine a higher affinity between the receptor and an NGF dimer.

mal physiology was made possible by generation of mice in which the normal genes are eliminated. The phenotypes of the mice carrying NGF, BDNF, or NT-3 gene disruptions are very similar to those of TrkA-, TrkB-, and TrkC-null mice, respectively, supporting the biochemical and cell biological findings that linked NGF to TrkA, BDNF to TrkB, and NT-3 to TrkC. It is noteworthy that the disruption of the receptors usually leads to a more severe phenotype, indicating that individual neurotrophins are more dispensable than receptors and that other neurotrophins may substitute for the genetically ablated one.

The mechanism of signal transduction by p75 remains largely unknown. NGF binding to p75 has been reported to

stimulate the generation of ceramide from sphingomyelin and to activate NF-κB, the latter in clear analogy to the p55 TNF-R. Trk receptors usually trigger most of the intracellular cascades classical of RTK. Therefore, why do these factors trigger differentiation rather than proliferation? First, Trk receptors are mainly expressed in cells of tissues whose ultimate fate is that of differentiation. Through the development of these tissues it is therefore likely that the expression of Trk receptors occurs at a stage in which the progenitor cells are already precommitted by other factors to enter a differentiation program. Second, the signal sent by the Trk receptor may have qualitative or quantitative differences with that triggered by classical mitogenic receptors

(Marshall, 1995). In PC12 pheochromocytoma cells, the activation of a kinase cascade (that of the MAP kinase) lasts longer in cells treated with NGF (which acts as a differentiation factor) than with EGF (which acts as a mitogenic factor). Qualitative differences relate to the activation of potential differentiation pathways instead of (or together with) growth pathways. Although Trk receptors may induce tyrosine phosphorylation of cellular substrates such as Shc or PLCγ (whose phosphorylation is also triggered by EGF), other substrates apparently exist, such as SNT, that are specifically phosphorylated by NGF but not by mitogenic growth factors.

The TGF-β Factors Participate in Growth Arrest, Morphogenesis, and Remodeling of Tissues

The TGF-β superfamily includes more than 30 isoforms that regulate several developmental and proliferative aspects in organisms that diverge from lower invertebrates to humans. In *Drosophila*, the TGF-β-related factors drive early embryonic processes such as dorsoventral axis formation, gut development, and proximal–distal patterning of adult appendages. In *Xenopus*, a TGF-β-related gene functions at one end of the fertilized egg directing the initial events responsible for the establishment of the body plan. In mammals, physiological actions of TGF-β include the control of sexual development, pituitary hormone produc-

tion, and the formation and conservation of bones and cartilages (Kingsley, 1994).

As secreted proteins, the TGF-β family of polypeptides are synthesized as precursor molecules containing a hydrophobic leader sequence at the N terminus, followed by a propeptide region. This propeptide region is cleaved at a dibasic motif to produce a mature, 100- to 150-amino acid C-terminal fragment. TGF-β factors belong to the cystine knot family of polypeptide factors, and they hetero- or homodimerize to generate the biologically active soluble form (Massagué, 1996).

Major advances in the knowledge of receptors and signal transduction pathways that are triggered by TGF-β polypeptides have been obtained by biochemical and genetic studies. Initial cross-linking experiments carried out in the 1980s identified three types of transmembrane proteins that bind TGF-β factors: the type I, II, and III receptors. Molecular cloning of these receptor subtypes has uncovered some unique structural and functional properties of this receptor system. Type III receptors are high M_r proteoglycans that bind TGF-β but do not have any known signaling capability. They have a large glycosylated extracellular domain and a relatively short cytosolic region devoid of any known signaling domain. The type III receptor (also called betaglycan because of the proteoglycan structure of the ectodomain) acts as a presenting accessory molecule, i.e., allows transfer and efficient binding of TGF-β to the type II receptor (TβRII) (Fig. 9). The effect of betaglycan on ligand binding is minimal in the case of TGF-β_1 and -β_3 but strong for TGF-β_2, an isoform that has low affinity for TβRII. In

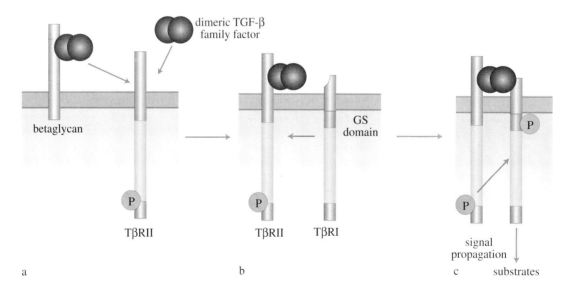

FIGURE 9 Schematic representation of signal propagation by TGF-β factors. The first step (a) is ligand binding to the ectodomain of the TβRII or presentation of the ligand to this receptor by the accessory molecule betaglycan. (b) This leads to association of TβRII with type I receptors and phosphorylation of the TβRI at the level of the GS domain. This activates the TβRI that transduces a signal by phosphorylation of specific substrates (c).

fact, cells that lack betaglycan, such as myoblasts and certain epithelial cells, are insensitive to TGF-β_2, although they express type I and II receptors and can be sensitized to TGF-β_2 by betaglycan expression. The type II and type I (TβRI) receptors are endowed with kinase activity. However, in clear divergence from the tyrosine kinase family of receptors, these two receptors are plasma membrane serine/threonine kinases.

Structurally, the receptors for type I and II can be subdivided into two families (Massagué, 1996). The type I receptors are more similar between the families than are the type II receptors. These receptors have 500–600 amino acids, with a relatively short ectodomain (e.g., 115 amino acids in the extracellular domain of the type II activin receptor) rich in cysteine residues. Some of these cysteines cluster in a membrane-proximal region, constituting a motif called the "cysteine box." Alternative splicing of the ectodomain has been implicated in the generation of receptor isoforms with different affinities for the ligand. The ectodomain is followed by a single hydrophobic transmembrane domain that precedes the cytosolic domain. The cytosolic domain contains a conserved region of approximately 30 amino acids N terminal to the kinase domain. This region, called the GS domain, contains the amino acid motif Ser-Gly-Ser-Gly-Ser-Gly and is essential for signal transduction.

Adequate signaling by TGF-β factors requires the presence of both type I and type II receptors. Type II receptors are required for high-affinity ligand binding, and type I receptors can only bind ligand when type II receptors are coexpressed. A different situation exists in the case of the bone morphogenic proteins (BMPs). The BMPs bind to both receptors with low affinity but synergize to create a high-affinity complex.

The mechanism of activation of the TβRI and TβRII receptor complex includes a first step of binding to type II receptors, followed by oligomerization with type I receptors and phosphorylation of the latter at a cluster of serine and threonine residues in the GS domain (Fig. 9). The fact that TβRII is the actual receptor subunit that binds TGF-β factors, and that TβRI acts as the principal transducer of the signal, led to the elaboration of a model in which both receptors closely interact in order to transduce the signal, with TβRII acting as the "receptor" for TGF-β factors and TβRI acting as a "signal transducer."

Once the receptor is activated, signals are transduced by intracellular proteins whose identity is beginning to be uncovered. Genetic studies in flies have identified a gene, *Mad,* that is required for decapentaplegic (dpp) activity (dpp is a fly analog of TGF-β). Also, in the worm *C. elegans* a search for genes with phenotypes analogous to daf-4 (another TGF-β factor of invertebrates) uncovered *sma-2, -3,* and *-4,* which code for proteins analogous to Mad. Sma-2 and daf-4 are cell autonomous and act on the same cell. In addition, daf-4 is unable to rescue sma-2 mutations, sug-

gesting that sma-2 is not required for daf-4 expression and indicating that sma-2 is in the signaling cascade of daf-4.

Using the TβRII cytoplasmic domain as a bait in the yeast two-hybrid system, a human protein called TRIP-1 was isolated. This protein interacts with the TβRII but not with type I, and it contains a WD motif, which is a consensus sequence present in multiple copies in diverse proteins and suspected to generate structural folds involved in protein–protein interactions. The association of TRIP-1 with TβRII requires the kinase activity and leads to TβRII-mediated TRIP-1 phosphorylation. Three other proteins have been identified using TβRI cytoplasmic domain as a bait: BMPR-II, the FK506/rapamycin-binding protein FKBP12, and the shared subunit of isoprenyltransferases FTα.

Biological Responses to TGF-β Factors

Cell biological and genetic studies have provided information regarding the role of TGF-β polypeptides in several aspects related to cell growth, development, and pathology (Massagué, 1996). TGF-β can elicit a wide variety of actions, including the control of cell adhesion, phenotype, and proliferation. In proliferation studies the response to TGF-β can be opposite depending on the experimental conditions. This has been elegantly illustrated in AKR-2B fibroblasts. TGF-β induces growth arrest in these cells but concomitantly stimulates PDGF-B expression. When these cells are placed in a mitogen-rich medium, the inhibitory effect of TGF-β is prevalent. However, when these cells are placed in a mitogen-free medium, the predominant response to TGF-β is proliferative due to stimulation of an autocrine loop by PDGF-B.

The predominant action of TGF-β, however, is that of inhibition of cell proliferation. This action is observed in many different cell types, including fibroblastic, lymphoid, neuronal, and hematopoietic. In certain cells TGF-β opposes the action of several well-known mitogens such as EGF or cytokines. This effect is mediated by a lengthening or an arrest of the cell cycle at a late G_1 point and does not directly affect other early signaling pathways triggered by growth factors. The degree of this inhibition depends on the cell type, and it is almost complete in certain lung fibroblasts.

The effect of TGF-β on cell adhesion depends on the effects of this factor on cell surface factors and secreted molecules that remodel the extracellular matrix. TGF-β usually increases cell adhesion by increasing deposition of the extracellular matrix, inhibition of its degradation, and modifications in cell surface adhesion receptors. Several extracellular matrix proteins, including fibronectin, collagens, and proteoglycans, are upregulated by TGF-β by a mechanism that includes increased transcription of the respective genes together with augmented mRNA stability. In addition to elevated synthesis of extracellular matrix proteins, TGF-β increases the levels of plasminogen activator inhibitor-1 and tissue inhibitor of metalloprotease, both inhibitors of enzymes involved in the degradation of the extracellular matrix. In addition, TGF-β also decreases the production of extracellular proteases such as collagenase and stromelysin.

Another group of target molecules in TGF-β actions that are involved in cell adhesion are the integrins. These molecules are heterodimers composed of an α and a β subunit. TGF-β increases the expression of individual subunits of the integrins. This is important because proper expression of cell adhesion receptors on the cell surface depends on the correct assembly of the heterodimeric cell adhesion receptors. In fact, if a subunit is present in limited amounts, the other subunits, even if they are present in excess, cannot be sorted to the cell surface and are degraded intracellularly. Therefore, increased expression of an integrin at the cell surface depends not only on an increased rate of biosynthesis but also on the balance between the α and the β subunits produced by the cell.

Much of the knowledge of the actions of TGF-β factors in development was derived from genetic studies of the inactivation of TGF-β factors in different animal models. In flies, the *dpp* gene product functions as a morphogen that dictates several developmental processes. In fact, the name decapentaplegic denotes the 15 or more defects found in the mutant animals. One of the prominent effects of *dpp* is its participation in the building of the dorsoventral body axis. Loss of function mutants of this gene lead to an extension of anterior body structures into the posterior region of the embryo. The importance of this gene in normal development is demonstrated by the fact that these mutations are lethal even in heterozygous animals, suggesting that the normal dosage of *dpp* is quite limiting in the normal embryo. Overproduction of *dpp,* on the other hand, leads to the opposite effect, i.e., dorsalization of the embryo.

Inactivation of the TGF-β locus in the mouse does not have any gross developmental effect. However, over time, infiltration of several organs by lymphocytes causes severe pathological processes leading to death. This is probably due to the immunosuppressive effect of TGF-β on lymphocyte growth and emphasizes the control of lymphocyte proliferation as one of the major functions of TGF-β in mammals.

References Cited

CAGAN, R. L., KRÄMER, H., HART, A. C., and ZIPURSKY, S. L. (1992). The bride of sevenless and sevenless interaction: Internalization of a transmembrane ligand. *Cell* **69,** 393–399.

CHAO, M. V. (1994). The p75 neurotrophin receptor. *J. Neurobiol.* **25,** 1373–1385.

CROSS, M. C., and DEXTER, T. M. (1991). Growth factors in development, transformation, and tumorigenesis. *Cell* **64,** 271–280.

HELDIN, C. (1995). Dimerization of cell surface receptors in signal transduction. *Cell* **80,** 213–223.

KAZLAUSKAS, A. (1994). Receptor tyrosine kinases and their targets. *Curr. Opin. Genet. Dev.* **4,** 5–14.

KINGSLEY, D. M. (1994). The TGF-beta superfamily: New members, new receptors, and new genetic tests of function in different organisms. *Genes Dev.* **8,** 133–146.

MARSHALL, C. J. (1995). Specificity of receptor tyrosine kinase signaling: Transient versus sustained extracellular signal-regulated kinase activation. *Cell* **80,** 179–185.

MASSAGUÉ, J. (1990). Transforming growth factor-alpha: A model for membrane-anchored growth factors. *J. Biol. Chem.* **265,** 21393–21396.

MASSAGUÉ, J. (1996). TGF-beta signaling: Receptors, transducers, and mad proteins. *Cell* **85,** 947–950.

MASSAGUÉ, J., and PANDIELLA, A. (1993). Membrane-anchored growth factors. *Annu. Rev. Biochem.* **62,** 515–541.

MCDONALD, N. Q., and CHAO, M. V. (1995). Structural determinants of neurotrophin action. *J. Biol. Chem.* **270,** 19669–19672.

NAGLICH, J. G., METHERALL, J. E., RUSSELL, D. W., and EIDELS, L. (1992). Expression cloning of diphtheria toxin receptor: Identity with a heparin-binding EGF-like growth factor precursor. *Cell* **69,** 1051–1061.

PAWSON, T. (1995). Protein modules and signaling networks. *Nature* **373,** 573–580.

PEREZ, C., ALBERT, I., DEFAY, K., ZACHARIADES, N., GOODING, L., and KRIEGLER, M. (1990). A nonsecretable cell surface mutant of tumor necrosis factor (TNF) kills by cell-to-cell contact. *Cell* **63,** 251–258.

RUBIN, G. M. (1991). Signal transduction and the fate of the R7 photoreceptor in *Drosophila. Trends Genet.* **7,** 372–377.

SNIDER, W. D. (1994). Functions of the neurotrophins during nervous system development: What the knockouts are teaching us. *Cell* **77,** 627–638.

ULLRICH, A., and SCHLESSINGER, J. (1990). Signal transduction by receptors with tyrosine kinase activity. *Cell* **61,** 203–212.

VAN DER GEER, P., HUNTER, T., and LINDBERG, R. A. (1994). Receptor protein tyrosine kinases and their signal transduction pathways. *Annu. Rev. Cell Biol.* **10,** 251–337.

ROSARIO RIZZUTO

*Department of Experimental and
Diagnostic Medicine,
Section of General Pathology
Via Borsari 46
44100 Ferrara, Italy*

Fluorescent and Chemiluminescent Molecules for Studying Dynamic Processes in Living Cells

Fluorescence microscopy, an experimental approach which allows the direct monitoring of biological processes in living cells, has undergone a major expansion in the past few years. This is mostly due to two factors: the technological advancement of the instrumentation, with the development of confocal microscopy and mathematical analysis of wide-field fluorescence microscopy ("digital imaging"), and the availability of novel probes which allow one to trace and dynamically follow cell structures or physiological parameters. In this article, I discuss two types of probes that, for the ease of use and the important applications, have had explosive success: the fluorescent Ca^{2+} indicators, which have extended to virtually all cell types in culture and in situ the measurement of this signaling ion, and the recombinant proteins, which, at least in principle, provide the great advantage of being selectively targeted to the cell compartment of interest upon addition of specific targeting signals to the polypeptide sequence. I discuss in detail two recombinant proteins: aequorin and green fluorescent protein.

Introduction

In the past few decades, an increasing number of researchers have undertaken the direct observation of the key processes of the life of a cell, such as movement, secretion, contraction, division, and simply death. At least three factors have contributed to the success of this experimental approach. The first was indirect: the explosive development of molecular biology, which allowed the "dissection" of the complex processes mentioned previously and the identification of the molecular actors. In many cases, the molecular analysis revealed complex scenarios, identifying different isoforms of the same protein with distinct tissue distribution and functional properties. The molecular reason for this complexity is obscure and awaits the direct analysis of the processes *in situ*. At the same time, molecular biology

techniques provided the possibility of recombinantly expressing a heterologous protein, i.e., of "adding" a specific protein to the molecular repertoire of a cell. It is now possible to endow a cell with a protein (e.g., a signaling molecule such as a kinase) that was not part of its repertoire to express a mutated form of an endogenous protein, thus verifying the effect of specific biochemical alterations, etc. Finally, with a more complex approach, it is possible to eliminate ("knockout") the gene of a specific protein and thus investigate directly the effect of its removal. In all cases, the direct monitoring of the process of interest is an essential part of these studies.

The second important factor was the technological advancement of optical and computer technology, which allowed the introduction of highly sophisticated microscopes and imaging systems. Indeed, the 1990s were characterized by the wide diffusion of confocal microscopy and the development of algorithms which allow the calculation, as in traditional wide-field fluorescence images, of the amount of signal originating from the out-of-focus portion of the sample and thus remove computationally the spurious haze. These two approaches allow one to obtain images with high contrast and resolution from complex samples, including whole organisms.

A third, decisive element for fully exploiting the new imaging technologies is the availability of probe which allows the labeling, within a living cell, of subcellular components of interest (organelles, proteins, etc.) or the measurement of relevant physiological parameters, such as the concentration of Ca^{2+}, which is an important intracellular second messenger. To this end, two groups of fluorescent molecules are particularly useful and have been increasingly used in the past few years: the fluorescent indicators of cell parameters, which are small organic molecules trappable in the cytoplasm, among which the Ca^{2+} indicators are the most important; and green fluorescent protein (GFP), a protein of the jellyfish *Aequorea victoria*. These probes, and in particular GFP, which has radically changed the perspectives of fluorescence microscopy in the past few years, are the main topics of this article. I also discuss the photoprotein aequorin, which is a chemiluminescent and not a fluorescent protein. However, at least historically, it is very interesting since it was the first recombinant protein to be employed in living cells for measuring an important physiological parameter (the concentration of Ca^{2+}), thus highlighting the extraordinary potential usefulness of this experimental approach.

Fluorescent Ca^{2+} Indicators

The first demonstration of the role of Ca^{2+} ions in controlling a cellular function (e.g., the cardiac muscle contraction) dates back more than a century ago; however, the extension of this concept — the general role of Ca^{2+} in cell signaling, which is today established knowledge — had to wait for the development of specific probes capable of measuring, in a wide variety of cell types, the intracellular Ca^{2+} concentration. In particular, there are two major methodological advancements. The first was the isolation of natural molecules, such as the photoprotein aequorin (Ridgway and Ashley, 1967), or the synthesis of chemical compounds, such as the metallochromic indicators arsenazo III and antipyrilazo III (Thomas, 1982), which allowed reliable measurements of [Ca^{2+}]. These indicators played an important role in the study of Ca^{2+} ions as intracellular second messengers. For example, the observation that various extracellular stimuli cause rhythmic oscillations of [Ca^{2+}], a phenomenon which has generated a very broad interest, was made in studies employing aequorin as the Ca^{2+} probe (Woods *et al.*, 1986). Studies with these probes, however, were associated with

the major experimental drawback that the indicator had to be introduced into the cells by microinjection because neither a protein nor a charged molecule, such as a metallochromic indicator, could permeate through the cellular membrane of a living cell. This obviously restricted this approach to the few cell types that were amenable to microinjection.

The scenario changed radically when R. Tsien and colleagues (1982) developed a new method for measuring intracellular Ca^{2+} concentration based on the use of fluorescent indicators, which can be trapped in the cytoplasm of the cells. The introduction of these indicators was a revolution in the study of Ca^{2+} homeostasis and they were the prototype of a variety of indicators for physiological parameters, such as pH (Rink *et al.*, 1982), the Na^+ concentration (Minta and Tsien, 1989), and Mg^{2+} concentration (Raju *et al.*, 1989).

The essential characteristics of a good indicator are its selectivity, the quality of the signal, and the ease of use. Ca^{2+} fluorescent indicators, among which the quin-2 molecule was the precursor, enumerate tens of different compounds which share the same start point: a Ca^{2+}-specific chelator, such as the EGTA or BAPTA (Fig. 1). The portion of the molecule that binds Ca^{2+} is a carboxylic backbone that is perfectly adapted to the dimension of the ion: this confers its specificity. A fluorophore group, associated with the carboxylic group, endows the molecule with fluorescent properties dependent on the binding of Ca^{2+} to the carboxylic cage. Due to their structure, particularly the presence of carboxylic groups, these indicators do not freely permeate across cellular membranes. To solve this problem the original molecule was modified by Tsien and coworkers by esterification of the charged groups in order to neutralize the positive charges. In this form, the molecule cannot bind Ca^{2+} but can permeate through the plasma membrane of a living cell and be trapped in the cytoplasm, where cellular enzymes (the esterases) hydrolyze the ester, yielding the active form of the indicator (which cannot diffuse back out). The key to the success of these Ca^{2+} indicators is their simplicity of use. Indeed, it is sufficient to incubate the cells for 20–30 minutes with the ester form of the indicator in the culture media in order to load a sufficient amount of the dye to proceed with Ca^{2+} measurements. To carry out the measurement it is necessary to remove spurious fluorescence by washing the cells with fresh medium, and the active fraction of the indicator, trapped in cytoplasm of living cells by the action of esterases, will be utilized in the measurements.

The New Generation of Fluorescent Indicators

Compared to the first fluorescent indicator developed by Tsien, the indicators developed in the following years

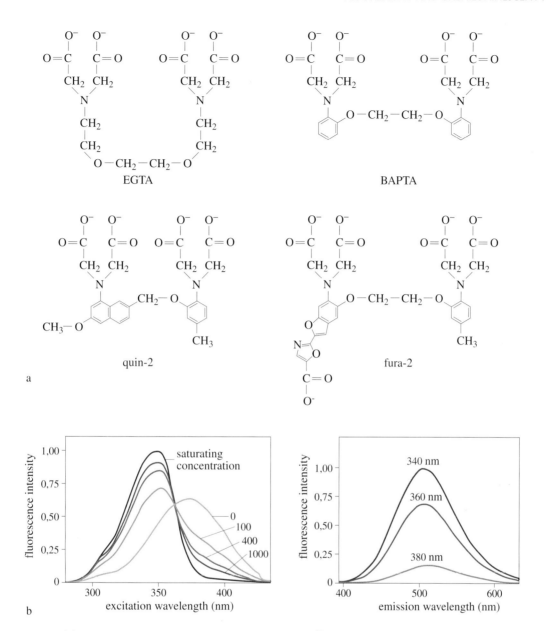

FIGURE I (a) Chemical structures of two chelators of Ca^{2+} ions, ethyleneglycol-bis(β-aminoethylene)-
N,N,N',N'-tetraacetic acid (EGTA) and 1,2-bis(2-aminophenoxy)ethylene-N,N,N',N'-tetraacetic acid (BAPTA)
and of the two derived fluorescent dyes quin-2 and fura-2. (b) Absorbance spectrum characteristics of the fluorescent indicator fura-2. (Left) The excitation spectrum in the presence of different nanomolar concentrations
of Ca^{2+} (indicated on the side of each curve), starting from the null concentration to the concentration at which
the fluorescent emission is maximal (saturating concentration). (Right) Changes in the emission intensity (peak
at 505 nm) at different excitation wavelengths (indicated above each curve) in the presence of a constant Ca^{2+}
concentration (400 nM). In both graphs, fluorescence intensity is normalized on the maximum peak.

(Grynkiewicz *et al.*, 1985) are chemically modified in order
to optimize their use in physiological studies.

First, it is possible to utilize smaller quantities of indicator because of their major molar fluorescence intensity,
thus reducing the Ca^{2+} buffering effect within the cytoplasm. The higher the intrinsic fluorescence of the indicator,
the smaller the amount of indicator needed for the measurement and hence the perturbation of Ca^{2+} intracellular

homeostasis. On average, although it was necessary to load
the cells with a concentration of quin-2 of approximately
0.5–1 mM, 10 times lower concentrations of fura-2, a widely
employed fluorescent indicator, are sufficient to accurately
carry out Ca^{2+} measurements.

A second important property of the new generation of
Ca^{2+} indicators is their lower affinity for Ca^{2+} ions (e.g.,
fura-2 has a dissociation constant $k_d = 224$ nM vs a $k^d =$

115 nM for quin-2). For this reason, these dyes are better suited for measuring Ca^{2+} concentrations in the range of 1–3 μM (values which are normally reached in the cytoplasm of stimulated cells). At these Ca^{2+} values, the fluorescence of quin-2 is completely saturated.

Finally, the binding of Ca^{2+} ions not only changes the quantity of emitted fluorescence but also induces an alteration in the spectrum (depending on the indicator, in the excitation or in the emission spectrum). In other words, the free form of the dye and the Ca^{2+} complex form are both fluorescent but they have different spectral characteristics. For example, if fura-2 is excited at a 340 nM wavelength, when the Ca^{2+} concentration increases the emission of the dye (at 505 nM) increases, but if the fura-2 is illuminated by a 380 nM wavelength the emission decreases (Fig. 1). The Ca^{2+} concentration can be calculated by the ratio of the intensity of fluorescence emitted upon illumination with the two wavelength; in this way, the measurement is independent of the quantity of indicator loaded into the cells which, in contrast, severely affects the measurements obtained by exciting the fluorophore at a single wavelength. This property, which represents a major methodological advancement, is of utmost importance in single cell imaging experiments. In these cases, there may be dishomogeneities in the loading of the dye due to the different thicknesses of the various portions of the cell.

The Ca^{2+}-dependent change of emission spectrum, typical of some new indicators (e.g. indo-1), is particularly helpful in confocal microscopy. In this case, the excitation of the sample is carried out by a laser beam. For these reasons, it is difficult to change the excitation wavelength, so it is inconvenient to use indicators such as fura-2 which are endowed with a Ca^{2+}-dependent variation in the excitation spectrum. Conversely, by using two different photomultipliers and appropriate filter sets which allow the collection of light at different wavelengths, with dyes such as indo-1 it is possible to carry out the same type of measurement (the ratio between Ca^{2+}-sensitive and Ca^{2+}-insensitive wavelengths) at the confocal microscope.

Another important advantage of some of the new-generation indicators, such as fluo-3 (Minta *et al.*, 1989), is that they are excited by visible light. On the contrary, quin-2 and fura-2 utilize UV light as an illumination source, which induces modification of biological macromolecules and thus may damage the cells or cause undesirable collateral effects. Moreover, upon UV illumination autofluorescence [due to endogenous fluorophores such NAD(P)H] is quite high, whereas it is much more modest with visible light.

Figure 2 shows a very recent application of fluorescent dyes in the measurement of Ca^{2+} concentration. The experiment consists of the measurement of variation in cytosolic Ca^{2+} concentration in astrocytes and neurons *in situ*,

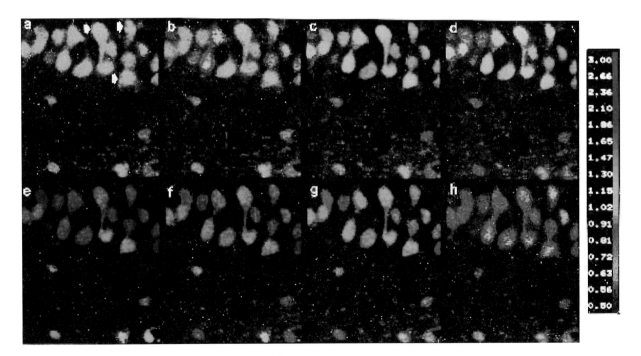

FIGURE 2 Measurements of the cytosolic Ca^{2+} concentration using the fluorescent dye indo-1 in brain slices of newborn rat. (a) The cells (pyramidal neurons of the hippocampal region CA1) are in resting conditions. (b–g) The cells' response to stimulation with 40 mM KCl. The acquisition time is 2 s. The relationship between pseudocolor and Ca^{2+} concentration (mM) is shown by the scale on the right (courtesy of G. Carmignoto and L. Pasti).

namely, in an intact section of cerebral tissue (in this case the CA1 hippocampal region). In the experiment shown, thin slices 200 μM thick were obtained by sectioning a neonatal rat brain with a vibratom and were loaded with the fluorescent indicator indo-1 using the same protocol described for the cells in culture. At the end of the incubation, the brain slice was transferred to the field of analysis of the confocal microscope which, in a complex sample such as a tissue portion, permits one to selectively analyze the fluorescent signal originating from single cells.

The cytosolic Ca^{2+} concentration changes, which in the case of indo-1 are translated in changes in the ratio between the emitted fluorescence at two different wavelength (405 and 485 nm), are expressed in pseudocolors. When a depolarizing stimulus was applied (40 mM KCl), the pyramidal cells showed a clear increase in the intracellular Ca^{2+} concentration due the entry of Ca^{2+} ions through voltage-dependent Ca^{2+} channels. Upon washout of KCl (thus eliminating the depolarization stimulus) the cytoplasmic Ca^{2+} concentration returned to the basal values.

The Protein Probes: Recombinant Aequorin, a Specifically Targeted Ca^{2+} Probe

The wide variety of molecular biology techniques, with the ensuing possibility of modifying and expressing exogenous proteins in living cells, has extensively expanded the number of applications of protein probes in cell biology. Two types of protein probes currently employed derive from the wide variety of bioluminescent organisms. The first is the group of chemiluminescent proteins, which emit light usually in response to changes in a physiological parameter, such as the concentration of ATP or Ca^{2+}. Since mammalian cells are not endowed with chemiluminescent proteins, the use of these probes in cell biology is usually associated with a low signal to noise ratio, as in the case of the photoprotein aequorin. Conversely, because the signal reflects only the light emitted from the recombinant protein, it is low and must be collected either by integrating the signal coming from at least 103 or 104 cells or, at the single cell level, using imaging analysis systems of high sensitivity and high cost.

The second group of protein probes, which has recently encountered increasing success, is that of fluorescent proteins. Among these proteins, green fluorescent protein of *Aequorea victoria* is the undisputed star and will be discussed in-depth here.

As previously discussed, aequorin and metallochromic dyes were gradually replaced in the study of Ca^{2+} homeostasis by the fluorescent dyes developed by Tsien and coworkers which can easily be used in virtually all cell types.

In particular, the new generation of dyes, as a result of the high fluorescent signal, allow, with traditional or confocal fluorescence microscopes, to image at the single cell level the changes in Ca^{2+} concentration induced by physiological stimuli. This methodological improvement has revealed an unexpected spatiotemporal complexity of this signaling mechanism.

The cloning of the aequorin cDNA (Inouye *et al.*, 1985), however, has shed new light on the use of this protein as a Ca^{2+} probe in living cells. First, the availability of the cDNA solves the problem of introducing this protein into the cells. Indeed, transfecting a cell with a plasmid which allows the recombinant expression of exogenous protein, such as aequorin, is a simple and effective procedure. By this means, it is possible to introduce aequorin in a wide variety of cell types, different both in morphology and in embryological origin, such as stable epithelial cell lines (HeLa and CHO), primary cultures of skeletal muscle (Brini *et al.*, 1997), neurons, and colangiocytes. After the cloning of the cDNA simplified the use of aequorin, its revaluation depended on two additional factors. The first is some favorable functional properties: the excellent signal to noise ratio, the wide spectrum of measurable Ca^{2+} concentrations, and its efficiency at a very low protein concentration, thus allowing the reduction of the buffering effect on the Ca^{2+} concentration (Brini *et al.*, 1995). The second is the great advantage provided, in principle, by protein probes, i.e., that of controlling in a selective way its intracellular localization. In fact, it is known that the intracellular sorting of the endogenous proteins depends, in most cases, on the presence of specific signaling information included within the primary sequence of the protein. These "targeting" sequences are recognized by specific sorting machineries and are necessary and sufficient for the correct delivery of the polypeptide. Using molecular biology techniques, if such specific targeting sequences are added to the sequence of the proteins of interest, the latter acquire the intracellular distribution determined by the signal sequence. In our case, by fusing the cDNA of different proteins to that of aequorin, we were able to add specific targeting information to the Ca^{2+}-sensitive photoprotein. Each chimeric protein, recombinantly expressed in cell cultures, maintains the ability of aequorin to measure Ca^{2+} concentration but is exclusively localized in the intracellular compartments specified by the targeting sequence. This approach has allowed the construction of Ca^{2+} probes which, in contrast to the fluorescent dyes, are exclusively localized in the intracellular district of interest.

Figure 3 is a schematic representation of the reaction of light emission of aequorin. Aequorin is an apoprotein of 21 kDa of different species of *Aequorea* which, in the active form, binds a prosthetic hydrophobic group, coelenterazine. When Ca^{2+} ions bind to the three high-affinity Ca^{2+} binding sites (homologous to the sites present in other Ca^{2+}-binding proteins, such as calmodulin), aequorin undergoes

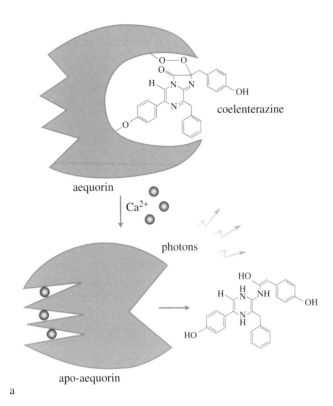

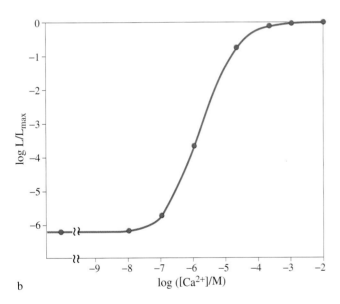

FIGURE 3 (a) Scheme of the chemiluminescent reaction of aequorin triggered by binding of Ca^{2+} ions (in the presence of oxygen and 2-mercaptoethanol) to the specific binding sites of the photoprotein. (b) Graphical representation of the changes in the rate of light emission of aequorin as a function of the different Ca^{2+} concentrations to which it is exposed. L, rate of photon emission at each value; L_{max}, rate of photon emission at saturating Ca^{2+} concentrations. The ratio L/L_{max} is 1 at saturating Ca^{2+} concentrations.

an irreversible reaction in which the prosthetic group is modified and a photon is emitted. The rate of the reaction depends on the $[Ca^{2+}]$ to which the photoprotein is exposed. This is apparent from the graph presented in Fig. 3, which shows the relationship between $[Ca^{2+}]$ and the fractional rate of aequorin consumption, expressed as the ratio between the emission of light at a defined Ca^{2+} concentration and the maximal rate of light emission at saturating $[Ca^{2+}]$. In the physiological range of $[Ca^{2+}]$ light emission is proportional to the second or third power of $[Ca^{2+}]$.

The Measurement of Ca^{2+} Concentration in Various Intracellular Compartments Using Aequorin Chimeras

Figure 4 shows the chimeric cDNA constructs prepared for targeting aequorin to the various intracellular compartments and some examples of measurements of Ca^{2+} concentration carried out with these probes.

Cytoplasm

The construct employed for measuring $[Ca^{2+}]$ in this compartment (cytAEQ) corresponds to the wild-type aequorin cDNA, with the addition of only a short (9-amino acid sequence (HA1) derived from hemagglutinin of influenza virus which is specifically recognized by a monoclonal antibody. This modification [i.e., the addition of a strong epitope to a protein with weak immunogenic properties (epitope tagging)] is useful for determining the recombinant protein in transfected cells by immunocytochemistry. In the corresponding trace, the cells transfected with cytAEQ were treated with a compound (histamine) which stimulates the cells by causing an increase in the cytoplasmic Ca^{2+} concentration. This occurs via the interaction with specific receptors which are coupled, via a trimeric G protein, to an effector enzyme (phospholipase c) that produces inositol 1,4,5-triphosphate (IP3), which in turn acts on channels localized in the endoplasmic reticulum and causes the release of Ca^{2+} from this compartment to the cytoplasm. In agreement with the data obtained with fluorescent Ca^{2+} indicators, the Ca^{2+} concentration in the cytoplasm, which at rest is 0.1 or 0.2 μM, increases upon addition of histamine to peak values of approximately 2.5 μM and then declines to a sustained plateau, which is maintained throughout agonist stimulation.

Mitochondria

Apart from the 13 polypeptides encoded by the mitochondrial genome, all other mitochondrial proteins are encoded by the nucleus, translated in cytosolic ribosomes, and then imported in the organelle (Hartl *et al.*, 1989). In most cases, the recognition as mitochondrial protein and the import into the organelle depend on a signal sequence rich in

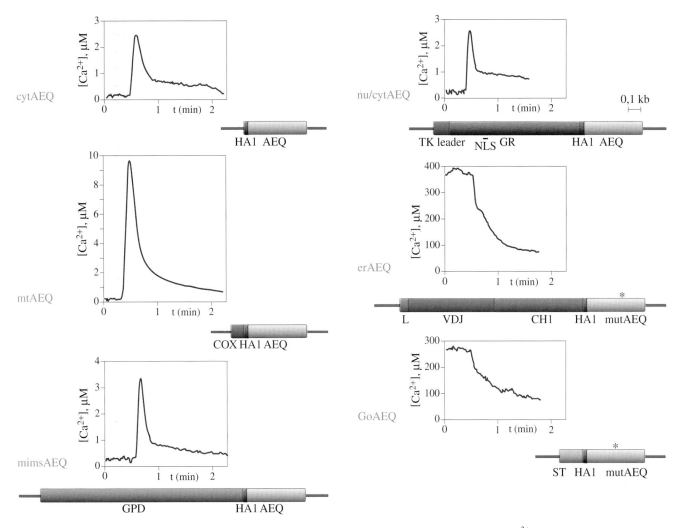

FIGURE 4 The cDNAs encoding the various aequorin chimeras and measurements of Ca^{2+} concentrations in HeLa cells. For each chimeric aequorin the trace of a typical experiment is shown on the top and the schematic map of the cDNA construct on the bottom. From top to bottom, left to right, cytosolic aequorin (cytAEQ), mitochondrial matrix aequorin (mtAEQ), mitochondrial intermembrane space aequorin (mimsAEQ), nuclear aequorin (nu/cytAEQ), endoplasmic reticulum aequorin (erAEQ), and Golgi apparatus aequorin (GoAEQ). NLS, nuclear localization sequence; GR, glucocorticoid hormone receptor; HA1, hemagglutinin epitope; GDP, glycerol-phosphate dehydrogenase; ST, sialyltransferase; COX, cytochrome c oxidase; TK, thymidylate kinase; VDJ, CH1, immunoglobulin fragments; L, leader sequence; AEQ, aequorin; mutAEQ, mutated aequorin (the asterisk indicates the mutation site).

basic residues which is located at the N terminus of the protein and is then removed after import by proteases of the mitochondrial matrix. In order to construct a "mitochondrial" aequorin, the cDNA of aequorin/HA1 was fused to the portion of the cDNA encoding the signal sequence and the first 6 amino acids of subunit VIII of cytochrome c oxidase, a mitochondrial protein. The encoded protein, denominated mitochondrial aequorin or mtAEQ, when expressed in mammalian cells is selectively localized in mitochondria (Rizzuto *et al.,* 1992). The Ca^{2+} measurements carried out with mtAEQ in HeLa cells, using an experimental protocol similar to that described previously, gave

a very surprising result. Upon stimulation with a physiological agonist such as histamine, mitochondria undergo changes in the matrix Ca^{2+} concentration which result in even higher concentrations than those of the cytoplasm (peak value approx 12 μM). The result is surprising because the mitochondria are endowed with low-affinity Ca^{2+} uptake systems, and thus at the cytoplasmic Ca^{2+} concentrations (both at rest and upon stimulation) mitochondrial uptake was expected to be negligible. As discussed later, the issue was solved by using GFP, which, appropriately targeted to the endoplasmic reticulum (ER) and mitochondria, allowed the demonstration of a close structural relationship between the ER (the organelle which acts as intracellular Ca^{2+} store) and mitochondria. The latter organelles, after stimulation, "sense" very high Ca^{2+} concentrations which meet the low affinity of their transport systems.

Mitochondrial Intermembrane Space

The mitochondria are enclosed by two membranes: The outer membrane is freely permeable to ions and small solutes, whereas the inner membrane is impermeable, thus allowing the creation of a proton electrochemical gradient by the respiratory chain. In order to measure the Ca^{2+} concentration between the two mitochondrial membranes, which was expected to differ substantially from that of the matrix and to be similar, if not identical, to that of the cytoplasm, aequorin was fused to a protein resident in that space (Rizzuto et al., 1998b). The protein we employed was glycerol phosphate dehydrogenase (GPD), a mitochondrial protein which is inserted in the inner membrane with a large C-terminal portion protruding in the intermembrane space. By appropriate fusion of the two cDNAs, a chimeric cDNA was constructed that encodes a protein (mimsAEQ) in which aequorin is added to the C terminus of GFP. The measurements of Ca^{2+} concentration with this probe showed that, in HeLa cells, upon stimulation with histamine the mitochondrial intermembrane space undergoes a Ca^{2+} change which is larger than that of the cytoplasm (peak value 3.5 vs 2.5 μM in the cytoplasm). This result, which is somewhat surprising considering that Ca^{2+} is released from the ER into the cytoplasm and then diffuses into the intermembrane space of mitochondria, suggests that the surface of mitochondria is not exposed to the bulk cytosolic $[Ca^{2+}]$ but rather to cytosolic microdomains in which very high Ca^{2+} concentrations are reached. This observation is in good agreement with the data obtained with the chimeric GFPs that showed a close contact between the mitochondria and the intracellular Ca^{2+} store, the ER.

Nucleus

Nuclear localization of proteins depends on a short targeting sequence with defined characteristics, which is denominated nuclear localization sequence (NLS). In distinction to the mitochondrial targeting sequence, NLS does not have a preferential localization within the protein and is not removed after the protein reaches its final destination (Garcia-Bustos et al., 1991). In some cases, nuclear localization is transient, e.g., the protein reaches the nucleus only in some phases of the life of the cell or in some physiological conditions. This happens because the protein may change its conformation and in some cases expose only the NLS, which is then recognized and allows the transport of the protein to the nucleus. For example, the NLS sequence of the glucocorticoid receptor (GR) is exposed and thus active only when the receptor binds the hormone; only in this case does the protein translocate to the nucleus and exert

its biological effect. Nuclear aequorin (nu/cytAEQ) was constructed by fusing the cDNAs of aequorin and GR, and it retains the intracellular sorting pattern of the latter polypeptide (Brini et al., 1993). Thus, it translocates to the nucleus only if the cells are treated with glucocorticoid hormones; otherwise, it is retained in the cytoplasm. In other words, the localization of this aequorin chimera depends on the experimental conditions. This is very useful because it allows the use of the same probe — thus avoiding any possible artifact due to differences in the Ca^{2+} affinities of two aequorin chimeras — to measure the Ca^{2+} concentrations of two compartments which were expected to have very similar Ca^{2+} concentrations, both at rest and upon stimulation. The results obtained indeed show that the changes in concentration which occur in the nucleus upon stimulation are virtually identical to those of the cytoplasm (the peak value after stimulation is approx. 2.5 μM), suggesting that the nuclear pores are open in vivo and the Ca^{2+} signal elicited in the cytoplasm diffuses to the nucleus, with no significant delay or decrease in amplitude.

Endoplasmic Reticulum

The proteins of the ER are retained in this compartment because they usually contain a double targeting signal. A hydrophobic sequence located at the N terminus of the protein causes its translation on membrane-bound ribosomes and its insertion in the ER. This signal and this mechanism are also shared by proteins which reside in other portions of the intracellular vesicular system (Golgi apparatus and secretory vesicles) and by secreted proteins (Sitia and Meldolesi, 1992). The retention of proteins within the ER depends on a second signal, which retrieves the proteins escaping into later compartments. The best characterized of these signals is the tetrapeptide KDEL, localized at the C terminus of the protein. In the case of aequorin, this latter signal could not be employed because the modification of the C terminus of the photoprotein drastically impairs its chemiluminesce properties. An alternative strategy was thus devised which takes advantage of a retention signal localized in a different portion of the protein. One of these cases is the signal that allows the heavy chain of immunoglobulins to be retained within the ER until assembly with the light chains occurs so that only mature immunoglobulins progress toward secretion (Sitia and Meldolesi, 1992). The map of the construct (Fig. 4) shows that the aequorin chimera localized in the ER (erAEQ) derives from the fusion of the cDNA of aequorin/HA1 with that of the heavy chain of immunoglobulins (Montero et al., 1995). The chimeric protein includes, from the N to the C terminus, the hydrophobic sequence that allows the insertion in the ER (leader sequence; L), the VDJ and CH1 portions of immunoglobulin (CH1 includes the ER retention signal), the HA1 epitope, and aequorin. In order to measure Ca^{2+} concentrations within the ER, however, a second problem needed to be solved. Indeed, although aequorin has a much broader dynamic range than fluorescent dyes, it does not allow one to measure reliably the high luminal Ca^{2+} concentration of the ER. It was thus necessary to reduce the affinity of aequorin. This was done using three approaches: the mutation of one of the three Ca^{2+}-binding sites (Kendall et al., 1992), the use of a modified prosthetic group, coelenterazine n (Barrero et al., 1997), and the substitution of Ca^{2+} within the cells with Sr^{2+}, a cation that mimics the behavior of Ca^{2+} but for which aequorin has a 50-fold lower affinity (Montero et al., 1995). The results obtained with mutated aequorin and coelenterazine n, using an experimental protocol similar to the one described previously, show in this compartment a radically different picture. In the lumen of the ER the resting Ca^{2+} concentration is approximately 400 μM (i.e., more than 1000 times higher than that in the cytoplasm) and, upon cell stimulation with histamine, it rapidly decreases due to the release of Ca^{2+} via the IP3-sensitive channels. These results demonstrate that the ER is the intracellular store of Ca^{2+} which is released upon agonist stimulation.

Golgi Apparatus

The targeting sequences responsible for the localization of the resident proteins of the Golgi apparatus are not as well-known as those of the compartments described previously. However, it is known that a transmembrane portion of sialyltransferase, an enzyme localized in the lumen of the Golgi apparatus, contains the targeting information and, when added to a heterologous protein, causes its retention in this organelle (Masibay et al., 1993). In order to construct an aequorin chimera localized in the Golgi apparatus, via appropriate fusion of their cDNAs the photoprotein was fused to the N-terminal portion of sialyltransferase, which included the transmembrane targeting segment. The Ca^{2+} measurements carried out with this chimera (GoAEQ) gave interesting, unexpected results (Pinton et al., 1998). As is apparent from Fig. 4, which shows an experiment using HeLa cells with a protocol similar to that described previously, the resting Ca^{2+} concentration in the lumen of the Golgi apparatus is very high (approximately 300 μM), comparable (albeit lower) to that of the ER. Upon stimulation with histamine, the Ca^{2+} concentration rapidly decreases, providing the novel notion that the Golgi apparatus shares with the ER the role of intracellular store of agonist-mobilizable Ca^{2+}.

The Protein Probes: Green Fluorescent Protein

GFP of A. victoria has recently become a widely employed tool in cell biology. It has been known for more than two decades that the luminous properties of some marine or-

ganisms depend not only on photoproteins (e.g., aequorin) that emit light upon enzymatic conversion of a cofactor but also on fluorescent proteins that modify the wavelength of the emitted light. However, the idea of using the latter proteins as biological tools is very recent. After the cloning of the cDNA of GFP in 1992 (Prasher et al., 1992), in an enlightening study M. Chalfie and coworkers (1994) showed that, as a result of the fluorescence of GFP, it was possible to identify neuronal cells in living animals, in the genome of which the cDNA of GFP was inserted under the control of a promoter active only in neurons. This elegant experiment not only demonstrated that the simple recombinant expression of GFP yields a strongly fluorescent protein but also provided the first clear example of biological application. The diffusion of this experimental approach was outstanding. Only a few years after the study by Chalfie and coworkers, hundreds of laboratories throughout the world utilized this protein for a wide variety of research applications (in 1997, more than 400 papers describing experiments using GFPs were published).

The reasons for the success of GFP are evident. First, as described previously, the fluorescence of GFP is not species specific and does not require the addition of cofactors to the cells transfected with the GFP cDNA. Its simple recombinant expression thus yields an intense fluorescent signal in bacteria, fungi, molds, plants, and mammalian cells. Moreover, the small size of GFP (approximately 27 kDa) allows one to fuse it with proteins of interest without interfering, in most cases, with their assembly or functional properties. Finally, the isolation of mutants with different spectral properties, as discussed later, significantly expands the number of applications.

Figure 5 shows the three-dimensional structure of GFP (Ormo et al., 1996). It is apparent that the protein forms a compact cylinder composed of 11 β-sheets, which surround an α-helix containing the fluorophore formed by the cyclization of three amino acids of the protein Ser(65)–Tyr–Gly. The cyclization is a slow process which occurs posttranslationally in the presence of molecular oxygen and thus accounts for the delay between the biosynthesis of the protein and the appearance of a fluorescent signal. This delay may represent a major problem in some applications of GFP (e.g., as a reporter of gene expression).

The fluorophore of GFP exists in two chemical forms which account for the two excitation peaks of the fluorescent protein (Fig. 5). The peak at 470 nm is due to the anionic form of the fluorophore, whereas the peak at 395 nm (UV light) is due to the neutral form. Both forms then emit green light (emission peak 510 nm). In wild-type GFP, the two forms of the fluorophore are in 6:1 ratio (neutral/ionized) and thus the excitation peak is higher at UV light. However, UV illumination causes, through the process of photoisomerization, the ionization of the fluorophore and thus the reduction of the fluorescent signal, with a parallel

a

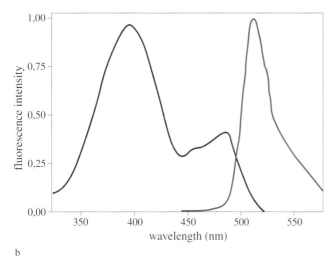

b

c

FIGURE 5 (a) Stereoscopic image of the three-dimensional structure of GFP. The β-sheets are shown in yellow and the α-helices in red (molecular coordinates from Ormo *et al.*, 1996). (b) Excitation (red) and emission (blue) spectra of native GFP. The fluorescence intensity is normalized to the emission spectrum peak. (c) Fluorescence image of HeLa cells transiently transfected with cDNA encoding native GFP.

increase in the signal detected upon illumination with blue light.

Figure 5c shows a fluorescence image of HeLa cells expressing the cDNA encoding wild-type GFP upon illumination with UV light. It is possible to observe an intense green fluorescence diffused to the whole cytoplasm as well as to the nucleus. This confirms the notion that GFP, a heterologous protein, does not possess any targeting signal recognized by mammalian cells. Also, the diffusion to the nucleus is not surprising given the small size of the GFP, below the cut-off for nuclear exclusion (approx 60 kDa).

GFP Mutants

Shortly after the first report on recombinant GFP and the understanding of its usefulness, mutants of GFP with stronger, or at least different, fluorescence properties started to be isolated. Recently, the solving of the three-dimensional structure (Ormo *et al.*, 1996) enhanced the possibility of generating useful mutants. The main classes of GFP mutants are briefly discussed in the following sections, with an emphasis on those which can be applied to cell biology studies.

Deleted GFPs

Regarding deleted GFPs, the results obtained have been largely unsuccessful. It was initially hoped that the size of GFP could be drastically reduced, thus reducing its steric hindrance and its potential interferences with proteins of interest in appropriately prepared chimeras. Progressive amino acid deletion from the two extremities of the polypeptide, however, showed that only seven and one amino acids can be eliminated from the carboxy and amino terminus, respectively, without interfering drastically with the fluorescence properties (Dopf and Horiagon, 1996). In other words, only the whole protein is functional, as could be predicted from its complex three-dimensional structure (Ormo *et al.*, 1996).

Bright Mutants

Enhancing the GFP signal is a major goal, especially for those applications in which the amount of protein or the time necessary for the formation of the chromophore may be a limiting factor. In principle, this goal can be achieved by two different approaches, i.e., either modifying the intrinsic fluorescence properties of the protein or acting on other parameters which are important in the production of a fluorescent compound.

Among the first type of mutations, the most successful is the substitution, within the fluorophore, of the serine residue at position 65 with a threonine (the S65T mutation) (Heim *et al.,* 1995). These mutations change the ratio between the two chemical forms of the chromophore, increasing the amount of the ionized form. The result is an increased efficiency of GFP as a fluorophore excited by blue light. This is important because these are the illumination conditions which are less damaging to living cells and in which GFP is more stable. In the case of the S65T mutant, the excitation peak at UV light is nearly absent, whereas that at blue light is 6-fold higher than for the wild-type protein (Fig. 6). Moreover, the rate of formation of the chromophore with this mutant is 4-fold higher than for wild-type GFP, and the active form of the protein is 3.5-fold more resistant to photobleaching. For these reasons, S65T has largely replaced wild-type GFP for most applications, thus becoming the mutant of choice for most experiments in living cells.

The other mutations do not modify the fluorophore, and thus the intrinsic fluorescence properties, but rather interfere with various steps of the process that leads to the formation of the fluorescent protein. The first series of mutations, now widely employed, are the silent mutations that, although they do not change the protein sequence, introduce in the coding region of the cDNA the most common (and efficient) codons of mammalian cells, which often do not coincide with those of *Aequorea* (Zolotukhin *et al.,* 1996). This procedure, denominated humanization of the cDNA, allows for equal amounts of mRNA (i.e., for the same transfection efficiency, time elapsed after transfection, etc.) to produce a larger amount of protein, thus anticipating the attainment of the detection threshold. A second group of mutations solves two major problems in the production of this heterologous protein. The first is the tendency of the protein, when overexpressed, to acquired a wrong conformation and precipitate in nonfluorescent inclusion bodies (Cormack *et al.,* 1996). The second is the relative inefficiency in the formation of the chromophore at 37°C (i.e., the experimental conditions of cell biology studies) when compared to 15–20°C, the temperature to which the native protein is normally exposed (Siemering *et al.,* 1996). Finally, in some plants, such as *Arabidopsis thaliana* (a widely employed organism in plant biology), a silent mutation that eliminates a cryptic splice site present in the GFP cDNA markedly increases the amount of protein produced (Haseloff and Amos, 1995).

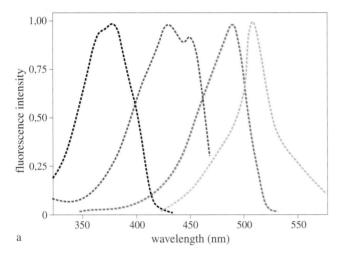

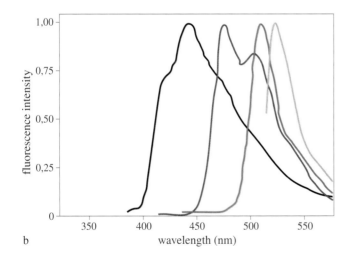

FIGURE 6 Excitation and emission spectra of GFP mutants with altered spectral characteristics. The fluorescent intensity values are normalized on the maximum peak. (a) Excitation spectra (dotted lines) of the Y66H,Y145F (dark blue), Y66W (light blue), S65T (dark green), and S65T,T203Y (light green) mutants. The letters in the names indicate the amino acids (F, phenylalanine; H, histidine; S, serine; T, threonine; W, tryptophan; Y, tyrosine). (b) Emission spectra (continuous lines) of the same mutants indicated in a.

The Mutant with Altered Excitation or Emission Wavelength

As discussed in more detail later, these mutants are extremely useful and considerably expand the possible applications of GFP. Indeed, they can be employed, in association with wild-type GFP or the S65T mutant, to reveal simultaneously two or more different proteins or structure of interest, to reveal the activity of different promoters, or to identify, in transgenic animals, cells of different embryological origin. In these applications it is necessary that the emitted light of the various mutants can be easily distinguished from that of the native protein. At least three spectral mutants of GFP fulfill this requirement (Heim and Tsien, 1996). The blue mutant, in which the spectral shift is due to a mutation in the fluorophore (Y66H), emits blue light (peak at 445 nm) when excited with UV light. The cyan mutant, due to a different mutation of the same amino acid (Y66W), has a bimodal excitation peak in the violet (433 and 453 nm) and emission peak between the blue and the green (475 and 501). The yellow mutant, due to the combination of the S65T mutation described previously with the T203Y mutation, when excited with green light (513 nm) emits yellow light (527 nm). Together with the mutations which actually cause the spectral shift, these mutants often include other mutations which stabilize the tertiary structure around the modified fluorophore (e.g., the blue mutant also contains the Y145F mutation).

··•

Applications

GFP as a Reporter of Gene Expression

The control of the spatial and temporal distribution of the expression of a gene rests usually in the information included upstream of the transcribed region. The understanding of these mechanisms is an intense field of study, and a common approach is that of fragmenting the putative informative region of the gene and verifying the capacity of the various portions to drive the expression of a reporter gene placed under their transcriptional control. In other words, the cDNA of the reporter protein is placed immediately downstream of the putative promoter; this construct is then transfected into cultured cells, and the quantity of expressed protein reveals the efficacy of the promoter. The protein normally employed for this task (chloramphenicol acetyltransferase β-galactosidase, or luciferase), which converts specific substrates into measurable products, depends on the interaction of the enzyme with the substrate and thus, in intact cells, on the diffusibility of the latter. Since the diffusion across biological membranes of the currently available substrates is poor, the cells must be fixed or permeabilized in order to carry out the experiments. With the exception of luciferase (which, given the low light output, needs sophisticated and expensive instrumentation), the reporter proteins currently available do not allow one to monitor dynamically the process of gene expression. Conversely, GFP, as a result of the autonomous formation of the fluorescent species and its strong intensity, allows analyses at the single cell level and, in some transparent organisms such as the nematode *Caenorhabditis elegans* (Chalfie *et al.*, 1994) also in transgenic animals. In contrast to luminescent proteins, however, GFP has to be revealed over a sizable endogenous fluorescence and thus it is necessary to express a relatively high amount of fluorescent product in order to detect a clear signal. Therefore,, the bright mutants described previously are particularly useful for this application, in which it is very important to detect early the activation of the promoter.

GFP as a Transfection Marker

The modification of the molecular repertoire of a cell by expressing specific proteins, through transfection experiments, is an effective approach for studying complex cellular processes such as the mechanisms which allow the cell to recognize extracellular ligands (hormones, growth factors, and neurotransmitters) and translate them into intracellular activation events (e.g., secretion, motility, differentiation, and growth). These are usually complex, often redundant signaling pathways which then converge onto common effectors. To reveal the significance of this complexity, and the specific role of proteins which are similar in structure and function, it is possible to express one of these proteins in its native form or as a mutant with altered biochemical properties and then verify the effect on cell function. This can be done two different ways: the selection, after cotransfection with a selectable marker, of cell clones which have inserted in the genome the exogenous gene and express it indefinitely, or the analysis shortly after transfection of the fraction of transfected cells which express the exogenous protein. The first procedure is more laborious and, since it analyzes the progeny of a single cell, it is associated with the major risk of emphasizing random, nonsignificant differences already present in the parent population. The second procedure analyzes a much more representative population of cells but is associated with a major experimental problem: Depending on the cell type and the experimental conditions, the transfected cells are comprise a fraction which varies between 3 and 40%. Therefore, functional studies must be carried out on this subset of cells, which must be distinguished from the others, which are unmodified with respect to the parental cells. Before the advent of GFP, the functional analysis (e.g., a single-cell Ca^{2+} measurement or an electrophysiological study) was carried out "blind," and then the cell was killed and the expression of the heterologous protein was verified a posteriori by immunocytochemistry. GFP solves this problem because, if cotransfected with the cDNA of interest, it plays the role of marker of transfection in living cells, which will be invariably positive for both the transfected gene of interest and GFP.

GFP as a Marker of Organelles and Other Cell Structures

The three-dimensional organization of the cell and its components is an important biological issue which the current systems for acquiring and analyzing images solve with high spatial resolution and image quality. GFP is very useful in these studies because, if the specific targeting sequences are known, it can be addressed selectively to an organelle or compartment of interest by constructing chimeric polypeptides with the appropriate localization signal. Since the construction of organelle targeted chimeras of the photoprotein aequorin was previously discussed, simply note that the strategies and the targeting sequences employed for addressing GFP to various intracellular compartments are the same as those used and described for aequorin.

I focus on an experimental result which provides a clear example of the usefulness of this approach. The measurements of mitochondrial Ca^{2+} concentration with targeted recombinant aequorin have revealed large increases of the Ca^{2+} concentration in the matrix that were unexpected based on the biochemical characteristics of the Ca^{2+} transport pathways The most reasonable explanation for this apparent discrepancy, as previously discussed, is that mitochondria are in close contact with the intracellular Ca^{2+} sources (i.e., the ER) and thus sense local domains of high Ca^{2+} concentration. Studies carried out with GFP allowed the direct demonstration of this hypothesis (Rizzuto *et al.,* 1998b). Mitochondria and the ER were labeled, in living cells, with two different GFP chimeras, each containing a localization signal for mitochondria and the ER, respectively. The two organelles could be distinguished because the two chimeras included GFP moieties with different spectral properties: Mitochondrial GFP, or mtBFP, included the blue mutant (Y66H), and the GFP of the ER, or erGFP, included the bright green mutant (S65T). By cotransfecting the two constructs in HeLa cells, it was possible to acquire the images of the two organelles separately, illuminating the sample first with UV light (and collecting the blue emission of mtBFP) and then with blue light (and collecting the green emission of erGFP). Then, in order to obtain the three-dimensional image of the two organelles, which was necessary to fully appreciate their spatial relationships, serial images were acquired through the vertical axis of the cell. To minimize distortions due to the movement of the cell and its internal components, a sophisticated instrument was employed which is based on a rapid motor of the plane of focus and a very sensitive camera for collecting the emitted light (and thus reducing the exposure time at each plane of focus to 10–20 ms). With this approach, it is possible to carry out the optical scanning of the sample in approximately 1 s, a time frame in which the cell movements are quite modest and do not significantly affect the data obtained. After image acquisition, specifically developed algorithms allow the removal of the fluorescence signal originating from the out-of-focus portions of the sample and then reconstruction of the correct three-dimensional image (Rizzuto *et al.,* 1998a). The final result is shown in Fig. 7. It represents a small portion of the cell and has an unprecedented resolution (80 nm), much higher than previously obtained with other instruments and probes. It is possible to appreciate that the two organelles form complex networks which, in limited domains, come into such close contact that the images of the two organelles cannot be resolved (Rizzuto *et al.,* 1998b).

GFP as a Marker of Specific Proteins

The distribution of a protein inside a cell and its dynamic variations are frequently investigated for at least two reasons. The first is that the spatial organization of a cell strictly depends on the distribution of its molecular components, and thus the understanding of the mechanisms that control

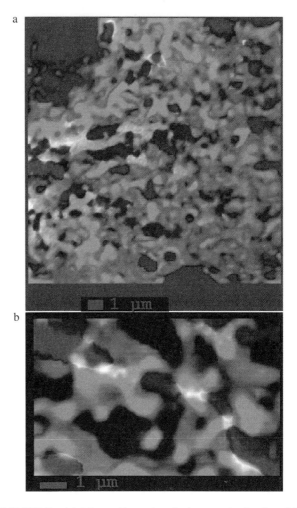

FIGURE 7 (a) Three-dimensional picture of mitochondria (in red) and endoplasmic reticulum (in green) visualized in HeLa cells with specifically targeted GFP chimeras. Where the two images are overlapped, the pixels appear white, showing the cellular domains where the two organelles are less than 80 nm distant. (b) A magnification of the picture shown in a, in which it is possible to appreciate the close relationship between the two organelles (reproduced with permission from Rizzuto *et al.,* 1998).

the intracellular targeting of the proteins is an important goal. Examples are the trafficking of proteins across the nuclear membrane or the progression of proteins along the secretory pathway, two topics of great interest in the past few years. In this type of study, it is very useful to be able to follow directly a native protein or a mutant in which the putative targeting sequences have been mutated. However, there is also another field of study that can take advantage of the possibility of dynamically following the localization of a protein: the study of the mechanisms which allow the intracellular decoding of extracellular signals. Indeed, it is known that, with different mechanisms, extracellular stimuli which differ substantially in chemical structure and

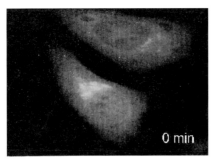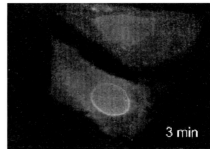

FIGURE 8 Intracellular distribution of a PKCδ/GFP chimera in HeLa cells before (left) treatment with 3 m*M* phorbol-12-miristic-13-acetate, a phorbol ester which induces translocation of the kinase to the plasma and nuclear membranes, and 3 min after treatment (right).

mode of action, such as steroid hormones and agonists which cause the breakdown of phosphatidylinositol biphosphate into diacylglycerol (DAG) and IP3, cause the intracellular translocation of signaling molecules. In particular, steroid hormones cause the translocation to the nucleus of their receptor, which can then act on the promoter of responsive genes, while the agonist-dependent increases in cytoplasmic Ca^{2+} concentration and DAG causes the translocation to the plasma membrane of the various isoforms of protein kinase C (PKC). The direct monitoring of the intracellular fate of these proteins thus allows one to dynamically follow the activation of a cell and to study the mechanisms through which it is triggered.

The experiment shown in Fig. 8 is a typical example of this application. In this study, the delta isoform of PKC has been fused to GFP. The transfection of this chimera endows a cell of a PKCδ moiety directly visible at the microscope analysis. The PKCδ/GFP chimera shows, in a resting cell, a cytoplasmic distribution. In agreement with this observation, if the plasma membrane is permeabilized with digitonin the chimera is completely released (data not shown). Upon stimulation with phorbol esters — cancer promoters which mimic the biological action of DAG, thus acting through the DAG/Ca^{2+} signaling pathway — PKCδ/GFP is almost entirely translocated to the plasma membrane, where it can act on specific effectors and cause the biological response. The PKCδ/GFP chimera allows one to follow, in real time and in a living cell, this important biological process.

References Cited

ALLEN, D. G., BLINKS, J. R., and PRENDERGAST, F. G. (1977). Aequorin luminescence: Relation of light emission to calcium concentration — A calcium-independent component. *Science* **195,** 996–998.

BARRERO, M. J., MONTERO, M., and ALVAREZ, J. (1997). Dynamics of $[Ca^{2+}]$ in the endoplasmic reticulum and cytoplasm of intact HeLa cells: A comparative study. *J. Biol. Chem.* **272,** 27694–27699.

BRINI, M., MURGIA, M., PASTI, L., PICARD, D., POZZAN, T., and RIZZUTO, R. (1993). Nuclear Ca^{2+} concentration measured with specifically targeted recombinant aequorin. *EMBO J.* **12,** 4813–4819.

BRINI, M., MARSAULT, R., BASTIANUTTO, C., ALVAREZ, J., POZZAN, T., and RIZZUTO, R. (1995). Transfected aequorin in the measurement of cytosolic Ca^{2+} concentration ($[Ca^{2+}]_c$): A critical evaluation. *J. Biol. Chem.* **270,** 9896–9903.

BRINI, M., DE GIORGI, F., MURGIA, M., MARSAULT, R., MASSIMINO, M. L., CANTINI, M., RIZZUTO, R., and POZZAN, T. (1997). Subcellular analysis of Ca^{2+} homeostasis in primary cultures of skeletal muscle myotubes. *Mol. Biol. Cell* **8,** 129–143.

CHALFIE, M., TU, Y., EUSKIRCHEN, G., WARD, W. W., and PRASHER, D. C. (1994). Green fluorescent protein as a marker for gene expression. *Science* **263,** 802–805.

CORMACK, B. P., VALDIVIA, R. H., and FALKOW, S. (1996). FACS — Optimized mutants of the green fluorescent protein (GFP). *Gene* **173,** 33–38.

DOPF, J., and HORIAGON, T. U. (1996). Deletion mapping of the *Aequorea victoria* green fluorescent protein. *Gene* **173,** 39–44.

GARCÍA-BUSTOS, J., HEITMAN, J., and HALL, M. N. (1991). Nuclear protein localization. *Biochim. Biophys. Acta* **1071,** 83–101.

GRYNKIEWICZ, G., POENIE, M., and TSIEN, R. Y. (1985). A new generation of Ca^{2+} indicators with greatly improved fluorescence properties. *J. Biol. Chem.* **260,** 3440–3450.

HARTL, F. U., PFANNER, N., NICHOLSON, D. W., and NEUPERT, W. (1989). Mitochondrial protein import. *Biochim. Biophys. Acta* **988,** 1–45.

HASELOFF, J., and AMOS, B. (1995). GFP in plants. *Trends Genet.* **11,** 328–329.

HEIM, R., and TSIEN, R. Y. (1996). Engineering green fluorescent protein for improved brightness, longer wavelengths and fluorescence resonance energy transfer. *Curr. Biol.* **6,** 178–182.

HEIM, R., CUBITT, A. B., and TSIEN, R. Y. (1995). Improved green fluorescence. *Nature* **373,** 663–664.

INOUYE, S., NOGUCHI, M., SAKAKI, Y., TAKAGI, Y., MIYATA, T., IWANAGA, S., MIYATA, T., and TSUJI, F. I. (1985). Cloning and sequence analysis of cDNA for the luminescent protein aequorin. *Proc. Natl. Acad. Sci. USA* **82,** 3154–3158.

KENDALL, J. M., SALA-NEWBY, G., GHALAUT, V., DORMER, R. L., and CAMPBELL, A. K. (1992). Engineering the Ca^{2+}-activated photoprotein aequorin with reduced affinity for calcium. *Biochem. Biophys. Res. Commun.* **187,** 1091–1097.

MASIBAY, A. S., BALAJI, P. V., BOEGGEMAN, E. E., and QASBA, P. K. (1993). Mutational analysis of the Golgi retention signal of bovine beta-1,4-galactosyltransferase. *J. Biol. Chem.* **268,** 9908–9916.

MINTA, A., and TSIEN, R. Y. (1989). Fluorescent indicators for cytosolic sodium. *J. Biol. Chem.* **264,** 19449–19457.

MINTA, A., KAO, J. P., and TSIEN, R. Y. (1989). Fluorescent indicators for cytosolic calcium based on rhodamine and fluorescein chromophores. *J. Biol. Chem.* **264,** 8171–8178.

MONTERO, M., BRINI, M., MARSAULT, R., ALVAREZ, J., SITIA, R., POZZAN, T., and RIZZUTO, R. (1995). Monitoring dynamic changes in free Ca^{2+} concentration in the endoplasmic reticulum of intact cells. *EMBO J.* **14,** 5467–5475.

ORMO, M., CUBITT, A. B., KALLIO, K., GROSS, L. A., TSIEN, R. Y., and REMINGTON, S. J. (1996). Crystal structure of the *Aequorea victoria* green fluorescent protein. *Science* **273,** 1392–1395.

PINTON, P., POZZAN, T., and RIZZUTO, R. (1998). The Golgi apparatus is an inositol 1,4,5-trisphosphate-sensitive Ca^{2+} store, with functional properties distinct from those of the endoplasmic reticulum. *EMBO J.* **18,** 5298–5308.

PRASHER, D. C., ECKENRODE, V. K., WARD, W. W., PRENDERGAST, F. G., and CORMIER, M. J. (1992). Primary structure of the *Aequorea victoria* green-fluorescent protein. *Gene* **111,** 229–233.

RAJU, B., MURPHY, E., LEVY, L. A., HALL, R. D., and LONDON, R. E. (1989). A fluorescent indicator for measuring cytosolic free magnesium. *Am. J. Physiol.* **256,** C540–C548.

RIDGWAY, E. B., and ASHLEY, C. C. (1967). Calcium transients in single muscle fibers. *Biochem. Biophys. Res. Commun.* **29,** 229–234.

RINK, T. J., TSIEN, R. Y., and POZZAN, T. (1982). Cytoplasmic pH and free Mg^{2+} in lymphocytes. *J. Cell Biol.* **95,** 189–196.

RIZZUTO, R., SIMPSON, A. W., BRINI, M., and POZZAN, T. (1992). Rapid changes of mitochondrial Ca^{2+} revealed by specifically targeted recombinant aequorin. *Nature* **358,** 325–328.

RIZZUTO, R., CARRINGTON, W., and TUFT, R. A. (1998a). Digital imaging microscopy of living cells. *Trends Cell Biol.* **8,** 288–292.

RIZZUTO, R., PINTON, P., CARRINGTON, W., FAY, F. S., FOGARTY, K. E., LIFSHITZ, L. M., TUFT, R. A., and POZZAN, T. (1998b). Close contacts with the endoplasmic reticulum as determinants of mitochondrial Ca^{2+} responses. *Science* **280,** 1763–1766.

SIEMERING, K. R., GOLBIK, R., SEVER, R., and HASELOFF, J. (1996). Mutations that suppress the thermosensitivity of green fluorescent protein. *Curr. Biol.* **6,** 1653–1663.

SITIA, R., and MELDOLESI, J. (1992). Endoplasmic reticulum: A dynamic patchwork of specialized subregions. *Mol. Biol. Cell* **3,** 1067–1072.

THOMAS, M. V. (1982). *Techniques in Calcium Research,* pp. 90–136. Academic Press, New York.

TSIEN, R. Y., POZZAN, T., and RINK, T. J. (1982). T cell mitogens cause early changes in cytoplasmic free Ca^{2+} and membrane potential in lymphocytes. *Nature* **295,** 68–71.

WOODS, N. M., CUTHBERTSON, K. S., and COBBOLD, P. H. (1986). Repetitive transient rises in cytoplasmic free calcium in hormone-stimulated hepatocytes. *Nature* **319,** 600–602.

ZOLOTUKHIN, S., POTTER, M., HAUSWIRTH, W. W., GUY, J., and MUZYCZKA, N. (1996). A "humanized" green fluorescent protein cDNA adapted for high-level expression in mammalian cells. *J. Virol.* **70,** 4646–4654.

General References

DE GIORGI, F., BRINI, M., BASTIANUTTO, C., MARSAULT, R., MONTERO, M., PIZZO, P., ROSSI, R., and RIZZUTO, R. (1996). Targeting aequorin and green fluorescent protein to intracellular organelles. *Gene* **173,** 113–117.

NUCCITELLI, R. (1994). A practical guide to the study of calcium in living cells. *Methods Cell Biol.* **40.**

PELHAM, H. R. (1989). Control of protein exit from the endoplasmic reticulum. *Annu. Rev. Cell Biol.* **5,** 1–23.

RIZZUTO, R., and FASOLATO, C. (1998). *Imaging Living Cells.* Springer-Verlag, New York.

RIZZUTO, R., BRINI, M., DE GIORGI, F., ROSSI, R., HEIM, R., TSIEN, R. Y., and POZZAN, T. (1996). Double labelling of subcellular structures with organelle-targeted GFP mutants *in vivo. Curr. Biol.* **6,** 183–188.

SULLIVAN, K. F., and KAY, S. A. (1998). Green fluorescent protein. *Methods Cell Biol.* In press.

GIULIO F. DRAETTA

Department of Experimental Oncology
European Institute of Oncology
Milan, Italy

Throughout evolution, the mechanisms responsible for control of cell proliferation have remained basically unchanged from the simplest eukaryotes to man. The simultaneous use of model systems, such as yeasts, the amphibian Xenopus laevis, and mammalian cell cultures, has made it possible since the late 1980s to clarify how the biochemical cascades responsible for activating cell proliferation, and the progression and completion of each division cycle, occur. Because of these control mechanisms, progression through the cycle takes place with extreme accuracy, in such a way as to preserve the integrity of the genome from one cell generation to the next. Alterations in these control mechanisms are frequently responsible for the onset of neoplasias.

The Cell Division Cycle: From Small Modifications to Drastic Changes

The Eukaryotic Cell Division Cycle

The progression of eukaryotic cells through the cell division cycle is controlled by molecular mechanisms necessary to ensure that all the intracellular components are duplicated before their physical separation into two daughter cells. To maintain the functional integrity of a living organism, whether unicellular or multicellular, it is fundamental for the reproduction of each cell to be controlled in such a way as to guarantee that its DNA, which contains all the information necessary for the cells to function, is replicated without error and is transmitted in complete form by the mother cell to the two daughter cells. Hence the existence in the eukaryotes of two distinct processes, DNA replication and subsequent mitosis, through which the DNA is first repli-

cated and then distributed to the daughter cells. The period between a mitosis and the ensuing DNA replication and that between the completion of the DNA replication and the ensuing mitosis are called G_1 and G_2, respectively (G = gap). The name S (synthesis) phase has been given to the period when the DNA replication takes place, whereas the phase that includes both mitosis and cytokinesis (i.e., the division of the nucleus and of the cytoplasm, respectively) has been called the M phase. The three phases G_1–S–G_2 constitute the interphase (or interkinesis).

The dividing process is organized in such a way as to permit the orderly succession of the various phases of the cycle. This takes place due to the existence of molecules which perform a control function without being directly involved in the execution of the single stages of the process, such as

DNA replication, mitosis, and cytokinesis. These molecules belong to a family of serine/threonine protein kinases, the cyclin-dependent kinases (Cdk), so-called because their enzymatic activity depends on their association with regulating proteins called cyclins. Signals of both stimulating and inhibiting type depend on these and determine whether or not a cell starts the dividing process and whether it completes it. The discovery and clarification of the mechanism of action of the Cdk is the result of the convergence of biochemical and genetic methodologies applied to a simple unicellular organism (yeast) and represents one of the major recent successes of the molecular biology of the cell (Fig. 1).

The G_1 Phase

The presence of a G_1 phase following mitosis and cytokinesis and preceding the beginning of DNA replication is used by all eukaryotic cells for growing and reaching the necessary size before starting the division process in such a way that, for that given cell type, maintaining its critical size will be guaranteed. In the simplest eukaryotes such as yeasts, as in multicellular organisms, reaching a certain critical mass is an essential signal for the onset of the dividing process. With the exception of the yeast *Schizosaccharomyces pombe,* whose cells — upon emerging from mitosis — have already reached an adequate size to enable them to

directly enter the S phase and thus skip the G_1 phase (Fantes and Nurse, 1977; Nurse, 1975), the majority of eukaryotic cells use the time spent in G_1 for cell growth. During this phase, the cells are generally sensitive to variations in their external environment and to the presence of nutrients and growth factors; in the absence of these substances there may be an interruption in their cycle and they may remain in a quiescent state called G_0.

DNA Replication

In the majority of cells, the DNA is replicated just once per cell cycle. The existence of one or more factors capable of inducing DNA replication was initially demonstrated by experiments of mammalian cell fusion (Rao and Johnson, 1970). When a cell in the G_1 phase is fused with a cell in the S phase, the nucleus in G_1 is induced to start the synthesis of its DNA immediately. This phenomenon can be explained by the fact that the replication of the DNA responds to cell factors present in the S phase which bring about its activation.

A detailed description of the biochemical mechanisms that determine the semiconservative replication of the DNA beyond the scope of this article. It is important, however, to consider the phenomena that permit the start of the DNA replication (Stillman, 1996). In yeast, in mammalian cells, and in cells of *X. laevis,* protein components have recently been identified which, in the G_1 phase, bond the DNA and contribute to the formation of the prereplicative complex. These include the proteins of the origin recognition complex (ORC), the protein Cdc6 (cell division cycle 6), and the MCM (mini chromosome maintenance) proteins, six of which have been identified in yeast. The ORC complex recognizes and becomes bonded at the sites of the onset of DNA replication, constituted by nucleotidic sequences from which the replication process stems. The ORC bond then fosters the interaction of the Cdc6 protein and the MCM proteins at these sites. These components are necessary to start DNA replication and are activated by phosphorylation events brought about through the Cdk.

Existence of a Licensing Factor for Entering Phase S

In the previously mentioned experiments of cellular fusion, when a cell in G_1 becomes bonded with one in S the nucleus in G_2, contrary to what has been observed for the nuclei in G_1, is not induced to replicate its DNA (Fig. 2). This suggests that many factors present in the nucleus have the capacity of "remembering" the DNA replication that has taken place. This phenomenon is called the rereplication block: The DNA replicated in phase S can be replicated again only after the cell has completed its mitosis. Nuclear permeabilization experiments have suggested that this block may be explained by the existence of a licensing factor bonding with the DNA during mitosis, in a barely

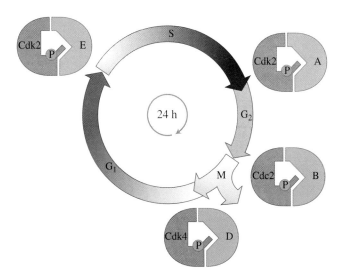

FIGURE 1 The passage of eukaryotic cells through the various phases of the cell division cycle depends on the activation of the cyclin–Cdk complexes. In mammalian cells, four types of complex are activated which are formed from four cyclins (in green) and three Cdks (in red). Cdk2 is able to associate with two cyclins, A and E, forming two complexes involved in the passages G_1–S and S–G_2. Cdc2 associates with cyclin B and is involved in the passage G_2–M, whereas the complex cyclin D–Cdk4 intervenes in the passage M–G_1. P indicates phosphorylation on a threonine residue which stabilizes the enzyme.

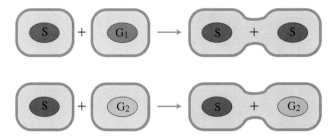

FIGURE 2 Induction of DNA synthesis in G_1 phase cells by cells in the S phase. When cells in S are fused with cells in G_1, the nuclei of the latter rapidly enter the S phase. In contrast, nuclei of cells in G_2 are resistant to the induction.

sufficient quantity to permit the replication of the DNA just once per cycle. According to this line of reasoning, given that the nuclear membrane is impermeable to this factor, whose synthesis takes place in the cytosol (cell sap), the replication is prevented until the passage of the cell through the subsequent mitosis, in which phase the disgregation of the nuclear envelope permits the association of another licensing factor with the sites of the starting of DNA replication. It has in fact been shown that the permeabilization of the nuclear membrane of cells in interphase can induce the rereplication of the DNA (Blow and Laskey, 1988; Coverley *et al.*, 1993; Leno *et al.*, 1992). The complex constituted by the ORC and MCM proteins could represent the licensing factor because, on the one hand, they are necessary for the DNA replication and, on the other hand, they are rendered inactive in each cycle by a cyclin–Cdk complex. It is necessary for the cell to complete the S phase and carry out mitosis before it is again able to accumulate the prereplicative complexes containing ORC and MCM on the DNA.

Duplication of the Centrosome

During interphase, parallel with the DNA replication, the duplication of the centrosome occurs, which is an organelle essential for the formation of the mitotic spindle and the secretion of the chromatids among the daughter cells. The centrosome is composed of centrioles and a protein matrix whose components have not been identified. Part of this protein matrix is the microtubule organizing center (MTOC), which allows the generation of microtubules. Its function is regulated during the cell cycle; in the interphase the microtubules generated by the MTOC are long and fairly stable, whereas during mitosis, probably as a consequence of the phosphorylation of the components of the MTOC by a cyclin–Cdk complex (e.g., cyclin B–Cdc2; see Fig. 1), an increase in the number of microtubules running from each MTOC and a shortening in their average length due to decreased stability occur (McNally, 1996; Verde *et al.*, 1990).

Mitosis and Transition from Anaphase to Metaphase

Dramatic morphological changes occur when the mitosis phase is reached: the dissolving of the nuclear envelope, the condensation of the chromatin, the reorganization of the network of microtubules with the formation of the mitotic spindle, and the fragmentation of the intracellular membranes. The purpose of these processes is to permit the secretion of the genetic material and of all the cell components among the daughter cells. Although the distribution of organelles, such as the mitochondria and the Golgi apparatus, occurs in a casual manner, the distribution of the DNA among the daughter cells must absolutely be quantitative. The accuracy of the process is guaranteed by a series of regulating circuits, known as checkpoints, which prevent entry into the mitosis phase unless the cell has completed its DNA replication, and during this phase they prevent the separation of the pairs of chromatids if these are not aligned on the metaphase plate (Murray, 1995; Rudner and Murray, 1996). Experimental evidence has demonstrated, for example, that the administration of DNA replication inhibitors or the physical moving, with micromanipulation techniques, of a single chromosome from the metaphase plate can block the progression of the cycle. Cytokinesis starts in parallel with mitosis; the first signs of this process are clearly seen with the formation of a ring of actin under the cell membrane. In this position the fission sulcus forms next, which is an invagination that will sink toward the center of the cell causing the separation of the two daughter cells.

Biochemical Cell Cycle Control Mechanisms

Cdc Mutations in Yeast

In the 1970s, studies commenced by L. Hartwell on the yeast *Saccharomyces cerevisiae* and by P. Nurse on *S. pombe* (Hartwell *et al.*, 1974; Nurse, 1975) led to the identification of a family of genes specifically involved in regulating the cell division cycle (Forsburg and Nurse, 1991; Pringle and Hartwell, 1981). The experiments carried out aimed to isolate mutants in which a blockage of cell division occurred; despite this, the yeast cells continued increasing in volume and metabolizing nutrients (Fig. 3). These results suggested the existence of specific cell division control mechanisms different from those that control growth. Great excitement was caused by the discovery that the genes identified in yeast and named *cdc* (cell division cycle) have homologs in all eukaryotic organisms, indicating that the regulation of the cell cycle is based on common mechanisms. These studies showed that the use of simple organisms, such as yeast, can lead to illuminating discoveries for the biology of more complex organisms, including man.

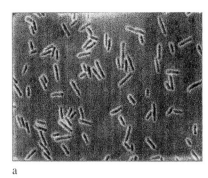 

a   b

FIGURE 3   Phase-contrast microscope photographs of cells of *Saccharomyces pombe.* (a) Normal cells in the exponential growth phase. (b) Cells carrying a mutation (inactivation) of the gene *Cdc25.* These cells are unidimensional and continue to grow without being able to divide (courtesy of G. Cottarel).

MPF and *cdc2*

A fundamental role in the cell division process is performed by the gene *cdc2,* which codifies a Cdk protein, the first one to be identified in yeast with the genetic methodologies just described. At the same time, biochemical studies conducted on a completely different system, that of the oocytes of *X. laevis,* led to the discovery of a cytosolic factor able to induce mitosis called MPF (maturation promoting factor), initially described as an activity capable of inducing meiotic maturation in oocytes of *X. laevis* blocked at the prophase of the first meiotic division. These oocytes are about 1 mm in diameter and are therefore large enough to be easily injected. The injection of cytoplasm taken from mitotic cells arrested in metaphase (both of *Xenopus* and of other organisms) induces the completion of the first meiotic division and progression until the metaphase of the second meiotic division. These experiments and successive ones demonstrated the existence, in the mitotic cells, of a factor able to induce mitosis in cells in interphase. From 1987 to 1990, using similar trials, the proteins present in the purified MPF were identified and it was demonstrated that these proteins are the homologs of the Cdc2 of *Xenopus* and of cyclin B of yeast (Dunphy *et al.,* 1988; Gautier *et al.,* 1988, 1990; Lohka *et al.,* 1988). It is important to observe that, although in yeast the protein Ccd2 is necessary for both phase S and mitosis to take place, in the higher eukaryotes it regulates only the entry to the mitosis phase, through the phosphorylation of the proteins associated in the nuclear envelope or with the microtubules (Nigg, 1991). Other Cdks are involved in the regulation of the other phases of the cycle (see Fig. 1).

Amplification of MPF and Entry to the Mitosis Phase

Studies in which the technique of microinjection has been used have demonstrated that in interphase (Fig. 4) the protein Cdc2 is present, but that it is inactive and can be activated following posttranslational modifications (e.g., dephosphorylation). *In vivo,* the Cdc2 protein is phos-

phorylated on residues of threonine (Thr14) and tyrosine (Tyr15), and this makes the enzyme inactive. By injecting small quantities of MPF into an oocyte in interphase, the onset of the mitosis phase and, at the same time, an accumulation of quantities of MPF exceeding 100 times that injected are achieved; the whole process takes place in the absence of protein synthesis. A latent form of MPF and an amplification mechanism must therefore be present which cause the appearance of active MPF in the oocyte injected. The explanation of this phenomenon lies in the capacity of MPF to phosphorylate and activate the protein phosphatase cdc25-C, which is able to determine an active MPF through dephosphorylation, thus triggering a self-amplification mechanism (Hoffmann *et al.,* 1993). This mechanism (Fig. 5) occurs in an extremely rapid transition which determines the onset of the mitosis phase (Murray, 1993).

Cdk: Its Structure and Function

Previously, it was shown how Cdk activity is fundamental for control of progression from one phase to the next in the cell cycle. Much effort has therefore been devoted to

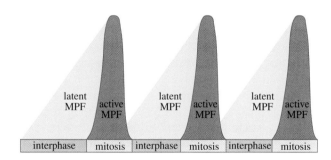

FIGURE 4   Differing concentrations of MPF during the cell cycle. During each cell cycle the MPF (i.e., the cyclin B–Cdc2 complex) accumulates and is kept inactive until the completion of DNA replication when, through a self-amplification mechanism, all the latent MPF is activated. At the end of mitosis, complete inactivation of the MPF is necessary for the return to interphase.

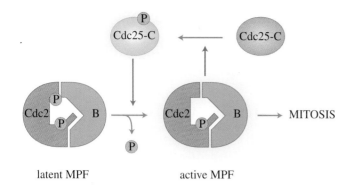

latent MPF   active MPF

FIGURE 5 During the interphase MPF is kept inactive by phosphorylation which prevents the catalytic activity of the kinase (Cdc2; in red). Entry into mitosis takes place through the activation of the complex by the phosphatase protein Cdc25-C. The cyclin B–Cdc2 complex is able to superactivate Cdc25-C, thereby generating a signal amplification mechanism which is essential for the onset of mitosis.

the study of these kinases for the purpose of establishing which relations between structure and function are at the basis of their regulation. These proteins constitute the catalytic subunit of multimeric complexes. Applying crystallographic analysis by X-rays, the three-dimensional structure of a Cdk, the protein Cdk2, was determined; two lobes are present, typical of the catalytic domains of protein kinases, which by means of a hinge movement close over the substrates, permitting the transfer of the phosphate groups from the ATP to residues of serine or threonine present in the protein substrate. Each Cdk, while maintaining the structural characteristics of every known protein kinase,

is inactive in the absence of the regulating subunit, cyclin, which is so named due to the fact that some of these proteins are expressed only in certain phases of the cell cycle. The bond with cyclin stabilizes the structure of the Cdk and permits its phosphorylation on a residue of threonine located in a region of the protein known as the T-loop in the vicinity of the active site of the kinase, that is, Thr161 in the human Cdc2 and Thr160 in Cdk2 (Fig. 6). This phosphorylation is not necessary for catalysis but contributes toward the stabilization of the enzyme through the formation of ionic bonds with the basic residues of the amino-terminal lobe of the Cdk and the cyclin (De Bondt *et al.,* 1993; Jeffrey *et al.,* 1995; Russo *et al.,* 1996b).

Regulation of the Cdks

As should be clear, the molecular mechanisms that control the activity of cyclin-dependent kinases are both complex and multiple. Enzymatic activation includes

1. Activation at the level of the transcription and translation of cyclin and Cdk
2. Formation of the cyclin–Cdk complex
3. Phosphorylation of the threonine of the T-loop by Cak (Cdk-activating kinase) (Fig. 7)

Enzymatic inactivation takes place through:

1. Proteolysis of the regulating subunit (i.e., cyclin)
2. Phosphorylation on the tyrosine and threonine of the Cdk by members of a family of protein kinases called Weel within the context of a preserved region of the enzyme that is involved in the bond of the substrate ATP, which is represented in the human Cdc2 by Thr14 and Tyr15. This phos-

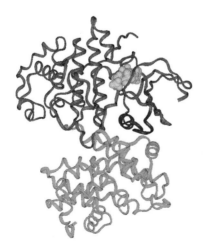
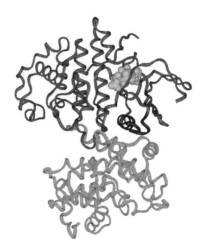

FIGURE 6 Model of the three-dimensional structure of the cyclin A–Cdk2 complex. The cyclin A is shown in green, and the Cdk2 is shown in red. The figure also indicates the region PSTAIRE (in blue), which is present in all members of the cyclin-dependent kinase family and involved in the bonding with the cyclins, and the T-loop (in purple) containing the phosphorylation site Thr160. A molecule of ATP (yellow) is visible in the active site (molecular coordinates from Jeffrey *et al.,* 1995).

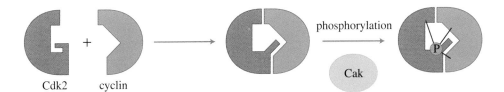

FIGURE 7 Phosphorylation of the T-loop by the kinase Cak. The binding of cyclin (in green) with the subunit Cdk (in brown) induces a structural reorganization of the region of this molecule defined as the T-loop. The complex is stabilized by phosphorylation on a threonine residue (Thr161 in the human Cdc2) by Cak, through the formation of ionic links (in violet) between basic residues of cyclin and the Cdk, and the threonine phosphate (modified from Russo *et al.*, 1996).

phorylation prevents the closure of the Cdk molecule on the substrates, thereby inhibiting catalytic activity.

3. Association with cyclin-dependent kinase inhibitors (Ckis), including two proteins, p16 and p21, so called because their molecular mass is 16 and 21 kDa, respectively, and proteins correlated with them.

Although the degradation of the cyclin brings about the irreversible inactivation of the Cdk and occurs at its emergence from a certain phase of the cycle, inactivation by means of phosphorylation (i.e., at the level of Thr14 and Tyr15) is reversible due to the dephosphorylation brought about by the protein kinases of the Cdc25 family. The third inactivation mechanism is due to the association of the cyclin–Cdk complexes with Ckis. Whereas the Ckis of the p21 class bond and inhibit the function of all known Cdk complexes, the protein p16 inhibits specifically the complexes that contain the D cyclins and the Cdk4 or Cdk6 proteins. The three-dimensional structure of the cyclin complex A–Cdk2, bonded with a fragment of p27, a protein correlated with p21, has enabled us to understand the mechanism through which these inhibitors exercise their action (Russo *et al.*, 1996a). In fact, it can be observed that the bond of the inhibitor modifies the structure of the Cdk2, physically obstructing access of the substrates to the active site, thereby preventing the interaction of the substrates with the cyclin. Regarding the interaction of the substrates with Cdk4, natural mutants of Cdk4 have been identified which are refractory to inhibition by p16, whereas they are inhibited by p21 and interact with cyclin D in the same way as with normal protein (Wölfel *et al.*, 1995). The residues of mutated amino acids have been located in the amino-terminal domain of Cdk4, far from the site of arrival of the enzyme. It has been hypothesized that the bond of p16 with Cdk4 impedes the enzymatic function by causing the molecule to stiffen, thus preventing it from bending over the catalytic site.

Checkpoint Controls

Control circuits allow a cell to divide in an orderly fashion, for example, preventing mitosis from taking place be-

fore the synthesis of the DNA has been completed or recognizing spontaneous errors or errors in response to cell damage, both in DNA replication and in the alignment of the chromosomes on the mitotic plate. These are caused by biochemical reactions which lead to the arrest of the cell division cycle in response to the activation of the checkpoint. During the cell division cycle, the existence of feedback controls guarantees that, should factors of disturbance intervene (e.g., damage to the DNA caused by ionizing radiations), the Cdk will be inactivated, thus stopping the division. This may be transitory if the cells have the capacity to repair the damage or permanent with the induction of apoptosis in the presence of serious and irreversible alterations.

Inactivation of Cdk complexes can occur through phosphorylation on the tyrosine of the Cdk subunit by a protein kinase of the Wee1 family or through the bond of Cki inhibitors, such as p16 and p21. The molecular cascades which lead to the inactivation of these inhibiting reactions in the higher eukaryotes are not properly known, whereas in yeast cells numerous genes involved in these processes have been identified (Elledge, 1996).

An example of well-characterized checkpoint control is represented by the inactivation of the cyclin B–Cdc2 complex during the S phase and the G_2 phase. Since this complex starts to accumulate during the course of the S phase, long before the onset of the mitosis phase, it is necessary to prevent its activation to allow the cell to complete the S phase. There are some unidentified signals that maintain high activity in kinases of the Wee1 family, and that of the Cdc25-C protein phosphatase is low, until completion of the DNA replication. It has been demonstrated that the administration of any pharmacological agent that blocks the S phase of cells in culture prevents the onset of the mitosis phase through the use of this inactivation mechanism (Fig. 8).

Degradation of Mitotic Cyclins

The hydrolytic degradation of many proteins takes place in the proteasome, a cylinder-shaped cytosolic enzymatic complex. Prior to being subjected to this process, the ma-

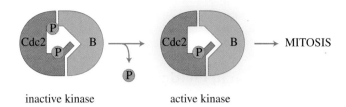

inactive kinase   active kinase

FIGURE 8 Factors which block DNA synthesis prevent entry of the cell into the mitosis phase by not permitting the passage of the cyclin B–Cdc2 complex from the inactive to the active form. This feedback control guarantees that the cell does not enter mitosis before the completion of DNA duplication.

jority of the proteins are modified by molecules of ubiquitin, a 7-kDa protein. Isopeptidic bonds are formed between the carboxy-terminal residue of the ubiquitin and the ε-aminic group of the target protein. The proteasome 26S, a multiprotein complex with enzymatic activity similar to that of chemotrypsin, recognizes the ubiquitinylated proteins, which are in fact polyubiquitinylated through the formation of polyubiquitin chains, and degrades them (Hochstrasser, 1995). The ubiquitinylation reactions start with the activation of the ubiquitin by the enzyme E_1, also called Uae (ubiquitin activating enzyme), which in the presence of ATP bonds the ubiquitin to its own active site and thus forming a thioester bond with the carboxy-terminal residue of the ubiquitin. The function of the enzyme E_1 is to transfer ubiquitin–thioester intermediates to the active sites of the enzymes E_2 or Ubc (ubiquitin conjugating enzyme). In the yeast *S. cerevisiae*, 13 Ubcs have been described which are able to recognize, in a specific manner, the substrates to be ubiquitinylated. In many cases there are additional components, the enzymes E_3, also known as ubiquitin ligases, enzymes that bond the ubiquitin to the protein. The induction of polyubiquitinylation and, consequently, of the degradation of a certain substrate takes place in response to the activation of the enzymes Ubc and E_3. In addition, posttranslational modifications in the substrates (i.e., phosphorylation and dephosphorylation) are in certain cases responsible for recognition of the substrate to be ubiquitinylated by the ubiquitinylating enzymes Ubc and E_3.

Studies conducted on the amphibian *X. laevis*, on yeast, and on many invertebrates have shown that the periodic degradation of the cyclin B during mitosis is necessary so that it is possible to leave this phase of the cycle and to enter the ensuing one (King *et al.*, 1996). In fact, the generation of a cyclin that cannot be degraded blocks the cell in mitosis. This periodicity in the degradation of the cyclin B is due to the presence of a sequence of amino acids called destruction box which, if conjugated with other proteins, confers this property on them. The degradation of cyclin during mitosis involves the ubiquitin system and requires a cascade of reactions using the enzyme E_1, the enzyme Ubc4 (or UbcX), and a multiprotein complex APC (anaphase promoting complex) or cyclosome. The function of this complex is to activate the degradation of cyclin in response to phosphorylation events depending on the activation of the cyclin B–Cdc2 complex and thus on actually entering the mitosis phase. Therefore, the activation of Cdc2 also causes its degradation, making progression possible through the cell division cycle (Fig. 9).

Degradation of Cdk Inhibitors during Phase G₁

In some cases, the activity of the Cki inhibitors is also regulated negatively and irreversibly through proteolysis. The activation of the degradation of some of these proteins takes place using a series of cellular components which were initially identified in the cells of the yeast *S. cerevisiae*, in which these components are the products of the genes *cdc34, cdk4, cdk53*, and *skp1 that* if inactivated result in a stoppage in phase G₁ caused by the failure of the cyclin–Cdk complexes to release the protein Sic1, a Cdk inhibitor.

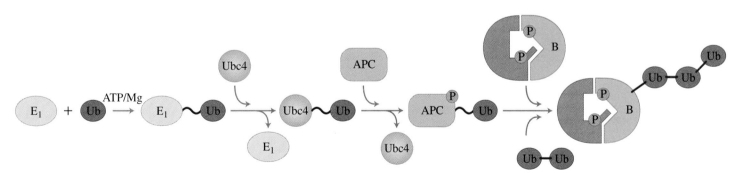

FIGURE 9 The process of ubiquitinylation of the cyclins, which involves various enzymes. The enzyme E_1 bonds the ubiquitin in the presence of ATP/Mg and transfers it to the enzyme Ubc4. The enzyme Ubc4, in turn, transfers the ubiquitin to the complex APC, which during mitosis is phosphorylated and causes the polyubiquitinylation of cyclin B (in green). One molecule of ubiquitin is bonded with others through isopeptidic links.

Ubiquitin is involved in the Cki degradation process during phase G_1. In fact, mutant yeast cells in the gene that codifies the ubiquitinylating enzyme Ubc3, the human homolog of Cdc34, which is normally blocked in the phase G_1, are able to start the DNA replication in the absence of the gene *sic1*, which codifies a Cki of yeast. This shows that the absence of the protein Sic1 makes the presence of the protein Ubc3, which is necessary for degradation and for the disappearance of the inhibitor Sic1, superfluous (Mendenhall, 1993; Nugroho and Mendenhall, 1994; Schwob *et al.*, 1994). Subsequently, it has been demonstrated that strains of yeast bearing mutations in the genes *cdc53, skp1,* and *cdc4* are also able to start DNA replication in the absence of Sic1. It is therefore probable that the proteins Cdc53, Skp1, and Cdk4 participate in the reactions of Sic1 degradation in phase G_1 of the cell cycle (Fig. 10).

It is interesting to note that, in human cells and in those of the nematode *Caenorhabditis elegans*, a homologous protein, Cdc53, to that of yeast is present. In particular, the cells of *C. elegans*, mutants of the protein family of which Cdc53 forms a part, show defects of the cell division cycle, some of which determine in the organisms affected an increase or a reduction in the number of cells (Kipreos *et al.*, 1996; Jackson, 1996). Studies of human cells have shown that the protein p27, a Cki present at high levels in quiescent cells and degraded in those stimulated to divide, is ubiquitinylated *in vivo* and the rate of the paired processes of ubiquitinylation and degradation of this protein is greater in growing cells than in quiescent ones. Moreover, *in vitro* the degradation of p27 requires the presence of Ubc3, the human homolog of Cdc34, or Ubc2 (Pagano *et al.*, 1995). Again, studies on the regulation of the cell cycle have made it possible to stress the convergence of the biochemical regulation mechanisms between such different organisms as man and yeast, revealing the preservation through their phylogenesis of the cell division cycle control mechanisms. The possibility of using different approaches for each of these organisms has enabled the rate of scientific discoveries in this field to increase dramatically.

Cell Division in Multicellular Organisms

The Cascade of Reactions Controlling Entry into the Cycle from State G_0

In multicellular organisms the systems of cell cycle control are more complex than those present in unicellular organisms, such as yeasts, because the divisions of each single cell have to take place respecting the requirements of the organism *in toto*. Once one division cycle has been completed, a cell can start a new one or become quiescent (i.e., it can enter the state G_0) and in certain cases can differentiate. This decision is reached in response to the signals coming from the external environment, such as nutrients, growth factors, contact with surrounding cells, and interactions with the extracellular matrix. During phase G_1 of the cycle, the cell has to pass a threshold known as the restriction point, past which it can complete the division process (Pardee, 1989). For cells of the higher eukaryotes, the restriction point in phase G_1 coincides with the phosphorylation of the protein Rb.

A series of mediators, through interaction with appropriate receptors located at the level of the plasmatic membrane or within the cell, coordinate cell division. Examples of these substances are also present in simple eukaryotes: In *S. cerevisiae*, during the conjugation process between the haploid cells, which necessarily have to stop in phase G_1, the cells of type a secrete a pheromone which in the α cells causes the arrest of the cycle in phase G_1, and those of type α secrete a pheromone that blocks the cycle of type a cells. In mammalian cells, stimulation with growth factors leads to the activation of a cascade of biochemical reactions that

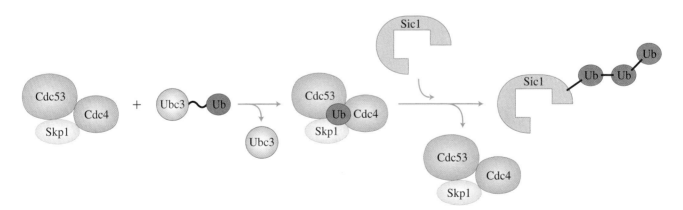

FIGURE 10 Degradation of the inhibitor Sic1 in the yeast *Saccharomyces cerevisiae* requires a complex containing the proteins Cdc53, Cdc4, and Skp1. The enzyme Ubc3 is responsible for the ubiquitinylation of Cdc4, which in turn transfers the ubiquitin to Sic1, initiating the process of degradation.

lead to the accumulation and activation of the cyclin complexes D–Cdk4 and E–Cdk2, both necessary for progression through phase G_1. The cyclin D_1 has in fact been identified (Matsushime *et al.,* 1991) as the product of a gene expressed in response to the activation of the macrophages by the growth factor colony-stimulating factor-1 (CSF-1). The removal of this factor causes the disappearance of the protein and the arrest of the cell cycle. Cyclin D_1 and two homologous cyclins, D_2 and D_3, are expressed in all the tissues and form complexes with the protein Cdk4 or with a very similar protein, Cdk6. For the activation of these complexes, the phosphorylation of the subunit Cdk is necessary, and in the protein Cdk4 this occurs at the level of Thr172. The active complex has as its principal substrate the protein Rb, codified by the gene *Rb,* which is altered in carriers of retinoblastoma, a tumor that starts in early infancy. Both for cyclin D_1 and in the case of cyclin E, it has been demonstrated that the superexpression of these proteins leads to a shortening of phase G_1 of the cycle, whereas inhibition of the expression or the activity thereof determines a stoppage in phase G_1. Various experimental approaches have shown that the protein Rb is the main substrate of the cyclin complex D–Cdk4. In fact, it has been observed that an increase in levels of the inhibitor p16 generated by means of the transfection of the expression carriers of the cDNA of p16 or the inhibition of the accumulation of cyclin D_1 generated through the expression of counter-sense DNA are able to block the proliferation of cells that express the protein Rb. These treatments are not effective in cells that express a mutated Rb or do not express it at all (Baldin *et al.,* 1993; Lukas *et al.,* 1994; Quelle *et al.,* 1993). It can thus be concluded from these experiments that the protein Rb is an inhibitor of the cell cycle and that it constitutes a substrate of the cyclin complexes D–Cdk4. The phosphorylation of Rb by Cdk4 abolishes its activity and by so doing

makes it possible for the cell cycle to progress. Instead, the cyclin complexes D–Cdk4 are not necessary for the division of cells in which the protein Rb is lacking; this is therefore the only important substrate for their activity. Through the formation of a complex with Cdk2, subsequently phosphorylated at the Thr160 level, the activation of cyclin E occurs. This complex is active in phase G_1, following the activation of the complexes containing cyclin D. Even if the cyclin complex E–Cdk2 *in vitro* can phosphorylate protein Rb, experiments analogous to those conducted with cyclin D have shown that the expression of cyclin E is also necessary in cells in which the function of protein Rb is lacking, suggesting that it recognizes other substrates in phase G_1 whose phosphorylation is necessary for the activation of phase S (Ohtsubo *et al.,* 1995).

In multicellular organisms, in the event of adherent cells, the activation of the cell division process, in addition to the presence in the medium of nutrients and growth factors, depends on interaction with the surrounding cells and with the extracellular matrix. Cells derived from the connective tissue and epithelium cells must necessarily grow on a solid support to which they adhere. If this is not the case, or if the cells arrive at a confluence (i.e., if there is contact inhibition), the Cdk complexes are inactivated and the cells exit from the proliferative cycle.

Regulation of Entry to Phase S

The inactivation of the protein Rb is necessary to trigger the cell division process. It is an important regulating element of the restriction point, present in cells in phase G_1 a few hours prior to the onset of phase S (Fig. 11). The protein Rb inhibits the function of the transcription factors of the family E2F, necessary for the expression of genes on which DNA duplication depends. The first factor of the family, E2F-1, has been isolated due to the capacity to bond

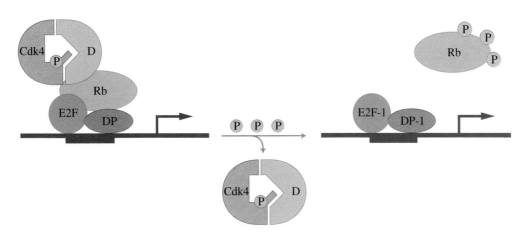

FIGURE 11 Phosphorylation of Rb by the cyclin D–Cdk4 complex causes the release of the transcription factor E2F and of DP and the start of the transcription of the genes necessary for entry into the S phase. The blue line represents a segment of DNA and the arrow a generic transcription initiation site.

the protein Rb and four other proteins similar to it are known: 2F-2, 2F-3, E2F-4, and E2F-5. The E2F proteins form heterodimeric complexes with the proteins DP-1 and DP-2. These complexes constitute the active form of the transcription factor. The dimerization of the DP proteins with 2F is necessary for the formation of a bond of high affinity with the DNA, for transcriptional activation, and for interaction with Rb. The factor E2F determines the transcription of genes whose expression is necessary for the induction of phase S. Furthermore, E2F regulates the transcription of Cdc2, cyclin A, and cyclin E. The protein Rb, in the absence of proliferative stimuli, prevents entry to phase S. It is responsible for the inactivation of the transcription factor E2F during the G_1 phase of the cell cycle. The inhibition is removed when Rb is phosphorylated by the Cdk complexes. In effect, the cell cycle could also be described as the Rb phosphorylation cycle (Weinberg, 1995): During the G_1, S, and M phases, various cyclin-dependent kinases bring about the phosphorylation of Rb. Phosphorylation in the G_1 phase by the cyclin complexes D–Cdk4 and E–Cdk2 determines entry to phase S. Subsequently, the protein remains phosphorylated until the completion of mitosis. It is thus to be presumed that the Cdk complexes containing cyclin A and cyclin B, which are active in phases S and G_2, respectively, and during mitosis, can also contribute to maintaining the protein Rb in phosphorylated form; these complexes are in fact able *in vitro* to phosphorylate the protein.

Stoppage of the Cell Cycle by Means of p53

The protein p53 plays a fundamental role in the control of the response of cells to alterations of the genome (Gottlieb and Oren, 1996). Under extreme conditions of stress, the cell may respond to cell damage by entering into apoptosis, or it may repair the damage suffered and resume normal growth. In mammalian cells, one of the checkpoint mechanisms of the cell cycle studied in greatest detail is that dependent on p53. In response to ionizing radiations or chemical compounds that damage the DNA, the intracellular level of p53 increases, causing the transcriptional activation of p53-dependent genes, one of which codifies the protein p21. Increasing the intracellular level of p21 results in a stoppage of the cell cycle, enabling the cells to repair the DNA (Fig. 12). Supporting this hypothesis, it has been demonstrated that the cells in which the gene that codifies p53 or p21 has been inactivated do not stop the cell division cycle in response to genomic damage by radiations or by drugs.

The increment in the cell levels of p53 is due to an increase in the half-life of the protein. Regulation of the accumulation of p53 occurs through the modulation of protein degradation caused by the reactions of ubiquitinylation. Comparing the half-life of p53 in cells during growth and in those in which because of ionizing radiations a stoppage of the division cycle has occurred, there is a 5–10 times increase. Normally, the half-life of p53 is very short and can be lengthened by the addition to cells in culture of inhibitors of the enzymatic activity of the proteasome. A shortening of the half-life of p53 prevents its intracellular accumulation and is caused by infections by the human papilloma virus, responsible for the beginning of carcinomas of the uterine neck. In infected cells, the protein E6 of the virus bonds p53, inducing conjugation with the ubiquitin and degradation by the proteasome (Scheffner *et al.*, 1991). p53 is also able to induce cell death: In the presence of high levels of stress, p53 can cause the entry of cells into apoptosis.

In approximately 50% of all tumors in humans, *p53* is absent or mutated. It has been demonstrated that alterations of the functions of *p53* are associated with greater resistance to antitumor medicines, which is to be expected given that the lack of *p53* permits the cells to continue to divide in the presence of cell damage.

Cell Cycle and Tumors

It has been proven that alterations in the control mechanisms of the cell division cycle are responsible for the uncontrolled growth and the genetic instability of cancer cells (Sherr, 1996). The absence of regulation of the activity of complexes containing cyclin D, due to superrepression of the regulating or catalytic subunit, or lack of inhibiting pro-

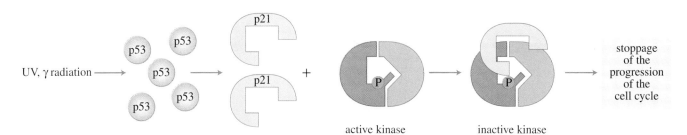

FIGURE 12 Cell damage due to chemical or physical agents, such as ultraviolet and γ rays, induces an accumulation of p53, which in turn induces the accumulation of p21. This unites with the cyclin–Cdk complexes and inhibits their activity, causing the stoppage of the progression of the cycle.

teins such as p16 makes cancer cells independent of growth factors, thus promoting the characteristic uncontrolled cell division.

In many human tumors, the following molecular alterations of Cdk regulation have been identified:

1. Genetic alterations that lead to an accumulation of cyclin and that are caused by genic amplification and translocations

2. Mutations that make the subunit Cdk nonsusceptible to inhibition by Cki inhibitors

3. Mutations and deletion of genes that codify Cki

4. Alterations of the enzymes that dephosphorylate and activate Cdk (e.g., Cdc25)

5. Alterations of Cki p27 degradation

6. Mutations and deletion of the genes that codify the Cdk substrates

All the molecular alterations described affect the Cdks which regulate the progression through phase G_1 of the cell cycle, during which the cells have to decide whether to continue the process of division or to stop in response to extracellular signals, such as the presence of growth factors or other molecules able to induce differentiation, physical contact with the surrounding cells, and interactions with the extracellular matrix. In each of the cases mentioned, it has been determined that the cells affected by cancer show an alteration in their entry to phase S, in which DNA duplication occurs.

The intrinsic toxicity of antitumor drugs currently available limits their use in treatment. These substances have been identified for their capacity to inhibit the growth of cells in culture or experimental tumors *in vivo*. The preferential cytotoxic activity of antitumor agents on cancer cells, compared with normal ones, probably depends on the fact that cancer cells are more sensitive to perturbations of the cell division cycle because they already have alterations of their cycle control mechanisms. Because their growth is independent of conditions in the external environment, cancer cells are very sensitive to the perturbations induced by antiproliferative agents since they are not able to react to them in an appropriate way. It would be advantageous to finalize a strategy to identify inhibitors of checkpoint functions with very low toxicity for normal cells, in the absence of further perturbations, as seems to occur in experimental models of cell death induction *in vitro*. It appears that it is already possible to treat tumors showing p53 alterations by using this type of approach.

New experimental strategies could therefore be based on the inhibition of the tumor proliferative capacity, by means of the specific inhibition of enzymes (altered in tumors) which control progression through the cell cycle, and on interference with checkpoint processes so as to cause the cells to plunge into apoptosis in response to the activation of mechanisms that already exist.

References Cited

BALDIN, V., LUKAS, J., MARCOTE, M. J., PAGANO, M., and DRAETTA, G. F. (1993). Cyclin D_1 is a nuclear protein required for cell cycle progression in G_1. *Genes Dev.,* **7,** 812–821.

BLOW, J. J., and LASKEY, R. A. (1988). A role for the nuclear envelope in controlling DNA replication within the cell cycle. *Nature,* **332,** 546–548.

COVERLEY, D., DOWNES, C. S., ROMANOWSKI, P., and LASKEY, R. A. (1993). Reversible effects of nuclear membrane permeabilization on DNA replication: evidence for a positive licensing factor. *J. Cell. Biol.,* **122,** 985–992.

DE BONDT, H. L., ROSENBLATT, J., JANCARIK, J., JONES, H. D., MORGAN, D. O., and KIM, S. H. (1993). Crystal structure of cyclin-dependent kinase 2. *Nature,* **363,** 595–602.

DUNPHY, W. G., BRIZUELA, L., BEACH, D., and NEWPORT, J. (1988). The *Xenopus* cdc2 protein is a component of MPF, a cytoplasmic regulator of mitosis. *Cell,* **54,** 423–431.

ELLEDGE, S. J. (1996). Cell cycle checkpoints: preventing an identity crisis. *Science,* **274,** 1664–1672.

FANTES, P., and NURSE, P. (1977). Control of cell size at division in fission yeast by a growth-modulated size control over nuclear division. *Exp. Cell. Res.,* **107,** 377–386.

FORSBURG, S. L., and NURSE, P. (1991). Cell cycle regulation in the yeasts *Saccharomyces cerevisiae* and *Schizosaccharomyces pombe. Ann. Rev. Cell. Biol.,* **7,** 227–256.

GAUTIER, J., NORBURY, C., LOHKA, M., NURSE, P., and MALLER, J. (1988). Purified maturation-promoting factor contains the product of a *Xenopus* homolog of the fission yeast cell cycle control gene cdc2+. *Cell,* **54,** 433–439.

GAUTIER, J., MINSHULL, J., LOHKA, M., GLOTZER, M., HUNT, T., and MALLER, J. L. (1990). Cyclin is a component of maturation-promoting factor from *Xenopus. Cell,* **60,** 487–494.

GOTTLIEB, T. M., and OREN, M. (1996). p53 in growth control and neoplasia. *Biochim. Biophys. Acta,* **1287,** 77–102.

HARTWELL, L. H., CULOTTI, J., PRINGLE, J. R., and REID, B. J. (1974). Genetic control of the cell division cycle in yeast, *Science,* **183,** 46–51.

HOCHSTRASSER, M. (1995). Ubiquitin, proteasomes, and the regulation of intracellular protein degradation. *Curr. Opin. Cell. Biol.,* **7,** 215–223.

HOFFMANN, I., CLARKE, P. R., MARCOTE, M. J., KARSENTI, E., and DRAETTA, G. F. (1993). Phosphorylation and activation of human cdc25-C by cdc2-cyclin B and its involvement in the self-amplification of MPF at mitosis. *EMBO J.,* **12,** 53–63.

JACKSON, P. K. (1996). Cull and destroy. *Curr. Biol.,* **6,** 1209–1212.

JEFFREY, P. D., RUSSO, A. A., POLYAK, K., GIBBS, E., HURWITZ, J., MASSAGUE, J., and PAVLETICH, N. P. (1995). Mechanism of Cdk activation revealed by the structure of a cyclin A-Cdk2 complex. *Nature,* **376,** 313–320.

KING, R. W., DESHAIES, R. J., PETERS, J. M., and KIRSCHNER, M. W. (1996). How proteolysis drives the cell cycle. *Science,* **274,** 1652–1659.

KIPREOS, E. T., LANDER, L. E., WING, J. P., HE, W. W., and HEDGECOCK, E. M. (1996). cul-l is required for cell cycle exit in *C. elegans* and identifies a novel gene family. *Cell,* **85,** 829–839.

LENO, G. H., DOWNES, C. S., and LASKEY, R. A. (1992). The nuclear membrane prevents replication of human G_2 nuclei but not G_1 nuclei in *Xenopus* egg extract. *Cell,* **69,** 151–158.

LOHKA, M. J., HAYES, M. K., and MALLER, J. L. (1988). Purification of maturation-promoting factor, an intracellular regulator of early mitotic events. *Proc. Natl. Acad. Sci. USA*, **85**, 3009–3013.

LUKAS, J., PAGANO, M., STASKOVA, Z., DRAETTA, G. F., and BARTEK, J. (1994). Cyclin D_1 protein oscillates and is essential for cell cycle progression in human tumour cell lines. *Oncogene*, **9**, 707–718.

MATSUSHIME, M., ROUSSEL, M., ASHMUN, R., and SHERR, C. J. (1991). Colony-stimulating factor 1 regulates novel cyclins during the G_1 phase of the cell cycle. *Cell*, **65**, 701–713.

MCNALLY, F. J. (1996). Modulation of microtubule dynamics during the cell cycle. *Curr. Opin. Cell. Biol.*, **8**, 23–29.

MENDENHALL, M. D. (1993). An inhibitor of p34CDC28 protein kinase activity from *Saccharomyces cerevisiae*. *Science*, **259**, 216–219.

MURRAY, A. (1993). Turning on mitosis. *Curr. Biol.* **3**, 291–293.

MURRAY, A. (1995). The genetics of cell cycle checkpoints. *Curr. Opin. Genet. Dev.*, **5**, 5–11.

NIGG, E. A. (1991). The substrates of the cdc2 kinase. *Semin. Cell. Biol.*, **2**, 261–270.

NUGROHO, T. T., and MENDENHALL, M. D. (1994). An inhibitor of yeast cyclin-dependent protein kinase plays an important role in ensuring the genomic integrity of daughter cells. *Mol. Cell. Biol.*, **14**, 3320–3328.

NURSE, P. (1975). Genetic control of cell size at cell division in yeast. *Nature*, **256**, 451–457.

OHTSUBO, M., THEODORAS, A. M., SCHUMACHER, J., ROBERTS, J. M., and PAGANO, M. (1995). Human cyclin E, a nuclear protein essential for the G_1-to-S phase transition. *Mol. Cell. Biol.*, **15**, 2612–2624.

PAGANO, M., TAM, S. W., THEODORAS, A. M., BEER-ROMERO, P., DEL SAL, G., CHAU, V., YEW, P. R., DRAETTA, G. F., and ROLFE, M. (1995). Role of the ubiquitin-proteasome pathway in regulating abundance of the cyclin-dependent kinase inhibitor p27. *Science*, **269**, 682–685.

PARDEE, A. B. (1989). G_1 events and regulation of cell proliferation. *Science*, **246**, 603–614.

PRINGLE, J. R., and HARTWELL, L. H. (1981). In *The molecular biology of the yeast* Saccharomyces: *life cycle and inheritance*, a c. di Strathern J. N., Jones, E. W., Broach, J. R., Cold Spring Harbor, Cold Spring Harbor Laboratory Press, pp. 97–142.

QUELLE, D. E., ASHMUN, R. A., SHURTLEFF, S. A., KATO, J. Y., BAR-SAGI, D., ROUSELL, M. F., and SHERR, C. J. (1993). Overexpression of mouse D-type eclins accelerates G_1 phase in rodent fibroblasts. *Genes Dev.*, **7**, 1559–1571.

RAO, P. N., and JOHNSON, R. T. (1970). Mammalian cell fusion: studies on the regulation of DNA synthesis and mitosis. *Nature*, **225**, 159–164.

RUDNER, A. D., and MURRAY, A. W. (1996). The spindle assembly checkpoint. *Curr. Opin. Cell. Biol.*, **8**, 773–780.

RUSSO, A. A., JEFFREY, P. D., PATTEN, A. K., MASSAGUE, J., and PAVLETICH, N. P. (1996a). Crystal structure of the p27Kipl cyclin-dependent-kinase inhibitor bound to the cyclin A-Cdk2 complex [see comments]. *Nature*, **382**, 325–331.

RUSSO, A. A., JEFFREY, P. D., and PAVLETICH, N. P. (1996b). Structural basis of cyclin-dependent kinase activation by phosphorylation. *Nat. Struct. Biol.*, **3**, 696–700.

SCHEFFNER, M., MÜNGER, K., BYRNE, J. C., and HOWLEY, P. M. (1991). The state of the p53 and retinoblastoma genes in human cervical carcinoma cell lines. *Proc. Natl. Acad. Sci. USA*, **88**, 5523–5527.

SCHWOB, E., BOHM, T., MENDENHALL, M. D., and NASMYTH, K. (1994). The B-type cyclin kinase inhibitor p40SIC1 controls the G_1 to transition in *S. cerevisiae*. *Cell*, **79**, 233–244.

SHERR, C. J. (1996). Cancer cell cycles. *Science*, **274**, 1672–1677.

STILLMAN, B. (1996). Cell cycle control of DNA replication. *Science*, **274**, 1659–1664.

VERDE, F., LABBE, J. C., DOREE, M., and KARSENTI, E. (1990). Regulation of microtubule dynamics by cdc2 protein kinase in cell-free extracts of *Xenopus* eggs. *Nature*, **343**, 233–238.

WEINBERG, R. A. (1995). The retinoblastoma protein and cell cycle control. *Cell*, **81**, 323–330.

WÖLFEL, T., HAUER, M., SCHNEIDER, J., SERRANO, M., WÖLFEL, C., KLEHMANN-HIEB, F., DE PLAEN, E., HANKELN, T., MEYER ZUM BÜSCHENFELDE, K., and BEACH, D. (1995). A p16INK4a-insensitive Cdk4 mutant targeted by cytolytic T lymphocytes in a human melanoma. *Science*, **269**, 1281–1284.

General References

GLOVER, D. M., and HUTCHISON, C., a c. di, *Cell cycle control*. Oxford-New York, Oxford University Press, 1995.

HOCHSTRASSER, M. (1996). Protein degradation or regulation: Ub the judge. *Cell*, **84**, 813–815.

LUKAS, J., BARTKOVA, J., ROHDE, M., STRAUSS, M., and BARTEK, J. (1995). Cyclin D_1 is dispensable for G_1 control in retinoblastoma gene-deficient cells, independently of Cdk4 activity. *Mol. Cell. Biol.*, **15**, 2600–2611.

MURRAY, A., and HUNT, T. *The cell cycle: an introduction*. Oxford-New York, Oxford University Press, 1994.

PAGANO, M., a c. di, *Cell cycle: materials and methods*. Heidelberg, Springer Verlag, 1995.

SHERR, C. J. (1993). Mammalian G_1 cyclins. *Cell*, **73**, 1059–1065.

SHERR, C. J., and ROBERTS, J. M. (1995). Inhibitors of mammalian G_1 cyclin-dependent kinases. *Genes. Dev.*, **9**, 1149–1163.

THOMAS G. COTTER

Tumour Biology Laboratory
Department of Biochemistry
University College
Lee Maltings, Prospect Row
Cork, Ireland

In multicellular organisms cell generation occurs via the process of mitosis and we know a considerable amount about the biology of this process. In contrast, we know very little about the mechanisms by which cells die. We do know that cell death occurs in a very controlled fashion with the dying cell playing an active part in its own demise. The cell death that takes place under physiological conditions is called apoptosis; it is regulated genetically and is important not only in physiological development but also in many disease situations. Exploring the underlying biology of apoptosis, in particular the characterization of cell suicide genes, has been a very active focus of research in the scientific world for the past 5 years.

Apoptosis: A Special Cell Death Program

Introduction

Man is a multicellular organism and cell division takes place by a process known as mitosis, as it does in almost all eukaryotic life forms. During mitosis the cell duplicates its DNA and divides to form two daughter cells, each exact replicas of their parent. Cell division by mitosis is a regular event in our bodies and, for example, is responsible for the replacement of the cells that line our gastrointestinal and respiratory tracts. This sort of cell replacement occurs every few days. Therefore, even though there is a considerable amount of mitosis constantly going on in multicellular organisms such as ourselves, there is no real change in the total number of cells in our bodies. This of course implies that for every cell that is generated via mitosis one must die

to balance the "cellular account book" of life. The question is how do cells die, and die they must if this balance is to be maintained. Regarding cell death, we now know that a cell can die by one of two processes — that of necrosis or apoptosis.

Necrosis is the classically defined form of cell death and results from severe trauma to the cell. A cell dying by necrosis undergoes a rapid and uncontrolled swelling and simply bursts. Its subcellular contents are spilled out into the extracellular milieu and this often results in an associated inflammatory response. The inflammatory response is characterized by an increased flow of blood to the area, an influx of white blood cells, and the release of various inflammatory mediator molecules. Cell death via necrosis is rarely seen under physiological conditions and cells dying by this process have no control over their fate. This lack of control is an important point and will be discussed later.

The second form of cell death is called apoptosis, and this process differs markedly from that of necrosis. First, it was discovered quite recently, in contrast to necrosis, which has been known for almost a century. Second, apoptosis is a genetically controlled process in direct contrast to necrosis which is not. This article is a review of the current understand of the cell biology and genetics of apoptosis, the role the process plays in normal physiology, and what happens when it becomes deregulated in diseases such as AIDS and cancer.

Apoptosis

In 1972 Andrew Wyllie and colleagues published a paper (Kerr *et al.,* 1972) that described the cell death process of apoptosis (Fig. 1). The term apoptosis is of Greek origin and is used to describe the falling of a petal from a flower or the loss of a leaf from a tree. In pronouncing the word correctly, the second "p" is silent. What may appear as a rather obscure origin for the derivation of the term apoptosis does in fact have some relevance to the features of the death process. As cells die by apoptosis they round up and literary fall off the supporting tissue structures on which they are growing, just like leaves falling from a tree in autumn. The process of apoptosis was identified 1972, but before that there were many examples of the process reported in the literature, albeit using different terminology. For ex-

ample, tingle bodies in the thymus, sunburn cells in the skin, and councilman bodies in the liver are all pre-1972 examples of apoptosis. Another term used to describe the phenomenon of apoptosis pre-1972 was shrinkage necrosis (Kerr, 1971). This is an accurate histological description of apoptosis since the outflow of water is one of the key features of apoptosis. In contrast, during necrosis there is cell swelling due to the inflow of water.

Although apoptosis was discovered in 1972, almost a decade passed before it caught the imagination of scientists. In the mid- to late 1980s immunologists and developmental biologists rediscovered the 1972 paper of Wyllie, Kerr, and Currie and began to explore the underlying biology of apoptosis in both immunological and development systems. This led to a slow increase in the number of publications in the area by the end of the 1980s. In the 1990s there has been an explosion in the number of papers devoted to cell death. In addition, there are at least a dozen books and two journals specifically devoted to the study of apoptosis. The reason for this rapid surge of interest is that scientists have realized that cell death via apoptosis is a fundamental biological process which we know very little about. In addition, the idea of a death process with "suicide genes" has stirred a "Faustian" interest in the subject. The quest to discover why and how a cell switches on its death program is one of the most active areas of biology in the 1990s and for good reason: If we can understand the process by which a cell kills itself, we may be able to interfere and either prolong the life of the cell or speed up its death. This is a desirable objective in many human disease conditions.

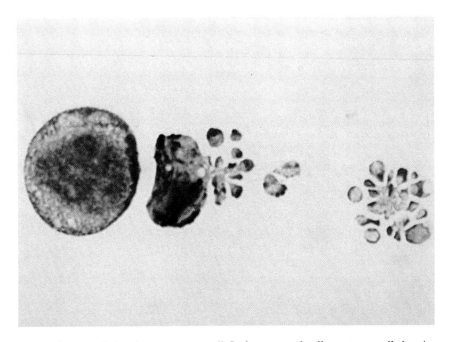

FIGURE I Stages of apoptosis in a human tumor cell. Left, a normal cell; center, a cell showing reduced volume and formation of the first apoptotic bodies; right (final stage), only a group of apoptotic bodies is present.

Cell Death in Development

Before discussing the cell biology of apoptosis it is worthwhile to ask the question as to why cells have developed a sophisticated mechanism to die and in what circumstances the death program is activated. There appears to be a variety of situations in which apoptosis is fundamentally important for the development and survival of multicellular organisms. One of these is the cell death which occurs under normal physiological processes. During fetal development, for example, there is a web of cells between the fingers and toes giving the hands and feet paddle-like features. As development progresses the fingers and toes are sculpted from this paddle, with the cells between the new digits dying via apoptosis (Fig. 2). This is a clear example of how the body uses the genetically programmed death process of apoptosis to form new anatomical structures. Another example of the role played by apoptosis is in the development of the neural system. During development the body makes a huge excess of neurons. These cells then compete in the newly forming nervous system to make connections with other cells. If they fail to make these connections they die via apoptosis. In fact, approximately 90% of all neurons formed during our lives die by apoptosis because they fail to make life-sustaining connections with their neighbors. This type of connect or die process is vitally important because it ensures that our nervous system, which so crucially depends on proper cell connections being made, is carefully put together to form a fully functioning neural network. A final example from the world of the development biologist and one which will be familiar to most people is the reabsorption of the tadpole tail during its transition into a frog. The cells of the tail are destroyed by apoptosis and reabsorbed (Lockshin and Zakeri, 1991).

There are additional examples from the field of immunology in which autoreactive cells of the immune system are deleted before they can cause harm to the body by spearheading an unwanted immune response against normal host tissue. However, there are many situations in which the highly regulated cell deletion of the immune system breaks down and the consequence is an array of crippling autoimmune diseases. Such diseases will be discussed later.

Caenorhabditis elegans and Apoptosis

A favorite model organism for the study of many developmental processes, in particular apoptosis, is the nematode worm *C. elegans.* The reason for this is that the organism is composed of 1090 cells and during development 131 die. These cells die in a programmed manner via apoptosis. The cell death primarily affects cells of the nervous system (105 cells); however, other cell types including those of muscle also die. Central to the death of these 131 cells are two key genes called *ced-3* and *ced-4.* These genes are selectively switched on when the program for cell death is activated in the 131 cells destined to die. If these genes are mutated such that they do not perform their normal function, then none of the 131 cells destined to die do so. Strangely, worms with the " extra" 131 cells appear normal. Seven genes control the engulfment and removal of the dead apoptotic cells. Mutations in these genes do not affect whether cells die or not, indicating that there is not a direct link between the *ced-3* and *-4* death-inducing genes and those involved in removing the dead cells.

The main reason for the cell death seen in *C. elegans* is the removal of unwanted cells, which has many consequences. For example, it may serve to remove cells that had a specific task during the early stages of development but have no further function in the adult organism. Second, it can generate sexual dimorphism. Many of the genes identified in the regulation of apoptosis in this organism have counterparts in higher organisms such as mammals, and what makes *C. elegans* so interesting is that it can be easily manipulated for genetic and development studies. Such studies are particularly difficult in more complex higher organisms.

Cell Stress and Apoptosis

In nature there is a constant struggle for survival as organisms are subjected to a variety of natural stresses during the course of their lives. This struggle even exists at the level of the cell in multicellular organisms, and strangely cell death, via apoptosis, may have an important role to play in the organisms' survival response in the face of stress. For example, when a cell is exposed to stress due to environmental or biological insult, it either survives or dies. This is a fact of cell life. If the cell experiences a low level of stress, then the synthesis of a group of stress or heat shock proteins can afford a level of protection to the cell, providing the stressing agent is removed after a short period. These stress proteins

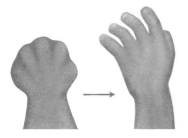

FIGURE 2   Early and late stages in the embryonal development of the human hand. Apoptosis is responsible for the death of the interdigital cells during embryonic tissue sculpting.

are found in almost all cell types from bacteria to humans and may be a primitive self-defense system. If, however, a cell experiences an insult which is not sufficient to kill it outright and to which stress/heat shock proteins cannot protect it, then the program for apoptosis is activated and the cell dies in a controlled fashion (Samali and Cotter, 1996). This is of clear benefit to multicellular organisms because when a cell dies via apoptosis its contents are not released to the cell's exterior as is seen in necrosis. Such a release would cause injury or even death to its near neighbors and this can have serious detrimental effects for the survival of other cells and the organism. In other words, by dying via apoptosis a cell is performing an altruistic act to save its cellular neighbors. Finally, if the cell experiences a high insult such that there is no time to activate the apoptosis program, then death occurs via necrosis with clear-cut damage and injury to neighboring cells due to the release of cell contents from the dying cell. Thus, a cell has two levels of defense to protect itself and the organism of which it is a part. In the first instance, with the help of stress proteins the cell survives, and in the second case the cell

dies via apoptosis but this does not affect neighboring cells or damage the organism (Fig. 3).

Biological Features of Apoptosis

One of the key biochemical hallmarks of apoptosis is the fragmentation of the cell's DNA into nucleosome fragments which can be seen in an agarose electrophoresis gel as a DNA ladder (Wyllie, 1980). This occurs due to the activation of an endonuclease enzymes which remains in an inactive state in the cell until the apoptosis program is switched on. The exact identification and characterization of this pivotal enzyme remains elusive and currently there are six candidate enzyme vying for the honor. It may well be that there are many enzymes capable of cleaving DNA in the manner seen in apoptosis, but which one is the key enzyme is unclear. The fragmentation also appears to occur in a two-stage process, with the DNA being cleaved into large pieces initially and then these pieces are broken down into nucleosome-size fragments at a later stage. It is likely that there

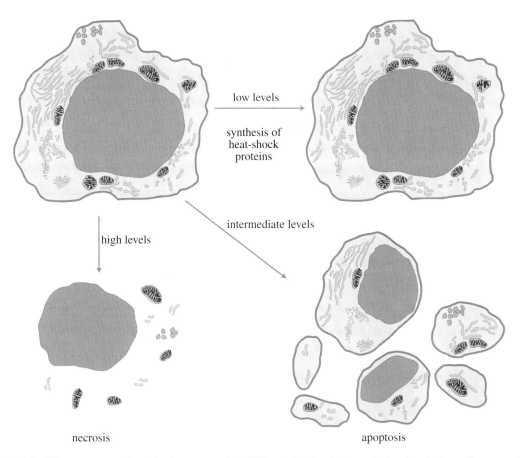

low levels

synthesis of
heat-shock
proteins

high levels

intermediate levels

necrosis

apoptosis

FIGURE 3 The responses of a cell when exposed to different levels of stress. At low levels the cell can produce heat shock or stress proteins which protect the cell. At intermediate levels, cell death occurs by apoptosis. At high levels of stress the cell dies by necrosis, with release of the cell contents into the extracellular milieu.

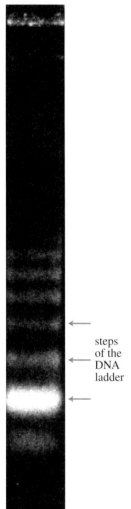

FIGURE 4 A DNA ladder formed of fragments of the size of one or a few nucleosomes, separated by electrophoresis on an agarose gel.

steps
of the
DNA
ladder

are two distinct enzyme systems that perform this process because some cells have only the ability to produce the large fragments, whereas other can produce both fragment types. Figure 4 shows a typical DNA fragmentation pattern seen in cells dying via apoptosis.

During apoptosis there are many dramatic changes in the morphology of the dying cell, including the extrusion of water from the cell leading to its shrinkage. Cells can shrink up to 50% during this process. Almost nothing is known about how or why the cell undergoes this shrinkage process. In addition to cell shrinkage, there is also a marked condensation of the cell's chromatin giving rise to the formation of pycnotic nuclei. In some cells, the nucleus then undergoes fragmentation inside the cell. The next clear morphological change that is seen is the break up of the cell into small sealed vesicles called apoptotic bodies. This is clearly seen in Fig. 1. The vesicles are then engulfed and destroyed by neighboring phagocytic cells. During this whole process there is no leakage of the cell's contents into the extra cel-

lular milieu since the integrity of the plasma membrane is maintained.

One reason why it took a very long time for scientists to discover apoptosis was not because the process rarely occurs but because apoptotic cells are very quickly removed under physiological conditions. The recognition and removal of apoptotic cells appear to be correlated with the expression of novel molecules on the surface of the apoptotic cells which allows them to be recognized by phagocytes. One of these molecules is the membrane lipid phosphatidyl serine. This molecule is normally located on the inside surface of the plasma membrane; however, during apoptosis the molecule flips to the outside and this acts as a recognition signal for phagocytosis (Fadok *et al.*, 1992). It is unclear how precisely the phosphatidyl serine is recognized by phagocytes (Fig. 5).

Genes and Apoptosis

Cohen and Duke (1984) determined that apoptosis is a genetically controlled process by demonstrating that inhibitors of macromolecular synthesis retard the process of apoptosis. These experiments indicated that the process of apoptosis was probably under some form of genetic control. Subsequent experiments over a period of years by many groups, including our own, identified two classes of genes that are involved in the regulation of apoptosis. In one class there are genes such as c-*myc* and *p53* which drive the process of apoptosis. In the second class there are genes such as *bcl-2* and *bcr-abl* that can inhibit cell death.

The gene c-*myc* is a type of "Janus" gene in that it can drive both cell proliferation and cell death (Fanidi *et al.*, 1992). Which direction a cell takes seems to depend on the situation in which a cell finds itself. For example, in the

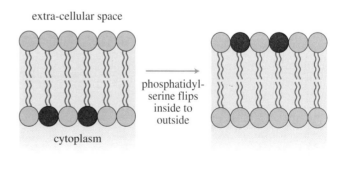

extra-cellular space

phosphatidyl-
serine flips
inside to
outside

cytoplasm

normal cell

apoptotic cell

FIGURE 5 The flipping of membrane lipid phosphatidylserine (in red) from the inside to the outside of the plasma membrane is characteristic of apoptosis. This event acts as a signal to enable the dying cells to be recognized by phagocytic cells.

presence of cell survival factors such as insulin-like growth factor-1, c-*myc* will drive proliferation when it is expressed. However, in the absence of such survival factors c-*myc* appears to trigger cell death via apoptosis. This could be part of a fail-safe mechanism to prevent the development of tumors. c-*myc* is a key gene in regulating cell proliferation, and thus any mutations that upregulate expression of c-*myc* have the potential to lead to uncontrolled cell proliferation and possible tumor development. However, c-*myc* by itself does not control cell proliferation. There must be adequate amounts of cell survival factors produced and their production is regulated by other genes in the cell that are only switched on when required. Thus, if there is a deregulated upregulation in c-*myc* expression, tumor development will not occur unless there is a similar deregulated upregulation in the expression of survival factors. In other words, at least two key genes have to be mutated for a tumor cell to develop, and this probably in part explains why the conversion of a normal cell to a tumor cell is such a rare event, considering the high rates of normal cell division that occur in our bodies (Fig. 6).

p53 is another key proapoptotic gene that will drive cell death in specific circumstances and it is also mutated in the vast majority of human cancers. Therefore, *p53* is part of a family of genes known as tumor suppressor genes because when they are active tumor development is suppressed. The main function of *p53* is to stop cells in G_1 phase of the cell cycle from continuing through the cell cycle following DNA damage. This stoppage allows the cell time to repair the DNA damage, and if the damage cannot be repaired then apoptosis is triggered (Yonish-Rouach *et al.*, 1991). Should *p53* fail to perform its duties (e.g., due to mutations), then cells with genetic damage can reenter the cell cycle, possibly giving rise to cancer. Mutated *p53* cannot trigger apoptosis. It is unclear how *p53* mediates its effects, but probably stops cells in G_1 by interacting with other genes, such *waf-1* and *gadd 45*, known to be involved in cell cycle

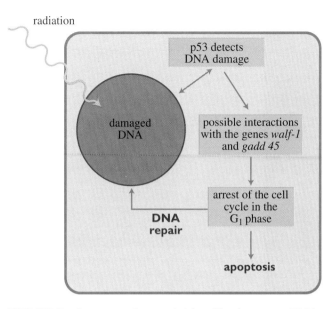

FIGURE 7 Sequence of events initiated by damage to DNA. The tumor suppressor gene *p53* detects DNA damage and halts the cell in the G_1 phase of the cell cycle to allow time for repair. The arrest in G_1 seems to be mediated by the interaction of *p53* with the products of the genes *waf-1* (*p21*) and *gadd45*. If the damage cannot be repaired, then apoptosis is triggered.

control (Fig. 7). Because of its cell cycle policing role, *p53* has been dubbed a guardian of the genome.

Just as there are proapoptotic genes such as c-*myc* and *p53*, likewise there is a class of genes that inhibit apoptosis and into this category are genes such as *bcl-2* and *bcr-abl*. *bcl-2* was the first oncogene to be identified as having a role in the control of apoptosis. Work emanating from the laboratory of Stanley Korsmeyer in St. Louis, Missouri, demonstrated that cells which expressed high levels of this gene were considerably more resistant to the induction of apoptosis. Cells transfected with the gene also showed resistance to cell death (McDonnell and Korsmeyer, 1991). Expression of the *bcl-2* gene was shown to be associated with follicular cell lymphoma, with the implication that this cancer may be due more to a lack of cell death than excess proliferation. In other words, for cells which fail to die at the end of their life span, their immortality contributes to tumor development. An additional consequence of elevated *bcl-2* expression is that there is an increased drug resistance in cells expressing this gene to undergo apoptosis induced by agents normally used to kill tumor cells. Therefore, there is now a scenario in which a gene, by inhibiting cell death, not only contributes to tumor development but also makes it more difficult to kill the cell with conventional anticancer drugs. *bcl-2* expression also appears to contribute to the development and progression of other cancers such as prostate.

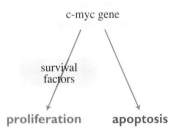

FIGURE 6 Functions of the oncogene c-*myc*, which can act both as a promoter of cell proliferation in the presence of cell survival factors and, in their absence, as an activator of apoptosis. For this reason, it is called a Janus gene after the two-faced Roman god.

Cancer of the prostate is the second or third most common cancer in the Western world. In the United States, it is the second most commonly diagnosed cancer in men. Patients with the disease usually experience difficulty in urination due to the enlargement of the prostate gland. Diagnosis of the disease is relatively straightforward and treatment consists of surgery and/or chemotherapy. Prostate cancer can be characterized as a two-phase disease with the initial phase being responsive to treatment. During this phase, the cancerous prostate cells grow in response to the growth-promoting properties of androgens such as testosterone. In fact, they are almost completely dependent on these male hormones for their survival. If the growth-promoting androgens are removed then the prostate tumor undergoes a dramatic reduction in size and the patient feels considerably better. This can be achieved by using drugs that antagonize the effects of the androgens or, alternatively, surgery in the form of castration can be used to remove the source of the hormones. Unfortunately, like many other cancers, prostate cancer invariably returns and the second phase is considerably more difficult to treat and is usually fatal because the returning tumor is independent of androgens and the cells do not die by apoptosis when this growth-promoting hormone is removed. Recent evidence suggests that an increase in *bcl-2* expression may be central to this change in tumor phenotype. The hormone-responsive tumor cells have little or no *bcl-2* expression but express the death-promoting gene *bax*. When the life-supporting androgens are removed either chemically or by castration, the cells rapidly die. However, *bcl-2* prevents this death and the tumor cells survive (McDonnell *et al.*, 1993). Thus, the expression of this gene may be involved in the transformation of a cancer which initially responds quite well to therapy to one which is resistant. This conversion of a tumor that is responsive to treatment to one that is resistant is not only seen for tumors of the prostate. Thus, the expression of antiapoptotic genes such as *bcl-2* appears to be of considerable significance in the development and treatment of many human cancers.

bcl-2 is just one member of an expanding family of apoptosis regulating genes, which includes its prodeath gene partner *bax*. Like *bcl-2, bax* is involved in the regulation of apoptosis, and whether a cell lives or dies may depend on the level of either of these two genes. An imbalance toward *bcl-2* and a cell lives; tip the scales toward *bax* and it dies. How *bax* drives apoptosis is unknown, as is how *bcl-2* promotes life. An attractive suggestion is that the main function of *bcl-2* is to keep *bax* in an inactive state (Oltvai *et al.*, 1993). It has also been suggested that *bcl-2* may act as an antioxidant; this idea is supported by the fact that many chemical antioxidants can also inhibit apoptosis in a manner similar to that of *bcl-2*. Finally, studies from my own laboratory and others have shown that the production of reac-

tive oxygen molecules is a feature of apoptosis induced by a variety of agents.

In addition to *bcl-2, bcr-abl* and v-*abl* have been shown to be potent inhibitors of apoptosis. Chronic myelogenous leukemia is characterized by the translocation of the c-*abl* gene from chromosome 7 to 22. This leads to the production of the Bcr-Abl fusion protein because the *abl* gene is aligned with the *bcr* gene, leading to an upregulated Abl kinase activity. For this type of cancer, it is interesting that there is a massive increase in the number of mature cells without any change in normal cell mitosis rates (Spiers, 1992). Further studies demonstrated that both *bcr-abl* and its viral counterpart v-*abl* are potent inhibitors of apoptosis. *bcr-abl* and v-*abl* do not seem to mediate their effects through *bcl-2* or *bax* but rather through an unknown effector mechanism. Chronic myelogenous leukemia is a biphasic disease in which the initial phase is characterized by the accumulation of large numbers of mature cells of the myeloid series. This appears to be due to a defect in the cell death process of apoptosis, probably mediated by the increased activity of the mutated Bcr-Abl kinase. The disease may remain in the chronic phase for a period of weeks or years but invariably it undergoes a transition to a more acute disease. The acute or accelerated phase of the disease is particularly difficult to treat and is generally fatal. At the biochemical level there is mutated expression of genes that drive proliferation during the acute phase of the disease, and this in combination with genes that prevent cell death provide the tumor cell with the necessary ingredients to survive and proliferate.

The following is one of the key biochemical questions: How does the elevated activity of the Abl kinase enzyme enable it to suppress cell death? The simple answer is we do not know. All we know is that the kinase catalyzes the phosphorylation of many protein targets in the cell's cytoplasm but we have yet to determine which one of these is important in the regulation of apoptosis.

Proteolytic Enzymes and Apoptosis

A role for proteolytic enzymes was suggested by many early observations, including the fact that protease inhibitors seemed to slow the onset of apoptosis. A second observation that stimulated a focus of research on protease in apoptosis was that *ced-3*, a cell death gene from *C. elegans*, had a sequence similarity with the inflammatory mediator human interleukin-1β converting enzyme (ICE). This enzyme was originally discovered as a component of macrophages which could cleave pro-interleukin-I into its active form. The enzyme belongs to a family of enzymes called cysteine protease (Martin and Green, 1995). With an established link between *ced-3* and ICE, many laboratories began to search not only for ICE-like enzymes but also for their substrates in apoptosis. Many such enzymes were revealed. A role for ICE in apoptosis has been suggested by many observations in addition to the homology link to *ced-3*. Purified ICE induced apoptosis when microinjected into many cell types and this could be inhibited by the ICE inhibitor Crm-A. Crm-A is a serpin-like protein produced by the cowpox virus and inhibits both ICE and apoptosis (Ray *et al.*, 1992). When Crm-A is overexpressed in cells, they are inhibited from undergoing apoptosis. These experiments provided convincing proof that ICE played a role in apoptosis. The one difficulty was that ICE knockout mice (mice that have been engineered to produce no ICE) exhibited a relatively normal phenotype during development. It may be that there are other enzymes capable of replacing ICE or that it does not have a key role in the apoptosis seen during development.

A variety of experiments have demonstrated that protease suspected of being involved in apoptosis can cleave poly(ADP-ribose) polymerase (PARP). PARP is a key enzyme in DNA repair and is cleaved into an 85-kDa fragment during apoptosis in most if not all systems examined to date. It probably makes sense for the cell to inactivate a key DNA repair enzyme if the cell's DNA is going to be fragmented during apoptosis. The cleavage of PARP is an early event in apoptosis and the loss of this function coincides with endonuclease activation. ICE was of course believed to be a key candidate enzyme for PARP cleavage. However, when the purified enzyme was added to PARP, nothing happened. Clearly, ICE was not directly responsible for the cleavage of PARP, but the fact that Crm-A blocked both apoptosis and PARP cleavage indicated a role in the system for ICE. Shortly after these experiments were reported an enzyme called cpp32 or YAMA (after the Hindu god of death) was shown to be capable of cleaving PARP (Tewari *et al.*, 1995). This enzyme is cleaved to produce a 12- and 17-kDa subunit during activation and this cleavage can be accomplished by ICE. The active form of YAMA/cpp32 was recently characterized and called apopain. This molecule has the capacity to cleave PARP.

Combining the previous pieces of information results in the following sequence of possible events. ICE or a related ICE family member cleaves cpp32/YAMA to produce an active apopain, which in turn cleaves and inactivates PARP. This then leads to the activation of the apoptotic endonuclease. The question remains as to how ICE is activated. A schematic drawing of these events is shown in Fig. 8.

Cell Death in the Immune System

AIDS is a disease characterized by the slow deterioration of the immune system. This appears to be due to the depletion of a group of T lymphocytes called CD4 cells (Cotter, 1993). The main function of these cells is to control and stimulate the immune response to infective organisms. The death of these cells in HIV patients appears to occur via apoptosis, although the cellular events which stimulate cell death are unclear. It is known that there is considerable apoptosis in CD4 cells that is not mediated directly by the AIDS virus but rather in an indirect manner. One line of evidence indicates that apoptosis occurs because of a cellular "short circuit" in the signals that would normally lead to proliferation of CD4 cells and an immune response. The following events appear to occur: One of the structural proteins of the virus is gp120, which is a key protein in allowing the virus to gain access to the CD4 lymphocyte. Once the virus has gained entry to the cell it incorporates into the genome and stays there until it is triggered to replicate. As part of the replicative process gp120 is produced and this protein can be released into the serum. This in turn can bind to other cells, and if these cells are triggered to mount an immune response they undergo apoptosis rather than proliferating, which they would do as part of a normal response. Thus, when CD4 cells receive a signal to proliferate they do the direct opposite when they are receiving signals through gp120, which is also bound to the cell at the same time. Although gp120 and stimulation of the CD4 cell through its T cell receptor may be key in triggering apoptosis, little is known about exactly how CD4 cell apoptosis

FIGURE 8 A possible sequence of proteolytic events leading to the cleavage of PARP, a critical enzyme in DNA repair which is inactivated during the early stages of apoptosis. The ICE protein, whose mode of action is unknown, acts on cpp322/YAMA transforming it into its active form (apopain) which cleaves and inhibits PARP; this then leads to the activation of the apoptotic endonuclease. Crm-A inhibits the action of ICE.

occurs. The actual death blow may be carried out through a cell surface molecule called Fas and its ligand (FasL). In the past few years, Fas has become a particularly important molecule in our understanding of the first few steps in the mechanism of apoptosis.

Fas is a member of the tumor necrosis family, so named because of its role in cell death. Fas is expressed on many cells in the body, and when it engages its natural ligand it induces apoptosis. Most of what is known about Fas derives from studies done on cells of the immune system in which Fas appears to be a key player in immune cell apoptosis (Nagata and Goldstein, 1995). In the thymus, which is one of the pivotal organs of the immune system responsible for the education of T lymphocytes, there is considerable natural apoptosis. In fact, approximately 99% of all thymocytes produced die within the thymus and they appear to do so via apoptosis. Thymocytes express a high level of Fas and there is accumulating evidence that the cell death seen in these cells is mediated through Fas. Mutations in the Fas molecule lead to lymphoproliferation, indicating that this molecule is pivotal in regulating the balance of lymphocyte proliferation and death. Fas also appears to have a role in the destruction of tumor cells by cytotoxic T lymphocytes. In other words, cytotoxic T lymphocytes kill their target cells by the induction of apoptosis and this appears to involve the Fas protein.

Fas is a transmembrane protein with a cytoplasmic tail that contains a "death domain." This is known to be true because mutations in this region of the molecule block the death-promoting activities of Fas. The following is a key question: How does Fas trigger the major morphological and biochemical features of apoptosis? This is a question for which we have only a partial answer and it may involve a lipid molecule called ceramide. Ceramide is normally a structural component of sphingomyelin which is found in cell membranes. When sphingomyelin is hydrolyzed by the enzyme sphingomyelinase ceramide is generated (Hannun and Obeid, 1995). When the Fas receptor is engaged there is an increase in the production of ceramide within cells and this seems to be a key molecule in the Fas signal cascade leading to apoptosis. In addition, ceramide by itself is a potent inducer of apoptosis in many cell types. How ceramide propagates the death signal is unclear, but it is known that it stabilizes mRNA and this may be important for apoptosis to occur.

Cell Death in the Nervous System

The death of cells in the nervous system has been recognized for almost a century. In 1949, Levi-Montalcini demonstrated extensive neuronal cell death during the development of the nervous system in the chick embryo. It was thought that the cell death seen was part of the complex sequence of developmental events leading to the construction of a nervous system. These early studies provided insight into the role cell death played in developmental processes but failed to ignite the imagination of scientists at the time. The reason for this may have been the concept that cell death was basically a passive event and as such unworthy of serious investigation. It was not until the seminal work of Wyllie and colleagues in the 1970s that interest in cell death in the nervous system was rekindled. Subsequent studies demonstrated that the cell death originally documented by Levi-Montalcini was in fact apoptosis. The dying neurons showed the classical morphological features of apoptosis and also fragmented their DNA into nucleosome-size pieces. The fact that neuronal cell apoptosis is an active process and dependent on RNA and protein synthesis was demonstrated by culturing neuronal cell lines in the absence of growth factors such as nerve growth factor in which apoptosis occurred, and this could be prevented by treating the cells with inhibitors of protein synthesis such as cycloheximide (Martin et al., 1992). It appears that during the developmental process a huge excess of neurons are produced. The cells will survive if they can make contact with neighboring cells and die if they do not. This connection with neighboring cells not only provides the cells with a lifeline but also helps construct the nervous system. In other words, if newly formed neurons do not connect with the neighboring neurons they die. Thus, there is a clear biological survival incentive for neurons to make nervous connections. The use of transgenic mice has given an even clearer insight into the role played by apoptosis in this system. For example, mice made transgenic for the antiapoptotic gene bcl-2 so that it is expressed in the nervous system have a brain size one and a half to twice the size of that of their nontransgenic counterparts. It is unknown whether there is any relationship between these animals with genetically altered brain size and intelligence.

One of the features of neurodegenerative diseases in humans is the death of cells in the central nervous systems. There is debate as to whether the cell death seen is apoptosis or not, but the balance of evidence is swaying in the direction of apoptosis. With this loss of cells there is a deterioration of brain function and this is manifested in a variety of ways. If the loss of brain cells can be prevented, we may be able treat diseases such as Alzheimer's and Parkinson's diseases. The ability to do this is the goal of many major pharmaceutical companies. Since it has been demonstrated that bcl-2 is a potent inhibitor of neuronal cell death, there may be scope in this area for developing therapies based on bcl-2. It may be possible to use gene therapy techniques to target bcl-2 to specific brain cells and prevent the cell death seen in diseases such as those described previously. An alternative avenue of action would be to design drugs that would cause an elevation of bcl-2 in target cells or even a drug that would mimic the life-preserving effects of bcl-2.

Prospects for Apoptosis-Based Therapies

In the preceding section, I briefly discussed the subject of altering the expression of apoptosis-regulating genes, and it is worthwhile to explore this concept as a potential future treatment for many human diseases. For example, as mentioned earlier, the chimeric *bcr-abl* gene is the driving force behind the development of a particular form of leukemia called CML. This appears to be due to the suppression of normal levels of apoptosis. We demonstrated that when we treated cells with an antisense oligonucleotide to block the synthesis of Bcr-Abl protein, the resistance to apoptosis was reversed (McGahon *et al.*, 1994). The principle of the antisense approach is to synthesize a short oligonucleotide about 20 bases long. This is usually targeted to the leading sequence of an mRNA of a particular protein. When the antisense enters the cell it binds to that one particular mRNA targeted like a piece of "genetic bubble gum" and prevents translation of the mRNA into protein. Thus, using this approach it should be possible to block the synthesis of any particular protein. In the case of our *bcr-abl* work we could reverse the resistance to apoptosis in CML using this approach. A similar strategy has also been used with *bcl-2* and follicular cell lymphomas. This particular cancer arises due to the abnormally increased expression of *bcl-2*, and when this is reversed using antisense the lymphoma cells undergo apoptosis.

The antisense approach is not the only genetic approach that can be used in an effort to alter the rate of apoptosis. For example, *p53* is a proapoptotic gene that is mutated and as a result nonfunctional in many human cancers. As a result, many tumor cells do not die by apoptosis when they otherwise would have done so. However, if a fully functioning *p53* is expressed in tumor cells lacking they rapidly undergo apoptosis. What appears to happen is that once *p53* is expressed in the tumor cell it detects cancer-causing DNA mutations and immediately triggers cell death via apoptosis. This can be demonstrated in the laboratory. Of course, the question is how does one get a normal *p53* into cells. There are many approaches that can be taken in this respect and perhaps the most promising is to use a viral vector to transport the *p53* gene into the target tumor cells. One such transport vector is the adenovirus, which can be genetically engineered to transport the *p53* gene into the cell and once there express the gene. When this occurs in the tumor cell the result is cell death. There are some significant scientific hurdles that need to be jumped before this approach can be turned into a viable cancer therapy, including targeting the gene to the desired cell. There is little or no information available on what might happen if the *p53* gene reached normal cells instead of the tumor cells. Such cells would then have an elevated expression of *p53* and this might make them supersensitive to apoptosis which could be undesirable. Second, once *p53* reaches its target cell, will it be expressed in sufficient amounts to trigger apoptosis? Once we have the answers to these key questions, we may be able to use this approach to tip the scales between mitosis and apoptosis toward the latter in tumor cells.

One of the interesting features of the adenovirus is that it carries a gene called *E1B* which functions as an anti-apoptotic gene. Of course, this would have to be deleted if this virus was to be used as a gene therapy transporter for *p53*. Luckily, this can be done quite easily. One might ask the question as to why a virus such as the adenovirus has such a gene in the first place and whether other viruses have similar genes. The answer is that many viruses have been documented as having genes that are capable of blocking apoptosis. In addition to the *E1B* gene of the adenovirus, the cowpox virus produces a protein called Crm-A, which is a protease inhibitor that blocks some of the key proteolytic events in the signaling pathway of apoptosis. We can rationally speculate why viruses have such genes as follows. The main life goal of any virus is to replicate and to do this it must infect a cell. It must then hijack the cell's replicative machinery to produce copies of itself. By producing anti-apoptotic proteins such as E1B, Crm-A, or others, the virus can prevent the cell from killing itself and allow it to continue producing copies of itself. The infected cell would like to undergo apoptosis because by doing so it will prevent the virus from replicating itself and spreading to other cells in the body. Unfortunately, many viruses and other microorganisms that need to replicate with cells have acquired the ability to keep cells alive while they replicate.

References Cited

COHEN, J. J., and DUKE, R. C. (1984). Glucocorticoid activation of a calcium-dependent endonuclease in thymocyte nuclei leads to cell death. *J. Immunol.* **132**, 38–42.

COTTER, T. G. (1993). Cell death in the immune system. *The Immunologist* **1**, 181–184.

FADOK, V. A., VOELKER, D. R., CAMPBELL, P. A., COHEN, J. J., BRATTON, D. L., and HENSON, P. M. (1992). Exposure of phosphatidylserine on the surface of apoptotic lymphocytes triggers specific recognition and removal by macrophages. *J. Immunol.* **148**, 2207–2216.

FANIDI, A., HARRINGTON, A. E., and EVAN, G. (1992). Cooperative interaction between c-*myc* and *bcl*-2 proto-oncogenes. *Nature* **359**, 554–556.

HANNUN, Y. A., and OBEID, L. M. (1995). Ceramide: An intracellular signal for apoptosis. *Trends Biochem. Sci.* **20**, 73–77.

KERR, J. F. R. (1971). Shrinkage necrosis: A distinct mode of cellular death. *J. Pathol.* **105**, 13–20.

KERR, J. F. R., WYLLIE, A. H., and CURRIE, A. R. (1972). Apoptosis: A basic biological phenomenon with wide-ranging implications in tissue kinetics. *Br. J. Cancer* **26**, 239–257.

LOCKSHIN, R. A., and ZAKERI, Z. (1991). In *Apoptosis, The Molecular Basis of Cell Death* (L. D. Tomei and F. O. Cope, Eds.), pp. 47–60. Cold Spring Harbor Laboratory Press, Cold Spring Harbor, NY.

MARTIN, D. P., ITO, A., HORIGOME, K., LAMPE, P. A., and JOHNSON, E. M. (1992). Biochemical characterization of programmed cell death in NGF-deprived sympathic neurons. *J. Neurobiol.* **23,** 1205–1220.

MARTIN, S. J., and GREEN, D. R. (1995). Protease activation during apoptosis: Death by a thousand cuts. *Cell* **82,** 349–352.

McDONNELL., T. J., and KORSMEYER, S. J. (1991). Progression from lymphoid hyperplasia to high-grade malignant lymphoma in mice transgenic for the t(14;18). *Nature* **349,** 254–256.

McDONNELL, T. J., TRONCOSO, P., BRISBAY, S., CHUNG, L., LOGOTHETIS, C., HSIEH, J. T., CAMBELL, M., and TU, S. M. (1993). In *Programmed Cell Death, the Cellular Molecular Biology of Apoptosis* (M. Lavin and D. Watters, Eds.), pp. 179–186. Academic Press, San Diego.

McGAHON, A., BISSONNETTE, R., SCHMITT, M., COTTER, K. M., GREEN, D. R., and COTTER, T. G. (1994). BCR-ABL maintains resistance of chronic myelogenous leukemia cells to apoptotic cell death. *Blood* **83,** 1179–1187.

NAGATA, S., and GOLDSTEIN, P. (1995). The Fas death factor. *Science* **267,** 1449–1456.

OLTVAI, Z. N., MILLIMAN, C. L., and KORSMEYER, S. J. (1993).

Bcl-2 heterdimerizes *in* vivo with a conserved homolog, Bax, that accelerates programmed cell death. Cell **74,** 609–619.

RAY, C. A., BLACK, R. A., KRONHEIM, S. R., GREENSTREET, T. A., SLEATH, P. R., SALVESEN, G. S., and PICKUP, D. J. (1992). Viral inhibition of inflammation: Cowpox virus encodes an inhibitor of the interleukin-1 beta converting enzyme. *Cell* **69,** 597–604.

SAMALI, A., and COTTER, T. G. (1996). Heat shock proteins increase resistance to apoptosis. *Exp. Cell Res.* **223,** 163–170.

SPIERS, A. S. D. (1992). In *Leukemia* (J. A. Whittaker, Ed.), pp. 434–467. Blackwell, Oxford.

TEWARI, M., QUAN, L. T., O'ROURKE, K., DESNOYERS, S, ZENG, Z., BEIDLER, D. R., POIRIER, G. G., SALVESEN, G. S., and DIXIT, V. M. (1995). Yama/CPP32beta, a mammalian homolog of CED-3, is a CrmA-inhibitable protease that cleaves the death substrate poly(ADP-ribose)polymerase. *Cell* **81,** 801–809.

WYLLIE, A. H. (1980). Cell death: The significance of apoptosis. *Int. Rev. Cytol.* **68,** 251–306.

YONISH-ROUACH, E., RESNITZKY, D., LOTEM, J., SACHS, L., KIMCHI, A., and OREN, M. (1991). Wild-type p53 induces apoptosis of myeloid leukemic cells that is inhibited by interleukin-6. *Nature* **352,** 345–347.

PAOLO M. COMOGLIO
CARLA BOCCACCIO

Institute for Cancer Research (IRCC)
University of Torino
Str. Prov. 142, Km. 3.95
10060 Candiolo, Torino, Italy

When Regulation Fails: How Cancer Arises

Cancer is a somatic genetic disease affecting a limited number of genes: oncogenes and tumor suppressor genes. The former control cell growth by stimulating cell division, and lesions occurring in these genes cause them to gain uncontrolled functions. The latter inhibit growth and if damaged become inactivated. Seldom are these lesions inherited; more often, they result from mutations induced by chemical carcinogens. Since these mutations affect somatic rather than germinal cells, they are not hereditary but give rise to disease only in individuals exposed to mutagenic activities. Cancer is a slowly evolving disease requiring systematic accumulation of damage to the genetic program controlling cell proliferation. When cells lose sensitivity to the signals regulating their interaction with the environment, tumor progression toward malignancy ensues. Conventional cancer therapy uses toxic drugs to ablate neoplastic growth, and the field of molecular medicine is developing a strategy for gene therapy that aims to replace or repair mutated genes in neoplastic cells.

Cancer Derives from an Accumulation of Genetic Lesions

The billions of different cells in a pluricellular organism live in a complex and interdependent community in which proliferation of every single element is strictly controlled. Normal cells give rise to daughter cells only when the balance between stimulatory and inhibitory signals tends toward cellular division. These signals consist of molecules secreted by other cells freely diffusing in the extracellular environment; thus, cells exposing appropriate "antennas," named receptors, on their surface can capture them. Cells can reciprocally control their activity via an extensive signaling network, and each tissue (i.e., an ensemble of similar cell types) can maintain its size, structure, and functions appropriate to the organisms' need.

Tumor, or neoplastic, cells are "anarchic elements." They violate the rules of obedience to the common signals, evade the control mechanisms for proliferation, and follow their own reproduction program. Moreover, they develop an even more dangerous ability: They break off from their original seat and, carried by the blood circulation system, they migrate toward other tissues, thus spreading to different organs to form secondary tumors or metastasis. The ability to invade tissues and form metastasis differentiates the malignant tumor from the benign form. Cancer cells spontaneously acquire growth autonomy and the ability to colonize and expand in different tissues until the organism succumbs to this sort of "parasite" generated within.

We know that in most cases tumor cells are descendants of one damaged common progenitor cell in which the genes are mutated. Genes are contained in the chromosomes in

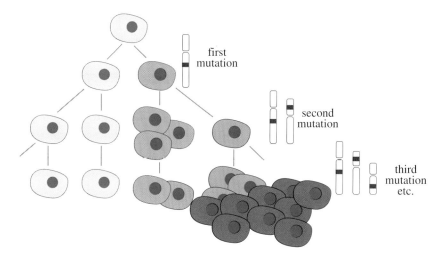

FIGURE 1 Schematic representation of clone selection in neoplastic cells. Malignant transformation occurs through an accumulation of mutations transmitted from the parent to the daughter cells. Each mutation, shown by a colored rectangle within the chromosome (in yellow), confers a selective advantage. The mutated cell expands its progeny at the expense of the others and the clone takes on progressively more malignant characteristics until a cancer is formed (red cells).

the nucleus; they are made up of DNA and encode the necessary instructions for synthesizing proteins, that is, the structural and functional components of the cell. Genetic damage can also occur many years before the tumor is clinically evident and confers to the cell the ability to divide without respecting external control signals. The cell transmits this property to the daughter cells together with the rest of its genetic patrimony.

According to the "clone selection" theory (Fig. 1) cell transformation, or transformation of a population of cells from normal to tumor, occurs gradually through the accumulation of numerous mutations transmitted from the mother cell to daughter cells. Each of these genetic lesions increases the proliferative autonomy of the cell that expands its progeny (i.e., the clone) at the expense of all the others. In other words, a miniature biological evolutionary process occurs in which the mutations produce clones able to rapidly and independently reproduce. It will be shown that the conditions of active proliferation predispose the cell to undergo further genetic lesions.

Studies on clinical progress in human tumors and experimental animal models confirm the theory of the tumors' gradual evolution. In fact, it is possible to recognize different stages, easily distinguishable by their morphology. The first is the appearance of a cell with increased proliferative ability, hyperplasia, and some morphological alteration, or dysplasia. Then a benign tumor is formed that grows rapidly but is unable to pass beyond the tissue confines. Transformation to the malignant form follows with the appearance of a so-called cancer *in situ,* exceeding the confines of the tissue without penetrating into the blood vessels. Then it develops into a cancer that is fully capable of invading

and metastasizing. As the tumor proceeds toward stages of higher aggressiveness, the number of genetic lesions increases. Figure 2 shows the accumulation of mutations characterizing the various phases of colon cancer (Fearon and Vogelstein, 1990).

Research carried out since the 1970s found that the lesions responsible for tumor onset affect two distinct gene groups: the protooncogenes and oncosuppressors. Under physiological conditions, the proteins encoded by these genes provide the necessary control for cell proliferation in response to extracellular signals. The oncogene-encoded proteins mediate proliferative signal transmission from the cell surface to the nucleus. Here they induce the key event in cell replication, i.e., the duplication of the genome. On the contrary, proteins encoded by oncosuppressor genes antagonize cell proliferation.

In neoplastic cells, genetic lesions of protooncogenes are mutations that cause either increased or uncontrolled protein activity (oncogene activation). Consequently, cell proliferation is hyperstimulated and could eventually function autonomously. On the contrary, oncosuppressor lesions responsible for tumor onset are inactivating mutations, provoking the loss of a "breaking mechanism" to cell proliferation (Bishop, 1991).

The accumulation of a sufficient number of mutations for normal cell transformation into an invasive tumor usually takes many years: This explains the proportional increase in malignant tumors with age. In theory, this provides time to recognize neoplastic growth at its initial stages and excise it before it becomes lethally malignant. However, all advanced stages of neoplastic transformation evolve very rapidly. Lesions tend to accumulate exponentially, and the

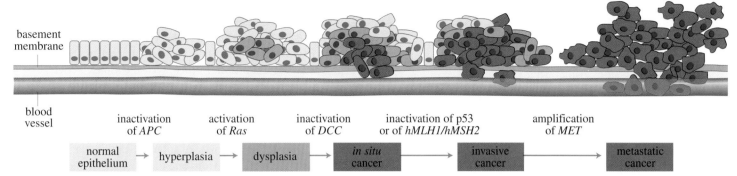

basement
membrane

blood
vessel

| inactivation of *APC* | activation of *Ras* | inactivation of *DCC* | inactivation of p53 or of *hMLH1/hMSH2* | amplification of *MET* |

| normal epithelium | → | hyperplasia | → | dysplasia | → | *in situ* cancer | → | invasive cancer | → | metastatic cancer |

FIGURE 2 Stages in the evolution of an epithelial tumor. Normal epithelial cells increase and undergo progressive behavioral and morphological alterations (indicated by the change in color of the predominant cell population). The passage from dysplasia to cancer is marked by the acquisition of the ability to cross the basement membrane (gray) and pass tissue confines. Cancer cells invade and exit from blood vessels to form secondary distant tumors (metastasis). (Bottom) Gene mutations that determine the passage between the various phases of colon cancer.

rate of spontaneous mutation does not justify this rapid accumulation. The neoplastic clone manifests "genetic instability": This phenomenon is due to lesions in genes that do not directly control cell proliferation. The products of these genes, on the other hand, control and repair DNA lesions that occur during the delicate phase of replication of the genetic material. Loss of defense mechanisms against genetic damage dramatically increases mutation onset, which accelerates the selection of aggressive tumor clones (Kinzler and Vogelstein, 1996).

Oncogenes Stimulate Cell Proliferation

Protooncogenes encode proteins that transmit signals for cell proliferation. These proteins mediate a cascade of biochemical events moving from the cell exterior toward the nucleus. The final event is the control of transcription of many genes (thus, protein synthesis), some of which induce DNA replication. This accompanies duplication of the whole cell structure and terminates with division into two daughter cells.

The cell cycle (i.e., the process of cell duplication) is divided into various stages:

G_1, characterized by gene activation by proliferative signals

S, in which DNA is duplicated

G_2, in which the other cell structures double

M (mitosis), in which the cell divides into two daughter cells via a complex and accurate portioning out of its genetic material contained in the chromosomes

As a rule, after stage M, cells enter a phase of quiescence, called G_0, in which there is no external proliferative signal.

The majority of the cells in humans are normally in this condition and, after the human has reached adult size, only in exceptional conditions can the cells reenter the proliferative cycle. Some cells, such as nerve cells, completely lose their ability to proliferate; this is why lesions in the nervous system are usually irreversible, and cell loss cannot be repaired by multiplication of the ones remaining intact. In most other organs, there are a small number of cells that are difficult to identify and isolate, called staminal cells. These never exit the reproduction cycle because their duty is to replace physiological or pathological loss of organ cells. Tissue activity is like life in a beehive: Worker bees (the normal cells) work hard (building combs, producing honey, etc.) but cannot reproduce. The queen bee (the staminal cells), whose only duty is to procreate, ensures species continuation. Proliferation mechanisms are physiologically activated in staminal cells. If a protooncogene undergoes a mutation that makes its protein active, the cell can easily proliferate ignoring either positive of inhibitory signals. Loss of these limiting factors is a crucial step in the transformation to neoplastic cells (Morrison *et al.*, 1997).

What are protooncogenes and what are the mutations that transform these physiological regulators of cell proliferation into activated oncogenes, that is, in those responsible for tumor onset? The genes that encode proteins involved in signal transduction of proliferative stimuli are potential oncogenes (Fig. 3). These proteins include the so-called growth factors that are mainly small soluble proteins, free to diffuse in the extracellular environment and circulate in the blood. Depending on their functions, growth factors can be divided into competence and progression factors. The first type recruit quiescent cells in the cell cycle or provoke transition from G_0 to G_1. Examples include epidermal growth factor (EGF), platelet-derived growth factor (PDGF), fibroblast growth factor (FGF). The second

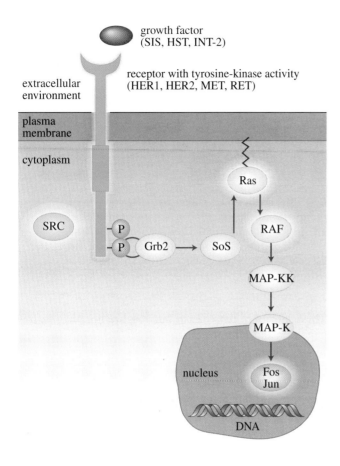

FIGURE 3 Some of the oncogene products involved in control of cell proliferation are classified according to their action mechanism: receptors with tyrosine kinase activity and enzymes with nonreceptor tyrosine kinase activity (green), growth factors (red), transducers (yellow), and transcription factors (purple). Transducers participate in chain reactions in which protein complexes are formed and phosphorylation reactions occur to determine transcription factor activation. The figure shows the molecules involved in activation of transcription factors Fos and Jun, starting from Grb2 association to a tyrosine phosphorylated receptor. Proteins are shown (yellow ring) that are constitutively activated after mutation of the oncogenes that encode them.

type are responsible for transition from G_1 to S, that is, passing through the restriction point of the cell cycle: for example, hepatocyte growth factor (HGF) and insulin-like growth factor (Cross and Dexter, 1991).

Biological activity of growth factors is mediated by protein receptors located in the plasma membrane, which is the target cell "casing." Receptors have an extracellular portion to which the factor binds with high affinity, a part that crosses the plasma membrane, and an intracellular part endowed with enzyme activity. In most cases, tyrosine kinase activity

catalyzes the transfer of phosphate groups from ATP to substrate tyrosine residues. Enzyme activity is tightly regulated and is inhibited when the receptor is not bound to its ligand. Factor binding to the extracellular part of the receptor removes inhibition. In turn, tyrosine kinase activity triggers a cascade of biochemical reactions within the cell that connect extracellular signal reception to its final destination, the genes in the nucleus. The cytoplasmic vehicles of the signals (i.e., the transducers) are also enzymes, endowed with complex regulatory subunits called adaptors. These "pass the word on" from one to the other via modifications such as phosphorylation of tyrosine and serine residues or the production of second messengers, in particular inositol-derived phospholipids. The final targets of this chain of signal transduction are the nuclear transcription factors. These proteins either stimulate or inhibit the functions of the enzyme complex RNA–polymerase II that transcribes in RNA messengers, which are the templates for protein synthesis (Van der Geer and Hunter, 1994).

The whole process in physiological conditions is accurately regulated by negative feedback mechanisms able, at any point, to interrupt the flow of signals from membrane to nucleus. Contrary to the protooncogenes described previously, oncogenes activated by a mutation encode altered proteins able to trigger proliferation but insensitive to the negative feedback mechanisms that inhibits them (Hunter, 1997).

In general, the alteration of the receptor's physical structure or the signal transducer is the result of a deletion, translocation, or point mutation. The first type are gross alterations of the gene. In the deletion, part of the DNA sequence is lost, whereas in translocation the gene is broken and joined to fragments of other genes. The point mutation is apparently a small genetic damage, but its consequences can be devastating for the product functionality. It can cause amino acid substitution that confers completely different properties to the protein. Otherwise, it can insert a signal that blocks protein synthesis, producing an incomplete protein. Oncogene activation may also be the consequence of gene amplification; instead of a double copy, the oncogene has numerous ones. Thus, messenger RNA transcription is increased and the protein accumulates in the cell. In other cases, transcription can be excessive even if the gene copy numbers are normal. This phenomenon may derive from anomaly in the promoter (i.e., the regulatory region of gene transcription) or hyperactivity in transcription factors that control it (Bishop, 1991).

Oncogene classification is based on product function (Fig. 3). Common classification divides them into groups starting with the proteins that act on the cell surface to those localized in the cytoplasm, and therefore the nucleus (Hunter, 1991, 1997). Class I includes oncogenes encoding growth factors, such as SIS encoding the PDGF-2 chain and HST and INT-2 encoding the FGF family molecules. The oncogenic potential of these proteins seems to lie in the formation of autocrine circuits. Under physiological conditions, cells exposing receptors for growth factors are distinct from the cells that produce the factors. However, the pathological condition under which the same cells produce both the factor and the receptor is called autocrine stimulation. This provokes the onset of proliferation and is very difficult to interrupt. Moreover, the combined production of growth factor and receptor and consequent meeting within the cell gives rise to a particularly dangerous stimulus. Some brain tumors called gliomas express a PDGF autocrine circuit.

Class II includes oncogenes that encode mutated forms of membrane receptors for growth factors. Among the first identified were the receptor family members binding EGF and similar factors: HER1 and HER2 (human EGF receptors 1 and 2), which are found in mammary tumors. A recent finding is the MET oncogene; it encodes the HGF receptor and is activated in colorectal, thyroid, ovary, and bone tumors (osteosarcoma). Activation of both the EGF and HGF receptors is a consequence of overexpression and accumulation of many receptors on the cell membrane. This situation of "overcrowding" causes receptor activation even in the absence of the growth factor outside the cell. Moreover, the MET oncogene is activated by point mutation found in hereditary and sporadic papillary kidney cancers. The RET oncogene, responsible for hereditary tumor syndromes (MEN2; multiple endocrine neoplasia type 2), encodes a receptor in which the catalytic activity shows alterations in substrate affinity caused by point mutations.

In addition to the tyrosine kinase receptors, oncogenes also encode tyrosine kinases completely within cytoplasm such as those in the SRC families or ABL, which is also active in the nucleus. These oncogenes belong to class III. ABL alteration occurs when the tyrosine kinase encoding part fuses to the *bcr* gene after a translocation that gives rise to the Philadelphia chromosome. This alteration is a marker of blood cells affected by chronic myeloid leukemia. In this case, lesions in oncogenes cause an increase in tyrosine kinase activity of their products.

Class IV includes the RAS family oncogenes found activated in a high percentage of tumors developing from various organs. The products of these oncogenes are monomeric proteins associated to the inner side of the plasma membrane. Their GTPase activity is similar to that of the so-called G protein. They receive an activating signal from the tyrosine kinase receptors, which in turn transmit the signal to the cytoplasm serine kinase encoded by RAF, the class V family prototype. Ras proteins isolated from tumors usually carry point mutations that slow down GTP hydrolysis, the role of which is to limit protein signaling.

Class VI encodes nuclear proteins, many of which are transcription factors. One representative example is MYC, encoding a protein that, in physiological conditions, is produced only after growth factor stimulation. In many forms

of cancer, particularly leukemia, the oncogene is overexpressed and protein levels are constantly high. Other transcription factors involved in tumor onset are Fos and Jun.

Oncosupressors Inhibit Cell Growth

In addition to oncogene activation, for a cell to transform from a normal to neoplastic there must be an accompanying oncosuppressor inactivation. Contrary to the previous examples, these genes encode proteins that mediate negative signals for cell proliferation. The signals include either soluble molecules (growth inhibitory factors) or molecules associated with the surface of contiguous cells or fixed structural components of the surrounding cell environment (i.e., the extracellular matrix). Contact with the surface of nearby cells, or with proteins produced by them, slows the cells tendency to proliferate. The cell captures these inhibitory messages with high-affinity receptors, analogous to that occurring in growth factor receptors. Also in this case, the signal is transduced to the cell nucleus via a cascade of molecular reactions. The nature and relationship of the proteins participating in these signaling pathways are not well-known compared to knowledge on oncogenes. In fact, it is technically more difficult to study a negative signal that a positive one. However, despite the difficulties, it is known that oncosuppressor alterations appear to have a fundamental role, particularly in the early stages of neoplastic transformation, and confer a hereditary predisposition to cancer. Their lesions, responsible for carcinogenesis, are "inactivating," contrary to those of the oncogenes that are "activating." Within each cell, every gene appears in two copies or alleles, each one inherited from each of the two parents. To achieve a pathological effect both alleles of the oncosuppressor must be inactivated. Double inactivation takes a long time to occur, but the possibility that it does occur notably increases if one of the two alleles is inherited in an already mutated form from one of the two parents. On the contrary, an activated oncogene is rarely transmitted from parent to child since it could irreparably damage intrauterine development (Levine, 1993).

Oncosuppressor classification does not follow precise criteria as does its oncogenic counterpart. However, one can distinguish genes that encode (i) extracellular factors and their receptors, (ii) cytoplasmic proteins, and (iii) nuclear proteins (Fig. 4). Transforming growth factor-β (TGF-β) is a soluble extracellular protein which, despite its name, is able to block cell proliferation. The receptor for this factor is inactive in some forms of colon cancer and has intracellular catalytic serine kinase activity (i.e., it can phophorylate serine residues). The *DPC4* gene is inactive in some pancreatic tumors and its product works downstream of the TGF-β. The protein encoded by the oncosuppressor *DCC* (deleted in colon carcinoma) lies across the plasma mem-

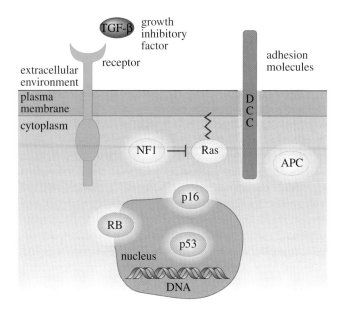

FIGURE 4 Some of the proteins encoded by oncosupressor genes (yellow ring) mediate inhibitory signals of cell proliferation. Proteins include soluble extracellular proteins (e.g., TGF-*β*) or receptors on the cell surface (e.g., DCC) that mediate adhesion by binding to proteins of the extracellular matrix. The other products of gene oncosuppressors shown here are cytoplasmic (yellow) or nuclear (purple) proteins found in mutated forms in human tumors. The interactions between these proteins are not clear; it is known that NF-1 is a negative regulator of *ras* oncogene activity.

brane in an analogous mode to growth factor receptors and binds structural elements of the extracellular matrix. A family of genes called GAS (growth arrest specific), specific for growth arrest, encode proteins, some of which are on the cell membrane, responsible for proliferative inhibitory signals.

APC and *NF1* products are representative of the cytoplasmic proteins encoded by oncosuppressors. *APC* is responsible for familial adenomatous polyposis of the colon, a relatively rare disease characterized by the early appearance of hundreds of benign tumors that successively turn malignant. Moreover, *APC* is mutated during onset of different, nonhereditary tumors of the gastrointestinal tract. The protein encoded by this gene acts as modulator for contact signaling between the cells and works downstream from the receptors known as cadherins, which are situated in the intercellular junctions. Alteration of *NF1* is implicated in Von Recklinghausen's neurofibromatosis, which is characterized by hereditary tumor development in the nervous system. This oncosuppressor's product inhibits *ras* oncogene activity.

Among the oncosuppressors encoding proteins in the nucleus, the better known ones are *p53, Rb, p16,* and *VHL*.

p53 is inactivated in Li–Fraumeni syndrome, a rare hereditary disease characterized by the onset of multiple cancers; however, more often it is mutated in many sporadic tumors. This gene is extremely interesting for understanding carcinogenesis mechanisms and is a promising candidate for molecular cancer therapy. p53 has been nicknamed "the guardian of the genome" because it monitors DNA integrity, becoming active as a transcription factor which promotes expression of genes that halt cell division and, in extreme cases, provoke cell death. Here, the danger of transmitting mutations to the daughter cells is avoided. *Rb* oncosuppressor inactivation is responsible for retinoblastoma, a tumor in the eye occurring in children. The corresponding Rb protein inhibits cell cycle progression. *p16* is inactivated in numerous types of nonhereditary cancers. This gene product inhibits factors that regulate the passage through the various stages of the cell cycle. *VHL*, a recently identified oncosuppressor, is responsible for the rare Von Hippel–Lindau syndrome, characterized by onset of numerous malignant tumors in kidney, adrenal gland, and blood vessels. Its product is a protein that regulates RNA polymerase functions. In the absence of *VHL*, excess mRNA is produced for synthesis of proteins that stimulate cell proliferation (Hunter, 1997; Levine, 1997).

The genes described previously that negatively regulate cell duplication have been collectively defined as "gatekeeper" gene. Recently, some of the genes defined as oncosuppressors, due to their familial transmission and type of lesion (i.e., the inactivation of both alleles), have been considered a separate group — the "caretaker" genes. These do not mediate proliferative stimuli or inhibiting signals; rather, they are involved in control of DNA integrity or repair, an action that the cell takes for protection against environmental assault. *BRCA1* and *BRCA2* belong to this group and are inactivated in familial breast cancer: Although they have a different structure, the proteins that they encode are both involved in controlling correct chromosome replication. *ATM* encodes a protein that controls DNA integrity; it is mutated in Ataxia–teleangectasia syndrome, which is characterized by nervous disorders, blood vessel malformation, and hypersensitivity to radiation-induced carcinogenesis. Among the oncosuppressors, *p53* has both caretaker and gatekeeper functions because it detects DNA mutations and inhibits cell proliferation (Enoch and Norbury, 1995; Kinzler and Vogelstein, 1997).

Other caretaker genes encode enzymes that repair lesions detected by proteins controlling DNA structure. A hereditary lesion in caretaker genes strongly predisposes cancer development. It is also true that when neoplastic transformation initiates because of oncogene and oncosuppressor mutations, a compromised caretaker gene triggers genetic instability causing a rapid evolution of the tumor to high stages of aggression.

The discovery of caretaker genes has shed light on mechanisms that generate and on those that, under physiological conditions, repair damage of genes regulating cell proliferation.

·····································

How Cancer Genes Break Down

Our genetic patrimony is subject to damage that threatens its integrity and risks destruction or distortion of the information for constructing and maintaining cells and thus the whole organism. The loss of a copy of a damaged protein is not, within limits, a dangerous event for the cell; rather, the larger part of macromolecular components are subject to physiological wear and continuous renewal. On the contrary, the genetic material is jealously guarded and synthesized *ex novo* only during cell replication. As is known, it contains a complex sequence of information conserved in double copy or even singly, as in the sex chromosomes. DNA is made up of two parallel and complementary filaments, one the mirrored copy of the other. During replication the two filaments split and each acts as a template for synthesis of its new partner.

To keep this precious heritage intact over time, DNA copying covers only its exact needs, therefore minimizing the risk of errors. Moreover, DNA has complex integrity monitoring and repair systems. The cell will attempt at all cost to avoid the transmission of erroneous copies of a gene to daughter cells.

The genome is in equilibrium between events that cause its alterations and "first-aid" systems for lesion repair. Mutations are, in part, spontaneous, that is, inherent to an imperfect DNA duplication system. This not only has negative consequences but also determines the appearance of new characteristics. If these make up a selective advantage, they will be successfully transmitted throughout the generations and are the basis for species evolution. However, some environmental situations accelerate the rate of spontaneous mutation to bring about inauspicious rather than positive consequences.

The first chemical carcinogens were recognized two centuries ago when English physician Percival Pott (1713–1788) associated scrotum tumors in chimney sweeps to prolonged exposure to the tar contained in soot. Later it was found that chemical colorants provoked bladder cancer in paint industry workers. From these substances, compounds inducing cancer in experimental animals were thus isolated, and subsequently many other natural or artificial molecules were found on the basis of their analogous chemical structures. The most common are the aromatic polycyclic hydrocarbons, aromatic amines, nitrosamines, and chlorohydrocarbons. These molecules are products of industrial synthesis or often residues of combustion of organic molecules in chimney as well as cigarette smoke. Inorganic molecules are also carcinogens, including beryllium, cadmium, cobalt, chrome, and nickel. Many synthesis molecules, in which the carcinogenic potential is known, have been abolished, such as the food colorants banned in the 1970s. However, many carcinogen composites are unknown or not under control by those who should guard environmental health. Others (e.g., those in tobacco) survive because commercial interest outweighs that of individual health. Apart from industrial products, some substances of vegetable origin are also carcinogens, such as cicasine and microfungine aphlatoxin B1. There is, however, a low exposure risk to these substances, at least in Western countries.

Identification of "carcinogens" with "mutagens" was not an immediate occurrence: There were many studies and much controversy before it was recognized that the target of the damaging activity of carcinogens is the DNA and that cancer onset is caused by its mutations. Understanding the action mechanism of a carcinogen is complicated by the fact that these molecules are not mutagens in the form in which they exist in the external environment. They become so only after chemical transformation occurring inside the body. Paradoxically, metabolic reactions are catalyzed by detoxification systems — chains of enzymes in the liver and other tissues exposed to external factors. They are modifications that aim to facilitate elimination of extraneous substances via excretory organs such as liver and kidney. However, before they reach their definitive form, these composites take on an intermediate oxidized form that is very reactive to other electrically charged molecules, including DNA. Carcinogens react with the bases — adenine, guanine, cytosine, and thymidine — which constitute the four-letter "alphabet" with which the genetic information is written. Every time a base is damaged, a "letter" is changed and the original meaning may be distorted: This is a mutation.

Not all carcinogens are chemical: Ionizing radiation emitted during nuclear fission and X-rays are also potent mutagens and carcinogens. Radiation is composed of electromagnetic waves such as those that compose light, but they are shorter in length and therefore have a higher energy content. They can ionize the molecules they hit and upset the electronic charge surrounding the atoms, transforming chemically inert molecules into reactive ones. Sunlight is also ionizing radiation; although at a lower energy level than X-rays, it is able to penetrate the skin surface in which it causes mutations. A very aggressive cancer called melanoma arises from melanocytes, the epidermis pigmentation cells. Radiation hits DNA either directly or indirectly through ionization of cell molecules. Like with chemical carcinogens, the DNA molecule undergoes structural alterations that modify the information contained in the gene.

Genetic alterations occurring in the oncogenes and oncosuppressors are not limited to the damage induced by chemical carcinogens and radiation. In fact, the discovery of oncogenes happened concomitantly with the discovery of oncogenic viruses that, for a brief period, provided hope for discovering etiological cancer agents. Later, it was discovered that viruses were responsible for only a small percentage of human tumors. Among the viruses closely associated to neoplastic growth are the T cell human leukemia virus, the hepatitis B virus, the Epstein–Barr virus, and the human papilloma virus. Viruses are microorganisms that, contrary to bacteria, are not able to replicate; they must first infect the cell by inserting their own genome and use the host DNA replication apparatus. In doing so, they damage the DNA structure as do chemical carcinogens, thus triggering the so-called insertional mutagenesis. In some animal species viruses can produce tumors by driving parts of mutated DNA to interfere with the genetic information of the host cell (Bishop, 1995).

Whatever the mutagenic agent, the lesions found in oncogenes and oncosuppressors can be either qualitative or quantitative. The smallest lesions are called point mutations. In very simple terms, when a single base of the DNA sequence is modified, the faithful copy of the filament is compromised and another "chance" one is inserted in place of the damaged base. Consequently, protein-building information is incorrect, which causes substitution of one amino acid for another during protein synthesis. Thus, dramatically different structural and functional properties may derive from this event. An example of activating point mutation is that affecting the *ras* oncogene: Substitution of an amino acid in the GTPase catalytic site abolishes inhibition of the proliferative signal. Inactivating point mutations are common in the oncosuppressor *p53*.

Sometimes, however, structural damage by mutagens is very evident: Deletions may occur, that is, "breaks" in the DNA filament causing production of truncated proteins. Usually, deletions are accompanied by a loss of function, which is particularly dangerous in the case of oncosuppressor genes. In other examples, the result may be the opposite — that of provoking oncogene activation. A representative case is that of the v-*erbB* oncogene, driven by an avian virus, encoding a truncated form of the EGF receptor, devoid of its extracellular domain and endowed with constitutively tyrosine kinase activity.

When there are multiple breaks in the DNA a sort of "sewing together" of different gene fragments may occur, followed by the production of chimeric proteins, with obvious functional alterations, as in the already mentioned case of *bcr-abl*. Anomalous breaks and sewing together are seen at a chromosome level, under the form of the so-called translocations.

In neoplastic cells, quantitative damage to genetic information is usually due to errors that accumulate during duplication of whole chromosomes. This often occurs after poor functioning of the genes that guard the integrity of the genome. These errors generate chromosome anomalies called "homogeneously staining regions," "abnormally banding regions," or even "double minutes." These anomalies correspond to regions of genetic amplification, that is, regions containing multiple copies of the same gene (Fig. 5). This phenomenon directly contributes to neoplas-

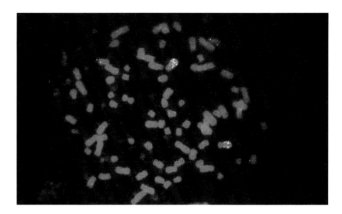

FIGURE 5 Microphotograph of a neoplastic cell in metaphase ($\times 1000$) in which the chromosome (orange) shows multiple copies (yellow) of the *MET* oncogene. Gene amplification causes an excessive production of the encoded protein with consequent loss of regulation of the proliferative signal (courtesy of G. Della Valle).

tic transformation because, as previously explained, an oncogene can be activated when its product is structurally correct but is synthesized in excess (Bishop, 1991).

The cell has evolved complex systems to defend itself from DNA damage. Even before artificial carcinogens contaminated our environment, it was necessary to repair lesions induced by exposure to ultraviolet light and endogenous products of oxidative metabolism, that is, the free radicals that behave like bland mutagens. The following are the main enzymes involved in DNA repair: nucleases, which are able to recognize alterations in DNA nucleotides and cleave them from the sequence in which they are inserted; DNA polymerases, which substitute the excised nucleotides with correct ones; and ligases, which reform bondings for making the DNA chain continuous. There are also hereditary genetic diseases in which the DNA repair enzymes are deficient and in which patients are prone to tumor development from childhood. The most common is xeroderma pigmentosum, characterized by hypersensitivity to ultraviolet light and causing serious damage to skin and eyes even after very short periods of exposure to the sun. Here, the nucleases are the faulty genes (Lehman, 1995). The importance of enzymes for postreplicative correction of the so-called mismatches is clear; that is, the coupling errors between bases juxtaposed on two parallel DNA filaments. Some genes encoding them, such as *hMSH2* and *hMLH1*, are inactivated in hereditary nonpolypose colorectal cancer (Kolodner, 1995).

Transformation, Cell Cycle, and Apoptosis

The target of positive or negative signals through which the oncogenes and oncosuppressors regulate proliferation is called the "cell clock" (Fig. 6). This biological clock scans

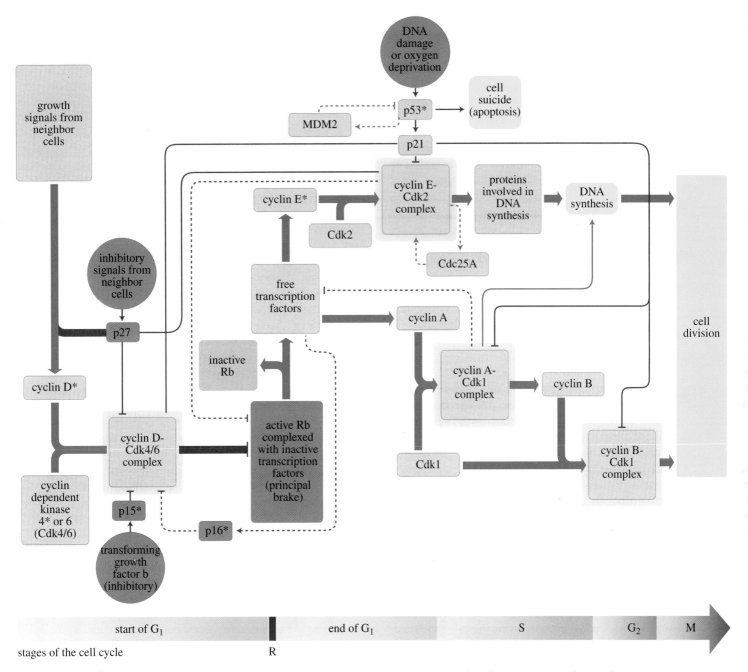

FIGURE 6 Schematic representation of the cell cycle. The stimulatory (gray) and inhibitory (orange) signals of proliferation converge in regulation of the so-called cell clock that scans cell replication times. The activity of enzyme complexes formed by cyclin and Cdk (yellow-framed squares) determines the passage between the G_1, S, G_2, and M phases of the cycle, indicated on the arrow at the bottom of the figure. R is the restriction point, a crucial moment for regulation of the cycle; when the cyclin–Cdk complex phosphorylates the Rb protein the R point can be surpassed, and the cell proceeds to DNA duplication. The dashed lines indicate feedback loops and the asterisks indicate that a mutation or a defect in gene regulation has been found in that protein in some human tumors (modified from Weinberg, 1996).

the replication times of the cell. The passage between the various phases in the cycle (G_1, S, G_2, and M) is caused by the activity of protein complexes formed by a regulatory subunit called cyclin and an enzyme called cyclin-dependent kinase (Cdk). There are different types of cyclin and

Cdk; they associate to form specific complexes to regulate the passage through each phase in the cycle. During G_1, growth factors stimulate cyclin D synthesis that binds to Cdk4 or Cdk6. The activated complex phosphorylates protein Rb, an inhibitor of transcriptional factors, on many serine residues. When Rb is phosphorylated, transcriptional factors are released and they stimulate expression of a series of genes, including cyclin E. Cyclin E, in turn, forms an active complex with Cdk2, responsible for transition from G_1 to S, in which DNA synthesis occurs. To complete duplication of genetic material and division into two daughter cells, a further two cyclin–kinase complexes intervene: cyclin A with cdk2 and cyclin B with Cdc2 (Sherr, 1993).

The protein complexes stimulating the passage through the various stages of the cell cycle are counterbalanced by proteins able to slow down or block the cyclin–kinase activity. These include p15, p16, and p27, which can block cyclin D-dependent Cdk activity, thus impeding passage through G_1. p27 also inhibits the cyclin E–cdk2 complex, specific for passage from G_1 to S. Proliferation inhibitory factors from the extracellular environment, such as TGF-β, induce expression of these three proteins. There is also an inhibitor active on all Cdks, called p21, that is able to block the cell in any stage of its cycle. p21 is induced at the transcriptional level by p53, the oncosuppressor protein guarding DNA integrity, especially during the delicate duplication stages (Sherr and Roberts, 1995).

How do oncogene and oncosuppressor alterations interfere in this complex cell regulatory mechanism? One of the critical moments is certainly the passage from G_1 to S. Activation of oncogenes mediating growth factor signal transduction can potentially overstimulate cyclin D synthesis and therefore activate the complex with Cdk4 or -6 to stimulate progression through G_1. In some tumors there is a spontaneous overproduction of cyclins D and E. However, there is no proof that cyclins alone can induce neoplastic transformation.

Regarding the oncosuppressors, *Rb* inactivation, occurring in a high percentage of tumors, causes constitutive activation of transcriptional factors that stimulate cyclin E synthesis and other factors responsible for DNA duplication. Inactivation of *p15* and *p16,* found in skin and breast tumors, provokes loss of two direct inhibitors of the passage between G_1 and S. p53 is crucial among the cell cycle regulators, inactivated in a very high number of cancers. This oncosuppressor can detect physical DNA damage and consequently block the cell cycle at any stage via p21 induction. It thus prevents the mutated cell from transmitting the lesion to its progeny, a fundamental task. p53 is also an essential regulator of programmed cell death or "apoptosis." Physiologically, this phenomenon occurs during growth and development of healthy tissue when definitive shape, size, and function are reached due to the constant balance between increase in cell number and loss of superfluous or damaging elements. In pathological conditions, destroying a damaged cell is a fine strategy for the organism as a whole in that the potential danger of a mutation is much more serious than the simple loss of a single cell. p53 induces a protein, Bax that, in turn, inhibits the Bcl-2 protein. The latter,

produced in abundance from a B leukemia cell line, in normal conditions slows the genetic programs leading to apoptosis. The conclusion of this complex series of activations and inhibitions is that mutated p53 in tumor cells is no longer able to stimulate apoptosis of damaged cells (Levine, 1997).

Invasion and Metastasis

Although consistent information is available about mechanisms leading to neoplastic transformation, little or nothing is known about genetic alterations that favor malignant progression to the successive stages. Particularly elusive are the genes permitting neoplastic cells to attract the blood vessels that nourish them; that is, to promote neoagiogenesis, to invade surrounding tissue, and to form metastasis.

A series of independent and interconnecting events are critical to the metastatic process. This includes detachment of cells from the primary tumor mass; invasion into local connective tissue, blood, and lymph vessels; arrest in the new location; extravasation and invasion of new territory; and colonization of the new site. Throughout all stages in invasive growth, angiogenesis (i.e., the formation of new vascular branches) and escape from immunitary defense are required. Finally, the migrant neoplastic cells, removed from their physiological context, must survive apoptosis.

Tissue invasion and metastasis dissemination occur through genetic programs in which expression is normally limited to a few developing embryonic cells. Stimulation of these programs in a neoplastic cell could occur through alterations in expression of genes active in limited phases of embryonic life but repressed in somatic cells of the adult organism. Regulation of these genes responds to extracellular cues that include migration (motogens), growth (mitogens), and differentiation (morphogens) signals and is mediated by the intracellular signal transduction apparatus and by specific transcription factors.

During differentiation, either embryonic or metastatic cells move within the intercellular matrix. Movement is the result of a series of coordinated modifications in the cytoskeletal and adhesive cell structures. Soluble molecules have been identified that stimulate cell motility and direct them to precise locations. These factors, called motogens, are secreted from the stroma cells (i.e., the fibroblasts and endothelial cells) or from the migrating cells infiltrating into tissue and acting on the surrounding parenchyma cells by means of a paracrine circuit. A neoplasm may aberrantly induce secretion of motogenic factors by the surrounding stroma or even produce these factors itself in an autocrine circuit. Consequently, stimulation of neoplastic cell movement and unleashing of the invasive properties ensue. An example is the scatter factor family, molecules secreted by stroma fibroblasts that stimulate dissociation of epithelial layers and invasion of extracellular matrixes, directing cell migration toward their concentration gradient. The first two scatter factors identified were HGF and MSP (macrophage stimulating protein). It is interesting to note that their receptors are encoded by two oncogenes, *MET* and *RON,* both of which are able to induce the whole invasive and migratory program in tissues (Comoglio and Boccaccio, 1996).

Deletion or inactivation of the genes for cadherins or other receptors mediating cellular adhesion frequently accompanies the metastatic process. In other situations, these receptors are inactivated as a consequence of posttranscriptional modifications, including alternative splicing or tyrosine phosphorylation of the receptor–catenin complex within the cell. Phosphorylation interferes with anchorage of this complex to the cytoskeleton and destabilizes intercellular adhesion. Cadherins are integral membrane proteins that mediate adhesion between same-type cells and are the main proteins responsible for epithelial tissue cohesion. The extracellular portion of these homophilic receptors binds specifically, in the presence of calcium ions, to an identical molecule on an adjacent cell. In this way, same-type cells recognize each other and anchor reciprocally. Cadherin E appears to be directly involved in the metastatic process. Manipulating expression to lower levels with genetic engineering techniques promotes epithelial cell dissociation in culture. On the contrary, introducing cadherin E in malignant cells suppresses the invasive capacity *in vitro.* Analysis of cadherin E expression levels in human tumors reveals that progression toward more aggressive malignant forms correlates with a general tendency to reduce its expression level.

To invade a tissue, the neoplastic cell adheres to the extracellular matrix and, after having modified it, uses it as support for migration. The adhesion molecules responsible for matrix interaction are the integrins, which are dimeric molecules consisting of two different subunits, α and β. These various α and β subunits combine to form at least 20 different integrins functioning as receptors for the main matrix components: fibronectin, laminin, and collagen. Specific expression of some types of integrin contributes positively or negatively to progression toward malignancy. In neoplastic cells posttranscriptional variants have been identified that behave as dominant-negative receptors, able to inhibit cellular adhesion (Bernstein and Loitta, 1994).

The invasive capacity is a crucial aspect of progression toward the metastatic phenotype. With invasion, the neoplastic cell commits the most serious transgression to the organism's "social order" by encroaching the barriers of the tissue boundaries, an unsurpassable border for epithelial cells. Under physiological conditions, separation of tissue compartments occurs precociously during embryonic development by specialized structures of the extracellular matrix called "basement membranes." These are efficient barriers for cell dissemination. To cross, the neoplastic cell has to partially destroy them. The surface of the cells able to invade and migrate has a rich complement of hydrolytic enzymes able to digest the various matrix components. The better known ones include uPA (urokinase-type plasminogen activator), plasmin, collagenases, metalloproteases, cathepsins, and glycosidases. The genes that encode these enzymes are subject to complex regulation, guaranteeing space–time delimitation of their lytic activity (Liotta *et al.,* 1991).

A primitive neoplasm and its metastasis can grow and simultaneously supply sufficient nourishment to the cells within the tumor mass by inducing its vascularization, that is, giving rise to neoangiogenesis. Genes that encode for neoangiogenic factors have been identified; the best known is VEGF (vascular endothelial growth factor). The nonregulated production of this factor by neoplastic epithelial cells is a relatively frequent occurrence.

Toward Neoplastic Cell Defeat

Cancer can be prevented by knowing the behavior patterns that expose genes to risks of mutation and hence avoiding them. For example, in 1990 in the European community, 140,000 cigarette smokers were diagnosed with lung cancer and only 15,000 nonsmokers were diagnosed (Smans *et al.,* 1993). However, the tendency to accumulate lesions is an inherent characteristic in the actual DNA nature, as is the price that the cell has to pay to keep its ability to adapt to the environment.

Cell aging automatically brings indisputable tendency to undergo genetic damage of varying forms, including those that bring about neoplastic transformation. As the average age of the population increases, tumor incidence is destined to increase proportionally. The deeper understanding of the causes of cancer causes one to think that it is not possible to eradicate cancer as was done for smallpox and polio and as sooner or later will be possible with AIDS.

The possible breakthroughs in the fight against cancer, however, are early diagnosis and cure. Early diagnosis is actually defined as secondary prevention because it greatly increases possibilities of success in cure. It consists of recognizing the tumor at its initial stage, before transformation in invasive cancer. When the neoplasm is well defined and has not metastasized, it is easy to attack with an available "arsenal of arms": surgical resection, chemotherapy, and, to a certain extent, radiotherapy. The situation becomes complicated when cancer is disseminated in the organism because, to provide a definitive cure, it would be necessary to kill all the neoplastic cells. Only a very few cells need to survive treatment because the extraordinary capacity to duplicate, which distinguishes neoplastic cells, would cause the tumor to reform. The largest limit in tumor therapies available today is their toxic effects on healthy tissue, particularly bone marrow and epithelial layers. Collateral effects can be serious and include, paradoxically, the risk of developing new tumors. In fact, chemotherapy agents are nearly always mutagens in that they attack the DNA. The ideal antitumor therapy should fulfill two basil requirements: the ability to wipe out all tumor cells and low toxic effects for healthy cells.

To recognize the molecular mechanisms responsible for tumor onset means finding the targets to which future therapy must turn. Treatment must be aimed at the anomaly responsible and specifically distinguishing the neoplastic cell from the healthy one. Many new-generation antitumor drugs are in advanced phases of experimentation and soon will be available to patients. Encouraging examples are the *ras* oncogene inhibitor which is activated in 20–30% of tumors. In order to make it work, the *ras* oncogene must be bound to a lipid sequence, a farnesyl, permitting it with associate to the intracellular side of the plasma membrane. *Ras* binding to farnesyl is catalyzed by the enzyme farnesyl-tranferase. In experimental animals, inhibitors of this enzyme can completely block pathological activity of the Ras protein and induce tumor regression. Specific inhibitors directed against oncogene-encoded receptors or cytoplasmic transducers are being studied. The classical approach is to develop molecules able to compete with their enzymatic function, such as tyrphostine, lavendustine, and chinazoline. Recently, more sophisticated approaches have been taken, such as the use of antisense inhibitors (i.e., fragments of genetic material able to interfere with RNA that drives oncogenic protein synthesis). Finally, further experimentation has been attempted with synthetic peptides that compete for binding to intracellular transducers to modulate growth factor receptor signal transduction (Gibbs and Oliff, 1994).

As already described, numerous genes are defective rather than hyperactive in tumor cells. This is true for onco-supressors and genes encoding DNA repair enzymes. These are destined to evoke an increasing clinical interest because they are inheritable and therefore responsible for predisposition to tumor onset. An ideal prevention method should systematically recognize these lesions and correct them before the tumor develops. Much attention has been given to the p53 and Rb proteins that inhibit progression in the cell cycle and mismatch repair genes due to the notable frequency with which they are found inactivated. In all these cases, therapy should attempt to revive the missing functions, restoring the cell with its lost or damaged genetic functions. This approach has been defined as "gene therapy" and is the most ambitious aim in molecular medicine. It consists of artificially reconstructing the necessary genetic information for protein synthesis and substituting this to the damaged gene with a sort of molecular surgery.

Currently it is relatively easy to reproduce the whole DNA sequence in a gene, equipped with regions for controlling its expression. However, there are difficulties in directing the DNA to the desired target. A second problem is that of stabilizing the presence and function of the exogenous gene inside the cell, which tends to reject it as a foreign body. Genetically modified viruses are being used to direct "therapeutic" genes to the tumor cells. Viruses are very potent natural genetic vectors because they have been selected to penetrate the cell with particularly high specificity, after they recognize their target with ligand receptor-type mechanisms. Moreover, they are equipped with devices that can be manipulated to allow integration of genetic material carried to the host cells, inserting it in a physiological context. Enormous technological difficulties have dampened the initial enthusiasm. Surprising hope has come from the AIDS virus, which suitably modified has proven to be one of the most potent instruments available (Mulligan, 1993).

References Cited

BERNSTEIN, L. R., and LIOTTA, L. A. (1994). Molecular mediators of interactions with extracellular matrix components in metastasis and angiogenesis. *Curr. Opin. Oncol.* **6,** 106–113.

BISHOP, J. M. (1991). Molecular themes in oncogenesis. *Cell* **64,** 235–248.

BISHOP, J. M. (1995). Cancer: The rise of the genetic paradigm. *Genes Dev.* **9,** 1309–1315.

COMOGLIO, P. M., and BOCCACCIO, C. (1996). The HGF receptor family: Unconventional signal transducers for invasive cell growth. *Genes Cells* **1,** 347–354.

CROSS, M., and DEXTER, T. M. (1991). Growth factors in development, transformation, tumorigenesis. *Cell* **64,** 271–280.

ENOCH, T., and NORBURY, C. (1995). Cellular responses to DNA damage: Cell cycle checkpoints, apoptosis, the roles of p53. *Trends Biochem. Sci.* **20,** 426–430.

FEARON, E. R., and VOGELSTEIN, B. (1990). A genetic model for colorectal tumorigenesis. *Cell* **61,** 759–767.

GIBBS, J. B., and OLIFF, A. (1994). Pharmaceutical research in molecular oncology. *Cell* **79,** 193–198.

HUNTER, T. (1991). Cooperation between oncogenes. *Cell* **64,** 249–270.

HUNTER, T. (1997). oncoprotein networks. *Cell* **88,** 333–346.

KINZLER, K. W., and VOGELSTEIN, B. (1996). Lessons from hereditary colorectal cancer. *Cell* **87,** 159–170.

KINZLER, K. W., and VOGELSTEIN, B. (1997). Cancer-susceptibility genes: Gatekeepers and caretakers. *Nature* **386,** 761.

KOLODNER, R. D. (1995). Mismatch repair: Mechanisms and relationship to cancer susceptibility. *Trends Biochem. Sci.* **20,** 397–401.

LEHMAN, A. (1995). Nucleotide excision repair and the link with transcription. *Trends Biochem. Sci.* **20,** 402–405.

LEVINE, A. J. (1993). The tumour suppressor genes. *Annu. Rev. Biochem.* **62,** 623–651.

LEVINE, A. J. (1997). p53, the cellular gatekeeper for growth and division. *Cell* **88,** 323–331.

LIOTTA, L. A., STEEG, P. S., and STETLER-STEVENSON, W. G. (1991). Cancer metastasis and angiogenesis: An imbalance of positive and negative regulation. *Cell* **64,** 327–336.

MORRISON, S. J., SHAH, N. M., and ANDERSON, D. J. (1997). Regulatory mechanisms in stem cell biology. *Cell* **88,** 287–298.

MULLIGAN, R. C. (1993). The basic science of gene therapy. *Science* **260,** 926–932.

SHERR, C. J. (1993). Mammalian G$_1$ cyclins. *Cell* **73,** 1059–1065.

SHERR, C. J., and ROBERTS, J. M. (1995). Inhibitors of mammalian G$_1$ cyclin-dependent kinases. *Genes Dev.* **9,** 1149–1163.

SMANS, M., BOYLE, P., and MUIR, C. S. (1993). *Cancer Mortality Altlas of the European Economic Community.* IARC, Lyon, France.

VAN DER GEER, P., and HUNTER, T. (1994). Receptor protein-tyrosine kinases and their signal transduction pathways. *Annu. Rev. Cell. Biol.* **10,** 251–337.

WEINBERG, R. (1996). *Scienze* **339,** 33.

General References

DE VITA, V. T., HELLMAN, S., and ROSENBERG, S. A. (Eds.) (2000). *Cancer: Principles and Practice of Oncology.* 6th ed. Lippincott, Williams and Wilkins. Philadelphia.

FRANKS, L. M., and TEICH, N. M. (Eds.) (1997). *Introduction to the cellular and molecular biology of cancer.* Oxford University Press. Oxford.

HANAHAN, D., and WEINBERG, R. A. (2000). The hallmarks of cancer. *Cell* **100,** 57–70.

LEWIN, B. (1999). *Genes VII.* 7th ed. Oxford University Press. Oxford.

SOUHAMI, R., TANNOCK, I., HOHENBERGER, P., and HORIOT, J.-C. (Eds.) (2001). *Oxford Textbook of Oncology.* 2nd ed. Oxford University Press. Oxford.

GUIDO TARONE

Department of Genetics, Biology, and Chemical Medicine
University of Torino
Torino, Italy

Adhesion and Recognition between Cells and the Organization of Tissues and Organs

In vertebrates there are three types of adhesive receptors responsible for the aggregation of cells in tissues and organs: the cadherins, the cell adhesion molecules, and the integrins. These receptors are connected to the cytoskeletal filaments inside the cells providing a continuum between the scaffolding structures inside and outside the cell. The analysis of these molecules led to the identification of many adhesive receptors that during development of the organism are differentially expressed in time and space, suggesting the existence of a recognition code based on the presence of a given combination of adhesive receptors at the cell surface. Upon cell–cell or cell–extracellular matrix interaction, adhesive receptors trigger intracellular reactions that modify both cytoskeleton organization and cellular response to differentiation and proliferation stimuli. Adhesion events between cells thus exert a crucial control on cell differentiation and proliferation necessary for a coordinated development of tissue within an healthy organism.

Introduction

The organization of multicellular organisms requires cell migration and aggregation to form tissue and organs during development. These morphogenetic processes require the ability of cells to recognize themselves and to establish adhesive interactions both with other cells and with the "extracellular matrix," a network of fibrous proteins and polymeric carbohydrates that hold the cells together in tissues but also provide a substrate for cell migration during development and tissue regeneration.

Aggregation of cells in tissues in multicellular organisms leads to a second important event consisting of cell differentiation and the acquisition of specialized functions in, for example, muscular, nervous, or epithelial tissues. Also, cell proliferation in tissues must be rigorously controlled. In

fact, whereas in unicellular organisms cell proliferation depends only on the availability of nutrients, in multicellular organisms control systems are present ensuring that a tissue reaches a given mass and maintains its size. As discussed in this article, both cell differentiation and proliferation are controlled in part by the interaction of a cell with other cells and by its ability to sense the surrounding environment. Research in the 1990s resulted in the definition of the molecular basis of such control mechanisms.

The Discovery of Adhesive Receptors

The ability of cells to recognize and adhere to each other was initially indicated by a series of experiments on cells isolated from different tissues such as liver and retina. When

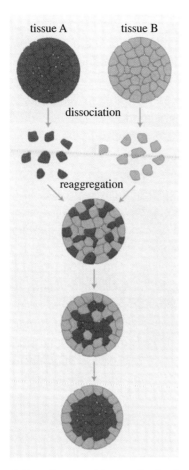

FIGURE 1 Cell–cell recognition and adhesion. Cells isolated from different tissues (A and B) by enzymatic dissociation are mixed together. After incubation in an appropriate medium, cells initially form a mixed aggregate, but soon A and B cells segregate forming distinct and homogeneous masses.

these two cell types are mixed in an appropriate physiological medium they sort out, forming homogeneous aggregates consisting of liver or retina cells. It is clear that liver cells are capable of recognizing and forming stable bonds with cells from the same tissue but not from the retina and vice versa (Fig. 1). Treatment of isolated cells with proteolytic enzymes, such as trypsin or pronase, that remove proteins from the cell surface prevents the formation of aggregates, indicating that the ability of cells to recognize and to adhere to each other requires the presence of membrane proteins on the cell surface, defined as "adhesive receptors." Analysis of these molecules resulted in the discovery that adhesive receptors can mediate cell–cell recognition via three different molecular mechanisms (Takeichi, 1990). The homophilic interaction involves a receptor that binds an identical molecule on the surface of a neighboring cell. The heterophilic interaction involves the binding of two different types of receptors on the surface of the interacting cells. The third type of interaction is mediated by an extracellular

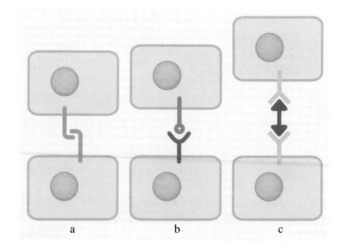

FIGURE 2 Mechanisms of cell–cell adhesion. (a) Homophilic adhesion: A receptor recognizes and binds to an identical molecule on the membrane of a neighboring cell. (b) Heterophylic adhesion: A receptor recognizes and binds to a different molecule on the membrane of a neighboring cell. (c) Adhesion through a soluble bivalent protein capable of making a bridge between two adhesive receptors on the membranes of neighboring cells.

molecule that binds to cell surface receptors bridging two cells together (Fig. 2).

Molecular analysis indicated that adhesive receptors that mediate recognition among cells can be classified into two major groups according to their structure and properties: the cadherins (Ca^{2+} adherins), which function only in the presence of Ca^{2+}, and the cell adhesion molecules (CAMs) belonging to the immunoglobulin superfamily that mediates adhesion in the absence of Ca^{2+}.

Cadherins

Cadherins include a dozen distinct molecules expressed on different cell types. E-cadherin is expressed very early during embryonic development and is responsible for the reaction of compaction (Fleming *et al.*, 1993), a process in which the cells of the outer layer of the morula, a spherical mass filled with cells, establish firm adhesion among each other, sealing the structure with a compact shell of cells. Due to this sealing reaction the fluid pumped inside allows the swelling of the structure and the formation of the blastula, in which the primitive embryo starts to develop. This is one of the earliest processes during development in which adhesive processes between cells allow the organization of primitive tissue structures.

E-cadherin molecules are also present in epithelial cells of adult tissues and are localized in restricted areas of the lateral plasma membrane, "adherent junctions," that represent specific sites of contact between neighboring cells. Different cadherins have been identified in different tissues. Among these, the P- and N-cadherins are present in placenta and nervous tissue, respectively, whereas VE- and M-cadherins are expressed specifically in vascular endothelium and muscle.

Cadherins are glycoproteins that span the plasma membrane and are exposed in both the extracellular and cytoplasmic space. The portion of the molecule that is exposed in the extracellular side contains five homologous repeated sequences capable of binding divalent cations such as Ca^{2+} (Fig. 3). In the most amino-terminal module several cadherins contain a three-amino acid sequence, His-Ala-Val, important in the homophilic recognition process and thus important for the adhesive function of these receptors.

It was demonstrated that cadherins mediate homophilic adhesion between cells by transfection of the corresponding gene in cells that normally do not express this molecule.

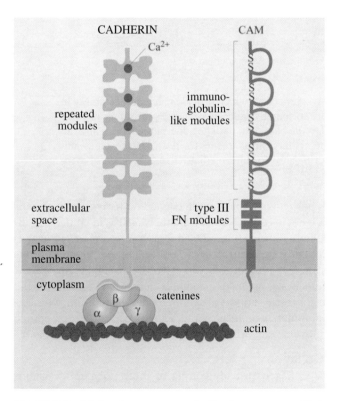

FIGURE 3 Molecular structure of adhesive receptors. Cadherins contain five repeated modules of homologous sequences in the region of the molecule exposed at the outer cell surface, stabilized by binding with Ca^{2+} ions. The more external one is capable of binding to the corresponding module of an identical cadherin on the membrane of a neighboring cell, thus allowing homophilic recognition. On the cytoplasmic side the cadherins bind a group of three proteins, α-, β-, and γ-catenins, that allow interaction with actin filaments. The CAMs are made up of a series of immunoglobulin-like and fibronectin (FN) type III structural modules. The former, present on opposed molecules, are able to bind reciprocally, allowing homophilic adhesion.

The cells that now express E-cadherin aggregate with each other in the presence of Ca^{2+} but do not interact with cells that do not express E-cadherin nor with cells that express P-cadherin. When E-cadherin-positive cells are mixed with N-cadherin- or P-cadherin-positive cells, aggregation occurs only among cells expressing the same type of cadherin. The homophilic binding among cadherins allows the automatic sorting of cells in homogeneous groups. Once cells interact, stable adhesion requires molecular functions associated with the cytoplasmic region of the cadherins. This region of the cadherin molecule interacts with the cytoskeletal network inside the cells, and this interaction is required to transfer to the whole cell the mechanical forces generated during adhesion. As discussed later, the same process applies to integrin-mediated adhesion.

Some cadherin molecules are connected to actin filaments of the cytoskeleton via bridging proteins known as α-, β-, and γ-catenins (Fig. 3) and participate in the organization of the so-called adherent junctions of epithelial cells (Fig. 4). Other types of cadherin molecules bind to cytokeratin intermediate filaments of the cytoskeleton to organize "desmosomes," which are specialized junctions with high mechanical resistance and thus important for providing cell–cell cohesion in epithelial tissues (Fig. 4).

Cell Adhesion Molecules

Another important system of adhesive receptors is represented by CAMs that mediate cell–cell interactions independently from Ca^{2+} ions. N-CAM is the prototype molecule of this group and was one of the first adhesive receptor to be identified (Edelman and Crossin, 1991). N-CAM is expressed on several cell types and mediates homophilic-type interactions. Its molecular structure consists of repeated immunoglobulin and fibronectin type III amino acid sequence modules (Fig. 3). N-CAM is coded for by a single gene that can generate different protein isoforms by an alternative splicing mechanism. One of these isoforms lacks the transmembrane and cytoplasmic portions and it is thus released from the membrane as a soluble form of N-CAM. The physiological role of such a form is unclear. A soluble N-CAM in the extracellular space can bind to N-CAM molecules on the cell membrane preventing cell–cell adhesion and thus acting as an antiadhesive regulator of cell–cell interactions. Another element of regulation in N-CAM function is glycosylation. In fact, N-CAM expressed on cells during early stages of embryonic development is bound to oligosaccharide chains rich in sialic acid, whereas at later stages the sialic acid is absent. Sialic acid is highly charged and its presence favors repulsion among N-CAM molecules, thereby reducing their adhesive functions. These modification of N-CAM structure can thus be utilized by the cell to regulate its adhesive functions according to specific needs. In fact, sometimes during embryonic development or tissue repair cells undergo cycles of migration and

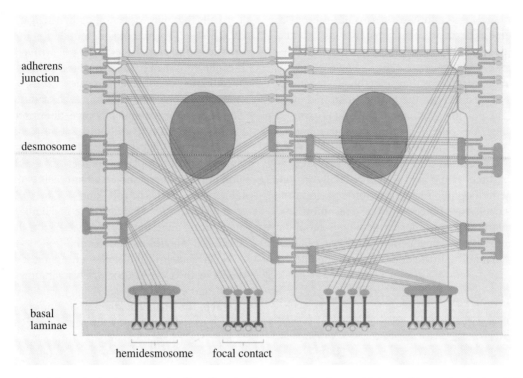

adherens
junction

desmosome

basal
laminae

hemidesmosome   focal contact

FIGURE 4 The principal systems of cell–cell and cell matrix junctions, which provide mechanical stability in epithelial tissue. The adherent junctions and desmosomes are formed by cadherins, whereas the integrins form hemidesmosomes and the focal contacts that anchor the cell to the basal lamina. The adherent junctions and the focal contacts are membrane-anchoring sites for contractile actomyosin filaments (in red). Desmosomes and hemidesmosomes, on the other hand, are membrane-anchoring sites for intermediate filaments made of cyto-keratins (in brown), responsible for the mechanical resistance.

arrest, implying that there is rupture and formation of adhesive contacts.

CAMs comprise a very large and heterogeneous group of adhesive receptors. Some molecules in this group also mediate heterophilic-type interactions, including ICAM-1 and VCAM, which bind to receptors of the integrin family and mediate the interaction of leukocytes to the blood vessel endothelium during the inflammatory response.

Whereas cadherins can organize specialized junctional structures between cells such as the adherent junctions and desmosomes (Fig. 4), CAMs mediate diffused, nonjunctional contact between cells. Due to this important difference the CAMs are more important in establishing transient adhesive interactions such as those occurring during embryogenesis. As tissues and organs form, stable contacts between cells are required and specialized junctions involving cadherins are formed (neural crest cells).

The Extracellular Matrix

As mentioned previously, in addition to cell–cell adhesion, interaction with the extracellular matrix is also required for the organization of cells in tissues. The extracellular matrix is particularly abundant in connective tissues in which cells are not in contact with each other, such as in epithelia. Collagens, fibronectins, and laminins are the major extracellular matrix proteins. All these proteins, due to their fibrous structure and the ability to establish multiple reciprocal interactions, organize a complex network important for the mechanical support of cells in tissues. In addition, extracellular matrix molecules interact with cell surface receptors, allowing anchoring cells to support the protein scaffold.

Collagens

Collagens usually consist of three polypeptide chains containing a high amount of the amino acid proline and coiled to form an elongated rigid rod (Fig. 5a). Twenty-five different polypeptide chains have been identified that can associate to form 15 collagen types. These different collagen molecules have distinct properties and tissue distributions.

In the case of fibrillar collagens, single trimeric molecules polymerize spontaneously to form long fibers extremely resistant to mechanical tension (Fig. 5). Collagen molecules have sites capable of interacting with other matrix molecules such as fibronectins (collagen types I and III), laminins (collagen type IV), and proteoglycans. Moreover, collagens have sites for binding to cell surface receptors and are thus capable of direct interaction with cells.

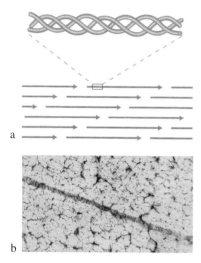

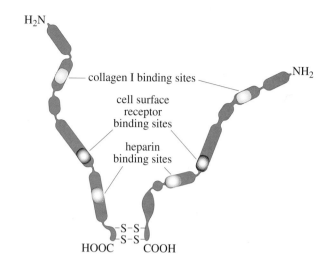

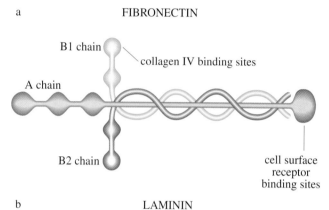

FIGURE 5 Structure of collagen. (a) (Top) The molecular structure of collagen composed of three polypeptides folded in a helix. (Bottom) The out-of-phase organization of the collagen molecules which generates long fibers endowed with high mechanical resistance. (b) Electron microscope image of a collagen fiber (courtesy of Rockefeller University Press).

Fibronectins and Laminins

The collagen fiber network and its interaction with the cell surface are further strengthened by molecules such as fibronectins and laminins (Fig. 6). Fibronectins consist of two fibrous polypeptides covalently linked by disulfide bonds in the carboxy-terminal region (Hynes and Yamada, 1982). Both polypeptides have three binding sites: a site for interaction with collagen, a site for a receptor present at the cell surface, and a binding site for heparin, a glycosylaminoglycan of the extracellular matrix. By a mechanism of alternative splicing, the fibronectin gene can code for different forms of the molecule that differ only in short segments of amino acid sequences. The various fibronectin isoforms are expressed at different stages and in different tissues, suggesting specific roles for each isoforms.

Laminins represent an important family of matrix proteins consisting of three polypeptide chains folded together to form a molecule with the typical Latin cross shape (Fig. 6). As in the case of collagens, there are several laminin polypeptide chains that associate in different combinations, forming at least 10 distinct laminin molecules expressed in different tissues (Timpl, 1996). Laminins contain binding sites for collagens and cell surface receptors. In association with collagen type IV, they form planar sheets known as basal laminae that function as anchoring substrates for epithelial cells (Fig. 4). Basal laminae also surround nerve and muscle cells, providing mechanical support for these tissues. In the kidney glomerulus, basal laminae perform a specific function: in addition to providing mechanical support, they separate the capillary blood

FIGURE 6 (a) Molecular structure of fibronectin, which consists of two very similar polypeptide chains bound together at one end by two disulfide bonds. On both chains, there are sites capable of binding respectively to collagens type I and III, to the cell surface receptor, and to heparin, an acidic polysaccharide of the matrix. (b) Molecular structure of laminin, which consists of three polypeptide chains, A, B1, and B2, folded to form a cross whose lateral arms bind to collagen type IV, whereas the lower end of the long arm binds cell surface receptors. Like collagens, the laminins also exist in different forms, characterized by their polypeptide subunits and by their specific tissue distribution.

vessel endothelium from the podocytes, thus functioning as blood filters. Fibronectins and laminins, forming cross-bridges between collagen fibers, stabilize the extracellular matrix network and, by binding to cell surface receptors, provide additional anchoring sites for the cells to the supporting collagen network.

Proteoglycans

The proteoglycans, proteins covalently linked to long acidic polysaccharides, are another important component of the extracellular matrix. These polysaccharide chains are capable of binding to fibronectins, laminins, and collagens, thus allowing their interaction with the matrix network. There are different types of polysaccharide chains, but all have the common property of being highly hydrophilic and rich in negative charges. Due to these properties proteoglycans bind high amounts of water, cations, and basic molecules such as growth factors. Because of the large water volume retained, they generate turgor necessary to resist mechanical compression.

In addition, there are several other molecules that are part of the extracellular matrix, such as tenascin, nidogen, vitronectin, trombospondin, and fibrinogen, that are not be discussed here. It is important to note that the extracellular matrices of different tissues differ in their physicochemical properties. The matrix of loose connective tissue, such as the derma, is rich in proteoglycans, whereas that of dense connective tissues, such as the tendon, consists predominantly of collagen fibers. In other cases, such as the bone, the matrix becomes a specialized rigid calcified structure due to deposition of insoluble calcium phosphate crystals between the collagen fibrils.

Integrins and Cell–Matrix Interaction

The Discovery of Integrins

Cells bind to the matrix using specific membrane receptors known as "integrins." These are glycoproteins that span the plasma membrane connecting the extracellular matrix with the cytoskeleton (Fig. 7), the intracellular filament network controlling the shape of the cell and its movement (Hynes, 1987). Integrins were discovered in the early 1980s by researchers interested in identifying the fibronectin receptor. At that time, it was known that cells in culture deposit a large amount of fibronectin on the culture plate surface, and by binding to it they attach to the culture dish, forming a cell monolayer. A few laboratories, including ours, identified a group of membrane glycoproteins with a molecular mass of approximately 140 kDa that allowed binding of the cell to the culture dish coated with fibronectin. At the same time the laboratory of Erkki Ruoslahti, investigating the fibronectin structure, identified the minimal amino acid sequence necessary for the adhesive activity of this molecule. This sequence, consisting of three amino acids, Arg-Gly-Asp, was sufficient, once adsorbed to the culture dish surface, to promote cell adhesion and thus was predicted to be recognized by the fibronectin receptor present on the cell surface. This prediction was determined to be correct and the membrane glycoproteins of 140 kDa described previously were found to bind the Arg-Gly-Asp tripeptide sequence. The fibronectin receptor was then characterized at the molecular level and shown to consist of two glycoprotein subunits, α and β, spanning the plasma membrane. At this point it became clear that the structure of the

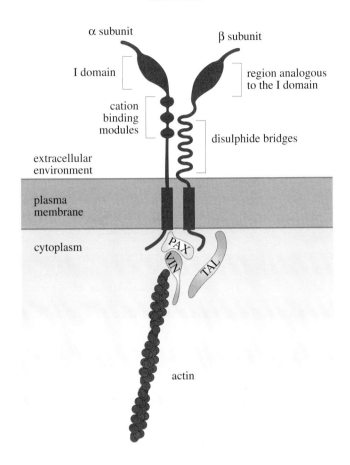

INTEGRIN

FIGURE 7 Molecular structure of integrins, which consist of two subunits, α and β. The α subunits are characterized by the presence of modules that bind divalent cations (such as Ca^{2+}) and by the I domain involved in ligand binding. Some α subunits lack the I domain and consist of two polypeptide chains linked by a disulfide bond. The β subunits are characterized by a region containing a high number of cysteine residues linked in disulfide bonds and by a segment homologous to the I domain of the α subunits. On the cytoplasmic side, integrins bind actin filaments via cytoskeletal proteins such as paxillin (PAX), talin (TAL), and vinculin (VIN).

fibronectin receptor was similar to that of membrane receptors identified in other cellular systems and involved in functions unrelated to cell adhesion to fibronectin. Among these were the fibrinogen receptor of platelets, two classes of leukocyte antigens defined as very late activation and leukocyte function antigen involved in cellular interaction during the immune response, as well as a group of recep-

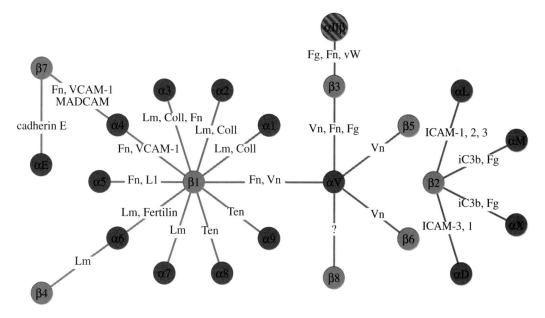

FIGURE 8 The integrins form a very large family of adhesive receptors. The schema summarizes all known combinations of α and β subunits. Each dimer is indicated by an α and by a β connected by a line. On the line, the receptor specificity of the corresponding heterodimer is indicated. Coll, collagen; Fg, fibrinogen; Fn, fibronectin; iC3b, complement factor; Lm, laminin; Vn, vitronectin; Ten, tenascin; MADCAM, mucosal addressing cell adhesion molecule; vW, von Willebrand factor.

tors identified in the fruit fly *Drosophila* and having an important role in the organization of the fly wing. Cloning of the genes of these receptors showed how all these molecules had a common evolutionary origin having highly correlated amino acid sequences. It was clear that these molecules were part of a large family of receptors for which the name of "integrins" was proposed (Hynes, 1987) to indicate their capacity to functionally integrate the extracellular matrix and the cytoskeleton across the plasma membrane.

Seventeen α and eight β subunits can associate with each other in different combinations forming 22 integrin heterodimers capable of binding different ligands (Fig. 8). The complexity of the integrin family is increased by the existence of different isoforms of α and β subunits generated by alternative splicing with specific functional activity and tissue distribution.

Molecular Structure and Function of Integrins

The α subunits are characterized by the presence in the amino-terminal portion exposed to the extracellular environment of three repeated sequences homologous to the EF hand structural domains capable of binding divalent cations. Divalent cations, such as Ca^{2+}, Mg^{2+}, and Mn^{2+}, are in fact required for binding of integrins to their ligands. Some integrin subunits also contain an amino acid sequence in this region of the molecule that is defined as I (inserted) domain, which is directly involved in binding to the ligand. α Subunits indeed define the ligand-binding specificity of

the heterodimers, and often integrin complexes that bind to different matrix proteins share common β subunits but have distinct α chains. β Subunits also participate in ligand binding with a portion of the molecule that is structurally related to the I domain present in α subunits. An important property of integrins is the low binding affinity for their ligands. This is in contrast with the high binding affinity typical of most hormone and neurotransmitter receptors. The low ligand binding affinity, however, is crucial for integrin function particularly during cell migration when a dynamic interaction with the extracellular matrix is required. Adhesion to the matrix become stable only when a high number of integrins concentrate in small areas of the membrane known as "focal contacts." In this way the binding strength is equal to the sum of the single interactions and can reach quite strong levels. When cell–matrix interaction reaches these levels, cells usually stop migrating and become stationary. A second important property of integrins is their ability to regulate their ligand-binding affinity state. Several stimuli can act on the cell via a series of intracellular reactions, inducing changes in ectodomain folding and shifting the integrin complex toward an "active" or an "inactive" conformation with respect to ligand binding. Due to this property a cell can modify its adhesive state by regulating integrin function. This can occur, for example, during cell division when the cell detaches transiently from the matrix and then attaches again when cell division is complete. The ability to regulate integrin activation state is thus an important way to control cell adhesion. In fact, while the interaction of hormones and neurotransmitters with the corresponding receptors is regulated by changes in the concentration of the ligand in the extracellular fluids, matrix proteins are constantly present in the extracellular environment and cell–matrix interactions can be regulated only by modifying the receptor binding capacity. Thus, when all integrin molecules at the cell surface are in the active state the cells are firmly adherent to the matrix with a strength identical to the sum of the strength of each interaction. By inactivating an increasingly number of integrin molecules, the cell can weaken the interaction with the matrix and due to the lower binding affinity the interactions can decrease and the cell is free to migrate on the matrix. These mechanisms are crucial in regulating cell motility not only during organogenesis in the embryo but also in the adult organism when damaged tissues need repair.

The portion of the integrin molecule facing the cytoplasmic side of the plasma membrane binds to cytoskeletal proteins of the actin contractile system. This connection system is most clearly appreciated in stationary cells in culture in which integrins are concentrated in small patches at the ventral surface of the cell in contact with the adhesive substratum and known as "focal contacts." The terminal ends of the actomyosin filament do not directly interact with integrins but rather form a complex bridging structure consisting of cytoskeletal proteins, such as talin, vinculin, paxillin, tensin, and α-actinin. Talin and vinculin likely play a pivotal role in this connecting structure. It has been proposed that talin binds to the β subunit integrin cytoplasmic domain at one end, and vinculin binds at the other end; vinculin then binds to F-actin. The molecular interaction at this level, however, needs to be completely clarified.

Binding to the cytoskeleton is a prerequisite for integrin adhesive activity and allows the mechanical forces generated during adhesion to be transmitted inside the cell during cell movement.

Like cadherins, integrins can also interact with intermediate filaments of the cytoskeleton. The integrin involved in this interaction, $\alpha_6\beta_4$, localizes at the hemidesmosomes, which are specialized adhesive structures in epithelial cells of the epidermis that are in contact with basal lamina (Fig. 4).

Both cadherins and integrins are capable of functioning as adhesive receptors only by interacting simultaneously with the extracellular ligand and with the cytoskeleton. Due to these interactions the mechanical forces acting on the cells within a tissue can be transmitted to an intracellular scaffold capable of providing adequate resistance to both compression and traction. The skin provides a clear example of how tissues must resist mechanical stress. In the absence of interaction with the cytoskeleton, the forces applied to the adhesive receptor could be sufficient to pull it out of the lipid bilayer.

Morphogenetic Role of Adhesive Receptors

The cell adhesion systems described previously function in a coordinated manner to control the ability of cells to interact with each other, organize into tissue, and move within tissue.

A classical example of these processes is the formation of the neural tube during embryonic development (Takeichi, 1995). The neural tube, the embryonal form of the central nervous system, forms by invagination of the ectoderm, the outer cell layer of the embryo that will give rise predominantly to skin and connected structures. Cells in the ectoderm express E-cadherin, but in the region in which the neural tube forms cells stop expressing E-cadherin and start to express N-cadherin. The differential expression of these receptors is likely to be required to segregate two cellular populations that will eventually form two different tissues (Fig. 9). Once formed, the neural tube will give rise in its more dorsal portion to a group of cells called "neural crest" cells. These are highly motile cells that detach from the neural tube and start to migrate in the ventral direction of the tube to form different structures in the cranial and trunk region of the embryo once migration is completed. These cells stop expressing N-cadherin when they detach from the tube and start to migrate. Migration occurs within the fibronectin-rich undifferentiated mesoderm using integrins as adhesive receptors, as indicated by the fact that microinjection of chicken embryos with monoclonal antibodies capable of interfering with fibronectin/receptor interaction leads to inhibition of neural crest cell migration *in vivo*. When cells stop migrating and start to aggregate, forming dorsal root ganglia, E-cadherin expression is turned on. These data indicate that expression of adhesive receptors at the cell surface is tightly regulated during development and suggest the existence of a mechanism of cell sorting based on the use of a recognition code defined by differential expression of specific cell surface receptors.

Not only are adhesive receptors important for tissue formation during embryonic development but also they play a pivotal role in maintaining tissue integrity in the adult individual. Some human pathologies, such as epidermolysis bullosa, clearly illustrate the important role of the adhesive molecules in maintaining the appropriate tissue organization and function. In people affected by epidermolysis bullosa epithelial tissues lining the internal lumen of the digestive tube and of the skin form severe blisters in response to even mild mechanical insults. This defect is due to the detachment of the epithelial cell layer from the underlying connective tissue (Fig. 4) and can be the consequence of genetic defects affecting any of the proteins involved in the cell–matrix adhesive system of the epithelial cells (Epstein, 1996). The defects can occurs in the cytokeratin intermediate filament proteins, in the $\alpha_6\beta_4$ integrin, or in the basement membrane extracellular matrix proteins such as laminin or collagen VII.

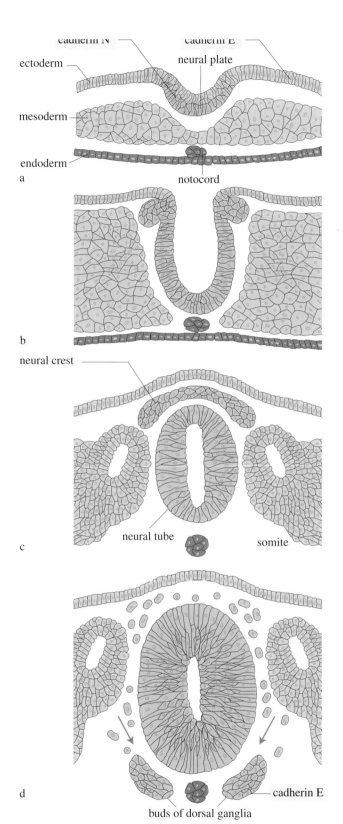

Pathologies involving cadherins have also been identified, proving the role of these molecules in tissues. In particular, the ability of some carcinoma cells to invade surrounding tissues and form metastasis requires the loss of expression of E-cadherin in carcinoma cells (Birchmeier and Behrens, 1994). This important function has been confirmed by the analysis of experimental tumors in mice in which abrogation of E-cadherin expression in tumor cells obtained by different experimental manipulation led to increased malignancy of tumor cells *in vivo.* At the same time, constitutive reexpression of E-cadherin by transfection in tumor cells that have lost it causes strong reduction in their metastatic capacity. Cadherins can thus function as oncosupressor genes. Their expression, in fact, by controlling cell–cell adhesion, prevents cell detachment from the primary tumor and migration in surrounding tissues to form metastasis.

Adhesive Receptors Generate Signals inside the Cell

The mechanical function of the extracellular matrix and the adhesive receptors in mediating cell recognition processes important in organizing and maintaining tissue organization have been discussed previously. Adhesive receptors, however, also play an important role in regulating cell proliferation and differentiation. One of the best examples of this property is represented by the epidermis. The epidermis is the outer layer of the skin and consists of epithelial cells, the keratinocytes, organized in layers piled on top of each other. The lower, or basal layer, consists of proliferating keratinocytes adherent to laminin of the basal lamina. Keratinocytes of the upper layers no longer proliferate and undergo a differentiation program that allows them to

FIGURE 9 Distribution of cadherins during development of the neural tube and the cells of the neural crest. The scheme represents a transverse section at trunk level of a chick embryo in the early phase of development. Initially, the embryo consists of ectoderm, mesoderm, and endoderm. In the central portion of the ectoderm a depression (neural plate) forms (a) that, upon further invagination (b), will form the neural tube (c), the embryonal form of the central nervous system. The ectoderm cells express cadherin E, but in the neural plate region cadherin E is substituted by cadherin N. In the region between the neural tube and the ectoderm, neural crest cells form. These cells emerge from the neural tube and migrate to a region between the neural tube and the lateral structures (somites) moving toward the ventral portion of the embryo where they will aggregate, forming various structures including the dorsal root ganglia of the nervous system (d). When they start to emerge from the neural tube and during migration, neural crest cells no longer express cadherin N; cadherin E, however, is expressed as soon as they aggregate into the dorsal root ganglia.

become mature keratinocytes. Because they are located in the upper layers, these cells are not in contact with the basal lamina but rather are held together by cell–cell adhesive junctions. If a cell of the upper layers is kept in contact with extracellular matrix proteins *in vitro* it stops differentiating (Adams and Watt, 1989), suggesting that the interactions with surrounding cells and the extracellular matrix generate information that regulates their differentiation in addition to controlling their positioning in the tissue.

This is possibly due to the fact that adhesive receptors, like hormone receptors, function as switches capable of activating intracellular reactions. The first demonstrations that adhesive receptors can generate intracellular signals on interaction with their ligands came from the laboratories of R. Hynes, R. Juliano, and J. Brugge, who showed that the interaction of integrins of the $\beta 1$ and $\beta 3$ class with their respective ligands fibronectin and fibrinogen leads to tyrosine phosphorylation of two groups of proteins in the molecular weight range of 120,000–130,000 and 60,000–70,000. Among these, two major proteins have been identified: p125FAK, a cytoplasmic tyrosine kinase that is specifically localized at focal adhesions, and paxillin, a protein functioning as an adaptor molecule and also localized at focal contacts (Clark and Brugge, 1995).

The phosphorylation of p125FAK and paxillin allows the organization of actin filaments in stress fibers during cell adhesion and migration as shown by experiments in which their phosphorylation was either inhibited or increased by using inhibitors of tyrosine kinases or tyrosine phosphatases (Defilippi *et al.*, 1994; Retta *et al.*, 1996).

In addition to activating tyrosine kinases, integrins also activate other pathways of intracellular signaling such as an increase in Ca^{2+} ion concentration, alkalinization of the cytoplasm, and phosphorylation of the mitogen activated protein (MAP) kinases (Clark and Brugge, 1995; Schwartz *et al.*, 1995; Wary *et al.*, 1996) (Fig. 10). All these signaling pathways are also activated by several receptors for growth factors. MAP kinases in particular are known to act in the nucleus of the cell by phosphorylating, and thereby activating, proteins such as Elk-1 that regulate a cascade of gene expression, including fos and cyclins D and A, responsible for induction of cell proliferation (Fig. 11). In particular, cyclins, by regulating the activity of cyclin-dependent kinases, control the onset of DNA synthesis and thus cell proliferation. The laboratory of R. Assoian has shown that the induction of cyclins D and A requires the combined action of growth factors and cell–matrix adhesion (Zhu *et al.*, 1996). In fact, growth factors acting on cells detached from the matrix do not induce cyclin synthesis. On the contrary, this induction occurs when they act on cells adherent to the appropriate extracellular matrix.

These experiments offer a molecular explanation to classical observations of S. Penman and A. Moscona in the early 1970s (Folkman and Moscona, 1978; Farmer *et al.*, 1978),

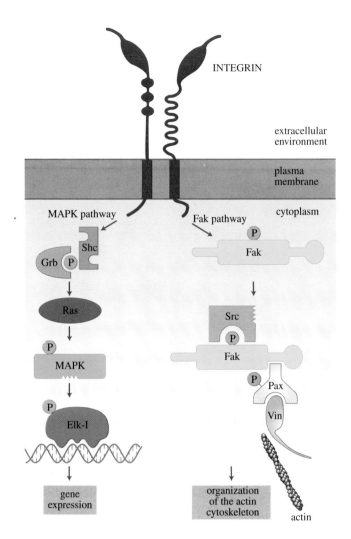

FIGURE 10 Schematic representation of the major intracellular signals triggered by integrins upon interaction with extracellular matrix proteins. (Right) The pathway leads to tyrosine phosphorylation of the FAK kinase. Once phosphorylated, FAK interacts with other kinases such as Src and with adaptor proteins such as paxillin (Pax) that in turn can interact with vinculin (Vin) and actin. In this way integrins can control the organization of the actin cytoskeleton necessary for the processes of cell adhesion and motility. (Left) The pathway that activates the MAPK and the transcription factor Elk-1, regulating gene expression and cell proliferation. This pathway requires the tyrosine phosphorylation of Shc, an adaptor protein, and of a series of intermediary proteins such as Grb and Ras that lead to activation of MAPK.

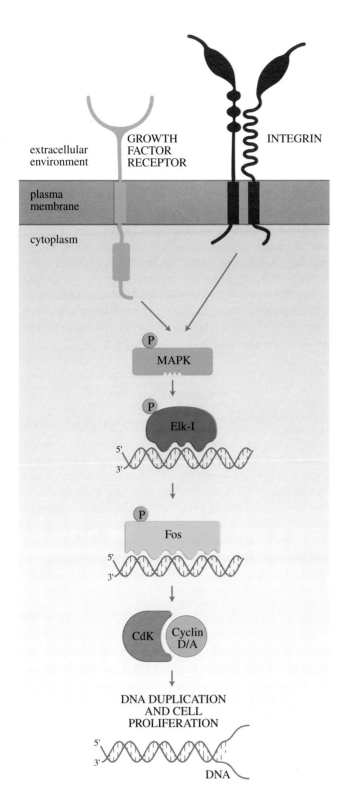

GROWTH
FACTOR
RECEPTOR

INTEGRIN

extracellular
environment

plasma
membrane

cytoplasm

P

MAPK

P

Elk-I

5'
3'

P

Fos

5'
3'

CdK

Cyclin
D/A

DNA DUPLICATION
AND CELL
PROLIFERATION

5'
3'

DNA

who showed that most mesechymal and epithelial cells proliferate *in vitro* only if allowed to adhere to the culture dish surface. If these cells are artificially maintained detached from the matrix, even in the presence of growth factors and appropriate nutrients, DNA synthesis does not occurs. Cell proliferation is thus controlled by the combination of at least two signals: one represented by soluble factors acting at long range, such as growth factors, and a second one acting locally and represented by insoluble factors of the extracellular matrix, such as fibronectins, laminins, and collagens, or cell–cell contacts. The molecular mechanisms of intracellular signaling of the two systems are similar since both growth factor receptors and integrins trigger the activation of MAP kinases. However, only when MAP kinases are simultaneously activated by both stimuli do they reach a threshold level sufficient to induce cyclin synthesis and cell proliferation (Fig. 11). These mechanisms explain how, in the epidermis, only keratinocytes of the basal layer that are in contact with the extracellular matrix are capable of proliferating in response to mitogenic stimuli.

The requirement of cell–matrix adhesion for proliferation allows a cell to respond to growth factors only when it is correctly positioned within a tissue. The importance of such a control mechanism is indicated by the pathological behavior of tumor cells. A property that more clearly defines a normal cell derived from a healthy tissue from a tumor cell is the fact that the latter one is capable of proliferating in the absence of adhesion to an extracellular matrix. This property, known since the early 1970s as "anchorage-independent growth," allows tumor cell growth in tissues different from those of origin and the formation of metastasis. The ability of tumor cells to grow in the absence of anchorage to the extracellular matrix is due to mutations in genes coding for molecules involved in the transmission of the mitogenic signal and leading to constitutive and sustained activation of MAP kinases and cyclins.

Cadherins are also capable of generating intracellular signals in response to cell–cell adhesion. In this case β-catenin plays a pivotal role. This cytoplasmic protein has two important functions. β-Catenin not only binds actin, bridging cadherins to the actin filament system at the cyto-

---

FIGURE 11 Synergy between growth factor receptors and integrins in the control of cell proliferation, which requires the presence of stimuli from both growth factors and extracellular matrix. The dual stimulation is necessary for the induction of the transcription factor Fos and of cyclins D and A. The intracellular signals generated by growth factor receptors and by integrins both lead to the activation of MAPK but are not sufficient individually to induce cell proliferation. It is thought that only the combination of the two signals leads to a threshold level of activation of MAPK sufficient to trigger the activation in cascade of Elk-1, Fos, the cyclins, and hence DNA synthesis.

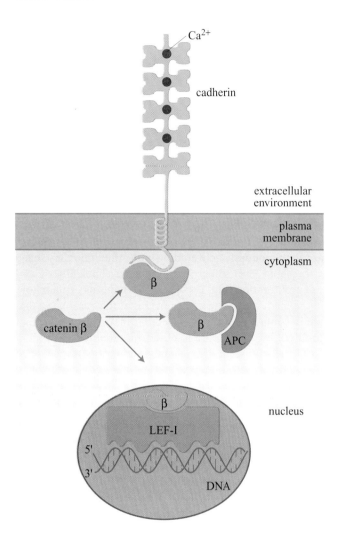

interaction with LEF-1 and thus its action on gene expression (Peifer, 1997). β-Catenin can also bind to a cytoplasmic protein known as adenomatous polyposis coli (APC), which by sequestering β-catenin regulates the level of the protein available for binding to cadherins and to LEF-1. The APC gene was initially discovered as a gene frequently lost or inactivated in colon carcinomas. It can thus be hypothesized that loss of APC protein leads to the accumulation of excess β-catenin available for the interaction with LEF-1; this could cause an imbalance of gene expression and cell–cell adhesion that is at the basis of neoplastic transformation. The level of β-catenin in the cytoplasm is also regulated by Wnt1, a secreted protein that plays a pivotal role in tissue organization during embryonic development and in tumorigenesis. Wnt1 binds to its cell surface receptors and triggers a cascade of reactions, involving molecules with unusual names such as disheveled and shaggy, and causes an increase in β-catenin levels in the cytoplasm. Thus, Wnt1 induces a cellular response similar to that induced by the inactivation of the APC protein.

Conclusion

In summary, adhesive receptors have a dual function in the organization of tissues: By providing the transmembrane connection between the cytoskeletal and extracellular structures they allow cell–cell and cell–matrix recognition and adhesion. At the same time, they play a crucial role in determining the ability of the cells to differentiate and proliferate within the tissue. This process is possible due to the ability of adhesive receptors to trigger biochemical reactions inside the cell that are capable of controlling gene expression. When this dual function is altered, the cell loses the ability to respond correctly to mitogenic stimuli and to coordinately interact with the surrounding environment necessary for tissue formation in healthy individual. A dramatic example of alteration of these mechanisms is represented by tumor cells that, having lost adhesion, become capable of escaping from the tissue of origin and proliferate in different tissues forming metastasis. A deeper knowledge of the functional properties of adhesive receptors can lead to a more rational approach for cancer treatment and to the development of technologies allowing the assembly of tissues *in vitro* to replace damaged or degenerating tissues.

FIGURE 12 Role of β-catenin in intracellular signaling by cadherins. The β-catenin, binding to the cadherin, allows interaction with actin and the adhesive function of this receptor; it can also bind the transcription factor LEF-1 and regulate its capacity to activate gene expression. Cell–cell adhesion mediated by cadherins can thus regulate gene expression by reducing the β-catenin available for interaction with LEF-1. The concentration of β-catenin free in the cytoplasm is further regulated by the APC protein that can bind the β-catenin and induce its degradation. The absence of the APC protein causes high levels of β-catenin and deregulation of gene expression, and this alteration is an essential step in the generation of some tumors of the colon.

References Cited

ADAMS, J. C., and WATT, F. M. (1989). Fibronectin inhibits the terminal differentiation of human Keratinocytes. *Nature* **340,** 307–309.

BIRCHMEIER, W., and BEHRENS, J. (1994). Cadherin expression in carcinomas: Role in the formation of cell junctions and the prevention of invasiveness. *Biochim. Biophys. Acta* **1198,** 11–26.

plasmic face of the plasma membrane, but also binds to the transcription factor LEF-1 and translocates to the nucleus to regulate gene expression (Fig. 12).

Cell–cell adhesion causes preferential binding of β-catenin to cadherins at the cell membrane preventing its

CLARK, E. A., and BRUGGE, J. S. (1995). Integrins and signal transduction pathways: The road taken. *Science* **268,** 233–239.

DEFILIPPI, P., BOZZO, C., VOLPE, G., ROMANO, G., VENTURINO, M., SILENGO, L., and TARONE, G. (1994). Integrin-mediated signal transduction in human endothelial cells: Analysis of tyrosine phosphorylation events. *Cell Adhesion Commun.* **2,** 75–86.

EDELMAN, G. M., and CROSSIN, K. L. (1991). Cell adhesion molecules: Implications for a molecular histology. *Annu. Rev. Biochem.* **60,** 155–190.

EPSTEIN, E. H., JR. (1996). The genetics of human skin diseases. *Curr. Opin. Genet. Dev.* **6,** 295–300.

FARMER, S. R., BEN-ZE'AV, A., BENECKE, B. J., and PENMAN, S. (1978). Altered translatability of messenger RNA from suspended anchorage-dependent fibroblasts: Reversal upon cell attachment to a surface. *Cell* **15,** 627–637.

FLEMING, T. P., JAVED, Q., COLLINS, J., and HAY, M. (1993). Biogenesis of structural intercellular junctions during cleavage in the mouse embryo. *J. Cell Sci. Suppl.* **17,** 119–125.

FOLKMAN, J., and MOSCONA, A. (1978). Role of cell shape in growth control. *Nature* **273,** 345–349.

HYNES, R. O. (1987). Integrins: A family of cell surface receptors. *Cell* **48,** 549–554.

HYNES, R., and YAMADA, K. M. (1982). Fibronectins: Multifunctional modular glycoproteins. *J. Cell Biol.* **95,** 369–377.

LISENMAYER, T. F. (1991). Collagen. In *Cell Biology of Extracellular Matrix* (E. D. Hay, Ed.), 2nd ed., pp. 7–44. Plenum, New York.

PEIFER, M. (1997). Beta-catenin as oncogene: The smoking gun. *Science* **275,** 1752–1753.

RETTA, S. F., BARRY, S. T., CRITCHLEY, D. R., DEFILIPPI, P., SILENGO, L., and TARONE, G. (1996). Focal adhesion and stress fiber formation is regulated by tyrosine phosphatase activity. *Exp. Cell Res.* **229,** 307–317.

SCHWARTZ, M. A., SCHALLER, M. D., and GINSBERG, M. H. (1995). Integrins: Emerging paradigms of signal transduction. *Annu. Rev. Cell Dev. Biol.* **11,** 549–599.

TAKEICHI, M. (1990). Cadherins: A molecular family important in selective cell–cell adhesion. *Annu. Rev. Biochem.* **59,** 237–252.

TAKEICHI, M. (1995). Morphogenetic roles of classic cadherins. *Curr. Opin. Cell Biol.* **7,** 619–627.

TIMPL, R. (1996). Macromolecular organization of basement membranes. *Curr. Opin. Cell Biol.* **8,** 618–624.

WARY, K. K., MAINIERO, F., ISAKOFF, S. J., MARCANTONIO, E., and GIANCOTTI, F. G. (1996). The adaptor protein Shc couples a class of integrins to the control of cell cycle progression. *Cell* **87,** 733–743.

ZHU, X., OHTSUBO, M., BOHMER, R. M., ROBERTS, J. M., and ASSOIAN, R. K. (1996). Adhesion-dependent cell cycle progression linked to the expression of cyclin D1, activation of cyclin E-cdk2, and phosphorylation of the retinoblastoma protein. *J. Cell Biol.* **133,** 391–403.

ROBERTO BRUZZONE

Pasteur Institute
Paris, France

PAOLO MEDA

University Medical Center
Geneva, Switzerland

Barriers between the Organism and the Environment

In vertebrates, epithelial tissues are destined to contact both the environment and the internal compartments of the body and form a barrier between them. All cavities and free surfaces of an organism are lined by an epithelial cell sheet, and all epithelial cells express several types of adhesion systems which provide a barrier to the movement of water and solutes between compartments as well as the structural architecture of the tissue. Cell adhesion may lead to the establishment of cell junctions, which are morphologically recognizable and functionally distinct structures. The elucidation of the molecular organization of the protein components implicated in adhesion and the formation of junctional complexes between cells has revealed that these structures are not only essential to protect the integrity of the organism but also serve as signaling centers necessary to coordinate cellular activity. The molecular basis of several human diseases has been linked to the functional perturbation of adhesion systems and cellular junctions.

Introduction

The development of multicellular organisms has necessitated the solution to complex problems of cellular architecture and organization. Cells have acquired properties that enable them to assemble in a selective fashion during early stages of embryonic development and form tissues which, in turn, associate into organs, functional units of a higher order of complexity.

In vertebrates, epithelial tissues are destined to both contact the environment and form a barrier, which is essential to preserve the integrity of the organism. All cavities and free surfaces of the organism are lined by an epithelial cell sheet, and all epithelial cells express several types of adhesion systems which prevent the movement of water and solutes between compartments and ensure the structural architecture of the tissue. Adhesion between epithelial cells is indispensable to avoid both the loss of fluids and water-soluble molecules from the interior milieu and its invasion by foreign substances and pathogens from the environment. Cells are also in contact with a variety of secreted macromolecules which constitute the extracellular matrix. Cell–matrix junctions are essential to ensure that nutrients diffusing from blood vessels in the underlying connective tissue can reach epithelial cells. In addition, cell-to-cell and cell-to-matrix adhesion systems promote the formation of a signaling network which controls the activity of groups of cells. Adhesion molecules are also able to transmit signals generated intra- or extracellularly to the nucleus, thereby coordinating gene transcription in neighboring cells of the same tissue. The proteins which participate in these signaling pathways are either distributed throughout the plasma membrane or become clustered in discrete regions to form well-defined intercellular junctions.

The normal differentiation of epithelial cells requires the sequential expression of several molecules that will ensure the development of the previously mentioned functions (Fig. 1). The ability to organize adhesion and signaling systems likely appeared relatively early during evolution and may have been of great importance in allowing the survival of the first metazoans. Animal species that have been able to organize specific areas of contact between cells into complex junctional structures have evolved, whereas other species, such as Porifera, which do not exhibit intercellular junctions have remained at the lower levels of the phylo-

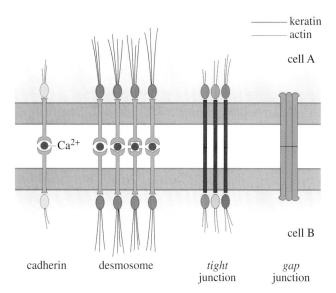

FIGURE I Schematic view of the organization of adhesion molecules and junctions between epithelial cells. The amino-terminal regions of homologous cadherins (in orange) interact in the extracellular space to ensure, in the presence of Ca^{2+} (red), the adhesion of adjacent cells. The carboxyl-terminal tails of each cadherin extend into the cytoplasm to link, via the interaction with various attachment proteins (in green), the actin microfilaments of two cells in contact. Cadherins may concentrate in the lateral portion of the cell membrane to form adherens junctions or desmosomes that link the intermediate filaments of keratin of the cells in contact. In tight junctions, integral membrane proteins (in dark red) link two adjacent cells and establish a barrier separating the apical from the basal and lateral portions of the plasma membrane. These proteins are connected through multiple intracellular attachment proteins (in brown) to actin microfilaments of the two interacting cells. In gap junctions, the integral proteins (in pink) that permit the exchange of ions and molecules between the cytoplasm of adjacent cells, belong to the connexin family.

genetic tree. It should also be noted that, in vertebrates, epithelial cell junctions are structurally complex and, as a result of the molecular diversity of their protein components, provide adhesion as well as barrier and signaling functions.

Cell–Cell Adhesion Molecules Belong to Two Classes

Cell adhesion is mediated by several molecular components and is more dependent on the number of adhesion molecules rather than their affinity, which is usually weak. Epithelial cells utilize more than one adhesion system and the predominant type varies in different tissues. In vertebrates, there are two distinct classes of cell adhesion molecules (CAMs), one which is functionally dependent on Ca^{2+} and another which is not. Cadherins operate in Ca^{2+}-dependent cell adhesion (Takeichi, 1991; Gumbiner, 1996), whereas the CAMs which belong to the immunoglobulin superfamily are implicated in Ca^{2+}-independent adhesion (Edelman and Crossin, 1991). Cadherins form a family of proteins that include at least 12 members encoded by distinct genes. Some cadherins are named according to the tissue in which they are more abundantly expressed. For example, E-cadherin is mostly present in epithelial cells, N-cadherin in nerve cells, and P-cadherin in the placenta. All cadherins are glycoproteins and share a common structure with one transmembrane domain, a short cytoplasmic carboxyl-terminal tail, and a large extracellular portion (Fig. 2). The extracellular part (EC) consists of five homologous regions (EC1–EC5) which contain Ca^{2+} binding sites (Shapiro *et al.*, 1995). The presence of extracellular Ca^{2+} is essential for the adhesive properties of cadherins. If Ca^{2+} is removed, the extracellular portion is cleaved by proteases and the function of cadherins is lost. Cadherin-based adhesion is defined as homophilic when it is mediated by identical proteins and as heterophilic when the cadherins involved differ. Alternatively, the binding between cells may be dependent on the presence of extracellular adapter proteins. Homophilic binding, the most prevalent form, has been demonstrated using mutant cell lines that do not possess intrinsic adhesion mechanisms. By transfecting these cell lines with the appropriate missing components, it has been possible to reconstitute the various steps of cadherin binding. For example, disaggregated cells that express different cadherins tend to associate on the basis of homophilic recognition systems (Takeichi, 1991; Gumbiner, 1996). X-ray crystallographic studies indicate that E-cadherin forms homodimers with the EC1 region oriented outwards, thus allowing the interaction with the EC1 domains of homologous dimers present on the adjacent cell (Shapiro *et al.*, 1995). This model proposes that the adhesive elements of neighboring

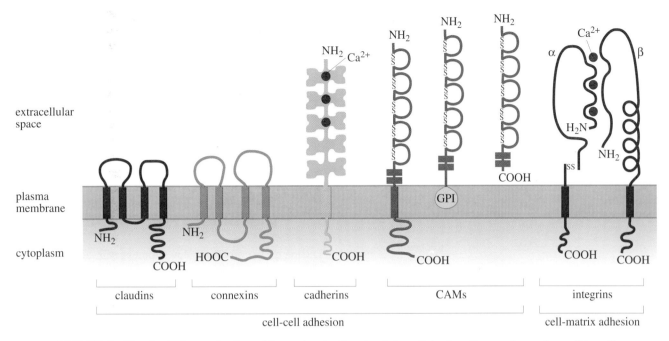

FIGURE 2 Topological organization of the molecules forming intercellular junctions and ensuring cell-to-cell and cell-to-matrix adhesion. Claudins and connexins feature four transmembrane regions, whereas cadherins, cell adhesion molecules (CAMs), and integrins are single-pass transmembrane proteins. In addition, two isoforms of neural-CAM (N-CAM) do not cross the cell membrane: One is attached to the membrane by a glycosylphosphatidylinositol anchor (GPI) and the other is secreted and incorporated into the extracellular matrix.

cells interact to form a sort of intercellular "zipper" (Fig. 3). It has not been directly demonstrated whether such a model applies to other recently identified cadherins (Huber *et al.,* 1996).

Most cell types express more than one cadherin in a pattern that is characteristic of each tissue. Cadherins are the main components of intercellular adhesion in embryonic tissues. In vertebrates, E-cadherin is the first to appear during early development and is involved in the compaction of the embryo at the eight-cell stage. This process permits cells in contact both to form organized intercellular junctions which are crucial to maintain a barrier between the organism and the environment and to participate in complex morphogenetic events. Thus, antibodies blocking the adhesive function of E-cadherin prevent embryo compaction. Cadherins also appear to be involved in the specification of the fate of cells derived from a given embryonic layer. For example, the formation of the neural tube that buds off the overlying neural ectoderm is accompanied by a switch in cadherin expression from E-cadherin in the neural ectoderm to N-cadherin in cells of the neural tube. The two cadherins seem unable to maintain adhesion between the cells of these two distinct tissues. Inactivation of E-cadherin by homologous recombination results in intrauterine death. E-cadherin mutant embryos exhibit cellular disaggregation

during early development and fail to form the trophoectoderm epithelium of the placenta (Takeichi, 1991; Gumbiner, 1996).

Continued expression and functional activity of E-cadherin are also crucial for epithelial cell adhesion in the adult organism. This adhesion, however, can be modulated to reduce the strength of cell cohesiveness in defined conditions, such as the cell migration which occurs during wound healing.

The molecules involved in Ca^{2+}-independent adhesion belong to the immunoglobulin superfamily of proteins. This family is characterized by the presence of sequence motifs that were first identified in antibodies and hence are termed Immunoglobulin (Ig)-like repeats. The most abundant and best studied member of this class of adhesion molecules is the neural cell adhesion molecule (N-CAM). N-CAM consists of at least 20 isoforms which result from the alternative RNA splicing of a single gene. Some of these isoforms are soluble, some are attached to the plasma membrane via a glycosyl-phosphatidylinositol anchor, but the majority are glycosylated integral membrane proteins (Fig. 2). N-CAM, as well as other adhesion molecules of the immunoglobulin superfamily (e.g., L1), participate in several steps of embryonic development such as the migration of nerve cells and growth of their processes. Ca^{2+}-independent adhesion

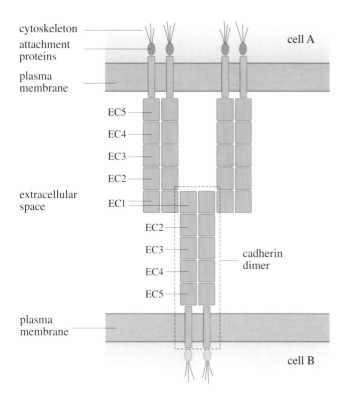

cytoskeleton

attachment proteins

plasma membrane

EC5

EC4

EC3

EC2

extracellular space

EC1

EC2

EC3

cadherin dimer

EC4

EC5

plasma membrane

cell A

cell B

FIGURE 3 "Zipper" model of cadherin-mediated cell adhesion. The functional unit of the model is a dimer of two cadherin molecules that project their EC1 regions outwards. In intercellular junctions, each dimer interacts with two analogous structures contributed by the adjacent cell. In this model, which assumes that only the EC1 regions bind to each other, the interdigitation of the dimers, similar to that of the elements of a zipper, provides the basis for cadherin-dependent cell adhesion. Although this model has been validated only for N-cadherin, it may also be prototypical of the cellular interactions that depend on other cadherins (adapted from Shapiro *et al.*, 1995).

molecules form mainly homophilic interactions but do not appear to be involved in the establishment of junctional complexes (Edelman and Crossin, 1991).

Formation of tissues is dependent on the presence of precursor cells whose progeny remain restricted to a defined region before migrating to their final destination. Such processes are accompanied by the selective expression of distinct adhesion molecules which contribute to the progressive organization of the tissue according to its final architectural plan. If cells from different embryonic organs are first dissociated and then mixed together, they will adhere preferentially according to the tissue of origin and will sort out to restore an arrangement similar to that observed in the native tissue (Edelman and Crossin, 1991). Cadherin-based intercellular adhesion is mechanically stronger than that mediated by adhesion molecules of the Ig superfamily and is mainly responsible for maintaining tissue integrity

and determining segregation between groups of cells. In contrast, Ca^{2+}-independent adhesion may participate in the fine modulation of specific cell–cell interactions. From a hierarchical standpoint, cadherins are more dominant. Thus, ectopic expression of cadherins causes an early and gross disruption in embryonic development, whereas the same experimental approach with N-CAM leads to less dramatic consequences (Takeichi, 1991; Gumbiner, 1996).

Epithelial Cells Are Joined by Several Types of Junctions with Distinct Functional Tasks

During early embryonic development cells adhere to each other without major reorganization of the cytoskeleton, which is a hallmark of most junctional complexes. Such cell-to-cell contacts ensure cell adhesion and can be dynamically modulated to comply with the needs of a developing organism which undergoes continuous remodeling. At later stages, epithelial cells establish more complex junctional structures that lead to the separation of the apical from the basolateral region of the plasma membrane. This segregation allows these two domains to acquire distinct functional and structural properties. Cell adhesion and intercellular junctions are therefore two facets of the same basic function. Although each cell utilizes a large variety of molecules to accomplish the function of adhesion to neighboring cells, intercellular junctions between epithelial cells are organized according to a common structural plan which includes three components: a membrane protein, an intracellular attachment protein, and a component of the cytoskeleton (Table 1 and Fig. 1). The integral membrane protein crosses the lipid bilayer of the plasma membrane, binds to molecules present in the extracellular space, and links them to cytoplasmic elements. In the extracellular space, adhesion molecules can interact with adjacent cells via proteins of the same type or of a different family. The cytoplasmic portion of an adhesion molecule can interact with a wide variety of adapter proteins that provide a structural and functional link between cell surface molecules and the cytoskeleton (Kirkpatrick and Peifer, 1995; Yamada and Geiger, 1997). Most cell types form several types of intercellular junctions which were originally classified on the basis of their morphological appearance and presumed function (Table 2). The continuous presence of these structures is crucial for cell homeostasis and for maintaining the integrity of an organism. In epithelial cells of vertebrates one can distinguish three functionally distinct intercellular junctions: occluding (tight) junctions, anchoring junctions (adherens junctions and desmosomes), and gap junctions (Table 1 and Fig. 1).

Tight junctions ensure the impermeability of a cell layer

	TABLE I		
	Proteins of Intercellular Junctions in Epithelia[a]		
JUNCTIONS	**INTEGRAL MEMBRANE PROTEINS**	**INTRACELLULAR ATTACHMENT PROTEINS**	**CYTOSKELETAL STRUCTURE (PROTEIN)**
Tight (zonula occludens)	Occludin, claudins	ZO-1, ZO-2, cingulin	Microfilaments (actin)
Adherens (zonula adherens)	Cadherins (E, N, P)	Catenins (α, β, γ), vinculin	Microfilaments (actin)
Desmosome	Cadherins (Dsc, Dsg)	Desmoplakin, γ-catenin	Intermediate filaments (keratins)
Gap	Connexins	20-1	α-Spectrin?

[a] Each intercellular junction is made of three types of proteins: integral membrane proteins that interact in the extracellular space, intracellular attachment proteins that associate the junction proteins to the cytoskeleton, and the cytoskeletal proteins. A few examples of the proteins involved are given in parentheses. Question mark indicates that the information is still incomplete. Dsc, desmocollin; Dsg, desmoglein.

by preventing the free diffusion of fluids, solutes, and cells from one side to another through the paracellular space. Moreover, tight junctions function as barriers to the movement of membrane proteins, thus maintaining a separation between structurally and functionally distinct regions of the plasma membrane of epithelial cells which become polarized. Adherens junctions and desmosomes confer adhesive properties to adjacent cells. This feature permits, on the one hand, the recognition between groups of cells during morphogenesis and provides, on the other hand, adult tissues with anchoring units that give strength and resistance to mechanical insults. Gap junctions are clusters of channels that allow the exchange of signals between cells without the need of diffusion in the extracellular space. Their function is to ensure metabolic cooperation and co-ordination of the activity of groups of cells, thereby improving homeostatic control.

Tight Junctions Form a Barrier

In most epithelia, cells are linked to prevent the leakage of fluids and soluble molecules through the paracellular route, i.e., the space between cells. This sealing belt is established by tight junctions (zonulae occludentes), which separate fluids of different chemical composition that bathe opposite sides of the cell (Table 2). The composition of these fluids depends, in turn, on the polarity of cells which express a unique set of proteins and lipids in distinct regions of the plasma membrane (Drubin and Nelson, 1996). In addition,

	TABLE 2	
	Functions of Intercellular Junctions and Adhesion Systems	
FUNCTION	**STRUCTURES/MOLECULES**	
Maintenance of membrane polarity	Tight junctions	
Sealing of intercellular space	Tight junctions	
Cell movements	Adherins junctions, integrins	
Direct intercellular communications	Gap junctions, cadherins, integrins	
Cell recognition	Cadherins	
Cell adhesion	Cadherins, tight junctions, desmosomes	
Adhesion to connective tissues	Hemidesmosomes, integrins, anchoring filaments, anchoring fibrils	

[a] Intercellular junctions are classified into three functionally distinct groups. The first group comprises the tight junctions that seal the extracellular space, thus preventing the free paracellular diffusion of molecules, and separate the apical from the basolateral region of the plasma membrane. The second group comprises adherens junctions and desmosomes that mechanically link adjacent cells via an interaction with the cytoskeleton as well as the hemidesmosomes that link the cells to the extracellular matrix. The third group comprises the gap junctions that allow the passage of electrical and biochemical signals between adjacent cells. During specific developmental stages or under particular conditions of cell growth and migration, the same junctional proteins may be used for cell recognition, adhesion, and communication, despite the absence of ultrastructurally detectable junctional structures.

the barrier formed by tight junctions maintains this asymmetrical distribution of proteins and lipids, preventing their movement to inappropriate membrane domains. In view of their key role in the regulation of the diffusion of molecules via the paracellular route, tight junctions are responsible for the development of transepithelial resistance. Thus, electrophysiological measurements of resistance between apical and basolateral spaces of an epithelial sheet represent a functional demonstration of the presence of tight junctions.

In electron micrographs, tight junctions appear as areas of focal contact where transmembrane junctional proteins adhere to occlude the extracellular space. The structural organization of these proteins is better appreciated after freeze-fracture electron microscopy (Fig. 4). Tight junctions are then seen as rows of intramembrane particles that form an anastomosing, belt-shaped network which separates the apical and basolateral domains of the plasma membrane. Tight junctions are found in most epithelia, regardless of their embryonic origin. The molecular orga-

nization of this type of junction comprises an extracellular region, a transmembrane, and a cytoplasmic portion (Kim, 1995; Kirkpatrick and Peifer, 1995). The extracellular and transmembrane regions are formed by integral membrane proteins which interact, in the cytoplasm, with other proteins, some of which have been identified and cloned (Table 1 and Fig. 2).

The first transmembrane protein identified in tight junctions is occludin, whose deduced amino acid sequence predicts an arrangement with four transmembrane domains, similar to the topology of other proteins localized in regions of close contact between the outer leaflets of two interacting plasma membranes. Another putative tight junction component is zonula occludens-1 (ZO-1), a 220-kDa protein which does not contain long hydrophobic stretches and therefore cannot be inserted into the plasma membrane. ZO-1 is expressed in two isoforms which are distinguished by the presence or absence of an 80-amino acid motif called the α region. The functional implications of this alterna-

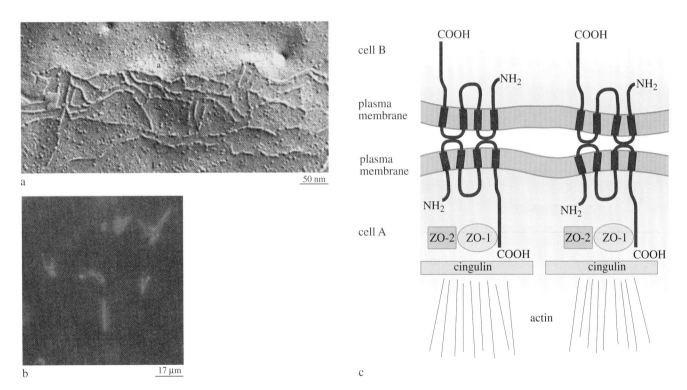

FIGURE 4 (a) Electron micrograph of a freeze-fractured replica of the plasma membrane, showing tight junctions recognizable for the fibrillar arrangement of the constituent proteins, organized in a belt-like distribution separating the basolateral (l) region of the cell membrane from the apical (a) portion, in which the density of protein particles is lower. (b) Fluorescence photograph, after incubation with antibodies against ZO-1, showing that the distribution of this protein, associated with tight junctions, is restricted to the apical membrane of epithelial cells. (c) Schematic drawing of the structure of a tight junction. Points of contact between two nearby cells are made by claudins (dark red), which are integral membrane proteins. Several cytoplasmic proteins, including ZO-1, ZO-2, and cingulin, concentrate at tight junctions where they interact with each other. The cingulin could form a scaffold underneath the membrane, thereby holding together other junctional proteins and allowing their connection with the microfilaments of the cytoskeleton, made of molecules of actin (red).

tive splicing are not obvious since there is no correlation between the expression of a given ZO-1 isoform and the structural organization or sealing properties of the resulting tight junctions. In addition to its localization at tight junctions, ZO-1 is also present in cells that are not believed to form tight junctions. In this case, the function of the protein is not clear. ZO-2 is a 150-kDa protein that tightly associates with and shows sequence similarity to ZO-1. ZO-2 is a component of the cytoplasmic domain of tight junctions and, contrary to ZO-1, is only found at these junctions. Cingulin is also an exclusive component of tight junctions. The derived amino acid sequence suggests a high content in α-helical structures, the absence of hydrophobic regions, and the possibility of supramolecular interactions. These characteristics could indicate that cingulin may provide a submembrane scaffold which links other proteins involved in tight junction formation and attaches them to the cytoskeleton.

The development of tight junctions reflects the constant necessity of epithelia to establish new boundaries in order to maintain the integrity of the organism while simultaneously permitting the tridimensional organization of cells. Recently, it has been shown that, in the frog embryo, tight junctions appear at the two-cell stage, when cingulin concentrates at the limit between apical and basolateral areas of the plasma membrane. In mammals, tight junctions are formed at embryonic compaction, a process that depends, in turn, on E-cadherin-mediated adhesion.

Tight junctions are functionally heterogeneous and their selective permeability can be dynamically modulated in different tissues (Kirkpatrick and Peifer, 1995). For example, intestinal tight junctions are more permeable to sodium ions than are renal junctions. This property is crucial for the kidney to transport salts vectorially, from the glomerular filtrate against the steep concentration gradients present across the junctional barrier, and prevent backward leakage of the reabsorbed salt via the paracellular pathway. Another role of tight junctions in the formation of compartments within an organism is illustrated by the blood–brain barrier of mammals. The exquisite low permeability of this structure has been attributed in part to the presence of high electrical resistance tight junctions which severely limit paracellular flux. The sealing properties of the blood–brain barrier can be transiently altered following pharmacological treatments, a finding that suggests the presence of a proteinaceous channel. By connecting the plasma membranes of adjacent cells, this channel could serve as a selective and adjustable filter which copes, in a timely fashion, with the needs of the paracellular transport pathway.

Several studies have exploited the sensitivity of tight junctions to extracellular Ca^{2+} concentrations to demonstrate that their assembly depends on the proper expression and functional competence of cadherins. Thus, removal of extracellular Ca^{2+}, which induces the dissociation of cad-

herins and the adherens junctions they form, is accompanied by a disruption of the cytoskeleton which in turn leads to a redistribution of tight junction proteins. The latter process requires protein phosphorylation since it can be blocked by several inhibitors of intracellular kinases. Moreover, antibodies directed against the extracellular region of cadherins block their association and disrupt the assembly of different types of junctions, including tight junctions. The opposite, however, is not true. Thus, disruption of tight junction assembly does not compromise cell–cell adhesion (Gumbiner, 1996). The barrier formed by tight junctions therefore appears to be under the dual control of cell adhesion molecules and intracellular signals, including the phosphorylation state of some of their protein components.

Adherens Junctions and Desmosomes Anchor Plasma Membranes

Adherens junctions connect the cytoskeleton of one cell to that of contacting cells. These structures confer strength to tissues and, not surprisingly, are most abundant in epithelia that are subjected to severe mechanical stress, such as the epidermis. From a structural and functional standpoint, all adherens junctions are composed of three sets of proteins. The first comprises adhesion proteins, which are integral membrane proteins with an extracellular, a transmembrane, and a cytoplasmic region. The extracellular region allows interaction with similar adhesion molecules expressed by the adjacent cell; the transmembrane segment ensures anchorage to the plasma membrane; and the cytoplasmic domain binds to intracellular attachment proteins, which link the adhesion proteins to the cytoskeleton. The second set of proteins comprises these attachment proteins, including the catenins (α, β, and γ), which connect the carboxyl-terminal tail of cadherins to actin filaments. β-Catenin plays a strategic role, linking cadherins to α-catenin, which in turn binds directly to actin. γ-Catenin can substitute for β-catenin in this complex (Kirkpatrick and Peifer, 1998). The third set of molecules includes various cytoskeletal proteins, whose identity depends on the type of junction (Table 1). These proteins maintain the tridimensional architecture of cells and establish a link between the plasma membrane and the nucleus as well as between neighboring cells (Kirkpatrick and Peifer, 1995; Garrod et al., 1996).

Adherens junctions of epithelial cells form an interrupted belt that is located toward the apical portion of the cell membrane, just underneath tight junctions. The adhesion molecules present in these structures belong to the cadherin family and are connected to the adapter proteins, catenins and vinculin, which in turn provide a link to actin filaments (Kirkpatrick and Peifer, 1995). The result of these interactions is that actin bundles both within and between cells are functionally linked in an intercellular network. The contraction of this network, which is dependent on myosin, enables several reorganizations of epithelia, such as the folding of epithelial sheets and the formation of complex structures during the early phases of morphogenesis (Table 2).

Desmosomes allow cells to share, and therefore tolerate, mechanical tension. They also consist of adhesion molecules of the cadherin family which are anchored, via adapter proteins (γ-catenin or desmoplakins), to cytoskeletal proteins forming the intermediate filaments (Garrod *et al.*, 1996). The molecular composition of these filaments varies according to the cell type. In epithelia, intermediate filaments are made of keratins, a multigene family of proteins which are expressed in characteristic type I/type II heteropolymeric pairs. Electron micrographs show that keratin filaments are connected to the dense cytoplasmic plaques of desmosomes (Fig. 5), therefore establishing an intercellular network that links and stabilizes the plasma membranes of adjacent cells. The adhesion receptors of desmosomes are desmocollins and desmogleins, which belong to the cadherin superfamily. Comparison of protein sequences between these molecules and classical cadherins shows a high degree of similarity in the extracellular domains, which are involved in the adhesive interactions, whereas the cytoplasmic tails, which are responsible for the binding to the junctional plaques, are quite different.

There are three isoforms of both desmogleins and desmocollins, which are encoded by separate genes and exhibit a distinct pattern of tissue expression (Garrod *et al.*, 1996). Some are ubiquitous, whereas others have a restricted expression pattern. Because each cell can produce several isoforms, it has been of interest to determine whether different desmosomal cadherins can participate in the organization of a single desmosome or whether they remain segregated in distinct structures. Available data suggest that desmosomes can be composed of more than one isoform of desmogleins and desmocollins. In addition, desmosomes

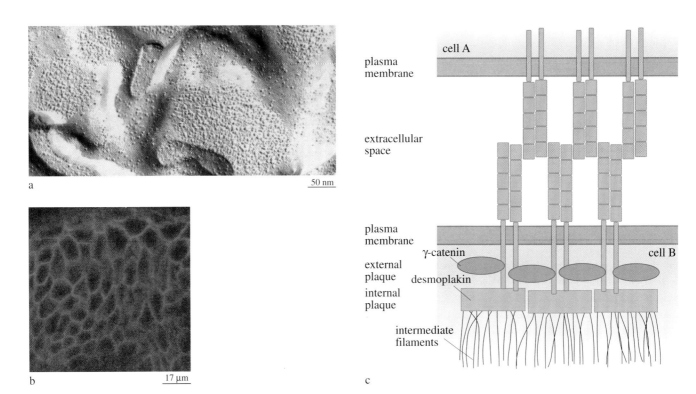

FIGURE 5 (a) Electron micrograph of the lateral membrane of an epithelial cell, in which desmosomes are seen as irregular aggregates of proteinaceous structures of various size and shape. This appearance results from the freeze fracture of the intermediate filaments of keratin, which associate with the cytoplasmic leaflet of the plasma membrane. (b) Fluorescence photograph of human epidermis after incubation with antibodies against desmoglein, one of the most abundant desmosomal cadherins. The staining shows that desmosomes are distributed throughout the entire lateral membrane of the epithelial cells in contact. (c) Schematic drawing of the structure of a desmosome. Desmogleins and desmocollins (orange) are integral membrane proteins of the cadherin family which, on the one hand, interact with each other in the extracellular space and, on the other hand, associate with γ-catenin in the external plaque and desmoplakins in the internal plaque. The latter are the intracellular attachment proteins to the intermediate filaments of the cytoskeleton which bind to the internal plaque of the desmosomal junction.

are dynamic structures that are constantly renewed and vary their molecular composition as a function of cell differentiation. Following the same building scheme of adherens junctions, desmosomal cadherins interact via their cytoplasmic carboxyl tail with the proteins of the junctional plaque, primarily desmoplakins and γ-catenin. Desmoplakins share sequence homology with proteins of intermediate filaments and are thought to connect them to desmosomes; γ-catenin shows a strong sequence similarity to β-catenin and binds to the central region of the cytoplasmic sequence of desmocollins and desmogleins (Kirkpatrick and Peifer, 1995; Garrod et al., 1996).

Desmosomes are assembled at blastocyst formation (32-cell stage) and contribute to the integrity of embryonic tissues at a time of complex structural modifications. The formation of desmosomes begins with the transcription, and subsequent protein biosynthesis, of desmocollin Dsc2 and is independent of physiological concentrations of extracellular Ca^{2+}. The assembly of desmosomes and adherens junctions is regulated in an interdependent fashion. Thus, the organization of a desmosome is secondary to the ability of cells to form adherens junctions. The role of the latter structures in desmosome assembly has been demonstrated in cell lines devoid of classical cadherins, which are unable to form desmosomes although they express the necessary molecular components. Transfection of either E-cadherin or P-cadherin into these cells is accompanied by the assembly of desmosomes, whereas blockade of cadherin-based adhesion is followed by the disruption of adherens junctions and desmosomes (Garrod et al., 1996). The formation of adherens junctions and the presence of strong cell–cell adhesion are prerequisites for the assembly of the other junctional complexes. This finding reinforces the concept of a hierarchical organization within junctional structures, which appears to be dominated by cadherin-based adhesion.

Gap Junctions Are Clusters of Intercellular Channels

Multicellular organisms must be able to coordinate the activity of groups of cells under different functional conditions. For example, the response to an external signal may be amplified by recruiting neighboring cells and, conversely, cells undergoing deleterious changes need to be isolated to preserve the integrity of the group. In either case, a system is required to permit cells to review and shape the functional state of their neighbors by the exchange of signaling molecules. Such a system of communication should also be rapidly modulated to continuously adapt to the changing needs of coupled cells. These features are met by the intercellular channels, specialized structures which are usually localized in distinct regions of the plasma membrane known as gap junctions. These channels mediate a unique form of communication, such that the exchange of molecules between cells is direct and does not involve transit through the extracellular space (Paul, 1995; Bruzzone et al., 1996). In freeze fractures of lateral plasma membranes, gap junctions appear as homogeneous aggregates of protein particles, each of which represents a channel between adjacent cells (Fig. 6). Intercellular channels are found in all metazoans, regardless of their position in the phylogenetic tree. In mammals, virtually all cell types express gap junction channels during either specific developmental stages or following terminal differentiation, with the exception of adult skeletal myocytes, most circulating blood cells, and spermatozoa.

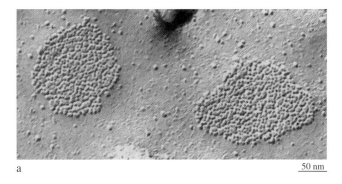
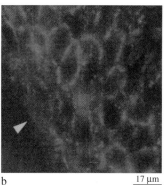

a   50 nm

b   17 μm

FIGURE 6 (a) Electron micrograph after freeze fracture of the lateral membrane in which gap junctions are recognizable as homogeneous aggregates of large protein particles. Each particle represents a channel for the cell-to-cell exchange of cytoplasmic ions and molecules (up to a molecular mass of approximately 1 kDa). (b) Fluorescence photograph of human epidermis after incubation with antibodies against connexin43, one of the constitutent proteins of the gap junction. The distribution of these is seen to be restricted to discrete domains of the lateral membrane of adjacent cells. Note the absence of these intercellular junctions on the basal region of the membrane of the epithelial cells of the first tissue layer (arrow) that is in contact with the underlying connective tissue.

Intercellular channels are structurally more complex than other ion channels because a complete cell-to-cell channel spans two plasma membranes and is formed by the association of two half channels, or connexons, contributed separately by each of the two participating cells. The interaction of two connexons in the extracellular space, or "gap," allows the establishment of communication between the cytoplasm of two adjacent cells. Connexons are made of structural proteins called connexins that comprise a multigene family. Connexins are distinguished on the basis of the molecular mass deduced from the cloned cDNA sequences, e.g., the 32-kDa protein of liver gap junctions is termed connexin32. The topological structure of a generic connexin predicts that it contains four hydrophobic regions, corresponding to transmembrane domains. This model also suggests the presence of two extracellular loops, whose location has been confirmed by several assays (Fig. 7). It is likely

that these extracellular portions are important for connexon pairing and participate, together with transmembrane domains, in the formation of the channel wall. The major cytoplasmic domains are unique in both sequence and length for each connexin and are thought to be involved in the regulation of channel permeability. Intercellular channels are defined as either homotypic, when connexons contributed by each cell are composed of the same connexin, or heterotypic, when the connexons differ. In addition, most tissues express more than one connexin and the same connexin is usually expressed by several cell types. Thus, a given cell can assemble homomeric connexons, made up of one connexin form, and different connexins may oligomerize into a heteromeric channel (Paul, 1995; Bruzzone et al., 1996).

The ubiquitous presence and the molecular heterogeneity of connexins have not facilitated the assignment of specific functions to intercellular channels and the identification of the relevant signals that cross them in different organs. It is well established that neurons, cardiac myocytes, and other electrically excitable cells use gap junction channels to generate synchronous electrical and mechanical responses. This type of signaling is faster than that achieved through chemical synaptic transmission and, as a result, is used by neuronal circuits that control escape responses in certain animal species. In addition, cell–cell communication is established early during embryogenesis and participates in controlling the activity of dividing cells. The functional blockade of connexin channels and the illicit communication between groups of cells that are usually segregated results in drastic alterations of morphogenesis. There is evidence that cadherin-based adhesion is required for the establishment of intercellular communication. For example, some cell lines are communication incompetent although they express connexins, unless they are transfected to induce E-cadherin synthesis. Signaling through connexins is also important to maintain adhesion. Thus, in at least one system, functional blockade of connexin-dependent cell-to-cell communication in a two-cell mouse embryo does not interfere initially with cell division but prevents compaction at the eight-cell stage, a process which is strictly dependent on the expression of E-cadherin. The available data suggest the existence of bidirectional feedback signals between cadherin-mediated adhesion and communication through connexins. This model also emphasizes the dual role of intercellular junctions, both in structural maintenance and in signaling (Kim, 1995; Bruzzone et al., 1996; Gumbiner, 1996).

The role of connexins in tissue homeostasis is being examined via the generation of transgenic animals, which allow the testing of specific working hypotheses (Nicholson and Bruzzone, 1997). For example, the abundance of gap junction channels between liver cells has long been thought to reflect a need to share and coordinate the metabolic

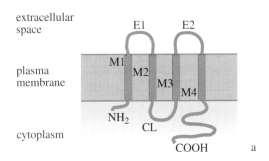

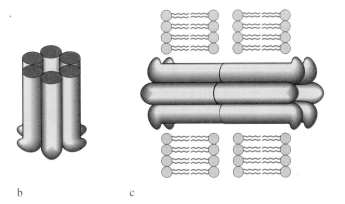

FIGURE 7 Schematic view of intercellular channels in gap junctions. (a) The connexins consist of four transmembrane regions (M1–M4), two extracellular regions (E1 and E2), a cytoplasmic loop (CL), and intercellular termini (both NH$_2$ and COOH). (b) Connexons, assembled by oligomerization of connexin molecules into a tubular hexameric structure, form the wall of half of a channel. (c) The intercellular channel is formed by a connexon, transported to the plasma membrane, and aligned with another connexon from a cell in contact. Clusters of intercellular channels constitutes a gap junction (adapted from Bruzzone and Ressot, 1997).

workload of a hepatic lobe. Connexin32 knockout animals show a great reduction in gap junction plaques and cell coupling in the liver, in which the highest levels of connexin-32 mRNA and protein are found. The amount of glucose released from hepatocytes of connexin $32^{-/-}$ after stimulation of sympathetic nerves is vastly reduced compared to that mobilized in wild-type animals. This phenotype suggests that connexin32-containing intercellular channels are required for the propagation of signals, triggered by the release of noradrenaline and other neurotransmitters at the sympathetic nerve endings, from the periportal area to the less innervated perivenous hepatocytes.

A different defect is produced by the ablation of the mouse connexin43 gene, which results in gross abnormalities of the development of the cardiac pulmonary outflow tract. The phenotype of connexin43 knockout mice is reminiscent of some forms of stenosis of the pulmonary artery occurring in the context of visceroatrial heterotaxia (Ivemark's syndrome), another disease which has been proposed to be related to connexin mutations. The causal relationship between connexin43 mutations and morphogenetic heart defects, however, has not been determined (Paul, 1995).

An exchange of signals through connexin channels is also critical to set in motion the cellular events required for oogenesis and ovulation. Knocking out connexin 37, which is expressed in oocytes, abolishes intercellular communication between the oocyte and the surrounding granulosa cells and results in female infertility. The absence of this signaling pathway, which orchestrates the cellular interactions necessary for oogenesis and ovulation, appears to result in this phenotype. Careful examination of the ovaries of connexin $37^{-/-}$ mice has revealed several abnormalities, including the inappropriate formation of corpora lutea of reduced size. Luteinization of granulosa cells normally occurs after expulsion of the oocyte from the follicle. The premature formation of corpora lutea in the knockout animals suggests that a signal transferred through connexin 37 channels from the oocyte to granulosa cells prevents this transformation. These findings suggest the existence of a bidirectional molecular dialogue between the oocyte and its supporting granulosa cells, although the identity of the transferred signals has not been determined (Simon et al., 1997). It is not known whether connexin defects are responsible for female infertility in humans, although the distribution of connexin37 in human ovarian follicles is similar to that in mouse. Moreover, inappropriate luteinization is a specific feature of spontaneous premature ovarian failure, a human disorder of unknown etiology.

This series of experiments demonstrates that junctional channels play an important role in the functional homeostasis of nonexcitable cells. Similar conclusions have been reached by studying secretory epithelia, particularly those that form the endocrine islets and exocrine acini of the

pancreas (Meda, 1996). Thus, stimulation of exocrine pancreatic secretion is accompanied by a decrease in gap junctional communication both in vitro and in vivo. In the case of the endocrine pancreas, however, gene expression, biosynthesis, and secretion of insulin are stimulated under conditions that promote the formation of intercellular channels. Moreover, cell lines characterized by abnormal insulin content and exocytosis do not express connexins. The restoration of cell coupling that follows transfection of such cell lines with a connexin is accompanied by a partial rescue of the secretory abnormality (Vozzi et al., 1995). Therefore, it appears that the influence of cell–cell communication on secretion is opposite in the exocrine and endocrine pancreas. This finding may be due, at least in part, to the expression of distinct and functionally incompatible connexins in the two tissues (Meda et al., 1993). This interpretation is supported by the initial observation of transgenic mice in which expression of connexin32, the main connexin of exocrine tissue, was forced in the insulin-producing cells, which normally express connexin43. The ectopic expression of connexin32 results in decreased hormonal secretion despite a major increase in gap junctional communication.

Gap Junction Channels Participate in the Elaboration of an Intercellular Language

The molecular complexity of intercellular channels bears important consequences on their function. Connexins have evolved to permit diverse languages of communication among cells. The first form of intercellular language is based on the realization that channels composed of different connexins are endowed with distinct biophysical and regulatory properties. It is likely that each connexin conveys a particular set of signals and that connexins may not be interchangeable. Although connexins do not form ion-selective channels in the classical sense, gap junction channels exhibit a certain degree of selectivity for cations and anions. Because gap junctions are permeable to cyclic nucleotides and inositol trisphosphate, which are negatively charged at normal intracellular pH, selectivity may allow to discriminate and differentially modulate the kinetics of intercellular propagation of second messengers. Given that changes in second messenger levels are usually short-lived, a decreased permeability of gap junction channels may result in the transfer of the signal to reduced numbers of cells. Size selectivity is also a connexin-specific feature. It had long been thought that connexins were unselectively permeable to molecules up to 1 kDa of molecular mass. The use of different fluorescent tracers has convincingly demonstrated that some connexins can discriminate between molecules of 400 and 600 Da, thus favoring the passage of a distinct set of signals. An example of the physiological implications of this selectivity is provided by cultured keratinocytes which, during their differentiation *in vitro,* switch the pattern of connexin expression. As a result, keratinocytes remain electrically coupled (i.e., they can exchange ions) but lose their initial permeability to larger molecules, such as certain fluorescent tracers. Thus, modulation of connexin expression is a feature that may operate to control the exchange of signals implicated in differentiation processes.

A second form of intercellular language is that of compatibility, that is the ability of connexins to discriminate among and functionally interact with different family members. Intercellular channels require the proper alignment and docking of two connexons in the extracellular gap. On the basis of the accepted models of connexin topology relative to the plasma membrane, the interconnexon docking implicates the two extracellular domains. Their similarity in primary sequence had led to speculation that interactions would favor the head-to-head pairing of connexons composed of a distinct connexin, thus promoting the unrestricted formation of heterotypic channels. In contrast, experimental data have shown that the ability to establish direct communication between adjacent cells is dependent, at least in part, on the ability of connexins to discriminate between different partners (White *et al.,* 1995). A consequence of the restricted ability of connexins for heterotypic interactions is that expression of different connexins in groups of cells may provide a powerful means to limit or even segregate, rather than facilitate, communication. This code of compatibility could explain the observation of "communication compartments" which are characterized by clusters of cells that are permeable to tracer molecules within, but not between, adjacent cell groups, even in the absence of obvious anatomical boundaries. Limits to the extent of intercellular communication would result in the heterogeneous dispatch of signals between neighboring cells, which could then lead to the activation of different gene expression programs. Incompatibility among connexins may also ensure that a given stimulus is more faithfully transmitted to a group of coupled cells without dissipation of the signal to the surrounding tissue. An interesting example of such compartmentalization is provided by the cardiovascular system, in which multiple connexins are expressed in heart, smooth muscle, and endothelial cells of the arterial wall. The Purkinje fibers of the cardiac conduction system express largely connexin40, whereas the working myocardium predominantly displays connexin43. These two connexins are incompatible and do not form heterotypic junctions in expression systems, thus minimizing the potential for undesired excitation of myocardial cells along the length of the conducting fibers.

The functional state of intercellular channels can be regulated by both covalent and noncovalent modifications of connexins which affect the channel structure. Such modulation amplifies the diversity of information that is transferred through gap junctions. This third form of intercellular language is called gating. Protein phosphorylation on serine, threonine, or tyrosine is a well-characterized mechanism by which extracellular signals can significantly alter the function of ion channels, and connexins are no exception to this rule. Connexin phosphorylation participates in the regulation of connexon oligomerization, interconnexon interactions, and the biophysical properties of intercellular channels. Activation of certain kinases, such as protein kinase A or protein kinase C, results in specific phosphorylation of gap junction proteins that is dependent on the type of kinase and connexin identity. This form of regulation results in a differential phosphorylation of distinct connexins under similar experimental conditions. Phosphorylation of connexins is associated with reduced channel permeability, which contributes to the elaboration of a connexin-specific language adapted to the functional needs of communicating cells (Fig. 8). An interesting example of gating is demonstrated in the transmission of the visual impulse along the retinal circuitry. The vertebrate retina presents a highly ordered architecture in which several types of neurons are organized in distinct layers. Gap junction channels, present between virtually all cell types, allow the lateral diffusion of signals and participate in light transmission along the vertical pathway which, from rod photoreceptors through bipolar cells, travels to the ganglion cells of the optic nerve which in turn relays impulses to the visual cortex. Signals from rods, which are active in twilight (scotopic vision), are transferred to ganglion cells via the intercellular channels which connect amacrine cells both together and to cone bipolar cells. This route amplifies the intensity of the specific signal by recruiting more cells and utilizes more efficiently the limited number of photons available in twilight. In normal light (photopic vision), intercellular channels between cone bipolar cells and amacrine cells are closed by an increase in cyclic GMP, thus preventing backward diffusion to amacrine cells and lateral spread of the light signal and maintaining the spatial detail carried by ganglion cells. The specificity of connexin phosphorylation may therefore explain why intercellular transfer of signals between amacrine cells, presumably through homotypic channels, is blocked by elevation of cyclic AMP levels (and activation of protein kinase A), whereas channels between amacrine and cone bipolar cells, probably heterotypic, are inhibited by cyclic GMP (and activation of protein kinase G) but are insensitive to cyclic AMP. In this system, connexin channels behave as a molecular switch that controls signal transmission between the cone and rod pathways that are associated with adaptation to different light conditions.

Cell–Matrix Adhesion Is Responsible for the Functional Integrity of Epithelia

The functional integrity of an epithelial sheet is strictly dependent on its anchorage to a connective tissue. Epithelial tissues lining cavities are not directly vascularized and de-

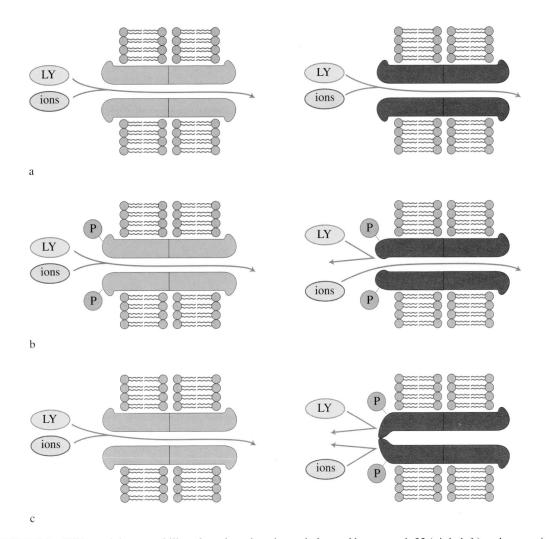

FIGURE 8 Differential permeability of gap junction channels formed by connexin32 (pink, left) and connexin-43 (red, right). (a) The channels are permeable to both the fluorescent tracer Lucifer yellow (LY) and ions. (b) After phosphorylation (P) by one isoform of protein kinase C (PKC), the passage of LY, but not of ions, is blocked in channels composed of connexin43, whereas the permeability of channels composed of connexin32 is unchanged. (c) In contrast to PKC, the tyrosine kinase encoded by the viral oncogene v-*src* is only able to phosphorylate connexin43. This phosphorylation results in a complete block of cell-to-cell communication (adapted from Bruzzone *et al.*, 1996).

rive their nutriment from the underlying blood vessels that vascularize connective tissues. Limited thickness and exposure to the environment render these epithelia particularly vulnerable to mechanical injuries which, by interrupting the adhesion between cells and between cells and the extracellular matrix, compromise their barrier function. The mechanical resistance of epithelia is considerably enhanced by their adhesion to thicker and stronger tissues which are found deeper in the organism. Moreover, adhesion to connective tissues determines the architecture of epithelia, both during morphogenetic development and in adult life, when environmental changes (i.e., the type or concentration of essential growth factors) modify the proliferation and differentiation of epithelial cells. Finally, contacts be-

tween epithelia and connective tissues modulate several specialized functions of epithelial cells by means of an intercellular signaling system that involves the interaction of extracellular molecules with their cognate receptors present on epithelial cells.

Adhesion of an epithelial cell to a connective tissue is ensured by the basal lamina, a layer of extracellular material produced by the epithelial cells (Fig. 9). The basal lamina consists of a superficial region, the lamina lucida, that is composed mainly of laminin and perlecan and a deeper portion, the lamina densa, in which proteoglycans and glycoproteins are intermingled with type IV collagen and fibronectin, an abundant component of the extracellular matrix which is synthesized by cells of connective tissues. The

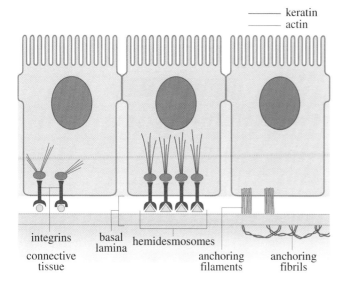

keratin
actin

integrins
basal lamina
hemidesmosomes
connective tissue
anchoring filaments
anchoring fibrils

FIGURE 9 Schematic view of the systems ensuring the adhesion of epithelial cells to the extracellular matrix. In the cell on the left, the integrins anchor the extracellular components of the basal lamina (yellow circles) to the actin microfilaments of the cytoplasm. The carboxyl termini of the integrins interact with these microfilaments via various intracellular attachment proteins (brown). In the cell in the center, a particular form of integrin ($\alpha_6\beta_4$) accumulates in the basal region of the cell membrane to form hemidesmosomes that link the glycoprotein components of the basal lamina (yellow triangles) to the intermediate filaments of keratin. In the cell on the right, the anchoring filaments link the basal plasma membrane to the first layer (lamina lucida; in yellow) of the basal lamina. The lamina lucida, in turn, is connected to the second layer (lamina densa; in dark brown). In stratified epithelia, the lamina densa is further attached via anchoring fibrils to collagen fibers of connective tissues.

combination and prevalence of each of these molecules, as well as other minor components, differ according to tissue type (Timpl, 1996; Yancey, 1995).

Although extracellular, the basal lamina performs at least three functions that are vital to the epithelial cells that produce it, and therefore it cannot be functionally separated from them. First, it regulates the passage of cells and molecules between connective and epithelial tissues. The basal lamina carries out this duty as a result of its tridimensional organization which is comparable to that of a sieve whose mesh discriminates molecules on the basis of size and electric charge (Timpl, 1996). A normal basal lamina prevents the passage of cells from either epithelia or connective tissue, with the notable exception of cells of the immune system which pass easily in both directions. The basal lamina also controls the passage of ions and soluble molecules which diffuse from capillaries and serve as nutrients for epithelia.

The second function of the basal lamina is to provide

epithelial cells with signals that are crucial for the control of their movements, growth, and differentiation. To this end, the basal lamina binds hormones and growth factors which are delivered via the systemic circulation, thus maintaining a high concentration of these molecules in the vicinity of the target epithelial cells. In addition, changes in the molecular composition of the basal lamina are sensed by integrins in the membrane of epithelial cells. This signaling system can be regarded as both intracellular, because integrins are connected in the cytoplasm to cytoskeletal proteins that modify several cellular functions, and intercellular, because all cells lying on the same basal lamina and expressing the same set of integrins will react in similar fashion to molecular changes in the basal lamina.

The third function of the basal lamina is to ensure the adhesion of epithelia to connective tissues through a series of layers. Thus, epithelial cells adhere to the lamina lucida, which in turn is attached to the lamina densa and the latter is bound to connective tissue. The first layer of this attachment system is composed of proteoglycans which are inserted in the basolateral membrane of epithelial cells and, more important, of integral membrane proteins, belonging to the integrin family, which act as receptors for individual components of the lamina lucida (Yancey, 1995; Fuchs, 1997).

Integrins are dimers of two glycoprotein chains, broadly divided into α and β subtypes, which associate in noncovalent fashion (Fig. 2). Approximately 20 different integrin dimers are known; they differ in the molecular nature of the α and β chains and it is likely that more will be identified. The majority of integrins are expressed by several cell types and most cells express multiple integrin (Schwartz *et al.*, 1995). Integrins that operate in cell–matrix adhesion usually contain the β_1 chain. Regardless of the subunit composition, α and β chains function as receptors that bind with their amino-terminal region to specific ligand molecules of the basal lamina. This binding is dependent on Mn^{2+} and Mg^{2+} ions, whereas dissociation is dependent on extracellular Ca^{2+} ions. Each cell utilizes several integrin dimers, each with its own affinity for specific extracellular molecules and, conversely, distinct components can be bound by several types of integrins. This adjustable system allows epithelial cells to adhere firmly to the basal lamina while preserving their ability to crawl along its surface (Sheppard, 1996).

The carboxyl-terminal tails of α and especially β integrin chains are connected to cytoplasmic microfilaments via adapter proteins, such as astalin or α-actinin (Table 3). This organization leads to a bidirectional system in which the cytoskeleton regulates the concentration of integrins in the plasma membrane, whereas the molecular composition of the basal lamina modulates, through integrin binding, several cellular functions, such as secretion of metalloproteases, apoptosis, the cell cycle, and gene expression (Ruoslahti and Reed, 1994).

In stratified epithelia such as the epidermis and some

TABLE 3

Proteins Ensuring the Adhesion of Epithelial Cells to the Extracellular Matrix[a]

MOLECULES/ JUNCTIONS	MATRIX PROTEINS	MEMBRANE PROTEINS	INTRA-CELLULAR ATTACHMENT PROTEINS	CYTOSKELETAL STRUCTURE (PROTEIN)
Integrins	Glycoproteins of the lamina lucida	Integrins	Talin, vinculin	Microfilaments (actin)
Hemidesmosomes	Glycoproteins of the lamina lucida	α_6/β_4 integrin, BPA1, BPA2	Desmoplakins	Intermediate filaments (keratins)
Anchoring filaments	Glycoproteins of lamina lucida and lamina densa	Kalinin	?	?
Anchoring fibrils	Type VII collagen	—	—	—

[a] Adhesion to the extracellular matrix is necessary for epithelial cell homeostasis, which depends on diffusion of nutrients from the capillaries of connective tissues. Epithelial cells may modulate their adhesive properties by changing the type and number of integrins, their binding sites to the components of extracellular matrix, or their interaction with cytoskeletal elements. BPA, bullous pemphigoid antigen.

areas of the digestive tract, which are subjected to considerable mechanical stress, integrins are complemented by three additional cell–matrix adhesion systems. The first consists of hemidesmosomes, which are regions of the plasma membrane in which epithelial cells cluster $\alpha_6\beta_4$ integrins in association with other proteins including, in the epidermis, bullous pemphigoid antigens (BPA) 1 and 2 (Yancey, 1995; Borradori and Sonnenberg, 1996). These molecules are connected through their carboxyl-terminal ends to the intermediate filaments of two types of keratins, K5 and K14, which form the backbone of epithelial cells in the basal layer. The second system is composed of anchoring filaments made of kalinin. This protein complex attaches the basal membrane of epithelial cells to the lamina lucida and the latter to the lamina densa (Yancey, 1995). Finally, the third system is composed of anchoring fibrils formed by type VII collagen which ensure the binding of the lamina densa, the deeper layer of the basal lamina, to the underlying connective tissue (Fig. 9).

Cell–Cell and Cell–Matrix Adhesion Complexes Provide a Dynamic Signaling System

Epithelial cells communicate among themselves not only through connexin channels but also by exchanging signals generated in other sites of contact with either cells or the extracellular matrix. There is increasing evidence that a sophisticated signal transduction machinery is deployed near junctional complexes and other sites of contact which act as centers to sort and dispatch cellular messages (Fig. 10). Complex and varied molecular interactions have been de-

scribed in recent years even though many of these protein–protein associations remain to be demonstrated *in vivo*. It is likely that the molecular composition of such signaling centers is dynamically regulated in order to adapt to specific functional requirements.

An interesting and unexpected discovery has been the recent finding of structural relationships between junctional proteins and tumor suppressor genes. Amino acid sequence analysis of ZO-1 and ZO-2 has revealed a significant similarity to the MAGUK (membrane-associated guanylate kinase) family of proteins, which include the disc large-1 tumor suppressor gene product (Dlg). This protein is required for control of cell proliferation in both epithelial tissues and imaginal discs of *Drosophila* (Kim, 1995). Dlg protein is specifically localized in the septate junctions of invertebrates, which are functionally equivalent to the tight junctions of vertebrates. Mutations in the *dlg* gene in *Drosophila* cause a complete loss of septate junctions from imaginal discs, in which *dlg* is normally expressed, and are accompanied by an overgrowth of these structures. Other domains share homology with motifs which further suggest a signaling role for MAGUK proteins. It is not known whether mutations in ZO-1 and ZO-2 are associated with abnormalities in cell proliferation and which second messenger systems are under the control of these tight junction-associated proteins.

There is compelling evidence that adherens junctions also function as signaling centers, exerting a role well beyond adhesion. A key element in the signaling pathway is β-catenin, a protein that binds directly to the carboxyl-terminal region of cadherins, linking them to the cytoskeleton via α-catenin (Table 1 and Fig. 10). β-Catenin can undergo phosphorylation, a modification which, under certain experimental conditions, leads to the inactivation of the

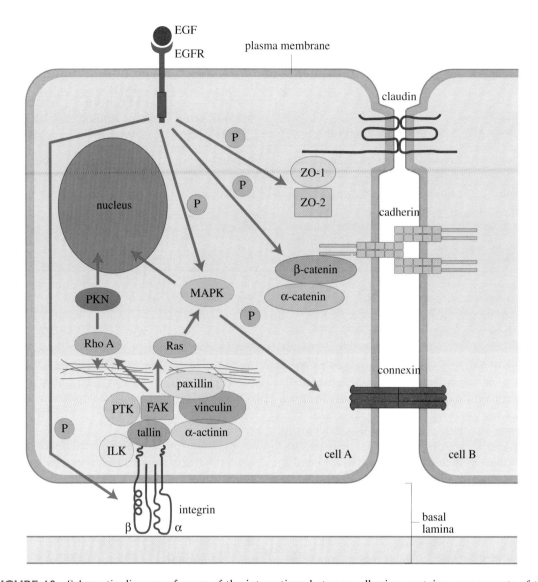

FIGURE 10 Schematic diagram of some of the interactions between adhesion proteins, components of the junction, and molecules involved in intracellular signaling. Junction complexes serve as integration and sorting centers for signals for individual cells and between neighboring cells. The activation of the EGF receptor (EGFR; dark green), which is associated with the direct or indirect phosphorylation (P) of a series of substrates, is an example of exogenous modulation of the functional state of adhesion molecules and of intercellular junctions. Phosphorylation of ZO-1 and ZO-2 affects tight junctions. Phosphorylation of β-catenin may be associated with the inactivation of cadherin/catenin adhesion systems. Phosphorylation of MAPK can affect connexins. Phosphorylation of integrins activates, by a complex system of proteins, the transcription pathways dependent on Ras and Rho proteins. Ras transfers membrane signals to the nucleus, whereas Rho affects both the organization of the actin cytoskeleton and, via PKN, the nucleus through cascade phosphorylations which converge to activate MAPK. EGFR, epidermal growth factor receptor; FAK, focal adhesion kinase; ILK, integrin-linked kinase; MAPK, mitogen-activated protein kinase; PKN, protein kinase N; PTK, proteins with tyrosine kinase activity; Ras and Rho A, members of the family of small GTP-binding proteins with GTPase activity.

cadherin–catenin adhesion system. Several tyrosine kinases, such as c-*src* and c-*yes,* accumulate in the vicinity of adherens junctions where they act on substrate junctional molecules, thereby modulating cellular adhesion and morphology (Adams, 1997; Yamada and Geiger, 1997). A direct

interaction has also been demonstrated between activated epidermal growth factor (EGF) receptor and β-catenin, which is a substrate for the intrinsic tyrosine kinase activity of the receptor. The crucial role of β-catenin has been confirmed by the discovery that it forms a complex with the product of the APC (adenomatous polyposis coli) gene, a tumor suppressor mutated in familial adenomatous polyposis. APC protein competes with E-cadherin to bind β-catenin, which then links it to the cytoskeleton. As is the case in junctional complexes, γ-catenin may substitute for β-catenin in APC–catenin complexes (Kirkpatrick and Peifer, 1995; Gumbiner, 1996). Together, these findings suggest that APC participates in the control of cell–cell interactions downstream of cadherin-dependent adhesion. Although there is no information on the precise role of the APC–β-catenin complex in cell proliferation, it is likely that shifts in the prevalence of this association may lead to the generation of selective signals. One possibility is that APC regulates the availability of β-catenin by preventing the association of the functionally inactive forms in junctional structures. In addition, it has been recently shown that both β-catenin and γ-catenin can modify gene expression by complexing lymphoid enhancer factor-1. This transcription factor, which is expressed at gastrulation, drives the translocation of catenins to the nucleus where these molecules bind to the promoter of the E-cadherin gene and inhibit its transcription (Huber *et al.*, 1996; Adams, 1997).

Another signaling system has been identified which involves cell–matrix interactions via integrins and their ability to participate in both intracellular and intercellular signaling (Clark and Brugge, 1995; Schwartz *et al.*, 1995). Thus, binding of integrins to their extracellular ligand transfers a signal which may alter attachment to the cytoskeleton, and therefore cell anchorage, whereas stimulation of intracellular messenger pathways may alter integrin affinity for proteins of the extracellular matrix and hence cell adhesion. Following occupancy of their binding sites by specific extracellular ligands, integrins acquire the ability to interact with cytoskeletal proteins, a step that in hemidesmosomes requires tyrosine phosphorylation of the β_4 subunit (Mainiero *et al.*, 1995). This phosphorylation event appears to be a prerequisite to activate Rho- and Ras-dependent signaling pathways (Fig. 10). These members of the superfamily of small GTPases transduce signals from tyrosine kinase receptors both to the nucleus (Ras) and to the cytoskeleton (Rho) and initiate a cascade of phosphorylations that converge in the activation of the MAPK (mitogen-activated protein kinase) family. The consequence of the increasing number of these protein–protein interactions is that integrins are likely to be involved in several distinct cellular functions, such as apoptosis, differentiation, and gene transcription. Spatial accumulation of these signaling molecules follows precise rules. Some proteins can associate after simple clustering of integrins in a region of the plasma membrane, others require tyrosine phosphorylation of integrins, whereas others accumulate in adhesion complexes only after integrin occupancy by extracellular ligands (Clark and Brugge, 1995; Yamada and Geiger, 1997). A central element in the coordination of this bidirectional transfer of information is focal adhesion kinase (FAK), a multifunctional tyrosine kinase which can interact directly with several β integrin subunits. FAK is also capable of complexing other molecules which control a series of synergistic interactions, eventually leading to a cascade of signals that are differentially dispatched to the appropriate cellular compartments.

Adhesion Systems and Human Diseases

The importance of the integrity of cell-to-cell and cell-to-matrix adhesion is emphasized by the functional perturbation of these systems in several human genetic and autoimmune disorders. During the past few years there has been considerable progress in our understanding of the molecular mechanisms of these diseases due to the identification both of the genes responsible for genetic disorders and of the blocking autoantibodies that cause acquired autoimmune diseases (Table 4). This information, together with the wealth of knowledge on cell adhesion molecules, has led to analysis of the consequences of functional perturbation of the proteins involved in these pathological conditions (Amagai, 1995; Yancey, 1995; Borradori and Sonnenberg, 1996; Garrod *et al.*, 1996; Adams, 1997; Fuchs, 1997; Nicholson and Bruzzone, 1997).

Genetic Disorders

The implications of the loss of the signaling role provided by junctional structures are well illustrated by one form of Charcot–Marie–Tooth disease (CMT). CMT is the most common inherited peripheral neuropathy, with an estimated incidence of about 1:2500, and all forms of Mendelian segregation have been demonstrated. From the clinical standpoint, the disease is characterized by a progressive weakness of the distal limb and intrinsic hand muscles, the development of pes cavus, absent or diminished deep tendon reflexes, and a variable sensory loss. The finding that the X-linked form of CMT was associated with mutations in the gene encoding connexin32 was totally unexpected because no connexins had been previously identified in the peripheral nervous system (Paul, 1995).

In vertebrates, most axons are protected by myelin, an insulating sheath which is formed by the spiral rolling of the Schwann cell membrane (Fig. 11). Each Schwann cell contributes a single myelin segment, which is separated from adjacent ones by nonmyelinated regions called nodes of Ranvier in which sodium and potassium channels, which are necessary for the propagation of action potentials, are clustered. Hence, the insulating properties of myelin control the propagation of action potentials between nodes, a phenomenon called saltatory conduction. Electron microscopic examination of myelin reveals the alternation of major dense lines, formed by the cytoplasmic sides of the

	TABLE 4	

Examples of Diseases Associated with Mutations or Dysfunction of Adhesion Proteins and Junctional Complexes[a]

GENETIC DISEASES

Disease	Mode of transmission	Structure (protein) affected
Charcot–Marie–Tooth	X-linked	Gap junctions (Cx32)
Nonsyndromic sensorineural deafness (DFNA3)	Autosomal dominant	Gap junctions (Cx26)
Nonsyndromic sensorineural deafness (DFNB1)	Autosomal recessive	Gap junctions (Cx26)
Zonular pulverulent cataract (CZP1)	Autosomal dominant	Gap junctions (Cx50)
Epidermolysis bullosa simplex	Autosomal dominant or recessive	Intermediate filaments (K5, K14)
Epidermolysis bullosa junctional	Autosomal recessive	Hemidesmosomes (laminin V, β_4 integrin) and anchoring filaments (kalinin)
Epidermolytic hyperkeratosis	Autosomal dominant	Intermediate filaments (K1, K9, K10)
Epidermolysis bullosa dystrophica	Autosomal dominant or recessive	Anchoring fibrils (type VII collagen)

AUTOIMMUNE DISEASES

Disease	Structure (antigen) affected	Antibody isotype
Pemphigus vulgaris	Desmosomes (Dsg3)	IgG
Bullous pemphigoid	Proteins of 180 and 230 kDa, desmoplakin I-like	IgG
Pemphigus foliaceous	Dsg1	IgG/IgA?
Paraneoplastic pemphigus	Proteins of 170 and 230 kDa, desmoplakins I and II	IgG
Pemphigoid	Hemidesmosomes (BPA1, BPA2)	?
Pustular dermatosis	Desmosomes (proteins of 105 and 115 kDa, Dsg1, Dsc1, Dsc2)	IgA
Acquired epidermolysis bullosa	Anchoring fibrils (type VII collagen)	IgG

OTHER DISEASES PUTATIVELY ASSOCIATED WITH ALTERATIONS OF ADHESION SYSTEMS

Disease	Structure (protein) affected
Cholestatic jaundice	Tight junctions (?)
Acute pancreatitis	Tight junctions (?)
Myocardial arrhythmia	Gap junctions (Cx43?)
Cardiac conduction system disorders	Gap junctions (Cx40?)
Carcinomas	Gap junctions (connexins), cadherins, catenins

[a] Diseases due to a dysfunction of intercellular junctions and other adhesion systems may result either from mutations of genes coding for structural proteins or from the presence of antibodies that bind specific proteins, thus altering their proper functioning. Blistering disorders of the skin are distinguished according to the plane of tissue cleavage, which depends on the nature of the affected protein. Blistering lesions may result from a cleavage occurring within the epidermis, in the upper dermis, or at the dermoepidermal junction. From both a genetic and a pathophysiological standpoint, several diseases may be included within each of these groups. K, keratin; Cx, connexin; Dsc, desmocollin; Dsg, desmoglein; BPA, bullous pemphigoid antigen.

plasma membrane, with double intraperiod lines which correspond to the fusion of the external leaflets of two adjacent layers of the plasma membrane. The remaining cytoplasm is confined to the paranodal loops, at the periphery of compact myelin, or to the Schmidt–Lanterman incisures, which are specialized regions traversing compact myelin. In

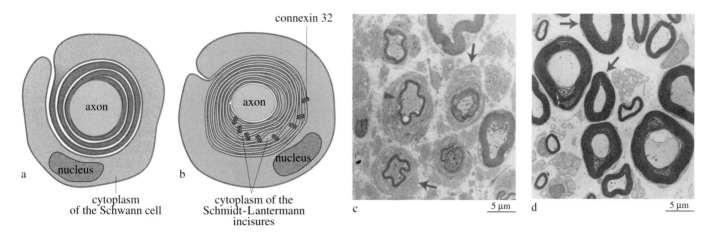

FIGURE 11   (a) Schwann cell of the peripheral nervous system which surrounds an axon in a spiral to form an insulating sheath of myelin. (b) Schwann cells express connexin32 (in pink) at paranodal regions and Schmidt–Lanterman incisures where some cytoplasm is retained (adapted from Paul, 1995). (c) Under the electron microscope, morphological analysis of a peripheral nerve of a patient affected by the X-linked form of Charcot–Marie–Tooth disease shows a decrease in the number of myelinated axons, most of which are surrounded by so-called "onion bulbs" (arrows). These structures, which result from the proliferation of Schwann cells around either demyelinated or thinly myelinated axons, reflect repeated episodes of demyelination and remyelination. Some of the myelin sheaths appear abnormally thin with respect to the diameter of the wrapped axon (arrowheads) (courtesy of J.-M. Vallat). (d) The biopsy of a normal peripheral nerve shows the characteristic appearance of thick and compact myelin (arrows) that surrounds all the axons in this field (courtesy of J.-M. Vallat).

Schwann cells, connexin32 is exquisitely localized in those areas of the myelin sheath containing cytoplasm (Paul, 1995). This distribution is incompatible with the formation of orthodox intercellular channels but suggests that connexin32 forms channels between turns of myelin, within an individual Schwann cell (Fig. 11). This cytoplasmic continuity would greatly shorten the diffusion pathway for the transfer of ions, second messengers, and metabolites from the perinuclear to the periaxonal region of Schwann cells, through the compact myelin wraps.

More than 100 different mutations in the gene encoding connexin32 have been reported. Some mutations are associated with the functional loss of connexin32 channels, thereby interfering with the intracellular exchange of messenger molecules and nutrients between different regions of the Schwann cell along the radial diffusion pathway (Bruzzone *et al.,* 1994). The precise sequence of events by which connexin 32 mutations lead to progressive demyelination in the peripheral nervous system remains to be elucidated. It should be noted that, although X-linked CMT has usually been classified among demyelinating neuropathies, several cases characterized by marked axonopathy have been described. The other two most frequent forms of CMT are caused by genetic defects in either the adhesion protein zero (P0) or in a gene encoding an integral membrane protein of 22 kDa with four transmembrane domains, peripheral myelin protein-22 (PMP22), whose function remains elusive. It is conceivable that all mutated proteins perturb the main function of myelin, that is, the maintenance of a barrier between axons and the surrounding tissues. In this special case, it appears that adhesion and communication systems are necessary for the proper functioning of distant areas of the same cell rather than for the coordination of the activity of neighboring cells. Recently, other connexin mutations have been established as the molecular cause of other genetic disorders. Thus, zonular pulverulent cataract and two forms of nonsyndromic sensorineural deafness are linked to mutations in connexin50 and connexin26, respectively (Table 4).

Autoimmune Diseases

Several components of cell–cell and cell–matrix adhesion complexes can become the targets of circulating autoantibodies. Some of these antibodies exhibit sufficient affinity to interfere with the function of critical adhesion proteins. Others can block the synthesis or the association in the plasma membrane of these proteins. In either case, the consequence is a severe dysfunction of cell-to-cell and cell-to-matrix adhesion systems which leads to the loss of the ability of epithelial tissues to act as a barrier between the organism and the environment. In some diseases, the functional and structural disruption of large epithelial areas results in the severe loss of body fluid and in bacterial infection, two complications that can be fatal in the absence of appropriate therapy (Amagai, 1995; Yancey, 1995).

This group of diseases is particularly evident in the epidermis, the largest epithelial tissue of the human organism, which acts as a barrier to the environment. To ensure such functional specialization, this tissue is formed by keratinocytes, cells which express high levels of most cell adhesion systems described in this article. Adhesion between keratinocytes is provided by several cadherin isoforms as well as by abundant desmosomes and gap junctions. In contrast, tight junctions are rare, if not absent, probably because the epidermis is one of the few epithelia located between a dry compartment (the environment) and a liquid one (the extracellular fluid). In the absence of these junctions, the epidermis uses two specific systems, which are very effective under normal conditions, to remain impermeable to fluids. First, keratinocytes secrete a lipid-rich substance which cements the intercellular spaces of the upper epidermal layers, thus preventing the paracellular diffusion of water and soluble molecules. The second system is formed during the program of terminal differentiation of keratinocytes, which consists of the cytoplasmic accumulation of high-molecular-weight keratins, the proteins forming intermediate filaments, and of specific proteins such as involucrin which thicken the plasma membrane. Together, these modifications of the superficial layers of the epidermis contribute to the maintenance of its role as a protective envelope and an effective barrier to water (Fuchs, 1997).

The epidermis is also the epithelial tissue most subject to continuous and intense mechanical stress. To ensure tissue integrity, keratinocytes rely not only on cell–cell adhesion but also on a wide variety of cell–matrix adhesion systems that anchor them to the connective tissue of the underlying dermis through a basal lamina of unique composition. These adhesion complexes involve integrins and hemidesmosomes which, together with the anchoring filaments formed by kalinin, tie keratinocytes of the basal epidermal layer to the lamina lucida of the basal lamina. More deeply, kalinin filaments provide an attachment for the lamina lucida to the lamina densa, which in turn is connected to the underlying stroma of the dermis through a network of anchoring fibrils which consist of type VII collagen (Yancey, 1995; Borradori and Sonnenberg, 1996; Fuchs, 1997). Thus, the epidermis is firmly attached to the upper dermis through a complex set of interactions involving cell adhesion molecules and specialized components of the extracellular matrix.

The variety and molecular complexity of these adhesion mechanisms account for the heterogeneity of human diseases caused by the autoimmune attack on the epidermis. From a theoretical standpoint, these diseases can be classified into two groups according to the site of the initial lesion. Either the interkeratinocyte or the keratinocyte-to-matrix adhesion system is targeted. Pemphigus is an example of a disorder of the first group (Fig. 12). In this disease, circulating autoantibodies, directed against either a cadherin-type of cell adhesion molecule or a cytoplasmic component of desmosomes, block the formation of these junctions which are constantly remodeled as keratinocytes migrate from the basal to the suprabasal epidermal layers under normal conditions. A decrease in the number of functional desmosomes, which are of paramount importance to provide adhesion between epidermal cells and confer upon them mechanical resistance, causes the dissociation of these cells and the ensuing loss of the epithelial barrier (Amagai, 1995). Keratinocytes that have lost contact with neighboring cells acquire a round shape as a consequence of the rearrangement of intermediate filaments, which are normally attached to the cytoplasmic plaques of desmosomes. In some cases, similar lesions can also be observed in other stratified epithelia, such as that which lines the oral cavity. In contrast, epithelial cells that are less dependent on desmosomes for adhesion, such as those in the basal epidermal layer, maintain an apparently normal shape and intercellular contacts as well as their attachment to the basal lamina.

Skin blistering diseases, which derive their name from the pathognomonic bullous lesions of the epidermis, are an example of the second group of cutaneous autoimmune disorders (Fig. 13), in which keratinocyte-to-extracellular matrix adhesion is attacked. In this group of diseases, circulating autoantibodies directed against one of the integrin subunits, a component of hemidesmosomes, anchoring filaments, or anchoring fibrils block their synthesis and/or proper assembly in either the basolateral membrane of keratinocytes or the basal lamina (Yancey, 1995; Fuchs, 1997). A decreased availability of these molecules, which normally provide a firm attachment between epidermis and dermis, is responsible for causing a cleavage at the dermoepidermal interface. In the case of autoantibodies against integrins or hemidesmosomes, the plane of cleavage occurs between keratinocytes of the basal layer and the lamina lucida portion of the basal lamina. Other bullous disorders are characterized by a plane of cleavage between the lamina lucida and lamina densa, when anchoring filaments are attacked, or between the lamina densa and the underlying dermis when anchoring fibrils represent the site of lesion. Under these conditions, cleavage results in the inadequate supply of nutrients to epithelial cells which become separated from the blood vessels present in the connective tissue of dermis. This defect, in turn, causes the rapid loss of a functional epithelial barrier, although cell–cell adhesion remains unaffected during the initial phases of the disease (Amagai, 1995; Yancey, 1995; Borradori and Sonnenberg, 1996; Fuchs, 1997).

Tumors

A crucial step of cancer progression is the ability of tumor cells to metastasize, that is, to colonize and destroy tissues distant from the original site of growth. To acquire greater mobility, epithelial cancer (carcinoma) cells must become loosely associated with neighboring cells and with the extracellular matrix. Loss of several components of

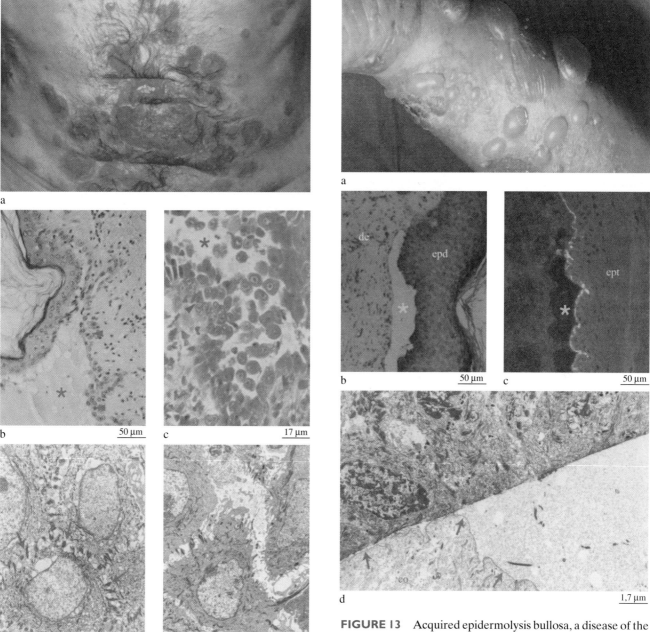

FIGURE 12 Pemphigus vulgaris, a disease of the cell–cell junction, complexes with dissociation of keratinocytes. (a) Photograph of extensive skin erosions, caused by the autoimmune attack, by antibodies directed against Dsg3, one of the cadherins that form the desmosome. (b and c) Histological analysis shows that these lesions are due to the detachment of epithelial cells in the suprabasal layers of the epidermis that split from those of the basal layer (arrowheads). The detached cells assume a round morphology and are separated by large spaces (asterisks) filled with inflammatory fluid. Noticeably, the cells of the basal layer of the epidermis maintain their normal attachment to the basal lamina and the connective tissue of the underlying dermis. (d) Under the electron microscope, the phenotypically normal regions of the skin show numerous desmosomes (arrows) which connect the keratinocytes of the suprabasal layers of epidermis normally. (e)In contrast, in the lesioned regions, desmosomes are markedly reduced in number (courtesy of D. Salomon).

FIGURE 13 Acquired epidermolysis bullosa, a disease of the cell–matrix junction, complexes with detachment of the epidermis. (a) Photograph of the epidermis of a patient affected by this disease. The autoimmune attack by antibodies directed against kalinin, one of the molecules of the anchoring filaments which attach the epidermis to the basal lamina, causes multiple bullous lesions. (b) Histological analysis shows that these lesions are due to the detachment of the entire epidermis (epd), which appears otherwise normal, from the underlying dermis (de) and from the consequent accumulation between the two tissues of inflammatory fluid (asterisk). (c) Fluorescence microscope photograph of the detached epidermis after incubation with antibodies directed against bullous pemphigoid antigen 1, a protein that allows for the adhesion of hemidesmosomes to the basal lamina. It can be seen (fluorescence) that the lamina lucida has retained its normal attachment to the basal membrane of epithelial cells (ept). (d) Under the electron microscope, examination of the side of a bullous lesion reveals that the lamina densa (arrows) has detached from the lamina lucida. Note that a normal adhesion is maintained between the lamina densa and the underlying connective tissue (co) and also between the keratinocytes of the epidermis (epd) (courtesy of D. Salomon).

adhesion and junctional systems is implicated in the invasiveness of solid tumors, whereas maintenance of cell–cell and cell–matrix junctions contributes to reducing the metastasizing potential of transformed cells. To escape from the tumor, carcinoma cells cross the basal lamina, as a result of the secretion of proteases which destroy some of its components. Then, cells enter into hematic or lymphatic circulation and colonize new territories, following the interaction with endothelial cells which line the blood vessels of the target organs. It is obvious that cell recognition and communication via adhesion molecules and connexins play a key role in these steps of tumor progression (Yamasaki and Naus, 1996).

Many studies have demonstrated a correlation between the loss of cadherin/catenin-dependent cell adhesion and the invasiveness of human cell lines derived from different tumors. Others have found a strong correlation between the adhesive function of cadherins and prognosis, as well as with certain clinical parameters such as the differentiation state of the tumor cells. For example, E-cadherin expression is reduced in basal cell carcinomas of the skin, in lobular breast carcinomas, and in colorectal adenocarcinomas. Recently, mutations in the gene encoding E-cadherin have been identified in 50% of cases of diffuse-type gastric tumors (Birchmeier, 1995). These mutations are either deletions of variable length of the protein or single amino acid substitutions destroying the Ca^{2+} binding sites of E-cadherin. In the first case, the functional consequences have not been determined, whereas the loss of the ability to bind Ca^{2+} should compromise the adhesion function of E-cadherin. Finally, the unexpected discovery of gain-of-function mutations of β-catenin in some forms of colon tumors indicates that this protein may also be considered an oncogene (Morin et al., 1997).

The concept that the maintenance of the integrity of adhesion complexes is necessary to control cell proliferation implies the existence of a messenger system that allows the coordination of cell growth. In parallel to the signaling code provided by cadherins and integrins, gap junction channels represent an ideal means to transfer the information needed to coordinate the growth rate of individual cells with that of their neighbors. Although a causal relationship has not been demonstrated, there is ample evidence of an inverse correlation between the extent of intercellular communication and the rate of cell growth, which points to the participation of gap junction channels in tumorigenesis. In particular, perturbation of junctional communication among tumor cells and between tumor and normal cells has been reported (Yamasaki and Naus, 1996). The loss of the latter form of communication, which has also been observed in vivo, may allow neoplastic cells to escape from the growth control of nontransformed cells, although it is likely that the inhibition of connexin channels is the result rather than the cause of the oncogenic transformation. Mutations in some connexin genes have been

found in cell lines and human tumors, but these events are rare and reflect a genetic polymorphism. In addition to communication defects, another argument in favor of the involvement of connexins in carcinogenesis has emerged from the demonstration that, in addition to tumor promoters such as phorbol esters, several oncogenes also inhibit cell communication. The latter include retroviral oncogenes such as v-src, DNA viral oncogenes such as SV40 (Simian virus) large T antigen and HPV (human papillomavirus) 16-E5, and cellular oncogenes such as c-src and c-erbB2. Connexins have also been regarded as tumor suppressor genes because they allow the passage of growth regulatory signals which inhibit cell proliferation. One model postulates that such signals are generated by a small number of scattered cells and then are passed to the rest of the cell population via gap junction channels. Evidence in support of this theory has come from the demonstration that transfection of connexins in transformed cells is generally accompanied by the restoration of cell coupling among neoplastic cells and a reduction of cell growth in vivo (Yamasaki and Naus, 1996; Vozzi et al., 1997).

Junctions and New Therapeutic Strategies

An important consequence of the relationship between connexins and cancer is the possibility of amplifying the efficacy of antitumor gene therapy due to the presence of intercellular communication. The strategy is based on the introduction in tumor cells of a gene which, once expressed, renders the transduced cells sensitive to a pharmacological treatment that kills them. For example, cells transfected with the herpes simplex thymidine kinase gene become sensitive to the antiviral agent ganciclovir. This guanosine derivative is phosphorylated by the viral enzyme and can then intercalate into DNA, thus causing the death of proliferating neoplastic cells. Unfortunately, the efficacy and specificity of such a treatment are hampered by the small percentage of cells that can be successfully transfected. In certain experimental models in vivo, only 10% of cells express the viral thymidine kinase, but the reduction of the tumor mass observed after ganciclovir treatment is dramatic enough to suggest an effect on the entire population of cells (Culver et al., 1992). Because the phosphorylated derivative of ganciclovir has a molecular mass (approximately 300 Da) that makes it permeable to connexin channels, these results have been interpreted as evidence that damaged cells can transfer toxic molecules to their neighbors most likely through connexins. Thus, a bystander tumor cytotoxicity has been observed in vitro only if tumor cells retained functional intercellular channels. Together, these data suggest that the presence of cell–cell communication could exert a dual negative regulation on tumor growth by contributing to inhibit cell proliferation and by extending the diffusion of toxic metabolites to neoplastic cells.

Another interesting therapeutic approach involving intercellular communication exploits a similar amplification mechanism mediated by connexin channels to extend the beneficial effect of the correction of a cellular defect by gene therapy. Cystic fibrosis is caused by a functional deficit that results from mutations in the gene encoding the cystic fibrosis transmembrane regulator (CFTR), a protein chiefly involved in transport of chloride ions. Transfection with wild-type CFTR of epithelial cells isolated from the upper respiratory airways of cystic fibrosis patients is followed by a restoration of ion transport, although only a small fraction of cultured cells express the wild-type protein. The most plausible explanation for these findings is, again, the possibility that normalized cells can exchange chloride ions with neighboring cells through gap junction channels, thus correcting the deficit of the entire population despite the presence of the mutated, inactive CFTR in the vast majority of cells (Johnson *et al.*, 1992).

···▼

Conclusions

Epithelial cells exploit multiple systems to recognize each other and adhere to both neighboring cells and the extracellular matrix. Each cell expresses a specific combinatorial set of adhesion molecules, and each cell also possesses receptors to sense various chemical signals. Similar to receptors that generate intracellular signals, adhesion molecules transduce signals within and between cells. Adhesion molecules are necessary, but not sufficient, to provide stability to the barrier they help form. Once they have ensured the functions of recognition, migration, and adhesion, which are essential to initiate the building plan of a tissue, adhesion molecules become necessary for both the formation of junctions and the interaction with the cytoskeleton which stabilizes tissue structure. As a result, epithelial cells become polarized and develop specialized functions in distinct membrane domains, whose maintenance is dependent on contact with other cells and the environment. Elucidation of the proteins implicated in these processes has allowed their manipulation *in vivo* and *in vitro* and the definition of the rules that underlie cell–cell recognition, adhesion, and the formation of intercellular junctions. Adhesion systems and junctional complexes play a pivotal role as structures that generate, receive, and integrate extracellular as well as intra- and intercellular signals. In addition, the abnormal function of junction-associated proteins is responsible for several human diseases. The integration of complementary experimental approaches is needed to clarify the molecular basis of these pathophysiological processes and to investigate feasible therapeutic solutions.

References Cited

ADAMS, J. C. (1997). Cell adhesion — Spreading frontiers, intricate insights. *Trends Cell Biol.* **7,** 107–110.

AMAGAI, M. (1995). Adhesion molecules. I: Keratinocyte–keratincyte interactions; Cadherins and pemphigus. *J. Invest. Dermatol.* **104,** 146–152.

BIRCHMEIER, W. (1995). E-cadherin as a tumor (invasion) suppressor gene. *BioEssays* **17,** 97–99.

BORRADORI, L., and SONNENBERG, A. (1996). Hemidesmosomes: Roles in adhesion, signaling, and human disease. *Curr. Opin. Cell Biol.* **8,** 647–656.

BRUZZONE, R., and RESSOT, C. (1997). *Eur. J. Neurosci.* **9,** 1–6.

BRUZZONE, R., WHITE, T. W., SCHERER, S. S., FISCHBECK, K. H., and PAUL, D. L. (1994). Null mutations of connexin 32 in patients with X-linked Charcot–Marie–Tooth disease. *Neuron* **13,** 1253–1260.

BRUZZONE, R., WHITE, T. W., and GOODENOUGH, D. A. (1996). The cellular Internet: On-line with connexins. *BioEssays* **18,** 709–718.

CLARK, E. A., and BRUGGE, J. S. (1995). Integrins and signal transduction pathways: The road taken. *Science* **268,** 233–239.

CULVER, K. W., RAM, Z., WALLBRIDGE, S., ISHII, H., OLDFIELD, E. H., and BLAESE, R. M. (1992). *In vivo* gene transfer with retroviral vector-producer cells for treatment of experimental brain tumors. *Science* **256,** 1550–1552.

DRUBIN, D. G., and NELSON, W. J. (1996). Origins of cell polarity. *Cell* **84,** 335–344.

EDELMAN, G. M., and CROSSIN, K. L. (1991). Cell adhesion molecules: Implications for a molecular histology. *Annu. Rev. Biochem.* **60,** 155–190.

FUCHS, E. (1997). Of mice and men: Genetic disorders of the cytoskeleton. *Mol. Biol. Cell* **8,** 189–203.

GARROD, D., CHIDGEY, M., and NORTH, A. (1996). Desmosomes: Differentiation, development, dynamics, and disease. *Curr. Opin. Cell Biol.* **8,** 670–678.

GUMBINER, B. M. (1996). Cell adhesion: The molecular basis of tissue architecture and morphogenesis. *Cell* **84,** 345–357.

HUBER, O., BIERKAMP, C., and KEMLER, R. (1996). Cadherins and catenins in development. *Curr. Opin. Cell Biol.* **8,** 685–691.

JOHNSON, L. G., OLSEN, J. C., SARKADI, B., MOORE, K. L., SWANSTROM, R., and BOUCHER, R. C. (1992). Efficiency of gene transfer for restoration of normal airway epithelial function in cystic fibrosis. *Nature Genet.* **2,** 21–25.

KIM, S. K. (1995). Tight junctions, membrane-associated guanylate kinases and cell signaling. *Curr. Opin. Cell Biol.* **7,** 641–649.

KIRKPATRICK, C., and PEIFER, M. (1995). Not just glue: Cell–cell junctions as cellular signaling centers. *Curr. Opin. Genet. Dev.* **5,** 56–65.

MAINIERO, F., PEPE, A., WARY, K. K., SPINARDI, L., MOHAMMADI, M., SCHLESSINGER, J., and GIANCOTTI, F. G. (1995). Signal transduction by the α_6/β_4 integrin: Distinct β_4 subunit sites mediate recruitment of Shc/Grb2 and association with the cytoskeleton of hemidesmosomes. *EMBO J.* **14,** 4470–4481.

MEDA, P. (1996). The role of gap junction membrane channels in secretion and hormonal action. *J. Bioenerg. Biomembr.* **28,** 369–378.

MEDA, P., PEPPER, M. S., TRAUB, O., WILLECKE, K., GROS, D., BEYER, E., NICHOLSON, B., PAUL, D., and ORCI, L. (1993). Differential expression of gap junction connexins in endocrine and exocrine glands. *Endocrinology* **133,** 2371–2378.

MORIN, P. J., SPARKS, A. B., KORINEK, V., BARKER, N., CLE-

VERS, H., VOGELSTEIN, B., and KINZLER, K. W. (1997). Activation of β-catenin-Tcf signaling in colon cancers by mutations in β-catenin or APC. *Science* **275,** 1787–1790.

NICHOLSON, S. M., and BRUZZONE, R. (1997). Gap junctions: Getting the message through. *Curr. Biol.* **7,** 340–344.

PAUL, D. L. (1995). New functions for gap junctions. *Curr. Opin. Cell Biol.* **7,** 665–672.

RUOSLAHTI, E., and REED, J. C. (1994). Anchorage dependence, integrins, and apoptosis. *Cell* **77,** 477–478.

SCHWARTZ, M. A., SCHALLER, M. D., and GINSBERG, M. H. (1995). Integrins: Emerging paradigms of signal transduction. *Annu. Rev. Cell Dev. Biol.* **11,** 549–599.

SHAPIRO, L., FANNON, A. M., KWONG, P. D., THOMPSON, A., LEHMANN, M. S., GRÜBEL, G., LEGRAND, J.-F., ALS-NIELSEN, J., COLMAN, D. R., and HENDRICKSON, W. A. (1995). Structural basis of cell–cell adhesion by cadherins. *Nature* **374,** 327–337.

SHEPPARD, D. (1996). Epithelial integrins. *BioEssays* **18,** 655–660.

SIMON, A. M., GOODENOUGH, D. A., LI, E., and PAUL, D. L. (1997). Female infertility in mice lacking connexin 37. *Nature* **385,** 525–529.

TAKEICHI, M. (1991). Cadherin cell adhesion receptors as a morphogenetic regulator. *Science* **251,** 1451–1455.

TIMPL, R. (1996). Macromolecular organization of basement membranes. *Curr. Opin. Cell Biol.* **8,** 618–624.

TSUKITA, S., and FURUSE, M. (2000). The structure and function of claudins, cell adhesion molecules at tight junctions. *Ann. N.Y. Acad. Sci.* **915,** 129–135.

VOZZI, C., ULLRICH, S., CHAROLLAIS, A., PHILIPPE, J., ORCI, L., and MEDA, P. (1995). Adequate connexin-mediated coupling is required for proper insulin production. *J. Cell Biol.* **131,** 1561–1572.

VOZZI, C., BOSCO, D., DUPONT, E., CHAROLLAIS, A., and MEDA, P. (1997). Hyperinsulinemia-induced hypoglycemia is

enhanced by overexpression of connexin 43. *Endocrinology* **138,** 2879–2885.

WHITE, T. W., PAUL, D. L., GOODENOUGH, D. A., and BRUZZONE, R. (1995). Functional analysis of selective interactions among rodent connexins. *Mol. Biol. Cell* **6,** 459–470.

YAMADA, K. M., and GEIGER, B. (1997). Molecular interactions in cell adhesion complexes. *Curr. Opin. Cell Biol.* **9,** 76–85.

YAMASAKI, H., and NAUS, C. C. (1996). Role of connexin genes in growth control. *Carcinogenesis* **17,** 1199–1213.

YANCEY, K. B. (1995). Adhesion molecules. II: Interactions of keratinocytes with epidermal basement membrane. *J. Invest. Dermatol.* **104,** 1008–1014.

General References

ALBERTS, B., BRAY, D., LEWIS, J., RAFF, M., ROBERTS, K., and WATSON, J. D. (1994). *Molecular Biology of the Cell,* 3rd ed. Garland, New York.

BISSELL, M. J., *et al.* (1993). Form and function in the epithelia. *Sem. Cell Biol.* **4,** 157–236.

CEREIJIDO, M., *et al.* (1992). *Tight Junctions.* CRC Press, Boca Raton, FL.

CITI, S., *et al.* (1994). *Molecular Mechanisms of Epithelial Cell Junctions: From Development to Disease.* Landes, Austin, TX.

EDELMAN, G. M., CUNNINGHAM, B. A., THIERY, J.-P., *et al.* (1990). *Morphoregulatory Molecules.* Wiley, New York.

HART, I., HOGG, N., *et al.* (1995). *Cell Adhesion and Cancer.* Cold Spring Harbor Laboratory Press, Cold Spring Harbor, NY.

MARSH, J., GOODE, J. A., *et al.* (1995). *Cell Adhesion and Human Disease.* Wiley, New York.

SOSINSKY, G. E. (1996). Molecular organization of gap junction membrane channels. *J. Bioenerg. Biomembr.* **28,** 297–309.

TRINKAUS, J. P. (1984). *Cells into Organs: The forces That Shape the Embryo,* 2nd ed. Prentice Hall, Englewood Cliffs, NJ.

EDOARDO BONCINELLI

Department of Biological Research and Technology
San Raffaele Scientific Institute
Milan, Italy

There are whole families of genes that regulate development. These genes may be divided into two categories: those that codify transcription factors and those that codify proteins that affect intercellular communication. The study of the genes that regulate development constitutes an article in itself, with future chapters day by day becoming more autonomous and substantial. In this article, genes of the first category will mainly be considered, with the emphasis on the genes that codify homeoproteins. Emphasis will also be placed on the historical process that led to the discovery of these genes and to clarifying their role first in Drosophila *(fruit fly) and then in vertebrates.*

The Genes That Control Development

Development and Its Various Phases

The development of a multicellular organism is characterized by a long series of biological events leading up from the zygote, that is, the fertilized egg cell, to the adult individual capable of fending for itself, interacting, and reproducing (Boncinelli, 1994). In particular, the development of animals will be considered here because they are the best known and most studied, although an increasing amount of knowledge is accumulating on the development of plants. The very first stages of development (Fig. 1) are common to the majority of animal species (Gilbert, 1994). Thus, immediately after the fertilization of the egg cell by the spermatozoon, a phase called segmentation or cleavage begins which consists of the rapid succession of a certain number of consecutive cell divisions. The effect of these cell divisions is to produce a few dozen cells, called blastomeres, which constitute the first anlage of the embryo. Generally, blastomeres are smaller than the zygote because the cell divisions in the segmentation phase have the effect of sharing among the various blastomeres the enormous mass of cytoplasm present in the zygote.

The zygote is a very large cell because the egg cell is in turn generally far larger than the other cells of the organism. The egg cell contains in its cytoplasm many molecules that are the depositories of precious information for guiding the earliest development of the future embryo as well as reserve nutritive substances that will enable the zygote to proceed at a quick rate through the first cell divisions without any energy sources of its own. The first cell divisions in

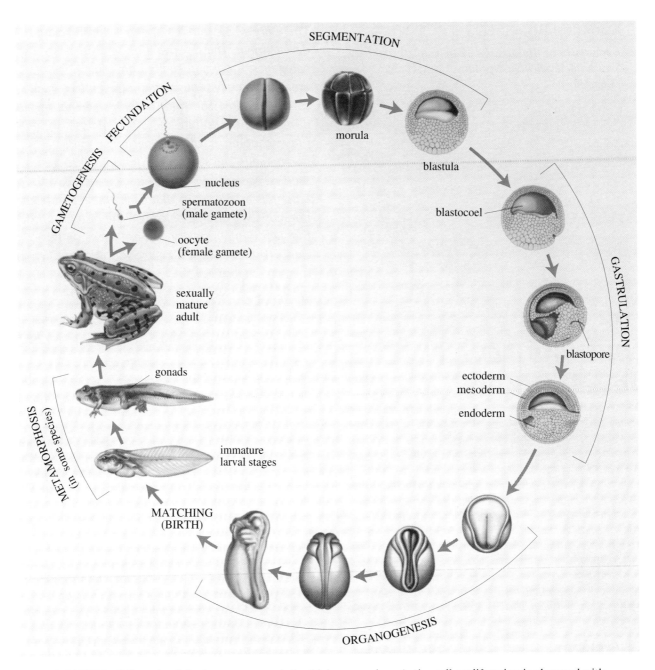

FIGURE I Life cycle of the frog, a representative higher organism. Active cell proliferation is observed without an increase in the size of the embryo during the segmentation phase. The fundamental architecture of the organism is defined during gastrulation, and structures and organs develop during the organogenesis phase. Metamorphosis is characteristic of some (indirect development) species, whereas it is absent in others (direct development species).

the segmentation phase generally take place without any overall growth in the mass of the embryo. The cytoplasmic material contained in the zygote simply subdivides in the cytoplasm of the first blastomeres, whereas each of these will obviously have its own nucleus and thus its own set of chromosomes. The complex of the very earliest blastomeres

sometimes takes the name of morula, whereas at the end of this stage the few cells present arrange themselves to form a little hollow ball, called blastula, which has a central cavity called the blastocoel.

At a certain moment the cells of the blastula move into action. Many of those originally outside of the blastula move inside, whereas many small cell layers become delaminated and form germ foliations. This basic process is called gastrulation. It is in this stage that the foundations are laid for the future individual's architecture. During gastrulation, in fact, the three fundamental embryonal layers called ectoderm, mesoderm, and endoderm can be identified. The external layer, the ectoderm, will later develop into the epidermis and the nervous system. The inner layer, the endoderm, will develop into the internal lining of the digestive tract and of the internal organs associated with this, such as the pancreas and the liver, whereas the intermediate layer, the mesoderm, will develop into the connective tissue of the cartilage and the bones, the muscles and such vital organs as the heart, the kidney and the gonads, as well as the blood cells. From the morphological standpoint, gastrulation transforms a hollow little ball into an elongated structure showing its anteroposterior and dorsoventral polarity and which is formed in turn by a bundle of parallel tubes. What emerges from gastrulation is generally a miniature organism already possessing many of the features of its definitive form.

In some species a series of transformations occur successively in the ectodermic layer, culminating in the formation of the neural tube. This phase is sometimes referred to as neurulation. At the same time as the formation of the neural tube, the clear structuring of the main organs of the future organism starts. This lengthy phase of more or less conspicuous morphogenetic events is generally known as organogenesis. At the end of this process, there will be an organism potentially capable of surviving which only has to grow slightly more and many of its parts need to mature. The cells composing its body and enabling it to function are called somatic cells or cells of the somatic line, whereas the cells that will produce the future gametes, thus contributing toward the perpetuation of the species, are called germ cells or cells of the germinal line.

The Biology of Development

Up to a certain point of their development, the embryos of the various organisms follow a relatively uniform ontogenetic pattern. Beyond these first stages, however, the variations of biological development become practically infinite. The embryos of organisms belonging to different taxonomic subdivisions, and at times also those belonging to the same subdivision, pursue different strategies in their development. It is thus extremely difficult to make a reasonably comprehensive treatment of this series of disparate events. There is no doubt, however, that all multicellular organisms develop, and that it would not be reasonable to think that they did so following radically different laws and mechanisms. Indeed, the concept of a profoundly unitary pattern of development mechanisms is constantly gaining accep-

tance. The recognition of this unitary process has led to the birth of a substantially new, autonomous discipline, developmental biology, which now coexists with embryology and its various more specialized branches. This new discipline (Boncinelli, 1994; Gilbert, 1994) aims directly at understanding the primary elements and the basic processes of the mechanism of development in the various systems, and attempts to enunciate, where possible, general principles of development. This is a tremendous task but also an exciting challenge. In tackling this task it can sometimes be of use to consider the chain of events occurring under the name of development as the overall outcome of three classes of processes regarded as the most elementary: growth, cell differentiation, and morphogenesis (Boncinelli, 1994).

The most evident parameter in an organism that is developing is the increase in its size. This phenomenon is often indicated with the term embryonic growth. Growth, which generally becomes evident at the end of gastrulation, is largely identified with the increase in the number of cells forming the embryo. A necessary condition for growth, and hence for the entire embryonic development, is cell proliferation which occurs in every phase of development and has to be controlled with extreme precision. Traditionally, however, this increase in the number of cells in the embryo is not considered an essential factor specific to growth because, even though it leads to an increase in the mass of the developing embryo, it indicates nothing more precise about its form and its structure.

On the contrary, the phenomenon of cellular differentiation is assuredly closer to the heart of the problem. This term is understood to mean the complex of events which, starting from an initial homogeneity, cause the cells of an embryo to become differentiated from one another. This process is considered so important that the term differentiation and development is often used as though it were a single thing. We have seen that all the cells of an adult organism derive from a single fertilized cell, the zygote. The first blastomeres deriving from cell divisions in the segmentation phase are generally very much like one another, whereas the cells of the adult definitely differ greatly among themselves and belong to hundreds of different histological types. During the different phases of development various groups of cells must therefore diversify, i.e., become differentiated. The complex of these diversifications is called cellular differentiation. The molecular and cellular mechanisms from which this process stems constitute the primary subject of study of modern molecular and cellular embryology.

Cellular differentiation as a whole can be regarded as consisting of two distinct processes or two distinct phases of the same process. The first phase is represented by everything that happens in the period between fertilization of the egg cell and the separation, or segregation, of the various cell lines. In the second phase, differentiation may ideally be broken down into a series of differentiations of specific cell lines, such as the blood cell line, the nerve cell line, and the line of muscle cells or cartilages. A process of terminal differentiation and maturity of the cells is completed in this phase.

The events that characterize the second phase of cellular differentiation are much easier to analyze than the events that characterize the first phase, and many of these specific processes of differentiation can be studied directly in the laboratory, separately from the system represented by the live animal. This often means cultivating *in vitro* specific cell lines that represent as closely as possible the process to be studied. Experiments of this type offer a wonderful opportunity for studying cellular and molecular events that accompany the differentiation of the various types of tissues. The majority of the knowledge that has accumulated in the past few decades on the mechanisms of gene regulation during development derives from this type of analysis.

Here, I diverge a moment to consider the general lesson to be learnt from studies of this type: Cellular differentiation is basically a question of differential genic expression. The nucleus of the cells of any tissue contains the same genetic complex, that is, exactly the same set of genes. What distinguishes a cell of the striated muscle from one of the liver, and these two from a nerve cell, is the fact that in the first cell a certain group of genes will be active (i.e., expressed) as opposed to another group of genes in the second cell and yet another group in the third one. In addition, if in the cells of a given tissue a gene is not active, its presence is absolutely useless and everything takes place as though said gene were not there. The problem of differentiation is thus transformed basically into that of control of genic expression. In this relatively simple way we can explain the majority of the biological phenomena observable in a higher organism. Naturally, this does not mean that the molecular details of this process are understood, and indeed it is the main contemporary biological problem.

The problem has in fact been shifted. The question is now the regulation of the activity of the various genes. This activity has to be very closely controlled not only cell by cell but also moment by moment. In fact, a cell, in response to given signals received from other cells or through the blood circulation, can also decide to suddenly "switch on" one of its passive genes or to silence another active one. What does it mean to say that a gene is active or "expressing," as it is termed in modern biological parlance? A gene may be said to be active only when it is able to synthesize its protein product, naturally in the right amount and at the right moment. For this to happen, however, several key passages are necessary: On the basis of the gene's DNA, an RNA molecule [the messenger RNA (mRNA)] has to be conveyed, that is, synthesized. This molecule of RNA must then leave the nucleus and migrate to the cytoplasm of the cell where,

finally, it must be translated, meaning that it has to act as a template or model for the synthesis of the corresponding protein.

Control of a gene's activity may be exercised at the level of each of these passages even though it is clear that the first is the most important. If, in fact, the mRNA molecule is not produced, all the other steps are absolutely useless. It is a fact that the majority of studies are conducted on the first passage in this chain, namely, on the synthesis of a mRNA molecule, a fundamental biological process called transcription. If a gene is transcribed, this does not necessarily mean that it is active, but if it is not transcribed it will certainly not be active. The transcription of a gene, and thus the first phase of its expression, is governed by the interaction of a certain number of nuclear protein factors called transcription factors or factors of transcription with specific regions of said gene called control regions or regulating regions.

The undoubted success of studies of this type, however, must not let one conclude that differentiation in general can be explained in this way. There is a whole chain of events that lead the cells, or rather a line of differentiation of the cells, from their relatively undifferentiated state as a blastomere to the almost terminally differentiated state of myoblast, erythroblast, or pre-B lymphocyte. Much study remains to be carried out on this chain of events, although there is no reason to think that the principles and the mechanisms implied in these early events are different than those forming the basis of terminal differentiation, which brings the erythroblast to the erythrocyte, the myoblast to the myotubule, and the pre-B lymphocyte to the adult B lymphocyte.

Can we say that the problem of the development of multicellular organisms is similar to the problem of cellular differentiation? As often occurs with questions of this type, the answer is first and foremost a semantic one. In this specific case it depends on how the term differentiation is defined. Biologists who tend to give an all-inclusive definition to this term would without hesitation give an affirmative answer to the question we have posed. Others prefer a narrower definition of the term and, consequently, they are inclined to think that development constitutes something more than cellular differentiation, even if this undoubtedly represents the basic cellular phenomenon of the whole process.

According to the logic of the second interpretation, the events and the processes for cellular differentiation therefore include almost everything under the concept of development, but something remains outside of this concept. What remains outside is essentially the concept of a precise arrangement and spatial organization of the cells belonging to the body's various cellular lines. In substance, the problem of the origin and the elaboration of the biological form remains outside the concept. This concept is often included in the more technical one of morphogenesis, understood as creation of forms, that is, of complex biological structures organized in space and time. This problem, known as pattern formation, can concern the formation of particular structures, such as a butterfly's wing, the hand of a primate, or the eye of a cephalopod, or of global structures, such as a limb or the overall body plan of an organism (Slack, 1991). Following this line of thought it is conceivable to envisage that in a hand, an eye, or an ear there is something more than an aggregate of tissues that have differentiated in the correct way and at the correct moment. One has the impression that in a complex structure another type of biological information different than differentiative or histological information is present: information which guides a given cell population toward its given histological destiny. In this connection we speak of positional information. The concept of positional information (Wolpert, 1969, 1989, 1993) — that is, of the type of biological information that confers on a cell of a developing organism a given, continuously reiterated positional direction — was introduced in the 1970s to give a less imprecise and less hazy scientific basis to the belief that an orderly biological structure represents something more than a complex of tissues. Since then, this concept of positional information has proved to be anything but imprecise and evanescent, and recently a great deal has been learned about the genetic control it exercises.

Genes and Development

All biological processes are controlled by genes, and development is certainly no exception. Indeed, the identification of gene families and networks that play a role in this process is a conquest of this century. The genetic and the genetic–molecular approaches have recently joined descriptive embryology and experimental embryology in contributing toward supplying an ever clearer and more precise picture of the whole ontogenetic phenomenon. Along this line, the fundamental conclusion has been reached that cellular differentiation as a whole is entirely explainable in terms of differential genetic expression. Part of this genetic program has been revealed and many genes have been identified and given a name, and in some cases they have been assigned a function.

In general, every gene is a development gene. Even the gene that controls the color of the body of the fruit fly (*Drosophila*) or the one that controls the color of the eyes of a human being can be considered as development genes because their action in some way influences an individual's development. It is clear, however, that these do not play a decisive role in regulating development processes, and their study does not bring us significantly closer to understanding the phenomenon of development. However, genes

do exist that control the presence or the absence of a limb, and genes also exist whose action is essential for the presence of motor neurons in the spinal cord. Genes of this type may correctly be termed genes that control development. It is advisable to use this term in reference to those genes that have at least one of the following two characteristics: their dysfunction causes conspicuous defects in the development of the individual possessing them or their behavior supplies practical and theoretical instruments essential for a better understanding of the overall development process.

The number of genes qualifying for this definition is probably about 100, whereas only 30 years ago there were only 1 or just a few. This expansion of our knowledge increased quickly in the 1980s and is progressing at such a frenzied rate that certain negative aspects ensue. For example, there is no room for an in-depth analysis of the results acquired and often even researchers feel frustration at their inability to keep up with the impetuous advances in knowledge. It is impossible to deal with these developments in detail; I limit discussion here to summarizing their principal concepts.

For this purpose, it is best to consider the genes in question as belonging to two major categories: that of regulator genes which codify the syntheses of nuclear located proteins, which act as transcription factors, and that of genes that codify diffusible or quasi-diffusible factors, among which are the growth factors, or membrane proteins, which regulate intercellular communication. In fact, the genes of these two categories interact continuously, especially during development. The expression of a regulator gene is almost always regulated on the basis of intercellular communication, whereas it is obvious that the various effectors of this communication are in turn controlled by the products of the regulator genes. Moreover, these two types of factors represent the basic elements of growth and in particular of the regulation of the genetic expression within each single cell which is the innermost essence of development. We can see what happens in the cells that originate from the first divisions of the zygote. To date, only two mechanisms are known (Fig. 2) that are capable of controlling the expression of genes in their very earliest phases of embryo development: location, with consequent segregation, of cytoplasmic determinants and induction (Horvitz and Herskowitz, 1992; Boncinelli, 1994).

From the standpoint of the single cell, we begin with the simple hypothesis that a gene of a given cell always expresses itself until different circumstances occur. These circumstances are unleashed by biological signals that can come from inside the cell or from the outside. In the first case, the cell behaves as though it were a deaf and blind entity whose evolution is simply the consequence of a preestablished plan written somewhere within it. In the second case, the state of the cell evolves on the basis of biological

SEGREGATION
OF CYTOPLASMIC
DETERMINANTS

INDUCTION

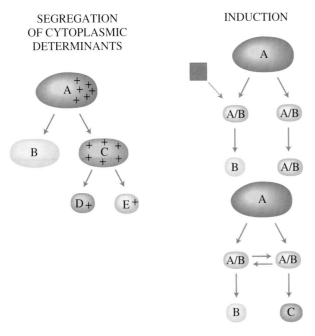

FIGURE 2 Generation of asymmetry among cells in the first stages of development: segregation of cytoplasmic determinants and induction. A cytoplasmic determinant (+) segregates into cell C and not cell B. The two daughter cells of cell A will therefore differ from each other as well as from the mother cell. In successive divisions the process may repeat and the determinant will be diluted. The two schematic drawings showing induction illustrate instead the inductive effect of a third cell on one of the two daughter cells (left) and the mutual inductive effect between the two daughter cells (right). It is assumed that without the induction effect the two daughter cells would be equal.

information from other cells or from the surrounding world in general. In each cell, at every moment, signals of both types are present. The two processes may be separated only for study purposes to better understand what is happening. According to the organisms, tissues, and stages, a signal from inside or one from outside may at times predominate (Boncinelli, 1994; Gilbert, 1994).

It is worthwhile to clarify this fundamental point. It is not difficult to understand how a signal from outside can reach the cell. This signal must sooner or later reach a molecule or a complex of molecules which enter the nucleus, occupy a specific region of the genome, and dictate their instructions to one or more genes. For each gene these instructions concern being "switched off" or "switched on" and, if it must be "on" and therefore express itself, the intensity with which it has to do this.

Understanding the mechanism of regulation from the inside appears somewhat more complex. We can think of information contained in a more or less concealed manner in the cytoplasm of certain cells and absent in those of oth-

ers. At the moment of the division of the embryo cell that contains it, this principle, called cytoplasmic determinant, can pass to only one of the two daughter cells, and from this to part of the progeny. The cytoplasmic determinant can act immediately or may be preserved in a latent state for cellular generations. The simplest way of viewing this principle is to consider a protein product sequestered in a given cellular compartment or a mRNA molecule that is translated only at the appropriate moment. Leaving this state of latency must be regulated by other endocellular determinants, by a change in the conditions outside of the cell, or simply by time. We can imagine, for instance, that the molecule that stops the expression of the principle is transmitted through the generations in such a way as to become progressively diluted. At the outset there is a quantity more than sufficient to keep the manifestation of the latent principle at bay. At each cell division its concentration diminishes until, at a certain moment, it will not be sufficient to carry out its task. Viewed in this way, it is not surprising that the location of certain cytoplasmic determinants should start in the egg cell, the asymmetrical original cell par excellence. Indeed, an extreme instance of influence of cytoplasmic determinants is provided by the location and the abundance of vitelline material (yolk) inside the egg. According to the various species, the egg cell can either contain little yolk and thus be almost symmetrical or contain much yolk. In the second case, the yolk will be located in a clearly asymmetrical manner at one end, which is known as the vegetal pole, of the egg cell. Diametrically opposed to this is the animal pole, where the nucleus is generally located (Gilbert, 1994).

Much of the study of development mechanisms currently concerns the identification of the genes belonging to one or the other of the two categories just mentioned. The known development genes are becoming numerous and their number is increasing. Fortunately, almost all of them can be grouped together in genic families on the basis of their structural similarity, which is very often (but incorrectly) called homology. Not always do genes belonging to the same family also perform similar tasks, although this does happen frequently.

This article examines these two categories of development genes and the transcriptional factors in detail. It is convenient to follow the history of the various discoveries with regard to the development genetics of *Drosophilia melanogaster* (the fruit fly). The impressive progress made by genetic technology has facilitated the study of this insect and constitutes one of the most important model systems of development biology this century. All of this would not have been so important, however, had it not been discovered that much of what is true for this insect is true also for many more complex organisms, including man. I will proceed with order and will present what has been learned about the genetic control of this insect's development, starting with the regulation of its main axes, the anteroposterior axis and the dorsoventral one.

Genes That Control the Development of the Fruit Fly: A Historical Outline

The discovery of the genes that control development is one of the most fascinating stories of this century and one of the most exciting ones in the epic of biology (Boncinelli, 1994). In particular, a family of genes has been discovered that controls the general setup and the organization of animals' bodies, which therefore decides where the head "should go" and the shoulders, the chest, and so forth. These are the genes that impose that the chest shall be between the head and the abdomen in all animals and not, for example, at one end of the body. These genes were originally discovered when studying the fruit fly and have since been found in every higher organism. On the basis of these discoveries we are beginning to understand how the structure of our own bodies is regulated, from the head down to the feet.

The story starts at the end of the 1950s: An American researcher, E. B. Lewis, possessing a pioneering spirit and considerable tenacity, made an analysis of many bizarre mutations (Lewis, 1964, 1978, 1985). This research was continued through the 1960s and 1970s by various geneticists, outstanding among them the Spaniard Antonio García-Bellido (García-Bellido *et al.,* 1973; García-Bellido, 1977). The genetic analysis of these phenomena culminated, in the early 1980s, with the monumental work by the German Christiane Nüsslein-Volhard and the Swiss Eric Wieschaus (Nüsslein-Volhard and Wieschaus, 1980) with the discovery of all the possible mutants of this type. At this point molecular biology made its debut and, toward the end of the 1980s, the situation evolved rapidly with a firework-like suc-

cession of discoveries in which genetic analysis proceeded in step with molecular studies. The rest is present-day history (Lawrence, 1992).

There are very many mutants observable in the fruit fly. The merit of E. Lewis was that he foresaw that some of these were special. The fruit fly normally has two wings, as do all Diptera, but after a mutation four-winged ones or wingless ones may be born (Fig. 3a), or they may be born with a pair of legs on their head in place of their antennae (Fig. 3b) or even in place of their nose. These are mutations regularly inherited and not irreproducible phenomena. Consequently, at the base of such mutations as these there must be some genes whose task is that of preventing such events from taking place and of ensuring that the legs, wings, and antennae are all in their proper place.

Upon closer inspection, these alterations reveal a common logic. What is transformed in these mutants is not the single anatomical structure but the whole segment of the body to which said structure belongs. The fly that is born with two pairs of wings is born in reality with two thoraxes, in which the wing is undoubtedly the most spectacular structure. The one born with legs in place of the antennae shows a total transformation of the region to which the antennae belong into that to which the legs normally belong. All this can be observed easily in a fly because it is a clearly segmented animal. Its body consists of a regular succession of 14 segments, 3 of which form the head, 3 the thorax, and 8 the abdomen. These segments are quite similar but not identical. The fly is a highly evolved insect deriving from an ancestor probably consisting of many almost identical body segments. In the course of their evolution these segments have progressively differentiated, but not to the point of not being able to preserve their affinity or, rather, their homology.

In particular, the fruit fly's thorax consists, in an antero-

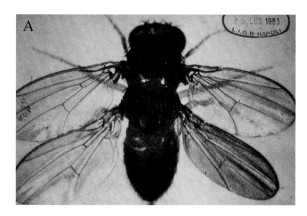

FIGURE 3 Examples of homeotic mutations in *Drosophila*. (a) Mutant *bithorax* which displays a transformation of the third thoracic segment into an excellent imitation of the second one. The most evident result is the presence of two pairs of wings instead of just a single pair. (b) Mutant *Antennapedia* showing the presence of a pair of legs at the top of the head instead of the usual antennae.

posterior direction, of three segments called T1–T3. Each of these thoracic segments has a pair of legs; insects, as known, have six legs. The second segment also has a pair of wings; the third one does not have wings but rather two membranous structures which represent a sort of advance in evolution of the second pair of wings characteristic of the standard insect which have been called the balancers. It is easy to observe that in the mutant fly, possessing four wings, the whole third segment of the thorax has been transformed into a copy of the second one. Consequently, the mutated insect has four wings, although this is not its only characteristic feature. The wingless insect shows a transformation of the second segment into a copy of the third one; consequently, it has no wings but four small balancers.

The peculiar aspect of these mutations, therefore, is the transformation of the identity of a whole segment of the body into that of another one. If mutations exist, there must also be genes responsible for the identity of the various segments: of the second segment of the thorax or of the third, of that of the antennae, or of that of the upper lip. These mutants enable us to have a glimpse of quite a new world — the world of the genes that issue a number of fundamental biological instructions, such as "Here there must be a leg and there a wing, here there must be an antenna and there the nose."

Toward the end of the nineteenth century mutations had been observed in which repeated structures, such as legs, wings, or phyllids, were transformed into other similar repeated structures. The name homeosis (from the Greek *homoios*, meaning "similar to" or "like") was given to the phenomenon of the transformation of a repeated structure, being a case of the transformation of "like into like" (Bateson, 1894). The importance of this type of transformation had already been realized but, as often occurs in the history of science, 100 years had to pass before this concept was fully developed. The mutants in question were also within this category and were therefore called homeotic mutants. The corresponding genes were called homeotic genes. The homeotic genes of the fruit fly are thus those genes that control the identity of the single body segments of this insect.

The importance of these genes in the hierarchy of the regulator genes was immediately clear. Furthermore, genetic analysis showed that at least some of these occurred, one close to the next, on one of the chromosomes of the fruit fly not in random order but one which reproduced the order of the body segments that each of these controls: The homeotic gene found farthest to the right on the chromosome controls the body segment at the front end, whereas the one farthest to the left controls the segment at the rear end. This fascinating, mysterious phenomenon was given the name of colinearity (Lewis, 1978, 1985).

For all these reasons the homeotic genes of the fruit fly were among the first to be isolated when molecular cloning became available. Compared with the results of purely genetic analysis, the molecular analysis of these genes led to a certain amount of confirmation and a certain number of novelties. The confirmations concern the correspondence between the regions of the body controlled by a given homeotic gene and the domain of expression of its genic products. The genic product of a gene that controls the first thoracic segment is effectively present in the anlage of the first thoracic segment and only there, and the genic product of a gene that controls the first abdominal segment is effectively present in the anlage of the first abdominal segment. Another confirmation concerns the phenomenon of colinearity. The eight principal homeotic genes of the fruit fly are effectively located, one against the other, in a limited physical region of chromosome 3 and their order reflects that of the body segments that each of these controls and in which their genic products can be identified.

Among the novelties, at least one has been of great importance. In the early months of 1984 it was discovered that these eight genes had a very similar structure and in particular contained a region of about 180 nucleotides practically identical in all eight. This region was called homeobox and the protein domain of approximately 60 amino acids that this codifies was called the homeodomain. Within a short time it was demonstrated that the homeodomain is a protein region capable of recognizing specific nucleotidic sequences on the DNA and of bonding therewith. The protein that contains a homeodomain is therefore a transcriptional factor capable of bonding with specific DNA sequences and regulating the expression of a certain number of genes. These eight regulator genes thus codify that same number of proteins belonging to a single family of transcriptional factors.

Today, various protein motives are known that are capable of bonding with specific DNA sequences, such as the so-called zinc fingers and the HLH (helix–loop–helix) domains, but the first protein motive of the sort was precisely the homeodomain, identified for the first time in the genic products of the homeotic genes of the fruit fly. The most important fact is that genes similar to the homeotic genes of the fruit fly are present in all higher organisms, including the frog, rat, and man.

General Background

The body of the fruit fly, like that of any other higher organism, has a major anteroposterior axis, a dorsoventral axis, and a bilateral right–left symmetry. Study of the development mutants of the fruit fly has taught us much about the genetic determination of these axes. Here, I discuss the genetic determination of the main axis, the anteroposterior one, which runs from the head to the tail. However, first it is necessary to note certain peculiarities of this insect's development. Immediately after fertilization, a series of nuclear divisions start in the zygote, without cytodieresis. The early embryo of this fly therefore appears as a syncytium and this feature clearly differentiates it from the majority of other systems and in particular from those of mammals. After approximately 12 nuclear divisions, the nuclei are largely at the periphery of the embryo which thus assumes the shape of a hollow ball called the blastoderm. This hollow ball does not yet possess cellular membranes between one nucleus and another and so is called syncytial blastoderm. Very soon, however, septa form, dividing the various nuclei and the ball changes from syncytial to multicellular. We then speak of cellular blastoderm. At this stage the embryo of the fruit fly possesses practically all its positional information: these first 6000 cells are the precursors of all the internal and external organs of the fruit fly and much of what will occur has already been determined. About 3 h have passed since fertilization. After this first phase all other phases are more or less the same as those in other systems. Gastrulation occurs and after a few hours the embryo is transformed into a small larva, which will subsequently turn into the winged adult animal. For the correct development of the anteroposterior axis of this insect the concerted action is required of three categories of genes (Fig. 4): maternal genes, segmentation genes, and homeotic genes (Lawrence, 1992).

In the egg of the fruit fly, the information necessary for the determination of the future front region and of the back region is already present. This information is introduced into the egg cell in the form of protein synthesized by many genes present in the gene complex of the mother, called maternal genes or genes of maternal effect. The result of the action of these genes is to establish, inside the egg cell, a gradient of an anteriorizing substance, whose maximum concentration in the anterior part decreases toward the posterior part, and that of a posteriorizing substance with a posterior maximum and a gradually decreasing trend toward the anterior part.

After fertilization, the zygotic genes come into play, i.e., those proper to the genetic patrimony of the future individual, and among these are the segmentation genes. On the basis of the positional information already present in the egg cell, the genes of this second category subdivide the future embryo into a certain number of potential segments along the anteroposterior axis. These segments must be of the correct number and in the correct order. This is a potential subdivision because at the moment of the action of these genes there is no possibility of distinguishing a cell from one region of the embryo from that of another. The segmentation genes act in three successive waves, subdividing first the embryo into three or four fairly large, partly overlapping areas and then into seven double strips and finally into 14 segments. Each of these waves is controlled by a particular class of segmentation genes. The first subdivision is the work of the *gap* or cardinal genes. The second

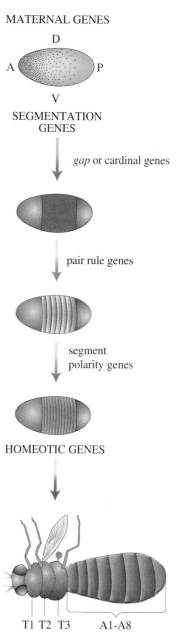

FIGURE 4 General scheme of the three categories of genes implicated in the specification of positional information along the anteroposterior axis of *Drosophila*. The maternal genes control the presence in the egg of substances necessary to specify, from the beginning, where the future regions will be localized. A, anterior; P, posterior; D, dorsal; V, ventral. The segmentation genes regulate the potential subdivision of the embryo into the correct number of segments. They act in three successive stages, controlled by three classes of segmentation genes. The *gap* or cardinal genes subdivide the embryo into at least three large areas. The equal module genes subdivide it into seven strips (and seven interstrips). The segmental polarity genes regulate the final 16 (14 + 2) segments. Finally, the homeotic genes give each single segment the appropriate biological structures; that is, they control the final biological identities of the different segments.

one is controlled by the pair rule genes and the third by the segment polarity genes.

After these various specifications have taken place, in the correct way and at the correct moment, the genes of the third category come into play, i.e., the homeotic *genes*. These control, along the anteroposterior axis, the final identity of the various segments which, at the end of the process, possess their own structures and organs. In this way, the second thoracic segment, T2, will be provided with a pair of wings, whereas the segment T3 will be wingless but have a pair of balancers. A homeotic gene exists, called *Ultrabithorax*, which controls the determination of the identity of the single segments T2 and T3. A single mutation of this gene can transform T3 into T2 or T2 into T3. The result will be a fly with four wings because T3 believes that it is T2 or a fly wholly without wings because T2 believes that it is T3. The interpretation of these phenotypes is quite simple. The gene in question determines at what point on the insect's anteroposterior axis T2 or T3 should be located; as a consequence of its decisions numerous other genes, known as executor genes or cytodifferentiating genes, implement successive operational plans. A homeotic gene resolves the dilemmas of wing/no wing and leg/no leg, whereas the other genes execute the corresponding project. Homeotic genes are therefore regulator genes of a higher order in the hierarchy capable of controlling the action of many other operator genes.

Maternal Genes of the Anteroposterior Axis

The action mechanisms of the three categories of genes, starting with the maternal ones, are discussed here (St. Johnston and Nüsslein-Volhard, 1992). The early embryo of the fruit fly can be ideally subdivided into five regions, three segmented and two nonsegmented. The three segmented regions occupy the central part of the embryo and consist of a region that later becomes the segmented portion of the head, another that will become the thorax, and a third that will become the abdomen. At the two ends of the embryo are the two nonsegmented regions: the anterior one, called acron, and the posterior one, called telson. The maternal genes that superintend the correct anteroposterior polarity of this insect belong to three different groups (Table 1): (i) those of the anterior group which provide for the location and the development of the anterior regions, (ii) those of the posterior group which supply the correct information for the location and the development of the posterior regions, and (iii) those of the terminal group which supply the correct information for the development of the extreme nonsegmented parts.

In a mutant of a gene of the anterior group, it is as though the information needed to form the head and the thorax had been lost. The segmented region of the mutant consists

TABLE I

Maternal Genes of Anteroposterior Polarity

GROUP	CONTROL GENES	CONTROLLED GENES
anterior	exuperantia swallow	bicoid
posterior	oskar vasa valois tudor staufen pumilio	nanos
terminal	trunk torso l(1)Nasrat l(1)pole hole	gene Y

only of an abdomen, whereas the acron is transformed into a telson. In a mutant of posterior type the damage is more limited: The abdomen is missing. In the mutants of the third group, both the acron and the telson are missing. The complementary nature of the three phenotypes is well illustrated by the constitution of the double mutants. An embryo deriving from a double-mutant mother for a gene of the anterior group, such as *bicoid*, and for a gene of the posterior group, such as *oskar*, is without the whole of the segmented region and consists essentially of two contrasting telsons. An embryo deriving from the combination of a mutation of a terminal gene, such as *torso*, with one of an anterior gene, such as *bicoid*, will consist of only an abdomen, whereas an embryo derived from the combination of a mutation in *torso* with one in *oskar* will be all head and thorax without terminal regions. A triple mutant will have no structural articulation at all.

For some of these mutants experiments have been conducted to transfer a certain quantity of the egg cell's cytoplasm. These experiments have demonstrated that in the fruit fly's egg two basic principles are present regulating its anteroposterior polarity: an anteriorizing principle and a posteriorizing principle. The normal product of the gene *bicoid*, i.e., the bicoid protein located in the anterior part, constitutes the anteriorizing principle operating in the egg cell of the fruit fly. The egg obtained from a mutant mother for the gene *bicoid* is lacking in its anteriorizing principle and can easily accept one from outside. To confirm these conclusions, it can be verified that the anterior plasma of a normal egg can induce an anteriority located in any region of a *bicoid* mutant embryo. If this is transplanted, for example, in a central region of the mutant egg, an embryo is obtained with a cephalic structure located in a central position and flanked by two opposed thoracic regions.

The gene *bicoid* has been isolated and the nature of its product has been analyzed. This is a transcriptional factor containing a homeodomain which thus acts to switch other genes on or off. The isolation of the gene has also provided a direct demonstration that its product is the anteriorizing principle operating in the fruit fly's egg. The results of these experiments confirm its identity as an anteriorizing factor.

Consider the mechanisms that lead to the location of this principle in the future anterior region of the embryo. The fruit fly's egg develops with the help of a certain number of accessory cells, called host cells, which are found at its anterior tip (Fig. 5). The mRNA of the gene *bicoid* is synthesized inside these cells and then passes to the developing ovocyte, in which it is located in the anterior part throughout the maturation phase of the egg cell. It seems that it remains anchored to the cytoskeleton of the anterior region of the egg as a result of the concerted action of the other maternal genes of the anterior group, such as *exuperantia* and *swallow*.

After fertilization, starting from this mRNA the corresponding protein is synthesized. In the embryo, at the syncytial blastoderm stage, a graduated distribution of the bicoid protein is observed, and its concentration, which is high in the anterior region, gradually diminishes in the posterior part. This occurs because the protein source is located at the anterior end of the embryo, whereas a uniform downgrading mechanism of the protein product operates throughout its length. When passing from the syncytial blastoderm to the cellular blastoderm the various nuclei bring with them information deriving from the concentration of bicoid protein to which they have been locally exposed and the cells that will enclose them will be influenced and their fate will be specified. The bicoid protein, with its concentration gradient along the anteroposterior axis of the early embryo of the fruit fly, is one of the few solidly documented instances of a morphogene. This term refers to a biological substance generally distributed according to a gradient and capable of influencing the specification and differentiation of the cells of a given region of the embryo on the basis of its local concentration.

The gene *bicoid*, with its mechanism of action, constitutes a real paradigm in the panorama of the genetic regulation of development. Far less is known, though, about the maternal genes of the posterior group. What is known, however, is sufficient to understand that the system of these posterior maternal genes operates with quite a different logic than that of the anterior maternal genes. It appears that the posteriorizing principle may be identified in the product of the gene *nanos*. This protein is not a transcriptional factor but carries out its action interfering with the work of the genes of the anterior group, in practice preventing the translation into protein of their RNA messengers. The group of terminal maternal genes is required for the formation of both the anterior and the posterior nonsegmented regions. The formation of the telson, for instance, does not depend on the genes of the posterior group but on those of the terminal group.

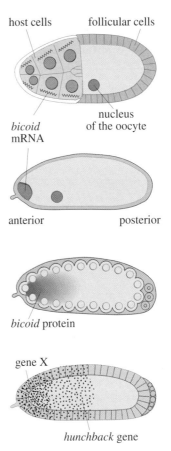

FIGURE 5 Molecular mechanism leading to the creation of the anteroposterior gradient of the protein product of the gene *bicoid* inside the fruit fly zygote. The protein is abundant in the anterior and becomes gradually more rarefied toward the posterior region because the messenger on which it is synthesized is localized and trapped in the anterior tip of the oocyte. This messenger was synthesized in the 15 host cells, found anterior to the oocyte, and subsequently transported to this region of the oocyte. Note that the oocyte is surrounded by other cells, called follicular cells, which control the dorsoventral polarity. After fecundation, the mRNA for the *bicoid* gene is translated and the anteroposterior gradient of the corresponding protein forms. As a consequence of this gradient, another anteroposterior gradient is established in the developing embryo which will last much longer — that of the cardinal gene *hunchback*. It is supposed that the anterior tip region is the zone of action of another unknown gene called X.

The Segmentation Genes

After the egg cell has been fertilized, producing the zygote, the segmentation genes come into play (Ingham and Martinez Arias, 1992). These genes can also be subdivided into

TABLE 2		
Segmentation Genes		
Gap or key cardinal genes	Giant	
	hunchback	
	huchebein	
	knirps	
	krüppel	
	tailless	
Paired module genes — primary	even-skipped	
	hairy	
	runt	
Paired module genes — secondary	fushi tarazu	
	paired	
	odd-skipped	
	odd-paired	
	sloppy-paired	
Segmental polarity genes	armadillo	
	cubitus interruptus	
	engrailed	
	fused	
	gooseberry	
	hedgehog	
	pathed	
	wingless	

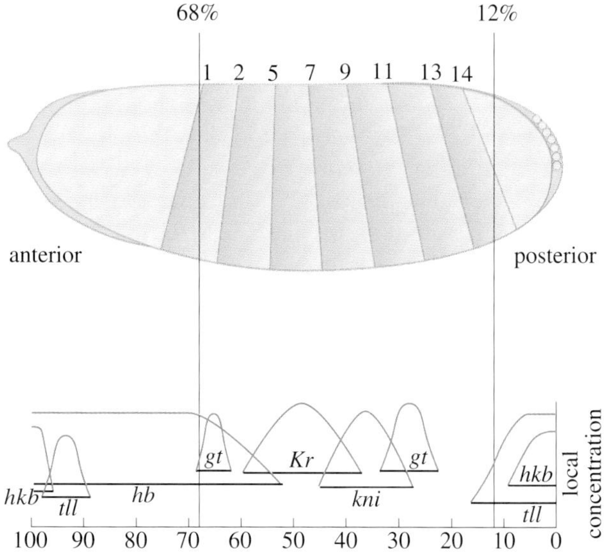

FIGURE 6 Distribution of the products of the various *gap* genes along the axis of the embryo. The total length is set at 100, starting from the posterior tip. The profiles of the various *gap* genes indicate the local concentration of the corresponding products. In addition to the three best known genes — *hunchback* (*hb*), *Krüppel* (*Kr*), and *Knirps* (*kni*) — the recently discovered cardinal genes *huckebein* (*hkb*), *tailless* (*tll*), and *giant* (*gt*) are indicated (see Table 2).

three groups: *gap* or key genes, pair rule genes, and segment polarity genes (Table 2). Mutations in the *gap* genes lead to larvae in which extensive regions are lacking at one of the two extremities or in a more central position. Mutations in the paired module genes produce larvae lacking a certain number of more restricted regions corresponding to the even segments (second, fourth, sixth, etc.) or to the odd ones (first, third, fifth, etc.). Mutations in the segmental polarity genes regard the identity of the anterior or the posterior compartment of each segment comprising the body of the larva.

This series of phenotypes obviously refers to a series of genes. The first *gap* genes studied were *hunchback, Krüppel,* and *knirps.* These genes are also called key genes because they truly play a key role: They have to subdivide the embryo into three major regions — the midanterior region, the central region, and the posterior region (in this order). It was subsequently determined that these three *gap* genes were not sufficient to account for the entire subdivision process of the embryo and three or four more key genes were identified and were given even more fanciful names.

Many of these *gap* genes have been isolated and it has been possible to determine their domains of expression. These coincide with embryo regions that subsequently give rise to regions of the body of the larva absent or significantly reduced in the respective mutants (Fig. 6). The first

expression of these *gap* genes is directly controlled by the maternal genes. It can be stated with a fair degree of approximation that where there is an abundance of anteriorizing factor the *hunchback* gene is activated, where there is a prevalence of posteriorizing factor the *knirps* gene is activated, and where there is a more or less equal concentration *Krüppel* is activated. Once they are activated, the *gap* genes begin to control each other. This dual system of control leads to a closely regulated expression of the various key genes in well-defined regions of the embryo, and this leads in turn to a subdivision of the embryo into four or five broad regions along the anteroposterior axis.

Many studies have been carried out on the mechanisms through which a given key gene expresses itself in a limited region of the embryo under the control of the product of the various material genes. The expression of the *hunchback* gene, for example, is controlled by the distribution of the *bicoid* product. It has been seen that the protein codified by *bicoid* directly activates *hunchback.* At the anterior end of the embryo there is a greater concentration of the *bicoid* product and it is there that *hunchback* is activated, whereas proceeding toward the posterior end the concentration of the *bicoid* product is too low for it to activate *hunchback.* Therefore, it can be understood that the continuous distribution of a molecule according to a gradient may translate into the localized expression of a second molecule.

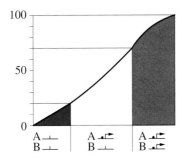

FIGURE 7 Conceptual model of the French flag. The gradient of concentration of a morphogene can be used for the threefold division of a territory if it is thought that the cells composing it can respond to the local concentration of the morphogene with a threshold response mechanism. In this case, it has been supposed that the response derives from two cellular genes (A and B) with two different activation thresholds. The activation of a gene is symbolized by arrows, whereas the solid circles indicate activator transcriptional factors switched on, directly or indirectly, by the morphogene.

This objective is achieved if the local concentration of the morphogene in question, in this case the bicoid protein, is interpreted with a threshold response mechanism. A given phenotype will appear only where the concentration of the morphogene is greater than a certain threshold value and possibly less than a second threshold value (Fig. 7). In the case of the pair of genes in question, a direct measurement of the entity of the response can be obtained which is represented by the extent of the target gene's domain of expression. In fact, as the bicoid protein concentration of a factor of 2 increases, an extension is observed in the domain of the expression of *hunchback* toward the posterior pole equal to 10% of the total length of the embryo. This extension is a discreet process: The posterior edge moves in leaps. In this way the gradual distribution of the morphogene is transformed into a total or zero response of the product controlled by it. The key genes are then subject to a dense network of reciprocal regulations which give rise to a tangle of domains of expression regulated in time and space. This network of domains in turn controls the activation of the even module genes.

Mutations caused by pair rule genes alter the pattern of alternate body segments: one yes and one no. The main role of these genes is to distribute the single cells of the embryo and to assign them to one of the 14 final segments. Mutant larvae in the gene *fushi tarazu*, for instance, die quickly because they lack typical parts of the odd abdominal segments, such as A1, A3, and A5; mutant larvae in the gene *even-skipped* instead lack the even segments.

Some of these genes have been isolated and the study of the distribution of their products during development has shown that they are mainly present in seven thin transversal strips of cells, very similar to belts, located in the central portion of the embryo. Each of these thus determines a particular striped pattern in the central part of the blastoderm (Fig. 8). The motives of the various genes are phased so as to supply the cells with a continuum of reference positional information. The gene *fushi tarazu,* for example, is expressed in seven strips three cells thick separated by empty regions five cells thick. Along the main axis of the embryo there will therefore be some cells that receive the information conveyed by *fishu tarazu* and other cells that do not receive it. It is not known what this information consists in, nor is the reason known why the segments where one of these genes is expressed are fated to disappear in the absence of their product.

At this point the segmentation genes of the third group — that is, those of segment polarity, the most famous of which is *engrailed* — also come into action. Their main role is to subdivide each segment into an anterior compartment and a posterior compartment. Thus, the 7 strip–interstrip complexes discussed previously become 14 and the primordium of the 14 segments constituting almost the whole of the segmented part of the fly appears. For example, the *engrailed* mutants show defects of the posterior compartments of all the segments. The current interpretation implies that the product of this gene in some way marks the cells of all the posterior compartments. Lacking this marking, the cells of the posterior compartments lose their identity.

The absence of the *engrailed* product does not influence the destiny of the cells of the anterior compartments but drastically alters that of the cells of the posterior compartments. In fact, those that do not contain it can, for example, cross the borders of the compartment. It is believed that ordinarily the cells of the posterior compartment possess membrane properties that do not allow them to mix with cells of the anterior compartment, but little or nothing is known about the mechanism that regulates this process.

The True Homeotic Genes

As the result of the action of the genes of the preceding categories, every single cell of the blastoderm knows its location along the anteroposterior axis of the embryo (Fig. 8). It is presumed that this takes place through some sort of appeal made by the cell: "Is gene 1 switched on? Yes. Is gene 2 switched on? No. Is gene 3 switched on? Yes." On the basis of the result of this appeal, a given homeotic gene can be activated in that cell. The expression of the appropriate homeotic gene will then determine the phenotype of that cell, which will contribute together with the others toward the overall identity of the segment to which it belongs.

A mutation of a homeotic gene consists of a more or less perfect transformation of structures belonging to one segment into those belonging to another segment. The homeotic genes (McGinnis and Krumlauf, 1992) are sometimes

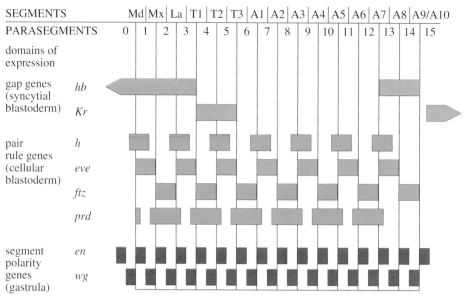

SEGMENTATION GENES

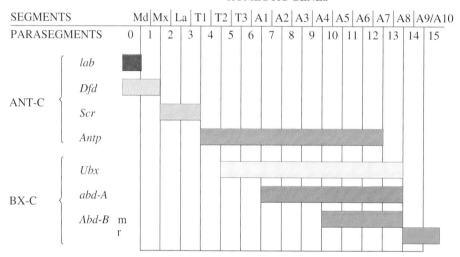

HOMEOTIC GENES

FIGURE 8 Diagram of the domains of expression of the various genes which regulate the anteroposterior axis of the fruit fly: (top) segmentation genes; (bottom) homeotic genes. For simplicity, only two cardinal genes [*hunchback* (*hb*) and *Krüppel* (*Kr*)], four equal module genes [*hairy* (*h*), *even-skipped* (*eve*), *fushi tarazu* (*ftz*), and *paired* (*prd*)], and two segmental polarity genes [*engrailed* (*en*) and *wingless* (*wg*)] are shown. The domains of expression of two different products of the gene *Abd-B* are indicated which are called morphogenetic (m) and regulative (r).

called selector genes because their action entails a choice among various alternative segmental identities. On the basis of this definition, there are many fruit fly mutations which may be termed homeotic, but the homeotic genes par excellence are the eight genes located inside two contiguous genic complexes (Fig. 9) called the bithorax complex (BX-C) and the Antennapedia complex (ANT-C). A

peculiar characteristic of these genes is that they are located next to each other in two relatively limited genomic regions. The situation currently found in such evolved insects as the fruit fly is considered atypical. The normal situation would be one in which all eight of these genes are aligned in a single complex called HOM-C.

In the BX-C complex three of these genes, *Ultrabithorax*,

FIGURE 9 Detail of two loci, ANT-C and BX-C, containing the homeotic genes and illustration of the concept of colinearity.

Abdominal-A, and *Abdominal-B*, are found in a line next to each other, whereas the genes *Labial, Proboscipedia, Sex combs reduced, Deformed,* and *Antennapedia* are found next to each other in the ANT-C complex. The two complexes take their names from the two best known homeotic mutants, *bithorax* and *Antennapedia*. Mutants of the first type present a more or less perfect transformation of part of the thoracic segment T3 into the homologous part of the segment T2, and from this a four-winged fly results. Mutants of the second type instead present a pair of legs in place of the antennae on top of the head.

Consider the two most important homeotic genes, i.e., *Ultrabithorax* and *Antennapedia. Ultrabithorax* controls the segment T3 plus the posterior compartment of T2 and the anterior one of A1. Prior to the physical isolation of this gene various homeotic mutations located in this region of the body were known, and these bear special, fanciful names such as *bithorax, postbithorax, anterobithorax, thoraxoid, Ultrabithorax,* and *Contrabithorax*. Each of these mutations has its own particular phenotype, but it was surprising when it was observed that they all in reality alter the same gene, *Ultrabithorax*.

The differences between the various mutations were clarified with the isolation of the gene *Ultrabithorax* (Bender *et al.*, 1983). Only the mutation called *Ultrabithorax* consists of a structural lesion of the protein produced by the gene *Ultrabithorax*. All the other mutations occur in noncodifying regulatory regions of this gene and this is the real reason for their bizarre conduct and their fascination. Only in the mutation called *Ultrabithorax* is the product of the gene lacking. In all the others the product is present and

functional, only it is distributed anomalously during development both in time and in space.

It was thus discovered that the various regulatory regions of this gene guide the expression of its genic product in one part of the body rather than in another. As a consequence of mutational events in these regulatory regions, the product of *Ultrabithorax* is expressed in regions of the body not pertaining to it, whereas at times it is absent in those regions in which it should be present. For example, in a *postbithorax* mutant the level of expression of *Ultrabithorax* in the posterior compartment of the future balancer is equal to that which would pertain to the posterior compartment of the future wing, whereas in an *anterobithorax* mutant the level of expression of *Ultrabithorax* in the anterior compartment of the future balancer is equal to that which would pertain to the anterior compartment of the future wing.

From the location of the regulatory mutations the location of the corresponding control regions was found. Currently, there is an almost complete map of these control regions and of the compartments in which they direct the expression of the gene. It is interesting that recessive mutations of this genic unit entail anteriorizing homeotic transformations: A posterior segment acquires many characteristics of a segment located in a more anterior position (e.g., the transformation of T3 into T2). On the contrary, dominant mutations cause posteriorizing homeotic transformations (e.g., T2 is transformed into T3).

A dominant mutation of the gene *Antennapedia*, the first one to be studied and the most famous, also called *Antennapedia*, transforms the antennae of the fly into two thoracic legs of type T2. This could indicate that the gene *Antennapedia* controls the identity of the segment bearing the antenna. However, this is not the case. In fact, this gene essentially controls the identity of the posterior compartment of T1 and the anterior compartment of T2. This is a case of one of the many strange spontaneous mutations of the fruit fly in which, in fact, the codifying region of *Antennapedia* has ended up under the control of a regulatory region of a gene expressed in the head; this represents a case of ectopic expression. The thoracic leg originates from the region of the antenna due to the imaginal disk of the antenna which is always ready to renounce its nature and transform into a leg.

Many researchers, lead by Walter Gehring, constructed a transgenic fly in which the product of *Antennapedia* was caused to express more or less everywhere in the animal's body. The result was that the fly displayed a transformation of its antennae into mesothoracic legs. In this case, the "redesign of an animal's body scheme" was correctly predicted (Schneuwly *et al.*, 1987). This experiment and others prove that although the product of the homeotic genes is necessary to specify the various segments of the fly's body, it is often also sufficient to carry out this function. A different direction of the expression of one of these genes can lead to a different specification of one or more somatic regions and this then leads to a series of new histological determinations.

The Homeodomain and Its Preservation

The newest element emerging from the molecular analysis of the homeotic genes is the presence of a homeodomain fairly well preserved in all eight of the different proteins codified by these eight genes. This proteic domain, within the homeotic genes *Antennapedia* and *Ultrabithorax* and in the segmentation gene *fushi tarazu,* was identified for the first time in 1984 by two different groups of researchers, the first one led by Walter Gehring at Basel and the second led by Matthew Scott in Colorado. Today, it is known that the homeodomain is present in many other regulator genes of the fruit fly and in all higher organisms, from the most elementary worms to the vertebrates and man.

The homeodomain, therefore, is a proteic domain capable of recognizing specific nucleotide sequences of the DNA. A protein that contains a homeodomain is therefore called a homeoprotein and is a nuclear protein with transcriptional factor properties. Each gene that contains a homeobox has, a priori, many probabilities of being a regulator gene that codifies a transcriptional factor. The presence of a homeobox is a clear indication in this sense and has been the guideline for the discovery of newer regulator genes in the most diverse organisms, from yeasts to man. On the basis of this criterion, whole families of regulator genes containing a homeobox (i.e., homeogenes) have been identified.

The homeodomains codified by the homeogenes are generally very similar to each other and are preserved from species to species, but naturally there are differences between them. On the basis of their primary sequence the various homeodomains can be subdivided into many classes. The first class includes the homeodomains most similar to the prototype sequence constituted by the homeodomain present in the homeotic genes of the fruit fly. The homeogenes containing a homeodomain of this class constitute a genic family, the family of Hox genes, present in vertebrates and studied particularly in the mouse and man (McGinnis and Krumlauf, 1992). These homeogenes constitute a genic family on the basis of at least two criteria: They are located beside one another in relatively compact genomic regions, and they collectively specify the identity of the various regions of the body along the anteroposterior axis. These properties place them in immediate relation with the homeotic genes of the fruit fly. It is believed that these homeogenes are homologs of the homeotic genes belonging to the complexes BX-C and ANT-C of the fruit fly or to the complex HOM-C which combines them.

The Hox Genes

There are approximately 40 mammal Hox homeogenes, about four times the number of fruit fly homeotic genes. They are arranged in four different compact chromosomic regions called genic complexes or Hox loci, each containing about 10 genes. The genic complex Hoxa contains 11 genes and in man is found on chromosome 7: Hoxb, Hoxc, and Hoxd each contain 9 genes and in man they are located on chromosomes 17, 12, and 2, respectively. All these genes are transcribed in the same direction, from 5′ to 3′ of their own locus. The four loci can be aligned in such a way that the corresponding genes of the loci are arranged on the same vertical. It is usually said that the Hox genes, aligned on the same vertical, belong to the same group or subfamily (Fig. 10). In terms of evolution this means that the four loci originated by successive duplications from the same primitive locus and were subsequently distributed onto different chromosomes. During this process some of the genes originally present were lost, whereas no new gene has yet appeared.

The most interesting characteristic is that the four loci can easily be aligned on the complex HOM-C of insects. It is in fact easy to identify the homologous vertebrate genes of *Abdominal-B, Deformed, proboscipedia, labial,* and, although not as easily, those of the others. Each of the four loci can therefore be considered homologous to the complex HOM-C which contains the homeotic genes of insects. It can be concluded that an ancestral gene complex, containing at least four or five of these genes, must have existed in a common ancestor of insects and vertebrates, whose lines of evolution separated more than half a billion years ago. Confirming this supposition, a Hox locus containing five correctly aligned homeogenes has recently been identified in *Caenorhabditis elegans,* a nematode far more distant from man than even the fruit fly. It is incredible that these genes have remained together for tens or even hundreds of millions of years and have remained aligned in the same order. This means that it must be a phenomenon of great biological significance or else a "frozen" accident that dies hard.

The Hox genes are all expressed during the embryonal development of vertebrates; some are expressed in certain embryonal tissues and others in different sites, but all are expressed in the cells of the embryo's central nervous system. The domain of expression of each single Hox gene extends considerably along the anteroposterior axis with a very clear-cut front edge and a far hazier posterior distribution. Current functional data suggest, however, that the region of the body controlled by each of these genes is in reality a limited belt of tissues adjacent to the front edge of their domain of expression. Now, the regions controlled by

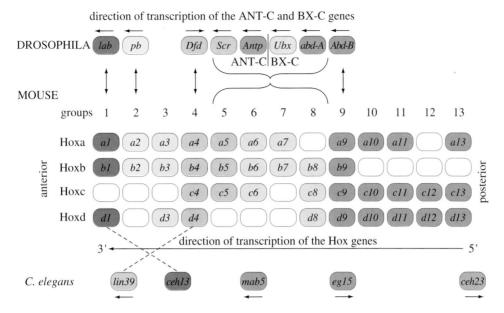

FIGURE 10 The genes HOM/Hox in the fruit fly, in vertebrates (mouse), and in the nematode *Caenorhabditis elegans*. The new enumeration of the Hox genes, which differs from that in many works prior to 1992, has been used. The genes on the left are expressed in the anterior region of the body and those on the right in the posterior region. The direction of transcription of the Hox genes in mammals is always the same, from right to left. The exact correspondence between the Hox genes of groups 5–8 and the *Drosophila* genes *Scr, Antp, Ubx,* and *abd-A* is not clear, whereas it is considered that the Hox genes of groups 9–13 all derive from a common ancestral gene very similar to *Abd-B*. The white boxes, although predicted by the scheme, do not correspond to any known Hox genes.

the various Hox genes also show the phenomenon of co-linearity observed for the homeotic genes of the fruit fly (Fig. 11). In this case, the rules of structural colinearity are even easier to enunciate because the Hox genes of the various loci are all transcribed in the same direction: Genes that are found toward the 3′ end of their respective loci are expressed further to the front than genes found gradually closer to the 5′ end of the same locus.

The scanning of the neural tube by the Hox genes is not uniform. There are regions that require less finish and regions that require a far more accurate distribution and specification. This is the case for the rhombencephalon or hindbrain, that part of the embryonal brain at the boundary between the brain proper and the spinal cord which subsequently gives way to the spinal bulb and to the medulla oblongata and from which almost all the cranial nerves will depart. In a very precise phase of the development of the rat and the chicken, it can be observed that the rhombencephalon is subdivided into eight morphological regions, called rhombomers, consisting of eight swellings separated by furrows.

It has been observed that the anterior edge of the domain of expression of some Hox genes coincides precisely with the border between some of these rhombomers being formed. The homeogenes located 3′ of the locus Hoxb, for example, scan the rhombencephalon in a regular, colinear manner (Fig. 11). The fifth-to-last gene is expressed in the spinal cord and the front edge of its expression coincides with the separation between rhombencephalon and spinal cord. The fourth-to-last gene is also expressed in the last two rhombomers, the seventh and the eighth; the third-to-last of the last four rhombomers is expressed from the fifth to the eighth; and the penultimate gene in the last six rhombomers is expressed from the third to the eighth. According to this pattern, the last gene at 3′, *Hoxa1*, should cover the whole of the rhombencephalon. For unknown reasons this does not occur: *Hoxa1* is expressed only in the fourth rhombomer and then along the spinal cord. What applies for the genes 3′ of the locus Hoxb applies also for the genes 3′ of the three other loci: The first four genes at 3′ of each of the four Hox loci subdivide into four regions of the rhombencephalon and the whole surrounding branchial region.

In the successive development phases, the roots of the cranial nerves and the dorsal spinal ganglia are positioned corresponding to a number of rhombomers. The rhombencephalon is therefore the sole region primarily segmented

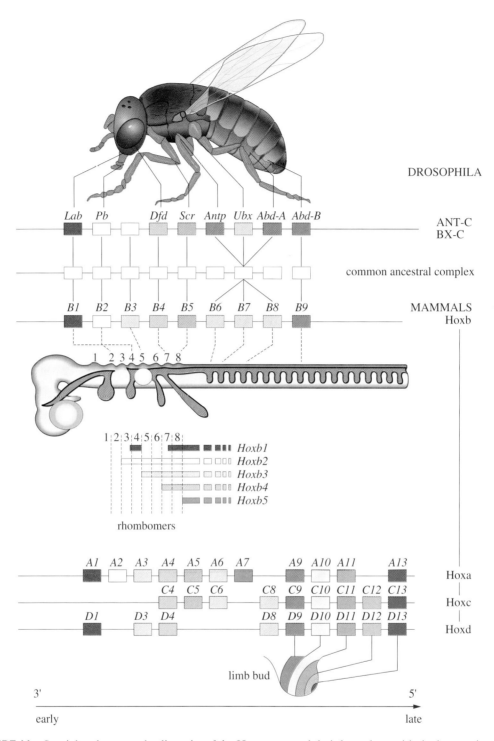

FIGURE 11 Spatial and temporal colinearity of the Hox genes and their homology with the homeotic genes of the fruit fly. The Hox genes of group 1 are the first to be activated during embryogenesis, whereas those of group 13 are the last. For simplicity, only the central nervous system of a mammalian embryo is shown with the principal cranic nerves and the rhombomers, indicated with numbers. The expression of the Hoxb genes in the eight rhombomers is shown in detail (middle). (Bottom) Knowledge of the expression of the genes *Hox9–d13* in the limb bud. Also shown is an anteroposterior colinearity.

of the embryonal nervous system of mammals and the expression of the Hox genes can be noted well before morphological signs of borders between one rhombomer and another appear. The Hox genes thus subdivide the region of the body corresponding to the rhombencephalon into distinct territories of subregions.

In the fruit fly the function of the homeotic genes and all other regulator genes can be studied by observing the respective mutants. In vertebrates, unfortunately, no spontaneous mutation concerning a gene of the Hox family has been identified. This could mean that a mutation of one of these genes has no effect or, probably, that it has such a disastrous effect that the animal literally cannot succeed in being born. Lacking spontaneous mutants, it has been sought to obtain more precise information on the function of these genes by having recourse to substitute genetics.

A certain number of transgenic mice bearing artificial mutations, by excess or insufficiency of function, of a number of Hox genes were produced. Artificial mutants by excess of function of a Hox gene showed defects of the axial skeleton and craniofacial malformations. Some transformations of the first cervical vertebrae of these mutants suggest transformations in a posterior direction as may effectively be expected of a homeotic mutation with increased function. For example, the first cervical vertebra appears as an imperfect copy of the second one.

Transgenic mice mutating through deficiency of function, destroying both copies of an endogenous gene, have also been obtained. Some conclusions of a general nature may be drawn from this type of research. First, the absence of the product of a given Hox gene produces generalized defects in different tissues and organs inside a relatively restricted region of the body, well defined and delimited along the anteroposterior axis. This region is located at the front end of the domain of expression containing the transcripts of said gene. Second, no clear overlapping can be observed between the belts of action of two contiguous Hox genes. Evidently, the overlapping of the domains of expression in embryos has no functional value, at least with regard to the most evident phenotypic features. It cannot be ruled out, however, that finer malformations, for instance, in the cellular architecture of the spinal cord, might have escaped observation. Third, as expected, a transformation of identity in the anterior direction is observed: For example, certain thoracic vertebrae could acquire characteristics of those further forward. Lastly, the mutants for two or more Hox genes that are on the same vertical show a far more severe phenotype than that of the single mutants. At least in part, the Hox genes that are on the same vertical, and which therefore belong to the same group, cooperate in regulating the same body region and in carrying out concurrent functions.

Can we term the transformations observed in transgenic mice mutating because of these genes homeotic? The answer is not univocal. At least in part, the identity of certain structures is modified in imitation of a more anterior structure of similar type, in the case of mutants due to deficiency of function, or of a more posterior structure of similar type, in the case of mutants due to excess of function, and this is particularly evident in the case of double mutants due to two genes belonging to the same group. In other cases, however, the transformation is not clearly of the homeotic type or, on the contrary, the suppression is observed of a given repeated body structure, as if segmentation genes were involved rather than homeotic genes. Certain differences thus exist between the mechanism of action of the Hox genes and that of the homeotic genes of the fruit fly. However, mammals are not clearly segmented organisms as are these insects. It can also be considered that the existence of four loci in place of just one serves to mitigate the strictly segmental effect of the various genes and that the continuity of our body regions derives from an overlapping of different superimposed and slightly staggered segmental sequences.

Recently, another type of colinearity displayed by genes belonging to various Hox genic complexes, the so-called temporal colinearity, has been brought to light. The Hox genes do not all become activated simultaneously during development. The first ones to be activated are those located 3' of the various loci, then those in a central position are activated, and finally those located 5' of their respective loci are activated. In the embryo of the rat, whose gestation lasts about 20 days, the genes located 3' are activated for the first time around the seventh day after conception, whereas those located 5' appear at least 2 days later. In the embryo of the toad, which takes 24 h to develop, 5 or 6 h pass between the appearance of the first Hox genes and that of the last ones.

The relation between topographic (or structural) colinearity and temporal colinearity of the Hox genes is not clear. Some believe that temporal colinearity is indeed the primary event and structural colinearity just a consequence of it. After all, development is nothing more than an orderly sequence of events in time which, after their temporal succession, are positioned in space.

The action of the genes belonging to the Hox complexes can be considered as follows: In a specific region of the nucleus of our cells there is a group of cells constituting as a complex a microrepresentation of our body or, in other words, of the sequence of the main parts of our body — a sort of homunculus dormant along the chromosome. At the appropriate moment all these genes swing into action, one after the other, at a measured rate. Their linear sequence on the chromosome takes place in the first instance as a temporal progression of the beginning of their action and subsequently as a spatial and positional succession of the various parts of the body. This microrepresentation of our body, this homunculus that we house in four of our chromosomes, simultaneously contains both a map and a clock.

Other Homeogenes

Mention must be made of at least three other genic families bearing a homeodomain: Pax, Emx, and Otx. The proteins codified by certain genes in the family *Pax* bear a homeodomain homologous to that contained in the gene *paired,* a fruit fly segmentation gene (Gruss and Walther, 1992). Again, in this case it is a question of regulator genes codifying transcriptional factors, with an important role in the development of vertebrates; as opposed to the case for the Hox genes, several spontaneous mutants for the various genes of this family are known. One of these is associated with the gene *Pax1* and corresponds to a spontaneous mutation known as *undulated.* Rats bearing this mutation have a malformation of the vertebral column which occurs distorted and with a zig-zag structure. Molecular analysis has shown that the product of *Pax1* is located in the embryo at the level of the various intervertebral disks throughout the length of the embryo. It can easily be understood that an alteration of the product of this gene or of its regulation can lead to a deformation of the vertebral column.

Many other members of the family are expressed along the whole anteroposterior axis and the majority of them are also active in the central nervous system. One of them, *Paxó,* has been found to be implicated in the murine mutation *small eye* and in the corresponding human mutation *aniridia,* whereas another, *Pax3,* is the basis of the mutation *splotch* in the rat and of the Waardenburg syndrome in man. In homozygote mice, through the mutation *small eye* it is noted that the eyes are absent as are many specific nasal structures, confirming the decisive role of the genes of this family in the development of various organs. However, in *splotch* rats malformations of the spinal column and of the heart are observed. Lastly, transgenic rats, in which both copies of the gene *Pax2* have been destroyed, display numerous defects in the brain and in the sense organs and are born without kidneys.

The Hox genes are expressed in the embryo in all the regions posterior to the rhombencephalon and the branchial arches. Their expression does not concern the head and even less the brain. Moreover, even the corresponding fruit fly genes operate mainly in the embryonal region corresponding to the trunk. There are two other families of vertebrate homeogenes, homologous to the same number of genes that regulate the development of the head in the fruit fly, which seem to regulate the development and the regionalization of the head and in particular the brain. These are the genes *Emx1* and *Emx2* and the genes *Otx1* and *Otx2,* which are expressed in extensive regions of the embryonal brain. Their limits of expression form quite an interesting picture (Fig. 12): The domains of expression of the four genes are concentric regions, one contained in the other (Simeone *et al.,* 1992).

The gene *Otx2* is expressed in all of the embryonal ros-

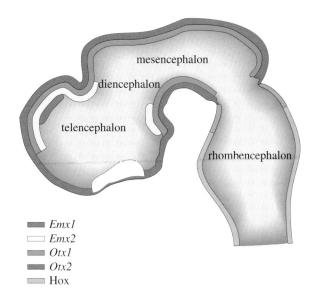

- ▬ *Emx1*
- ▭ *Emx2*
- ▬ *Otx1*
- ▬ *Otx2*
- ▬ Hox

FIGURE 12 Expression of the Emx and Otx genes of the brain of a vertebrate embryo. The domains of expression of the Hox genes are shown separately from those of the Otx genes.

tral brain, including the prosencephalon (forebrain) and the mesencephalon (midbrain). This gene appears to be expressed in an attempt to distinguish this region of our central nervous system from the posterior ones, composed of the rhombencephalon (hindbrain) and the spinal cord, where the Hox genes are active. The domain of expression of another gene, *Otx1,* is contained in that of *Otx2,* which in turn contains that of *Emx2.* Finally, *Emx2* contains the domain of expression of *Emx1,* whose activity is essentially confined to the dorsal telencephalon (endbrain), including the future cerebral cortex.

The first gene to be expressed during development is *Otx2,* whereas the last one is *Emx1.* It can therefore be assumed that the specification of the various regions of the brain by these genes occurs in successive steps, in increasingly more restricted regions, and that only at the end of this process should the cerebral cortex region receive its precise biological instructions laying down and regulating its structure and function. It can be imagined that the genic products of *Otx2* specify the regions destined to become the rostral brain. On the basis of this specification the products of *Otx1* then specify a more limited subregion therein. Within this, the products of *Emx2* then come into play, delimiting the future proencephalon. Finally, the products of *Emx1* which inform the cells of the region corresponding to the dorsal telencephalon which have to organize themselves to produce the cerebral cortex are activated.

Again, we observe an incredible preservation of the structure of corresponding genes, present in such different species as the fruit fly and man, and the even more incredible preservation of the location of their genic products.

These genes expressed much further forward in the fruit fly are also expressed much further forward in vertebrates, with the addition of the influence of the brain, or rather of the most noble part of the brain, the cerebral cortex, which is the seat of memory and abstract thought. *Emx1* is expressed exclusively there, whereas its closest homolog, *Emx2,* is expressed not only in the cerebral cortex but also in many of the anatomical structures linked to olfaction, such as the olfactory bulb and the olfactory epithelium.

Two of the four genes have recently, for different reasons, attracted the interest of researchers. The gene demarcates the region corresponding to the anterior portion of the brain in vertebrates from the very earliest phases of their embryonal development. This gene is in reality activated early. It is switched on when the blastocyst of the mouse is implanted in the maternal uterus and before the blastula of the frog begins gastrulation. Its genic products are in place from the very first moment, when an antero-posterior polarity is established in the embryo, and from the outset they mark the future position of the head and the part of the brain furthest forward. All this indicates that *Otx2* plays a primary role in the early determination of the head and the rostral brain. This assumption seems to be borne out by experiments on the manipulation of its expression conducted both on the mouse and on the frog. Embryos of mice deprived of this gene cease developing and degenerate halfway through gestation, have no head, and show very grave defects in all their axial structures. Embryos of frogs in which this gene is switched off have no brain, whereas in embryos in which this gene is expressed more than necessary there is an expansion of the head and the brain at the expense of the trunk and the tail, which are very reduced.

Emx2 seems to control the multiplication of the nerve cells which will form the cerebral cortex. These must be neither too numerous nor too few, and any dysfunction of this gene, such as has been observed in a congenital human illness called schizoencephalia, gravely compromises the intelligence if not the whole complex of the higher functions coordinated by the cerebral cortex.

These new homeogenes expressed in the brain complete the picture of the genes that control the development and the regionalization of the central nervous system and the entire body of vertebrates, from the head to the feet. These genes thus contain the secret of the form and the structure of our bodies. The identification of these genes and many others has enabled us to gain a better understanding of the genetic control of embryonal development and is a good omen for a complete understanding of this fascinating biological process in the future. To be sure, the road is still a long one. Many phenomena exist for which solutions are not known, but if the rate of discoveries and technical innovations continues at the current pace, there is no doubt that many questions, including that of the embryonal and postembryonal development of the brain and of the cerebral cortex in particular, will soon have answers.

References Cited

BATESON, W. (1894). *Materials for the Study of Variation.* Macmillan, London.

BENDER, W. AKAM, M., KARCH, F., BEACHY, P., PFEIFER, M., SPIERER, P., LEWIS, E., and HOGNESS, D. (1983). Molecular genetics of the bithorax complex in *Drosophila melanogaster. Science* **221,** 23–29.

BONCINELLI, E. (1994). *Biologia della Sviluppo.* La Nuova Italia Scientifica, Rome.

GARCÍA-BELLIDO, A. (1977). Homeotic and atavic mutations in insects. *Am. Zool.* **17,** 613–629.

GARCÍA-BELLIDO, A., RIPOLL, P., and MORATA, G. (1973). Developmental compartmentalization of the wing disk of *Drosophila. Nat. New Biol.* **245,** 251–253.

GILBERT, S. F. (1994). *Developmental Biology,* 4th ed. Sinauer, New York.

GRUSS, P., and WALTHER, C. (1992). Pax in development. *Cell* **69,** 719–722.

HORVITZ, H. R., and HERSKOWITZ, I. (1992). Mechanisms of asymmetric cell division: Two Bs or not two Bs, that is the question. Boundaries and fields in early embryos. *Cell* **68,** 237–255.

INGHAM, P. W., and MARTINEZ ARIAS, A. (1992). Boundaries and fields in early embryos. *Cell* **68,** 221–235.

LAWRENCE, P. A. (1992). *The Making of a Fly.* Blackwell, Oxford.

LEWIS, E. B. (1964). *The Chromosomes in Development* (M. Locke, Ed.). Academic Press, New York.

LEWIS, E. B. (1978). A gene complex controlling segmentation in *Drosophila. Nature* **276,** 565–570.

LEWIS, E. B. (1985). Regulation of the genes of the bithorax complex in *Drosophila. Cold Spring Harbor Symp. Quant. Biol.* **50,** 155–164.

MCGINNIS, W., and KRUMLAUF, R. (1992). Homeobox genes and axial patterning. *Cell* **68,** 283–302.

NÜSSLEIN-VOLHARD, C., and WIESCHAUS, E. (1980). Mutations affecting segment number and polarity in *Drosophila. Nature* **287,** 795–801.

SCHNEUWLY, S., KLEMENZ, R., and GEHRING, W. J. (1987). Redesigning the body plan of *Drosophila* by ectopic expression of the homeotic gene *Antennapedia. Nature* **325,** 816–818.

SIMEONE, A., ACAMPORA, D., GULISANO, M., STORNAIUOLO, A., and BONCINELLI, E. (1992). Nested expression domains of four homeobox genes in developing rostral brain. *Nature* **358,** 687–690.

SLACK, J. M. W. (1991). *From Egg to Embryo.* Cambridge Univ. Press, Cambridge, UK.

ST. JOHNSTON, D., and NÜSSLEIN-VOLHARD, C. (1992). The origin of pattern and polarity in the *Drosophila* embryo. *Cell* **68,** 201–219.

WOLPERT, L. (1969). Positional information and the spatial pattern of cellular differentiation. *J. Theor. Biol.* **25,** 1–47.

WOLPERT, L. (1993). *Il Trionfo dell'embrione.* Sperling & Kupfer, Milan.

General References

BONCINELLI, E. (1998). *I Nostri Geni.* Einaudi, Torino, Italy.

WILKINS, A. S. (1992). *Genetic Analysis of Animal Development,* 2nd ed. Wiley-Liss, New York.

MARK E. FORTINI

Department of Genetics
University of Pennsylvania School of Medicine
Stellar-Chance Laboratories
Philadelphia, Pennsylvania 19104-6069, USA

Cell Signaling and Developmental Patterning in the Insect Retina

A central question in developmental biology is how a three-dimensional multicellular structure is assembled using information encoded in the nucleotide sequence of genomic DNA. Pattern formation in the insect retina is one such process that has been examined at the molecular level using the powerful methods of genetic analysis. Cellular patterning and fate specification in this tissue require the cooperative action of several different biochemical pathways, involving diffusible factors that organize large cellular territories and membrane-bound factors that mediate interactions between neighboring cells. These factors are conserved in other organisms, including humans, and have been implicated in a wide variety of normal developmental events as well as aberrant developmental processes such as cancer. The biological principles that have emerged from studies on the relatively simple insect retina are thus applicable to cellular patterning as it occurs in many different tissues and animal body plans.

An Eye for an Eye: Developmental Biology and the Insect Retina

The renowned Dutch naturalist Anton van Leeuwenhoek, an early inventor of the microscope, was fascinated by the new scientific perspectives afforded by his apparatus. Leeuwenhoek scrutinized all manner of objects and animals under his microscope lens, and he seems to have been particularly astonished by the wondrous perfection of the tiny insect eye. Writing more three centuries ago, Leeuwenhoek marveled at the exquisite detail of the fly retina and commented upon its scientific implications:

If we now see that provident Nature works in all Creatures, from the biggest to the smallest, almost in one and the same way, and if we remember that each round protuberance of the Eye is composed of many superposed scale-like parts, as I have said about the Eye of the Dragon-Fly, and still is provided with its perfect roundness, which most ingenious structure exceeds anything we see on the Earth with the naked Eye, we must say again: Away with the opinion of Aristotle and all those who still follow him and want to maintain that flying Creatures or any other living Creature is generated from rotten matter and who are trying to obscure the Truth with their writings. (Leeuwenhoek, 1695/1979, p. 251)

Leeuwenhoek would doubtless be gratified to know that the theory of spontaneous generation has long since been discredited, but he might be surprised by the considerable attention currently being devoted to the beautifully patterned insect compound eye by modern scientists. In approximately the past 15 years, developmental biologists have discovered that the retina of one insect in particular, the fruit fly *Drosophila melanogaster,* is a most favorable system for studies on the development of multicellular organisms. These studies have sought to answer several questions that are relevant to the development of many different animals, including humans: How is a complex array of differentiated cells generated from an initially homogeneous reservoir of undifferentiated cells? Are cell fates determined primarily by intrinsic cellular factors or by extrinsic cues? What are the molecular mechanisms that underlie pattern formation and cell fate acquisition during development?

Its sheer beauty aside, the *Drosophila* compound eye has several more prosaic advantages for developmental studies. First, the adult compound eye is a remarkably regular hexagonal array of approximately 750 essentially identical unit eyes, termed ommatidia, each capped by its own tiny lens (Fig. 1). Each ommatidium is composed of 20 different cells, including 8 neuronal photoreceptors (R1–R8) and 12 nonneuronal accessory cells. Moreover, all 20 cells are uniquely identifiable by their position within the ommatidium, their morphology, and their physiological function. In essence, the different cell types are sorted out by Nature into a convenient grid pattern in the adult compound eye and are not jumbled together as in many other tissues and organs (Fig. 2). Subtle defects or experimentally induced perturbations in cellular form or identity are thus displayed several hundred times in a single eye for the benefit of the biologist, whose poor eyesight or limited patience might otherwise allow an interesting pattern defect to slip by unnoticed. Second, the developing cellular architecture of the eye has been traced from the late larval stages to the adult stage using high-resolution electron microscopy, providing a wealth of detailed knowledge needed to interpret the rapidly accumulating genetic and molecular data. Third, the compound eye of *D. melanogaster* is located on the anterior end of one of the premier experimental model organisms currently in use, with a repertoire of sophisticated genetic and molecular techniques. Because the fruit fly eye is completely dispensable for survival in a laboratory environment, it is relatively simple to isolate mutants that affect only eye development, and special genetic methods may be employed to recover mutations in genes that are essential for development of the embryo and the eye. Intensively studied for more than three-fourths of a century, the fruit fly boasts thousands of mutants and genetic variants, thousands of researchers worldwide, and its own recently completed genome sequencing project. More than any other factor, the critical mass of manpower and research material devoted to *D. melanogaster* has been responsible for the

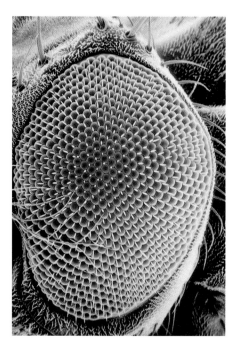

FIGURE I Scanning electron microscope image of the surface of the eye of the fruit fly *Drosophila melanogaster.* The anterior part of the eye is on the left and the dorsal part at the top. The lenses of the different ommatidia form a precise hexagonal array, with small interommatidial bristles present between the lenses. Each ommatidia is a separate functional visual system that samples the light from a small sector of the visual field. Because the compound eyes encompass a large portion of the adult head capsule, the fly has a much larger field of vision and binocular view than humans or most other animals. These features are believed to be evolutionary adaptations that allow the fly to perform high-speed airborne maneuvers while pursuing other flies or evading predators.

rapid rise to prominence of the insect compound eye in current developmental biology.

This article focuses on a few of the intercellular signaling pathways that have emerged during that past few years as regulatory mechanisms of cell patterning and cell fate specification during *Drosophila* eye development. Following a brief description of the developmental origins of the compound eye, three patterning events that occur at distinct stages of eye development are considered in more detail (Fig. 3). The first event is a wave of morphogenesis that is responsible for large-scale patterning of the entire ommatidial array and that is driven by secreted factors that may diffuse over a sizable portion of the developing eye epithelium. The second event is the arrangement of cells behind this wave into appropriately sized, evenly spaced clusters of cells that will eventually differentiate into mature ommatidia. This event involves the activities of short-range diffusible signals as well as local (nondiffusible) signals that are exchanged between initially equivalent cells. The third

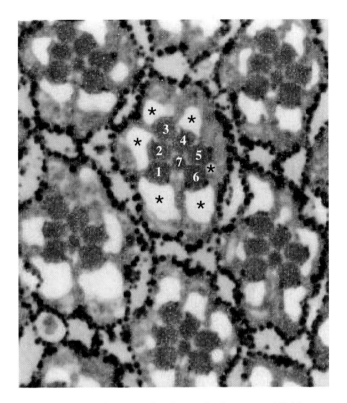

FIGURE 2 A tissue section beneath the ommatidial lenses reveals the internal arrangement of cells in the *Drosophila* retina. The central portion of each ommatidium is occupied by the photoreceptor cells R1–R8, each of which possesses a tubular membrane-dense structure called the rhabdomere. This acts as the waveguide for incident light and contains the rhodopsin molecules that absorb light to trigger the first step in the phototransduction cascade. In this section, at the level of the apical plane, the trapezoidal arrangement is clearly visible for the rhabdomeres of photoreceptor cells R1–R7; these are marked with the appropriate numbers in one of the ommatidia. The R8 cell rhabdomere is located beneath the R7 rhabdomere in underlying tissue sections. The lighter colored areas surrounding the rhabdomeres (asterisks) are the photoreceptor cell bodies, which project their rhabdomeres into the central cavity on stalk-like cellular extensions. The darker hexagonal lattice separating each group of photoreceptors is composed of pigment cells, which extend throughout the depth of the retina and optically insulate each ommatidium from stray light.

event is the inductive recruitment of one particular ommatidial cell type into the growing cell cluster. This recruitment, which is thought to be representative of the cellular interactions that add the many different cell types to the assembling ommatidial unit, is triggered by a local inductive signal transmitted between nonequivalent cells. These three patterning events illustrate how cells use long-range and short-range diffusible factors together with local signals to organize surrounding tissue regions and influence the cell fate choices of neighboring cells. Remarkably, the molecu-

long-range
diffusible
signal

short-range
diffusible
signal

local
inductive
signal

FIGURE 3 The three general classes of cell signaling that are responsible for developmental patterning in the *Drosophila* compound eye: signaling mediated by long-range diffusible substances (top), signaling mediated by short-range diffusible substances (middle), and local inductive signaling that occurs between neighboring cells in physical contact with each other (bottom). In all three cases, an extracellular cue emanating from the signaling cell (yellow) is detected by other cells that express the appropriate cell surface receptor for the signal. In response to the signal, the receiving cells select a particular ontogenetic program specifying their fate and differentiate accordingly (red). Cells that are competent to respond to the signal but that are not exposed to it may remain undifferentiated or select an alternative ontogenetic pathway by default (white). In the case of long-range and short-range diffusible substances, the diffusion of the signal from a defined cellular source presumably produces a concentration gradient within the tissue, with the possibility of different cellular responses elicited by different concentrations of the signal. In the case of an inductive signal, the signaling event requires direct physical contact between the two cells and does not involve nonneighboring cells.

lar components of these signaling pathways are highly conserved in other species, including humans, and have been associated with both normal and cancerous developmental processes in these organisms. These findings suggest that similar molecular and genetic programs may underlie seemingly unrelated developmental events in different organisms, vindicating the selection of relatively simple model organisms, such as the fruit fly *D. melanogaster* and the nematode *Caenorhabditis elegans*, to address questions of general significance in developmental biology.

Structure and Origins of the *Drosophila* Compound Eye

The *Drosophila* compound eye has been shaped by evolution to fulfill its purpose as a miniature optical apparatus, and fortunately evolution has left us with a biological structure of extraordinary precision and relative simplicity. Each of the approximately 750 ommatidia of a single adult eye contains 20 identifiable cell types, all of which are well characterized morphologically and whose developmental origins are understood in detail. Of the 20 ommatidial cells, 8 are neuronal photoreceptor cells, 4 are lens-secreting cone cells, and 8 are pigment cells that form a sheath around the photoreceptor cell group (Fig. 4). All these cells exhibit highly elongated cellular shapes and are bundled tightly together to form a single thin, tapered ommatidium that traverses the retina. The photoreceptor cells are immediately recognizable in tissue sections by their rhabdomeres, which are membrane-dense tubular structures that function as waveguides for the incoming light and that contain the opsin molecules that absorb light energy. The rhabdomeres project into the central cavity of the ommatidia on cellular extensions of the photoreceptors. By several criteria, the eight photoreceptors, R1–R8, may be divided into three general classes: R1–R6, R7, and R8. The R1–R6 photoreceptors contain large-diameter rhabdomeres that occupy peripheral locations in the rhabdomere ensemble, all express the same opsin and display the same spectral sensitivity, and their axons all terminate in the same optic lobe region of the fly brain. The R7 and R8 photoreceptors each contain small-diameter rhabdomeres that are located in the central position of the rhabdomere set, with the R7 rhabdomere situated atop the R8 rhabdomere such that the two form a continuous waveguide. The R7 and R8 cells also contain distinct opsins and thus exhibit characteristic spectral sensitivities, and they project axons to two different regions of an optic lobe not innervated by R1–R6. The four cone cells are responsible for secreting the lens apparatus of the ommatidium, and they have most of their cell mass located just under the lens, although long cellular processes from these cells extend the length of the retina. The pigment cells, classified as primary, secondary, or tertiary pigment cells based on their ommatidial location and cellular shape, are located peripheral to the photoreceptor cell complement of the ommatidium and form the interlocking hexagonal latticework of the compound eye. This latticework provides structural integrity to the retina and the dense pigment granules present in these cells optically insulate each individual ommatidium from stray light not entering through its own lens.

The *Drosophila* eye, like most other tissues of the adult fly or imago, develops from a small group of progenitor cells that are set aside in the early embryo (Garcia-Bellido and Merriam, 1969; Wieschaus and Gehring, 1976). These cells proliferate but do not begin differentiation, giving rise to a monolayer epithelium of several thousand cells, termed the eye imaginal disc, by the late larval stages prior to pupariation. Ommatidial assembly and patterning does not occur

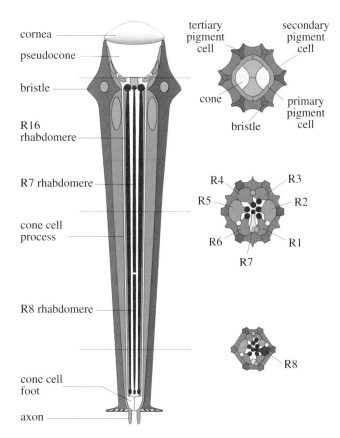

FIGURE 4 Cellular architecture of a single *Drosophila* adult ommatidium. (Left) A schematic longitudinal view of the ommatidium; (right) cross sections at three different levels (indicated by dotted lines). At the apical surface of the ommatidium is the lens apparatus consisting of the corneal lens and the underlying pseudocone which gather and focus light on the tips of the photoreceptor rhabdomeres. The lens is secreted by the cone cells (yellow), which form a pocket beneath the pseudocone and extend long processes to the base of the ommatidium. The photoreceptors and their light-absorbing rhabdomeres also extend nearly the complete length of the ommatidium and display highly elongated cell profiles. The photoreceptor complement of each ommatidium is nestled within a precisely constructed, opaque shield composed of three different classes of pigment cells (primary, secondary, and tertiary). Axons from the photoreceptor cells extend from the base of the ommatidium and project to different synaptic target regions of the insect brain. Sensory inputs from the ommatidia are relayed to the brain by these axons and their subservient interneurons and used by the insect to derive visual information about its environment (modified from Wolff and Ready, 1993).

synchronously over the entire retinal field; rather, columns of ommatidia emerge from within a visible indentation, termed the morphogenetic furrow, that sweeps across the eye disc from posterior to anterior (Fig. 5). Since a single column emerges from the furrow every 1.5 to 2 h, an eye

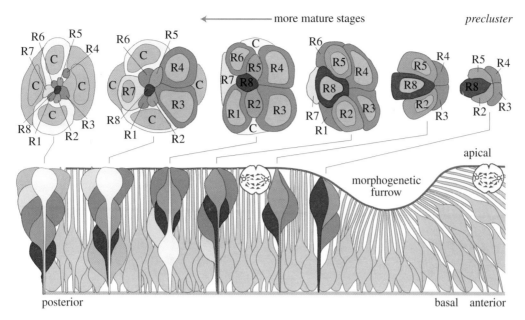

FIGURE 5 Ommatidial assembly sequence during larval eye disc development. A longitudinal section of a developing eye imaginal disc is depicted (with the anterior part of the disc to the right). Above the disc, cross sections of ommatidia are shown which, from right to left, show progressively more mature stages. The first overt signs of ommatidial assembly occur within the morphogenetic furrow, which sweeps across the developing eye disc tissue during the late larval stages of the insect life cycle. Within the furrow, small groups of cells assemble into arcs and then into five-cell preclusters. These preclusters contain the precursor cells for the R8 photoreceptor cell and for the R2–R5 and R3–R4 pairs, which are assigned their respective cell fates in that invariant sequence. A round of cell division posterior to the furrow, denoted by a symbol depicting a mitotically active cell in the center of the disc, generates additional cells which contribute to the next photoreceptor recruitments, namely, those of the R1–R6 pair and finally the R7 cell. Once the complete set of eight photoreceptors has been determined, the four cone cells (C) are assembled around them in the posterior disc regions. Pigment and bristle cells (not shown) are specified at later stages in development (modified from Wolff and Ready, 1993).

imaginal disc removed from a late-stage larva for histological examination contains a neatly arranged developmental series of progressively more mature ommatidia extending posteriorly from the furrow (Wolff and Ready, 1993).

As undifferentiated cells emerge from the furrow, they are organized into recognizable groups of 10–15 cells, one of which adopts the R8 photoreceptor cell fate. The R8 precursor cell is the founder cell of the ommatidium, around which the remaining seven photoreceptor precursors are assembled in a fixed order with the pairwise additions of R2 and R5, R3 and R4, R1 and R6, and finally the R7 precursor cell (Wolff and Ready, 1993). Recruitment and neural differentiation of each photoreceptor cell type has been visualized by staining larval eye discs for neuronal proteins, revealing that the process occupies about 24 h or 12 ommatidial columns (Fig. 6). Addition of the 12 accessory cells of each ommatidium, the cone cells, pigment cells, and bristle cells, occurs during the pupal stages. The cone cells secrete the corneal lenses that form the exterior surface of the adult eye, and the pigment cells comprise the interlocking honey-

comb support structure of the adult retina and optically insulate the ommatidia from stray light.

Long-Range Diffusible Factors Drive the Anterior Progression of the Morphogenetic Furrow

The morphogenetic furrow reflects a transient shortening of the long columnar cells of the eye disc epithelium and is accompanied by constriction of apical (topmost) membrane surfaces, basal (downward) migration of nuclei, and synchronization of the cell cycle. The importance of these events for the imminent differentiation of cells within the furrow is unclear, but it has been suggested that apical membrane constriction may coordinate the onset of ommatidial assembly by dramatically concentrating cell surface receptors and ligands present on the apical membranes of the cells (Wolff and Ready, 1993). Anterior progression of the

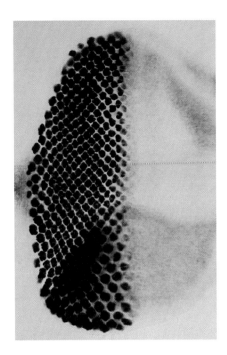

FIGURE 6 Micrograph illustrating ommatidial maturation in the *Drosophila* larval eye imaginal disc visualized with antibodies. In this case, an antibody was used that recognizes the ELAV protein, an RNA-binding protein expressed in the nuclei of all newly born neurons. The morphogenetic furrow is located approximately in the middle of the disc, runs from top to bottom, and is moving to the right (anterior). As nascent ommatidia emerge from the furrow, a few cells show positive staining for ELAV. These cells correspond to the R8, R2, and R5 precursor cells. Columns of progressively older ommatidia extend to the left (posterior) and display increasingly more staining as additional cells are recruited as neurons until all eight photoreceptors are specified for each ommatidium. The cells ahead of the furrow are unstained since they have not yet been recruited into ommatidia.

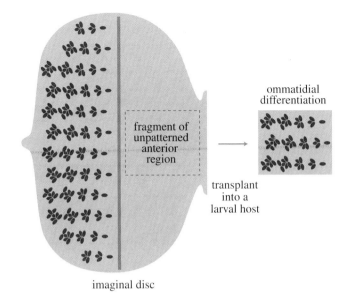

FIGURE 7 Scheme showing how fragments dissected out of the unpatterned region of a normal eye imaginal disc and cultured in a larval host are able to form a new furrow and initiate ommatidial differentiation. These studies provided the first clues that the regions of the disc ahead of the furrow, although morphologically unpatterned, possess information that can be used to drive cellular patterning events and organogenesis.

furrow was first considered to result from local interactions that add new ommatidia to the growing eye field, using the most recently formed ommatidial column as a template for the next (Ready *et al.*, 1976). However, doubts were cast on this idea by transplantation experiments in which anterior disc fragments, lacking ommatidial clusters and furrow, were cultured in larval hosts and found to be capable of generating a new furrow and ommatidia (Lebovitz and Ready, 1986). These experiments indicated that the morphogenetic furrow can reinitiate *de novo*, and that the information necessary to do so is mapped onto the anterior, unpatterned regions of the disc (Fig. 7).

Within the past few years, much progress has been made in uncovering the molecular mechanism behind furrow progression (Heberlein and Moses, 1995). As is often the case with *Drosophila*, the first crucial clues came from the analysis of mutants, in this case flies with incomplete, poorly

formed eyes (Fig. 8). When developing eye imaginal discs from these mutants were stained for neuronal proteins, mature ommatidial clusters were observed immediately posterior to the furrow, in contrast to the nascent preclusters that emerge from the furrow in wild-type eye discs (Heberlein *et al.*, 1993; Ma *et al.*, 1993). This result implies that the mutant animals exhibit a specific defect in furrow movement, and that ommatidial assembly otherwise proceeds normally behind the halted furrow.

A second important realization, namely, that diffusible substances are involved in this process, came from the study of genetic mosaics. Mosaics are animals in which patches or clones of mutant tissue are produced and allowed to develop in an otherwise wild-type tissue background or vice versa, allowing the effects of the mutant or wild-type cells upon the surrounding tissue to be assessed (Fig. 9). When a clone of wild-type tissue was generated in a mutant eye disc with a halted furrow, the furrow continued moving through the clonal region, showing that the wild-type cells produce a factor that keeps their segment of the furrow moving and that the surrounding mutant cells do not produce a diffusible inhibitor capable of interfering with the furrow in the clone (Heberlein *et al.*, 1993). More significant, mutant eye disc regions, without a functional furrow of their own, developed relatively normally if located near the clone, showing that the wild-type cells of the clone must produce a diffusible molecule that stimulates furrow movement (Fig. 10).

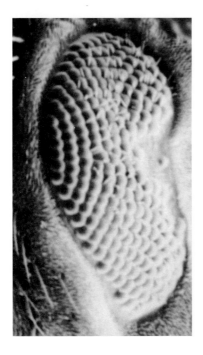

FIGURE 8 Scanning electron micrograph of the eye of a recessive viable mutant of the *hedgehog* gene, which is involved in the early phases of the formation of the regular structure of the compound eye in *Drosophila*. The exterior phenotype is characterized by an anterior scar and by a much smaller number of ommatidia than is observed in wild-type flies. This adult eye phenotype is caused by a premature halt in the anterior progress of the morphogenetic furrow during larval development. The eye is oriented with anterior to the right and dorsal at the top (reproduced with permission from Heberlein *et al.*, 1993).

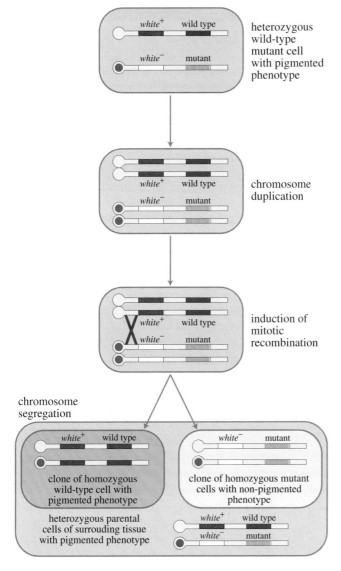

FIGURE 9 Generation of a clone of mutant tissue in a heterozygous wild-type tissue background, a technique used in *Drosophila* to assess the functions of essential gene products in the postembryonic development of particular tissues and to examine the role of the gene in interactions between different cells or cell groups during development. In this example, the animal is heterozygous for the mutation of interest as well as the linked marker gene *white* (top), which is necessary for the production of pigment granules in pigment and photoreceptor cells. The mutation (*white⁻*) causes the absence of pigment and thus allows the identification of homozygous recessive mutant cells. Mitotic recombination is experimentally induced in dividing cell populations after chromosome duplication (middle) and leads to the formation of hybrid chromosomes composed of nonidentical chromatids. Segregation of these chromatids (distinguished with dotted and undotted centromeres), followed by expansion of the daughter cell populations, produces clones of unpigmented, homozygous mutant tissue (lighter rectangle at bottom right) accompanied by homozygous wild-type pigmented twin spots (darker rectangle at bottom left) on a background of parental heterozygous pigmented tissue (large rectangle at bottom). The recessive mutant phenotype of the gene of interest may thus be studied within the unpigmented tissue patch, and the potential effects of this mutant clone on nearby tissue may be examined in the surrounding pigmented eye tissue.

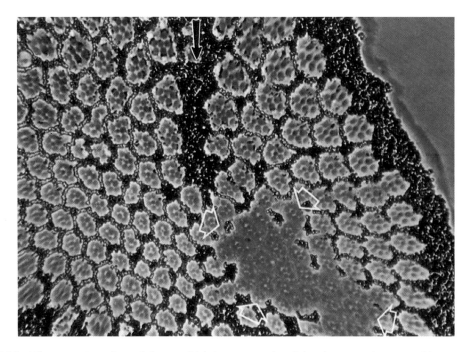

FIGURE 10 Tissue section of an adult eye which is composed mainly of mutant tissue, unable to support furrow progression, marked with a darkly colored lattice of pigment cells. The position at which the furrow would normally stop in this disc preparation is indicated by the arrow. A clone of unpigmented but otherwise wild-type tissue (arrowheads) in the furrow-stop mutant disc rescues the anterior progress of the morphogenetic furrow through the clonal territory and also through a considerable area of surrounding mutant tissue. The ability of the wild-type clone to support furrow progression through its own territory and adjacent mutant cell areas provides direct evidence for the existence of a diffusible, long-range substance that drives furrow movement (reproduced with permission from Heberlein *et al.*, 1993).

Among the mutations first identified as capable of stopping the furrow were certain rare, viable mutations in *hedgehog*, a gene that encodes a secreted protein that is thus a good candidate for the diffusible signal (Heberlein *et al.*, 1993; Ma *et al.*, 1993). The *hedgehog* gene is expressed in differentiating photoreceptors that have already emerged from the furrow, and if neural differentiation is prevented by genetic means Hedgehog protein does not accumulate and furrow movement is impaired (Heberlein *et al.*, 1993). Conversely, if *hedgehog* is expressed ectopically in cells anterior to the normal furrow, new morphogenetic furrows appear and cause precocious neural differentiation of these anterior eye regions (Fig. 11; Heberlein *et al.*, 1995). Hedgehog is thought to exert its effects by inducing the expression of another diffusible factor, the product of the *decapentaplegic* gene, in cells within the morphogenetic furrow (Heberlein *et al.*, 1993; Ma *et al.*, 1993). The *decapentaplegic* gene product is a member of the transforming growth factor-β (TGF-β) family (Padgett *et al.*, 1987), other members of which play key roles in vertebrate development. Expression of *decapentaplegic* in the furrow is abolished or greatly reduced in mutants that stop the furrow (Fig. 12) and induced in the secondary furrows caused by ectopic Hedgehog protein (Heberlein *et al.*, 1993, 1995). Both *hedgehog*

and *decapentaplegic* are essential genes; therefore, loss-of-function mutations that abolish activity of these genes cause early lethality and can only be studied in the compound eye by inducing homozygous mutant clones to grow in an animal with otherwise normal tissue. Using this technique, the long-range patterning abilities of the diffusible products of *hedgehog* and *decapentaplegic* are apparent in clones of loss-of-function mutations in either gene, which can grow to encompass large areas (>20 ommatidia) with few or no obvious defects presumably due to rescue of the mutant cells by the surrounding wild-type tissue (Fig. 13; Heberlein *et al.*, 1993; Ma *et al.*, 1993). By one estimate, the patterning effects of the morphogenetic furrow may extend anteriorly from the furrow over a distance of approximately 30% of the entire retinal field (Ma *et al.*, 1993).

Recent work has uncovered additional components of the mechanism by which neural differentiation behind the furrow, acting through *hedgehog* and *decapentaplegic*, influences events ahead of the furrow. Secondary furrows such as those caused by ectopic Hedgehog are also generated by clones of loss-of-function mutations in at least two other genes, *patched* and the gene encoding the major catalytic subunit of protein kinase A (Pan and Rubin, 1995; Ma and Moses, 1995; Strutt *et al.*, 1995; Wehrli and Tomlin-

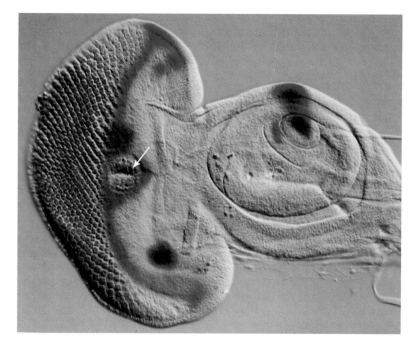

FIGURE 11 Imaginal disc of a fly in which molecular genetic techniques were utilized to express the *hedgehog* gene in a group of cells ahead of the moving furrow, in a region in which it would not normally be expressed. Ectopic expression of the Hedgehog protein in these cells triggers a series of events that reflect the propagation of a second, circular furrow (arrow). One event is the induction of the *decapentaplegic* gene, a downstream target of Hedgehog signaling in this tissue. In this histological preparation, *decapentaplegic* expression is visualized by the blue staining produced by a *decapentaplegic-lacZ* marker gene. A second event is the assembly and neural differentiation of ommatidia within the area encompassed by the new furrow, as shown by the photoreceptor clusters stained brown in this sample. This eye imaginal disc (left) and associated antennal imaginal disc (right) are oriented with the anterior at right (courtesy of U. Heberlein).

son, 1995). Protein kinase A is a member of an evolutionarily conserved family of proteins that phosphorylate and thereby regulate the activities of specific substrate proteins. These clones also display classic properties of an organizing center in that they can affect the growth and patterning of nearby wild-type tissues. The exact relationship of these proteins to one another is unclear. Patched may be the receptor for Hedgehog, although some data argue against this idea (Perrimon, 1995). It is unknown whether protein kinase A is repressed directly by Hedgehog or if it acts in a

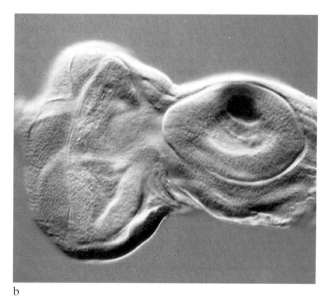

a

b

FIGURE 12 Discs from flies carrying a *decapentaplegic-lacZ* marker gene that allows the pattern of *decapentaplegic* gene expression to be visualized by a blue staining product in histochemical studies. High levels of expression of the *Drosophila decapentaplegic* gene are detected in the morphogenetic furrow in a wild-type disc (a) but not in a disc from a furrow-stop mutant such as *hedgehog* (b). These observations have implicated the *decapentaplegic* gene in the genetic hierarchy controlling morphogenetic furrow progression in the *Drosophila* eye (reproduced with permission from Heberlein *et al.*, 1993).

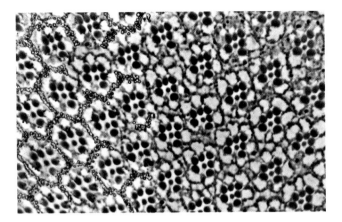

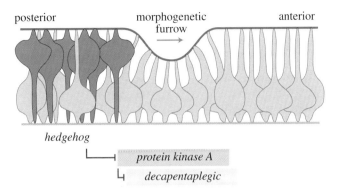

posterior   morphogenetic   anterior
furrow

hedgehog

protein kinase A

decapentaplegic

FIGURE 13 Tissue section of an adult eye in which a large clone of Decapentaplegic-deficient ommatidia (unpigmented area at right) has been able to develop normally, presumably because the mutant cells are rescued by the long-range signaling activity of Decapentaplegic protein produced in the neighboring wild-type disc regions (yellow pigmented area at left). Analysis of such clones suggests that the Decapentaplegic protein, like other secreted proteins with important developmental roles, is capable of exerting its effects on cells far from that in which it was initially synthesized (reproduced with permission from Heberlein *et al.,* 1993).

FIGURE 14 Schematic representation of the genetic mechanism underlying morphogenetic furrow progression in the developing *Drosophila* eye. The disc is shown in side view, with the morphogenetic furrow indicated at the top. The furrow is progressing from posterior (left) to anterior (right), such that differentiating ommatidia (colored) are depicted posterior to the furrow at left. Beneath the disc diagram, colored rectangles indicate the relative time of action and genetic relationships between genes involved in furrow progression. As indicated in this schematic, neural differentiation of the ommatidial cells posterior to the furrow induces expression of the *hedgehog* gene. Anterior diffusion of secreted Hedgehog protein represses protein kinase A activity, leading directly or indirectly to the induction of the *decapentaplegic* gene. The genetic relationships depicted here are not meant to imply molecular pathways or biochemical interactions (reproduced with permission from Heberlein and Moses, 1995).

distinct, antagonistic pathway (Kalderon, 1995; Perrimon, 1995). It is also doubtful that all the effects of these signaling molecules are mediated solely by Decapentaplegic since secondary morphogenetic furrows have not been reported as a consequence of ectopic *decapentaplegic* expression and may thus require additional Hedgehog-dependent factors. A similar regulatory hierarchy involving Hedgehog, Patched, protein kinase A, and Decapentaplegic operates in the developing wings and legs of *Drosophila,* and both *hedgehog* and *patched* are required at the same stage of embryonic development (Perrimon, 1995). Furrow propagation may thus be directed by a common signaling mechanism that is used, with some modifications, in a variety of developmental contexts (Fig. 14).

Since the Hedgehog signal emanates from differentiating neurons posterior to the furrow, it clearly cannot account for the initiation of furrow movement at the posterior disc margin prior to the onset of neural differentiation. What factors enable the furrow to form and start moving in the absence of Hedgehog? Another gene important for embryonic patterning, termed *wingless,* is expressed along most of the dorsal margin of the disc, except in the posterior disc, and acts to restrict furrow initiation to the posterior margin (Ma and Moses, 1995; Treisman and Rubin, 1995; Heberlein and Moses, 1995). In the absence of *wingless* function, a secondary furrow is triggered along the ventral margin and moves across the disc perpendicular to the

normal furrow (Fig. 15), a phenotype that is also observed occasionally in eye discs with ectopic *hedgehog* expression (Ma and Moses, 1995; Treisman and Rubin, 1995; Heberlein *et al.,* 1995). The *wingless* gene encodes a member of the Wnt protein family, a group of secreted factors that are thought to act locally (Nusse and Varmus, 1992). Wingless may help align the normal furrow with the posterior margin by blocking the formation of Hedgehog-producing cells at the lateral margins, although the biochemical mechanism by which Wingless exerts its inhibitory effects on the furrow is unknown.

The discovery that a transient wave of morphogenesis in the *Drosophila* eye is driven by diffusible factors of the Hedgehog and TGF-β protein families has provided further evidence for the importance of these signaling molecules in animal development and the value of the *Drosophila* eye as an experimental system. Because proteins that interact with Hedgehog and TGF-β in other developmental contexts, such as *Drosophila* embryonic development and mammalian limb bud formation, may also function in the morphogenetic furrow, further studies involving the eye imaginal disc may provide information about such common signaling components. Since there is a working knowledge

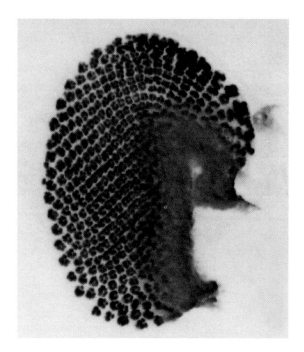

FIGURE 15 Imaginal disc in which removal of the *wingless* function derepresses furrow initiation along the dorsal margin of the disc, resulting in a secondary furrow that moves orthogonally to the normal furrow. This observation leads to the conclusion that, in contrast to Hedgehog and Decapentaplegic, which promote furrow movement, the Wingless protein acts in a defined sector of the eye imaginal disc to repress furrow progression. A normal function of Wingless is therefore to ensure that the furrow progresses in a single direction as a linear depression in the disc surface. In this sample, neural differentiation of ommatidia is revealed by antibody staining for the Hairy protein (black) and Scabrous protein (brown); the disc is oriented with the anterior at right and dorsal at top (reproduced with permission from Ma and Moses, 1995).

of furrow behavior and some of the genes that control it, rapid progress in this field can be expected. For example, furrow initiation at the posterior margin has been shown to require the nuclear protein encoded by the *dachshund* gene (Mardon *et al.*, 1994) and other nuclear proteins have been implicated in regulating the speed of furrow propagation (Brown *et al.*, 1995). These nuclear proteins are excellent candidates for regulators of the genetic program controlling furrow progression and cellular patterning within the furrow since they are transcription factors that are thought to bind directly to target genes and thereby control the activity of these genes. Perhaps most interesting will be the identification of the downstream targets of Hedgehog and Decapentaplegic — those genes and molecules that encode the patterning information mapped onto anterior disc regions in advance of the furrow and used to instruct cells to enter the furrow. One such candidate gene, *shortsighted*, has

been isolated on the basis of its *hedgehog*-dependent expression in a stripe of cells located just ahead of the furrow (Treisman *et al.*, 1995). The *shortsighted* gene interacts genetically with *decapentaplegic* and, remarkably, is homologous to a mouse gene that is transcriptionally induced in response to TGF-β.

Short-Range Diffusible Signals and Local Interactions Establish the Evenly Spaced Ommatidial Array

As cells emerge from the posterior edge of the morphogenetic furrow, they are organized into evenly spaced ommatidial preclusters within a few columns. The first morphological signs of cellular patterning are visible in these first few columns (Wolff and Ready, 1993). Imaging of the apical cell profiles with lead sulfide reveals that within the furrow, groups of approximately 10–15 cells assemble at regular intervals and are transformed into arcs of 7–9 cells each by the next column or two (Fig. 16). Each arc collapses in on itself, giving rise to an anterior quintet of cells that comprises the ommatidial precluster (R8, R2, R5, R3, and R4 precursor cells) and one or two so-called mystery cells located just posterior to the precluster. The mystery cells are expelled from the developing ommatidium at this stage and reenter the populations of uncommitted epithelial cells (Fig. 17).

Although much remains to be learned about the complex interplay of cells emerging from the furrow, the process of establishing the regularly spaced groups of cells is thought to involve the activities of short-range diffusible factors, including the product of the *scabrous* gene as well as the *Drosophila* epidermal growth factor receptor (EGFR) homolog and its ligands. Local interactions between cells in each group, mediated by the so-called neurogenic genes *Notch* and *Delta*, are then believed to lead to the selection of a single R8 precursor cell from among many competent candidate cells of the group. The R8 precursor cell is the founder cell of each ommatidium, around which are recruited the remaining cell types by a series of position-dependent inductive events (Tomlinson and Ready, 1987a; Jarman *et al.*, 1994).

The role of the *scabrous* gene in establishing the ommatidial spacing pattern was first appreciated from studies of loss-of-function *scabrous* mutants. These flies display an irregular eye surface and fused ommatidia due to excessive numbers of furrow cells entering the R8 differentiation pathway (Baker *et al.*, 1990). The *scabrous* gene encodes a protein related to the extracellular proteins fibrinogen and tenascin, and Scabrous protein is secreted by *Drosophila* tissue culture cells (Hu *et al.*, 1995; Baker *et al.*, 1990). In wild-type *Drosophila* eye imaginal discs, Scabrous protein

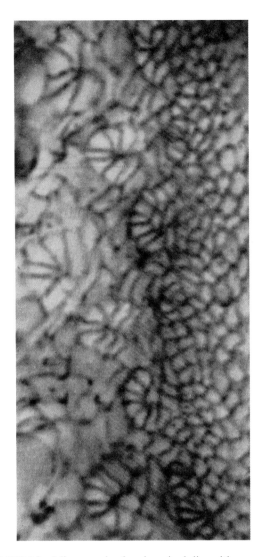

FIGURE 16 Micrograph of an imaginal disc with a segment of the furrow, which is proceeding from posterior (left) to anterior (right). The appearance of the initial ommatidial structure is visible within and immediately posterior to the morphogenetic furrow in eye imaginal discs treated with lead sulfide, which stains the apical cell profiles in the disc epithelium. Toward the right, arcs of 7–10 ommatidial precursor cells are seen emerging from the posterior edge of the furrow. These arcs then close in on themselves and resolve into discrete ommatidial preclusters containing 5–7 cells over the next few columns, as seen to the left (reproduced with permission from Baker and Zitron, 1995).

is detected in regularly spaced groups of 5–10 R8 candidate cells at the leading edge of the furrow. Within each group, Scabrous protein becomes restricted to a single R8 precursor cell over the next four columns. In *scabrous* mutants, however, apparently nonfunctional Scabrous protein is observed in a disorganized stripe of R8 candidate cells in the anterior furrow that resolves into a disorganized pattern of

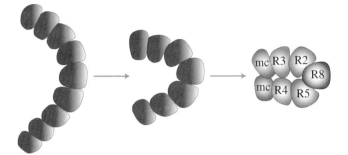

FIGURE 17 Schematic of ommatidial precluster formation immediately posterior to the morphogenetic furrow in the developing *Drosophila* eye. An arc of nine cells (left) as it first appears during its formation in the morphogenetic furrow is shown. As the arc closes in on itself (center), a few cells may be lost from the structure. By the time the arc has completely closed to form a group of cells (right), this precluster typically contains five precursor cells and one or two extra cells, termed mystery cells (mc). These events occur over a period of several hours and occupy only the first few ommatidial rows posterior to the furrow. In later rows, the mystery cells are excluded from the precluster and rejoin the population of uncommitted cells in the eye imaginal disc. The remaining five cells of the precluster are destined to become the photoreceptors R8, R2, R5, R3, and R4.

mostly single R8 cells (Fig. 18). Thus, *scabrous* expression precedes the organization of furrow cells into arcs and ommatidial preclusters, and Scabrous is needed to establish the initial spaced array of ommatidial preclusters (Baker and Zitron, 1995; Baker *et al.*, 1990). These observations have led to the proposal that Scabrous acts as a short-range diffusible factor that inhibits anterior furrow cells closest to the *scabrous*-expressing cell groups from initiating ommatidial formation, thus ensuring that each new ommatidial column arises as a periodic array, precisely out of register with the preceding template column (Baker and Zitron, 1995). No biochemical data exist on the diffusion distance of Scabrous protein, but it cannot be far because R8 precursor cells are separated by only approximately eight cell diameters within each column and by only approximately four to six cell diameters from R8 precursor cells in the two more anterior columns (Wolff and Ready, 1993; Baker *et al.*, 1990). The cell surface receptors for Scabrous and other proteins needed to interpret the signal have not been identified.

The earliest stages of ommatidial assembly also appear to be regulated by the *Drosophila* homolog of the epidermal growth factor receptor (EGFR) and its two ligands, which are encoded by the *spitz* and *argos* genes (Fig. 19). The role of EGFR in eye development is complex and poorly understood, affecting not only cellular differentiation but also cell division and cell death (Xu and Rubin,

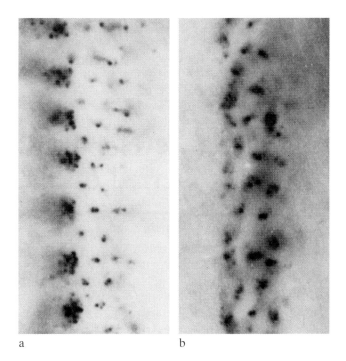

a   b

FIGURE 18 Comparison of the distribution of the Scabrous protein in a wild-type eye disc (a) and in a *scabrous* mutant eye disc (b), in which nonfunctional Scabrous protein is produced. In both panels, the morphogenetic furrow runs vertically slightly to the right of center and the anterior of the disc is at the right. In the wild-type disc, a regular array of *scabrous*-expressing groups of cells is seen in the furrow and isolated *scabrous*-expressing cells are observed in more posterior disc regions. In contrast, the *scabrous* mutant disc displays a disordered zone of poorly grouped *scabrous*-expressing cells within the furrow which do not resolve into single *scabrous*-expressing cells in more posterior rows. In the absence of Scabrous protein function, the correct formation of regularly spaced preclusters is severely compromised and the result in the adult is an uneven, irregular compound eye structure (reproduced with permission from Baker and Zitron, 1995).

1993; Baker *et al.,* 1990). Nevertheless, it is clear that the number of cells that are recruited into the nascent ommatidia is somehow regulated by the activity of the *Drosophila* EGFR homolog. Clones of cells completely lacking EGFR do not survive to contribute to adult eye structures (Xu and Rubin, 1993; Baker and Rubin, 1989) and few ommatidia develop in the eyes of animals homozygous for a hypermorphic allele of EGFR (Baker and Rubin, 1989). The reduced number of ommatidia in these eyes arise further apart from one another than in wild-type eyes, and *scabrous* expression is abolished in the intervening undifferentiating regions of the morphogenetic furrow, suggesting that EGFR functions at an earlier stage than Scabrous in ommatidial spacing processes (Baker *et al.,* 1990; Baker and Rubin, 1989). Unexpectedly, wild-type cells located near

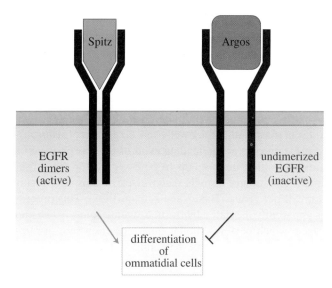

FIGURE 19 Schematic representation of the regulation of ommatidial differentiation in *Drosophila* by the EGFR homolog. In *Drosophila* EGFR is required throughout eye development for the proper cell fate specification of numerous different cell types that contribute to the final adult structure. Like other tyrosine kinases, the *Drosophila* EGFR is thought to require dimerization for its activation, with dimerization controlled by extracellular ligands present on the surfaces of neighboring cells. The activity of EGFR is stimulated by its ligand Spitz and negatively regulated by its ligand Argos so that these ligands have opposing effects on ommatidial differentiation.

the edge of an EGFR-deficient clone often differentiate inappropriately as supernumerary photoreceptor cells, demonstrating an apparently nonautonomous effect of the mutant tissue on the surrounding wild-type regions (Xu and Rubin, 1993).

These nonautonomous effects of EGFR were unexpected since the *Drosophila* EGFR homolog is a transmembrane tyrosine kinase thought to act as a receptor for extracellular signals (Schejter *et al.,* 1986). Presumably, autonomous activity of EGFR directly affects the differentiation of eye disc cells, which in turn alters their signaling properties and leads to indirect, nonautonomous effects on neighboring cells. Many of the molecular components of EGFR signaling have been determined in *Drosophila*. Extracellularly, EGFR interacts with its ligands Spitz and Argos, an agonist and an antagonist of EGFR activity, respectively (Schweitzer *et al.,* 1995). Clonal studies performed in the developing eye indicate that Argos, like Scabrous, is a short-range diffusible signaling molecule able to act over a distance of several cell diameters (Freeman *et al.,* 1992). Ommatidia in animals homozygous for *argos* loss-of-function mutations contain excess photoreceptor cells and cone cells, indicating that Argos, like EGFR, is normally required at multiple stages of ommatidial development (Freeman *et al.*, 1992; Kretzschmar *et al.*, 1992; Okano *et al.*, 1992). Activation of the EGFR homolog in the *Drosophila* eye initiates a well-characterized intracellular signal transduction pathway involving proteins of the Ras, Raf, and MAP kinase families (Hafen *et al.*, 1993).

Although Scabrous and the EGFR homolog play critical roles in establishing the evenly spaced groups of *scabrous*-expressing cells at the leading edge of the morphogenetic furrow, neither gene seems to be essential for the subsequent selection of a single *scabrous*-expressing R8 precursor cell per group (Baker and Zitron, 1995). This process is apparently mediated by the neurogenic genes *Notch* and *Delta*, which perform similar functions elsewhere in *Drosophila* development, most notably in the segregation of neuroblasts from surrounding epidermoblasts during formation of the embryonic central nervous system and the specification of sensory organ precursors during development of the adult peripheral nervous system (Campos-Ortega, 1993). These patterning events are all examples of lateral specification, the specification of cell fates by local interactions between initially equivalent cells (Fig. 20).

Notch and *Delta* encode a large transmembrane receptor protein and its membrane-anchored ligand, respectively, and homologs or close relatives of these two proteins have been implicated in a wide variety of developmental processes in several other organisms including humans (Artavanis-Tsakonas *et al.*, 1995). When temperature-sensitive mutations in *Notch* and *Delta* are used to eliminate their activity in the morphogenetic furrow, groups of *scabrous*-

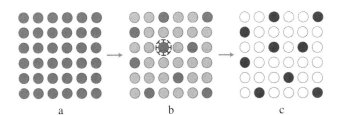

FIGURE 20 Hypothetical model for the generation of a pattern of spatially intermingled, distinct cell types from a field of initially equivalent cells by lateral signaling. Prior to lateral signaling, all cells possess intrinsic signaling and receiving activities in equal measures (a). These activities are negatively coupled to each another, such that lateral signaling autonomously reinforces the signaling activity of a cell and inhibits the signaling activity of neighboring cells (b). According to this model, slight random fluctuations in signaling and receiving activities that arise among the initial cellular territory would become amplified over time and ultimately give rise to a spaced pattern of different cell types (c). This model can account for some features of ommatidial precluster formation in the morphogenetic furrow and might explain how the size of individual ommatidial fields is regulated at the earliest stages of their development (modified from Wigglesworth, 1940).

expressing cells emerge at regular intervals along the anterior furrow as in wild type, but *scabrous* expression persists in many cells of each group instead of rapidly becoming restricted to single R8 precursor cells as the groups mature (Fig. 21; Baker and Zitron, 1995; Parks *et al.*, 1995). Local interactions among equivalent R8-competent cells, mediated by Notch and Delta proteins, thus refine the rough pattern of spaced cell clusters generated by Scabrous and EGFR activities into a precise array of isolated R8 founder cells.

The mechanism by which the binding of the ligand Delta by the Notch receptor protein leads to the selection of a single R8 precursor cell from among many eligible candidates is unclear. In the nematode *C. elegans*, a similar lateral specification event between two neighboring cells is accompanied by reciprocal modulations in the expression levels of a Notch-like protein and its ligand in the two cells (Wilkinson *et al.*, 1994). In the morphogenetic furrow, Notch and Delta protein levels are also modulated in a periodic manner, but the significance of these changes in cell fate decisions is not obvious (Baker and Zitron, 1995; Parks *et al.*, 1995). Recent studies have led to the identification of proteins that bind to the intracellular domain of the *Drosophila* Notch receptor and influence its signaling activity (Fig. 22), including an evolutionarily conserved transcription factor and a novel cytoplasmic protein (Artavanis-Tsakonas *et al.*, 1995). Both of these proteins function in *Drosophila* eye development, but their role in refining the R8 ommatidial founder cell array has yet to be assessed.

A likely target of the Notch pathway in the eye is the *atonal* gene, which is related to the four members of the *achaete–scute* gene complex. The *achaete–scute* genes are located in close proximity to one another in the genome and encode related proteins belonging to the same family of DNA-binding transcription factors, as does *atonal* (Jarman *et al.*, 1993). During development of the embryonic central nervous system and adult peripheral nervous system, *achaete–scute* genes are expressed first in clusters of cells that then require Notch activity to restrict expression to isolated neural precursors (Campos-Ortega, 1993). In an analogous manner, *atonal* is expressed in evenly spaced groups of cells early in the furrow and in single R8 precursor cells thereafter, thus closely paralleling *scabrous* expression during eye development (Jarman *et al.*, 1994). In a viable *atonal* mutant, R8 ommatidial founder cells fail to arise, leading to subsequent failures in neural differentiation and morphogenetic furrow progression (Jarman *et al.*, 1995).

Local Inductive Events between Neighboring Cells Assemble Ommatidia around the R8 Founder Cells

Once the array of R8 founder cells has been established, the remaining seven photoreceptor cell precursors are added

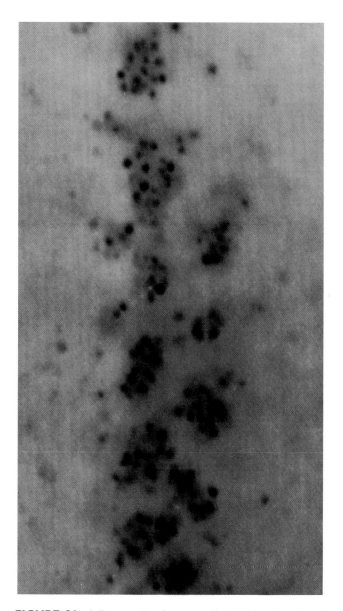

FIGURE 21 Micrograph of an eye disc lacking functional Notch protein, in which the pattern of early ommatidial formation in the morphogenetic furrow is revealed by Scabrous protein expression. Groups of Scabrous-expressing cells emerge from the furrow at regular intervals (center) but fail to resolve into a corresponding pattern of single isolated R8 precursor cells over the next few columns (right). These results suggest that Notch activity is not required for the initial appearance of evenly spaced cell groups in the furrow, but that it is needed for lateral signaling events within the group that lead to the specification of a single R8 founder cell per group, around which the future ommatidium is constructed (reproduced with permission from Baker and Zitron, 1995).

to the ommatidia in a precise spatiotemporal sequence, beginning with R2 and R5, followed by R3 and R4, then R1 and R6, and finally R7. Accessory cells are then assembled around the photoreceptor core to complete the ommatidial

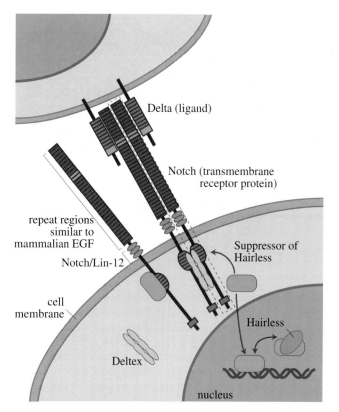

FIGURE 22 Schematic representation of the molecular events involved in signaling by the transmembrane receptor Notch during the development of the *Drosophila* eye. Binding of the Notch protein (in dashed box) to its ligand Delta is thought to initiate intracellular events in the Notch-expressing cell that lead to the transmission of a signal to target genes in the nucleus. Extracellular ligand binding could influence the interaction of Notch with its intracellular partners, Deltex and Suppressor of Hairless, and thus control their activity. Suppressor of Hairless is a DNA-binding protein that directly regulates the transcription of downstream target genes in concert with its nuclear binding partner Hairless. Close homologs of most of these proteins, including Notch, Delta, Suppressor of Hairless, and Deltex, have been identified in several other species including man, and this molecular pathway is believed to regulate similar developmental processes in many different organisms (modified from Artavanis-Tsakonas *et al.,* 1995).

cell complement. Until recently, ommatidial assembly was considered to proceed by a clonal mechanism, whereby one or a few stem cells would divide to produce all the cells used for a single ommatidium (Wolff and Ready, 1993). For example, a single photoreceptor stem cell could divide three times to generate the eight photoreceptor cell precursors R1–R8, and the other stem cells would give rise to the accessory cells in a similar fashion (Bernard, 1937). Comparatively little evidence exists for this model in any insect eye, however, and mosaic studies have definitively excluded such clonal mechanisms for the *Drosophila* ommatidium. When clones of marked cells are induced in an otherwise wild-

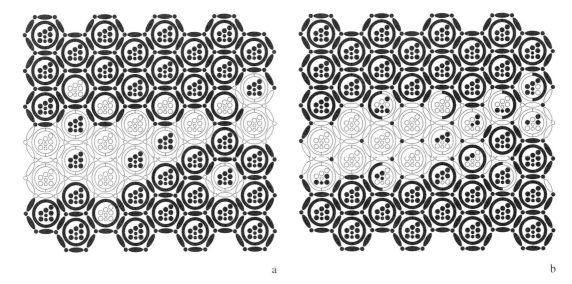

a   b

FIGURE 23 Potential lineage relationships between ommatidial cells examined in mosaic *Drosophila* retinae. Unpigmented mutant cells (white) arise within a mutant clone induced during larval development and intermingle with surrounding wild-type pigmented cells (red) along the edges of the clone. (a) A hypothetical mosaic clone illustrating the pigmentation pattern expected if all eight photoreceptors (trapezoidal set of circles) of a single ommatidium are derived from one founder cell and the pigment cells are derived from a different founder cell. Primary, secondary, and tertiary pigment cells are represented respectively by semicircular arcs, ovals, and small circles at the vertices of the ovals. Within a single ommatidium, all photoreceptors are of the same genotype, as are both primary pigment cells, but such relationships among secondary and tertiary pigment cells are less apparent due to their lattice-like arrangement. (b) A mosaic clone in which there are no strict lineage relationships among any of the cell types. Note that photoreceptor and pigment cells of different genotype are intermingled within individual ommatidia. This pigmentation pattern is not a hypothetical example; it represents an actual clone (reported by Ready *et al.*, 1976) showing that in the *Drosophila* retina there are no fixed lineage relationships among the ommatidial cell types. In both schematic clones, the anterior is at the right.

type eye, the border between the clone and the surrounding tissue contains normally constructed ommatidia composed of both marked and unmarked cells (Fig. 23). Such intermingling has been documented for the photoreceptors, pigment cells, and cone cells of border ommatidia, demonstrating that these cell types are not strictly related by lineage as required for a clonal mechanism of development (Wolff and Ready, 1991; Lawrence and Green, 1979; Ready *et al.*, 1976). In contrast to these cell types, the four cells that comprise the interommatidial mechanosensory bristle group are clonally derived from a single progenitor cell (Wolff and Ready, 1991; Cagan and Ready, 1989a).

The absence of strict lineage relationships among the cells of a single ommatidium implies that cellular identities are instead determined by extrinsic cues. Unlike furrow progression or R8 founder cell specification, ommatidial assembly is thought to be directed primarily by a series of strictly local interactions between already committed cells of the maturing cluster and their immediate, uncommitted neighbors (Tomlinson and Ready, 1987a). This type of event is termed an inductive interaction, which is defined as an interaction between different types of cells that results in one cell instructing the other to select a certain fate. Any

requirement for the activities of short-range or long-range diffusible factors produced by nearby ommatidia in this process is apparently negligible because isolated ommatidia that arise far from any other ommatidia in the retinae of hypermorphic EGFR homolog mutants are quite normal in structure and orientation (Fig. 24; Baker and Rubin, 1989). Additional evidence for local inductive events within the growing ommatidium has come, not surprisingly, from mutational analyses in *Drosophila*. In particular, the induction of the R7 photoreceptor cell, the last photoreceptor cell to join the ommatidial cluster, has been the subject of intensive study. The specification of the R7 precursor cell is thought to be typical of the numerous inductive events needed to build a complete ommatidium, and it illustrates how multiple developmental mechanisms cooperate to limit the outcome of these events to a single, uniquely positioned cell in the ommatidium.

Genetic screens for flies with aberrant visual behavior led to the identification of two loci, *sevenless* and *bride of sevenless*, that share a remarkably specific mutant phenotype — the absence of only the UV-sensitive R7 cell in every ommatidium of the eye (Reinke and Zipursky, 1988; Harris *et al.*, 1976). Null mutations in either gene are homozy-

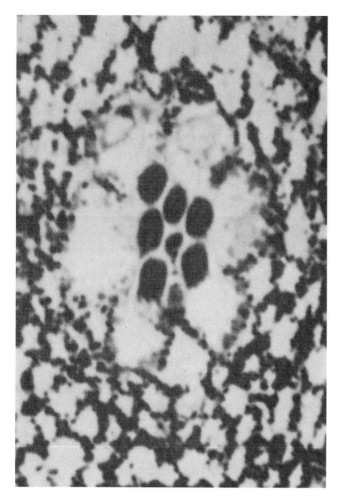

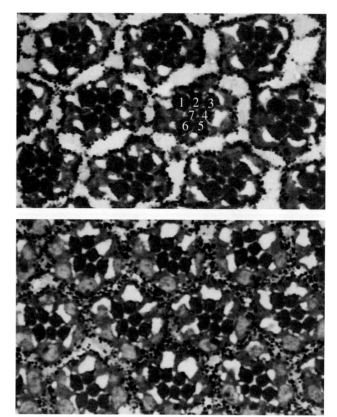

FIGURE 24 Micrograph of the eye of a *Drosophila* mutant bearing a hyperactive form of the EGFR homolog. As can be seen, ommatidia can develop normally even in the absence of nearby ommatidia. This observation implies that, unlike morphogenetic furrow progression and ommatidial precluster formation (which rely on secreted factors from nearby disc territories or neighboring cells), specification of cell types within an already formed ommatidial cell cluster occurs through local cell interactions within the group (courtesy of N. E. Baker and G. M. Rubin).

FIGURE 25 Micrographs of ommatidia from (top) a wild-type and (bottom) a *sevenless* mutant fly. Although in every ommatidium of the wild-type fly there is a central R7 cell with a small rhabdomere, the ommatidium of a mutant *sevenless* fly contains the normal six photoreceptors R1–R6 but is completely missing the R7 photoreceptor. The R8 photoreceptor is present in both wild-type and *sevenless* flies and is located in more basal tissue sections not included in these micrographs.

gous viable and display no detectable defects other than missing R7 cells, even though both genes are expressed in other tissues in addition to the eye imaginal discs (Fig. 25). Developmental analyses have revealed that in *sevenless* and *bride of sevenless* mutants, the uncommitted cell that finds itself in contact with the R1, R6, and R8 precursor cells is transformed from an R7 precursor into a cone cell precursor (Reinke and Zipursky, 1988; Tomlinson and Ready, 1986, 1987b). Mosaic studies have again been instrumental in clarifying the roles of these genes in eye development. In ommatidia of mixed cellular genotype that arise along the border between *sevenless* mutant and wild-type tissues,

functional *sevenless* gene activity is needed only in the R7 precursor cell for its proper formation, indicating that the *sevenless* gene product acts autonomously in the presumptive R7 cell to receive or interpret the inductive signal (Tomlinson and Ready, 1987b). In contrast, *bride of sevenless* is required only in the R8 cell for the nonautonomous specification of the R7 cell in mosaic ommatidia, and thus it is needed to produce or transmit the inductive signal (Reinke and Zipursky, 1988).

Consistent with these mosaic data, molecular studies have revealed that *sevenless* encodes a transmembrane receptor tyrosine kinase (Hafen *et al.*, 1987) and *bride of sevenless* encodes the putative membrane-anchored ligand for the Sevenless receptor (Krämer *et al.*, 1991; Hart *et al.*, 1990). Receptor tyrosine kinases span the cell membrane and phosphorylate substrate proteins in response to an extracellular signal, and many members of this large protein family play prominent roles in developmental signaling and cancer progression (Ullrich and Schlessinger, 1990). In larval eye imaginal discs, Sevenless protein is detected transiently at the apical microvillar membrane surface of a subset of ommatidial cells (Tomlinson *et al.,* 1987; Banerjee *et al.,* 1987). Sevenless accumulates first in the R3 and R4 precursor cells and the mystery cells, and weakly in the R1 and R6 precursor cells, and later in the R7 precursor cell and cone cell precursors (Fig. 26). Transient expression of *sevenless* precedes the overt differentiation of each of these

FIGURE 27 Fluorescence micrograph showing the distribution of the Boss protein (in green). Boss accumulates at high levels in each ommatidium specifically on the apical microvilli of the R8 cell precursors and in vesicles within neighboring R7 cell precursors, giving rise to an evenly spaced pattern posterior to the morphogenetic furrow in wild-type eye imaginal discs. The limited distribution of Boss results in Sevenless activation in only one cell per ommatidium, such that only a single R7 cell is normally produced. Indiscriminate expression of Boss under experimental conditions can force many additional cells into an inappropriate pathway of R7 development. In this disc preparation, the anterior is at the top. The anti-Boss antibodies were kindly provided by H. Krämer and S. L. Zipursky.

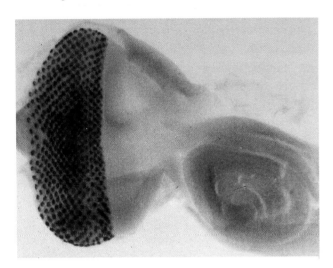

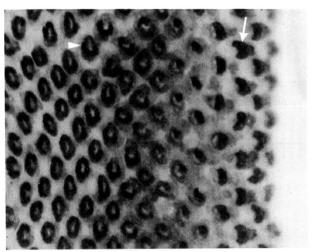

FIGURE 26 Micrographs at (top) low and (bottom) high resolution showing the distribution of the Sevenless protein in the developing *Drosophila* eye. In a, it can be seen that the Sevenless protein (darker areas) accumulates on the apical microvillar surface of the eye disc epithelium posterior to the morphogenetic furrow. The micrograph in b shows the apical profiles of R3 and R4 (arrow) cell precursors and of R7 cell and cone cell precursors (arrowhead), which accumulate high levels of Sevenless in a dynamic spatiotemporal pattern. Under certain conditions, the R3, R4, and cone cell precursors all have the ability to be misrouted into the R7 cell fate. The discs are oriented with the anterior at right. Sevenless antibodies kindly provided by G. M. Rubin.

cell types, and no expression is seen in the R2, R5, and R8 precursor cells. In contrast, the Bride of sevenless protein accumulates specifically on the apical microvilli of the R8 precursor cell throughout this period of eye development (Fig. 27; Krämer *et al.,* 1991). Direct binding of the Sevenless receptor protein to its putative ligand Bride of sevenless has been inferred from the colocalization of both proteins within large intracellular membranous structures, termed multivesicular bodies, that are detected in the R7 precursor cell and in *sevenless*-expressing tissue culture cells that have been aggregated with *bride of sevenless*-expressing cells (Krämer *et al.,* 1991). The functional significance of these multivesicular bodies is not entirely clear; they may be sites for sorting of internalized receptor–ligand complexes, analogous to the late endosomal compartment of mammalian cells, or they may play a more direct role in R7 cell induction (Zipursky and Rubin, 1994).

How important are the spatially restricted distributions of the Sevenless and Bride of sevenless proteins for the eventual specification of a single R7 cell at a precise position within the ommatidium? This question has been addressed by placing the expression of each protein under the control of a heat shock promoter, allowing ubiquitous expression throughout the eye imaginal disc. In transgenic flies carrying these engineered hybrid genes, exposure of the flies to high temperature for a defined period of time at the appropriate developmental stage results in the accumulation of the desired protein in all cells of the fly. Surprisingly, ubiquitous expression of Sevenless using this approach has no apparent developmental consequences other than the rescue of R7 cells in *sevenless* mutant animals (Bowtell *et al.,* 1989; Basler and Hafen, 1989). Ubiquitous expression of the ligand Bride of sevenless, however, results in the transformation of cone cell precursors into R7 cells, producing adult ommatidia with several extra R7 cells (Van Vactor *et al.,* 1991). In these flies, ectopic activation of the Sevenless receptor tyrosine kinase in uncommitted cone cell precursors that express Sevenless but normally do not contact the ligand-bearing R8 cell causes them to adopt the R7 cell fate. Thus, the spatially restricted presentation of the ligand by only the R8 precursor cell is normally a critical regulatory feature of this inductive interaction. Restricted localization of both receptor and ligand to single cells in such a developmental context would be unlikely because it would imply that both the signaling cell and the receiving cell have already been uniquely specified prior to the receptor–ligand interaction. Spatially restricted distributions of cell surface receptors and their ligands are likely to play major roles in many developmental processes, including other aspects of compound eye development. As discussed earlier, the localized expression of *wingless* along the eye disc margin is crucial for the proper dorsoventral alignment of the morphogenetic furrow.

Genetic screens have elucidated many of the cytoplasmic and nuclear events that follow ligand-induced activation of the Sevenless receptor (Fig. 28). Stimulation of the intracellular tyrosine kinase activity of Sevenless leads to the subsequent activation of Ras1, a small GTPase of the evolutionarily conserved $p21^{ras}$ superfamily, via the putative guanine nucleotide exchange protein encoded by *Son of sevenless* and the receptor-binding protein encoded by the *downstream of receptor kinase* gene (Zipursky and Rubin, 1994). Ras1 activity is inhibited by the GTPase-activating protein Gap1; this inhibition is overcome by Sevenless activity in the R7 precursor cell. Ras1 activation in turn triggers a signaling cascade of cytoplasmic kinases of the Raf, MAPK kinase, and MAP kinase families (Zipursky and Rubin, 1994). Many nuclear factors have been implicated as potential targets of this kinase cascade, including DNA-binding proteins related to the mammalian ETS and Jun transcriptional regulators (Dickson, 1995).

The signaling cascade downstream of Sevenless bears a strong resemblance to those operating downstream of many receptor tyrosine kinases in other organisms, including mammals, and it also functions downstream of other receptor tyrosine kinases in *Drosophila* (Hafen *et al.,* 1993). In contrast to mutations in *sevenless* and *bride of sevenless,* mutations in most other components of the pathway have

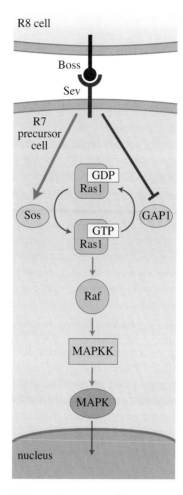

FIGURE 28 Schematic representation of the signaling pathway involving the Sevenless receptor tyrosine kinase during induction of the R7 photoreceptor precursor in *Drosophila.* Sevenless protein (Sev) in the R7 cell precursor is activated by binding to its ligand Bride of Sevenless (Boss), presented by the adjacent R8 cell. This interaction results in the activation of Ras1, a small GTPase, via stimulation of the activity of its guanine nucleotide exchange factor, termed Son of Sevenless (Sos), and inhibition of its GTPase-activating protein, termed Gap1. Ras1 activation leads to the activation of a cascade of protein kinases whose first member is Raf. The Raf protein regulates the kinase activities of downstream components of this cascade — MAP kinase kinase (MAPKK) and MAP kinase (MAPK) — which phosphorylate and thereby regulate transcription factors in the nucleus. The general features of this signaling pathway have been conserved in multicellular organisms ranging from simple worms and flies to humans, and this pathway appears to represent a very widespread mechanism for intercellular signaling mediated by receptor tyrosine kinase proteins.

widespread effects on embryonic and postembryonic tissue development in the fly. Several components, including the Ras1, Son of sevenless, and Downstream of receptor kinase proteins, are required for the development of all eight photoreceptor cells, raising the possibility that the sequential recruitment of different photoreceptor cell precursors into the maturing ommatidium may be effected by a battery of cell-type specific receptor–ligand pairs that activate a common downstream signaling pathway (Simon *et al.*, 1991). Moreover, the EGFR homolog, another receptor tyrosine kinase active in ommatidial assembly, also utilizes the Ras1 pathway. These observations led to the notion that this pathway acts as a general biochemical switch with little intrinsic information content instead of a specialized instructive signal dedicated to R7 cell development (Zipursky and Rubin, 1994; Dickson and Hafen, 1993). A major challenge for future studies in *Drosophila* and in mammalian cells is to determine how different physiological responses are elicited by the stimulation of different receptor tyrosine kinases that apparently use the same downstream effector pathway.

During studies on the molecular events downstream of Sevenless activation, it was found that expression of constitutively activated Sevenless, Ras1, or Raf under *sevenless* gene regulatory sequences resulted in a cell fate transformation opposite to that seen in *sevenless* mutants, namely, transformation of cone cells into extra R7 cells (Fig. 29; Fortini *et al.*, 1992; Dickson *et al.*, 1992; Basler *et al.*, 1991). These studies suggest that the eye disc cells that normally become the cone cell and R7 cell precursors comprise a group of equivalent immature cells having just two developmental options, with the final outcome dictated by the activity of the Ras1 switch (Fig. 30). However, the R3 and R4 precursor cells also express *sevenless,* but they were not affected in these experiments nor are they induced to differentiate as R7 cells by their R8 neighbors in wild-type flies. An attractive hypothesis is that the Ras1 pathway is switched on in these cells in a Sevenless-independent manner, as required for R1–R6 photoreceptor cell develop-

ment, but that prior commitment of these cells to alternative photoreceptor cell fates prevents them from differentiating as R7 cells. Prior commitment of R1–R6 cells constitutes a second mechanism, in addition to spatial localization of the Bride of sevenless ligand, for ensuring the fidelity of R7 cell specification within the precisely arranged ommatidium.

Evidence supporting this view has come from studies on the *seven-up* and *rough* genes, which encode putative transcription factors required for development of R1–R6 photoreceptor cell subpopulations (Mlodzik *et al.*, 1990; Saint *et al.*, 1988; Tomlinson *et al.*, 1988). In mutants lacking *seven-up* or *rough* function, R1–R6 cell commitment is impaired, causing some of these cells to assume an R7 cell fate instead (Heberlein *et al.*, 1991; Van Vactor *et al.*, 1991; Mlodzik *et al.*, 1990). Conversely, ectopic expression of *rough* in the R7 precursor cell results in the opposite cell fate transformation, turning the R7 cell into an R1–R6-like cell (Fig. 31; Kimmel *et al.*, 1990; Basler *et al.*, 1990). Significantly, this transformation requires wild-type *sevenless* gene activity, reinforcing the notion that the Sevenless–Ras1 pathway operates as a general switch used to select a developmental program from among the available options. Exactly which programs are available presumably depends on the prior history of the cell in question and the activity of its transcriptional regulators, such as the homeodomain protein encoded by the *rough* gene.

A third mechanism contributes to R7 cell fate specification by regulating the general competence of cells to respond to developmental cues. This mechanism involves the action of the Notch transmembrane receptor, whose role in the lateral specification of R8 precursor cells has already been described. Studies in different organisms have shown that Notch and its ligands act not only in lateral interactions between initially equivalent cells but also in inductive interactions between different cell types (Artavanis-Tsakonas, 1995). In the *Drosophila* eye, as in the embryonic central nervous system, Notch activity is required at multiple times in development for the proper execution of numerous cell fate decisions (Cagan and Ready, 1989b). Transient expression of a constitutively activated intracellular domain of the Notch receptor under *sevenless* gene regulatory elements results in the transformation of R1–6 precursor cells into ectopic R7 cells and the transformation of the true R7 precursor cell into a nonneuronal cone cell (Fig. 32; Fortini *et al.*, 1993). These cell fate transformations closely resemble those seen in *seven-up, rough, sevenless,* and *bride of sevenless* mutants, as if inappropriate Notch activation renders the ommatidial precursors insensitive to their normal inductive cues. Cell fate changes have also been obtained with similarly activated Notch proteins in vertebrates, indicating that the control of cellular differentiation is a general and conserved function of the Notch signaling pathway (Artavanis-Tsakonas, 1995).

Unlike *hedgehog, decapentaplegic, bride of sevenless,*

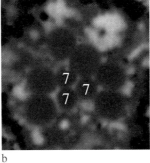

a   b

FIGURE 29 Activation of Ras1, a component of the Sevenless pathway, results in supernumerary R7 cells in the adult ommatidium. (a) A wild-type ommatidium with labeled photoreceptor elements contains a single R7 cell. (b) A transgenic fly bearing a *sevenless* promoter-activated Ras1 construct contains an ommatidium with three central R7 rhabdomeres, indicating that two cone cell precursors have been forced to adopt the R7 cell fate (reproduced with permission from Fortini *et al.*, 1992).

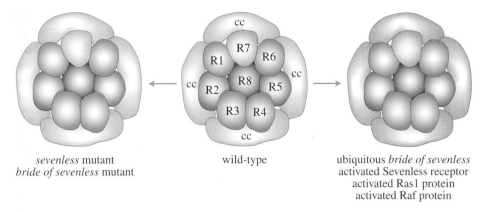

sevenless mutant
bride of sevenless mutant

wild-type

ubiquitous *bride of sevenless*
activated Sevenless receptor
activated Ras1 protein
activated Raf protein

FIGURE 30 In particular mutant flies or under certain experimental conditions, the precursors of the cone cells (cc) can be misrouted into the cell fate of the R7 precursors and vice versa. The *sevenless* and *bride of sevenless* mutations both cause the transformation of R7 precursors into cone cells (left, purple). Conversely, spatially unrestricted expression of the Bride of Sevenless protein or expression of constitutively activated Sevenless, Ras1, or Raf causes the transformation of cone cell precursors into R7 cells (right, yellow). A complete failure in Sevenless signaling causes these cells to all adopt the cone cell fate, whereas indiscriminate Sevenless signaling causes the opposite effect of driving all these cells into the R7 photoreceptor fate.

and *rough*, whose activities in eye development depend critically on their restricted expression patterns, *Notch* is expressed ubiquitously throughout the eye imaginal disc during development (Parks *et al.*, 1995; Kooh *et al.*, 1993). Moreover, overexpression of wild-type Notch protein under *sevenless* gene regulatory sequences has no apparent effect on eye development, indicating that the activation state of Notch, and not just its expression level, is crucial for local inductive events such as R7 cell fate specification (For-

tini *et al.*, 1993). Notch activity in the eye is presumably regulated by its ligand Delta, which exhibits a complex and dynamic subcellular distribution in cells of the eye imaginal disc (Parks *et al.*, 1995; Kooh *et al.*, 1993). Additional research is needed to clarify the relationship of Notch signaling activity to the other two known mechanisms involved in R7 cell induction, namely, the spatially restricted activation of the Sevenless receptor tyrosine kinase and the prior cell fate commitment of the R1–R6 photoreceptor cell classes.

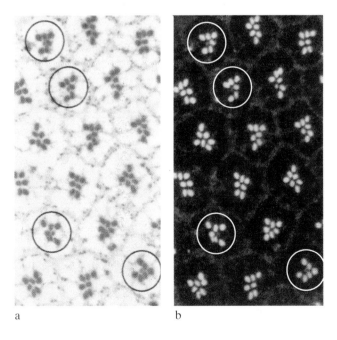

a   b

FIGURE 31 Micrographs showing (a) brightfield and (b) fluorescence views of the retinal field of adult ommatidia from a transgenic fly bearing a construct with the *rough* gene under the control of the promoter of *sevenless*. With the exception of a few unaffected wild-type ommatidia (circled), the others contain a cell similar to that of the R1–R6 group in the position normally occupied by the R7 cell, as revealed in b by immunostaining for the R1–R6-specific opsin protein. Although the wild-type ommatidia have six cells which stain positively in immunofluorescence, the other ommatidia have seven positive cells, one of which was generated from an R7 precursor. Unlike the perturbations in Sevenless signaling, which invariably induce the precursors of the R7 cells to transform into cone cells or vice versa, altered expression (in an R7 precursor) of the Rough protein reprograms the fate of this cell and converts it into that of one of the R1–R6 group. This cell-fate transformation indicates that the Rough protein functions as part of the biochemical code that controls photoreceptor subtype identity using a developmental mechanism distinct from Sevenless signaling. Consistent with this notion, the R1–R6-like cell that occupies the R7 position in these transgenic flies requires *sevenless* gene function for its formation, regardless of its altered cellular identity. The Rough protein belongs to a large class of evolutionarily related transcription factors bearing the homeodomain DNA-binding motif. This class of proteins has a prominent role in developmental patterning and cell-type specification in diverse organisms (reproduced with permission from Kimmel *et al.*, 1990).

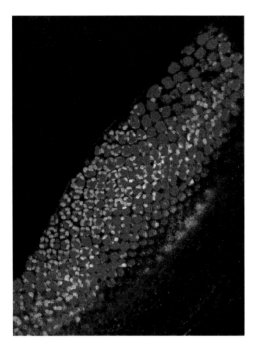

FIGURE 32 An image obtained by laser scanning confocal microscopy of an eye disc from a transgenic larva of *Drosophila* in which the expression of the intracellular domain of Notch is under the control of the *sevenless* gene. The image reveals that R7 photoreceptor cell precursors expressing activated Notch (green) are prevented from adopting the appropriate neuronal fate (red). This result implies that during normal R7 cell induction, the Notch pathway must be repressed in the presumptive R7 precursor cell in order for it to respond properly to the Sevenless pathway signal. In this micrograph, the disc is oriented with the anterior at right (reproduced with permission from Artavanis-Tsakonas *et al.*, 1995).

Conclusion

The experimental advantages of the *Drosophila* compound eye have allowed rapid progress in understanding the development of this highly ordered biological structure. The picture of insect eye development that has emerged is one in which numerous signaling events cooperate to transform an initially unpatterned monolayer epithelium into a remarkably precise assembly of differentiated cells. At the molecular level, long-range and short-range diffusible factors, such as Hedgehog, Decapentaplegic, Scabrous, and Argos, organize the immature disc epithelium into progressively more restricted cellular territories. Local interactions between neighboring cells within these territories, mediated by molecules such as Notch, Delta, Sevenless, and Bride of sevenless, are responsible for the final specification of individual cell types. The precision of the adult retina, essential for its physiological function, is thus achieved by sequential signaling events that operate within a context provided by prior patterning.

Perhaps most reassuring to those who study *Drosophila* eye development, not to mention the government agencies and private philanthropies that fund the research, is that the molecular mechanisms uncovered to date are evolutionarily conserved and appear to play key roles in mammalian development. Numerous developmental processes in mammals require the activities of the mammalian Sonic hedgehog, Wnt, and Decapentaplegic-like TGF-β proteins (Ingham, 1994; Jessell and Melton, 1992). Mammalian tyrosine kinases and their associated Ras proteins have been implicated in normal development and oncogenesis in many organisms, including humans (Ullrich and Schlessinger, 1990). Cancerous growth is often the consequence of a failure to regulate properly the signaling activities of these pathways. For example, an altered human Notch receptor is associated with a particular form of leukemia (Ellisen *et al.*, 1991), but Notch may act normally during vertebrate neurogenesis in a manner similar to its role in R8 cell specification in the *Drosophila* eye (Austin *et al.*, 1995). Several other aspects

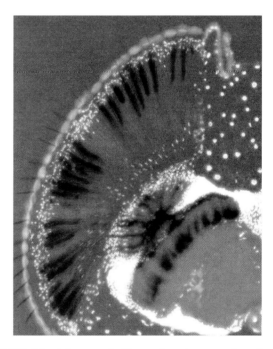

FIGURE 33 Micrograph illustrating the axonal projection pattern of one class of R7 cells to a specialized synaptic target within the medulla, an optic lobe of the brain that receives inputs from the R7 and R8 photoreceptors but not from the R1–R6 class of photoreceptors. The photoreceptor cell bodies and their axons are visualized by histochemical staining for β-galactosidase in transgenic flies carrying a construct with a rhodopsin promoter–*lacZ* gene promoter fusion. As revealed by the blue staining, axons from a subset of R7 photoreceptor cells in the adult *Drosophila* retina extend into the optic lobes of the brain and make synaptic connections with interneurons at a specific target layer of the medulla. Nuclei are labeled with a fluorescent dye (white), and the adult retina is oriented with the anterior at the top (reproduced with permission from Fortini and Rubin, 1990).

of *Drosophila* eye development not considered here, such as cell death, cell cycle regulation, and axon pathfinding, also share molecular details with mammalian development (Fig. 33). In this respect, it is sobering to recall that Anton van Leeuwenhoek, perhaps the first person to examine an insect under a microscope lens, prefaced his remarks on the insect compound eye with the assertion that "provident Nature works in all Creatures, from the biggest to the smallest, almost in one and the same way" (Leeuwenhoek, 1695/1979, p. 251). Research on the molecular genetics of *Drosophila* eye development has proven him correct and has demonstrated the value of using simple model organisms to address questions of general significance in developmental biology.

Acknowledgments

I thank Nick Baker, Ulrike Heberlein, Kevin Moses, Donald Ready, Gerry Rubin, and Tanya Wolff for providing photomicrographs and other illustration material. I am grateful to Helmut Krämer and Larry Zipursky for antibodies and to all my colleagues in the Rubin and Artavanis-Tsakonas laboratories for advice and productive discussions.

References Cited

ARTAVANIS-TSAKONAS, S., MATSUNO, K. and FORTINI, M. E. (1995). Notch signaling. *Science* **268**, 225–232.

AUSTIN, C. P., FELDMAN, D. E., IDA, J. A., JR., and CEPKO, C. L. (1995). Vertebrate retinal ganglion cells are selected from competent progenitors by the action of *Notch. Development* **121**, 3637–3650.

BAKER, N. E., and RUBIN, G. M. (1989). Effect on eye development of dominant mutations in *Drosophila* homologue of the EGF receptor. *Nature* **340**, 150–153.

BAKER, N. E., and ZITRON, A. E. (1995). *Drosophila* eye development: *Notch* and *Delta* amplify a neurogenic pattern conferred on the morphogenetic furrow by *scabrous. Mech. Dev.* **49**, 173–189.

BAKER, N. E., MLODZIK, M., and RUBIN, G. M. (1990). Spacing differentiation in the developing *Drosophila* eye: A fibrinogen-related lateral inhibitor encoded by *scabrous. Science* **250**, 1370–1377.

BANERJEE, U., RENFRANZ, P. J., POLLOCK, J. A., and BENZER, S. (1987). Molecular characterization and expression of *sevenless,* a gene involved in neuronal pattern formation in the *Drosophila* eye. *Cell* **49**, 281–291.

BASLER, K., and HAFEN, E. (1989). Ubiquitous expression of *sevenless:* Position-dependent specification of cell fate. *Science* **243**, 931–934.

BASLER, K., YEN, D., TOMLINSON, A., and HAFEN, E. (1990). Reprogramming cell fate in the developing *Drosophila* retina: Transformation of R7 cells by ectopic expression of *rough. Genes Dev.* **4**, 728–739.

BASLER, K., CHRISTEN, B., and HAFEN, E. (1991). Ligand-independent activation of the *sevenless* receptor tyrosine kinase changes the fate of cells in the developing *Drosophila* eye. *Cell* **64**, 1069–1081.

BERNARD, F. (1937). Recherches sur la morphogénèse des yeux composés d'arthropodes. *Bull. Biol. France Belgium* **23**(Suppl.), 3–162.

BOWTELL, D. D. L., SIMON, M. A., and RUBIN, G. M. (1989). Ommatidia in the developing *Drosophila* eye require and can respond to *sevenless* for only a restricted period. *Cell* **56**, 931–936.

BROWN, N. L., SATTLER, C. A., PADDOCK, S. W., and CARROLL, S. B. (1995). Hairy and Emc negatively regulate morphogenetic furrow progression in the *Drosophila* eye. *Cell* **80**, 879–887.

CAGAN, R. L., and READY, D. F. (1989a). The emergence of order in the *Drosophila* pupal retina. *Dev. Biol.* **136**, 346–362.

CAGAN, R. L., and READY, D. F. (1989b). *Notch* is required for successive cell decisions in the developing *Drosophila* retina. *Genes Dev.* **3**, 1099–1112.

CAMPOS-ORTEGA, J. A. (1993). Early neurogenesis in *Drosophila melanogaster*. In *The Development of Drosophila melanogaster,* Vol. 2, pp. 1091–1129 Cold Spring Harbor Laboratory Press, Cold Spring Harbor, NY.

DICKSON, B. (1995). Nuclear factors in sevenless signalling. *Trends Genet.* **11**, 106–111.

DICKSON, B., and HAFEN, E. (1993). Genetic dissection of eye development in *Drosophila*. In *The Development of Drosophila melanogaster,* Vol. 2, pp. 1327–1362. Cold Spring Harbor Laboratory Press, Cold Spring Harbor, NY.

DICKSON, B., SPRENGER, F., MORRISON, D., and HAFEN, E. (1992). Raf functions downstream of Ras1 in the Sevenless signal transduction pathway. *Nature* **360**, 600–603.

ELLISEN, L. W., BIRD, J., WEST, D. C., SORENG, A. L., REYNOLDS, T. C., SMITH, S. D., and SKLAR, J. (1991). *TAN-1,* the human homolog of the *Drosophila Notch* gene, is broken by chromosomal translocations in T lymphoblastic neoplasms. *Cell* **66**, 649–661.

FORTINI, M. E., and RUBIN, G. M. (1990). Analysis of *cis*-acting requirements of the *Rh3* and *Rh4* genes reveals a bipartite organization to rhodopsin promoters in *Drosophila melanogaster*. *Genes Dev.* **4**, 444–463.

FORTINI, M. E., SIMON, M. A., and RUBIN, G. M. (1992). Signalling by the *sevenless* protein tyrosine kinase is mimicked by Ras1 activation. *Nature* **355**, 559–561.

FORTINI, M. E., REBAY, I., CARON, L. A., and ARTAVANIS-TSAKONAS, S. (1993). An activated Notch receptor blocks cell-fate commitment in the developing *Drosophila* eye. *Nature* **365**, 555–557.

FREEMAN, M., KLÄMBT, C., GOODMAN, C. S., and RUBIN, G. M. (1992). The *argos* gene encodes a diffusible factor that regulates cell fate decisions in the *Drosophila* eye. *Cell* **69**, 963–975.

GARCIA-BELLIDO, A., and MERRIAM, J. R. (1969). Cell lineage of the imaginal discs in *Drosophila* gynandromorphs. *J. Exp. Zool.* **170**, 61–76.

HAFEN, E., BASLER, K., EDSTROEM, J.-E., and RUBIN, G. M. (1987). *Sevenless,* a cell-specific homeotic gene of *Drosophila,* encodes a putative transmembrane receptor with a tyrosine kinase domain. *Science* **236**, 55–63.

HAFEN, E., DICKSON, B., RAABE, T., BRUNNER, D., OELLERS, N., and VAN DER STRATEN, A. (1993). Genetic analysis of the sevenless signal transduction pathway of *Drosophila. Dev. Suppl.,* 41–46.

HARRIS, W. A., STARK, W. S., and WALKER, J. A. (1976). Genetic dissection of the photoreceptor system in the compound eye of *Drosophila melanogaster. J. Physiol. (London)* **256,** 415–439.

HART, A. C., KRÄMER, H., VAN VACTOR, D. L., JR., PAIDHUN-GAT, M., and ZIPURSKY, S. L. (1990). Induction of cell fate in the *Drosophila* retina: The bride of sevenless protein is predicted to contain a large extracellular domain and seven transmembrane segments. *Genes Dev.* **4,** 1835–1847.

HEBERLEIN, U., and MOSES, K. (1995). Mechanisms of *Drosophila* retinal morphogenesis: The virtues of being progressive. *Cell* **81,** 987–990.

HEBERLEIN, U., MLODZIK, M., and RUBIN, G. M. (1991). Cell-fate determination in the developing *Drosophila* eye: Role of the *rough* gene. *Development* **112,** 703–712.

HEBERLEIN, U., WOLFF, T., and RUBIN, G. M. (1993). The TGFβ homolog *dpp* and the segment polarity gene *hedgehog* are required for propagation of a morphogenetic wave in the *Drosophila* retina. *Cell* **75,** 913–926.

HEBERLEIN, U., SINGH, C. M., LUK, A. Y., and DONOHOE, T. J. (1995). Growth and differentiation in the *Drosophila* eye coordinated by *hedgehog. Nature* **373,** 709–711.

HU, X., LEE, E.-C., and BAKER, N. E. (1995). Molecular analysis of *scabrous* mutant alleles from *Drosophila melanogaster* indicates a secreted protein with two functional domains. *Genetics* **141,** 607–617.

INGHAM, P. W. (1994). *Hedgehog* points the way. *Curr. Biol.* **4,** 347–350.

JARMAN, A. P., GRAU, Y., JAN, L. Y., and JAN, Y. N. (1993). *atonal* is a proneural gene that directs chordotonal organ formation in the *Drosophila* peripheral nervous system. *Cell* **73,** 1307–1321.

JARMAN, A. P., GRELL, E. H., ACKERMAN, L., JAN, L. Y., and JAN, Y. N. (1994). *atonal* is the proneural gene for *Drosophila* photoreceptors. *Nature* **369,** 398–400.

JARMAN, A. P., SUN, Y., JAN, L. Y., and JAN, Y. N. (1995). Role of the proneural gene, *atonal,* in formation of *Drosophila* chordotonal organs and photoreceptors. *Development* **121,** 2019–2030.

JESSELL, T. M., and MELTON, D. A. (1992). Diffusible factors in vertebrate embryonic induction. *Cell* **68,** 257–270.

KALDERON, D. (1995). Morphogenetic signalling. Responses to *hedgehog. Curr. Biol.* **5,** 580–582.

KIMMEL, B. E., HEBERLEIN, U., and RUBIN, G. M. (1990). The homeodomain protein *rough* is expressed in a subset of cells in the developing *Drosophila* eye where it can specify photoreceptor cell subtype. *Genes Dev.* **4,** 712–727.

KOOH, P. J., FEHON, R. G., and MUSKAVITCH, M. A. T. (1993). Implications of dynamic patterns of Delta and Notch expression for cellular interactions during *Drosophila* development. *Development* **117,** 493–507.

KRÄMER, H., CAGAN, R. L., and ZIPURSKY, S. L. (1991). Interaction of *bride of sevenless* membrane-bound ligand and the *sevenless* tyrosine-kinase receptor. *Nature* **352,** 207–212.

KRETZSCHMAR, D., BRUNNER, A., WIERSDORFF, V., PFLUG-FELDER, G. O., HEISENBERG, M., and SCHNEUWLY, S. (1992). *giant lens,* a gene involved in cell determination and axon guidance in the visual system of *Drosophila melanogaster. EMBO J.* **11,** 2531–2539.

LAWRENCE, P. A., and GREEN, S. M. (1979). Cell lineage in the developing retina of *Drosophila. Dev. Biol.* **71,** 142–152.

LEBOVITZ, R. M., and READY, D. F. (1986). Ommatidial develop-

ment in *Drosophila* eye disc fragments. *Dev. Biol.* **117,** 663–671.

MA, C., and MOSES, K. (1995). *wingless* and *patched* are negative regulators of the morphogenetic furrow and can affect tissue polarity in the developing *Drosophila* compound eye. *Development* **121,** 2279–2289.

MA, C., ZHOU, Y., BEACHY, P. A., and MOSES, K. (1993). The segment polarity gene *hedgehog* is required for progression of the morphogenetic furrow in the developing *Drosophila* eye. *Cell* **75,** 927–938.

MARDON, G., SOLOMON, N. M., and RUBIN, G. M. (1994). *dachshund* encodes a nuclear protein required for normal eye and leg development in *Drosophila. Development* **120,** 3473–3486.

MLODZIK, M., HIROMI, Y., WEBER, U., GOODMAN, C. S., and RUBIN, G. M. (1990). The *Drosophila seven-up* gene, a member of the steroid receptor gene superfamily, controls photoreceptor cell fates. *Cell* **60,** 211–224.

NUSSE, R., and VARMUS, H. E. (1992). *Wnt* genes. *Cell* **69,** 1073–1087.

OKANO, H., HAYASHI, S., TANIMURA, T., SAWAMOTO, K., YOSHI-KAWA, S., WATANABE, J., IWASAKI, M., HIROSE, S., MIKO-SHIBA, K., and MONTELL, C. (1992). Regulation of *Drosophila* neural development by a putative secreted protein. *Differentiation* **52,** 1–11.

PADGETT, R. W., ST. JOHNSTON, R. D., and GELBART, W. M. (1987). A transcript from a *Drosophila* pattern gene predicts a protein homologous to the transforming growth factor-β family. *Nature* **325,** 81–84.

PAN, D., and RUBIN, G. M. (1995). cAMP-dependent protein kinase and *hedgehog* act antagonistically in regulating *decapentaplegic* transcription in *Drosophila* imaginal discs. *Cell* **80,** 543–552.

PARKS, A. L., TURNER, F. R., and MUSKAVITCH, M. A. T. (1995). Relationships between complex Delta expression and the specification of retinal cell fates during *Drosophila* eye development. *Mech. Dev.* **50,** 201–216.

PERRIMON, N. (1995). Hedgehog and beyond. *Cell* **80,** 517–520.

READY, D. F., HANSON, T. E., and BENZER, S. (1976). Development of the *Drosophila* retina, a neurocrystalline lattice. *Dev. Biol.* **53,** 217–240.

REINKE, R., and ZIPURSKY, S. L. (1988). Cell–cell interaction in the *Drosophila* retina: The *bride of sevenless* gene is required in photoreceptor cell R8 for R7 cell development. *Cell* **55,** 321–330.

SAINT, R., KALIONIS, B., LOCKETT, T. J., and ELIZUR, A. (1988). Pattern formation in the developing eye of *Drosophila melanogaster* is regulated by the homeobox gene, *rough. Nature* **334,** 151–154.

SCHEJTER, E. D., SEGAL, D., GLAZER, L., and SHILO, B.-Z. (1986). Alternative 5′ exons and tissue-specific expression of the *Drosophila* EGF receptor homolog transcripts. *Cell* **46,** 1091–1101.

SCHWEITZER, R., HOWES, R., SMITH, R., SHILO, B.-Z., and FREE-MAN, M. (1995). Inhibition of *Drosophila* EGF receptor activation by the secreted protein Argos. *Nature* **376,** 699–702.

SIMON, M. A., BOWTELL, D. D. L., DODSON, G. S., LAVERTY, T. R., and RUBIN, G. M. (1991). Ras1 and a putative guanine nucleotide exchange factor perform crucial steps in signaling by the Sevenless protein tyrosine kinase. *Cell* **67,** 701–716.

STRUTT, D. I., WIERSDORFF, V., and MLODZIK, M. (1995). Regulation of furrow progression in the *Drosophila* eye by cAMP-dependent protein kinase A. *Nature* **373,** 705–709.

TOMLINSON, A., and READY, D. F. (1986). *Sevenless:* A cell specific homeotic mutation of the *Drosophila* eye. *Science* **231,** 400–402.

TOMLINSON, A., and READY, D. F. (1987a). Neuronal differentiation in the *Drosophila* ommatidium. *Dev. Biol.* **120,** 366–376.

TOMLINSON, A., and READY, D. F. (1987b). Cell fate in the *Drosophila* ommatidium. *Dev. Biol.* **123,** 264–275.

TOMLINSON, A., BOWTELL, D. D. L., HAFEN, E., and RUBIN, G. M. (1987). Localization of the *sevenless* protein, a putative receptor for positional information, in the eye imaginal disc of *Drosophila. Cell* **51,** 143–150.

TOMLINSON, A., KIMMEL, B. E., and RUBIN, G. M. (1988). *rough,* a *Drosophila* homeobox gene required in photoreceptors R2 and R5 for inductive interactions in the developing eye. *Cell* **55,** 771–784.

TREISMAN, J. E., and RUBIN, G. M. (1995). *wingless* inhibits morphogenetic furrow movement in the *Drosophila* eye disc. *Development* **121,** 3519–3527.

TREISMAN, J. E., LAI, Z.-C., and RUBIN, G. M. (1995). *shortsighted* acts in the *decapentaplegic* pathway in *Drosophila* eye development and has homology to a mouse TGF-β-responsive gene. *Development* **121,** 2835–2845.

ULLRICH, A., and SCHLESSINGER, J. (1990). Signal transduction by receptors with tyrosine kinase activity. *Cell* **61,** 203–212.

VAN LEEUWENHOEK, A. (1695/1979). Letter of 18 May 1695. In *The Collected Letters of Antoni van Leeuwenhoek* (L. C. Palm, Ed.), Vol. 10, p. 251. Swets and Zeitlinger, Lisse,.

VAN VACTOR, D. L., JR., CAGAN, R. L., KRÄMER, H., and ZIPURSKY, S. L. (1991). Induction in the developing compound eye of *Drosophila:* Multiple mechanisms restrict R7 induction to a single retinal precursor cell. *Cell* **67,** 1145–1155.

WEHRLI, M., and TOMLINSON, A. (1995). Epithelial planar polarity in the developing *Drosophila* eye. *Development* **121,** 2451–2459.

WIESCHAUS, E., and GEHRING, W. (1976). Clonal analysis of primordial disc cells in the early embryo of *Drosophila melanogaster. Dev. Biol.* **50,** 249–263.

WIGGLESWORTH, V. B. (1940). Local and general factors in the development of "pattern" in *Rhodnius prolixus* (Hemiptera). *J. Exp. Biol.* **17,** 180–200.

WILKINSON, H. A., FITZGERALD, K., and GREENWALD, I. (1994). Reciprocal changes in expression of the receptor *lin-12* and its ligand *lag-2* prior to commitment in a *C. elegans* cell fate decision. *Cell* **79,** 1187–1198.

WOLFF, T., and READY, D. F. (1991). Cell death in normal and rough eye mutants of *Drosophila. Development* **113,** 825–839.

WOLFF, T., and READY, D. F. (1993). Pattern formation in the *Drosophila* retina. In *The Development of Drosophila melanogaster,* Vol. 2, pp. 1277–1325. Cold Spring Harbor Laboratory Press, Cold Spring Harbor, NY.

XU, T., and RUBIN, G. M. (1993). Analysis of genetic mosaics in developing and adult *Drosophila* tissues. *Development* **117,** 1223–1237.

ZIPURSKY, S. L., and RUBIN, G. M. (1994). Determination of neuronal cell fate: Lessons from the R7 neuron of *Drosophila. Annu. Rev. Neurosci.* **17,** 373–397.

PART TWO

The Defenses of the Organism

Edited by Antonio Lanzavecchia, Bernard Malissen, and Roberto Sitia

INTRODUCTION

ANTONIO LANZAVECCHIA

BERNARD MALISSEN

ROBERTO SITIA

In unicellular organisms, survival of the species is based essentially on the rapidity of division and diversification. The more effective these processes are, the more frequent will mutants occur in the populations that are capable of adapting to the new environment and eluding the attack of a hostile species. By means of a classical Darwinian selection process, microorganisms resistant to antibiotics are selected in this way.

Accompanying the development of multicellular organisms is the evolution of ever more sophisticated defense mechanisms. The bigger mammals, whose cycle of reproduction stretches over decades, cannot in fact rely on elementary selective mechanisms but have to elaborate specific strategies to defend themselves against constant attack by pathogens of all types. "The Defenses of the Organism" examines the logic and the functioning of the immune system, the main system of defense that organisms possess.

Our journey through the immune system starts with Benvenuto Pernis, who accompanies us on a short historical excursus, summarizing the main stages of immunological thought, its evolution, and the moments that Thomas Kuhn would define "of revolution" in the discipline. One such moment — the triumph of selective theories over instructive theories — is described by one of the protagonists, Gustav J. V. Nossal, in the second chapter in the first section.

A classic allegory of the immune system concerns a comparison with military defense systems. As does every army, the immune system has to deploy troops that are specialized but well coordinated, sophisticated weapons, and systems of information and counterinformation engaged in continuous battle against often hardly visible enemies. The second section summarizes the main characteristics of the cells and the molecules of the immune system. Fritz Melchers outlines the development and the specialization of the lymphocytes, the most specialized corps in the immune army. Roy Mariuzza shows the structure of the weapons used by the lymphocytes, the antibodies and the receptors for the antigen of T lymphocytes (T cells), both in repose (i.e., free) and at work (linked with their respective antigens). Cristina Rada discusses the mechanisms at the basis of diversity generation, the problem which dominated immunology in the 1960s and 1970s. Jonathan Sprent and David F. Tough, summarize our knowledge of another key topic: immunologic memory.

The efficiency of an army requires the constant collaboration and coordination of the corps that form it. Similarly, the cells of the immune system have to interact constantly with one another and exchange information for the purpose of identifying and neutralizing the enemy, without causing too much damage to the battlefield. The first problem, identifying the target, is crucial for the functioning of the defense mechanisms. The third section is largely dedicated to this fundamental requisite of the immune response. Any excesses or inaccuracies in these mechanisms could unleash responses against components of the organism itself, and thereby cause serious autoimmunity diseases. The T lymphocytes' system is able to recognize potentially dangerous extraneous antigens and to generate an effector response to eliminate them. The principles on which the efficiency, specificity, and flexibility of this system are based are discussed by Antonio Lanzavecchia. Elena A. Armandola, Harald Kropshofer, Anne B. Vogt, and Günter J. Hämmerling discuss how the cells responsible for presenting the antigen carry out their task. It will be shown how elaborate systems guarantee that the immune system is promptly informed of the presence of enemies inside or outside of the cell and how at the same time the lymphocytes share the tasks against these two basic classes of pathogens. The role of these mechanisms in the organism is described by Ralph M. Steinman, whereas John C. Cambier and Idan Tamir describe the mechanisms through which the danger signals generated on the surface of the lymphocytes are transferred to the nucleus and generate the appropriate response, whether this be cell activation or tolerance.

The fourth section stresses the fact that the work of the cells in the immune system is assuredly not a sedentary one but one requiring considerable mobility. In our organism there is an intense traffic of leukocytes. How this is regulated in molecular terms is described by Alberto Mantovani. Soluble factors, cytokines, play a central role in directing the different leukocyte populations to the various sites and in modulating the intensity and type of their response. It is no surprise, therefore, that — as described by M. Raffaela Zocchi and Anna Rubartelli — the production and release of cytokynes are subject to extremely tight regulation imposed at many different levels.

It is also not surprising that many pathogens smuggle themselves into the intercellular communication mechanisms for the purpose of sabotaging them or using them to their own advantage. The fifth section begins with an enlightening example of this counterinformation or "red-herring" activity, the "genetic piracy" of certain viruses, described by Mauro S. Malnati and Paolo Lusso. Lorenzo Moretta and Maria Cristina Mingari take stock of the way in which the delicate balance between response and tolerance is maintained in normal subjects, describing many pathological situations resulting from the breakdown of this balance. An overall vision of the arsenal of the immune system, i.e., of its effector mechanisms against viruses, is provided by Pietro Pala, Tracy Hussel, and Peter J. M. Openshaw. There are two chapters dedicated to the mechanisms of sabotaging the immune system: Jacques A. Louis, Pascal Launois, Reza Behin, Yasmine Belkaid, and Geneviève Milon illustrate an emblematic case — that of the protozoon *Leishmania* — analyzing the strategies adopted by this microorganism to elude the immune response, whereas Emmanuel J. H. J. Wiertz discusses the sabotage of the antigen presentation mechanisms operated by the cytomegaloviruses to dodge the immune defenses. To conclude this section, Guido Poli discusses the battle royal being waged between the human race and the first large-scale retrovirus pathogen for man, the AIDS virus, capable of becoming integrated with our chromosomes and being transmitted as a transgene to our offspring. In his chapter, he summarizes the brief but tumultuous history of the past 15 years, during which victories and defeats have often been — and continue to be — the two faces of one and the same medal.

Lastly, the sixth section illustrates many present and future applications based on the manipulation of the immune system. Antonello Covacci and Rino Rappuoli describe the various stages of vaccination, perhaps the greatest success of medicine regarding the impact on the length and the quality of our life, illustrating the current frontiers of the field, primarily the one regarding tumors. Michael J. Owen describes the development of transgenic animals, one of the most incisive experimental systems for the analysis of genetic diseases and for the targeted development of new therapeutic protocols. Silvia Biocca and Antonino Cattaneo analyze the more or less infinite possibilities of generating differences that characterize the immune system, exploiting the fitting comparison of Borges's Library of Babel. The creation of an artificial immune system and the design of mathematical models capable of describing how they work, discussed by Ulrich Behn, Franco Celada, and Philip E. Seiden, are far more than just possible ideas for science fiction movies.

We realize that the chapters are written in a variety of styles, and that some will not be easy reading. However, would it have been correct to reduce this diversity in a volume on the immune system, whose logic is based precisely on diversity?

And now, happy reading.

Benvenuto Pernis

Departments of Microbiology and Medicine
Columbia University
New York, NY 10032, USA

In this article, concepts that have influenced the progress of immunology in this century, from the discovery of antibodies to molecular biology, are discussed.

Conceptual Milestones in Immunology

Introduction

This article is neither a summary of the immunology dealt with in the other articles of this encyclopedia nor a description of the history of immunology, which would require a book by itself (Silverstein, 1989). It is instead a description, inevitably biased, of the evolution of the basic concepts that have dominated Immunology during this century, particularly the second half. Indeed, this century has witnessed the transformation of this branch of biology from a derivate of microbiology with mostly applied purposes into an independent discipline, itself subdivided into different sections, with an independent conceptual basis and a fascinating capacity for progress.

I have been fortunate to live long enough and to be involved in immunology closely enough to be able to witness directly the evolution of this discipline in the second half of the century. It has been an evolution that has been in part determined by some well-defined discoveries of high relevance (that might be defined as quantal steps) but even more has been marked by the emergence of new areas of investigation. These new areas have been generated in part by these discoveries, in part they have followed the introduction of new experimental methods, and in part they have been the consequence of the interactions of immunology with other disciplines, such as genetics, molecular biology, and cellular biology. The result has been an uninterrupted fruitful expansion.

This article, therefore, is intended as an overview of the conceptual background of immunology in an effort to understand how the current state of this discipline, with its highly dynamic condition, has been reached.

Serology in the Late 1940s

For medical students graduating in the late 1940s, immunology was essentially the study of a reaction of vertebrates to bacterial products, mostly bacterial toxins (Behring and Kitasato, 1890), culminating in the production of specific serum proteins which, in the absence of a better denomination, were called antibodies. The field was dominated by applicative purposes consisting of the use of these "antibodies" for the diagnosis of infectious diseases and for the treatment of some of these, particularly those marked by the production of soluble toxins such as diphtheria or tetanus. Conversely, the bacterial products that induced the production of antibodies were called "antigens." The potential to use these antigens to induce a state of active immunity, specific and protective against the bacterial toxins and bacterial infections, was well-known and overshadowed the theoretical problem of the biological mechanisms that are involved in the production of antibodies. However, curiosity was generated by the fact that specific antibodies could also be produced against nonbacterial products, such as the erythrocytes from another species, and even against small chemicals (haptens) covalently bound to protein vectors.

Gradually, however, there was a growth of interest in the chemical nature of antibodies. First, it was realized that antibodies were included in the fraction of serum proteins called "euglobulins" because they could be precipitated with relatively low concentrations of ammonium sulfate. Toward the end of the 1930s, Arne Tiselius at Uppsala introduced a method, electrophoresis, to separate soluble proteins in an electric field. Soon after, Elvin Kabat brought serum rich in antibodies to Tiselius's laboratory; this showed a prominent peak in the electophoretic fraction called "gamma." The excess gamma globulins disappeared after precipitation of the antibodies with their specific antigen; antibodies were therefore shown to be gamma globulins (Tiselius and Kabat, 1939). Later, studies by Williams and Grabar (1955) with immunoelectrophoresis showed that antibodies could also be found among the serum molecules with alpha or beta mobility; however, these were always antigenically related with the gamma globulins.

In the same period the reactions between antigens and antibodies were the object of quantitative studies. These studies were denominated "immunochemistry" (Kabat and Mayer, 1961) and generated much new knowledge, including the fact that antibody molecules had at least two combining sites for the same determinants on the antigen and were therefore bivalent and symmetrical.

In 1942, Landsteiner and Chase showed that another form of immunity, the so-called "delayed hypersensitivity," did not depend on serum antibodies but could be transferred only with blood cells. These cell-bound antibodies were in fact the specific receptors of T cells, as was shown many years thereafter.

Somatic Selection in the Production of Antibodies

The realization that one could induce an apparently endless number of antibodies with different specificities spurred a growing interest in the biochemical and cellular basis of the production of these molecules. The first general theory capable of accounting for the cellular mechanisms of antibody production was formulated by Paul Ehrlich (1900). In his "side-chain" theory, Ehrlich viewed antibodies as side chains located on cellular membranes (Fig. 1). These side chains were heterogeneous and each possessed a combining

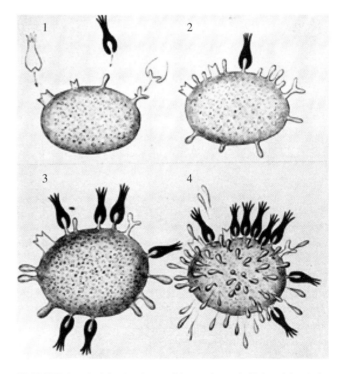

FIGURE 1 Original scheme illustrating Ehrlich's side chain theory of antibody production. Ehrlich thought that antibodies were receptors located on the cell membrane (1) which, after interaction with the specific antigen(in black, 2 and 3) left the cell (4) to diffuse in the body fluids where they would neutralize toxins (reproduced with permission from Ehrlich, 1900).

site for a different antigen. Upon reaction of a given chain with an antigen the cell was stimulated to produce more of the same kind, which were then released and appeared in the serum as specific antibodies. Considering what is now know of B lymphocytes and their membrane immunoglobulins, which are subsequently secreted after stimulation by antigens and maturation of the lymphocyte to a plasma cell, Ehrlich's theory was not far off the mark, although it postulated that one cell would carry on its membrane not only one kind of receptor but also more or less all possible side chains for all possible antigens. This was the weak point that progressively discredited Ehrlich's theory, the first theory that explained immunity as the result of a process of somatic selection, because immunologists considered it highly unlikely that one single cell could have on its membrane preformed antibodies directed against each one of the many different antigens that were progressively being found active as immunogens. This supported "instructive" theories that considered a direct role of the antigens in the generation of antibody specificity. Antibodies were thought to be proteins with an incomplete tridimensional folding that was only acquired after contact with antigens. The antigen molecules were "templates" on which the final folding of antibodies occurred (Fig. 2). Clearly this required that the antibody-producing cells be the same cells that could phagocytose antigens, namely, the macrophages. These views (Breinl and Haurowitz, 1930; Pauling, 1940) were dominant in immunology for a relatively long time, but eventually they

conflicted with the demonstration that the tridimensional structure of proteins is determined by their amino acid sequence, which is coded for by the sequence of DNA bases in their genes. Therefore, the information for the structure of each antibody could not reside in the antigen but had to be searched for at the gene level.

Niels Jerne was the first to realize this fact, and in his "Natural Selection Theory of Antibody Formation" (1955) he openly rejected the instructive views. Jerne was interested in the so-called "natural" antibodies — those antibodies that can be found in the serum in the absence of a deliberate or obvious immunization process. In particular, he was studying the antibacteriophage antibodies, for which he had worked out a very sensitive phage-neutralization test. The fact that these antibodies were as a rule found in normal serum suggested to Jerne that preformed antibodies against all antigens might exist and that, upon administration of the antigen, they would form complexes that were a signal for the cells to produce multiple copies of that particular antibody. The biochemical steps involved in this copying process were not specified, and this was an obvious flaw in Jerne's theory. For example, he did not consider the possibility that natural antibodies might actually have been the consequence of immune stimulation by antigens, e.g., bacteriophages present normally in the digestive tract. Despite these flaws, Jerne's theory had the merit to drastically divert the views on antibody formation from an "instructive," or Lamarckian, concept to a "selective," or Darwinian, concept. Thus, Jerne's 1955 publication marked a conceptual milestone in the progress of immunology. I remember asking Niels Jerne what had prompted him to entertain selective views to explain the basis of antibody formation and receiving the startling answer that sometimes in science it is necessary to consider concepts that are the exact opposite of the prevalent ones.

Selective concepts were clearly superior to instruction to explain antibody formation; however, a selection at the level of antibody molecules was not satisfactory. It was necessary, instead, to envisage selection at the level of the antibody-producing cells, complete with their genes and structures for the synthesis and secretion of proteins.

This concept was conceived and developed by F. M. Burnet: Cells with the potential capacity for antibody production are subject to somatic mutations of the genes that code for antibodies. These are then expressed on their membranes as receptors for antigens generating many different cellular clones, each potentially capable of reacting with a different antigen or antigen determinant. This diversity of clones, at the level of the receptors for antigens, is the basis for the specificity of the immune response, which occurs through the selective proliferation and maturation of the clones capable of interacting with a given antigen (Fig. 3). From the basic concept, which has been subse-

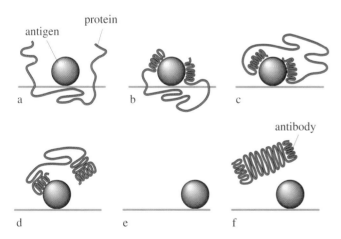

FIGURE 2 Diagram illustrating the direct role of the antigen in determining the specificity of antibodies (after P. Pauling). Following this theory, a protein that still has not undergone a tridimensional folding interacts (a) with the antigen on the cell surface. The folding then proceeds in contact with the antigen (b–e) so that the antibody molecule takes up a configuration specific for the antigen. Only when the process is completed does the antibody separate from the antigen (f) (reproduced with permission from Pauling, 1940).

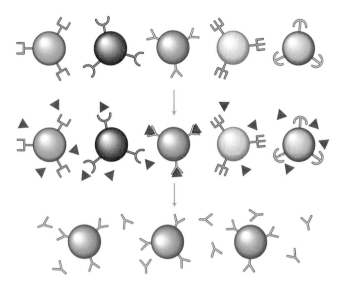

FIGURE 3 The clonal selection theory. Many different cellular clones express different antibodies. There is one kind of antibody for each clone (a). When antigen interacts with a specific membrane receptor (b) it stimulates the proliferation of the clone and the secretion of the corresponding antibodies (c). Note that this figure corresponds almost exactly to the series of events that occur in an immune response to a T-independent antigen (such as lipopolysaccharide).

quently proven to be in principle correct, a series of consequences were derived to explain immunological memory, immunological tolerance, autoimmunity, and the capacity of the immune system to distinguish self from not-self.

Today, 40 years after its writing as the text for a series of lectures held at Vanderbilt University in Nashville, Tennessee, it is a pleasure to read the book by Burnet, *Clonal Selection Theory of Acquired Immunity* (1959), and to note the clearness and far-sightedness of it. It is very enlightening, after so much time, to follow the original thoughts and to realize their connection with the experimental evidence then available and their vast relevance for the research that followed. In essence, *Clonal Selection* is more than a milestone on the road of immunology because it can be considered a true conceptual revolution of this discipline. Actually, upon scrutiny, its aftermath still influences the progress of research in immunology and related areas. For instance, the last chapter of Burnet's book extends the concept of clonal selection to the general field of tumors, which we now know to be monoclonal, and insists on the validity of somatic cell selection, a concept that has proven its validity in hematology and will doubtlessly be extended in other fields.

Despite some understandable resistance among those immunologists who had favored instructive concepts, the clonal selection theory of acquired immunity was quickly accepted and within 1 year was favored by geneticists. In fact, J. Lederberg (1959) concluded that "the problem is solved, the rest is technology." As will be discussed later, immunological "technology" has proven to be a vast field and far from free of conceptual relevance.

Immunochemistry after Clonal Selection

The conceptual revolution brought about by the clonal selection theory did not slow the progress of immunochemistry but rather fostered it. An important link between clonal selection and immunochemistry was the evidence that the proteins appearing in large amounts in the serum of humans or mice affected by plasma cell tumors (plasmocytomas) were monoclonal immunoglobulins related to normal antibodies. The chemical study of these clonal products was clearly much more informative than that of the heterogeneous mixture of immunoglobulins present in normal, and even in hyperimmune, serum. Therefore, the study of plasmacytoma proteins became a guide for the investigation of the structure of normal antibodies. It was thus established that antibody molecules were made of the covalent combination, through disulfide bonds, of heavy chains (45 kDa) with light chains (20 kDa). In most cases, each molecule included two heavy and two light chains in a symmetrical structure that was in agreement with the functional bivalence of antibodies. Many plasmacytomas produced an excess of light chains that could appear in the urine as "Bence–Jonce" protein. It was further established that there were two types of light chains, "kappa" and "lambda"; only one type appeared in a given molecule of plasmocytoma product. The heavy chains instead could be of one of six different kinds: gamma one, gamma two, gamma three, alpha, epsilon, and mu (to which a type delta was later added). On the basis of the heavy chain, seven different isotypes, or classes, of immunoglobulins were identified in human serum. Much of this work was performed in the laboratory of H. Kunkel at the Rockfeller Institute in New York. It was noted that the molecules with mu chains tended to form pentamers of molecular weight close to 1 million kDa that were called macroglobulins; antibodies with this molecular weight tended to dominate early in the immune response. Antibodies with alpha heavy chains were instead dominant in secretions and those with epsilon chains were highly cytophilic for mast cells and were mainly responsible for allergic reactions of the immediate type.

The light chains were the first for which a complete amino acid sequence was established. From the comparison of Bence–Jones proteins obtained from different plasmocytoma patients, it was apparent that these light chains included a variable part, different for every patient, and a constant part that was identical for all proteins of a given type, kappa or lambda. These monoclonal proteins, therefore, showed a chemical diversity, as was expected on the basis of the genetic diversity of clones generated at the somatic level. A closer analysis of this chemical diversity

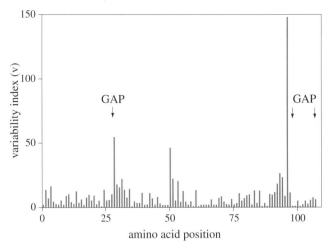

FIGURE 4 Histogram showing the variability of a set of variable regions of immunoglobulin light chains. Variability, reported on the y axis, is measured by the variability index v = a/b, where a is the number of different amino acids at a given position and b is the frequency of the most common amino acid at that position. GAP indicates the positions where insertions or deletions of amino acids have been noted. The histogram reports the variability indexes obtained comparing numerous human and mouse kappa light chains. Three hypervariable regions stand out, corresponding to positions 24–34, 50–56, and 89–97 (reproduced with permission from Wu and Kabat, 1970).

showed that it was prominent in some defined segments of the chain, called "hypervariables" (Fig. 4), and that these were therefore good candidates for those parts of the molecule capable of establishing specific contacts with the antigen (Wu and Kabat, 1970). This assumption was supported later by X-ray crystallographic analysis.

Cellular Immunology: The Cell That Produces Immunoglobulins

Clonal selection changed the emphasis of cellular immunology from the study of nonspecific phagocytes to that of specific cells (lymphocytes). A basic fact contributed by the new cellular immunology was the discovery of two different populations of lymphocytes approximately equivalent in number and gross morphology but marked by different membrane molecules (Raff, 1970). One population had immunoglobulin molecules on their membranes, easily identified with fluorochrome-labeled anti-immunoglobulin antibodies, whereas the other population lacked immunoglobulins and instead, in mice, had a different marker that was denominated Thy-1 to indicate the thymic origin of these cells. In humans the two populations of lymphocytes were respectively marked by membrane immunoglobulins or by a receptor capable of binding sheep erythrocytes.

This second population could therefore be identified under the microscope by its capacity to form rosettes with sheep erythrocytes. The receptor for sheep erythrocytes has been subsequently identified and called CD2, but its function has not been clarified.

The discovery of the two main populations of lymphocytes, denominated B and T to indicate their respective origin from the bone marrow or the thymus, was certainly a milestone of cellular immunology. The study of B cells developed, at least initially, at a more rapid pace probably because their specific products, the immunoglobulins, were well-known and could be recognized by a series of antibodies that identified their isotypes, allotypes, and idiotypes. These antibodies, employed in immunofluorescence microscopy that allowed the study of the immunoglobulins produced by single cells, revealed a series of facts that together resulted in a satisfactory understanding of the physiology of B cells.

It was established that single B cells used the genes of only one chromosome to code for the immunoglobulin chains that they expressed (allelic exclusion; Pernis *et al.,* 1965), whereas they could, for a short time, express two different heavy chain isotypes while performing the so-called "class switch." Furthermore, it became clear that, as predicted by theory, one B cell expressed membrane immunoglobulins endowed with one single combining site for antigen, and that this specificity was preserved in the antibodies that it secreted after maturation (one cell, one antibody; Nossal and Lederberg, 1958). The functional significance of these facts was fairly obvious, but their genetic basis was not; therefore, the definition of the phenotype of B cells preceded that of their genotype by about 10 years.

Another observation of considerable functional relevance was made at the same time, namely, that surface immunoglobulins were mobile in the plane of the membrane and, if cross-linked with polyvalent ligands, formed small patches which subsequently were internalized in endosomes and/or formed polar caps. This observation was made at the same time that it was being realized that cellular membranes in general were made of a double layer of phospholipids in which membrane proteins are embedded.

The lipid bilayer is fluid at body temperature and the membrane proteins can move laterally, forming the so-called "fluid mosaic." This lateral movement of receptors allows their aggregation after interaction with ligands, a phenomenon that supports the reception of signals derived from the contact with extracellular molecules. Furthermore, the inclusion in the receptor aggregates of other membrane molecules allows the modulation of the signal and can influence the reaction of the cell in different ways. Thus, the mobility of membrane proteins can be considered a basic fact in immunology and in cell biology in general.

The study of B cells was bound to explain a fact that had been apparent from the beginning of the studies on antibody formation, namely, that there was an increase in the affinity of antibodies concomitant with ongoing immunization. In terms of clonal selection, Burnet had foreseen that "the combination of frequent small mutations with an efficient positive selection could quickly determine an increase of the complementarity between antigen and antibody." Today, we know that this in fact happens due to a combination of frequent point mutations in the DNA coding for the variable segments of the immunoglobulin chains with selection by antigen under the guidance of T cells. This occurs in the germinal centers of lymphoid follicles; these structures are therefore the site of immunological events that conform, in an elegant way, to the general rules of cellular immunology.

Cellular Immunology: T Cells

The molecular basis of the specificity of T cells [i.e., their clonotypic receptor (TCR)] was not easily identified. The TCR was for a relatively long time "elusive" probably because there is not a soluble counterpart, as there is for the immunoglobulin receptors of B cells. Furthermore, unlike the B receptors, the TCR did not show the capacity to bind antigen in an easily demonstrable interaction. The first successes in the identification of the TCR were obtained with the use of molecular biology through the isolation of their cDNA in mouse (Hedrick *et al.,* 1984) or human cells (Yanagi *et al.,* 1984), respectively. Progress was rapid, both at the level of genes and at the level of proteins. At either level, both the similarities and the differences between the TCR and the clonal receptor of B cells (BCR) appeared clear. The similarities included the multichain structure (two for the TCR) and the presence in each chain of a variable and a constant region. Structural differences included a monovalent TCR, instead of an at least bivalent BCR, and the lack of a flexible hinge in the middle of the TCR molecule. However, the basic difference between TCR and BCR was determined to be in the mechanism of antigen recognition — a difference that could be elucidated only later with the discovery of the detailed structure of histocompatibility antigens and their role in the presentation of antigen to the TCR. It is clear, however, that the identification of the TCR was a milestone in immunology, comparable to the discovery of immunoglobulins.

After the identification of the molecular basis of T clonal diversity, the study of the function of these cells proceeded more quickly. The selection of T cell clones in the thymus, foreseen by Burnet, was studied in detail by cell transfer. It was found that T cell clones were subject not only to negative but also to positive selection. It appears reasonable, but it has not been definitely established, that positive selection precedes the negative one and that it occurs mainly in the cortex of the thymus. Positive selection chooses those clones that have receptors capable of interaction with an-

tigens presented by the histocompatibility molecules of the same organism and is therefore responsible for the phenomenon called histocompatibility restriction of T lymphocytes. Negative selection in the thymus, mainly in the medulla of the organ, aims instead to eliminate potentially autoreactive T cells.

The selection of T cells in the thymus is a basic fact that partially explains the tolerance that the immune system displays toward self-antigens and therefore its ability to distinguish self from nonself. This distinction, certainly a fundamental property of the immune system, must be grounded in lymphocyte selection. It is not certain, however, that negative selection only occurs in the thymus. Actually, T cell clones with potential autoreactivity are present in the peripheral lymphoid tissues and must be kept in check by "peripheral tolerance." The molecular foundation of the peripheral tolerance is certainly quite complex; for instance, in the mouse at least 14 different genetic loci have been shown to be involved in the control of autoimmune diabetes. Some molecules implicated in peripheral tolerance have been shown to be functional in the transmission of negative, or self-destructive, signals to T lymphocytes, including CD95, or Fas, and its ligand, Fas ligand, and the CTLA-4 receptor. The inactivation or the genetic knockout of these molecules starts generalized autoimmune reactions by T cells, which is proof of the role of these receptors in tolerance.

It is easy to predict that the field of peripheral tolerance will include some of the next milestones in immunology. These will likely include the function of some known T cell molecules or the demonstration of some novel ones and the elucidation of molecular sequences involved in the transduction of signals in these cells. It is hoped that the understanding of these molecular events and their regulation will provide the key to understanding the logic of peripheral tolerance, a process that has rules well understood by T lymphocytes but not by those who are studying these cells.

Histocompatibility Molecules

This field, originated to study the genetic basis of the rejection of tissues transplanted among individuals of the same species, has reached a milestone that has included the understanding of the immunological function of major histocompatibility complexes (MHCs). It has been shown that MHC molecules work as structures functional in the presentation of fragments (peptides) of protein antigens to the receptors of T cells.

The problem faced by the immune system in discriminating proteins, which are sometimes quite similar in overall structure, is not an easy one. Actually, the surface of a globular protein molecule is chemically a complex mosaic of atoms in which the potential number of different antigenic determinants is ill defined. B cells solve the problem of discriminating between protein antigens with the production of antibodies of increasing affinity and specificity produced by cells whose immunoglobulin genes mutate during the immune response in the course of selective processes beginning under the guidance of T cells. However, the genes of T cell receptors do not mutate while the cells are responding to a given antigen. Thus, T cells must solve the problem of an unequivocal discrimination between protein antigens, the only ones to which they respond, in a different way. The strategy followed by T cells is not that of recognizing surface determinants of proteins but instead recognizing amino acid sequences of fragments (peptides) of these antigens. These peptides are produced by a process of protein fragmentation occurring inside antigen-presenting cells and are then presented to T cell receptors after insertion in a given pocket of the MHC molecules. The elucidation of the tridimensional structure of MHC molecules (Bjorkman *et al.,* 1987), complete with inserted peptides, was certainly another milestone in the progress of our knowledge of histocompatibility and of antigen recognition by the receptors of T cells.

The study of the processing of protein antigens and the presentation of these peptides through MHC molecules of class I or class II, with all its interesting details and clear functional implications, has been one of the most interesting and satisfactory sections of cellular immunology.

....•

The Accessory Receptors

In addition to clonal receptors that are responsible for their specificity, both T and B lymphocytes express on their membranes nonclonal, so-called "accessory" receptors that modulate their function. Some of these are only present on the membranes of T lymphocytes, others are only found in B cells, and many are expressed by both B and T lymphocytes or by lymphocytes and other cells of the immune or hemopoietic systems.

The definition of the structure and reactivity of some of these accessory receptors has allowed a better understanding of the function of some subpopulations of T or B lymphocytes. This is best explained by discussing some examples. The CD4 receptor is a monomer that interacts with a determinant present on the nonpolymorphic region of the beta chain of class II MHC molecules. As a consequence, T lymphocytes that express the CD4 accessory receptor are geared to recognize antigenic peptides presented by class II MHC. Instead, the CD8 accessory receptor is a heterodimer that reacts with a determinant in the alpha 3 (nonpolymorphic) region of the heavy chain of class I MHC molecules. Accordingly, T lymphocytes that express CD8 molecules are directed toward the recognition of peptides presented by class I MHC. In the thymus, immature T lymphocytes go through a phase in which each cell expresses

both CD4 and CD8 accessory receptors; these cells are apparently still undecided. Only very few mature lymphocytes possess both CD4 and CD8 molecules; therefore, there are two main subpopulations marked by one or another coreceptor. This dichotomy of T cells is associated with a well-defined difference of function: Most CD4 lymphocytes do not kill the cells that present the corresponding antigens presented on class II MHC but rather provide help through the secretion of appropriate lymphokines. CD8 lymphocytes, however, are cytotoxic for the cells that present to them antigenic peptides bound to their class I MHC.

The logic for the existence of two different lymphocyte subpopulations, as well as two different classes of MHC molecules, is to be found in the existence of two different ways of antigen processing and presentation by target cells. The MHC class II route works essentially for the processing of extracellular antigens, whereas the class I route is employed preferentially for antigens generated inside the cells, such as viruses. It is clear that in this latter instance the killing of target cells by cytotoxic CD8 lymphocytes that recognize viral peptides presented by class I MHC molecules is instrumental in the fight against the spread of a virus within the infected organism.

It is evident that the expression of accessory receptors CD4 and CD8 fits well with the functional dichotomy of lymphocyte populations and that this is an instance in which the study of nonclonal receptors of T lymphocytes has provided a milestone in the progress of immunology. Another instance is certainly that of CD3. The epsilon, gamma, and delta chains of this complex that have an extracellular region associate with the alpha and beta chains of the clonotypic receptor and with the dimer of another chain, zeta, which is mostly intracellular. The interaction of the clonal receptor with the specific antigen, presented by the MHC molecules, induces an aggregation of the TCR–CD3 complexes, which can be identified with the "reception" of the antigenic signal. The different CD3 chains, particularly the zeta and epsilon chains, then start the "transduction" of the signal which goes through a complex series of molecular events that end in the transcription of the genes involved in the activation of T lymphocytes. The function of the CD3 accessory complex, which is expressed by all subpopulations of T lymphocytes, appears as a milestone located at the beginning of a road in the transduction of signals from the receptors to the nucleus of T lymphocytes. Of this road, we know many bends and checkpoints, but its full extent, with all its potential for meaningful regulation of cell function, is unknown.

Another example that shows the relevance of accessory receptors for the understanding of the function of T and B lymphocytes is that of the CD40 molecules, present on the membrane of B lymphocytes, and of their corresponding ligands (CD40-L), expressed by T cells. The interaction between these two groups of molecules is part of a chain of events that occur as part of the cooperation between T and B lymphocytes in the production of antibodies directed against protein antigens.

This chain starts with the interaction between an intact protein antigen and the BCR (the membrane immunoglobulins). The protein antigen is then internalized by B cells and, in their endosomes, fragmented into peptides, some of which are then presented by the class II MHC molecules that reach the surface of the B cell. A specific CD4[+] T cell interacts with this presented peptide through its clonotypic receptor and, in addition, engages its CD4 coreceptor with a determinant in the nonpolymorphic region of the beta chain of class II MHC molecules. Following these events the T cell is stimulated through its CD3 complex and secretes the cytokine interleukin-4 (IL-4) that interacts with its corresponding receptors on B and T cells. The activated T cell then expresses on its membrane CD40-L, which fits with the CD-40 molecules present on B cells, thus establishing a firm conjugate of T and B cells essential for the progression of the antibody response to the protein antigen. Note that the cooperation between B and T cells occurs between two cells that are both endowed with clonotypic receptors specific for the same protein but with different determinants of the protein, namely, a surface configuration for the BCR and a presented peptide for the TCR. In the T and B cell cooperation, in addition to the clonotypic receptors of both cells and to the MHC molecules, at least four accessory receptors and at least one cytokine are involved.

Where are the milestones in the study of this complex chain of events? Obviously, the whole chain, with its complex molecular events occurring at the level of the membranes and in the cytoplasm of B and T cells, must be considered as a unit.

I have discussed in-depth T and B cell cooperation in response to a protein antigen to emphasize that the study of the structure, and even the function, of one given molecule should not be conceptually separated from an effort to consider the whole of a complex function in which it is involved. This does not detract, of course, from the necessity to study the individual molecular components of a complex function; I note only that the details and the whole are both worth attention. For instance, 30 accessory molecules on the membranes of T cells have been identified to date, although not all are present at the same time and on the same cells. Those of B cells are even more numerous. For many of these, we ignore the function; we must make a continued effort dedicated to their biochemical and genetic study so we can gain a satisfactory understanding of lymphocyte physiology.

The Genes of Clonotypic Receptors

The discovery of restriction enzymes in the early 1970s provided a great impulse for the study of DNA sequences; in fact, it opened the field of molecular biology. This impulse soon affected immunology with the study of immunoglobulin genes. An important step forward in the study of the genes was provided by the work of Susumo Tonegawa which showed, in a monoclonal plasma cell line, that in mature B cells the distribution of immunoglobulin genes in a set of restriction fragments was different from that in embryo DNA or in the DNA of nonlymphoid cells (Hozumi and Tonegawa, 1976). In other words, Tonegawa showed that during the maturation of B cells these genes had undergone a process of somatic rearrangement. I vividly remember the enthusiasm generated at the Basel institute of Immunology by this discovery. I also remember trying without success (and wrongly) to dampen this enthusiasm by pointing out that it was already known from the formal genetics of rabbit immunoglobulin allotypes that the genes coding for the variable and constant regions of immunoglobulin heavy chains were separate in the germline and that therefore they had to be joined by a somatic process. This does not change the fact that Tonegawa's work was a conceptual, as well as an experimental, milestone in the progress of immunology.

The discovery of the somatic arrangement of immunoglobulin genes was quickly followed by other findings at the DNA level, such as the definition of the J and D segments, the chromosomal localization in various species of the genes for the different immunoglobulin chains, and the existence in the 3′ position of the constant region genes of extensions necessary for the membrane insertion of immunoglobulins. Subsequently, repetitive sequences located in the noncoding regions 5′ of each heavy chain constant region gene (with the exception of the delta) were identified. These support recombination events that can place a DNA segment coding for a given VDJ adjacent to one or another constant region gene; this recombination provides the genetic basis for the so-called class switch of B cells. Finally, it was shown that in the variable region of the immunoglobulin chain genes, there was an accumulation of point mutations that paralleled the increase of antibody affinity during an immune response, exactly as predicted by Burnet.

When the genes coding for the chains of T receptors were identified, it was apparent that similar events of redistribution of the DNA segments coding for the variable and constant regions occurred. Notable differences between the B and T genes were the absence in the latter of repetitive sequences for the class switch and the lack of accumulation of point mutations in the variable regions during a peripheral immune response.

The considerable amount of knowledge collected in a relatively short time on immunoglobulin genes provided a satisfactory explanation of many facets of the phenotype of B cells. Basic was the elucidation of the mechanism for the somatic generation of antibody gene diversity. The pattern included the existence of multiple variable region genes that could be assembled with a constant region gene with the addition of a J and D (for heavy chains) segment, with possible additional variability at the junction. Furthermore, the dynamic state of the immunoglobulin genes explained allelic exclusion, the existence of membrane and secreted immunoglobulins with the same specificity, the class switch, and the increase of antibody affinity during the immune response.

Likewise, for T receptors, the genetic basis of the generation of diversity and of allelic exclusion was elucidated. Of course, the genes of the T receptors did not show the features related to the control of secretion or those connected with the class switch, and not even the point mutations responsible for the increase of affinity, since all these events happen to immunoglobulins but not to T receptors. On the whole, the genes for immunoglobulins and T receptors, with their somatic dynamics strictly limited to B and T cells, are perhaps one of the best known aspects of immunology, in which the connections between structure and function are most satisfactory. The biochemical machinery responsible for the cuts of DNA at precise times and locations and that which handles the repairs were studied later. It was thus established that these events were complex and that they required not only defined sequences in the noncoding regions flanking the segments subject to recombination but also a series of enzymes. Two of these enzyme systems, RAG-1 and RAG-2, are involved in the act of DNA cutting and are expressed exclusively in lymphoid cells during the process of receptor diversity generation. Knocking out the genes that code for these enzymes completely blocks the maturation of both B and T lymphocytes. Together with enzymes involved in the repair of the breaks of double-stranded DNA, many of which are also present in nonlymphoid cells, RAG-1 and RAG-2 form an enzymatic complex that as a whole has been called "recombinase." The emergence of the recombinase system about 500 million years ago has been an essential step in the evolution of an immune system.

I now compare the somatic mutation and selection processes that support acquired immunity with natural selection in the evolution of species. That is, I am comparing Burnet with Darwin. It is obvious that the roles of the "recombinase" system and especially the "ad hoc" structures of the antibody and TCR genes and their flanking sequences are peculiar to the process of somatic selection in the immune system. Although the difference between the two systems might not be as extreme as it appears because the recombinase system must have progressively evolved, Darwin might say that the immune system is cheating at the game of natural selection because it relies on an ad hoc setup of genes and enzymes to focalize mutations in a defined set of genes and cells and at a definite time. However, Burnet might answer that the system that generates the diversity of clonal receptors operates at random, like the germ-like mutations that drive the evolution of species. Actually, the sequences in the receptor genes of a given B or T cell that will generate its specificity are not predictable; one only knows that, eventually, the receptor chains will conform to one of the 1 billion or more different specificities that provide the so-called repertoire of the immune system. Furthermore, Burnet would note that the numerous point mutations in the immunoglobulin genes, under persistent clonal selection by antigen, that provide a progressive increase of antibody affinity during a given immune response are as Darwinian as one can expect from a biological system that must solve its problems within 1 or 2 weeks and not in millions of years. In the following sections, I discuss recent areas of immunology that were not predictable on the basis of clonal selection alone.

The Cytokines

This rapidly developing area of immunology includes the discovery of a series of relatively small molecules secreted by the lymphoid cells, or by accessory cells of the immune system, together with the receptors for them. These molecules, with a relatively limited life and diffusion space outside of the cell that secretes them, are communication signals between different cells. Therefore, many of them have also been denominated "interleukins." The interleukins contribute to change the immune system from a series of independent cellular clones to a system integrated within itself and connected with the other systems of the organism.

To date, we know of the existence of about 40 cytokines, which can be subdivided into four major groups according to their main functions: (i) support of natural immunity, (ii) regulation of the development of lymphoid cells, (iii) control of inflammation, and (iv) stimulation of hemopoiesis.

The discovery of each one of these cytokines has been an important contribution to immunology, but with regard to their conceptual relevance perhaps three cytokines belonging to the second group are outstanding: IL-2, IL-4, and IL-12. IL-2 is produced and secreted by T cells, mainly by CD4$^+$ cells with the Th1 phenotype, at the beginning of their stimulation. Simultaneously, these cells also synthesize and express on their membranes the chains of the receptor for IL-2. The result is that, to progress in the activation process, the cell must use the very IL-2 molecules that it has secreted (autocrine stimulation). At the same time, in a kind of cross-fertilization, the IL-2 molecules produced by nearby T cells are bound by the receptors of other lymphocytes (paracrine stimulation). It should be noted that the paracrine stimulation occurs only among those cells that express IL-2 receptors and therefore, within a given space and in a given time, have recognized an antigenic determinant with their TCR. This explains why, despite the paracrine stimulation by IL-2, the response of the T cells remains strictly antigen dependent. Nevertheless, it may be surprising that a cell that has recognized antigen must, to proceed in its activation, use a molecule that itself (or one of its neighbors) has produced, secreted, and bound back on its membrane receptors. Why can't the cell be short-circuited and proceed in the activation process independently from the secreted lymphokine? One may think that dependence from secreted IL-2 puts all the cells that recognize antigen in a given area on an equal standing. Actually, it is relevant for an optimal protective effect that there may be a polyclonal response directed against the higher possible number of determinants of a protein antigen and even to the higher possible number of antigens present in one infectious agent. The role of IL-2 as a molecule for communication between T lymphocytes can be considered as one of the elements favoring a complete response, with the engagement of the highest possible number of clones available, rather than an oligoclonal reaction. In addition to such considerations of the physiology of this interleukin, I now discuss the most interesting knowledge so far obtained on the regulation of the IL-2 gene. The transcription of the gene requires signals originating at the cell membrane both from the TCR–CD3 complex and from the CD28 coreceptor. In the absence of the engagement of this second molecule with its ligand (B7) present on the antigen-presenting cell, the T lymphocyte is not activated but rather placed in a state of specific nonreactivity called "anergy." If one also considers the data on the regulation of the expression of the three different chains that may form part of an IL-2 receptor and those on the transduction of the signals that originate from the IL-2 receptors and eventually induce the activation of the genetic program for the proliferation of T cells, the molecular biology of IL-2 alone would comprise a chapter. Furthermore, if the numerous studies on IL-2 that are relevant for medicine were considered, one would see how wide is a field that at the beginning appeared only as an interesting, but limited, observation in immunology.

The same can be said for IL-4. This molecule is produced and secreted mainly by activated Th2 cells. It is quite interesting that it then directs the differentiation of other T lymphocytes toward the Th2 phenotype in a kind of self-amplification circuit. IL-4, as already discussed, is part of the channels through which T cells communicate with B lymphocytes. It is involved in inducing the maturation and the class switch of B cells in the T-dependent response to protein antigens; in particular, it favors the switch of B cells to IgE production and therefore the appearance of allergies of the immediate type.

Another molecule of both theoretical and practical interest is IL-12 (Trinchieri, 1995). This interleukin is produced by macrophages and B cells. It favors the differentiation of CD4 T cells toward the Th1 phenotype and stimulates the maturation of CD8 T lymphocytes to the stage of active cytotoxic cells. However, perhaps the most relevant activity of IL-12, the one that resulted in its discovery, is that of stimulating the function of natural killer (NK) cells, which in turn, after activation, can produce IL-12. This is another instance of self-stimulation of a given kind of cells, mediated by a cytokine.

Natural Killer Cells

These lymphoid cells are cousins of T lymphocytes with the same precursors and they share some membrane markers and a similar morphology. However, there is a basic difference between NK cells and T lymphocytes: In NK cells the genes coding for the T cell receptor chains (and those for the immunoglobulins) are not recombined but instead are in the germline position. As a consequence, the NK cells are not clonally diverse at that level.

Some properties of these cells have been known for quite a while, including their ability to mediate the so-called antibody-dependent cytotoxicity that derives from their built-in cytotoxic properties combined with the expression of a receptor, CD16, for the constant part (Fc) of the immunoglobulins. This receptor focuses the cytotoxic activity of NK cells on cellular targets labeled by antibodies of the IgG class. Furthermore, the NK cells were known to be able to recognize and kill in a "natural" way many tumor and/or virus-infected cells. The molecular basis of this natural ability was not known, but it was quite clear that there was no specificity for cellular targets of a given tumor or for cells infected by a given virus. This apparent lack of specificity separated the NK cells from classical antigen-specific immunocytes and therefore tended to reduce interest in them as an aspect of study. Only recently has it been realized that there is a logic in the apparent folly of NK cells. It has in fact been discovered that NK cells, at a difference from their cousins the CD8 cytotoxic T lymphocytes, do not kill target cells marked by non-self-antigenic determinants but instead kill on the basis of the absence in the target of some self-molecules (Karre et al., 1986). In fact, NK cells detect in their targets the lack of the presence of class I MHC molecules. This property is related to the expression by NK cells of a series of membrane receptors that recognize one or another of the isotypes of class I MHC and, in some instances, even one allele at a given locus (Moretta et al., 1994). These receptors inhibit or completely block what appears to be the innate tendency of NK cells to kill. Therefore, the natural target of NK cells is any other cell that, for different reasons, has lost the capacity to express class I MHC molecules, which are normally present on the membranes of all nucleated cells in the organism. The loss of membrane expression of class I MHC is one of the more frequent events through which tumor cells can escape the so-called "immune surveillance." In parallel many viruses have found a way to prevent the expression of class I MHC on the membrane of infected cells. For the virus, this provides the obvious advantage of protecting the infected cells from the kill by CD8 T lymphocytes specific for determinants derived from viral antigens. Following this logic, the NK cells complete the protective cytotoxic function of the immune system by killing those targets that the specific cytotoxic T lymphocytes cannot identify because of the lack of the molecules necessary for the presentation of the antigens. This function, which is linked to the expression of a series of receptors specific for class I MHC molecules, places the NK cell in a position closer to that of the immunocytes proper rather than to that of the accessory cells of the immune system, such as the macrophages. Therefore, despite the fact that NK cells do not express clonal receptors for antigens, the demonstration of their capacity to identify "holes" in the histocompatibility system (the "missing self") and the identification of the receptors through which this is accomplished certainly can be viewed as a conceptual milestone in immunology.

Consequently, the interest of immunologists in NK cells has grown and the pace of the study of these cells has increased. It had been found that there are receptors for class I MHCs which activate NK cells instead of inhibiting them, and the biochemistry of the activation process is being studied. Finally, it has been realized that a minority of T cells can express, in addition to their clonal antigen-specific receptor, the same class I MHC-restricted receptors that the NK cells express, both inhibitory and activating. This observation, in addition to being a further demonstration of the relationship between NK and T cells, opens a new field to study the possible modulation of T lymphocyte function by NK receptors. We await with curiosity what will result from the study of these TINK (T cells with inhibitory NK receptors) and TANK (T cells with activating NK receptors) cells.

The Growth of Methodology and the Future of Immunology

It is appropriate to recall the already quoted opinion of J. Lederberg concerning clonal selection as the basis of the immune response. Lederberg said in 1959, "The problem is solved, the rest is technology." Even if we can share his opinion, at least in part, we cannot fail to be impressed by the vast expansion of immunological technology in the past 40 years and to recognize its impact on the evolution of basic concepts in immunology. The work in the laboratory of Cesar Milstein on the fusion of immonoglobulin-producing cells, initiated to study the mechanism of allelic exclusion, brought to the hybridoma methodology for the production of monoclonal antibodies. This methodology proved to be a powerful tool in almost all areas of immunology. When conjugated with fluorochromes and coupled with the use of the fluorescence-activated cell sorter (FACS), monoclonal antibodies allowed the identification and quantitative study of many lymphocyte membrane molecules and the identification, on this basis, of different lymphocyte subpopulations as well as their development and differential functions.

Advances in the techniques of culture and cloning of lymphocytes, as well as the establishment of continuous lymphocyte lines, has allowed the *in vitro* study of many aspects of the immune response. Perhaps the largest impact of the progress of technology on the advance of immunology came from the widespread use of procedures originated in other branches of biology, including the cloning of genes and the study of their function in transfected cells or, alternatively, of the function of cells in which genes have been knocked out. Recently, the investigation of the role of a given gene in immune function has been extended to the whole animal with the study of transgenic mice. In these genetically manipulated strains of mice a given gene has been introduced in the germline, often associated with a defined promoter allowing its expression only in a given time or after a given stimulation. Conversely, mouse strains in which a given gene had been knocked out by recombination in embryonic stem cells have been very useful. In addition to genetics, biochemistry has also contributed methods employed in immunological studies, notably studies of signal transduction and DNA transcription in cells of the immune system.

The progress of technology has revealed many new facts which, while proving the variety and complexity of biological events involved in the immune reactions, have been difficult to categorize into clear comprehensive views. It has therefore not been easy to place conceptual milestones in the progress of recent immunology. Therefore, we might ask ourselves the following question: How much do we know today of the whole of the biological phenomena that we call immunology? The answer can only be given in relative terms: not enough. This last statement is grounded in our persistent inability to cure the more widespread and serious diseases of the immune system, such as multiple sclerosis, rheumatoid arthritis, and juvenile diabetes and our woefully incomplete knowledge of the immunopathogenesis of HIV infection. It is likely that our inadequate capacity to solve these practical problems resides in our inefficient knowledge of the details of peripheral tolerance. It is obvious that, as long as medical problems of this magnitude, and many others as well, have not been solved, immunologists cannot claim to have completed their task. It appears that much time will have to elapse before immunologists will be able to say, with Niels Jerne, that they are "waiting for the end." However, will there be an end for immunology? It is reasonable that we shall experience, for immunology as for other branches of the study of life, a progressive blurring of the boundaries with other disciplines. If so, it will almost be impossible to claim to have placed the final conceptual milestones of immunology. Therefore, for the foreseeable future, immunologists will continue to exist and will be kept quite busy with the double motivation of theoretical curiosity and practical necessity.

References Cited

BEHRING (VON), E., and KITASATO, S. (1890). Uber das Zustande-kommen der Diphterie-Immunitat under der Tetanus-Immunitat bei Thieren. *Deutsche Med. Wochenschrift* **16,** 1113.

BJORKMAN, P., SAPER, B., SAMROUI, W., BENNET, J., STROMINGER, J., and WILEY, D. (1987). Structure of the human class I histocompatibility antigen HLA-A2. *Nature* **329,** 506.

BREINL, F., and HAUROVITZ, F. (1930). Chemische Untersuchung des Prazipitates aus Hamoglobin und anti-Hamoglobin Serum und Bemerkungen uber die Natur der Antikorper. *Z. Physiol. Chem.* **192,** 45.

BURNET, F. M. (1959). *The Clonal Selection Theory of Acquired Immunity.* Cambridge Univ. Press, Cambridge, UK.

EHRLICH, P. (1900). The Croonian lecture. *Proc. R. Soc. London (Biol.)* **66,** 424.

HEDRICK, S., COHEN, D., NIELSEN, E., and DAVIS, M. (1984). Isolation of cDNA clones encoding T-cell-specific membrane-associated proteins. *Nature* **308,** 149.

HOZUMI, N., and TONEGAWA, S. (1976). Evidence for somatic rearrangement of immunoglobulin genes coding for variable and constant regions. *Proc. Natl. Acad. Sci. USA* **73,** 3628.

JERNE, N. K. (1955). The natural selection theory of antibody formation. *Proc. Natl. Acad. Sci. USA* **41,** 849.

KABAT, E., and MAYER, M. (1961). *Experimental Immunochemistry,* 2nd ed. Charles C. Thomas, Springfield, IL.

KARRE, K., LJUNGGREN, H. G., PIONTEK, G., and KIESSLING, R. (1986). Selective rejection of H-2-deficient lymphoma variants suggests alternative immune defense strategy. *Nature* **319,** 675.

LANDSTEINER, K., and CHASE, M. (1942). Experiments on transfer of cutaneous sensitivity to simple compounds. *Proc. Soc. Exp. Biol. Med.* **49,** 688.

LEDERBERG, J. (1959). Genes and antibodies. *Science* **129,** 1669.

MORETTA, L., CICCONE, E., MINGARI, M. C., BIASSONI, R., and MORETTA, A. (1994). Human NK cells: Origin, clonality, specificity and receptors. *Adv. Immunol.* **55,** 341.

NOSSAL, G., and LEDERBERG, J. (1958). Antibody production by single cells. *Nature* **181,** 1419.

PAULING, L. (1940). A theory of the structure and process of formation of antibodies. *J. Am. Chem. Soc.* **62,** 2643.

PERNIS, B., CHIAPPINO, G., KELUS, A., and GELL, P. (1965). Cellular localization of Immunoglobulins with different allotypic specificities in rabbit lymphoid tissues. *J. Exp. Med.* **122,** 853.

RAFF, M. C. (1970). Two distinct populations of peripheral lymphocytes in mice distinguished by immunofluorescence. *Immunology* **19,** 637.

SILVERSTEIN, A. M. (1989). *A History of Immunology.* Academic Press, San Diego.

TISELIUS, A., and KABAT, E. A. (1939). Electrophoretic study of immune sera and purified antibody preparations. *J. Exp. Med.* **69,** 119.

TRINCHIERI, G. (1995). Interleukin-12: A proinflammatory cytokine with immunoregulatory functions that bridge innate resistance and antigen-specific adaptive immunity. *Annu. Rev. Immunol.* **13,** 251.

WILLIAMS, C., and GRABAR, P. (1955). Immunoelectrophoretic studies on serum proteins I. The antigens of human serum. *J. Immunol.* **74,** 158.

WU, T. T. E., and KABAT, E. A. (1970). An analysis of the sequence of the variable regions of Bence–Jones proteins and myeloma light chains and their implications for antibody complementarity. *J. Exp. Med.* **132,** 211.

YANAGI, Y., YOSHIKAI, Y., LEGGET, K., CLARK, S., ALEKSANDER, I., and MAK, T. (1984). A human T cell-specific cDNA clone encodes a protein having extensive homology to immunoglobulin chains. *Nature* **308,** 145.

General References

BABBIT, B., ALLEN, P., MATSUEDA, G., HABER, E., and UNANUE, E. (1985). Binding of immunogenic peptides to histocompatibility molecules. *Nature* **317,** 359.

LOOR, F., FORNI, L., and PERNIS, B. (1972). The dynamic state of the lymphocyte membrane: Factors affecting the distribution and turnover of surface immunoglobulins. *Eur. J. Immunol.* **2,** 203.

MILSTEIN, C. (1993). From the structure of antibodies to the diversification of the immune response. *Scand. J. Immunol.* **37,** 386.

NOSSAL, G., WARNER, N., and LEWIS, H. (1971). Incidence of cells simultaneously secreting IgM and IgG antibodies to sheep erythrocytes. *Cell. Immunol.* **2,** 41.

TAYLOR, R., DUFFUS, P., RAFF, M., and DE PETRIS, S. (1971). Redistribution and pinocytosis of lymphocyte surface immunoglobulin molecules induced by anti-immunoglobulin antibody. *Nature New Biol.* **233,** 225.

Gᴜꜱᴛᴀᴠ J. V. Nᴏꜱꜱᴀʟ

Department of Pathology
The University of Melbourne
Parkville, Victoria, 3052, Australia

Lymphocyte Selection by Antigen

The concept of lymphocyte selection by antigen required a new paradigm of immune recognition. Previously, it had been thought that antibodies assume their correct shape and specificity through being molded against a template of antigen. The selective theories of immune activation recognize that genes for antibodies and T cell receptors must preexist in the unimmunized animal. Evolution has solved the puzzle of how to mount specific immune responses against a vast diversity of antigens by a unique somatic minigene shuffling process which results in each lymphocyte carrying a receptor of only one specificity. For immune activation, the antigen stimulates only those lymphocytes with a corresponding, preformed receptor. To ensure that the body is not destroyed by autoimmune processes, mechanisms of negative lymphocyte selection by antigen also exist, predominantly because of a susceptibility of immature lymphocytes to deletion on an encounter with a relevant antigen. This article surveys the history and the current status of lymphocyte selection by antigen.

The Nature of Bodily Defense

Defense against pathogenic microorganisms is critical to all living species. For the animal kingdom, defense mechanisms are divided into innate and immunological. Innate processes are many and varied, comprising such actions as cilia beating rhythmically to rid mucous surfaces of particulate matter, digestive enzymes such as lysozyme degrading pathogens, scavenger cells of various sorts possessing the capacity to eat and digest microorganisms, and physical barriers such as skin making it difficult for microbes to penetrate into the body. These are just a few examples, and clearly the innate mechanisms will vary as one moves up the phylogenetic tree.

Immunity has one key difference from innate processes: It displays memory. The critical element of immunological defense is that certain infectious diseases can only strike once. The host suffers from an infection, and either dies or survives, following which the same pathogen cannot take hold again in that individual. This kind of immunity depends critically on white blood cells called lymphocytes, which appeared for the first time in the lowest vertebrates, the cartilaginous fishes. Thus, adaptive immune responses are confined to vertebrates and become progressively more efficient, reaching their highest sophistication in mammals. This article poses the following question: How do lymphocytes react to the foreign molecules of invading microorganisms in order to mount the protective immune response?

A generic term for substances capable of inducing immunity is the word "antigen." It must therefore be asked, How do antigens "speak" to lymphocytes?

The Antibody Problem: How to Create Infinite Recognition Capability

In 1796, Edward Jenner was struck by the observation that milkmaids with cowpox sores on their fingers or elsewhere seemed insusceptible to smallpox. A century and a quarter before the discovery of viruses, he used cowpox, a close relative of the smallpox virus, as the first live attenuated vaccine. This was important empirical science, but a major step forward was taken when Louis Pasteur established the microbial origin of infectious diseases and undertook the careful, planned attenuation of what we now know to have been both bacteria and viruses through repeated passages of the infectious agent either in living tissue or in culture. Frequently, passaged microbes lost their virulence and could be used as vaccines. In other words, the state of immunity could be induced without subjecting the animal or person to the risk of the disease.

In 1891, von Behring and Kitasato discovered antibodies. They found that the serum of immune individuals contained specific substances capable of reacting with and neutralizing the harmfulness of pathogens. Later it was shown that antibodies physically associated with antigens, causing the formation of a lattice resulting in precipitation or aggregation of the antigen in question. The next stunning finding was that not just microbial antigens could cause antibody formation. In fact, it proved possible to inject chemical substances made in the test tube into experimental animals and to elicit specific antibody. Moreover, minuscule chemical changes to an artificial antigen could somehow be recognized by the immune system, eliciting different antibodies. As Karl Landsteiner showed, the specificity of immunological reactions was of an extraordinary degree. Within nature, there are many thousands of different pathogens, each composed of many thousands of molecules. Artificially, the number of small organic molecules which the organic chemist can synthesize is virtually unlimited. Another interesting fact is that animals can make antibodies to the proteins of other animals. The immune system of the rabbit has no difficulty in distinguishing egg white (ovalbumin) from human albumin or mouse albumin. Therefore, somehow we must explain how an absolutely huge number of different antibodies can be made by every individual.

A straightforward concept to explain antibody diversity was the direct template hypothesis (Breinl and Haurowitz, 1930). This viewed antigen entering a cell, perhaps the macrophage scavenger cell specialized for taking up microbes, and acting as a template against which the protein-synthetic machinery then fashioned an exactly corresponding anti-

body molecule. The antibody simply copied the shape of the antigen much as a molded product shapes itself around the corresponding dye. When Linus Pauling embraced the theory, it became the standard way of viewing antibody production and went almost unchallenged for a quarter of a century.

Selective Theories of Antibody Formation

Burnet and Fenner (1949) wrote an influential monograph which discussed many of the aspects that the direct template theory failed to explain, including the exponential kinetics of the appearance of antibody, the increase in the affinity of antibodies with increasing time after immunization, the well-known booster effect in which a second injection of antigen causes more antibody production than the first, and the observation that healthy animals and people do not form antibody against their own bodily proteins and carbohydrates. Other problems regarding the direct template hypothesis arose in the 1950s. Watson and Crick's demonstration of the structure of DNA launched an avalanche of work on the principles of genetic coding, and soon after the Crick dogma was enunciated: Information flows from DNA to RNA to protein and not in the reverse direction. Anfinsen found that the way a protein folds into its three-dimensional configuration depends in large measure on its amino acid sequence. Therefore, how could an antigen "instruct" a cell to make a particular antibody? The recently developed fluorescent antibody technique showed that antibody was made by plasma cells, not by antigen-capturing macrophages. A new hypothesis was urgently required.

Jerne (1955) took a major step in his natural selection theory of antibody formation. He revived an idea first suggested by Paul Ehrlich, namely, that antibodies were a natural product of a cell. The role of antigen was merely to cause overproduction of the particular natural antibody capable of uniting with the antigen in question. He argued that, even if there were 10^{11} possible sorts of antibodies, 1 ml of normal serum could still contain 10^6 molecules of each specificity. Antigen could unite with these and foster their selective synthesis. Though the way in which so many patterns emerged was unclear, and the template notion was preserved, this widely read paper did focus minds on the possibility that the antibody response was a selective, not an instructive, phenomenon. Talmage (1957) deserves much credit for pointing out that the entity to be selected could be a cell—Ehrlich indeed had conceived of antibodies as specific receptors or "side chains" on cell surfaces—and when the antibody that the cell is synthesizing matches the invading antigen, that cell is selected for multiplication. Burnet (1957) was prompted by Talmage's article to publish his own views (Fig. 1). They were similar to Talmage's, with three elaborations. Burnet was more forceful than Talmage

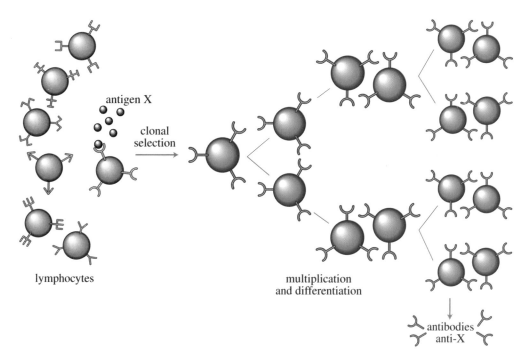

FIGURE 1 Schematic representation of the clonal selection theory proposed by Burnet to explain antibody formation. Each lymphocyte has a different type of receptor (represented by the different symbols on the cell surface); the specific interaction with a determined antigen X (in red) determines the formation of a clone of cells which produce the same antibody (clonal selection and multiplication). The first formulation of this theory did not include costimulatory influences such as T cell help to B cells.

in promoting the idea that each lymphocyte had only one kind of receptor and that stimulating the "right" receptor would create a clone of cells, each making the same unique antibody. He also postulated that somatic mutation of the genes coding for antibodies would create variants, some of which might be making antibody of higher affinity for the antigen. If so, these mutants might possess a survival advantage in the face of decreasing antigen concentration. Thus, the affinity of the antibody could increase with time. In a way, the lymphocytes of the body were seen as a Darwinian microcosm. Also, Burnet foresaw how the clonal selection theory could explain that "self" antigens were not recognized. Perhaps, in an immature animal, an encounter between antigen and a cell capable of reacting with it would cause deletion rather than selection of the cell in question (Fig. 2). Thus, animals are tolerant of their own antigens. If, for some reason, "forbidden" antiself clones were allowed to survive, then autoimmune disease might supervene. Burnet expanded his views into a full-length monograph and vigorously promoted the theory during the next few years. Lederberg (1959) was an early supporter and added the valuable insight that the transition from a capacity to be rendered tolerant to a capacity to be positively selected by antigen had to be a transition made by every lymphocyte, not by the animal as a whole. "Self" antigens are always present and are thus able to "catch" each lymphocyte

during a "window" of tolerance susceptibility. Foreign antigens enter unexpectedly and find some cells within this window, but many more have already passed it. These immunocompetent cells are clonally selected and make antibody which soon rids the body of the foreign invader. Deleting a few lymphocytes inappropriately was viewed as a small price to pay for achieving the vital task of self–nonself discrimination.

It is worthwhile to discuss the extraordinary range of affinities with which antibodies react with antigen (Fig. 3). Association constants of 10^{-5} M or even lower are not uncommon, whereas some very special antibodies display a K_a of 10^{-10} or even 10^{-11} M. In most recognition systems between biological macromolecules, eons of evolution have determined an optimal affinity. Thus, enzyme–substrate, hormone–receptor, transcription factor–DNA, or virus–receptor recognition occur at a defined, predetermined strength. The immune system is different. It does not "know" what it will "need" to recognize next. It must somehow create recognition units to fit any conceivable antigen more or less well. As discussed later, it uses special tricks to improve on the strength of the original recognition. However, this great variation in affinities creates confusion for academic immunologists. One scientist may be working with a system operationally geared to detect only high-affinity antibodies or cells. This scientist may find a particu-

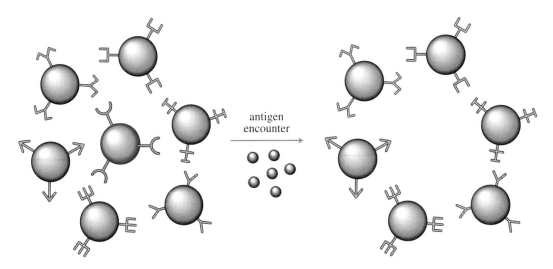

antigen
encounter

FIGURE 2 Immunological tolerance to self-antigens is due to a negative selection of self-reactive lymphocytes by the antigen. Initially, only the elimination or killing of the potentially self-reactive cells (clonal deletion) was envisaged. Later research showed that the antigen could induce a refractory state (clonal anergy) without the actual disappearance of the cell.

lar effect, e.g., that the lymphocyte population has been rendered tolerant to the antigen in question. A second scientist may be in a position to detect antibodies of much lower affinity. The relevant cells may not have been affected by the particular experimental protocol and, because they are present in greater numbers, lead the investigator to an opposite conclusion. This dilemma explains many of the discrepancies in the immunological literature.

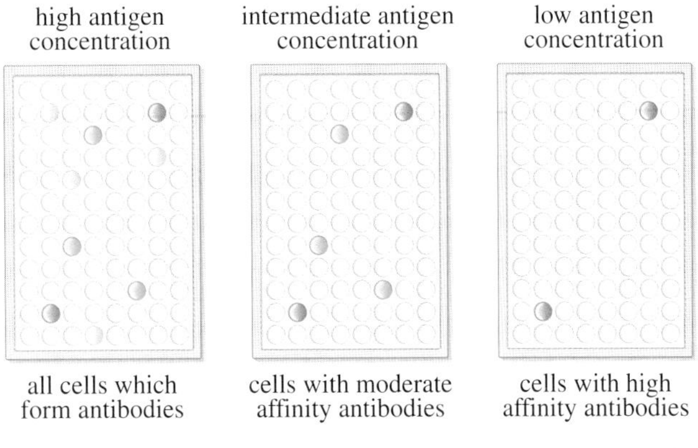

high antigen concentration	intermediate antigen concentration	low antigen concentration
all cells which form antibodies	cells with moderate affinity antibodies	cells with high affinity antibodies

FIGURE 3 Immune recognition can vary in affinity. The diagram portrays the antibody products of a population of lymphocytes assayed under various conditions using an ELISA assay in which antibodies bind to antigen fixed to a microtiter plate and are then revealed by a color-generating enzyme reaction (green). If a high concentration of antigen is used (left), many cells score positive (all those that produce antibodies). If a low concentration is used (right), few score positive (those producing high-affinity antibodies). With an intermediate concentration (middle), an intermediate result is obtained.

The Cellular Basis of Antibody Formation

It must first be considered how lymphocytes come to be the cells that need to be considered for selective stimulation by antigen. It is of interest that early researchers such as Pasteur, von Behring, and even Ehrlich did not concern themselves greatly with the cellular source of antibodies. At the turn of the century, Pfeiffer and Marx demonstrated that spleen, lymph nodes, and bone marrow were important sites of antibody production; this was confirmed by Alexis Carrel in some of the first tissue culture experiments ever conducted. McMaster and Hudack in the 1930s showed that lymph nodes draining the site at which an antigen had been injected were particularly active, and Ehrich and Harris carefully documented both the great enlargement of such lymph nodes and the increase in the number of lymphocytes that poured out from them into the efferent lymphatic vessels, which connected a lymph node with nodes further up the draining chain and, finally, with the venous blood. Because cells harvested from the efferent lymphatic vessels could form antibody *in vitro*, it was not surprising that these authors concluded that lymphocytes were the antibody-forming cells. Moreover, lymphocyte populations harvested from one rabbit during the course of an immune response could be injected into a second, unimmunized rabbit and could transfer the capacity to form antibodies.

Despite this excellent American research, conclusions of a different sort were being reached in Europe in the 1940s. Bing and Plum made the association between high immunoglobulin levels in certain patients and the appearance in their tissues of accumulations of very characteristic cells.

These so-called plasma cells had an eccentric nucleus, a bulky cytoplasm which stained with dyes taken up by RNA, and a prominent Golgi region (a hallmark of cells specialized for large-scale protein synthesis). Bjorneboe and Gormsen noted plentiful plasma cells in heavily immunized experimental animals. Fagraeus (1948) performed two sets of very important experiments. She described the sequential appearance after immunization of what appeared to be a lineage of cells: first, large, rapidly dividing plasmablasts, then medium-sized "immature" plasma cells, and finally smaller, nondividing cells with typical plasma cell morphology. Furthermore, she noted that most of the plasma cells were found in the red pulp of the spleen, not the white pulp in which most of the lymphocytes lived. On dissection of splenic fragments into portions containing mainly red or white pulp, it was found that the former made far more antibody. She concluded that the plasma cell was the predominant antibody-forming cell. This view was supported by Coons' group (Leduc *et al.*, 1955), who devised the "sandwich" immunofluorescence assay, a clever way of staining for specific antibody, and showed plasma cells to be strongly positive.

Where does the lymphocyte come into play? The key finding reconciling the prior lymphocyte transfer experiments with the previous results was made by Gowans, who showed that, on appropriate antigenic stimulation, a small lymphocyte could turn into a blast cell, which then began to proliferate rapidly. This suggested that lymphocytes must be the population of cells on which Burnet's clonal selection depended.

Proof of Clonal Selection

When enunciated in 1957, the clonal selection theory was regarded as fairly outrageous. It occurred to me that perhaps it could be quite readily refuted. If an animal were simultaneously immunized with two or three different antigens, and single cells from the draining lymph node could be examined for antibody production, each cell might perhaps produce two or even three different antibodies. Joshua Lederberg joined me in this research (Nossal and Lederberg, 1958), and we decided to use immunobilization of motile *Salmonella* bacteria by antiflagellar antibody as our assay method. Single cells were incubated in tiny microdroplets under mineral oil, and after 4 h the tissue culture fluid which had been nourishing the single cell was tested for antibody content by micromanipulating into each droplet 5–10 motile bacteria. Even the first experiments, later refined and extended by several years' work, showed that one cell made only one antibody.

This research did not reveal why one cell only made one antibody. To resolve this question, one would have to go back to the starting lymphocyte population. However, Gordon Ada and I thought we should first put the nail in the coffin of the direct template hypothesis. We injected rats with small amounts of a strong antigen that had been rendered highly radioactive and later examined the antigen content of single antibody-forming cells by autoradiography. A method which would readily have detected four molecules of antigen failed to find any trace of antigen within the cells. Bearing in mind that each plasma cell has thousands of separate antibody factories, associated with the polyribosomes of the endoplasmic reticulum, it was clear that antigen, acting as a template, was simply not there.

Figure 4 summarizes the process performed to provide a formal proof of the clonal selection theory (Nossal and Pike, 1976). It was necessary to find a way to purify the lymphocytes from an unimmunized mouse which had the capacity to bind to a particular antigen. This was done by layering 10^8 spleen cells on gelatin which was coupled to the antigen. The vast majority of unbound cells were carefully washed away. The tiny fraction which adhered was recovered by melting the gel in the presence of collagenase. Subsequently, the single, antigen-specific cells were cultured in specialized microcultures and found to be capable of making only the kind of antibody capable of reacting with the fractionating antigen.

Molecular biology has played a large role in validating selective theories of antibody formation and in demonstrating how the diversity of antibodies can be accommodated within the Crick dogma. It was determined that the genes coding for antibodies are divided into variable (V) and constant (C) portions. Furthermore, the V genes for the heavy chains of immunoglobulins are made up of three sets of "minigenes" (V, D, and J), and the V genes for the light chains are composed of two sets of minigenes (V and J). A unique series of gene translocations selects out particular gene combinations for each lymphocyte as it develops, such that each cell possesses its own particular assembled gene combination. Furthermore, each cell possesses receptors of a single specificity. The lymphocyte population as a whole constitutes an extensive repertoire of multiple specificities, among which the antigen must select the best fitting.

Interacting Cells in Immune Responses

It is known that the lymphocytes capable of forming antibodies represent only one kind of lymphocyte. Figure 5 shows the main interacting cells of the immune system (see also Table 1). The sequence of lymphocyte activation, clonal proliferation, and differentiation is a complex cascade of events. Antigen capture by an antigen presenting cell (APC) is the first important step. The two main types of APC are dendritic cells (DCs) and macrophages. Cells

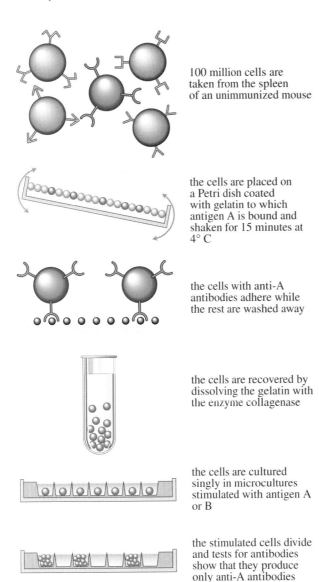

100 million cells are taken from the spleen of an unimmunized mouse

the cells are placed on a Petri dish coated with gelatin to which antigen A is bound and shaken for 15 minutes at 4° C

the cells with anti-A antibodies adhere while the rest are washed away

the cells are recovered by dissolving the gelatin with the enzyme collagenase

the cells are cultured singly in microcultures stimulated with antigen A or B

the stimulated cells divide and tests for antibodies show that they produce only anti-A antibodies

FIGURE 4 Experimental proof of the clonal selection theory of antibody formation. Cells from unimmunized mice were prepared on the basis of their capacity to recognize an antigen. *In vitro*, they could only form antibody corresponding to that antigen.

of the antibody-forming series (B cells) can also act as APCs and this form of presentation serves to augment immune responses. The APC is by no means only a passive magnet for antigens. It serves other very important functions in lymphocyte selection and activation. Special endocytic vacuoles called endosomes are responsible for digesting protein antigens into shorter peptide fragments. This processing step is a prelude to the foreign peptide being inserted into a groove on a "self" molecule belonging to the major histocompatibility complex (MHC), namely, a class II MHC molecule. The MHC–peptide complex is then transported to the surface of the APC. Here, it can act as a specific part-

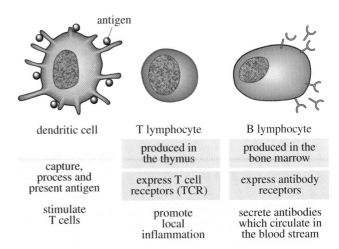

FIGURE 5 Dendritic cells, T lymphocytes, and B lymphocytes work together in the immune response. The principal characteristics of these cells are shown.

ner for a suitable T cell receptor (TCR) molecule on the surface of a T lymphocyte. T lymphocytes are formed in the thymus and are so named to distinguish them from the B lymphocytes, which are formed in the bone marrow and which are responsible for antibody formation. T lymphocytes, also called T cells, do not form antibody. They recognize the peptide–MHC complexes with very fine specificity using their TCR, which consists of an α and a β chain. The overall design of the TCR is similar to that of antibody molecules, and the same kind of somatic minigene translocation events form the final receptor so that a T cell repertoire is set up that is ready to mount immune responses against the whole universe of foreign peptides. There is another trick that T cells have learned. How APCs present peptides from foreign antigens was previously discussed. However, T cells can also "see" peptides from altered (and thus perhaps dangerous) self proteins and peptides from the inner portions of viruses. They can do so because a certain type of T cell, CD8$^+$, is designed to recognize peptides in the groove of not class II but class I MHC molecules. Class I MHC molecules pick up intracytoplasmic peptides such as fragments of viral molecules or housekeeping gene products. Peptide-loaded class I molecules are brought to the surface of the cell and displayed there. CD8$^+$ T cells can thus "patrol" the body and find virus-infected cells or cells displaying mutated gene products. Because CD8$^+$ cells possess powerful cytotoxic properties, this surveillance function can rid the body of virus-infected cells frequently before the virus has completed it life cycle; and it can also destroy aberrant cells, for example, malignant or premalignant cells.

An additional feature of APCs is that they display molecules on their surface which subserve costimulatory functions. Cross-linking of the TCR on a T cell is not sufficient to cause full activation of several cognate interactions which mediate costimulation; that between B7.1/B7.2 and CD28 is one of the more important (Hathcock *et al.,* 1993). When the APC has managed to stimulate the T cell, the latter emerges from its resting state, enlarges, and begins to divide. At the same time, it secretes hormone-like factors known as lymphokines. The pattern of lymphokine secretion by different T cells will be discussed later. Suffice it to say here that lymphokines mediate the proinflammatory functions of T cells and are thus responsible for most of the phenomena of delayed-type hypersensitivity; they are responsible for the "help" which T cells give to B cells in the process of antibody formation, and they can, in some circumstances, also negatively regulate immune responses, serving to suppress excessive inflammation. T cells, via the cytokines they secrete, are key regulators of immunological processes. The "helper" function, the proinflammatory cytokine release, the cytotoxic killing of specific target cells, and the suppressor function all act in a highly localized manner. In fact, in many cases actual cell contact is required. This is in contrast to the function of B cells, the antibody product of which can be effective a long distance from the site of secretion.

The basic paradigms of immune recognition are summarized in Table 1.

....•

The Controversial Question of T Cell Subsets

In addition to certain less prominent types of T cells, such as $\gamma\delta$ T cells, CD-1-restricted T cells, or NK1.1$^+$ T cells (which will not be considered further), there exist among the "standard" $\alpha\beta$ TCR-positive T cells different subsets. One division is essentially absolute. T cells emerge from the thymus as CD4$^+$ or CD8$^+$ cells. As a general rule, the former play the major part in helper function to B cells, in cell-mediated immune responses such as delayed-type hypersensitivity, and in lymphokine secretion. The latter are specialized for

cytotoxic killing of other cells, for example, virus-infected cells. However, there is some overlap in these functions. CD8$^+$ cells certainly secrete some lymphokines, and CD4$^+$ cells can be demonstrated to have cytotoxic potential. The most important difference is that CD4$^+$ cells see antigenic peptides in association with MHC class II molecules, whereas CD8$^+$ cells react with peptides attached to the groove of MHC class I. As noted previously, class II-associated peptides are derived from proteins which the APC has pinocytosed and fragmented within endosomes, whereas class I-associated peptides are derived from the cytosol of that cell, e.g., fragments of self proteins or viral proteins (York and Rock, 1996). The downregulatory function of T cells was initially ascribed to CD8$^+$ "suppressor" T cells. Recently, circumstances have been described in which the suppressive function has clearly been ascribable to CD4$^+$ T cells (Weiner *et al.,* 1994). It is likely that both CD4$^+$ and CD8$^+$ cells can contribute to suppression, depending on experimental circumstances.

A hypothesis which has gained great currency because it provides certain helpful simplifications is that CD4$^+$ T cells can be divided into T-helper type 1 (Th1) or Th2 cells (Mosmann and Coffman, 1989). This is based on the discovery that a majority of cloned T cell lines, at least in the mouse, secrete either interleukin-2, interferon-γ, and lymphotoxin or IL-4, -5, -6, and -10. Th1 cells, the former type, are seen as guiding immune responses down a proinflammatory pathway, whereas Th2 cells set the system up for an antibody response. When more physiological sources of T cells were examined, the Th1/Th2 categorization did not hold (Kelso *et al.,* 1991). T cells secreted a varied mixture of cytokines. Rather, the Th1-like or Th2-like pattern of secretion emerged as the immune response matured.

One striking feature of both *in vitro* and *in vivo* T cell responses is the presence of positive feedback loops and cross-inhibition. Thus, IL-4 is a strong promoter of the development of Th2-type cytokine responses and thus of more IL-4 production. At the same time, IL-4 inhibits the production of Th1-type cytokines (Gross *et al.,* 1993). In contrast, IL-12 promotes Th1-type T cell development and inhibits its Th2-type cytokine production (Heinzel *et al.,* 1993). It can be seen that T cell responses can be "locked in" to a certain direction which, once established, is difficult to change. Our knowledge of how an antigen influences a deviation to one or another pattern is quite scant. Certainly the nature of the adjuvant used is important, as is the molecular form of the antigen and route whereby it is introduced into the body. It is hoped that we will learn enough about T cell regulation to be able to devise vaccine formulations selective for the type of immune response which offers the most protection.

A new concept which is gradually emerging is that of a Th3-type cytokine pattern, with the immunoinhibitory molecule transforming growth factor-β as a dominant com-

---

TABLE I

The Basic Paradigms of Immune Recognition

- System doesn't know what it must recognize: "infinite" capability.
- Recognition depends on T cell and antibody receptors. The rule: One cell, one receptor.
- Receptor genes are created by a unique process of somatic minigene assembly.
- The lymphocytes, each with a different receptor, constitute a repertoire. Self antigens can punch holes in the repertoire.

ponent. This is perhaps the modern version of the suppressor T cell.

The Question of B Cell Subsets

The question of B cell subsets has been much less discussed but is also important. B cells exit from the bone marrow and the specificity of their immunoglobulin receptors is different as a result of the differing somatic gene selection and assembly events, but initially they are a homogeneous population in functional terms. The subsets arise through different kinds of lymphocyte selection by antigen. There is a primary "decision" to be made: whether to become an antibody-forming cell (AFC) or to engage in a series of divisions which result in more B cells of similar specificity, i.e., "memory" cells. In both cases, there is a secondary decision: Should the cell continue to make the same chemical type of antibody, or should it switch to another type? The B cell emerges from the marrow carrying receptors of two chemical kinds or isotypes, namely, IgM and IgD. However, through a second unique kind of gene translocation event, the B cell can "decide" to switch to making a different isotype with different biological properties without undergoing a change in antibody specificity. There are six other isotypes, namely, IgA and IgE and four different sorts of IgG. IgA, for example, is prominently found in mucosal sites, acting as a kind of barrier cream. IgE is prominently associated with allergic phenomena and with defense against parasites. The IgGs are quantitatively the most abundant antibodies and are very good at neutralizing toxins and viruses. Although a single B cell clone can make multiple isotypes, an individual cell will only transiently do so. The main drivers of the isotype switch are T cell-derived cytokines. For example, IL-4 is essential for a switch to IgG1 and IgE. Other cytokines drive other switches. Having undergone an isotype switch, both the antibody-secreting cell and the memory cell remain faithful to one particular isotype, making it until they die.

To gain a better understanding of how lymphocyte selection by antigen can result in these diverse responses, it is necessary to briefly examine the microarchitecture of lymphoid tissues.

Dual Pathway of *in Vivo* B Cell Selection

Both lymph nodes and spleen are designed to trap and retain antigens and to put these in the closest contact with large numbers of lymphocytes. Antigens reach lymph nodes largely via the afferent lymphatic vessels which drain the different tissues. They reach the spleen via the blood circulation. Some of the antigen is already within antigen-capturing cells (e.g., DC), which pick up antigen in the tis-

sues or the blood. The lymphoid tissues are divided into portions in which lymphocytes traffic and develop and other portions in which macrophages abound and mature antibody-forming cells called plasma cells take up residence. In the lymph nodes, these two portions are called the cortex and the medulla; in the spleen, they are called the white pulp and the red pulp. I now consider the events taking place in the white pulp of the spleen after injection of an antigen. It is important to recall that lymphocytes circulate extensively through lymphoid tissues, journeying restlessly from blood to lymph nodes and spleen and back again, thus giving them the maximum chance of meeting trapped antigen. The first place the recirculating lymphocytes reach in the spleen is in the white pulp. The central portion of the white pulp, the so-called periarteriolar lymphocyte sheath (PALS), is particularly rich in T lymphocytes and is surrounded by an area which contains rounded nests of densely packed lymphocytes termed lymphoid follicles. These follicles have a very high proportion of B lymphocytes. Both the PALS and the follicles also contain numerous DCs, which are referred to respectively as interdigitating DC (IDC) and follicular DC (FDC).

When a T cell of appropriate specificity encounters an IDC in the PALS, it is activated and begins to divide. In turn, the activated T cell is now in a position to help a B cell in its pathway of clonal expansion and differentiation. The B cell is normally specific for a different part of the antigen than the T cell. It comes close to the T cell either because it has pinocytosed and processed antigen which the T cell can recognize or as part of a three-cell cluster, namely, DC, T cell, and B cell. Considering that the cells of the appropriate specificity have to find one another, immune responses take time to initiate. The activated B cell can now develop along one of two different pathways (Jacob *et al.,* 1991). Either it can grow into an expanding focus of AFC or it can migrate toward the lymphoid follicles and create a germinal center. Evidence has been presented (Jacob and Kelsoe, 1992) that a single B cell can have progeny which initiate both events. Figure 6 illustrates the separate cellular collections of the two proliferative cascades.

The AFC foci typically develop toward the edge of the white pulp and some of the AFCs eventually reach the red pulp in smaller clusters. In a primary T-dependent immune response, AFC foci reach their maximum size at approximately Day 7 or 8, when they may consist of scores or hundreds of cells. This extrafollicular B cell proliferation occurs without concomitant Ig V gene somatic mutation. The AFCs which comprise the foci can switch isotype, frequently from a synthesis of IgM to IgG1. It is possible to find foci with approximately half the B cells forming IgM and the other half IgG1. Eventually, the B cells in the foci die by apoptosis, which can be markedly retarded by the transgenic insertion of the gene *bcl-2*, the product of which inhibits apoptosis (Smith *et al.,* 1996).

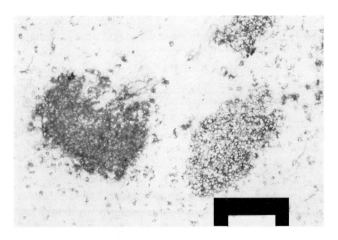

FIGURE 6 Micrograph of a germinal center (in blue) and of an AFC focus (in red). B cells, following T-dependent antigenic stimulation, can form a germinal center or grow into an AFC focus. The colorations were obtained by histochemical methods.

The events leading to the formation of a germinal center are schematically summarized in Fig. 7. A critical player in their development is the FDC. This cell has a remarkable capacity to capture antigen and retain it on its surface for many months (Nossal and Ada, 1971; Szakal *et al.*, 1989)

without processing and in an immunogenic form. However, antigen only binds to FDC in two circumstances: if antigen is complexed to antibody because of Fc receptors that FDCs possess, or if the antigen has activated the alternate complement pathway because FDCs also have C3 receptors. With a soluble protein antigen, follicular localization is delayed for several days until some antibody has been formed. With a viral or bacterial antigen, it can begin within a few hours.

A second key group of cells within germinal centers are CD4$^+$ T lymphocytes. Their role within this largely B cell microenvironment is not clear, but mice lacking T cells have greatly diminished germinal center formation. Presumably, the CD4$^+$ cells and the cytokines they produce help to maintain the growth drive of the B cells. As activated B cells accumulate in lymphoid follicles, and begin to divide rapidly, unknown processes also initiate IgV gene somatic mutation. This begins approximately 6 days after antigen injection and occurs at an extraordinarily high rate estimated at approximately 10^{-3} mutations/base pair/generation or nearly one mutation per cell division. Many of these mutations will have no effect on the affinity of the mutated antibody to the antigen in question, and many others will lower the affinity. However, some will increase the affinity. It is important to be able to select the higher affinity variants. This is where the FDC-bound antigen plays a role. Mutated

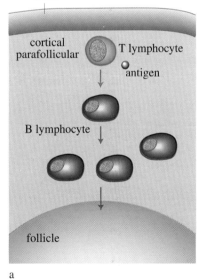
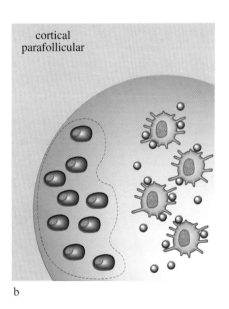
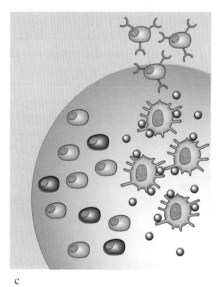

a   b   c

FIGURE 7 (a) The B lymphocyte, with T lymphocyte help, is activated by antigen and begins to divide. Some of the progeny migrate into the lymphoid follicle. (b) In the follicle, the dendritic cells hold the antigen on their surface and the B lymphocytes, dividing rapidly, create a germinal center. (c) Deep in the germinal center, far from the antigen, hypermutations of the Ig V genes occur in the cells. After expressing the mutated receptor, these cells migrate toward the antigen source. Positively selected cells (light green) either leave the center to form antibodies or B memory cells or iteratively repeat the mutation-selection process. The cells which do not receive a positive signal binding to the antigen die and are phagocytosed by the macrophages.

B cells migrate to the vicinity of the FDCs. Only those with Ig receptors of higher affinity can displace previously synthesized antibody from the immune complex on the FDCs. Such cells will be selectively stimulated, and it appears that the process can be iterative. Whereas some positively selected, higher affinity B cells leave the germinal center, others recommence V gene hypermutation followed by further testing and selection. As a result, B cells can carry 50 or more mutations and their antibody product is very significantly different from the germline gene product initially carried by the ancestors of these B cells. A secondary B cell repertoire is thus created that is heavily influenced by the antigenic history of the individual concerned. Germinal centers reach their maximal size 2 or 3 weeks after antigen, but can become even larger if new antigen reaches them.

B cells which leave the germinal center can become either AFCs or memory recirculating B cells. In the latter case, they will usually be isotype switched, relatively long lived, and, according to some evidence, more readily stimulated than primary B cells if antigen is reencountered. In the former case, the life span is variable. Most AFCs live only a few days, but a significant minority can live for weeks or months. The high-affinity antibodies which they produce constitute a significant component of the complex phenomenon of immunological memory. The various B cell subsets arising from the diverse effects of antigenic stimulation differ in certain molecular markers and can be separated in reasonably pure form by flow cytometry.

Negative Selection of Lymphocytes by Antigen

Immunological tolerance, namely, the fact that animals do not react to their own antigens, is discussed elsewhere in this volume, but no account of lymphocyte selection by antigen would be complete without reference to the fact that antigen can negatively select lymphocytes as well as positively select them for activation and immune effector function. Special mention is made here of the primary lymphoid organs, the thymus and the bone marrow, in which extensive elimination of self-reactive cells occurs. An excellent example of negative selection by clone elimination is the classical experiment of von Boehmer's group (Kisielow *et al.*, 1988) schematically represented in Fig. 8. Mice were rendered transgenic for a TCR with specificity for the male transplantation antigen H-Y. The majority of thymocytes carried the transgenic receptor. In the female, the thymus developed normally and reached normal size. In the male, in which many cells in the thymus expressed the target antigen, the thymus was only 5% of its normal size. All the transgenic receptor-positive anti-H-Y T cells had been eliminated. The only cells capable of surviving were those which

had "broken through" and had expressed an endogenous, not a transgenic, TCR or those which had downmodulated the CD8 molecule, which acts as a coreceptor in this system. A conceptually similar process of B cell elimination occurs in the bone marrow, in which B cells with reactivity to, for example, a common cell surface antigen, die locally. Two additional phenomena should be noted for the B cell. For a limited period of time, an immature, potentially self-reactive B cell is given a second chance. Having formed a self-reactive receptor, it can rearrange the light-chain V genes on the chromosome not used on the first occasion to determine if now, in combination with the same heavy-chain V gene, a receptor results which is no longer self-reactive. This process is known as receptor editing (Tiegs *et al.*, 1993). Second, the B cell can be negatively signaled by antigen without being eliminated. This refractory state, initiated by less extensive receptor cross-linking, is known as clonal anergy (Nossal and Pike, 1980). In fact, induction of anergy is not confined to immature B cells or the bone marrow. As illustrated in Fig. 9, there is a differential sensitivity of B cells to anergy induction dependent on the degree of maturity (Nossal, 1983). Obviously, pre-B cells not yet expressing the rearranged IgM receptor are blind to antigen. The most sensitive stage occurs when the IgM receptor first

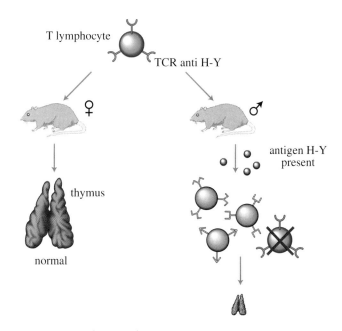

FIGURE 8 The experiment of von Boehmer, which is an example of negative selection. TCR H-Y transgenic mice express a TCR with specificity for a peptide of the male antigen H-Y. In the female, in which the antigen is not present, the thymus develops normally; in the male, in which the antigen is present, all the transgenic T cells positive for the H-Y receptor die and the thymus reaches only 5% of its normal size. The only cells that survive are those which express an endogenous TCR or cells with reduced levels of CD8.

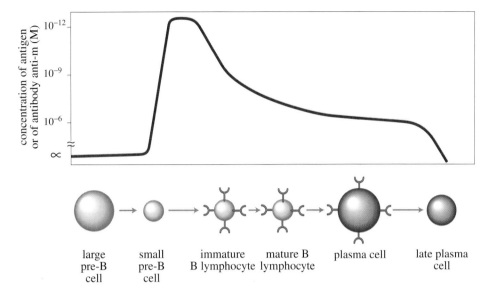

large pre-B cell   small pre-B cell   immature B lymphocyte   mature B lymphocyte   plasma cell   late plasma cell

FIGURE 9 Differential sensitivity of B cells to inactivation by antigen as a function of differentiation status. Pre-B cells, which do not yet display Ig receptors for antigen, are insensitive, as are late-stage plasma cells which have lost their Ig receptors. All other B cell stages can be functionally inactivated but at different antigen concentrations.

emerges. As the cell gains immunocompetence, it can still be negatively signaled by antigen, provided costimulatory influences are absent. Even the actual AFC can be inhibited in its secretory process, admittedly only by quite high antigen concentrations. The AFC becomes refractory to the action of antigen only quite late in its life history when it loses its Ig receptors. The state of anergy is potentially reversible in the B cell if antigen is withdrawn. In the continued presence of antigen, which is the physiological situation for self-antigens, the anergic B cell suffers a shortened life span. In some ways, therefore, clonal anergy shades into clonal deletion. Clonal anergy has also been induced in T cells in experimental circumstances. It is not clear what role it plays in physiological self-tolerance.

Tolerance within the Secondary B Cell Repertoire

Given the extent of V gene hypermutation within germinal centers, and the large number of self-antigens, occasionally a B cell, appropriately stimulated by a foreign antigen, fortuitously mutates to acquire specificity for some self-antigen. Fortunately, evolution has designed a mechanism to eliminate such cells. Within the germinal center, B cells go through a "second window" of tolerance susceptibility. If they encounter free antigen, not associated with an FDC, they are particularly susceptible to apoptotic cell death (Pulendran *et al.*, 1995; Shokat and Goodnow, 1995). Figure 10

shows a typical example in which such cells have died and are being engulfed by macrophages within the germinal center. Thus, in health, the secondary or postantigenic B cell repertoire is "purged" of self-reactive cells. Clearly, this mechanism is not perfect because autoimmune diseases do occur. In these cases, autoantibodies which appear are usually heavily mutated. The basis of this failure of regulation is not understood.

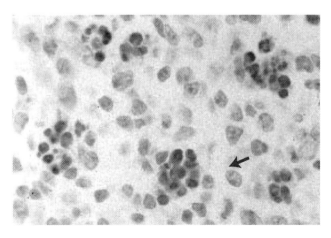

FIGURE 10 Micrograph of apoptotic cells colored brown by the immunohistological TUNEL technique. Most of the cells are grouped in clusters because they have been taken up by macrophages (arrow). B lymphocytes in the germinal centers are susceptible to apoptosis if they encounter free antigen.

T Cell Tolerance for Antigens Not Present in the Thymus

Negative selection of self-reactive T cells in the thymus is subject to two constraints. First, there must obviously be some affinity cutoff point for clone elimination and thus low-affinity antiself T cells with affinity below the threshold can escape. Second, there are self-antigens which are not widely distributed in the body but rather confined to particular tissues. These may not gain entry into the thymic microenvironment to a sufficient degree to provoke thymic negative selection. Thus, cells against such self-antigens seed out from the thymus into the peripheral lymphoid tissues. Such antiself cells may simply ignore their target antigen. For example, migration of unprimed T cells into bodily organs is quite limited because the cells tend to home primarily to lymphoid tissues. Furthermore, even if an encounter were to take place, factors such as low affinity or the absence of costimulatory influences may prevent a significant autoimmune response. It has recently been shown that antigen-induced T cells are subject to apoptotic death after a few cycles of division and this feedback loop may limit the degree of damage which an occasional self-reactive T cell reaching its target can actually inflict (Lenardo et al., 1995). Many examples of such peripheral autoreactive T cell deletion have been reported. Finally, immunoregulatory or suppressor T cells can develop in some experimental tolerance situations and may play a role in self-tolerance as well. Therefore, it is clear that many backup mechanisms support central thymic deletion, thus resulting in the normal tolerant state.

Conclusions

From origins as rather outrageous ad hoc speculations, selective theories of lymphocyte activation have advanced greatly. The "one cell, one receptor" rule has been amply validated both through cellular experimentation and through the revelation of the extraordinary molecular events underlying the creation of somatically assembled Ig and TCR V genes. Two elements not foreshadowed in the original formulations are now the subject of a great deal of experimental work, namely, costimulatory influences other than receptor cross-linking which are required for lymphocyte activation and methods other than clonal deletion that underlie self–nonself discrimination. However, the broad rules of lymphocyte selection by antigen, both positive and negative, are well understood. The focus of immunology has therefore moved to another level, namely, that of immunoregulation. The mechanisms which guide immune responses in various directions (e.g., predominantly antibody formation versus predominantly T cell responses) or into one of the various antibody isotypes or a particular kind of T cell response are only partially understood. However, it is these choices which medical science seeks to influence, whether it be for the design of an effective vaccine, a treatment for autoimmune disease, or a more perfect result in organ transplantation. There is a real sense of excitement because these issues are being addressed by the most powerful tools of the new biology. At the same time, we must not forget the pioneers who enabled us to reach the current level of understanding of lymphocyte selection by antigen.

References Cited

BREINL, F., and HAUROWITZ, F. (1930). Chemische Untersuchung des Präzipitates aus Hämoglobin und Anti-Hämoglobin-Serum und Bemerkungen über die Natur der Anti-körper. Z. Physiol. Chem. **192**, 45–57.

BURNET, F. M. (1957). A modification of Jerne's theory of antibody production using the concept of clonal selection. Austral. J. Sci. **20**, 67–69.

BURNET, F. M., and FENNER, F. (1949). The Production of Antibodies, 2nd ed. Macmillan, Melbourne.

FAGRAEUS, A. (1948). The plasma cellular reaction and its relation to the formation of antibodies in vitro. J. Immunol. **58**, 1–13.

GROSS, A., BEN-SASSON, S. Z., and PAUL, W. E. (1993). Anti-IL-4 diminishes in vivo priming for antigen-specific IL-4 production by T cells. J. Immunol. **150**, 2112–2120.

HATHCOCK, K. S., LASZLO, G., DICKLER, H. B., BRADSHAW, J., LINSLEY, P., and HODES, R. J. (1993). Identification of an alternative CTLA-4 ligand costimulatory for T cell activation. Science **262**, 905–907.

HEINZEL, F. P., SCHOENHAUT, D. S., RERKO, R. M., et al. (1993). Recombinant interleukin-12 cures mice infected with Leishmania major. J. Exp. Med. **177**, 1505–1509.

JACOB, J., and KELSOE, G. (1992). In situ studies of the primary immune response to (4-hydroxy-3-nitrophenyl)acetyl. II. A common clonal origin for periarteriolar lymphoid sheath-associated foci and germinal centers. J. Exp. Med. **176**, 679–687.

JACOB, J., KASSIR, R., and KELSOE, G. (1991). In situ studies of the primary immune response to (4-hydroxy-3-nitrophenyl)-acetyl. I. The architecture and dynamics of responding cell populations. J. Exp. Med. **173**, 1165–1175.

JERNE, N. K. (1955). The natural-selection theory of antibody formation. Proc. Natl. Acad. Sci. USA **41**, 849–857.

KELSO, A., TROUTT, A. B., MARASKOVSKY, E., et al. (1991). Heterogeneity in lymphokine profiles of CD4[+] and CD8[+] T cells and clones activated in vivo and in vitro. Immunol. Rev. **123**, 85–113.

KISIELOW, P., BLÜTHMANN, H., STAERZ, U. D., STEINMETZ, M., and VON BOEHMER, H. (1988). Tolerance in T-cell-receptor transgenic mice involves deletion of nonmature CD4[+]8[+] thymocytes. Nature **333**, 742–746.

LEDERBERG, J. (1959). Genes and antibodies: Do antigens bear instructions for antibody specificity or do they select cell lines that arise by mutation? Science **129**, 1649–1653.

LEDUC, E. H., COONS, A. H., and CONNOLLY, J. M. (1955). Studies on antibody production. II. The primary and secondary responses in the popliteal lymph node of the rabbit. *J. Exp. Med.* **102,** 61–104.

LENARDO, M. J., BOEHME, S. A., CHEN, L., COMBADIERE, B., FISHER, G., FREEDMAN, M., MCFARLAND, H., PELFREY, C., and ZHENG, L. (1995). Autocrine feedback death and the regulation of mature T lymphocyte antigen responses. *Int. Rev. Immunol.* **13,** 115–134.

MOSMANN, T. R., and COFFMAN, R. L. (1989). TH1 and TH2 cells: Different patterns of lymphokine secretion lead to different functional properties. *Annu. Rev. Immunol.* **7,** 145–173.

NOSSAL, G. J. V. (1983). Cellular mechanisms of immunologic tolerance. *Annu. Rev. Immunol.* **1,** 33–62.

NOSSAL, G. J. V., and ADA, G. L. (1971). *Antigen, Lymphoid Cells and the Immune Response.* Academic Press, New York.

NOSSAL, G. J. V., and LEDERBERG, J. (1958). Antibody production by single cells. *Nature* **181,** 1419–1420.

NOSSAL, G. J. V., and PIKE, B. L. (1976). Single cell studies on the antibody-forming potential of fractionated, hapten-specific B lymphocytes. *Immunology* **30,** 189–202.

NOSSAL, G. J. V., and PIKE, B. L. (1980). Clonal anergy: Persistence in tolerant mice of antigen-binding B lymphocytes incapable of responding to antigen or mitogen. *Proc. Natl. Acad. Sci. USA* **77,** 1602–1606.

PULENDRAN, B., KANNOURAKIS, G., NOURI, S., SMITH, K. G. C., and NOSSAL, G. J. V. (1995). Soluble antigen can cause enhanced apoptosis of germinal-centre B cells. *Nature* **375,** 331–334.

SHOKAT, K. M., and GOODNOW, C. C. (1995). Antigen-induced B-cell death and elimination during germinal-centre immune responses. *Nature* **375,** 334–338.

SMITH, K. G. C., HEWITSON, T. D., NOSSAL, G. J. V., and TARLINTON, D. M. (1996). The phenotype and fate of the antibody-forming cells of the splenic foci. *Eur. J. Immunol.* **26,** 444–448.

SZAKAL, A. K., KOSCO, M. H., and TEW, J. G. (1989). Microanatomy of lymphoid tissue during humoral immune responses: Structure function relationships. *Annu. Rev. Immunol.* **7,** 91–109.

TALMAGE, D. W. (1957). Allergy and immunology. *Annu. Rev. Med.* **8,** 239–256.

TIEGS, S. L., RUSSELL, D. M., and NEMAZEE, D. (1993). Receptor editing in self-reactive bone marrow B cells. *J. Exp. Med.* **177,** 1009–1020.

WEINER, H. L., FRIEDMAN, A., MILLER, A., *et al.* (1994). Oral tolerance: Immunologic mechanisms and treatment of animal and human organ-specific autoimmune diseases by oral administration of autoantigens. *Annu. Rev. Immunol.* **12,** 809–837.

YORK, I. A., and ROCK, K. L. (1996). Antigen processing and presentation by the class I major histocompatibility complex. *Annu. Rev. Immunol.* **14,** 369–396.

FRITZ MELCHERS

Basel Institute for Immunology
Basel, Switzerland

Lymphocyte Development

During embryonic development, lymphocytes derive from the multipotent hematopoietic staminal cell, which has its center in the bony medulla. They are continually regenerated throughout their entire life. T-lymphocytes mature in the thymus, while B-lymphocytes develop within the bone marrow. The development of the two cellular lines takes place thanks to a selection of a function address and a series of rearrangements of the genes that codify the specific receptors of the antigens. In particular, the development begins with a small number of progenitors and then expands through proliferation and differentiation, generating cells that acquire the receptor structures specific to the antigen, at the conclusion of the maturation cycle. In both primary lymphatic organs (the thymus and bone marrow) and in secondary lymphatic organs (for example, the lymph nodes) a phenomenon of negative selection takes place for those lymphatic receptors that are potentially able to bring about processes of auto reactivity. At the same time, the cells may be selected for entrance in the peripheral lymphatic organs, prepared for codification to become T- and B-lymphocytes capable of responding to the antigens.

All Lymphocytes Originate from Pluripotent Hematopoietic Stem Cells

A single cell, the pluripotent hematopoietic stem cell (HSC), is capable of generating all cell lineages of blood and, hence, also all the lymphocytes of the immune system (Morrison *et al.*, 19997). A HSC is the ancestor of the 2×10^{12} lymphocytes of an adult human or the 2×10^9 lymphocytes of an adult mouse.

HSC must be capable of self-renewal. Self-renewal upon division is maintained as long as at least one of the daughter cells retains the potential of the original cell while the other daughter cell might differentiate further along the pathways of myeloid/lymphoid cell differentiation. It is not clear whether the immune system is built up from one or from many HSCs. HSCs are likely to be retained in their state of development by contact with an environment of stromal cells which also determine the size of the HSC compartment. The HSC compartment may enter a resting state once it has been filled and may remain in that state while more differentiated cells are required.

HSCs have not been grown in tissue culture. Hence, it is not known how often this pluripotent cell can divide and maintain its state of development and how fast is loses its state as an HSC. In humans and mice, HSCs have been found throughout life, although their numbers appear to decrease with age. It has been possible to estimate the frequencies of HSC since a single cell is capable of populating the appropriate environment of a developing or a deficient (usually lethally irradiated) host with HSCs and their descendants upon adoptive transfer. However, the determined state of an HSC appears to be less stable when it is removed from its natural environment during this adoptive transfer since secondary adoptive transfer from the transplanted host into a new host is often unsuccessful (Spangrude *et al.*, 1995).

The Compartments of the Immune System

Approximately half of all lymphocytes in an adult are found disseminated as single cells in the epithelia of skin, tongue, and the respiratory, reproductive, and gastrointestinal tracts in which they are thought to form first lines of defense and in which they might ensure the integrity of these epithelia (Rocha *et al.*, 1992, 1995; Boismenu and Havran, 1994). These intraepithelial lymphocytes (IELs) are mostly T lymphocytes, and many of them express γ/δ T cell receptors (TCRs), although α/β TCR$^+$ IELs are also found.

The other half of the lymphocytes in the immune system are organized in extrafollicular and follicular structures in the secondary lymphoid organs, such as spleen or lymph nodes. Two-thirds of them are T lymphocytes, most of which express α/β TCRs, whereas one-third are B lymphocytes, most of which express IgM as B cell receptor for antigen (Rolink and Melchers, 1993; Spits, 1994; Kisielow and von Boehmer, 1995; Melchers *et al.*, 1995; Shortman and Wu, 1996; Burrows and Cooper, 1997; Fehling and von Boehmer, 1997; Papavasiliou *et al.*, 1997).

Approximately 10% of all T and B lineage cells are found in the primary lymphoid organs of an adult (i.e., in the thymus and bone marrow, respectively) as progenitors, precursors, and immature lymphocytes. They all have a short life span, with half-lives on the order of 2–4 days. The majority (90%) of all lymphocytes are located in the peripheral lymphoid organs or circulate from blood to lymph and back. Most of them are long lived in the adult, with half-lives of 6 weeks and longer. The mature, peripheral lymphocytes are subdivided into many phenotypically and functionally distinct subpopulations. The major T lymphocyte subpopulations are CD4$^+$ α/β TCR$^+$ helper T cells, CD8$^+$ α/β TCR$^+$ cytolytic T cells, γ/δ TCR$^+$ cytolytic T cells, α/β or $\gamma\delta$ TCR$^+$ IEL T cells, B$_1$ IgM$^+$ B cells, and conventional IgM$^+$ B cells. Many numerically smaller populations of T and B cells are known, including those with previous experience with a foreign antigen that are considered to carry memory for the previous encounter.

It is evident that many lymphocytes die every day and are replaced by newly generated and newly selected cells. With increasing age, this regenerative capacity of the immune system decreases while the relative representation of long-lived cells increases. It is not known, or only incompletely analyzed, how the different phenotypic and functional subpopulations of lymphocytes in the immune system turn over during life. Nevertheless, it is clear that this regenerative process never stops completely. However, it is not known whether HSCs or more committed, less pluripotent stem cells (i.e., lymphoid and T and B lineage committed cells) are the sources of this continuous regeneration during adulthood.

Embryonic Development

The primary germ layers of a mouse embryo — endoderm, ectoderm, and mesoderm — are formed on Day 6 or 7 of embryogenesis. Thereafter, the mesoderm moves in two directions, anterior and posterior. The posterior part develops, among others, progenitors of cardiac muscle cells, endothelial cells, and hematopoietic cells (Fig. 1). The putative common progenitors of the endothelial and the hematopoietic cells are called hemangioblasts. It is thought that embryonic development, at Day 10 of gestation of the mouse and, hence, at a cellular stage close to hemangioblasts, generates the first pluripotent HSC. The intraembryonic splanchnopleura in the aorta–gonad–mesonephros region initiates HSC formation and also expands HSCs (Cumano *et al.*, 1996; Medvinsky and Dzierzak, 1996). It can not be excluded that HSCs occur later in more than one place in the embryo since, early in development, mesenchyme, and later HSCs, are found in the yolk sac, the fetal liver, the spleen, and the bone marrow. In developing bone mesenchymal cells become located in subendosteal areas where they are thought to develop on one side into a layer of osteoblasts and osteoclasts and on the other side into HSC.

In vitro commitment of primitive ectoderm to mesoderm and mesenchyme, and its subsequent development to HSCs, can be studied with embryonic stem (ES) cells. Several ES cell lines exist which grow in tissue culture and which can be induced to develop into embryoid bodies which resemble at least parts of the proper morphological organization of the embryo. Further *in vitro* culture of the disintegrated embryoid bodies can develop erythroid, myeloid, and lymphoid cells (Potocnik *et al.*, 1994).

ES cells can be genetically altered *in vitro* by the heterologous, random integration of a transfected gene or by the homologous integration of a mutated form of an endogenous gene. *In vitro* development from ES cells can be compared with the development of the same ES cells *in vivo*. For this, ES cells are injected into mouse blastocysts, and the mixed blastocysts are reimplanted into a foster mother. The offspring are chimeric. Some cells are from the original blastocyst, and others are from the injected ES cells. Whenever germ cells are generated from ES origin, the chimeric mouse can be a founder of a new mouse strain, carrying the integrated ES-derived mutations (Thomas and Capecchi, 1987; Torres and Kühn, 1997).

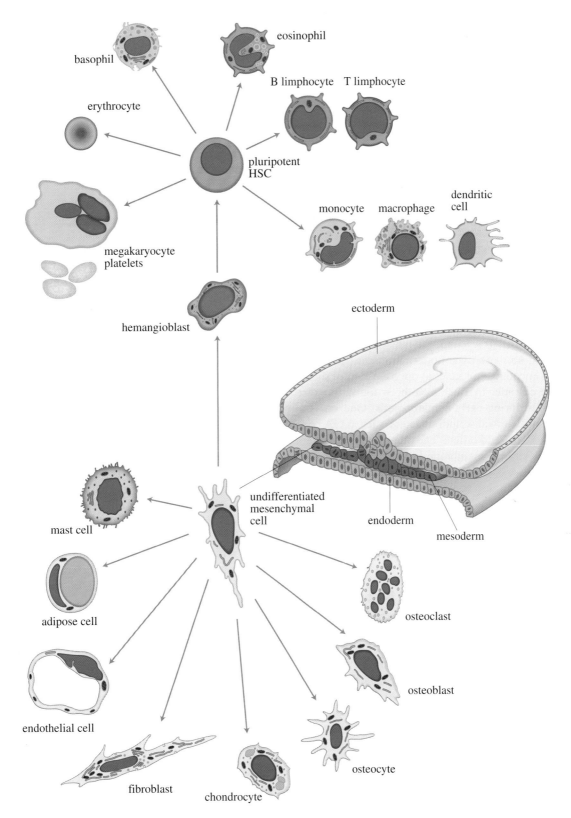

FIGURE 1 During embryonic development three germ layers are recognized: endoderm, ectoderm, and mesoderm. The mesoderm gives rise to nondifferentiated mesenchymal cells, the putative common precursors of hemopoietic (top) and endothelial cells (bottom). Development of the hemangioblast leads to the HSC, the pluripotent hemopoietic stem cell, which differentiates into the different blood cells. The nondifferentiated mesenchymal cell also gives rise to the other cell populations).

The method of injection of ES cells into blastocysts can also be used to study the effects of embryonally or neonatally lethal mutations in the hematopoietic pathway of differentiation. In this so-called $RAG^{-/-}$ blastocyst complementation assay blastocysts of $RAG-1^{-/-}$ or $RAG-2^{-/-}$ mice are injected with ES cells carrying a lethal mutation introduced by homologous recombination. Since the $RAG^{-/-}$ mutation abolishes the development of T and B lymphocytes, lymphocytes developing in these complemental mice must derive from the mutated ES cells (Chen, 1996).

Two strategies are used to identify rare early progenitors for hematopoietic cell lineages. In one strategy, one works backwards against the direction of development. In this strategy, a set of markers and functions of a given cellular stage are used to search for its immediate precursors. It assumes that the precursor already expresses some of the markers or some of the transcription factors regulating the expression of the markers of the unknown cell. It furthermore depends on the capacity of the as yet unidentified precursor to develop either *in vitro* or, upon transplantation into a deficient host, *in vivo* into the later determined cell.

The second strategy uses modern transgenetics. Two sets of genes encoding molecules at opposite ends of signal transduction cascades in a cell have been found to exert major controls during early hematopoiesis. One set is a collection of nuclear transcription factors, such as GATAs, PU-l, Ikaros, Pax 5, and others. The other set is composed of families of surface membrane-located and intracellular tyrosine kinases, such as Flk-2, c-kit, and others. Targeted disruption of these genes introduces genetic lesions, disrupting the hematopoietic cell development at defined stages (Singh, 1996; Georgopoulos, 1997). Analyses of these lesions allow us to draw the first contours of fate maps of erythroid, myeloid, and lymphoid cell development (Fig. 2).

Waves of Embryonic and Neonatal Lymphocyte Development

The pioneering studies of Le Douarin and Jotereau (Le Douarin and Jotereau, 1975; Jotereau and Le Douarin, 1982) have shown that the avian fetal thymus emits three discrete waves of T cells — two before and one after birth. The waves are started by the immigration of lymphoid progenitor cells into a thymus that is receptive for colonization at two times before and one time after birth (Fig. 3). Jotereau *et al.* (1987) demonstrated comparable waves of T cell development in the mouse. Again, a first wave between Days 10 and 13 and a second after Day 13 occur in the embryo, whereas the third wave was detectable in the neonate from Day 7 of life onwards. The third wave likely begins a continual generation of T cells throughout life. This continuous generation increases until puberty and then declines

for the rest of the life, although it never ceases completely (Fig. 3).

In the human embryo T lymphoid progenitors have been detected between Weeks 6 and 8 of gestation, well before the formation of the thymus anlage at Week 8 (Fig. 3). Fetal liver is one source of progenitors of these types. Hence, cells with T lymphoid markers and functions can develop outside the thymus (Phillips *et al.*, 1992; Poggi *et al.*, 1993; Sanchez *et al.*, 1993). The human embryo has a functional immune system with T and B cells at Week 10 of gestation, well before birth. The same is true for the fetal lamb. In contrast, chicken and mice are born at approximately the time when the first antigen-reactive mature lymphocytes appear in the peripheral lymphoid organs. This exemplifies that the development of T and B cells is independent of the timing of the transition from embryonic to perinatal existence of an organism, although the embryo is shielded by the mother from stimulation by foreign antigens (except the mother's antibodies made in response to such foreign antigens). Exposure to foreign antigen does influence the repertoires of mature lymphocytes after the organism is born.

Early T Cell Development

The repertoires of T cells arising early in ontogeny has been analyzed in more detail in the mouse (Rocha *et al.*, 1992; Kisielow and von Boehmer, 1995). Before birth, two waves of almost monoclonal γ/δ TCR^+ T cells are generated in the thymus (Fig. 3). Athymic mice do not make them, nor does it appear possible to regenerate them by HSC transplantation into an adult mouse. T cells of the first wave have the $V_{\gamma 3}$ segment rearranged, with no N-region diversity, to $J_{\gamma 1}$ which is expressed with $C_{\gamma 1}$. The δ chain of this γ/δ TCR is $V_{\delta 1}$, $J_{\delta 2}$, C_{δ}. In the second wave, $V_{\gamma 4}$ is used in the same combination, with the same δ chain, again without N-region diversity (Fig. 3). Since the other, nonproductively rearranged γ and δ alleles of the TCR loci show heterogeneity in rearrangements, the two waves of T cells appear to be positively selected by an unknown self-antigen. The first wave populates the epidermal layer of the skin and the second the epithelia of the tongue and the female reproductive tract. It is not clear how general these waves of populations of γ/δ TCR^+ T cells are for other species. Later in life, the populations of γ/δ TCR^+ T cells found in these epithelia are much more diversified in their TCR repertoires.

The two early waves are followed by the generation of more diverse γ/δ TCR^+ T cells which also have N regions inserted in their V(D)J joints. They populate the mucosal epithelia of the intestine, and they are also found in secondary lymphoid organs as well as circulating in the blood. In some individuals, large parts of these recirculating γ/δ TCR^+ T cells appear to be oligoclonal. Positive selection

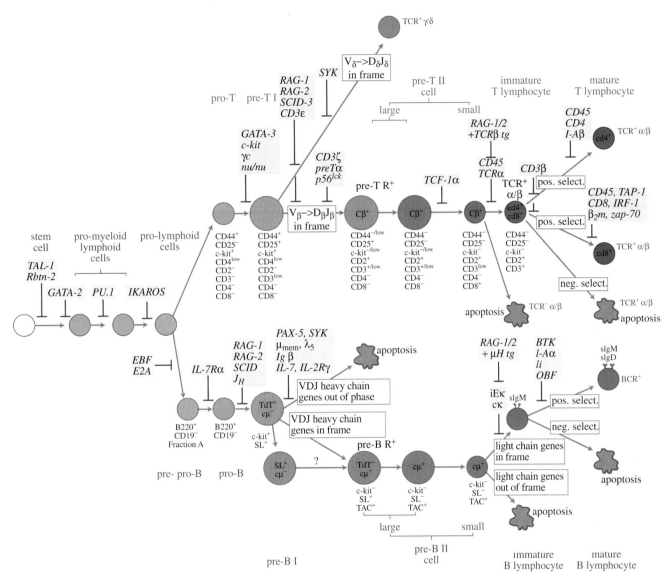

FIGURE 2 Schematic representation of the two directions of development of the lymphoid lineage, T (in blue) and B (in green). The pluripotent stem cell (HSC, in white) gives rise to progeny which in the bone marrow generate the B lineage and in the microenvironment of the thymus develop into the T lineage. It is possible to analyze the early stages of development by identification of markers which characterize a determined stage or one can identify specific functions (transgenic analysis), which permit the description of the various stages of hemopoiesis. For example, some nuclear transcription factors are indicated (yellow rectangles) which are active in the various stages of evolution of the HSC. From the moment when the prolymphoid (pro-T and pro-B) lines take different directions of differentiation it is possible, step by step, to describe the morphofunctional characteristics of the cells of the two lymphoid lines. The CD markers indicate a series of monoclonal antibodies which describe, by the absence or the presence of the molecule recognized on the lymphocyte, the stages of maturation. Through the study of some congenital diseases (such as severe combined immunodeficiency disease) it is possible to understand the genetic defects that are at the origin of the anomalous, or missing, differentiation. For example, the genes *Rag-1* and *Rag-2,* active during some of the rearrangement phases of T maturation, can condition determined steps which follow up to the mature T cell which expresses the antigen receptor (TCR, red labels). In the B differentiation steps one can observe a gradual appearance of the molecular structures that form the membrane immunoglobulin (such as $c\mu+$, cytoplasmatic μ chain) not yet expressed on the membrane. Activation of the light chain gene implies the final assembly and hence expression of the B lymphocyte receptor (BCR) in a mature form on the surface of the lymphocyte; the VDJ genes contribute to the variability of the heavy chain.

T LYMPHOCYTE DEVELOPMENT

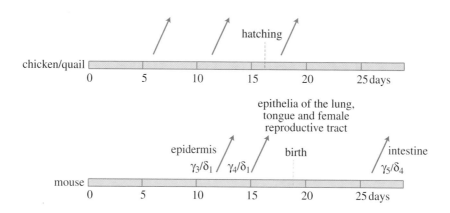

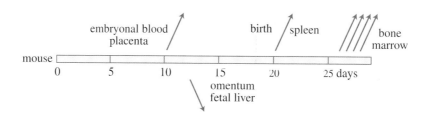

B LYMPHOCYTE DEVELOPMENT

FIGURE 3 Waves of T lymphocyte (top) and B lymphocyte (bottom) development during pre- and postnatal life of birds and mice. The avian fetal thymus emits three discrete waves (red arrows) of T cells, two before and one after hatching. In the mouse, the waves of T cell development are comparable to those in birds. Two waves occur before birth, the first between Days 10 and 13 and the second after Day 13, whereas a third wave occurs in the neonate from Day 7 onwards. The two embryonal waves of T cells are almost monoclonal γ/δ TCR$^+$: The first populates the epidermis and the second the epithelium of the tongue, lung, and the female reproductive tract. In the following wave, the γ/δ TCR$^+$ T cells are more heterogeneous and populate the mucosal epithelia of the intestine. In the mouse a first wave of B lymphocyte progenitors occurs in the embryonal blood and in the placenta between Days 10 and 13 of gestation. Between Day 13 and birth a second wave of B cells develops in the fetal liver and omentum. After birth, first the spleen and then the bone marrow generate B cells. In the adult the bone marrow appears to be the major site for the continuous production of B cells.

could be a consequence of exposure to either infections agents, such as mycobacteria, or self-antigens.

In transgenic mice expressing a γ/δ TCR characteristic of one given epithelium, T cells with this transgenic TCR can be found in other epithelia. Thus, these TCRs are not used for migration and homing to a specific site in the body. Since γ/δ TCR$^+$ T cells have been found on occasion to express the rearrangement-active genes RAG-1 and RAG-2 in the epithelia, it cannot be excluded that a first rearrangement occurs in the thymus, which is then followed by a second VJ rearrangement or V replacement in the second site, i.e., in the epithelia.

The generation of α/β TCR$^+$ T cells also begins before birth of a mouse and appears to lead to a diverse repertoire,

with TCR-lacking N regions before birth; N-region insertions at the V(D)J joints of TCRα and -β chains are found after birth (Bogue *et al.*, 1992). The α/β TCR$^+$ T cells which populate the extrafollicular and follicular areas of peripheral lymphoid organs express CD8 molecules as α/β heterodimers whenever they have differentiated to cytolytic T cells. On the other hand, α/β TCR$^+$ T cells, which populate the epithelia as IELs, express CD8 as α/α homodimers (Rocha *et al.*, 1992, 1995). Even more significant, they express the FcRI-γ subunit of the Fc receptor as a signal transducing molecule in the CD3 complex associated with γ/δ or α/β TCRs instead of the ζ subunit, which is used in the CD3 complex of the α/β TCR$^+$ extrafollicular and follicular T.

Genetically, athymic nude mice and human patients with

di George's syndrome (which effects a similar thymic dysplasia as that of nude mice) lack the α/β TCR$^+$ (CD8 α/β^+) extrafollicular and follicular T cells. However, they, as well as the TCRβ-deficient (TCR$\beta^{-/-}$) mice, produce γ/δ TCR$^+$ T cells and populate epithelia with IEL T cells in normal numbers. In the case of the nude mice (but not the TCR$\beta^{-/-}$ mice), some of the IELs express α/β TCR. This strongly argues for extrathymic ways to develop T cells. Although some RAG-1$^+$/RAG-2$^+$ T lineage cells have been found in epithelia, the extrathymic sites of T cell development and the environmental stromal cells which induce this development need to be found. In summary, IEL T cells can be developed inside and outside the thymus, whereas extrafollicular and follicular T cells appear to need the thymus for their development. Development in the thymus is described in more detail later.

Mutant mice which are deficient in the TCF-1 gene produce the first embryonal waves of α/β TCR$^+$ T cells but are deficient in the generation of α/β TCR$^+$ T cells in the adult (Clevers and Grosschedl, 1996). This indicates that embryonic waves of T cell development and the continuous generation of T cells throughout adult life appear to follow at least partially different genetic programs.

Early B Cell Development

B cell development during embryonic and neonated life of mice occurs in waves and at different sites (Melchers, 1979; Marcos *et al.*, 1991) (Fig. 3). Between Days 10 and 13 of gestation, a first wave of progenitors committed to B lineage development are found circulating in embryonic blood and in sites which are vascularized by the embryonic blood circulation (e.g., the placenta). Thereafter, between Day 13 and birth, a second wave of B cells develop in fetal liver and omentum. After birth, first spleen and then bone marrow generate B cells. During adult life, bone marrow appears to be the major site for the continuous production of B cells. This development in bone marrow is described in more detail later.

Fetal liver is also the major site of embryonic B cell development in humans (approximately Week 8 of gestation), rodents, and ruminants. In sheep and rabbits, ileal Peyer patches have been found to be a site of primary lymphocyte development. The stage of progenitors in which this development is started remains to be investigated (Reynaud *et al.*, 1995; Knight and Winstead, 1997).

In contrast to TCR, in which both α and β and γ and δ chains have N regions inserted at their V(D)J joints during neonatal and adult lymphopoiesis of the mouse, respectively, only IgH chains but not L chains carry N-region insertions. In the first waves of embryonic B cell development, particularly in fetal liver, H chains are made in which no N

regions are inserted at the V(D)J joints. Consequently, precursor B cells of the early embryonic types do not express the enzyme terminal desoxynucleotidyltransferase (TdT).

Early embryonic B cell development, as is the case for embryonic T cell development, can be distinguished from neonatal and adult B cell development in bone marrow by the differential expression MHC class II molecules and a myosin light chain-like gene. In addition, a defecting *ets-1* gene leaves embryonic development intact but abolishes development in the adult mouse, and a defective *OBF* gene does the reverse (Clevers and Grosschedl, 1996). Again, it can be concluded that embryonic and adult B cell development follow at least partially different genetic programs.

The early repertoires of B lineage cells in all B lymphopoietic sites show a strong bias for overrepresentation of the most 3'-located V$_H$ segments within the H chain gene locus (Coutinho *et al.*, 1992). In the mouse, these are the V$_H$ 7183 and V$_H$ Q52 families, in rabbit the V$_H$1 segment, and in human the V$_H$3 family. Later in the developmental pathway of immature and mature B cells, these families become underrepresented in the productively rearranged H chain loci, indicating selection of B cells by the specificity of certain μH chains during development, either as a positive or a negative influence. In the mouse, the most prominent member of the early overrepresentation and the later suppression within the B cell repertoire is the V$_H$81x gene segment. A possible scenario for this early preference and later suppression of the representation of these V$_H$ gene segments will be discussed later, and B cell development in bone marrow is described in more detail.

Lymphoid Development *in Vitro*

Early T and B cell development can also be observed in tissue culture. Thus, single-cell suspension cultures of fetal thymus and fetal liver from Day 13 onwards will go through the same stages of development from early progenitors to immature T or B cells, although without any proliferative expansion (Melchers *et al.*, 1995; Rolink *et al.*, 1995).

Precursor B cells of mouse at the stage of c-kit$^+$ CD25$^-$ B220$^+$ CD19$^+$ D$_H$J$_H$-rearranged pro/pre-B I cells can be cloned on preadipocytic stromal cells in the presence of exogenously added recombinant mouse IL-7, and they will expand by cell division every 18–24 h to continuously growing cell lines. Neither human precursor B cells at a comparable stage of differentiation nor precursor T cells of either mouse or human at comparable stages of differentiation are capable of this extensive proliferation at a stable state of differentiation.

Removal of IL-7 from the continuously growing pre-B I cell cultures, or culture of *ex vivo* isolated pre-B I cells in the absence of IL-7, induces pre-B I cells to differentiate without division. This differentiation encompasses a loss of clonability on stromal cells in the presence of IL-7, induction of apoptosis, upregulation of the rearrangement-active proteins and V_H to $D_H J_H$, as well as V_L to J_L rearrangements in and out of frame so that $sIgM^-$ and $sIgM^+$ immature B cells are generated.

Development from early precursors can also be followed in organ cultures of fetal thymus (FTOC) (Ceredig *et al.,* 1983) and fetal liver (FLOC) (Owen *et al.,* 1974). The remarkable difference of FTOC and FLOC in comparison to single-cell suspension cultures is the capacity of precursor cells not only to differentiate in a normally timed manner to immature cells but also to expand by proliferation. In the case of FTOC, the differentiation from $CD4^+$ $CD8^+$ immature to $CD4^+$ or $CD8^+$ mature T cells can also be observed; hence, these culture systems are similar to *in vivo* differentiation, proliferation, and cellular selection conditions. Since these processes can be positively and negatively influenced in FTOC and FLOC by ligands, mAbs, cytokines, and other molecules, they allow more in-depth study of some of the molecular mechanisms underlying T and B lymphocyte differentiation *in vitro*.

Commitment to the Lineages of α/α TCR$^+$ and γ/δ TCR$^+$ T Cells and Ig$^+$ B Cells

The descendants of a pluripotent stem cell can commit to becoming erythrocytes, granulocytes, mast cells, monocytes or macrophages, megakaryocytes or platelets, or T or B lymphocytes. Commitment to all lymphoid lineages, not just those developed in the thymus, must include the activation of the machinery which rearranges gene segments of the TCR and immunoglobulin (Ig) gene loci (Fig. 4). The RAG-1 and RAG-2 gene products are indispensable for these rearrangements. Targeted disruption of either gene results in the complete inability of all progenitors of lymphoid cells to rearrange any of the TCR or Ig gene segments. Nevertheless, the mutated RAG loci still allow the development of early progenitors of both T and B lineage cells in which sterile transcription of the TCRβ locus (in the thymus) or the Ig heavy (H) chain locus (in the bone marrow) occurs (Fig. 2).

Early pro/pre- and pre-I lymphocytes of both T and B lineages from both wild-type and RAG mutant mice express other components of the rearrangement machinery. It appears that this machinery is composed of both ubiquitously used and lymphoid-specific parts. In addition to RAG-1 and RAG-2, TdT also appears to be lymphoid specific.

It is likely that for all lineage commitments, cells have to migrate to a specialized environment in which a special program of gene expression is initiated as a step toward commitment to a given lineage (Fig. 5).

Thymus: The Primary Lymphoid Organ for T Cell Development

The most impressive case of a special environment in which lymphoid development is induced is the thymus (Fig. 5). The fact that some progenitor cells are attracted to migrate to the thymus is the earliest sign of the commitment of such cells to become T cells. Even before they arrive at the periphery of the thymus, a T cell progenitor expresses molecules on its surface which are thought to mediate its extravazation through the endothelial cell layers into the microenvironment of the thymus.

Close contacts with a stroma of specialized epithelial cells in the thymus are thought to further commit T cell precursors along the lineage, leading to α/β TCR- or γ/δ TCR-expressing T cells. It is likely that this contact induces the opening of those chromosomal regions which harbor the TCRβ, -γ, and -δ chain genes so that these loci can be transcribed. The transcribed RNAs are initially sterile (i.e., they are not translated into useful protein) and do not contain the information for the variable regions because the relevant V-region genes have not yet been rearranged.

It is likely that the early T lineage progenitors are capable of entering either the γ/δ TCR or the α/β TCR T lineage pathway of differentiation. Although single cells at such an early stage of development have not been analyzed in detail, it is likely that they activate and transcribe sterily both the TCRβ and the -δ gene loci. Entry into the γ/δ TCR or α/β TCR T lineage pathway could well depend on which of the two initially activated loci is first productively V(D)J rearranged, i.e., whether they first make TCRδ or TCRβ chains.

Bone Marrow: The Primary Lymphoid Organ of the Adult for B Cell Development

In the adult mouse or man, pluripotent stem cells do not have to migrate far to become B cells. During adult life, the bone marrow is the major site of B lymphopoiesis, and bone marrow also harbors the pluripotent stem cells (Fig. 5). Nevertheless, it is believed that the cells must migrate from an environment which keeps them in the pluripotent state to another environment in which they are induced to become B cells. As with cells in the thymus committed to becoming T cells, it is possible that the commitment of cells to enter the B lineage begins with the expression of adhesion molecules which bind to the B lineage inductive stromal cells, even before the cells have arrived there.

Close contact of migrating progenitors with the B lineage inductive stromal cells is expected to open the chromo-

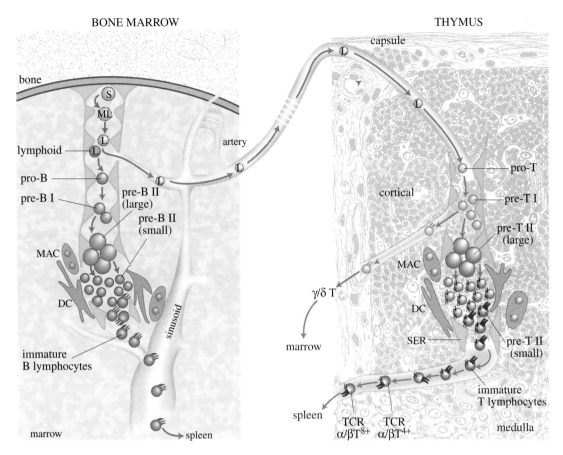

FIGURE 4 Lymphocyte development in bone marrow and thymus. Myeloid–lymphoid (ML) and then lymphoid (L) progenitors develop from the pluripotent HSC (S), near the endosteum in the bone marrow. These progenitors can either remain in the marrow to commit to B lineage development (green) or migrate to the thymus to become T lineage committed (blue). Pro-, pre-I, large pre-II, small pre-II, and immature, antigen receptor-expressing lymphocytes are generated in close contact with a reticular stromal cell compartment (pink). It is thought that these cells have important functions in the interactions which occur at the different stages of development. In the medulla of the thymus, radio-sensitive dendritic cells (DC, violet) are involved in negative selection, whereas radio-resistant epithelial stromal cells (SERs) are mediating positive selection. Dying cells are taken up and degraded by macrophages (MAC, brown). Similar negative, and possibly positive, selection also occurs in bone marrow with immature B cells. Selected immature T and B cells emigrate from the primary lymphoid organs to the spleen and to other peripheral, secondary lymphoid sites.

somal region which harbors the H chain gene locus. Subsequently, sterile transcription from the H chain locus is initiated. The rearrangement machinery is then activated and ready to rearrange the relevant gene segments.

The population of bone marrow B lineage cells does not express CD19. It expresses the tyrosine kinase Flk-2, and some of the cells express surrogate L chain; others express Nk1.1 and are progenitors of natural killer cells. When cultured in the appropriate environment of stromal cells and cytokines, these cells develop into a B lineage state in which further contact with stroma and exposure to IL-7 will induce their continued proliferation. These later CD19$^+$ cells

have downregulated Flk-2 expression while they continue to express c-kit. They are called pro/pre-B I cells. The marrow of a young mouse has approximately 5×10^6 such cells. They are capable of long-term proliferation on stromal cells in the presence of IL-7.

The thymus of a young mouse has approximately 5×10^6 T lineage cells at two comparably early stages of differentiation. These cells, called pro/pre-T and pre-T I cells by analogy to B lineage cells, are Thy1$^+$ CD4$^-$ CD8$^-$ CD2$^-$. The pro/pre-T cells are CD3$^-$ CD25$^-$ CD44$^+$, whereas the pre-T I cells are CD3$^{-/low}$, CD25$^+$, CD44$^-$ (Fig. 4).

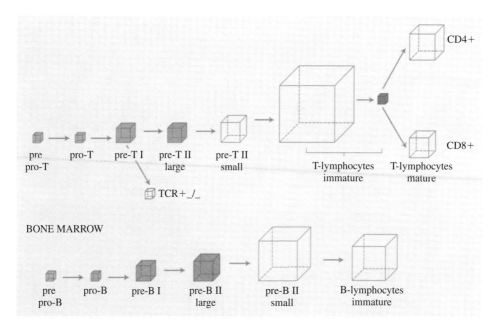

FIGURE 5 Pools of precursors and immature cells of the T lymphocyte (from gray to blue) and B lymphocyte (from gray to green) lineages in the thymus (top) and in the bone marrow (bottom), respectively, of a young mouse. Shaded cubes indicate cells in active division, whereas open ones indicate resting cells. The sizes of the boxes are proportional to the number of cells that the different T and B lineage subpopulations occupy in the primary lymphoid organs. The volume of the pro-/pre-cells is equivalent to 5×10^5 cells.

Ordered Rearrangements of the Ig and TCR Alleles

After T and B cells have become committed to a lineage of lymphocytes, they begin to rearrange TCR and Ig gene loci, respectively, and to follow a program of cellular development so that TCR$^+$ or Ig$^+$ cells which are resting and ready to respond to a foreign antigen emerge in the periphery.

The developmental stages through which the intrathymic α/β TCR T lineage and intrabone marrow B lineage cells pass, and the rules which they follow, show striking similarities (Figs. 4 and 5). They are described together so that their similarities — and their differences — are more visible. Both T lineage-committed cells in the thymus and B lineage-committed cells in bone marrow rearrange their antigen receptor-encoding gene segments in an ordered fashion. Both begin with D segment to J segment rearrangements on both alleles, T lineage cells at the TCRβ locus and probably also at the TCRδ^+ locus, and B lineage cells at the H chain locus. Secondary D to J rearrangements at DJ-rearranged loci are possible, particularly in B lineage cells in which approximately 15 D segments and 4 J segments are located in the same orientation within the H chain locus. The pre-B I cells in bone marrow which have undergone D_H to J_H rearrangement express B220, CD19, the ty-

rosine kinase c-kit, the surrogate L chain, together with two glycoproteins as pro-B receptor and are reactive to stromal cells and the cytokine IL-7 (Fig. 4). In human bone marrow, a comparable population has been found. It constitutes 10% of all B lineage cells. This population expresses CD19, CD10, CD34, human V_{preB}, and TdT.

In vitro stromal cells and IL-7 maintain mouse pre-B I cells in their $D_H J_H$-rearranged state of differentiation. In fact, IL-7 inhibits further differentiation, just as removal of IL-7 induces it. These findings suggest that the same results could occur *in vivo*. In bone marrow, pre-B I cells are expected to fill a compartment which is limited by many specialized stromal cells. Once filled, any further cell division will be asymmetric, leaving one daughter cell in contact with stroma and IL-7 as a pre-B I cell, whereas the other daughter cell loses contact to stroma and IL-7 and thus differentiates into pre-B II cells. The cells are known to divide twice a day; hence, 5×10^6 pre-B I cells generate 1×10^7 pre-B II cells per day by two asymmetric divisions (Fig. 6).

Approximately 5×10^6 pro/preT-I and large preT-II cells are expected to fill compartments in thymus specified by stromal cells from which the α/β TCR$^+$ and γ/δ TCR$^+$ T cells are generated (Fig. 6). The pro/preT-I cells are expected to enter D to J rearrangements of the TCRδ and -β loci in a fashion similar to that of the H chain loci in pro/pre B-I cells.

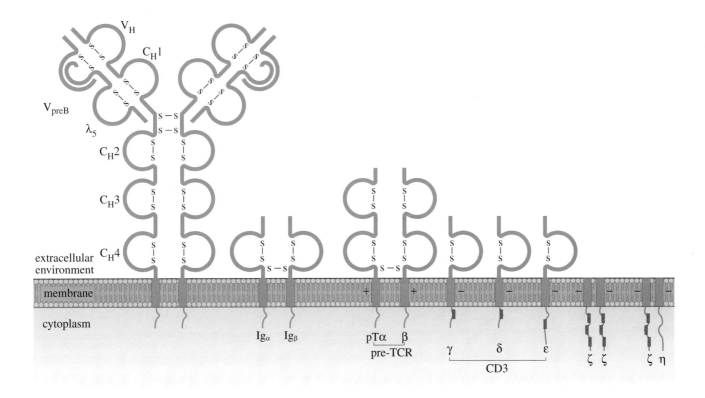

FIGURE 6 The prelymphocyte receptors. (Left) The pre-B cell receptor complex expressed on large pre B-II cells. In mature lymphocytes the IgM, Igα, and Igβ associate. The different loops of the molecules are held together by disulfide bridges. In IgM, the two proteins, V_{pre-B} and λ_5, associate with each other to form the surrogate L chain. V_H is the variable domain of the heavy chain, and C_H1–C_H4 are the heavy chain constant domains. (Right) The components of the receptor complex expressed on the large pre T-II cells. The pre-TCR is made up of α and β chains which, in mature lymphocytes, associate with the CD3γ, -δ, and -ε chains; the ζ and ν chains are present as homodimers ζζ or heterodimers ζν, probably associated with the TCR or with other CD3 chains. The + and − symbols refer to the polar residues present in the transmembrane regions which are involved in the association of the chains. The red rectangles represent the recognition and activation domains containing conserved sequences.

The Surrogate L and α Chains

One of the earliest molecules to signify commitment for B lineage differentiation is the surrogate light (L) chain (Melchers *et al.*, 1993). Surrogate L chain is encoded by two genes, V_{preB} and λ_5, producing two proteins that associate with each other to form an L chain-like structure (Fig. 7). The surrogate L chain genes are located on mouse chromosome 22 and human chromosome 2. Surrogate L chain is selectively expressed in three early progenitor states of B lymphopoiesis and in no other cell in the body (Fig. 4). In the two earliest progenitor states — that is, on cells which express the B lineage-specific molecule, B220, as well as the tyrosine kinase c-kit and which either do or do not express CD19 — surrogate L chain appears to be associated

with two glycoproteins, gp130 and gp46. The function of the early, surface membrane-bound Ig complex has not been elucidated.

One of the earliest molecules to signify commitment for T lineage differentiation is the surrogate TCRα chain, called pre-Tα chain (Fehling and von Boehmer, 1997) (Fig. 7). At least pre-T I cells express the pTα gene (Fig. 4). pTα is located on mouse chromosome 17, telomeric to the major histocompatibility complex (MHC), the human version of pTα on the short arm of chromosome 6, again linked to MHC. It is suspected that pTα protein has a partner polypeptide with which it associates in order to form a surrogate TCRα chain and to appear on the surface of pre-T cells Ig association with TCRβ chain, but such a second component has not been found. The pTα protein could have a function analogous to the λ_5 protein on the surrogate

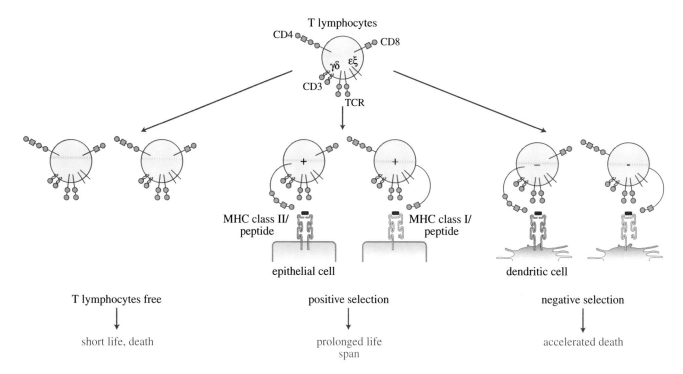

FIGURE 7 The T cells in the thymus express both CD4 and CD8. When they bind to the MHC class I/peptide complex they use the CD8 molecules and downregulate expression of CD4, whereas when they bind to MHC class II/peptide complex they use the CD4 molecules and downregulate expression of CD8. The lymphocytes which bind to epithelial cells (yellow) undergo positive selection, exit the thymus, and survive; those that bind to dendritic cells (violet) undergo negative selection and die.

L chain. The missing partner of pTα could then have V_{preB}-analogous properties.

V to DJ Rearrangements: The Formation of Prelymphocyte Receptor

The second ordered recombination event occurs between a V segment and a previously formed DJ rearrangement on the same chromosome, i.e., V_β to $D_\beta J_\beta$ or V_δ to $D_\delta J_\delta$ or V_H to $D_H J_H$. V to DJ rearrangements can occur in and out of the reading frame set by the DJ rearrangement at the TCRβ, TCRδ, or IgH chain loci, respectively. The productively rearranged forms of the genes are expressed as proteins and transported to the surface. TCRδ chain appears on the surface as soon as a productive V_γ to J_γ rearrangement has been made, and γ/δ TCR can be made. TCRβ chains appear on the surface because they associate with the surrogate α chain, pTα, to form pre-T receptors. Likewise, IgH chains do so with surrogate L chain to form pre-B receptors. These three receptors appear on the surface to be associated with the appropriate signal transducing complexes, i.e., Igα and -β for the pre-B receptor and CD3 for

the preT receptor. Surface expression of the prelymphocyte receptors is crucial for further development. Surface pre-B receptor-positive cells enter a wave of proliferative expansion in bone marrow, as do pre-T receptor-positive cells in the thymus. They divide three to seven times before they fall into a resting, small pre-B II and pre-T II, respectively, stage of development (Figs. 5 and 6).

When genes encoding components of the surrogate chains are inactivated by targeted mutation, i.e., in λ_5-deficient and pre-Tα-deficient mice, this proliferative expansion is abolished in B and T lineage cells, respectively, because the prelymphocyte receptors cannot be formed and deposited on the surface. Membrane deposition of the pre-B lymphocyte receptors appears mandatory for effecting proliferative expansion since in the B lineage expansion does not occur in mice in which the transmembrane portion of the μH chain is defective. In all these mutants the large, cycling precursor cells of type II are not formed, and all subsequent immature and mature compartments are either drastically reduced in size (in the surrogate chain mutants) or totally absent (in the transmembrane mutant).

Targeted disruption of the pTα chain gene does not affect the intrathymic generation of the γ/δ TCR-expressing T cells (Fehling and von Boehmer, 1997). However, in these

α/β TCR thymocyte-deficient mice, there are only a small number of precursors of γ/δ TCR-expressing T cells and these do not expand into the compartment normally held by α/β TCR T lineage precursors.

···•

Do Prelymphocyte Receptors Secure Isotype and Allelic Exclusion?

In pre-T II cells successful expression of either of the two TCR isotypes, γ/δ TCR or α/β TCR, appears to preclude any further rearrangements of the other isotype so that no double-expressing γ/δ TCR$^+$/α/β TCR$^+$ T cells are found in the system. This isotype exclusion in pre-T cells might follow the same rules as allelic exclusion of productive V(D)J rearrangements at the two alleles of the TCRβ and the IgH chain loci. It is not known whether TCRγ chain rearrangements actually precede TCRδ chain rearrangements. If this were the case, then γ chains could take the place of surrogate L and pre-Tα chains for the detection and amplification of cells with successful, i.e., productive, V(D)J rearrangements of the TCRδ locus.

The cells which carry productively rearranged TCRβ or IgH loci are identified by markers. For the B lineage, the markers are c-kit$^-$ and CD 25$^+$, which selectively express the tyrosine kinases syk, lck, and fyn and are called large pre-B II cells; they are large because they are in cell cycle. For the T lineage they are CD4$^-$, CD8$^-$, and CD25$^-$ and they express lck. In analogy to B cell development, I call them large pre-T II cells. Syk-deficient mice arrest B cell development at the transition from pre-B I to large pre-B II cells; lck-defective mice do so in T cell development at the transition from pre-T I to large pre-T II cells. In the latter mice, the development of γ/δ TCR$^+$ T cells is not affected. As a consequence of the proliferative expansion, pre-B II and pre-T II cells both have positively selected productively VDJ-rearranged IgH and TCRβ loci, respectively, over unproductively rearranged loci, at least on one allele, since more than 99% of them express TCRβ and IgH chains, respectively, in their cytoplasm.

Polymerase chain reaction (PCR) analyses of the rearrangement status of the two H chain alleles in single pre-B cells have shown that half of the cells have one allele productively and the other allele nonproductively VDJ rearranged. The other half of the cells have one productively VDJ-rearranged allele, with the other only DJ rearranged. Hence, when VDJ-rearranged μH chain-expressing pre-B cells first appear in development, they are already allelically excluded. This secures from the beginning that one B cell makes only one μH chain of an antigen-specific receptor, an essential prerequisite for the validity of the clonal selection hypothesis. Although a similar single-cell PCR analysis has not been done for the constitution of TCRβ alleles

in pre-T II cells, it can be expected that they in fact are allelically excluded as well.

Two mechanisms could operate to secure allelic exclusion. In half of all cells at the transition from pre I to pre II cells, the first V to DJ rearrangement could be nonproductive and the second productive. In the other half, the first could be productive. In that cell, the μH chain or TCRβ chain produced from the first allele should signal the inhibition of V to DJ rearrangement at the second allele. How could it do that? Since these cells express pre-T and pre-B receptors, respectively, when they have made TCRβ or IgH chains, the prereceptors could signal. What would this signal change in the cell? An analysis of the expression of the rearrangement machinery, in particular of RAG-1 and RAG-2, has shown that the expression of these essential components is downregulated in the prereceptor-expressing cells (Fig. 4). The RAG genes are reactivated at later stages of development when L chain genes need to be rearranged, and a similar analysis in T lineage cells shows the same down- and upregulation of the RAG genes. Before the RAG genes are reactivated, the DJ-rearranged allele should be closed and made inaccessible to the rearrangement machinery. The same developmental stages of B lineage cells in man show the same modulated expression of the RAG genes.

•···

Rearrangements of IgL and TCRα Chain Gene Loci: The Generation of Immature sIgM$^+$ or α/β TCR$^+$ Lymphocytes

In order to be able to express sIgM and α/β TCR, the precursors of the immature lymphocytes have to enter a third stage of ordered gene rearrangements: V segments of the TCRα, Igκ L, and Igλ L loci are rearranged to the J segments of these loci, again in or out of frame. The T lineage and B lineage cells in these rearranged loci are found at the transition of the large T cells cycling to the small resting CD4$^+$8$^+$ α/β TCR-immature T cells, and at the transition of the large, cycling pre-B II to the small resting c-kit$^-$CD25$^+$ pre-B II cells and the resting, surface IgM-expressing immature B cells (Figs. 2, 4, and 5). Comparable B lineage cells in the human are the CD19$^+$10$^+$34$^-$ V$_{preB}^-$ cells which express μH chain in the cytoplasm or, together with L chains, as Ig on the surface.

In the bone marrow of a young mouse, 5×10^6 proliferating pre-B I cells originally generate 1×10^7 large, cycling pre-B receptor-expressing pre-B II cells which expand by proliferation and feed into a pool of 4×10^7 resting pre-B II cells and 2×10^7 resting immature/surface Ig-expressing lymphocytes (Fig. 6). This resting pool of cells turns over as a consequence of programmed cell death (apoptosis) with a half-life of 2–4 days. Most of the cells generated this way die without ever leaving the bone marrow. One thousand times as many cells of each compartment are found in human bone marrow, all with strikingly similar properties. In the thymus of a young mouse, 5×10^6 proliferating pre-T I cells generate 1 or 2×10^7 large, cycling, pre-T receptor-expressing pre-T II cells which feed into a pool of resting surface α/β TCR negative and later surface α/β TCR-expressing, CD4$^+$8$^+$ double-positive immature thymocytes. Again, most of them die by apoptosis, with a half-life of 2–4 days.

Hence, although the majority of all immature cells in bone marrow and thymus never leave these organs and die *in situ*, they must be generated in large numbers in order to secure the accumulation of the peripheral, mature lymphocyte pool. In surrogate chain-deficient mice, peripheral, normal mature B and T lymphocytes are made, although at a very reduced rate. It takes these deficient mice over one-half of a year to accumulate only half of their normal numbers of peripheral mature cells.

Secondary Rearrangements in IgL and TCRα Loci

Pre-B II cells, small pre-T II cells, immature B cells, and immature T cells continue to express the rearrangement machinery (Fig. 4). Since the IgL and TCRα loci remain accessible, secondary rearrangements can occur in these cells. They increase the chances of converting a nonproductively rearranged TCRα or IgL chain locus into a productive one, thus rescuing the cell for later selection. They may also change the expression of the antigen receptor of an immature T or B cell, either by giving it a new specificity with a new chain or by losing the expression of the receptor through the loss of the productive and the gain of a nonproductive rearrangement (Nemazee, 1993; Radic and Weigert, 1994; Goodnow *et al.*, 1995). The rearrangement machinery is inactivated as B cells transit from the immature to the mature stage, but nothing is known of the molecular modes by which this downregulation occurs.

Lymphocyte Development without Ig and TCR Expression

Even when pro-T or pro-B cells cannot express TCR and Ig chains, respectively, as is the case in RAG-defective mice, these cells can be induced to differentiate to the phenotypic stages of CD4$^+$ CD8$^+$ CD25$^-$ immature T cells to c-kit$^-$ CD25$^-$ small pre-B II/immature B cells without expressing TCR or BCR. Antibodies specific for CD3 or Igα stimulate this cellular expansion *in vivo* of T and B lineage cells, respectively, as does the constitutive, T cell-specific expression of lck. This demonstrates that early lymphocyte differ-

entiation does not require the expression of TCR or Ig, although it is needed for proper filling of the later, more mature compartments with sufficient numbers of normally functional (i.e., antigen-receptor positive) cells.

Normally, transcription of the Ig or TCR gene loci precedes their rearrangement. In RAG-deficient mice, induction of transcription of TCRα and IgL chain loci is seen in cells differentiating from pre-B I or pre-T I cells to pre-B II-like or pre-T II-like cells *in vitro*. Stimulation of differentiating pre-B cells with IL-4 and a CD40-specific antibody induces sμ–sε switching on the H chain locus. This suggests that lymphocyte differentiation to fully mature stages of cells depends on cell–cell contacts and cytokines but not on the expression of the antigen-specific receptors (Rolink *et al.*, 1996; Fehling and von Boehmer, 1997).

Arrest of Differentiation and Death of Autoreactive Immature Lymphocytes

The diversity of antigen receptors, generated by a wide variety of VDJ and VJ recombinations and further diversified by N regions inserted into their joints, is so great that the originally generated repertoires of lymphocytes, each with one particular VDJ/VJ pair, must include autoantigen-recognizing cells. Autoantigens are recognized first where lymphocytes are made: in the thymus and in the bone marrow. This recognition results in an arrest of differentiation of the autoreactive cell and its accelerated death (Nemazee, 1993; Radic and Weigert, 1994; Goodnow *et al.*, 1995; Kisielow and von Boehmer, 1995). The molecular modes by which death is effected are not known. Since macrophages have been found in these primary lymphoid organs, it is possible that at least some of the death of cells is in fact achieved by phagocytosis of the self-reactive cells, possibly after antigenic encounter and consequent phenotypic changes. Normally, these cells might try to change their antigen receptors through secondary rearrangements at the TCRα or IgL loci, respectively (i.e., they might "edit" their receptors in order to evade self-recognition). However, if the autoantigen-recognizing receptor is the only one present in the mouse (i.e., in a rearrangement-deficient mouse carrying either a transgenic TCR or Ig with specificity for an autoantigen), then the cells cannot evade self-recognition through secondary gene rearrangement. In this case, all cells are arrested in their development. For B lineage cells at this stage of development we know practically nothing of the molecular and cellular requirements for this autoantigen presentation.

It appears that it is possible to interfere with the arrest. The injection of a competitive form of an autoantigen can release cells from this arrest (Andersson *et al.*, 1995). It allows their development into autoantibody-secreting plasma cells. This finding suggests a possible scenario for the induction of B cell-mediated autoimmune diseases such as lupus erythematosus.

The arrest of differentiation, also called negative selection, represents the first line of tolerance induction, in which autoantigen-reactive cells presented with autoantigens in the primary lymphoid organs are induced to die. Secondary rearrangements of the IgL chain or the TCR chain loci often result in the generation of immature lymphocytes which express — either transiently or stably — more than one L chain or TCRα chain. At this point in development, therefore, the rule that one receptor makes only one antigen-specific receptor is frequently not obeyed. The "edited" cells join the pool of immature lymphocytes.

Antigen Presentation to Immature T Cells in the Thymus

Much of the detailed knowledge of the genes and the molecular processes that control antigen presentation to T cells has come from studies of the antigen-presenting cells of the peripheral lymphoid system (Koopman *et al.,* 1997; Pieters, 1997). Antigen is taken up by specialized antigen-presenting cells (APCs). It is processed, in the case of proteins, by proteolytic digestion into peptides. The peptides are loaded into the binding clefts of MHC class I and class II molecules, and the resulting complexes are then transported to the cell surface where they are presented to T cells.

Although antigen uptake, processing, and presentation have not been studied in detail in the thymus, we expect that autoantigens (often liberated by the death of T lineage precursors) are handled and presented in similar ways.

The processing and presentation of antigenic fragments by MHC class I and MHC class II molecules follow different routes. After their synthesis, class II molecules travel inside the cell through the endoplasmic reticulum (ER), first into Golgi and post-Golgi compartments and then into endosomal and lysosomal compartments where they become loaded with peptides. They are then deposited on the cell surface. Two chaperones have been characterized which protect the peptide binding site before loading: the invariant chain (Ii) protects the groove, and the HLA-DM molecules and HLA-DO uncover the groove and load the peptides.

MHC class I molecules, on the other hand, are loaded with peptides processed from proteins before they ever leave the ER. The degradation of proteins into peptides occurs in special intracellular protease complexes called proteasomes, located within the cytoplasm, from where the peptides are transported with special transporters to the loading sites in the ER. Access to the proteasomes, and thereby to class I loading, is via the cytoplasm; consequently, proteins which are not synthesized do not get there unless they use special tricks.

Immature α/β TCR$^+$ T cells that recognize their auto-antigens with high avidity are arrested in their differentiation and will rapidly die by apoptosis unless they have had a chance to change the specificity of their receptor by secondary rearrangements of their TCRα genes. This allows them to edit their receptors. If this lowers the avidity for the autoantigen, they may survive, but only for a short time. Secondary rearrangements of the IgL chain or the TCR chain loci often result in the generation of immature lymphocytes which express — either transiently or stably — more than one L chain or TCRα chain. At this point in development, the rule that one cell makes one antigen-specific receptor is frequently not obeyed. The edited cells join the pool of immature lymphocytes.

Positive Selection of Immature T Cells in the Thymus

In the thymic pool of immature T cells, cells with TCRs of lower avidity for the peptide/MHC complexes can still be found. In fact, these immature T cells might essentially recognize MHC molecules alone: A single, soluble α/β TCR that is specific for certain peptide complexed with a particular MHC molecule can bind to this MHC molecule even when it carries many different unrelated peptides; evidently it does so to simply stabilize the conformation of the MHC molecule so that it will be recognizable by the TCR. It remains a hotly debated issue whether the germline-encoded V segments of TCRα and/or -β loci are selected in evolution for their capacity to bind MHC molecules. This should not be the case for the V segments of loci which encode antigen receptors that do not have to recognize peptide/MHC complexes in order to become functional, i.e., V_H, V_λ, V_δ, and V_γ. This recognition with lower avidity changes the life of the α/β TCR T cell. If its TCR fits MHC class I, it uses the CD8 molecules coexpressed with CD4 molecules on immature α/β TCR$^+$ T cells to make additional contacts with MHC class I molecules. This signals the cell to downregulate CD4, exit the thymus, and become long-lived and mature in the periphery. Conversely, if the TCR fits MHC class II, the cell uses its CD4 molecules to make additional contacts to MHC class II molecules, downregulates CD8, exits the thymus, and becomes long-lived and mature. The mature CD8$^+$, class I-restricted T cells are precursors of killer T cells, whereas CD4$^+$, class II-restricted T cells are precursors of helper T cells. Hence, all the peripheral α/β TCR$^+$ cells which have gone through the processes of negative and positive selection should be tolerant to autoantigens and should recognize antigens only in the context of MHC. Positive selection of helper T cells is done on epithelial, radio-resistant cells, and it does not work when DCs are the only APCs capable of expressing MHC class II molecules.

It has been estimated that only 2 or 3 million cells are selected each day from the pools of immature lymphocytes, from which more than 20 million cells leave the sites of primary lymphoid development in the cortex of the thymus and in the B cell areas of bone marrow. Immature T cells then appear in the medulla of the thymus and in the extrafollicular regions of the spleen. The selection into the mature lymphocyte pools is thought not to involve cell proliferation and might or might not change the life expectancy of the cells, which initially is short.

Positive Selection of Immature B Cells

No genetic factors are known which restrict the repertoire of mature, peripheral B cells, nor is the repertoire skewed toward the recognition of foreign antigens in the context of some of the body's own molecules. Hence, it is not clear whether immature B cells undergo positive selection via their antigen receptors before they become responsive and begin to proliferate and to differentiate. However, double-mutant $CD40^{-/-}/xid$ ($btk^{-/-}$) mice are incapable of selecting a long-lived B cell repertoire that can mediate T cell-independent and T cell-dependent responses. Nevertheless, both mutant strains develop B lineage precursors and immature B cells in the bone marrow in normal numbers and with normal phenotypes. This makes the positive selection of peripheral, long-lived B cells at least plausible; nevertheless, it does not exclude the likelihood that longevity is induced only by experience with a foreign invader and after the proliferative expansion of a mature, resting B cell.

In summary, a small number of progenitors and precursor ($\sim5 \times 10^6$) expand in the primary lymphoid organs by proliferation to 50×10^6 immature, antigen receptor-expressing cells, all of which have short life spans. Only 2 or 3×10^6 of them leave the sites of primary lymphoid development in the cortex of the thymus and in the B cell areas of bone marrow. Immature T cells then appear in the medulla from where they exit to appear in the T cell-rich zones of the extrafollicular regions of the spleen. Immature B cells exit the bone marrow to enter the B cell-rich zones of the spleen. In the extrafollicular regions they are resting, mature, antigen-sensitive cells, ready to enter an immune response against a foreign antigen.

Acknowledgments

I thank Dr. Antonio Lanzavecchia for critical reading of this article. The Basel Institute for Immunology was founded and is supported by F. Hoffmann-La Roche Ltd., Basel, Switzerland.

References Cited

ANDERSSON, J., MELCHERS, F., and ROLINK, A. (1995). Stimulation by T cell independent antigens can relieve the arrest of differentiation of immature autoreactive B cells in the bone marrow. *Scand. J. Immunol.* **42**, 21–33.

BOGUE, M., GILFILLAN, S., BENOIST, C., and MATHIS, D. (1992). Regulation of N-region diversity in antigen receptors through thymocyte differentiation and thymus ontogeny. *Proc. Natl. Acad. Sci. USA* **89**, 11011–11015.

BOISMENU, R., and HAVRAN, W. L. (1994). Modulation of epithelial cell growth by intraepithelial gamma delta T cells. *Science* **266**, 1253–1255.

BURROWS, P. D., and COOPER, M. D. (1997). B cell development and differentiation. *Curr. Opin. Immunol.* **9**, 239–244.

CEREDIG, R., McDONALD, H. R., and JENKINSON, F. J. (1983). Flow microfluorometric analysis of mouse thymus development *in vivo* and *in vitro*. *Eur. J. Immunol.* **13**, 185–190.

CHEN, J. (1996). Analysis of gene function in lymphocytes by RAG-2-deficient blastocyst complementation. *Adv. Immunol.* **62**, 31–59.

CLEVERS, H. C., and GROSSCHEDL, R. (1996). Transcriptional control of lymphoid development: lessons from gene targeting. *Immunol. Today* **17**, 336–343.

CUMANO, A., DIETERLEN LIEVRE, F., and GODIN, I. (1996). Lymphoid potential, probed before circulation in mouse, is restricted to caudal intraembryonic splanchnopleura. *Cell* **86**, 907–916.

FEHLING, H. J., and VON BOEHMER, H. (1997). Early a/b T cell development in the thymus of normal and genetically altered mice. *Curr. Opin. Immunol.* **9**, 263–275.

GEORGOPOULOS, K. (1997). Transcription factors required for lymphoid lineage commitment. *Curr. Opin. Immunol.* **9**, 222–227.

GOODNOW, C. C., CYSTER, J. G., HARTLEY, S. Z., BELL, S. E., COOKE, M. P., HEALY, J. L., AKKARAJU, S., RATHMELL, J. C., POGUE, S. L., and SHOKAT, K. P. (1995). *Adv. Immunol.*, 279–368.

JOTEREAU, F. V., and LE DOUARIN, N. M. (1982). Demonstration of a cyclic renewal of the lymphocyte precursor cells in the quail thymus during embryonic and perinatal life. *J. Immunol.* **129**, 1869–1877.

JOTEREAU, F. V., HENZE, F., SALOMON-VIC, V., and GASCAN, H. (1987). Cell kinetics in the fetal mouse thymus: Precursor cell input, proliferation and emigration. *J. Immunol.* **138**, 1026–1030.

KISIELOW, P., and VON BOEHMER, H. (1995). Development and selection of T cells: Facts and puzzles. *Adv. Immunol.* **58**, 87–209.

KNIGHT, K. L., and WINSTEAD, C. R. (1997). Generation of antibody diversity in rabbits. *Curr. Opin. Immunol.* **9**, 228–232.

KOOPMAN, J.-O., HÄMMERLING, G. J., and MOMBURG, F. (1997). Generation, intra-cellular transport and loading of peptides associated with MHC class I molecules. *Curr. Opin. Immunol.* **9**, 80–88.

LE DOUARIN, N. M., and JOTEREAU, F. V. (1975). Tracing of cells of the avian thymus through embryonic life in interspecific chimeras. *J. Exp. Med.* **142**, 17–40.

MARCOS, M. A., GUTIERREZ, J. C., HUETZ, F., MARTINEZ, C., and DIETERLEN-LIEVRE, F. (1991). Waves of B-lymphopoiesis in the establishment of the mouse B-cell compartment. *Scand. J. Immunol.* **34**, 129–135.

MEDVINSKY, A., and DZIERZAK, E. (1996). Definitive hematopoiesis is autonomously initiated by the AGM region. *Cell* **86,** 897–906.

MELCHERS, F. (1979). Three waves of B lymphocyte development during embryonic development in the mouse. In *Cell Lineage, Stem Cells and Cell Determination* (N. le Douarin, Ed.), INSERM Symp. No. 10, pp. 281–289. Elsevier, Amsterdam.

MELCHERS, F., ROLINK, A., GRAWUNDER, U., WINKLER, T. H., KARASUYAMA, H., GHIA, P., and ANDERSSON, J. (1995). Positive and negative selection events during B lymphopoiesis. *Curr. Opin. Immunol.* **7,** 214–227.

MORRISON, S. J., WRIGHT, D. E., CHESHIER, S. H., and WEISSMAN, I. L. (1997). Hematopoietic stem cells: Challenges to expectations. *Curr. Opin. Immunol.* **9,** 216–221.

NEMAZEE, D. (1993). Promotion and prevention of autoimmunity by B lymphocytes. *Curr. Opin. Immunol.* **5,** 866–872.

OWEN, J. J. T., COOPER, M. D., and RAFF, M. C. (1974). *In vitro* generation of B-lymphocytes in mouse fetal liver, a mammalian bursa equivalent. *Nature* **249,** 361–363.

PAPAVASILIOU, F., JANKOVIC, M., GONG, S., and NUSSENZWEIG, M. C. (1997). Control of immunoglobulin gene rearrangements in developing B cells. *Curr. Opin. Immunol.* **9,** 233–238.

PHILLIPS, J. H., HORI, T., NAGLER, A., BHAT, N., SPITS, H., and LANIER, L. L. (1992). Ontogeny of human natural killer (NK) cells: Fetal NK cells mediate cytolytic function and express cytoplasmic CD3 epsilon,delta proteins. *J. Exp. Med.* **175,** 1055–1066.

PIETERS, J. (1997). MHC class II restricted antigen presentation. *Curr. Opin. Immunol.* **9,** 89–96.

POGGI, A., SARGIACOMO, M., BIASSONI, R., PELLA, N., SIVORI, S., REVELLO, V., COSTA, P., VALTIERI, M., RUSSO, G., and MINGARI, M. C. (1993). Extrathymic differentiation of T lymphocytes and natural killer cells from human embryonic liver precursors. *Proc. Natl. Acad. Sci. USA* **90,** 4465–4469.

POTOCNIK, A. J., NIELSEN, P. J., and EICHMANN, K. (1994). *In vitro* generation of lymphoid precursors from embryonic stem cells. *EMBO J.* **13,** 5274–5283.

RADIC, M. Z., and WEIGERT, M. (1994). Genetic and structural evidence for antigen selection of anti-DNA antibodies. *Annu. Rev. Immunol.* **12,** 487–520.

REYNAUD, C. A., GARCIA, C., HEIN, W. R., and WEILL, J. C. (1995). Hypermutation generating the sheep immunoglobulin repertoire is an antigen-independent process. *Cell* **80,** 115–125.

ROCHA, B., VASSALLI, P., GUY-GRAND, D., and NOV, S. (1992). The extrathymic T-cell development pathway. *Immunol. Today* **13,** 449–454.

ROCHA, B., GUY-GRAND, D., and VASSALLI, P. (1995). Extrathymic T cell differentiation. *Curr. Opin. Immunol.* **7,** 235–242.

ROLINK, A., and MELCHERS, F. (1993). Generation and regeneration of cells of the B-lymphocyte lineage. *Curr. Opin. Immunol.* **5,** 207–217.

ROLINK, A. G., MELCHERS, F., and ANDERSSON, J. (1996). The SCID but not the *RAG-2* gene product is required for Sm-Se-heavy chain class switching. *Immunity* **5,** 319–330.

SANCHEZ, M. J., GUTIERREZ-RAMOS, J. C., FERNANDEZ, E., LEONARDO, E., LOZANO, J., MARTINEZ, C., and TORIBIO, M. L. (1993). Putative prethymic T cell precursors within the early human embryonic liver: A molecular and functional analysis. *J. Exp. Med.* **177,** 19–33.

SHORTMAN, K., and WU, L. (1996). Early T lymphocyte progenitors. *Annu. Rev. Immunol.* **14,** 29–47.

SINGH, H. (1996). Gene targeting reveals a hierarchy of transcription factors regulating specification of lymphoid cell fates. *Curr. Opin. Immunol.* **8,** 160–165.

SPANGRUDE, G. J., BROOKS, D. M., and TUMAS, D. B. (1995). Long-term repopulation of irradiated mice with limiting numbers of purified hematopoietic stem cells: *In vivo* expansion of stem cell phenotype but not function. *Blood* **85,** 1006–1016.

THOMAS, K. R., and CAPECCHI, M. R. (1987). Site-directed mutagenesis by gene targeting in mouse embryo-derived stem cells. *Cell* **51,** 503–512.

TORRES, R. M., and KÜHN, R. (1997). *Laboratory Protocols for Conditional Gene Targeting.* Oxford Univ. Press, Oxford.

General References

COUTINHO, A., FREITAS, A. A., HOLMBERG, D., and GRANDIEN, A. (1992). Expression and selection of murine antibody repertoires. *Int. Rev. Immunol.* **8,** 173–187.

MELCHERS, F., KARASUYAMA, H., HAASNER, D., BAUER, S., KUDO, A., SAKAGUCHI, N., JAMESON, B., and ROLINK, A. (1993). The surrogate light chain in B-cell development. *Immunol. Today* **14,** 60–68.

ROLINK, A., GHIA, P., GRAWUNDER, U., HAASNER, D., KARASUYAMA, H., KALBERER, C., WINKLER, T., and MELCHERS, F. (1995). *In vitro* analyses of mechanisms of B-cell development. *Sem. Immunol.* **7,** 155–167.

SPITS, H. (1994). Early stages in human and mouse T cell development. *Curr. Opin. Immunol.* **6,** 212–221.

Roy A. Mariuzza

Center for Advanced Research in Biotechnology
University of Maryland
Biotechnology Institute
Rockville, Maryland, USA

Antigen recognition by the immune system is mediated by two classes of molecules, antibodies and T cell receptors (TCRs). Antibodies recognize antigens in their native form, whereas TCRs recognize antigens only as peptide fragments bound to the major histocompatibility complex (MHC) molecule. In addition. TCRs interact with a class of disease-causing molecules known as superantigens (SAGs). X-ray crystallography has revealed the three-dimensional structure of antigen–antibody, TCR–peptide/MHC, and TCR–SAG complexes.

Structural Basis of Antigen Recognition by Immune System Receptors

Introduction

The binding of foreign antigens to complementary structures on the surface of B and T lymphocytes represents the initial step in the sequence of events leading to activation of the immune system. Two distinct classes of molecules, both members of the immunoglobulin supergene family, are responsible for antigen recognition: antibodies, which function both as antigen receptors on B lymphocytes and as circulating effector molecules, and T cell receptors (TCRs), which occur only in membrane-bound form. Although antibodies recognize antigen in intact form, TCRs recognize antigen only in the form of peptides bound to molecules of the major histocompatibility complex (MHC). TCRs also interact with a class of disease-causing and immunostimu-

latory proteins of bacterial or viral origin known as "superantigens" (SAGs). Here, the current knowledge of the three-dimensional structure of antigen–antibody, TCR–peptide/MHC, and TCR–SAG complexes is reviewed.

Three-Dimensional Structure of Antigen–Antibody Complexes

Immunoglobulin G (IgG) molecules are homodimers composed of two identical polypeptide chains spanning approximately 250 amino acids (the light or L chain) and 450 amino acids (the heavy or H chain). Each L chain contains two domains formed by two antiparallel β-sheets, whereas each H chain consists of four such domains. These β-sheet domains are structurally very similar and have been termed the "immunoglobulin fold." The N-terminal variable (V) domains of the L and H chains each contain three loops which link the β-strands and which are highly variable in length and sequence among different antibodies. These so-called hypervariable or complementarity-determining regions (CDRs) form a structure at the end of the immunoglobulin molecule which is complementary to antigen. The primary paradigm of antigen–antibody interactions is that the three-dimensional structure of the six CDR loops recognize and bind a specific antigenic surface (epitope) which is determined by the three-dimensional structure of the antigen. Interactions of antibodies with protein antigens, the most common and diversified antigens encountered by the immune system, occur over large sterically and electrostatically complementary areas. That is, hydrophobic patches of the antigen surface are recognized and interact with hydrophobic patches on the surface of the antigen binding site of the antibody, atoms of polar character in turn interact with atoms of opposite charge in the antibody, and proton donors and acceptors are involved in hydrogen bonds.

The most extensively studied model for the interaction of antigen and antibody is the mouse anti-lysozyme antibody D1.3, an IgG1,κ produced in the secondary response to hen egg white lysozyme (HEL). The structural features of this antibody were established initially with the X-ray crystal structure of the Fab fragment complexed with HEL at 2.8 Å resolution (Fig. 1) (Amit *et al.,* 1986), followed by high-resolution crystal structures of the free D1.3 Fv fragment (a heterodimer consisting of only the V_L and V_H domains) (Bhat *et al.,* 1990) and the Fv–HEL complex (Bhat *et al.,* 1994). Sixteen HEL residues form the discontinuous, conformational epitope recognized by D1.3. Seventeen amino acids from the antibody contact the epitope. The

H chain contributes 10 of these residues and the L chain 7. All six CDRs and, in addition, two residues that belong to the framework regions (FRs) participate in contacts with the antigen. The H chain makes more contacts with the antigen than does the L chain, and in particular its CDR3 loop makes many more contacts than any other CDR; V_L CDR2 contributes the least to antigen binding. Many antibody side chains that contact the antigen (9 of 17) are aromatic, presenting large areas of hydrophobic surface to the antigen. Some of these side chains (His30 and Tyr50 in V_L and Tyr101 in V_H) also contribute to hydrogen bonding with the antigen through their polar atoms. Therefore, hydrogen bonds and van der Waals interactions describe the chemical nature of antigen–antibody contacts. Approximately 750 Å^2 of the solvent-accessible surface of HEL and approximately 700 Å^2 of antibody D1.3 are buried by complex formation.

Somatic recombination and imprecise joining of the V_L–J_L and V_H–D–J_H genetic segments of the L and H chains, respectively, generate sequence diversity affecting position 96 in κ chains and CDR3 in H chains. The structural model of the D1.3–HEL complex allows an evaluation of the contribution of that diversity to antigen binding. V_L residue Arg96 is relatively distant (>4 Å) from the antigen, as are the J_κ-encoded residues. The residues encoded by the D segment of V_H CDR3 (Arg99, Asp100, Tyr101, and Arg102) are involved in very specific contacts with HEL. In contrast, neither J_H residues nor those that could originate from imprecise joining at the D_H–J_H junction contribute directly to the contacts made by D1.3 to its antigen.

The overall picture of the antigen–antibody interface in the D1.3–HEL complex is that of two irregular, flat surfaces with protuberances and depressions that fit into the complementary features of the other (Fig. 1). Small conformational changes in the antibody upon complex formation are involved in producing the observed complementarity between the interacting surfaces, consistent with an "induced-fit" mechanism for antigen–antibody recognition. These changes include adjustments in side chain conformation as well as a small displacement of the V_L and V_H domains relative to their positions in the free Fv. The latter movement brings contacting residues closer to the antigen, optimizing the fit between the two proteins. Significant rearrangements of individual CDR loops have been demonstrated for anti-DNA and anti-peptide antibodies (Davies and Padlan, 1992; Wilson and Stanfield, 1993).

A comparison of the structure of the D1.3–HEL complex with those of other antigen–antibody complexes in which the antigen is a protein reveals common features such as similar areas of interaction, similar number of contacting residues, small conformational changes in the antigen and antibody, and participation of all the CDRs and one or two residues of the framework regions in the contacts with antigen (Davies *et al.,* 1990; Braden and Poljak, 1995). However,

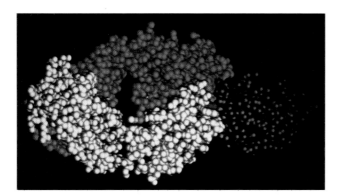

FIGURE I Model of the three-dimensional structure of the complex between the Fab fragment of antibody D1.3 and HEL. The light chain is in blue and the heavy chain in yellow; the antigen, hen egg white lysozyme (HEL), is in green. Glutamine 121, the residue of HEL critical for complex stabilization, is indicated in red.

some significant differences have been noted, such as the presence of salt bridges in the complex between antibody HyHEL-5 and HEL, which is absent in the D1.3–HEL complex. Another difference is that in the D1.3–HEL complex the largest number of contacts are made by V_H CDR3, whereas in HyHEL-5 it is V_L CDR1 and V_H CDR2 that make the most contacts with the antigen.

Although the crystal structure of the D1.3–HEL complex provides detailed information on the molecular architecture of the antigen–antibody interface, it does not indicate the actual functional contribution of individual residues to complex stabilization. To answer this question, alanine-scanning mutagenesis has been carried out for all residues of D1.3 and HEL in contact in the crystal structure. The affinities of the mutant proteins were then measured in order to determine the contribution of individual residues to complex formation (Dall'Acqua et al., 1996). Figure 2 shows the relative loss in binding free energy (DDG) for individual alanine substitutions in D1.3 and HEL mapped onto the three-dimensional structure of each protein. As can be seen, the energetics of binding to HEL are dominated by only 3 of 13 contact residues tested (DDG > 2.5 kcal/mol): V_L Trp92, V_H Asp100, and V_H Tyr101. These form a patch at the center of the interface and are surrounded by residues whose apparent contributions are much less pronounced (DDG < 1.5 kcal/mol). Similarly, significant decreases in binding to D1.3 were only

observed for substitutions at 4 of 12 contact positions of HEL: Gln121, Ile124, Arg125, and Asp119. These residues form a contiguous patch located at the periphery of the surface contacted by the antibody (Fig. 2). Therefore, formation of the D1.3–HEL complex is mediated by only a few productive interactions which dominate the energetics of association. This is similar to findings in the case of human growth hormone binding to its receptor (Clackson and Wells, 1995) and demonstrates that "functional epitopes" (defined as those residues that contribute the most to binding) are not necessarily equivalent to "structural epitopes" (those residues making direct contacts in the three-dimensional structure).

Three-Dimensional Structure of TCR–Peptide/MHC Complexes

Antigen recognition by T cells is mediated by TCRs, which are disulfide-linked heterodimers composed of α and β (or γ and δ) chains consisting of variable (V) and constant (C) regions analogous to those of antibodies. The interaction of TCRs with self-peptide/MHC ligands is responsible for the developmental fate of T cells (Jameson et al., 1995). Thymocytes (precursor T cells that clonally express TCRs on their surface) develop their repertoire and mature to T cells in the thymus before entering the periphery. Thy-

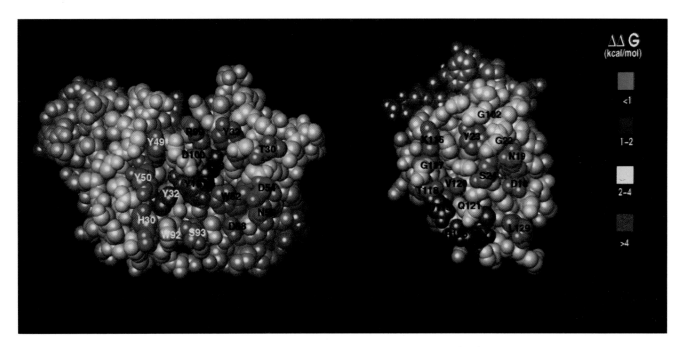

FIGURE 2 Model of the three-dimensional structure of the surface of D1.3 (left) in contact with HEL and of the surface of HEL (right) in contact with D1.3. The residues are color coded according to the loss of binding free energy when substituted by alanine as indicated on the right. Glycine residues on HEL in contact with D1.3 are labeled at their α-carbon atoms. D, aspartic acid; G, glycine; K, lysine; L, leucine; N, asparagine; Q, glutamine; R, arginine; S, serine; T, threonine; V, valine; W, tryptophan; Y, tyrosine.

mocytes bearing TCRs with moderate affinity ($K_D = 10–100 \mu M$) for self-peptide/MHC ligands undergo "positive selection" and are rescued from apoptosis (Alam *et al.*, 1996). On the other hand, thymocytes whose TCRs bind too tightly to self-peptide/MHC are "negatively selected" and, like those with no affinity for self-peptide/MHC, die without reaching the periphery (or become anergized). In this way, most potentially dangerous autoreactive T cells are eliminated during thymic development. Surprisingly, the affinity difference between positively and negatively selecting T cell ligands is relatively small, only approximately 3- to 10-fold. Thus, TCRs are believed to scan many different self-peptide/MHC ligands for those that provide the appropriate degree (neither too much nor too little) of complex stabilization required for positive selection. This screening process generates a pool of mature T cells in the periphery that mostly react strongly only with foreign peptides presented by self-MHC molecules.

X-ray crystallographic studies have shown that the TCRs resemble antibody Fab fragments in their overall structure (Bentley *et al.*, 1995; Fields *et al.*, 1995; Garcia *et al.*, 1996; Garboczi *et al.*, 1996) (Fig. 3). The TCR V region is composed of immunoglobulin-like V_α and V_β domains connected by short peptide segments to the C_α and C_β domains of the C region. The relative orientation of V_α and V_β domains is similar to that of V_L and V_H domains. Likewise, the individual CDRs (except for V_α CDR2) are in nearly equivalent positions in TCRs and antibodies. There are, however, three important structural differences between TCRs and antibodies. The first is a switch in a β-strand from the inner sheet in antibody V_L and V_H domains to the outer sheet in V_α, resulting in a repositioning of the V_α CDR2 loop. The second is the C_α domain, which has only a partial immunoglobulin fold, with the outer three β-strands being replaced by a loop, a short α-helix, and a β-strand, respectively. The third is that the V_β and C_β domains are in much more intimate contact than the corresponding domains in antibodies: The total buried surface area between V_β and C_β is approximately 800 Å^2 compared to 200–350 Å^2 between V_H and $C_H 1$, depending on the elbow angle.

The current understanding of TCR–peptide/MHC interactions is based on the crystal structures of two TCR–peptide/MHC class I complexes (Garboczi *et al.*, 1996; Garcia *et al.*, 1998): (i) the complex between TCR 2C and a self-peptide (dEV8) derived from a mitochondrial respiratory protein bound to the self-MHC class I molecule H-2K^b (Fig. 3) and (ii) the complex between TCR A6 and a viral peptide (Tax) derived from HTLV-1 bound to the HLA-A2 MHC class I molecule (see Fig. 6b). A common view of peptide/MHC recognition by TCRs has emerged from a comparison of these two structures. In both complexes, the TCR fits diagonally across the peptide/MHC ligand. The total surface area buried in the interface is similar in the two TCR–peptide/MHC complexes, approximately 1800 Å^2,

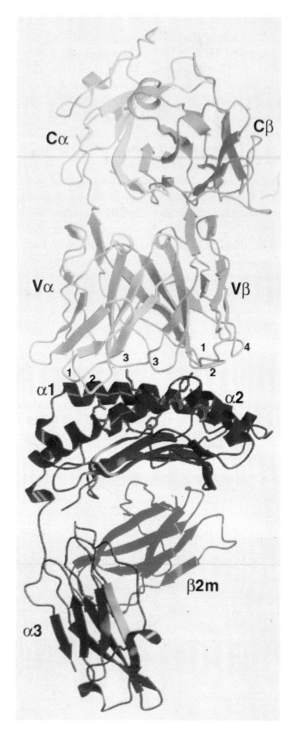

FIGURE 3 Structure of the 2C TCR–dEV8/H-2K^b complex. The TCR α chain is shown in yellow and the β chain is in gray; MHC class I α chain is in green, the dEV8 peptide in red, and β_2m in blue. The numbers 1–3 indicate the CDRs of the V_α and V_β domains and 4 indicates the hypervariable region (HV4) of the V_β chain.

TABLE I

Contacts in TCR–Peptide/MHC Interfaces

	2C TCR– dEV8/H-2K[b]	A6 TCR– Tax/HLA-A2
CDR1α	P1, P2, α1, α2	P1, P2, P4, α1, α2
CDR2α	α2	α2
CDR3α	P4, α1	P4, P5, α1
HV4α		α2
CDR1β	P4, P7, α2	P8
CDR2β	α1	
CDR3β	P6, α2	P5, P6, P7, P8, α2

Note. P denotes the position of the residue in the peptide, starting from the N terminus; α1 and α2 are the two domains of the MHC class I molecule that together form the peptide-binding groove.

with the peptide much more deeply buried in the MHC than in the TCR. The TCR α chain covers the N-terminal residues of the bound peptide and the β chain the C-terminal residues (Table 1). CDR1α and CDR1β lie between the α-helices of the MHC class I heavy chain and simultaneously contact the peptide and MHC molecule. CDR2α and CDR2β lie directly over the α2 and α1 helices, respectively, interacting exclusively with MHC. CDR3α and CDR3β are thus positioned to "read out" the contents of the peptide-binding groove. This arrangement appears to take full advantage of TCR sequence diversity. CDR1 and CDR2 are encoded by the V_α and V_β germline genes. However, CDR3α and CDR3β, which show much greater variability and are positioned to interact with the bound peptide, are encoded by recombination events at the V_α–J_α and V_β–D_β–J_β junctions. It should be noted, however, that the CDR1s also contact peptide and that the CDR3s also contact MHC (Table 1), such that peptide/MHC recognition is mediated by the particular combination of CDRs unique to each TCR binding site. The other two regions of TCR variability, hypervariable region 4α (HV4α) and HV4β, have only minor participation in complex formation. In the A6 TCR-Tax/HLA-A2 complex, CDR1β and CDR2β barely interact with peptide/MHC; CDR3β, which is unusually long, contributes the largest buried surface area of all the CDRs. In the 2C TCR–dEV8/H-2K[b] complex, however, CDR3β is positioned over a largely empty pocket and makes only limited contacts with peptide/MHC. Based on the number of contacts and buried surface area, V_α appears to be dominant in orienting the TCR relative to peptide/MHC in both complexes.

The generality of the binding mode seen in the 2C TCR–dEV8/H-2K[b] and A6 TCR–Tax/HLA-A2 complexes is based on several considerations (Garboczi *et al.*, 1996). The diagonal binding appears to be the consequence of structural features shared by both MHC class I and class II molecules. The top surface of these proteins is not flat but rather has two protuberances near the N termini of the α-helical regions that form the peptide-binding groove. In order for the fairly flat TCR combining site to fit at a low enough point on the MHC to contact the length of the peptide, a diagonal orientation is required. Large translations along the peptide-binding groove or rotations from the diagonal orientation would disrupt the fit between TCR and peptide/MHC.

Despite the overall shape complementarity, the interface between TCR and peptide/MHC appears to be significantly less tightly packed than those of other protein–protein (including antigen–antibody) complexes (Garcia *et al.*, 1998). In the case of the 2C TCR–dEV8/H-2K[b] complex, the poor fit is mostly the result of a limited number of TCR contacts with bound peptide rather than of TCR contacts with MHC. In particular, there is a large, empty cavity at the center of the interface between the CDR3 loops of the α and β chains of the TCR. This relatively poor complementarity is consistent with the surprisingly weak affinities of TCRs for peptide/MHC (10^{-5}–10^{-6} M compared to 10^{-8}–10^{-10} M for antigen–antibody interactions) which have led to serial triggering (Valitutti *et al.*, 1995) and kinetic proofreading (Rabinowitz *et al.*, 1996) models of T cell activation. In these models, weak affinities and fast off rates are in fact necessary to enable a single peptide/MHC complex to serially engage many TCRs. Indeed, ligands with very high affinity for the TCR are predicted to be actually less effective at triggering T cells. It is noteworthy in this respect that no evidence for somatic hypermutation of TCR genes has been found among the many α and β chains that have been sequenced, suggesting that there is no selective advantage to generating high-affinity TCRs by improving interface complementarity, in contrast to the case for antibodies.

An important finding in the case of the 2C TCR–dEV8/H-2K[b] complex (Garcia *et al.*, 1998) is the relatively large structural rearrangements in the TCR combining site associated with ligand binding. As shown in Fig. 4, these rearrangements primarily involve the CDR1 and CDR3 loops of the V_α domain, which are displaced by 4–6 Å relative to their positions in the unliganded A6 TCR structure. Some of these changes are necessary to avoid steric clashes with peptide/MHC, whereas others probably serve to maximize productive interactions with the ligand. This structural flexibility, or "plasticity," would allow a single TCR to adopt multiple conformations, thereby enabling it to interact with different peptides presented by the same MHC molecule. This in turn could help account for the well-known ability of many T cells to cross-react with both self- and foreign peptide/MHC complexes. It is important to emphasize, however, that the conformational changes seen in the A6 TCR are localized to the combining site and are not transmitted

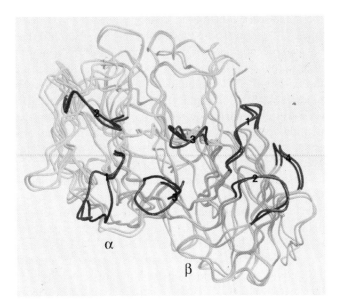

FIGURE 4 Superposition of free and MHC/peptide-bound TCR 2C structures shows conformational differences in the CDR loops of the two forms. The numbers 1–3 indicate the CDRs of the V_α and V_β domains and 4 indicates the hypervariable region (HV4) of the V_β chain, shown in green in the free structure and red in the bound structure.

to the C regions. Thus, as in the case of the TCR A6–Tax/HLA-A2 complex (Garboczi et al., 1996), it is unlikely that changes in TCR conformation upon ligand binding are responsible for initiating T cell signaling. Rather, mechanisms based on ligand-induced TCR oligomerization (Reich et al., 1997) appear more likely to account for T cell activation by peptide/MHC. Conformational adjustments of the magnitude seen in the 2C TCR–dEV8/H-2K^b complex have been observed in antigen–antibody complexes in which the antigen is a peptide or DNA (Davies and Padlan, 1992; Wilson and Stanfield, 1993) but not with protein antigens.

Three-Dimensional Structure of TCR–Superantigen Complexes

The structurally and immunologically best characterized group of SAGs are the *Staphylococcus aureus* enterotoxins, which cause both toxic shock syndrome and food poisoning (Marrack and Kappler, 1990). Some *S. aureus* isolates also produce toxic shock syndrome toxin-1, which has been implicated in the majority of cases of menstrual toxic shock. In addition, mouse mammary tumor viruses encode endogenous SAGs that enable these retroviruses to exploit the host immune system for their transmission (Kotzin et al., 1993). It has also been proposed that SAGs derived from bacteria, mycoplasma, or viruses may initiate autoimmune

disease by activating T cells specific for self-antigens. For example, an analysis of pancreatic islet-infiltrating T cells from patients with insulin-dependent diabetes mellitus (IDDM) revealed preferential expression of the $V_\beta 7$ gene segment but no selection for particular V_α segments or V_β–D_β–J_β junctional sequences (Conrad et al., 1994). This led to the proposal that a SAG associated with pancreatic islets may be involved in the pathogenesis of IDDM. Conrad et al. (1997) isolated a novel human endogenous retrovirus from supernatants of IDDM islets and showed that the envelope gene encodes a MHC class II-dependent SAG specific for $V_\beta 7$.

The three-dimensional structure of the complex between a TCR β chain (mouse $V_\beta 8.2$–$J_\beta 2.1$–$C_\beta 1$) and the SAG staphylococcal enterotoxin C3 (SEC3) has been determined (Fields et al., 1996). The complex is formed through contacts between the V_β domain and the small and large domains of SEC3 (Fig. 5). The solvent-excluded area of the interface is 1300 Å^2, which is similar in size to antigen–antibody and TCR–peptide/MHC interfaces. The V_β residues in contact with SEC3 are Asn28 and Asn30 of CDR1; Tyr50, Gly51, Ala52, Gly53, Ser54, and Thr55 of CDR2; Lys57 and Lys66 of FR3; and Pro70 and Gln72 of HV4. The CDR1, CDR2, FR3, and HV4 regions account for 30, 47, 13, and 10%, respectively, of the total contacts to the SAG.

As shown in Fig. 5, there are no direct contacts between V_β CDR3 and SEC3. This provides a simple explanation for the finding that bacterial and viral SAGs stimulate T cells bearing particular V_β elements without obvious selection for V_β CDR3 length or amino acid sequence (Marrack and Kappler, 1990). However, this does not rule out the possibility that, in certain cases, V_β CDR3 may modulate SAG reactivity. For example, when the V_β domain of TCR A6 (Garboczi et al., 1996) is superposed onto the 14.3.d V_β domain in the $V_\beta C_\beta$–SEC3 structure, residues 97–100 of A6 $V\beta$ CDR3, located at the tip of this long protruding loop, are found to make extensive contacts to SEC3 residues 99–104. Similar interactions may explain the observed influence of V_β CDR3 residues on T cell reactivity toward mouse retroviral Mtv-9 SAG (Ciurli et al., 1998).

The SEC3 residues in contact with V_β are Asn60, Tyr90, Val91 (small domain) and Gly19, Thr20, Asn23, Tyr26, Phe176, and Gln210 (large domain). These mutations are located in the cleft between the small and large domains, which forms the interface with the β chain in the crystal structure (Fig. 5). A sequence alignment of bacterial SAGs revealed that SEA, SED, and SEE, which do not activate $V_\beta 8.2$-bearing T cells, differ from SEB and SEC at nearly all V_β-contacting positions. On the other hand, SAGs which bind the 14.3.d β chain with an affinity similar to that of SEC3 ($K_D = 6.2\,\mu M$) retain several key contacting residues, in particular Asn60, Tyr90, and Gln210.

The crystal structure enables us to understand how SEC3 can stimulate T cells expressing V_β domains from many dif-

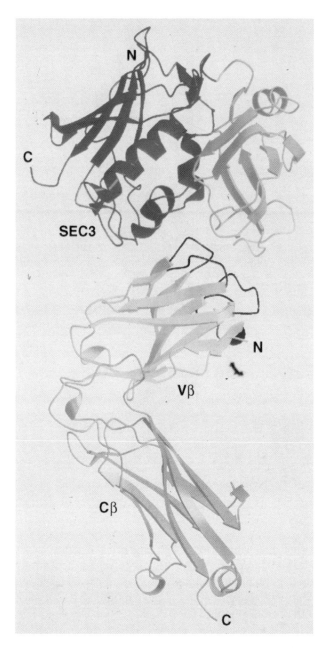

FIGURE 5 Structure of the complex formed by the binding of a $V_\beta C_\beta$ TCR (bottom) to SEC3 (top). The V_β and C_β domains of the TCR are shown in yellow and orange, respectively, and the large and small domains of SEC3 are shown in blue and violet, respectively. Inside the V domain, CDR1, CDR2, CDR3, and HV4 are highlighted in different colors.

ferent families since all the hydrogen bonds between SEC3 and V_β are formed between SEC3 side chains and V_β backbone atoms. A similar mode of binding, involving a major role for backbone hydrogen bonds, has been observed in peptide/MHC complexes (Fremont et al., 1992). In agreement with this mechanism, the three-dimensional structures of other V_β domains reveal that the positions of back-

bone atoms in the SAG-binding region of V_βs reactive with SEC3 are very similar to those of 14.3.d V_β but significantly different in V_βs that do not bind this SAG. For example, SEC3 activates T cells bearing mouse $V_\beta 8$ and human $V_\beta 12$ but not mouse $V_\beta 2$ (Kotzin et al., 1993). When human $V_\beta 12.3$ (Garboczi et al., 1996) is superposed onto mouse $V_\beta 8.2$, the root mean square (rms) difference in α-carbon positions for residues in the SEC3 binding site is only 0.9 Å. However, when mouse $V_\beta 2.3$ (Housset et al., 1997) is superposed onto mouse $V_\beta 8.2$, the rms difference is 3.0 Å. This difference is mainly attributable to a strand switch in $V_\beta 2.3$ relative to other V_β domains of known structure: In $V_\beta 2.3$ the c″ strand is hydrogen bonded to the d strand of the adjacent (outer) β-sheet, whereas in other V_βs the c″ strand is associated with the c′ strand in the same (inner) sheet.

No gross conformational changes occur in either the 14.3.d β chain or SEC3 upon complex formation, although many changes in the side chain conformation of interface residues are observed; these could indicate an "induced fit" mechanism for TCR–SAG recognition, as described by Garcia et al. (1998) for the 2C TCR–peptide/MHC complex. In the β–SEC3 complex, the close association between the V_β and C_β domains found in the structure of the free β chain and in associated $\alpha\beta$ TCR heterodimers is preserved. Similarly, no large conformational changes were noted upon formation of the complex between HLA-DR1 and SEB (Jardetzky et al., 1994). Therefore, excluding possible cooperative effects, these results suggest that formation of the TCR–SAG–MHC ternary complex does not involve major domain movements or rearrangements in the polypeptide backbones of the interacting species. Similarly, except for rearrangements in certain of the CDR loops, no gross conformational changes in the TCR were observed in the crystal structures of the TCR–peptide/MHC class I complexes reported by Garboczi et al. (1996) and Garcia et al. (1998).

Several models for the interaction of T cells with SAGs have been proposed (Kotzin et al., 1993). These models differ in the extent to which the TCR interacts with MHC in the TCR–SAG–MHC complex. The available crystal structures allow us to clearly discriminate among these models, at least for the staphylococcal enterotoxins. We have constructed a model of the TCR–SAG–MHC complex by least-squares superposition of the following crystal structures: (i) the $V_\beta C_\beta$–SEC3 complex, (ii) the SEB–peptide/HLA-DR1 complex (Jardetzky et al., 1994), and (iii) the A6 TCR $\alpha\beta$ heterodimer (Garboczi et al., 1996) (Fig. 6a). This model has several important features. The SAG is well positioned to bridge the antigen-presenting cell (APC) and the T cell. The V_α domain makes no direct contacts with the SAG, consistent with the ability of the β chain alone to bind SAGs (Malchiodi et al., 1995). The MHC molecule is oriented with its α chain over the β chain of the TCR and its β chain over the TCR α chain. However, in contrast to

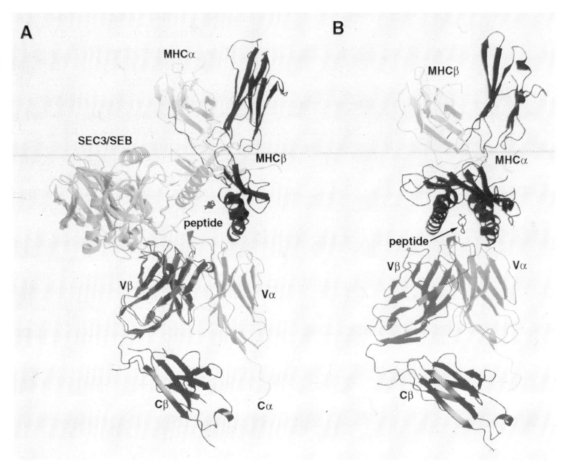

FIGURE 6 (a) Structure of the TCR–peptide/MHC complex. (b) Hypothetical model of the interaction between this complex and the superantigen SEC3/SEB.

the conventional TCR–peptide/MHC complex (Fig. 6b), there are no direct contacts between the TCR β chain and the MHC α or β chain. On the other hand, the MHC $\beta1$ domain interacts with the V_α domain of the TCR. This can account for evidence of a functional interaction between the class II β chain and the TCR α chain in recognizing bacterial SAGs as well as for evidence that the ability of certain V_α regions to interact more favorably than others with different class II alleles is responsible for the preferential expression of these V_αs among SAG-reactive T cells. It should be noted, however, that the extent of the interaction between V_α and the MHC $\beta1$ domain will largely depend on the relative orientation of V_α and V_β domains in the TCR heterodimer and that a comparison of different TCRs has revealed significant variability in the geometry of V_α/V_β association (Garcia et al., 1998). The MHC molecule is only partially engaged by the TCR in our model, with the SAG acting as a wedge between the TCR β chain and the MHC α chain (Fig. 6a). As a consequence, the antigenic peptide is effectively removed from the combining site of the TCR. Therefore, the SAG is able to circumvent the

normal mechanism for T cell triggering by specific peptide/MIIC complexes. The result is polyclonal activation of whole populations of T cells expressing particular V_β elements, largely irrespective of the peptide/MHC specificity of the corresponding TCRs.

References

ALAM, S. M., et al. (1996). T-cell receptor affinity and thymocyte positive selection. Nature **381,** 616–620.

AMIT, A. G., MARIUZZA, R. A., PHILLIPS, S. E. V., and POLJAK, R. J. (1986). Three-dimensional structure of an antigen–antibody complex at 2.8 Å resolution. Science **233,** 747–753.

BENTLEY, G. A., BOULOT, G., KARJALAINEN, K., and MARIUZZA, R. A. (1995). Crystal structure of the beta chain of a T cell antigen receptor. Science **267,** 1984–1987.

BHAT et al. (1990). Small rearrangements in structures of Fv and Fab fragments of antibody D1.3 on antigen binding. Nature **347,** 483–485.

BHAT et al. (1994). Bound water molecules and conformational stabilization help mediate an antigen–antibody association. Proc. Natl. Acad. Sci. USA **91,** 1089–1093.

BRADEN, B. C., and POLJAK, R. J. (1995). Structural features of the reactions between antibodies and protein antigens. *FASEB J.* **9**, 9–16.

CIURLI, C., POSNETT, D. N., SEKALY, R.-P., and DENIS, F. (1998). Highly biased CDR3 usage in restricted sets of beta chain variable regions during viral superantigen 9 response. *J. Exp. Med.* **187**, 253–258.

CLACKSON, T., and WELLS, J. A. (1995). A hot spot of binding energy in a hormone–receptor interface. *Science* **267**, 383–386.

CONRAD, B., *et al.* (1994). Evidence for superantigen involvement in insulin dependent diabetes mellitus aetiology. *Nature* **371**, 351–355.

CONRAD, B., *et al.* (1997). A human endogenous retroviral superantigen as candidate autoimmune gene in type I diabetes. *Cell* **90**, 303–313.

DALL'ACQUA, W., GOLDMAN, E. R., EISENSTEIN, E., and MARIUZZA, R. A. (1996). A mutational analysis of the binding of two different proteins to the same antibody. *Biochemistry* **35**, 9667–9676.

DAVIES, D. R., and PADLAN, E. A. (1992). Twisting into shape. *Curr. Biol.* **2**, 254–256.

DAVIES, D. R., *et al.* (1990). Antigen–antibody complexes. *Annu. Rev. Biochem.* **59**, 439–473.

FIELDS, B. A., *et al.* (1995). Crystal structure of the $V\alpha$ domain of a T cell antigen receptor. *Science* **270**, 1821–1824.

FIELDS, B. A., *et al.* (1996). Crystal structure of the beta chain of a T-cell receptor complexed with a superantigen. *Nature* **384**, 188–192.

FREMONT, D. H., *et al.* (1992). Crystal structures of two viral peptides in complex with murine MHC class I H-2Kb. *Science* **257**, 919–927.

GARBOCZI, D. N., *et al.* (1996). Structure of the complex between human T-cell receptor, viral peptide and HLA-A2. *Nature* **384**, 134–141.

GARCIA, K. C., *et al.* (1996). An alpha/beta T cell receptor structure at 2.5 A and its orientation in the TCR–MHC complex. *Science* **274**, 209–219.

GARCIA, K. C., *et al.* (1998). Structural basis of plasticity in T cell receptor recognition of self peptide–MHC complexes. *Science* **279**, 1166–1172.

HOUSSET, D., *et al.* (1997). The three-dimensional structure of a T-cell antigen receptor $V\alpha V\beta$ heterodimer reveals a novel arrangement of the $V\beta$ domain. *EMBO J.* **16**, 4205–4216.

JAMESON, S. C., HOGQUIST, K. A., and BEVAN, M. J. (1995). Positive selection of thymocytes. *Annu. Rev. Immunol.* **13**, 93–125.

JARDETZKY, T. S., *et al.* (1994). Three-dimensional structure of a human class II major histocompatibility complex molecule complexed with superantigen. *Nature* **368**, 711–718.

KOTZIN, B. L., LEUNG, D. Y. M., KAPPLER, J., and MARRACK, P. (1993). Superantigens and their potential role in human disease. *Adv. Immunol.* **54**, 99–166.

MALCHIODI, E. L., *et al.* (1995). Superantigen binding to a T cell receptor beta chain of known three-dimensional structure. *J. Exp. Med.* **182**, 1833–1845.

MARRACK, P., and KAPPLER, J. (1990). The staphylococcal enterotoxins and their relatives. *Science* **248**, 705–711.

RABINOWITZ, J. D., *et al.* (1996). Kinetic discrimination in T-cell activation. *Proc. Natl. Acad. Sci. USA* **93**, 1401–1405.

REICH, Z., *et al.* (1997). Ligand-specific oligomerization of T-cell receptor molecules. *Nature* **387**, 617–620.

VALITUTTI, S., *et al.* (1995). Serial triggering of many T-cell receptors by a few peptide–MHC complexes. *Nature* **375**, 148–151.

WILSON, I. A., and STANFIELD, R. L. (1993). Antigen–antibody interaction. *Curr. Opin. Struct. Biol.* **3**, 113–118.

CRISTINA RADA

MRC Laboratory of Molecular Biology
Cambridge CB2 2QH, United Kingdom

The Generation of Diversity

The fundamental property of the mammalian immune system is its unique ability to recognize almost any possible molecular structure and prevent it from potentially damaging the organism. The front line of the specific immune defense is the production of antibodies whose role is to neutralize pathogens and to prevent their spread. Thus, antibodies must be as diverse as their potential targets. In order to generate this diversity, the immune system has evolved a two-step strategy. First, it produces a vast repertoire of molecular structures by combinatorial assembly of a small number of gene segments. This is the first approximation or "primary repertoire" which must provide the initial recognition of any possible antigen, although it might do so with low efficiency. The second step is the refinement of the primary repertoire against a particular antigen by somatic hypermutation of the antibody genes and the subsequent Darwinian selection of high-affinity mutants. This article describes in general terms how the diversity of the primary repertoire is generated and the known features of the mechanism underlying the further diversification and maturation of the immune response by somatic hypermutation are discussed.

Introduction

Genetic variation and in some cases sexual reproduction combine to produce variations in the progeny in living organisms, and we know from the work of Charles Darwin that natural selection operates on those differences to modify and occasionally generate new species. There are approximately 2 million living species on Earth today. However, this represents less than 1% of the estimated total number of species that have ever existed. The information necessary to design and operate complex organisms, such as oak trees, viruses, or humans, is contained within the sequence of the DNA molecule (occasionally RNA). This information is transmitted from generation to generation and although the fidelity of the newly synthesized copies of

the parental DNA is very high, mistakes (or mutations) occur which result in new varieties of the same genes or new genes. Life as we know it is therefore defined by constant change and the potential to generate an infinite degree of variation. The differences in genes between both individuals and species are translated in the structural diversity of the chemical molecules that make the organism.

After the discovery in the nineteenth century of the nature of infectious diseases by Robert Koch, it became clear that any defense mechanism which would protect its host from microbiological attack would need to be able to recognize and neutralize a myriad of different and ever-changing molecular structures. The pioneering work of Jenner in 1796 on the protective effect of vaccination established that such a defense mechanism did indeed exist. Jenner showed that exposure of humans to cowpox conferred "immunity" against the human disease smallpox. The first step in understanding the molecular basis of immune defenses came in 1890 with the discovery by Behring and Kitasato of antibodies in the serum of vaccinated individuals as substances that could specifically bind the pathogen. It was later discovered that the quality of an antiserum improved with time and repeated exposure. In 1903, the experiments of Rudolf Kraus showed that the protective activity of an antitoxin serum to lethal doses of a toxin was significantly improved after donor rabbits had been repeatedly exposed to sublethal doses of the toxin. However, it was not until 1951 that Jerne actually determined in a modern way that the affinity of the antibodies to a specific antigen increased with time following immunization. It was clear since the 1930s, through the work of Landsteiner, that organisms could produce specific antibodies against virtually any foreign substance, including man-made chemicals. The question remaining was how it was possible to genetically encode a molecule that was able to recognize virtually any other molecular structure with enough affinity to neutralize the pathogen and with the exquisite specificity which would minimize interaction with anything but the intended target. The question of the generation of diversity (GOD) has been central to immunology since its beginning. For some time, two conflicting theories, instruction versus selection championed by Pauling (1940) and Jerne (1955), respectively, wrestled to explain the specificity of antibodies. The instruction theory postulated that specificity was obtained by the folding of the antibody around an antigen template, whereas the selective theory predicted that antigen would simply be bound by one of a repertoire of preexisting antibody structures. In the early 1960s, the understanding that the primary amino acid sequence of a protein was responsible for its three-dimensional structure tilted the balance in favor of the selection camp. The initial work of Anfinsen and the specific application of the principle to antibody molecules by Haber and Tanford dashed any remaining hopes for instructionalists. By chemically denaturing antibodies of a known specificity and allowing renaturation in the absence of antigen, the original specificity of the antibody was restored, clearly demonstrating that the structural information was entirely contained in the primary amino acid sequence of the molecule. The realization that each B cell clone produced antibody of only a single defined specificity led Burnet in 1957 and others to formulate the clonal selection theory, which explained how the specificity would be achieved by selection and expansion of the best preexisting antibody-producing clones by antigen. The final answer to the question of diversity had to wait for the development of molecular biology and molecular genetics. Although the analysis of the protein sequences of antibodies produced by myeloma tumors had demonstrated the heterogeneity of antibody proteins, immunologists tried to explain the

question of diversity by either of two mutually exclusive hypothesis. One theory assumed that all information used to produce the diversity was encoded in the germline, whereas the other suggested that it was newly generated in each individual by somatic mechanisms. The solution was as simple and as elegant as anyone could have dreamt. The immune system in fact employed a small number of genes in a combinatorial way to generate new genes. This process required a specialized new genetic mechanism of gene rearrangement which is used in both T and B cell antigen receptors. The first conclusive evidence for DNA rearrangement of the immunoglobulin genes during B cell development was presented in 1976 by Tonegawa. By comparing the configuration of the V and the C segments encoding light chains in embryos (prior to antibody production) and in myeloma cells, he demonstrated that the V and C segments were brought together in the myeloma cells which secreted antibodies. As in macroevolutionary processes, this strategy generates molecular diversity by coupling genetic mechanisms with a process of selection in order to identify and expand useful variants. The advent of DNA–RNA hybridization techniques established that the number of genes available to make antibody molecules was limited, whereas the advances in mRNA sequencing together with monoclonal antibody technology subsequently demonstrated that somatic variability was an important contributor to the diversity of antibody molecules. The germline versus somatic diversity controversy vanished as it became clear that both parts make an essential contribution.

The rapid progress in molecular biology has dispelled the enigma of GOD but not the fascination.

Diversity of the Primary Repertoire

Genetic Configuration of Antibody Genes

Immunoglobulin molecules are assembled from two heavy and two light chains. Both types of chain are constructed from modular immunoglobulin domains. Each module is encoded by a distinct genetic element and although the basic structure of each module is the same (two β-sheets joined together by a disulfide bond), this modular structure allows functional specialization of the different parts of the molecule. The primary function of the antibody molecule, recognition of other molecules, is carried out by the variable domains (V_L and V_H) at the amino-terminal end of both heavy and light chains. The effector functions of the immunoglobulin as antigen receptor on the surface of B cells or as secreted antibody, reside at the carboxy-terminal end, which is composed of one or several constant domains (C_L and C_{H1}–C_{H4}). The variable regions of heavy and light chain associate to form the antigen-combining site of the molecule and it is this region that displays a staggering degree of variability. As discussed pre-

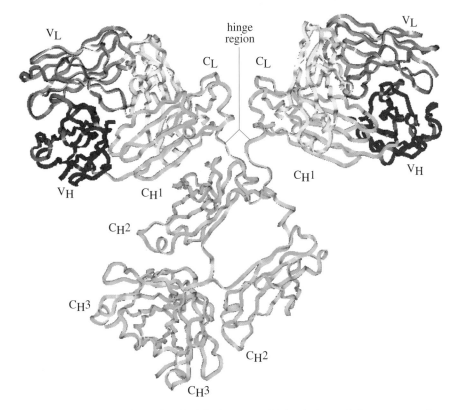

FIGURE I Structure of an antibody molecule. The heavy and light chains and the corresponding variable and constant regions are highlighted. The hypervariable loops, which join the β-sheet strands (colored ribbons) and form the antigen binding site, are highlighted in red. The transmembrane domain and cytoplasmic tail of the membrane-bound antibody are not shown. The average size of a single immunoglobulin fold is $40 \times 25 \times 25$ Å. A whole antibody molecule is 20–30 nm, although the dimensions vary depending on the angle of the hinge region (dashed lines) and the relative positions of the V and C domains at the tip of the molecule. Different isotypes have different amounts of flexibility at the hinge region, for example, IgD-type immunoglobulins are more flexible than IgM-type immunoglobulins. For comparison, the average size of an animal cell is 10–30 μm, whereas a bacterium such as *Escherichia coli* is 3–5 μm.

viously, this high degree of variability is essential for the ability of antibodies to recognize almost any possible molecular structure and it is achieved with remarkable genetic economy (Fig. 1).

The genes which encode the immunoglobulin molecules are organized in clusters. The immunoglobulin (Ig) gene loci have a complex organization which varies slightly in different species, but the complexity reveals a common overall structure (exemplified by the human loci depicted in Fig. 2), which reflects the common evolutionary origin of the antibody genes. The best characterized Ig loci are the human and mouse loci, and the human genes have been almost completely mapped and sequenced. The genes encoding the Ig heavy chains are clustered in chromosome 14 in humans (chromosome 12 in mice). A copy of each C_H region gene is located together with regulatory elements, such as enhancers and recombination signals, at one end of the locus, whereas several families of V genes extend over the other end of the locus. Six J_H segments and 25 D_H seg-

ments are located in between. To date, 111 V_H genes have been identified in man, 51 of which are known to be functional. The light chain can be encoded by two different loci, the kappa locus, which is located on chromosome 2, and the lambda locus, which is located on chromosome 22 (chromosomes 6 and 16 in mice, respectively). In both human and mice, lambda chains are less common than kappa chains, as are the number of V_λ versus V_κ gene segments. Seventy-six V_κ genes have been identified in man, with at least 40 of these being functional. The kappa locus is a simplified version of the heavy chain locus. Many V_κ genes are located at one end, followed by 5 J_κ segments and a single C_κ gene. The lambda locus shows a slightly different configuration with 7 C_λ genes, each of which is preceded by a corresponding J_λ segment. Three of the $J_\lambda C_\lambda$ tandem units are known to be nonfunctional pseudogenes. The total number of V_λ genes has not been established, but at least 31 are known to be functional. In mice, the $J_\lambda C_\lambda$ genes are organized into two clusters with 2 $J_\lambda C_\lambda$ units each and the corresponding

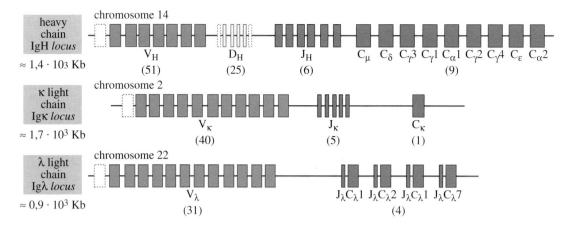

FIGURE 2 Schematic representation of the human heavy and light chain loci. Only the functional C regions are represented (not to scale). The numbers in parentheses indicate the number of functional genes. The size of the each locus is shown to the right.

V_λ segment or segments are located upstream. Three of the $J_\lambda C_\lambda$ units are known to be functional. The number of V_λ elements seems to vary within both different laboratory mouse strains and wild mice.

The number and organization of the gene fragments in each locus are variable not only between different species but also within the same species. This is a reflection of the evolutionary dynamics of the immunoglobulin genes, in which expansion by duplication and contraction by deletion constantly reshape the gene pool. In addition, a certain degree of polymorphism is found among the same genes in different individuals and a process of gene conversion seems to contribute to the diversification of the V segments in the germline of any given species.

Combinatorial Diversity by Gene Rearrangement

The diversity of antibody molecules is accomplished by the assorted use of the variable regions encoded in the germline and the combination of a single heavy and a light chain to form the binding site. In humans, at least 10^4 different antigen binding sites can be generated depending on which V genes are utilized (Table 1).

A complete antibody gene is assembled by joining the

TABLE I				

Combinatorial Diversity of Human Antibody Genes[a]

COMBINED ELEMENTS				PRODUCT
V_H 51	$\times$	D_H 25 = 1275	D segment reading frames $\times 3$	$V_H D_H$ = 3,825
VD_H 3,825	$\times$	J_H 6		VDJ_H = 22,950
V_κ 40	$\times$	J_κ 5		VJ_κ = 200
V_λ 31	$\times$	JC_λ 4		VJC_λ = 124
H 22,950	$\times$	λ 124		= 2,845,800 (2.8×10^6)
H 22,950	$\times$	κ 200		$+7.4 \times 10^6$ = 459,000,000 (4.5×10^6)

[a] Additional diversity (joining diversity) is introduced by the imprecise joint of the V and (D)J segments. In the heavy chain further diversity is achieved by the addition of untemplated nucleotides to the ends of the coding sequences before the joint is completed.

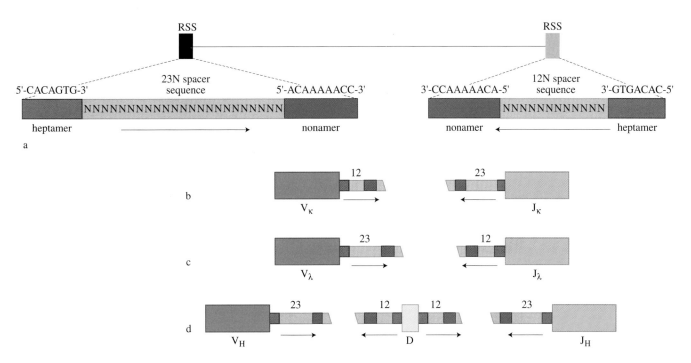

FIGURE 3 Signals involved in the control of V(D)J rearrangements. The rearrangement of the V segment to the C segment is directed by specific signals flanking the coding sequences. Such signals are recognized by the specialized V(D)J recombinase which includes the products of the *Rag1* and *Rag2* genes. Each recombination signal sequence (RSS) (gray boxes) is composed of (a) a conserved heptamer sequence 5'-CACAGTG-3' and a nonamer sequence 5'-ACAAAAACC-3' (blue box) joined by a nonconserved spacer sequence either 23 or 12 nucleotides long. The spacer length is meticulously conserved (positioning the heptamer/nonamer signals on the same face of the DNA double-helix structure where they can be bound by RAG1 and RAG2) and it determines the directionality of the rearrangement. A gene segment can only be joined to a segment flanked by a RSS with the opposite spacer length (the 12/23 rule). Recombination occurs in a hierarchical manner such that, at any given time, only one allele of the same locus is accessible to recombination. The 12/23 rule ensures that the right segments are joined together. Coding segments are represented by green, yellow, and orange boxes. Details of the RSS are highlighted, such as the heptamer (pink boxes) and the nonamer (blue boxes). The arrows represent the direction of the joint.

necessary V and C modules with the intervening J or DJ segments to encode a heavy and a light chain. The joining of the genetic units is accomplished by specialized machinery which promotes orderly rearrangements in the DNA (Fig. 3). This occurs during the early stages of lymphocyte differentiation and is tightly linked to the program of B cell development. During rearrangement a V segment which can be located a very long distance away in the same chromosome is brought to the proximity of the C region genes and the DNA is broken and rejoined in a new configuration.

The process of V(D)J recombination contributes an additional source of diversity because prior to the ligation of the recombined segments the ends of the coding sequences are usually the target of different modifications. These include deletions, N-region additions, P-nucleotide addition, and homology-mediated joining. The result is the addition or the removal of several nucleotides at the junctions between the V and J segments or the V, D, and J segments,

modifying the ends of the coding segments and making the joining very imprecise. In the process of N-region addition, several nucleotides are enzymatically added to the ends of rearranging segments by an enzyme, terminal deoxynucleotidyl transferase (TdT), which unlike DNA polymerases does not require a DNA template. Palindromic P nucleotides are usually one or two nucleotides palindromic to the terminal nucleotides of the coding segments and occur as a consequence of the resolution of intermediary hairpin structures during the recombination process (Fig. 4). Occasionally, overlapping homologous nucleotides at the joining end of the recombining segments can direct the join, possibly by utilizing the machinery of double-strand break DNA repair, and are partially retained in the join region providing additional variation.

The joining diversity is more extensive in heavy chains since an extra D segment is involved in the joining process. Furthermore, N-region additions are almost exclusively

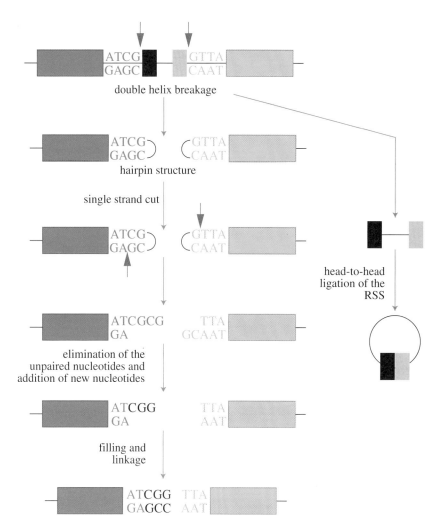

FIGURE 4 Schematic representation of the molecular events involved in V(D)J rearrangement. The coding segments are represented by colored boxes, whereas the recombination signal sequences (RSS) are represented by gray or black boxes. Recombination occurs at the border between the heptamer sequence and the coding sequence. The first step is the cleavage of both DNA strands adjacent to the RSS and the formation of a hairpin structure at both ends of the coding segments. The hairpin structure is resolved by a single-strand nick close to the apex, which can subsequently result in the addition of P nucleotides. The ends can be trimmed and N nucleotides added (red) before filling-in and ligation. Alternatively, when homology is present at both ends, these can be aligned and ligated. The intervening sequences are circularized or inverted by straight ligation of the RSS without any modification of the end prior to ligation. Depending on the relative orientation of the recombined segments, the DNA between them might be deleted.

confined to the developmental stage at which active rearrangement of the heavy chains occurs. The TdT enzyme is switched off by the time kappa and lambda rearrangements start; thus, N additions are very rarely found in light chains.

Together, the potential diversity generated by joining variability coupled with combinatorial variability is approximately on the order of 10^{10}. If we assume that an Ig molecule can be encoded by 100 kb of DNA (including a heavy chain, a light chain, and all of the regulatory elements), in the absence of the combinatorial and rearrangement processes approximately 10^{15} bp would be required to store the information necessary to achieve the same degree of diversity. The size of a haploid human genome is approximately 3×10^9 bp, whereas the approximate size of the heavy and light chain loci is ~2% of one average chromosome (~0.1% of the total human genome). It thus appears that the process of combinatorial assortment and V(D)J recombination is able to generate vast antibody diversity with only a relatively small genetic investment.

This genetic diversity directly translates into the chemical diversity of the antibody combining site. Three complementary determining regions (CDRs) supported by the rigid structure of the β-sheets which form the framework (FW) are responsible for the antibody specificity (Fig. 1). Indeed, the CDRs from one antibody molecule can be transplanted into another antibody, thus transferring the specificity for antigen binding. Provided some minor changes are made to certain FW residues, the new grafted molecule can attain the same antigen specificity as that of the original antibody without any significant loss of affinity. Thus, although small changes in the FW regions can affect the affinity of the antibody, the FW may essentially be considered as being interchangeable, whereas specificity is principally confined to the hypervariable CDR loops.

Despite the genetic diversity of V genes, they share a high degree of structural similarity. The folding patterns of the CDRs (with the exception of the V_H CDR3s) may be grouped into a restricted set of canonical structures. In human heavy chains there are 3 CDR1 and 6 CDR2 canonical structures, although only 8 of the 18 potential combinatorial forms are found in the germline. The question of whether this merely reflects historically contingent evolutionary events or whether certain combinations have been negatively selected remains unclear.

The final picture of the variability of the antigen combining site may therefore be simplified (Fig. 5). The highest degree of variability is afforded by the CDR3 of the heavy chain and the CDR3 of the light chain, which occupy the central part of the antibody binding site and represent the variability generated by somatic V(D)J recombination. This is flanked by the CDR1 and -2 of either the heavy or the light chain, which represent variability encoded in the germline.

Alternative mechanisms to V(D)J recombination are used by other species to achieve the same goal of diversification of the germline. The very limited joining diversity seen in birds is increased during embryogenesis by a gene conversion mechanism (Fig. 6) which transfers segments of a pool of V pseudogenes to a single functional V. Similarly, the primary repertoire of the sheep is diversified in the ileal Peyer's patches by extensive point mutations, both before and for a short time after birth. In the case of rabbits, the limited number of V_H genes utilized in V(D)J rearrangement are likely to be further diversified by both gene conversion and somatic point mutation shortly after birth. The limited germline diversity in this case is compensated by high levels of N addition before birth.

Somatic Diversification of the Germline: The Actual Repertoire

Not all of the theoretically possible Ig molecule variants are found in the actual repertoires of the real world. Certain heavy–light chain combinations are structurally incompatible, as are some of the structures resulting from V(D)J recombination. Furthermore, certain constraints are imposed by the nature of the recombination process. For example, the V and D genes most proximal to the J segments are preferentially expressed. Only one in three of the possible coding frames of D segments leads to a functional rearrangement. Biases also exist in the joining of some V regions to certain J segments. In addition, the N-residue addition mechanism has a preference for the insertion of G and C residues. Furthermore, short deletions, with or without P nucleotides, are more common than either large or no deletions. Finally, the coding sequence of the V region can determine the recombination efficiency of that particular V. Thus, not all of the different possible antibody molecules will actually be present in the primary repertoire, and of those which are present the frequency of each may vary considerably.

In addition, certain physiological limitations shape the actual expression of the repertoire. The recombination process is hierarchically organized so as to ensure that only a single antibody species is expressed on each cell. This phenomenon is known as "allelic exclusion" and it is achieved by the orderly rearrangement of heavy and light chains in a sequential manner. After a functional heavy chain is produced, it is then expressed on the surface of the developing B cell in conjunction with an invariant surrogate light chain. This provisional receptor is able to deliver a signal to induce the initiation of rearrangement on the kappa lo-

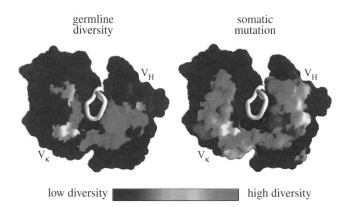

germline diversity

somatic mutation

low diversity ▬▬▬▬ high diversity

FIGURE 5 Representation of the germline diversity and the diversity introduced during somatic hypermutation within the antigen combining site of a model antibody (modified from Tomlinson *et al.*, 1996). Diversity is primarily limited to the antigen combining site. Although the germline-encoded diversity tends to concentrate within the center of the antigen binding site, the diversity generated by somatic hypermutation appears to target the periphery of the antigen binding site and therefore complement the germline-encoded variability. The gray structure in the center of the model represents the CDR3 loop of the heavy chain. The end of the light chain CDR3 lies at the center of the binding site and is not shown. Both CDR3s are excluded from the analysis and represent the junctional variability introduced during rearrangement.

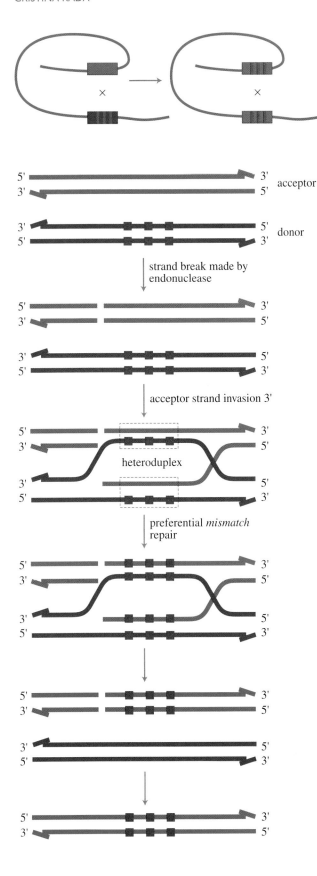

5' ———————————————————▶ 3'
3' ◀——————————————————— 5' acceptor

3' ◀——————■■■——————— 5' donor
5' ———————————————————▶ 3'

strand break made by
endonuclease

5' —————— ——————————▶ 3'
3' ◀——————————————————— 5'

3' ◀——————■■■——————— 5'
5' ———————————————————▶ 3'

acceptor strand invasion 3'

5' —————— ——————————▶ 3'
3' ◀——————————————————— 5'

heteroduplex

3' ◀——————■■■——————— 5'
5' ———————————————————▶ 3'

preferential *mismatch*
repair

5' —————— ——————————▶ 3'
3' ◀——————————————————— 5'

3' ◀——————■■■——————— 5'
5' ———————————————————▶ 3'

5' ——————■■■——————▶ 3'
3' ◀——————————————————— 5'

3' ◀——————————————————— 5'
5' ———————————————————▶ 3'

5' ——————■■■——————▶ 3'
3' ◀——————■■■——————— 5'

cus and to suppresses further rearrangements on the heavy chain locus. Lambda chains are usually rearranged after kappa rearrangements have failed to produce a functional light chain. Presumably this is the reason why fewer lambda-bearing antibodies are found in mice and to a lesser extent in humans compared with kappa-bearing antibodies. In other species such as sheep, light chain rearrangements probably follow the opposite order since lambda chain antibodies are the predominant combination.

The deletion of autoreactive B cells, or alternatively the short life span of autoreactive B cells rendered anergic, is probably one of the most drastic constraints on the shape of the expressed repertoire. The imperative of avoiding antibodies which can react with proteins or other structures within the individual organism eliminates from the repertoire a vast amount of possible variants by deletion of the self-reacting B cell clones before they leave the bone marrow. A mechanism known as "receptor editing" can further modify the primary repertoire by inducing secondary rearrangements of the light chain despite the presence of a functional receptor in immature B cells. These rearrangements are understood in the context of tolerance. For instance, an autoreactive B cell undergoing editing of its antibody receptor will be given a second chance to try an alternative V gene for the light chain in the hope that the new heavy–light chain combination will eliminate the autoreactive specificity and thus spare the cell from apoptosis (programmed cell death).

The function of the recombining sequence (RS) signals (also known as kappa-deleting element in humans) downstream of C_κ, which mediate the V(D)J recombinase-dependent deletion of the C region, may be interpreted in

FIGURE 6 Schematic overview of the gene conversion process. Gene conversion involves a nonreciprocal homologous recombination event which results in templated "transfer" of sequences from a homologous donor DNA segment. The details of the molecular process are better understood in the *Saccharomyces cerevisiae* mating-type locus. The functional Mat a locus "switches" to Mat α by acquiring sequences from a silent distant homologous donor locus (HMLα or HMLa). Recombination is induced by the HO endonuclease which makes a double-strand break in the acceptor Mat a locus. The 3' single-strand end of the cleaved DNA segment invades the homologous donor gene (red) forming a heteroduplex paired region. Mismatches (X) are then corrected by a highly preferential form of mismatch repair which corrects the invading donor strand. Repair might be directed by the free 3' end of the acceptor strand as in bacterial mismatch repair, in which a nick in the DNA determines which strand undergoes repair. A similar mechanism is involved in the transfer of sequences from nonfunctional pseudogenes to the functional single rearranged V gene in chickens and possibly the diversification of the germline V genes within a species.

the same receptor editing context. Although RS rearrangements appear in a large proportion of cells with lambda rearrangements, they are not a prerequisite for rearrangement in the lambda locus. Thus, recombination between the downstream recombination signal sequence and usually a J segment results in the deletion of the C_κ in the functionally rearranged light chain which would eliminate the autoreactive heavy–light combination and allow a new rearrangement to occur.

A stochastic element is also involved in shaping the actual repertoire since, at any given time, the number of B cells in the organism is less than the potential diversity. Time is one of the variables that determines the particular specificities which are present in the repertoire. Two additional examples of the temporal constraints depend on the age of the individual. First, in fetal lymphocytes the TdT polymerase responsible for the N additions is not expressed; therefore, newborns have a more restricted joining diversity. Second, although the total number of B cells remains relatively constant throughout the life of the individual, the proportion of memory B cells in the periphery increases with age, resulting in a relative bias of the available repertoire toward past experience.

Initiation of T Cell-Dependent Immune Responses: The Formation of Germinal Centers

In general, protein-derived antigens are dealt with by the immune system in a regulated fashion. This requires the activation of T cells, which in turn provide help to develop the B cell antibody response. These so-called T cell-dependent B cell responses lead to the further diversification of antibodies by the introduction of somatic point mutations into the variable regions of the Ig genes which encode the antibody receptor. Regulation of this process requires the interaction of several cells, which ensures that newly generated self-reacting antibodies are not produced. This is important since the final stages of differentiation in this pathway produce the repertoire of recirculating memory cells which are both more frequent and more readily activated.

Following activation by a T cell-dependent antigen, B cells do not immediately proliferate or secrete antibody. Instead, they migrate to the draining secondary lymphoid areas, and the first stop is the T cell areas. Once there, T helper cells interact with the incoming B cells via their antigen receptors (TCR) and the costimulatory molecules (such as CD28). Both antigen presented by the major histocompatibility complex (MHC) class II molecules and the costimulatory molecules B7.1 and B7.2 on the surface of the B cell lead to the activation of T cells which can then provide the help signal by upregulating the surface expression

of CD40 ligand. The CD40–CD40 ligand interaction and cytokine secretion eventually lead to the full B cell activation, proliferation, and differentiation.

The differentiation of B cells involves migration to the follicles to form the germinal centers. B cells remaining in the foci next to the T cell areas proliferate and differentiate into plasma cell-secreting antibody of the IgM class. These foci are short lived and most disappear by the time germinal centers appear. The antibody produced at this stage is essential not only for the initial immune response to the pathogen but also for the germinal center reaction and affinity maturation. It allows fixation of antigen in the form of antigen–antibody complexes on the follicular dendritic cells and also removes antigen from the circulation, thereby restricting the available antigen to that localized within the germinal center.

The activated B cells which migrate to the follicles initiate the germinal center reaction, which is first detectable 5 or 6 days after primary immunization. Germinal centers have a characteristic architecture which corresponds to their functional compartmentalization (Fig. 7). The dark zone, light zone, and the follicular mantle can be distinguished morphologically, and in humans the B cell subpopulations in each zone may be discriminated on the bases of their unique combination of surface makers. The follicle is organized around the follicular dendritic cells (FDCs), which are specialized cells with a delicate network of processes that are lined with antigen–antibody complexes. Their role is central to the function of the germinal center since they act as a reservoir of native antigen and are essential for selection during the affinity maturation process. They are localized in the basal light zone, which forms the interface between the dark zone, in which very rapidly proliferating B cells are undergoing somatic hypermutation, and the light zone, in which selection of high-affinity B cells and apoptosis occur.

Germinal center cells are oligoclonal and prone to cell death. This phenotype correlates with high expression of the apoptosis-inducing proteins Fas, p53, c-Myc, and Bax and the absence of apoptosis-preventing proteins, such as Bcl-2. Several signals are required to prevent apoptosis of GC B cells. First, after proliferation and hypermutation, the B cell is thought to interact with the FDC to retrieve antigen. Only antigen-specific B cells with receptors of the same or higher affinity than the that of antibodies on the FDC will be able to "snatch" antigen and present it via MHC class II molecules. Next, B cells must interact with specific CD40 ligand-expressing T cells. The specific interaction with T cells and the cytokines provided by it will switch off the apoptotic pathway. This involves reexpression of Bcl-2 and depending on the type of cytokines involved and allows the positively selected high-affinity B cells to leave the GC and to differentiate into plasma cells or recirculating memory cells.

PERIPHERAL LYMPHOID AREAS

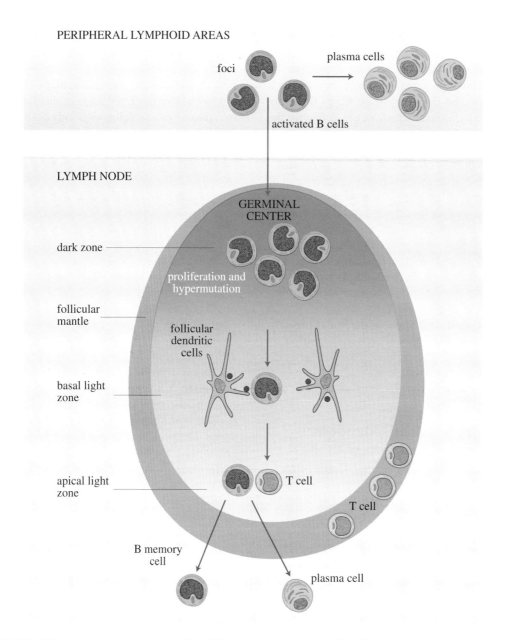

FIGURE 7 Schematic representation of the differentiation pathways of B cells in germinal centers during a T cell-dependent immune response. After activation in the peripheral lymphoid areas, B cells from the foci migrate to the B cell follicles. Typically, between 4 and 10 cells initiate a germinal center in human lymph nodes. Only cells with sufficient initial affinity for the antigen will enter the dark zone and undergo fast proliferation and somatic hypermutation. The majority of these cells die by apoptosis, whereas the few cells able to interact with antigen on the follicular dendritic cells receive a survival signal and migrate through to the apical light zone where the majority of the germinal center T cells are localized. The selected high-affinity cells will be able to present antigen and interact with the T cells to confirm the survival signal, undergo class switch, and migrate out of the follicles to become terminally differentiated plasma cells secreting high-affinity, isotype-switched antibodies. Alternatively, they will differentiate into antigen-specific, resting, recirculating memory cells with high-affinity receptors.

The germinal center acts as a regulated organ of selection by restricting the availability of antigen and setting strict requirements of specificity and affinity on the B cells which are allowed into the memory population. The fierce competition for antigen and thus survival results in a very efficient and fast mechanism of focusing the immunoglobulin repertoire on a particular pathogen by creating an expanded secondary repertoire of high-affinity antibodies. It could be argued that the principal evolutionary pressure on this stringent selection process is more the requirement for the immune system to ensure fast clearance of the pathogen and prevention of its spread before the organism is killed than the need to delete autoreactive mutants.

Somatic Hypermutation

General Features

The mutations that accumulate in the V region of immunoglobulin genes have been studied during the development of immune responses to several antigens, in particular to very simple molecules (haptens) coupled to carrier proteins. The defining features of the hypermutation process have been established from the analysis of monoclonal antibodies which are generated at different stages of the immune response. A variable number of single-point mutations (from 1 to 24 or more) are usually seen along the rearranged V_H and/or V_L. The number of mutations in individual genes can vary, but the overall mutation rate has been consistently calculated to be approximately 1 in 1000 mutations per base pair per cell division. This frequency is significantly higher that the mutation frequencies observed in somatic tissues or in B cell lines in culture. The mutations occur preferentially on the rearranged allele but are also present in nonfunctional rearrangements. Most of the mutations in functional genes are found in the CDRs, although they also appear in framework regions and in noncoding areas. The nature and the distribution of the mutations observed are greatly dependent on antigen selection. A single point mutation (characteristically in the CDR) can be responsible for a 100-fold increase in the affinity of an antibody and, due to selection and expansion, will be found very frequently after the initial stage of the immune response. These selected hot spots will appear frequently even if the initial mutation was a rare one.

The first evidence of hypermutation can be detected approximately 4 or 5 days following immunization. The number of cells undergoing hypermutation appears to peak at approximately 14–24 days. This approximately correlates with the time window of the germinal center reaction. Mutations accumulate in a stepwise fashion, and it is possible to trace the genealogy of individual clones within single germinal centers during a specific immune response. The

overall frequency of mutations observed increases with time and successive immunizations, as does the average affinity of the secreted antibodies. This process is known as affinity maturation.

The analysis of hypermutation has advanced greatly as a result of the cloning and sequencing of the majority of the germline V segments in humans and mice. The technological aid provided by the polymerase chain reaction and the use of transgenic mice carrying immunoglobulin minigenes have made it possible to analyze mutations in the absence of the bias imposed by antigen selection and to accumulate large databases of mutations. This has facilitated the characterization of some of the intrinsic properties of the mechanism and the dissection of the DNA elements that control the recruitment and targeting of hypermutation to the V segment. The cellular signals that control the activation and regulation of hypermutation remain unknown, as does the molecular machinery responsible for introducing the mutations.

The Target of Hypermutation

Mutations are targeted to the rearranged V segment and surrounding area and are only very rarely found within the C region. Mutations reported in the mouse lambda C region have been attributed to the shorter distance between the J and C segments in this locus compared with the kappa locus. The bulk of the mutations are found in a domain which starts downstream from the promoter and extends downstream from the J segment into the noncoding J–C intron (Fig. 8). Mutations are found approximately 2 kb from the promoter, but it appears that their frequency declines toward the 3′ end. The 5′ border of hypermutation has been finely mapped in mouse kappa light chains and is revealed by a sharp increase in the frequency of mutation at approxi-

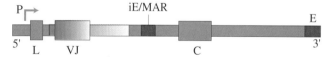

FIGURE 8 The mutation domain in immunoglobulin kappa chains. The main target area of the hypermutation mechanism includes the VJ region and part of the J–C intron. A sharp 5′ border is located in the middle of the leader intron. The color gradient (from red to pink) indicates the declining mutation frequency seen toward the 3′ end of the mutation domain. The extent and precise details of this decline have not been determined. The absence of mutations in the C region of kappa chains suggests a possible role for the MAR and intron enhancer regulatory elements as boundaries for hypermutation since they are located between the mutation domain and the unmutated C region. At the 5′ end, the promoter region and the proteins associated with the transcription initiation complex may contribute to define the border.

mately 200 bp from the initiation of the transcription site. The location of this border is independent of the primary sequence and is only related to the distance from the transcription initiation site. Occasional mutations are observed upstream of the border and the promoter, but their frequency is significantly lower than the frequency observed within the hypermutation domain. Thus, for all practical purposes, the transcriptional promoter defines the 5′ border of the mutational domain. The 3′ border is less well defined and although it is possible that other regulatory elements present in the J–C intron such as the matrix attachment regions (MARs) or the enhancers may have a role in defining the boundaries of the mutation domain, it is also possible that the distance from the initiation of transcription may ultimately determine the extent of the hypermutation domain. The precise details of the mutation frequency and distribution of mutations within the domain might eventually reveal interesting features about the mechanism. For instance, a slow decline in the mutation frequency toward the 3′ end might suggest a mechanism in which an enzyme with limited procesivity (as is the case with some polymerases) is loaded at the 5′ end.

The target domain has no particular features which are required to attract the hypermutation machinery. Replacement of the V segment in transgenic constructs has no significant effect on the frequency of hypermutation targeting nor on the location or size of the targeted domain. Several heterologous DNA, including bacterial and human genes, have been successfully placed in lieu of the murine physiological target, and even a transgenic construct with a complete replacement of the mutation domain (encompassing the promoter and immediately adjacent 5′ region, the leader, the VJ segment, and part of the J–C intron) is an efficient target for hypermutation.

Distribution of Mutations, Hot Spots, and Clusters

As mentioned previously, some individual nucleotide positions appear to attract an increased number of mutations. These mutational hot spots tend to be located within the CDR regions and reflect the antigenic selection for improved antibody affinity. Significantly, a few hot spots have been observed in parts of the antibody which do not affect antigen binding or which result in silent changes. Analysis of "passenger" rearranged V segments, either in disabled transgenic constructs or in endogenous nonproductive rearrangements and nonfunctional Ig chains, have demonstrated that hot spots are intrinsic features of the hypermutation machinery. Not only are deletions and insertions extremely rare but also some nucleotide positions are consistently preferred targets for mutation. Even heterologous DNA sequences bear the same bias for hot spots. Large databases of mutations are required to provide statistically significant information on hot spots. Analysis of many mu-

tations which have accumulated in the same V gene in response to many different antigens (as is the case for cells isolated from the germinal centers of Peyer's patches in the gut), or mutations accumulated in passenger transgenes, reveals that the majority of the hot spots occur within a sequence consensus. The primary sequence, RGYW [where R stands for purine (A or G), Y stands for pyrimidine (C or T), and W stands for A or T] is found in a large proportion of the hot spots observed to date. Significantly, the same consensus is applicable to hot spots in human, mouse, or sheep V genes as well as to heterologous DNA sequences targeted by the hypermutation machinery. The sequence TA(G/C/T) also appears to be a preferential target for mutation. However, not all mutations or even hot spots lie within the consensus and not all potential consensus sequences are actually targeted by the somatic hypermutation machinery. Some other features in addition to the primary sequence context must be involved in defining which position becomes a hot spot. The importance of the consensus primary sequence is clearly demonstrated by the main hot spot in the VκOx1 gene (the second base of the Ser31 A**GT**; Fig. 9). The AGT serine codon at amino acid position 31 of the VκOx1 light chain was replaced by a silent change (a TCA serine codon) in a transgenic mouse Ig construct. This change destroyed the consensus primary sequence at that position without any changes in the protein encoded by the transgenic light chain. As a result, the Ser31 position not only ceased to be the preferred target for hypermutation but also became a cold spot (Goyenechea and Milstein, 1996). Significantly, this silent change also resulted in the emergence of new hot spots, suggesting the possibility that long-range sequence interactions could be important in defining the position of mutational hot spots.

A slight bias for clustered substitutions has been observed which might reflect evolutionary pressures to remove hot spot targets from the FW regions that form the structural scaffold of the antibody and to concentrate potential mutations in the antigen binding site which is principally encoded by the CDR regions. Thus, the ratio of possible replacement/silent mutations is higher in the CDR regions than in the FW. In addition, analysis of the distribution of the two types of serine codon usage [AGC or AGT versus TC(A/C/T/G)] in human immunoglobulin V genes revealed a significant bias in favor of AG serine codons (which can be part of the consensus RGYW) within the CDRs, whereas the FWs were biased in favor of the TC serines. A similar analysis of human T cell receptor genes did not reveal a bias for AG serines in the antigen binding site encoding regions (Wagner et al., 1995).

Although selection pressures during evolution might have favored the concentration of hot spots within the CDR regions (CDR1 most prominently in kappa chains and all three CDRs in lambda chains), a high frequency of muta-

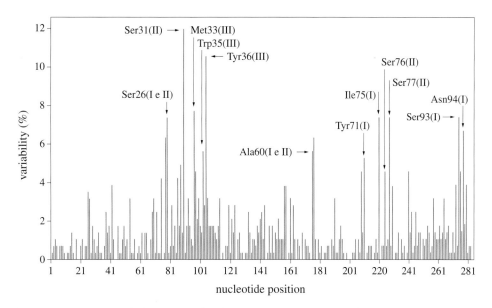

FIGURE 9 Hot spots of somatic hypermutation in a mouse kappa light chain V gene. A database of 284 mutated VκOx1 sequences obtained from Peyer's patches germinal center cells, spleen, and memory B cells in transgenic and nontransgenic mice is shown. The y axis represents the proportion of sequences with a mutation at that position. The x axis represents each nucleotide position along the V segment. There are 1136 mutations in the database. The main hot spots are highlighted. The area with the greatest density of hot spots, including the main Ser31 hot spot, coincides with the CDR1 region.

tions and hot spots are also found in noncoding regions such as the J–C intron and heterologous DNA targets. This suggests that the hot spot bias is a preexisting intrinsic feature of the hypermutation machinery.

An additional bias is observed in the nucleotide substitution preferences of the hypermutation machinery. Transitions are more frequently observed than would be predicted by a random distribution. Individual bases are also nonrandomly targeted in that A residues on the coding strand are more likely to be mutated than T and G residues more than C. Given the base-pairing nature of DNA, this asymmetry implies that the mutation machinery has the ability to discriminate between the coding and the noncoding strand; a feature which is shared with some forms of DNA repair.

DNA Sequences and Regulatory Elements Involved in Recruiting and Targeting Hypermutation

The hypermutation machinery can target transgenic immunoglobulin constructs integrated in the genome outside the immunoglobulin loci. Although the integration site may influence the observed mutation frequency perhaps by altering the accessibility of the transgene to the hypermutation machinery, it is possible to identify the *cis*-acting elements which are required to recruit hypermutation by deletion analysis of modified immunoglobulin transgenes. This has been accomplished in the kappa light chains since

the regulation of the kappa locus is probably less complex than that of the heavy chain locus. It thus appears that the control elements which are required to promote efficient transcription are also involved in the recruitment of hypermutation (Fig. 10). Transgenic kappa chains with deleted MAR, intron enhancer, or 3′ enhancer, all of which are involved in regulation of transcription of kappa chains, have impaired ability to recruit hypermutation. All deletions that affect the mutation frequency also reduce transgene expression. However, although germinal center cells which

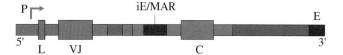

FIGURE 10 *Cis*-acting DNA sequences required to recruit somatic hypermutation to the mutation domain in kappa chains. An active promoter is required (red arrow), as is the presence of the MAR/iE (matrix attachment region and intron enhancer) and 3′E (3′ enhancer) shown in red. The mutation target domain is shown in green. The requirements of the areas in gray, including the C region and part of the J–C intron, have not been tested. The target domain can be replaced by heterologous sequences. The position of the enhancers can vary with respect to the promoter and still efficiently target hypermutation to the transcribed domain immediately downstream of the promoter. Areas in blue are not required.

express low levels of the mutant transgene have mostly unmutated transgenes, cells which achieve higher levels of expression are also capable of accumulating mutations in the transgene. Interestingly, replacement of the Ig kappa promoter by a different active promoter has little or no impact on the recruitment or the frequency of hypermutation. Thus, the initiation of transcription and the targeting of hypermutation appear to be linked. Furthermore, the placement of an immunoglobulin promoter immediately upstream of the C region in a transgene construct targets the mutation to the C region, directly linking hypermutation to transcription initiation. The relative position and orientation of the MAR and the enhancers, as is the case for enhancement of transcription, have no obvious effect on the recruitment of hypermutation.

The regulation of hypermutation in heavy chains may require elements which have not been identified. Although rearranged IgH transgenes can to a certain extent accumulate mutations, full hypermutation is only achieved after *trans*-switching into the endogenous IgH locus. Transgenic mice containing partial human IgH loci (transloci) have confirmed the idea that sequences in addition to the μ enhancer are required to target and recruit hypermutation in heavy chains. New technology which enables the manipulation of large DNA fragments is currently available and should eventually help to ascertain which elements control hypermutation in the heavy chain locus. It is likely that the elements regulating transcription in the IgH and Igλ loci will also control hypermutation, as appears to be the case in the kappa locus.

Although transcription is clearly required for the recruitment of hypermutation, a different element of regulation is likely to be involved in addition to the differentiation state of the B cell since mutations do not accumulate in all well-expressed transgenes, even in cells in which the endogenous loci are being effectively targeted.

Recently, the methylation status of the immunoglobulin locus has been implicated as an additional control in the accessibility of the locus to transcription. Active demethylation of the kappa locus precedes transcription and rearrangement. Interestingly, there is a correlation between expression, demethylation, and mutation frequency in kappa transgenes. Whereas highly mutated transgenes are not only well expressed but also fully demethylated, poorly expressed and poorly mutated transgenes tend to remain completely or partially methylated. Methylation offers a simple way of modifying individual genes in a clonally stable manner and can also affect the regulation of transcription; however, its role in controlling hypermutation needs further investigation.

Molecular Mechanism of Somatic Hypermutation

The original model described by Brenner and Milstein (1966) to explain the molecular basis of somatic hypermutation invoked a form of error-prone DNA repair. The link between hypermutation and transcription has reinforced the idea that a form of error-prone DNA repair analogous to transcription-coupled repair could be involved in hypermutation. Several features of the hypermutation mechanism are similar to features observed in transcription-coupled repair. Both are linked to transcription and both appear to be able to discriminate between the two strands. A 5′ boundary downstream of the promoter has been described in both — a boundary between fast and slow repair in the case of transcription-coupled repair and a boundary between low and high frequency of mutations in the case of hypermutation. Nonetheless, no defect in somatic hypermutation has been detected in most of the complementation groups of the nucleotide excision repair pathway in humans. Thus, although some of the factors involved in hypermutation might be similar or phylogenetically related to the DNA repair proteins, it is likely that somatic hypermutation is effected by a specific and distinct machinery.

Recently, it was proposed that a hypermutation factor, which would be expressed in B cells at a particular stage in their differentiation, would be physically linked to the transcription complex including RNA polymerase II. The hypermutation factor would promote stalling of the RNA pol II complex during elongation of the immunoglobulin transcript and thus promote spurious repair of the transcribed strand, which would introduce misincorporations (Peters and Storb, 1996). This model would not necessarily require actual damage to the DNA to trigger repair; however, it still requires that the misincorporated nucleotides are not corrected by the mismatch repair machinery before DNA replication. Also, a model (Fig. 11) including a hypermutating priming factor (HPF) which is recruited to the transcription complex in a clonally stable manner and which would induce local error-prone DNA repair would account for the observed incorporation of multiple mutations during clonal expansion within the germinal center.

In any case, the model must account for the observed features of hypermutation, including the hot spots, the observed high frequency of mutation, substitution bias, and strand discrimination. The link to transcription and clonality and the decline in the observed frequency of mutations

FIGURE 11 A model linking somatic hypermutation and clonality. (described in Goyenechea *et al.*, 1997). Recruitment of a hypermutation priming factor (HPF) is favored by the correct assembly of all the transcription regulatory elements but retains a certain "unpredictability" (thin arrow). Similarly, the HPF may occasionally be assembled in the absence of some of the regulatory elements as long as a transcription initiation complex is present. The essential feature of this model is the ability to maintain clonality not only within the cell but also at the level of individual genes.

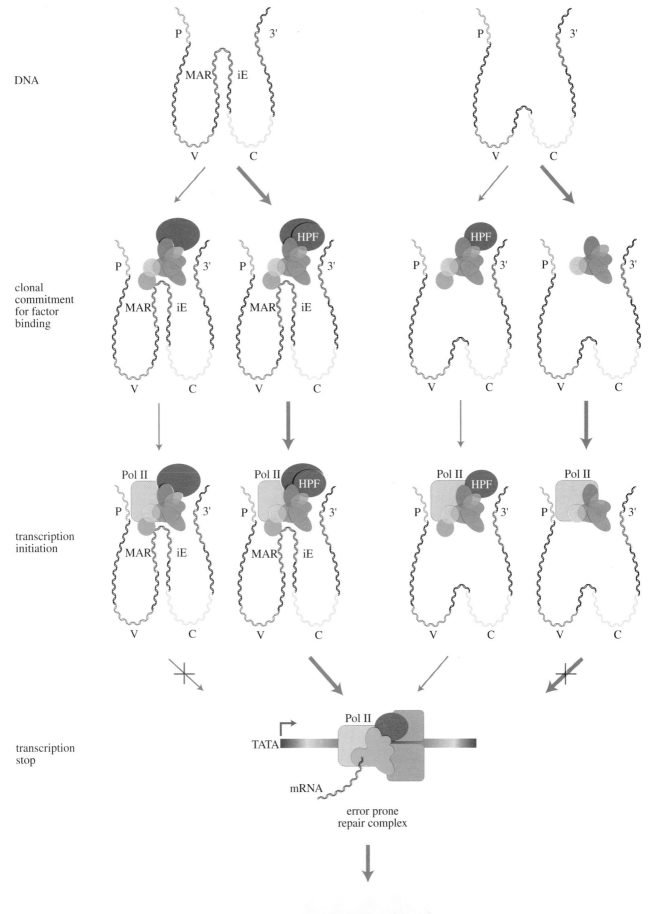

DNA

clonal
commitment
for factor
binding

transcription
initiation

transcription
stop

error prone
repair complex

HYPERMUTATION

with the distance from the initiation of transcription (well before the end of the transcription unit), together with the rare mutations upstream of the transcribed region, are all features which should provide clues regarding the nature of the molecular basis of hypermutation. By analogy with the mechanism of transcription, several steps could be involved in this process. First, a hypermutation factor must be loaded onto the transcription initiation complex at the promoter (a step which would require the interaction of the promoter and enhancers). This factor would be stable once bound to the RNA pol II complex and could be responsible for the introduction of nicks or damage onto the DNA. A second factor would be recruited which would be involved in the error-prone repair of the initial damage. The mismatches introduced during this error-prone repair would have to be fixed either by local inhibition of mismatch repair or by further employing an error-prone mismatch repair mechanism. It is likely that the whole process would be carried out by a complex association of proteins in analogous fashion to the "spliceosome," "rybosome," "transcriptiosome," or "repairosome."

Although the models discussed here (and some others which have not been discussed) are consistent with the available experimental evidence, they constitute nothing more than useful working hypotheses. Biochemical analysis of the somatic hypermutation mechanism is still awaiting the characterization of an appropriate *in vitro* experimental system or a genetic defect in which hypermutation is abolished or significantly altered. Recently, defects in a protein involved in the normal DNA mismatch repair, the MSH2 protein (the analog of the bacterial mutS protein which is involved in the initial recognition of base-pair mismatches), has been shown to affect the mutation pattern of hypermutation, raising the possibility that some of the pathways involved in DNA repair are directly or indirectly involved in somatic hypermutation. The unraveling of the molecular mechanisms involved in the generation of diversity has proven very fruitful in the understanding of some of the molecular events involved in cancer and disease as well as basic genetic mechanisms such as recombination. A few more surprises are to be expected.

References

BRENNER, S., and MILSTEIN, C. (1966). Origin of antibody variation. *Nature* **211,** 242–246.

GOYENECHEA, B., and MILSTEIN, C. (1996). Modifying the sequence of an immunoglobulin V-gene alters the resulting pattern of hypermutation. *Proc. Natl. Acad. Sci. USA* **93,** 13979–13984.

GOYENECHEA, B., *et al.* (1997). Cells strongly expressing Igk transgenes show clonal recruitment of hypermutation: A role for both MAR and the enhancers. *EMBO J.* **16,** 3987–3994.

PETERS, A., and STORB, U. (1996). Somatic hypermutation of immunoglobulin genes is linked to transcription initiation. *Immunity* **4,** 57–65.

TOMLINSON, I. M., *et al.* (1996). The imprint of somatic hypermutation on the repertoire of human germline V genes. *J. Mol. Biol.* **256,** 813–817.

WAGNER, S. D., *et al.* (1995). Codon bias targets mutation. *Nature* **376,** 732.

General References

The anatomy of antigen specific immune responses (1997). *Immunol. Rev.* **156.**

COOK, G. P., and TOMLINSON, I. M. (1995). The human immunoglobulin VH repertoire. *Immunol. Today* **16,** 237–242.

GELLERT, M. (1992). Molecular analysis of V(D)J recombination. *Annu. Rev. Immunol.* **22,** 425–446.

HONJO, T., and ALT, F. W. (1995). *Immunoglobulin Genes,* 2nd ed. Academic Press, London.

JANEWAY, C. A., and TRAVERS, P. (Eds.) (1997). *Immunobiology: The Immune System in Health and Disease,* 3rd ed. Current Biology/Garland, London/New York.

MILSTEIN, C. (1990). Antibodies: A paradigm for the biology of molecular recognition: The Croonian Lecture, 1989. *Proc. R. Soc. London* **239,** 1–16.

MILSTEIN, C., and NEUBERGER, M. S. (1996). Maturation of the immune-response. *Adv. Prot. Chem.* **49,** 451–485.

PAUL, W. E. (Eds.) (1993). *Fundamental Immunology,* 3rd ed. Raven Press, New York.

SILVERSTEIN, A. M. (1989). *History of Immunology.* Academic Press, London.

Somatic hypermutation of immunoglobulin genes (1998). *Immunol. Rev.* **162.**

WEIL, J. C. (Ed.) (1996). *Sem. Immunol.* **8.**

JONATHAN SPRENT
DAVID TOUGH

Department of Immunology
The Scripps Research Institute
La Jolla, California 92037, USA

Primary immune responses to infectious agents cause marked clonal expansion and differentiation of antigen specific T and B cells into effector cells. Many of the responding lymphocytes are eliminated at the end of the response, but some cells survive and differentiate into long lived memory cells. Immunological memory leads to a more efficient secondary response to the pathogen concerned and reflects both affinity maturation and an increase in precursor frequency. The factors controlling the production and survival of memory cells are discussed.

Immunological Memory

Introduction

The fact that prior exposure to an infectious agent often leads to lifelong protection against subsequent infection has been known for centuries. Indeed, this observation is probably largely responsible for the development of modern immunology. It is well accepted that resistance to secondary infection reflects a long-lasting state of immunological memory carried by T and B lymphocytes. This article provides an overview of the properties of memory cells and the factors controlling the generation and survival of these cells.

Memory responses to antigen typically occur more rapidly than primary responses and are more intense, thus causing rapid elimination of the pathogen of concern (Cerottini and MacDonald, 1989; Gray, 1993; MacKay, 1993;

Sprent, 1994; Zinkernagel *et al.,* 1996). The accelerated tempo of memory responses is largely attributable to an increase in the precursor frequency of antigen-reactive lymphocytes. In addition, memory cells show phenotypic differences from naive cells, which heightens their sensitivity to antigen. To understand the properties and functions of memory cells, it is first important to consider the features of naive lymphocytes and the fate of cells participating in the primary response.

The Primary Immune Response and the Fate of Effector Cells

In young animals the majority of T and B cells are immunologically naive (Sprent and Tough, 1994). These cells are generated in the primary lymphoid organs — the thymus for T cells and the bone marrow for B cells — and migrate to the secondary lymphoid organs, i.e., the spleen, lymph nodes (LNs), and Peyer's patches. T and B cells in these sites are not sessile but form a pool of recirculating lymphocytes that migrate continuously from one organ to another via blood and lymph (Sprent, 1993b).

Naive T and B lymphocytes are small resting cells and display a characteristic surface phenotype (Swain *et al.*, 1991; Sprent and Tough, 1994). Typical naive T cells express low to intermediate levels of CD44 (CD44^{lo}) and high levels of the LN homing receptor CD62L (L-selectin) (CD62L^{hi}). Naive T cells also express high levels of the A, B, and C isoforms of the CD45 molecule (CD45RA/B/C^{hi}). The majority of naive T cells express high levels of the $\alpha\beta$ T cell receptor (TCR) and comprise a 1:2 mixture of CD4$^-$8$^+$ and CD4$^+$8$^-$ cells; these two subsets of $\alpha\beta$ T cells recognize antigen (peptides) bound to major histocompatibility complex (MHC) class I and class II molecules, respectively. The features of the minor subset of $\gamma\delta$ T cells expressing $\gamma\delta$ TCRs will not be discussed since the antigen specificity of these cells is unclear. At the B cell level, typical naive B cells express high levels of both IgM and IgD and, like T cells, are CD44^{lo} and CD62L^{hi}.

The life span of naive T and B cells has long been controversial (Sprent, 1993a; Sprent and Tough, 1994). Many immunologists view naive cells as intrinsically short-lived cells that die within a few weeks unless exposed to specific antigen. However, in animal models it is well established that typical T and B lymphocytes with a naive phenotype exclude DNA precursors such as [^{3}H]thymidine and bromodeoxyuridine (BrdU) for prolonged periods, indicating that the cells have a slow turnover (Fig. 1). In addition, murine T and B cells survive almost indefinitely (>1 year) when transferred to immunoincompetent (SCID) hosts, with many of the cells retaining a naive phenotype (Table 1). In man, chromosome marker studies on patients exposed to irradiation have indicated that naive phenotype T cells can remain in interphase for a period of many years (Michie *et al.*, 1992).

Although aging is associated with a gradual decline in the proportion of naive phenotype cells, these cells do not disappear and are still apparent in advanced old age (Sprent and Tough, 1994). Some of the naive phenotype cells present in old age may represent revertants from memory phenotype cells, but others are probably the direct descendants of cells generated in young life. *De novo* generation of naive lymphocytes is most prominent before the age of puberty, but this process may continue at a limited rate throughout life. Hence, some of the naive phenotype cells formed in old age could be newly formed. Direct evidence, however, is lacking.

Recirculation of lymphocytes is a device to mobilize antigen-reactive cells from throughout the body and cause

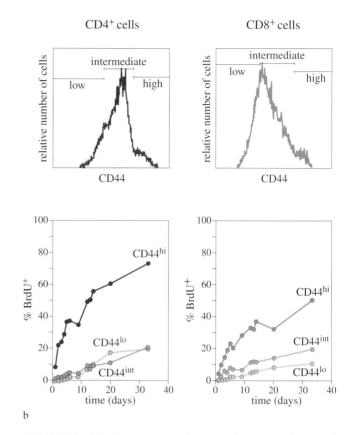

b

FIGURE I Results of an experiment to determine the rate of turnover of T cells in young adult mice. The mice were thymectomized and given BrdU (a precursor which is incorporated into DNA providing a marker of the formation of new cells) in the drinking water (0.8 mg/ml). Cells extracted from the lymph nodes at determined intervals were stained to measure the expression of CD44, a marker that is upregulated to high levels on activated and memory T cells. (a) Relative numbers of CD4$^+$ and CD8$^+$ expressing different levels of CD44, defined on an arbitrary scale. It is apparent that most of the T cells express CD44 at a low or intermediate level; such staining is typical of naive T cells. (b) Percentage of CD44^{lo}, CD44^{int}, and CD44^{hi} T cells that incorporated BrdU during the time shown. It can be seen that the turnover of naive phenotype T cells was much slower than that for memory phenotype cells (CD44^{hi}) (modified from Sprent and Tough, 1994).

these cells to accumulate in sites of antigen concentration, namely, the spleen and the draining LN (Sprent, 1993b). In these sites, antigens are engulfed by professional antigen presenting cells (APCs), such as macrophages and dendritic cells, and displayed on the surface of these cells as peptides bound to MHC class I and II molecules. Dendritic cells are concentrated in the T-dependent areas of the lymphoid tissues and are thus strategically positioned to make contact with recirculating T cells. Confrontation with antigen on APCs causes T cells to be withdrawn from the recirculating lymphocyte pool and become temporarily se-

TABLE I

Number and Type of T and B Cells in the Thoracic Duct Taken from Mice Suffering from Severe Combined Immunodeficiency (SCID) Injected with 10^7 BALB/c LN Cells

MICE EXAMINED	AVERAGE NUMBER OF CELLS COUNTED IN 48 HOURS	SURFACE MARKERS PRESENT AT 24–48 HOURS (%)[1]		
		CD4[2]	CD8	IgM[+]
normal BALB/c	223.4	39 (65)	16	50
C.B.-17SCID not injected	0.1	nr (nr)	nr	nr
C.B.-17SCID injected one month before sample	9.9	65 (nr)	23	7
C.B.-17SCID injected two months before sample	11.8	68 (nr)	17	10
C.B.-17SCID injected ten months before sample	20.2	64 (56)	22	12[3]

1. Measured with FACS (Fluorescence-Activated Cell Sorter).
2. In parentheses, an indication is given of the percentage of CD4 cells that are CDE62L[hi].
3. The tests conducted with the antibody anti-μ of the specific allotype have indicated that 85–95% of μ cells came from the donor.
nr = not found.

questered, thus initiating the immune response. This process also applies to B cells, although sequestration of B cells probably reflects contact with free antigen rather than the processed antigen found on APC.

As discussed in detail elsewhere in this volume, contact with antigen causes antigen-specific T and B cells to proliferate and differentiate into effector cells (Zinkernagel *et al.*, 1996). These latter cells leave the lymphoid organs after several days and percolate throughout the body via the bloodstream. Because of the expression of a new cohort of homing molecules (Picker and Butcher, 1992), effector cells are able to penetrate the walls of capillary blood vessels and thus make direct contact with virtually any cell type expressing the antigen in question, e.g., with epithelial cells in the lung in the case of influenza infection.

The issue of how effector cells destroy infected cells is dealt with elsewhere in this volume (see also Zinkernagel *et al.*, 1996; Kagi *et al.*, 1996). In the context of memory cells and their functions, it is important to note that the mechanisms used to eliminate pathogens vary considerably depending on the pathogen concerned. In the case of extracellular bacteria, it is well established that humoral immunity (i.e., production of antibody by B cells) is of crucial importance. Since antibody production generally requires the T helper function of CD4[+] T cells, the elimination of extracellular bacteria thus requires the combined actions of two different types of effector cells, namely, antigen-specific B cells and CD4[+] T helper cells. These cell types also control responses to certain viruses, especially cytopathic viruses such as vaccinia or influenza viruses in mice. These and other cytopathic viruses destroy the cells in which they replicate. Eliminating these viruses thus requires soluble mediators such as neutralizing antibodies from B cells and antiviral cytokines such as type I interferon (IFN) from macrophages and IFN-γ and tumor necrosis factor from CD4[+] cells. Since cytopathic viruses destroy cells, the activity of cytotoxic T cells (CTLs) (i.e., CD8[+] cells) is unimportant.

In contrast, CTLs play a crucial role in eliminating noncytopathic viruses such as lymphocytic choriomeningitis virus (LCMV) in mice (Zinkernagel *et al.*, 1996; Kagi *et al.*, 1996). These viruses replicate in cells without causing damage and therefore, once inside a cell, are protected from the effects of antibodies and other soluble mediators. Thus, eliminating noncytopathic viruses during the stage of viral replication entails destroying the infected cell, i.e., by the action of CD8[+] (and sometimes CD4[+]) CTLs. These cells kill their target cells via a perforin-dependent mechanism. A similar process may also operate for the destruction of intracellular bacteria such as *Listeria monocytogenes*. Here, the infected cells are largely MHC class II[+] macrophages and related cells. Via surface expansion of class II-associated peptides, the infected cells act as targets for CD4[+] cells and are destroyed by local release of toxic cytokines or via direct CTL activity.

After the elimination of pathogens, effector cells become redundant and most of these cells die rapidly (Sprent and Webb, 1995). The widescale destruction of effector cells at the end of the immune response presumably reflects a pressure to preserve the primary repertoire of naive lymphocytes. Thus, if effector cells were allowed to survive indefinitely the lymphoid system would rapidly become dominated by a restricted repertoire of sensitized T cells; responsiveness to new pathogens would then be impeded. This problem is avoided by the destruction of effector cells at the end of the immune responses.

The mechanisms involved in the elimination of effector cells after the primary response are complex. Some of the cells die locally in the lymphoid tissues via apoptosis and rapidly undergo phagocytosis by macrophages and related cells. Other effector cells are cleared by the liver after entering the bloodstream. In addition, a sizable proportion of T cells die only after mediating their effector functions in distal sites, in particular in Peyer's patches and the lamina propria of the gut. How effector cells die in these various sites is controversial, although "exhaustion," expression of "death" molecules such as Fas, and loss of contact with growth factors such as IL-2 are probably of prime importance (Sprent and Webb, 1995; Zinkernagel et al., 1996).

In general, the extent of effector cell elimination correlates with the intensity of the primary response (Sprent and Webb, 1995). Thus, a strong primary response associated with marked clonal expansion of the responding T cells is followed by extensive elimination of these cells at the end of the response. This process of expansion followed by deletion is most marked for high-affinity T cells and suggests that T cell elimination may reflect a process of exhaustion. As discussed later, T cell exhaustion can be extreme in certain conditions of priming and result in complete T cell elimination and tolerance. Whether exhaustion reflects hyperstimulation of T cells or simply strong upregulation of death molecules such as Fas is unclear.

Widescale death of effector cells also applies to B cells (Sprent, 1994; MacLennan, 1994). Guided by the helper function of CD4$^+$ cells, antigen-specific B cells differentiate rapidly into antibody-forming blast cells and then to fully differentiated plasma cells. Like effector T cells, plasma cells disseminate throughout the body during the primary response and lodge in large numbers in the lamina propria of the gut and also in the bone marrow. Most plasma cells are short lived and die within a few days, although some cells can survive for several weeks. Why plasma cells have a short life span is unclear, although, as for T cells, loss of contact with growth factors and/or expression of death molecules may be important. For both T and B cells, the elimination of effector cells at the end of the primary response is extensive but is generally not complete (Sprent and Webb, 1995). This is obviously essential because if all the responding cells died the end result would be tolerance rather than immunity. In practice, a significant proportion of the responding cells survive for prolonged periods, thus leading to a state of specific memory.

The Generation of Memory Cells

The fact that most effector cells have only a brief life span implies that memory cells have special properties which enable these cells to resist the onset of apoptosis. In considering this problem, the obvious question is whether memory cells are the direct progeny of effector cells or are descended from an early branch point of the initial response. Despite much speculation over many years, this question has yet to be resolved. Nevertheless, it is clear from studies with TCR transgenic mice that prolonged exposure to very high doses of antigen prevents the induction of T memory cells (Zinkernagel et al., 1996). Under these extreme conditions, the responding T cells undergo extensive division but then disappear en masse, thus leading to a state of tolerance (Fig. 2). Such widescale near-complete elimination of T cells at the end of the primary response is also seen in the response to superantigens (Sprent and Webb, 1995).

For superantigens, the T cells surviving the primary response appear to be a subset of low-affinity T cells. It might seem to follow, therefore, that the memory cells generated during typical viral infections are drawn from a precursor pool of low-affinity cells, i.e., from cells that are less susceptible to exhaustion than high-affinity cells. This idea is unlikely, however, because memory cells typically show higher affinity for antigen than their naive precursors (Cerottini and MacDonald, 1989; McHeyzer-Williams and Davis, 1995).

An alternative possibility is that memory cells are derived from precursors that for some reason make only limited contact with antigen during the primary response. For example, during influenza infection the primary response is most intense in the draining LN in the mediastinum. This means that, despite the constant mobility of naive lymphocytes, the cells percolating through distal LN at the time of infection are less likely to make prolonged contact with antigen than the cells initially present in the draining LN. Cell death through exhaustion would then be limited and thus favor survival of the responder cells and differentiation into memory cells. This scenario begs the question of how lymphocytes can respond to antigen without succumbing to the lethal effects of exhaustion and/or committing suicide through expression of death molecules such as Fas. Although this paradox is not fully understood, there is increasing evidence that the upregulation of death molecules following T cell activation can be countered by the expression of antiapoptotic molecules such as Bcl-2 and Bcl-X$_L$ (Cory, 1995; Boise et al., 1995). Hence, it is highly likely that the successful generation of memory cells hinges on timely upregulation of antiapoptotic molecules. The factors controlling the upregulation of survival molecules such

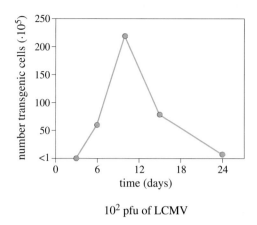
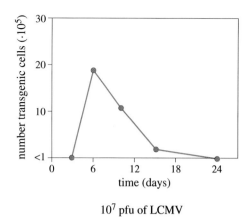

10^2 pfu of LCMV 10^7 pfu of LCMV

FIGURE 2 Expansion and elimination of T cells transgenic for TCR reactive to lymphocytic choriomeningitis virus (LCMV) in hosts infected with either a small or a large dose of live LCMV. A very small dose (10^3) of transgenic T cells is transferred into normal C57BL/6 mice together with either 10^2 (left) or 10^7 (right) plaque-forming units (pfu) of LCMV. At various intervals of time, total numbers of clonotype-positive CD8$^+$ T cells were measured in the spleens of the infected mice. The virus was completely eliminated in the case of 10^2 pfu of LCMV but remained at high levels with 10^7 pfu. The results can be summarized as follows: The expansion of the injected T cells was enormous in both groups but was clearly higher with the lower dose of virus; in both groups the expansion of the injected T cells was followed by rapid disappearance of the proliferating cells; and the successive elimination of the responding T cells was complete with the high dose of virus (suggestive of insurgence of tolerance by exhaustion of the T cells) but incomplete with the low dose (suggestive of the appearance of memory after elimination of the virus) (modified from Moskophidis *et al.*, 1993).

as Bcl-2 and Bcl-X$_L$ are unclear. One possibility is that expression of these molecules is favored by gradual (rather than abrupt) loss of contact with antigen during the later stages of the primary response. As discussed later, this idea could explain why memory is generally associated with affinity maturation, i.e., with preferential survival of high-affinity cells.

Since most T cells are not subject to somatic hypermutation, the simplest explanation for affinity maturation is that high-affinity cells compete more effectively for antigen than low-affinity cells and thus undergo greater clonal expansion during the primary response. As discussed earlier, however, T cells undergoing strong clonal expansion are prone to subsequent strong clonal elimination at the end of the response. How can one explain affinity maturation? The rate at which T cells lose contact with antigen could be the key. In the case of low-affinity cells, one can envisage that the decline in antigen concentration at the end of the primary response causes these cells to rapidly lose contact with antigen, thus leaving insufficient time for the cells to upregulate antiapoptotic molecules. Low-affinity cells then die quickly through apoptosis. The situation with high-affinity cells may be different. As mentioned earlier, the strong initial proliferative response of high-affinity cells leads to exhaustion and eliminates many of these cells just after the peak of the primary response. Some of these high-affinity cells survive, however, and because of their enhanced sensitivity to antigen are able to maintain contact with antigen for prolonged periods during the declining stages of the

primary response and thereafter. Such chronic exposure to low concentrations of antigen may facilitate upregulation of antiapoptotic molecules and thus lead to preferential survival of high-affinity memory T cells at the end of the response (Fig. 3). It should be emphasized that this scenario is largely speculative, and the precise conditions controlling the expression of death and survival molecules on lymphocytes during and after the primary response are still poorly understood.

As for T cells, the relationship of memory B cells to the effector cells generated during the primary response is controversial. Since fully differentiated plasma cells are thought to be end cells, the possibility that memory B cells are derived from plasma cells seems unlikely. As for T cells, memory B cells could be the progeny of cells that evade contact with high concentrations of antigen. However, there is also evidence that memory B cells are the descendants of a subset of B cells that are distinct from the typical B cells participating in the primary response (Klinman, 1994). According to this idea, naive B cells comprise two separate subsets of cells, one controlling primary responses and the other committed to the formation of memory cells. Although this dichotomy of B cells is supported experimentally, the factors controlling the production and differentiation of these two B cell subsets and the relationship of the cells are still largely obscure.

Memory B cells are generated in a unique site in the lymphoid tissues, namely, germinal centers. The differentiation of antigen-stimulated B cells in germinal centers is

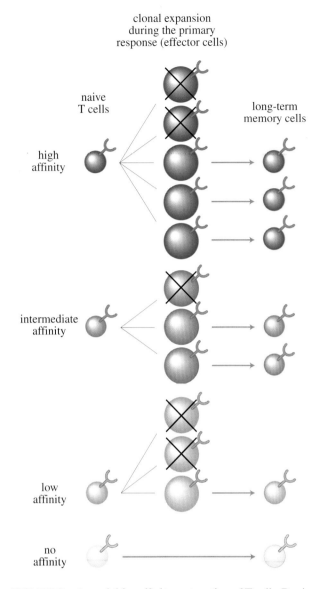

clonal expansion
during the primary
response (effector cells)

naive
T cells

long-term
memory cells

high
affinity

intermediate
affinity

low
affinity

no
affinity

FIGURE 3 A model for affinity maturation of T cells. During the primary response T cells undergo clonal expansion, with the extent of expansion being controlled by the intrinsic TCR affinity encountered on APC. The abundance of antigen during the early stages of the immune response induces both high- and low-affinity T cells to proliferate extensively and differentiate into effector cells. Most effector cells die rapidly through increased expression of death genes such as *fas,* loss of contact with life-sustaining cytokines (IL-2), or via exhaustion (hyperstimulation). Reduction of the levels of antigen, which occurs during the late stages of the response, is followed by competition for antigen, leading to preferential survival of high-affinity T cells and elimination of low-affinity cells. By this process of affinity maturation, the pool of long-term memory T cells is selected toward high-affinity cells (reproduced with permission from Sprent, 1995).

highly complex and is discussed in detail elsewhere (MacLennan, 1994). In brief, contact with antigen in the form of antigen–antibody complexes on the surface of follicular dendritic cells (a unique type of APC found only in germinal centers) drives B cells to proliferate and undergo somatic hypermutation of Ig-combining sites (Fig. 4). This

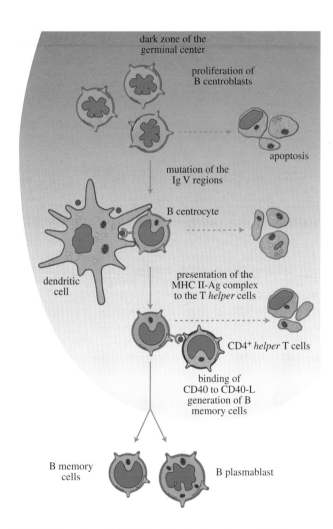

dark zone of the
germinal center

proliferation of
B centroblasts

apoptosis

mutation of the
Ig V regions

B centrocyte

dendritic
cell

presentation of the
MHC II-Ag complex
to the T *helper* cells

CD4[+] *helper* T cells

binding of
CD40 to CD40-L
generation of B
memory cells

B memory
cells

B plasmablast

FIGURE 4 A model for B cell selection in germinal centers of the lymph nodes. B cells proliferate as centroblasts in the dark zone of the germinal center, and mutations are introduced into the variable regions (V) of their immunoglobulin (Ig) genes. These cells give rise to centrocytes, which express the mutated Ig (in blue) on their surface. The centrocytes bind specific antigen (Ag; in red) from the surface of follicular dendritic cells via surface Ig. Processed antigen is then expressed on the cell surface bound to MHC class II molecules. The centrocytes thus present the antigen to CD4[+] T helper cells which express CD40-L. Ligation of CD40 on centrocytes by T cell CD40-L then leads to the generation of memory B cells. Further signals may induce differentiation of these cells to antibody-secreting plasmablasts. Note that failure of the cells to pass from one stage to another will lead to apoptosis. Thus, selection of B cells with high affinity for antigen may occur at earlier stages (modified from MacLennan, 1994).

process favors high-affinity mutants and results in wide-scale local death of low-affinity mutants and cells that fail to mutate. The extensive attrition of cells involved in the germinal center reaction thus leads to preferential survival of high-affinity B cells which then enter the recirculating lymphocyte pool as long-lived memory B cells. As for T cells, affinity maturation of memory B cells may hinge on a process of gradual weaning from contact with antigen, leading to effective upregulation of antiapoptotic molecules.

Surface Phenotype of Memory Cells

Whereas naive lymphocyte are resting cells, memory cells generally display an activated or semiactivated phenotype and thus show a partial resemblance to effector cells (Swain *et al.*, 1991; MacKay, 1993; Sprent, 1994; Zinkernagel *et al.*, 1996).

For T cells, typical effector cells show high expression of CD44 (CD44^{hi}) and downregulation of CD62L (CD62L^{lo}); effector cells also show variable downregulation of CD45RA/B/C and upregulation of CD45RO. Reflecting their activated state, effector T cells are generally substantially larger than naive cells and display a variety of activation markers such as IL-2 receptors. As measured by BrdU incorporation, effector cells also show a high rate of proliferation. For memory cells, the phenotype of these cells depends on the time at which these cells are examined relative to the primary response. When tested within the first few weeks of the response, continuous contact with antigen in waning concentrations maintains many early memory cells in cell cycle and thus in an overtly activated state. Indeed, early memory cells generally display direct effector function, implying that early memory is controlled by the residual effector cells generated during the later stages of the immune response. Within 1 or 2 months of infection, sensitized T cells with overt effector function become less prominent. Contact with specific antigen is greatly reduced and the surviving T cells revert toward a resting phenotype associated with upregulation of CD45RA/B/C isoforms and reexpression of CD62L. Nevertheless, the reversion of activated memory cells to resting cells is generally not complete because the cells tend to remain CD44^{hi} and CD45RO^{hi}. As discussed later, the retention of this semiactivated phenotype may reflect the fact that long-term memory cells continue to make contact with antigen, albeit at low levels. Alternatively, the upregulation of markers such as CD44 may be permanent and unrelated to persistent contact with antigen.

As for T cells, the B cells participating in the primary response are predominantly activated effector cells at the height of the response but then progress toward a resting phenotype when antigen becomes limiting. As discussed earlier, memory B cells occur as high-affinity mutants in germinal centers and then leave these sites to enter the re-circulating lymphocyte pool. Typical memory B cells show a switch in Ig isotype. Thus, whereas naive B cells are IgM$^+$ IgD$^+$, memory B cells express other isotypes, i.e., IgG, IgA, or IgE. Another distinguishing feature of memory B cells is that these cells express only low levels of the heat-stable antigen (HSA). Typical naive B cells, in contrast, are HSA^{hi}. Like T cells, long-term memory B cells show a relatively slow rate of turnover but express a semiactivated phenotype for such markers as CD44.

Specificity and Function of Memory Cells

As discussed previously, T and B memory cells tend to be enriched in high-affinity cells relative to their naive precursors. This process of affinity maturation makes memory cells highly sensitive to low concentrations of antigen, thus enabling memory cells to make potent responses during the early stages of secondary infection. In addition, the precursor frequency of antigen-specific cells is generally substantially higher than that for the primary response. The combined effects of affinity maturation and increased precursor frequency thus make secondary responses much more intense than primary responses and thereby lead to more effective elimination of the pathogen concerned.

The types of effector function generated in secondary responses are usually the same as those generated in the primary response. Nevertheless, the particular subsets of lymphocytes participating in secondary responses can be of crucial importance. As emphasized by others (Zinkernagel *et al.*, 1996), efficient secondary responses are highly dependent on humoral immunity. Likewise, successful vaccines for pathogens almost invariably lead to prolonged production of specific antibody. As discussed earlier, antibody production is especially important for infection with extracellular bacteria and cytopathic viruses.

Since antibody production is T cell dependent, humoral immunity depends on successful priming of both B cells and CD4$^+$ T helper cells. This is an important point because the particular conditions encountered during T cell priming can dictate the type of function the cells subsequently display at the level of memory cells (Seder and Paul, 1994). Several different factors, including antigen density, exposure to cytokines, and the range of costimulatory molecules displayed on APC, can have marked effects on T cell function. The type of cytokines encountered during priming is particularly important. Thus, priming in the presence of IL-12 skews T cells toward a Th1 phenotype. These cells induce preferential production of IgG$_{2a}$ antibody by B cells and synthesize a restricted pattern of cytokines dominated by IFN-γ. These inflammatory cytokines are important for delayed-type hypersensitivity reactions and hence for responses to intracellular bacteria, such as *Listeria* and *Leishmania*. Conversely, priming in the presence of IL-4 induces Th2 function. These cells fail to produce inflammatory cytokines and so are not useful for responses to intracellular bacteria. Moreover, Th2 cells can inhibit the function of Th1 cells and thereby have a suppressive effect. However, Th2 cells are especially effective as T helper cells and induce strong production of IgM and IgG_1 antibodies. These antibodies have efficient neutralizing activity and are thus crucial for responses to extracellular pathogens.

The observation that the functions displayed by memory cells can be determined by the conditions encountered during initial priming has obvious implications for designing appropriate vaccines. This topic is attracting much attention and is likely to dominate the discussion of memory cells in the future.

Factors Influencing the Survival of Memory Cells

In the past, memory cells were viewed as intrinsically long-lived cells which could survive for the lifetime of the host without requiring contact with specific antigen. This notion was challenged by observations that memory cells survived poorly on adoptive transfer unless coinjected with specific antigen (Gray, 1993).

To explain this finding, it was suggested that the survival of memory cells in long-term primed hosts reflected persistent contact with residual antigen. In this respect it is well accepted that antigen can survive for prolonged periods after the primary response in the form of immune complexes on the follicular dendritic cells in germinal centers. These complexes could thus serve as a reservoir of antigen

and induce continuous low-level stimulation of memory cells. However, it has been shown that long-term memory at the T cell level can be induced in mice lacking B cells (Di Rosa and Matzinger, 1996). Hence, the survival of memory T cells does not appear to require the presence of immune complexes. An alternative possibility is that antigen may persist in the form of stable peptide–MHC complexes (Zinkernagel et al., 1996). This idea seems unlikely because it is difficult to envisage that peptide–MHC complexes could remain stable for protracted periods, i.e., for years in man. Another possibility is that the elimination of pathogens at the end of the primary response is incomplete and leads to subclinical infection in sequestered sites (Zinkernagel et al., 1996). This notion is clearly valid for certain viruses such as herpes simplex viruses and perhaps measles virus. However, for other viruses, such as influenza, the elimination of virus seems to be complete.

The notion that the maintenance of memory requires persistent contact with antigen has been challenged by several reports documenting the fact that purified memory cells from LCMV-infected mice can survive for prolonged periods on adoptive transfer in the apparent complete absence of specific antigen (Matzinger, 1994) (Fig. 5). These findings seem to provide prima facie evidence that primed lymphocytes do not need chronic stimulation for their survival. However, the previously mentioned experiments only prove that specific antigen is not required. In this respect, some workers have suggested that memory cell survival could reflect cross-reactive contact with various environmental antigens (Beverley, 1990). The problem with this notion is that unless the diversity of environmental anti-

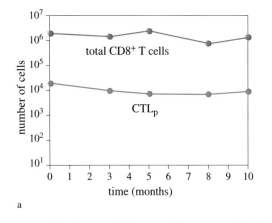
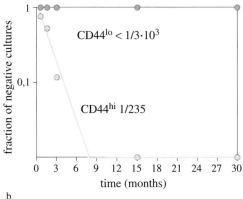

FIGURE 5 Persistence of virus-specific memory $CD8^+$ T cells in the absence of specific antigen. $CD8^+$ T cells were purified from mice immune to LCMV 90 days after infection: 2×10^6 of these cells were injected intravenously into mice which had never been infected with LCMV. The input population contained 10^4 LCMV-specific CTL precursors (CTLp) and was free of any LCMV genome or antigen. At the indicated times post-transfer, total donor $CD8^+$ cells were counted and LCMV-specific CTL precursors (CTLp) were measured. (a) The number of CTLps remained constant for up to 10 months from the transfer. (b) The fraction of LCMV-specific CTLps present in the $CD44^{lo}$ and $CD44^{hi}$ populations is shown. It is evident that 5 months after the transfer the CTLps present are represented only by $CD44^{hi}$ cells (modified from Lau et al., 1994).

gens is extreme, which seems unlikely, memory cell survival would be a hit-or-miss affair and favor T cells with broad cross-reactivity.

An alternative possibility is that memory cells require chronic stimulation, but receive it via cytokines rather than through the TCR. This idea deserves consideration because certain cytokines (i.e., type I IFN and IFN-γ) are able to stimulate T cells *in vivo* in the apparent absence of TCR signaling (Tough *et al.*, 1996). Viruses are powerful producers of IFNs via macrophage activation, and local production of these cytokines probably accounts for the prominent "bystander" proliferation of T cells found during the height of the primary response to viruses. Bystander proliferation of T cells *in vivo* is largely restricted to the CD44^hi subset of CD8^+ cells and occurs when mice are injected with Poly I:C, an inducer of type I IFN, or with purified type I IFN or IFN-γ (Tough *et al.*, 1996). Similar findings apply when mice are injected with gram-negative bacteria (D. Tough and H, Sprent, unpublished data); this effect seems to reflect IFN production elicited by lipopolysaccharide, a component of the bacterial cell wall.

The fact that viruses and bacteria both elicit strong production of IFNs *in vivo* implies that bystander stimulation of T cells may be a common occurrence and operate throughout life. In the case of type I IFN, it is of interest that the CD44^hi CD8^+ cells stimulated after Poly I:C injection do not die. Instead, the cells proliferate for a brief period (several days) and then differentiate into resting cells (Tough *et al.*, 1996) (Fig. 6). These cells survive for prolonged periods and display the same CD44^hi CD62L^hi CD45RB^hi phenotype as long-lived memory cells. IFN production may also promote cell survival during the primary response. Thus, at least for responses to MHC alloantigens in mice, the widescale elimination of T cells at the end of

the primary response is reduced when antigen is coinjected with type I IFN (Poly I:C; Tough *et al.*, 1996). This adjuvant effect of IFN could explain why memory cell generation to viruses is generally highly efficient.

The finding that T cell stimulation via IFNs augments cell survival suggests that intermittent contact with IFNs released during various infections could provide a nonspecific boost for preexisting memory cells. This mechanism could thus play an important role in maintaining the longevity of memory.

Conclusions

Despite evidence that memory cells can survive in the apparent absence of specific antigen in experimental models, it is highly likely that memory cells are normally subject to intermittent stimulation. Such stimulation could reflect TCR signaling (contact with deposits of specific antigen or cross-reactive environmental antigens) or exposure to cytokines or both. In addition to promoting cell survival, maintaining at least a proportion of memory cells in an activated or semiactivated state could be important for ensuring that memory responses can occur as rapidly as possible. Thus, if memory cells reverted fully to resting cells, the generation of secondary responses would be qualitatively similar to primary responses (except for an increase in precursor frequency) and thus require the same sequential process of T cell induction in the lymphoid tissues followed by effector cell migration into the bloodstream. Conversely, maintaining a proportion of memory cells in an activated state would enable the immune system to continuously monitor the portals of antigen entry (e.g., the lungs) for a return of the pathogen concerned. Activated (or semi-

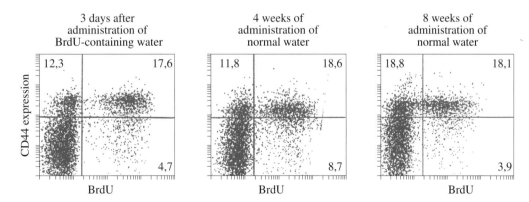

FIGURE 6 Phenotype and longevity of CD8^+ T cells proliferating in response to type I interferon. Adult thymectomized B6 mice were injected with Poly I:C, a potent inducer of type I (α/β) interferon, and given BrdU in their drinking water for 3 days. Subsequently, the mice were given normal drinking water for 4 or 8 weeks. Dot plots obtained after flow cytometric analysis of stained cells show CD44 expression with respect to BrdU labeling of the CD8^+ T cells. The numbers within each quadrant refer to the percentage of stained cells detected (reproduced with permission from Tough *et al.*, 1996).

activated) cells would then be in a position to attack the pathogen peripherally at a very early stage of reinfection. Continuous production of specific antibody would intensify this process.

References Cited

BEVERLY, P. C. L. (1990). Is T cell memory maintained by cross-reactive stimulation? *Immunol. Today* **11,** 203–205.

BOISE, L. H., MINN, A. J., NOEL, P. J., JUNE, C. H., ACCAVITTI, M. A., LINDSTEN, T., and THOMPSON, C. B. (1995). CD28 co-stimulation can promote T cell survival by enhancing the expression of Bcl-xL. *Immunity* **3,** 87–98.

CEROTTINI, J.-C., and MACDONALD, H. R. (1989). The cellular basis of T-cell memory. *Annu. Rev. Immunol.* **7,** 77–89.

GRAY, D. (1993). Immunological memory. *Annu. Rev. Immunol.* **11,** 49–77.

CORY, S. (1995). Regulation of lymphocyte survival by the *blc-2* gene family. *Annu. Rev. Immunol.* **13,** 513–544.

DI ROSA, F., and MATZINGER, P. (1996). Long-lasting CD8 T cell memory in the absence of CD4 T cells or B cells. *J. Exp. Med.* **183,** 2153–2163.

KAGI, D., LEDERMANN, B., BURKI, K., ZINKERNAGEL, R. M., and HENGARTNER, H. (1996). Molecular mechanisms of lymphocyte-mediated cytotoxicity and their role in immunological protection and pathogenesis *in vivo*. *Annu. Rev. Immunol.* **14,** 207–232.

KLINMAN, N. R. (1994). Selection in the expression of functionally distinct B-cell subsets. *Curr. Opin. Immunol.* **6,** 420–424.

LAU, L. L., JAMIESON, B. D., SOMASUNDARAM, R., and AHMED, R. (1994). Cytotoxic T cell memory without antigen. *Nature* **369,** 648–652.

MACKAY, C. R. (1993). Immunological memory. *Annu. Rev. Immunol.* **53,** 217–265.

MACLENNAN, I. C. M. (1994). Germinal centers. *Annu. Rev. Immunol.* **12,** 117–139.

MATZINGER, P. (1994). Memories are made of this? *Nature* **369,** 605–606.

MCHEYZER-WILLIAMS, M. G., and DAVIS, M. M. (1995). Antigen-specific development of primary and memory T cells *in vivo*. *Science* **268,** 106–111.

MICHIE, C. A., MCLEAN, A., ALCOCK, C., and BEVERLEY, P. C. (1992). Lifespan of human lymphocyte subsets defined by CD45 isoforms. *Nature* **360,** 264–265.

MOSKOPHIDIS, D., LECHNER, F., PIRCHER, H., and ZINKERNAGEL, R. (1993). Virus persistence in acutely infected immunocompetent mice by exhaustion of antiviral cytotoxic effector T cells. *Nature* **362,** 758–761.

PICKER, L. J., and BUTCHER, E. C. (1992). Physiological and molecular mechanisms of lymphocyte homing. *Annu. Rev. Immunol.* **10,** 561–591.

SEDER, R. A., and PAUL, W. E. (1994). Acquisition of lymphokine-producing phenotype by CD4$^+$ T cells. *Annu. Rev. Immunol.* **12,** 635–673.

SPRENT, J. (1993a). Lifespans of naive, memory and effector lymphocytes. *Curr. Opin. Immunol.* **5,** 433–438.

SPRENT, J. (1993b). T lymphocytes and the thymus. In *Fundamental Immunology* (W. E. Paul, Ed.), pp. 75–110. Raven Press, New York.

SPRENT, J. (1994). T and B memory cells. *Cell* **76,** 315–322.

SPRENT, J. (1995). Immunological memory. *The Immunologist* **3,** 212–215.

SPRENT, J., and TOUGH, D. F. (1994). Lymphocyte life-span and memory. *Science* **265,** 1395–1400.

SPRENT, J., and WEBB, S. R. (1995). Intrathymic and extrathymic clonal deletion of T cells. *Curr. Opin. Immunol.* **7,** 196–205.

SPRENT, J., SCHAEFER, M., HURD, M., SURH, C. D., and RON, Y. (1991). Mature murine B and T cells transferred to SCID mice can survive indefinitely and many maintain a virgin phenotype. *J. Exp. Med.* **174,** 717–728.

SWAIN, S. L., BRADLEY, L. M., CROFT, M., TONKONOGY, S., ATKINS, G., WEINBERG, A. D., DUNCAN, D. D., HEDRICK, S. M., DUTTON, R. W., and HUSTON, G. (1991). Helper T-cell subsets: Phenotype, function and the role of lymphokines in regulating their development. *Immunol. Rev.* **123,** 115–144.

TOUGH, D. F., BORROW, P., and SPRENT, J. (1996). Induction of bystander T cell proliferation by viruses and type I interferon *in vivo*. *Science* **272,** 1947–1950.

ZINKERNAGEL, R. M., BACHMANN, M. F., KUNDIG, T. E., OEHEN, S., PIRCHET, H., and HENGARTNER, H. (1996). On immunological memory. *Annu. Rev. Immunol.* **14,** 333–368.

ANTONIO LANZAVECCHIA

Basel Institute for Immunology
Basel, Switzerland

This essay offers a complete picture of the recognition mechanism of the antigen by the T-lymphocytes, which is distinguished by its extraordinary sensitivity, specificity, and flexibility. It begins with a discussion of the quantitative aspects of the presentation of the antigen in various cells. This is followed by an examination of the T-cell activation mechanism, placing special emphasis on the activation dynamics of TCR and on the role of the accessory molecules. Finally, there is a presentation of a summary of the recognition mechanism of the antigen and the activation of the T-lymphocytes, which illustrates the capacity of the immune system to generate appropriate responses according to the nature of the antigen that is found.

Specificity, Sensitivity, and Flexibility of T Cell Antigen Recognition

The Remarkable Efficiency, Specificity, and Flexibility of T Cell Antigen Recognition

Since the discovery of major histocompatibility complex (MHC) restriction (Zinkernagel and Doherty, 1974), the molecular basis of T cell antigen recognition has been unveiled and is now resolved at the level of a crystal structure (Bjorkman *et al.,* 1987; Garcia *et al.,* 1996). The molecular basis is the interaction of the T cell receptor (TCR) with a MHC molecule carrying a peptide derived from antigen breakdown. This interaction is responsible for the selection of T cells in the thymus and for their activation in the peripheral lymphoid tissues, resulting in the induction of specific immune responses. TCR and peptide–MHC interact at the level of the functional synapse between the T cell and an antigen-presenting cell (APC) where peptide–MHC complexes engage and trigger TCRs resulting in the delivery of activation signals to the T cells.

Sensitivity, specificity, and flexibility are the hallmarks of T cell antigen recognition. T lymphocytes are sensitive since they can detect few peptide–MHC complexes, are specific because they can discriminate between minor details of these complexes, and are flexible in the sense that they can give different types of functional responses to antigen, depending on the type of APC on which the antigen is presented. How can all this be accomplished?

There are three important questions that must be addressed. First, the APC should be able to present enough complexes to trigger T cells. However, how is this accomplished given that APCs have a limited number of MHC molecules which are exposed to a large variety of peptides derived from degradation of self-proteins? Second, T cells should be able to detect few complexes of MHC and antigenic peptide among a vast excess of nonspecific, but similar complexes of the same MHC molecules bound to self-peptides. How can such a high degree of sensitivity be matched with the unique specificity of recognition required? Third, there are different types of foreign antigens and the immune system must be able to mount the appropriate type of response. For instance, it must be able to raise cytotoxic T cells (CTLs) in response to a virus, to raise inflammatory T helper 1 responses to intracellular bacteria, to mount T helper 2 responses to parasites, and, at the same time, be able to avoid inflammatory responses to food antigens and maintain tolerance to those antigens expressed in peripheral tissues outside the thymus. How does the immune system classify the type of antigen in order to mount the most appropriate response? Clearly the immune system must have developed ways to optimize loading of antigenic peptides and efficiently detect the specific complexes as well as means to discriminate between infectious and innocuous antigens.

In this article, I first discuss quantitative aspects of antigen presentation in different types of APCs. Second, I will review the mechanism of T cell activation with particular emphasis on the kinetics of TCR triggering and the role played by accessory and costimulatory molecules. Finally, I will offer a synthetic view of antigen presentation and T lymphocyte activation that may account for the capacity of the immune system to mount appropriate types of effector T cell responses, depending on the context in which the antigen is recognized.

······························●

MHC Molecules Are Disposable Receptors for Peptides

MHC class I and class II molecules sample peptides in distinct intracellular compartments and present the resulting complexes on the cell surface for scrutiny by cytotoxic and helper T lymphocytes, respectively (Germain, 1994). MHC class I molecules bind peptides in the endoplasmic reticulum (ER). These peptides are generated in the cytosol by breakdown of cytosolic proteins and are subsequently translocated to the ER by specific transporters. Once formed, the MHC class I–peptide complexes reach the cell surface, where they are scrutinized by class I-restricted cytotoxic T cells. In this way, the class I pathway provides the immune system with the capacity to survey all proteins which are synthesized within the cell as well as those exogenous proteins that may eventually gain access to the cytosol. In contrast to MHC class I, MHC class II molecules bind peptides in the endocytic compartment. These pep-

tides are derived from the degradation of proteins taken up by fluid phase or by receptor-mediated endocytosis. The process of peptide loading is regulated by the invariant chain (Ii) that targets the newly synthesized MHC class II molecules to the loading compartment. Here, the Ii is removed by proteases, whereas DM and DO molecules catalyze the release of the Ii fragments from the groove of MHC class II molecules and promote binding of high-affinity peptides. These complexes are then transported to the cell surface, where they are scrutinized by class II-restricted T helper cells. In this way, the class II pathway provides the immune system with the capacity to recognize protein antigens which have been selectively captured from the extracellular milieu as well as any other protein which is present in the endocytic compartment.

When binding of antigenic peptides to MHC molecules was recognized as the initial critical step in the immune response, immunologists were faced with the problem of how foreign antigens could effectively compete for binding with an overwhelming excess of self-peptides. It was suggested that, upon encounter with a foreign antigen, self-peptides already bound to MHC molecules would be exchanged for foreign peptides derived from incoming antigens. This mechanism would provide an APC with a high capacity for peptide binding but at the same time would not guarantee the persistence of these complexes because at subsequent time points the reaction would be reversed.

This popular notion of peptide exchange had to be reassessed when it was shown that antigenic peptides are preferentially loaded on newly synthesized rather than preformed molecules (Davidson et al., 1991). In living cells the peptide–MHC class II complexes were found to be very stable with half-lives of approximately 30 h (Lanzavecchia et al., 1992). In addition, the half-life of the complexes was the same as the half-life of the class II molecules to which the peptide was bound, indicating that in living cells peptide binding to MHC class II molecules is essentially irreversible. It was also shown that strong peptide binders can actually increase the life expectancy of MHC molecules (Nelson et al., 1994). Finally, recent work has shown that APCs have specific chaperons (DM and DO molecules) that can edit peptide binding, leading to selection of stable complexes (Kropshofer et al., 1997).

The selection of stable complexes also applies to MHC class I molecules. It has been shown that, in the absence of peptides, empty class I molecules are rapidly removed from the cell surface and degraded (Ljunggren et al., 1990), whereas they can be stabilized by addition of an appropriate binding peptide. Thus, the emerging paradigm is that both MHC class I and class II molecules behave as disposable receptors for peptides and that there is a natural selection for stable complexes in living cells. The concept of complex stability makes biological sense because it allows the APC to accumulate antigenic complexes and retain them for periods of time sufficient to encounter T cells. These findings emphasize the critical need for new synthesis of MHC molecules to sustain peptide loading. As discussed later, the biosynthesis of class II molecules is precisely regulated to match this need.

Specialized Mechanisms of Antigen Capture Determine the Relative Efficiency of Different APCs

Since T cells recognize only peptide–MHC complexes, what ultimately matters is not the concentration of free antigen but rather the capacity of APCs to capture and process antigen and load antigenic peptides onto newly synthesized MHC molecules. Antigen capture and MHC biosynthesis are differentially regulated in different APC types to match their specific needs.

B lymphocytes have low levels of fluid-phase endocytoses but are very efficient at capturing and internalizing antigens that bind to their membrane immunoglobulin (mIg) receptor (Lanzavecchia, 1990). This ensures that, upon encounter with antigen, only the specific B cells will present antigen to specific T cells and, in turn, will be stimulated by the T cells to proliferate and differentiate to APCs. It is interesting that Fc receptors present on B cells do not internalize antigen–antibody complexes, illustrating that in B cells uptake is restricted to those antigens which bind specifically to mIg receptors.

Although for B cells it makes sense to present only the antigen for which they have specific receptors, for dendritic cells (DCs) the need is exactly the opposite. As discussed elsewhere in this encyclopedia, DCs must capture any incoming antigen in peripheral tissue and present it to T cells in secondary lymphoid organs. Therefore, DCs must use a broad strategy for antigen capture. We have shown that immature DCs, such as those that are present in peripheral tissues, have an extraordinarily high and diverse antigen-capturing capacity (Sallusto et al., 1995). At least three distinct mechanisms are used. The first is macropinocytosis, which in DCs is constitutive and allows uptake of large volumes of fluid (as much as the cell's own volume every 2 h) and concentrates the macrosolutes in the MHC class II compartment. The second mechanism is provided by the mannose receptor, a lectin that mediates uptake of mannosylated and fucosylated antigens. Finally, DCs express a form of the Fc receptor that allows efficient internalization and presentation of antigen–antibody complexes. As a consequence, immature DCs can present antigens, taken up in the fluid phase at concentrations of ~10^{-10} M, and are therefore as efficient as antigen-specific B cells that take up antigen via mIg. Presentation of antigen can be boosted even further (~100-fold) in DCs if the antigen is mannosy-

lated or complexed with IgG antibodies, allowing uptake via the mannose or the Fc receptors.

DCs also have the peculiar capacity of shunting antigens from the endocytic compartment to the cytosol. This allows presentation of endocytosed or phagocytosed antigen on MHC class I molecules via the classical cytosolic pathway (Norbury et al., 1997; Albert et al., 1998). This pathway may be particularly important to generate MHC class I-restricted cytotoxic responses to those antigens that are not directly synthesized within DCs, for instance, tumor antigens and antigens from viruses that infect only tissue cells.

Dendritic Cell Maturation Optimizes Presentation of Infectious Antigens on MHC Molecules

A breakthrough in understanding DC physiology was made when it was found that DCs can change their properties when they move from the peripheral tissues to the secondary lymphoid organs (Steinman, 1991). This process, defined as maturation, probably occurs at a low rate constitutively but can be remarkably increased by inflammation. Since DCs are very rare in tissues and sensitive to isolation procedures, it has been difficult to study the details of this maturation process and to identify the stimuli that induce it. Therefore, we developed an in vitro model to study the life cycle of human DCs (Sallusto and Lanzavecchia, 1994) and to identify the signals that modulate their function (Fig. 1). We found that the process of DC maturation can be triggered either by inflammatory cytokines, such as interleukin-1 (IL-1) and tumor necrosis factor-α (TNF-α), or by recognition of patterns characteristic of infectious agents, such as lipopolysaccharide (LPS) or double-stranded RNA (dsRNA). Another powerful stimulus for DC maturation is provided by T helper cells via CD40L–CD40 interaction. All these stimuli lead to a complete switch in the properties of DCs. On the one hand, the antigen-capturing capacity is progressively lost due to downregulation of macropinocytosis, Fc, and mannose receptors. On the other hand, costimulatory molecules are upregulated and production of IL-12 is stimulated, which together lead to a higher T cell stimulatory capacity. In addition, the maturation process results in a change in migratory capacities as exemplified by the switch in chemokine receptor expression which is key to the migration of DCs from inflammatory tissues to the secondary lymphoid organs (Sallusto et al., 1998).

The maturation process also optimizes antigen presentation by promoting the accumulation of peptide–MHC complexes (Cella et al., 1997). As shown in Fig. 2, MHC class II molecules are synthesized at a high rate by immature DCs and are loaded with peptides, and the complexes are trans-

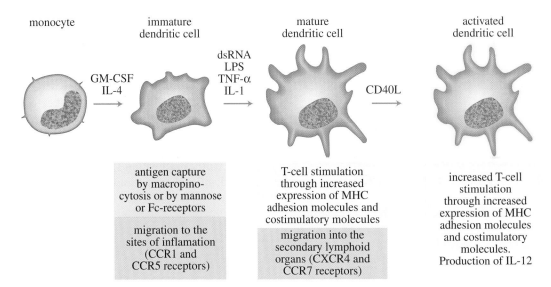

FIGURE I   The origin, differentiation, and maturation of dendritic cells (DCs). Human peripheral blood monocytes represent the immediate precursors of DCs. When cultured with GM-CSF and IL-4, these differentiate without proliferation into typical immature DCs. These cells are characterized by a high capacity for antigen capture through macropinocytosis or by means of receptors for the Fc portion of IgG and for mannose. They have low T cell stimulatory capacity and express CCR1 and CCR5 receptors, which are specific for inflammatory chemokines, that allow them to migrate to inflammatory sites. Inflammatory cytokines (TNF-α and IL-1) and bacterial and viral products (LPS and dsRNA) induce DC maturation. Mature DCs lose antigen-capturing capacity, upregulate adhesion and costimulatory molecules, and express chemokine receptors (CXCR4 and CCR7) that drive their migration to secondary lymphoid organs. CD40L, expressed by T helper lymphocytes, induces the DCs to produce IL-12 and further activates them.

ported to the cell surface. However, in immature DCs the newly synthesized molecules have a short life expectancy since they turnover rapidly with half-lives of approximately 10 h. When DC maturation is induced, a striking change occurs. The rate of MHC class II synthesis is increased about fourfold, whereas the half-life progressively shifts from 10 to more than 100 h. The increased rate of MHC class II synthesis endows maturing DCs with a higher capacity for antigen presentation, whereas the shift to high stability results in accumulation of the newly formed peptide–MHC complexes which persist for several days. It is interesting that approximately 40 h after induction of maturation, MHC class II synthesis is shut off, and DCs completely lose the capacity to present new antigens. Thus, DCs can be considered as disposable APCs in the sense that they can be used only once soon after induction of maturation. This mechanism maximizes the capacity to present those antigens which are encountered in a short window during the maturation process. These will be typically infectious antigens which trigger DC maturation either directly via LPS or dsRNA or indirectly via induction of inflammatory cytokines.

A particular challenge for an APC is the presentation of a cytopathic virus on MHC class I molecules (Fig. 2). In this case the cell should make a difficult compromise because it must synthesize enough viral proteins to present viral antigens but at the same time must resist the cytopathic effect of the virus. In the case of influenza virus, DCs are particularly capable of making this difficult compromise. The immature DCs are very susceptible to influenza virus infection and consequently produce high levels of viral proteins. However, they promptly respond to viral infection by initiating the maturation process. This leads to the production of type I interferon that rapidly builds up a resistance to the cytopathic effect of the virus by inducing expression of protective proteins such as MxA. At the same time, maturation induces a 10-fold upregulation of MHC class I synthesis which results in production of large numbers of complexes loaded with viral peptides. Interestingly, the half-life of MHC class I–peptide complexes is short, approximately 5–10 h in both mature and immature DCs. This relative instability of class I complexes is compensated by a sustained synthesis that allows peptide loading to proceed for long periods of time.

The difference in stability between class I and class II molecules makes sense if we consider the different functions of these two systems of peptide presentation. MHC class II molecules should present antigens which are transiently encountered in the surrounding environment of a DC, so it is important that DCs make a maximum effort to load antigenic peptides over a short period of exposure to the antigen and then retain them as stable complexes. On

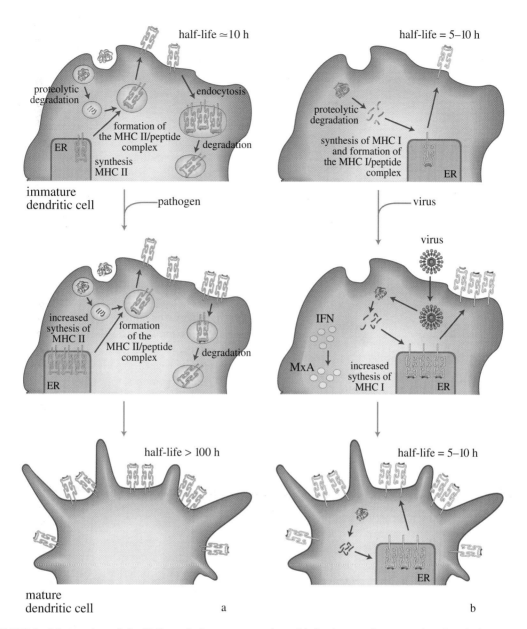

FIGURE 2 Maturation of the DCs optimizes presentation of infectious antigens on class I and class II MHC molecules. (a) Presentation of exogenous antigens on MHC class II molecules. Immature DCs efficiently capture antigens (green) and load the processed peptides onto newly synthesized MHC class II molecules. Because of the high endocytic activity the peptide–MHC class II complexes are rapidly removed from the membrane and degraded (half-lives ~10 h). A pathogen (red) triggers DC maturation either directly (e.g., via LPS) or indirectly via stimulation of TNF-α production. The resulting increased levels of MHC class II biosynthesis allow a more efficient formation of peptide–MHC complexes. Many of these complexes contain peptides derived from infectious antigens. Together with a decrease in the endocytic activity, degradation of MHC class II molecules progressively decreases. Later (after approximately 2 days), MHC class II synthesis is shut off and the mature DCs retain long-lived complexes (half-lives > 100 h) that have been formed during maturation. (b) Presentation of viral antigens on MHC class I molecules. When infected with influenza virus, DCs produce large amounts of viral proteins, upregulate synthesis of MHC class I molecules approximately 10-fold, and produce type I interferon (IFN). IFN induces a resistance to the cytopathic effects of the virus through the production of protective proteins such as MxA. In mature DCs, MHC class I synthesis remains sustained, whereas the half-life of the surface MHC–peptide complexes remains short (5–10 h).

the other hand, MHC class I molecules should present endogenously synthesized antigens; therefore, it is instrumental that these complexes be continuously generated inside the cell to allow expression of all possible viral proteins for the time that the cell remains infected.

The High-Sensitivity, Low-Affinity Paradox of TCR Antigen Recognition

T lymphocytes are equipped with approximately 30,000 TCRs which, with few exceptions, are all identical, displaying the same specificity. They also carry many molecules that contribute to TCR triggering and T cell stimulation. These are the CD4 or CD8 coreceptors which bind to non-polymorphic regions of MHC class II and class I molecules, respectively (Janeway, 1992) and the CD28 molecule which binds to two ligands, B7.1 and B7.2, expressed on professional APCs (Lenschow *et al.*, 1996). This equipment, linked to several distinct signal transduction pathways, allows T lymphocytes to be activated by recognizing few peptide–MHC complexes. It has been estimated that 100 complexes on a professional APC are sufficient to activate a T helper cell (Demotz *et al.*, 1990; Harding and Unanue, 1990), whereas 1 complex may be sufficient to induce killing of a target cell by a cytotoxic T cell (Sykulev *et al.*, 1996).

Given the high sensitivity and specificity of T cell antigen recognition, researchers in the field were surprised when the affinity of the TCR–peptide MHC interaction was found to be low (Matsui *et al.*, 1991; Weber *et al.*, 1992). In particular, the kinetics of the interaction was characterized by very fast off rates on the order of a few seconds. This was perceived as a paradox because it was known that in T lymphocytes signaling needs to be sustained for much longer periods, on the order of hours, to lead to T cell activation (Crabtree, 1989). How is it possible that a low-affinity interaction could sustain signaling in T cells for such a long period of time? One possibility is that multimerization of TCRs and coreceptors would lead to the formation of stable signaling complexes capable of sustaining the signaling process. Indeed, there is evidence for redistribution and enrichment of TCRs as well as other molecules in the area of contact between a T cell and an APC (Monks *et al.*, 1997). These microscopical images, however, represent a snapshot of the triggering process and do not account for the extraordinary dynamic nature of these interactions.

Sustained Signaling by Serial TCR Triggering

Many observations in our laboratory originally suggested that in living cells the TCR–peptide–MHC interaction is a very dynamic process. T cells conjugated with APCs carrying the antigen undergo a sustained signaling that can be measured by increases in intracellular calcium or tyrosine phosphorylation. This signaling, which otherwise lasts for several hours, can be terminated within 1 min by either dissociating T cells from the APCs, by adding antibodies that mask the MHC molecules recognized by the TCR, or by blocking the T cell's actin cytoskeleton (Valitutti *et al.*, 1995a). These findings are incompatible with the notion of stable signaling complexes and rather suggest that sustained signaling requires a continuous engagement of TCRs by peptide–MHC.

A breakthrough in the field was reached when it was found that in T cells that carry two distinct TCRs only those that are specifically triggered are downregulated in a dose- and time-dependent fashion. This observation provided the basis to estimate the kinetics and number of triggered TCRs by simply measuring TCR downregulation (Valitutti *et al.*, 1995b). Using this method it was estimated that ~100 peptide–MHC complexes displayed on the surface of an APC can trigger and downregulate as many as 20,000 TCRs in a few hours. Together with the requirement for continuous TCR engagements, the striking difference between the number of ligands on the APC and the number of TCRs triggered provided the basis for the TCR serial triggering model (Fig. 3; Valitutti and Lanzavecchia, 1997).

According to this model the fast kinetics of interaction allows a single peptide–MHC complex to engage a TCR, trigger it, and subsequently dissociate from it, becoming available for a new round of TCR engagement and triggering. The triggered TCRs provide a quantal amount of signal to the T cells but are rapidly downregulated and degraded in lysosomes. A central feature of this model is that only an optimal kinetics is compatible with efficient TCR triggering. The kinetics must be such to simultaneously allow triggering of engaged receptors and dissociation from the ligand once the TCR has been triggered in order to allow for new engagements to occur. The estimated half-life of the TCR–peptide–MHC interaction is ~10 s, which is consistent with this model. Deviation from this optimal kinetics is expected to result in a weaker stimulus. On the one hand, high-affinity ligands characterized by very slow off rates, although capable of triggering TCRs, will be able to engage only one or very few with time because of their failure to dissociate. Therefore, they will behave as weak agonists. On the other hand, ligands with off rates faster than optimal, although capable of engaging many TCRs, will trigger them only rarely or not at all and will therefore be inefficient at stimulating T cells. In the latter case, CD4 or CD8 coreceptors can increase the frequency of triggering by stabilizing the TCR–peptide–MHC interaction.

The fact that small deviations from an optimal kinetics result in a sharp decrease in the efficiency of TCR triggering explains the exquisite specificity of the T cells account-

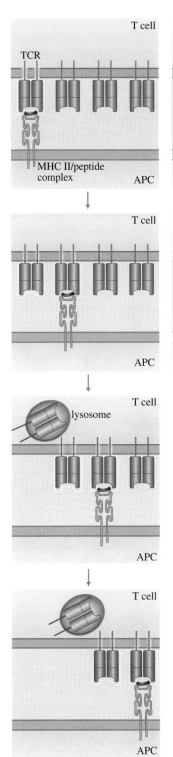

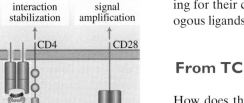

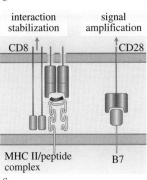

a

ing for their capacity to discriminate between high homologous ligands.

From TCR Triggering to T Cell Activation

How does the extent and kinetics of TCR triggering determine T cell response? We found that T cells appear to "count" the number of triggered TCRs since they are activated to proliferate and produce cytokines only when an appropriate threshold has been reached (Viola and Lanzavecchia, 1996). This threshold is tunable by costimulation and consequently differs according to the nature of the APC. When APCs express high levels of costimulatory molecules, the threshold is low since triggering of 1000 TCRs or less is sufficient for T cell to be activated. In contrast, when the APCs lack costimulatory molecules the threshold is much higher since up to 8000 TCRs need to be triggered to induce a comparable response. Such high levels of TCR occupancy require very high doses of antigen and are therefore unlikely to be reached under physiological conditions. This tunable T cell activation threshold thus confers a great advantage to professional APCs such as DCs that display high levels of B7 molecules and can therefore activate T cells even by displaying a relatively small number of complexes.

A different requirement for priming of naive T cells and elicitation of effector functions has been recognized as a fundamental property of the immune system (Swain *et al.*, 1996). Naive T cells have selective requirements for priming since they are activated only by DCs which express high levels of both antigens and costimulatory molecules. On the contrary, effector T cells can respond to low doses of antigen displayed by various APCs, including those that lack costimulatory molecules.

An example is provided by cytotoxic T lymphocytes. Although priming of naive CD8 T cells requires high doses of antigen, costimulation, or T cell help, lysis of target cells by effector CTLs can be triggered by a single peptide–MHC

FIGURE 3 The sustained signaling necessary for T cell activation is maintained by a dynamic process of serial triggering of the TCR. (a) During the interaction between T cells and APCs every MHC–peptide complex triggers many TCRs (typically 100 complexes trigger up to 20,000 TCRs). The activated TCRs are internalized and degraded in the lysosomes. This process of serial triggering depends on an appropriate kinetics of TCR–ligand interaction and can be regulated by CD4/CD8 coreceptors. (b) Binding of the CD28 molecule expressed on the T cells to the B7 molecule expressed on the APCs increases the signal transduced by activated TCRs, allowing T cells to become committed more rapidly and at lower thresholds.

class I complex on essentially any target cell, even in the absence of costimulation (Valitutti *et al.*, 1996).

What is the significance of the tunable threshold? Do T cells really "count" the number of triggered TCRs or does this number reflect a cumulative amount of signal delivered to the cell? We have shown that rather than absolute numbers of TCRs triggered, what matters is the duration of stimulation (Iezzi *et al.*, 1998). In order to be committed to proliferation, naive T cells require a continuous TCR stimulation for at least 12 h in the presence of costimulation and even for a longer time (up to 30 h) in its absence. In contrast, effector T cells require a much shorter stimulation since they are rapidly committed after only 1 h of stimulation. Strikingly, if stimulation is prolonged, effector T cells undergo activation-induced cell death.

The differential requirements for activation of naive and effector T cells explain the different roles played by professional and nonprofessional APCs in the course of the immune response (Fig. 4). The long time of commitment of naive T cells determines the requirement for APCs that express high levels of adhesion and costimulatory molecules. On the one hand, adhesion molecules allow formation of stable APC–T cell conjugates that are essential to deliver the sustained TCR triggering. On the other hand, costimulation, by lowering the time required for commitment, increases the possibility that this time threshold is reached. In contrast, nonprofessional APCs have low levels of adhesion molecules and therefore engage naive T cells only transiently. In addition, they lack costimulation, which further increases the requirement for longer contact times.

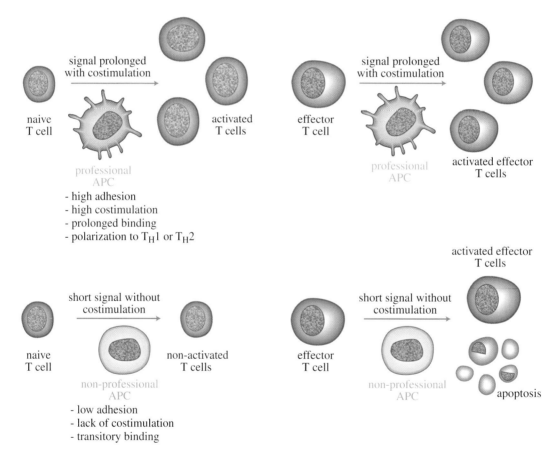

FIGURE 4 Roles of professional and nonprofessional APCs. Naive T cells have strict requirements for activation. They need signaling sustained for at least 10 h in the presence of costimulation and up to 30 h in its absence. These needs are met only by DCs, which express high levels of adhesion and costimulatory molecules. In contrast, nonprofessional APCs can only interact transiently with naive T cells. They may bind and trigger TCRs but, in the absence of costimulation, the signal obtained is not strong enough and not sustained enough to lead to T cell activation. In addition to the requirements for a full T cell commitment, DCs can also deliver signals essential for T cell polarization to type 1 or type 2 effector cells. Effector cells require shorter times for commitment and can consequently be activated by nonprofessional APCs to deliver their effector function. It is possible, however, that continuous stimulation by nonprofessional APCs may lead to activation-induced cell death; instead, stimulation by professional APCs, through costimulatory molecules, sends the T effector cells an antiapoptotic signal.

Therefore, it is unlikely that commitment is reached when antigen is presented by nonprofessional APCs.

Conversely, the short time of commitment of effector T cells decreases the need for adhesion and costimulation, allowing a productive interaction with nonprofessional APCs as well. However, in this case a prolonged stimulation may result in induction of cell death, thus explaining the rapid exhaustion of effector cells by massive antigenic stimulation *in vivo* (Moskophidis *et al.*, 1993).

A Kinetic View of Antigen Presentation and T Cell Activation

In this section, I provide a synthetic view of the whole process of T cell antigen recognition to emphasize some of the key features. The system is geared to allow efficient recognition of pathogens, which is independent from the capacity of APCs to increase synthesis of MHC molecules and express costimulatory molecules. As exemplified in Fig. 5, the availability of antigen in the endosomal or cytosolic compartments is the first limiting factor. As previously discussed, this step is regulated in APCs at the level of antigen capture (for exogenous antigens) or at the level of antigen synthesis (for endogenous viral proteins). In both cases, high amounts of antigens are delivered in the loading compartments. The second critical factor is the availability of newly synthesized MHC molecules to be loaded with antigenic peptides. In DCs, synthesis of MHC molecules is regulated in response to pathogens allowing an increased binding capacity at the same time as the processed antigen becomes available. The stability of MHC–peptide interac-

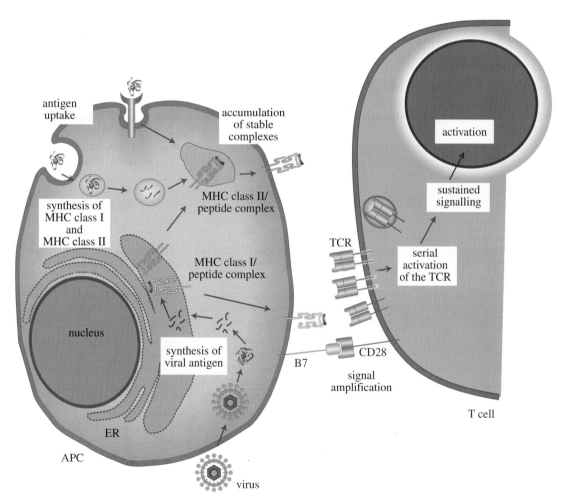

FIGURE 5 Critical steps in antigen recognition by T cells. Uptake of soluble antigen by macropinocytosis or through receptors or synthesis of viral antigens provides antigenic peptides in the appropriate cell compartments. The synthesis of MHC molecules is boosted by infectious stimuli and promotes the accumulation of stable complexes containing antigenic peptides. Serial triggering of TCRs by MHC class I–peptide or MHC class II–peptide complexes leads to a sustained signaling process resulting in T cell activation. The interaction between CD28 and B7, which is induced in APCs by pathogens, amplifies the signal delivered by each triggered TCR. Therefore, B7 represents a recognition code that allows the generation of an effective response to pathogens.

tion allows accumulation of complexes and their retention for a period of time sufficient for the APC to meet specific T cells. This stability is also important because it allows a single MHC–peptide complex to engage and trigger many TCRs, thus maintaining a sustained signal required for T cell activation. Finally, the signaling process can be tuned by costimulatory molecules (CD28) if they engage their ligands (B7.1 and B7.2) on APCs. B7 is indeed the hallmark of professional APCs and confers to these cells a selective advantage in the capacity to trigger T cells. The fact that the expression of B7, as well as other costimulatory molecules, is regulated by pathogens or T cell help facilitates the immune response toward infectious antigens.

References Cited

ALBERT, M. L., SAUTER, B., and BHARDWAJ, N. (1998). Dendritic cells acquire antigen from apoptotic cells and induce class I-restricted CTLs. *Nature* **392**, 86–89.

BJORKMAN, P. J., SAPER, M. A., SAMRAOUI, B., BENNETT, W. S., STROMINGER, J. L., and WILEY, D. C. (1987). Structure of the human class I histocompatibility antigen, HLA-A2. *Nature* **329**, 506–512.

CELLA, M., ENGERING, A., PINET, V., PIETERS, J., and LANZAVECCHIA, A. (1997). Inflammatory stimuli induce accumulation of MHC class II complexes on dendritic cells. *Nature* **388**, 782–787.

CRABTREE, G. R. (1989). Contingent genetic regulatory events in T lymphocyte activation. *Science* **243**, 355–361.

DAVIDSON, H. W., REID, P. A., LANZAVECCHIA, A., and WATTS, C. (1991). Processed antigen binds to newly synthesized MHC class II molecules in antigen-specific B lymphocytes. *Cell* **67**, 105–116.

DEMOTZ, S., GREY, H. M., and SETTE, A. (1990). The minimal number of class II MHC–antigen complexes needed for T cell activation. *Science* **249**, 1028–1030.

GARCIA, K. C., DEGANO, M., STANFIELD, R. L., BRUNMARK, A., JACKSON, M. R., PETERSON, P. A., TEYTON, L., and WILSON, I. A. (1996). An alphabeta T cell receptor structure at 2.5 A and its orientation in the TCR–MHC complex. *Science* **274**, 209–219.

GERMAIN, R. N. (1994). MHC-dependent antigen processing and peptide presentation: Providing ligands for T lymphocyte activation. *Cell* **76**, 287–299.

HARDING, C. V., and UNANUE, E. R. (1990). Quantitation of antigen-presenting cell MHC class II/peptide complexes necessary for T-cell stimulation. *Nature* 574–576.

IEZZI, G., KARJALAINEN, K., and LANZAVECCHIA, A. (1998). The duration of antigenic stimulation determines the fate of naive and effector T cells. *Immunity* **8**, 89–95.

JANEWAY, C. A., JR. (1992). The T cell receptor as a multicomponent signalling machine: CD4/CD8 coreceptors and CD45 in T cell activation. *Annu. Rev. Immunol.* **10**, 645–674.

KROPSHOFER, H., HAMMERLING, G. J., and VOGT, A. B. (1997). How HLA-DM edits the MHC class II peptide repertoire: Survival of the fittest? *Immunol. Today* **18**, 77–82.

LANZAVECCHIA, A. (1990). Receptor-mediated antigen uptake and its effect on antigen presentation to class II-restricted T lymphocytes. *Annu. Rev. Immunol.* **8**, 773–793.

LANZAVECCHIA, A., REID, P. A., and WATTS, C. (1992). Irreversible association of peptides with class II MIIC molecules in living cells. *Nature* **357**, 249–252.

LENSCHOW, D. J., WALUNAS, T. L., and BLUESTONE, J. A. (1996). CD28/B7 system of T cell costimulation. *Annu. Rev. Immunol.* **14**, 233–258.

LJUNGGREN, H. G., STAM, N. J., OHLEN, C., NEEFJES, J. J., HOGLUND, P., HEEMELS, M. T., BASTIN, J., SCHUMACHER, T. N., TOWNSEND, A., KARRE, K., et al. (1990). Empty MHC class I molecules come out in the cold. *Nature* **346**, 476–480.

NELSON, C. A., PETZOLD, S. J., and UNANUE, E. R. (1994). Peptides determine the lifespan of MHC class II molecules in the antigen-presenting cell. *Nature* **371**, 250–252.

NORBURY, C. C., CHAMBERS, B. J., PRESCOTT, A. R., LJUNGGREN, H. G., and WATTS, C. (1997). Constitutive macropinocytosis allows TAP-dependent major histocompatibility complex class I presentation of exogenous soluble antigen by bone marrow-derived dendritic cells. *Eur. J. Immunol.* **27**, 280–288.

MATSUI, K., BONIFACE, J. J., REAY, P. A., SCHILD, H., FAZEKAS-DE-ST-GROTH, B., and DAVIS, M. M. (1991). Low affinity interaction of peptide–MHC complexes with T cell receptors. *Science* **254**, 1788–1791.

MONKS, C. R., KUPFER, H., TAMIR, I., BARLOW, A., and KUPFER, A. (1997). Selective modulation of protein kinase C-theta during T-cell activation. *Nature* **385**, 83–86.

MOSKOPHIDIS, D., LECHNER, F., PIRCHER, H., and ZINKERNAGEL, R. M. (1993). Virus persistence in acutely infected immunocompetent mice by exhaustion of antiviral cytotoxic effector T cells. *Nature* **362**, 758–761.

SALLUSTO, F., and LANZAVECCHIA, A. (1994). Efficient presentation of soluble antigen by cultured human dendritic cells is maintained by granulocyte/macrophage colony-stimulating factor plus interleukin 4 and downregulated by tumor necrosis factor alpha. *J. Exp. Med.* **179**, 1109–1118.

SALLUSTO, F., CELLA, M., DANIELI, C., and LANZAVECCHIA, A. (1995). Dendritic cells use macropinocytosis and the mannose receptor to concentrate macromolecules in the major histocompatibility complex class II compartment: Downregulation by cytokines and bacterial products. *J. Exp. Med.* **182**, 389–400.

SALLUSTO, F., SCHAERLI, P., LOETSCHER, P., SCHANIEL, C., LENIG, D., MACKAY, C. R., QIN, S., and LANZAVECCHIA, A. (1998). Rapid and coordinated switch in chemokine receptor expression during dendritic cell maturation. *Eur. J. Immunol.* **28**, 2760–2769.

STEINMAN, R. M. (1991). The dendritic cell system and its role in immunogenicity. *Annu. Rev. Immunol.* **9**, 271–296.

SWAIN, S. L., CROFT, M., DUBEY, C., HAYNES, L., ROGERS, P., ZHANG, X., and BRADLEY, L. M. (1996). From naive to memory T cells. *Immunol. Rev.* **150**, 143–167.

SYKULEV, Y., JOO, M., VTURINA, I., TSOMIDES, T. J., and EISEN, H. N. (1996). Evidence that a single peptide–MHC complex on a target cell can elicit a cytolytic T cell response. *Immunity* **4**, 565–571.

VALITUTTI, S., and LANZAVECCHIA, A. (1997). Serial triggering of T-cell receptors: A basis for the sensitivity and specificity of T cell antigen recognition. *Immunol. Today* **18**, 299–304.

VALITUTTI, S., DESSING, M., AKTORIES, K., GALLATI, H., and LANZAVECCHIA, A. (1995a). Sustained signaling leading to T cell activation results from prolonged T cell receptor occupancy. Role of T cell actin cytoskeleton. *J. Exp. Med.* **181,** 577–584.

VALITUTTI, S., MULLER, S., CELLA, M., PADOVAN, E., and LANZAVECCHIA, A. (1995b). Serial triggering of many T-cell receptors by a few peptide–MHC complexes. *Nature* **375,** 148–151.

VALITUTTI, S., MULLER, S., DESSING, M., and LANZAVECCHIA, A. (1996). Different responses are elicited in cytotoxix T lymphocytes by different levels of T cell receptor occupancy. *J. Exp. Med.* **183,** 1917–1921.

VIOLA, A., and LANZAVECCHIA, A. (1996). T cell activation de-termined by T cell receptor number and tunable thresholds. *Science* **273,** 104–106.

WEBER, S., TRAUNECKER, A., OLIVERI, F., GERHARD, W., and KARJALAINEN, K. (1992). Specific low-affinity recognition of major histocompatibility complex plus peptide by soluble T-cell receptor. *Nature* **356,** 793–796.

ZINKERNAGEL, R. M., and DOHERTY, P. C. (1974). Restriction of *in vitro* T cell-mediated cytotoxicity in lymphocytic chorio-meningitis within a syngeneic or semiallogeneic system. *Nature* **248,** 701–702.

Elena A. Armandola
Harald Kropshofer
Anne B. Vogt
Günter J. Hämmerling

German Cancer Research Center
Department of Molecular Immunology
Im Neuenheimer Feld 280
69120 Heidelberg, Germany

The Cell Biology of Antigen Presentation

The immune system has developed multifaceted strategies to fight the invasion of pathogens. Pathogens lingering and replicating in the cytosol of a cell are dealt with by the cytosolic proteolytic machinery and their proteins are digested into peptides. These peptides are then translocated by a specialized transporter into the lumen of the endoplasmic reticulum where they bind to major histocompatibility complex (MHC) class I molecules and are presented to cytotoxic T lymphocytes. This process allows the T cells to continuously screen the cytosol for foreign antigens, not only from pathogens but also from tumors, and to eliminate the potential source of danger. Microbes dwelling in intracellular vesicles or their products, such as toxins, are disposed of by a different pathway, namely, the endosomal/lysosomal proteolytic system. The resulting peptides are loaded in endosomal/lysosomal compartments onto MHC class II molecules with the help of accessory molecules and are presented to T helper lymphocytes. These cells will produce cytokines which will both boost the T cell response and drive the production of antibodies that are important for the elimination of the pathogen. Although the two pathways are different, both exploit part of the machinery already present in the cell, but some highly specialized molecules and systems have also evolved to optimize the ability of an individual to fight pathogens.

Introduction

Microorganisms and viruses are a daily challenge for the immune system. The variety of pathogens and the different strategies they have developed to invade host cells require the host defense system to be able to recognize them regardless of their antigenic variability and/or cellular location. Adoptive immunity has evolved to meet this task.

T cells play a central role as effectors of the immune response against pathogens. They recognize and respond to antigens expressed by pathogens only when the latter are presented as peptides on the cell surface in the context of the two main classes of major histocompatibility complex (MHC) molecules, MHC class I and MHC class II.

The pathways through which antigenic peptides are generated and meet the different MHC molecules on their way to the cell surface as well as the cascade of events triggered by recognition of these peptides by $CD8^+$ cytotoxic or $CD4^+$ helper T cells are quite different. These differences are not only fundamental for maximizing the chances of the immune system to cope with the different invasion strategies of pathogens but also represent an interesting example of how, during evolution, basic cellular systems and mechanisms have differentiated to varying extents and have been adapted to perform very specialized tasks.

Some pathogens (e.g., viruses and some bacteria) localize and replicate in the cytosol, whereas others (e.g., some bacteria) propagate in vesicular compartments to which they gain access via the endocytic/phagocytic pathway. The different cellular compartments in which pathogens and their products localize required the development of different antigen processing and presentation pathways (Fig. 1).

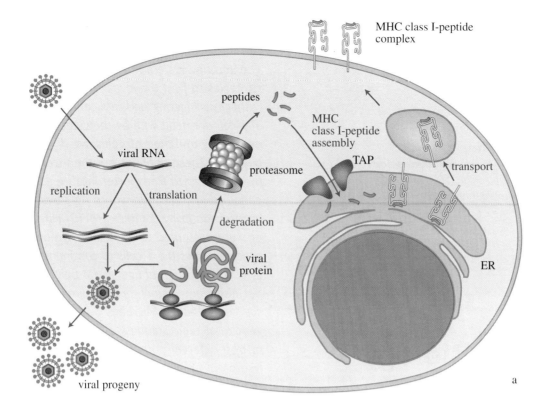

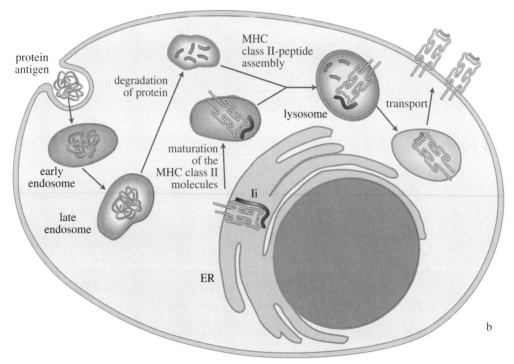

FIGURE I Antigen presentation pathways. (a) Virus and some bacteria are found in the cytosol, in which they give rise to new progeny. The antigenic proteins of these pathogens are degraded into peptides by the proteasome and enter the endoplasmic reticulum (ER) via the TAP transporter. MHC class I molecules (light yellow) are assembled in the ER, in which they bind antigenic peptides. After assembly the MHC class I/peptide complex leaves the ER and arrives, via the vesicles, on the cell surface. (b) Extracellular antigens and some bacteria are found in vesicles of endocytic or phagocytic origin. Passing from the early endosome to the late endosome, the endocytic vesicle environment becomes progressively more acidic and endosomal/lysosomal proteases are activated, degrading the proteins contained in the vesicles. After synthesis, MHC class II molecules (dark yellow) are routed to endosomal/lysosomal vesicles by the targeting signal of the invariant chain (Ii, in blue), which is progressively degraded. When the MHC class II-containing vesicles meet antigenic peptide-containing vesicles, the MHC class II/peptide complex is formed which then is directed to the cell surface.

Cytosolic pathogens are degraded by the major proteolytic machinery present in the cytosol, the proteasome, and the antigenic peptides generated are then transported across the endoplasmic reticulum (ER) membrane to newly synthesized MHC class I molecules. Peptide loaded on MHC class I molecules will be recognized by CD8[+] cytotoxic T lymphocytes. This recognition will lead to the activation of the lymphocyte and the elimination of the infected cell.

Antigens derived from pathogens dwelling in vesicular compartments will be degraded by the proteases present in the endosomal/lysosomal compartment. They will not be brought to the ER but will meet newly synthesized MHC class II molecules, which are heading toward the surface, in specialized endosomal/lysosomal compartments. Peptides loaded on MHC class II molecules will be recognized by CD4[+] helper T lymphocytes whose activation will lead to the production of cytokines which, in turn, can stimulate the antigen presenting cell (APC) or stimulate antibody production by B cells.

The MHC

The MHC is a region of the genome containing several highly polymorphic genes, and it was first identified through the analysis of the pattern of rejection or acceptance of tissues transplanted among different strains of mice. It was established that the recognition of a graft as "foreign" has a genetic basis. The genes responsible for causing graft rejection were therefore called "histocompatibility genes." Individuals expressing the same set of MHC molecules accept tissue grafts from one another, whereas individuals who differ in their MHC genes reject the grafts.

Because tissue grafting is not a naturally occurring event, the MHC system must have developed and evolved for some function other than monitoring the grafting of a foreign tissue. It became clear that the products of the MHC genes are fundamental in determining the ability of an individual to recognize foreign antigens, such as those expressed by invading pathogens, and that the immune system can respond to a foreign antigen only when the latter is displayed in the context of "self" MHC molecules. This phenomenon is called MHC restriction and is the basis of immune recognition by T cells (Zinkernagel and Doherty, 1974).

After the discovery of the MHC, it was soon recognized that multiple loci were present in the complex. The genes at the different loci code for proteins with a very similar structure which are codominantly expressed. The MHC class I molecules are encoded by genes at three different loci, called HLA-A, -B, and -C for man and H2-K, -D, and -L for mouse. The main MHC class II loci are named HLA-DR, -DQ, and -DP for man and I-A and -E for mouse (Fig. 2). The genes of the MHC are located on human chromosome 6 and on mouse chromosome 17. The MHC loci are highly polymorphic; that is, many alternative forms (alleles) of each gene exist. For example, more than 100 allelic variants have been identified in the population for the *HLA-B* locus and for the *HLA-DRβ* locus. Each individual carries a different combination of alleles of the genes encoding the different classes of MHC molecules. Therefore, in an outbreak population, the allelic variability created by the coexpression of the genes at these loci is enormous, and the chances of two individuals being identical at all loci are minimal, except in the case of identical twins.

The advantage of this diversity is that the spectrum of peptides that can be bound by the MHC molecules of one individual is greatly increased, and thereby the number and kind of pathogens to which T cells can respond are increased. Alternatively, grafting of tissues from one individual to another, although it has become a powerful tool to treat irreversible organ damage, is very problematic because of the strong and rapid immune responses mounted against the graft.

As for all genes, the MHC genes undergo selective pressure from the environment, and in some populations it has

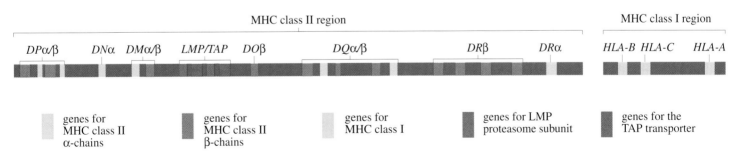

FIGURE 2 Organization of the MHC class I and class II genes. The MHC region is located on human chromosome 6 and contains the genes coding for the α and β chains of MHC class II (*HLA-DPα/β, -DNα, -DMα/β, -DOβ, -DQα/β, -DRβ,* and *-DRα*) and the genes *HLA-A, -B,* and *-C*, which encode the α chain of MHC class I. This region also contains the genes for the TAP subunits and the LMP subunits of the proteasome, but the gene encoding β2-microglobulin (β2m), part of the MHC class I molecule, is located on chromosome 15.

been found that certain alleles are overrepresented with respect to what is expected from the classical rules of inheritance. For example, the *HLA-B53* allele has been found to be commonly expressed in the West African population. In this region malaria is endemic and it has been found that *HLA-B53*$^+$ individuals can recover from a potentially fatal form of the disease. The *HLA-B53* allele seems to confer a selective advantage to individuals expressing it, probably by presenting a specific peptide antigen from the malaria parasite which can more effectively stimulate the immune system (Hill *et al.*, 1992).

Since the recognition of certain environmental antigens is often achieved best in combination with a particular antigen-presenting MHC allele, it is not surprising that immune recognition of endogenous tissue antigens is also often dominated by certain HLA alleles. This is the reason why many autoimmune diseases, such as diabetes, ankylosing spondylitis, autoimmune thyroiditis, and rheumatoid arthritis, are found to be associated with the expression of certain HLA alleles. Significant associations were found, for example, between the expression of *HLA-B27* and the development of ankylosing spondylitis or between the expression of an *HLA-DQ* allele with certain polymorphic amino acids at position 57 in its β chain and the susceptibility to insulin-dependent diabetes mellitus (Morel *et al.*, 1988). Association of certain MHC alleles with protection from the development of certain diseases has also been found.

Structure of MHC Molecules

MHC class I molecules are cell surface glycoproteins consisting of two polypeptide chains: an (or heavy) chain encoded by genes in the MHC and a small β chain, β$_2$-microglobulin (β$_2$m) whose gene lies outside the MHC (human chromosome 15 and mouse chromosome 2) and displays little or no polymorphism. The α chain is a transmembrane protein, carries N-linked glycans, and has a molecular weight of 40–43 kDa. The β$_2$m chain (12 kDa) does not span the membrane but noncovalently pairs with the α chain to form a heterodimer.

The molecular structure of MHC class I molecules has been elucidated by X-ray crystallography (Bjorkman *et al.*, 1987) (Fig. 3). The structures of both human and mouse class I molecules have been determined and show a great degree of structural similarity.

The α chain contains three domains, α$_1$, α$_2$, and α$_3$, of which the α$_3$ domain is proximal to the cell membrane and pairs with β$_2$m. The amino acid sequence of the α$_3$ domain is highly conserved among all class I molecules. The α$_3$ domain and β$_2$m fold to form immunoglobulin (Ig)-like disulfide-linked domains. Ig domains are regions homolo-

gous to the constant domains of immunoglobulin which form compact globular structures held together by disulfide bonds. They are the building blocks of Igs and many other molecules playing a role in cell–cell interactions. These molecules are grouped in the Ig superfamily and are thought to have evolved from one ancestral gene. In addition to MHC molecules, the T cell receptor, Fc receptors, costimulatory molecules, such as CD4, CD8, B7, and CD28, and several integrins, such as ICAM-1 and VCAM-1, belong to the family.

In MHC molecules the Ig-like domains α$_3$ and β$_2$m form the scaffold onto which the α$_1$ and α$_2$ membrane distal domains of the heavy chain rest. The α$_1$ and α$_2$ domains fold to form a groove that accommodates the antigen-derived peptide (Fig. 3a). Eight antiparallel β-pleated sheets form the base of the peptide-binding groove, whereas the walls are made of two α-helices. The ends of the groove are closed by interaction among the side chains of amino acids of the α-helices and the outer β-pleated sheets, limiting the access to the groove to peptides of a restricted length. Polymorphic amino acid residues are frequently found in the peptide-binding groove and form pockets which accommodate and interact with the side chains of the peptide residues. These interactions are involved in the stabilization and the allele specificity of peptide binding.

MHC class I and class II molecules are evolutionarily related, as shown by their high degree of sequence homology and by their structural features. MHC class II molecules are also composed of two subunits, the α and the β chain, both of which are encoded by MHC genes. The α chain (33–35 kDa) bears two N-linked oligosaccharides, whereas the

---

FIGURE 3 (a) Crystallographic structures of MHC class I (left) and class II (right) molecules. The transmembrane and cytoplasmic domains are not visualized (they should extend from the bottom of the figure). Peptides binding in the groove are shown in red. (b) Three-dimensional structure (seen from above on the left and from the side on the right) of the peptide-binding groove of MHC class I (above) and class II (below) molecules with bound peptides (HIV-reverse transcriptase for class I and hemagglutinin A of influenza virus for class II). Asterisks indicate pockets in the grooves in which MHC class I molecules bind the peptide in an extended conformation. The ends of the groove are closed and both extremities of the peptide bind in the cleft. Also, the peptide which binds to MHC class II molecules is in an extended conformation, but the ends of the groove are open and the extremities of the peptide can extend outside the cleft. Peptide side chains that interact with pockets in the groove are underlined. A, alanine; E, glutamic acid; H, histidine; K, lysine; I, isoleucine; L, leucine; N, asparagine; P, proline; Q, glutamine; T, threonine; V, valine; Y, tyrosine (reproduced with permission from Stern and Wiley, 1994).

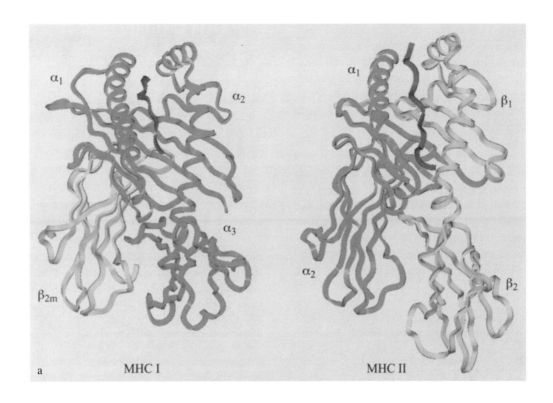

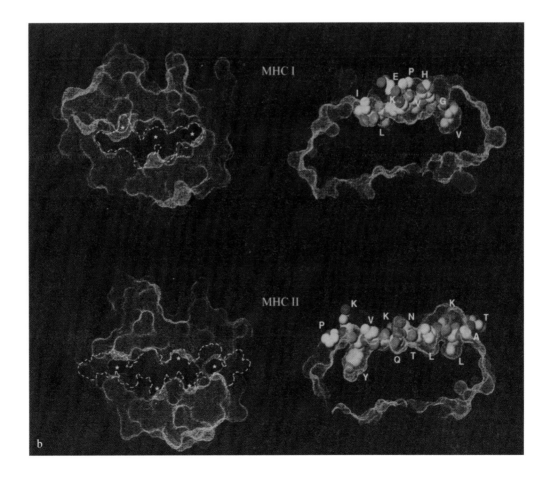

β chain (26–30 kDa) has only one N-linked glycan. Each subunit consists of two extracellular domains, a transmembrane region and a short cytoplasmic tail. The membrane-proximal regions, α_2 and β_2, fold to form typical Ig-like domains, whereas each of the membrane-distal domains, α_1 and β_1, constitute one-half of the cleft which functions as peptide-binding groove.

Determination of the crystal structure of the human MHC class II molecule HLA-DR1 confirmed the overall structural similarities between MHC class I and class II molecules: The floor of the groove in class II molecules is also composed of eight antiparallel β-pleated sheets and its walls are formed by two α-helices. However, in contrast to the class I molecules peptide-binding groove, both ends of the class II molecule cleft are open, allowing longer peptides to bind with their N and C termini protruding from either end of the cleft (Fig. 3b). As in the case of class I molecules, polymorphic residues are found mainly in the peptide-binding groove and form allele-specific pockets which accommodate specific peptide side chains. For both MHC class I and class II molecules, hydrogen bonds between the peptide backbone and the groove determine the binding affinity of a peptide for the MHC heterodimers. Each MHC allelic variant is able to bind a set of peptides, each with different affinity, and the resulting complexes display a wide range of stabilities. The peptide bound to the groove can be viewed as the third subunit of MHC molecules since α/β dimers devoid of peptide are inefficiently brought to the cell surface and are mainly unstable and undergo functional inactivation and/or denaturation.

Structure of MHC-Associated Peptides

Abundant and important information on the nature of the MHC–peptide interaction has been derived from the characterization of peptides that bind to MHC molecules *in vivo*. Peptides could be isolated and their length and sequence determined. It was shown that the peptides which bind to MHC class I molecules are mainly 8–10 amino acids long. Alignment of the amino acid sequences of different peptides eluted from various MHC class I alleles showed a predominance of one or few amino acids at certain positions of peptides binding to the same allele. These residues are called "anchor residues." The binding of the peptide to the groove of the MHC molecule is stabilized both by contact of its NH_2- and COOH-terminal residues with nonpolymorphic residues in the groove and by the interaction of the side chains of the anchor residues with the polymorphic residues located within pockets of the MHC class I groove (Fig. 4). Different anchor residues on the peptides are important for binding to different MHC class I alleles.

Study of MHC class II-associated peptides demonstrated that, unlike class I-associated peptides, the peptide

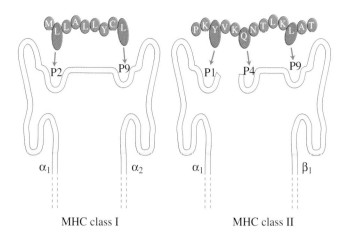

FIGURE 4 Interaction of peptides with MHC molecules. Peptides contain anchor residues (in red) which tightly bind into pockets (P1, P2, P4, and P9) of the groove of MHC molecules; those of class I are formed from the α_1 and α_2 domains and those of class II from the α_1 and β_1 domains. The polymorphic residues of MHC molecules form pockets in which specific residues of the peptide have to fit. These residues are conserved among peptides binding to the same MHC allele. M, methionine; L, leucine; A, alanine; Y, tyrosine; C, cysteine; P, proline; K, lysine; V, valine; Q, glutamine; N, asparagine; T, threonine.

length is highly variable (12–30 residues), with an average of 15–18 residues. Moreover, from numerous MHC class II alleles nested sets of self-peptides derived from a single protein were isolated, extending from a central core region of 11–13 residues toward the N and C termini. The existence of these N- and C-terminal truncation variants indicates that processing by amino- and carboxypeptidases can occur either prior to or after binding of the peptides to MHC class II molecules.

Sequence alignment of self-peptides eluted from different MHC class II alleles also indicates the existence of allele-specific binding motifs. The binding motifs are created not only by the sort and relative position of anchor residues fitting into the respective pockets of the groove but also by biases introduced by the antigen processing machinery (e.g., specificity of proteases). For example, proline residues are frequently found at the N termini of class II-binding peptides. This may be due to the fact that prolines act as stop signals for aminopeptidases and prevent further N-terminal proteolysis.

There has been much interest in developing methods for identifying peptide binding motifs for the various MHC alleles, with the goal of predicting from which regions of a protein antigenic peptides might be generated. This has important implications, for example, for peptide vaccine development.

The generation of antigenic peptides and their ability to bind MHC molecules does not strictly depend on the peptide binding motifs but is also influenced by the amino acids surrounding a potential antigenic peptide in the protein. Flanking sequences influence antigen processing and sometime determine if an antigenic peptide will be generated or not. It has also been observed that synthetic peptides designed with the appropriate anchor motifs and predicted to bind MHC molecules may not bind at all or only to a limited extent; however, peptides with mutations at predicted critical residues can sometimes bind very well to MHC molecules provided that the overall conformation of the peptides allows them to fit into the groove. Although the general features regulating the binding of a peptide to MHC molecules are clear, additional factors can enter into play which make the prediction of a potential antigenic site in a protein difficult.

MHC Class I Pathway of Antigen Presentation

MHC class I molecules are expressed by all nucleated cells, although their level of expression varies in different cell types. Cells of the immune system express high levels of class I molecules, whereas liver cells, for example, show a lower expression.

The level of MHC expression influences T cell activation and it can be regulated by immune and inflammatory stimuli. Cytokines can modulate the rate of transcription of MHC class I genes. On almost all cell types, for example, interferons (IFNs)-α, -β, and -γ and tumor necrosis factor (TNF) markedly increase the level of class I expression. This phenomenon is not limited to MHC class I genes, but extends to other molecules of the antigen processing pathway (e.g., components of the proteolytic machinery and peptide transporters) and to MHC class II genes. The expression of the different proteins within a pathway is coordinately regulated by cytokines, providing an important amplification mechanism for T cell responses.

Antigen Processing in the MHC Class I Pathway

MHC class I molecules are synthesized in the ER, in which they need to be loaded with an antigenic peptide to fold correctly before they are transported to the cell surface. The antigenic peptides presented by MHC class I molecules are mostly derived from cytosolic proteins, raising two fundamental questions: How are peptides generated and how are they loaded onto MHC class I molecules in the ER?

An example of antigens which are processed and loaded as peptides onto MHC class I molecules are viral proteins. A virus infects a cell and takes over its protein synthesis machinery. Many of the viral proteins are then present in the cytosol, which is the major site of protein degradation in

a cell. A key player in cytosolic protein degradation is the multicatalytic proteinase called proteasome (Coux *et al.*, 1996). This is a very large (ca. 700 kDa) proteolytic complex, also called 20S proteasome because of its sedimentation properties, composed of several α and β subunits. The β-type subunits contain the proteolytically active sites of the complex. Twenty-eight low-molecular-weight (20–30 kDa) subunits (14 α and 14 β) form heptameric ring-like structures. Four heptameric rings are stacked to form the main barrel-shaped body of the proteasome, with the two central rings being formed by β-type subunits and the two outer rings by α-type subunits (Fig. 5). The rings surround a central cavity to which unfolded proteins must access to be proteolysed (Groll *et al.*, 1997).

The 20S proteasome is present *in vivo* in at least two forms. It can bind to an IFN-γ-inducible regulator (PA28), which influences the efficiency and specificity of the 20S proteasome proteolytic activity, or it can constitute the core of the so-called 26S proteasome, a 1500- to 2000-kDa complex resulting from the assembly of other regulatory subunits at both ends of the 20S core. The 26S proteasome is responsible for the breakdown of proteins targeted for degradation by ubiquitination of their lysine residues.

It is of interest that the genes encoding two of the 20S proteasome subunits (LMP2 and LMP7) have been localized within the MHC. The expression of these subunits, similar to the expression of MHC class I, is enhanced by IFN-γ, which is frequently produced during infections. Thus, the immune system appears to respond to an infection by upregulation of various components of the antigen processing

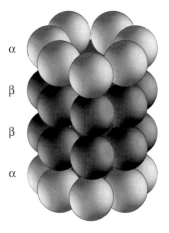

FIGURE 5 The cylindrical structure of the 20S proteasome is formed by 28 subunits arranged in four stacked heptameric rings. The α subunits form the two external rings, and the β subunits form the two central rings and contain the proteolytically active sites. At the ends of the proteasome other subunits can assemble that can regulate the activity of the 20S proteasome or form the 26S proteasome, which is responsible for the breakdown of ubiquitinylated proteins.

machinery. The incorporation of LMP2 and LMP7 in the 20S proteasome modifies its proteolytic specificity so that peptides suitable for MHC class I binding are generated more efficiently.

The incorporation of the LMP subunits in the proteasome provides a means by which the immune system can exploit a preexisting cellular system for performing a very specialized task. It will be shown that this is a common occurrence along the antigen presentation pathway. Once viral proteins are degraded by the proteasome in the cytosol to the corresponding peptides, the latter are still unaccessible to newly synthesized MHC class I molecules which wait in the ER for their peptide cargo to complete assembly and then move to the cell surface.

Cells have evolved sophisticated and specialized systems to transport a variety of compounds across membrane barriers. One of these systems has been dedicated to delivering peptides to the ER lumen for MHC class I loading.

Transport of Antigenic Peptides into the Endoplasmic Reticulum

The translocation of antigenic peptides, generated in the cytosol, into the ER is achieved through the action of very specialized transporter molecules [transporter associated with antigen presentation (TAP)] belonging to the large family of ATP-binding cassette (ABC) transporters. All ABC transporters exploit the energy derived from ATP hydrolysis for their function, but the various members of the family translocate a great variety of compounds across membranes. ABC transporters are found in all organisms from bacteria to primates. The oligopeptide permease of *Salmonella typhimurium*, for example, translocates peptides of two to five amino acids, whereas the STE6 transporter of *Saccharomyces cerevisiae* is involved in the secretion of the mating pheromone α-factor. In humans, the multidrug resistance transporters, which translocate hydrophobic drugs out of the cell, and the cystic fibrosis transmembrane conductance regulator transporter, a chloride channel, also belong to this family of transporters (Higgins 1992).

The TAP complex is formed by two subunits, TAP1 and TAP2 (Fig. 6), each containing an NH_2-terminal hydrophobic multimembrane-spanning domain and a COOH-terminal hydrophilic cytosolic domain, harboring the ATP binding site. The expression of TAP1 and TAP2 is upregulated by IFN-γ. The TAP transporters were identified by studying mutant cell lines in which MHC class I molecules were synthesized but not efficiently transported to the cell surface. The defect in these cells was identified as the lack of expression of the TAP transporter complex which resulted in an impaired supply of peptides to the ER, leading to a large decrease in the expression of MHC class I molecules at the cell surface. The importance of the TAP transporters in the antigen processing and presentation pathway

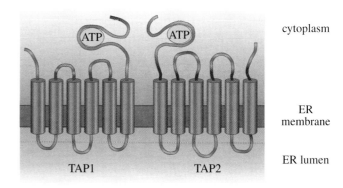

FIGURE 6 The TAP transporter is formed from two subunits, TAP1 and TAP2. These are multimembrane spanning molecules residing on the ER membrane. They carry an ATP binding site on their cytoplasmic portion. ATP hydrolysis drives the translocation of peptides (preferably of 8–12 amino acids) from the cytosol to the ER.

is underscored by the fact that individuals with a defect in the TAP genes develop an immunodeficiency syndrome and suffer from recurrent bacterial infections (de la Salle *et al.,* 1994).

TAP has been shown to reside in the ER membrane and to translocate peptides, preferably 8–12 amino acids long, across the membrane into the ER lumen. The size of the peptides preferentially transported by TAP is very similar to the size requirements of MHC class I molecules. It has been found, however, that longer peptides (up to 40 amino acids long) can also be transported by TAP, although with a much lower efficiency (Momburg *et al.,* 1994). These longer peptides cannot normally bind to class I molecules, but an exception has been reported for *HLA-B27*, which has been found to bind peptides of up to 33 amino acids in a TAP-dependent fashion. Generally, however, longer peptides are either further trimmed in the ER by ER-resident proteases (a process that does not seem to be very efficient) or can be reshuttled out of the ER back into the cytosol to be further degraded to a length that would make them better substrates to undergo a new cycle of transport by TAP.

The TAP transporter operates a sort of preselection of the peptide repertoire offered to MHC class I molecules for loading in regard to both size and sequence. Peptide transporters seem to have adapted to the requirements of MHC class I molecules in the various species or to have coevolved with them.

Assembly of MHC Class I Molecules in the ER and Their Transport to the Cell Surface

Proper assembly of MHC class I molecules occurs through sequential steps which involve several additional proteins, some of which act as chaperones for MHC molecules (Fig. 7). In human cells, shortly after synthesis, the α

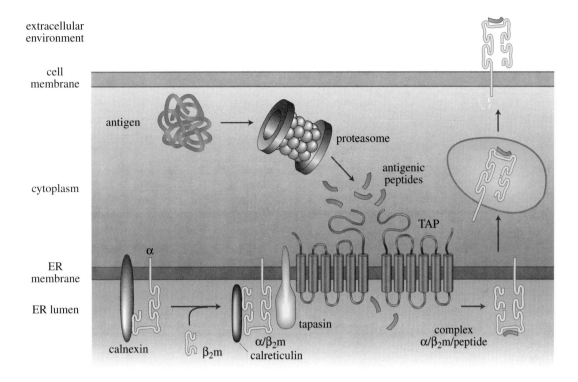

FIGURE 7 MHC class I assembly in the ER. The MHC class I heavy chain (α) assembles in the ER with the membrane-bound chaperone calnexin (red). When β_2m joins the complex, calnexin is released and substituted by calreticulin (purple), which delivers α/β_2m complexes to the TAP transporter and tapasin (orange). The antigenic peptide (green) is delivered to the ER by TAP and binds class I molecules, and the trimeric complex α/β_2m/peptide is released and transported, via vesicles, to the cell surface.

chain complexes with the ER-resident, membrane-bound protein calnexin, which retains the partially folded α chain in the ER. The association of β_2m with the α chain releases the heterodimer from calnexin, and the α chain/β_2m pair interacts with another chaperone molecule, calreticulin. To a certain extent these chaperones are interchangeable. In fact, it has been observed that in cells lacking calnexin, calreticulin can take over its function.

The complex α/β_2m/calreticulin subsequently moves to form a larger multimolecular complex with the TAP transporter. To this multimolecular complex also belongs the protein tapasin, whose function is under investigation but which seems to be essential for proper peptide loading of MHC class I molecules. Until the proper peptide is loaded onto them, class I molecules rely on the interaction with one or more molecules acting as chaperones to guarantee their survival. These interactions are fundamental for avoiding misfolding and aggregation of the α/β_2m dimer. The loading of a peptide onto the MHC class I molecule releases it from the interaction with TAP and tapasin, and the properly folded trimolecular complex (α/β_2m/peptide) can now start its journey toward the cell surface.

Peptide loading is the limiting step in MHC class I assembly and it depends on the supply of peptides to the ER.

Here, an array of empty class I molecules are maintained, partially folded, ready to be loaded and released to the cell surface. This allows the cell to quickly respond at any given time to an antigenic challenge (Koopmann *et al.*, 1997).

The MHC Class II Pathway of Antigen Presentation

Unlike MHC class I molecules, class II molecules are constitutively expressed by only a few cell types, such as B cells, dendritic cells, macrophages, Langerhans cells, and epithelial cells of the thymus, which are all specialized APCs. However, as is the case for class I expression, class II expression can be induced by some cytokines (e.g., IFN-γ), for example, in activated T cells, in astrocytes of the brain, and in keratinocytes.

Assembly of Class II Molecules in the ER

Newly synthesized class II α and β chains are cotranslationally inserted into the ER lumen and glycosylated. The newly synthesized chains transiently associate with the ER-resident chaperone calnexin, probably to assist subunit assembly. Calnexin is also involved in the first stages of folding of class I, indicating the importance of chaperones in preventing misfolding and aggregation before all the components (α chain, β chain, and peptide) are correctly assembled. In contrast to class I, to which antigenic peptides are delivered in the ER, MHC class II molecules are delivered to the compartments in which antigenic peptides are produced. Peptide loading of class II molecules occurs in endosomal/lysosomal compartments, which they reach empty of antigenic peptides but are chaperoned by a specialized accessory molecule, the invariant chain (Ii).

Shortly after biosynthesis, Ii replaces calnexin in the association with class II α and β chain and assists class II molecules in their further maturation. In contrast to class II α and β chains, Ii is a nonpolymorphic glycoprotein which possesses a noncleavable signal–anchor sequence near the N terminus so that the C terminus becomes inserted into the lumen of the ER. The most abundant variant of Ii is the so-called p33 form. Alternative splicing gives rise to a p41 form. There are several functionally important regions in the Ii molecule (Fig. 8a), including the region between amino acids 163 and 183 which is essential for trimerization of Ii (Fig. 8a). The trimers form a scaffold used by α and β chains for proper folding and interchain pairing, resulting in a nonameric $(\alpha\beta)_3$Ii$_3$ complex (Roche *et al.*, 1991) (Fig. 8b). Although class II molecules can form dimers in the absence of Ii, a significant portion of them form large aggregates in the ER and only a few such class II dimers reach the cell surface. Hence, Ii functions as a class II-specific chaperone that optimizes the efficiency of dimer formation during the initial stages of folding and guarantees the survival of α/β heterodimers until peptide loading.

The nonameric complex $(\alpha\beta)_3$Ii$_3$, is sorted from the ER to the Golgi complex and it is hypothesized that three adjacent cytosolic Ii tails may form a motif that is recognized by a cytosolic protein complex which sorts the complex to the endocytic pathway. The ER harbors a multitude of different chaperones which can at times replace each other. As mentioned previously, calreticulin binds to newly synthesized class I heavy chains in the absence of calnexin and, likewise, in Ii-negative cells α/β dimers have been found associated to the ER resident proteins GRP94 and ERp72

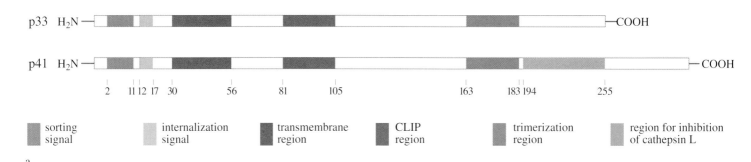

a

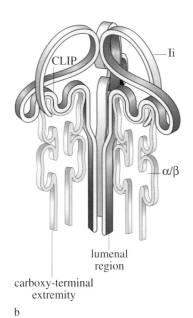

b

FIGURE 8 The invariant chain and MHC class II molecules. (a) Schematic representation of the p33 and p41 forms of human invariant chain. The specialized functions are reported for the various regions. Numbers refer to amino acid positions. (b) Hypothetical three-dimensional structure of the nonameric $(\alpha\beta)_3$Ii$_3$ complex. According to the current model, Ii uses a lumenal region and the transmembrane region for trimerization. The extended CLIP region and the carboxy-terminal sequence interact with the α/β class II dimers.

for prolonged periods of time. However, such Ii-negative cells are severely impaired in their ability to shuttle class II molecules to the cell surface or to load them with antigenic peptides, indicating that Ii cannot be replaced by other chaperones. Ii is therefore a highly specialized chaperone whose structure has evolved to fulfill the needs of MHC class II molecules.

The sites and modes of Ii interaction with MHC class II molecules have been extensively investigated. Ii interacts with the groove of class II molecules as well as with other sites. The groove is occupied by a region of Ii denominated CLIP (class II-associated invariant chain peptide) (Riberdy *et al.*, 1992), which binds to all the different class II alleles although with different affinities due to the polymorphism of the class II molecules. Whether the CLIP region in the context of intact Ii also interacts with the binding groove and adopts the same conformation as the CLIP peptide in the crystal is not known. Elucidation of the crystallographic structure of the class II α/β/Ii complex will provide more definitive insight.

Therefore, the binding of Ii to class II molecules in the ER is not only important for proper folding of the α/β dimer but also substitutes for antigenic peptides in an environment in which the conditions for loading of class II molecules are not optimal (e.g., neutral pH is optimal for MHC class I but not for class II peptide loading). However, it also protects the groove from premature loading until the α/β

dimers reach the endosomal/lysosomal compartments, in which MHC class II molecules meet antigenic peptides and loading efficiently occurs at low pH.

Sorting of MHC Class II Molecules to the Endocytic Pathway

Membrane proteins synthesized in the ER and transported to the Golgi complex follow a default pathway to the cell surface, and deviations from it require the presence of specific sorting signals. Signal-dependent sorting within the secretory pathway occurs in the *trans*-Golgi network (TGN). Two pathways have been described by which proteins may enter the endocytic pathway: either directly from the TGN by using transport vesicles or by transit to the cell surface followed by rapid internalization and delivery to endocytic compartments. Both pathways employ the same signals (Table 1).

Endosomal sorting signals are located in the cytoplasmic tail of proteins and consist of leucine-based (L$\varnothing$ and $\varnothing$L) and tyrosine-based (Y$xx\varnothing$ or NxxY) motifs, where x is any amino acid and $\varnothing$ is one of the hydrophobic amino acids Leu, Ile, Val, Phe, Met, or Ala. The N-terminal tail of Ii contains two leucine-based sorting signals, which also individually function as internalization signals: Leu-Ile at position 7 and 8 and Met-Leu at position 16 and 17. Typical sorting signals are presented in Table 1.

The endocytic pathway is commonly divided into early

TABLE I				
Endosomal/Lysosomal Sorting Signals				
TYPE OF SIGNAL		**SORTED PROTEIN**	**SIGNAL SEQUENCE**	**DESTINATION**
Tyrosine	Y$xx\varnothing$	Transferrin receptor	$(x)_{17}$-LS**Y**TR**F**-*TM*	Early endosomes/PM
		Cation-independent mannose-6-P receptor	*TM*-$(x)_{42}$-AA**Y**RG**V**	Late endosomes
		LAMP-1	*TM*-$(x)_5$-AG**Y**QT**I**	Lysosomes
		Acid phosphatase	*TM*-$(x)_6$-PG**Y**RH**V**	Lysosomes
		HLA-DMβ	*TM*-$(x)_6$-SS**Y**TP**L**	MIIC
	NxxY	LDL receptor	*TM*-$(x)_{12}$-F**D**NP**VY**	Early endosomes
		IGF1 receptor	*TM*-$(x)_{16}$-S**V**NP**EY**	Early endosomes
		Mannose receptor	*TM*-$(x)_{12}$-F**E**NT**LY**	Phagosomes
Leucine	L$\varnothing$	Cation-independent mannose-6-P receptor	*TM*-$(x)_{151}$-DDSDF**LL**HV	Late endosomes/TGN
		Invariant chain	MDDQRD**LI**S-$(x)_{22}$-*TM*	Endosomes
		LIMP III	*TM*-$(x)_5$-DERAP**LI**RT	Lysosomes
	$\varnothing$L	Invariant chain	NNEQ**LPM**L-$(x)_{13}$-*TM*	Endosomes
		HLA-DR1β	*TM*-$(x)_7$-GLQPTG**FLS**	MIIC/PM

Note. Abbreviations used: Y$xx\varnothing$, signal with tyrosine at the first and a hydrophobic residue at the fourth position (x stands for any amino acid); NxxY, signal with asparagine at the first and tyrosine at the fourth position (x stands for any amino acid); L$\varnothing$ or $\varnothing$L, dileucine signal involving one leucine paired with another aliphatic or aromatic residue; PM, plasma membrane; TM, transmembrane region; TGN, *trans*-Golgi network.

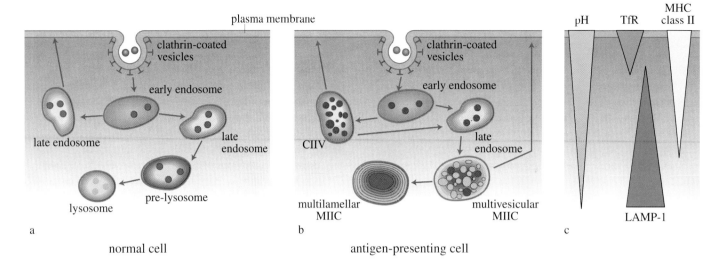

FIGURE 9 Compartments of the endocytic pathway in normal cells (a) and in antigen presenting cells (b). Normally, endocytosed proteins are unidirectionally sorted to lysosomes via clathrin-coated vesicles, early endosomes, late endosomes, and prelysosomes. In parallel, some proteins recycle to the cell surface via early and/or late endosomes. In antigen presenting cells morphologically distinct compartments are involved: Multivesicular and multilamellar MHC class II compartments (MIICs) correspond to lysosomal organelles, whereas class II vesicles (CIIVs), similar to late endosomes because of their buoyant density, are thought to be part of the recycling pathway. The triangles (c) represent the endocytic pH gradient, the distribution of the marker protein transferrin receptor (TfR), the distribution of the lysosome-associated membrane protein 1 (LAMP-1), and that of MHC class II molecules.

endosomes, recycling endosomes or CURL (compartments of uncoupling of receptors and ligands), late endosomes, and lysosomes. In APCs the compartments in which class II molecules accumulate, such as late endosomes and lysosome-like organelles, are referred to as CIIV (class II MHC vesicles) (Amigorena *et al.*, 1994) and MIIC (MHC class II compartment) (Peters *et al.*, 1991) (Fig. 9). Depending on the cell types examined, each of these compartments displays considerable heterogeneity and the reciprocal interactions of the compartments are not clear.

The route taken by class II molecules to reach the endocytic pathway is fundamental in determining the efficiency and quality of antigen presentation: Direct delivery of nascent class II molecules from the TGN to a late endocytic compartment would limit the environment in which loading takes place to vesicles in which the pH is low and proteolysis efficient, preventing those antigens that are highly susceptible to proteolytic degradation from having access to class II molecules. Transport of α/β/Ii complexes to the cell surface followed by internalization allows the colocalization of internalized antigen and class II molecules along the endocytic pathway. A gradually declining pH, together with a concomitant increase in the concentration of proteases, may yield a large spectrum of processed antigenic fragments to be sampled by class II molecules. It appears that α/β/Ii complexes can reach endosomes by both of the previously described pathways. The majority of α/β/Ii com-

plexes are thought to traffic directly to endosomal compartments, but a portion of newly synthesized class II molecules reach the cell surface very rapidly and slowly internalized. This is reminiscent of other membrane proteins, including the lysosome-associated membrane protein (LAMP1) or the cation-independent mannose-6-phosphate receptor (CI-M6P receptor).

Endosomal Proteolysis of Ii and Peptide Loading

The removal of Ii is necessary for peptide loading of class II molecules to occur. During transport of α/β/Ii complexes through the endocytic pathway, Ii is sequentially degraded starting from its C-terminal domain (Fig. 10).

Intermediate breakdown products of Ii still contain the N-terminal cytoplasmic tail responsible for sorting and retention. Thus, the complete proteolytic breakdown of Ii is required not only to generate an empty peptide-binding groove but also to free class II dimers from the Ii-mediated endosomal retention mechanism and to allow their transport to more acidic lysosome-like compartments.

Proteases of the cathepsin family are thought to be responsible for the stepwise degradation of Ii and generation of α/β/CLIP complexes. The aspartyl proteases cathepsin D and E are thought to be involved in the early steps of Ii breakdown. The cysteine protease cathepsin B can catalyze the final steps of Ii proteolysis *in vitro* (Fig. 10), and the en-

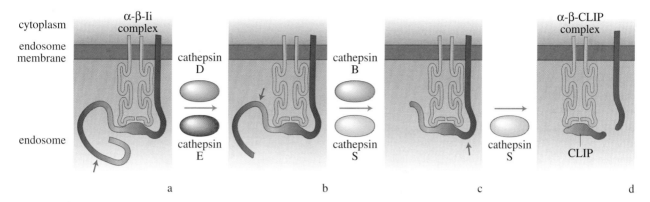

FIGURE 10 Stepwise proteolysis of the invariant chain (Ii) in the endosomes. (a) The Ii (in blue) assembles with α/β class II complexes in the ER, in which it prevents premature peptide binding and directs the complex to the endosomal compartment. Here, Ii is first cleaved by aspartic acid proteases, such as cathepsin D and E. (b–d) Successively Ii is then cleaved by thiol proteases, such as cathepsin B and S, and finally by cathepsin S to generate α/β/CLIP. Red arrows indicate the cleavage sites.

doproteinase cathepsin S was recently reported to have a key role in Ii processing. Cathepsin S is highly expressed in class II-positive cells and inducible by IFN-γ, which is a characteristic of most other proteins involved in MHC class I- or class II-mediated antigen presentation. Specific inhibition of cathepsin S in B cells results in accumulation of a class II-associated Ii fragment. In addition, only cathepsin S is able to digest α/β/Ii complexes in such a way that α/β/CLIP complexes are generated. Therefore, cathepsin S appears to be essential for the final step of Ii processing. The compartment(s) in which Ii processing takes place is a matter of debate. There are indications that the processing of Ii occurs in organelles different from those where antigenic peptide loading has been observed. There are also indications that the same proteases involved in Ii degradation also contribute to processing of antigenic proteins. Cathepsin L is one of these proteases. Interestingly, cathepsin L is strongly inhibited by an Ii-derived fragment encoded by the extra exon of the p41 form (Fig. 8a). Hence, another role of Ii may be to function as an endosomal protease inhibitor, modulating antigen processing in endosomes.

The majority of class II molecules are transported to compartments exhibiting lysosomal characteristics (Peters *et al.,* 1991). These organelles were found in human B cells and designated MHC class II compartments (MIIC), are 200–400 nm in diameter, are mildly acidic, and they have multivesicular or multilamellar morphology, contain typical lysosomal marker proteins such as β-hexosaminidase, LAMP-1, and CD63, are negative for Ii, and, most important, are highly enriched for class II molecules (Fig. 9). Exogenous protein antigens have access to these compartments, suggesting that the MIIC is a site at which the assembly of class II–peptide complexes occurs. B cells, dendritic cells, and activated macrophages contain MIIC-like compartments (Table 2).

Cell type-specific differences in MHC class II loading compartments may exist because class II-containing vesicles, called CIIV (Amigorena *et al.,* 1994), have been identified in mouse cells which differ from the described MIICs mainly by three criteria: They have a lower buoyant density, contain the early endosomal marker transferrin receptor but not the lysosomal membrane protein lgpA, and are devoid of Ii. It is not clear whether the CIIV found in mouse cells is analogous to the MIIC of human B cells or whether human cells possess a CIIV equivalent that has not been detected.

An attractive hypothesis is that in APCs a system of connected endocytic compartments has developed whose microenvironment is optimal for loading of class II molecules

TABLE 2		
Endocytic MHC Class II-Loading Compartments		
CRITERIA	**MIIC**	**CIIV**
Morphology	Multivesicular or multilamellar	Multivesicular
Density	High	Low
Markers		
TfR	−	+
CI-M6P receptor	±	+
Igp A	++	−
MHC class II	+++	+++
Ii	±	−
APC type	B cells Dendritic cells Macrophages	B cell lymphoma A20

with antigenic peptides. Examples of specialization of endocytic compartments can be found in different cell types: Muscle cells and adipocytes store glucose transporter molecules in specialized endosomes. Similarly, nerve cells and epithelial cells deliver proteins from the TGN to either the apical or basolateral surface via distinct endosomes. However, vesicles of the MIIC type have also been described in cells that do not express class II molecules; therefore, it is equally possible that the class II-enriched loading compartments of APCs are not unique to these cells but are part of the endocytic machinery normally present in almost any other eukaryotic cell type and that are exploited by APCs for a special task.

Mechanism of Protein Uptake by APCs

Receptor-mediated endocytosis, macropinocytosis, and phagocytosis contribute to various extents to antigen capture in different APCs (Fig. 11). Several parameters influence the efficacy of antigen uptake: (i) the number and specificity of respective cell surface receptors, (ii) the overall endocytic activity of the APC, and (iii) the nature of the antigen (e.g., soluble or particulate).

B Cells. Antigen-specific B cells present antigen at about 1000-fold lower concentrations than nonspecific (naive) B cells. This is mainly due to efficient active endocytosis via the antigen-specific B cell receptor (BCR) present on the plasma membrane of activated B cells, whereas naive B cells passively internalize antigen via fluid-phase pinocytosis or by other receptors with lower affinity for the antigen (Fig. 11, left). The BCR consists of a membrane immunoglobulin (mIg) associated with the disulfide-linked heterodimer Igα/Igβ. The cytoplasmic regions of Igα and Igβ provide two functions: After binding of the antigen to the mIg they induce signaling events mediated by tyrosine-based activation motifs (ITAMs) which initiate a cascade leading to B cell activation, and they mediate the delivery of BCR-associated antigen to processing compartments of the endocytic pathway, in which the antigen and the BCR components are degraded.

Dendritic Cells. Dendritic cells (DCs) are the most powerful class II-restricted APCs. They are able to efficiently internalize almost any soluble antigen and express high levels of class II molecules. Antigen uptake and T cell stimulatory capacities are tightly regulated during the maturation of DCs. In their immature state, when they reside in nonlymphoid tissues in which they encounter foreign antigens, DCs have a high antigen capture and processing capability but do not efficiently stimulate T cells. Inflammatory mediators such as bacterial lipopolysaccharide, tumor necrosis factor-α, or IL-1 promote DC maturation and their homing to the T cell areas of secondary lymphoid organs. These mature DCs lose their high capacity to internalize antigen but express high amounts of class II molecules on their cell surface and efficiently stimulate T cells.

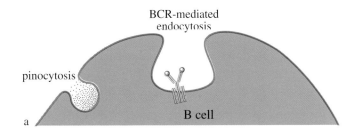

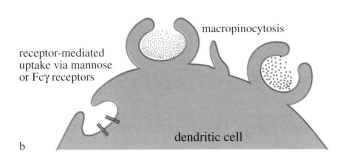

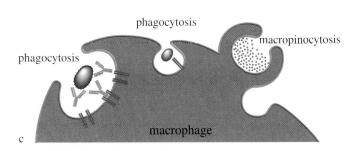

FIGURE 11 Mechanisms of antigen capture used by B cells, dendritic cells, and macrophages to ingest different forms of antigen (in red). (a) Active endocytosis in the B cell is mediated by the specific receptor (BCR, in yellow) present on the plasma membrane of activated B cell. Naive B cells internalize the antigen by fluid-phase pinocytosis. (b) Immature dendritic cells can capture antigen by constitutive macropinocytosis, mediated by membrane ruffling, or by endocytosis mediated by the mannose receptor or mediated by various types of receptor for Fcγ (in blue). (c) Macrophages are the primary phagocytic cells and thus play a major role in the defense against microbial infection. They also possess various types of Fcγ receptors (in dark blue) on the plasma membrane, which permit them to capture and degrade microorganisms coated with antibody (in light blue) and use macropinocytosis for aspecific capture of particulate material, although to a lesser extent than dendritic cells.

Immature DCs can capture antigen by at least three different mechanisms. The first is constitutive macropinocytosis mediated by membrane ruffling and fluid-phase endocytosis via large vesicles. This enables DCs to take up almost half of the cell's volume per hour. The content of the vesicles is concentrated and finally accumulates in lysosome-

TABLE 3

Receptors Involved in Antigen Capture

APC TYPE	RECEPTOR	ANTIGENIC LIGAND
B cells	B cell receptor (BCR) Ig + (Igα/β)$_2$	Soluble antigens
Dendritic cells	Mannose receptor	Mannosylated or fucosylated antigens
	FcγRII	AntigenIgG complexes
	DEC-205 (mouse)	Broad specificity
Macrophages	FcγRI	Particulate
	FcγRII	AntigenIgG complexes
	Mannose receptor	Mannosylated or fucosylated antigens

like compartments in which class II molecules are found. The second important pathway involves endocytosis via the mannose receptor. This receptor takes up mannosylated or fucosylated antigens typical of prokaryotic glycoproteins, releases them at low pH in early endosomal compartments, and recycles to the plasma membrane (Table 3). Finally, DCs use different sorts of Fc receptors to internalize antigen–antibody complexes and to target these immunocomplexes to lysosomal compartments for processing (Fig. 11, center). Inflammatory stimuli can influence the efficiency of antigen uptake through the different mechanisms.

Macrophages. Macrophages are the primary phagocytic APCs (DCs have only low and B cells almost no phagocytic function). Phagocytosis plays a major role in the defense against microbial infection: Phagocytosed pathogens are killed and their antigenic fragments are processed and presented by MHC class II molecules. Macrophages also use macropinocytosis for the unspecific uptake of particulate material, albeit to a lesser extent than DCs. They also express the mannose receptor on their surface and are thus able to internalize microbes that express mannosylated surface glycoproteins by receptor-mediated phagocytosis. Most important, macrophages express different variants of Fcγ receptors on the plasma membrane which allow the uptake and degradation of antibody-coated microorganisms. The internalized microorganisms are then encapsulated in acidified vesicles, the phagosomes (Fig. 11c). Phagosomes fuse with class II MHC-positive lysosomal compartments, in which lysosomal enzymes can destroy the internalized microorganisms and the antigenic peptides are loaded onto class II molecules.

HLA-DM and Peptide Loading

The interaction with Ii is essential in the initial stages of class II molecules folding and migration to the endocytic

compartment. Once Ii is degraded, the CLIP fragment remains bound to the groove and needs to be removed for antigenic peptides to be loaded. Because peptide loading of α/β/CLIP complexes occurs much faster *in vivo* than *in vitro*, it was hypothesized that another accessory molecules may be necessary.

Two new genes in the MHC class II region, *HLA-DMA* and *DMB*, were identified through the analysis of mutant B cell lines unable to present exogenous protein antigen and displaying abnormally high amounts of CLIP associated with their surface class II molecules. These genes differ from the classical class II genes in that the sequence homology with known class I and class II genes is lower than 30% and that they display only limited polymorphism. Their gene products, the DMα and DMβ chains, form a heterodimer, HLA-DM (DM), that does not bind peptides and is not expressed on the cell surface but is sorted to MIICs by virtue of a sorting signal in the cytosolic tail of the DMβ chain. Due to these characteristics DM is regarded as a nonclassical class II molecule.

DM accelerates the dissociation of CLIP from class II molecules and facilitates loading of α/β dimers with peptide (Sloan *et al.,* 1995). It acts like a catalyst and one DM molecule catalyzes loading of 3–12 DR/CLIP complexes (depending on the class II allele) per minute at mildly acidic pH (5.0). As such, DM is classified into the category of low-capacity catalysts, such as lysozyme or protein disulfide isomerase. Members of the Hsp70 chaperone family also catalyze folding of their substrate proteins with comparably low turnover. The amount of DM in the MIIC loading compartment of B cells is about one-fifth the DR molecules supporting the view that DM acts as a catalyst of loading. DM binds efficiently to αβ dimers associated with CLIP but poorly to DR molecules carrying the normal set of self-peptides. DM is not, however, absolutely required for the removal of CLIP because presentation of exogenous peptide antigens also can take place in DM-deficient cells. A set of self-peptides different from CLIP is in fact found associated to HLA-DR3, DR4, and I-A^k in DM-negative mutants. Moreover, protein antigens are successfully presented by DM-negative cells in the context of several class II alleles, such as A^k, A^d, DR1, or DR4. Release of CLIP in the absence of DM can be explained by the low intrinsic stability of CLIP at endosomal/lysosomal pH for some class II alleles. The self-release of CLIP is facilitated by its N-terminal nine residues (Kropshofer *et al.,* 1995). These residues are likely to interact with an effector site outside the peptide groove, thereby enhancing the dissociation of the CLIP core from the groove. The self-release of CLIP raises the possibility that class II loading may also be accomplished in endocytic compartments devoid of DM as early or recycling endosomes. Peptide loading in early endosomes has actually been demonstrated in DM-deficient cells after antigenic protein uptake via the transferrin receptor.

The activity of DM is not limited to dissociating class II/ CLIP complexes but extends to class II/peptide complexes. The intrinsic kinetic stability of the class II/peptide complexes determines whether DM will be able to promote peptide release. For example, CLIP is released by DM because it has a high off-rate, whereas the DR1-bound HA (307–319) peptide from influenza hemagglutinin is DM resistant because it binds with high intrinsic stability to DR1. The crystallographic analysis of the DR1/HA complex demonstrates that two sorts of interactions contribute to the high intrinsic stability: (i) HA has an optimized set of anchor residues fitting into the specificity pockets of the DR1 cleft and (ii) it has an appropriate length for the formation of hydrogen bonds between its backbone and residues lining the DR1 groove. Replacement of anchor residues in a peptide or changes in length influence its DM sensitivity by changing its intrinsic stability. From these findings a kinetic proofreading model has been deduced to describe the activity of DM (Fig. 12).

The groove of α/β dimers is assumed to oscillate between an open and a closed state at mildly acidic pH, with DM preferentially binding to the open conformer, thereby increasing the half-life of this conformer. A plausible model is that binding of an optimal peptide will cross-link both subunits of the groove and keep it closed, thereby counteracting the DM-mediated opening (Fig. 12). In contrast, low-stability ligands dissociate from the open conformer and DM will remain associated with and thus chaperone the empty α/β dimer. In conclusion, DM has the capacity to skew the class II-associated peptide repertoire seen by different T cells toward kinetically long-lived α/β/peptide complexes. DM thus reduces the diversity of peptides exposed on the cell surface and appears to be involved in the selection of immunodominant epitopes in APCs (Fig. 13).

DM as a Chaperone

As long as there are no appropriate ligands associated with α/β dimers, the latter tend to aggregate due to the exposure of hydrophobic surfaces and subsequently will be degraded. In the ER and during the transport to the endocytic pathway, Ii protects class II molecules. What is the fate of α/β dimers after the loss of Ii in the endosomal/lysosomal system? Peptides present in lysosomal compartments are expected to take over the task of Ii by binding into the groove and protecting class II molecules from aggregation or unfolding. However, the availability of appropriate peptides depends on several factors, e.g., the structural requirements for peptide binding to the respective class II allele, the biological half-life of each peptide, and the presence of processing proteases. *In vivo,* not all these conditions seem to be fulfilled. In B cells 10–20% of the class II molecules in lysosomal compartments do not contain peptides and are associated with DM. DM preserves the functionality of empty class II molecules. Incubation for less than 30 min at pH 5 in the absence of peptide is sufficient to inactivate DR1/CLIP complexes, whereas in the presence of DM most of the DR1 molecules remain functional. DM dissociates from DR$\alpha\beta$ complexes when peptide is present. Thus, DM qualifies as a molecular chaperone that prevents aggregation of transiently empty class II molecules as long as the peptide binding groove is not stably occupied with peptide. DM might chaperone the class II groove by partially reaching into the groove and mimicking a peptide or it may bind outside the groove and stabilize its structure so that the collapse of the groove is prevented. In lysosomal compartments DM is usually found associated with a considerable fraction of the class II molecules. Therefore, under normal conditions, there seems to be a shortage of peptides in these compartments. The scarcity of self-peptides may offer the advantage that foreign peptides do not have to compete against ligands derived from self-proteins and may have easy access to the cohort of empty class II molecules chaperoned by DM. Thus, the existence of empty class II/DM complexes seems to enable APCs to rapidly respond to a new antigenic challenge. This situation is reminiscent of the MHC class I pathway, in which a sizable fraction of unloaded class I molecules is engaged in complexes with TAP and tapasin and appears to wait for appropriate peptides.

In addition to DM, other players are likely to be involved in the loading process. For example, the products of two

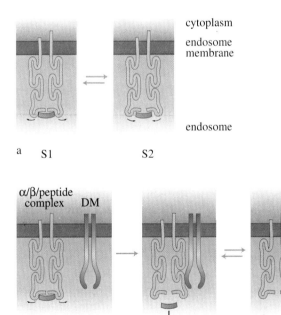

a S1   S2

α/β/peptide complex   DM

b

FIGURE 12 Conformational dynamics in the peptide-binding groove of MHC class II molecules. (a) According to this model the groove oscillates between a closed (S1) and an open state (S2). (b) The DM heterodimer (in purple) stabilizes the open conformer.

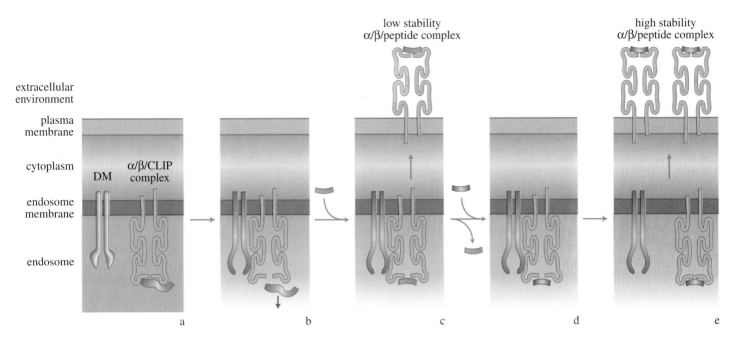

FIGURE 13   Selective editing of peptides by DM. (a, b) In the MHC class II loading compartments, DM associates with the $\alpha\beta$ dimer, permitting the release of CLIP (in blue) or (c) of other low-stability peptides (in green) from the $\alpha\beta$ complexes until (d) a peptide with high stability (in orange) is captured. As a result, the α/β/peptide repertoire on the cell surface is dominated by high-stability complexes (e).

other nonclassical MHC class II genes, HLA-DNA and -DOB, have recently been described from a heterodimer, designated HLA-DO (DO). DO binds tightly to DM in the ER and depends on this interaction to exit the ER (Liljedahl *et al.*, 1996). DO is not expressed at the cell surface but is sorted to lysosome-like compartments, in which it remains complexed with DM. Its evolutionary conservation and selective expression in APCs (B cells and dendritic cells) are indicative of DO having a role in antigen presentation, possibly acting as a regulator of DM.

Routes to the Cell Surface

The mechanism through which α/β/peptide complexes are targeted to the cell surface is poorly understood. It is unclear whether there is a default route from the loading compartments to the plasma membrane or whether some other sorting machinery is involved. One hypothesis is that class II molecules are retained in the endocytic pathway as long as they are associated with Ii or DM, both of which carry endosomal/lysosomal targeting signals. Proteolytic removal of the Ii sorting signal would allow α/β/CLIP complexes to travel freely to the surface if DM is not present. This has been shown to occur in DM-negative cells. Class II/CLIP complexes, however, are mainly retained in endosomal/lysosomal compartments and can go to the cell surface only after dissociation from DM, induced by loading with stably binding peptides. Most class II/peptide complexes on the cell surface are stable and thus have a long half-life.

Transport of loaded class II molecules to the cell surface could involve either a direct fusion of MIICs or CIIVs with the plasma membrane or vesicle-mediated trafficking. In human B cells and in a melanoma cell line, multivesicular MIICs have been shown to fuse with the plasma membrane, but this does not rule out the possibility that shuttling vesicles are also used to transport mature class II molecules to the cell surface.

Strategies Devised by Pathogens to Evade Immune Recognition

The sophisticated ways developed by the immune system to recognize and eliminate virus-infected cells are matched by strategies that viruses and microorganisms have evolved to escape immune surveillance. Thus, several pathogens encode proteins that can interfere with the complement cascade, inhibit apoptosis, act as cytokines or cytokine antagonists, or regulate antigen presentation.

Viruses, whose antigens are mainly presented by MHC class I molecules, directly affect these molecules, thus impairing antigen presentation to cytotoxic T lymphocytes (CTLs), whereas microorganisms endocytosed or phagocytosed by cells often modify the environment of intracellular vesicles or their trafficking to create a protective niche that allows their survival and escape from the MHC class II antigen presentation machinery (Spriggs, 1996).

MHC class I molecules are one of the viral targets through which elimination of virus-infected cells is circumvented either by affecting the ability of class I molecules to present virus-derived peptides to CTLs or by exploiting their function as inhibitors of natural killer (NK) cell-mediated lysis. For example, herpesviruses (e.g., herpes simplex, cytomegalovirus, and Epstein–Barr virus) are among the most successful viruses in immune evasion and have been extensively studied. They employ a variety of strategies to downregulate MHC class I expression on the surface of infected cells, thereby escaping recognition. Various viral proteins have been identified which interfere with different steps of class I synthesis and maturation. Two gene products, US2 and US11, of human cytomegalovirus (hCMV) bind in the ER to newly synthesized class I molecules and promote their translocation to the cytosol, in which they are degraded by the proteasome. Another protein of CMV (US3) binds $\alpha/\beta_2 m$ complexes and retains them in the ER. Different proteins expressed by herpes simplex virus (HSV) and hCMV affect the supply of antigenic peptides to the ER. These proteins (i.e., ICP47 for HSV and US6 for hCMV) bind to the TAP transporters and block their activity. This results in a lower expression of class I molecules on the cell surface.

Although the downregulation of class I is clearly advantageous to the virus because it prevents recognition by CTLs, cells lacking class I are susceptible to lysis by NK cells. Also, herpesviruses have developed strategies to bypass the host defense system. hCMV and mCMV express, in the late phase of infection, a protein called UL18 which displays a high homology to MHC class I and has been shown to interact with NK cells' inhibitory receptors. The interaction blocks NK recognition and killing of the infected cell.

Other well-studied examples of MHC class I regulation by viruses involve adenoviruses. Adenovirus 12 expresses a protein, E1A, which negatively regulates MHC class I gene transcription by inducing a decrease in the production of the p50 subunit of the transcription factor NF-κB. This factor plays an important role in class I gene transcription and lower levels of p50 result in decreased expression of MHC class I on the cell surface. Class I expression can also be more directly affected by adenovirus by blocking transport of the molecules to the surface through the protein 19K, the product of gene *E3*. 19K is a transmembrane protein with an ER retention signal and it interacts with the class I α chain replacing $\beta_2 m$ blocking transport of class I to the surface.

Although much less frequently, the function of MHC class II molecules can also be downregulated by viruses. One strain of HSV, for example, downregulates IFN-γ and IL-6 expression, indirectly affecting class II expression. Another example is the BZLF2 protein of Epstein–Barr virus

(EBV) which interacts directly with the β chain of HLA-DR on the cell surface and blocks its antigen presenting ability.

Antigen presentation by class II molecules, however, is more frequently affected by subversion of the antigen processing pathway rather than by direct impairment of class II molecule functions. The two basic mechanisms by which this is achieved are the deacidification of endosomal organelles and the uncoupling of pathogen-containing vesicles from the endocytic pathway so that they cannot develop into lysosomes. Several bacteria, such as salmonella, legionella, mycobacterium, and chlamydia, create an intracellular niche in which they can thrive, protected from the action of endosomal/lysosomal enzymes and separated from normal vesicular trafficking. The molecular mechanisms through which this occurs are not fully understood.

As the details of pathogen–host interactions are unveiled, a better understanding of the molecular mechanisms involved in the immune defense process will also be gained. Several immune modulators have microbial counterparts (e.g., viral cytokine homologs), and the effects of the microbial protein on the host cell can clarify the function of the corresponding host proteins. Therefore, pathogens represent valuable tools for studying normal immune mechanisms and for developing new strategies to selectively modulate the immune system to fight specific infections.

References Cited

AMIGORENA, S., DRAKE, J. R., WEBSTER, P., and MELLMAN, I. (1994). Transient accumulation of new class II MHC molecules in a novel endocytic compartment in B lymphocytes. *Nature* **369,** 113–120.

BJORKMAN, P. J., SAPER, M. A., SAMRAOUI, B., BENNET, W. S., STROMINGER, J. L., and WILEY, D. C. (1987). Structure of the human class I histocompatibility antigen, HLA-A2. *Nature* **329,** 506–512.

COUX, O., TANAKA, K., and GOLDBERG A. L. (1996). Structure and functions of the 20S and 26S proteasomes. *Annu. Rev. Biochem.* **65,** 801–847.

DE LA SALLE, H., HANAU, D., FRICKER, D., URLACHER, A., KELLY, A., SALAMERO, J., POWIS, S. H., DONATO, L., BAUSINGER, H., LAFORET, M., *et al.* (1994). *Nature* **265,** 237–241.

GROLL, M., DITZEL, L., LOWE, J., STOCK, D., BOCHTLER, M., BARTUNIK, H. D., and HUBER, R. (1997). Structure of 20S proteasome from yeast at 2.4 Å resolution. *Nature* **386,** 463–471.

HIGGINS, C. F. (1992). ABC transporters: From microorganisms to man. *Annu. Rev. Cell Biol.* **8,** 67–113.

HILL, A. V., ELVIN, J., WILLIS, A. C., AIDOO, M., ALLSOPP, C. E. M., GOTCH, F. M., GAO, X. M., TAKIGUCHI, M., GREENWOOD, B. M., TOWNSEND, A. R. M., MCMICHAEL, A. J., and WHITTLE, H. C. (1992). Molecular analysis of the association of HLA-B53 and resistance to severe malaria. *Nature* **360,** 434–440.

KOOPMANN, J. O., HÄMMERLING, G. J., and MOMBURG, F. (1997). Generation, intracellular transport and loading of peptides associated with MHC class I molecules. *Curr. Opin. Immunol.* **9,** 80–88.

KROPSHOFER, H., VOGT, A. B., STERN, L. J., and HAEMMERLING, G. J. (1995). Self-release of CLIP in peptide loading of HLA-DR molecules. *Science* **270,** 1357–1359.

LILJEDAHL, M., KUWANA, T., FUNG-LEUNG, W. P., JACKSON, M. R., PETERSON, P. A., and KARLSSON, L. (1996). HLA-DO is a lysosomal resident which requires association with HLA-DM for efficient intracellular transport. *EMBO J.* **15,** 4817–4824.

MOMBURG. F., ROELSE, J., HOWARD, J. C., BUTCHER, G. W., HÄMMERLING, G. J., and NEEFJES, J. J. (1994). Selectivity of MHC-encoded peptide transporters from human, mouse and rat. *Nature* **367,** 648–651.

MOREL, P. A., DORMAN, J. S., TODD, J. A., McDEVITT, H. O., and TRUCCO, M. (1988). Aspartic acid at position 57 of the HLA-DQ beta chain protects against type I diabetes: A family study. *Proc. Natl. Acad. Sci. USA* **85,** 8111–8115.

PETERS, P. J., NEEFJES, J. J., ORSCHOT, V., PLOEGH, H. L., and GEUZE, H. L. (1991). Segregation of MHC class II molecules from MHC class I molecules in the Golgi complex for transport to lysosomal compartments. *Nature* **349,** 669–676.

RIBERDY, J. M., NEWCOMB, J. R., SURMAN, M. J., BARBOSA, J. A., and CRESSWELL, P. (1992). HLA-DR molecules from an antigen-processing mutant cell line are associated with invariant chain peptides. *Nature* **360,** 474–476.

ROCHE, P. A., MARKS, M. S., and CRESSWELL, P. (1991). Formation of a nine-subunit complex by HLA class II glycoproteins and the invariant chain. *Nature* **354,** 392–394.

SLOAN, V. S., CAMERON, P., PORTER, G., GAMMON, M., AMAYA, M., MELLINS, E., and ZOLLER, D. M. (1995). Mediation by HLA-DM of dissociation of peptides from HLA-DR. *Nature* **375,** 802–806.

SPRIGGS, M. K. (1996). One step ahead of the game: Viral immunomodulatory molecules. *Annu. Rev. Immunol.* **14,** 101–130.

STERN, L. J., and WILEY, D. C. (1994). *Behring Institute Mitteilungen* **94,** 1–10.

ZINKERNAGEL, R. M., and DOHERTY, P. C. (1974). Restriction of *in vivo* T cell-mediated cytotoxicity in lymphocytic choriomeningitis within a syngenic or semiallogenic system. Nature **248,** 701–702.

General References

HILL, A., JUGOVIC, P., YORK, I., RUSS, G., BENNINK, J., YEWDELL, J., PLOEGH, H., and JOHNSON, D. (1995). Herpes simplex virus turns off the TAP to evade host immunity. *Nature* **375,** 411–415.

KROPSHOFER, H., HÄMMERLING, G. J., and VOGT, A. B. (1997). How HLA-DM edits the MHC class II peptide repertoire: Survival of the fittest? *Immunol. Today* **18,** 77–82.

MADDEN, D. R. (1995). The three-dimensional structure of peptide–MHC complexes. *Annu. Rev. Immunol.* **13,** 587–622.

NEEFJES, J. J., and MOMBURG, F. (1993). Cell biology of antigen presentation. *Curr. Opin. Immunol.* **5,** 27–34.

RAMMENSEE, H. G. (1995). Chemistry of peptides associated with MHC class I and class II molecules. *Curr. Opin. Immunol.* **7,** 85–96.

WATTS, C. (1997). Capture and processing of exogenous antigens for presentation on MHC molecules. *Annu. Rev. Immunol.* **15,** 821–850.

RALPH M. STEINMAN

Laboratory of Cellular Physiology and Immunology
Rockefeller University
New York, NY 10021, USA

GEROLD SCHULER

Department of Dermatology
University of Erlangen
D-91054 Erlangen, Germany

KAYO INABA

Laboratory of Immunobiology,
Graduate School of Biostudies
Kyoto University
Kyoto 606-01, Japan

Antigen Presentation in Vivo

A system of dendritic cells is specialized for presenting antigens in vivo. These cells are found at body surfaces and in the interstitial spaces of most solid organs except brain parenchyma. Dendritic cells efficiently capture and present antigens on major histocompatibility complex class I and II products. Receptors for antigen and distinct endocytic compartments are being identified in dendritic cells, which also express very high levels of adhesion and costimulatory molecules and produce large amounts of IL-12. All of these features, coupled with the capacity to migrate and localize to the T cell areas of lymphoid tissues, allow dendritic cells to act as natural adjuvants for helper and killer T cell responses in vivo. Dendritic cells are abundant and accessible at the surfaces of mucosal-associated lymphoid tissues, and they can actively support the replication of HIV-1. The study of vaccines, immune therapy, transplantation, and autoimmunity should benefit from additional emphasis on antigen-presenting cells in vivo.

Introduction: Distinctions between *in Vitro* and *in Vivo* Analyses of Antigen Presentation

The biology of T cells is rooted in incisive observations *in vivo*, i.e., in experimental animals and several clinical conditions (Table 1). *In vivo* studies, for example, identified the major histocompatibility complex (MHC) and its biological roles and distinguished B and T cells and their functions. To pursue these findings, tissue culture approaches were used. Subsets of lymphocytes and other white cell types were discovered. Monoclonal antibody and DNA probes were produced for more than 200 white cell products. The steps required for antigen presentation on MHC class I and II molecules were outlined. Critical steps in the signaling

pathways for lymphocyte development and function were unraveled.

In a real sense, however, the study of immunity *in vivo* has reawakened. The major stimuli are the ability to selectively alter genes that contribute to immune function and the need to prevent or treat many diseases that have long been known to involve the immune system, particularly T cells. Many principles in T cell biology must now be placed into an *in vivo* context, and what has been learned from T cells in culture needs to be translated into better vaccines and better treatments for allergy, transplantation, autoimmunity, AIDS, and malignancy.

Antigens are presented in an unusual way to T cells. The antigens are fragmented to peptides that bind to MHC products within other cells, antigen presenting cells (APCs).

TABLE I

Some Milestones in T Cell Immunology: *In Vivo* Studies of Patients and Experimental Animals

The MHC in transplantation and immune responsiveness

Lymphocytes as the cellular mediators of antigen-specific immunity

The thymus and bursa as central lymphoid organs

Separate B cell and T cell systems for humoral and cell-mediated immunity

MHC restriction during viral immunity and T cell development

The MHC–peptide complexes then emerge at the APC surface for T cell recognition. As a result, T lymphocytes recognize other cells that present peptides from infecting organisms, from self constituents to which the T cells have not been tolerated, and even from malignant tumors. These principles of antigen presentation have been worked out largely with T cell clones and T–T hybridomas *in vitro*. However, the homogeneous T cell lines that have been useful for unraveling the principles of antigen recognition *in vitro* differ in several ways from the T cells that emerge from the thymus and mediate immune responsiveness *in vivo* (Fig. 1).

In cloned T cell populations, all the T cells have a single specificity, whereas immunologically naive populations contain only approximately 1/100,000 cells that are specific for any one peptide within a pathogen or tumor. T cell clones and T–T hybridomas are chronically stimulated large cells and therefore express many surface and regulatory molecules that are not found in primary, quiescent, small T cells. Cloned T cells are typically stimulated in stationary culture systems, whereas immune responses must occur within the

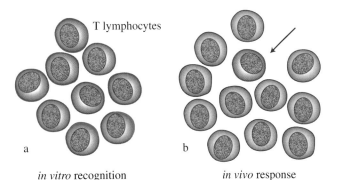

in vitro recognition   *in vivo* response

FIGURE I Two levels for studying antigen presentation to T lymphocytes. (a) *In vitro,* chronically stimulated, homogeneous, stationary clones are used to study the capacity of T cells to recognize MHC–peptide complexes. (b) *In vivo,* responses to presented antigens must be generated from quiescent, infrequent, circulating T cells (arrow).

confines of complex animals in which T cells are constantly circulating and must gain access to antigen on APCs. How are rare, quiescent, circulating T cells identified and activated by antigen *in vivo*? The dendritic cell (DC) system of APCs is specialized to capture antigens and stimulate antigen-specific T cells *in vivo*. DCs are also involved in tolerance, but this function has been explored less extensively and is not discussed here.

Primary Immune Responses Typically Begin in Lymphoid Organs That Drain the Site of Antigen Deposition

Lymphoid organs are designed to bring lymphocytes, antigens, and different subsets of APCs together for the purpose of generating immunity. Some of the critical early experiments that demonstrated this function of lymphoid organs were carried out by colleagues of Peyton Rous, the discoverer of the Rous sarcoma virus. Rous had additional interests, one of which was the distribution of injected antigens and proteins *in vivo*. His colleagues McMaster and Hudak injected colored dyes into the skin and tracked their movement. They found that the vital dyes collected into thin channels or "streamers," which are the afferent lymphatics (Fig. 2). Within minutes the lymphatics carried the dye to the local or draining lymph node. This finding reflects the fact that one function of lymphatics is to retrieve the plasma proteins that normally leak across capillaries and to recycle these proteins in lymph fluid back to the bloodstream. If lymphatics are blocked, the tissues will swell. A second function is that antigens and DCs access the afferent lymphatics and thereby the local or draining lymph node (Fig. 2).

When these investigators administered microorganisms, they discovered that antibodies to an injected antigen appeared first in the draining lymph node and later in the serum. They proved this by injecting two different antigens into the right and left sides of the neck. The draining cervical nodes only formed antibodies to the organism that had been injected on the corresponding side of the neck, indicating that antibody had to be made locally and not derived from some other site. Therefore, following the deposition of an antigen in one part of the body the antigen travels to the lymph node that drains that site, and an immune response develops locally. These scientists did not know that lymphocytes were the mediators of immunity, nor were they aware of the distinction between B and T cells, but their findings apply to many types of T cell responses (Table 2).

The most studied experimental immune response is the priming of CD4$^+$ helper T cells. One injects an antigen beneath the skin together with an adjuvant, usually complete Freunds' adjuvant (CFA). Five or more days after injecting an antigen with an adjuvant, the draining lymph node con-

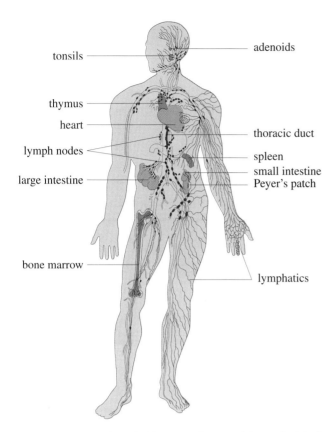

tonsils

adenoids

thymus

heart

lymph nodes

thoracic duct

spleen

small intestine
Peyer's patch

large intestine

bone marrow

lymphatics

FIGURE 2 The lymphoid system in man (shown in blue). Most parts of the body, including organs such as the heart and the intestine, have lymphatics that carry lymph, antigens, and APCs toward the lymph nodes. The lymphatics from the intestine and lower parts of the body come together to form the large thoracic duct in the chest.

tains primed CD4$^+$ T cells. This is demonstrated by boosting the lymph node suspension with the immunizing protein in culture and searching for DNA synthesis in the responding T cells. If the antigen fetuin is given in the left paws of mice with CFA, cells from the lymph node behind the left knee (the popliteal lymph node) and left arm (the axillary and brachial nodes) (Fig. 2) will proliferate specifically in response to fetuin (Table 3). If the mice simultane-

---

TABLE 2

Generation of Immunity in Lymphoid Organs That Drain the Site of Antigen Deposition

Antibody formation

CD4$^+$ T helper cells (DNA synthesis assay)

CD8$^+$ T killer cells (^{51}Cr-release assay)

Adoptive transfer of immune cells to naive recipients

TABLE 3

T Cell Priming Occurs in the Draining Lymph Node[a]

ANTIGEN USED TO STIMULATE DNA SYNTHESIS	DNA SYNTHESIS [CPM ^{3}H-TdR $\times$ 10^{-3}] BY CELLS FROM	
	Right Node, Primed to Lysozyme + CFA	Left Node, Primed to Fetuin + CFA
None	1.5	1.5
Fetuin	4.7	**130.1**
Lysozyme	**79.0**	4.9
PPD	**121.8**	**124.3**

[a] Groups of four mice were primed in the right paws with lysozyme and complete Freunds' adjuvant (CFA) and in the left paws with fetuin and CFA. Seven days later, the draining lymph nodes from individual mice were removed, and the lymph node suspensions were challenged with the indicated antigens [i.e., fetuin, lysozyme, or purified protein derivative (PPD; an antigen in the mycobacteria that is a component of CFA)]. On the third day of the challenge or booster culture, DNA synthesis was measured as cpm [^{3}H]thymidine uptake. The data are the means of the four mice, where the standard errors were less than 20% of the means. Large responses (in bold) are seen in the lymph node draining the site of antigen deposition. cpm, counts per minute.

ously receive a second antigen, lysozyme with CFA in the right paws, cells from nodes on the right side respond to lysozyme but not fetuin (Table 3). Cells from the lymph nodes on both sides of the mouse are primed to purified protein derivative since this antigen is part of the CFA that was given as the adjuvant to prime the T cells. Additional studies can be carried out to prove that the primed cells are CD4$^+$ T cells and are recognizing antigen presented in the context of MHC class II products.

If one injects a virus or an immunogenic tumor, then the draining lymph node also generates specific, CD8$^+$ cytolytic T lymphocytes (CTLs). The frequency of antigen-specific CTL precursors (CTLp) increases, and when cells from the primed animal are boosted with antigen in culture antigen-specific "effector" CTLs develop. Again, the draining lymph node is the site at which the frequency of CD8$^+$ CTLp increases and T cells are primed to become killers.

If one measures T cell priming by the adoptive transfer of cells from immunized to naive recipients *in vivo*, then cells from lymph nodes or spleen that drain the site of antigen deposition can transfer CD4$^+$ or CD8$^+$ immunity. We will begin the discussion of antigen presentation *in vivo* by discussing mechanisms underlying the fundamental phenomenon that T cell-mediated immunity develops in lymphoid organs that drain the site of antigen deposition. We will emphasize lymph nodes and spleen, for which most of the work has been done. Later we will consider the distinct features of a much larger reservoir of lymphoid tissue, mucosal-associated lymphoid tissues (MALT). Finally, we will summarize some changes that occur at the level of APCs during T cell memory and T cell-mediated diseases.

The Microscopic Anatomy of Peripheral Lymphoid Organs

Lymphocyte Recirculation

The early investigators of the immune response knew that lymphoid organs were collections of lymphocytes, but they did not know that lymphocytes were the critical mediators of immunity. This was established at Oxford University, where Florey, the codiscoverer of penicillin, had had a long interest in the lymphatic circulation. Building on this experience, Gowans cannulated the thoracic duct (Fig. 2) to obtain lymph that was rich in lymphocytes (in fact, the term "lymphocyte" originates from the fact that these are the cells that predominate in thoracic duct lymph). All the lymph and lymphatics that leave the lymph node below the diaphragm eventually access the thoracic duct. Using highly enriched thoracic duct lymphocytes, Gowans could induce immunity and even reverse immune tolerance. The thoracic duct lymphocytes initially contained small and large cells, but the large ones disappeared upon overnight culture. The cultured, small lymphocytes were sufficient to transfer immunity. It also became apparent at this time (the early 1960s) that large lymphocytes developed from small lymphocytes upon stimulation with mitogens or antigens. Therefore, small lymphocytes are the mediators of immunity and "transform" into large lymphocytes or lymphoblasts upon successful encounter with antigen.

Gowans then took advantage of the fact that small lymphocytes could be radiolabeled by incorporating [³H]uridine into cellular RNA. Radioisotopes of this sort had become available at approximately this time. Gowans could then use autoradiography to search for the labeled cells following injection into the bloodstream. He found that small lymphocytes continually traffic across special vessels, called high endothelial venules (HEVs), into the lymph node; the cells reemerge in the thoracic duct approximately 3–6 h later (Fig. 3). The endothelium in most venules is flat or cuboidal, but in lymph nodes the cells are much thicker or "high."

Next, it was shown that within 24 h of giving an antigen to an animal, the thoracic duct lymphocytes become selectively depleted of reactivity to that antigen. Approximately 3–5 days later, numerous enlarged lymphocytes or lymphoblasts reappear in the thoracic duct presumably as a result of activation in the lymphoid tissue. Therefore, lymphocytes continually recirculate through lymphoid organs, as if in constant search for their corresponding antigen. The

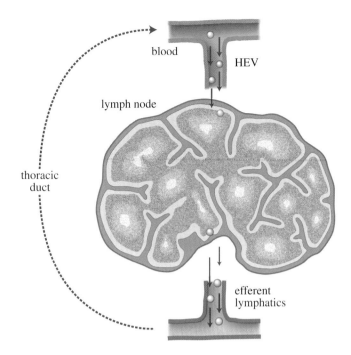

FIGURE 3 The recirculation of lymphocytes, from blood to lymph node via the high endothelial venules (HEV) and back to blood via the thoracic duct.

presence of antigen in a lymphoid organ efficiently selects the rare clones of circulating lymphocytes that are antigen specific.

B Cell and T Cell Areas

As criteria became available to distinguish the subsets of B and T lymphocytes, it was found that B and T cells circulate through distinct regions of the lymphoid organ, the B cell and T cell areas (Fig. 4). In all lymphoid tissues, the areas that are enriched in B cells are called follicles. The T cell-rich regions have different names depending on the organ, e.g., the deep cortex of lymph node, the periarterial sheaths of spleen, and the interfollicular regions of Peyer's patch or MALT. The T cell areas contain most of the HEVs through which lymphocytes recirculate.

A good way to demonstrate the existence of distinct B and T cell compartments *in vivo* is to stain a section of lymphoid tissue with monoclonal antibodies that are specific for each cell (Fig. 5a). However, around the HEVs in the T cell areas are cuffs of B cells which must make their way from the HEV to the B cell follicle (Fig. 5a). The molecular basis for the migration of B and T cells through select B and T compartments is being established. Important roles also could be played by different kinds of extracellular matrix and by distinct types of APCs in each region.

Two Types of Dendritic Cells in the B Cell and T Cell Areas

B cell and T cell areas contain distinct types of APCs termed follicular dendritic cells (FDCs) and DCs, respec-

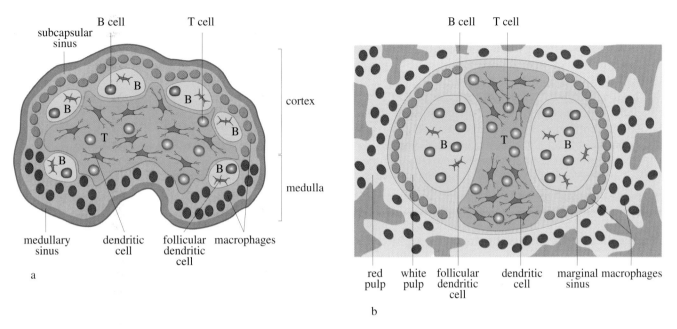

FIGURE 4 Schematic section of two lymphoid organs, a lymph node (a) and the spleen (b), showing the separate compartments for B (in green) and T (in blue) cell recirculation. The follicular dendritic cells and the dendritic cells are found in the B and T cell areas, respectively. The regions are also shown where the two types of macrophages (see text) that clear the blood and lymph of particulates and antigens are found.

tively (Fig. 4). FDCs retain native antigens as immune complexes extracellularly. The antigens are presented to B cells. During presentation by FDCs, high-affinity memory cells are selected that form as a result of somatic mutation in the proliferating or germinal center regions of the follicles. T cell area DCs display processed antigens as MHC–peptide complexes. These are presented to T cells during the early stages of an immune response.

The easiest way to display these networks of DCs is to stain sections with appropriate monoclonals. An interesting example involves antibodies to antigen-presenting MHC class II molecules or to the invariant chain that is required

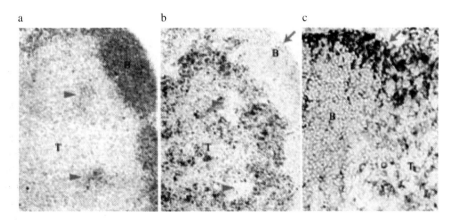

FIGURE 5 Identification of the lymphocyte compartments and APCs by staining tissue sections with monoclonal antibodies and immunocytochemistry. (a) Staining for B lymphocytes (anti-B220, in brown) and T lymphocytes (anti-CD4, in blue) marks the follicles, which are rich in B cells, and the deep cortex, which is rich in T lymphocytes. Note the B cells surrounding HEVs in the T cell areas (arrowheads). (b) Staining of an adjacent section with anti-MHC class II (in brown) and anti-CD4 (in blue) shows the much more intense brown staining of DCs in the T cell areas (mouse T cells are MHC class II negative) than in B cells in the follicles and macrophages in the subcapsular sinus (arrow). (c) A section in which the invariant chain, an polypeptide essential for antigen presentation, is stained in blue, whereas macrophages in the subcapsular sinus are in brown (arrow). Note that the invariant chain is more abundant in the DCs in the T cell area than in follicular B cells. Micrographs provided by Dr. Maggi Pack.

for presentation on MHC II. These antibodies stain B cells in the follicles and large, stellate DCs in the T cell areas (Figs. 5b and 5c). DCs in the T areas stand out following immunocytochemical staining because of their very high levels of expression of MHC II and invariant chain.

This long-standing view of distinct types of APCs in B cell and T cell areas is changing. Antigen-processing DCs have been identified in the germinal centers of human tonsil (Grouard *et al.*, 1996). These "GCDCs" express the CD11c integrin that is often abundant on DCs (Witmer-Pack *et al.*, 1993) and CD4. GCDCs probably present antigen to T cells that in turn are required to help B cells undergo the germinal center reaction including somatic mutation. GCDCs also may have a transporting role in germinal center function, transporting immune complexes and even memory T and B cells into the germinal center.

Two Types of Macrophage-Rich Regions for Antigen Clearance or Scavenging

Macrophages are also an abundant component of lymph nodes and spleen (Fig. 4). Macrophages are the major site for antigen deposition and catabolism *in vivo*. Scavenging and digestion are activities of macrophages in all tissues, not just lymphoid organs. Clearance is especially prominent in the liver in which sinusoidal-lining macrophages, called Kupffer cells, clear the portal circulation of antigens and other gut-derived substrates.

In lymph node and spleen, macrophages are not only abundant but also localized to discrete regions (Fig. 4). In lymph node, macrophages line the subcapsular sinus, which is where afferent lymphatics enter the node, and the medullary sinuses, which is where efferent lymphatics exit the node. Together, these pools of macrophages can substantially clear the lymph of antigens. In spleen, macrophages line both sides of the marginal sinus, which is where splenic arteries terminate, and the red pulp sinuses, which are a sponge of vessels that leave the spleen. Again, phagocytes can filter the blood for antigens, altered proteins, and damaged cells. MALTs do not contain the large collections of macrophages that are evident in lymph node and spleen. Instead, numerous macrophages are associated with the vast areas of mucosal epithelium.

There appear to be two types of macrophages that in turn are localized to defined regions of lymph node and spleen. Most macrophages are found in the lymph node medulla and splenic red pulp (Fig. 4). In the mouse these cells express high levels of a seven-membrane-spanning molecule called F4/80 but lower levels of a sialoadhesin called SER-4. In contrast, macrophages in the subcapsular sinus of lymph node and marginal sinus of spleen express little F4/80 but are particularly rich in SER-4 (Fig. 5c). The SER-4-rich, marginal sinus macrophages disappear in osteopetrotic mice, which have a genetic deficiency in macrophage colony-stimulating factor (M-CSF), but the F4/80-rich macrophages remain numerous. The functional significance of

the two types of M-CSF-dependent and -independent macrophages in anatomically discrete sites is not apparent. They could have distinct receptors for clearing different antigens from the blood and lymph or distinct cytokine and other secretory products.

Generation of Primary T Cell Responses *in Vitro*

Discovery of Dendritic Cells

Before we consider how lymphocytes and antigens come together to generate specific immunity, we discuss some tissue culture experiments that were essential for discovering DCs and their role in adaptive immunity. The primary mixed leukocyte reaction (MLR) and the primary antibody response were the *in vitro* assays used to identify the distinct accessory function of DCs. In the MLR, cells from one individual stimulate the T cells from an MHC-mismatched (or allogeneic) individual to proliferate, secrete lymphokines, and develop killer cells. This T cell reaction mimics the early stages of the powerful response against transplants that develops across an MHC barrier *in vivo*. The MLR therefore is commonly used as a method of tissue typing prospective donors in transplantation. The absence of an MLR means that the donor and recipient are closely matched at the MHC.

In the primary antibody response, mixtures of B and T lymphocytes require accessory nonlymphocytes to produce antibody to a "nominal" antigen in a fully syngeneic system. When the accessory cells for antibody responses were examined, DCs were recognized because of their many differences from macrophages (Steinman and Cohn, 1974). At the time, macrophages were the main nonlymphocytes that had been identified in lymphoid organs. The DCs had a peculiar morphology that had never been seen in macrophages. This was highlighted by the presence of many fine processes (dendrites) when the cells were attached to glass or plastic and large motile sheet-like, lamellipodia (veils) when the cells were nonadherent (Fig. 6a). The mature DCs did not actively phagocytose particles and lacked Fc and C3 receptors for immune complexes. DCs also lacked an antimicrobial apparatus, e.g., myeloperoxidase, lysozyme, and numerous lysosomes. In contrast, macrophages from different organs and species were adherent, actively phagocytic, and rich in immune complex receptors and antimicrobial functions. DCs could be isolated from the white pulp or lymphoid nodules of spleen, whereas phagocytes derived from the red pulp and various inflammatory sites. Later it was found that DCs lacked receptors for M-CSF but had abundant receptors for granulocyte-macrophage colony-stimulating factor (GM-CSF).

Methods were developed to enrich the trace subset of DCs from mouse spleen and other organs. When the DCs were tested for functional activity, the results could not

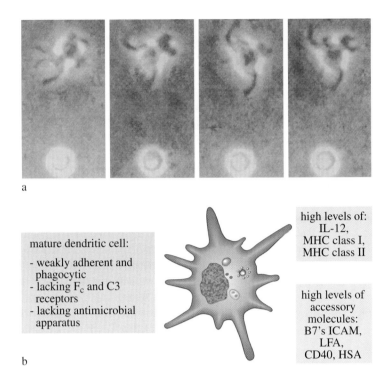

a

mature dendritic cell:

- weakly adherent and phagocytic
- lacking F_c and C3 receptors
- lacking antimicrobial apparatus

b

high levels of:
IL-12,
MHC class I,
MHC class II

high levels of accessory molecules:
B7's ICAM,
LFA,
CD40, HSA

FIGURE 6 Features of mature dendritic cells. (a) Phase-contrast micrographs of a typical, nonadherent live DC, with large motile lamellipodia, shown at 15-s intervals; below is a small lymphocyte. (b) Some general features of mature DCs in most tissues of mammalian species.

have been foreseen (Steinman and Witmer, 1978; Inaba *et al.*, 1983; Inaba and Steinman, 1985). Very small numbers of DCs were sufficient to initiate both the MLR and the primary antibody response (DC:T cell ratios of 1:100 or less;

Fig. 7). It had been thought that all cells that expressed MHC class II products (then known as Ia antigens in the mouse and HLA-DR in humans) were strong MLR stimulators, but this was being studied at stimulator to responder

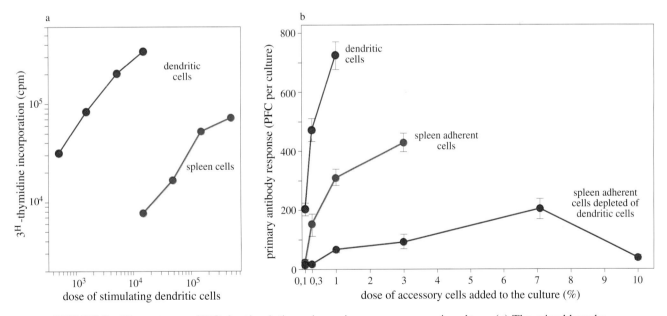

FIGURE 7 The potency of DCs in stimulating primary immune responses in culture. (a) The mixed lymphocyte reaction (expressed as uptake of tritiated thymidine) in microcultures of 300,000 T lymphocytes. Allogeneic T cells respond much more efficiently than bulk spleen cells. cpm, counts per minute. (b) Primary antibody responses [expressed as plaque forming cells (PFC)] of mixtures of B and T cells in the presence of increasing doses of accessory cells such as DCs (in violet), bulk spleen adherent cells (in blue), and spleen adherent cells selectively depleted of DCs with 33D1 monoclonal antibody (in red) [reproduced with permission from (a) Steinman and Witmer (1978)and (b) Inaba *et al.* (1983)].

ratios of 1 to 1. When DCs are fully mature, a stimulator responder ratio of 1:1000 or less will induce strong MLRs. Purified B cells, or macrophages that are induced to express MHC class II with lymphokines (Steinman *et al.,* 1980), are much weaker or inactive as MLR stimulators. Therefore, suspensions of lymphoid cells had a trace DC component that was responsible for initiating the two available assay systems for primary immune responses.

Recent criteria have verified that DCs are efficient APCs for immunologically naive T cells. For example, in the MLR, DCs are powerful stimulators of T cells in neonatal cord blood (Sallusto and Lanzavecchia, 1995). It is thought that cord blood T cells have not had a chance to encounter antigen, so they are operationally naive or unprimed. DCs also stimulate primary responses to protein antigens. The frequency of protein-specific T cells is usually too low to detect primary responses in culture, but primary responses can be studied through the use of T cells from a T cell receptor (TCR) transgenic mouse. In these mice, most of the lymphocytes express a receptor for a peptide presented in the context of MHC class I or II products. To rule out the possibility that the TCR transgenic cells have been primed by a cross-reacting antigen in the environment, the cells can be further selected for their naive status by removing those with a memory phenotype, e.g., low levels of the selectin CD62L or high levels of CD44. Again, DCs prove to be powerful APCs for primary responses by peptide-specific, TCR transgenic T cells (Seder *et al.,* 1992).

Experiments would soon reveal that instead of being active scavenging cells, DCs had very high levels of antigen-presenting MHC products and many different accessory molecules that are needed for T cell binding and costimu-

lation. The cell physiology of DCs segregated the scavenging and clearance functions of phagocytes from the requirements for presentation and stimulation to T cells.

Afferent and Efferent Limbs during Immune Responses *in Vitro*

Presentation during immune responses *in vitro* occurs in two stages, termed afferent and efferent limbs. In the afferent limb, T cells are activated by antigens presented on DCs, yielding large numbers of T lymphoblasts; in the efferent limb, these activated T cells respond vigorously to antigens presented on other cells that are capable of specialized effector functions. For example, when DCs prime helper T cells from H-2^b mice *in vitro*, the activated T blasts will help H-2^b B cells to make antibody (Inaba and Steinman, 1985). The activated T blasts will not help H-2^k B cells, even if additional H-2^b DCs are added to the cultures. Therefore, DCs prime the helper T cells that go on to help B cells or macrophages in an MHC-restricted fashion (Fig. 8).

In the MLR and antibody responses, the growth and function of CD4$^+$ helper cells are typically monitored. DCs also are potent APCs for the afferent limb of CD8$^+$ killer cell responses. In the efferent limb, killer cells that have been activated by DCs can lyse other targets, such as macrophages and tumor cells that are presenting antigen. In other words, antigen presentation allows any cell that expresses MHC molecules to present MHC–peptide complexes and be recognized by T cells (usually activated T cells), whereas the accessory roles of antigen-presenting DCs initiate CD4$^+$ and CD8$^+$ responses from quiescent T cells.

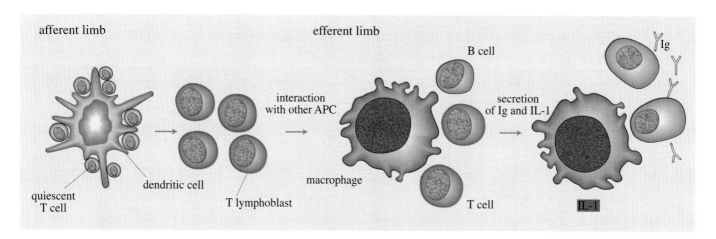

FIGURE 8 Two stages of an immune response *in vitro*. Dendritic cells activate quiescent T lymphocytes in the afferent limb to form T lymphoblasts, which then interact with other APCs in the efferent limb to elicit effector functions such as secretion of immunoglobulins by B cells and IL-1 secretion by macrophages. The T cells can be of either helper or cytolytic types.

Localization of Dendritic Cells to T Cell Areas *in Vivo*

Shortly after the time that MHC class II-rich DCs were recognized as potent APCs for T cell responses *in vitro*, it became evident that similar cells populate the T cell areas of peripheral lymphoid organs. When tissue sections are stained for MHC class II, as in Figs. 5 and 9, large stellate cells with abundant MHC class II are noted. In mice, T cells do not express MHC II, and there are relatively few B cells or macrophages in the T cell area (Figs. 5 and 9). T cell area DCs are also rich in the invariant chain (Figs. 5 and 9), a polypeptide that is essential for presentation on MHC class II. These and other criteria established that the MHC II-rich cells in the T cell areas are the *in vivo* counterpart of DCs (Witmer-Pack *et al.,* 1993).

The Potency of Dendritic Cells as APCs *in Vitro:* Some Underlying Mechanisms

Antigen Handling

Recent evidence indicates that DCs have specialized mechanisms for capturing and ferrying antigens so that peptides are optimally presented on MHC class II. This is a change from traditional points of view. At the level of antigen uptake, it has been thought that an immunologically nonspecific cell such as a DC could only capture antigens by bulk or nonspecific mechanisms. At the level of antigen processing, it has been reasoned that immunologically rele-

vant peptides are generated within the digestive lysosomes. The impetus for a change in thinking has been the availability of larger numbers of DCs so that cellular and molecular approaches are more feasible with this formerly trace APC. The DCs are generated from proliferating and nonproliferating progenitors using appropriate cytokines. These methods will not be described here.

Both primary and progenitor-derived DCs can have large numbers of MIICs, which are endocytic compartments that are rich in MHC class II products (Sallusto and Lanzavecchia, 1995; Nijman *et al.,* 1995). Of interest is a subset of nonlysosomal MIICs in which peptide–MHC class II complexes are destined for the cell surface for presentation to T cells. These "CIIVs" can be distinguished from standard lysosomes by cell fractionation. A role for nonlysosomal CIIVs in antigen presentation makes sense. Lysosomes are typically the site for complete protein digestion and little membrane recycling to the cell surface, whereas antigen presentation requires that peptides (not amino acids) be secured and affixed to surface-seeking MHC II products.

DCs also have multilectin receptors that can enhance the binding, uptake, and presentation of antigens. Two such receptors have been identified on DCs, and both mediate adsorptive uptake via coated pits (Fig. 10). These receptors are termed the macrophage mannose receptor (Sallusto and Lanzavecchia, 1995) and DEC-205 (Jiang *et al.,* 1995). Both are large, similarly configured molecules of 175 and 205 kDa, respectively. The mannose receptor has 8 contiguous, calcium-dependent or C-type lectin domains. This

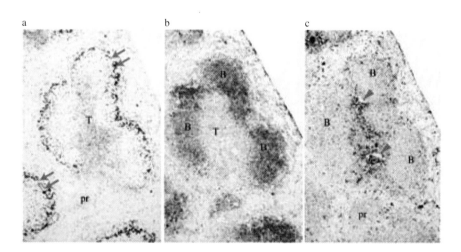

FIGURE 9 Abundance of MHC class II products and the invariant chain on DCs in the T cell areas of the spleen, the organ in which DCs were discovered. (a) A section stained for the SER-4 sialoadhesin of marginal zone macrophages (in brown, double arrows) and for CD4[+] T lymphocytes (in blue). Most of the section contains a large, white pulp nodule surrounded by red pulp (pr). (b) An adjacent section in which the CD4[+] T lymphocytes are stained in blue and the B lymphocytes in brown. (c) Another adjacent section stained for T lymphocytes in blue and the invariant chain in brown. The latter is an essential polypeptide for presentation on MHC class II and is much more abundant on T cell area DCs than B cells in the follicles and macrophages in the marginal zone and red pulp. The central arteries of the white pulp are marked with arrowheads.

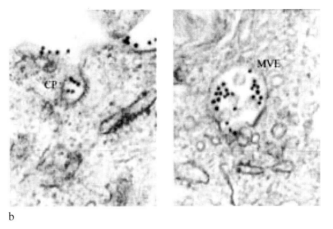

FIGURE 10 (a) Multilectin-type receptors for antigen uptake and presentation. The two receptors expressed on DCs are the macrophage mannose receptor (MMR) and the decalectin DEC-205. The structure of this receptor family is characterized by many contiguous C-type lectin domains (colored shapes), with each domain having distinct binding sequences for polysaccharides. (b) Micrograph showing antibodies to DEC-205 (black granules) which rapidly enter into the DCs via coated pits (CP; after 1 min) and multivesicular endosomes (MVE; after 5–20 min).

receptor can bind mannosylated proteins that are present on the surface of many microbes. The mannose receptor was first identified on lung macrophages, but it is expressed at high levels on some populations of DCs. DEC-205 has 10 contiguous C-type lectin domains, so it is a DECAlectin. DEC-205 is expressed at high levels on DCs in the T cell areas *in vivo*, but it is also found on some epithelia and brain endothelium. This multilectin very likely recognizes glycans, but these have yet to be identified. Antibodies to DEC-205 are rapidly internalized via coated pits, delivered to intracellular endosomes (Fig. 10b), and efficiently presented to Ig-specific T cells (Jiang *et al.*, 1995).

The glycans that are recognized by DEC-205 are not known but might be found on both foreign and self-glycoproteins. If DEC-205 recognizes self components, this could be important in the self-tolerizing function of DCs during negative selection in the thymus. The role of DCs in tolerance is not emphasized in this article. Presumably, the DCs that mediate negative selection would have to be able to present small amounts of a large panel of self-antigens.

Given the abundance of CIIVs, as well as receptors for uptake, it is likely that DCs capture and process antigens and self-constituents with high efficiency. This efficiency could have several consequences for antigen presentation *in vivo*, in which DCs clearly express very high levels of MHC II and invariant chain (Figs. 5 and 9). For example, DCs may be able to initiate a response when the concentrations of antigen are very low, and DCs may generate prolonged responses and memory when the amounts of MHC II peptide complexes are high and expressed for long periods. By modifying antigens with appropriate glycans, it might be possible to enhance targeting to DCs *in vivo*.

Cell Surface, Adhesion, and Costimulatory Molecules

In addition to antigen presentation, other accessory molecules, often termed "signal two," are needed to activate T cells. Many of the accessory molecules that are currently known to be expressed by DCs are not qualitatively different from other APCs. These include CD40, ICAM-1/CD54, LFA-3/CD58, B7-2/CD86, and heat-stable antigen. It is possible that unique accessory molecules will be identified on DCs. Alternatively, several adhesive and costimulatory molecules may need to work in concert to induce immunity *in vivo* and DCs, because they have several-fold higher levels of each, become much more efficient in activating T cells.

An estimate of the combined efficacy of accessory molecules on DCs relative to other APCs has been obtained by studying cells that are presenting known amounts of an MHC class II-binding superantigen. When DCs are presenting the same amount of superantigen as B cells, the T cell response per APC is increased at least 30-fold (Bhardwaj *et al.,* 1993).

As discussed later, the DCs that are normally resident in skin and other tissues are "immature" in that they are not potent stimulators of T cell responses. The immature DCs have low levels of CD40, ICAM-1, LFA-3, and B7-2. When the cells are cultured for 1 day, the amount of these accessory molecules quickly increases. Such high levels are not seen when B cells or macrophages are placed in culture (Inaba *et al.,* 1994) (Fig. 11). The upregulation is selective since other DC surface products, such as CD44 and CD45, and some accessory molecules, such as ICAM-3 and heat-stable antigen, are abundant on immature DCs. DCs in many sites are immature but are poised to upregulate CD40, CD54, CD58, and CD86 rapidly and to high levels. When the controls for the expression of DC accessory molecules are identified, we will understand how to regulate the efficacy of these potent APCs and thus the onset of T cell-dependent immunity *in vivo* (Schuler and Steinman, 1985; Romani *et al.,* 1989).

Cytokines

Originally it was thought that IL-1, which is an abundant product of macrophages and other cell types, was an essential lymphocyte-activating factor. After DCs were identified, however, it became apparent that neutralizing antibodies to IL-1 did not block T cell activation. Nonetheless, there is evidence that IL-1 can enhance DC function and IL-1β seems to be required for the induction of contact sensitivity, possibly via DCs in the skin.

The role of chemokines in the biology of DCs is under scrutiny. DCs can respond to select chemokines such as RANTES, MIP-1α, and MIP-1β. The response of DCs to chemokines has added meaning given the capacity of DCs to support HIV-1 replication and the recently discovered roles for chemokine receptors as coreceptors for HIV-1 entry.

DCs are known to make very high levels of a particular cytokine, IL-12 (Koch *et al.,* 1996; Cella *et al.,* 1996). The IL-12 heterodimer induces interferon-γ (IFN-γ) production from T cells and NK cells. When CTL responses to DCs are weak, IL-12 enhances CTL formation. DCs make large amounts of IL-12 upon interacting with T cells, particularly primed CD4$^+$ helpers. Activated T cells, especially CD4$^+$ cells, express CD40 ligand. Triggering via CD40 is one pathway for inducing high levels of production of biologically active IL-12 from DCs.

DCs may be able to induce both Th1 and Th2 types of responses. The "default pathway" may be Th1. The release of IL-12, induced by the expression of CD40 ligand on activated T cells, would skew DC-mediated responses to the Th1 type of immune response, with extensive IFN-γ production. The alternative Th2-type response develops if exogenous IL-4 is present (Seder *et al.,* 1992). This IL-4 might be produced, for example, by mast cells or by the NK1.1 subset of T cells.

Gaps in the Molecular Understanding of DCs

Once DCs were separated from other cell types, especially macrophages, their unusual potency as accessory cells and other distinctive properties became evident. However, it only recently has become feasible to approach DC function in-depth at a molecular level. This stems from the fact that large numbers of DCs can be generated with cytokines. DCs are being subjected to many technologies that can identify products that are expressed in a cell-restricted fashion.

It is evident that DCs also share with other cells many of the molecules that lead to strong T cell binding and costimulation, e.g., ICAMs, LFAs, and B7s. Nonetheless, these shared molecules can be expressed particularly quickly and

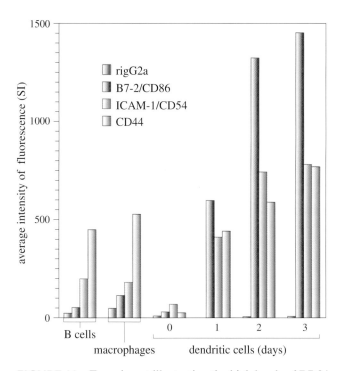

FIGURE 11 Experiment illustrating the high levels of B7-2/CD86 (in red) and ICAM-1/CD54 (in green) expressed on DCs relative to B cells and macrophages. The histogram also shows the rapid upregulation of these molecules from the moment the cells are placed in culture until the third day.

at high levels on DCs. Therefore, DCs may be regulated by distinct mechanisms and are poised to rapidly upregulate several APC functions. Some of the most distinctive physiological features of DCs relate to their maturation and homing properties *in vivo*. Next, we discuss antigen presentation *in vivo* and consider other important features of DCs as APCs.

Three Cellular Pathways and Subcellular Fates for Antigen *in Vivo*

Macrophages for Catabolism within Lysosomes

The bulk of the antigen that can be visualized shortly after injection into an animal is in macrophages. The antigens are delivered to lysosomes, which are powerful digestive organelles (Steinman and Cohn, 1972). An antigen, because it is delivered to lysosomes of macrophages, is degraded to the level of building blocks (i.e., amino acids and monosaccharides; Fig. 12) that move across the lysosome membrane. Larger peptides and saccharides do not permeate the membrane and exert an osmotic pressure that swells the cell.

To monitor sequestration in lysosomes *in vivo*, it is best to inject an indigestible material such as colloidal carbon. In the case of spleen, the black colloid is sequestered in macrophages of the marginal zone and red pulp (Fig. 13a). A small amount is retained on FDCs. Little if any uptake is observed in DCs, which can be selectively marked with monoclonal antibodies such as the CD11c integrin shown in Fig. 13 (Witmer-Pack *et al.*, 1993).

FDCs for Surface Retention of Native Antigen in Immune Complexes

A second fate for an injected antigen is to totally escape digestion and be retained intact on the surface of FDCs. Antigens (in relatively small amounts) can be retained in a native form as immune complexes attached to complement receptors on FDCs or perhaps Fc receptors (Fig. 4). It is likely that any small particle and antigen that fixes complement, directly or via antibody, can be retained on the surface of FDCs (Chen *et al.*, 1978). FDCs were originally called "dendritic macrophages," but the cells are fundamentally different from phagocytes. FDCs retain immune complexes intact for months in an extracellular location, whereas macrophages rapidly internalize and degrade these complexes.

Dendritic Cells and B Cells for Antigen Presentation via MIIVs

A third fate for an injected antigen is to be processed into peptides that are affixed to and presented by MHC molecules, especially the MHC products of DCs (Fig. 12). This fate of antigen, in contrast to the interaction of antigens

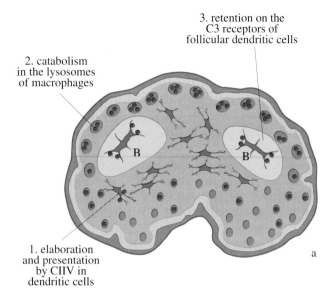

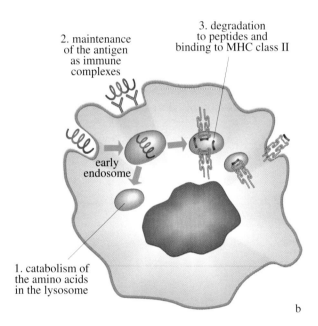

FIGURE 12 Three possible fates of antigen in different cell types at the level of an organ such as the lymph node (a) and at the level of subcellular pathways (b).

with macrophages and FDCs, is difficult to visualize for two reasons: The processed antigen is no longer in its native state, and the amount of antigen that is retained as peptide–MHC complexes may be small. For example, a macrophage can easily accumulate and degrade 10,000–100,000 protein molecules per hour following an exposure to 100 μg/ml of antigen (Steinman and Cohn, 1972), whereas many APCs only present 100 peptide molecules as peptide–MHC complexes following much greater exposures to antigen.

Functional assays, instead of direct visualization, have

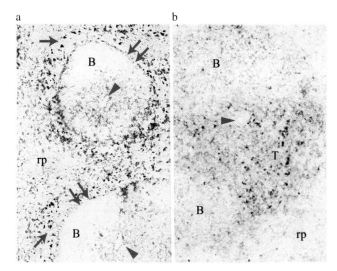

FIGURE 13 Monitoring antigen accumulation and processing in the spleen. (a) Endocytic macrophages take up black colloidal carbon, yielding a ring of black stain in the marginal metallophils on the inner side (double arrows) and on the outer side (single arrows) of the marginal sinus. The DCs, marked in brown with a monoclonal antibody against the CD11c integrin, do not accumulate the black colloidal carbon. The nonphagocytic DCs are abundant at the edge of the marginal zone, where T cells enter the white pulp, and surrounding the central arteries of the white pulp (arrowheads). (b) Formation of an endogenous MHC–peptide complex in a splenic white pulp nodule is visualized using the Y-Ae monoclonal to I-A^{b+} and I-Eα peptide. Note the intensely stained DCs in the T cell area (T) which surrounds the central artery of the spleen (arrowhead), the less intensely stained B cell areas (B), and the macrophages, essentially negative, in the marginal zone and in the red pulp (rp).

therefore been used to detect the presence of MHC–peptide complexes on APCs. A cell that is bearing MHC–peptide complexes and has other requisite "second signals" will stimulate the corresponding antigen-specific T cells without the further addition of antigen. If one injects protein antigens and then isolates different types of APCs, DCs seem to be the major cell type that have processed the antigen into a form that can be presented to T cells or T–T hybridomas (Bujdoso *et al.,* 1989; Crowley *et al.,* 1990; Liu and MacPherson, 1993). This is also the case for self-antigens that are being handled chronically *in vivo.* Therefore, although it is difficult to visualize the accumulation of antigens in DCs, functional assays indicate that these APCs are a major repository of peptide–MHC complexes *in vivo.*

Recently, monoclonal antibodies that directly identify MHC–peptide complexes were identified. The first such monoclonal, Y-Ae, recognizes a complex formed from an endogenous peptide (from the I-Eα chain) presented on an I-A^b class II molecule (Rudensky *et al.,* 1991). The Y-Ae antibody stains DCs more strongly than B cells (Fig. 13b).

The results with this one antibody to MHC–peptide complex show directly that DCs and B cells are forming endogenous MHC–peptide complexes *in vivo.*

Summary

There are three known fates of antigen *in vivo* — scavenging or destruction, retention as immune complexes, and presentation as peptides on MHC products — and for each, select subsets of cells and subcellular pathways predominate. Antigen scavenging and destruction primarily involves macrophages and delivery to lysosomes. The latter are end-stage digestive organelles that degrade proteins and monosaccharides to the level of their building blocks. Retention of native antigen involves FDCs and binding to receptors, particularly C3 receptors. These are not overtly internalized into the cell. Antigen processing and presentation as peptides on MHC products seems most efficient in DCs and B cells *in vivo.* Efficiency most likely is enhanced by receptors that target antigen to a specific subset of endosomal vacuoles, CIIVs (class II MHC-rich vacuoles).

Dendritic Cells Are Nature's Adjuvant, Able to Prime CD4 Helpers and CD8 Killers *in Vivo*

DCs that carry a foreign antigen will immunize T cells *in vivo* if injected into naive mice or rats. The induction of cell-mediated immunity *in vivo* with DCs was first achieved in transplantation experiments in which a single injection of MHC-incompatible DCs elicited the rejection of tolerated transplants. In these and other experiments, the DC was the only adjuvant that was used; hence the term nature's adjuvant to describe this system of APCs. This is not meant to imply that other "noncellular" adjuvants such as CFA operate to stimulate DCs, but this is possible.

When DCs were pulsed with foreign proteins and injected into the paws of syngeneic mice or rats, CD4$^+$ T cells in draining lymph nodes were primed specifically to the foreign protein that has been pulsed onto the DC (Inaba *et al.,* 1990; Liu and MacPherson, 1993). As in the case of priming with antigens in an adjuvant such as CFA, which was described in Table 3, the priming with antigen-bearing DCs was restricted to the draining lymph node, i.e., the nodes that drain afferent lymph from the site at which the DCs were injected (Fig. 14).

It was possible that in these experiments, DCs were simply delivering antigen for the recipient animal's APCs to present. However, this was ruled out in experiments in which DCs were of MHC type A, and the recipients were MHC A$\times$ B (Inaba *et al.,* 1990; Liu and MacPherson, 1993) (Table 4). F$_1$ T cells contain separate clones that recognize antigen in the context of either MHC A or MHC B. If DCs were priming T cells directly, the primed T cells should only

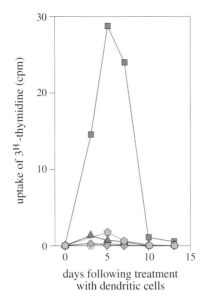

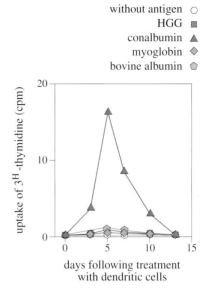

without antigen ○
HGG ■
conalbumin ▲
myoglobin ◆
bovine albumin ⬠

FIGURE 14 Priming *in vivo* of CD4⁺ T lymphocytes by APCs (expressed as the incorporation of tritiated thymidine during DNA synthesis). The APCs prime the T cells against a foreign protein; priming occurs in the lymph nodes that drain the site of DC injection and is specific for the protein pulsed onto the DCs: human gamma globulin, HGG (left) and conalbumin (right). cpm, counts per minute (reproduced with permission from Inaba *et al.*, 1990).

respond to peptides presented on MHC A; if the DCs were delivering antigen to host APCs, the primed T cells should respond to antigen on either MHC A or MHC B APCs. The experiments showed that DCs prime T cells directly since

DCs of MHC A only primed F_1 T cells that were restricted to MHC A.

CD8⁺ killer cells can also be primed by antigen bearing DCs *in vivo*. When mice are primed with DCs bearing ap-

TABLE 4

Primed F_1 T Cells Are Restricted to the Parental Strain of the Dendritic Cells Used in Priming[a]

MHC OF SPLEEN APCs USED TO BOOST F_1 T CELLS *IN VITRO*	DNA SYNTHESIS BY F_1 T CELLS			
	Primed with Ia^k DCs		Primed with Ia^d DCs	
	−Ag	+Ag	−Ag	+Ag
Experiment 1				
A × B	0.6	**11.4**	0.3	**11.8**
A	0.5	**22.4**	0.1	0.1
B	0.3	2.7	0.2	**22.9**
Experiment 2				
A × B	1.1	**25.0**	3.4	**67.1**
A	1.8	**26.2**	2.9	3.4
B	2.0	2.2	2.5	**69.0**

[a] T cells from F_1 mice were primed with protein antigen (Ag) pulsed onto parental strain H-2^d or H-2^k DCs. Seven days later, T cell priming in the draining lymph node was measured by challenging the T cells with antigen and APCs from either parent or from F_1 mice. Priming was measured as DNA synthesis in these challenge or boosting cultures. Note that the F_1 T cells were primed to recognize antigen in association with the MHC of the injected DCs (in bold) (from Inaba *et al.*, 1990).

propriate peptides from an experimental tumor, the animals develop specific CTLs and are resistant to a challenge with the specific tumor (Celluzzi *et al.,* 1996). The DCs can also induce resistance to tumors that are already growing in the animal. It will be important to test if DCs can be used to induce protective, $CD8^+$-dependent immunity to infectious agents, such as viruses and protozoa including plasmodia, the causative agent of malaria. Many investigators wish to use autologous DCs to try to increase the T cell-mediated resistance of patients who are chronically infected with a virus or who are not making an immune response to a malignancy. Although it is not established whether DCs have absolutely essential roles in initiating T cell responses *in vivo,* it is clear that DCs represent a specialized, physiologic, and more potent pathway.

Dendritic Cells Traffic via the Afferent Lymph or Bloodstream to the T Cell Areas of Draining Lymphoid Tissues

One property of DCs that helps explain their immunizing function *in vivo* is that DCs can migrate to the T cell areas of the draining lymphoid organ. This homing capacity places DCs in the path of recirculating T lymphocytes so that rare antigen-specific T cell clones can be plucked from the circulation and activated. Homing has been visualized using DCs labeled with a nontoxic dye that fluoresces blue (Austyn *et al.,* 1988). One day after injecting the DCs into the blood or into the footpad, sections are made of the spleen or draining lymph node, respectively. When the sections are double labeled with specific monoclonals (e.g., antibodies to T cells that fluoresce green), it is evident that the blue DCs home to the green T cell areas (Fig. 15a. Another approach is to apply monoclonals that see a polymorphic epitope that is expressed by the injected DCs and not the host. For example, when MHC mismatched DCs (labeled with a red cytochemical reaction) are injected, they home to the T cell areas (labeled with a brown cytochemical reaction) of the recipient lymph node (Fig. 15b). These homing experiments can utilize DCs that are obtained from any of a number of tissues, e.g., spleen, skin, and DCs grown from bone marrow progenitors. Macrophages, in contrast to DCs, home to the macrophage-rich regions of the lymphoid tissues.

It also is clear that DCs normally traffic in blood and afferent lymph so that there is a steady-state movement of DCs from nonlymphoid to lymphoid organs. DCs in organ cultures of skin begin to migrate into afferent lymphatics. Afferent lymphatics can be cannulated in larger animals, such as sheep or humans, and DCs routinely are found trafficking at 10,000 cells/h or more. After an injection of an antigen intramuscularly, the DCs in the lymph carry the antigen in a form that is stimulatory for specific T cells (Buj-

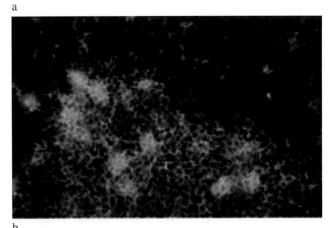

a

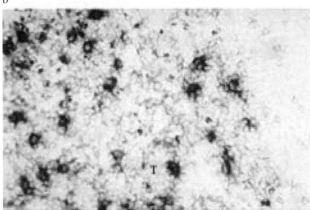

b

FIGURE 15 Homing of DCs to the T lymphocyte areas. (a) Micrograph of a spleen section; the DCs are shown in blue and the T cells in green (courtesy of J. M. Austyn, J. W. Kupiec-Weglinski, D. W. Hankins, and P. J. Morris). (b) Micrograph of a lymph node section; the DCs are shown in red and the T cells in brown.

doso *et al.,* 1989; Liu and MacPherson, 1993). To do this experiment in smaller animals, investigators surgically remove certain draining lymph nodes. The afferent lymphatics then reconnect to efferent lymphatics, which in turn drain directly into the thoracic duct. In this way, cannulation of the thoracic duct (which normally lacks DCs) provides the investigator with access to DCs in "pseudoafferent" lymph. Again, the DCs are found to carry antigens and particulates that have been injected into the site that is drained by the corresponding lymphatics. Therefore, if the mesenteric lymph nodes are removed, DCs that drain the intestine move directly into the thoracic duct and carry antigens that are deposited in the intestinal lumen (Liu and MacPherson, 1993). If the celiac lymph nodes are removed, liver DCs move into the thoracic duct and carry particulates that are injected into the bloodstream (Matsuno *et al.,* 1996).

The DCs within the T cell areas of lymphoid tissues can express high levels of the costimulator molecule B7-2 or

CD86 (Inaba *et al.,* 1994), but there may be heterogeneity in these APCs. The existence and development of distinct functional subsets of DCs is an area of active research but is not considered here.

The Onset of T Cell-Dependent Immunity *in Vivo:* Maturation and Migration of Dendritic Cells

Distinct Afferent and Efferent Limbs to the Immune Response *in Vivo*

The findings in the preceding sections lead to a working hypothesis whereby T cell-mediated immune responses *in vivo* develop in successive afferent and efferent limbs (Fig. 16). The afferent limb begins when DCs capture and process antigens and are induced to mature, e.g., to express high levels of accessory molecules for T cell stimulation. The DCs then migrate via the blood or lymph to the T cell areas of spleen or lymph node. In the case of antigens that gain direct access to the T cell areas, these could be presented by local lymphoid DCs. The DCs that present antigen or superantigen in the T cell areas select the corresponding antigen-reactive T cell clones and induce the T cells to divide and differentiate (to secrete lymphokines or to become killers). For the efferent limb, activated T cells or T blasts exit the lymphoid organ via the efferent lymphatics. These T cells, because they have been activated, are able to egress the vasculature wherever it is inflamed and expresses lymphocyte adhesion molecules. For example, stimulated T cells have active β2 and β1 integrins that can bind to the corresponding ICAM and VCAM on inflamed endothelium. Since the site of antigen deposition is likely to have inflamed vessels, the antigen-activated T cells would be mobilized to the appropriate place to interact with APCs that carry out the efferent limb of immunity. For example, T blasts could activate macrophages to kill microbes in an infectious focus, help B cells to make antibodies at mucosal surfaces, or kill targets that are infected with a virus.

The two phases to immune responses *in vivo* correspond to the two phases that occur during immune responses *in vitro* (Fig. 8). The afferent limb depends on DCs that activate antigen-specific T cells in the draining lymphoid organ; the efferent limb depends on T blasts which work together with other APCs to eliminate antigens in an inflamed tissue. In fact, cell-mediated immune responses such as contact sensitivity and graft rejection have long been known to be associated with lymphocyte blastogenesis (enlargement and entry into cell cycle) in the T cell areas of the draining lymph nodes. It is also known that during an immune response, many T blasts emerge into the efferent lymphatics and thoracic duct lymph. These blasts are cleared rapidly from the circulation. Therefore, the physiology of immune

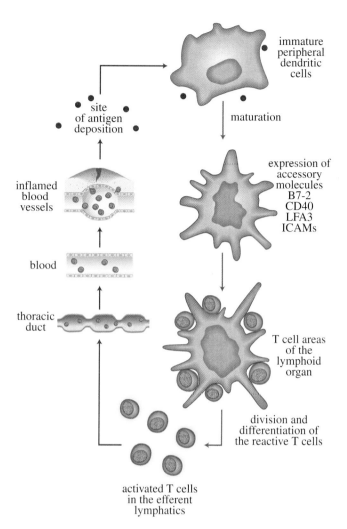

FIGURE 16 Afferent and efferent limbs of T cell-mediated immunity *in vivo.* Antigen presentation is required for both limbs. The afferent limb (blue arrows) response involves small T lymphocytes that are activated by DCs usually in the lymphoid organ during primary responses. The efferent limb (red arrows) response depends on T blasts and endothelial cells of the inflamed vessels that interact to allow the T blasts to leave the bloodstream and activate other APCs to generate effector responses.

responses *in vivo* may be as dependent on the active contributions of DCs and vascular endothelium as on antigen recognition and T cell activation.

Maturation of DCs: *In Vitro* Studies

As described previously, the afferent limb *in vivo* is oversimplified in that the onset of cell-mediated immunity *in vivo* likely involves a series of changes in DCs termed maturation. DCs in most peripheral tissues are immature in the sense that they are not powerfully immunostimulatory. As first recognized for skin DCs, called Langerhans cells, immature DCs express much lower levels of antigen-presenting MHC products and T cell adhesion and costimulator molecules. DCs in skin, for example, express low levels of cell surface MHC II and barely detectable levels of CD40, CD54 or ICAM-1, CD58 or LFA-3, and CD86 or B7-2. DCs in organs other than skin may have even less MHC II and accessory molecules. When the DCs are placed in culture, however, the cells mature. Cell size and processes increase, the levels of surface MHC products increase 10-fold or more, and all the previously mentioned T cell interaction molecules are expressed at high levels (Schuler and Steinman, 1985; Inaba *et al.,* 1994). This maturation, an obvious control point for the onset of T cell-mediated immunity, does not occur if the DCs are purified away from the surrounding cells. The latter are producing cytokines that mediate maturation, for example, GM-CSF (Witmer-Pack *et al.,* 1987) (Fig. 17).

Antigen-capture mechanisms are also regulated during maturation. There are clear-cut examples in which immature DCs (i) are able to capture exogenous proteins and some particulates, (ii) synthesize high levels of the invariant chain that is required for presentation on MHC II, and (iii) have many acidic endocytic vesicles and MIICs (Romani *et al.,* 1989). These properties can all be extinguished in DCs that have very high levels of accessory molecules and T cell-stimulating capacity. Much of this information is derived from studies on epidermal dendritic cells or Langerhans cells that undergo the previously discussed changes when placed in culture. Another illustrative situation occurs with immature DCs in human blood. The development of these cells in GM-CSF plus IL-4 apparently "freezes" the DC in the antigen-capture mode (Sallusto and Lanzavecchia, 1995). The cells show remarkable pinocytic activity, especially with macropinocytic vesicles, and the internalized substrates are targeted to MHC II-rich vesicles. If the cells are then exposed to cytokines that can be found in monocyte-conditioned media, further maturation of cell surface accessory molecules occurs and the levels of endocytic activity and MIIVs decrease markedly.

Similar events can occur *in vivo* (Matsuno *et al.,* 1996). If colloidal carbon or latex particles are administered intravenously, immature DCs in the liver internalize the particles and move into the afferent lymph for homing to the celiac lymph nodes. When isolated from the lymph, the maturing DCs no longer take up the particles. This regulation of the antigen uptake and processing functions of DCs

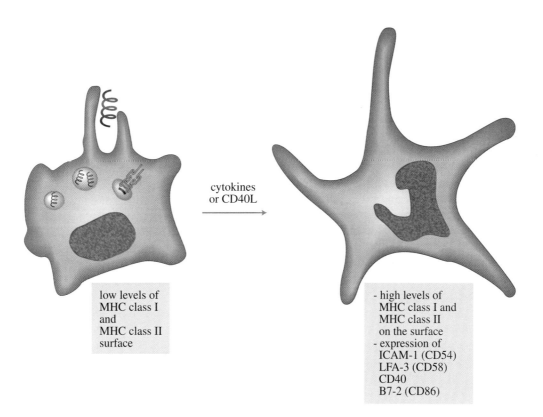

cytokines
or CD40L

low levels of
MHC class I
and
MHC class II
surface

- high levels of
  MHC class I and
  MHC class II
  on the surface
- expression of
  ICAM-1 (CD54)
  LFA-3 (CD58)
  CD40
  B7-2 (CD86)

FIGURE 17 Maturation of DCs occurs in the presence of cytokines or CD40L; it includes cell enlargement and process formation, downregulation of the MIIC system of endocytic vacuoles, and upregulation of several accessory molecules. Examples are to be found in Pierre *et al., Nature* **388,** 787–792 (1997), Inaba *et al., J. Exp. Med.* **191,** 927–936 (2000), and Turley *et al., Science* **288,** 522–527 (2000).

means that DCs primarily display peptides that are acquired at the site of antigen deposition.

The Coupling of Maturation and Migration *in Vivo*

The skin is the classic organ for studying the two most powerful and reliable T cell responses *in vivo* — transplant rejection and contact allergy. In both instances, afferent lymphatics must be intact for T cell priming to occur. During skin transplantation and contact allergy, epidermal DCs begin to mature and migrate into the lymph within the first day of stimulation. Normally, epidermal DCs are sessile, immature cells. However, if skin is transplanted (even to a syngeneic host), or if a contact allergen is applied (but not a nonsensitizing chemical), the DCs start to move into the dermis and afferent lymphatics. Simultaneously, the expression of MHC molecules and B7-2/CD86 increases dramatically (Larsen *et al.,* 1990). The cues for these changes toward immunogenicity have yet to be worked out, but tumor necrosis factor-α (TNF-α) has been implicated. Likewise, when lipopolysaccharide (LPS) is given to rats, there is a marked efflux of DCs into mesenteric lymph; this migration is TNF-α dependent (MacPherson *et al.,* 1995).

Although skin is the best studied organ, DC maturation and migration to the lymph nodes also occur in organs other than skin such as the lung and the spleen (Holt *et al.,* 1993). Following administration of LPS, the DCs in the splenic marginal zone mature and move into the periarterial T cell areas (De Smedt *et al.,* 1996).

One way to analyze the regulation of DC function is to identify transcriptional controls. Members of the important NF-κB or rel family have recently been studied. DCs express very high levels of all NF-κB/rel family proteins, i.e., rel-A, rel-B, rel-C, p50, and p65. Many of these are active when electromobility shift assays are performed on nuclear extracts from DCs, particularly p50 and rel-B (Granelli-Piperno *et al.,* 1995). Curiously, mature DCs are missing another group of otherwise ubiquitous, transcriptional control proteins called Sp1.

Localization of Responding T Cells *in Vivo*

Early Histologic Studies

In the early 1960s, it became apparent that lymphocytes were the mediators of immunity, and that large lymphocytes or blasts represented lymphocytes that had been activated by antigens. Histologic studies were carried out on lymphoid tissues undergoing cell-mediated immune responses. When draining lymph nodes were stimulated by application of contact allergens or transplants, two of the most powerful stimuli for T cell responses *in vivo*, numerous lymphoblasts developed in the DC-rich T cell areas.

T-Dependent Antibody Responses

The antibody response is more complicated since several types of APCs may be involved at different stages of the response. Three stages can be distinguished and all occur in different parts of the lymphoid organ. Early during a T-dependent antibody response, the responding B cells are found at the junction of T cell and B cell areas. One interpretation of this finding is that helper T cells are primed by antigens presented on DCs, and then the activated helper cells interact with clonally selected B cells, much as occurs *in vitro* (Inaba and Steinman, 1985) (Fig. 8).

Later in the response, by Days 5–7, plasma cells accumulate in the macrophage-rich red pulp and medulla of a stimulated spleen and lymph node. Plasma cells are considered to be the end stage of antibody-forming cell development. The cells contain abundant Ig within a well-developed rough endoplasmic reticulum. It is possible that the macrophage contributes a cytokine such as IL-6 or IL-10 that is important in B cell maturation. In any case, plasma cells are rare in MALTs, which lack the macrophage-rich regions of lymph node and spleen. Instead, the plasma cells are scattered beneath the mucosal epithelium, especially the lamina propria of the gut, where macrophages are again abundant. Long-lived plasma cells are also abundant in the bone marrow.

The third stage of the antibody response involves germinal center formation within the B cell follicles. This becomes readily evident at about Day 7 and leads to the affinity maturation of memory B cells. Germinal center development typically is driven by helper T cells, which in turn may be activated by the newly recognized germinal center DC (Grouard *et al.,* 1996).

Observations on TCR Transgenic T Cells

In all of the previously discussed studies, antigen-specific T cells were not identified directly. TCR transgenic mice provide large numbers of antigen-specific T cells that can be stimulated with defined antigens *in vivo* and identified with anti-TCR antibodies. By using adoptive transfer of such TCR transgenic T cells, immunologists can now use exciting new methods to monitor and visualize antigen-specific T cells *in situ* (Kearney *et al.,* 1994). When the corresponding antigen is given subcutaneously with adjuvant, the antigen-specific TCR transgenic cells accumulate first in the paracortical, DC-rich region of the draining lymph node as proposed in Fig. 16. After proliferation for several days, the T cells move to the follicles. The T cells then disappear slowly but remain hypersensitive to antigen restimulation, a property of memory cells. In contrast, when the antigen is given by the intravenous route, the T cells proliferate less extensively in the paracortex, do not move into the follicles, and are hyporesponsive to antigen restimulation. These findings show that the form and route of antigen administration influence the type of T cell response *in vivo*. The differences may entail different APCs (e.g., B cells in the case of intravenous antigens and DCs for antigens in adjuvants), or the function of APCs may be influenced by adjuvants.

Alterations of T Cell Immunity during Memory

Enhanced responses to recall immunization ("memory") are characteristic of both T cells and B cells. Memory B and T cells have undergone clonal expansion in the classical sense of Burnet and Lederberg. However, each type of lymphocyte also undergoes qualitative improvements in function but with different tactics. The B cell shapes a higher affinity receptor through somatic mutation and affinity maturation in germinal centers. Also, the B cell antigen receptor undergoes isotype switching with associated changes in biological function. Memory T cells do not alter their receptor for antigen, either by somatic mutation or by isotype switching, but instead change their adhesion, costimulatory, and homing molecules (Fig. 18a).

Some of these distinct features of memory T cells can facilitate their interaction with DCs in nonlymphoid tissues. There is a marked reduction in CD62L, a selectin that nor-

mally guides the naive T cell across the HEV and into the lymphoid organs. There is an upregulation in the adhesive function of integrins, especially the $\beta 2$ integrin LFA-1/CD11a, the $\beta 1$ integrin VLA-4/CD49d, and the $\alpha E\beta 7$ and $\alpha 4\beta 7$ mucosal homing integrins. Such changes in homing molecules alter the recirculating properties of the T cell, allowing memory T cells to traffic through nonlymphoid extravascular sites such as skin and the gut. In fact, memory T cells are the main T cells in afferent lymph, which drains the extravascular spaces of most nonlymphoid organs except the brain. As a result of traffic through nonlymphoid tissues, memory T cells gain more rapid access to DCs, which is the reciprocal of a primary response in which DCs migrate to the naive T cells that recirculate through lymphoid organs. In fact, within 1 day memory cells aggregate with local DCs in the classical delayed-type hypersensitivity skin test for T cell memory. Better T cell access to DCs may help to sustain memory or to accelerate the immune response of primed individuals.

Less consideration has been given to the possibility that DCs may contribute to the generation of B and T cell memory. DCs and memory T cells are enriched in afferent lymph relative to blood. In fact, memory T cells often bind to DCs in the afferent lymph (Pope *et al.*, 1994) (Fig. 18b). It is possible that these DC–T cell conjugates are able to traffic to the B cell follicles (rather than the expected T cell areas) to

- expression of CD58 (LFA-3)
- high levels of CD44
- loss of CD62b (selectin L)
- change of CD45 isoform
  (expression of CD45RO);
  loss of CD45RA and CD45RB

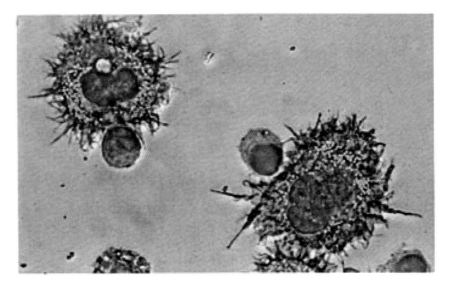

FIGURE 18 (a) Distinctive surface features of memory T cells including surface markers. (b) Micrograph of T memory cells (in blue) bound to DCs.

help initiate the germinal center reaction that underlies B cell memory.

Another possibility is that the maintenance of memory resembles positive selection in the thymus and involves the recognition of MHC products, which do not bear cognate peptides, on DCs within nonlymphoid organs. Nonspecific peptides may provide a sufficient signal to keep the memory cell alive, much like thymic epithelium is thought to rescue self-restricted thymocytes from programmed cell death. The constant traffic of memory cells through extravascular tissues and lymph would permit frequent contact with MHC products on DCs (Fig. 18b), and these might serve to positively select or maintain the viability of the memory lymphocyte.

Mucosal Immunity *in Vivo*: Positive and Negative Regulation at the Level of DCs

Mucosal surfaces are the principal sites through which antigens gain access to the body, including DCs. The area of respiratory and gastrointestinal epithelium is staggering, literally "football fields" in size in humans. Vaccine design is increasingly directed to mucosal immunization.

The term MALT refers to the organized collections of lymphocytes that resemble lymphoid organs in having B cell follicles and T cell areas, as well as the associated FDCs and DCs. A general feature of MALT is that antigen gains direct access to the lymphoid organs without a need for afferent lymphatics.

The Peyer's patch of the intestine (gut-associated lymphoid tissue) is the best studied, but there are lymphoid nodules in other mucosae, e.g., in the airways. The epithelium covering these lymphoid collections, particularly in the gut, contain special cells called M cells because of their unusual "microfolds" (Fig. 19a). Antigens and particulates cross the epithelium via these M cells, which are the functional equivalent of afferent lymphatics. DCs seem to be present in substantial numbers beneath the M cells and very likely are the first cells to capture antigens (Kelsall and Strober, 1996). These DCs might then move into the T cell areas of the MALT or may traffic via the lymphatics to draining lymph nodes, e.g., in the gut to the mesenteric lymph nodes and in the lung to the peribronchial and mediastinal lymph nodes.

Other types of MALT are the lymphoid organs in the pharynx. Best known are the tonsils in the oral pharynx and the adenoid at the back of the nose. These and the many other lymphoid collections in the pharynx are collectively known as Waldeyer's ring. Tonsils are not found in rodents, but they are prominent in humans. The mucosal surfaces covering pharyngeal lymphoid organs function comparably to a network of afferent lymphatics, allowing direct access of antigens to the lymphoid tissue (Fig. 19b). It is possible that some parts of the adenoid and tonsil surface epithe-

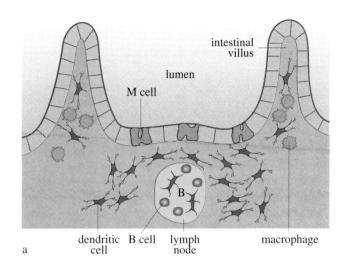

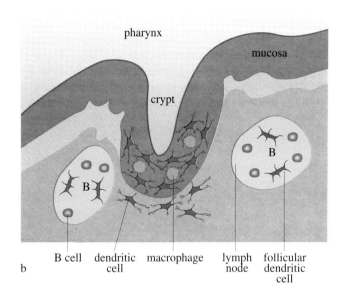

FIGURE 19 Schematic transverse sections of mucosal-associated lymphoid organs. (a) The Peyer's patch of the intestine, showing the position of the M cells, from which antigens cross the intestinal epithelium, and the mass of DCs below. (b) The mucosal surface of the tonsil which functions in a similar way to a network of afferent lymphatics, allowing direct access of antigens to the lymphoid tissue. The dendritic cells are particularly abundant around the crypts.

lium have antigen-transporting M cells, but other parts are a particularly rich source of DCs, many of which lie along crypts or plications (Frankel *et al.,* 1996).

In addition to organized MALT, DCs are found elsewhere on mucosal surfaces. DCs have been identified in the lamina propria of the gastrointestinal tract and the alveolar septae and airway epithelium of the lung (Holt *et al.,* 1987; Maric *et al.,* 1996). These DCs appear to be immature. Maturation into potent DCs can be inhibited by mucosal macrophages, as in the lamina propria of the gut or the alveolar spaces of the lung. The inhibitory mechanism is uncertain, but NO, transforming growth factor-β, and IL-10 are among many candidate mediators. As a result, DCs do not induce immunity to the numerous antigenic proteins that continually access the airway and gut. In contrast to mucosal surfaces, skin lacks many typical macrophages and therefore may be a particularly immunogenic route. Nevertheless, the processes that apply to skin and lymph nodes can occur in MALT. For example, if active DCs bearing an antigen are injected into an airway, the cells gain access to the draining lymph nodes in the chest, and the T cells are primed.

Can Immune Responses Be Manipulated Clinically at the Level of APCs?

Many major diseases, including diseases that are prevalent in the young, involve the immune system, especially T cells (Table 5). Therapeutic approaches in clinical immunology have concentrated on antigens and T cells, and APCs have been overlooked. For example, approaches to vaccines favor different formulations of antigens and noncellular adjuvants, and therapies of ongoing disease focus on suppression of T cell function or on adoptive transfer of T cells. Are DCs a physiologic pathway to *in vivo* immunity that must now be emphasized?

HIV-1 Infection

DCs may contribute in a major way to the pathogenesis of HIV-1 infection, i.e., to support HIV-1 replication and the death of memory-type CD4$^+$ T cells. DCs express both CD4 and chemokine receptors for HIV-1 binding and entry. In the past, it was difficult to document productive infection of DCs in HIV-1-infected individuals or of normal DCs exposed to HIV-1 *in vitro*. Recently, evidence has been obtained that certain DCs, especially at mucosal surfaces, can amplify HIV-1 infection, but T cells also seem essential. A

TABLE 5

Can Unknowns in Clinical Immunology Be Studied at the Level of APCs?

Rejection of transplants

Resistance to microbial infections and vaccine design

Resistance to tumors

Contact allergy and other delayed-type hypersensitivity reactions

Diseases attributed to autoimmunity, e.g., multiple sclerosis,
  juvenile diabetes, rheumatoid arthritis, psoriasis

AIDS

---

critical finding was made *in vitro* when HIV-1 was applied to tight conjugates between DCs and memory T cells derived from skin (Pope *et al.*, 1994). The skin was used as an accessible model for many of the mucosal surfaces through which HIV-1 gains access during sexual and perinatal transmission. That is, many of these mucosal surfaces are keratinized epithelia in which numerous DCs are located. HIV-1 replicated actively in cutaneous and mucosal DC–T cell conjugates in the apparent absence of cognate antigen since the infected cells were not in cell cycle (Fig. 20a). HIV-1 caused the DCs and T cells to fuse, creating heterokaryons or syncytia (Fig. 20b, left) that were sites for viral replication and cell death. One important feature of the syncytia, which may explain the high levels of viral gag protein synthesized there, is that DCs contribute high levels of the NF-κB proteins, whereas T cells contribute Sp1 proteins (Granelli-Piperno *et al.*, 1995). Together these two types of factors can upregulate transcription via the HIV-1 promoter.

The mucosal surfaces of the adenoid and tonsil *in vivo* contain HIV-1-infected syncytia that are derived from DCs (Frankel *et al.*, 1996). These mucosa normally have regions with activated DCs and memory-type T cells, i.e., cells that resemble the permissive mixtures of DCs and T cells that are isolated from human skin. Even in asymptomatic, HIV-1-infected individuals, cells containing intracellular HIV-1 protein are found within and just beneath the mucosa epithelium overlying the lymphoid tissue proper (Fig. 20b, right). Cells with intracellular viral protein are difficult to identify in the peripheral lymph nodes and skin. Instead, HIV-1 protein in lymph node is primarily found on FDCs which presumably retain HIV-1 virions and antigen as immune complexes extracellularly.

Therefore, the long-standing difficulty in identifying active sites for HIV-1 replication *in vivo* can be viewed as follows. HIV-1 replication is driven by DC–memory T cell conjugates, especially at mucosal surfaces. These infected cells then elicit a strong CD8$^+$ response (CTLs, IFN-γ, and chemokines) that limits the spread of the infection. HIV-1 is not really a latent infection but rather a battleground in which DCs control both sides of the conflict.

Tolerance in Autoimmunity and Transplantation

For candidate autoimmune diseases (insulin-dependent diabetes, multiple sclerosis, rheumatoid arthritis, psoriasis, etc.), there is optimism that investigators will soon be able to identify critical self-antigens. For example, the enzyme glutamic acid decarboxylase may be an important autoantigen in juvenile diabetes. Researchers will have to learn to use the emerging information on clinical autoantigens to induce a desired state of tolerance.

One approach to inducing tolerance might be to present critical autoantigenic peptides on a nonimmunogenic APC, perhaps a small B cell. DCs should not be overlooked, how-

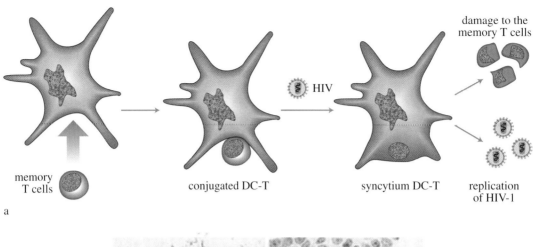

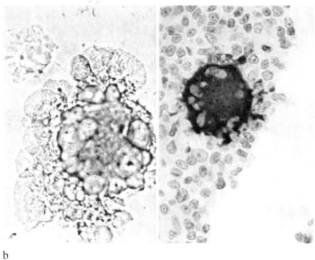

FIGURE 20 (a) DCs conjugated with T cells support the replication of HIV-1 virus, especially within multi-nucleated syncytia. (b) Micrographs of infected syncytia (brown staining for the HIV-1 gag protein) *in vitro* (left) and *in vivo* (right) from adenoid mucosa. Micrographs courtesy of Dr. Melissa Pope & Sara Frankel.

ever. For example DCs may kill antigen-responding T cells (Suss and Shortman, 1996) and contribute to peripheral (postthymic) tolerance.

If therapy for autoimmune disease is to begin at an early age in genetically predisposed individuals, as for insulin-dependent diabetes or pemphigus vulgaris, then the thymic DC may be an additional site to target the antigen to induce central tolerance. DCs within the medulla are likely to be critical in inducing clonal deletion to self-antigens.

The issue of tolerance also surfaces in the field of transplantation. Patients still suffer from problems of chronic rejection. In some studies, only 10–20% of patients with kidney transplants are able to survive 10 years, despite immunosuppression. One mechanism for chronic graft rejection may be the chronic presentation of polymorphic (especially MHC-encoded), graft-derived peptides on host DCs. If so, the induction of tolerance, especially to graft-derived peptides, would provide a means for blocking chronic re-

jection. Another strategy for inducing tolerance is to learn to induce chimerism between donor and host, especially at the level of DCs. In this way, host T cells might regard donor MHC molecules as "self."

Active Immunotherapy and Vaccines

Resistance to tumors, like that to some prevalent infectious diseases (e.g., influenza and malaria), may best be mediated by CD8+ killer cells. Whereas it is straightforward to elicit recall responses from primed human CD4+ T cells in tissue culture, this is not true for CD8+ responses. CD8+ killers are typically generated by chronic stimulation and the retrieval of a few T cell clones. A major obstacle to obtaining CD8+ responses in culture may be the need to target peptides for presentation on MHC class I molecules of DCs. Recently, it was shown that influenza infects DCs and elicits recall CD8+ responses, both CD8+ T cell proliferation and killing. The DC is so effective that nonreplicating forms of influenza are efficiently presented by human DCs (Bender *et al.,* 1995). The virus is treated with heat or UV light in such a way that nucleic acid synthesis cannot occur, but adsorptive uptake and endosomal fusion via the viral hemagglutinin is not blocked. In this way, virions can gain access to the cytoplasm and to the efficient antigen-presentation capacities of DCs. It is possible that many enveloped viruses can gain access to the cytoplasm of DCs, thereby leading to successful presentation without being able to replicate there.

IL-12 is another recently recognized component in the design of vaccines for strong T cell-mediated immunity. This cytokine skews NK and T cell responses toward high levels of IFN-γ production and, as a result, strong activation of macrophages with antimicrobial properties. The capacity of DCs to produce IL-12 is remarkably high (Cella *et al.,* 1996; Koch *et al.,* 1996); therefore, targeting antigens via DCs should yield stronger vaccines for cell-mediated immunity.

The success in generating large numbers of DCs *ex vivo* with cytokines is encouraging several groups to use these APCs for boosting host resistance *in vivo* (Fig. 21). A further modification would be to modify the DCs by genetic engineering with the relevant viral or tumor protein. Presumably, the DCs in a cancer patient are not presenting tumor antigens since there is little evidence that an immune response is occurring, e.g., expanded numbers of tumor-specific CTLp. The goal is to reinfuse autologous, antigen-bearing DCs to induce an active immune response, particularly a killer cell response, against tumor cells or chronically infected cells. This new form of active immunotherapy takes advantage of many of the findings discussed in this article.

Vaccination is of course the success story in immunology. However, most successful vaccines operate by inducing protective or neutralizing antibodies. Vaccines are still lacking for many diseases in which protection requires cell-mediated immunity more than antibodies. For example, CD4$^+$ T cells that generate high levels of IFN-γ, and CD8$^+$ T cells that have killer activity and make appropriate cytokines and chemokines, could have important protective roles in AIDS, malaria, and influenza. One strategy that has been emphasized is to identify peptides as subunit vaccines. Although safe, peptide vaccines have two major obstacles. One is MHC restriction since most peptides will only bind to a restricted number of MHC types. The other is the lack of targeting of peptides to specialized DCs. In the future, it might be more effective for vaccines to target proteins to DCs, e.g., via genetic manipulation or with nonreplicating forms of an infectious agent. Then, "nature's adjuvant" would tailor the peptides to the individual's MHC and provide the many other specializations that would enhance immunity *in vivo*. Recently identified specific steps in DC uptake, maturation, and migration provide some guideposts for such experiments.

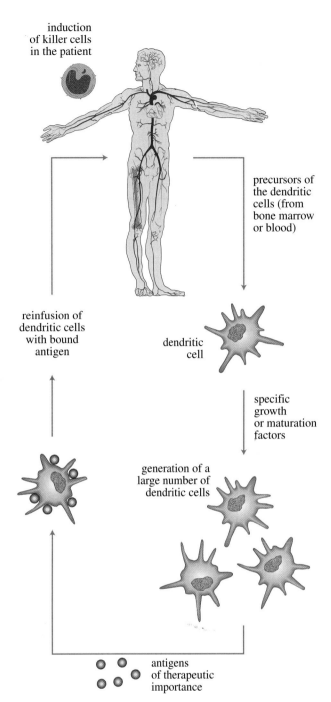

FIGURE 21 A possible strategy for the use of DCs for active immunotherapy *in vivo:* the generation of large numbers of DCs from their progenitors, the charging of these DCs *ex vivo* with antigens of clinical importance, and the strong accessory function of the DCs upon reinfusion into autologous recipients.

Manuscript originally submitted for publication in 1996.

References Cited

AUSTYN, J. M., KUPIEC-WEGLINSKI, J. W., HANKINS, D. F., and MORRIS, P. J. (1988). Migration patterns of dendritic cells in the mouse. Homing to T cell-dependent areas of spleen, and binding within marginal zone. *J. Exp. Med.* **167,** 646–651.

BENDER, A., BUI, L. K., FELDMAN, M. A. V., LARSSON, M., and BHARDWAJ, N. (1995). Inactivated influenza virus, when presented on dendritic cells, elicits human CD8⁺ cytolytic T cell responses. *J. Exp. Med.* **182,** 1663–1671.

BHARDWAJ, N., YOUNG, J. W., NISANIAN, A. J., BAGGERS, J., and STEINMAN, R. M. (1993). Small amounts of superantigen, when presented on dendritic cells, are sufficient to initiate T cell responses. *J. Exp. Med.* **178,** 633–642.

BUJDOSO, R., HOPKINS, J., DUTIA, B. M., YOUNG, P., and MC-CONNELL, I. (1989). Characterization of sheep afferent lymph dendritic cells and their role in antigen carriage. *J. Exp. Med.* **170,** 1285–1302.

CELLA, M., SCHEIDEGGER, D., PALMER-LEHMANN, K., LANE, P., LANZAVECCHIA, A., and ALBER, G. (1996). Ligation of CD40 on dendritic cells triggers production of high levels of Interleukin-12 and enhances T cell stimulatory capacity: T-T help via APC activation. *J. Exp. Med.* **184,** 747–752.

CELLUZZI, C. M., MAYORDOMO, J. I., STORKUS, W. J., LOTZE, M. T., and FALO, L. D., JR. (1996). Peptide-pulsed dendritic cells induce antigen-specific, CTL-mediated protective tumor immunity. *J. Exp. Med.* **183,** 283–287.

CHEN, I. L., FRANK, A. M., ADAMS, J. C., and STEINMAN, R. M. (1978). Distribution of horseradish peroxidase [HRP]–anti HRP immune complexes in mouse spleen, with special reference to follicular dendritic cells. *J. Cell Biol.* **79,** 184–199.

CROWLEY, M., INABA, K., and STEINMAN, R. M. (1990). Dendritic cells are the principal cells in mouse spleen bearing immunogenic fragments of foreign proteins. *J. Exp. Med.* **172,** 383–386.

DE SMEDT, T., PAJAK, B., MURAILLE, E., HEINEN, E., DE BAETSELIER, P., URBAIN, J., LEO, O., and MOSER, M. (1996). Positive and negative regulation of dendritic cell function by lipopolysaccharide *in vivo. J. Exp. Med.* **184,** 1413–1424.

FRANKEL, S. S., WENIG, B. M., BURKE, A. P., MANNAN, P., THOMPSON, L. D. R., ABBONDANZO, S. L., NELSON, A. M., POPE, M., and STEINMAN, R. M. (1996). Replication of HIV-1 in dendritic cell-derived syncytia at the mucosal surface of the adenoid. *Science* **272,** 115–117.

GRANELLI-PIPERNO, A., POPE, M., INABA, K., and STEINMAN, R. M. (1995). Coexpression of REL and SP1 transcription factors in HIV-1 induced, dendritic cell–T cell syncytia. *Proc. Natl. Acad. Sci. USA* **92,** 10944–10948.

GROUARD, G., DURAND, I., BANCHEREAU, J., and LIU, Y.-J. (1996). Dendritic cells capable of stimulating T cells in germinal centers. *Nature* **384,** 364–367.

HOLT, P. G., SCHON-HEGRAD, M. A., and OLIVER, J. (1987). MHC class II antigen-bearing dendritic cells in pulmonary tissues of the rat. Regulation of antigen presentation activity by endogenous macrophage populations. *J. Exp. Med.* **167,** 262–274.

HOLT, P. G., OLIVER, J., BILYK, N., MCMENAMIN, C., MCMENAMIN, P. G., KRAAL, G., and THEPEN, T. (1993). Downregulation of the antigen presenting cell function[s] of pulmonary dendritic cells *in vivo* by resident alveolar macrophages. *J. Exp. Med.* **177,** 397–407.

INABA, K., and STEINMAN, R. M. (1985). Protein-specific helper T lymphocyte formation initiated by dendritic cells. *Science* **229,** 475–479.

INABA, K., STEINMAN, R. M., VAN VOORHIS, W. C., and MURAMATSU, S. (1983). Dendritic cells are critical accessory cells for thymus-dependent antibody responses in mouse and man. *Proc. Natl. Acad. Sci. USA* **80,** 6041–6045.

INABA, K., METLAY, J. P., CROWLEY, M. T., and STEINMAN, R. M. (1990). Dendritic cells pulsed with protein antigens *in vitro* can prime antigen-specific, MHC-restricted T cells *in situ. J. Exp. Med.* **172,** 631–640.

INABA, K., WITMER-PACK, M., INABA, M., HATHCOCK, K. S., SAKUTA, H., AZUMA, M., YAGITA, H., OKUMURA, K., LINSLEY, P. S., IKEHARA, S., MURAMATSU, S., HODES, R. J., and STEINMAN, R. M. (1994). The tissue distribution of the B7-2 costimulator in mice: Abundant expression on dendritic cells *in situ* and during maturation *in vitro. J. Exp. Med.* **180,** 1849–1860.

JIANG, W., SWIGGARD, W. J., HEUFLER, C., PENG, M., MIRZA, A., STEINMAN, R. M., and NUSSENZWEIG, M. C. (1995). The receptor DEC-205 expressed by dendritic cells and thymic epithelial cells is involved in antigen processing. *Nature* **375,** 151–155.

KEARNEY, E. R., PAPE, K. A., LOH, D. Y., and JENKINS, M. K. (1994). Visualization of peptide-specific T cell immunity and peripheral tolerance induction *in vivo. Immunology* **1,** 327–339.

KELSALL, B. L., and STROBER, W. (1996). Distinct populations of dendritic cells are present in the subepithelial dome and T cell regions of the murine Peyer's patch. *J. Exp. Med.* **183,** 237–247.

KOCH, F., STANZL, U., JENNEWIEN, P., JANKE, K., HEUFLER, C., KÄMPGEN, E., ROMANI, N., and SCHULER, G. (1996). High level IL-12 production by murine dendritic cells: Upregulation via MHC class II and CD40 molecules and downregulated by IL-4 and IL-10. *J. Exp. Med.* **184,** 741–747.

LARSEN, C. P., STEINMAN, R. M., WITMER-PACK, M., HANKINS, D. F., MORRIS, P. J., and AUSTYN, J. M. (1990). Migration and maturation of Langerhans cells in skin transplants and explants. *J. Exp. Med.* **172,** 1483–1493.

LIU, L. M., and MACPHERSON, G. G. (1993). Antigen acquisition by dendritic cells: Intestinal dendritic cells acquire antigen administered orally and can prime naive T cells *in vivo. J. Exp. Med.* **177,** 1299–1307.

MACPHERSON, G. G., JENKINS, C. D., STEIN, M. J., and EDWARDS, C. (1995). Endotoxin-mediated dendritic cell release from the intestine: Characterization of released dendritic cells and TNF dependence. *J. Immunol.* **154,** 1317–1322.

MARIC, I., HOLT, P. G., PERDUE, M. H., and BIENENSTOCK, J. (1996). Class II MHC antigen [ia]-bearing dendritic cells in the epithelium of the rat intestine. *J. Immunol.* **156,** 1408–1414.

MATSUNO, K., EZAKI, T., KUDO, S., and UEHARA, Y. (1996). A life stage of particle-laden rat dendritic cells *in vivo:* Their terminal division, active phagocytosis and translocation from the liver to hepatic lymph. *J. Exp. Med.* **183,** 1865–1878.

NIJMAN, H. W., KLEIJMEER, M. J., OSSEVOORT, M. A., OORSCHOT, V. M. J., VIERBOOM, M. P. M., VAN DE KEUR, M., KENEMANS, P., KAST, W. M., GEUZE, H. J., and MELIEF, C. J. M. (1995). Antigen capture and MHC class II compartments of freshly isolated and cultured human blood dendritic cells. *J. Exp. Med.* **182,** 163–174.

POPE, M., BETJES, M. G. H., ROMANI, N., HIRMAND, H., CAMERON, P. U., HOFFMAN, L., GEZELTER, S., SCHULER, G., and STEINMAN, R. M. (1994). Conjugates of dendritic cells and memory T lymphocytes from skin facilitate productive infection with HIV-1. *Cell* **78,** 389–398.

ROMANI, N., KOIDE, S., CROWLEY, M., WITMER-PACK, M., LIVINGSTONE, A. M., FATHMAN, C. G., INABA, K., and STEINMAN, R. M. (1989). Presentation of exogenous protein antigens by dendritic cells to T cell clones: Intact protein is presented best by immature, epidermal Langerhans cells. *J. Exp. Med.* **169,** 1169–1178.

RUDENSKY, A. Y., RATH, S., PRESTON-HURLBURT, P., MURPHY, D. B., and JANEWAY, C. A., JR. (1991). On the complexity of self. *Nature* **353,** 660–662.

SALLUSTO, F., and LANZAVECCHIA, A. (1995). Dendritic cells use macropinocytosis and the mannose receptor to concentrate antigen in the MHC class II compartment. Downregulation by cytokines and bacterial products. *J. Exp. Med.* **182,** 389–400.

SCHULER, G., and STEINMAN, R. M. (1985). Murine epidermal Langerhans cells mature into potent immunostimulatory dendritic cells *in vitro. J. Exp. Med.* **161,** 526–546.

SEDER, R. A., PAUL, W. E., DAVIS, M. M., and FAZEKAS DE ST. GROTH, B. (1992). The presence of interleukin 4 during *in vitro* priming determines the lymphokine-producing potential of CD4$^+$ T cells from T cell receptor transgenic mice. *J. Exp. Med.* **176,** 1091–1098.

STEINMAN, R. M., and COHN, Z. A. (1972). The interaction of soluble horseradish peroxidase with mouse peritoneal macrophages *in vitro. J. Cell Biol.* **55,** 186–204.

STEINMAN, R. M., and COHN, Z. A. (1974). Identification of a novel cell type in peripheral lymphoid organs of mice. II. Functional properties *in vitro. J. Exp. Med.* **139,** 380–397.

STEINMAN, R. M., and WITMER, M. D. (1978). Lymphoid dendritic cells are potent stimulators of the primary mixed leukocyte reaction in mice. *Proc. Natl. Acad. Sci. USA* **75,** 5132–5136.

STEINMAN, R. M., NOGUEIRA, N., WITMER, M. D., TYDINGS, J. D., and MELLMAN, I. S. (1980). Lymphokine enhances the expression and synthesis of Ia antigen on cultured mouse peritoneal macrophages. *J. Exp. Med.* **152,** 1248–1261.

SUSS, G., and SHORTMAN, K. (1996). A subclass of dendritic cells kills CD4 T cells via Fas/Fas-ligand induced apoptosis. *J. Exp. Med.* **183,** 1789–1796.

WITMER-PACK, M. D., OLIVIER, W., VALINSKY, J., SCHULER, G., and STEINMAN, R. M. (1987). Granulocyte/macrophage colony-stimulating factor is essential for the viability and function of cultured murine epidermal Langerhans cells. *J. Exp. Med.* **166,** 1484–1498.

WITMER-PACK, M. D., CROWLEY, M. T., INABA, K., and STEINMAN, R. M. (1993). Macrophages, but not dendritic cells, accumulate colloidal carbon following administration *in situ. J. Cell Sci.* **105,** 965–973.

General References

CAUX *et al.* (1996). In *Blood Cells* (T. Whetton and J. Gordon, Eds.). Plenum, London.

LIU, Y.-J., GROUARD, G., DE BOUTEILLER, O., and BANCHEREAU, J. (1996). *Int. Rev. Cytol.* **166,** 139–179.

MOLL, H. (1995). *The Immune Functions of Epidermal Langerhans Cells.* Landes, Austin, TX.

SCHULER, G. (1991). *Epidermal Langerhans Cells.* CRC Press, Boca Raton, FL.

STEINMAN, R. M. (1991). The dendritic cell system and its role in immunogenicity. *Annu. Rev. Immunol.* **9,** 271–296.

WILLIAMS, I. A., EGNER, W., and HART, D. N. J. (1994). Isolation and function of human dedritic cells. *Int. Rev. Cytol.* **153,** 41–104.

JOHN C. CAMBIER
IDAN TAMIR

Division of Basic Sciences
Department of Pediatrics
National Jewish Medical and Research Center
Denver, Colorado 80206, USA

Signal Transduction and Cell Fate Decisions in the Immune System

Lymphocyte development, activation, and differentiation are regulated by a large number of ligand–receptor pairs, the most central of which are receptors for antigen and antigenic peptide-complexed major histocompatibility antigens expressed by B cells and T cells, respectively. These receptors determine the ultimate specificity of the response by transducing the initial signals that activate precursors of antigen-specific effectors or inactivate autoreactive cells. Consistent with this apparent dichotomy in signal transduction consequence, recent findings indicate that these receptors do not function as binary switches. Rather, antigen receptor ligation can result in a variety of immediate biological consequences depending on antigen structure and context and the differentiative stage of the responding cell. "Tuning" of receptor signaling to drive distinct responses is accomplished in part by coligation, or concurrent ligation, of other receptors with the antigen receptor and by changing expression and/or use of receptor-accessory molecules and signal transduction intermediaries as a function of cell differentiative stage. Here, we discuss the biochemical basis of B cell antigen receptor signal transduction and review several specific mechanisms by which responses through these receptors are modified by coreceptors and accessory molecules.

Introduction

Lymphocyte development and generation of immune responses are precisely orchestrated processes involving many regulatory mechanisms. The complexity of this regulation, which appears unparalleled in other organ systems, has evolved, in part, because of very strong selective pressures placed on the immune system. The system must respond swiftly using a variety of strategies to eliminate potential pathogens that seek to overcome the host while avoiding potentially devastating immunity to autologous antigens. The former requires the development and deployment of a variety of specialized effector mechanisms from the immune system's armamentarium. The latter requires that mechanisms must exist to purge the system of cells that bear autoreactive antigen receptors so that they cannot be activated later by exogenous cross-reactive antigens. Although generation of tolerance to self versus immunity reflects the most overt of cell fate decisions made by cells of the immune system, many component decisions underlie these processes. For example, B cells may respond to antigen by undergoing apoptosis, editing antigen receptors to alter specificity, proliferation, or priming to better present antigen to T cells and receive T cell signals which induce B cell proliferation and differentiation to become antibody-secreting plasma cells.

Transduction of signals with distinct consequences through receptors of unchanging specificity requires that signals be interpreted or integrated differently depending on the circumstance in which they are received. Evidence indicates that three major mechanisms exist by which antigen receptor signals are modified to generate specialized responses. First is the use of coreceptors, which are receptors that when ligated simultaneously, often coligated, with the antigen receptor transduce signals that modify or complement those transduced by the receptor. The second involves the changing expression and use of signal transduction effectors as cells differentiate. The consequence is a change in the response to receptor ligation depending on the cell's developmental or differentiative stage. Third, signal strength as determined by antigen affinity and valency can, by determining quantity, quality, and duration of the signal, modify biologic outcome. Multiple examples of these types of regulation exist in both T and B lymphocyte biology. Prominent examples in B cells include antigen receptor signal amplification and attenuation by coligation of the type II complement receptor (CR2) and the receptor for immunoglobulin G constant regions (FcγRIIB1), respectively. Antigen receptor signaling can be modified by alterations in expression of accessory or effector molecules, including CD22, CD45, and CD19, that increase as B cells mature. These effectors can exert both negative (CD22) and positive (CD45 and CD19) effects on receptor signaling. Finally, strength of signal effects are exemplified by clonal deletion tolerance induced by cell-associated antigen versus clonal anergy induced by soluble autologous antigens.

The molecular bases by which lymphocytes transduce and integrate the multitude of signals that regulate their function, and the mechanisms by which these signals are modified to initiate alternative cell fate decisions, is a fascinating and important area of study. Here, we review some current examples of these processes focusing on signal transduction by the B cell antigen receptor and its modification by coreceptors and accessory/effector molecules.

The Molecular Basis of Antigen Receptor Signal Transduction

Beginning at the transition from the pre-B to immature stage, B cells express specific receptors that function to internalize antigen for subsequent processing and presentation to T cells and transduce signals that lead to multiple, often alternative responses. The receptor is a multisubunit complex composed of a membrane-bound form of immunoglobulin (mIg), noncovalently associated with disulfide-linked heterodimers of CD79a and CD79b (Fig. 1) (Gold and DeFranco, 1994; Cambier, 1995; Pao et al., 1997a). Immunoglobulins of all isotypes can function as B cell antigen receptor (BCR) components. Whereas immature B cells express only mIgM-containing receptors, mature B cells coexpress mIgM and mIgD and memory B cells express the isotype (i.e., IgG, IgA, or IgE) that their differentiated daughter cells will secrete. Evidence indicates that immu-

noglobulin isotype content may affect the output of BCR signaling. For example, a variety of studies conducted during the past two decades suggest that mIgM and mIgD transduce partially distinct signals, but the biochemical basis of this difference has not been defined. Membrane-associated immunoglobulins differ from their secreted counterparts in containing a short spacer sequence at the normal heavy chain C terminus, a single transmembrane-spanning region, and a short [3 (mIgM and mIgD) to 28 (mIgG) residues] cytoplasmic tail. Although the mIg tail contributes to signaling (Weiser et al., 1997), CD79a and CD79b function as the receptor's primary transducers (Gold and DeFranco, 1994; Cambier, 1995; Pao et al., 1997a). They are members of the immunoglobulin superfamily, containing a single extracellular immunoglobulin-like domain, a single transmembrane-spanning region, and cytoplasmic tails of 48 (CD79b) and 61 (CD79a) amino acid residues. The cytoplasmic domains of both CD79a and CD79b contain the sequence motif $YX_2LX_7YX_2L$, termed the immunoreceptor tyrosine-based activation motif (ITAM), that functions as the receptor's interface with cytoplasmic effectors of signal transduction. As will be discussed later, both the amino acid content and residue spacing are critical for receptor binding and activation of SH2-containing accessory and effector molecules.

Resting BCRs associate with protein tyrosine kinases (PTKs) of the Src family by an interaction involving the receptor's ITAM and a domain found in the N terminus of the PTKs (Fig. 1). The kinases bind selectively to the CD79a ITAM via a site which involves the DCSM sequence found in CD79a but not in CD79b. The binding of nonphosphorylated ITAMs to the kinase N-terminal region, although of low affinity, is stabilized in vivo by concurrent kinase association with the adjacent plasma membrane via N-terminal myristate and palmitate moieties. The N-terminal sequence of Src family PTKs, which is responsible for their association with the resting receptor, differs extensively among Src family members and is thus termed "unique." In addition to this domain, Src family members share four structurally homologous domains, namely, the catalytic domain (termed SH1 for src homology 1), SH2 and SH3 (src homology 2 and 3, respectively), and a C-terminal tail containing a conserved tyrosyl residue (Brown and Cooper, 1996). SH2 and SH3 domains are found in many proteins and are involved in mediating protein–protein interactions. SH2 domains bind to phosphotyrosyl residues of the same or other proteins with specificity that is governed by the phosphotyrosyl-flanking residues. SH3 domains interact with proline-rich domains found within many adaptor proteins (Brown and Cooper, 1996). The role(s) of these interactions in BCR signaling will be discussed later.

The resting BCR-associated Src family PTKs exist in a dynamic equilibrium between inactive and partially active states. The transition between these states was shown to

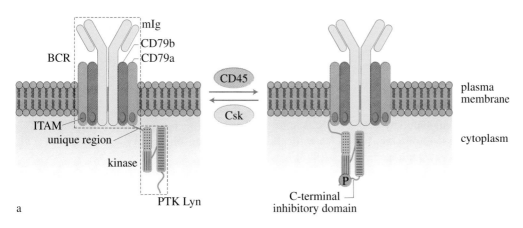

a

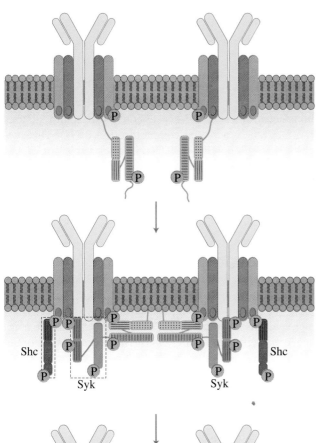

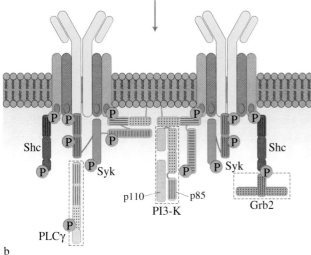

b

FIGURE 1 Autonomous signaling by the B cell antigen receptor (BCR). This receptor, consisting of mIg, CD79a, and CD79b, is constitutively associated with Src family tyrosine kinases (PTK, in green), usually PTK Lyn, via kinase unique region interactions with the ITAM sequences (blue) of CD79a and CD79b. The horizontally striped regions represent the SH1 domains, vertical striped regions represent SH2 domains, and dotted regions represent SH3 domains. (a) The kinases usually exists in two states — partially active, nonphosphorylated kinase and repressed, C-terminal phosphorylated (P) kinase — in an equilibrium determined by the relative activities of Csk and CD45. (b) BCR aggregation, by clustering nonphosphorylated receptor-associated kinase molecules, causes phosphorylation of the tyrosine residues in the ITAM regions and in the kinases. ITAM phosphorylation leads to recruitment of additional effectors, such as Shc (red) and Syk (blue), that associate via their SH2 domains. These in turn become phosphorylated and, together with PTK Lyn, recruit secondary effectors, such as PLCγ (light blue), PI3-K (light brown), and GRB2 (pink). Dynamic interactions within this extended complex lead to effector activation and downstream signal transduction.

correlate with the phosphorylation of a single tyrosyl residue found at the kinase's C terminus (Songyang *et al.*, 1995; Brown and Cooper, 1996; Pao *et al.*, 1997a; Pao and Cambier, 1997). The phosphorylation of this residue, which is conserved among Src-family members, is mediated by a cytosolic PTK named Csk (for C-terminal src-family kinase) and results in an enzymatically-inactive state. Dephosphorylation of this residue, which is mediated by the transmembranal protein tyrosine phosphatase (PTP) CD45, leads to a partially active kinase (Fig. 1). Several studies have suggested that kinase inactivation by C-terminal tyrosyl phosphorylation is a result of its intramolecular interaction with the kinase's SH2 domain (Brown and Cooper, 1996). The recently determined three-dimensional structures of two Src family members, c-Src and Hck, solved by X-ray crystallography, have actually demonstrated such an interaction.

These studies have also shed light on another facet of Src regulation involving its SH3 domain. It has been known for some time that the kinase's SH3 domain plays a role in regulating its activity; however, the mechanism was obscure since Src family kinases lack a proline-rich domain which, by binding its own SH3, could be responsible for this effect. The three-dimensional structure reveals, however, an unexpected intramolecular interaction between Src's SH3 and the back of its catalytic domain, largely mediated by the linker segment connecting the SH2 and catalytic domains (Fig. 1). The linker segment, which does not contain the consensus SH3-binding PXXP motif, nonetheless assumes a left-handed polyproline-II helical conformation that occupies the SH3 ligand binding site. Thus, via intramolecular interactions, both SH2 and SH3 domains contribute to the stabilization of a closed conformation which apparently does not allow the catalytic domain to assume an enzymatically active state. Therefore, destabilization of this conformation by either C-terminal dephosphorylation or competitor binding to Src's SH2 and/or SH3 domains should result in an increase in Src's catalytic activity. Consistent with this hypothesis, a recent study has shown a 50-fold increase in the activity of Hck, a Src family member, upon interaction with the HIV Nef protein, which is a high-affinity ligand of Hck's SH3 domain (Moarefi *et al.*, 1997).

Additional regulation of the activity of Src family kinases is achieved by phosphorylation of another conserved tyrosyl residue (Y416 for c-Src) found within the activation segment of the catalytic domain. Phosphorylation of this tyrosine is believed to result in a conformational change in the catalytic domain leading to its activation. Specifically, this event is believed to involve the interaction of the formed phosphotyrosyl residue with positively charged residues buried within the catalytic domain. By relieving steric hindrance, this interaction is assumed to allow a glutamate residue within the amino-terminal lobe of the catalytic domain to contact and reposition a lysine residue which is involved in coordinating the phosphate group of ATP (Ya-

maguchi and Hendrickson, 1996). The phosphorylation at Y416 was shown to follow first-order kinetics with respect to the kinase and is thus believed to be mediated by Src family PTKs. Thus, the close apposition of receptor-associated Src family molecules during receptor aggregation should allow for their efficient transphosphorylation (Moarefi *et al.*, 1997).

Consistent with the previously discussed observations, BCR aggregation leads to rapid tyrosyl phosphorylation of several Src family PTK members, including Lyn, Fyn, Lck, Blk, and Fgr, of which at least some have been shown to associate with the resting receptor (Cambier, 1995; Pao *et al.*, 1997a). This phosphorylation putatively occurs at the activation segment of these kinases, accounting in part for the observed increase in their enzymatic activity. Studies using B cells from Lyn knockout mice and Lck-negative T cell lines demonstrate dependence of receptor-mediated tyrosyl phosphorylation events (of ITAM and most other cellular substrates) on the expression of Src family kinases (A. DeFranco, personal communication, 1997). This suggests that this family of PTKs is responsible for ITAM tyrosyl phosphorylation and downstream signaling and is the most proximal kinase family activated upon receptor aggregation.

The expression of CD45 has been shown to be required for effective receptor aggregation-induced tyrosyl phosphorylation of Src family kinases, suggesting that kinase transphosphorylation is dependent on dephosphorylation of the C-terminal inhibitory motif (Pao and Cambier, 1997). However, tyrosyl phosphorylation of many other intracellular substrates, including receptor subunits, upon BCR aggregation was not compromised in CD45-negative cells, suggesting either that other kinases may compensate for Src family kinase activity or that the activity of C-terminally phosphorylated Src family kinases toward the ITAM is intact. In this respect, even the C-terminal tyrosyl phosphorylated, and thus repressed, kinase was shown to be capable of limited *in vitro* transphosphorylation, suggesting that Src family kinase activity is not completely inhibited by this covalent modification (Moarefi *et al.*, 1997).

Tyrosyl phosphorylation of Src family PTKs upon BCR aggregation is accompanied by the phosphorylation of conserved ITAM tyrosines (Cambier, 1995; Pao *et al.*, 1997a). These tyrosyl residues are within the optimal context, with respect to their flanking residues, for phosphorylation by Src family kinases (Songyang *et al.*, 1995; Schmitz *et al.*, 1996); however, only the N-terminal ITAM tyrosyl residue (Y182) of CD79a fully meets Src family kinase substrate recognition requirements, as determined *in vitro* for both Lyn and Blk (Schmitz *et al.*, 1996). Consistent with this, it has been shown that the presence of this residue within the BCR is critical for the tyrosyl phosphorylation of both CD79a and CD79b upon receptor aggregation and comprises the major *in vivo* tyrosyl phosphorylation site within the BCR.

Studies employing baculovirus-expressed Src family kinases and bacterially expressed CD79a/b further confirmed this observation, showing an approximately fourfold higher *in vitro* phosphorylation of CD79a at position Y182 than at Y193 (Pao *et al.*, 1997a). The results of these studies suggest that only a small fraction of the aggregated receptors are phosphorylated on all conserved ITAM tyrosines. The possible implications with respect to BCR signaling are discussed later.

Upon their phosphorylation, ITAM tyrosines have been shown to act as a scaffold, recruiting effectors such as Shc, Syk, and additional Src family kinases to the receptor via their SH2 domains (Pao *et al.*, 1997a). Binding of Src family kinases to the tyrosyl-phosphorylated ITAM further increases their specific activity, and this may account for induced phosphorylation of additional substrates such as Syk. Syk association with the doubly tyrosyl-phosphorylated ITAM (ppITAM) was shown to involve both of Syk's tandem SH2 domains binding to the two, properly spaced, phosphotyrosyl residues of the ppITAM (Kurosaki *et al.*, 1994, 1995). This binding was shown to increase Syk's specific activity and lead to its tyrosyl phosphorylation (Kurosaki *et al.*, 1994, 1995). The latter may be mediated by receptor-associated Src family PTKs, Syk transphosphorylation, or both, resulting in a further increase in Syk's activity and the creation of binding sites for other SH2-containing effectors. Thus, both Src and Syk classes of tyrosine kinases are activated as a consequence of their interaction with the phosphorylated ITAM. It is important to note, however, that the asymmetrical nature of ITAM phosphorylation dictates that only a fraction (probably <10%) of the aggregated receptors can bind and activate Syk; the majority will bind to single SH2 domain-containing signaling and/or adaptor molecules. Analysis of sequence requirements for the recognition of tyrosyl-phosphorylated peptides by different SH2-containing proteins revealed sensitivity to amino acids at positions −2 to 3 with respect to the phosphotyrosyl residue. Thus, phosphorylation of CD79a at Y182 creates a putative binding site for the SH2 domain of the adaptor protein Shc, whereas phosphorylation at Y193 is optimal for Src family kinase SH2 binding (Songyang *et al.*, 1994). Consistent with this result, Shc has also been shown to undergo tyrosyl phosphorylation in response to BCR aggregation and to bind *in vitro*, via its SH2 domain, to the ppITAM of CD79a/CD79b.

Available evidence from *in vitro* binding and coprecipitation studies indicates that the activated BCR, with its associated Syk and Src family kinases, is able to recruit an additional "layer" of effectors to the receptor (Pleiman *et al.*, 1994). These include phospholipase Cγ (PLCγ), which binds to phosphorylated Syk via its SH2 domain, and phosphatidylinositol 3-kinase (PI3-K), which binds to the SH3 domain of Src family kinases via a proline-rich sequence within its p85 subunit. The latter interaction may also lead to additional activation of Src family PTKs, as discussed previously. Conversely, the binding of PI3-K to Src family SH3 domains was shown to activate its enzymatic activity (Pleiman *et al.*, 1994), specifically the phosphorylation of phosphatidylinositides at the D3 position of the inositide ring. Upon association with the receptor, PLCγ undergoes tyrosyl phosphorylation, which leads to its activation and the hydrolysis of phosphatidylinositol-4,5-P_2 (PIP$_2$) producing two soluble products, phosphatidylinositol-1,4,5-P_3 (IP$_3$) and diacylglycerol. These soluble second messengers have been implicated in calcium ion mobilization and protein kinase C activation, respectively. The recruitment of Shc to the tyrosyl-phosphorylated receptor suggests a link between BCR activation and the $p21^{ras}$ signaling pathway. Shc was shown in various cell types, including B cells, to bind the adaptor molecule Grb2, which further mediates binding to SOS and $p21^{ras}$. The guanine nucleotide exchange activity of SOS converts $p21^{ras}$ from an inactive GDP-bound state to an active GTP-bound one.

The apparent linkage of specific ITAM phosphotyrosyl residues to specific downstream signaling pathways (e.g., CD79a tyrosine 182 to Shc) and the asymmetry of ITAM tyrosyl phosphorylation suggest a mechanism by which strength of signal may affect cell fate decisions. Thus, differences in affinity and/or avidity of the interaction between the B cell surface immunoglobulin and the antigen may determine BCR subunit tyrosyl phosphorylation pattern, which may in turn lead to different signaling pathways being activated with different biological outcomes. An example of such regulation was provided by a study of the biochemical events initiated upon T cell receptor (TCR) stimulation by antigen presenting cell-carried, peptide-occupied major immunohistocompatibility complex (MHC) molecules (Madrenas *et al.*, 1995). Although agonist peptide-occupied MHC molecules elicited "normal" tyrosyl phosphorylation patterns of TCR subunits and subsequent phosphorylation and activation of ZAP-70 (which shares structural homology with Syk), single amino acid changes within either the peptide or the MHC sequence were shown to lead to partial agonism or antagonism resulting in a change in TCR subunit tyrosyl phosphorylation pattern and a lack of ZAP-70 phosphorylation and activation. Thus, small changes in the affinity of TCR–(peptide–MHC) interaction may lead to distinct patterns of receptor ITAM phosphorylation, leading to altered recruitment and activation of signaling molecules and a modified biological response. In this respect, receptor density within the aggregate and/or aggregate lifetime, both dependent on the on- versus off-rates of receptor–ligand interactions, may determine the biochemical pathways initiated by such an aggregate. One possible way of rationalizing this correlation involves a model which has been termed "kinetic proofreading." This model suggests that the sequence of biochemical events initiated upon receptor aggregation introduces a time lag between initial receptor aggregate formation and the full-fledged biochemical response. Hence, the biochemical output of receptor aggregation, and thus the elicited cellular response, would depend on the extent of signal propagation before aggregate dissociation. It is possible that CD79a/b ITAM tyrosyl residues are also differentially phosphorylated as a function of cluster lifetime, thus providing different signals as a function of the time that has passed from receptor aggregation onset. In addition, coaggregation of specific coreceptors and/or accessory molecules with the BCR was shown in several cases to have a profound effect on the biochemical detail of the signaling process and thus on the cellular response. The best studied of these cases are discussed in the following section.

Accessory Function of CD19 in BCR Signaling

CD19 is a highly conserved, B lymphocyte-specific transmembranal glycoprotein that is expressed from the pro-B to the plasma cell stage. Its extracellular portion contains two immunoglobulin-like domains, and its 242-amino acid cytoplasmic tail contains nine potential tyrosine phosphorylation sites. It has been shown to coprecipitate with mIgM in Daudi B cells, to co-cap with aggregated BCR, and to undergo tyrosyl phosphorylation in response to BCR aggregation (Carter et al., 1997). Recent studies indicate that CD19 associates with BCR via the 17-amino acid juxtamembrane sequence in its intracellular domain. Thus, CD19 is either preassociated with the BCR or associates with it following BCR aggregation, and it serves as a substrate of BCR-activated PTKs. As noted previously, CD19 is expressed on pro- and pre- B cells that lack mIg and its aggregation in this context also leads to tyrosyl phosphorylation and activation of members of the Src family of PTKs. Evidence indicates that CD19 expression increases during B cell progress from the immature to mature phenotype. These and other coprecipitation data suggest that CD19 is directly associated with src family kinases. CD19 also occurs as a component of the type 2 complement receptor (CR2) complex and is involved in signaling via this complex, as will be discussed later. Finally, mice in which the CD19 is ablated by homologous recombination are defective in BCR-mediated activation of proliferation, germinal center formation, and generation of antibody responses, indicating that CD19 plays an important positive role in B cell activation and differentiation.

Recent evidence indicates that BCR-mediated activation of certain downstream pathways is dependent on CD19 expression. Tyrosyl phosphorylation of CD19 in response to BCR aggregation occurs predominantly at two residues, Y484 and Y515, located at the distal portion of the CD19 tail within YXXM motifs (Buhl et al., 1997). These phos-

photyrosyl residues have been shown to mediate the association of the p85 subunit of PI3-K with CD19 (Tuveson et al., 1993). The phosphorylation of these tyrosyl residues was found to be critical for BCR-mediated PI3-K activation but not required for other signaling events, such as the tyrosyl phosphorylation of CD79a, CD79b, Lyn, Syk, PLCγ1, and PLCγ2 (Buhl et al., 1997). Although BCR-mediated PLCγ tyrosyl phosphorylation is independent of CD19, phosphoinositide hydrolysis following BCR ligation is much reduced in the absence of CD19 expression. Recent evidence from our laboratory and others indicates that tyrosyl phosphorylation of PLCγ in the absence of CD19 is not sufficient by itself to induce effective phosphoinositide hydrolysis and calcium mobilization (Buhl et al., 1997; Pao and Cambier, 1997). Studies involving PI3-K inhibitors and B cells from CD19 knockout mice indicate that PI3-K activation is required for receptor-mediated phosphoinositide hydrolysis and Ca^{2+} mobilization (Buhl et al., 1997; Hippen et al., 1997).

Although the mechanism by which PI3-K complements PLCγ activity is not known, several lines of evidence suggest a key role for phosphatidylinositol-3,4,5-trisphosphate (PtdIns3,4,5P3), the product of phosphatidylinositol-4,5-bisphosphate (PtdIns4,5P2) phosphorylation by PI3-K. PtdIns3,4,5P3 was shown to bind to SH2 domains and to modulate PI3-K interaction with tyrosine phosphorylated proteins. In addition, phosphoinositides were shown to interact with another protein module known as the pleckstrin homology (PH) domain. This domain is found in the sequence of Bruton's tyrosine kinase (Btk), a PTK which was shown to be required for BCR-mediated Ca^{2+} mobilization, particularly for extracellular influx. Interestingly, Btk was shown to selectively interact with PtdIns3,4,5P3-carrying vesicles via its PH domain. Thus, Btk may be the complementary effector required for PLCγ-mediated phosphoinositide hydrolysis. Finally, it is tempting to speculate that in the active receptor complex, PI3-K simultaneously binds both to the Src family kinase SH3 domain and to the phosphotyrosyl residues of CD19, bridging these effectors (Fig. 2). It is also possible that this multipoint interaction is necessary for optimal PI3-K activation (Pleiman et al., 1994; Buhl et al., 1997).

In summary (Fig. 2), aggregation of the BCR leads to the assembly of an extended complex composed of CD79a, CD79b, CD19, and effectors which include proximal components of multiple downstream signal propagative pathways known to be activated by BCR aggregation. The dynamic inter- and intraprotein interactions that occur within the complex presumably trigger the activation of pathways that transduce BCR signals to the nucleus: the PLCγ–phosphoinositide hydrolysis–Ca^{2+} mobilization pathway, the Shc–Ras–MAPK pathway, and the PI3-K pathway. The importance of CD19 in amplifying and modulating the signal initiated by BCR aggregation is further exemplified

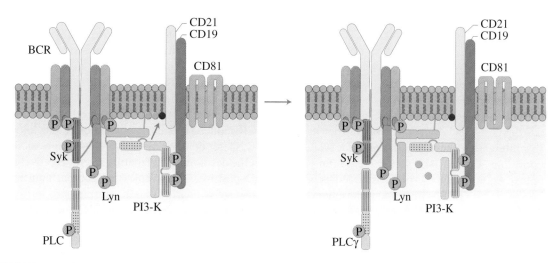

FIGURE 2 Accessory function of CD19 in BCR signal transduction. BCR signaling is modified qualitatively by CD19, which is phosphorylated upon BCR aggregation, and functions to recruit and activate PI3-K. CD19 expression and PIP$_3$ (purple circles) generation by PI3-K are necessary for the hydrolysis of PIP$_2$ (pink circles) mediated by PLCγ and for the subsequent mobilization of Ca^{2+} induced by BCR aggregation.

by its presence in membranal protein complexes (e.g., CR2) that act as BCR coreceptors.

The Type 2 Complement Receptor– An Amplifying BCR Coreceptor

One of the most biologically potent B cell antigen receptor signal modifiers is the receptor for iC3b and C3dg fragments of the complement C3 component. These fragments are generated upon complement fixation by immune complexes and certain biological surfaces. The fragments bind covalently to the fixing conjugate via an ester linkage and couple the complex to the CR2 complement receptor. Classical studies by Pepys demonstrated that mice depleted of complement using cobra venom factor make diminished responses to antigen (Pepys, 1974; Fearon and Carter, 1995). Subsequently, genetically determined deficiencies in C3a, C4, and C2 (involved in C3 activation) were found to correlate with reduced immune responsiveness. The role of C3dg and its receptor in the immune response was most elegantly demonstrated in studies using anti-hen egg lysozyme (HEL) immunoglobulin transgenic mice (Dempsey et al., 1996). It was shown that antigen HEL, conjugated with C3dg to a molar ratio of 1:3, is as much as 10,000-fold more immunogenic than HEL alone. Limited analyses conducted thus far of the CR2 signal transduction cascade have demonstrated that the increased immunogenicity of C3dg–antigen conjugates is correlated with an increased ability to induce phosphoinositide hydrolysis and calcium mobilization in responding cells (Dempsey et al., 1996).

CR2 is a multisubunit complex composed of CD21, CD19, Leu13, and CD81 (Fearon and Carter, 1995; Tedder

et al., 1997). The CD21 component of the complex is responsible for C3dg binding. CD19 interacts with CD81, a tetraspan family member, via its extracellular domain and with CD21 via both its extracellular and its transmembrane domains. CD21 is composed of an extracellular domain containing multiple short consensus repeats consisting of compactly folded units of 60–70 residues, a single transmembrane-spanning domain, and a 34-amino acid cytoplasmic tail. Consistent with the relatively small size of this tail, all detectable signaling functions of the CR2 complex appear to be vested in CD81 and CD19. CD81 contains all structural information for transducing signals leading to homotypic aggregation. Studies using chimeric receptors indicate that other signaling functions, most notably the enhanced Ca^{2+} response to BCR ligations, reside within the cytoplasmic tail of CD19. Genetic ablation of CD21 and CD19 has similar effects on the immune system. B cell activation is impaired, antibody responses are greatly inhibited, and germinal center formation is much reduced.

It appears that CD19 plays two distinct roles in BCR signal transduction: as an accessory molecule and as a coreceptor (upon CR2 coaggregation with the BCR by C3dg-containing immune complexes). The nature of the additional signal(s) generated upon BCR–CR2 co-cross-linking is unclear. Clearly, the increment could be purely quantitative; aggregation of the BCR with CR2 may simply lead to a more efficient tyrosyl phosphorylation of CD19 within the YXXM motifs and hence to a more efficient recruitment and activation of PI3-K. Based on our previous studies of CD19 accessory function, such an increase in PI3-kinase recruitment and activation would be expected to lead to enhanced BCR-mediated phosphoinositide hydrolysis and Ca^{2+} mobilization (Buhl et al., 1997; Hippen *et al.,* 1997). This is precisely what is seen empirically in both the anti-HEL–C3dg and the antibody-mediated CR2–BCR coaggregation models. Alternatively, or additionally, CR2–BCR coaggregation may lead to a qualitatively distinct signaling response. For example, coaggregation may lead to phosphorylation of CD19 tyrosines in addition to Y484 and Y515. It has been reported that Fyn binds phosphorylated CD19 via its SH2 domain. It was further suggested that this interaction occurs via the phosphorylated tyrosines at positions 405 and 445, which are within the YEND/E context. Additionally, CD19 tyrosine 391 is embedded in the sequence YEEP, which upon phosphorylation is predicted to bind to the multifunctional transducer molecule Vav (Songyang *et al.,* 1993). CD19 tyrosine 391 is reportedly required for optimal Vav activation following BCR coaggregation with CR2 (D. Fearon, personal communication). It is noteworthy, however, that no phosphorylation of these residues was detectable following BCR aggregation under conditions which induce strong phosphorylation of Y484/515 (A. M. Buhl, unpublished observation).

In conclusion, it is clear that CR2 coaggregation with BCR tremendously enhances the B cell response to low doses of antigen. The molecular basis of this amplification is likely to lie in quantitative and/or qualitative differences in signal transduction imparted by the cytoplasmic tail of CD19 when it is coaggregated with the BCR. Likely mechanisms include enhanced phosphorylation of Y484 and Y515, increasing PI3-K activation (quantitative), or phosphorylation of other tyrosines that are not phosphorylated upon BCR aggregation alone which activate new pathways (qualitative). As discussed later, receptor-coupled or recruited PTPs and phosphoinositide phosphatases appear to selectively regulate BCR activation of these pathways.

CD45 Is Required for Participation of Src Family Kinases in BCR Signaling

Cells of the hematopoietic lineage express alternately spliced forms of the membrane PTP CD45 (Thomas, 1989). B lineage cells express the full-length CD45 gene product, B220 (or CD45R), beginning at the pre-pro-B stage (fraction A). Expression increases gradually as B cells mature, becoming maximal at the mature B cell stage. Differentiation to antibody-secreting plasma cells is accompanied by loss of CD45 expression; thus, most myelomas (tumors of this stage) are CD45-negative. CD45 contains two protein tyrosine phosphatase domains in its cytoplasmic tail. Although variable use of three extracellular domain exons, depending on cell lineage and differentiation stage, implies ligand specificity, no specific CD45 ligand has been described to date.

Many studies have implicated CD45 in regulation of antigen receptor signaling. Among the first were studies of the antigen receptor-bearing mouse myeloma, J558Lμm3. CD45 "reconstitution" by transfection was found to be necessary for antigen-induced Ca^{2+} mobilization in this cell line (Justement *et al.,* 1991) . Analysis of isolated clones expressing varying levels of CD45 revealed a direct correlation between CD45 expression level and the magnitude of BCR-induced calcium mobilization. Using knockout mice, it was later demonstrated that CD45 expression is required for the development of mature splenic B cells (IgM-positive, IgD-positive). The splenic B cell compartment in the knockout mice is nearly twice its normal size but contains primarily immature (IgM-positive, IgD-negative) cells. These B cells failed to proliferate normally in response to BCR stimulation and exhibited defects in immune responsiveness. Studies of B cell function in CD45 knockout mice expressing anti-HEL immunoglobulin transgenes indicate that the CD45 functions to increase B cell sensitivity to antigen stimulation. Thus, CD45 plays a positive role in BCR signal transduction.

The molecular basis of CD45 function was discussed previously. To recapitulate, CD45 is known to specifically dephosphorylate a negative regulatory phosphotyrosyl residue within the C termini of Src family kinases, enhancing their ability to phosphorylate many substrates. Recent studies in the CD45-negative J558Lμm3 cell line indicate that C-terminally phosphorylated Src family kinases are competent to associate with BCR via their N-terminal unique regions (Pao and Cambier, 1997). However, in this system receptor-associated Src family kinases are not activated in response to receptor aggregation nor do they bind to tyrosyl-phosphorylated ITAMs. This is presumably the consequence of competition with C-terminal phosphotyrosyl residues for their SH2 domains. Interestingly, despite the inability to induce Src family kinase activation in CD45-negative J558Lμm3, BCR ligation in these cells leads to the tyrosyl phosphorylation of CD79a, CD79b, and many downstream effectors (Pao and Cambier, 1997; Pao *et al.,* 1997b).

Failed Src family kinase activation following BCR ligation in CD45-negative J558Lμm3 has a selective effect on signaling pathway activation. In these cells, BCR ligation does not lead to Shc phosphorylation nor to p21^{ras} activation (Pao *et al.,* 1997b). This is consistent with the previously demonstrated function of Shc, which by binding Grb2–SOS complexes acts as a linker to the ras pathway. Importantly, BCR-mediated phosphorylation of many other substrates, including Syk, CD79a, CD79b, PLCγ, Vav, Cbl, and HS1, proceeds normally in CD45-negative J558Lμm3 (Pao *et al.,* 1997b). Despite the apparently normal tyrosyl phosphorylation of PLCγ1 and PLCγ2, IP$_3$ production and mechanistically linked calcium mobilization are blocked in CD45-negative cells (Pao *et al.,* 1997b). This indicates that receptor-mediated phosphoinositide hydrolysis is dependent on a CD45-regulated effector that complements PLCγ function in IP$_3$ production. Although no direct evidence implicates any particular effector, obvious candidates for this function include CD19, Btk, and PI3-K, all of which are required for antigen induction of calcium mobilization. As discussed earlier, these effectors may act sequentially. Thus, phosphoinositide hydrolysis by phosphorylated PLCγ may depend on the translocation of Btk to and function at the plasma membrane. The dependence of Btk and PI3-K activation on CD45 expression is yet to be determined.

In conclusion, CD45 modulates BCR signaling in a qualitative fashion, determining the participation of Src family kinases and, perhaps via Src family kinases, the activation of downstream pathways including the p21^{ras}–MAPK cascade, phosphoinositide hydrolysis, and calcium mobilization. Variations in CD45 expression during B cell development may serve to vary the biologic output of BCR stimulation by selective modulation of these pathways.

FcγRIIB1, an Inhibitory BCR Coreceptor

FcγRIIB1 is a low-affinity receptor for IgG that is expressed throughout B cell development and maturation up to the plasma cell stage (Ravetch and Anderson, 1990; Amigorena *et al.,* 1992). This receptor was implicated in inhibitory signaling in early experiments which demonstrated that immune complexes containing IgG antibodies are not immunogenic. Later studies revealed that specific coaggregation of the FcγRIIB1 with the BCR leads to inhibition of BCR-mediated proliferative and blastogenesis responses (Phillips and Parker, 1983). This inhibition requires receptor coaggregation, i.e., independent FcγRIIB1 aggregation does not inhibit the BCR response. Negative cooperativity of BCR and FcγRIIB1 provides a mechanism for preventing the participation of newly arisen antigen-specific B cells in the immune response once the response has matured to produce high-affinity IgG antibodies. Genetic validation of this concept has been provided by studies of FcγRIIB1 knockout mice that exhibit as much as a 10-fold enhanced production of antibodies late in the immune response.

Analysis of the molecular mode of FcγRIIB1 action was first reported by Bijsterbosch and Klaus (1985) and Wilson *et al.* (1987). Bijsterbosch and Klaus found that FcγRIIB1 coaggregation led to premature termination of BCR-mediated phosphoinositide hydrolysis. Consistent with an FcγRIIB1 effect on the PLCγ activity, Wilson *et al.* showed that FcγRIIB1 coaggregation profoundly inhibited the late phase of the BCR-mediated calcium mobilization response. These findings indicated that the FcγRIIB1 signal impinges on BCR signaling pathways at a point or points proximal to the receptor.

Recent studies have sought to define this mechanism more precisely. A specific sequence in the FcγRIIB1 cytoplasmic tail was found to be necessary and sufficient for inhibitory function (Amigorena *et al.,* 1992). This sequence contains a tyrosyl residue that is phosphorylated upon FcγRIIB1 coaggregation with the BCR and is essential for receptor function. This tyrosyl residue conforms to the consensus sequence I/VXYXXL that has been termed immunoreceptor tyrosine-based inhibitory motif (ITIM) and shown to be involved in negative signaling by a variety of receptors (D'Ambrosio *et al.,* 1995; Vivier and Daeron, 1997). Using synthetic phosphopeptides, D'Ambrosio *et al.* showed that phosphorylation of this tyrosyl residue is required for binding of three potential effector proteins — the protein tyrosine phosphatases SHP-1 and SHP-2 and the phosphatidylinositol 4,5-bisphosphate 5'-phosphatase SHIP (D'Ambrosio *et al.,* 1995; Ono *et al.,* 1996). The latter was further shown to undergo tyrosyl phosphorylation upon BCR–FcγRIIB1 coaggregation. D'Ambrosio *et al.* further showed that inhibitory signaling by FcγRIIB1 is defective in moth-eaten mice, which lack functional SHP-1, implicating SHP-1 as an essential mediator. It is noteworthy that in some circumstances SHP-1 is not necessary for FcγRIIB1 inhibitory signaling. For example, FcγRIIB1-mediated inhibition of FcγRI activation of mast cells proceeds normally in bone marrow-derived mast cells from moth-eaten mice. The inescapable conclusion is that some degree of redundancy exists in FcγRIIB1 effector function. It should be noted, however, that no direct evidence implicates SHP-2 or SHIP in inhibitory signaling.

Based on early findings implicating protein tyrosine phosphatases, Keiner *et al.* (1997) and Hippen *et al.* (1997) explored the effect of FcγRIIB1 coaggregation on BCR-mediated protein tyrosine phosphorylation. Both groups found that tyrosyl phosphorylation of most substrates is unaffected by FcγRIIB1–BCR coligation. However, CD19 is an exception. BCR-mediated CD19 tyrosyl phosphorylation proceeds normally for the first few seconds following receptor coaggregation. However, after 15 s (when BCR-mediated phosphorylation of CD19 is normally maximal), its phosphorylation is reduced by >80%. By 45 s following stimulation, CD19 phosphorylation is at baseline levels, whereas it remains maximal in cells stimulated through BCR alone. Both groups showed that dephosphorylation of CD19 was accompanied by failure to recruit and activate PI3-kinase.

Together with studies of CD19 function discussed earlier, these findings suggest a working model for the mechanism of FcγRIIB1 inhibition of BCR signaling (Fig. 3). FcγRIIB1 coaggregation with BCR leads to "normal" initiation of signaling via BCR. However, tyrosine kinases activated via the BCR phosphorylate colocalized FcγRIIB1, triggering recruitment of SHP-1. In turn, SHP1 dephosphorylates CD19, reversing the recruitment and activation of PI3-kinase. The consequence is the premature termination of phosphoinositide hydrolysis and calcium mobiliza-

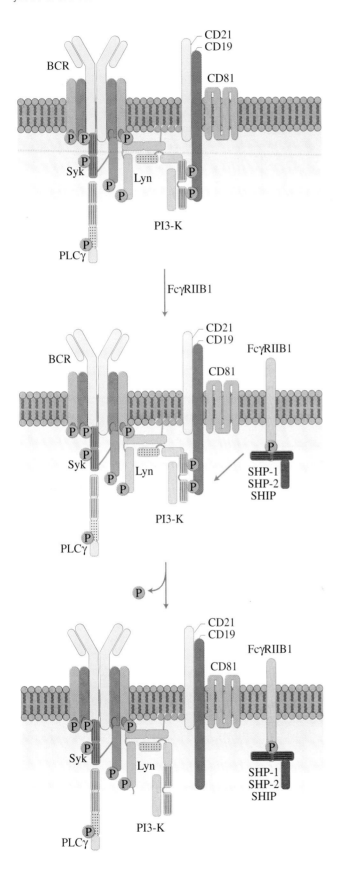

tion. This provides a mechanism for qualitative modulation of the BCR signal, blocking PI3-K and PLCγ pathways while leaving tyrosine kinase pathways intact. It has been suggested (Ono *et al.*, 1996) that SHIP mediates inhibitory signaling by degradation of Ins1,3,4,5P4, a potential mediator of extracellular calcium influx following BCR aggregation. This might explain the observed selective effect of FcγRIIB1 coaggregation on the calcium influx response. It also seems possible that SHIP may act by dephosphorylating PtdIns3,4,5P3 and thus may have the same ultimate effect as failed PI3-K activaton — reduction of PtdIns3,4,5P3 levels below those normally seen following BCR signaling. Regardless of which mechanism is operative, the consequence of FcγRIIB1 coaggregation with BCR is a qualitative change in the biologic output of BCR signaling when FcγRIIB1 is involved. In this context, it has been shown that coaggregation of FcγRIIB1 with BCR diverts the cell to an apoptotic pathway.

Although the previously discussed model is attractive, some pertinent data exist that are not accommodated. For example, in a recent series of studies it was shown that under certain conditions FcγRIIB1 coaggregation with BCR partially blocks PLCγ phosphorylation and p21^{ras} activation. Furthermore, it is unclear how inhibitory signaling is mediated in the SHP-1 negative mast cell models. Is signaling inhibition in this situation mediated by SHP-2 and/or SHIP? Activation of SHIP, because it stimulates the Akt kinase pathway, would be expected to promote viability rather than death. Finally, CD19 is only expressed in B cells, but FcγRIIB1 mediates inhibitory signaling in mast cells, T cells, and B cells. Do mast cells and T cells contain a CD19 analog that integrates PI3-K and is an FcγRIIB1 target?

CD22 Attenuation of BCR Signal Transduction

CD22 is a B lineage-restricted cell surface protein that functions as a lectin, binding certain α-2,6 sialoglycoproteins via a recognition site in the two most distal of its seven extracellular immunoglobulin-like domains. This molecule contains a single transmembrane-spanning domain and a cytoplasmic tail with six tyrosines. CD22 associates with

FIGURE 3 BCR signaling is negatively regulated by coaggregation with FcγRIIB1. FcγRIIB1 is phosphorylated by BCR-activated tyrosine kinases and recruits phosphatases, including SHIP, SHP-1, and SHP-2 (in red), that attenuate the BCR signal. These proteins act on CD19, which is dephosphorylated and inhibits PI3-K activation and downstream events.

the BCR, possibly via recognition of α-2,6 sialic acid moieties associated with the IgM substructure. This possibility is suggested by findings that CD22 binds secretory IgM via its lectin domains. CD22 is expressed at low levels on immature B cells, increases precipitously at the immature to mature transition, and remains at high levels thereafter. Recently, several groups prepared CD22 knockout mice and analyzed their phenotype. B cells develop normally in these mice but mature B cells exhibit enhanced calcium mobilization in response to BCR ligation. These B cells make enhanced BCR-mediated proliferative responses to suboptimal doses of anti-receptor antibodies and appear to exhibit an unusually short life span consistent with constitutive activation. Finally, CD22 knockout mice express unusually high levels of serum immunoglobulins. All these features are consistent with hyperstimulability, indicating that CD22 is a negative regulator of BCR signaling.

Analysis of CD22 signaling functions has demonstrated that CD22 is tyrosyl phosphorylated upon BCR aggregation and associates with multiple effectors of signal transduction, including Lyn, Syk, PLCγ isoforms, PI3-K, and SHP-1. In view of the similar hyperresponsive phenotypes of B cells from CD22 knockout mice and SHP-1 knockout mice, it is tempting to speculate that inhibitory CD22 signal is mediated by SHP-1 (Cyster and Goodnow, 1995). SHP-1 binds avidly to three phosphotyrosyl-containing sequences derived from CD22 and appears to utilize CD22 as a substrate. Although the possible role of CD19 as a SHP-1 substrate in this model has not been explored, evidence implicating CD19 dephosphorylation by SHP-1 in the FcγRIIB1 model makes it a logical candidate. It should be noted, however, that in natural killer cells, killer inhibitory receptor (KIR)-activated SHP-1 appears to mediate its effect by dephosphorylation of TCRζ chains, ZAP-70, and PLCγ. Substrate specificity may depend on the proximity and orientation of the phosphatase and its targeting structure (i.e., CD22) relative to potential substrates.

In conclusion, CD22 appears to function as an inhibitory receptor via recruitment of SHP-1. The mechanisms by which SHP-1 may mediate this effect are unknown. CD22 signaling is probably important in determining the sensitivity of B cells to antigen. The increase in expression of CD22 during maturation suggests that it may play a unique role in the function of the antibody repertoire of the animal. Reduced CD22 expression by immature B cells may make these cells hypersensitive to antigen at the time in development when antigen binding induces tolerance. Thus, reduced CD22 expression may ensure that autoreactive cells, even those bearing low-affinity receptors for antigen, are purged from the repertoire. Conversely, increased expression via mature cells may ensure that only cells with high-affinity receptors are activated to participate in an immune response.

References Cited

AMIGORENA, S., BONNEROT, C., DRAKE, J. R., CHOQUET, D., HUNZIKER, W., GUILLET, J.-G., WEBSTER, P., SAUTES, C., MELLMAN, I., and FRIDMAN, W. H. (1992). Cytoplasmic domain heterogeneity and functions of IgG Fc receptors in B lymphocytes. *Science* **256,** 1808–1812.

BIJSTERBOSCH, M. K., and KLAUS, G. G. B. (1985). Crosslinking of surface immunoglobulin and Fc receptors on B lymphocytes inhibits stimulation of inositol phosphate breakdown via the antigen receptors. *J. Exp. Med.* **162,** 1825–1827.

BROWN, M. T., and COOPER, J. A. (1996). Regulation, substrates and functions of Src. *Biochem. Biophys. Acta* **1287,** 121–149.

BUHL, A. M., PLEIMAN, C., RICKERT, R. C., and CAMBIER, J. C. (1997). CD19 is required for B cell antigen receptor mediated activation of PI3-kinase. *Immunity*.

CAMBIER, J. (1995). Antigen and Fc receptor signaling. The awesome power of the immunoreceptor tyrosine-based activation motif (ITAM). *J. Immunol.* **155,** 3281–3285.

CARTER, R. H., DOODY, G. M., BOLEN, J. B., and FEARON, D. T. (1997). Membrane IgM-induced tyrosine phosphorylation of CD19 requires a CD19 domain that mediates association with components of the B cell antigen receptor complex. *J. Immunol.* **158,** 3062–3069.

CYSTER, J. G., and GOODNOW, C. C. (1995). Protein tyrosine phosphatase 1C negatively regulates antigen receptor signaling in B lymphocytes and determines thresholds for negative selection. *Immunity* **2,** 13–24.

D'AMBROSIO, D., HIPPEN, K. L., MINSKOFF, S. A., MELLMAN, I., PANI, G., SIMINOVITCH, K. A., and CAMBIER, J. A. (1995). Recruitment and activation of PTP1C in negative regulation of antigen receptor signaling by FcγRIIB1. *Science* **268,** 293–297.

DEMPSEY, P. W., ALLISON, M. E., AKKARAJU, S., GOODNOW, C. C., and FEARON, D. T. (1996). C3d of complement as a molecular adjuvant: Bridging innate and acquired immunity. *Science* **271,** 348–350.

FEARON, D. T., and CARTER, R. H. (1995). The CD19/CR2/TAPA-2 complex of B lymphocytes: Linking natural to acquired immunity. *Annu. Rev. Immunol.* **13,** 127–149.

GOLD, M. R., and DeFRANCO, A. L. (1994). Biochemistry of B lymphocyte activation. *Adv. Immunol.* **55,** 221–295.

HIPPEN, K. L., BUHL, A. M., D'AMBROSIO, D., NAKAMURA, K., PERSIN, C., and CAMBIER, J. C. (1997). FcγRIIB1 inhibition of BCR mediated phosphoinositide hydrolysis and Ca^{2+} mobilization is integrated by CD19 dephosphorylation. *Immunity* **7,** 49–58.

JUSTEMENT, L. B., CAMPBELL, K. S., CHIEN, N. C., and CAMBIER, J. C. (1991). Regulation of B cell antigen receptor signal transduction and phosphorylation by CD45. *Science* **252,** 1839–1842.

KIENER, P. A., LIOUBIN, M. N., ROHRSCHNEIDER, L. R., LEDBETTER, J. A., NADLER, S. G., and DIEGEL, M. L. (1997). Coligation of the antigen and Fc receptors gives rise to the selective modulation of intracellular signaling in B cells. *J. Biol. Chem.* **272,** 3838–3844.

KUROSAKI, T., TAKATA, M., YAMANASHI, Y., INAZU, T., TANIGUCHI, T., YAMAMOTO, T., and YAMAMURA, H. (1994). Syk activation by the src-family tyrosine kinase in the B cell receptor signaling. *J. Exp. Med.* **179,** 1725–1729.

KUROSAKI, T., JOHNSON, S. A., PAO, L., SADA, K., YAMAMURA, H., and CAMBIER, J. C. (1995). Role of the syk autophosphorylation site and SH2 domains in B cell antigen receptor signaling. *J. Exp. Med.* **182,** 1815–1823.

MADRENAS, J., WANGE, R. L., WANG, J. L., ISAKOV, N., SAMELSON, L. E., and GERMAIN, R. N. (1995). Zeta phosphorylation without ZAP-70 activation induced by TCR antagonists and partial agonists. *Science* **267,** 515–518.

MOAREFI, I., LeFEVRE-BERNT, M., SICHERI, F., HUSE, M., LEE, C.-H., KURIYAN, J., and MILLER, W. T. (1997). Activation of the Src-family tyrosine kinase Hck by SH3 domain displacement. *Nature* **385,** 650–653.

ONO, M., BOLLAND, S., TEMPST, P., and RAVETCH, J. V. (1996). Role of the inositol phosphatase SHIP in negative regulation of the immune system by the receptor Fc(gamma)RIIB. *Nature* **383,** 263–266.

PAO, L. I., and CAMBIER, J. C. (1997). Syk, but not Lyn, recruitment to B cell antigen receptor and activation following stimulation of CD45⁻ B cells. *J. Immunol.* **158,** 2663–2669.

PAO, L., CARBONE, A. M., and CAMBIER, J. C. (1997a). *Antigen Receptor Structure and Signaling in B Cells.* Wiley, New York.

PAO, L. I., BEDZYCK, B. W., PERSON, C., and CAMBIER, J. C. (1997b). Molecular targets of CD45 in B cell antigen receptor signal transduction. *J. Immunol.* **158,** 1116–1124.

PEPYS, M. B. (1974). Role of complement in induction of antibody production *in vivo*. Effect of cobra venom factor and other C3 reactive reagents on thymus-dependent and thymus-independent antibody response. *J. Exp. Med.* **140,** 126–145.

PHILLIPS, N. E., and PARKER, D. C. (1983). Fc-dependent inhibition of mouse B cell activation by whole anti-μ antibodies. *J. Immunol.* **130,** 602–606.

PLEIMAN, C. M., HERTZ, W. M., and CAMBIER, J. C. (1994). Activation of phosphatidylinositol-3′ kinase by Src-family kinase SH3 domain binding to the p85 subunit. *Science* **263,** 1609–1612.

RAVETCH, J. V., and ANDERSON, C. L. (1990). *FcγR Family: Proteins, Transcripts and Genes.* ASM Press, Washington, DC.

SCHMITZ, R., BAUMANN, G., and GRAM, H. (1996). Catalytic specificity of phosphotyrosine kinases Blk, Lyn, c-Src and Syk as assessed by phage display. *J. Mol. Biol.* **260,** 664–677.

SONGYANG, Z., SHOELSON, S. E., CHAUDHURI, M., GISH, G., PAWSON, T., HASER, W. G., KING, F., ROBERTS, T., RATNOFSKY, S., LECHLEIDER, R. J., *et al.* (1993). SH2 domains recognize specific phosphopeptide sequences. *Cell* **72,** 767–778.

SONGYANG, Z., BLECHNER, S., HOAGLAND, N., HOEKSTRA, M. F., PIWNICA WORMS, H., and CANTLEY, L. C. (1994). Use of an oriented peptide library to determine the optimal substrates of protein kinases. *Curr. Biol.* **4,** 973–982.

SONGYANG, Z., CARRAWAY, K. L. I., ECK, M. E., HARRISON, S. C., FELDMAN, R. A., MOHAMMADI, M., SCHLESSINGER, J., HUBBARD, S. R., SMITH, D. P., ENG, C., *et al.* (1995). Catalytic specificity of protein-tyrosine kinases is critical for selective signalling. *Nature* **373,** 536–539.

TEDDER, T. F., INAOKI, M., and SATO, S. (1997). The CD19–CD21 complex regulates signal transduction thresholds governing humoral immunity and autoimmunity. *Immunity* **6,** 107–118.

THOMAS, M. L. (1989). The leukocyte common antigen family. *Annu. Rev. Immunol.* **7,** 339.

TUVESON, D. A., CARTER, R. H., SOLTOFF, S. P., and FEARON, D. T. (1993). CD19 of B cells as a surrogate kinase insert region to bind phosphatidylinositol 3-kinase. *Science* **260,** 986–989.

VIVIER, E., and DAERON, M. (1997). Immunoreceptor tyrosine-based inhibition motifs. *Immunol. Today* **18,** 286–291.

WEISER, P., MULLER, R., BRAUN, R., and RETH, M. (1997). Endosomal targeting by the cytoplasmic tail of membrane immunoglobulin. *Science* **276,** 407–409.

WILSON, H. A., GREENBLATT, D., TAYLOR, C. W., PUTNEY, J. W., TSIEN, R. Y., FINKELMAN, F. D., and CHUSED, T. M. (1987). The B lymphocyte calcium response is diminished by membrane immunoglobulin cross-linkage to the Fcγ receptor. *J. Immunol.* **138,** 1712–1718.

YAMAGUCHI, H., and HENDRICKSON, W. A. (1996). Structural basis for the activation of human lymphocyte kinase Lck upon tyrosine phosphorylation. *Nature* **384,** 484–489.

General References

COHEN, G. B., and BALTIMORE, D. (1995). Modular binding domains in signal transduction proteins. *Cell* **80,** 237–248.

FRANKE, T. F., KAPLAN, D. R., and CANTLEY, L. C. (1997). PI3K: Downstream AKTion blocks apoptosis. *Cell* **88,** 435–437.

PAWSON, T. (1995). Protein modules and signalling networks. *Nature* **373,** 573–579.

ALBERTO MANTOVANI

Department of Immunology and Cell Biology
Mario Negri Institute of Pharmacology Research
Milan, Italy

Section of Pathology and Immunology
Department of Biotechnology
University of Brescia
Brescia, Italy

The recruitment of leukocytes from the blood compartment constitutes a multistep process which involves primary and secondary inflammatory cytokines as well as adhesion molecules expressed on leukocytes and endothelial cells. The properties of interleukin-1, a primary prototypic inflammatory mediator characterized by peculiar pathways of regulation, and of chemokines, an emerging superfamily of secondary mediators, are analyzed. These mediators offer new paradigms to understand diverse pathologies, as illustrated by the role of chemokine receptors as fusion cofactors for HIV, and provide tools and targets for the development of novel therapeutic strategies.

Molecular Regulation of Leukocyte Recruitment

Introduction

The extravasation of leukocytes from the blood compartment and their accumulation in tissues represents an essential determinant of inflammatory and immune processes. Leukocyte recruitment and accumulation from the blood into tissues is a multistep process which involves a first step of rolling and adhesion to vascular endothelium followed by the transendothelial migration and passage through basement membrane (Fig. 1). The process of extravasation from the blood is essentially controlled by the properties of vascular endothelium and chemotactic signals coming from tissues, able to induce directional migration of leukocytes (Springer, 1994; Carlos and Harlan, 1994; Butcher and Picker, 1996; Mantovani *et al.*, 1997). In general, the process of leukocyte recruitment can be considerably selective, causing the preferential accumulation of one or another white blood cell population. For instance, at sites of allergic reactions eosinophils are a prominent cell population. B cells associated with mucosal tissues are enriched in cells producing IgA. In general, the selectivity of leukocyte recruitment is not determined by a single molecule, selective for one or another cell population, but rather is the end result of a combination of molecular determinants (adhesion molecules and chemotactic factors). Thus, the current paradigm is that of an area code model in which the address is generated by a combination of numbers and letters.

The identification of vascular endothelium as a crucial component, capable of playing an active role in thrombotic, inflammatory, and immune processes, is relatively recent.

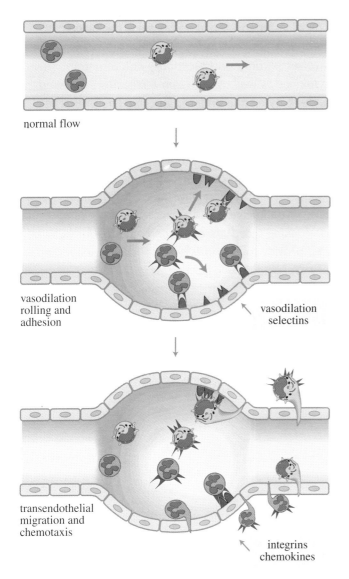

normal flow

vasodilation
rolling and
adhesion

vasodilation
selectins

transendothelial
migration and
chemotaxis

integrins
chemokines

FIGURE I Leukocyte recruitment as a multistep process. As an example, monocytes (beige) and neutrophils (pink) are shown. The first stage consists of vasodilation, stimulated by vasodilators and selectins, followed by rolling and adhesion to the vascular endothelium and, finally, by migration through the endothelium and the basement membrane. The process of extravasation from the blood flow is essentially controlled by the properties of the vascular endothelium and by tissue-derived chemotactic signals (integrins and chemokines) able to induce directional migration of leukocytes. Leukocyte recruitment in general is very selective, causing the preferential accumulation in different areas of one or another white blood cell population.

Traditionally, the endothelial lining of blood vessels was considered a passive element, endowed with negative properties (e.g., that of being nonthrombogenic). In the past few years there has been increased interest in the biology of vascular endothelium and studies have identified the vessel wall as a primary, active element in the processes of hemo-

stasis and thrombosis, inflammation, and immunity (Mantovani *et al.,* 1992a, 1997).

The second determinant involved in leukocyte recruitment is constituted by factors of tissue origin, capable of inducing directional migration of white blood cells, in response to a concentration gradient (chemotaxis). Classical chemotactic signals are represented by formyl peptides of bacterial origin, the prototype of which is the formylated tripeptide f-Met-Leu-Phe, and products of the complement cascade, in particular C5a.

These classical chemotactic signals obviously did not explain completely the recruitment of leukocytes in situations such as delayed-type hypersensitivity reaction and chronic inflammation in which there is no infectious process nor activation of the complement cascade. Moreover, classical understanding of leukocyte recruitment did not explain the selectivity of the accumulation of one or another leukocyte population under a variety of pathophysiological conditions. The identification of cytokines capable of inducing directional migration of the different leukocyte populations now provides the elements to understand the molecular basis for the regulation of recruitment of leukocytes from the blood compartment as well as to design experimental approaches for the search for agonists and antagonists.

Finally, it should be noted that the properties of blood flow are extremely important for allowing leukocyte recruitment (Mantovani *et al.,* 1992a, 1997). Indeed, sites of inflammation are characterized by vasodilation, with slowing of blood flow, and these rheological alterations are permissive for the action of the other two factors at play (chemotactic factors and adhesive molecules). It is important to observe that the blood flow alterations are controlled in part by endothelial cells which, when exposed to inflammatory signals such as interleukin-1 (IL-1), produce nitric oxide (NO), prostaglandin I2, and endothelin.

Cytokines are polypeptide mediators which play a central role in immunity and inflammation. Schematically, we can distinguish primary inflammatory mediators [IL-1, tumor necrosis factor (TNF), and IL-6] produced by macrophages, which set in motion the cytokine cascade, and secondary mediators, such as chemokines and adhesion molecules, which are induced by primary mediators and act further downstream. In this article, the circuits of molecular regulation of IL-1, chosen as a prototypic primary mediator (Dinarello, 1996), are first discussed. In the second part of this article, the role of the vessel wall, with analysis of the secondary mediators adhesion molecules and chemokines, is discussed.

Primary Cytokines: The IL-I System

IL-1 is the term used to identify two polypeptides, α and β, which are among the most active and multifunctional mediators described in immunology and cell biology. The

TABLE I

The Overlapping Spectrum of Action of IL-1, TNF, and IL-6

	IL-1	TNF	IL-6
Inflammation	+++	+++	−
Toxicity	++	+++	−
Hemopoiesis	+++	+	+++
Immunity	+	+	+++

spectrum of action of IL-1 includes cells of hematopoietic origin, from bone marrow precursors to differentiated leukocytes; the vessel wall; and cells of mesenchymal, nervous, and epithelial origin. The occupancy of a few, perhaps one, receptors per cell is sufficient to activate a response. The spectrum of action of IL-1 overlaps to a large extent with that of TNF and other cytokines such as IL-6 (Table 1). IL-1 is in fact a complex system with peculiar circuits of negative regulation (Fig. 2).

The identification of IL-1 is the result of two independent lines of research. Studies on the pathogenesis of fever led to the identification of soluble products of leukocyte origin as possible endogenous mediators. In 1943, Menkins

suggested that leukocytes release a pyrogenic substance called pyrexin. Atkins and Wood in 1955 identified a circulating pyrogenic factor in the rabbit and called it endogenous pyrogen. In different line of work focused on immunology, Gray and Waksman in 1973 identified a lymphocyte activating factor (LAF). This discovery was followed by the description of diverse monokine activities presumably mediated by the same factor. As predicted by the physicochemical and biological characterization of LAF/EP, molecular cloning in 1984 identified two genes (α and β) whose products were responsible for the immunologic activities as well as for the inflammatory and pyrogenic potential previously described. In the late 1970s different laboratories described the existence of IL-1 inhibitors. An IL-1 inhibitor (IL-1 receptor antagonist; IL-1ra) has been cloned and identified as a member of the IL-1 family. IL-1 is thus a prototype of pleiotropic inflammatory cytokines, belonging to a complex system constituted by IL-1α and -β in their diverse molecular forms, by IL-1ra, and by receptor structures (Table 2).

Structure

IL-1α and -β in their secreted form have a molecular weight of 17 kDa but different isoelectric point (pH 5.2 for IL-1α and pH 7 for IL-1β) (Table 2). The two forms are coded by distinct genes and show modest homology.

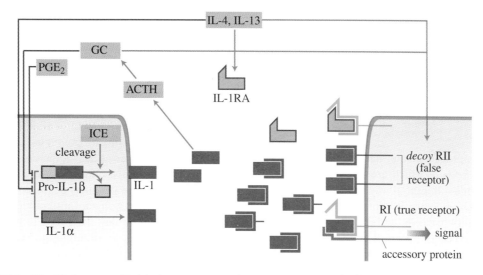

FIGURE 2 The IL-1 system. IL-1 belongs to a complex system composed of IL-1α and -β in their diverse molecular forms (red), the antagonist IL-1ra, and at least two receptor structures, a true type I receptor (green) which cooperates with the accessory protein (blue) and a decoy type II receptor (red). The precursor pro-IL-1β is processed by a proteolytic enzyme, ICE, whose activity is in turn subject to proteolytic regulation. The true receptor is able to transduce the signal following binding with IL-1 and activates the cells which express it, whereas the accessory protein increases the affinity of the receptor for IL-1 and probably cooperates in signal transduction. The false *decoy* receptor binds and sequesters IL-1, thus reducing the generation of activatory signals. Antiinflammatory/immunosuppressive signals such as glucocorticoids (GC) and some cytokines (IL-4 and IL-13) act in a coordinated way on different components of the system, blocking production of IL-1, increasing that of the antagonist, and stimulating the expression and release of the type II receptor, whereas prostaglandins (PGE$_2$) act only as inhibitors of IL-1α and -β.

TABLE 2

Molecules of the IL-1 Family

MOLECULE	MOLECULAR WEIGHT (kDa)	SIGNAL PEPTIDE	PRODUCING CELLS	ACTIVITY
IL-1α	17	No	Diverse, monocytes in particular	Agonist
IL-1β	17	No	Diverse, monocytes in particular	Agonist
Receptor antagonist (IL-1ra)	17–24	Yes/no[a]	Monocytes, fibroblasts, keratinocytes	Antagonist

[a]IL-1ra is produced in three isoforms, two of which are intracellular.

Sequence homology between α and β (human) is 26%, whereas at the level of nucleotide sequence the homology is 45%. Although mature extracellular IL-1 has a molecular weight of 17 kDa, IL-1 is produced as a 31-kDa precursor which is processed to its mature form. IL-1α and -β lack a signal sequence and the mechanism of secretion of these proteins remains unclear. Pro-IL-1β is processed by a proteolytic enzyme, IL-1 converting enzyme (ICE) (now called caspase), which in turn needs to be activated. ICE belongs to a family of cysteine protease involved in the regulation of apoptosis. ICE is also probably involved in the transport and secretion of mature IL-1β. IL-1ra is a protein of 17–20 kDa with homology to IL-1β. Three isoforms of IL-1ra have been cloned — one secreted and two intracellular (Dinarello, 1996; Muzio et al., 1995).

Receptors

Two distinct receptor molecules have been cloned (Colotta et al., 1994), both of which are able to bind IL-1α and -β, expressed in a differential way in diverse cell types (Fig. 3). The 80-kDa receptor is expressed in T lymphocytes, fibroblasts, epithelial cells, and endothelial cells. The 80-kDa receptor, also called type I, is a signal transducing molecule and activates cells. An accessory protein augments the affinity of the type I receptor for IL-1 and probably cooperates in signal transduction. The 60-kDa receptor (type II) is not a signaling receptor. It binds and sequesters IL-1 (decoy receptor) (Colotta et al., 1993, 1994, 1995; Re et al., 1994, 1996). The decoy receptor is predominantly expressed on myelomonocytic cells and B cells. IL-1ra binds preferentially the type I receptor with an affinity similar to that of IL-1 without eliciting any biological response. It is interesting to note that immunosuppressive anti-inflammatory molecules (glucocorticoids and cytokines such as IL-4 and

IL-13) act in a coordinated way on different components of the system, blocking the production of IL-1, augmenting that of IL-1ra, and stimulating expression and release of the decoy receptor (Colotta et al., 1993, 1994; Re et al., 1994).

Producing Cells and Regulation

IL-1 was originally described as a monokine, i.e., a product of cells belonging to the differentiation pathway of monocytes and macrophages. Subsequent studies revealed that virtually every cell type is able to produce IL-1 when appropriately stimulated, although monocytes–macrophages are quantitatively the major source of this cytokine. Cells able to produce IL-1 include T and B cells, natural killer (NK) cells, fibroblasts, synovial cells and mesengial cells, keratinocytes, and endothelial cells. A variety of stimuli are able to induce production of IL-1. Bacterial products, in particular endotoxin, are potent inducers of IL-1, representing one of the essential mediators of the response to these microorganisms. Other inducers of IL-1 relevant in pathophysiology are products of the complement cascade (C5a in particular), antigen–antibody complexes, and cytokines such as TNF and IL-1. IL-1ra has been identified in molecular terms due to induction by antigen–antibody complexes. It is not completely clear whether and to what extent stimuli and producing cells are the same for IL-1 and IL-1ra.

Biological Effects of IL-1

IL-1 is the prototype of pleiotropic cytokines, molecules active on a variety of cells and organs (Table 3). It is possible to summarize the biological activity of IL-1 distinguishing those related to actions on the immune and hemopoietic systems and those related to actions relevant for inflammation.

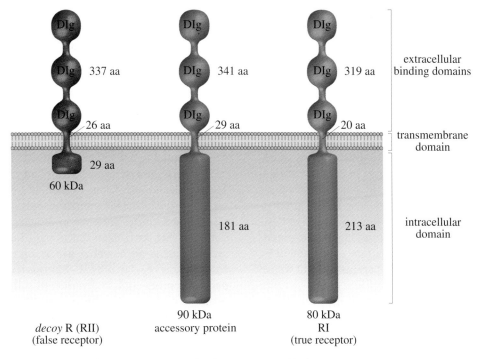

extracellular
binding domains

transmembrane
domain

intracellular
domain

decoy R (RII)
(false receptor)

90 kDa
accessory protein

80 kDa
RI
(true receptor)

FIGURE 3 Structure of the IL-1 receptors. The *decoy* receptor binds IL-1 but not the antagonist IL-1ra. In contrast, the true receptor RI binds both IL-1, in cooperation with the accessory protein, to generate the response (see Fig. 2), and IL-1ra. In this latter case, signal transduction in blocked.

Immune and Hemopoietic System

As indicated by the acronym LAF, IL-1 augments the blastogenic response of T lymphocytes to mitogens such as concanavalin A and phytohemagglutinin. These biological properties of IL-1 provide a basis for biological assays to measure IL-1. IL-1 induces the production of IL-6 in a variety of cell types including T lymphocytes, and it is probable that the action at this level is mediated at least in part by the induction of this secondary cytokine. The role actually played by IL-1 (possibly via IL-6) in antigen presentation to T cells has long been debated. Although in my opinion an essential and necessary role of IL-1 in antigen presentation has not been clearly established, available information suggests that this cytokine can be important in the induction of the response in virgin T cells, i.e., cells that have never come into contact with antigen. IL-1 also acts on the proliferation and differentiation of B cells. Induction of IL-6, a potent regulator of B cell function, represents an important pathway through which IL-1 acts on B cells. IL-1 acts on hematopoietic bone marrow precursors with colony stimulating factor (CSF)-like activities and as such has also been identified (hemopoietin 1). IL-1 acts on immature hematopoietic precursors. In addition to having CSF activity per se, IL-1 induces the production of various CSFs (G, GM, and M) by a variety of cell types. The CSF activity of IL-1 is probably a component of its radioprotective function.

Nonhematopoietic Cells

Neurons involved in thermoregulation in the hypothalamus are the target for the first identified activity of IL-1, the induction of fever. The pyrogenic activity of IL-1 is me-

---

TABLE 3

Main Biological Activities of IL-1

CELLS/ORGANS	MAIN EFFECTS
Central nervous system	Fever (via prostaglandins), anorexia, fatigue
Liver	Acute phase proteins (in part via IL-6)
Fibroblasts, synovial cells	Collagenase, prostaglandins
Bone	Resorption
Blood vessels	Vasodilation (via PGI_2 and NO); prothrombotic state; leukocyte recruitment
Bone marrow	Stimulation of precursors (in part via CSF)
T and B lymphocytes	Costimulation (in part via IL-6)
Neoplastic cells (some)	Inhibition of growth

diated by the induction of prostaglandins and blocked by the administration of cyclooxygenase inhibitors. Other effects on the central nervous system of IL-1 are the induction of slow wave sleep and anorexia. IL-1 stimulates the production of corticotropin releasing factor (CRF), which in turns activates the release of ACTH and glucocorticoids. Since glucocorticoids are important inhibitors of the synthesis of IL-1, this represents a negative inhibitory circuit which controls the production of IL-1. A second mechanism of inhibitory feedback of the IL-1 synthesis is represented by prostaglandins, particularly PGE_2. The stimuli that induce the production of IL-1 by macrophages [e.g., lipopolysaccharide (LPS)], and IL-1, also induce production of PGE_2, which in turn inhibits the production of IL-1. A further inhibitory mechanism whose therapeutic potential has been the subject of investigation is represented by $\omega 3$ fatty acids, which are present in various fish and whose mode of action has not been defined.

IL-1 is probably the main mediator of the acute phase response, i.e., the complex alterations of hematological and cellular parameters with which the organism responds to an emergency situation such as that represented by systemic exposure to bacteria or their products, in particular endotoxins. IL-1 induces in the liver the production of C reactive protein, fibrinogen, antiproteases (e.g., $\alpha 1$ antitrypsin), and A amyloid. For some of these proteins (A amyloid), IL-1 acts directly on hepatocytes, whereas for other (fibrinogen) IL-1 acts indirectly via stimulation of the production of IL-6. IL-1 also induces other alterations associated with the acute phase response, such as the reduction in blood levels of zinc and iron. It is of interest that the latter is part of a kind of "nutritional immunity" since iron is an essential element for many bacteria.

IL-1 induces a complex reprogramming of the function of vascular endothelium. The identification of the effects of IL-1 at this level has profoundly modified our understanding of the role played by the vascular lining of blood vessels in immunological, inflammatory, and thrombotic processes. In fact, endothelial cells have long been considered a passive lining of blood vessels, endowed essentially with nega-

tive properties (e.g., that of being nonthrombogenic). Endothelial cells are considered major players in a variety of pathophysiological conditions. IL-1 alters the hemostatic and inflammatory properties of endothelial cells, inducing synthesis of thromboplastin (procoagulant activity) and platelet-activating factor (PAF), inhibiting the thrombomodulin- and protein C-dependent anticoagulation pathway, and inhibiting thrombus dissolution via the induction of the inhibitor of plasminogen activator. In addition, vascular endothelium when exposed to IL-1 produces prostacyclin (PGI_2) and NO, which are potent vasodilatory agents. These vascular alterations explain, at least in part, the vasodilation at sites of inflammation and coagulation changes associated with the presence in the blood of bacterial endotoxins (Schwartzmann reactions) (Table 4). Leukocyte recruitment is an essential aspect of inflammatory processes and is regulated by IL-1 at two levels. First, IL-1 induces the production in a variety of cell types, and in particular in vascular cells of chemotactic cytokines, molecules able to induce directional migration of leukocytes. In addition, IL-1 induces the expression, increased or *de novo*, on the luminal surface of vascular endothelium of adhesion molecules which are recognized by different leukocyte populations. The appearance of adhesion molecules on the surface of endothelial cells and the production of chemotactic cytokines underlie leukocyte recruitment induced by IL-1 in tissues.

Secondary Mediators: Adhesion Molecules on Leukocytes and Vascular Endothelium

Throughout their natural history, leukocytes have undergone a close interaction with vascular endothelium. During embryogenesis hematopoietic cells and vascular endothelia share a common origin in hematopoietic blood islands. Subsequently, during the entire life of the organism, endothelial cells provide a crucial constituent of the hematopoietic microenvironment, producing conspicuous amounts of cytokines with CSF activity: These mediators regulate

---

TABLE 4

Endothelial Products That Regulate Leukocyte Recruitment

FACTOR	MOLECULE	SITE OF ACTION	ROLE
Adhesion molecules	ICAM-1, VCAM-1, E and P selectin	Postcapillary venules	Capture of circulating leukocytes
Chemotactic factors	IL-8, Gro, MCP-1	Capillaries	Adhesion, migration
Vasodilators	PGI_2, NO	Smooth muscle cells	Permissive for the action of other factors

the proliferation and differentiation of hematopoietic precursors. Once released in the circulation, leukocytes must interact with cells of the vessel wall which regulate their extravasation from the blood compartment in a variety of ways. Lastly, in tissues the development and expression of immune and inflammatory reactions require the active participation of endothelial cells which regulate and in turn are regulated by the various leukocyte populations.

The interaction between vascular endothelium and leukocytes involves two communication systems in close reciprocal connection. A first communication pathway is provided by the physical interaction of the two cell types, mediated by receptors and counterreceptors expressed by a regulated fashion. A second pathway of communication is provided by soluble polypeptide (cytokines) or lipids (e.g., PAF) mediators which influence in a bidirectional way the two cell types.

In the analysis of the interaction between vascular endothelium and leukocytes, one is stricken by the similarities between cells which belong to the mononuclear phagocyte system and vascular endothelium. These similarities include expression of common membrane structure and some functional properties (the production of some cytokines, accessory function, and response to hematopoietic cytokines). These similarities and interrelationships recall the old term now largely abandoned of "reticuloendothelial system," which was coined by pathologists who detected this strict symbiotic relationship between phagocytes and endothelial cells.

Vascular endothelium plays a central role both in innate immunity and in specific immunity. The role of endothelium in specific immunity is due, at least in part, to its role in the regulation of lymphocyte recirculation and its capacity to act as an accessory cell in the induction of immune response. T and B lymphocytes, unlike NK cells, polymorphonuclear leukocytes, and mononuclear phagocytes, have the capacity to recirculate; in other words, once they have extravasated, lymphoid cells enter into the lymphatic system and through the thoracic duct reenter into the bloodstream. Lymphocyte recirculation allows a general function of surveillance of all organs and tissues and localization at well-defined anatomical sites of the lymphoid population with specialized function.

The most characteristic anatomical substrate of lymphocyte recirculation is high endothelium. High endothelium is constituted by cuboid cells and was initially described in postcapillary venules [high endothelial venules (HEVs)] of lymph nodes. It is mainly at this level that, via the diapedesis, the exit of lymphocytes from the blood compartment occurs. HEVs augment in lymph nodes during immune responses and are present in tissues involved in cell-mediated immune reactions. HEVs are virtually absent in athymic or thymectomized animals. To date, it has not been possible to culture *in vitro* high endothelium and

to identify unequivocally mediators responsible for this peculiar morphology.

The most dramatic example of lymphocyte recirculation and selective localization of a lymphoid population at a given site is provided by B cells which produce IgA. These cells tend to localize in a preferential way to mucosal tissues, homing to what is generally known as mucosa-associated lymphoid tissue. This article is focused mainly on adhesion molecules involved in the interaction between cells of innate immunity and endothelium.

Endothelium and Inflammation

Endothelial Adhesion Molecules and Leukocyte Counterreceptors. In general, the adhesion of leukocytes to endothelium is a physiological process. Indeed, in normal conditions an appreciable proportion (up to 80%, although with considerable variability) of the intravascular pool of neutrophils is marginated; these cells remain adherent to the vascular endothelium in the microcirculation of the spleen, lungs, and other organs. This marginated pool is in a rapid equilibrium with the pool of circulating cells and, following acute stimuli such as adrenaline release, marginated cells can enter into the circulation. Marginated neutrophils can also migrate through the vessel wall but, in this case, they cannot recirculate and are eliminated. Lymphocytes adhere and transmigrate normally at the level of specialized sections of the vascular tree such as postcapillary venules in lymphoid organs. Lymphocytes can return to the circulation through the lymphatic system after a few hours.

In addition to this type of interaction, present under normal conditions, leukocytes adhere to vessel walls and infiltrate tissues following immune and inflammatory reactions. Endothelial cells contribute to these processes essentially by releasing chemotactic substances, which are able to attract and activate leukocytes at sites of inflammation, by expressing on their membrane adhesion molecules for leukocytes, and by producing mediators (PGI_2 and NO) which cause vasodilation (Table 4). Here, I discuss adhesion molecules expressed on endothelial cells and recognized by corresponding leukocyte counterreceptors.

Adhesion Molecule Present on Endothelial Cells. In the absence of stimuli vascular endothelium expresses certain adhesion molecules, in particular intercellular adhesion molecule 1 (ICAM-1) and ICAM-2. These provide an "adhesive tone" which allows the extravasation of leukocytes under physiological conditions and helps the initial recruitment during inflammation.

Following the interaction with diverse stimuli such as IL-1 and TNF, vascular endothelium expresses on its surface adhesion molecules for polymorphonuclear leukocytes, monocytes, and lymphocytes. Some of these molecules are synthesized *de novo;* in other words, they appear after a few hours and persist for a relatively long time (up

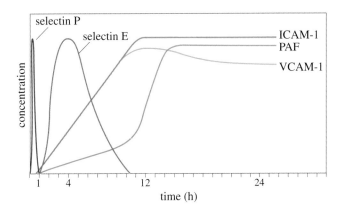

FIGURE 4 Kinetics of expression of endothelial adhesion molecules; concentrations are expressed as a function of the time after stimulation. ICAM-1, intercellular adhesion molecules; VCAM, vascular cell adhesion molecules; PAF, platelet-activating factor.

to 24–48 h). Other adhesion molecules are only transported from the cytoplasm of endothelial cells to the plasma membrane and appear on the endothelial surface a few minutes after stimulation; these molecules also disappear from the surface relatively rapidly (Fig. 4). This sequence reflects the need to have an initial rapid recruitment which is subsequently stabilized by diverse molecules and followed by transendothelial migration of leukocytes.

The following are molecules that are induced slowly and which have been characterized to date on endothelial cells:

1. Endothelial leukocyte adhesion molecule-1 (ELAM-1), now called E selectin
2. ICAM-1
3. Vascular cell adhesion molecule-1 (VCAM-1)

To date, only one adhesion molecule preexisting and rapidly induced on the surface of endothelial cells has been identified. It is called GMP140 or platelet activation-dependent granule external membrane protein or P selectin.

Structurally, E and P selectin belong to the same family of molecules, called selectins or leukocyte endothelium cell adhesion molecules. E and P selectin have homology in terms of structure and function. E selectin is expressed only on endothelial cells, it has a molecular weight of about 110 kDa, and it is an internal membrane protein (it has a transmembrane portion, a cytoplasmic portion, and an extracellular portion). E selectin is not expressed under normal conditions but only after activation of endothelial cells by inflammatory stimuli such as IL-1 and TNF. The expression of E selectin reaches a peak at 4 h and declines thereafter to unmeasurable values within 8–10 h. This molecules is recognized essentially by polymorphonuclear leukocytes, monocytes, and memory T cells. E selectin is found *in vivo*

essentially at postcapillary venules under inflammatory conditions. Moreover, it is also observed in lymph node venules under diverse pathological conditions such as reactive lymphoadenitis and lymphomas. In contrast, P selectin is always present in endothelial cells but, under nonactivated conditions, it is intracellular, located in small organs called Weibel and Palade bodies. When endothelial cells are activated by stimuli such as histamine and thrombin, P selectin is transported from Weibel and Palade bodies to the external surface of endothelial cells in contact with blood where it is available for interaction with circulating cells. As mentioned previously, this process is extremely rapid (it reaches a peak of expression in 10–15 min) and is reversible after approximately 1 h. P selectin in recognized only by polymorphonuclear leukocytes. All selectins have a domain homologous to lectins, of fundamental importance for their adhesive properties.

The E and P selectin counterreceptors have been identified. They are two sugars already known (Lewis x and a) because they have been identified as blood group antigens; these moieties are bound to lipid or proteins. In general, for these adhesive processes to occur the recognition of two cells must occurs due to a sugar–protein interaction rather than an interaction between two proteins as is generally the rule.

ICAM-1 and VCAM-1 are induced slowly on endothelial cells. These two molecules have some structural homology and are different from selectins. They belong to the immunoglobulin superfamily, which in addition to including antibodies also includes a series of adhesive molecules. These two molecules are expressed, although at low levels, on resting endothelial cells, but they augment several-fold upon activation of endothelial cells with inflammatory stimuli, such as IL-1 and TNF. Both are recognized by monocytes and lymphocytes, whereas polymorphonuclear cells only bind ICAM-1. Counterreceptors for these two molecules have been identified. ICAM-1 and VCAM-1 counterreceptors are two proteins which belong to the integrin family. LFA-1 or CD11a/CD18 is the ICAM-1 counterreceptor, whereas $\alpha_4\beta_1$ or VLA4 is the receptor for VCAM-1. ICAM-1 can also be recognized by another integrin known as MAC-1 or CD11b/CD18.

Under resting conditions endothelial cells also express a receptor known as ICAM-2, a molecule similar to ICAM-1. This protein is not regulated by inflammatory signals and is recognized by the same counterreceptor as ICAM-1 (LFA-1). The molecular structure of the main adhesion receptors responsible for the interaction of endothelial cells with leukocytes is schematized in Fig. 5.

Finally, one must consider that a phospholipid, PAF, which is synthesized by activated endothelial cells, can promote the adhesion of leukocytes. This phospholipid is not present in resting endothelial cells, but inflammatory sig-

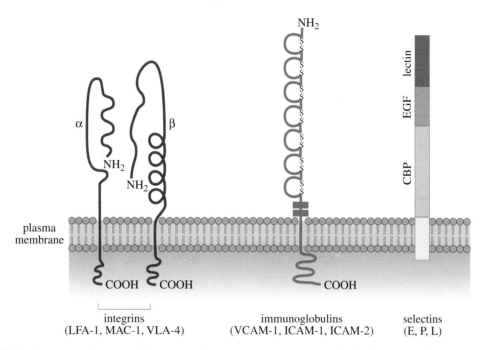

FIGURE 5 Molecular structure of the main adhesion receptors responsible for the interaction between endothelial cells and leukocytes. EGF, area of homology with epithelial growth factor; CBP, area of homology with complement-binding proteins.

nals such as histamine or cytokines induce its production. PAF remains mainly associated to the plasma membrane and is released only in minute amounts. PAF can also be defined as a chemotactic agent able to activate polymorphonuclear leukocytes and monocytes. It is of interest that, in contrast to most chemotactic agents, PAF remains cell associated: In this way it is protected from inactivation and can reach relatively high local concentrations.

Leukocyte Receptors for Adhesion Molecules on Endothelium. Specific receptors on leukocytes interact with endothelial cell adhesion molecules. In particular, all leukocytes express LFA-1 (receptor for ICAM-1 and ICAM-2 on endothelial cells), monocytes and lymphocytes express $\alpha_4\beta_1$ (receptor for VCAM-1), polymorphonuclear leukocytes and monocytes express the carbohydrate sialil-Lewis x (sLex) (receptor for E selectin) and MAC-1 (which can bind ICAM-1), and polymorphonuclear leukocytes express Lex (receptor for P selectin).

Because of this connection one can make some general statements which go beyond the understanding of specific molecules. All adhesion receptors or leukocytes are always present on the cell membrane; however, some need cells to be exposed to specific chemotactic signals (such as leukotrienes, PAF, and chemokines) to be able to bind. The LFA-1 and MAC-1 integrins belong to this group. It is likely that during inflammatory responses endothelial cells are first activated and leukocytes are recruited to the site

following the interaction with P selectin, ELAM-1, and VCAM-1, which they can recognize without activation. Chemotactic factors released locally activate adhesion molecules present on leukocytes, such as LFA-1 and MAC-1, and this leads to stabilization of adhesion and activation and transendothelial migration of cells.

A molecule known as leukocyte adhesion molecule 1 (LAM-1) or L selectin is also important for the interaction with vascular endothelium. L selectin expressed on circulating phagocytes and on lymphocytes interacts with sugar moieties present on endothelial cells. L selectin is shed by leukocytes following passage through endothelial cells into tissues. It is of interest that there is a relatively rare congenital pathology, called leukocyte adhesion deficiency (LAD-1), characterized by the absence of adhesion receptors (LAF-1 and MAC-1) on leukocytes. These patients lack a normal defense response to bacterial infections. This observation indicates how adhesion processes described herein can be essential for normal host defenses. A second syndrome of congenital immunodeficiency, LAD-2, characterized by a defect in the production of sLex, has also been identified.

Adhesion Molecules on Endothelial Cells and Pathogens. As discussed previously, vascular endothelium expresses adhesion molecules in a regulated way and these are used for the recruitment of professional migrants such as circulating leukocytes. It was recently shown that

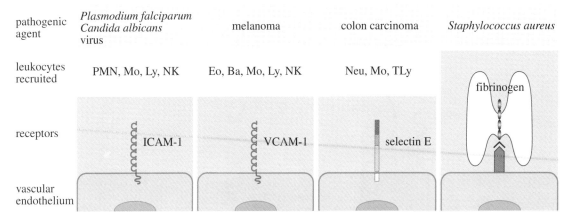

| pathogenic agent | *Plasmodium falciparum* *Candida albicans* virus | melanoma | colon carcinoma | *Staphylococcus aureus* |

FIGURE 6 Endothelial adhesion molecules recognized by pathogens for invasion of the tissues. PMN, polymorphonuclear leukocytes; Mo, monocytes; Ly, lymphocytes; TLy, T lymphocytes; Ba, basophils; Eo, eosinophils; NK, natural killer cells.

diverse pathogens use the same molecular tools to interact which endothelial cells and undergo dissemination (Fig. 6). Certain viruses (rhinovirus), bacteria (*Klebsiella pneumoniae*), fungi (*Candida albicans*), and protozoa (red blood cells infected by *Plasmodium falciparum*) interact with ICAM-1. In the case of *P. falciparum*, the interaction with endothelial cells is central for the pathogenesis of cerebral malaria, characterized by the arrest of red blood cells in cerebral microcirculation and production of inflammatory cytokines. In addition, tumor cells utilize the same molecules for dissemination and implantation at distant anatomical sites. In fact, melanomas (unique in this respect as nonhematopoietic elements) express the VLA-4 integrin which recognizes VCAM-1, whereas colon carcinomas express sugars recognized by E selectin. Endothelial molecules recognized by tumor cells of different hystogenetic origin remain to be defined. Finally, *Staphylococcus aureus* interacts with vascular endothelium due to the intermediary function of fibrinogen. Indeed, *S. aureus* binds fibrinogen and this in turn is recognized by a specific endothelial receptor; the interaction with vascular endothelium is probably crucial for the pathogenesis of dissemination and implantation of this germ at distant sites such as that observed in septic endocarditis.

Secondary Mediators: Chemokines

General Characteristics

The term chemokines (chemotactic cytokines) identifies a superfamily of proteins with chemotactic activity (Oppenheim *et al.,* 1991; Baggiolini *et al.,* 1994). The name chemokine was officially adopted at the Third Interna-

tional Workshop on Cytokines and substitutes for a series of alternative names previously used, such as intercrine, SIS (small inducible secreted cytokines), and scy (small cytokines). Chemokines are a family of more than 20 proteins which share some common structural features. They are basic proteins of relatively small size (60–80 amino acids) with a molecular weight usually between 8 and 10 kDa. They are characterized by the presence of four cysteines which give origin to two disulfate bridges between the first and the third and the second and the fourth. This structural feature provides a three-dimensional structure crucial for the interaction with receptors. The destruction of disulfide bridges causes a loss of biological activity. The relative position of the first two cysteines identifies two distinct families (Table 5). In the first, the cisteine tandem is interrupted by an extra amino acid that is different in different molecules so that this is called Cys-X-Cys or α chemokine family. The second family is characterized by an uninterrupted cysteine tandem and is called Cys-Cys or β (Oppenheim *et al.,* 1991; Baggiolini *et al.,* 1994). Members of the chemokine α family are localized on chromosome 2 in humans, whereas β chemokines are localized on chromosome 17—with the exception of Rantes, which is localized on chromosome 4. α Chemokines show a level of amino acidic homology of 25–50%. For β chemokines the homology is usually 30–70%, and the homology is not more than 40% among members of the two families. Recently, a protein with chemotactic activity with a certain degree of homology with β chemokine was identified: This chemokine shows peculiar properties, such as the presence of only two Cys residues (the first and the third), and it is localized on chromosome 1. This protein, called lymphotactin, represents a third family of chemokines, the Cys or γ chemokines (Table 5).

TABLE 5

TABLE 5

Members of the Chemokine Superfamily and Their Cellular Targets

FAMILY	CYTOKINE	Neu	Eo	Ba	Mo	T	B	NK	DC
C-X-C (α)	IL-8	++	±	±	–	±	–	±	–
	NAP-2	++	–	–	–	nt	nt	nt	nt
	Gro (α, β, γ)	++	–	–	–	nt	nt	nt	nt
	IP-10	–	nt	nt	+	+	–	nt	–
C-C (β)	MCP-1	–	–	++	++	++	nt	++	–
	MCP-2	–	+	+	+	+	nt	++	–
	MCP-3	–	++	++	+	+	nt	++	++
	Rantes	–	++	++	+	++	+	++	++
	MIP-1α	±	+	+	+	+	+	++	++
	MIP-1β	–	–	–	+	+	+	–	nt
	Eotaxin	–	+	nt	–	nt	nt	nt	nt
C (γ)	Lymphotactin	–	nt	nt	–	+	nt	nt	nt

Note. Abbreviations used: Neu, neutrophils; Eo, eosinophils; Ba, basophils; Mo, monocytes; T, T lymphocytes; B, B lymphocytes; NK, natural killer cells; DC, monocyte chemotactic protein (MCP).

Producing and Target Cells

α Chemokines include 14 proteins, of which the best characterized are IL-8, Gro (three forms, α, β, and γ), IP10, NAP-2 (neutrophil activating protein-2), and ENA-78. β Chemokines include at least 12 proteins, including monocyte chemotactic protein (MCP)-1, MCP-2, MCP-3, LD78/MIP-1α, MIP-1β, Rantes, and eotaxin. As it can be seen, nomenclature is extremely confusing and this is in part due to the fact that often different laboratories have independently identified the same molecules using different approaches. As shown in Table 5, α chemokines are also active on neutrophils and T lymphocytes and are inactive on monocytes. β Chemokines, in contrast, show a wider spectrum of action and are inactive on neutrophils (Oppenheim *et al.,* 1991; Baggiolini *et al.,* 1994). Information on lymphotactin is limited; however, data suggest this protein is chemotactic for T cells but not for monocytes and neutrophils.

IL-8 is the most extensively studied member of the α chemokine family. This cytokine was initially identified as a chemotactic factor for neutrophils produced by peripheral blood monocytes stimulated with LPS and was originally called NAP-1. It is now clear that IL-8 is produced by diverse cell types, such as fibroblasts, epithelial cells, nervous cells, and certain types of transformed cells, and by cells with barrier function such as endothelial cells, mesothelial cells, and smooth muscle cells–mesangial cells. IL-8 is produced as a 99-amino acid precursor, 20 of which are the signal peptide. The mature form can be further processed by proteases at the NH_2 terminus giving origin to pro-

teins with different dimensions (69–79 amino acids). The 77-amino acids form is produced mainly by fibroblasts and endothelial cells, whereas monocytes produce mainly a 72-amino acids protein. *In vitro* 72-amino acids IL-8 is biologically more active that the 77 form. The action of proteases is also crucial for the production of NAP-2. This protein is produced following secretion and proteolytic digestion of a precursor (platelet basic protein) present in α granules of platelets.

α Chemokines essentially have chemotactic activity (directional migration in response to a concentration gradient) both *in vitro* and *in vivo*. Recent *in vitro* studies have shown that IL-8 is able to bind to the surface of endothelial cells of postcapillary venules by interacting with glycosaminoglycans. IL-8 immobilized on these substrates shows an augmented capacity to activate neutrophils. In the presence of priming signals such as TNF-α or IL-1, α chemokines are also able to induce activation of other biological responses such as oxidative burst and degranulation. These functions play an important role in the development of inflammatory reactions.

MCP-1 is the most extensively studied cytokine among those belonging to the β chemokine family. It is composed of 76 amino acids and is glycosylated. MCP-1 is produced by different cell types and in particular by endothelial cells, fibroblasts, and monocytes following cellular activation (Fig. 7). Diverse types of transformed cells, such as sarcomas, gliomas, and carcinomas, produce MCP-1 constitutively (Mantovani *et al.,* 1992b). MCP-1 is a chemotac-

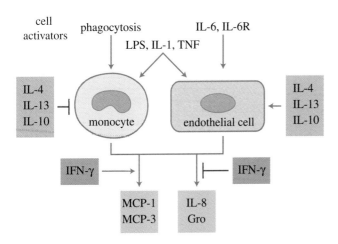

FIGURE 7 Regulation of the production, following cell activation, of certain chemokines in monocytes and vascular endothelium cells. Interferon-γ (IFN-γ) increases the production of MCP and inhibits that of IL-8 and Gro. Some interleukins (IL-4, IL-13, and IL-10) act in opposing directions on the two types of cells, stimulating the endothelial cells and inhibiting the monocytes.

tic stimulus for monocytes, T lymphocytes, NK cells, basophils, and dendritic cells. In basophils, it also acts as a sectetagogue inducing histamine release. This effect is augmented by pretreatment of cells with IL-3, IL-5, or granulocyte macrophage-colony stimulating factor. MCP-1, in contrast, is inactive on eosinophils and is the only stimulus to date able to discriminate between these two cellular populations. MCP-3 and eotaxin are potent stimuli for eosinophils. These proteins are thus important candidates for the regulation of allergic reactions. MCP-1, MCP-2,

MCP-3, Rantes, MIP-1α, and MIP-1β are also chemotactic for T lymphocytes, although they show some differences: Rantes is active on both resting and activated cells, whereas the other chemokines are active mainly on preactivated cells. Rantes and MCP-1 act preferentially on memory T cells (CD45RO), whereas MIP-1α and MIP-1β are preferentially active on CD8+ and CD4+ T lymphocytes, respectively. MCP-3 and other β chemokines are also active on dendritic cells (Sozzani *et al.,* 1994, 1995).

Activities Other Than Chemotaxis

Although chemokines' main activity is that of inducing directional migration of leukocytes, recent evidence suggests that some of these molecules can also act as regulators of cell growth. Thus, Gro and IL-8 stimulate the proliferation of certain melanomas *in vitro.* MIP-1α has a peculiar activity in that it inhibits selectively and reversibly the proliferative capacity of immature hematopoietic precursors. On the basis of this unique activity, this member of the β chemokine family is undergoing investigation as a candidate for the protection of bone marrow precursors from chemotherapy toxicity.

Receptors

α and β chemokines act via membrane receptors which belong to the family of rhodopsin-type receptors, characterized by seven transmembrane domains (Fig. 8). These receptors are associated with heterotrimeric G proteins whose activation causes an increase in intracellular calcium concentrations and activation of phospholipid metabolism caused by phospholipases C, D, and A2. Currently, five receptors for β chemokines (CCR1–5) and four for α chemokines (CXCR1–4) are known. Among the first, CCR1 was cloned first; this is a shared receptor for MIP-1α, Rantes,

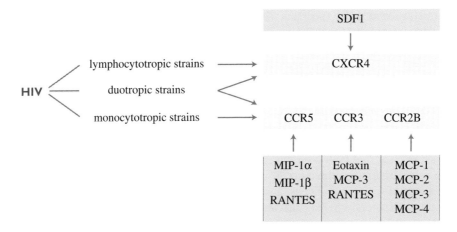

FIGURE 8 Chemokine receptors as cofactors for HIV in AIDS. Different strains of HIV use, in addition to CD4, the receptors for the chemokines (C-X-C and C-C; in light green) as their entry gateway. The chemokines specific for the various receptors are indicated in the dark green boxes.

and MCP-3. CCR2 is expressed in two forms, A and B, obtained by differential splicing, and it binds both MCP-1 and MCP-3. Finally, CCR3 was recently cloned as a shared receptor for Rantes, MCP-3, and eotaxin. Chemokine receptors thus obey two rules: A single receptor generally binds more than one chemokine and the same chemokine binds more than one receptor.

Chemokine and Virus

It is of great interest to determine whether certain viruses encode proteins which show a high level of homology with chemokine receptors. To date, two viral proteins homologous to chemokine receptors have been identified. US28 is a protein coded by cytomegalovirus which shows at the extracellular NH_2 terminus 60% homology with CCR1. ECRF3 is encoded by herpesvirus saimiri and has a high level of homology (44% at the NH_2 terminus) with the IL-8 receptor CXCR2. Both proteins are able to bind the respective cytokines and to transduce signals. However, it is not clear why in evolution the virus found it advantageous to capture these two genes. Finally, on the surface of red blood cells, there is a transmembrane protein able to bind chemokines belonging to both the α family (IL-8, Gro, and platelet factor-4) and the β family (MCP-1 and Rantes). This protein is identical to the Duffy antigen, which is the receptor through which *Plasmodium vivax* invades erythrocytes. This receptor binds an IL-8 sequence different from that recognized by the CXCR1 and -2 receptors. Therefore, it is possible to hypothesize that one could use mutants of IL-8 as antimalarial pharmacological agents.

Chemokine receptors have recently emerged as essential factors for the fusion of HIV with target cells (Moore *et al.,* 1996). In particular, CCR5 is involved in the infection by monocytotropic strain, whereas CXCR4, also known as fusin, is used by lymphocytotropic strains. Chemokines such as Rantes, which occupies these receptors, block HIV infection *in vitro,* and subjects who are homozygous for an allele of CCR5 which results in lack of surface expression of the protein on the plasma membrane (D32) are resistant to HIV *in vivo.*

Chemokines and Tumors

Several lines of evidence suggest that chemokines play a crucial role in the formation of leukocyte infiltrate of neoplastic tissues (Mantovani *et al.,* 1992b). Initial studies conducted by this group showed that the amount of macrophages which infiltrate tumors can correlate directly with the capacity of tumor cells to release *in vitro* factors with chemotactic activity for monocytes. Recently, gene transfer experiments have shown in a direct way the role of chemokines in the recruitment of leukocytes in tumor tissues. Several tumor lines, human and murine, produce *in vitro* various chemokines and in some cases they have provided

a source of these molecules, invaluable for their initial purification. Human melanomas and Kaposi's sarcoma, characterized by a prominent infiltrate of phagocytic cells, produce MCP-1 and IL-8 *in vitro.* In some cases, such as in Kaposi's sarcoma and in ovarian carcinoma, the production of MCP-1 by tumor cells has been shown *in vivo* via *in situ* hybridization and immunohistochemistry of tumor lesions. High levels of MCP-1 and IL-8 have been found in the ascites of patients with ovarian carcinoma and in the pleural exudate associated with lung carcinoma. Collectively, these observations suggest that chemokines are actually important molecules for the process of leukocyte recruitment in neoplastic tissues.

As already mentioned, tumor cells are strong producers of chemokines. In most cases, these molecules have not shown important activities on cell growth *in vitro.* Under certain conditions, it was shown that these molecules can act as growth factors. This is the case for Groa-MGSA, a C-X-C chemokine initially identified and purified as an autocrine growth factor produced by a human melanoma cell line and for this reason called MGSA (melanoma growth stimulatory activity). Consistent with these observations, certain human melanoma cell lines have been shown to be dependent for growth on IL-8 produced in an autocrine fashion. In certain cases, the mitogenic activity of IL-8 on some human tumor lines was shown to be mediated by induction of Groa-MGSA. Finally, IL-8 and other members of the C-X-C family which have the "ELR" motif at the NH_2 terminus are chemotactic for endothelial cells and capable of inducing angiogenesis.

To study the effect of chemokines on tumor growth, cells of malignant melanoma transfected with the MCP-1 gene have been injected *in vivo.* The tumors obtained showed an increased macrophage infiltration, and they were still tumorigenic but expressed a slower growth rate compared to controls. This effect was also evident in athymic mice and was likely due to a change in macrophage infiltration. Similar studies conducted in Chinese hamster ovary cells injected in allogeneic mice have shown an inhibition of tumor growth under conditions of overproduction of MCP-1 (C-C) or IL-8 and MIP-1β (C-X-C). In all these cases, chemokines were not able to influence cellular growth *in vitro.*

Leukocyte recruitment in tumors coexists with the state of systemic immunosuppression and defective capacity to mount inflammatory responses at different anatomical sites. It is possible that the local augmented production of chemokines may contribute to the setting up of systemic inhibition of macrophage function. MCP-1 is able to inhibit the generation of NO by the macrophage cell line J774. Moreover, different chemotactic factors, including chemokines, induce rapid release of the type II decoy receptor for IL-1 (Colotta *et al.,* 1995) and of the p75 TNF receptor. This dual action could contribute to maintaining inflammatory response at the local site and inhibit the systemic action of proinflammatory cytokines.

·································

Concluding Remarks

The extravasation of leukocytes from the blood compartment into tissue is a crucial determinant of a variety of physiological and pathological conditions. The identification of chemokines and their spectrum of action has provided a new and more in-depth level of understanding of the molecular basis of leukocyte recruitment and of its selectivity.

One can now identify elements (chemotactic cytokines and adhesion molecules) which together allow the recruitment, selective and with different kinetics, of one versus another leukocyte population. Although studies in human and animal pathology are still in the initial phase, available information is consistent with the hypothesis that these cytokines are involved in the pathogenesis of a variety of disease conditions ranging from atherosclerosis to acute and chronic inflammation and allergic diseases. It is therefore not surprising that a variety of groups are actively involved in investigating this emerging superfamily of inflammatory cytokines searching for antagonists or to exploit some of the agonist properties of these molecules.

References Cited

BAGGIOLINI, M., DEWALD, B., and MOSER, B. (1994). Interleukin-8 and related chemotactic cytokines-CXC and CC chemokines. *Adv. Immunol.*, **55**, 97–179.

BUTCHER, E. C. and PICKER, L. J. (1996). Lymphocyte homing and homeostasis. *Science*, **272**, 60–66.

CARLOS, T. M., and HARLAN, J. M. (1994). Leukocyte-endothelial adhesion molecules. *Blood*, **84**, 2068–2101.

COLOTTA, F., RE, F., MUZIO, M., BERTINI, R., POLENTARUTTI, N., SIRONI, M., GIRI, J. G., DOWER, S. K., SIMS, J. E., and MANTOVANI, A. (1993). Interleukin-1 type II receptor: a decoy target for Il-1 that is regulated by IL-4. *Science*, **261**, 472–475.

COLOTTA, F., DOWER, S. K., SIMS, J. E., and MANTOVANI, A. (1994). The type II 'decoy' receptor: novel regulatory pathway for interleukin-1. *Immunol. Today*, **15**, 562–566.

COLOTTA, F., ORLANDO, S., FADLON, E. J., SOZZANI, S., MATTEUCCI, C., and MANTOVANI, A. (1995). Chemoattractans induce rapid release of the interleukin 1 type II decoy receptor in human polymorphonuclear cells. *J. Exp. Med.*, **181**, 2181–2186.

DINARELLO, C. A. (1996). Biological basis for Il-1 in disease. *Blood*, **87**, 2095–2147.

MANTOVANI, A., BOTTAZZI, B., COLOTTA, F., SOZZANI, S., and RUCO, L. (1992a). The origin and function of tumor-associated macro-phages. *Immunol. Today*, **13**, 265–270.

MANTOVANI, A., BUSSOLINO, F., and DEJANA, E. (1992b). Cytokine regulation of endothelial cell function. *FASEB J.*, **6**, 2591–2599.

MANTOVANI, A., BUSSOLINO, F., and INTRONA, M. (1997). Cytokine regulation of endothelial cell function: from molecular level to the bed side. *Immunol. Today*, **18**, 231–240.

MOORE, J. P., and KOUP, R. A. (1996). Chemoattractans attract HIV researchers. *J. Exp. Med.*, **184**, 311–313.

MUZIO, M., POLENTARUTTI, N., SIRONI, M., POLI, G., DE GIOIA, L., INTRONA, M., MANTOVANI, A., and COLOTTA, F. (1995). Cloning and characterization of a new isoform of the interleukin-1 receptor antagonist. *J. Exp. Med.*, **182**, 623–628.

OPPENHEIM, J. J., ZACHARIAE, C. O., MUKAIDA, N., and MATSUSHIMA, K. (1991). Properties of the novel proinflammatory supergene 'intercrine' cytokine family. *Annu. Rev. Immunol.*, **9**, 617–648.

RE, F., MUZIO, M., DE ROSSI, M., POLENTARUTTI, N., GIRI, J. G., MANTOVANI, A., and COLOTTA, F. (1994). the type II 'receptor' as a decoy target for IL-1 in polymorphonuclear leukocytes: characterization on of induction by dexamethasone and ligand binding properties of the released decoy receptor. *J. Exp. Med.*, **179**, 739–743.

RE, F. *et al.* (1996). Inhibition of interleukin-1 responsiveness by type II receptor gene transfer: a surface 'receptor' with anti-interleukin-1 function. *J. Exp. Med.*, **183**, 1841–1850.

SOZZANI, S., ZHOU, D., LOCATI, M., RIEPPI, M., PROOST, P., MAGAZIN, M., VITA, N., VAN DAMME, J., and MANTOVANI, A. (1994). Receptors and transduction pathways for monocyte chemotactic protein-2 and monocyte chemotactic protein-3: similarities and differences with MCP-1. *J. Immunol.*, **152**, 3615–3622.

SOZZANI, S., SALLUSTO, F., LUINI, W., ZHOU, D., PIEMONTI, L., ALLAVENA, P., VAN DAMME, J., VALITUTTI, S., LANZAVECCHIA, A., and MANTOVANI, A. (1995). Migration of dendritic cells in response to formyl peptides, C5a and a distinct set of chemokines. *J. Immunol.*, **155**, 3292–3295.

SPRINGER, T. A. (1994). Traffic signal for lymphocyte recirculation and leucocyte emigration: the multistep paradigm. *Cell*, **76**, 301–314.

M. Raffaella Zocchi

Laboratory of Tumor Immunology
San Raffaele Scientific Institute
Milan, Italy

Anna Rubartelli

Protein Biology Unit
National Institute of Cancer Research
Genova, Italy

Nonclassical Mechanisms of Secretion in the Physiopathology of the Immune System

Intercellular communications are fundamental for many biological processes involved in the survival and reproduction of living organisms. Secretory proteins are among the most important messengers in this network of information. Proteins with a hydrophobic signal sequence are released in the extracellular environment through a constitutive or regulated pathway of secretion (classical secretion). However, increasing evidence shows that nonclassical mechanisms must exist for the secretion of proteins lacking the typical secretory signal peptide, many of which are involved in the pathogenesis of infectious, autoimmune, and neoplastic diseases. In this article, many molecular mechanisms possibly involved in nonclassical secretion will be discussed; in addition, the role of some leaderless proteins in the regulation of the immune response and the possible physiopathological implications will be analyzed.

Introduction

Protein Secretion: A Way for Intercellular Communications

In all living organisms, cells must intercommunicate in order to guarantee fundamental processes, such as growth, differentiation, and development; similarly, information must be exchanged among the various organelles of each cell. Cells are surrounded by a plasma membrane, which separates the cytoplasm from the extracellular space and possess a system of internal membranes that surround all the different organelles. These membranes, with their lipidic bilayer, are able to guarantee isolation and, as a result of their proteic structures, allow communication among the different compartments. Indeed, messages allowing in-

tercellular communications are represented by molecules which specifically interact with proteic systems, such as channels or receptors, that are able to transduce the message (Alberts *et al.,* 1991). Cells have a private life but, at the same time, they belong to a society. As an individual, each cell possesses a characteristic pattern of gene expression; as a member of a society, it can be influenced by the external environment and can selectively transmit information to other cells.

Signal Sequences

Among the molecules acting as information messengers, many are proteins. The synthesis of all proteins begins on the free cytosolic polyribosomes: The fate of these new synthesized proteins depends on the presence of specific sig-

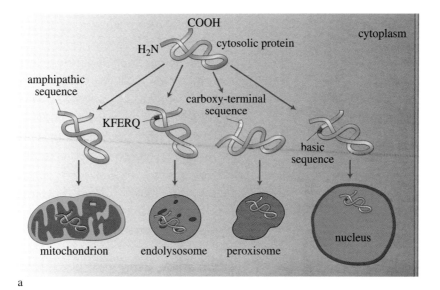

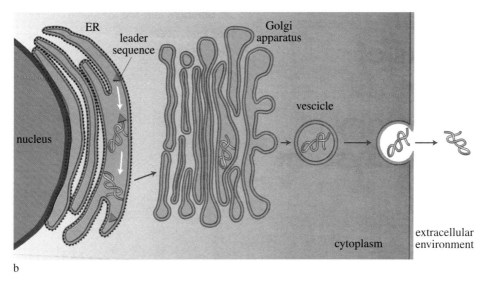

FIGURE I   Signal sequences. (a) An N-terminal amphipathic sequence directs proteins to the mitochondria and is then removed in the organelle. The KFERQ sequence targets cytosolic proteins to endolysosomes under stress conditions and a basic sequence characterizes the proteins destined for the nucleus. Both of these sequences lack a specific localization in the protein sequence and remain as part of the protein even within the organelles. This is true also for the C-terminal sequence responsible for protein targeting to the peroxisomes. (b) The N-terminal sequence (leader sequence) of proteins addressed to the endoplasmic reticulum is synthesized in the cytosol and translocates to the lumen of the reticulum. In this compartment, biosynthesis of the protein sequence is completed, whereas the leader sequence is cleaved by a specific peptidase. The protein then passes via the Golgi apparatus to the secretory vesicles and to the extracellular environment by fusion of these with the plasma membrane.

nal peptides on their amino acidic sequence. The signal peptides (also termed targeting sequences or leader sequences) are composed of 15–60 amino acid residues, situated at the extremities or inside the molecule (Schatz and Dobberstein, 1996), which are able to address the nascent polypeptide to the different organelles (Fig. 1).

The signal sequences facilitate the specific interaction between the proteins and their appropriate translocation complex located on the target membrane. This interaction induces a modification of the translocation complex which consequently allows translocation of the protein through the membrane. The contact between a signal sequence and its translocation complex can be direct or mediated by chaperonines (cytosolic transporter proteins). In the case of mitochondrial and secretory proteins, the signal peptides are eliminated by specific endopeptidases located in the lumen of the endoplasmic reticulum (ER) or mitochondria after translocation of the target membrane (Schatz and Dobberstein, 1996). Cytosolic proteins without any characteristic signal will stay in the cytosol, where they will carry out their functions, whereas those which possess a signal sequence will be transported following the signals to their specific destinations. Proteins destined to mitochondria, chloroplasts, and peroxisomes have to overcome a membrane after completion of their synthesis, whereas translocation of proteins at the ER membrane occurs cotranslationally and the synthesis of these proteins is completed inside the ER. The signal sequence which directs the proteins to the ER is a hydrophobic region of 15–30 residues at the N terminus of the protein which is called "leader peptide." The newly synthesized protein in the ER lumen can be retained inside the ER if a characteristic C-terminal sequence (KDEL) is present or it can go toward the Golgi apparatus the protein can stop or go through it and then be sorted to lysosomes, endosomes, or, upon fusion of the secretory vesicles with the plasma membrane, to the extracellular space (Fig. 2).

The Secretory Apparatus as a Post Office

If we consider the secretory apparatus as a post office, we can compare the leader peptide to the post code which addresses the secretory proteins to the ER membrane. Like a post office, which can be the addressee of missives, the ER

may represent the final destination for some proteins: In this case, the amino-acidic sequence KDEL, which corresponds to the address, will signal the protein to stop in the ER. In the same way, other sequences on a protein translocated into the ER may address it to different subcompartments of the exocytotic apparatus: When none of these sequences are present, the protein goes on by default to the extracellular environment (Fig. 2).

Not all the proteins will be secreted with the same speed. Indeed, speed and frequency of secretion are regulated on the basis of cellular or extracellular necessities. For instance, albumin is secreted with an almost constant kinetics for maintaining oncotic blood pressure (constitutive secretion); in contrast, some neuropeptides accumulate in neuronal vesicles and then are rapidly released during synaptic transmission (regulated secretion). Often, the basal level of secretion is regulated by external request: This is the case for many proteic hormones, such as insulin, the secretion of which is regulated by the sugar concentration in the blood.

The modality of secretion can also be modified during pathological conditions. For example, in normal conditions, the secretion of proteins of the extracellular matrix has a slow rhythm because the producing cells have to maintain the homeostasis, substituting the aged components undergoing degradation; however, in the presence of tissue lesions (wounds or inflammation), the rhythm of secretion accelerates in order to repair the damage. In the same way, the enzymes stored inside the leukocyte granules in normal

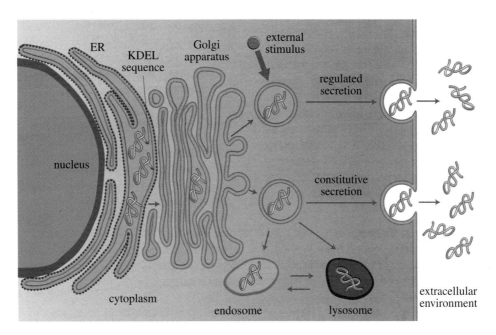

FIGURE 2 The classical secretory pathway. A protein which finishes its synthesis inside the ER can be retained due to its C-terminal sequence KDEL or transported by vesiculation to the Golgi apparatus and from here to lysosomes and endosomes or to the extracellular environment after fusion of the secretory vesicle membranes with the plasma membrane. Secretion can be constitutive or regulated: In the latter case, the protein accumulates in secretory vesicles until exocytosis is induced by an extracellular stimulus.

conditions will be released during the antibacterial response, acting as mediators of inflammation.

Returning to the similitude between secretion and post office, constitutive secretion is comparable to the ordinary mail. In the case of the regulated secretion, cells can choose among different possibilities: Secretion of hormones resembles the express mail; that of neurotransmitters, involved in the urgent communications, can be compared to telegrams; and the release of the enzymes of leukocyte granules is comparable to the delivery of Christmas packages, which is rapid and intense but limited to a restricted period.

The Secretion of Leaderless Proteins

Mail is not the only way for sending messages: There are also alternative systems such as pony express or carrier pigeons. Similarly, the passage through the ER and Golgi is not the only possibility for a secretory protein to go outside the cell. In fact, a family of proteins exist which are secreted and carry out an extracellular function, even if they lack a secretory signal sequence; they are called leaderless secretory proteins and have been described in bacteria (de Lima Pimenta *et al.*, 1997), yeast (Kuchler and Egner, 1997), and the higher eukaryotes (Rubartelli and Sitia, 1997). In higher eukaryotes, these proteins share some structural features, such as the molecular mass of 12–45 kDa, the absence of N-linked glycosylation even if potential sites of glycosylation are present, and the presence of free cysteines not engaged in disulfide bridges. In addition, these proteins are usually monomeric. These characteristics suggest that leaderless proteins do not pass in the ER, in which N-linked glycosylation and formation of disulfide bridges occur; indeed, their secretion is not inhibited by drugs such as brefeldin A and monensin, which are able to specifically block protein transit through the ER–Golgi route. Moreover, their presence outside the cell is not the consequence of a passive release of cytoplasmic content, as may occur in the case of cell death; in fact, the presence of leaderless proteins in the extracellular environment is selective and does not correlate with the presence of other cytoplasmic proteins such as lactate dehydrogenase (LDH), which represents a marker for cell lysis (Rubartelli and Sitia, 1997).

Mechanisms of Leaderless Secretion

On the basis of the studies carried out to date, it is conceivable that the secretion of leaderless proteins involves a sequential series of events. First, the protein has to be recognized: A given leaderless protein must somehow be selected among myriad cytosolic macromolecules. However, it has not been possible to evince a common motif on the sequences of these proteins that could represent a signal sequence: This obviously does not exclude that such a motif(s) exists, and the search for it is ongoing. Biochemical modifications such as myristylation, farnesylation, and acetylation can play a role in the leaderless secretion by making

the proteins more hydrophobic, thus improving their affinity for the membranes. Nevertheless, these modifications do not represent a signal for the secretion because other cytosolic proteins undergo myristylation or acetylation but are not secreted.

Once recognized, the proteins to be secreted must be transported to the extracellular space. This can be accomplished by either of two mechanisms: translocation or vesiculation (Fig. 3).

Vesiculation. This mechanism implies that a given protein must concentrate into patches beneath localized regions of the plasma membrane in order to be secreted. The membrane then evaginates, forming a bleb, and releases extracellularly vesicles enriched in that protein; by a similar mechanism, reticulocytes get rid of transferrin receptors from their membrane. In this case, however, transferrin receptor-enriched vesicles are generated by a two-step process. First, the plasma membrane invaginates with generation of multivesicular bodies, which are multilamellar complexes containing vesicles called exosomes. Then, the exosomes are released upon the fusion of the membrane of the multivesicular bodies with the plasma membrane (Johnstone *et al.*, 1987). The lectin L-14/galectin1 is an example of a leaderless secretory protein that is secreted by vesiculation (Cooper and Barondes, 1990). Immunofluorescence analysis revealed that, after synthesis, this protein accumulates in a region beneath the membrane. The increased concentration of L-14 at the submembrane level somehow induces evagination of the membrane, vesicle formation, and release of L-14-containing vesicles in the extracellular environment. The membrane of these vesicles probably undergoes a precox lysis, thus allowing the rapid solubilization of the protein.

Translocation. For other leaderless proteins translocation has been proposed as a way to leave the cell. The requirement of unfolding is a common feature in the classical models of translocation: Proteins must unfold in order to get a "translocation competent conformation." Unfolding is assisted by cellular chaperonines; since the newly synthesized proteins tend to assume their three-dimensional structure, which is energetically favorable, the action of chaperonines appears to be an energy-dependent process. The general thesis that proteins must be in a loose conformation in order to translocate a membrane has some exceptions. For instance, a peroxisomal protein was shown to be imported into peroxisomes as an heterotrimer (McNew and Goodman, 1994). Indeed, in the case of peroxisomes, the import of folded proteins is quite a common event: This mechanism seems to be mediated by specific sequences which are recognized by membrane receptors, but the transporter(s) involved has not been identified. Although the necessity of unfolding for the secretion of leaderless proteins has been demonstrated in prokaryotes, a clear demonstration of this hypothesis in higher eukaryotes is still lacking.

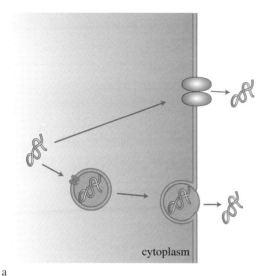

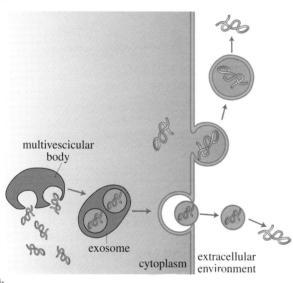

FIGURE 3 Models of leaderless secretion: translocation (a) or vesiculation (b). (a) Translocation of a leaderless protein (green ribbon) can occur through transporters (in blue) on the plasma membrane (top) or on the membrane of intracellular vesicles (bottom); the protein will be released after fusion of the vesicle membrane with the plasma membrane. (b) Vesiculation can occur by evagination of a vesicle enriched in a given leaderless protein at the plasma membrane (top) or by the formation of exosomes inside multivesicular bodies (bottom). These bodies are generated from endolysosomal vesicles by membrane invagination and the formation of exosomes enriched in the leaderless protein. The exosomes, released by exocytosis from the multivesicular bodies in which they are contained, liberate the soluble protein when their membranes dissolve.

Translocation Implies the Presence of Membrane Transporters

In yeast and bacteria (deLima Pimenta *et al.*, 1997; Kuchler and Egner, 1997) the secretion of leaderless proteins is often dependent on membrane proteins belonging to the

family of ATP-binding cassette (ABC) transporters. This family of transporters includes the multidrug resistance protein (MDR), responsible for the resistance to several drugs developed by many tumors; the cystic fibrosis gene product; and the ER proteins Tap-1 and Tap-2, which are implicated in the translocation of antigenic peptides from the cytosol to the ER lumen, where they associate with major histocompatibility complex (MHC) class I molecules (Hamon *et al.*, 1997a). Functional ABC transporters show a symmetrical four-domain structure composed by two membrane-anchoring domains and two cytoplasmic ATP-binding domains, repeated in tandem. In the general model, the membrane-anchoring domains are composed by six hydrophobic transmembrane regions: Variants of the basal structure differ in the number of transmembrane sequences or in the fact that the ABC transporter may be a dimer instead of a monomer (Hamon *et al.*, 1997a). In yeast, five classes of ABC transporters have been described: For four of them, the mammalian homolog has been identified. A sixth class, characterized by an additional transmembrane domain located between the two anchor regions flanked by two regulatory sequences, seems to be restricted to mammals: its prototype is ABC1 (Fig. 4; Hamon *et al.*, 1997a). Recently, a new model for leaderless secretion has been proposed in yeast that involves the product of two genes, *NCE-1* and *NCE-2* (nonclassical export) which do not belong to the ABC family (Cleves *et al.*, 1996). Because it has been observed in yeast that overexpression of some endogenous proteins whose intracellular accumulation may be toxic is coupled to a positive regulation of *NCE-1* and *NCE-2* expression (Cleves *et al.*, 1996), it is possible that this mechanism of nonclassical export is involved in maintaining the cellular homeostasis.

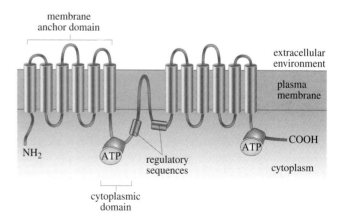

FIGURE 4 Schematic structure of an ABC1 transporter. The structural constants of a functional ABC transporter are represented by four domains: two membrane-anchoring domains (composed of six hydrophobic sequences) and two cytoplasmic ATP-binding domains sequentially repeated. The ABC1 transporter represented in the figure has an additional transmembrane domain between the two anchorage domains, flanked by two regulatory sequences.

TABLE I

Leaderless Proteins

PROTEIN	kDa	FUNCTION
Annexin1/lipocortin	35	Control of exocytosis; differentiation of myelocytes
Ciliary neurotrophic factor (CNTF)	24	Neurotrophic factor
Endothelial/monocyte activating polypeptide II (EMAP II)	20	Activating factor for endothelium and monocytes
Coagulation factor XIII, α chain	83	Coagulation
FGF-1	17	Differentiating and growth factor
FGF-2	17	Differentiating and growth factor
β-Galactoside-binding protein	15	Antiproliferating factor
Glia activating factor (FGF-9)	30	Glial activating factor
ICE (?)	45–20/10	Caspase, IL-1β converting enzyme
IL-1β	35–17	Proinflammatory and immunoregulatory cytokine
IL-1RA	20	IL-1 receptor antagonist
IL-18	24–18	γ-Interferon-inducing cytokine
Mammary-derived growth inhibitor	15	Antiproliferative factor
Parathymosine	15	Thymic hormone
Platelet-derived endothelial cell growth factor (PD-ECGF)	45	Endothelial cell growth factor
Prothymosine	15	Thymic hormone
Tat (HIV-1)	15	Viral transactivating factor; endothelial cell growth factor; immunomodulating factor
TRX/ADF	12	

Mammalian Leaderless Secretory Proteins

A number of leaderless secretory proteins is listed in Table 1. Some of them, such as the fibroblast growth factor (FGF) family of proteins and interleukin-1β (IL-1β), carry out extracellular functions only, whereas others, such as thioredoxin (TRX), high mobility group 1, human immunodeficiency virus-1 (HIV-1) Tat protein, and transglutaminase also have an intracellular role. Many leaderless secretory proteins are involved in the control of cellular proliferation and act in an autocrine or paracrine manner; some of them are mediators of inflammation, immunoresponse, or both. In the following sections, IL-1β and IL-18, two leaderless cytokines which play a role in the regulation of the immune response, and the HIV-1 protein Tat, an example of a heterologous leaderless protein with immunosuppressive function, are briefly discussed.

··

Interleukin-1β

Biological Activity and Physiopathological Implications

IL-1 is a multifunctional cytokine, produced mainly by activated mononuclear phagocytes. IL-1β is an important mediator of inflammation and plays a central role in the regulation of the immune response; it stimulates the synthesis of IL-6 and its own synthesis in mononucleated phagocytes and in vascular endothelium. Its action on endothelial cells promotes the coagulation and stimulates the expression of leukocyte adhesion molecules. At high concentration, IL-1 acts as an endogenous pyrogen, causing fever and inducing the synthesis of plasma proteins characteristic of the acute phase by liver. Its excessive production or deregulation contributes to many different pathologies (Dinarello, 1996), such as septic shock, disseminated intravasal coagulation, acute myeloid leukemia, in which there is an autocrine stimulation by IL-1, and myeloma, in which the production of IL-1 by neoplastic plasma cells is implicated in the bone reabsorption characteristic of this disease. IL-1 is also an important mediator of allograft rejection and of different chronic syndromes such as Alzheimer's disease, diabetes, some vasculitis, and other autoimmune pathologies such as rheumatoid arthritis. IL-1 seems to be involved in degenerative processes of the cartilage directly by inhibiting proteoglycan synthesis and indirectly by activating the C3 fraction of the complement and promoting bone reabsorption. All these effects seem to be key events in the induction of the inflammatory reaction which leads to the formation of synovitis, synovial pannus, and then to the erosive lesions which characterize rheumatoid arthritis. High levels of IL-1 have been identified in the sinovial fluid of patients with gout or pseudogout: In these cases, IL-1 secretion by tissutal macrophages seems to be induced by the phagocytosis of microcrystals. The capacity of stimulating collagen I and II production by fibroblast also emphasizes the role of IL-1 as an important mediator in the pathogenesis of different disorders of the connective tissue, especially scleroderma. Finally, a role for IL-1 in the pathogenesis of the immunocomplexes-mediated alveolitis has been hypothesized: The underlying mechanism is only partially understood and seems to correlate with the induction of free radical production and the activation of complement.

Biosynthesis and Regulation of IL-1 Activity

Two forms of IL-1 exist, IL-1α and IL-1β, which are encoded by two different genes; even if they have a low sequence homology (30%), they interact with the same receptors expressed on the surface of target cells (Dinarello, 1996). Despite their extracellular function, both lack the secretory signal sequence. IL-1α and IL-1β are synthesized in the cytosol as precursor proteins of apparent molecular mass of 35 kDa (pro-IL-1), the proteolysis of which generates the mature forms of 17 kDa. IL-1α is active in the precursor form, whereas IL-1β becomes active only after proteolytic maturation by the cytosolic enzyme interleukin-1β converting enzyme (ICE). The active form of ICE is a heterodimer composed by two subunits, p20 and p10, which result from processing of a single 45-kDa precursor protein. ICE does not display bioactivity in the cytosol, in which it is present in the precursor form; indeed, even if the cytosolic ICE colocalizes with pro-IL-1β within the cytosol, it does not induce maturation of the cytokine as demonstrated by the absence of extracellular 17-kDa IL-1β (Rubartelli, 1996). Recently, it has been proposed that the maturation of IL-1β may occur during the translocation of its precursor through a transmembrane channel, a hypothetical transporter of unknown nature, to which ICE could be associated. This hypothesis is based on immunoelectron microscopy studies which demonstrate the presence of ICE on the plasma membrane (Singer et al., 1995). Of the two forms of IL-1, IL-1β is the most represented extracellularly and is responsible of most of the biological activities of the cytokine in humans (Dinarello, 1996). Since IL-1β has a very high biological activity, its production must be finely tuned in order to avoid tissue damage in the course of acute or chronic inflammation. Indeed, many of the inflammatory processes are autolimited by mechanisms of negative feedback which control the activity of IL-1β and by those of others cytokines (Dinarello, 1996). Regarding other cytokines involved in inflammation and in immune response, such as IL-6, tumor necrosis factor-α (TNF-α), interferon-γ (IFN-γ), IL-1β production is regu-

lated at the transcriptional and translational levels; moreover, several specific inhibitors of IL-1 activity have been identified. Among these, IL-1 receptor antagonist (IL-1RA), which competes with IL-1 for the binding to IL-1 receptor (Dinarello, 1996), seems to play the major role. In addition, IL-1β activity is regulated at posttranslational levels by modulation of processing and secretion. For instance, the rate of IL-1β secretion in basal conditions is very low but can be considerably increased by extracellular ATP. Exogenous ATP or the ATP produced by monocytes stimulated by lypopolysaccharide (LPS) interacts with the purinergic P2Z receptors expressed on the surface of activated monocytes (Ferrari et al., 1997). It has been proposed that the stimulation of the purinergic receptor by ATP results in induction of K$^+$ efflux from the cell, with consequent depletion of intracellular potassium (Ferrari et al., 1997). This could in turn induce the activation of ICE, which is responsible for the maturation of IL-1β, with a consequent increase in IL-1β secretion (Walev et al., 1995).

IL-1β: A Secretory Mechanism Alternative to the ER–Golgi Route

When in 1985 the gene coding for IL-1β was cloned, the problem concerning the secretion of this leaderless protein was set aside by claiming that IL-1β is released by cell lysis. It was assumed that activated monocytes, which produce a large amount of intracellular IL-1β, after migration to the inflammation site could degenerate and then passively release IL-1β in the extracellular environment. This hypothesis was then definitively discarded on the basis of two observations. First, the secreted biologically active form of IL-1β is the mature 17-kDa form, whereas intracellular IL-1β is the 35-kDa precursor; when cell lysis is induced in vitro, the precursor form of 35 kDa is found in the supernatant, in which mature IL-1β is undetectable. This means that the secretion of IL-1β is associated with its proteolytic maturation, a process which needs living cells. Second, IL-1β secretion is selective because its presence in the extracellular environment does not correlate with the presence of other cytosolic proteins, such as LDH or IL-1α, which are kept inside the cell. The possibility that IL-1β is secreted by the classical mechanism of secretion has been evaluated and excluded. Indeed, drugs able to inhibit at different levels the protein transport along the ER–Golgi classical secretory route, such as brefeldin A and monensin, block the secretion of leader sequence-bearing cytokines, such as IL-6 and TNF-α, but not IL-1β. In contrast, these drugs potentiate the secretion of IL-1β (Rubartelli et al., 1990). Thus, an alternative mechanism of secretion for IL-1β seems to exist which is independent of the ER–Golgi route.

It has been reported that IL-1β secretion by monocytes is concomitant with apoptosis (Hogquist et al., 1991). This hypothesis was supported by the fact that ICE, which is able to induce IL-1β maturation, is also involved in the triggering of the apoptotic process. Further analyses have clarified that IL-1β secretion also occurs in cells which do not present either early or late signs of apoptosis. Thus, although it is possible that IL-1β is released during apoptosis, apoptosis per se is not necessary for the process of IL-1β secretion (Rubartelli *et al.,* 1993).

Possible Role of ABCI in IL-1β Secretion

Since the secretion of many leaderless proteins in yeasts and bacteria occurs through ABC transporters, it has been proposed that a mammalian ABC homolog is implicated in the secretion of IL-1β (Hamon *et al.,* 1997a). The role of ABC transporters in IL-1β secretion was investigated using drugs able to block the activity of different ABC transporters. The results demonstrated that blockers of the ABC1 transporter activity are able to inhibit the release of IL-1β (Hamon *et al.,* 1997b), at variance with blockers of MDR or of other ABC transporters. Structural studies on ABC1 have shown its homology with a protein coded by the gene *ced-7* of *Caenorhabditis elegans;* functional studies have demonstrated that this transporter is involved in the phagocytosis of apoptotic cells by macrophages (Hamon *et al.,* 1997a). Experiments on oocytes of *Xenopus laevis* microinjected with ABC1 demonstrated that this protein is an electroneutral anionic transporter whose function is inhibited by drugs such as glybenclamide, bromosulfophthalein, and diisothiocyanatostylbene disulfonic acid. These drugs are able to block IL-1β secretion at the same doses at which they block the anionic transport (Hamon *et al.,* 1997b). For these reasons, ABC1 is a major candidate for mediating IL-1β secretion.

A Model for IL-1β Secretion

The fact that inhibitors of ABC1 transporter are able to block the secretion of IL-1β suggests that the release of this cytokine occurs with a mechanism of translocation rather than of vesiculation. The translocation can occur directly at the plasma membrane or at an internal membrane. Studies in our laboratory have demonstrated that even if IL-1β has a predominant cytosolic localization in the producing cell, it is partially contained within intracellular vesicles which cofractionate with late endocytic vesicles; these vesicles share with lysosomes some biochemical and structural features. Since conditions which stimulate IL-1β secretion, such as high extracellular ATP levels, are able to induce the release of vesicular IL-1β, we have proposed a model of secretion which predicts the exocytosis of the IL-1β-containing endolysomes (Andrei *et al.,* 1999). According to this model, IL-1β translocates from the cytosol to the lumen of these organelles in its immature form. Upon extracellular stimuli, the membrane of these vesicles fuses with the plasma membrane and IL-1β is released in the mature form (Fig. 5). A similar mechanism of secretion has been proposed in *Dictyostelium discoideum* for the exter-

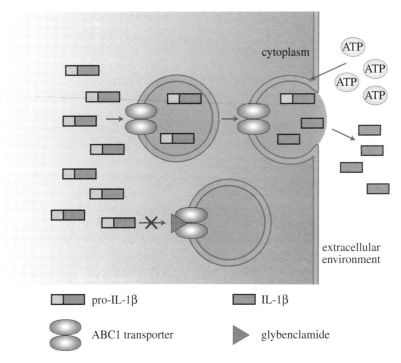

FIGURE 5 Model of IL-1β secretion. In this model a fraction of cytosolic pro-IL-1β translocates inside lysosomal-related vesicles through a transporter present on their membranes. Translocation is inhibited by glybenclamide, an ABC1 transporter inhibitor. In a second step mature IL-1β is released extracellularly by exocytosis of these vesicles. The basal level of exocytosis is low, but it is induced by extracellular stimuli such as high ATP concentrations.

nalization of DdCAD-1, a leaderless membrane protein. This protein translocates from the cytosol into acid vesicles, similar to lysosomes, whose exocytosis is regulated by extracellular osmotic conditions (Sesaki *et al.,* 1997). Although the major function of lysosomes is degradation of their content, in some cell types such as macrophages they behave like secretory organelles, as demonstrated by the regulated secretion of hydrolytic lysosomal enzymes present in these cells. Moreover, the specialized secretory granules present in different cell types — such as the azurophyl granules of granulocytes, the specific granules of mastocytes, and the lytic granules of T lymphocytes — all derive from lysosomes (Stinchcombe and Griffiths, 1999). This regulated endolysosomal exocytosis gives the macrophage the potential to exert a regulatory influence in the course of inflammation, infection, and induction of immune response by modulating the release of IL-1β and lysosomal hydrolases in its surrounding environment.

Interleukin-18

Biological Activity

Recently, the gene coding for a new cytokine, IL-18, has been cloned (Ushio *et al.,* 1996) and the cytokine called interferon-γ inducing factor (IGIF) due to its capacity to induce the production of IFN-γ by T helper 1 (T$_H$1) CD4$^+$ lymphocytes, CD8$^+$ cytotoxic T lymphocytes, and natural killer (NK) cells. Thus, IL-18 belongs to the restricted group of cytokines (including IL-12, TNF-α, and IL-2) which are able to regulate INF-γ production. One of the most important effects of IFN-γ is the activation of macrophages, with a consequent increase of antibacterial activity, antigen presentation, and cytolitic activity of T lymphocytes. Moreover, IFN-γ contributes to the expansion of the T$_H$1 cell subpopulation and to the activation of endothelial cells. Thus, it is evident that IL-18 plays a crucial role in the regulation of cell-mediated inflammatory responses. In modulating INF-γ production, IL-18 seems to act synergically with IL-12, which apparently induces the expression of IL-18 receptor on target cells, and with TNF-α. It can be speculated that IL-18 participates in the "polarization" of the T helper system, promoting the differentiation of T$_H$0 lymphocytes in T$_H$1, which produce INF-γ and IL-2, rather than in T$_H$2, which produce different cytokines (IL-4 and IL-10). An IL-18 receptor with a structure similar to the IL-1 receptor has been described (Torigoe *et al.,* 1997); nevertheless, its expression and the regulation of its functions in the immunocompetent cells are still under investigation. Cells deputated to the IL-18 production are not fully characterized. In mouse, the gene coding for IL-18 has been cloned from Kupffer liver cells, and cells belonging to the monocyte–macrophage lineage seem to be the major producers. Nevertheless, the gene is also expressed in other cell types, such as keratinocytes and osteoblasts. Recently,

it was proposed that IL-18 expression is associated with the active stage of autoimmune insulin-dependent diabetes in mice (Rothe *et al.,* 1997).

Structure and Biosynthesis

IL-18 is structurally correlated to IL-1β since both present a characteristic β-sheet conformation, whereas the α-helix structure predominates in the other cytokines. Moreover, like IL-1β, IL-18 also lacks a secretory signal sequence, is not glycosylated, and is synthesized as an inactive precursor, pro-IL-18, displaying a molecular weight of 24 kDa. The same caspase ICE, responsible for IL-1β maturation, processes pro-IL-18, yielding to the IL-18 mature form of 18 kDa, which is biologically active (Ushio *et al.,* 1996). These characteristics suggest that, like IL-1β, IL-18 is secreted by an alternative mechanism which is different from the classical route; however, the modalities of secretion of this newly discovered cytokine are still under investigation.

Possible Role of IL-18 as a "First-Aid" Cytokine in the Immune Response

The biological activity of IL-18 and the fact that it is produced by cells of the monocyte–macrophage lineage suggest that this cytokine plays an important role in the immune response. Recent experiments in our laboratory have shown that dendritic cells, which are able to present the antigens to T lymphocytes, constitutively produce IL-18. The contact with antigen-specific T lymphocytes induces a rapid release of preformed IL-18, whereas its synthesis is inhibited (Gardella *et al.,* 2000). Preliminary results suggest that MHC molecules are implicated in the signal, transduced from the T lymphocytes to the dendritic cells, which leads to the modulation of IL-18 secretion. This suggests that, in the presence of pathogens, IL-18 could represent an early warning for the immune system, followed by the recruitment of T$_H$ lymphocytes and their differentiation into the T$_H$1 subpopulation. The decrease of IL-18 synthesis that follows the interaction between dendritic cells which present the antigen and antigen-specific T lymphocytes could represent a mechanism of negative feedback aimed at preventing an excessive amplification of the response.

HIV-1 Tat Protein

Structure and Biosynthesis

Tat is an 86- to 104-amino acid protein encoded by two exons, the first of which contributes the N-terminal 72 residues. Its sequence can be subdivided into several regions: a cysteine-rich domain, a core region, a highly conserved basic domain which contains the nuclear localization sequence, and a glutamine-rich domain. The last three regions are all involved in the binding of the viral RNA. Despite many efforts, Tat has not been crystallized. On the basis of nuclear magnetic resonance and molecular dynamics studies, a structural model has been proposed with two rigid (the core region and the glutamine-rich domain) and two highly flexible domains (the cysteine-rich region and the basic domain). The sequence coded by the second exon did not yield a defined structure. Interestingly, this region contains the tripeptide RGD that, as discussed later, mediates some extracellular functions of Tat. In infected cells, Tat localizes in the nucleus, but it is also present in the cytoplasm and it can be released extracellularly in the absence of cell death or membrane permeability modifications. Since the release of Tat protein is not inhibited by drugs blocking the classical secretory route, such as brefeldin A, it is possible that Tat belongs to the leaderless secretory protein family and is secreted through an alternative route. Once Tat is secreted, it can be taken up by bystander healthy or infected cells and interfere with their growth/differentiation processes (Goldstein, 1996; Rubartelli et al., 1998).

Biological Activity

Tat is a transactivating factor involved in the process of viral replication. In addition to acting on viral genes, Tat is also able to drive the transcription of many cellular genes, such as those codifying for IL-2, IL-6, and TNF-α. In addition to its endogenous role of transcription factor, Tat has been reported to exert many biological activities on several cell types. Tat is an angiogenic factor involved in the development of Kaposi's sarcoma; furthermore, it exerts important immunosuppressive effects, such as the inhibition of NK cell activity and antigen-specific or nonspecific T lymphocyte proliferation, the induction of apoptosis in T cells, and the block of several dendritic cell functions including apoptotic body phagocytosis and IL-12 secretion (Rubartelli et al., 1998). Several different mechanisms have been proposed to explain the extracellular functions of Tat:

• Low-affinity interactions with eparan sulfates and high-affinity interactions with vascular endothelial growth factor (VEGF) through its basic domain with a consequent angiogenic activity
• Interaction with integrins through the RGD domain, leading to inhibitory effects, such as on apoptotic body engulfment by dendritic cells, or activating effects, such as induction of endothelial cell spreading and of leukocyte chemotaxis
• Interaction with the dipeptidyl peptidase CD26, which is expressed on the membrane of T lymphocytes, followed by the block of cell proliferation
• Interaction with functional L-type Ca^{2+} channels with consequent functional block of some cell types, including NK cells and dendritic cells, due to the inhibition of Ca^{2+} entry and of the following signal transduction cascade: Tat can enter the cytoplasm of target cells from the extracellular space, reach the nucleus, and induce the expression of different genes. The capacity to penetrate the cell by translocation across the plasma membrane or the endosomal

membrane, if the internalization occurs through endocytosis, is a quite rare and peculiar feature that Tat shares with a few bacterial toxins and some other leaderless proteins, such as FGF-1 and FGF-2. Whatever the extracellular mechanism of Tat action, it seems clear that this protein is crucial in determining the gravity of the AIDS. In fact, there is an inverse correlation between the appearance of natural anti-Tat antibodies and the progression of the disease. Therefore, the functional block of exogenous Tat, obtained by an active immunization, might reduce the immunosuppressive effects of this protein, supporting the antiviral response. However, more information on the molecular structure of Tat is needed in order to design successful vaccination strategies (Rubartelli et al., 1998).

Advantages of a Leaderless Secretory Pathway

Why do cells, which present an efficient secretory system, utilize an alternative pathway of secretion for some proteins? In fact, returning to the post office similitude, the leaderless secretion would represent a slow mail service such as the carrier pigeon service. Nevertheless, two features of leaderless secretory proteins balance for their inefficient secretion: (i) Most of them are characterized by a high biological activity and (ii) they act in an autocrine or paracrine way. The slow and inefficient secretion may thus become an advantage by self-limiting the protein activity in a small environment and allowing the control of potentially toxic effects. In the following sections, other hypotheses on the necessity or the advantages of the nonclassical secretion are discussed.

Leaderless Secretion as Safety Valve

In yeast, even proteins needed for the cell survival are tolerated only at low levels of expression and become toxic when present at high levels (Cleves et al., 1996): Nonclassical secretion might impede the cytosolic accumulation of these proteins by inducing their prompt secretion, thus acting as a safety valve. A similar mechanism could be utilized in mammals: An example is the mytochondrial sulfotransferase rhodanese, which in physiological conditions accumulates in mitochondria but when overexpressed is efficiently secreted even if it lacks the secretory signal sequence (Rubartelli and Sitia, 1997).

Prevention of Intracellular Autocriny through the Compartmentalization of Receptor and Ligand

Many mammalian leaderless proteins are active as autocrine growth factors on several cell types (Rubartelli and Sitia, 1997); thus, along their way to the extracellular milieu they should avoid the interaction with their own receptors, which might result in early activation. This kind of intracellular stimulation would escape any feedback regulation, such as the downmodulation of membrane receptors. The impossibility of regulating the cascade of events triggered by the interaction of a growth factor with its receptors might lead to uncontrolled proliferation and cellular transformation. This was indeed demonstrated in different experimental systems (Rubartelli and Sitia, 1997). IL-3 is a classical secretory protein endowed with the signal sequence which addresses its synthesis to the ER; if the IL-3 gene is deprived of this sequence, IL-3 accumulates in the cytosol and it is not secreted. In contrast, if a KDEL sequence, which is specific for maintaining the protein into the ER compartment, is engineered at the carboxy terminus of the protein, and cells expressing the IL-3 receptor are transfected by this modified IL-3 gene, cell growth becomes independent of the presence in the medium of this interleukin. This means that the receptor is already functional in the ER and is able to transduce to the nucleus the proliferative signal initiated by the interaction with the ligand. By the same technique it has been possible to demonstrate that *v-sis,* the oncogene which codes for the polypeptide homolog to the B chain of platelet-derived growth factor (PDGF), is able to activate the PDGF receptor at the ER level; this event, in the absence of a negative regulation, induces cell transformation. In the case of the leaderless protein FGF-2, the insertion of a secretory leader sequence at the N terminus induces cell transformation in the cell line expressing the FGF-2 receptor. This implies that FGF-2, when forced through the ER–Golgi, binds to its receptor and induces its uncontrolled activation. Since the autocrine stimulation mediated by FGF-2 is a physiological mechanism shared by several cell types both during embryonic and fetal life (during which it mediates differentiation of different organs and tissues) and during the adult life (FGF-2 is an important angiogenic factor), it appears that compartmentalization of receptor and ligand represents an efficient mechanism for controlling both proliferation and differentiation.

Prevention of Misfolding by the Exclusion from Oxidizing or Sugar-Rich Environments

Many leaderless proteins bear free sulfhydryl groups which must be maintained in the reduced state in order to guarantee a correct folding and hence the bioactivity of the protein. The ER lumen presents some peculiar characteristics, such as a higher concentration of Ca^{2+} and a more oxidating condition compared to those of the cytosol. The latter is essentially due to the presence of reduced glutathione and cysteine transporters, which favor the oxidation of sulfhydryl groups. As a consequence, proteins secreted through the ER–Golgi classical route present many intra- and interchain disulfide bridges. For the proteins which present free thiols, the transit through an oxidizing milieu, such as the ER lumen, might result in either retention or secretion in a nonfunctional folding. The following are examples of the

proteins for which cysteine oxidation might result in alteration of functions: TRX, whose redox potential depends on its reduced sulfhydryl groups; Tat, which contains seven cysteines; FGF-1, which is active only in the reduced state; and ICE, the convertase implicated in IL-1β maturation, that bears a reduced sulfide in its active site. Another example of how avoiding the classical route of secretion may represent an advantage for protein structure is given by lectins: These proteins, involved in cell to cell adhesion through specific binding with membrane carbohydrates, might take advantage from avoiding the Golgi compartment, which is rich in sugars and may stack lectins, thus impairing their transport.

Proteins Might Have Both Intra- and Extracellular Functions

Among the proteins which are actively secreted by nonclassical mechanisms, some have a function inside the cell beside the extracellular one. For example, TRX is a cytosolic oxide reductase which acts as a growth factor outside the cell (Rubartelli and Sitia, 1997); Tat is a transcription factor with a nuclear localization sequence, but it is also able to exert several extracellular functions (Goldstein, 1996; Rubartelli *et al.,* 1998); and the presence of an intracellular form of IL-1 receptor antagonist suggests that IL-1 protein can also play a function inside the cell (Rubartelli and Sitia, 1997). It is possible that, on the basis of its physiological and developmental state, the cell is able to address a cytosolic protein toward an alternative extracellular function: The nonclassical secretory route would guarantee the possibility of this double function.

Efficient Posttranscriptional Regulation of Secretion

Another advantage of nonclassical secretion is the possibility of increasing rapidly the bioactive extracellular protein by inducing a secretory switch of the preformed intracellular protein. Examples of leaderless secretory proteins whose extracellular availability is regulated by rapid modification of their rate of secretion are IL-1β and IL-18.

According to the model proposed for IL-1β secretion, a fraction of the protein translocates from the cytosol into lysosomal-related vesicles. Exocytosis of these vesicles, with consequent secretion of IL-1β, is low in LPS-activated monocytes, but it is strongly induced by extracellular conditions, such as high ATP levels or interaction with immunocomplexes, which mimic *in vitro* conditions occurring *in vivo* at the site of inflammation. Similarly, an extracellular pH lower than 7 enhances IL-1β maturation and thus the availability of the bioactive form. The induction of secretion is very rapid, and equally fast is the return to the basal condition when the inducing stimuli are removed. In the case of IL-18, we recently observed that this cytokine is constitutively produced by dendritic cells but barely secreted; secretion is rapidly induced by the contact with T lymphocytes during the antigenic response. Since IL-1β is an important mediator of inflammation and IL-18 is an enhancer of T helper responses, the importance of a fine and efficient regulation of their secretion is evident.

References Cited

ANDREI, C., DAZZI, C., LOTTI, L., TORRISI, M. R., CHIMINI, G., and RUBARTELLI, A. (1999). The secretory route of the leaderless protein interleukin-1β involves exocytosis of endolysosome-related vesicles. *Mol. Biol. Cell* **10**, 1463–1475.

CLEVES, A. E., COOPER, D. N., BARONDES, S. H., and KELLY, R. B. (1996). A new pathway for protein export in *Saccharomyces cerevisiae. J. Cell Biol.* **133**, 1017–1026.

COOPER, D. N., and BARONDES, S. H. (1990). Evidence for export of a muscle lectin from cytosol to extracellular matrix and for a novel secretory mechanism. *J. Cell Biol.* **110**, 1681–1691.

DE LIMA PIMENTA, A., BLIGHT, M. A., CHERVEAUX, C., and HOLLAND, I. B. (1997). Protein secretion in gram negative bacteria. In *Unusual Secretory Pathways: From Bacteria to Man* (K. Kuchler, A. Rubartelli, and B. Holland, Eds.). Chapman & Hall Landes Bioscience, New York/Austin, TX.

DINARELLO, C. A. (1996). Biological basis for interleukin 1 in disease. *Blood* **87**, 2095–2147.

FERRARI, D., CHIOZZI, P., FALZONI, S., HANAU, S., and DI VIRGILIO, F. (1997). Purinergic modulation of interleukin-1β release from microglial cells stimulated with bacterial endotoxin. *J. Exp. Med.* **185**, 579–582.

GARDELLA, S., ANDREI, C., POGGI, A., ZOCCHI, M. R., and RUBARTELLI, A. (2000). Control of interleukin-18 secretion by dendritic cells: role of Ca^{2+} influxes. *FEBS Lett.* **481**, 245–248.

GOLDSTEIN, G. (1996). HIV-1 Tat protein as a potential AIDS vaccine. *Nature Med.* **2**, 960–964.

HAMON, Y., LUCIANI, M. F., BECQ, F., VERRIER, B., RUBARTELLI, A., and CHIMINI, G. (1997a). Interleukin-1β secretion is impaired by inhibitors of the Atp binding cassette transporter, ABC1. *Blood* **90**, 2911–2915.

HAMON, Y., LUCIANI, M. F., and CHIMINI, G. (1997b). Mammalian ABC transporters and leaderless secretion: Facts and speculations. In *Unusual Secretory Pathways: From Bacteria to Man* (K. Kuchler, A. Rubartelli, and B. Holland, Eds.). Chapman & Hall Landes Bioscience, New York/Austin, TX.

HOGQUIST, K. A., NETT, M. A., UNANUE, E. R., and CHAPLIN, D. D. (1991). Interleukin 1 is processed and released during apoptosis. *Proc. Natl. Acad. Sci. USA.* **88**, 8485–8489.

JOHNSTONE, R. M., ADAM, M., HAMMOND, J. R., ORR, L., and TURBIDE, C. (1987). Vesicle formation during reticulocyte maturation: Association of plasma membrane activities with released vesicles (exosomes). *J. Biol. Chem.* **262**, 9412–9420.

KUCHLER, K., and EGNER, R. (1997). Unusual protein secretion and translocation pathways in yeast: Implication of ABC transporters. In *Unusual Secretory Pathways: From Bacteria to Man* (K. Kuchler, A. Rubartelli, and B. Holland, Eds.). Chapman & Hall Landes Bioscience, New York/Austin, TX.

McNEW, J. A., and GOODMAN, J. M. (1994). An oligomeric protein is imported into peroxisomes *in vivo. J. Cell Biol.* **127**, 1245–1257.

ROTHE, H., JENKINS, N. A., COPELAND, N. G., and KOLB, H. (1997). Active stage of autoimmune diabetes is associated with the expression of a novel cytokine, IGIF, which is located near Idd2. *J. Clin. Invest.* **99**, 469–474.

RUBARTELLI, A., and SITIA, R. (1997). Secretion of mammalian proteins that lack a signal sequence. In *Unusual Secretory Pathways: From Bacteria to Man* (K. Kuchler, A. Rubartelli, and B. Holland, Chapman & Hall Landes Bioscience, New York/Austin, TX.

RUBARTELLI, A., COZZOLINO, F., TALIO, M., and SITIA, R. (1990). A novel secretory pathway for interleukin 1β, a protein lacking a signal sequence. *EMBO J.* **9**, 1503–1510.

RUBARTELLI, A., BAJETTO, A., ALLAVENA, G., COZZOLINO, F., and SITIA, R. (1993). Post-translational regulation of Interleukin 1β secretion. *Cytokine* **5**, 117–124.

RUBARTELLI, A., POGGI, A., SITIA, R., and ZOCCHI, M. R. (1998). HIV-1 Tat: A Polypeptide for all seasons. *Immunol. Today* **19**, 543–545.

SCHATZ, G, and DOBBERSTEIN, B. (1996). Common principles of protein translocation across membranes. *Science* **271**, 1519–1526.

SESAKI, H., WONG, E. F., and SIU, C. H. (1997). The cell adhesion molecule DdCAD-1 in Dictyostelium is targeted to the cell surface by a nonclassical transport pathway involving contractile vacuoles. *J. Cell Biol.* **138**, 939–951.

SINGER, I. I., SCOTT, S., CHIN, J., BAYNE, E. K., LIMJUCO, G., WEIDNER, J., MILLER, D. K., CHAPMAN, K., and KOSTURA, M. J. (1995). The interleukin-1 beta converting enzyme (ICE) is localized on the external cell surface membranes and in the cytoplasmic ground substance of human monocytes by immunoelectron microscopy. *J. Exp. Med.* **182**, 1447–1459.

STINCHCOMBE, J. C., and GRIFFITHS, G. M. (1999). Regulated secretion from hemopoietic cells. *J. Cell Biol.* **147**, 1–6.

TORIGOE, K., *et al.* (1997). Purification and characterization of the human interleukin-18 receptor. *J. Biol. Chem.* **272**, 25737–25742.

USHIO, S., *et al.* (1996). Cloning of the cDNA for human IFN-γ-inducing factor, expression in *Escherichia coli,* and studies on the biologic activities of the protein. *J. Immunol.* **156**, 4274–4279.

WALEV, I., RESKE, K., PALMER, M., VALEVA, A., and BHAKDI, S. (1995). Potassium-inhibited processing of IL-1 β in human monocytes. *EMBO J.* **14**, 1607–1614.

General References

ALBERTS, B., BRAY, D., LEWIS, J., RAFF, M., ROBERTS, K., and WATSON, J. D. (Eds.) (1995). *Biologia Molecolare della Cellula.* Zanichelli, Bologna.

KUCHLER, K., RUBARTELLI, A., and HOLLAND, B. (Eds.) (1997). *Unusual Secretory Pathways: From Bacteria to Man.* Chapman & Hall Landes Bioscience, New York/Austin, TX.

Mauro S. Malnati
Paolo Lusso

Unit of Human Virology
San Raffaele Science Institute
Milan, Italy

Viruses and Molecular Mimicry: The First "Cybernetic Pirates"

Several viruses have copied and integrated in their genome many genes derived from the infected host, successfully utilizing them for propagation of their own progeny. This process of molecular mimicry represents one of the most fascinating mechanisms by which the parallel evolution between the virus and its host has been achieved. Retroviruses, herpesviruses, and poxviruses are three large families that have developed strategies of adaptation to the host based on this form of "cybernetics piracy" ante litteram. However, these families differ in the choice of genes utilized and in the number of cellular targets engaged. In this article, two complex models of molecular mimicry, the result of different evolutionary strategies, will be discussed; both represent examples of the sophistication reached by large DNA viruses in the manipulation and adaptation of cellular DNA in the attempt to conserve their genetic record over time.

Introduction

Despite the current wealth of information on the genomic structure of viruses, some of which have been entirely sequenced and their molecular organization and regulation of gene expression have been defined in detail, comprehension of the mechanisms by which viral infections induce disease in the host is still limited. This issue deals with a very complex system in which the different viral functions that contribute to the determination of the pathologic process are added to the functions of the host (innate and specific immunity) which render it susceptible or resistant to disease. Because of the extraordinary progress achieved in recent years in the fields of molecular biology and gene manipulation, several experimental models have been developed to study the function of single viral genes; however, it is clear that the numerous factors determining the virulence and pathogenicity of the virus, on the one hand, and the host's defense mechanisms, on the other hand, create a complex network of interactions that are barely reproducible even in the most sophisticated experimental model. A further level of complexity is represented by the fact that each specific virus–host interaction is often a unique and unrepeatable system. Within the babel of these multiple interactions, however, it is possible to identify strategies that are common to different families resulting from precise evolutionary choices. One of these strategies is based on the ability of the virus to acquire genetic information from the host: At least three large families of viruses have been able to copy and incorporate in their genome genes that play a key role in the regulation of cell growth and the immune response. This form of "cybernetics piracy" ante litteram, described for the first time in the 1970s by J. M. Bishop and H. Varmus in retroviruses (a family of RNA viruses), has also been widely used by two families of DNA viruses: Herpesviridae and Poxviridae.

Although the strategy of incorporation of exogenous genes may be interpreted as a response to a common evolutionary pressure, different modus operandi (way of action) are used by each viral family. Retroviruses have gained long-term survival by relying on their capacity to copy themselves through retrotranscription, a process that generates a more stable form of genetic information (DNA instead of RNA) able to recombine directly with the DNA of the host and to persistently integrate in the cellular genome. To keep the process of retrotranscription efficient, retroviruses have been obliged to maintain the size of their genome almost unchanged, a constraint which has greatly limited their freedom to withdraw exogenous genes from the host's cell. Having the chance in the majority of instances to make only one choice, the retroviruses have incorporated cellular factors involved in the induction of cell proliferation: the so-called protooncogenes. These factors, controlled by precise mechanisms of regulation of cellular genes, have become viral oncogenes capable of determining neoplastic transformation in infected cells. In this way, retroviruses have obtained maximal efficiency in the processes of recombination and integration in the host's genome which strictly depend on the cell's activation state and on cellular proliferation.

Different from retroviruses, large DNA viruses have not developed strategies for their recombination into the genome of the host, most likely because of their large size and the complexity of their genome. These features represent an objective obstacle to the maintenance of the integrity and functions of their genetic program in the instance that a casual recombination with cellular DNA might occur. Furthermore, in the case of the poxviruses, which replicate exclusively within the cytoplasm of the infected cell, a physical barrier between the genetic material of the virus and that of the cell exists. Therefore, transfer of genetic information from the host has followed different pathways, for example, through the formation of molecular hybrids between viral DNA and cellular messenger RNA. This kind of strategy, which is less efficient than the direct recombination and integration performed by retroviruses, most likely occurred due to a strong selective pressure imposed by the host's complex programs of antiviral resistance. To control and rapidly eliminate viruses which represent a serious threat for an individual's survival, the host has been obliged to set up cellular programs able to catalyze a rapid and efficient inflammatory reaction (innate or natural immune response), ensuring in the meantime the survival of the cells proximal to the site of infection and the effectors

of the specific immune response in the difficult milieu created by an acute inflammatory process. In addition to imposing an adaptive pressure, the host also preselects the genetic material on which the viruses could operate their strategic choices: the transcripts of messenger RNA induced in the cells in response to infection. Therefore, not casual appears the choice made by DNA viruses to incorporate cellular genes involved in the regulation of cell survival or of the immune system, either natural or specific. In the never-ending "chess game" played by the host and the virus, the progressive incorporation of cellular genes by the viruses has represented a true strategic program aimed to continuously elude, move after move, the cellular mechanisms arranged by the host to limit viral infection.

Herpesviruses and poxviruses have faced the biologic problem of their own persistence and propagation and have established two distinct strategies that are reflected in the different way they have carried out the "molecular robbery" of the host's material. Herpesviruses, which persist in a single infected individual in a latent form for the length of the individual's life, have mainly favored the long-term survival of the host cell, copying or inducing the transcription of genes involved in the regulation of cellular proliferation and survival. On the contrary, poxviruses, which are not able to establish a persistent infection in a single individual but remain in balance with the host population, continuously "jumping" from one individual to the other, have primarily invested in being invisible to the immune system, copying and adapting cellular genes involved in the regulation of the acute inflammatory response. Herpesviruses appear to be "gentle thieves" which attempt to not perturb the homeostasis between the infected cell and the organism, whereas poxviruses are "trained commandos" able to penetrate the programs of intercellular communication of the immune system in order to sabotage them. This article provides several examples of "cybernetics piracy" that poxviruses and herpesviruses have enacted during the course of a slow adaptive evolution, parallel to that of their natural hosts. Many excellent reviews have been written which the reader may consult for further information (Murphy, 1993, 1994a,b).

Molecular Mimicry among Poxviruses

Poxviruses are a family of DNA viruses that infect both vertebrates and invertebrates, among which the most famous member for the human species is the variola virus, the etiologic agent of smallpox. These viruses are characterized by a large and complex virion (Fig. 1) that contains the enzymes for the synthesis of the messenger RNA and a double-stranded linear DNA genome between 130 and 300 kb in length, which is able to replicate exclusively in the cytoplasm of infected cells. The general organization of

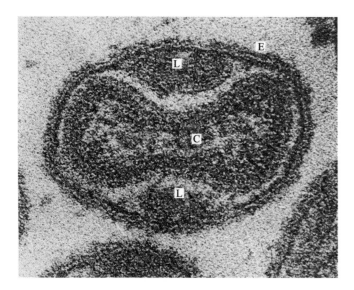

FIGURE 1 Electron micrograph of a mature virion of vaccinia virus. The core (C), lateral bodies (L), and outer pericapsid membrane (E) are distinguishable (reproduced with permission from *Fields Virology*, 1990).

these viruses and, in particular, the genes that encode structural proteins and those essential for viral replication are highly conserved and clustered around the central region of the genome (Turner and Moyer, 1990). At the periphery are two regions encompassing terminal sequences inverted and repeated which represent a preferential site for recombination events between host and virus (Fig. 2).

The large genome of poxviruses has allowed them to host inside the inverted and repeated terminal regions many genes (in the case of vaccinia virus, at least 55) which are not strictly required for replication and propagation in cellular cultures (Fig. 2). These genes, sometimes acquired from the host in a double copy (one for each terminal region), represent important virulence factors.

Strategies to Survive: Invisible and Promiscuous

Viruses represent an immunogenic stimulus for the host in a manner which is proportional to their complexity. Among the DNA viruses, poxviruses are the largest and the most complex: The host is therefore capable of rapidly mounting a specific immune response in order to limit viral replication and to properly protect itself from potential reinfection episodes. Hence, poxviruses have been obliged to elaborate specific strategies to indefinitely propagate themselves aimed at limiting the efficiency of the host's antiviral activity, represented by the inflammatory response or the specific and innate immune response. Furthermore, the ability of poxviruses to replicate exclusively within the cytoplasm of the host's cell most likely has prevented the evolution of a mechanism of latency, which instead has occurred for another family of large DNA viruses, the herpesviruses. Poxviruses have followed two main strategies in their interaction with the host. The first has been to gain the broadest host spectrum (indeed, poxviruses are common etiologic agents of zoonosis) in order to ensure a large pool of individuals susceptible to the infection. The second has been to adopt many genes, mainly of cellular origin, that are able to interfere at different levels with the mediators and effectors of the innate and adaptive responses, with the specific function of rendering the actively replicating virus "invisible."

The biologic compromise between propagation of the viral progeny and the survival of the infected host has been achieved for the majority of the members of the Poxviridae family through the "localization" of the infection by neutralizing the efficacy of the anti-inflammatory response at the site of entry and first replication of the pathogen, such as the exposed skin or, seldom, the mucosal tissues. In this manner, poxviruses have obtained an efficacious strategy of

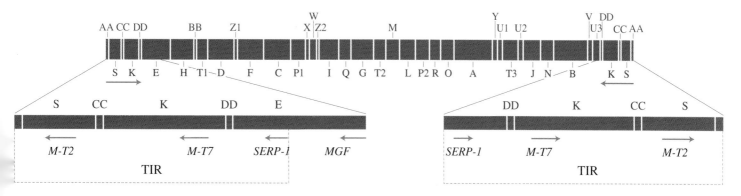

FIGURE 2 Schematic representation of the genome of myxoma virus. The letters in red indicate the restriction sites recognized by the enzyme *Bam*HI; the letters in blue indicate the different regions of the genome. In the lower part of the figure the terminal inverted and repeated (TIR) sequences are shown magnified, as are some of the genes described among the virulence factors of poxviruses; the arrows indicate the direction of transcription.

transmission relying on high levels of viral replication in the exposed primary lesion concomitant with a low pathogenetic impact on the individual and a negligible biologic disadvantage. Only few poxviruses are exceptions to this general model, notably the variola virus (completely eradicated as a result of the vaccination programs of the World Health Organization) and the myxoma virus. The fact that the myxoma virus has a dichotomous behavior causing a chronic asymptomatic infection in the American rabbit, its natural host, whereas infection of most European rabbits is lethal makes it one of the most fascinating models in which to study the molecular mechanisms responsible for the pathogenicity of viral infections.

The Virus of Myxomatosis: Paradigm of Pathogenesis by Poxviruses

Myxomatosis was described as a pathogenic entity in Uruguay at the end of nineteenth century when European rabbits imported into the New World were suddenly wiped out by an unknown disease, later called myxomatosis (Fenner and Meyers, 1978). In its natural host, the North American rabbit (*Sylvilagus bachmani*), and in the rabbit of South America tropical forests (*Sylvilagus brasiliensis*) myxoma virus causes a persistent infection which is almost asymptomatic and induces only a minimal immune response (McFadden, 1994). Instead, in the European rabbit (*Orictolagus cuniculus*), which belongs to a distinct but related genus, myxoma virus provokes the appearance of extensive dermal lesions at the site of primary infection, followed by a rapid dissemination through the lymphoreticular system with onset of multiple lesions in secondary sites both internal and external, such as the nose, the ear, and the eyes (Strayer, 1989). This sequence of events is accompanied by a profound immune dysfunction that favors the establishment of supervening secondary infections by gram-negative bacteria, in particular of the respiratory tract and of the conjunctiva, with mortality rates close to 100% within 1 or 2 weeks. In the early 1950s the extreme virulence of this pathogen was utilized to try to eradicate the population of European rabbits introduced a century earlier in Australia. In fact, in the absence of natural predators the population of wild rabbits had grown to such proportions to be considered a serious ecologic threat (Fenner and Meyers, 1978). This strategy represented the first form of "ethnic cleansing" with biologic weapons performed against vertebrates. The use of myxoma virus resulted in a rapid reduction of the rabbit population, which in a few years gradually acquired a certain degree of resistance to the virus and started to repopulate the ecosystem. Therefore, although in Australia today myxoma virus is enzootic, the rabbit population has reached almost the same level as that before the intro-

duction of the virus. In addition to the poor result of this strategy, this "field" experiment with myxoma virus has eased the study of the molecular mechanisms determining the pathogenicity of poxviruses. Several virulence factors have been identified and cloned and, in some instances, the direct effect of these factors on the host's immune system has been demonstrated.

Virulence Mechanisms of Poxviruses

Cellular Growth Factors

Several poxviruses have acquired genes that belong to the superfamily of epithelial and transforming growth factors [*epidermal growth factor* (EGF) and transforming growth factor (TGF), respectively] (Turner and Moyer, 1990; McFadden *et al.*, 1994). In particular, in the myxoma virus and in other related leporipoxviruses, notably Shope fibroma virus and malignant rabbit fibroma virus, these viral growth factors are encoded by single-copy genes located close to the terminal inverted repeated sequences in the genome (Upton *et al.*, 1987). The currently characterized products of these genes have maintained the six cysteine residues critical for correct folding of the receptor-binding domain: Indeed, these proteins are able to activate the cellular receptor of EGF (Lin *et al.*, 1988; Chang *et al.*, 1990). In the myxoma virus, the pathogenetic role of a gene belonging to this superfamily, called myxoma growth factor (MGF), has been studied using recombinant viruses in which MGF has been disrupted or replaced by another growth factor, such as rat TGF (Opgenorth *et al.*, 1992, 1993). Rabbits infected with the MGF-disrupted myxoma virus exhibited a marked decrease of epithelial hyperplasia and metaplasia at the level of the conjunctiva and respiratory tract (Opgenorth *et al.*, 1992). On the contrary, recombinant viruses containing rat TGF regained wild-type capacity of epithelial metaplastic transformation (Opgenorth *et al.*, 1993), demonstrating that MGF is a virulence factor equivalent to TGF, which acts through activation of the EGF receptor *in vivo*. For myxoma virus, the selective advantage provided by the expression of growth factors such as MGF might be represented by the positive regulation of cell growth through mitogenic stimulation (McFadden *et al.*, 1994). MGF, by mimicking a ligand for the ubiquitously expressed EGF receptor, would increase the pools of nucleotides and other precursors, normally depressed in quiescent cells, thus improving the efficiency of viral replication and increasing viral spread and infectivity. Recently, a poxviral homolog of vascular endothelial growth factor, the Orf virus, was described for sheep poxviruses (Lyttle *et al.*, 1994). It is conceivable that the myxoma virus might encode a related growth factor since one of the characteristic histological features of myxomatosis is the disregulated proliferation of endothelial cells near the viral lesions.

Proteinase Inhibitors

Many poxvirus genes encode proteins belonging to a family of proteins, the serpins, that act as inhibitors of cellular serine proteinases (Buller and Palumbo, 1991; Turner and Moyer, 1990). Among the orthopoxviruses, such as vaccinia and cowpox, three such genes have been identified and designated *SP-1*, *SP-2*, and *SP-3* (Law and Smith, 1992; Turner and Moyer, 1992). Two of these genes have been cloned in the myxoma virus: *SERP-1* and *SERP-2* (Upton et al., 1990; Petit et al., 1996). Serpins function as pseudosubstrates for the target serine proteinases, forming stable complexes that remove the proteinase from the pool of active enzyme (Potempa et al., 1994). The active site of these inhibitors contains the P1–P1′ amino acid residues that form, in the natural substrate, the peptide bond cleaved by the protease. These residues determine in part the specificity of the proteases inhibited by the viral serpins. Among the latter, *SP-2* and *SERP-2* show the highest degree of genetic identity, equivalent to 35% (Petit et al., 1996); both are intracellular, not glycosylated, and expressed at an early stage of infection (Palumbo et al., 1989). Furthermore, both are able to inhibit *in vitro* an atypical cysteine proteinase, the interleukin-1β converting enzyme (ICE) (Ray et al., 1992; Petit et al., 1996). This protease catalyzes a crucial reaction in the functional maturation of interleukin-1 (IL-1), which is the conversion of the inactive intracellular precursor, IL-1α, into IL-1β, the secreted, biologically active form. The gene *SP-2* is an important virulence factor, the deletion of which causes a marked attenuation of the virus (Thompson et al., 1993). It has been hypothesized that *SP-2* determines an early sequestration of the convertase, resulting in the inactivation of secretion of IL-1β by the infected cell and consequent reduction of the number and function of leukocytes infiltrating the infection site (Thompson et al., 1993; Petit et al., 1996). The protein SP-2 is also capable of inactivating *in vitro* granzyme B, a serine proteinase present in the granules of cytotoxic T lymphocytes and involved in the Ca^{2+}-dependent apoptotic pathway (Shi et al., 1992; Quan et al., 1995). Recently, ICE has been demonstrated to also be involved in the processes of cellular apoptosis mediated by the interaction of Fas with its ligand, both of which are antagonized *in vitro* by SP-2 (Gagliardini et al., 1994; Tewari et al., 1995). The hypothesis that, as a result of serpins, poxviruses have developed the ability to dampen apoptosis endogenously, initiated inside the infected cell, or exogenously as a result of the interaction of the infected cell with the cellular effectors of the immune response is very attractive. Among the other viral serpins, only *SERP-1* has been demonstrated to effectively act as a virulence factor in that deletion of both copies of the gene results in an attenuation of disease in infected animals by virtue of a more effective leukocyte infiltration into the sites of viral replication (Macen et al., 1993). *In vitro, SERP-1* has been shown to inhibit several extracellular serine proteinases, such as plasmin, urokinase, tissue plasminogen activator, and at least one of the components of the complement cascade, C1s (Lomas et al., 1993). However, it is unclear whether these proteinases also represent the true biologic target of *SERP-1 in vivo* (Lomas et al., 1993).

Inhibitors of Complement Activation

One of the major secreted proteins expressed by some orthopoxviruses consists of four, tandem, inexact copies of a 60-amino acid sequence known as short consensus repeat (SRC). SCR is normally present in many proteins that regulate complement activation. The protein encoded by the vaccinia virus (VCP; vaccinia complement control protein) is the most closely related to the human SCR which binds the C4b complement factor. This VCP inhibits the classical and alternative pathways of complement activation through its ability to bind and inactivate both C4b and C3b (Kotwal et al., 1990). At least *in vitro*, VCP prevents complement-enhanced neutralization of vaccinia virus by antibody (Isaacs et al., 1992). Viruses with deleted VCP are severely attenuated *in vivo*, causing lesions which are smaller and heal more rapidly (Isaacs et al., 1992).

Soluble Interferon Receptors

The α, β, and γ interferons (IFNs) form a heterogeneous family of cytokines initially discovered and defined by their ability to induce resistance to viral infections (Isaacs and Lindermann, 1957). This biologic function is achieved through activation of several cellular processes in which the different types of interferons play different regulatory roles. In general, type I IFNs (α and β) interfere directly with viral replication within the infected cell and reduce the susceptibility to infection of bystander cells, whereas type II IFN (γ) performs its antiviral activity by modulation of the immune effector cells (Callard and Gearing, 1994). Therefore, it is not surprising that many viruses have evolved strategies aimed at neutralizing the IFN action (McNair and Kerr, 1992). Also, poxviruses represent an interesting study model in that they encode at least four distinct proteins able to counteract at different levels the action of IFNs. Two of these proteins, VVE3L and VVK3L in vaccinia virus (Beattie et al., 1991; Chang et al., 1992), block the action of IFNs within the infected cell. The other two proteins, homologs of IFN receptors, are responsible instead for a form of "soluble sabotage" by virtue of their high-affinity binding to IFNs resulting in the neutralization of their biologic action. The first discovered was the *T7* gene of myxoma virus, which encodes a viral protein that binds IFN-γ (McFadden et al., 1995). Its protein product, MT7, has a molecular mass of 37 kDa and is the most abundant protein expressed in the infected cells. This protein has a significant homology to the α chain of the human and murine IFN-γ receptor. However, binding experiments with

IFN-γ from several animal species demonstrated a marked species specificity: MT7 selectively binds only rabbit IFN-γ, with an affinity comparable to that of its cognate cellular receptor (Mossman *et al.*, 1995, 1996), completely inhibiting its biologic activity. *In vivo,* MT7 is a virulence factor crucial for progression of myxomatosis. Rabbits infected with myxoma virus in which both copies of the *T7* gene have been disrupted survive to infection in 100% of the cases and present a vigorous and efficacious cell-mediated immune response, with an increase in the number of specific antimyxoma virus lymphocytes in secondary lymphoid organs and the appearance of a leukocyte infiltrate in the dermal tissue surrounding the lesion (Mossan *et al.*, 1996). However, the mechanism of action of MT7 is not completely understood. On the one hand, it is reasonable to predict that the protein binds and sequesters the IFN-γ produced by T lymphocytes and natural killer (NK) cells, thereby inhibiting the cell-mediated specific response; on the other hand, it is difficult to understand which role may be played by the hindrance of tissue redistribution of leukocytes in proximity of the infection sites. In this regard, it has been observed that MT7 is able to bind several chemokines *in vitro* (Lalani *et al.*, 1997). This interaction involves the carboxy terminus of chemokines, which is also responsible for their interaction with the glycosaminoglycans of cell membrane and extracellular matrix. In analogy to another protein of myxoma virus (MT2), MT7 could actually exhibit two distinct functions: one as a selective "trap" for IFN-γ, with consequent blockade of the effector immune response, and another as "flypaper" for chemokines, preventing the recruitment of the cellular effectors to the infection site. However, it should be noted that the perturbing activity of MT7 on cellular migration has not been directly tested in chemotaxis bioassays.

The second kind of soluble receptor for IFN, the B18R protein, was identified for the first time in vaccinia virus (Symons *et al.*, 1995). The gene that encodes this protein is present in a single copy and is conserved in several strains of orthopoxviruses. The protein belongs to the immunoglobulin superfamily and is structurally related to type I cytokine receptors, such as the receptors for IL-1 and IL-6. B18R is therefore unique because all the eukaryotic cytokine receptors previously described belong to class II cytokine receptors (Callard and Gearing, 1994). This protein efficiently binds both type I IFNs of different animal species, with greater affinity for the human and rabbit ones, and is able to inhibit their function in several biologic assays (Symons *et al.*, 1995). The deletion of this gene determines an attenuation of viral virulence *in vivo.* Therefore, B18R also seems to act as an efficacious "trap" for IFN.

Soluble Receptors for Other Cytokines

Many cellular cytokine receptors, particularly those with single membrane-spanning domains, have been found to also exist in secreted soluble forms produced by either proteolytic cleavage from the cell surface or alternative splicing of the messenger RNA (Rose-John and Heinrich, 1994). The biologic function of these soluble forms is to sequester bioactive ligands away from the cell surface receptors. Poxviruses have utilized this natural control strategy, already present in the host, by incorporating in their genome at least two cellular genes that encode receptors of inflammatory cytokines: the IL-1 receptor and the *p75* receptor for TNF-α. The viral homolog of the IL-1 receptor is encoded by the *B15R* gene, identified initially in vaccinia virus (Alcami and Smith, 1992) and subsequently in other orthopoxviruses. Culture supernatants of cells infected with myxoma virus and other leporipoxviruses contain a protein with analogous biologic function, the gene of which is unknown (McFadden *et al.*, 1995). *B15R* is transcribed late in the viral replicative cycle (Spriggs *et al.*, 1992). Surprisingly, in the model of intranasal infection in mice, deletion of the *B15R* gene does not significantly modify viral pathogenicity (Spriggs *et al.*, 1992). In addition to the difficult interpretation of the role played by this protein *in vivo*, B15R represents an attractive paradigm of poxviruses' mimicry strategy to circumvent the host's immune response. They have indeed selected the most appropriate cellular receptor to act as a competitor (type II receptor binds IL-1β with 10-fold higher affinity than the inactive precursor IL-1α or the antagonist IL-1RA), potentiating its already high specificity and transforming it into an "exclusive trap" for IL-β, the only biologically active form.

The viral homolog of the type II receptor for TNF-α is the first known example of a viral cytokine receptor. It was casually discovered at the beginning of the 1990s in the course of a database search for sequences related to the type II TNF-α receptor. The sequence with highest homology (i.e., 38% identity) was a gene of Shope virus, the causal agent of benign fibroma in the rabbit (Smith *et al.*, 1990). The homology was concentrated at the amino terminus of the protein, particularly in the four repeated domains rich in cysteines [cysteine-rich domain (CRD)] responsible for TNF-α binding. The viral gene, denominated *T2*, had been cloned a few years earlier and characterized as a typical early gene of poxviruses (Upton *et al.*, 1990). The *T2* gene, also present in the myxoma virus and called *MT2* (Upton *et al.*, 1991), has been found in several other poxviruses, including variola and vaccinia virus (Shchelkunov *et al.*, 1993), although in the latter both copies of the gene have been rendered inactive by mutations (Howard *et al.*, 1991). In the myxoma virus the experimental inactivation of both *MT2* copies results in a marked attenuation of the pathogenic effects, with a reduction of lethality >70% and a complete healing of the infected animals (Upton *et al.*, 1991), accompanied by the appearance of an efficient specific immune response. *MT2* is therefore an important virulence factor for myxoma virus and probably for other poxviruses in that vaccinia virus, which is highly attenuated by repeated *in vitro* passages, lacks functional copies of the *MT2* analogous gene. The viral protein is secreted either as a monomer (55–59 kDa) or as a homodimer (90 kDa) and, similar to the cellular receptor, shows high-affinity binding to TNF-α of rabbit (dissociation constant 170–195 pM) but not to that of other species (Schreiber *et al.*, 1996), resulting in the blockade of its biologic activity *in vitro* (Schreiber and McFadden, 1994). The binding of *MT2* to TNFα is mediated by the first three CRD domains of the protein, whereas the first 24 residues of the carboxy terminus favor its exocytosis from the infected cell (Schreiber *et al.*, 1996).

A new feature of *MT2* has recently been discovered, namely, the ability to interfere with the apoptotic processes induced by viral infection in lymphoid cells (Macen *et al.*, 1996). Replication of an *MT2* knockout virus in CD4$^+$ lymphocytes is usually abortive due to a rapid induction of apoptotic processes. The new function has been mapped to the amino terminus of the protein. The inhibition of apoptosis requires structural integrity of only the first two CRDs of the protein, which is different from its capacity to bind TNF-α (McFadden *et al.*, 1997). Furthermore, the exogenous addition of soluble protein is not able to prevent the rapid induction of cell death (McFadden *et al.*, 1997). These two observations allow the exclusion of TNF-α in the pathogenesis of apoptosis. The capacity to prolong survival of the infected lymphocyte may represent an important element in the pathogenesis of infection by myxoma virus: In fact, the dissemination of the virus to lymphatic organs represents a crucial step in the progression of the disease. It is reasonable to hypothesize that lymphocytes, temporarily surviving the infection, are the major carriers of viral dissemination.

Molecular Mimicry among Herpesviruses

The Herpesviridae family comprises large doublestranded DNA viruses which infect virtually all known animal species. The majority of herpesviruses are extremely efficient in their relation with their host species in that they are capable of establishing an infection which is almost ubiquitous. A distinctive feature of herpesviruses is the architecture of the mature virion, which has a diameter of approximately 150–300 nm and contains a nucleocapsid with icosadeltahedron symmetry formed by 162 capsomers and an external trilaminar envelope with glycoprotein spikes projecting from it (Fig. 3). The genome of herpesviruses has a large size and can vary between 120 and 230 kb due to the different lengths of repeated sequences set either at the extremities or internally to the coding segment. By virtue of their size and complexity, the genome of herpesviruses is particularly suitable to incorporation of exogenous genes copied from the host's DNA.

From the biologic point of view, herpesviruses represent the most classic model of viruses that establish a latent in-

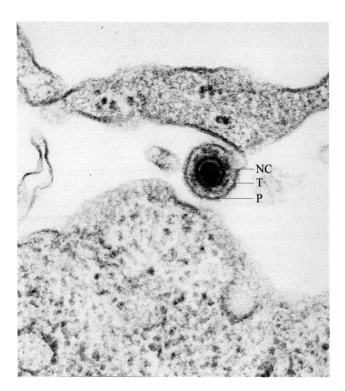

FIGURE 3 Electron micrograph of a mature HHV-6 virion. The nucleocapsid with icosadeltahedron symmetry (NC), the viral tegument (T), and the outer pericapsid membrane (P) can be distinguished.

fection, a strategy that allows them to persist in a single individual for the entire length of the individual's life. After the acute primary infection, herpesviruses induce a chronic persistent infection characterized by sequential cycles of reactivation and authentic virologic latency. The latent infection is established only in certain cell types and in specific anatomical sites, distinct for each herpesvirus. Integration of the viral genome into the cellular DNA occurs quite rarely; in fact, the genome usually persists in an episomal form with a minimal expression of viral genes, indispensable for maintaining the latent state. The latent infection constitutes a minor immunogenic stimulus for the infected host, for which many herpesviruses have further limited developing a specific strategy to interfere with antigen presentation mediated through the major histocompatibility complex (MHC). On the contrary, the acute primary infection as well as the phases of reactivation are characterized by the onset of the lytic viral cycle, with rapid destruction of target cells and induction of the host's defensive responses of both the inflammatory and the specific type.

Many herpesviruses belonging to the subfamilies β and γ herpetovirinae cunningly used the strategy of molecular mimicry to favor their own persistence in the host. The most pressing problem these viruses faced in order to survive was the natural limits of the life span of their target cells. In fact, different from α herpetovirinae, which establish a latent infection in cells with an almost unlimited life length, such as the neurons of spinal ganglia, β and γ herpetovirinae have instead a tropism for cells that, although long living, are predetermined to death, such as epithelial, mesenchymal, and lymphoid cells. Without altering the cells' natural life span, these viruses could therefore exist only for a limited time unless repeated events ensuring the induction of the lytic infective cycle occurred, thereby expanding the infection with involvement of younger, new target cells. To deal with these potential difficulties, herpesviruses have "chosen" to copy and incorporate in their genome a series of genes crucial for the proliferation and long-term survival of the cell. Several examples of this peculiar and efficacious strategy of persistence will be presented in the following sections.

Poxviruses have generally adopted the mimicry strategy in order to "sabotage" the host's antiviral response, copying molecules of the immune response, such as cytokine receptors, and mutating them appropriately to obtain soluble molecular decoys capable of blocking the inflammatory reaction during the course of acute infection. On the contrary, herpesviruses have incorporated the cytokine genes but have maintained them functional in order to exploit the durable proliferative or anti-inflammatory effects of these factors. Obviously, exceptions to this rule exist for both groups of viruses. A recent example is a soluble homolog of the receptor for macrophage differentiating factor colony-stimulating factor-1, encoded by the Epstein–Barr virus (EBV) which neutralizes the cellular cytokine (Strockbine et al., 1998).

Cellular Cytokine Homologs

A γ herpesvirus of the genus *Rhadinovirus*, the human herpesvirus 8 (HHV-8), has acquired in its genome several genes that are important virulence factors (Fig. 4). This virus was identified for the first time in 1995 in Kaposi's sarcoma (KS) tissue by a molecular technique called repre-

sentational difference analysis, based on repeated cycles of genetic subtraction after DNA amplification by the polymerase chain reaction. In addition to its involvement in KS etiology, HHV-8, which can infect B lymphocytes *in vivo* and thus establishing a latent infection, appears to play a role in the pathogenesis of a rare subtype of B cell lymphoma, defined body cavity lymphoma, with extranodal localization specific for thoracic or abdominal cavities. Analysis of the HHV-8 genome has revealed that the virus encodes a protein homologous to cellular IL-6, a cytokine that enhances the growth of B lymphocytes by inhibiting apoptosis. This protein has been implicated in the pathogenesis of many human lymphoproliferative disorders such as multiple myeloma. Considering the whole molecule, the degree of identity between the viral (vIL-6) and the cellular IL-6 is 24.7%, and there is even higher homology in the region that recognizes the IL-6 receptor. Unlike its cellular counterpart, the gene that encodes viral IL-6 is not spliced and therefore it has been presumably acquired from the messenger RNA of the host cell. It was demonstrated that the protein encoded by vIL-6 is functional by *in vitro* proliferation studies utilizing murine plasmacytoma B9 cells (Moore et al., 1996). Only seldom has expression of vIL-6 been observed in the KS tissue, which is a pleomorphic tumor of nonlymphoid origin characterized by a marked endothelial component. On the contrary, constitutive expression of vIL-6 has been demonstrated in primary cells of body cavity B lymphoma as well as in B cell lines derived from this lymphoproliferative disorder (Moore et al., 1996). Secretion of vIL-6 would therefore constitute an autocrine mechanism to support the expansion and long-term survival of B lymphoid cells. It is also curious that although HHV-8 is present in cell lines of neoplastic origin, a transforming capacity for HHV-8 has not been demonstrated.

Another example of a gene incorporated into the genome of herpesvirus and homologous to one that encodes cellular cytokines is the gene *BCRF-1* of EBV (Hsu et al., 1990), a γ herpesvirus implicated in the pathogenesis of

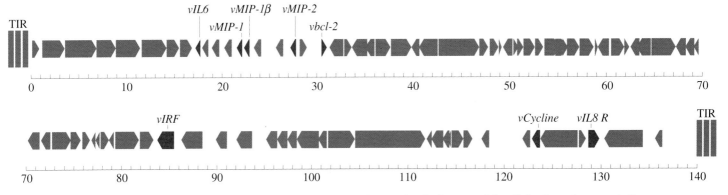

FIGURE 4 Schematic representation of the genome of HHV-8. Viral genes with cellular homologs are indicated in purple. At both the initial and terminal extremities, terminal inverted and repeated (TIR) sequences are indicated. The numbers denote the size of the genome in kilobases.

human tumors of lymphoid and epithelial origin, such as Burkitt's lymphoma and nasopharyngeal carcinoma, respectively. This gene encodes a homolog of IL-10 (Hsu *et al.*, 1990). Although functional data on this gene do not exist *in vivo,* its protein product appears to mimic *in vitro* all the antiproliferative and anti-inflammatory effects of human IL-10, including the negative regulation of the expression of the cytokines IL-12 and IFN-γ and of lymphocyte proliferation. Furthermore, both human and viral IL-10 exert a negative effect on the expression of class II MHC molecules and on the peptide transporter TAP (Zeidler *et al.*, 1997). This gene could represent a virulence factor more similar to those encoded by poxviruses in that its secretion in the proximity of the infection site would allow the virus to reduce the efficacy of the host's protective immune response.

Viral Oncogenes

HHV-8 and the monkey's herpesvirus Saimiri (HVS), a transforming virus which causes T cell lymphomas in New World primates, contain a gene that presents sequence homology with the genes of cellular cyclins, in particular type D cyclins (Nicholas *et al.*, 1992; Chang *et al.*, 1996). Cyclins are essential for cellular proliferation and division. In particular, cyclin D1 is a protooncogene implicated in the pathogenesis of certain parathyroid and liver tumors. The cyclins of HVS and HHV-8 have been demonstrated to be functionally active in *in vitro* tests (Jung *et al.*, 1994; Chang *et al.*, 1996). Their mechanism of action involves a kinase activity that determines the phosphorylation, and consequent inactivation, of the product of the antioncogene RB. Therefore, the viral cyclin, similar to its cellular counterpart, prevents the cell cycle arrest mediated by RB. It is interesting that the viral cyclin D is expressed in KS tissue as well as in body cavity B cell lymphoma. Therefore, it is hypothesized that in both tumors this viral protein may allow the infected cells to overcome the checkpoints that limit uncontrolled cell growth.

EBV appears to have concentrated maximum effort in an attempt to warrant itself a way of escape from apoptosis of the host cell, which would extinguish the infection. To this aim, it has developed two mechanisms, which are apparently redundant: the induction of expression of the cellular protooncogene *bcl-2* and the incorporation in its own genome of a homolog of *bcl-2,* called *BHRF1* (Henderson *et al.*, 1993). This latter gene, which is involved in the pathogenesis of follicular lymphomas and of other human neoplasias, has a specific antiapoptotic activity that is believed to be mediated by the formation of heterodimers with proteins, such as Bax, involved in the apoptosis mechanism. The *BHRF1* gene is actively expressed in cells of B lymphomas positive for EBV. Different from EBV, HHV-8 has chosen only the strategy of directly incorporating the gene *bcl-2* in its own genome (Sarid *et al.*, 1997). *In vitro* studies

have demonstrated that the viral Bcl-2 protein is able to prevent cytotoxicity induced by Bax and to form dimers with the cellular Bcl-2 protein (Sarid *et al.*, 1997).

Another viral oncogene is the *K9* gene of HHV-8 which has a remarkable degree of similarity to interferon regulatory factors (IRFs), proteins initially identified as transcriptional factors involved in the regulation, either positive (IRF-1 or IRF-3) or negative (IRF-2), of IFN-β expression (Lamphier and Taniguchi, 1994). It has subsequently been demonstrated that IRF-1 and IRF-2 have antioncogenic and oncogenic properties, respectively. The locus in which the *IRF-1* gene is placed is in fact involved in cytogenetic abnormalities identified in many human leukemias and preleukemias, whereas the dysregulated expression of IRF-2 induces transformation of murine fibroblasts *in vitro* (Lamphier and Taniguchi, 1994). Studies with knockout mice have revealed that *IRF-1* is indispensable for physiologic induction of apoptosis (Tanaka *et al.*, 1994) and plays an essential role in the development of cellular immune response, particularly for the function of NK cells (Duncan *et al.*, 1996) and for antigen presentation to CD8$^+$ T lymphocytes (White *et al.*, 1996). Recently, the IRF homolog present in HHV-8 was functionally characterized: Following stable transfection of murine fibroblasts, neoplastic transformation was observed, with tumor induction in immunosuppressed mice (Gao *et al.*, 1997). Therefore, the *K9* gene behaves as a true viral oncogene. Furthermore, an inhibitory effect of signal transduction induced by IFN-β has been documented, probably limiting the antiviral activity of this cytokine to the area of the infection.

Histocompatibility Antigens

With a strategy similar to the one pursued by poxviruses, cytomegalovirus, either human (HCMV) or murine (MCMV), has acquired in its own genome a homolog of MHC class I antigens. Because MHC class I antigens play a fundamental role in the specific response of CD8$^+$ T lymphocytes and NK cells, this kind of molecular mimicry could allow the virus the elusion of the host's cell-mediated immune control. In MCMV the disruption of gene *m144,* homologous to MHC, determines an effective *in vivo* reduction of the replicative capacity of the virus, which can be restored upon depletion of NK cells. The product of this gene appears to act as a molecular decoy, mainly for NK recognition. Conversely, it is not clear whether the HCMV gene, homologous to MHC *UL18,* is able to function similarly to its murine counterpart.

Viral Chemokines

Chemokines are small basic polypeptides with chemoattractant activity that are physiologically involved in the maturation, recycling, and recruitment to the inflammation sites of hematopoietic cells, particularly leukocytes. The chemokine superfamily has been subdivided into families based on a peculiar structural motif, formed by two cysteine residues, which is found near the amino terminus of the mature protein. The two main families are the α chemokines, or C–X–C, which present a single amino acid residue between the two cysteines, and the β chemokines, or C–C, with two contiguous cysteines (Baggiolini *et al.*, 1994). Recently, other chemokines have been identified belonging to the γ family, including leukotactin, which is characterized by a single cysteine, and fractalkine or neurotactin, which are characterized by the C–X₃–C motif, with three interposed residues (Bazan *et al.*, 1997). The latter is also the first example of a chemokine that is anchored to the cell membrane through a mucin stalk (Bazan *et al.*, 1997). In recent years, chemokines have gained a central role in virology with the discovery of a specific antiviral activity against immunodeficiency lentiretroviruses of humans (HIV-1 and HIV-2), monkeys (SIV), and cats (FIV). Three chemokines of the C–C family, notably RANTES (regulated on activation, normal T cell expressed and secreted) and MIP-1 (macrophage inflammatory protein-1) -α and -β, have been identified as the major HIV-suppressive factors produced by CD8$^+$ T lymphocytes (Cocchi *et al.*, 1995). A few months after this discovery, the physiologic basis of this peculiar antiviral activity was elucidated with the identification of a molecule, homologous to a chemokine receptor, which was at that time still an "orphan" of a known ligand (Feng *et al.*, 1996). This molecule, subsequently denominated CXCR4, was recognized as the specific receptor of the chemokine SDF-1 (stromal cell-derived factor-1) and as an essential element of the membrane receptor complex for one of the two principal biologic subtypes of HIV, CXCR4$^+$. CXCR4 belongs to the superfamily of receptors with seven transmembrane domains (7TM) coupled to G proteins. Subsequently, it has been demonstrated that other chemokine receptors play an important role as coreceptors for lentiviruses' infection, and how the differential use of these molecules constitutes the main physiologic basis of the biologic diversity among HIV strains present in the population has been elucidated. In particular, the molecule CCR5, a receptor specific for RANTES, MIP-1α, and MIP-1β, is the fundamental coreceptor for viral strains belonging to the second principal biologic subtype of HIV, CXCR4$^-$. The latter strains have the highest prevalence in the population and are most frequently implicated in viral transmission through the sexual route. These strains are not able to grow in T cell lines, which usually do not express CCR5. The crucial role played by CCR5 in viral transmission is also demonstrated by the fact that individuals resistant to HIV infection, albeit repeatedly exposed to the virus, are characterized by a homozygous genetic defect (CCR5/D32) in the gene that encodes CCR5. Chemokine receptors are involved in the fusion process that occurs between the viral and cellular membranes and which allows the injection of the viral core into the cell cytoplasm. Chemokines thus

block infection by preventing the use of the coreceptors by the virus, either by directly engaging the receptors or by negative regulation of their expression on the cell surface.

It has been observed that many herpesviruses encode homologs of chemokine genes. In particular, HHV-8 has acquired in its genome three genes that encode structural and functional proteins analogous to cellular chemokines. Two of these present more than 35% homology with the MIP-1α gene and have therefore been denominated *vMIP-I* (*K6*) and *vMIP-II* (*K4*). The third gene (*K4.1*) presents a certain degree of homology with MIP-1β as well as with the gene that encodes another C–C chemokine, MCP-1 (macrophage chemoattractant protein-1). It has not been completely elucidated whether *vMIP-I* and *vMIP-II* are expressed in body cavity B cell lymphoma; however, treatment with a phorbol ester (TPA) enhances its expression (Moore *et al.*, 1996). Instead, no expression of these genes has been observed in KS spindle cells, which are thought by many investigators to be the neoplastic component of the tumor (Moore *et al.*, 1996). However, both *vMIP-I* and *vMIP-II* have shown angiogenic activity in the chick chorioallantoic membrane assay (Boshoff *et al.*, 1997), a result that suggests a possible pathogenetic mechanism of HHV-8 in KS, a tumor characterized by a remarkable neoangiogenetic component. It is not clear whether vMIP-I is able to activate cellular chemokine receptors, whereas for vMIP- II the data are contradictory. One study reported that this protein, obtained by chemical synthesis, is functionally active in that it is able to chemoattract the eosinophil granulocytes (Boshoff *et al.*, 1997). In contrast, in another study a broad-spectrum antagonistic action was observed on several cellular receptors, including CCR1, CCR2, CCR3, CCR5, and CXCR4, as well as on the viral receptor US28 of CMV (Kledal *et al.*, 1997). It can be hypothesized that the antagonistic activity may represent a virulence factor for HHV-8 in that it could limit the recruitment of monocytes and lymphocytes to the infection site. More difficult to interpret is the possible agonistic activity for CCR3 because the recruitment of eosinophils would have local inflammatory effects that could counteract viral replication or persistence. However, the recent finding that the CCR3 receptor is selectively expressed on type 2 T helper cells (T$_H$2) may offer a key of interpretation in that a T$_H$2 polarized response could negatively affect the host's control of viral infection.

The study of the effects of HHV-8 chemokines on HIV infection is of particular importance since KS develops with elevated frequency in HIV-infected individuals, probably due to an overlap of the transmission routes of the two viruses. However, the results obtained to date are contradictory. Using feline cells transfected with the *vMIP-I* gene, a sharp reduction in their infectability by HIV-1 CCR5-dependent strains has been observed (Moore *et al.*, 1996); on the contrary, experiments performed with the protein obtained by chemical synthesis have demonstrated no inhibitory effect (Boshoff *et al.*, 1996). In regard to vMIP-II, one study demonstrated inhibition of HIV-1 infection mediated by CCR3 but not by CCR5 or CXCR4 and also a weak inhibitory effect on peripheral blood mononuclear cells (Boshoff *et al.*, 1997). Another study demonstrated that vMIP-II, either synthetic or recombinant, is a broad-spectrum potent inhibitor of HIV-1 able to block CXCR4$^-$ strains (through CCR5) as well as CXCR4$^+$ strains (through CCR3 or CXCR4) (Kledal *et al.*, 1997). A possible reason that might explain these discrepant results is the different amino-terminal sequence of the proteins produced in different laboratories (GDTLGA or LGA). In fact, it is known that the amino-terminal region of chemokines may determine their receptor specificity. The peculiar form of viral interaction between HHV-8 and HIV is an extremely interesting study model. Indeed, when HHV-8 and HIV coinfect the same individual, a condition of viral interference never hitherto described would occur that would be mediated not by the classic mechanism of receptor competition but indirectly through secretion of interfering soluble factors. However, the real clinical relevance of this viral interaction needs to be evaluated.

Although among herpesviruses the presence of chemokine gene homologs has been demonstrated only for HHV-8, in the genome of HHV-6 a gene exists which contains four cysteine residues, two of which form the C–C structural motif proximal to the amino terminus. HHV-6 is a β herpesvirus, highly cytopathic for CD4$^+$ T lymphocytes, that causes the early childhood disease exanthema subitum and that in adults may play a role in immunodeficiencies of viral or iatrogenic origin. However, the amino acid sequence predicted for this protein does not appear compatible with a classic chemokine structure.

Chemokine Receptors

A common feature of some members of β and γ herpetovirinae is the presence in their genome of genes that encode 7TM-type receptors coupled to G proteins. Among these, at least a few have been demonstrated to be functionally active chemokine receptors. However, an example of 7TM-type viral gene also exists among poxviruses — the *K2R* gene of the pig poxvirus (Massung *et al.*, 1993). Among herpesviruses, the first gene identified and studied was *US28* of HCMV, which indeed possesses two other genes homologous to those encoding 7TM receptors (Chee *et al.*, 1990). This receptor has a remarkable genetic homology with CCR1, the first receptor identified for C–C chemokines, which binds RANTES, MIP-1α, and MCP-3 (Neote *et al.*, 1993). The protein encoded by the *US28* gene, expressed late in the viral replicative cycle, is functionally active and behaves as a promiscuous receptor in that it binds chemokines of the C–C family and of the C–X–C family. Among the 7TM cellular receptors, the only receptor with

an analogous behavior is the so-called Duffy antigen, which is expressed on the membrane of red blood cells, binds chemokines belonging to different families, and acts as a receptor for the malarial parasite *Plasmodium vivax* (Horuk *et al.*, 1993). The functional significance of US28 for the success of CMV infection is unknown. Because chemokines exert a costimulatory effect on cellular activation, it can be hypothesized that expression of US28 gives the infected cell a proliferative advantage upon binding of chemokines that are broadly diffused in the organism. However, a precise functional characterization of this molecule in the physiology of CMV infection has not been achieved; therefore, other hypotheses cannot be excluded, namely, that it could have a decoy effect, sequestering chemokines near the site of infection and thereby limiting the infiltration of cells of the inflammatory and specific immune response. Two other β herpesviruses which contain in their genome 7TM-type genes are HHV-6 and HHV-7. However, the function of their protein products has not been studied. The protein products of 7TM-type genes encoded by HHV-8 and by the monkey HVS, two γ herpesviruses homologous and genetically colinear, are functional. Similar to the products of US28 of CMV, the 7TM receptor of HVS (Nicholas *et al.*, 1992) and HHV-8 (Arvanikatis *et al.*, 1997) also belong to the category of promiscuous receptors because they bind C–C chemokines (such as RANTES and I-309) and members of the C–X–C family (notably IL-8 and PF-4). The genes present in the two herpesviruses are located in the same genomic position (K74) and display the maximum degree of homology with the IL-8-specific receptor (CXCR1). Data obtained by *in vitro* experiments suggest that the receptor K74 of HHV-8 may play a role in cellular proliferation and therefore in the prolongation of the infected cell's life span. In fact, different from cellular receptors which can be activated only on high-affinity binding of a specific ligand, the K74 molecule of HHV-8 is constitutively active and transmits a continuous intracellular signal activating the inositol phosphate and protein kinase C cascade (Arvanikatis *et al.*, 1997). Upon transfection of the *K74* gene into murine fibroblasts, an enhancement of their proliferation has been observed. Therefore, *K74* could effectively act as a viral oncogene and participate to the neoplastic process induced by HHV-8 and HVS. The hypothesis that the 7TM receptor may play a role in HHV-8-induced neoplasia is supported by the finding that the *K74* gene is expressed in KS tissue and in cells of body cavity B cell lymphoma (Cesarman *et al.*, 1996). Unlike HHV-8 and HVS, another γ herpesvirus, EBV, lacks 7TM-type genes acquired from the host during its evolution. However, it is noteworthy that this virus, which has a transforming effect on B lymphocytes *in vitro*, is a potent inducer of expression of a 7TM cellular receptor, known as EBI-1 (EBV induced chemokine receptor-1) or CCR7, which is transactivated by the *EBNA-2* gene (Burgstahler *et al.,* 1995). In an analogous manner, other cellular genes that have homologs in HHV-8, such as those that encode the cytokine IL-6, the protooncogene *bcl-2*, cyclin D, and the chemokine *SDF-1* (S. Polo *et al.,* unpublished data), are also transactivated and induced by EBV infection. Thus, members of the γ herpetovirinae subfamily have developed different strategies to obtain the same result: to prevent apoptosis and the arrest of the cell cycle and to limit the efficacy of the host's antiviral immune response. EBV did not directly incorporate the host's genes in its genome because it possesses transactivators (such as EBNA/LMP) able to induce the expression of their cellular homologs. Instead, HHV-8, lacking these transactivators, has been obliged to incorporate these cellular genes in its genome (Table 1).

It has also been evaluated whether certain 7TM-type receptors encoded by herpesviruses may act as coreceptors for HIV. It has been documented that the US28 molecule of HCMV can be utilized by many HIV strains to penetrate inside the cell (Pleskoff *et al.,* 1997), although its efficiency as a coreceptor is low (Rucker *et al.,* 1997). No activity was demonstrated for the receptors encoded by HHV-6 (Rucker *et al.,* 1997). The significance of this peculiar interaction between CMV and HIV is unknown, although it is conceivable that a synergistic effect may occur between the two viruses in coinfected patients: CMV could indeed contribute to the expansion of the pool of cells infectable by HIV, thus acting as a cofactor for HIV spread in the body.

Conclusions

The phenomenon of molecular mimicry is of particular importance in the complex network of interactions between virus and host. The continuous process of reciprocal adaptation of the two species is one of the hallmarks of viral biology, intrinsic to the essence of these sophisticated intracellular parasites. A virus is, as the Nobel laureate Manfred Eigen artfully wrote, "a genetic program that carries from one cell to the other the simple message: reproduce me!" It is indeed a biologic entity uncapable of reproducing and spreading itself in the absence of other living organisms to which it transfers its genetic record. Consequently, the host's survival is the indispensable condition for the viral survival. Eigen's dictat "reproduce me!" is therefore transformed into "reproduce me and survive!", being enriched by the fine strategies that viruses enact in order to better preserve the host's physical integrity. Survival may refer to individuals, as in the case of herpesviruses that indefinitely persist in a single host, or to species, as for poxviruses that continuously "jump" from one individual to another and from species to species; nevertheless, it remains the main objective of the adaptation process. It is also by virtue of cybernetics piracy that large DNA viruses have established a compromise with the superior organisms, genetically advantaged because they are predisposed to "learn" from the external world. The plasticity of molecular mimicry has helped these viruses to overcome the trial of time, building

TABLE I

Comparison of Strategies between HHV-8 and EBV

GENE PRODUCT	FUNCTION	EBV	HHV-8
bcl-2	Oncogene — apoptosis block	induced and incorporated	incorporated
IL-6	Cytokine — B lymphocyte proliferation	induced	incorporated
IL-10	Cytokine — Negative regulation of IFN-γ, IL-12, MHC-II	incorporated	
IRF	Oncogene — Block of apoptosis, IFN-β function		incorporated
Cyclin D	Oncogene — Block of antioncogene RB	induced	incorporated
Chemokines	Chemokine antagonists — Block of chemotaxis	induced	incorporated
Chemokine receptors	Oncogene — Anti-chemotaxis decoy	induced	incorporated

a difficult but long-lasting balance. "Copy to survive" is a sort of "Zelig complex" that has granted the virus the easiest way to acquire new genetic information, "learning" directly from its own host how to face the challenge of a dynamic and clever ecosystem.

Acknowledgment.

We thank Mrs. Stefania Laus for precious editorial assistance.

References Cited

ALCAMI, A., and SMITH, G. (1992). A soluble receptor of interleukin-1β encoded by vaccinia virus: a novel mechanism of virus modulation of the host response to infection. *Cell*, **71**, 153–167.

ARVANITAKIS, L., GERAS-RAAKA, E., VARMA, A., GERSHENGORN, M. C., and CESARMAN, E. (1997). Human herpesvirus KSHV encodes a constitutively active G-protein-coupled receptor linked to cell proliferation. *Nature*, **385**, 347–350.

BAGGIOLINI, M., DEWALD, B., and MOSER, B. (1994). Interleukin-8 and related chemotactic cytokines-CXC and CC chemokines. *Adv. Immunol.*, **55**, 97–179.

BAZAN, J. F., BACON, K. B., HARDIMAN, G., WANG, W., SOO, K., ROSSI, D., GREAVES, D. R., ZLOTNIK, A., and SCHALL, T. J. (1997). A new class of membrane-bound chemokine with a CX3C motif. *Nature*, **385**, 640–644.

BEATTIE, E., TARTAGLIA, J., and PAOLETTI, E. (1991). Vaccinia-virus encoded eIF-2α homolog abrogates the antiviral effect interferon. *Virology*, **183**, 419–422.

BOSHOFF, C. *et al.* (1997). Angiogenic and HIV-inhibitory functions of KSHV-encoded chemokines. *Science*, **278**, 290–294.

BULLER, R. M., and PALUMBO, G. J. (1991). Poxvirus pathogenesis. *Microbiol. Rev.*, **55**, 80–122.

BURGSTAHLER, R., KEMPKES, B., STEUBE, K., and LIPP, M. (1995). Expression of the chemokine receptor BLR2/EB1I is specifically transactivated by Epstein-Barr virus nuclear antigen 2. *Biochem. Biophys. Res. Commun.*, **215**, 737–743.

CALLARD, R. E., and GEARING, A. J. H. (1994). *The cytokine factsbook*. Londra, Academic Press.

CESARMAN, E., NADOR, R. G., BAI, F. BOHENZKY, R. A., RUSSO, J. J., MOORE, P. S., CHANG, Y., and KNOWLES, D. M. (1996). Kaposi's sarcoma-associated herpesvirus contains G protein-coupled receptor and cyclin D homologs which are expressed in Kaposi's sarcoma and malignant lymphoma. *J. Virol.*, **70**, 8218–8223.

CHANG, Y., CESARMAH, E., PESSIN, M. S., LEE, F., CULPEPPER, J., KNOWLES, D. M., and MOORE, P. S. (1994). Identification of herpesvirus-like DNA sequences in AIDS-associates Kaposi's sarcoma. *Science*, **266**, 1865–1869.

CHANG, W., MACAULAY, C., HU, S. L., TAM, J. P., and McFADDEN, G. (1990). Tumorgenic poxviruses: characterization of the expression of an epidermal growth factor gene in Shope fibroma virus. *Virology*, **179**, 926–930.

CHANG, Y., MOORE, P. S., TALBOT, S. J., BOSHOFF, C. H., ZARKOWSKA, T., GODDEN-KENT, D., POTERSON, H., WEISS, R. A., and MITTNACHT, S. (1996). Cyclin encoded by KS herpesvirus. *Nature*, **382**, 410.

CHANG, H. W., WATSON, J. C., and JACOBS, B. L. (1992). The E3L gene of vaccinia virus encodes an inhibitor of the interferon-induced, double-stranded RNA-dependent protein kinase. *Proc. Natl. Acad. Sci. USA*, **89**, 4825–4829.

CHEE, M. S., SATCHWELL, S. C., PREDDIE, E., WESTON, K. M., and BARRELL, B. G. (1990). Human cytomegalovirus encodes three G protein-coupled receptor homologues. *Nature*, **344**, 774–777.

COCCHI, F., DEVICO, A. L., GARZINO-DEMO, A., ARYA, S. K., GALLO, R. C., and LUSSO, P. (1995). Identification of RANTES, MIP-1α, MIP-1β as the major HIV-suppressive factors produced by CD8+ T cells. *Science*, **270**, 1811–1815.

DUNCAN, G. S., MITTRUCKER, H. W., KAGI, D., MATSUYAMA, T., and MAK, T. W. (1996). The transcription factor interferon regulatory factor-1 is essential for natural killer cell function *in vivo*. *J. Exp. Med.*, **184**, 2043–2048.

FENG, Y., BRODER, C. C., KENNEDY, P. E., and BERGER, E. A. (1996). HIV-1 entry cofactor: functional cDNA cloning of a seven-transmembrane, G protein-coupled receptor. *Science*, **272**, 872–877.

FENNER, F., and MYERS, K. (1978). Myxoma virus and myxomatosis in retrospect: the first quarter century of a new disease. In *Virus and Environment* a c. di Kurstok E., Maramorosh K., New York, Academic Press, pp. 539–570.

GAGLIARDINI, V., FERNANDEZ, P. A., LEE, R. K., DREXLER, H. C., ROTELLO, R. J., FISHMAN, M. C., and YUAN, J. (1994). Prevention of vertebrate neuronal death by the crmA gene. *Science*, **263**, 826–828.

GAO, S. J., BOSHOFF, C., JAYACHANDRA, S., WEISS, R. A., CHANG, Y., and MOORE, P. S. (1997). KSHV ORF K9 (vIRF) is an oncogene which inhibits the interferon signaling pathway. *Oncogene*, **15**, 1979–1985.

HENDERSON, S., HUEN, D., ROWE, M., DAWSON, C., JOHNSON, G., and RICKINSON, A. (1993). Epstein-Barr virus-coded BHRF1 protein, a viral homologue of Bcl-2, protects human B cells from programmed cell death. *Proc. Natl. Acad. Sci. USA*, **90**, 8479–8483.

HORUK, R., CHITNIS, C. E., DARBONNE, W. C., COLBY, T. J., RYBICKI, A., HADLEY, T. J., and MILLER, L. H. (1993). A receptor for the malarial parasite *Plasmodium vivax*: the erythrocyte chemokine receptor. *Science*, **261**, 1182–1184.

HOWARD, S. T., CHAN, Y. S., and SMITH, G. L. (1991). Vaccinia virus homologues of the Shope fibroma virus inverted terminal repeat protein and a discontinuous ORF related to the tumor necrosis factor receptor family. *Virology*, **180**, 633–647.

HSU, D. H., DE WALL MALEFYT, R., FIORENTINO D. F., DANG, M. N., VIEIRA, P., DE VRIES, J. SPITS, H., MOSMANN, T. R., and MOORE, K. W. (1990). Expression of interleukin-10 activity by Epstein-Barr virus protein BCRF1. *Science*, **250**, 830–832.

ISAACS, A., and LINDERMANN, J. (1957). Virus interference. I. The interferon. *Proc. R. Soc. Lond.*, **147**, 258–267.

ISAACS, S. N., KOTWAL, G. J., and MOSS, B. (1992). Vaccinia virus complement-control protein prevents antibody-dependent complement-enhanced neutralization of infectivity and contributes to virulence. *Proc. Natl. Acad. Sci. USA*, **89**, 628–632.

JUNG, J. U., STAGER, M., and DESROSERS, R. C. (1994). Virus-encoded cyclin. *Mol. Cell Biol.*, **14**, 7235–7244.

KLEDAL, T. N. *et al.* (1997). A broad-spectrum chemokine antagonist encoded by Kaposi's sarcoma-associated herpesvirus. *Science*, **277**, 1656–1659.

KOTWAL, G. J., ISAACS, S. N., McKENZIE, R., FRANK, M. M., and MOSS, B. (1990). Inhibition of the complement cascade by the major secretory protein of vaccinia virus. *Science*, **250**, 827–830.

KOTWAL, G. J., and MOSS, B. (1988). Vaccinia virus encodes a secretory polypeptide structurally related to complement control proteins. *Nature*, **335**, 176–178.

LALANI, A. S., GRAHAM, K., MOSSMAN, K., RAJARATHNAM, K., CLARK-LEWIS, J., KELVIN, D., and McFADDEN, G. (1997). The purified myxoma virus gamma interferon receptor homolog M-T7 interacts with the heparin-binding domains of chemokines. *J. Virol.*, **71**, 4356–4363.

LAMPHIER, M., and TANIGUCHI, T. (1994). The transcription factors IRF-1 and IRF-2. *The immunologist*, **2/5**, 167–171.

LAW, K. M., and SMITH, G. L. (1992). A vaccinia serine protease inhibitor which prevents virus-indiced cell fusion. *J. Gen. Virol.*, **73**, 549–557.

LIN, Y. Z., CAPORASO, G., CHANG, P. Y., KE, X. H., and TAM, J. P. (1988). Synthesis of a biological active tumor growth factor from the predicted DNA sequence of Shope fibroma virus. *Biochemistry*, **27**, 5640–5645.

LOMAS, D. A., EVANS, D. L., UPTON, C., McFADDEN, G., and CARRELL, R. W. (1993). Inhibition of plasmin, urokinase, tissue plasminogen activator, and C1s by a myxoma virus serine proteinase inhibitor, *J. Biol. Chem.*, **268**, 516–521.

LYTTLE, D. J., FRASER, K. M., FLEMING, S. B., MERCER, A. A., and ROBINSON, A. J. (1994). Homologues of vascular endothelial growth factor are encoded by the poxvirus, orf virus. *J. Virol.*, **68**, 89–92.

MACEN, J. L., GRAHAM, K. A., LU, S. F., SCHREIBER, M., BOSHKOV, L. K., and McFADDEN, G. (1996). Expression of the myxoma virus tumor necrosis factor receptor homologue and M11L genes is required to prevent virus-induced apoptosis in infected rabbit T lymphocytes. *Virology*, **218**, 232–237.

MACEN, J. L., UPTON, C., NATION, N., and McFADDEN, G. (1993). SERP1, a scrine proteinase inhibitor encoded by myxoma virus, is a secreted glycoprotein that interferes with inflammation. *Virology*, **195**, 348–363.

McFADDEN, G. (1994). Poxviruses: rabbit, hare, squirrel, and swine. In *Encyclopedia of Virology*, a c. di Webster R.G.A.G., Sanders Scientific Publications, pp. 1153–1160.

McFADDEN, G., GRAHAM, K., ELLISON, K., BARRY, M., MACEN, J., SCHREIBER, M., MOSSMAN, K., NASH, P., LALANI, A., and EVERETT, H. (1995). Interruption of cytokine networks by poxviruses: lessons from myxoma virus. *J. Leuk. Biol.*, **57**, 731–737.

McFADDEN, G., GRAHAM, K., and OPGENORTH, A. (1995). Poxvirus growth factors. In *Viroceptors, Virokines and Related Modulators Encoded by DNA Viruses*, a c. di McFadden G., Austin R. G. Landes Co.

McFADDEN, G., SCHREIBER, M., and SEDGER, L. (1997). Myxoma T2 protein as a model for poxvirus TNF receptor homologs. *J. Neuroimmunol.*, **72**, 119–126.

McNAIR, A. N., and KERR, I. M. (1992). Viral inhibition of the interferon system. *Pharmacol. Ther.*, **56**, 79–95.

MASSUNG, R. F., JAYARAMA, V., and MOYER, R. W. (1993). DNA sequence analysis of conserved and unique regions of swinepox virus: identification of genetic elements supporting phenotypic

observations including a novel G protein-coupled receptor homologue. *Virology*, **197**, 511–528.

MOORE, P. S., BOSHOFF, C., WEISS, R. A., and CHANG, Y. (1996). Molecular mimicry of human cytokine and cytokine response pathway genes by KSHV. *Science*, **274**, 1739–1744.

MOSSMAN, K., NATION, P., MACEN, J., GARBUTT, M., LUCAS, A., and McFADDEN, G. (1996). Myxoma virus M-T7, a secreted homolog of the interferon-γ receptor is a critical virulence factor for the development of myxomatosis in European rabbits. *Virology*, **215**, 17–30.

MOSSMAN, K., UPTON, C., and McFADDEN, G. (1995). The myxoma virus-soluble interferon-γ receptor homologue, M-T7, inhibits interferon-γ in a species specific manner. *J. Biol. Chem.*, **270**, 3031–3038.

MURPHY, P. M. (1993). Molecular mimicry and the generation of host defense protein diversity. *Cell*, **72**, 823–826.

MURPHY, P. M. (1994a). Viral imitation of host defense proteins. *JAMA*, **272**, 1948–1952.

MURPHY, P. M. (1994b). Molecular piracy of chemokine receptors by herpesviruses. *Inf. Agents Dis.*, **3**, 137–154.

NEOTE, K., DI GREGORIO, D., MAK, J. Y., HORUK, R., and SCHALL, T. J. (1993). Molecular cloning, functional expression, and signaling characteristics of a C-C chemokine receptor. *Cell*, **72**, 415–425.

NICHOLAS, J., CAMERON, K. R., and HONESS, R. W. (1992). Herpesvirus saimiri encodes homologues of G protein-coupled receptors and cyclins. *Nature*, **355**, 362–365.

OPGENORTH, A., GRAHAM, K., NATION, N., STRAYER, D., and McFADDEN, G. (1992). Deletion analysis of two tandemly arranged virulence genes in myxoma virus, M11L and myxoma growth factor. *J. Virol.*, **66**, 4720–4731.

OPGENORTH, A., NATION, N., GRAHAM, K., and McFADDEN, G. (1993). Transforming growth factor alpha, Shope fibroma growth factor and vaccinia growth factor can replace myxoma growth factor in the induction of myxomatosis in rabbits. *Virology*, **192**, 701–709.

PALUMBO, G. J., PICKUP, D. J., FREDERICKSON, T. N., McINTYRE, L. J., and BULLER, R. M. (1989). Inhibition of an inflammatory response is mediated by a 38-kDa protein of cowpox virus. *Virology*, **172**, 262–273.

PETIT, F., BERTAGNOLI, S., GELFI, J., FASSY, F., BOUCRAUT-BARALON, C., and MILON, A. (1996). Characterization of a myxoma virus-encoded serpin-like protein with activity against interleukin-1β converting enxyme. *J. Virol*, **70**, 5860–5866.

PLESKOFF, O., TREBOUTE, C., BRELOT, A., HEVEKER, N., SEMAN, M., and ALIZON, M. (1997). Identification of a chemokine receptor encoded by human cytomegalovirus as a cofactor HIV-1 entry. *Science*, **276**, 1874–1878.

POTEMPA, J., KORZUS, E., and TRAVIS, J. (1994). The serpin superfamily of proteinase inhibitors: structure, function, and regulation. *J. Biol. Chem.*, **268**, 516–521.

QUAN, L. T., CAPUTO, A., BLEAKLEY, R. C., PICKUP, D. J., and SALVESEN, G. (1995). Granzyme B is inhibited by the cowpox virus serpin cytokine response modifier A. *J. Biol. Chem.*, **270**, 10.377–10.379.

RAY, C. A., BLACK, R. A., KRONHEIM, S. R., GREENSTREET, T. A., SLEATH, P. R., SALVESEN, G. S., and PICKUP, D. J. (1992). Viral inhibition of inflammation: cowpox virus encodes an inhibitor of the interleukin-1β converting enzyme. *Cell*, **69**, 597–604.

ROSE-JOHN, S., and HEINRICH, P. C. (1994). Soluble receptors for cytokines and growth factors: generation and biological function. *Biochem. J.*, **300**, 281–290.

RUCKER, J. *et al.* (1997). Utilization of chemokine receptors, orphan receptors, and herpesvirus-encoded receptors by diverse human and simian immunodeficiency viruses. *J. Virol.*, **71**, 8999–9007.

SALLUSTO, F., MACKAY, C. R., and LANZAVECCHIA, A. (1997). Selective expression of the eotaxin receptor CCR3 by human T helper 2 cells. *Science,* **277**, 2005–2007.

SARID, R., SATO, T., BOHENZKY, R. A., RUSSO, J. J., and CHANG, Y. (1997). Kaposi's sarcoma-associated herpesvirus encodes a functional bcl-2 homologue. *Nat. Med.*, **3**, 293–298.

SCHREIBER, M., and McFADDEN, G. (1994). The myxoma virus TNF-receptor homologue (T′) inhibits tumor necrosis factor-α in a species-specific fashion. *Virology*, **204**, 692–705.

SCHREIBER, M., RAJARATHNAM, K., and McFADDEN, G. (1996). Myxoma virus T2 protein. (TNF) receptor homolog, is secreted as a monomer and dimer that each bind rabbit TNFα but the dimer is more potent TNF inhibitor. *J. Biol. Chem.*, **271**, 13.333–13.341.

SCHCHELKUNOV, S. N., BLINOV, V. M., and SANDAKHCHIEV, L. S. (1993). Genes of variola and vaccinia viruses necessary to overcome the host protective mechanisms. *FEBS Lett.*, **319**, 80–83.

SHI, L., KAM, C. M., POWERS, J. C., AEBERSOLD, R., and GREENBERG, A. H. (1992). Purification of three cytotoxic lymphocyte granule serine proteases that induce apoptosis through distinct substrate and target cell interactions. *J. Exp. Med.*, **176**, 1521–1529.

SMITH, C. A., DAVIS, T., ANDERSON, D., SOLAM, L., BECKMANH, M. P., JERZY, R., DOWER, S. K., COSMAN, D., and GOODWIN, R. G. (1990). A receptor for tumor necrosis factor defines an unusual family of cellular and viral proteins. *Science*, **248**, 1019–1023.

SPRIGGS, M. K., HRUBY, D. E., MALISZEWSKI, C. R., PICKUP, D. J., SIMS, J. E., BULLER, R. M., and VANSLYKE, J. (1992). Vaccinia and cowpox viruses encode a novel secreted interleukin-1-binding protein. *Cell*, **71**, 145–152.

STRAYER, D. S. (1989). Poxviruses. In *Virus-Induced Immunosuppression*, a c. di Specter S., Bendinelli M., Fiedman H., Plenum Press, New York, pp. 125–151.

STROCKBINE, L. D., COHEN, J. I., FARRAH, T., LYMAN, S. D., WAGENER, F., DuBOSE, R. F., ARMITAGE, R. J., and SPRIGGS, M. K. (1998). The Epstein-Barr virus BARF1 gene encodes a novel, soluble colony-stimulating factor-1 receptor. *J. Virol.*, **72**, 4015–4021.

SYMONS, J. A., ALCAMI, A., and SMITH, G. L. (1995). Vaccinia virus encodes a soluble type I interferon receptor of novel structure and broad species specificity. *Cell*, **81**, 551–560.

TANAKA, N., ISHIHARA, M., KITAGAWA, M., HARADA, H., KIMURA, T., MATSUYAMA T., LAMPIER, M. S., AIZAWA, S., MAK, T. W., and TANIGUCHI, T. (1994). Cellular commitment to oncogene-induced transformation or apoptosis is dependent on the transcription factor IRF-1. *Cell*, **77**, 829–839.

TEWARI, M., QUAN, L. T., O'ROURKE, K., DESNOYERS, S., ZENG, S., BEIDLERE, D. R., POIRIER, G. G., SALVESEN, G. S., and DIXIT, V. M. (1995). Yama CPP32/beta, a mammalian homolog of ced-3, is a CrmA inhibitable protease that cleaves the death substrate poly (ADP-ribose) polymerase. *Cell*, **81**, 801–809.

THOMPSON, J. P., TURNER, P. C., ALI, A. N., CRENSHAW, B. C., and MOYER, R. W. (1993). The effects of serpin gene mutations on the distinctive pathobiology of cowpox and rabbitpox virus following intranasal inoculation of Balb/c mice. *Virology*, **197**, 328–338.

TURNER, P. C., and MOYER, R. W. (1990). The molecular pathogenesis of poxviruses. *Curr. Top Microbiol. Immunol.*, **163**, 125–152.

TURNER, P. C., and MOYER, R. W. (1992). An orthopoxvirus serpin-like gene controls the ability of infected cells to fuse. *J. Virol.*, **66**, 2076–2085.

UPTON, C., MACEN, J. L., and McFADDEN, G. (1987). Mapping and sequencing of a gene from myxoma virus that is related to those encoding epidermal growth factor and transforming growth factor alpha. *J. Virol.*, **61**, 1271–1275.

UPTON, C., MACEN, J. L., SCHREIBER, M., and McFADDEN, G. (1991). Myxoma virus expresses a secreted protein with homology to the tumor necrosis factor receptor gene family that contributes to viral virulence. *Virology*, **184**, 370–382.

UPTON, C., MACEN, J. L., WISHART, D. S., and McFADDEN, G. (1990). Myxoma virus and malignant rabbit fibroma virus encode a serpin-like protein important for virus virulence. *Virology*, **179**, 618–631.

WHITE, L. C., WRIGHT, K. L., FELIX, N. J., RUFFNER, H., REIS, L. F., PINE, R., and TING, J. P. (1996). Regulation of LMP2 and TAP1 genes by IRF-1 explains the paucity of CD8$^+$ T cells in IRF-1 -/- mice. *Immunity*, **5**, 365–376.

ZEIDLER, R., EISSNER, G., MEISSNER, P., UEBEL, S., TAMPE, R., LAZIS, S., and HAMMERSCHMIDT, W. (1997). Downregulation of TAP1 in B lymphocytes by cellular and Epstein-Barr virus-encoded interleukin-10. *Blood*, **90**, 2390–2397.

LORENZO MORETTA
MARIA CRISTINA MINGARI

*National Institute for Cancer Research and the
Center for Advanced Biotechnologies*

*Institute of General Pathology
University of Genoa
Genoa, Italy*

Immunological Defenses and Immunopathology

The principal role of the immune system is to defend the host against infections. The important players in the defense against extracellular bacteria and viral reinfections are the antibodies. On the other hand, the defense against endocellular illnesses, both viral and bacterial, is entrusted to more complex mechanisms that involve T-cytotoxic lymphocytes and helpers, macrophages and histocompatibility antigens, specialized molecules, which are able to "inform" the immune system of the presence of infective agents inside the cells. In several cases, however, the immune response directed towards pathogenic agents may cause pathology, involving the same mechanisms that usually constitute protection. This may occur because the microbic antigens are excessive or because they are not eliminated, causing an inflammatory reaction which is powerful and long; another reason may be that the antigen, which is in itself innocuous, such as allergens or a non-cytopathic virus, causes a useless and dangerous immune reaction; finally, it may be because microbic antigens that are structurally similar to host molecules bring about a reaction directed against them.

The Immune System Is Responsible for Protection against Pathogens

The function of the immune system is to protect the host against infecting agents; it therefore has to be able to identify them rapidly and to respond in an appropriate manner, triggering the effector mechanisms most likely to neutralize and eliminate them. The disastrous effect of infections in immunodeficient patients, with defects in terms of quality or quantity in one or more of their defense mechanisms, highlights the central role of the immunological response in the control of infections. It is therefore implicit that pathogenic agents have exercised the main selective pressure on the protective systems of organs, and that the immune system of current vertebrates is the result of an evolutionary process beginning with primitive defense mechanisms. Given that infectious agents display a considerable degree of variability, especially in their protein component, in order to recognize them and to combat them effectively the most highly evolved immune systems have developed mechanisms able to form a practically unlimited number of receptors. The problem of the generation of receptor diversity has been resolved by T lymphocytes and by B lymphocytes in a substantially analogous way — through a process of somatic rearrangement of gene segments. This solution implies the insertion of nucleotides in the joints between segments, apart from a process of somatic mutation in the gene segments which encode the variable portions of the anti-

---

TABLE I

Several Examples of Immune Response Which Cause Immune Pathology

IMMUNE MECHANISM	EXAMPLES
Response to pathogenic agents or their toxic products; the damage to the host reflects an excessive response due to the fact that the antigen is difficult to eliminate or due to the excess of the antigen.	Response mediated by T-lymphocytes and macrophages to tuberculosis or leprosy microbacteria. Antibody response to streptococcus products.
Several antigens, including allergens, have a type of use response due to the defense against certain pathogenic agents, i.e., helminthes, but not against the allergens themselves.	Allergic diseases mediated by IgE.
Using conventional mechanisms, the immune system responds to microorganisms that are not in themselves pathogenic. The tissue damage is due to the immune response of the host.	Illnesses from non cytopathic viruses: lymphocyte choriomeningitis virus, hepatitis-B virus, etc.
Response to microbic components that are structurally similar to components of the host (molecular mimesis). The anergy condition in respect to the self-components is exceeded and the immune response is directed against the host.	Rheumatic fever

---

bodies. The evolution of parallel recognition mechanisms in T and B lymphocytes can probably be regarded as the result of the different selective pressures exerted by the pathogens typical of different habitats. It is clear that the antibodies are particularly effective in recognizing extracellular pathogens and in triggering the effector mechanisms able to eliminate them, such as lysis brought about by the complement or the elimination by specialized cells possessing receptors for the Fc portion of the antibodies IgG (Table 1). Defense against pathogens that live within the cells — such as many bacteria, microbacteria, and certain parasites and viruses — cannot be mediated by antibodies. In these cases, to be able to identify an infected cell it was necessary to develop a mechanism that could convey the antigens of the pathogenic agent to the surface and expose them there (Kaufmann, 1993). The histocompatibility antigens are able to convey and expose peptides derived from the degradation of endocellular pathogens on the surface of the cell (York and Rock, 1996). T lymphocytes recognize molecules of self histocompatibility linked to the non-self-peptide, with the consequent activation of their functional program. In the case of infected macrophages, the T helper lymphocyte produces cytokines, especially the γ interferon (IFN-γ), which stimulate the macrophage and induce it to destroy the pathogens that live inside it. In the case of viral infections, the T cytolytic lymphocytes are able both to kill the infected cells and to produce IFN-γ, with the consequent inhibition of the virus replication (Kägi *et al.*, 1996).

It should be emphasized that the various pathogens occupy specific endocellular compartments. For example, the endocellular bacteria reside in the interior of endosomic vesicles and synthesize their proteins in this compartment

(Kaufmann, 1993); on the other hand, viruses use the protein biosynthesis apparatus of the cell infected by them. This difficulty has led to the evolution of two distinct classes of histocompatibility antigens which are specialized in bonding peptides originating in various cell compartments and in conveying them to the cell surface with different mechanisms. Class I histocompatible antigens convey peptides generated at the cytoplasmic level, whereas class II histocompatible antigens convey peptides originating in the endosomic and lysosomic vesicles. In the first case, T cytolytic lymphocytes CD8$^+$ come into play and directly destroy the infected cells; in the second case, T helper lymphocytes CD4$^+$ which, through the production of cytokines, activate the macrophages, inducing them to destroy the bacteria that live in their vesicles.

The simple interaction between the antigen receptors expressed by T lymphocytes and the major histocompatibility complex (MHC) and the peptide complex is not sufficient to induce an immune response by virgin T lymphocytes; that is, those T lymphocytes not yet in contact with the antigen. To obtain an effective response that involves proliferation and functional maturing of the antigen-specific cells, a second signal is necessary due to the interaction between the costimulating receptors present on the membrane of T lymphocytes and the molecules to which they are bonded (ligands), expressed on the cells that present the antigenic peptide associated with MHC molecules. In the absence of a costimulating signal, the simple interaction between the T cell receptor (TCR) with the MHC–peptide complex induces a state of anergy in which the cell is unable to respond to the stimuli or to cell death. On the other hand, effector T cells that have already come into contact with the antigen and which have given rise to a functional differentiation are able to be activated directly by the interaction between the TCR and the MHC–peptide complex without any need for costimulating signals. An important receptor molecule with costimulating activity expressed by the virgin T cells is CD28. Two main ligands have been identified for CD28: B7-1 and B7-2. B7-1 is expressed on B lymphocytes on the dendritic cells and on the activated monocytes, whereas B7-2 is also expressed on nonactivated monocytes (Lenschow *et al.,* 1996). It should be noted that other molecules — for example, adhesion molecules such as leukocyte function antigen-1 and its ligand intracellular adhesion molecule-1 (ICAM-1) or ICAM-2 — are able to increase the activation of T lymphocytes mediated by the TCR. However, as opposed to the costimulating molecules, the latter are unable to sustain the proliferation and differentiation of virgin T cells and to prevent their anergy. The need for a double signal to induce a response by virgin T cells is a general rule: It can serve as a safety mechanism to avoid responses caused by the activation of lymphocytes that express specific receptors for self-antigens. This safety mechanism is essential in the case of B lymphocytes, whose receptors rapidly come into contact with somatic mutations and can therefore easily generate self-reactive cells. On the other hand, another control mechanism of B lymphocytes is provided by the fact that their activation depends on T helper lymphocytes whose receptors are not affected by somatic mutations. Moreover, selection at the thymus level eliminates many potentially self-reactive lymphocytes, although it does not eliminate specific T lymphocytes for proteins synthesized selectively in certain peripheral tissues. For example, proteins contained in thyroid cells but absent in the thymus will not pass through the thymic "screen" during clonal selection. It has been demonstrated that T lymphocytes with receptors specific for peptides of peripheral (and therefore self-reactive) antigens are not normally activated due to the lack of costimulating molecules at the level of the target tissue. Contact with these antigens causes a state of anergy or cell death. It is therefore clear that discrimination between self and non-self is conditioned not only by the interaction between TCR and antigen but also by the existence or otherwise of effective costimuli (Lenschow *et al.,* 1996). As a corollary, it follows that regulation of the expression of costimulating signals has the same relevance as that of recognition of the antigen in the discrimination between self and non-self. The functional limitations of virgin T cells in the response to antigens apply not only for self-antigens but also for extraneous ones. It should be stressed that, to induce an immune response against purified proteins, adjuvants have to be used which contain bacterial products such as the components of the mycobacterium of tuberculosis or of *Bordetella pertussis.* This necessity is explained by the fact that various bacterial products are able to activate the cells that present the antigen and to induce the expression of the ligands for the costimulating molecules expressed by T lymphocytes. An immunological response against a given antigen can therefore easily be induced only during the course of an infection. This is important because, in this way, the immune system is able to respond effectively only to non-self components of infecting agents and not to harmless proteins and to self-proteins.

As the immune system has evolved to identify and combat pathogenic agents, it is not surprising that in the more primitive defense systems receptors have been selected which are able to recognize structures typical of pathogens that are absent in the host. Some of these structures have been preserved during evolution and are present in the more complex immune systems of vertebrates. These receptors not only supply a nonclonal form of defense against pathogenic agents but also have adapted to the immune system and foster the triggering of the immune response mediated by T and B lymphocytes. An example is provided by the receptors, which are present on the monocytes and on the macrophages, for the lipopolysaccharide (LPS) of gram-negative bacteria. LPS is the most powerful stimulator of macrophages and is considerably more effective than IFN-γ. Released by gram-negative bacteria, LPS stimulates the production of proinflammatory cytokines, such as IL-1, tumor necrosis factor-α (TNF-α), and IL-6, all of which are molecules that create a series of immediate direct and indirect effects (Ulevitch and Tobias, 1995) and that are able to limit the infection and to recall neutrophilic granulocytes able to eliminate the bacteria. In addition to these nonimmunomediated effects, LPS promotes the triggering of an immunological response as a result of the induction of the expression of ligands (e.g., B7-1) for costimulating molecules (e.g., CD28) of T lymphocytes.

A peculiar defense mechanism is mediated by natural killer (NK) cells. These NK cells, contrary to T cytolytic lymphocytes, express in a constitutive manner a cytolytic machinery and receptor structures able to interact with the membrane structures expressed by the majority of nucleated cells (Moretta *et al.,* 1994). Consequently, the NK cells can attack and destroy other cells, including normal ones. To prevent this from happening, the protection mechanism adopted by normal cells is used which foresees the expression of class I MHC molecules; the NK cells recognize the MHC molecules because they have inhibitor receptors able to inhibit their cytolytic function (Moretta *et al.,* 1996). In the case of particular viral infections caused, for example, by herpesvirus, the cell is unable to express class I MHC antigens and therefore is open to attack by NK cells. The latter function in a specular manner with respect to T cytolytic lymphocytes: Although these recognize cells that express class I MHC molecules, link to antigenic peptides, and destroy them, the NK cells are able to identify and kill cells that do not express class I molecules (Moretta *et al.,* 1996). It is clear that the two defense systems act in a complementary manner and, although both use different mechanisms, both contribute toward eliminating cells infected by many viruses.

From Normal Immune Responses to Immunopathological Responses

On the basis of the previous discussion and to have a clearer understanding of certain immunopathological reactions, namely, those defense reactions that become transformed into an injury for the host, it is necessary to bear the following considerations in mind: (i) The immune system originally evolved from a primitive defense system against pathogenic agents using more effective mechanisms; (ii) the primitive effector cells and mechanisms became modified and adapted to a more complex immune system, for example, acquiring receptors for the Fc fragment of IgGs or producing active cytokines on the lymphocytes; (iii) defense reactions exist in which these mechanisms play a prevalent role; (iv) in the course of certain infections, immunopathological reactions are to be regarded as the subproduct of the response necessary for the survival of the host; and (v) some of the effector mechanisms of specific immunity, such as phagocytic cells, cytokines, or the inflammatory cells and complement, are not specific for the antigen or for the pathogenic agent. Hence, many responses which are useful are often accompanied by a local injury, and sometimes even a systemic injury to the tissues of the host. The elimination of the pathogenic agent generally means the termination of the injury. In situations in which, due to the characteristics of the pathogenic agent or the quality of the host's response, control of the reaction is not achieved or a response is activated against self-antigens, a state of illness on an immunological basis is determined. Also, primitive defense mechanisms, such as the recognition of products typical of pathogens absent in the host such as LPS, can cause a pathological reaction. In this case, the injury depends on the quantity of LPS released by the bacteria and the entity of the stimulation of the microphages. A major part of the effect is in fact mediated by the TNF-α, which may have a harmful systemic action if released in large quantities. Concerning the more properly immunological pathologies, those pathological situations characterized by excessive or uncontrolled immunological reactions are referred to as hypersensitivity illnesses. As already mentioned, hypersensitivity often coexists with the protective response. The term self-immunity is instead applied to those pathological situations in which the immunity response is directed against self-antigens. In this article exemplary situations will be considered that illustrate how immunity protection mechanisms can lead to an immune pathology. An initial analysis will be made of many general mechanisms, a result of which immunohumoral and cell-mediated responses can cause an illness. Then, many pathological situations will be identified within the framework of allergic and self-immune pathology or the pathology of response against pathogenic agents.

Main Types of Immune Pathologies

Immune pathologies are often defined on the basis of the pathogenic mechanisms responsible for the injury to the tissues. Antibodies can in fact come into play, with their specific biological characteristics, able to recruit and activate components of the complement in addition to cells responsible for the inflammatory response and for the tissue damage, such as leukocytes and macrophages. The antibodies of concern may be specific for extraneous antigens such as IgE, which is specific for allergens, and IgG, which is specific for bacterial products, or for self-antigens. In IgE-mediated allergies, the IgE antibodies specific for allergens bond together with specific receptors for thc fragment Fc of the antibodies, which are present on the mast cells and on the basophils, where they remain at length acting as real receptors. When they again encounter the allergen, the IgE antibodies induce the release of potent mediators of inflammation contained in the granules of the cells. In some cases, the antibodies specific for cell antigens are bonded with the target cells and can induce their destruction as a result of the activation of the complement or the intervention of scavenger (phagocyte) cells. In other cases, they are directed against soluble antigens (e.g., bacterial or viral products) and form immune complexes circulating in the blood. The immune complexes are then deposited in the filter zones (e.g., the renal glomerulus), causing damage through mechanisms which will be analyzed in detail later.

Tissue damage can derive from mechanisms mediated by cells which implicate the intervention of T lymphocytes since these can act either directly, as in the case of cytolytic T lymphocytes, or through the activation of the macrophages. In all these cases, we speak of delayed hypersensitivity.

Normal immune responses, directed against microbic antigens, can be harmful and give rise to immune pathologies for various reasons. Many pathogenic agents are difficult to eliminate; therefore, there is persistent stimulation of the immune system which amplifies and makes chronic the effector mechanisms. The inflammatory response gives rise in this case to an often appreciable injury to the tissues, as in the case of infections caused by infections of tuberculosis or leprosy microbacteria. In other cases, the immune response is directed against microorganisms which would not cause an illness as noncytopathic viruses; the injury to the tissue is therefore due solely to the immune response and not to the direct effect of the infecting agent. Extraneous antigens can also have an antigenic determinant in common with self-components. The immune response induced against the pathogenic agent, made possible as a result of the mechanisms described previously, reacts in cross form with self-antigens, which are incapable of inducing an immune response because of the lack of costimulating signals, thus causing tissue injury as in the case of rheumatic fever (Fig. 1).

Pathology Determined by the Response to Bacterial LPS

Recall that primitive defense mechanisms still exist in vertebrates which, although they do not use very sophisticated receptor systems of recognition and have a nonclonal distribution (the receptors are equal on all the cells of a particular type, for example, on all the macrophagic cells), make it possible to discriminate between self and non-self components typical of certain pathogenic agents, such as LPS (Ulevitch and Tobias, 1995). The effect occurs to a large extent as a result of the production of proinflammatory cytokines, principally TNF-α. A prompt response to the LPS released in the course of gram-negative bacterial infections is important to limit the infection and to promote the arrival of cells able to eliminate the bacteria. Nevertheless, high LPS concentrations can cause serious tissue damage, shock, disseminated intravascular coagulation, and death.

When produced locally, in response to a limited quantity of LPS, the effects of TNF-α are of an autocrinal and paracrinal type since they mainly govern the function of the monocytes, the granulocytes, and the endothelial cells. Its action is in fact exercised (i) on monocytes and macrophages which produce cytokines, with a positive feedback

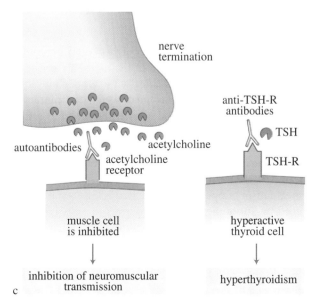

FIGURE 1 Immunopathology due to antibodies directed against cellular antigens. (a) Complement-mediated cell lysis. The antibodies (in blue) bind to an antigen (in red) on the cell surface and activate complement (C). Cell lysis is due to the activation of the last fractions of the complement cascade which form the membrane attack complex (MAC). (b) Opsonizing autoantibodies: phagocytosis or antibody-mediated cell cytotoxicity. Antibodies directed against cellular antigens (in red), which do not fix complement effectively, cover the cells, making them recognizable by the macrophages (M) or by the NK cells. The former destroy the opsonized cells by phagocytosis and the latter by antibody-mediated cytotoxicity mechanisms. (c) Antibodies directed against hormone receptors or neurotransmitters can cause grave illnesses. (Left) Autoantibodies directed against the acetylcholine receptor can inhibit neuromuscular transmission, preventing the receptor from binding with the acetylcholine. (Right) Antibodies against the thyroid stimulating hormone receptor (TSH-R) can stimulate the thyroid cells by mimicking a TSH effect: In this case, the antibody can have an agonist effect and induce an activation not controlled by the thyroid cells with consequent hyperfunction of the thyroid (hyperthyroidism).

mechanism, and (ii) on the cells of the endothelium to promote the production of chemotaxic cytokines, such as IL-8, which are able to recruit neutrophil leukocytes which play a central role in antibacterial defenses. The effect of TNF-α on neutrophil leukocytes is also due to the expression of adhesion molecules on the endothelial cells which favor the recruitment and the transfer of the neutrophils from the blood to the tissues. These effects, induced by the interaction of monocytes and macrophages with the LPS, are very important for a prompt response against bacteria so as to prevent their rapid multiplication. The positive result of the reaction to an infection also depends on the production of procoagulant substances by the macrophages and the endothelial cells. The formation of blood clots at the level of the microcircle in the site of the infection can prevent the spread of bacteria. However, if the TNF-α is released in large quantities there are systemic effects of varied entity which, in extreme cases, can lead to death. TNF-α causes pyrogenic metabolic effects due to its action at the level of the hypothalamus, with the consequent alteration of the hypothalamic thermostat and the inducement of fever with deleterious effects mainly on the circle. By activating the coagulation system, TNF-α induces in the most serious cases a widespread intravascular coagulation. In addition, it has an inhibitory effect on cardiac contractility: By reducing the flow of blood it induces a relaxation of the smooth musculature of the vessels, diminishing their tone and the production of vasodilating substances such as prostacyclin. The combination of widespread intravascular coagulation with increased permeability of the vessels and the negative effect on cardiac contractility explains the serious effects due to the massive release of LPS. Hence, at the level of innate defense mechanisms it is possible to determine the ways in which a defensive response against pathogenic agents can become transformed into a harmful response for the host.

Pathology Induced by Microbic Products with Superantigenic Activity

Certain microbic products, mostly toxins produced by staphylococci and enterotoxins, are able to activate significant amounts of T lymphocytes. The molecular basis of this effect has been clarified. These substances, termed superantigens, are able to bond specifically with given V_β segments of the TCR and to activate selectively all the T lymphocytes expressed by the important V_β segments (Marrack and Kappler, 1990, 1994; Acha-Orbea and McDonald, 1995). The superantigens differ from the classic antigens not only in these properties but also in the fact that they bond directly with the β chain of the TCR without going through any processing. The term superantigen denotes the number of lymphocytes activated — far more than the number activated by a conventional protein antigen. The activation of many T lymphocytes brings about the production of conspicuous quantities of cytokines

which, in extreme cases, can induce damage in many aspects, similar to septic shock due to LPS. It is possible to obtain the release of both lymphotoxin, which produces effects similar to TNF-α, and IFN-γ, which induces the production of TNF-α by the macrophages. As opposed to the septic shock, which is linked to the infection of the host by gram-negative bacteria, the toxic shock due to staphylococcal enterotoxin usually occurs through the contamination of food with the thermostable toxins. Enterotoxins are a very frequent cause of food poisoning.

Immune Pathology Caused by Antibodies

The harm caused by the deposition of antibodies in immune complex form was the first form of immune illness recognized. Antibody-mediated illnesses are classified into two main categories according to whether the antibodies are directed against cell or tissue antigens or recognize soluble antigens. A third situation occurs with antibodies of the IgE class which, once produced, become affixed, through their Fc fragment, to high-affinity receptors expressed on the mast cell membrane and basophilic membranes. The type of tissue damage generated in these three situations differs greatly, as discussed in the following section.

Damage Due to Antibodies Directed against Cells

In the case of antibodies directed against cells, the damage is mainly, if not exclusively, limited to the cells or the tissues affected. They are generally autoantibodies or antibodies produced against extraneous antigens which cross-react with self-antigens. The mechanisms that induce pathological effects depend on the class of the antibodies and, therefore, on their biological characteristics. Complement-mediated lysis occurs especially with IgM antibodies and with certain IgG antibodies. The activation of the complement can lead to osmotic lysis of the cells (Fig. 1) by the membrane attack complex. In the case of IgG antibodies, damage mainly occurs to the cells covered by antibodies due to the scavenging action of the macrophages or to the cytolytic action of the NK cells. In the case of autoimmune hemolytic anemias, for instance, which are characterized by autoantibodies directed against the patient's red blood cells, the elimination of the red blood cells occurs following phagocytosis by the macrophages of the spleen and the liver: The increased destruction of the red blood cells, over a protracted period, is the cause of anemia. An alternative mechanism of opsonized nucleated cell lysis (i.e., cells covered by antibodies) is mediated by NK cells (Fig. 1) with antibody-dependent cellular cytotoxicity.

In particular cases, the cell antigen recognized is a receptor structure, such as the receptor for the thyrotropin of the thyroid cells, and therefore interference can occur in the normal functions of the cells in both a positive and a negative sense. Often, the damage caused to normal cell functioning is clinically more evident than the damage caused by the activation of the complement or by the action of phagocytes. Two classic examples are illustrated in Fig. 1. An important consequence of the bond between antibodies and tissue antigens is the recruitment of neutrophil leukocytes and, in part, monocytes. This recruitment occurs primarily due to the chemotaxic effect of two small soluble molecules released in the course of the activation of the complement: C3a and C5a. Both the neutrophils and the monocytes and macrophages express membrane receptors specific for the Fc fragment of the IgG. They are thus able to interact with the IgG immune complexes and to release hydrolytic enzymes contained in the cytoplasmatic granules (Fig. 2). As will be discussed later, this mechanism takes on a preponderant role in the case of the depositing of immune complexes.

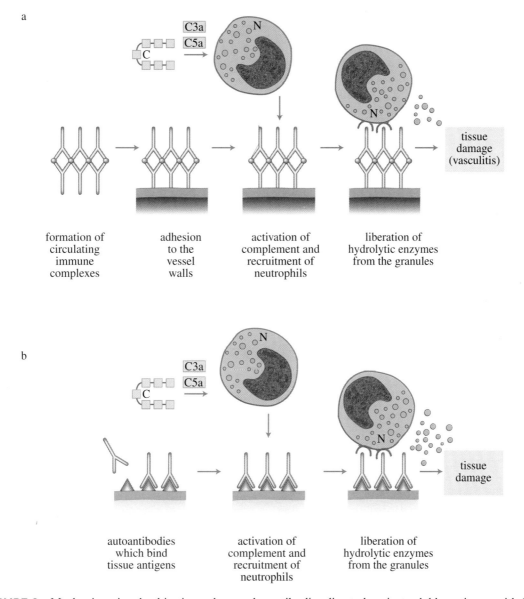

FIGURE 2 Mechanisms involved in tissue damage by antibodies directed against soluble antigens, with formation of immune complexes, or against tissue antigens. (a) In the case of antibodies (in blue) directed against soluble antigens (in red), immune complexes are formed in the blood. These are deposited, for example, at the level of the basement membrane of the renal glomerulus. The activation of complement (C), and in particular the chemotactic effect of C3a and C5a, released during the course of activation, causes the recruitment of neutrophils (N) with release of hydrolytic enzymes from their granules, leading to tissue damage. (b) In this case, the antibodies bind specifically to the tissue since they are directed against tissue antigens. The mechanisms that cause the damage are basically those illustrated in Fig. 2a. However, if the tissue antigen recognized is expressed on the cellular membrane, other effector mechanisms may come into play such as those illustrated in Fig. 1.

Antibodies Directed against Soluble Antigens and Formation of Immune Complexes. Under normal conditions, the formation of immune complexes facilitates the elimination of the antigen due to the neutralization of its harmful effects, as in the case of toxins and the action of scavengers which remove the immune complexes. In many cases, however, the formation of immune complexes may be responsible for serious damage to the host: The extent of the reaction depends both on their quantity and on their distribution in the organism. According to their deposit site, it is also possible to observe different immunopathological reactions. In the case of immune complexes that are not very soluble which are deposited in the tissues close to the antigen's entry site, a localized reaction develops. If, instead, the immune complexes are soluble and are formed in the blood, they are deposited in "filter zones" situated mostly on the basement membrane of the renal glomerulus, on the synovial membrane of the joints, on the choroid plexuses of the brain, and on the walls of the vessels. In any case, tissue injury is developed at the level of the site of deposit, brought about by small soluble molecules stemming from the activation of the complement, mainly C3a and C5a. These molecules induce the degranulation of the mast cells and the basophils with the release of histamine and a localized increase in vessel permeability. The C3a and C5a fragments also have a chemotaxic activity and principally attract the neutrophils which accumulate at the deposit site of the immune complexes. The neutrophils try to englobe them using the receptors for the Fc fragment of the IgG and for the C3b fraction of the complement, which are linked to the deposit site of the immune complexes. Given that the latter and the Cb3 are bonded with the basal membrane, this englobement is ineffective. The activated neutrophils will release lytic enzymes contained in the granules, oxygen radicals, and lipidic mediators, which are the elements mainly responsible for tissue damage (Fig. 2). Furthermore, the activation of the complement can induce the aggregation of the platelets and cause the release of coagulation factors, with the consequent formation of microclots. The formation of circulating immune complexes and of generalized pathological manifestations, including fever, edemas and erythemas, generalized vasculitis, arthritis, and glomerular nephritis, occurs in various pathological conditions, including autoimmune diseases such as lupus erythematosus and rheumatoid arthritis and infectious diseases such as poststreptococcic glomerulonephritis, viral hepatitis, mononucleosis, and malaria. Also, certain reactions against medicines, including sulfamides and penicillin, are characterized by the formation of immune complexes. In the case of infectious diseases, the immune complexes are formed of antibodies and various bacterial, viral, or parasitic antigens. In the case of autoimmune diseases, the self-antigens recognized are represented by various proteins, by DNA, and even by antibodies which, in this case, behave as antigens and are recognized by other antibodies, as occurs for rheumatoid factors typical of rheumatoid arthritis.

Poststreptococcic glomerulonephritis represents an exemplary pathological condition. Soluble antigens (e.g., toxins of the streptococcus) are found in the blood after acute streptococcic infections. The production of antibodies against these streptococcic antigens leads to the formation of immune complexes which are deposited at the level of the basal membrane of the renal glomerulus, a typical blood filter zone. The complement is unable to dissolve the glomerular basal membrane because of its chemical composition. Also, the component C3 is activated with the formation of C3a, which is soluble and able to attract the neutrophil leukocytes, and C3b, which is able to bond at the site where the immune complexes are deposited. The leukocytes thus recruited become bonded with the immune complexes by means of their receptors for the Fc fragment of the IgG antibodies and with the C3b by means of specific receptors. The result of the ensuing activation is the liberation of the content of the granules with consequent damage to the structures of the glomerulus. If the damage extends to a high percentage of glomeruli, the result could be a serious functional compromising of the kidneys with ensuing renal failure. Therefore, an immune reaction induced by the streptococci and by their toxic products can degenerate into a serious pathology for the host.

Hypersensitivity Reactions Mediated by IgE Antibodies. The immune reactions induced by some types of antigens, commonly called allergens, constitute one of the most rapid and most powerful effector mechanisms of the immune response. The antibodies of the IgE class, produced in response to allergens, bond together with high-affinity receptors, which are expressed on the mast cells of the tissues and on the circulating basophils. When the allergen bonds with the IgE adhering to these cells, there is a rapid release of various mediators, responsible for a sudden increase in vascular permeability, of vasodilation, of contraction of the musculature of the bronchi, and of the intestine and localized inflammation. This reaction is a typical example of immediate hypersensitivity, starting rapidly (within a few minutes) after the encounter with the allergen. In extreme cases, in which there is a very intense release of mediators by the basophils and the mast cells, a generalized reaction, anaphylaxis, occurs, that is characterized by accentuated bronchospasm with possible asphyxia, circulatory collapse due to the increased vascular permeability, widespread edemas, and shock.

The classic sequence of events typical of immediate hypersensitivity includes (i) the "sensitization" phase, with the production of IgE by B lymphocytes that have come into contact with the allergen; (ii) the bonding of the IgE with the mast cells and the basophils; (iii) the interaction of the allergen, following its subsequent introduction, with the IgE adhering to the mast cells; and (iv) the consequent cellular activation with the immediate release of the preformed mediators accumulated in the granules (Fig. 3). The reason why such a potent system evolved that is apparently harmful to the organism is not clear. It is possible that it was selected mainly for defense against parasites, helminths, and larvae of insects, probably as the initial part of a more complex response leading to the formation of inflammatory infiltrates rich in eosinophilia, particularly active in defense against the previously mentioned pathogenic agents.

Nevertheless, in addition to playing an important role in defense against parasites, the eosinophilia are also responsible for the induction of the tissue lesions typical of allergic diseases, such as allergic asthma. The reason why the most common allergens are able to induce similar immune responses to those of certain pathogenic agents, whereas the introduction into the organism of other antigens through the same paths as those of the allergens does not induce an allergic response is unknown. It may be that the capacity of an antigen to function as an allergen is an intrinsic property of the antigen. Another possibility is that the allergens are associated with substances of adjuvant type able to direct the immune response toward the production of IgE. Both antigens of certain parasites and normal allergens could induce allergic responses because both are dependent on adjuvant substances able to induce a polarized response with the activation of T_H2 lymphocytes.

Mediators of IgE-Mediated Reactions. The effects and clinical manifestations of hypersensitivity reactions mediated by IgE reflect principally the biological effects of the mediators released during the degranulation of mast cells and basophils. These mediators act both on the nearby tissues and on other effector cells, including the eosinophils, neutrophils, monocytes, platelets, and endothelial cells, which are recruited and activated. The mediators therefore act both directly and indirectly with a consequent amplification of the effector mechanisms. The release of the mediators in response to an infection by parasites triggers a defense mechanism useful for the host. The rapid contrac-

FIGURE 3 The main events of IgE antibody-mediated hypersensitivity. (a, b) B lymphocytes specific for the allergen are activated and produce IgE antibodies which are then fixed, through their Fc fragment, to the high-affinity receptors (FcR) present on mast cells and basophils (sensitization stage). (c, d) With a successive contact with the allergen, this interacts with the specific IgE adhering to the surface of the mast cells and the basophils, inducing the aggregation of the IgE on the membrane and the signal that leads to cellular degranulation. (e) The granules release their content consisting of powerful mediators which act mainly on the vessels, resulting in an increase in vessel permeability and the formation of edemas, and on the smooth musculature, with spasm at the bronchial level or hypermotility at the intestinal level.

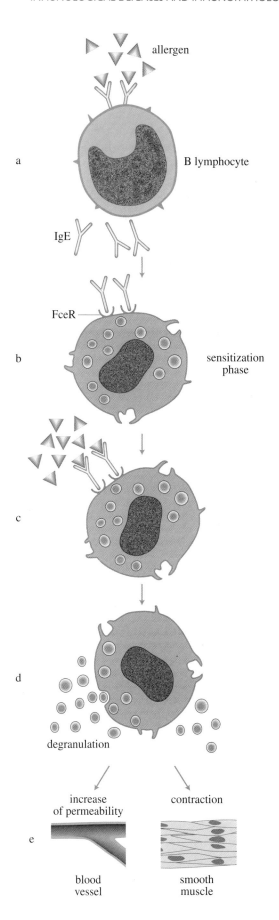

tion of the smooth musculature, for example, can cause the expulsion of the parasites or the larvae that infest the intestine, whereas the localized contraction of the smooth musculature at the postcapillary level causes a dilation of the capillary bed. This effect, together with the increase in vascular permeability, increases the inflow of plasma, antibodies, and inflammatory cells or lymphocytes necessary for the elimination of the parasite. In the case in which the production of mediators occurs in response to allergens, with antigens innocuous for the host, an inappropriate reaction occurs whose mechanism is similar to that described previously, with negative effects for the host. As illustrated in Table 2, some mediators are preformed; that is, they are synthesized and accumulated in the granules of the mast cells and the basophils prior to their degranulation. Aside from heparin, which prevents the blood from coagulating,

some powerful mediators are included, such as histamine, which act rapidly on the smooth musculature and on the vessels and are responsible for the immediate effects. The secondary mediators are synthesized following the activation of the mast cells and the basophils. Some of them derive from the enzymatic breakdown of the membrane phospholipids, as in the case of the leukotrienes and the prostaglandins; others, including many cytokines, can be synthesized *ex novo*. Some cytokines are able to contribute toward determining the pathological manifestations of the IgE-mediated hypersensitivity reactions. An example is provided by TNF-α and IL-1, both products of the mast cells and which can contribute to the shock. Moreover, both TNF-α and IL-1 induce an endothelial activation that determines the expression of adhesion molecules and the production of chemotaxic factors, fostering the accumu-

TABLE 2	
Principal Mediators Involved in Hypersensitivity Mediated by IgE	
PRE-SHAPED MEDIATORS IN THE GRAINS OF MAST CELLS AND BASOPHILS*	**PRINCIPAL EFFECTS (IMMEDIATE)**
Histamine	Increase in the permeability of vessels, contraction of smooth muscles.
Serotonin	Increase in the permeability of vessels, contraction of smooth muscles.
Proteases	Genesis of damage in the tissue and the base membrane of the vessels. Formation of complement degradation products.
Chemioactive factor for eosinophils	Recall of eosinophils.
SECONDARY MEDIATORS, DERIVED FROM A METABOLIC ACTIVITY**	**PRINCIPAL EFFECTS (DELAYED)**
Leucotrienes	Increase in the permeability of vessels, contraction of smooth muscles of the lungs.
Prostaglandins	Increase in the permeability of vessels, contraction of smooth muscles; aggregation of the platelets.
Platelet Activating Factor (PAF)	Aggregation of the platelets and their degranulation, contraction of smooth muscles of the lungs.
TNF-α. IL-1	Increase in the permeability of vessels, induction of procoagulant activity, expression of adesoine molecules, reclusion of inflammatory cells.

*The pre-shaped and accumulated mediators in the grains of mastcells and basophiles are released rapidly when the allergen interacts with the specific IgE, which adheres to the surface of those cells. The effects are thus immediate.

**The secondary mediators must be synthesized by the cells; for this reason they have a delayed effect.

lation of neutrophils, eosinophils, monocytes, and macrophages characterizing the delayed phase. As noted previously, the eosinophils that are accumulated during the course of the delayed phase contribute toward the development of chronic inflammations such as that of the bronchial mucosa, which may cause chronic bronchial asthma (Table 2).

Regulation of the IgE Response: The Role of T Lymphocytes and Cytokines. As mentioned previously, there is no definitive answer regarding the reason why many antigens behave as allergens. However, in the past few years we have obtained sufficient information to understand the mechanisms that determine the production of IgE rather than antibodies of other classes, such as IgA or IgG. First, we know that a genetic predisposition exists — this is evident in test animals: Only some species, in fact, respond to particular antigens with the production of IgE and sometimes, within one species (e.g., mice), some strains are characterized by their greater capacity to produce IgE. As a consequence, these animals can defend themselves effectively from infections by certain parasites but they more easily develop allergic responses. In man, the outcome of hypersensitivity reactions mediated by IgE antibodies is influenced by genetic components.

Other factors that determine the typology of the response include the dose of antigen because high doses can induce a temporary production of IgE followed by the production of IgG, the use of particular adjuvants, and other less important factors. It is clear that, for all events, a common mechanism exists through which various factors act which are determined by the regulation exercised by the T helper lymphocytes belonging to the subpopulations T_H1 and T_H2 (Romagnani, 1994). The T_H1 cells reduce the response, whereas the T_H2 cells increase it. Some species, or some individuals of the same species, respond to antigens with mainly T_H2 type cells. Furthermore, many allergens or substances associated with them have the capacity to stimulate the T_H2 cells more effectively. The cytokines produced by T_H2 cells, mostly IL-3–IL-5, are able to mediate different aspects of the IgE-mediated hypersensitivity response: IL-3 and IL-4 induce a proliferation of the mast cells, and IL-4 is responsible for the production of IgE. In other words, the antibodies produced against particular antigens will be mainly IgE instead of IgA or IgG. Lastly, IL-5 is responsible for the proliferation, maturation, accumulation, and activation of the eosinophils. In contrast, cytokines produced by T_H1 cells, principally IFN-γ, have an inhibitory effect on the IgE-mediated response (Romagnani, 1994). In addition to the effect of IL-4, a second signal seems to play an important role in the production of IgE. This signal involves the molecule CD40 expressed on B lymphocytes, producers of antibodies, and its ligand, CD40L, expressed on T_H2 lymphocytes (Foy *et al.*, 1996).

Cross-Reaction between Microbic Products and Antigens of the Host: Molecular Mimesis

Some pathogenic agents contain amino acid sequences similar to those of self-antigens. In the event of infection with these pathogens, it is possible that tolerance to self-antigens may disappear. This mechanism can explain the autoimmune reactions directed against the cardiac muscle and valves which occur in rheumatic fever, following streptococcic infections. Rheumatic fever is a rare complication of these infections which appear 3 or 4 weeks after the outbreak. Its pathogenesis is explained on the basis of a cross-reaction between carbohydrates of the streptococcus and components of the cardiac muscle. The fact that there is no correlation between the gravity of the infection and the outcome of rheumatic fever and that only a small percentage of individuals are affected by the disease suggests that individual factors of predisposition are also involved and that the disappearance of tolerance is not sufficient to explain its outcome. In this context, it is important to emphasize that autoimmune reactions, for example, with the production of autoantibodies, are relatively frequent after various infections but in general are limited in time. The outcome of real autoimmune pathologies seldom occurs. The majority of antigens that promote antibody cross-reactions are of glucidic or lipidic origin, although cross-reactive antibodies also specific for protein components have been identified. A typical example is an antigen of *Klebsiella pneumoniae* which is structurally similar to the allele HLA-B27. This structural analogy seems to be at the origin of the high risk for HLA-B27$^+$ subjects of developing ankylosing spondylosis after Klebsiella infections.

Immune Pathology Caused by Cells

As mentioned previously, in many cases tissue damage is caused by the activation of effector mechanisms mediated by some types of cells, mainly lymphocytes and macrophages. A typical example is provided by the reactions of delayed hypersensitivity, which are manifest when the antigen activates T lymphocytes, mostly belonging to the subpopulation T_H1.

Delayed Hypersensitivity Reactions

The activation of T helper cells, under certain conditions, can trigger a delayed hypersensitivity reaction. Such situations occur mainly in the case of pathogenic agents that infect the macrophages, stimulating them, or which activate NK cells that are able to produce great quantities of IFN-γ, cytokines which function as powerful activators of macrophages. The pathogenic agents that infect the macrophages are efficient in inducing delayed hypersensitivity and it is difficult to eliminate them. This situation determines a chronic activation of T helper lymphocytes, which recognize antigenic peptides, derived from the partial degradation of the pathogenic agent, presented by the macrophages in association with class II histocompatibility antigens. The macrophages activated produce IL-12, a very important for directing a prevalent response of the T_H1 lymphocytes (Trinchieri, 1995). The T_H1 cells, activated by the specific recognition of antigenic peptides, in turn produce two other important cytokines: IL-2, which induces a proliferation of the T_H1 lymphocytes, and IFN-γ, which induces a further activation of macrophages. The latter thus increase the synthesis of class II histocompatibility antigens, which are more efficient in stimulating the T_H1 lymphocytes. Furthermore, they stimulate the macrophages to produce numerous cytokines, the overall effect of which is to attract more macrophages to the reaction site, to activate them, and to promote their more efficient phagocytosis by means of increasing the concentration of lytic enzymes. The increases in cytolytic activity can contribute to the elimination of the pathogen. However, the protracted stimulation of the T lymphocytes, and indirectly of the macrophages, caused by the continued production of IFN-γ can cause an escape of lytic enzymes into the surrounding tissues, with consequent damage. An alteration of the normal architecture of the tissues can also occur following the formation of a granulomatosis type of reaction. The granuloma is the morphological expression of a defense mechanism able to circumscribe and sequester a pathogen. It consists of activated macrophages which take on the aspect of epithelial cells, which are large cells with very extensive cytoplasm, and giant cells with more than one nucleus, resulting from the fusion of many macrophagic cells. Visible near the macrophages that occupy the center of the granuloma is an accumulation of lymphocytes and a fibrous-type reaction. The formation of fibrous tissue is the result of the production by the activated macrophage of many cytokines, mainly platelet-derived growth factor, fibroblast growth factor, and transforming growth factor-β, which are able to induce a proliferation of the fibroblasts and the production of collagen.

It also must be emphasized that retarded-type reactions are fundamental for effective protection against endocellular bacteria and parasites. The antibodies are unable to reach these microorganisms inside the macrophages. For instance, mice whose gene of IFN-γ — whose macrophages are not therefore activated — has been knocked out, die after being inoculated with not highly virulent strains of microbacteria that are normally easily eliminated. Moreover, AIDS sufferers succumb to "opportunist infections" generally due to endocellular bacteria and other pathogens, such as mushrooms, that are normally controlled by mac-

rophages. In these patients, the destruction of T CD4$^+$ helper cells leads to insufficient production of IFN-γ, thus making it impossible to activate the macrophages.

The prevalence of the tissue damage caused by necrosis or by disarrangement following granulomatose reactions may have harmful consequences for the host. This occurs especially in the case of endocellular pathogens, which are eliminated with particular difficulty and which consequently induce a chronic stimulation of T_H1 lymphocytes and hyperactivation of the macrophages. Tissue damage due to necrosis may be detrimental. In the case of pulmonary infections caused by tuberculosis microbacteria in man, the necrotic tissue, which is rich in lytic enzymes, causes the destruction of the walls of the alveoli and even penetrates the bronchi with the resultant spread of the microbacteria through the airways to other parts of the lungs. Furthermore, the perforation of vessels may give rise to hemorrhages that may be more or less serious according to the size of the vessels, may allow the microbacteria to spread through the bloodstream to other organs, or both. In this way various organs are damaged and their functionality is seriously compromised. Complications due to infection by tuberculosis microbacteria occur in man in approximately 10% of cases. Tuberculosis, which had drastically diminished in the more developed countries as a result of the use of antibiotics and effective chemotherapeutic agents, recently returned to the limelight in the 1990s as the consequence both of mass migrations from countries in which the disease is endemic and of the selection of microbacterial strains resistant to the antibiotics usually effective against tuberculosis.

Delayed-type hypersensitivity reactions can also occur against inappropriate antigens such as the antigens of poison ivy or substances such as nickel, hair dyes, and cosmetics, which are able to induce reactions of cutaneous hypersensitivity such as contact dermatitis. In this case, there are small molecules which become fixed to cutaneous proteins, forming complexes. Captured by the Langherans cells or by macrophages, these complexes cause the formation of class II antigens, which in turn activate the response to the T CD4$^+$ cells which differentiate vis-à-vis the subpopulation T_H1, developing a typical reaction of delayed hypersensitivity. The accumulation and activation of the macrophages leads to the release of lytic enzymes with a reaction (dermatitis) characterized by redness and the formation of papules. In contrast to the infective pathology previously discussed, the damage caused to the tissues by the immune response is far greater than any possible positive effects.

Immunopathological Response to Noncytopathic Viruses

Virus–host relations are often quite complex and result from the fact that viruses have adopted various strategies to dodge the host's defense mechanisms and that the host is the outcome of a selection targeted for protection against the harmful effects of the viruses (Alcami and Smith, 1992; Ray *et al.*, 1992). This is clear in the case of cytopathic viruses, such as the influenza virus, the variola virus, and the hepatitis viruses. For the survival of the species, it became necessary to evolve particularly efficient defense systems (Zinkernagel and Hengartner, 1994), including both non-specific mechanisms, especially interferons and NK cell-mediated responses, and specific mechanisms, antibodies, or T cytolytic lymphocytes. In the case of noncytopathic viruses, the situation is very different. The viruses and the host may coexist independently of the effectiveness of the immune response (Zinkernagel and Hengartner, 1994). In some noncytopathic viral infections, it is the immune response against the virus that actually causes damage to the host's tissues. A classic example is the lymphocytic choriomeningitis virus which, in the mouse, induces an inflammation of the meninges, mostly at the level of the spinal cord. The virus infects the meninx cells, but does so without causing any direct damage. However, the cells infected by the virus induce a response by specific T cytolytic lymphocytes, which are responsible for the damage. In mice affected by immunodeficiency, or in which the T lymphocytes have been eliminated, no lesions are caused. Hence, this is a paradoxical situation: Instead of being more liable to an infective pathology, immunodeficient animals are resistant. An exemplary case in man is the hepatitis B virus (HBV). With HBV infections corresponding to a low viral load, a rapid, effective immune response is triggered which involves limited damage to the hepatic cells, the elimination of the virus, and cure. In the course of more extended infections or if the immune response develops more slowly, or in cases in which both situations occur, there may be different pathological frames — from acute fulminating hepatitis, in which the massive destruction of the cells of the liver by cytolytic T lymphocytes occurs, to a subacute or chronic course of the disease due to a progressive destruction of the infected hepatic cells by the T lymphocytes. Also, infection by HBV of a subject with a compromised or immature immune system can evolve into a situation in which the virus and the host coexist in harmony. In fact, the virus does not cause damage of the infected cells and, moreover, the host is unable to make an immune response. Conditions of this type are frequent in persons treated with immunosuppressant drugs such as cyclosporin, which is used in transplants to prevent rejection, or chemotherapeutic agents used in treatment of leukemia or tumors. These patients, although smitten by the virus, do not develop hepatitis, and in the long term they may develop other diseases such as immune complex pathologies caused by the production of antibodies against the virus and tumors of the liver as a result of the fact that the HBV can be integrated in the genome of the liver cells and play a part in their neoplastic transformation.

It should be emphasized that, even in the pathogenesis of viral infections of acquired immunodeficiency (HIV), the host's immunopathological reaction plays a more important role than the direct cytopathic effect of the virus. In other words, the progressive destruction of the CD4$^+$ T lymphocytes and progression toward the full-scale disease reflect immunopathological mechanisms directed against the virus and its products rather than the direct destruction of infected cells (Zinkernagel and Hengartner, 1994; Pantaleo and Fauci, 1995).

References Cited

Acha-Orbea, H., and McDonald, H. R. (1995). Superantigens of mouse mammary tumor virus. *Annu. Rev. Immunol.,* **13,** 459–486.

Alcami, A., and Smith, G. L. (1992). A soluble receptor for interleukin-1 beta encoded by vaccinia virus: a novel mechanism of virus modulation of the host response to infection. *Cell,* **71,** 153–167.

Foy, T. M., Aruffo, A., Bajorath, J., Bühlmann, J. E., and Noelle, R. J. (1996). Immune regulation by CD40 and its ligand GP39. *Annu. Rev. Immunol.,* **14,** 591–617.

Kägi, D., Ledermann, B., Bürki, K., Zinkernagel, R. M., and Hengartner, H. (1996). Molecular mechanisms of lymphocyte-mediated cytotoxicity and their role in immunological protection and pathogenesis *in vivo. Annu. Rev. Immunol.,* **14,** 207–232.

Kaufmann, S. H. (1993). Immunity to intracellular bacteria. *Annu. Rev. Immunol.,* **11,** 129–163.

Lenschow, D. J., Walunas, T. L., and Bluestone, J. A. (1996). CD28/B7 system of T cell costimulation. *Annu. Rev. Immunol.,* **14,** 233–258.

Marrack, P., and Kappler, J. (1990). The staphylococcal enterotoxins and their relatives. *Science,* **248,** 1066.

Marrack, P., Kappler, J. (1994). Subversion of the immune system by pathogens. *Cell,* **76,** 332.

Moretta, A., Bottino, C., Vitale, M., Pende, D., Biassoni, R., Mingari, M. C., and Moretta, L. (1996). Receptors for HLA class-I molecules in human natural killer cells. *Annu. Rev. Immunol.,* **14,** 619–648.

Moretta, L., Ciccone, E., Mingari, M. C., Biassoni, R., and Moretta, A. (1994). Human natural killer cells: origin, clonality, specificity and receptors. *Adv. Immunol.,* **55,** 341–380.

Pantaleo, G., and Fauci, A. S. (1995). New concepts in the immunopathogenesis of HIV infection. *Annu. Rev. Immunol.,* **13,** 487–512.

Ray, C. A., Black, R. A., Kronheim, S. R., Greenstreet, T. A., Sleath, P. R., Salvesen, G. S., and Pickup, D. J. (1992). Viral inhibition of inflammation: cowpox virus encodes an inhibitor of the interleukin-1 beta converting enzyme. *Cell,* **69,** 597–604.

Romagnani, S. (1994). Lymphokine production by human T cells in disease states. *Annu. Rev. Immunol.,* **12,** 227–257.

Trinchieri, G. (1995). Interleukin-12: a proinflammatory cytokine with immunoregulatory functions that bridge innate resistance and antigen-specific adaptive immunity. *Annu. Rev. Immunol.,* **13,** 251–276.

ULEVITCH, R. J., and TOBIAS, P. S. (1995). Receptor-dependent mechanisms of cell stimulation by bacterial endotoxin. *Annu. Rev. Immunol.*, **13,** 437–457.

YORK, I. A., and ROCK, K. L. (1996). Antigen processing and presentation by the class I major histocompatibility complex. *Annu. Rev. Immunol.*, **14,** 369–396.

ZINKERNAGEL, R. M., and HENGARTNER, H. (1994). T-cell-mediated immunopathology versus direct cytolysis by virus: implications for HIV and AIDS. *Immunol. Today,* **15,** 262–268.

Bibliografia generale

ABBAS, A. K., LICHTMAN, A. H., and POBER, J. S. *Cellular and molecular immunology.* 2ª ed., Filadelfia, W. B. Saunders Company, 1994.

KUBY, J. *Immunologia.* Torino, UTET, 1995.

MALE, D., COOKE, A., OWEN, M., TROWSDALE, J., and CHAMPION, B. *Advanced immunology.* 3ª ed., Londra-Baltimora, Mosby, 1996.

ROMAGNANI, S., and MORETTA, L. *Immunologia e immunopatologia.* Torino, UTET, 1989.

Pietro Pala
Tracy Hussell
Peter J. M. Openshaw

Department of Respiratory Medicine
National Heart and Lung Institute
Imperial College of Science,
Technology and Medicine
St Mary's Hospital
Norfolk Place
London W2 1PG, United Kingdom

Defenses against Viral Infections

As the immune system has evolved to combat infections, so too have pathogens evolved to escape it. The form and structure of pathogens and the immune system are the products of this constant struggle for survival. During this coevolution, both pathogens and the immune system have found each other's strengths and weaknesses. This competition has resulted in an immune system with multiple layers of defenses. These include passive obstacles to pathogen invasion, active chemical compounds with antimicrobial activity, sophisticated strategies to identify non-self components (antibodies that selectively recognize specific molecular features of pathogens, cells that broadly react with foreign materials, and specific effector lymphocytes that can discriminate between closely related viruses), and an array of destructive devices to break down the molecular structure of pathogens or to stop their replication in infected tissues. Pathogens are met first by nonadaptive and later by adaptive immune defenses. These merge seamlessly and multiple processes can occur in parallel, both independently and in cooperative fashion. Many of the ways in which the immune system and pathogens interact are understood at the molecular and even atomic level. Despite these advances, infections still constitute major worldwide threats. The challenge for the future is to apply our knowledge of immunology to defeating these continued threats to health.

Introduction

Throughout our lives, the immune system defends us against repeated attacks by potentially lethal pathogens while tolerating microbes that offer no immediate threat. Our own cells are outnumbered by approximately 10:1 by microbes that thrive on our skin and in our bowels. It has been estimated that our immune system handles and eliminates an average of nine respiratory viral infections in the first year of life alone (Denny, 1995), each of which presents a new panel of antigens to be recognized by the immune system. During each year that life continues, further novel antigens are encountered. The effector mechanisms that eliminate these infections form an intricate network, each component of which links to others. The main components of this sequential system of immune defenses are summarized in

Fig. 1 and Table 1. Although this network is highly complex, the molecular basis on which individual components function is, to a large degree, understood at the molecular level.

Mucosal routes of infection are the preferred mode of entry for most viral pathogens. The threat of respiratory and intestinal infections to world health is inevitable. Breathing, drinking, eating, and sexual contact are largely unavoidable processes. These functions necessitate the exchange of materials with the environment or with other hosts. Any well-adapted pathogen is prone to exploit these processes to gain access to the internal milieu.

The skin and mucosae are not only ideal sites of invasion but also well suited to spreading pathogens to other hosts. Irritation of the nasal mucosa by common cold viruses causes mucus to be secreted, which contains abundant in-

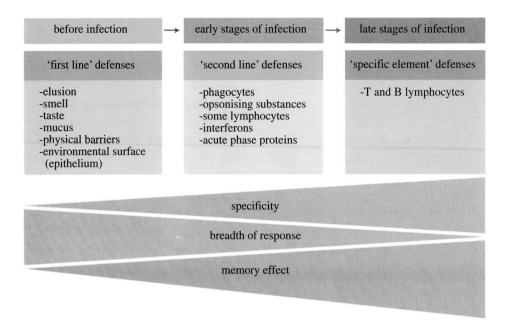

FIGURE 1 The sequential actions of the immune system. Immune defenses start to act before infection occurs. These first-line defenses are designed to reduce tissue invasion and depend on physical properties that many infectious agents have in common. As infection occurs, the responses become more specific. T and B lymphocytes, which are normally active during later phases of infection, have exquisite specificity; they can also be overaggressive and have the potential to make disease worse. The first- and second-line defenses are sometimes called innate, whereas T and B cell immunity is termed acquired. The specificity, level of response, and the memory effects of the cells involved in the various phases of the infection vary with the progression of the infection.

fectious virus. Sneezing and coughing cause secretions to be expelled. Particles of mucus settle on the mucous membranes of new hosts. The cycle of infection does not require the virus to spread to other tissues outside the respiratory tract. Transmission may indeed be most effective if the infected person is not made so ill as to prevent normal social contact. Agents that cause diarrhea and vomiting similarly assist their own spread by these effects, which are rarely lethal to the host. It can be argued that pathogens causing common colds and acute gastrointestinal symptoms have reached an ideal stage of stable evolution. To become either more or less pathogenic would be disadvantageous to the pathogen. Pathogens that do cause appreciable mortality may arguably be unstable "incomers" that have not yet evolved to a state of stable equilibrium with the host.

The World Health Organization estimates indicate that approximately 14 million people die each year from infections that are transmitted via the respiratory tract compared to 5 million from diseases transmitted by the gastrointestinal route. Most of these deaths occur in childhood. Viral infections of childhood are of key importance to childhood development and patterns of adult disease throughout the world. It is hoped that further advances in knowledge of defenses against infectious diseases will make them amenable to rational prevention and treatment, thereby reducing their toll on human health.

Defenses at the Surface

Viruses require living cells to replicate, and it is remarkable that almost all the cells that are visible on the skin are dead. Papillomaviruses require broken skin to enter and form a wart, and HIV cannot infect through intact skin. Although the surface of the respiratory and gastrointestinal tract contains live cells, most will die and be shed from the body within 24 h. This "wall of death" is undoubtedly the most important single defense against invasive infection.

Dead cells, however, are not the only defense at the surface. Nonpathogenic bacteria and fungi (the normal body flora) bathe the skin, upper airway, and digestive tract and compete with potentially harmful bacteria for attachment sites and nutrients. Reducing the density of this normal flora with antibiotics can increase the invasive potential of nonsusceptible microbes. This competitive occupancy is an important barrier to invasive infection.

In the respiratory tract, there is constant flow of mucus from the deep parts of the lung and from the nose to the pharynx, where mucus is normally swallowed or (in some cases) expectorated. This mucus sheet comprises islands of thick "gel" in a sea of mucus "sol." Mucus contains antibacterial and antiviral enzymes such as lysozyme (present in saliva, sweat, and tears). Sialic acid mucoproteins also adsorb onto viruses such as influenza A, reducing the ability of viruses to bind mucosal receptors. The skin also pro-

TABLE I			
Summary of Defenses against Pathogens			
EFFECTOR	**MECHANISM**	**REGULATION**	**COMMENTS**
Type I interferon	Activated by infection of host cells	Double-stranded RNA in most nucleated cells causes IFN-α and -β production	Viruses are powerful inducers and are often inhibited by interferons (see Fig. 2)
Macrophages	Engulfment of particles and initiation of inflammation	Opsonized particles are engulfed and digested and cytokines produced	Dengue virus, HIV, mycobacteria, and *Leishmania* all thrive in macrophages
NK cells	Kill virus-infected cells	Activated by decreased class I MHC expression and inhibited by KIR. Bind through the CD16 (IgG receptor) to antibody-coated cells	Cytomegalovirus evades NK killing by providing a class I analog (see Fig. 3)
Complement	Punches holes in membranes, contributes to neutralization by antibodies	Classical, alternative, and lectin pathway	HSV expresses complement receptors that enable it to infect certain cell types
Th	Cytokine and chemokine production: Class II restricted killing; enhancement of other responses	Recognition of exogenous (soluble) proteins in association with MHC class II molecules. Differentiate into cells which produce specific combinations of cytokines (see Fig. 6)	Any virus tends to stimulate Th responses. May be damaging in various infections, including measles and RSV. Depleted by HIV infection
CTLs	Kill virus-infected cells by perforin, granzyme, and Fas ligand. Also potent sources of TNF and IFN-γ (see Fig. 7)	Recognition of endogenously synthesized peptides together with MHC class I molecules	Probably important in clearing/controlling many infections (e.g., influenza, RSV, Epstein–Barr virus, and human papilloma virus). May have pathogenic role in hepatitis B and HIV infections
Neutralizing antibodies	Prevention of binding/entry	At mucosal surfaces (IgA) and in serum (IgM and IgG) (see Fig. 4).	Influenza and RSV infections can be prevented by specific, high-titer neutralizing antibodies
$\gamma\delta$T cells	Kill stressed cells. Cytokine production	Recognition of generic stress signals (such as MIC-A and MIC-B) induced in intestinal epithelial cells	Probable importance in metabolically independent intracellular infections

duces sebaceous secretions containing fatty acids, which are powerful antimicrobial agents. In the gut, the alternating low and high pH as peristalsis moves material from the stomach to the small intestine, the detergent and enzymatic properties of bile, and the other gut peptidases eliminate almost all infectious agents that enter the intestinal tract.

In the mucous membranes, a complex, extensive, and highly specialized immune system has evolved. The presence of foreign materials on the mucosal surfaces necessitates careful regulation of immune responses in order to avoid bystander tissue damage. The function of these superficial defenses is to protect against invasion by potential pathogens while tolerating nonpathogenic materials. By contrast, the internal immune system normally functions in

a sterile environment and reacts more vigorously to antigenic material.

Successful elimination of antigen from mucosal sites without bystander tissue damage depends on balanced immunological responses. If noninflammatory immune mechanisms are unsuccessful at eliminating the antigen, proinflammatory mechanisms will become dominant and immune damage will result. Such immune damage is responsible for some of the clinical manifestations that occur during viral infections.

Innate Immune Responses

Macrophages. Macrophages (so-called because they are "big cells that eat") are a key component of the innate immune system. They engulf viruses, and other pathogens which have been neutralized by circulating antibodies and need to be destroyed without damaging surrounding tissues. All macrophages are derived from bone marrow cells which become blood monocytes. These differentiate into macrophages in the lung, lymph node, spleen, skin, central nervous system (to form microglia), and liver (where they form Kupffer cells). Although their primary function is to engulf particulate matter, engulfment is improved if the particles are coated in antibody or complement (both of which function to "opsonize" the particle). This process of opsonization has been likened to adding sauce or gravy to food, making it more palatable and attractive. Once the macrophage has engulfed the particle, it digests the contents of the endocytic vesicle. Digestion products from the engulfed particle are then transported to the cell surface in association with major histocompatibility complex (MHC). If this peptide is not naturally occurring in the body, helper T cells are activated.

The process of engulfment starts with binding of opsonized material to Fc receptors on the macrophage membrane, followed by endocytosis into an internal vesicle called phagosome. This fuses with a vesicle containing lysozyme and other degradative enzymes, called lysosome; the resulting vesicle is called phagolysosome. In the phagolysosome, a combination of low pH maintained by ion pumps, acid hydrolases, hydrogen peroxide, reactive oxygen radicals, and nitric oxide break down the phagolysosome's content. Occasionally, the particle to be destroyed is too large to be phagocytosed (as in the case of a parasite). This results in the release of the lysosomal enzymes to the extracellular space, with the possibility of causing local damage to host cells. Macrophages are prominent in granulomas, inflammatory processes in which multiple effector cells (monocytes, lymphocytes, and polymorphonuclear cells) attracted by chemotactic factors released after tissue invasion by foreign particulate material, including pathogens, form a barrier around it. Typical examples are seen in tuberculosis, in which live mycobacteria can persist for decades inside a granuloma.

Although generally effective, many important microbes have developed methods to circumvent the antimicrobial effects of macrophages. First, certain viruses (notably Dengue and HIV) actually thrive within macrophages and indeed may use antiviral antibody to enhance their ability to infect macrophages. The macrophage may become ineffective and even act as a Trojan horse carrying the infection to other sites within the body. Some nonviral pathogens (e.g., *Leishmania* and many mycobacteria) are obligate intracellular pathogens and can replicate only once within cells.

The macrophage is not only efficient at digesting antigen and presenting components of foreign proteins. It is also a powerful producer of certain cytokines, including interleukins-1 and -6. An infected macrophage will therefore alert neighboring cells, including B cells, to the presence of foreign, potentially infectious material.

Natural Killer Cells and Interferons. A second component of the innate immune response is the natural killer (NK) cell system, which usually is linked with the actions of interferons (IFNs). The actions of IFNs are summarized in Fig. 2. Virus-induced IFN-α/β (type I interferon) is produced by a wide variety of virus-infected cells and enhances NK cell-mediated cytotoxicity. Vaccinia virus produces a secreted analog of the type I interferon receptor which inhibits the antiviral effect of IFN. The existence of this viral evasion mechanism suggests an important role for IFN in resistance to viruses. In addition to being released by NK cells, IFN-γ is released by activated T lymphocytes. It also induces an antiviral state (e.g., inducing the Mx gene in mice, which results in resistance to influenza virus) and stimulates antigen processing and presentation in a similar way to type I IFN. IFN-γ is produced by T cells infiltrating neuronal ganglia infected by herpes simplex virus and is thought to be involved in switching the virus from a lytic to a latent replication cycle.

NK cells differ from cytotoxic T cell in that they do not require the thymus for their differentiation. The mechanisms by which NK cells recognize and eliminate pathogens have remained largely mysterious until recently (Biron, 1997). NK cells are recruited within the first day or two of infection and divide and differentiate locally. They have important antiviral effects, in part because they kill infected cells by cytolytic granule formation and in part because they produce interferon-α, -β, and -γ. Many advances have rekindled interest in NK cells, particularly the characterization of many surface receptors displayed on NK cells and on the surface of potential target cells which regulate NK cell function. The first of these is the interaction between MHC class I on normal target cells and the killer inhibitory receptors (KIRs) on NK cells. Engagement in class I sends a negative signal to the NK cell inhibiting its lytic action. Altered or absent class I cannot stimulate this inhibitory signal. The NK cell is then activated by a signal through a second receptor, causing the NK cell to release its contents (Fig. 3). This directional release of granules causes apoptosis in the target cell. Any cells containing a virus that downregulates class I are prone to be killed by this mechanism.

Certain viruses, such as human and murine cytomegalovirus, counteract this mechanism by encoding an MHC class I homolog. Furthermore, normal uninfected cells respond to local release of interferons by upregulation of class I expression. This makes normally responsive cells resistant to NK killing. If the cell is infected with the virus, this normal upregulation may fail, causing the cell to be more susceptible to NK killing. This is one example of how killing inhibitory receptors operate, but the NK cell may test a variety of KIRs before a final decision is reached and

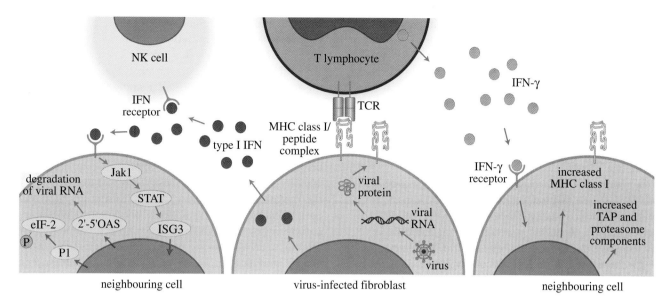

FIGURE 2 How interferons work. Viral nucleic acid, typically double-stranded RNA, and proteins introduced by viral infection cause production and release of type I interferon (IFN-α and -β). By binding to receptors on neighboring cells, IFN signals through the Jak1 cascade (tyrosine kinases of the Janus family), STAT (signal transducer and activator of transcription), and ISG3 (a gene induced by IFN), causing the production of many proteins, including the antiviral protein, 2′,5′-oligo-adenylate synthetase, which activates an endoribonuclease that degrades viral RNA, and the P1 kinase, which phosphorylates eIF-2 (a protein synthesis initiation factor) rendering it unavailable for viral replication. In addition, type I IFNs activate the NK cells. In the virus-infected cells, viral peptides are transported to the cell surface by MHC class I, recognized by the TCR of the lymphocytes which are activated and produce IFN-γ. This is recognized by the neighboring cells and expression of MHC class I, β2 microglobulin, TAP (transporter associated with antigen processing), and the proteasome components is enhanced, making viral infection more detectable by killer cells.

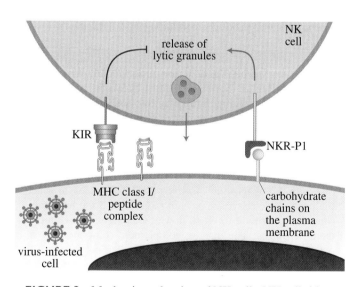

FIGURE 3 Mechanism of action of NK cells. NK cells identify virus-infected cells using two main families of receptors. Lectin-type receptors, such as NKR-P1, bind to carbohydrates on the cell membrane and send a signal for release of lytic granules by the NK cell. In contrast, killer inhibitory receptors (KIRs) of the immunoglobulin superfamily bind to MHC class I molecules and send a signal that inhibits the lytic action of the NK cells and prevents killing of the target cells.

the cytotoxic granules are deployed. Once specific local T cell responses are activated, additional mechanisms turn off NK cell functions. The activity of these potentially damaging cells is therefore of short duration but bridges the interval between initial infection and the development of specific immunity.

Acquired Immune Responses

Although innate effector responses are mobilized early and have little or no specificity for particular pathogens, acquired immune responses are delayed but highly specific. Antigen-specific lymphocytes proliferate and differentiate during this lag period, reaching numbers that are sufficient to form an effective defense. If a similar or identical pathogen has been previously encountered, the lag phase is shortened and the potency of the immune response is greatly enhanced. This phenomenon is termed immune memory.

At mucosal sites, lymphoid cells are concentrated into specialized aggregations called mucosal-associated lymphoid tissue. When these cells migrate, they preferentially seed back to mucosal sites. This "homing" behavior depends on the expression of surface receptors that specifically recognize vascular addressins displayed on high endothelial venules.

B Cells. One of the main cell types involved in acquired immunity is the B lymphocyte. B cells are so called because they develop in the bursa of Fabricius in chickens. In this species, bursectomy eliminates B cells and antibody production from the developing chick. There is no anatomic equivalent of the bursa in mammals. Mature B cells carry immunoglobulin on their surface, which acts as an antigen receptor. B cells are distributed by the blood and lymph to the lymphoid tissues, particularly the lymph nodes and spleen.

In mammals, the development of B cells starts with stem cells in the bone marrow, originally derived from the liver. These cells form a lineage through the pro-B cell, the pre-B cell, onto B cells proper, and finally to plasma cells. In order to move up this ladder of maturation, B cells respond to developmental stimuli (particularly IL-7 produced by stromal cells in the bone marrow). During ontogenesis, the genes that encode antibody molecules are recombined to produce unique light and heavy chains that make up the specific antigen receptor. The plasma cell is the final developmental stage, during which abundant antibody is secreted.

There are two main types of B cell response, the T cell dependent and the T cell independent. It has been proposed that the T cell-independent pathway is mainly responsive to repetitive antigens of the type displayed on the surface of microbes, and that this pathway is responsible for inducing potent antiviral responses. Such responses are more rapid, require no adjuvant, preferentially induce neutralizing responses, and are long-lived compared to the T cell-dependent responses (Bachmann and Zinkernagel, 1996). In contrast, T cell-dependent B cell responses (such as that to haptenated proteins or sheep red blood cells) depend on costimulation of B cells via B7.1/2 interactions with CD28 and CD40 interactions with CD40 ligand. Release of cytokines from T cells directed at B cells induces B cell proliferation, antibody production, and isotype switching.

Immunoglobulin moves on and off all proteins with which it makes contact and remains longer with a protein that has a surface with matching features which most exactly complement the surface shape, charge distribution, and hydrophobicity of the immunoglobulin. Where the spatial fit is very exact, the immunoglobulin remains fixed for a long enough period of time to trigger other immune responses. Therefore, there must be enough different immunoglobulins to recognize almost every possible existing and potential pathogenic protein. At the same time, strong self-recognition must be avoided.

The way in which this diversity of immunoglobulin is produced was one of the most intractable puzzles of immunology that was solved only a few years ago. Generation of diversity is achieved by combining many different gene segments that encode polypeptides (exons) that rearrange during development to produce hybrid heavy and light chains. There are many variable germline elements that can re-

combine and have a remarkable high rate of recombinatorial inaccuracy. Point mutations are frequent, and different heavy and light chains can associate to produce at least 10^{16} different shapes. This variability is concentrated in the antigen-binding domains of immunoglobulin. Although this native variability is not as great as that possessed by the T cell receptor, B cells have another very useful way of attaining specificity: They can undergo "somatic mutation." This is achieved by allowing errors to occur in DNA transcription, which occur in approximately 1 of every 1000 base-pair couplings. This rate is 100 times as fast as the mutation rate in RNA viruses, raising an interesting parallel between the capacities of the immune system and pathogens to microevolve. B cells with errors that improve binding to an antigen are rewarded by greater stimulation, growth, and division. However, if a good fit is achieved, the B cell expands and differentiates further. Cells that make antibodies that bind strongly to foreign particles proliferate, producing more of the antibody that finds an appropriate target.

The second component of immunoglobulin is the crystalizable fragment or Fc. For a given subclass of antibody this component is constant and faces away from the antigen. It interacts with complement or Fc receptors on cells, thus determining the effect of antigen engagement by the Fab component. The different types of immunoglobulin are named after the Fc component, and each has a distinct character. As an immune response develops, the B cells mature and undergo isotype switching. The same Fab encoded by VH genes will be successively associated with different CH gene products, producing IgM, IgG, etc (Fig. 4).

FIGURE 4 Mechanism of action of antibodies. (a) Neutralization: Antibodies that bind to the virus attachment proteins can block virus binding and entry into susceptible cells. (b) Opsonization: Immune complexes of virus particles and antibodies can attach through the antibody Fc moiety to cells bearing Fc receptors, such as macrophages, which engulf them by phagocytosis into the phagosomes. Lysosomes, containing digestive enzymes such as lysozyme and acid hydrolase, then fuse with the phagosomes forming phagolysosomes, in which the proteolytic enzymes, oxygen radicals, and the acid environment contribute to break down the contents. (c) Complement activation: The inactive complement component, composed of the subunits C1q, C1r, and C1s, can be activated by binding to a single IgM molecule or to at least two IgG molecules that have undergone a change in conformation after binding to antigen. This in turn activates C1q, which activates C1r, which activates C1s, starting the classical complement activation pathway. (d) Mast cell degranulation: Most IgEs are not freely circulating but are found bound to high-affinity Fc receptors on mast cells. Engagement of the surface-bound IgE causes the release of a variety of molecules, including histamine. (e) NK cell arming: Antibodies bind to CD16 (Fc receptors on NK cells), allowing these nonspecific effectors to acquire the antigen specificity of the antibody.

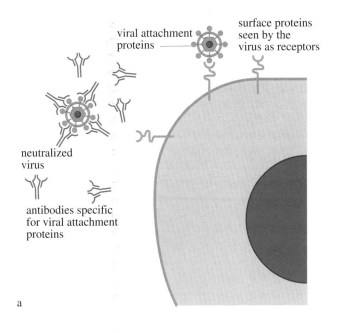

viral attachment proteins

surface proteins seen by the virus as receptors

neutralized virus

antibodies specific for viral attachment proteins

a

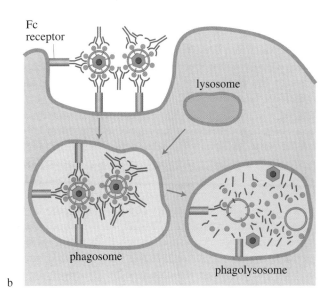

Fc receptor

lysosome

phagosome

phagolysosome

b

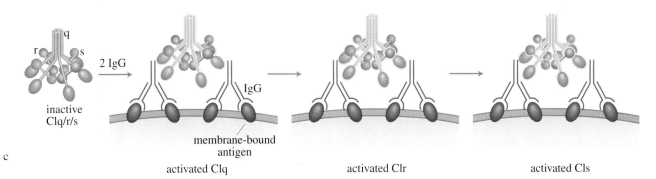

q

r

s

inactive Clq/r/s

2 IgG

IgG

membrane-bound antigen

activated Clq

activated Clr

activated Cls

c

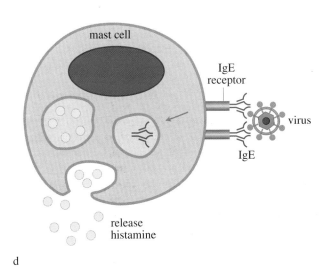

mast cell

IgE receptor

virus

IgE

release histamine

d

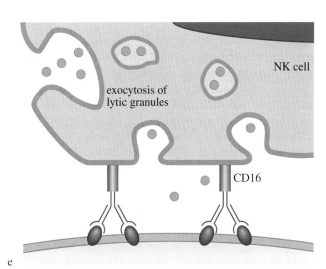

exocytosis of lytic granules

NK cell

CD16

e

IgG is the major serum immunoglobulin and is the main antibody that is found after immunization with most antigens. It is not actively secreted into the respiratory tract but can leak out from the circulation when the mucosa is inflamed. B cells that produce IgG often lie within the deeper layers of the mucosa. IgG is very potent at opsonizing antigen for engulfment by polymorphonuclear leukocytes (PMNs) and macrophages. Attachment is followed by engulfment (phagocytosis) and fusion of the internalized phagosome with lysosomes (vesicles containing degradative enzymes). Proteolytic enzymes, oxygen radicals, and acidification of the phagolysosome contribute to break down the phagolysosome contents. Monomeric IgM can be joined by J chains which bind secretory components on the basolateral surface of enterocytes. This released antibody complexes antigen, which is then excluded from the body's interior (immune exclusion). It is a very potent opsonic antibody and is present at the time of the first encounter of B cells with antigen. It is also a strong activator of complement components, which punch holes in cells. Inactive complement component (C1qrs) can be activated by binding to a single IgM molecule (or to at least two IgG molecules) that have undergone a change in conformation after binding to antigen. This in turn activates C1q, which activates C1r, which activates C1s, thus starting the "classical" complement activation pathway (Fig. 5). IgA is a specialized immunoglobulin present at low concentrations in serum, but it is very abundant in mucosal secretions. It is actively secreted into the lumen when bound by a secretory component. Examples of antiviral actions include prevention of poliovirus infection at the intestinal mucosa and prevention of respiratory tract infection by common cold viruses, including respiratory syncytial virus and influenza A. IgD is present at very low levels in serum, but it is present on the surface of many B cells early during differentiation. IgE is also present at very low concentrations in serum, but it is important because the Fc component binds very strongly to mast cells, basophils, and eosinophils. Engagement of this surface-bound IgE causes release of histamine, bradykinin, enzymes, leukotrienes, platelet-activating factor, etc. It is thought to be particularly important in the protection against infections by worms, but it also triggers allergic reactions in hay fever and asthma sufferers.

Antibody to viruses can be neutralizing, in which case antibody binding stops the virus from being infectious by physically blocking the sites on surface proteins critical to virus attachment and entry. Alternatively, it can bind other sites that do not interfere with virus binding. This nonneutralizing antibody can opsonize virus directly or lead to complement binding. Finally, some viruses seem to infect better in the presence of antibody, using opsonization to get into cells that are unable to destroy the virus. This is called infection-enhancing antibody.

No matter the type, antibody is often ineffective at preventing viral infections of the mucosal surface because it has the wrong specificity or is present in the wrong place or at the wrong time. If B cells have not encountered a given viral antigen before, those of the correct specificity will take time to expand. For example, the immune system may not have ever encountered the strain of influenza circulating in a particular year and antibody to other strains may be ineffective against the surface proteins of the new strain. One recent example was the 1997 Hong Kong outbreak of avian influenza H5N1. Some other species of cold viruses

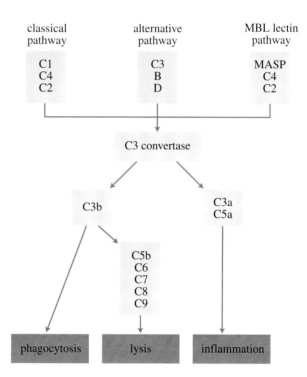

FIGURE 5 Complement is a collection of circulating factors that were originally identified as capable of "complementing" the antimicrobial activity of antibodies. The system is based on enzyme-based amplification of a trigger signal provided by the presence of pathogens. This signal generates membrane attack complexes that form pores on cell membranes and opsonize the pathogens. Three main activation pathways exist: the classical pathway, triggered by the interaction of the C1, C2, and C4 complement components; the alternative pathway, triggered by exposure on the pathogen surface of C3 convertase, produced by the interaction of the complement components C3, B, and D; and the lectin pathway, triggered by binding of mannose-binding lectins (MBL, MASP, C2, and C4) to carbohydrates on viruses or bacteria. The three pathways converge in the activation of C3 convertase, an enzyme which cleaves its substrates into functionally active fragments. This generates C3a and C5a (small fragments of C3 and C5 with inflammatory properties); C3b (large fragment of C3 which opsonizes the pathogen binding it to the complement receptor on macrophages); and C5b, which together with the fragments C6–C9 initiates the formation of the membrane attack complex, leading to cell lysis.

(i.e., rhinoviruses) are able to have 1 of approximately 100 different "coats," making the immune system treat each infection as a newcomer. In the case of respiratory syncytial virus, serum antibody at normal levels is ineffective against reinfection of the nose. Local IgA in the nose is effective but only lasts approximately 9 months after infection. Since outbreaks occur about once a year, this antibody is of little use in subsequent outbreaks.

Processing and Presentation of Viral Antigens to T Cells. T cells recognize antigens that have been "processed," that is, broken down into peptide fragments. These fragments are presented on the surface of normal host cells bound in specific clefts on surface proteins called MHCs. If the cells are infected with a virus, the novel endogenous peptides are sampled by the class I MHC and carried to the cell surface. In contrast, some specialized antigen-presenting cells (APCs) take up proteins from the extracellular space and digest these exogenous proteins for presentation by class II MHC. When T cells meet these novel MHC–peptide complexes they bind to them using the T cell receptor (TCR), the binding of which is stabilized by CD4 (which helper T cells use to bind to class II) or CD8 (which cytotoxic T lymphocytes (CTL) use to bind to class I). Individual mature T lymphocytes recognize a very limited number of antigens. They do this by matching the shape of the tip of the T cell receptor that projects from the surface of the T cell with a complementary shape of antigen presented by other host cells.

The most important of the specialized APCs are macrophages, dendritic cells (so named because of the branch-like projections that they cast out between other cells), and B lymphocytes. How the antigen enters APCs is important for deciding what type of T cell is stimulated by the antigen. Exogenous (soluble) antigen is taken up by APCs, particularly when opsonized by antibody or complement or when bound to antibody on the surface of B cells. Once internalized, the antigen is broken down in phagolysosomes. The peptides that result from this digestion meet with newly synthesized (or recycled) class II MHCs which are, until this point, accompanied by a protein called invariant chain. The invariant chain makes close contact with MHC II and projects into the cleft on the tip of the class II protein, which forms the binding site for peptides produced from digested antigen. Once the invariant chain moves off, the peptide binding site quickly takes up whatever peptides are available and which fit into the cleft. If no foreign antigenic peptides are available, MHC II takes up normal host cell peptides produced by the natural turnover of proteins in this compartment of the cell. Each MHC II protein can potentially hold one of several thousand different peptides. Although this seems a very large number, it is small compared to the potential number of peptides that are present in any given cell. The peptides are usually between 15 and 19 amino acids long and can project slightly from one end of the MHC II cleft. It is thought that about 100–200 MHC proteins need to be occupied by appropriate peptide for recognition to occur. This is normally approximately 0.1–10% of the MHC proteins present on the surface of an APC.

The pathway that presents endogenous antigens has some important differences from that presenting exogenous antigen. Endogenous antigens are synthesized within the cell and then broken down into peptides of 8–10 amino acids. These shorter peptides meet class I MHC protein as it is made in the rough endoplasmic reticulum, and the association of peptide with MHC I helps to stabilize the association with β2 microglobulin. This trimolecular complex is moved to the Golgi apparatus, where the proteins are glycosylated before display on the cell surface. These peptides fit into a cleft on the tip of MHC I in almost the same way that peptides fit into MHC II, except that the MHC I cleft is closed at both ends. Therefore, each MHC I has a more limited choice of peptides with adequate fit so that only a few hundred different peptides are displayed within MHC I clefts on the surface of typical cells. This more limited range is to some extent compensated by a much wider range of cells which bear MHC I. Most nucleated cells bear MHC I, including all the cells of the mucosal epithelium, APCs, fibroblasts, and vascular endothelium. In contrast, MHC II is only present in significant amounts on cells that are professional APCs. The two MHC proteins therefore function in different ways. Class I MHC shows what proteins the cell is making and invites cytotoxic T cells to kill the cell displaying the abnormal MHC–peptide complex. The class II MHC is present only on cells that are specialist "news reporters" that sample the proteins present in the extracellular fluid and trigger a coordinated immune response by helper T cells.

The ability of T cells to discriminate self from non-self develops in the thymus. After leaving the bone marrow, cells destined to become T cells migrate into the thymus. Here, they undergo a series of changes as they pass from the cortex to the medulla. The thymus is effectively a test bed designed to select T cells that have low reactivity to MHC displaying self-antigens but still capable of recognizing foreign peptides presented by self-MHC. Soon after arrival, new immigrants from the bone marrow display CD4 and CD8. When both these proteins are displayed along with TCR, the strength of interaction with APCs in the thymic stroma determines the fate of the developing T cell. Those that bind too strongly or too weakly are eliminated or go into a state of semipermanent inactivity (anergy). This process of thymic selection is very rigorous, and most developing T cells die within the thymus by apoptosis. Those T cells selected for being of future potential value to the host multiply and pass out from the thymus. They then circulate through the lymph nodes, spleen, liver, gut, lung, skin, and other sites searching for novel MHC–peptide complexes on the surface of host cells.

Once in the periphery, the opportunities for further change in the T cells' surface proteins are very few. These T cells are "committed" in that they have lost the ability to express other coreceptors and cannot change their TCR. They remain quiescent until they contact a cell that bears the characteristics that they were programmed to find and only then go into an activated state. The process by which T cells reach the APCs is not one of pure chance, although frequency must play a part when only very few lymphocytes have a T cell receptor that fits closely to any given MHC–peptide complex. At sites where infection is occurring, inflammation causes an increase in blood flow, more active extravazation of lymphocytes into the tissues, and an activation of APC that increases the amount of MHC protein on their surface. Once the APC meets a T cell with the correct T cell receptor, the T cell is driven into a state of activity and cell division. The process of lymphocyte development is therefore a finely controlled balance between growth, inactivity, activity, and death. All these features are essential for a properly functioning lymphoid system.

The development of lymphocytes to some extent parallels the developmental and adaptive ability of the pathogens that the immune system has grown to meet. The potential diversity of the TCR is immense. It has been calculated that the TCR genes are capable of joining up in different ways to generate approximately 10^{16} different TCR proteins, ensuring that one or more TCRs will be a near-perfect fit to any foreign protein. Many of these possibilities are rejected by the thymus because there is either a too strong or too weak interaction with normal host cells. In each of us, there has therefore been a microevolutionary race run by our lymphocytes before we were born which resulted in approximately 95% of the possible lymphocytes dying by apoptosis. Throughout life, lymphocytes continue to be selected, tolerated, or deleted. A relatively restricted range of T cells survive in the elderly, perhaps in part explaining the limited capacity of the immune system to respond to infections in old age.

Helper T Cells.

The presence of CD4 on the surface of a T cell causes greatly enhanced binding of the TCR to MHC II that bears the correct peptide. Once the virgin thymic emigrant makes contact with an appropriate APC, it proliferates and starts to make chemical signals called cytokines. Important examples of cytokines made by these helper T cells are IL-1, IL-3–IL-5, IL-10, and interferon-γ. These chemicals signal to other cells to prepare to attack the invading pathogen. IL-3 is a key factor that causes stem cells to grow and divide, making more B and T cells. IL-2 has a key role in T cell activation, whereas IL-4 and -6 activate B cells. IL-3 (in combination with granulocyte macrophage colony-stimulating factor is potent in promoting the growth and proliferation of phagocytic cells (including eosinophils, PMN, basophils, and macrophages). Helper T cells play a central role in making a precise iden-

tification of the invading pathogen and switching on the host "alarm system" which brings in other key players in immune defense.

Helper T cells can be classified into different types that produce characteristic cytokine combinations – some directing cell-mediated immune responses and others enhancing antibody production by B cells (Mosmann and Sad, 1996; Fiorentino et al., 1989; Street et al., 1990). Some antigens and modes of priming differentially induce two major functional subsets, Th1 and Th2 (Fig. 6). Th1 cells make type 1 cytokines, which include tumor necrosis factor-α (TNF-α) and IFN-γ. These have potent antiviral effects. Type 2 cytokines include IL-4, IL-5, and IL-10 – cytokines prominent in antihelminthic and allergic responses. Populations of cells making type 1 and type 2 cytokines are mutually inhibitory. Although polarized cytokine production has been seen in T cell clones from several species (both primate and rodent), the degree of polarization in natural T cells is only now becoming clear.

Cytotoxic T Cells.

The endogenous processing pathway leads primarily to activation of T cells that bear CD8 (which mostly function as cytotoxic T cells). Again, the process of engaging the correct T cell is essentially one of finding a T cell receptor with exactly the correct shape to fit over the peptide–MHC complex. Once this union has been made, the presence of CD8 on the surface of the T cell locks the T cell onto the APC that bears the appropriate peptide. The response of CD8$^+$ T cells is again one of activation (Fig. 7).

Although they also produce cytokines (mainly type 1), the main result of this activation is that the cells become cytotoxic. The activated CTL reorients its cytoskeleton and secretory apparatus toward the target cell, and granules containing perforin and granzymes are released in the membrane pocket formed between the two cells. Perforin undergoes a Ca^{2+}-dependent polymerization and inserts into the lipid bilayer of the target cell, forming polyperforin-delimited pores. The resulting inflow of water and ions causes the target cell to burst. The cell then breaks down its own DNA and RNA and dies. The T cell detaches from the dying cell, undergoes cell division, and moves off to search for other cells to engage and kill. It is important that cytotoxic T cells are very particular in their choice of antigen. Uncontrolled killing of cells mistakenly recognized as bearing virus would be disastrous.

Another cytotoxic mechanism (the main one used by CTLs that do not have granules) consists of activation-induced expression of the Fas ligand (FasL). FasL binds to Fas receptors present on the target cell membrane, bringing them into close proximity. This triggers the apoptosis (programmed cell death) mechanism, resulting in cellular DNA fragmentation and death. CTLs detach from the target cell after delivery of the lethal hit and can kill multiple cells in rapid succession.

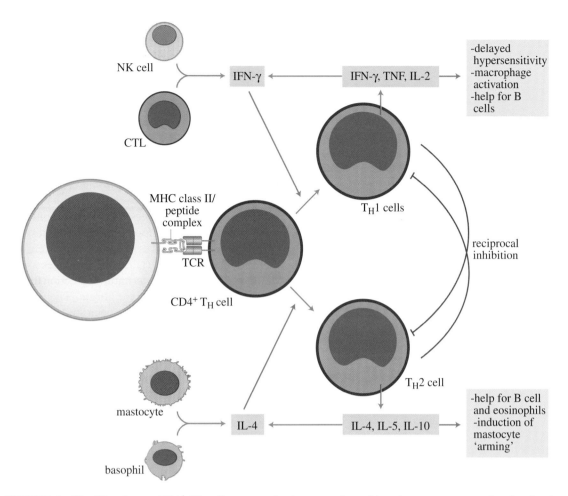

FIGURE 6 The Th subsets. CD4$^+$ Th cells can synthesize several cytokines in response to antigenic stimulation. According to the type of cytokine expressed, Th cells are divided into two main subsets, Th1 and Th2. Th1-type cells are characterized by production of IFN-γ, TNF, and IL-2. They are involved in delayed-type hypersensitivity responses, cause macrophage activation, and provide help to B cells producing antibodies capable of binding complement. IFN-γ, produced by the Th1 cells, NK cells, and CTL, induces the Th cells to differentiate into Th1. Th2-type cells specialize in IL-4, IL-5, and IL-10 production. They are able to support B cells producing most classes of antibodies, resulting in mast cell arming, support eosinophils, and can inhibit cytokine release from macrophages. These cells, like mastocytes and basophils, produce IL-4 which induces the Th cells to differentiate into Th2. The Th1 and Th2 cells inhibit each other reciprocally. Th1-type cells are most capable of hampering replication of many viruses through the activity of IFN-γ, TNF, and CTL. Th2-type cells are less effective and possibly detrimental if dominantly induced, as in RSV infection. Regulation of the Th1 and Th2 responses is complex and not fully understood.

Examples of Immune Responses to Particular Pathogens

Respiratory Syncytial Virus. Pulmonary infections are the leading causes of morbidity and mortality in children. Respiratory syncytial virus (RSV) is a major worldwide childhood respiratory pathogen and an unsolved challenge for vaccine development (Heilman, 1990). Clinically, infection is characterized by symptoms and signs of bronchial narrowing, and many children who recover from bronchiolitis are subsequently diagnosed as asthmatic. Immune defenses against RSV are interesting for several reasons.

First, reinfection occurs throughout life despite apparently strong immune responses to primary infection. Second, vaccination is capable of enhancing the severity of disease. Third, there are apparent links between childhood infections and the later development of asthma.

RSV belongs to the genus *Pneumovirus* in the family Paramyxoviridae, bearing close similarity to Marburg, measles, canine distemper virus, mumps, and parainfluenza viruses. Electron microscopy shows irregularly shaped and often clumped virions, with a lipid envelope bearing surface glycoproteins G, F, and SH. The nucleocapsid contains

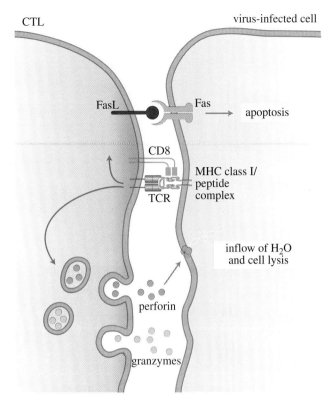

FIGURE 7 CTL effector mechanisms. Cytotoxic T lymphocytes bind to cells presenting peptide fragments (in green, bound to MHC) of viral proteins generated during infection. The binding of TCR to MHC class I–peptide complex activates the CTL. The interaction is further strengthened by accessory, non-antigen-specific molecules such as CD8, which binds to a nonpolymorphic part of MHC class I. The activated CTLs release granules containing perforin (in orange) and granzymes (in pink) into the space between the two cells. The perforin inserts itself into the lipid bilayer of the target cell, forming polyperforin delimited pores; the resulting inflow of water causes the target cell to burst. Another cytotoxic mechanism of activated CTLs is represented by the expression of Fas ligands (FasL); binding of FasL to Fas on the target cell triggers apoptosis.

a single-strand negative-sense RNA genome of 5×10^6 kDa which is nonsegmented. There are 10 genes, with 12 potential gene products. Sequential transcription occurs from $3'$ to $5'$; the first genes to be transcribed are 1c and 1b, which encode nonstructural proteins of unknown function. Then follows the nucleoprotein (N), which is relatively well conserved between natural isolates. The phosphoprotein (P) and small hydrophobic (SH) are transcribed next, followed by the attachment protein (G) and the fusion protein (F). These two proteins are the major surface glycoproteins, against which neutralizing antibody is directed and which show the most natural variability between different natural isolates of RSV. The last two proteins are the M2 (second

matrix or 22-kDa protein) and the large protein L, the latter being the RNA polymerase. Both are thought to be relatively well conserved.

Studies in man and in animal models have shown the dual role of antiviral T cells in eliminating virus and causing enhanced disease. This immunopathological paradox is more clearly understood for RSV disease than for that caused by any other common human infection. The first evidence that specific immunity could be harmful was derived in the 1960s, when children were vaccinated with formalin-inactivated RSV. Vaccine recipients developed strong serological responses but were not protected against infection. Most vaccinees who subsequently became infected with RSV developed severe lower respiratory tract disease, and some died as a result. Postmortem examination of formalin-inactivated (FI) RSV vaccinees who developed enhanced pathology upon natural infection documented an eosinophilic inflammatory infiltrate. The reasons for vaccine-augmented disease have been studied, but no safe and effective vaccine has been produced. Immune responses to RSV have recently been reviewed (Openshaw, 1995).

Human natural immunity following infection is short lived and reinfections occur throughout life. After the disappointing results of early trials with FI-RSV, alternative vaccine candidates were proposed that attempt to induce protective responses in the respiratory mucosa. These include attenuated RSV strains, peptide or subunit vaccines formulated in adjuvants that boost mucosal responses, recombinant adenovirus, or vaccinia viruses expressing RSV proteins F and/or G. To date, it is unclear whether any of these will be effective in humans.

In the mouse model of RSV disease, the majority of lymphocytes recovered during the first 5 days of primary infection are CD4$^-$ (null) in phenotype. Most have characteristics of NK cells, the activity of which peaks at about this time. During elimination of the virus from the lungs (Days 6–9), the main single subset is CD8$^+$, although CD4$^+$ cells are also found. In this model, induction of distinct forms of immunopathology can be explained by production of type 2 cytokines from T cells that recognize the major surface glycoprotein G and type 1 cytokines by T cells specific to the fusion protein and second matrix proteins. The most convincing proof that disease augmentation is caused by T cells is derived from studies of passive transfer of immunity. Mice infected with RSV alone show only mild illness, but disease severity is increased dramatically by injection of T cells that recognize RSV. Th1 cells or CD8 CTL can cause lung hemorrhage (reminiscent of shock lung), whereas CD4 T cells making type 2 cytokines cause lung eosinophilia (Alwan *et al.*, 1994). In contrast, antibody transfer never causes enhanced disease *in vivo* but can sometimes protect against infection.

Various studies have suggested a link between RSV bronchiolitis and asthma. Sigurs *et al.* (1995) performed a prospective cohort study of RSV bronchiolitis in infancy and found it to be a major risk factor for the development of asthma and atopic disease. It is not clear from these studies that bronchiolitis operates as an independent risk factor, but it did seem to be an additional factor that operates independently of family history of atopy.

The type of immune response that occurs upon initial encounter with RSV depends on several factors, including genetic influences, immunological status, the presence of ongoing immune responses to other antigens, and the presence of memory of similar antigens.

HIV and AIDS.

Infection with HIV is generally not cleared by the immune system. Upon entry into the body, HIV rapidly establishes reservoirs of virus that are effectively outside the reach of immune clearance mechanisms: (i) latent viral infection of quiescent $CD4^+$ T cells — these will generate new virus upon T cell activation; (ii) antibody- and complement-coated virions trapped in the follicular dendritic cell network in lymph nodes — these virions are not neutralized (on the contrary, they are highly infectious and concentrated at obligatory recirculation sites for $CD4^+$ T cells); and (iii) escape variants that have lost antigenic epitopes. In addition, virus genome instability is high and virus variation occurs throughout the infection and at different anatomical sites, providing an ever-changing target for immune effector mechanisms. As a result, many immunological processes are activated during the infection, although they fail to eradicate the virus.

Following entry of HIV into the body, infection rapidly spreads to cells expressing CD4 and the chemokine receptors CCR5 or CXCR4 (Hill *et al.*, 1997). Cell-mediated and antibody responses follow the first massive wave of viremia which results from viral replication. The number of circulating $CD4^+$ T cells decreases at the peak of viremia, consistent with their being the main host cell for HIV replication. The peak of CTL responses is followed by a decrease in viremia and return of $CD4^+$ cells to nearly normal levels, suggesting that CTLs are capable of at least temporarily controlling HIV. Neutralizing antibody responses appear later, slowly climbing to a plateau level. This "acute" phase of the infection lasts approximately 3 months and is followed by a "chronic" phase, during which CTL responses remain at a high level, antibody levels are high, circulating virus is low, and $CD4^+$ T cells slowly decline. This phase can last several years. It is characterized by constant high turnover of virus, which in turn erodes the number of circulating $CD4^+$ cells (Ho *et al.*, 1995). Eventually, a catastrophic transition to AIDS occurs in a majority of infected individuals, when the number of circulating $CD4^+$ T cells decreases below a critical threshold, CTL and antibodies become depleted, and the host eventually succumbs to opportunistic infections or AIDS-related malignancies.

Resistance to infection or to progression to AIDS is seen in individuals with genetic variants of the viral co-receptor molecules (Liu *et al.*, 1996). This is not due to immune effector mechanisms, even if the chemokine receptors involved have a role in immune responses, but rather simply reflects poor binding to cells. Other factors involved in slow progression to AIDS could be unidentified protective responses and "weak" infectious inocula, or this could simply represent one end of a spectrum of progression rates.

CD4$^+$ Cells. The disappearance of $CD4^+$ cells is probably due to a combination of direct viral cytopathic effects and killing by HIV-specific CTLs that recognize HIV peptides expressed with class I molecules on the membrane of infected $CD4^+$ cells. Even before a decrease in $CD4^+$ cell number is evident, immune responses to some other stimuli appear reduced. With time, responses to alloantigens and eventually mitogens are lost. Earlier loss of Th1-type responses (IFN-γ and IL-2 production) and preferential retention of Th2-type responses (IL-4 and IL-10 production) during progression to AIDS has been reported. This might be due to the fact that Th1 cells are more susceptible to HIV killing than Th2 cells or to HIV-mediated positive induction of Th2 cells. $CD4^+$ cytotoxic cells (recognizing gp120 peptides presented by class II molecules) have also been reported, but their importance is not clear.

CD8$^+$ Cells. Levels of circulating HIV-specific CTL are extremely high (approximately 1% of lymphocytes) throughout the chronic phase (Moss *et al.*, 1995), to the extent that it is possible to assay their activity directly in *ex vivo* cytotoxicity assays. Multiple CTL clones recognizing different viral antigens are usually present and complex dynamics of clonal expansion of $CD8^+$ cells and virus variant load have been described (Nowak *et al.*, 1995).

In addition to cytotoxic activity, CTL may release IFN-γ (direct antiviral effect) and chemokines such as RANTES, MIP-1α, and MIP-1β, which can compete with HIV for the coreceptors used to infect $CD4^+$ cells. This competition hinders entry of free virus into new host cells, although it is not a direct effector mechanism and is unable to eliminate the infection.

Antibodies. Antibody responses develop more slowly than CTL, perhaps due to dependence on help from $CD4^+$ cells. The main neutralizing antibodies are directed against the viral envelope glycoprotein gp120, particularly against a very variable loop region called V3. The V3 loop has a ratio of coding/noncoding mutations which suggests it is under a selective pressure, possibly a combination of antibody neutralization and the availability of many alternative coreceptors. The exact mechanism of neutralization is not clear. Not all anti-V3 loop antibodies are neutralizing. Neutralizing antibodies recognizing the V3 loop do not prevent binding to CD4 but could be blocking a postbinding step of virus entry. Another possible mechanism is inhibition of cleavage of the envelope glycoprotein (a protease-mediated step required for infectivity, which turns the gp160 env precursor into gp120 and gp41). Other neutralizing antibodies recognize the CD4-binding domain on gp120 and can cross-neutralize viral strains with different V3 loops. Other regions of gp120 are also targets of neutralizing antibodies, which could be acting by conformational effects or by interfering with binding to coreceptors.

In experimental systems in chimpanzees, high levels of neutralizing antibodies recognizing the envelope glycoprotein can protect against HIV infection, but these levels cannot be sustained. Furthermore, virus envelope variation can easily side-step such protection. In clinical trials of envelope-based vaccines, the antibodies induced could not neutralize primary isolates, showing the limitations of strain-specific neutralization.

Antibody-Directed Cellular Cytotoxicity. Antibody-directed cellular cytotoxicity (ADCC) mediated by antibodies recognizing gp120 and gp41 envelope glycoproteins and NK cells has been described. Its role in controlling HIV replication depends on the activity of NK cells because antibodies are present at high titers during most of the chronic phase. This activity may be less useful than CTL cytotoxicity, or it may even be detrimental since it could cause lysis of uninfected CD4$^+$ T cells that have bound released gp120 or B cells which express membrane antibodies recognizing gp120.

Enhancing Antibodies. Circulating complexes of HIV and antibodies are infectious and are not destroyed by macrophages. In fact, antibodies can enhance HIV infectivity. This occurs through direct binding of immune complexes to cells expressing the Fc receptor (mainly CD16 on NK cells and FcR1 on monocytes) or (for complement-mediated enhancement) to cells expressing CR2. Interestingly, enhancement is demonstrable with neutralizing sera that have been diluted past the titer required for neutralization. Thus, waning antibody levels with disease progression might precipitate progression to AIDS. Also, antibodies that neutralize one strain of HIV may enhance another. This effect may contribute to the selection of resistant variants and highlights a potential problem for vaccination aimed at induction of antibody-mediated immunity.

NK Cells. The activity of NK cells appears to decline with progression to AIDS, but addition of IL-12 to *in vitro* assays of NK function can restore activity.

Can the Immune System Protect against HIV? Although the unaided immune system seems at best only capable of slowing the progression to AIDS, it might still be possible to counteract the variation and kinetic advantage of HIV by preventive immunization. Several reports have described HIV-specific cellular responses in HIV-exposed, seronegative individuals, although it has been difficult to show that clearance of a virulent HIV inoculum had indeed occurred. Protection against HIV infection has been achieved in vaccinated chimpanzees and in macaques against infection with the related simian immunodeficiency virus (Daniel *et al.,* 1992). The recent success of antiviral combination therapy in lowering the viral load in HIV-infected individuals suggests that early, aggressive antiviral treatment after infection might prevent immune depletion and allow solid protective immunity time to develop and eventually control residual viral replication (Autran *et al.,* 1997).

Herpes Simplex Virus Infections

Herpes simplex virus (HSV) has a linear double-stranded DNA encoding 70 or more viral proteins. The genome is contained in a core (the capsid), that is surrounded by an amorphous material termed "tegument," which is covered by a lipid envelope studded with glycoproteins. There are two types, HSV-1 and HSV-2, with the latter usually associated with genital infection.

HSV causes a wide spectrum of diseases, ranging from common herpes labialis or cold sore, gingivostomatitis, genital herpes, keratoconjunctivitis, visceral infections of the immunocompromised host, and rare herpes encephalitis. In the United States, by 50 years of age 60–80% of the population has been infected with HSV-1 and 20% with HSV-2.

HSV infects by contact with mucosal surfaces or abraded skin. Infection is followed by replication in epithelial cells at the site of entry. At this time, HSV enters the local endings of sensory neurons and is transported by retrograde axonal flow to dorsal root ganglia, where it undergoes limited transient replication. Neuronal infection is usually controlled (although lethal central nervous system infection can occur in immunocompromised individuals) and the virus becomes latent, i.e., the genome persists in neurons as an episome, whereas no viral proteins are synthesized and only latency-associated transcripts indicate viral gene expression. Stressors such as UV exposure, emotional stress, fever, other infections (e.g., pneumococcal pneumonia), immunosuppression, or section of a peripheral nerve may be followed by reappearance of virus in the sites innervated by the infected neurons. Here, the virus enters a lytic replicative cycle in epithelial cells, producing vesicles and ulcers. Transmission to another individuals occurs at this stage.

Individual variation in frequency of reactivations is large, from 1 reactivation every few years to 10–20 reactivations per year. Frequency of recurrent (reactivated) infection might depend on initial virus load as well as host factors, including the immune response. HSV could potentially spread from its mucocutaneous entry site through sensory neurons to cause a fatal encephalitis, but most hosts resist disseminated HSV disease. A functional immune system is essential to this defense.

Infection in the naive host causes local inflammatory responses at the mucocutaneous site of replication. Local production of type I interferon (IFN-α and IFN-β) and NK cells are required for early containment of epithelial spread of the virus. In addition to direct antiviral effects, interferons increase expression of MHC class I and viral antigen presentation in keratinocytes, providing a better target for CTL. However, these initial effector mechanisms limit but do not prevent infection of neurons and establishment of latency. Clearance of the local infection requires T lymphocytes.

Experiments in the mouse model of HSV infection have revealed the complexities of interaction between genetic factors in the host and CD4$^+$ and CD8$^+$ cell responses. Generally, clearance of HSV from the primary site of infection depends on CD4$^+$ T cells, whereas CD8$^+$ T cells restrict neuronal spread. Development of HSV-specific CD8$^+$ T cells depends on CD4$^+$ cells in certain, but not all, strains of mice. Selective whole animal depletion of CD4$^+$ or CD8$^+$ T cells has limited effects on clearance of HSV from the primary lesion, whereas treatment of mice with antibodies that neutralize IFN-γ (which is secreted by both T cell subsets) extends the duration of local infection.

CD8$^+$ T cells infiltrate neuronal ganglia infected by HSV and appear to be essential to control neuronal infection since treatment with anti-CD8 antibodies results in greater neuronal destruction (Simmons and Tscharke, 1992). The mechanism of action of HSV-specific CD8$^+$ T cells in the infected ganglia is nonlytic: In fact, they increase the number of neurons that contain viral antigen. This may be because expression of class I molecules, normally absent in neural tissue, is blocked at a posttranscriptional level in neurons. However, CD8$^+$ cells recognize viral peptides produced in an infected cell and presented by that cell's class I molecules. Infection enhances class I expression in the ganglia, but only in satellite and Schwann cells (Pereira *et al.,* 1994). Unlike neurons, satellite cells are not permissive for HSV replication, but abortive infection may still occur and allow presentation of viral products synthesized in the initial part of the replicative cycle. Thus, the sequence of events would be as follows: Lytic infection of neurons releases virus which infects satellite cells abortively but stimulates their expression of class I molecules. These present viral peptides to infiltrating CD8$^+$ cells attracted by chemokines released during the initial stages of infection. The activated CD8$^+$ T cells then release IFN-γ, which acts on newly infected neurons and switches the virus replicative cycle from a lytic mode to the establishment of latency. MHC class I expression disappears during the latent phase, when minimal transcription of viral genes occurs and the state of immune activation declines.

Latency can persist for months or years, until appropriate stimuli cause reactivation of the latent virus. Viral protein synthesis then resumes and infectious virus travels back along peripheral sensory fibers to reappear at the mu-

cocutaneous site that they innervate. Here, viral replication may cause clinically evident cytopathic effect, appearing as clear vesicles and ulcers. In approximately 50% of reactivations, however, virus shedding is asymptomatic, thus facilitating transmission.

Various acquired immune mechanisms are primed at this stage so that recurrences are shorter lived than primary lesions. Neutralizing antibodies against envelope glycoproteins (mainly gD and gB) limit the spread of infection by free virus. HSV has evolved countermeasures against antibodies and complement. The envelope glycoproteins gE and gI are actually Fc receptors that will bind monomeric IgG, whereas gE on its own can bind aggregated IgG. This appears to protect virions from neutralization. Similarly, gC has partial homology to the complement receptor CR1 and binds C3b. This may actually help HSV to attach to certain cell types.

Recurrent infection lesions are infiltrated by mononuclear cells comprising both CD4$^+$ and CD8$^+$ lymphocytes. CD4$^+$ cells predominate in early lesions. HSV-specific CD4$^+$ and CD8$^+$ T lymphocytes are present at low frequency in PBMC. Cloning of T cells isolated from patients with recurrent lesions has shown that several different HSV antigens are recognized by CTL, both structural and nonstructural virus products. These studies used lymphoblastoid B cell lines infected with HSV as target cells. When fibroblast cell lines were used as targets, CTL-mediated lysis appeared less efficient. This led to the discovery that HSV contains a gene (referred to as α47 and coding the ICP47 product) whose expression can retain MHC class I molecules inside the cell by interacting with the transporter associated with antigen processing and inhibiting peptide transport to the endoplasmic reticulum (Tomazin *et al.,*; York et al., 1994). Thus, HSV can avoid or limit CTL-mediated clearance.

CD4$^+$ T cell clones isolated from HSV lesions are specific for multiple viral antigens, including the abundant tegument component VP16 and envelope glycoproteins gB, gC, and gD. The different roles of Th1 and Th2 cells have been studied in mouse models of herpetic stromal keratitis (HSK) (Thomas and Rouse, 1997). Eye infection by HSV is a major infectious cause of blindness in the developed world, and HSK is thought to be a T cell-mediated immunopathology since T cell-deficient animals are protected from HSK and HSK develops after virus titers are already declining but HSV-specific T cell responses are developing in the cornea.

Since mouse strain- and virus strain-related genetic factors are involved in HSK, it is not surprising that different effector cells appear to cause HSK in different model systems. In the BALB/c mouse infected with KOS or RE strains of HSV-1, HSK is mediated by TNF and IFN-γ secreting CD4$^+$ cells, but in A/J mice CD8$^+$ CTLs are the effectors of HSK. Anti-IFN-γ antibodies abrogate HSK, and infiltration of the cornea by Langerhans cells (as a result of inflammation) is required for HSK because MHC class II is not expressed by corneal cells. Other mechanisms involving antibody and complement-mediated damage, and antibody and NK cell-mediated ADCC, have been described. To highlight the complexity of these mechanisms, HSK can also be induced by a long-term CD4$^+$ clone specific for gD which produces IL-4 (i.e., a Th2 clone).

No effective vaccine against HSV exists. Neutralizing antibodies specific for envelope glycoproteins gD and gB can protect and improve recurrent disease in the guinea pig model and are candidates for vaccine development. Other promising approaches include attenuated or genetically "crippled" live virus and DNA immunization. A vaccine that completely prevents HSV infection may be unattainable, but induction of combined antibody and cell-mediated immunity at the mucosal level would probably limit HSV initial replication and the establishment of latent infection. Such a vaccine would result in milder or asymptomatic recurrent disease.

Conclusions

The immune response is essential in defense against viral infections but is capable of causing harm to the host. Without any immune response, the viruses would spread in an uncontrolled way. This, it can be assumed, would be bad for the host. When viral proteins are synthesized, they reduce the ability of the cell to synthesize normal cellular proteins. It is notable, however, that some viruses are not particularly damaging to cells in the absence of an immune response. It can be argued that the immune response is not only good and bad for the host but also good and bad for the virus. The inflammation which results from the immune response is often helpful to the virus in reaching a new host. Since the virus can only survive by spread to new hosts, the immune response is an essential assistant to many viruses.

Acknowledgments

We thank our current and past coworkers in the respiratory unit for their invaluable contributions and The Wellcome Trust for their continued support.

References Cited

ALWAN, W. H., KOZLOWSKA, W. J., and OPENSHAW, P. J. M. (1994). Distinct types of lung disease caused by functional subsets of antiviral T cells. *J. Exp. Med.* **179**, 81–89.

AUTRAN, B., CARCELAIN, G., LI, T. S., BLANC, C., MATHEZ, D., TUBIANA, R., KATLAMA, C., DEBRE, P., and LEIBOWITCH, J. (1997). Positive effects of combined antiretroviral therapy on CD4$^+$ T cell homeostasis and function in advanced HIV disease. *Science* **277**, 112–116.

BACHMANN, M. F., and ZINKERNAGEL, R. M. (1996). The influence of virus structure on antibody responses and virus serotype formation. *Immunol. Today* **17**, 553–558.

BIRON, C. A. (1997). Activation and function of natural killer cell responses during viral infections. *Curr. Opin. Immunol.* **9**, 24–34.

DANIEL, M. D., KIRCHHOFF, F., CZAJAK, S. C., SEHGAL, P. K., and DESROSIERS, R. C. (1992). Protective effects of a live attenuated SIV vaccine with a deletion in the nef gene. *Science* **258**, 1938–1941.

DENNY, F. W., JR. (1995). The clinical impact of human respiratory virus infections. *Am. J. Respir. Crit. Care Med.* **152**(Suppl.), S4–S12.

FIORENTINO, D. F., BOND, M. W., and MOSMANN, T. R. (1989). Two types of mouse T helper cell. IV. Th2 clones secrete a factor that inhibits cytokine production by Th1 clones. *J. Exp. Med.* **170**, 2081–2095.

HEILMAN, C. A. (1990). Respiratory syncytial and parainfluenza viruses. *J. Infect. Dis.* **161**, 402–406.

HILL, C. M., DENG, H., UNUTMAZ, D., KEWALRAMANI, V. N., BASTIANI, L., GORNY, M. K., ZOLLA, P. S., and LITTMAN, D. R. (1997). Envelope glycoproteins from human immunodeficiency virus types 1 and 2 and simian immunodeficiency virus can use human CCR5 as a coreceptor for viral entry and make direct CD4-dependent interactions with this chemokine receptor. *J. Virol.* **71**, 6296–6304.

HO, D. D., NEUMANN, A. U., PERELSON, A. S., CHEN, W., LEONARD, J. M., and MARKOWITZ, M. (1995). Rapid turnover of plasma virions and CD4 lymphocytes in HIV-1 infection. *Nature* **373**, 123–126.

LIU, R., PAXTON, W. A., CHOE, S., CERADINI, D., MARTIN, S. R., HORUK, R., MACDONALD, M. E., STUHLMANN, H., KOUP, R. A., and LANDAU, N. R. (1996). Homozygous defect in HIV-1 coreceptor accounts for resistance of some multiply-exposed individuals to HIV-1 infection. *Cell* **86**, 367–377.

MOSMANN, T. R., and SAD, S. (1996). The expanding universe of T-cell subsets: Th1, Th2 and more. *Immunol. Today* **17**(3), 138–146.

MOSS, P. A., ROWLAND, J. S., FRODSHAM, P. M., MCADAM, S., GIANGRANDE, P., MCMICHAEL, A. J., and BELL, J. I. (1995). Persistent high frequency of human immunodeficiency virus-specific cytotoxic T cells in peripheral blood of infected donors. *Proc. Natl. Acad. Sci. USA* **92**, 5773–5777.

NOWAK, M. A., MAY, R. M., PHILLIPS, R. E., ROWLAND-JONES, S., LALLOO, D. G., MCADAM, S., KLENERMAN, P., KÖPPE, B., SIGMUND, K., BANGHAM, C. R. M., and MCMICHAEL, A. J. (1995). Antigenic oscillations and shifting immunodominance in HIV-1 infections. *Nature (London)* **375**, 606–611.

OPENSHAW, P. J. M. (1995). Immunopathological mechanisms in respiratory syncytial virus disease. *Sem. Immunopathol.* **17**, 187–201.

PEREIRA, R. A., TSCHARKE, D. C., and SIMMONS, A. (1994). Up-regulation of class I major histocompatibility complex gene expression in primary sensory neurons, satellite cells, and Schwann cells of mice in response to acute but not latent herpes simplex virus infection in vivo. *J. Exp. Med.* **180**, 841–850.

SIGURS, N., BJARNASON, R., SIGURBERGSSON, F., KJELLMAN, B., and BJÖRKSTÉN, B. (1995). Asthma and immunoglobulin E antibodies after respiratory syncytial virus bronchiolitis: A prospective cohort study with matched controls. *Pediatrics* **95,** 500–505.

SIMMONS, A., and TSCHARKE, D. C. (1992). Anti-CD8 impairs clearance of herpes simplex virus from the nervous system: Implications for the fate of virally infected neurons. *J. Exp. Med.* **175,** 1337–1344.

STREET, N. E., SCHUMACHER, J. H., FONG, T. A. T., BASS, H., FIORENTINO, D. F., LEVERAH, J. A., and MOSMANN, T. R. (1990). Heterogeneity of mouse helper T cells: Evidence from bulk cultures and limiting dilution cloning for precursors of Th1 and Th2 cells. *J. Immunol.* **144,** 1629–1639.

THOMAS, J., and ROUSE, B. T. (1997). Immunopathogenesis of herpetic ocular disease. *Immunol. Res.* **16,** 375–386.

YORK, I. A., ROOP, C., ANDREWS, D. W., RIDDELL, S. R., GRAHAM, F. L., and JOHNSON, D. C. (1994). A cytosolic herpes simplex virus protein inhibits antigen presentation to $CD8^+$ T lymphocytes. *Cell* **77,** 525–535.

General References

FIELDS, B. N., KNIPE, D. M., HOWLEY, P. M., CHANOCK, R. M., MELNICK, J. L., MONATH, T. P., ROIZMAN, B., and STRAUS, S. E. (1996). *Fields Virology,* 3rd ed. Lippincott-Raven, Philadelphia.

JANEWAY, C. A., JR., and TRAVERS, P. (1997). *Immunobiology,* 3rd ed. Current Biology/Garland, London/New York.

MIMS, C. A. (1995). *Mims' Pathogenesis of Infectious Disease.* Academic Press, London.

PAUL, W. E. (1993). *Fundamental Immunology,* 3rd ed. Raven Press, New York.

Pascal Launois
Reza Behin
Jacques A. Louis

World Health Organization
Immunology Research and Training Center
Institute of Biochemistry
University of Lausanne
CH-1066 Epalinges, Switzerland

Yasmine Belkaid
Geneviève Milon

Cellular Immunophysiology Unit
Institut Pasteur
Paris, France

Some microorganisms (parasites) depend on other organisms for achieving their life cycle. Although within their vertebrate hosts, parasites are confronted by both innate and adaptive immune responses, they have developed means to subvert immune surveillance, allowing them to complete their life cycle. Using the example of an intracellular protozoan parasite, Leishmania major, *this article will illustrate particular strategies used by these microorganisms to subvert immune regulatory/effector mechanisms.*

How Do Parasites Subvert the Immune Surveillance? The Example of *Leishmania*

Introduction

In many developing countries, parasitism represents an enormous public health burden and constitutes an important threat to socioeconomic development. During the past 20 years, important advances have been made in the understanding of the mechanisms of slowly acquired immunity against several parasitic diseases as well as of the factors contributing to chronic immunopathological processes. Unfortunately, and despite these significant and important scientific advances, the goal of developing efficacious immunoprophylactic measures that can be applied in parasite-endemic areas has not been reached. Difficulty in attaining this goal stems, at least in part, from the crucial feature developed by many parasites, once they have invaded their host, to survive and grow even in presence of vigorous in-

nate and adaptive immune responses. In addition, the strategy developed by the individual parasite to overcome the innate immune reactivity and/or to escape or subvert the adaptive immune responses mounted by its host is unique to a given natural host–parasite pair, reflecting the coadaptation of the two genomes.

Thus, rather than enumerating the various distinct strategies developed by individual parasites allowing them to survive even when facing strong immune responses, this article will focus on recent results characterizing the immunological host factors which determine resistance and susceptibility to parasitism with an intracellular parasite, *Leishmania* spp.

Leishmania are protozoan parasites capable of achieving their life cycle in human beings and several other mammalian species. Delivered within the dermis of their mam-

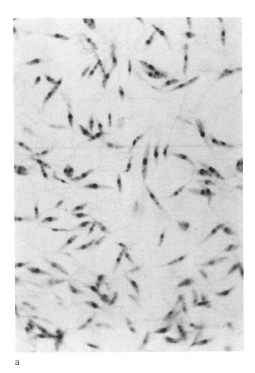
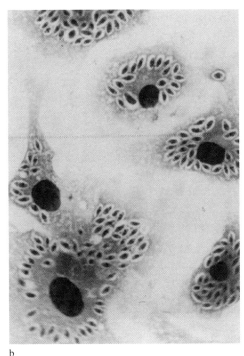

a   b

FIGURE 1   Developmental stages of *Leishmania* spp. (a) Promastigotes, the flagellate form which develops within the vector (sand fly). (b) Amastigotes, the aflagellate form present in mouse macrophages.

malian hosts in their flagellate form (promastigotes) by the bite of sand fly vectors, the parasites are rapidly taken up by resident dermal mononuclear phagocytes in which they transform into their aflagellate (amastigotes) forms (Fig. 1).

In human beings, infection with *Leishmania* can lead either to a clinically silent parasitic process or to a spectrum of clinical manifestations, depending on the species of the parasite and the genetic makeup of the host. The clinical features of the main categories of human leishmaniasis include (i) the cutaneous leishmaniasis characterized by cutaneous lesions at the site of inoculation that will eventually heal spontaneously; (ii) the mucocutaneous leishmaniasis thought to be due to metastasis of the microorganisms to mucosal sites from previous cutaneous lesions and leading to important necrotic damage of the soft tissues of the face and the buccopharyngeal cavity; and (iii) visceral leishmaniasis with multiple parasite-loaded tissues, particularly the bone marrow, spleen, and liver, with these last two organs becoming massively enlarged (Sacks *et al.,* 1993).

Strong evidence exists that indicates that resolution of infection, either spontaneous or following chemotherapy, is accompanied by a solid/persistent state of immunity to reinfection. Considering the high cost of drug therapy, particularly for developing countries, vaccination might thus represent a potent and economically feasible means by which to ultimately control leishmaniasis. The rational design of such vaccine depends not only on the identification of parasite molecules which are the targets of protective

immune responses but also, importantly, on a precise understanding of the cellular and molecular parameters leading to the exclusive development and maintenance of the protective immune responses.

In recent years, the study of a murine model of parasitism with *Leishmania major* has greatly contributed to the elucidation of immune mechanisms that either restrict or favor survival of these parasites in their hosts. Therefore, in this article we will summarize the recent data from several laboratories, including our own, pertaining to the characterization of the cellular and molecular parameters of the specific immune responses that parasitism with *L. major* can induce in mice and their effect on the ultimate fate of parasites. The applicability of results obtained in this experimental model to the human situation will also be briefly discussed.

The Murine Model of Parasitism with *L. major*

Almost the entire spectrum of clinical manifestations seen in human beings infected with *Leishmania* can be reproduced in mice after experimental inoculation of *L. major.* Mice from most inbred strains (e.g., C3H/He, CBA/Ca, B10.D2, C57BL/6, and 129/Sv/Ev) are resistant, developing small lesions that heal spontaneously and become immune to reinfection. In contrast, mice from a few geneti-

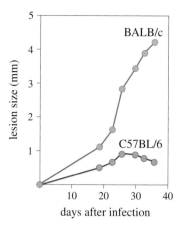

FIGURE 2 Development of lesions induced by *Leishmania major*. Susceptible mice (BALB/c) and resistant mice (C57BL/6) were infected with *L. major* given subcutaneously into the left hind foot pad. Foot pad swelling was monitored by use of a caliper and lesion size was determined by subtracting the thickness of the uninfected foot pad from the thickness of the infected foot pad.

cally susceptible strains (e.g., BALB/c) develop severe cutaneous lesions that never heal and do not become immune to reinfection (Fig. 2). Studies using this murine model of parasitism have significantly contributed to a precise understanding of some of the complex interactions between the host immune system and this obligate parasite of macrophages.

Resistance and Susceptibility of Mice to *L. major*: The Central Role of Macrophage Activation and Deactivation

Results from studies in both natural (human patients) and experimental (mice) hosts have unequivocally demonstrated that healing of lesions requires the development of T cell-mediated immune responses capable of activating macrophages harboring parasites to a microbicidal state. This activation of macrophages results in the synthesis of the inducible nitric oxide synthase (iNOS) leading to the L-arginine-dependent production of reactive nitrogen radicals toxic for the intracellular *Leishmania* (Assreuy *et al.*, 1994). It requires soluble factors (cytokines) produced by specifically activated T cells. The key cytokine produced by activated T cells responsible for macrophage activation has been demonstrated to be interferon-γ (IFN-γ), which acts in synergy with tumor necrosis factor α (TNF-α) endogenously produced by the macrophages (Green *et al.*, 1990). Accordingly, mice from a genetically resistant background but lacking either IFN-γ or the IFN-γ receptor gene fail to resolve their lesions (Swihart *et al.*, 1995; Wang *et al.*, 1994). In contrast, another cytokine, interleukin-4 (IL-4), which can also be released by activated T cells, exerts macrophage-deactivating function since it inhibits IFN-γ-trig-

gered activation of macrophages (Liew, 1989). Thus, IL-4 transgenic mice on a resistant genetic background fail to control lesions induced by *L. major* (Leal *et al.*, 1993). Other cytokines are also able to downregulate the anti-*Leishmania* effector function of macrophages, including IL-10, IL-13, and transforming growth factor-β, which are able to counter the IFN-γ-triggered induction of the iNOS (Moore *et al.*, 1993; Zurawski and De Vries, 1994).

Resistance and Susceptibility of Mice to *L. major*: The Crucial Role of Two Functionally Distinct Parasite-Specific CD4 T Cell Subsets

It has long been known that resistance to infection with *L. major* in mice is critically dependent on the activity of T cells (Mitchell *et al.*, 1981). Using monoclonal antibodies (mAb) that discriminate T cells of different subpopulations, resolution of lesions by genetically resistant mice was demonstrated to result from the expansion and activity of T cells from the CD4 subset. Strikingly, clear evidence was also obtained that the activity of CD4 T cells expanding in parasitized mice from genetically susceptible strains was an important factor in disease progression (Titus *et al.*, 1987). *Leishmania major* resides within the phagolysosomal compartment of macrophages, where protein processing for the major histocompatibility complex (MHC) class II pathway of presentation occurs. This particular localization accounts for the preferential induction of parasite-reactive CD4 T cells whose receptors [T cell receptors (TCRs)] recognize peptides loaded on MHC class II molecules.

It had been established from *in vitro* studies that CD4 T cell precursors can differentiate in two functionally distinct subsets, distinguishable by the mutually exclusive pattern of cytokines that they produce following reactivation (Mosmann and Coffmann, 1989). CD4 Th1 cells produce IFN-γ and IL-2, whereas Th2 cells produce IL-4, IL-10, and IL-13.

The dissection of the effector mechanisms developing slowly following infection of mice with *L. major* and accounting for either susceptibility or resistance revealed that these opposite functions were indeed exerted by two distinct subsets of CD4 T lymphocytes. The first clear correlation between disease manifestations and the pathway of CD4 T cell precursors differentiation *in vivo* was provided by results from studies using this murine model of infection with *L. major*. Indeed, it was clearly shown that, following infection with *L. major*, resistant mice mount a polarized CD4 Th1 cell response, in contrast to susceptible mice which mount a polarized CD4 Th2 cell response (Sadick *et al.*, 1989). The crucial importance of the CD4 Th1 cell response in resistance to *L. major* has been directly demonstrated by observations showing that the adoptive transfer of differentiated parasite-specific CD4 Th1 cell lines to BALB/c mice carrying the severe combined immunodeficiency (SCID) mutation render these otherwise extremely susceptible mice resistant to infection (Holaday *et al.*, 1991). In contrast, reconstitution of SCID mice with differentiated Th2 cell lines led to the development of even larger lesions than those in BALB/c mice. These results clearly show that parasite-reactive Th1 and Th2 cells are, by themselves, able to respectively interrupt the development of lesions or favor disease progression. The anti-*Leishmania* effector function of specific CD4 Th1 cells was demonstrated to be mediated by the IFN-γ released by the cells following activation, rendering the majority of macrophages parasiticidal. The pro-*Leishmania* effector function of specific CD4 Th2 lymphocytes appears to be mediated by the IL-4, IL-10, and IL-13 released by the cells following activation, thus inhibiting the macrophage parasiticidal activity.

Subsequently, it was possible to address the main issue of the kinetics/nature of the signaling molecules underlying the commitment of CD4 T cell precursors along one of the two pathways of differentiation *in vivo*. The strategies used relied on different experimental procedures: *in vivo* manipulations of mice such as selective depletion of circulating immunocompetent cells, neutralization of cytokines, or studies with mice expressing a transgene or harboring targeted mutations (e.g., gene disruption). Here, we focus on the kinetics/nature of the signaling molecules acting, at very early time points following infection, within the lymphoid organs draining the cutaneous site of parasite delivery. Indeed, mainly naive T cells constitutively recirculating within these lymphoid organs are expected to be responsive to differentiating/polarizing signals within the T cell–dependent area (Abbas *et al.*, 1996; Fearon and Locksley, 1996).

Early Signals, within the Lymphoid Organs, Instructing the Functional Polarization of CD4 T Cell Precursors

The process of T cell activation/differentiation could result from several different signals acting within the T cell–dependent area. Among the different potential T cell–polarizing signals presented in Table 1, cytokines have been best studied.

TABLE I

Signals That Could Contribute to the Functional Differentiation of CD4 T Cell Precursors

Route of administration of antigen
Type of cells presenting epitopes to CD4 T cells
Nature of costimulatory signals
  Transmembrane molecules
  Cytokines
Genetic polymorphisms
Dose of antigens

IL-12 is a cytokine produced by leukocytes, i.e., dendritic cells and mononuclear phagocytes, both of which have important functions for T cell activation (Trinchieri, 1996). Following infection with *L. major*, the development of Th1 T lymphocytes is regulated by IL-12: Administration of IL-12 to susceptible BALB/c mice during the first 5 days of infection inhibited the development of a Th2 response (Heinzel *et al.*, 1993; Sypeck *et al.*, 1993). Conversely, treatment of C57BL/6 mice with anti-IL-12 neutralizing antibodies prevented Th1 cell commitment in these otherwise resistant mice (Sypeck *et al.*, 1993). In addition, IL-12 knockout mice on a genetically resistant background developed a Th2 response and were susceptible to infection (Mattner *et al.*, 1996).

IL-4 was first shown to be produced by Th2 lymphocytes and later by other leukocytes, such as mast cells, basophils, and eosinophils, with these three populations being constitutively resident (mast cells) in the extravascular space or recruitable from the blood within this space. IL-4 appears to be required for CD4 Th2 cell development in susceptible BALB/c mice: Indeed, the injection of anti-IL-4 mAb at the initiation of infection prevents the differentiation and expansion of Th2 cells and leads to the development of a protective Th1 response (Sadick *et al.*, 1990).

In view of these results indicating the crucial role of IL-4 and IL-12 in CD4 T cell differentiation, experiments have been performed to directly compare the kinetics of production of these two cytokines within the lymph nodes of resistant and susceptible mice. IL-4 mRNA transcripts were monitored in susceptible and resistant mice from Day 0 to Day 10 after *L. major* inoculation. In contrast to resistant mice, susceptible mice exhibited a peak of IL-4 mRNA transcripts in draining lymph node as soon as 16 h following *L. major* infection (Launois *et al.*, 1995) (Fig. 3). Following this early burst of IL-4 transcripts, IL-4 mRNA transcription returned to baseline level until Day 4, when high and sustained levels of IL-4 transcripts were again detectable, reflecting the development of the Th2 polarization of parasite-reactive CD4 T lymphocytes.

The treatment of BALB/c mice with exogenous IL-12 the day before the infection completely suppressed the 16-h IL-4 burst (Launois *et al.*, 1995). Conversely, in C57BL/6 mice treated with anti-IL-12 neutralizing antibodies the day before *L. major* inoculation, increased IL-4 transcription was observed in their lymph nodes within the first 24 h (unpublished data). In the same vein, genetically resistant mice lacking the IL-12 gene exhibited an early burst of IL-4 mRNA expression in response to *L. major* (Mattner *et al.*, 1996). Together, these results suggest that in resistant mice, IL-12 downregulates this rapid IL-4 response (Fig. 4). Regarding the timing of IL-12 p40 and p35 production, too few studies are available and these provide conflicting results. Many laboratories are currently addressing this issue by monitoring both the IL-12 transcripts and the protein. The difficulties in assessing production of biologically ac-

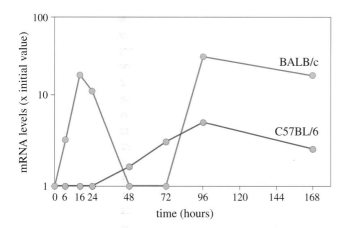

FIGURE 3 Comparison of the kinetics of IL-4 mRNA expression in the popliteal lymph nodes of susceptible mice (BALB/c) and resistant mice (C57BL/6) following injection of 3×10^6 stationary-phase *L. major* promastigotes. At various times following injection, mice were sacrificed, RNA was extracted from their popliteal lymph nodes, and the relative levels of IL-4 mRNA were determined. In the susceptible mice, a peak of IL-4 mRNA appears after only 16 h from infection (modified from Launois *et al.*, 1995).

tive IL-12 might be due to its heterodimeric nature and the complex regulation of the production of each monomer (Trinchieri, 1996). Thus, mainly data on the cellular origin of IL-4 produced during the early stages of the *L. major*-driven process will be summarized.

All available data strongly indicate that CD4 lymphocytes are responsible for the early IL-4 burst in *L. major*-infected susceptible BALB/c mice (Launois *et al.*, 1995). Within the $\alpha\beta$ TCR CD4 population, a minor subset has been identified which can rapidly produce large amounts of

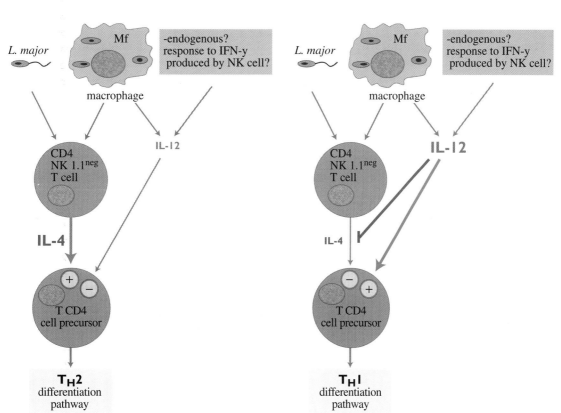

FIGURE 4 In susceptible mice CD4 T cell precursors reactive to *L. major* differentiate toward the Th2 functional phenotype in response to IL-4 rapidly produced by NK.1.1⁻ CD4 T cells following infection (directly or macrophage mediated). This IL-4 burst rapidly renders CD4 T cell precursors unresponsive to IL-12. In resistant mice, the early IL-4 response to *L. major* is prevented by IL-12, either constitutively present in a sufficient amount before infection or rapidly induced after infection. In the absence of IL-4, *L. major*-reactive CD4 T cell precursors maintain their responsiveness to IL-12 and differentiate toward the Th1 functional phenotype.

IL-4 following anti-CD3 stimulation (Yoshimoto and Paul, 1995). The cells within this minor subset express the NK1.1 transmembrane molecule and utilize a biased TCR repertoire (Vα14 paired with Vβ2, -7, or -8). These cells have been proposed as the source of IL-4 instructing the Th2 differentiation of *Leishmania*-reactive MHC class II-restricted CD4 T lymphocytes. Nevertheless, recent results have shown that this minor subpopulation does not contribute to the burst of IL-4 seen in lymph nodes of susceptible BALB/c mice as early as 16 h after *L. major* inoculation (Launois *et al.,* 1995). Furthermore, in BALB/c mice lacking NK1.1[+] CD4 T lymphocytes, a Th2 CD4 response is still developing following *L. major* infection (Brown *et al.,* 1996). An important issue, which remains to be investigated, concerns the stage of differentiation (naive, experienced, or memory?) of the conventional CD4 T lymphocytes triggered by *L. major*-derived products to produce the first burst of IL-4.

Recently, based on results from *in vitro* studies using naive CD4 T lymphocytes (reactive to an ovalbumin peptide) from either BALB/c or B10.D2 mice, a model to explain resistance/susceptibility to *L. major* has been proposed and its relevance is currently being experimentally tested in many laboratories. This model focuses on the different capacities of BALB/c mice and B10.D2 T lymphocytes to undergo IL-12-induced Th1 development: Whereas CD4 T lymphocytes from the B10.D2 background maintained IL-12 responsiveness in culture and exhibited a Th1 developmental phenotype, identically treated BALB/c CD4 T lymphocytes exhibited a Th2 developmental phenotype while losing IL-12 responsiveness as measured by IFN-γ production (Güler *et al.,* 1996). One of the most recent results available within this context indicates that a single locus on chromosome 11 might be critical in controlling the maintenance of IL-12 responsiveness and therefore the subsequent Th1/Th2 response (Gorham *et al.,* 1996). Of note, this locus is proposed as a modifier of the immune response (Gorham *et al.,* 1996). Thus, its more or less rapid contribution is expected to depend on the intrinsic properties of the immunogenic organisms or molecules delivered within the lymphoid organs in which naive T cells are polarized once they have been engaged through their TCRs. In agreement with this model, we have shown that soon after inoculation of *L. major,* parasite-specific CD4 cells from BALB/c mice no longer respond to IL-12 as assessed through IFN-γ production. However, the induction of this IL-12 unresponsiveness is totally dependent on the IL-4 produced during a critical temporal window, namely, the first 48 h following *L. major* infection (Launois *et al.,* 1997). Thus, these results favor the hypothesis that the genetically determined susceptibility to *L. major* and the rapid and sustained Th2 developmental commitment of parasite-reactive BALB/c CD4 lymphocytes are primarily the result of a rapid overproduction of IL-4, which in turn leads to IL-12 unresponsiveness (Fig. 4).

Early Leukocyte Trafficking from the Parasite-Loaded Dermis to the Lymphoid Organ

The parasites are delivered within the extravascular compartment of the dermis. The dermis is a site at which leukocytes reside and/or transiently migrate. Indeed, whereas macrophages, mast cells, and dendritic leukocytes are anchored to the extracellular matrix, T lymphocytes expressing cutaneous antigen are known to be able to leave blood capillary vessels and to migrate within the extracellular matrix before they enter the lymphatic vessels and the draining lymph node (Belkaid *et al.,* 1996). It is expected that leukocyte residency/trafficking will be transiently modified once parasites have been inoculated within the dermis. For example, Langherans dendritic cells and $\gamma\delta$-expressing T cells will be mobilized from the epidermis to the dermis, whereas neutrophils, natural killer cells, and monocytes will be recruited to this site from the blood. Only recently have methods derived from the "skin explant cultures" been developed to monitor this leukocyte trafficking (Belkaid *et al.,* 1996). Using this approach, important information has been obtained on the nature and kinetics of those leukocytes delivered from the parasite-loaded dermis to the draining lymph node. It is noteworthy that a larger number of CD4 T cells have been observed within the dermis of normal BALB/c mice compared to normal C57BL/6 mice. Thus, it is tempting to speculate that some of these CD4 cells, once activated by *Leishmania*-derived molecules, will enter the lymphatic vessels and the lymph nodes, in which they contribute to the burst of IL-4 observed 16 h after *L. major* inoculation.

Studies aimed at analyzing the flux of leukocytes, their lineages, and their signaling molecules (transmembrane and secreted) are only beginning and no doubt will provide a more precise characterization of the cellular origin of the signals underlying CD4 T cell differentiation.

Conclusions

Using genetically different mice as experimental hosts of *L. major,* it has been possible to document how these parasites can subvert some immune regulatory/effector molecules very early after their delivery within the dermis. BALB/c mice represent hosts in which *L. major* rapidly creates a microenvironment permissive to its optimal growth and dissemination. The characterization of the host genes "subverted" by these intracellular parasites is of prime importance and expected to lead to the rational design of immune intervention that will impede the uncontrolled parasitism and the subsequent pathogenic processes. This goal could be difficult to achieve, however, if the "genetic" difference results in low or high production of the crucial signals instructing CD4 cell differentiation. Nevertheless, the murine model of infection with *L. major* has shed light on some of the rules instructing differentiation of T cell subsets *in vivo*. It is hoped that the advances made using this model of infection will contribute to the development of means by which to control human leishmaniasis, which is a public health burden in many parts of the world.

Acknowledgments

Work from the authors' laboratories was supported by grants from the Swiss National Science Foundation, the European Union, the Roche Research Foundation, the World Health Organization, and the Institut Pasteur.

References Cited

ABBAS, A. K., MURPHY, K. M., and SHER, A. (1996). Functional diversity of helper T lymphocytes. *Nature* **383,** 787–793.

ASSREUY, J., CUNHA, F. Q., EPPERLEIN, M., NOROTHON-DUTRA, A., O'DONNELL, C. A., LIEW, F. Y., and MONCADA, S. (1994). Production of nitric oxide and superoxide by activated macrophages and killing of *Leishmania major. Eur. J. Immunol.* **24,** 672–676.

BELKAID, Y., JOUIN, H., and MILON, G. (1996). A method to recover, enumerate and identify lyphomyeloid cells present in an inflammatory dermal site: A study in laboratory mice. *J. Immunol. Methods* **199,** 5–25.

BROWN, D. R., FOWELL, D. J., CORY, D. B., WYNN, T. A., MOSKOWITZ, N. H., CHEEVER, A. W., LOCKSLEY, R. M., and REINER, S. L. (1996). $\beta2$-Microglobulin-dependent NK1.1⁺ T cells are not essential for T helper cell 2 immune responses. *J. Exp. Med.* **184,** 1295–1304.

FEARON, D. T., and LOCKSLEY, R. M. (1996). The instructive role of innate immunity in the acquired immune response. *Science* **272,** 50–53.

GORHAM, J. D., GÜLER, M. L., STEEN, R. G., MCKEY, A. J., DALY, M. J., FREDERICK, K., DIETRICH, W. F., and MURPHY, K. M. (1996). Genetic mapping of a murine locus controlling development of T helper 1/T helper 2 type responses. *Proc. Natl. Acad. Sci. USA* **93,** 12467–12472.

GREEN, S. J., CRAWFORD, R. M., HOCKMEYER, J. T., MELTZER, M. S., and NACY, C. A. (1990). *Leishmania major* amastigotes initiate the L-arginine dependent killing mechanism in IFN-γ stimulated macrophages by induction of tumor necrosis factor-α. *J. Immunol.* **145,** 4290–4297.

GÜLER, M. L., GORHAM, J. D., HSIEH, C. S., MCKEY, A. J., STEEN, R. G., DIETRICH, W. F., and MURPHY, K. M. (1996). Genetic susceptibility to *Leishmania*: IL-12 responsiveness in Th1 cell development. *Science* **271,** 984–987.

HEINZEL, F. P., SCHOENHAUT, D. S., RERKO, R. M., ROSSER, L. E., and GATELY, M. K. (1993). Recombinant interleukin 12 cures mice infected with *Leishmania major. J. Exp. Med.* **177,** 1505–1509.

HOLADAY, B. J., SADICK, M. D., WANG, Z., REINER, S. L., HEINZEL, F. P., PARSLOW, T. G., and LOCKSLEY, R. M. (1991). Reconstitution of *Leishmania* immunity in severe combined immunodeficient mice using Th1- and Th1-like cell lines. *J. Immunol.* **147,** 1653–1658.

LAUNOIS, P., OHTEKI, T., SWIHART, K., MACDONALD, H. R., and LOUIS, J. A. (1995). In susceptible mice, *Leishmania major* induce very rapidly interleukin-4 production by CD4⁺ T cells which are NK1.1⁻. *Eur. J. Immunol.* **25,** 3298–3307.

LAUNOIS, P., SWIHART, K. G., MILON, G., and LOUIS, J. A. (1997). Early production of IL-4 in susceptible mice infected with *Leishmania major* rapidly induces IL-12 unresponsiveness. *J. Immunol.* **158,** 3317–3324.

LEAL, L. M. CC. C., MOSS, D. W., KUHN, R., MULLER, W., and LIEW, F. Y. (1993). Interleukin-4 transgenic mice of resistant background are susceptible to *Leishmania major* infection. *Eur. J. Immunol.* **23,** 566–569.

LIEW, F. Y. (1989). Functional heterogeneity of CD4⁺ T cells in leishmaniasis. *Immunol. Today* **10,** 40–45.

MATTNER, F., MAGRAM, J., FERRANTE, J., LAUNOIS, P., DI PADOVA, K., BEHIN, R., GATELY, M. K., LOUIS, J. A., and ALBER, G. (1996). Genetically resistant mice lacking interleukin-12 are susceptible to infection with *Leishmania major* and mount a polarized Th2 cell response. *Eur. J. Immunol.* **26,** 1553–1559.

MITCHELL, G. F., CURTIS, J. M., SCOLLAY, R. G., and HANDMANN, E. (1981). Resistance and abrogation of resistance to cutaneous leishmaniasis in reconstituted BALB/c nude mice. *Aust. J. Exp. Biol. Med.* **59,** 539–544.

MOORE, K. W., O'GARRA, A., DE WAAL MALEFYT, T. R., VIERA, P., and MOSMANN, T. R. (1993). Interleukin-10. *Annu. Rev. Immunol.* **11,** 165–190.

MOSMANN, T. R., and COFFMANN, R. L. (1989). Th1 and Th2 cells: Different patterns of lymphokine secretion lead to different functional properties. *Annu. Rev. Immunol.* **7,** 145–173.

SACKS, D. L., LOUIS, J. A., and WRITH, D. F. (1993). Leishmaniasis. In *Immunology and Molecular Biology of Parasitic Infections,* pp. 71–86. Blackwell, Oxford.

SADICK, M. D., HEINZEL, F. P., HOLADAY, B. J., COFFMANN, L., and LOCKSLEY, R. M. (1989). Reciprocal expression of Interferon γ or interleukin 4 during the resolution or progression of murine leishmaniasis. Evidence for expansion of distinct helper T cell subsets. *J. Exp. Med.* **169,** 59–72.

SADICK, M. D., HEINZEL, F. P., HOLADAY, B. J., PU, R. T., DAWKINS, R. S., and LOCKSLEY, R. M. (1990). Cure of murine leishmaniasis with ant-interleukin 4 monoclonal antibody. Evidence for a T cell-dependent, interferon gamma-independent mechanism. *J. Exp. Med.* **171,** 115–127.

SWIHART, K., FRUTH, U., MESSMER, N., HUG, K., BEHIN, R., HUANG, S., DEL GIUDICE, G., AGUET, M., and LOUIS, J. (1995). Mice from a genetically resistant background lacking the interferon γ receptor are susceptible to infection with *Leishmania major* but mount a polarized T helper cell 1-type CD4⁺ T cell response. *J. Exp. Med.* **181,** 961–971.

SYPECK, J. P., CHUNG, C. L., MAYOR, S. E. H., SUBRAMANAYAM, J. M., GOLDMAN, S. J., SIEBURTH, D. S., WOLF, S. F., and SCHAUB, R. G. (1993). Resolution of cutaneous leishmaniasis: Interleukin 12 initiates a protective T helper type 1 immune response. *J. Exp. Med.* **177,** 1793–1802.

TITUS, R. G., MILON, G., MARCHAL, G., VASSALLI, P., CEROTTINI, J. C., and LOUIS, J. (1987). Involvement of specific Lyt-2⁺ T cells in the immunological control of experimentally induced murine cutaneous leishmaniasis. *Eur. J. Immunol.* **17,** 1429–1433.

TRINCHIERI, G. (1996). Interleukin-12: A proinflammatory cytokine with immunoregulatory functions that bridge innate resistance and antigen-specific adaptive immunity. *Annu. Rev. Immunol.* **13,** 251–276.

WANG, Z.-E., REINER, S. L., ZHENG, S., DALTON, D. K., and LOCKSLEY, R. M. (1994). CD4$^+$ effector cells default to the Th2 pathway in interferon γ-deficient mice infected with *Leishmania major. J. Exp. Med.* **179,** 1367–1371.

YOSHIMOTO, T., and PAUL, W. E. (1994). CD4^{pos} NK1.1^{pos} T cells promptly produce interleukin 4 in response to *in vivo* challenge with anti-CD3. *J. Exp. Med.* **179,** 1285–1295.

ZURAWSKI, G., and DE VRIES, J. E. (1994). Interleukin-13, an interleukin-4 like cytokine that acts on monocytes and B cells, but not on T cells. *Immunol. Today* **15,** 19–26.

General References

BOGDAN, C., GESSNER, A., SOLBACH, W., and RÖLLINGHOF, M. (1996). Invasion, control and persistence of *Leishmania major* parasites. *Curr. Opin. Immunol.* **8,** 517–525.

REINER, S. L., and LOCKSLEY, R. M. (1995). The regulation of immunity to *Leishmania major. Annu. Rev. Immunol.* **13,** 151–177.

SCOTT, P. (1996). Th cell development and regulation in experimental cutaneous leishmaniasis. *Chem. Immunol.* **63,** 98–114.

EMMANUEL J. H. J. WIERTZ

*Laboratory of Vaccine Development
and Immune Mechanisms
National Institute of Public Health
and the Environment
P.O. Box 1
3720 BA Bilthoven
The Netherlands*

In this article, immune evasion by viruses is reviewed. Recent studies will be discussed that reveal novel mechanisms by which viruses evade detection and elimination by the host immune system. In particular, evasion mechanisms of five persistent viruses are considered: herpes simplex virus, human cytomegalovirus, mouse cytomegalovirus, Epstein–Barr virus, and adenovirus. Unraveling the strategies used by viruses to survive within the host could identify new targets for antiviral drugs and for improved vaccines. Identification of the mechanisms that underlie these strategies might also reveal new, fundamental features of biology that occur in uninfected cells and are exploited by the viruses.

How Viruses Elude the Immune System

Introduction

Recently, remarkable escape strategies have been uncovered for many viruses. Rather than focusing on frequently described phenomena such as antigenic variation, some of the newly discovered "stealth strategies" will be discussed in this article. In particular, viruses that cause persistent infections, such as herpesviruses, seem to be masters at manipulating the host immune response and, in this respect, resemble the famous Trojan horse.

Herpesviruses are thought to have cospeciated with their natural hosts. This may explain the interesting coexistence that occurs for many herpesviruses, characterized by a life-long infection, usually without causing serious disease of the host. Throughout the life of the host new virus is produced, allowing inoculation of new individuals. During millions of years of coevolvement with the host immune system, herpesviruses appear to have acquired escape mechanisms that allow even slowly replicating viruses such as human cytomegalovirus (HCMV) to establish a persistent infection. An example of a typical herpes viron is shown in Fig. 1.

Cytotoxic T lymphocytes (CTLs) play an important role in the elimination of virus-infected cells, which they recognize by the presence of virus-derived peptides in the antigen binding site of major histocompatibility complex (MHC) class I molecules occurring on the surface of the infected cells. Viral coevolution with the host immune system is likely to favor the emergence of viruses that can elude

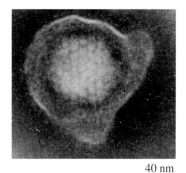

<u>40 nm</u>

FIGURE 1 Electron micrograph of an enveloped herpes virion negatively stained with 3% potassium phosphotungstate. A nucleocapsid, surrounded by a tegument of globular material and an external envelope, can be distinguished (courtesy of P. Roholl).

recognition by CTLs (Maudsley and Pound, 1991; Rinaldo, 1994; Zinkernagel, 1996). Many of the escape mechanisms involve downregulation of MHC class I molecules. This may be caused by attenuation of the genes encoding the class I molecules but also posttranslational mechanisms

have been described. Viruses appear to manipulate not only peptide loading of the MHC class I molecules but also the synthesis, assembly, and surface expression of the MHC-encoded glycoproteins — the very molecules that signal the presence of a virus-infected cell to the immune system.

Recognition of Viral Proteins by Cytotoxic T Cells

Viral mRNAs are translated using the normal translation machinery of the cell. Like other proteins, cytosolic viral gene products are degraded, as part of normal protein turnover, by cellular proteases, including a multisubunit catalytic protease complex known as the proteasome (discussed later) (Fig. 2). Degradation of viral proteins in the cytosol leads to the formation of polypeptides that might then be transported to the lumen of the endoplasmic reticulum (ER) by a dedicated transporter complex: the transporter associated with antigen presentation (TAP) (Fig. 2). In the ER, these peptides assemble with the membrane-bound heavy chain of the class I MHC molecules and the light chain, β_2 microglobulin (β_2m). The binding of peptide sta-

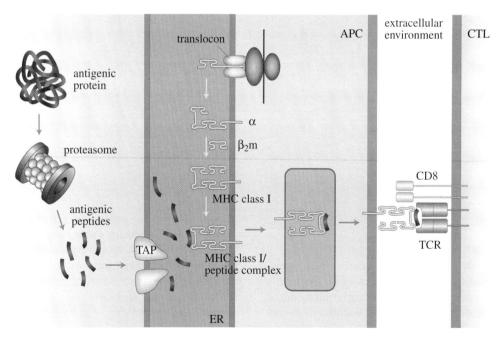

FIGURE 2 Processing and presentation of antigen by MHC class I molecules in antigen presenting cells (APC) and antigen recognition by cytotoxic T lymphocytes (CTL). (Left, from the top down) An antigenic protein is degraded by the proteasome and the resulting peptides are translocated to the ER by the TAP transporter. (Center, from the top down) The nascent heavy α chain is translated from mRNA (red line) and introduced into the ER via the translocation complex (translocon) where it associates with the light chain (β_2m). Successively, the peptides are loaded onto these molecules forming the MHC class I–peptide complex. (Right) Properly assembled complexes are transported to the cell surface where they are available for interaction with T cell receptors (TCR) and accessory molecules such as CD8.

bilizes the interactions of the MHC class I heavy and light chains and, as a result, a trimolecular complex of MHC class I, β_2m, and peptide is formed (Fig. 2) (Heemels and Ploegh, 1995; Townsend and Bodmer, 1989). These MHC complexes, loaded with peptides, are then transported from the ER, through the Golgi apparatus, to the plasma membrane (Fig. 2). At the cell surface, they are available for surveillance by CTLs. Upon recognition of the MHC–peptide complex on the plasma membrane of the virus-infected cell, the CTLs deliver a lethal hit, resulting in the lysis of the infected cell.

The Proteasome

The 26S proteasomal complex is a large ($\sim$1500 kDa) proteolytic complex (Coux *et al.*, 1996) that contains numerous subunits, some of which have not been characterized. The core is formed by the 20S proteasome ($\sim$700 kDa) which consists of 14 α and 14 β subunits (Fig. 3). Proteins are targeted for degradation by the attachment of chains of ubiquitin, a 8.4-kDa protein, to them. Initially, the C-terminal Gly residue of ubiquitin is covalently linked to the M-amino group of Lys side chains in the protein through an isopeptide bond. Further ubiquitin moieties are added to the primary ubiquitin and the resulting branched chains of ubiquitin target the protein for ATP-dependent destruction by the 26S proteasomal complex. Nonubiquitinated proteins can also be degraded by the proteasome. The proteasome is responsible for the normal turnover of proteins in the cytosol.

Proteasomal degradation leads to the formation of peptides which can be transported into the ER where they bind MHC class I molecules; in virally infected cells, peptides derived from viral proteins are generated by the proteasome and ultimately displayed by class I molecules at the cell surface (Fig. 2). Recently, several very effective and specific inhibitors of proteasomal degradation have been described, including peptide aldehydes and the antibiotic lactacystin. Leucyl-leucyl-norleucinal, the calpain-I inhibitor that was cocrystallized with the proteasome, is also highly effective, although it also affects other cellular proteases. Blocking proteasomal function by these drugs interferes with MHC class I-restricted antigen presentation and illustrates the important role the proteasomal complex plays in antigen processing.

A Herpes Simplex Virus-Encoded Protein Inhibits Peptide Transport by TAP

The herpes simplex viruses (HSV)-1 and HSV-2 are closely related viruses that infect the oral and genital mucosa, respectively. HSV-1 can maintain latency in the trigeminal and cervical ganglia and HSV-2 in the sacral ganglia. The molecular basis of the establishment of HSV latency is unclear. In the latently infected neurons, the virus is periodically reactivated and virus is transferred by axonal transport to cells innervated by the infected neurons. HSV is relatively insensitive to antiviral antibodies. It is therefore likely that T cells play a role in controlling viral infection and reactivation.

To escape immediate elimination by the host immune system, the virus has acquired a mechanism to reduce the expression of peptide–MHC complexes. As a result, within a few hours of infection with HSV, the infected cells are resistant to lysis by HSV-specific CTLs. The MHC class I molecules are synthesized normally but are retained in the ER (York *et al.*, 1994). This resistance to lysis by CTLs appears to depend on a small (9-kDa) cytosolic protein, ICP47, encoded by HSV.

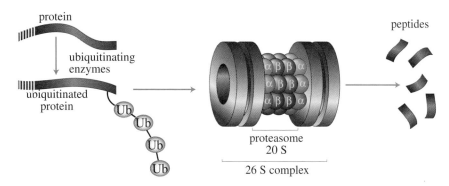

FIGURE 3 Structure and function of the 26S proteasomal complex, a large proteolytic complex of approximately 1500 kDa that contains numerous subunits, some of which have not been characterized. The proteolytic core is formed by the 20S proteasome which consists of 14 α and 14 β subunits. The 26S proteasomal complex degrades ubiquitinated proteins.

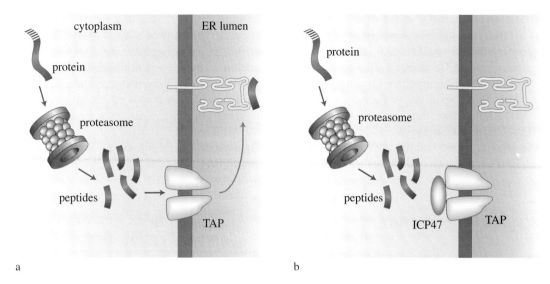

FIGURE 4   Herpes simplex virus interferes with peptide transport by TAP. (a) In a normal cell the transfer of antigenic peptides to the MHC class I molecules occurs via TAP. (b) In a cell infected with HSV-1 or -2, ICP47, a small cytosolic protein coded by this virus, competes with antigenic peptides preventing normal transport.

Since class I molecules require bound peptides and β_2m to form a stable trimolecular complex, the observation that the ER-retained MHC class I molecules from ICP47-expressing cells were unstable in detergent lysates suggested that the class I molecules lacked antigenic peptide. Since ICP47 is a cytosolic protein, it was possible that this HSV protein interfered with the generation of peptides or with the function of TAP, which moves peptides from the cytosol to the ER (Fig. 2b). Indeed, the use of a semipermeabilized cell system demonstrated a strong inhibition of peptide translocation of peptides into the ER in the presence of ICP47 (Früh *et al.*, 1995; Hill *et al.*, 1995) (Fig. 4). Furthermore, a physical association of ICP47 with the human, but not rodent, TAP complex could be demonstrated. The transporter consists of a heterodimer of TAP-1 and TAP-2, each containing six to eight membrane-spanning regions and an ATP-binding domain. Competition experiments have suggested that the TAP complex contains a single peptide binding site. ICP47 and peptides seem to compete for binding to TAP. This prevents the transport of peptides, making cytosol-derived peptides unavailable for binding to and display by class I molecules. Interestingly, the inhibitory effect of ICP47 can be mimicked with a 35-residue synthetic peptide corresponding to amino acid residues 1–35 of ICP47. The active site within this sequence is currently being mapped using a series of truncated synthetic peptides derived from ICP47 (Galocha *et al.*, 1997).

Since the only known role of the TAP complex is to transport peptides, inhibition of this function is unlikely to disrupt aspects of cellular physiology other than antigen presentation. Inhibitors of TAP, based on active sequences

in ICP47, might thus be immunosuppressive. Conversely, small molecules that prevent the TAP–ICP47 interaction might improve the immunogenicity of HSV-infected cells.

Immune Evasion by HCMV Involves Multiple Mechanisms

The study of the effects of cytomegaloviruses on the MHC class I-restricted antigen presentation pathway has yielded an embarrassment of riches. The HCMV encodes five and possibly six different glycoproteins, each interfering in a different way with elimination of the virus by the host immune system (Fig. 5). Most likely, it is the concerted action of these glycoproteins that allows HCMV to escape from elimination by the host immune system during acute and perhaps also persistent infection. Prime targets of these CMV glycoproteins are MHC class I glycoproteins. Recently, several novel links in the multistep process of immune evasion by HCMV have been discovered.

The HCMV-Encoded US2 and US11 Gene Products Destroy Newly Synthesized MHC Class I Molecules

The US region of the HCMV genome encodes most if not all glycoproteins known to play a role in downregulation of MHC class I expression. Using defined deletion mutants, the early gene product US11 was first identified as a major cause of the long observed instability of MHC class I molecules in HCMV-infected cells (Jones *et al.*, 1995; Wiertz *et al.*, 1996a). Another early gene product, US2, was

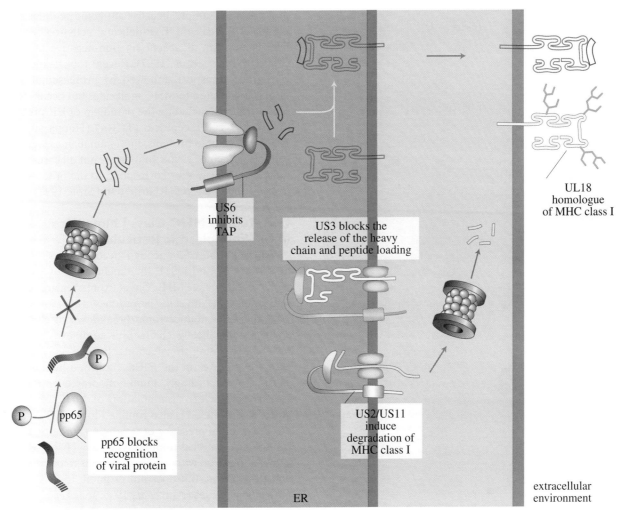

FIGURE 5 Schematic summary of the escape strategies adopted by HCMV to elude the host immune system. The pp65 protein interferes with recognition of viral proteins by the proteasome. US6 blocks the peptide transporter TAP. Two proteins, US2 and US11, induce rapid degradation of MHC class I molecules by a novel protein breakdown pathway. US3 causes the retention of the heavy chain of MHC class I in the ER and retards the loading of peptide. UL18, a homolog of MHC class I, acts as a molecular decoy for natural killer cells.

found to cause class I degradation in a manner indistinguishable from that caused by US11 (Wiertz *et al.*, 1996b). Simultaneous expression of two gene products with a similar function is unexpected, especially in view of the absence of significant similarity between US2 and US11.

Using cell lines stably transfected with US11 and US2, some of the highly unusual characteristics of what appears to be a novel protein breakdown pathway have been characterized (Fig. 6) (Wiertz *et al.*, 1996a,b). In the transfectants, as in HCMV-infected cells, the class I heavy chains are degraded with a half-life of <1 min. Surprisingly, inhibitors of the cytosolic proteasome block the degradation and, in the presence of these inhibitors, a breakdown intermediate has been identified. Further characterization re-

vealed that the observed intermediate is the class I heavy chain that has lost its single N-linked glycan. Unexpectedly, the removal of the sugar moiety involves an N-glycanase.

The precursor–product relationship, established by pulse-chase experiments, indicates that the deglycosylated intermediate is derived from fully glycosylated class I heavy chains. Since the proteasome and an N-glycanase are involved in the degradation, it is likely that the process is occurring in the cytosol. Indeed, the cytosolic localization of the breakdown intermediate has been confirmed by subcellular fractionation. Apparently, in the presence of the US11 glycoprotein the class I molecules are translocated into the ER normally but are then "dislocated" into the cytosol where they are degraded by the proteasome. We pro-

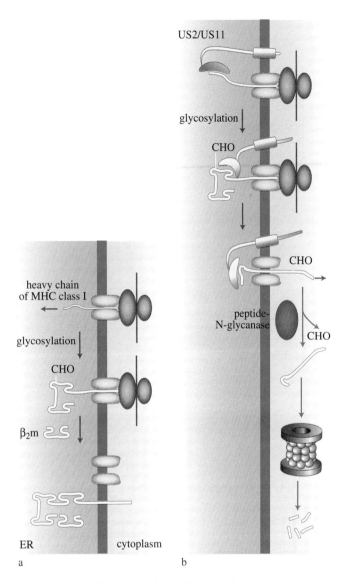

FIGURE 6 HCMV-encoded US2 and US11 induce destruction of MHC class I molecules. (a) In normal cells nascent class I heavy chains are translocated to the ER via the translocon (in green) and glycosylated by an enzymatic complex in the ER lumen. (b) In cells infected with HCMV, US2 or US11 induce the completed heavy chains to return by the inverse path back to the cytosol, where they are deglycosylated by a peptide-N-glycanase and degraded by the proteasome.

pose a model in which the class I heavy chains do not leave the translocation complex but slide back to the cytosol as soon as the translation of the heavy chain is completed and the product has been released from the ribosome (Wiertz *et al.*, 1996b) (Fig. 6).

This novel degradation pathway appears to be operational in the absence of viral gene products and is used to dispose class I heavy chains that fail to fold and assemble

properly, for example, because antigenic peptides are lacking (as is the case in TAP-deficient cells) or because the light chain, β_2m, is absent (in β_2m-negative Daudi cells) (Fig. 7) (Hughes *et al.*, 1997). Recent reports indicate that retrograde transport, followed by proteasomal degradation, is not limited to MHC molecules but occurs for many other proteins, including the paradigm of ER protein degradation, T cell receptor-α (TCRα) (Huppa and Ploegh, 1997) and soluble proteins, in mammalian as well as yeast cells (Bonifacino, 1996). In many if not all cases, this degradation pathway may correspond to the process referred to as ER degradation (Klausner and Sitia, 1990).

US3 Retains MHC Class I Molecules in the Endoplasmic Reticulum

Although the US3-encoded glycoprotein shows homology with US2 and US11, the mechanisms by which these molecules influence MHC class I expression are fundamentally different (Ahn *et al.*, 1996; Jones *et al.*, 1996). Whereas US2 and US11 induce rapid breakdown of class I heavy chains, stable heavy chain–β_2m complexes are formed in US3-expressing transfectants. Instead, US3 retains the class I complexes in the ER as indicated by their EndoH sensitivity. Accordingly, immunofluorescence microscopy reveals perinuclear accumulation of MHC class I molecules, whereas US3 is also confined to the ER. When cells are lysed in the presence of digitonin (but not NP40) US3 is found in a physical complex with class I heavy chains and β_2m. Association with US3 occurs prior to peptide loading and slows down, but does not prohibit, the acquisition of antigenic peptides (Ahn *et al.*, 1996; Jones *et al.*, 1996).

A careful analysis of the sequence of expression of US2, US3, and US11 indicates that US3 is present at immediate early times during infection, whereas US11 and US2 are expressed at early and late times (Ahn *et al.*, 1996). The half-lives of US3 and US11 are estimated to be 3 and 6 h ip, respectively, indicating that expression of US3 and US11/US2 overlaps briefly. It is tempting to propose a model in which US3, being the first nonregulatory immediate early protein expressed, retains stable MHC class I heterodimers in the ER, thereby rendering them susceptible to destruction by US11/US2 expressed at later time points. However, experiments with US11 or US2 transfected cells suggest that the majority of class I molecules are attacked as free heavy chains: Rapid degradation occurs immediately after completion of the polypeptide chain, prior to association with β_2m (Wiertz *et al.*, 1996a,b). In contrast, US3 appears to interact specifically with heavy chain–β_2m complexes (Ahn *et al.*, 1996; Jones *et al.*, 1996). Nevertheless, the interplay between US3 and US2/US11 may apply to the minor population of MHC class I molecules that escapes immediate destruction by US11 and US2 and forms a complex with β_2m.

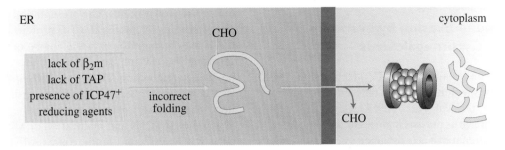

FIGURE 7 The degradation pathway that is used by the HCMV-encoded US2 and US11 to destroy MHC class I molecules appears to be operational in the absence of viral gene products. This novel breakdown pathway is used to dispose of class I heavy chains that fail to fold and assemble properly — for example, because antigenic peptides are lacking, as in TAP-deficient cells or where transport is inhibited by ICP47, or because the light chain β_2m is absent.

ER retention of MHC class I molecules is not unique for HCMV: The adenovirus E3-19k gene product possesses similar functions and murine cytomegalovirus also retains class I molecules in the ER. US3 and E3-19k are not homologous; both are type I membrane proteins and contain high mannose-type N-linked glycans.

US6 Blocks Peptide Transport by TAP

Although ER retention of class I molecules, and also degradation by US2, is mediated via a direct physical interaction with class I heavy chains, this is not a prerequisite for class I downregulation. Other viral gene products have been shown to act indirectly and block functions required for assembly of MHC class I–peptide complexes. Such a mechanism has been described for the HSV-1- and -2-encoded ICP47 (vide supra). Data indicate that HCMV also encodes a protein, US6, capable of inhibiting TAP function (Ahn *et al.*, 1997; Hengel *et al.*, 1997). Most likely, inhibition by US6 involves a novel type of interaction with TAP since US6 is a glycoprotein and does not demonstrate any homology with the cytosolic ICP47 (Fig. 4).

UL18 Acts as a Decoy for Natural Killer Cells

Natural killer (NK) cells carry triggering receptors and inhibitory receptors. Activation of the triggering receptor by a target cell will result in its destruction unless the inhibitory receptor detects an MHC class I molecule. Thus, cells that have lost cell surface expression of MHC class I are recognized and destroyed by NK cells. Interestingly, in addition to the class I retaining function, the murine CMV (MCMV) encodes a class I homolog, *m144*, which appears to act as a molecular decoy *in vivo* (Farrell *et al.*, 1997). A recombinant MCMV in which the *m144* gene has been disrupted demonstrates severely restricted replication compared with wild-type MCMV (Farrell *et al.*, 1997). *In vivo* depletion studies show that NK cells are responsible for the observed attenuation of the infection. HCMV also encodes a class I homolog, UL18, which is capable of inhibiting the function of NK cells *in vitro* (Reyburn *et al.*, 1997).

Phosphorylation of the HCMV Immediate Early (I-E) Protein by pp65 Inhibits T Cell Recognition of the I-E Protein

During the I-E phase of viral gene expression, an essential viral transcription factor, the 72-kDa I-E protein, is produced. Since this I-E gene product is expressed abundantly prior to synthesis of most glycoproteins known to be involved in immune suppression, a CTL response against I-E would be expected. However, very few I-E-specific CTLs are detected in seropositive individuals. Data suggest that a CMV matrix protein, pp65, might be responsible for this apparent immune suppression (Gilbert *et al.*, 1996). pp65 catalyzes phophorylation of I-E. The modified I-E gene product fails to activate CTLs *in vitro* (Fig. 8).

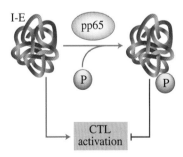

FIGURE 8 The HCMV-encoded pp65 interferes with T cell recognition of an HCMV immediate early (I-E) protein characteristic of the early phases of infection. pp65 catalyzes phosphorylation of the I-E protein that, thus modified, fails to activate CTLs *in vitro*.

Immune Evasion by Murine Cytomegalovirus

Like its human counterpart, the MCMV affects antigen presentation in many ways but does so via different mechanisms (Thale *et al.,* 1995; Del Val *et al.,* 1989; Ziegler *et al.,* 1997). After infection, the virus persists in mice for life. The infection can be asymptomatic or accompanied by disease. CTLs generated by mice against MCMV are essential for controlling the infection and for resistance to lethal doses of virus. In addition, NK cells play a role in controlling the infection.

Mice infected with MCMV generate CTLs that recog-

nize many viral proteins. The presentation of epitopes from these proteins, including the I-E protein pp89, to specific CTLs may be monitored in the course of the infection by exposing MCMV-infected cells to CTLs at different time points after infection. Although pp89-specific CTLs efficiently recognize the protein within the first hour of infection, no response is observed during the next few hours. When the synthesis and maturation of MHC class I molecules was investigated in the context of viral infection, an early gene was identified, *M152,* that causes retention of class I molecules in the ER. Upon expression of *M152,* recognition of simultaneously expressed MCMV proteins is inhibited.

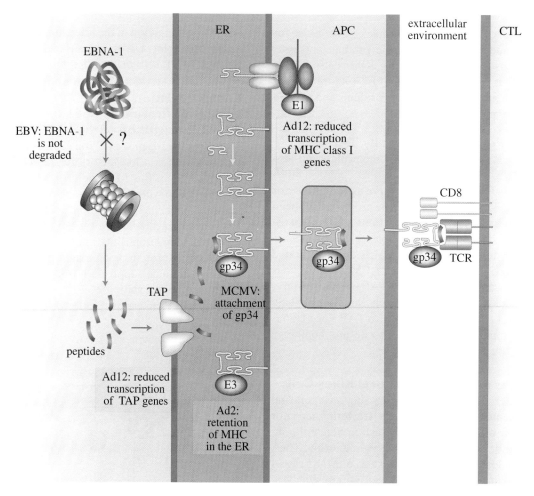

FIGURE 9   Interference by MCMV (in pink), EBV (in yellow), and adenoviruses Ad2 and Ad12 (in green) in the MHC class I pathway. The EBV-encoded EBNA-1 protein evades CTL recognition perhaps because it fails to be degraded by the proteasome. During infection with MCMV and with the adenoviruses, retention of the MHC class I molecules is observed in the ER. In the case of Ad2, retention is caused by an *E3*-encoded 19-kDa protein, whereas the *E1* gene products of Ad12 interfere with transcription of MHC class I and TAP genes. In addition, MCMV can produce other proteins, such as gp34, which associate stably with the MHC class I complex and probably interfere with the recognition of such complexes on the cell surface by the CTLs.

Like HCMV, MCMV also encodes a MHC class I homolog, m144, that acts as a molecular decoy for NK cells *in vivo*, thus inhibiting lysis of infected cells by NK cells (Farrell *et al.*, 1997). In addition to these known evasion strategies, MCMV may express additional genes that might affect the ability of the host immune response to detect infected cells. For example, a 34-kDa, membrane-bound glycoprotein, gp34, has been described that stably associates with MHC class I molecules (Kleijnen *et al.*, 1997). The gp34 protein binds to class I molecules in the ER and then moves to the cell surface in association with these molecules (Fig. 9). Only folded class I molecules, containing both heavy chain and β_2m, are bound by gp34. Whether gp34 interferes with the host immune response (e.g., by interfering with CTL recognition at the cell surface) is unclear.

Escape Mechanisms of Epstein–Barr Virus

The Epstein–Barr virus (EBV), a human gamma herpesvirus, occurs in all human communities as a widespread infection. EBV has potent growth-transforming activity in B lymphocytes, causes infectious mononucleosis, and shows a strong association with many malignancies, including Burkitt's lymphoma, nasopharyngeal carcinoma, lymphomas in immunocompromised individuals, and Hodgkin's lymphoma (Rickinson and Kieff, 1996; Khanna *et al.*, 1995; Rickinson and Moss, 1997).

During infection with EBV, a CTL response is observed against a variety of EBV-encoded proteins (Rickinson and Kieff, 1996; Khanna *et al.*, 1995; Rickinson and Moss, 1997; Kieff, 1996). This CTL response consists of a classic antiviral MHC class I-restricted CD8[+] T cell response, although cytotoxic class II-restricted CD4[+] T cells have also been described. Nevertheless, the virus persists for life as a latent infection in all infected individuals. This suggests that EBV has acquired highly effective escape mechanisms that allow the virus to elude the host immune system.

The life cycle of EBV can be divided into a latent and a lytic or productive phase of infection. During latency, expression of EBV-encoded proteins is limited to the EBV-encoded nuclear antigens EBNA-1–6; the latent membrane proteins LMP-1, 2α, and 2β; and two small polyadenylated RNAs, EBER-1 and -2. After activation of the lytic phase, more than 80 EBV gene products are synthesized. For several latent and lytic antigens interesting phenomena have been described that might play a role in immune evasion by EBV.

T Cell Recognition of EBV Latent Antigens

Contrary to all other latent antigens, EBNA-1 fails to activate MHC class I-restricted CTLs and possibly also class II-restricted T cells. Possible explanations for this surprising observation will be discussed later. CTL reactivities against the other latent antigens are highly skewed toward EBNA-3A, -3B, and -3C (also called EBNA-3, -4, and -6, respectively) (Steven *et al.*, 1996). Moreover, the response is directed toward only a limited number of epitopes within the EBNA-3 gene products, whereas the repertoire of T cell receptor genes involved in recognition of some of the immunodominant epitopes is limited to similar, highly conserved TCR$\alpha\beta$ genes. This skewing not only involves memory T cell responses but also is observed during the acute primary infection, which suggests that the EBNA-3 proteins might in some way be more accessible to the MHC class I processing and presentation pathway than other latent antigens.

Several EBV-encoded antigens demonstrate an unusual behavior during processing by the antigen presenting cell. For example, the EBV latent antigen LMP-2 contains epitopes that can be recognized by MHC class I-restricted T cell clones in a TAP-independent fashion (Lee *et al.*, 1996; Khanna *et al.*, 1996). Recognition of LMP-2 by another T cell clone requires a functional TAP transporter (Lee *et al.*, 1996; Khanna *et al.*, 1996). These differences in processing may be related to the localization of the epitopes in this multiple membrane-spanning protein: The TAP-independent epitopes are in putative transmembrane domains, whereas the TAP-dependent epitope is in a predicted cytosolic portion (Lee *et al.*, 1996; Khanna *et al.*, 1996).

EBNA-1

EBNA-1 has the highly unusual property: It fails to activate CTLs (Rickinson and Kieff, 1996; Kieff, 1996; Trivedi *et al.*, 1991; Levitskaya *et al.*, 1995). All other latent antigens, including EBNA-2–EBNA-6, LMP-1, LMP-2α, and LMP-2β, induce potent CTL responses. EBNA-1 expression is required for maintenance of the viral episome during latency. EBNA-1 is the only latent antigen consequently expressed during infectious mononucleosis and in all EBV-associated malignancies. Thus, limitation of EBV gene expression to EBNA-1 in long-lived nonreplicating B cells and in tumor cells may play a role in immune evasion by these cells.

The mechanism by which EBNA-1 prohibits CTL recognition remains an enigma. The phenomenon seems to be based on intrinsic properties of the protein, does not involve other EBV-encoded gene products, and is not limited to humans. For example, syngeneic murine tumor cells transfected with EBNA-1 are not eliminated, whereas the same tumor cells transfected with LMP-1 are rejected efficiently (Trivedi *et al.*, 1991). No *trans*-acting suppression has been observed. On the other hand, the inhibitory effect can be transferred to other antigens. Introduction of an EBNA-4-derived CTL epitope within the EBNA-1 sequence abrogates recognition by human T cell clones specific for the EBNA-4 epitope (Levitskaya *et al.,* 1995). EBNA-1 contains a long stretch of repeated glycine and alanine residues. The response of the EBNA-4-specific CTLs is restored when the EBNA-1-encoded Gly–Ala repeat is deleted in the EBNA-1/EBNA-4 hybrid protein. These experiments suggest that the repetitive Gly–Ala sequence is prohibitive for CTL recognition. However, the level at which CTL recognition of EBNA-1 is inhibited is unknown.

The nonresponsiveness of CTLs to EBNA-1 is probably not due to a lack of MHC class I binding sequences within EBNA-1. A computer search predicts that EBNA-1 contains class I binding motifs to many human leukocyte antigen (HLA) alleles. Several epitopes have been identified that can bind class I molecules in an *in vitro* stabilization assay performed with TAP-deficient cells. Also, it is unlikely that the lack of immunogenicity is related to limitations of the T cell receptor. More likely, the generation of recognizable epitopes is blocked due to interference with antigen processing and presentation. Possible mechanisms include sequestration in a compartment from which no antigenic peptides can be generated and binding to a protein that prohibits entry into the processing pathway. Furthermore, EBNA-1 might interfere with proteolytic degradation by the proteasome (Fig. 9). This multicatalytic complex represents the major nonlysosomal proteolytic system in the cytosol and in the nucleus. As a by-product of protein degradation, the proteasome generates small peptides that may serve as antigens for cytotoxic T cells. In terms of immune escape, the failure of EBNA-1 to be degraded by the proteasome would act as a two-edged sword. The generation of potential targets for CTL recognition would be prohibited, whereas stability of EBNA-1 in the nucleus would be increased.

An observation that is interesting in the context of possible protease resistance is the binding of EBNA-1 to palindromic recognition sequences within the plasmid origin of replication (ori-P). The DNA recognition function has been mapped to a C-terminal domain of approximately 18 kDa. Upon DNA binding this region acquires protease resistance. In addition to the effect of the Gly–Ala repeat, the DNA binding property of EBNA-1 might contribute to the lack of recognition of T cell epitopes within the EBNA-1 sequence.

EBV Lytic Antigens

MHC class II molecules usually present epitopes that are derived from soluble, exogenous antigens processed via the endocytic route (Fig. 10 for a comparison of MHC class I and class II presentation pathways). However, sometimes proteins that do not enter the cell via endocytosis but are endogenously synthesized by the antigen presenting cell manage to enter the endocytic pathway and are presented in the context of MHC class II molecules. An example of this unusual antigen presentation route is the endogenously synthesized, EBV-encoded gp340, which can be recognized by CD4[+] T cells in the context of MHC class II molecules (Lee *et al.,* 1993). Usually, presentation via the class II-restricted endocytic pathway can be blocked with chloroquine: This drug inhibits proteolytic degradation of endocytosed proteins into small peptides. However, the class II-restricted presentation observed for gp340 is an exception: Processing appears to occur via a chloroquine-resistant pathway and requires a functional signal peptide within the gp340 sequence (Fig. 10).

During the EBV lytic cycle a glycoprotein, gp42, the product of the BZLF2 open reading frame, is expressed that plays a role in the infection of B cells by binding to a cell type-specific receptor. When the specificity of the interaction of the BZLF2 gene product with B cells was investigated its extracellular domain was found to bind to J chains of MHC class II HLA-DR molecules (Spriggs *et al.,* 1996). Interestingly, a soluble form of BZLF2 consisting of its extracellular domain linked to the Fc portion of IgG1 was able to inhibit antigen-driven proliferative responses of MHC class II-restricted T cells as well as class II-restricted cytotoxicity in mixed lymphocyte cultures. It remains to be established whether the expression of BZLF2 influences class II-restricted T cell responses in the course of productive EBV infection. Also, it is not known at what stage during the biosynthesis of MHC class II molecules the association with the BZLF2 glycoprotein occurs.

Another EBV gene product with interesting immunomodulatory properties is BCRF1. This protein demonstrates strong homology to human IL-10, with almost 90% identity at the amino acid level (Rickinson and Kieff, 1996; Kieff, 1996). Like IL-10, BCRF1 downregulates gamma interferon production and MHC class II expression on monocytes. Although the protein was found to induce local anergy to tumors, enhancement of CTL and NK cell responses has been reported as well. *In vitro* studies indicate that BCRF1 promotes virus-induced transformation of human B cells. BCRF1 is expressed late in EBV replication and some data suggest that the protein could also play a role in latent infection (Rickinson and Kieff, 1996; Kieff, 1996).

Immune Escape by EBV-Associated Tumors

The EBV-associated Burkitt's lymphoma (BL) is a classical example of a tumor that evades CTL recognition (Rickinson and Kieff, 1996; Rickinson and Moss, 1997; Kieff, 1996). This has been attributed to cell phenotype-related downregulation of certain MHC class I alleles, low expression of cell adhesion molecules and costimulatory molecules, and downregulation of all transformation-associated viral antigens except EBNA-1. Moreover, the peptide transporter proteins TAP1 and/or TAP2 appear to be

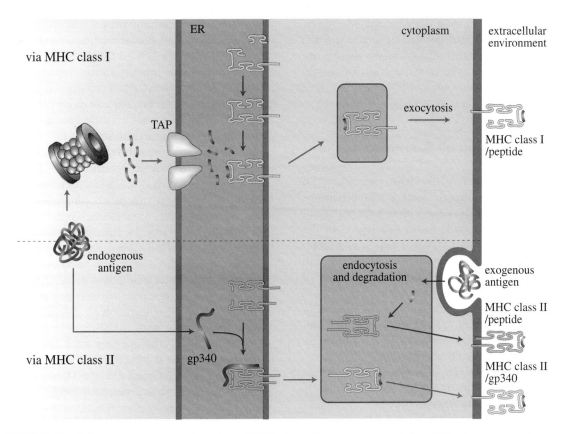

FIGURE 10 Schematic representation of the MHC class I (blue arrows) and class II (red arrows) restricted antigen presentation pathways. Usually, endogenously synthesized antigens enter the class I pathway, whereas exogenous antigens enter that of the class II. Occasionally, endogenously produced antigens are presented in the context of class II molecules, as in the case of the EBV-encoded gp340. Translocation of this protein, or part of it, to the ER via its signal peptide and premature binding to MHC class II in the ER is a possible explanation for the class II-restricted presentation of gp340.

downregulated consistently in many BL cell lines. It should be emphasized that multiple risk factors are involved in the pathogenesis of EBV-associated malignancies. In BL these include, in addition to infection and immortalization of B lymphocytes by EBV, other infections such as endemic malaria or AIDS, reciprocal translocations between chromosome 8 near the locus of the c-*myc* oncogene, either the Ig heavy chain locus on chromosome 14 or the light chain locus on chromosome 2 or 22, and other known and unknown factors.

The only latent gene invariably expressed in EBV-positive BL cells is EBNA-1. Studies on latent gene expression in nasopharyngeal carcinoma (NPC) cells indicated that EBNA-1 is present in nearly all cases, whereas LMP-1 and LMP-2α are detected in approximately 75% of the EBV-positive tumors (Rickinson and Kieff, 1996; Rickinson and Moss, 1997; Kieff, 1996). In EBV-associated Hodgkin's disease (HD) the pattern of latent gene expression usually resembles that seen in NPC. Due to its consistent expression

in all of these EBV-associated malignancies, EBNA-1 represents an important target for future research.

The immunoblastic lymphomas of the immunosuppressed comprise a special category of EBV-associated malignancies. All cases of posttransplant lymphomas and most cases of AIDS-related lymphomas appear to be EBV genome positive, consistently expressing latent and occasionally replicative genomes (Rickinson and Kieff, 1996). Usually, the full spectrum of latent antigens is expressed, but more heterogeneous expression has been observed as well. The failure to control latent (or active) EBV infection due to inhibition of T cell responses by immunosuppression is the major underlying factor of what appears to be an uncontrolled proliferation of EBV-infected B cells. Hence, the risk of lymphoproliferative disease may be reduced by increasing the level of EBV-specific immune surveillance, for example, by vaccination or by adoptive transfer. Since EBV gene expression is not limited to a small number of genes, as is the case in BL, NPC, and HD, all latent phase proteins should be included in an EBV vaccine. Moreover, if lytic antigens would be incorporated in such a vaccine, protection against productive infection might also be obtained. However, in order to be effective a vaccine should overcome the mechanisms by which EBV latent and lytic cycle antigens attenuate T cell recognition.

Immune Evasion by Adenoviruses

Adenoviruses have acquired two distinct strategies to avoid detection by host CTLs (Burgert, 1996). Both mechanisms reduce the availability of class I molecules on the surface of infected cells. Genes encoded in the *E1a* region in the genome of adenovirus strain 12 (Ad12) were shown to reduce the expression of class I molecules by interfering with the transcription of class I genes (Fig. 9). Expression of TAP molecules is also reduced by this mechanism. It is not known, however, whether peptide transport is reduced to the extend that assembly of MHC class I–peptide complexes is affected. Class I expression in cells is regulated by several proteins, including NF-κB, which bind to an element in the class I promoter. The *E1a* gene product inhibits the processing of precursors of NF-κB and consequently reduces the level of these two transcriptional activators.

The E3 region of the virus encodes a 19-kDa glycoprotein, E3-19k, which binds to several human and mouse class I alleles. In cells expressing the gene for E3-19k, there is reduced expression of class I molecules on the cell surface, although the synthesis of class I heavy or light chains remains unaltered. Examination of the maturation status of N-linked glycans on class I molecules recovered from such cells revealed that their egress from the ER had been blocked (Fig. 9). This retention in the ER caused by E3-19k binding was sufficient to reduce the availability of MHC molecules on the cell surface and consequently abrogate the lysis of infected cells by CTLs. The sequence responsible for retaining E3-19k in the ER has been identified and the disruption of the retention signal resulted in E3-19k molecules that matured to the cell surface and could not retain class I molecules in the ER.

Possible Applications of Viral Escape Mechanisms in Vaccine and Drug Development

Elucidation of the mechanisms by which viruses elude the immune system will facilitate the development of vaccines and drugs to treat infection with such viruses. For example, the genes of HCMV that encode the glycoproteins that cause retention (US3) and destruction (US2 and US11) of MHC class I molecules can be deleted from the viral genome. Likewise, in HSV-1 and -2 the ICP47 gene can be

eliminated. The resulting viruses will no longer interfere with MHC class I expression and, as a result, presentation to CTLs of peptides derived from other viral proteins will be restored. These mutant viruses represent interesting vaccine candidates. Instead of deleting the immunomodulatory molecules entirely, one can modify their active sites such that these proteins lose their destructive properties. Thus, these proteins can also serve as targets for a protective CTL response elicited by the vaccine. It is difficult to predict the efficacy of such vaccines. Theoretically, virus-infected cells in which 100% of the MHC class I molecules have been destroyed cannot be eliminated by CTLs induced by the vaccine. Most likely, however, immunization will shift the balance of the virus–host equilibrium toward a more rapid and efficient control of the infection.

Alternatively, the mechanisms used by the viruses described in this article could also be used to suppress an unwanted immune response, for example, in the case of autoimmune diseases or rejection of allogeneic tissue transplants. Interference with MHC class I-restricted antigen presentation in cells that are targets for a harmful attack by CTLs might rescue those cells from elimination. Possible applications of ICP47 and US2/US11 for the purpose of immune suppression are currently being investigated in several laboratories.

Another potential application of immunomodulatory viral gene products is in the field of gene therapy. One of the major problems encountered in gene therapy with viral vectors is the elimination of successfully transduced cells by the host immune system. Antigens derived from the gene therapy virus are presented at the cell surface in the context of MHC class I molecules and the infected cells are destroyed by CTLs. As a possible solution, the HCMV-encoded US2/US11 molecules or the HSV-derived ICP47 could be included in the gene therapy virus to downregulate MHC class I expression in infected cells, thereby rescuing the cells from elimination by CTLs.

Once the domains within US2 and US11 that determine the specificity for MHC class I have been identified, ligand specificity could be altered to target other molecules for destruction. For example, breakdown of HIV gp120 might be induced by reengineered US2/US11, engaging newly synthesized gp120 molecules. For this purpose, US2/US11 should be equipped with a natural ligand of gp120, i.e., the N-terminal domain of CD4 (Fig. 11).

Evidence is accumulating suggesting that ER to cytosol transport is not limited to US2/US11-induced degradation of class I molecules. Instead, this pathway may be a hitherto unknown route used by the cell to dispose misfolded proteins, thus providing an alternative to protein breakdown in the ER lumen. A better understanding of this pathway may allow the development of drugs inhibiting dislocation. Such inhibitors may have therapeutic applications. For example, in the case of cystic fibrosis, the *corpus de-*

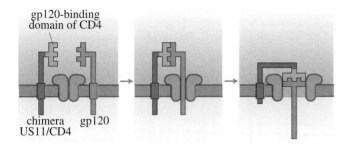

gp120-binding
domain of CD4

chimera   gp120
US11/CD4

FIGURE 11 Selective destruction of membrane proteins by dislocation, one of the possible therapeutic applications of viral evasion mechanisms. US11 is used by HCMV to downregulate expression of MHC class I molecules. A chimeric US11 glycoprotein may be used to target other proteins for destruction, for example, the HIV-encoded gp120. In this case, the specificity-determining domain of the US11 molecule would be replaced with the gp120-binding domain of CD4, the natural ligand of gp120.

licti, the mutant cystic fibrosis transmembrane regulator (CFTR) molecule is degraded. However, it is known that the most frequently occurring CFTR mutations yield variants that are still functional if they can reach the cell surface. Thus, if degradation could be inhibited by a drug, perhaps sufficient CFTR molecules could be rescued to restore function and cure the disease.

References Cited

AHN, K., ANGULO, A., GHAZAL, P., PETERSON, P. A., YANG, Y., and FRÜH, K. (1996). Human cytomegalovirus inhibits antigen presentation by a sequential multistep process. *Proc. Natl. Acad. Sci. USA* **93,** 10990–10995.

AHN, K., GRUHLER, A., GALOCHA, B., JONES, T. R., WIERTZ, E. J. H. J., PLOEGH, H. L., PETERSON, P. A., YANG, Y., and FRÜH, K. (1997). The ER-luminal domain of the HCMV glycoprotein US6 inhibits peptide translocation by TAP. *Immunity* **6,** 613–621.

BONIFACINO, J. S. (1996). Reversal of fortune for nascent proteins. *Nature* **384,** 405–406.

BURGERT, H. G. (1996). Subversion of the MHC class I antigen-presentation pathway by adenoviruses and herpes simplex viruses. *Trends Microbiol.* **4,** 107–112.

COUX, O., TANAKA, K., and GOLDBERG, A. L. (1996). Structure and functions of the 20S and 26S proteasomes. *Annu. Rev. Biochem.* **65,** 801–847.

DEL VAL, M., MUNCH, K., REDDEHASE, M. J., and KOSZINOWSKI, U. H. (1989). Presentation of CMV early immediate early antigen to cytolytic T lymphocytes is selectively prevented by viral genes expressed in the early phase. *Cell* **58,** 305–315.

FARRELL, H. E., VALLY, H., LYNCH, D. M., FLEMING, P., SHELLAM, G. R., SCALZO, A. A., and DAVIS-POYNTER, N. J. (1997). Inhibition of natural killer cells by a cytomegalovirus MHC class I homologue *in vivo. Nature* **386,** 510–514.

FRÜH, K., AHN, K., DJABALLAH, H., SEMPÉ, P., VAN ENDERT, P. M., TAMPÉ, R., PETERSON, P. A., and YANG, Y. (1995). A viral inhibitor of peptide transporters for antigen presentation. *Nature* **375,** 415–418.

GALOCHA, B., HILL, A., BARNETT, B. C., DOLAN, A., RAIMONDI, A., COOK, R. F., BRUMER, J., MCGEOCH, D. J., and PLOEGH, H. L. (1997). The active site of ICP47, a herpes simplex virus-encoded inhibitor of the major histocompatibility complex (MHC)-encoded peptide transporter associated with antigen processing (TAP), maps to the NH2-terminal 35 residues. *J. Exp. Med.* **185,** 1565–1572.

GILBERT, M. J., RIDDELL, S. R., PLACHTER, B., and GREENBERG, P. D. (1996). Cytomegalovirus selectively blocks antigen processing of its immediate-early gene product. *Nature* **384,** 720–722.

HEEMELS, M.-T., and PLOEGH, H. L. (1995). Generation, translocation and presentation of MHC class I-restricted peptides. *Annu. Rev. Biochem.* **64,** 463–491.

HENGEL, H., KOOPMAN, J. O., FLOHR, T., MURANYI, W., GOULMY, E., HÄMMERLING, G. J., KOSZINOWSKI, U. H., and MOMBURG, F. (1997). A viral ER resident glycoprotein inactivates the MHC encoded peptide transporter. *Immunity* **6,** 623–632.

HILL, A., JUGOVIC, P., YORK, I., RUSS, G., BENNINK, J., YEWDELL, J., PLOEGH, H. L., and JOHNSON, D. (1995). Herpes simplex virus turns off the TAP to evade host immunity. *Nature* **375,** 411–415.

HUGHES, E. A., HAMMOND, C., and CRESSWELL, P. (1997). Misfolded major histocompatibility complex class I heavy chains are translocated into the cytoplasm and degraded by the proteasome. *Proc. Natl. Acad. Sci. USA* **94,** 1896–1901.

HUPPA, J. B., and PLOEGH, H. L. (1997). The alpha chain of the T cell antigen receptor is degraded in the cytosol. *Immunity* **7,** 113–122.

JONES, T. R., HANSON, L. K., SUN, L., SLATER, J. S., STENBERG, R. M., and CAMPBELL, A. E. (1995). Multiple independent loci within the human cytomegalovirus unique short region downregulate expression of major histocompatibility complex class I heavy chains. *J. Virol.* **69,** 4830–4841.

JONES, T. R., WIERTZ, E. J. H. J., SUN, L., FISH, K. N., NELSON, J. A., and PLOEGH, H. L. (1996). Human cytomegalovirus US3 impairs transport and maturation of MHC Class I heavy chains. *Proc. Natl. Acad. Sci. USA* **93,** 11327–11333.

KHANNA, R., BURROWS, S. R., and MOSS, D. J. (1995). Immune regulation in Epstein–Barr virus-associated disease. *Microbiol. Rev.* **59,** 387–405.

KHANNA, R., BURROWS, S. R., MOSS, D. J., and SILINS, S. L. (1996). Peptide transporter (TAP-1 and TAP-2)-independent endogenous processing of Epstein–Barr virus (EBV) latent membrane protein 2A: Implications for cytotoxic T-lymphocyte control of EBV-associated malignancies. *J. Virol.* **70,** 5357–5362.

KIEFF, E. (1996). Epstein–Barr virus and its replication. In *Fields Virology* (B. N. Fields, D. M. Knipe, and P. M. Howley, Eds.), pp. 2343–2396. Lippincott-Raven, Philadelphia.

KLAUSNER, R. D., and SITIA, R. (1990). Protein degradation in the endoplasmic reticulum. *Cell* **62,** 611–614.

KLEIJNEN, M. F., HUPPA, J. B., LUCIN, P., MUKHERJEE, S., FARREL, H., CAMPBELL, A. E., KOSZINOWSKI, U. H., HILL, A. B., and PLOEGH, H. L. (1997). A mouse cytomegalovirus glycoprotein, gp34, forms a complex with folded class I MHC molecules in the ER which is not retained but is transported to the cell surface. *EMBO J.* **16,** 685–694.

LEE, S. P., WALLACE, L. E., MACKETT, M., ARRAND, J. R., SEARLE, P. F., ROWE, M., and RICKINSON, A. B. (1993). MHC class II-restricted presentation of endogenously synthesized antigen: Epstein–Barr virus transformed B cell lines can present the viral glycoprotein gp340 by two distinct pathways. *Int. Immunol.* **5,** 451–460.

LEE, S. P., THOMAS, W. A., BLAKE, N. W., and RICKINSON, A. B. (1996). Transporter (TAP)-independent processing of a multiple membrane-spanning protein, the Epstein–Barr virus latent membrane protein 2. *Eur. J. Immunol.* **26,** 1875–1883.

LEVITSKAYA, J., CORAM, M., LEVITSKY, V., IMREH, S., STEIGERWALD-MULLEN, P. M., KLEIN, G., KURILLA, M. G., and MASUCCI, M. G. (1995). Inhibition of antigen processing by the internal repeat region of the Epstein–Barr virus nuclear antigen-1. *Nature* **375,** 685–688.

MAUDSLEY, D. J., and POUND, J. D. (1991). Modulation of MHC antigen expression by viruses and oncogenes. *Immunol. Today* **12,** 429–431.

REYBURN, H. T., MANDELBOIM, O., VALÉS-GÓMEZ, M., DAVIS, D. M., PAZMANY, L., and STROMINGER, J. L. (1997). The class I MHC homologue of human cytomegalovirus inhibits attack by natural killer cells. *Nature* **386,** 514–517.

RICKINSON, A. B., and KIEFF, E. (1996). Epstein–Barr virus. In *Fields Virology* (B. N. Fields, D. M. Knipe, and P. M. Howley, Eds.), pp. 2397–2446. Lippincott-Raven, Philadelphia.

RICKINSON, A. B., and MOSS, D. J. (1997). Human cytotoxic T lymphocyte responses to Epstein–Barr virus infection. *Annu. Rev. Immunol.* **15,** 405–431.

RINALDO, C. R. J. R. (1994). Modulation of major histocompatibility complex antigen expression by viral infection. *Am. J. Pathol.* **144,** 637–650.

SPRIGGS, M. K., ARMITAGE, R. J., COMEAU, M. R., STROCKBINE, L., FARRAH, T., MACDUFF, B., ULRICH, D., ALDERSON, M. R., MULLBERG, J., and COHEN, J. I. (1996). The extracellular domain of the Epstein–Barr virus BZLF2 protein binds the HLA-DR beta chain and inhibits antigen presentation. *J. Virol.* **70,** 5557–5563.

STEVEN, N. M., LEESE, A. M., ANNELS, N. E., LEE, S. P., and RICKINSON, A. B. (1996). Epitope focusing in the primary cytotoxic T cell response to Epstein–Barr virus and its relationship to T cell memory. *J. Exp. Med.* **184,** 1801–1813.

THALE, R., SZEPAN, U., HENGEL, H., GEGINAT, G., LUCIN, P., and KOSZINOWSKI, U. H. (1995). Identification of the mouse cytomegalovirus genomic region affecting major histocompatibility complex class I molecule transport. *J. Virol.* **69,** 6098–6105.

TOWNSEND, A., and BODMER, H. (1989). Antigen recognition by class I restricted T-lymphocytes. *Annu. Rev. Immunol.* **7,** 601–624.

TRIVEDI, P., MASUCCI, M. G., WINBERG, G., and KLEIN, G. (1991). The Epstein–Barr-virus-encoded membrane protein LMP but not the nuclear antigen EBNA-1 induces rejection of transfected murine mammary carcinoma cells. *Int. J. Cancer* **48,** 794–800.

WIERTZ, E. J. H. J., JONES, T. R., SUN, L., BOGYO, M., GEUZE, H. J., and PLOEGH, H. L. (1996a). The human cytomegalovirus US11 gene product dislocates MHC class I heavy chains from the ER to the cytosol. *Cell* **84,** 769–779.

WIERTZ, E. J. H. J., TORTORELLA, D., BOGYO, M., YU, J., MOTHES, W., JONES, T. R., RAPOPORT, T. A., and PLOEGH, H. L. (1996b). Sec61-mediated transfer of a membrane protein from the endoplasmic reticulum to the proteasome for destruction. *Nature* **384,** 432–438.

YORK, I. A., ROOP, C., ANDREWS, D. W., RIDDELL, S. R., GRAHAM, F. L., and JOHNSON, D. C. (1994). A cytosolic herpes simplex virus protein inhibits antigen presentation to CD8$^+$ T lymphocytes. *Cell* **77,** 525–535.

ZIEGLER, H., THÄLE, R., LUCIN, P., MURANYI, W., FLOHR, T., HENGEL, H., FARRELL, H., RAWLINSON, W., and KOSZINOWSKI, U. H. (1997). A mouse cytomegalovirus glycoprotein retains MHC class I complexes in the ERGIC/cis-Golgi compartments. *Immunity* **6,** 57–66.

ZINKERNAGEL, R. M. (1996). Immunology taught by viruses. *Science* **271,** 173–178.

GUIDO POLI

AIDS Immunopathogenesis Unit
San Raffaele Scientific Institute
Milan, Italy

A Retrovirus Is Forever: The Lesson of HIV

HIV is the causative agent of AIDS, one of the most important plagues of this century. The anatomical dissection of its genome, completed in a very short time, has provided important indications for the development of pharmacological agents increasingly targeted to the different steps of the HIV life cycle. In this regard, the recent identification of the mechanism of viral entry into immune cells has generated new perspectives. However, the optimism that has accompanied the success of new therapies based on combinations of different antiretroviral agents has been diminished by the rediscovery of the fundamental property of any retrovirus: the capacity of latently infecting cells as a proviral DNA, the eradication of which represents the new frontier of biomedical research in this area. Substantial progress in the field of human genetics has allowed the discovery of genes conferring either resistance to HIV infection or the development of a more benign course of disease, thus indicating new strategies for therapy or prevention.

Introduction

At the beginning of the 1980s, the Centers for Disease Control of Atlanta, Georgia, noted in their bulletin *Morbidity and Mortality Weekly Report* the unusual emergence of deaths caused by *Pneumocystis carinii* pneumonia (PCP), an agent broadly present in the general population but unable to cause relevant diseases unless people also had important defects in their immune system, such as those observed in organ transplanted patients or in individuals with advanced tumors. The second characteristic of this epidemic was that all the individuals were male homosexuals, mostly living in the gay communities of San Francisco and New York City. Rapidly, the definition of "gay syndrome" was made to indicate both the particular type of sexual re-

lationship of the infected individuals and the fact that its etiology was unknown. Indeed, initially the cause of this new disease was thought to be the use or abuse of chemicals containing nitrites, frequently used as sexual boosters, rather than an infectious agent. However, shortly thereafter, PCP was described in intravenous drug users (IVDUs)—mostly, but not exclusively, males. In addition, other diseases, including tumors typical of immunocompromised individuals such as Kaposi's sarcoma and B cell lymphomas, were described in this population. All these diseases are considered opportunistic in that they are caused by infectious agents usually unable to harm individuals with an intact immune system. Although sexual promiscuity was quite common among IVDUs, the most relevant element of this population was undoubtedly the use of heavy drugs, such as heroin, as well as the habit of exchanging syringes and needles within the same groups of individuals gathering in "shooting galleries." The mixing of heroin and blood of different individuals represented a form of bonding within the group and a way of potentiating the effect of the drug. Finally, some patients affected by severe hemophilia developed the same immunodeficiency syndrome observed in homosexuals and IVDUs. These patients were receiving frequent transfusions of concentrates of clotting factor VIII (the absence of which is the cause of hemophilia) obtained by pools of several donors' plasma. This last observation indicated an infectious rather than toxic agent as the cause of the new clinical entity of unknown etiology that was quickly defined "acquired immunodeficiency syndrome" (AIDS) to underscore that it was a disease occurring in individuals that had been healthy and immunologically mature for a part of their lives. The acronym AIDS was unconsciously projected to become one of the most characterizing words of this century, symbolizing the potential of modern science to rapidly conquer knowledge that was impossible to obtain only a decade before and, at the same time, the limits in discovering rapid and effective solutions to emerging problems of public health. For these reasons, AIDS has definitively subverted the superficial rather than overoptimistic view matured during the 1960s and 1970s that infectious diseases would no longer be a major threat for humankind. This prejudice provoked several negative consequences, including the severe decrease in funding for research on infectious diseases, involving the most important biomedical centers in the world such as the National Institutes of Health (NIH) in Bethesda, Maryland.

Today, we know several features of AIDS. We know in great detail the anatomy of its etiologic agent, a virus belonging to the family of retroviruses named human immunodeficiency virus (HIV), and we have realized that no one on Earth can be considered immune from this infection. We know HIV's modalities of transmission, and we observe the paradox that this preventable disease continues to spread in vast areas of the world where poverty and ignorance prevail. By knowing the etiologic agent, we have developed therapeutic strategies to contrast its replication that are partially effective and that clearly indicate that the war against AIDS can be won. Also, with profound sadness, we are aware that the best therapeutic cocktails that can block, at least transiently, the disease progression today and that can even promote a partial recovery of the compromised immune defenses will unlikely be available for areas of the planet accounting for more than 90% of the HIV epidemic, areas that are afflicted with poverty, diseases, and the lack of the most elementary health supports such as clean water (Fig. 1). The best hope for this large portion of humanity is the development of a vaccine.

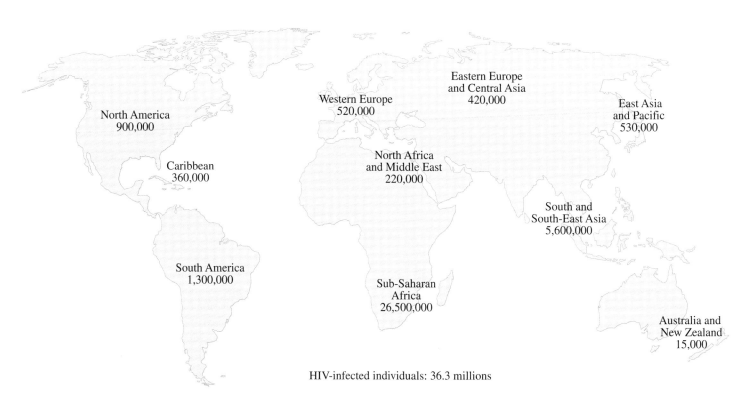

FIGURE I Cumulative distribution of HIV-infected individuals (adults and children) living with HIV/AIDS. It is clear that the most afflicted areas belong to the Third World (modified from UNAIDS/WHO, 2000).

In the following sections, the most solid notions regarding AIDS, with particular emphasis on its etiologic agent, the characteristics of its replication in the cells of the human immune system, and the interaction between the virus and the host will be discussed.

Anatomy and Physiology of HIV

HIV is the most important pathogenic human retrovirus identified to date. Indeed, HIV is responsible for the death of more than 30 million people in almost every area of the earth (Fig. 1) and it can infect the human *in utero* until old age. This virus is transmitted horizontally through sexual intercourse and via inoculation of infected blood or other body fluids. HIV infects the cells of the human immune system which express the CD4 molecule on their cell surface, including a subset of CD4$^+$ T lymphocytes and mononuclear phagocytes. It is possible that HIV can also infect cells lacking expression of CD4, but this hypothesis, with the exception of the astrocytes of the central nervous system (CNS), has not been proven *in vivo*. The interaction between HIV and CD4$^+$ cells results in their infection. *In vitro,* HIV infection of CD4$^+$ lymphocytes is followed by the replication of the virus and by cell death, thus providing a simple but quite convincing model of the main feature typically observed in HIV-infected individuals: a slow but inexorable decline of circulating CD4$^+$ T cells. The reduction of T lymphocytes in blood is a marker that predicts accurately the onset of the symptoms of the terminal phase of HIV infection, which is properly referred to as AIDS. As

A. S. Fauci of the NIH illustrated at the first World Conference on AIDS held in Atlanta in 1985, the CD4$^+$ T lymphocytes are "the orchestra director of the immune system"; consequently, their death caused directly by *in vitro* HIV infection appeared the most convincing explanation for the depletion of T lymphocytes typical in these individuals (Fauci, 1996).

Replication of the Viral Genome

The selectivity of HIV infection depends on the characteristics of its gp120 Env, a glycoprotein (gp) of the viral envelope with a mass of 1.2×10^5 kDa, which adapts to the CD4 molecule like a key in its lock. Binding of gp120 Env to CD4, comparable in terms of affinity to that of a hormone to its own receptor (10^{-9} M), allows a second glycoprotein of the viral envelope (gp41), capable of causing the fusion of biological membranes, to drastically change its conformation and to insert itself into the membrane of the target cell. The consequence is the injection of the content of the viral particle (or virion) in the cell cytoplasm. The injected components of the virions are made of a ribonucleoprotein complex formed by some Gag proteins, two copies of genomic viral RNA, and the key enzyme for this family of viruses: a RNA-dependent DNA polymerase or reverse transcriptase (RT) (Fig. 2). During the transfer of the preintegration complex from the cytoplasm to the nucleus, a process independent from ATP consumption, the RT begins to synthesize a DNA chain complementary to the genomic RNA, literally jumping both within a single strand and between the two strands of RNA (Fig. 3). In addition, the RT possesses another enzymatic function in a distinct domain of the protein, named RNAse-H, that

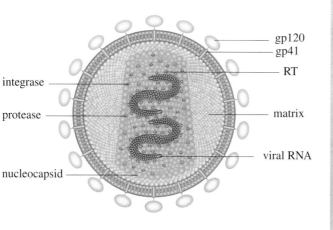
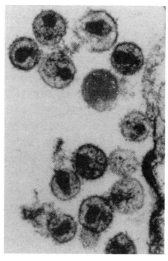

integrase

protease

nucleocapsid

a

gp120
gp41
RT

matrix

viral RNA

b

FIGURE 2 (a) Structure of HIV. The virus is made of a lipid bilayer (in gray), belonging to the cell from which the virion originated, in which the glycoproteins (gp) of the viral envelope, gp 120 and gp41, are inserted. Inside, there is a ribonucleoprotein complex containing two strands of genomic RNA and the enzymes reverse transcriptase (RT), integrase, and protease. b. Ultramicrograph of HIV particles.

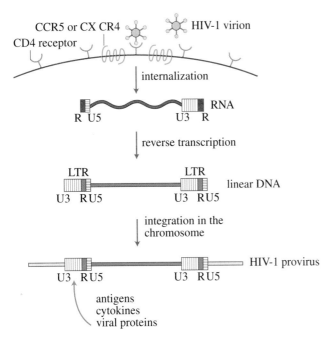

FIGURE 3 Early phase of HIV infection. The virion enters into the host cell and its RNA is reverse transcribed in linear double-helix DNA, which is then integrated in the host DNA (lighter blue), thus becoming a provirus. At the ends of the linear DNA repeated sequences (LTR) are present, subdivided into U3, R, and U5 regions. Antigens and cytokines can modulate either positively or negatively the levels of viral transcription, in synergy or in competition with viral proteins such as Tat and Rev (modified from Cullen and Greene, 1989).

progressively degrades the original RNA molecule forming the DNA–RNA hybrid, thus allowing the completion of the DNA double helix. HIV DNA is synthesized abundantly and in different molecular forms, including circular elements with one or two long terminal repeats (LTRs) and one linear form with two identical LTRs at the 5′ and 3′ ends. It is this last form that is randomly integrated into host chromosomes, as a "provirus," by the activity of another enzyme encoded by the *pol* gene, which is called integrase. It is from this very moment that "a retrovirus is forever" because, once integrated in the cellular genome, it will persist permanently for the entire life span of the infected cell and of the entire organism (Vogt, 1997). From this point of view, the biological strategy of retroviruses, and of HIV in particular, is much more subtle and dangerous for humankind than that of class 4 pathogens (e.g., filoviruses such as the Ebola virus) that are so cytopathic and aggressive that they burn out together with the infected host before having enough time to spread to several individuals.

The RT activity does not make just a carbon copy of RNA into DNA. During this process a sophisticated transposition of asymmetrically distributed nucleic acid se-

quences to the 5′ and 3′ ends of the viral genome occurs. The final result (Fig. 4) is the presence of identical LTRs at the two ends of the provirus. These sequences, further subdivided into U3, R, and U5 regions, although variable in their composition are common to all retroviruses and encompass the coding portion of the viral genome.

The Proviral Genome and Its Transcription

Conventionally, the HIV genome is subdivided into structural (*gag*, *pol*, and *env*) and, in the case of lentiviruses (the subfamily of retroviruses to which HIV belongs), regulatory genes that are crucial in order to understand both the pathogenic elements of the virus and the fundamental differences among retroviruses (Fig. 4). HIV-related retroviruses, discovered before HIV and named human T lymphotropic virus-I (HTLV-I) and HTLV-II, cause a moderate state of immunodeficiency *in vivo* and are associated with the pathogenesis of neurological diseases and a rare form of T cell leukemia of adults (ATLs). These viruses *in vitro* cause the neoplastic transformation and therefore they immortalize rather than kill cells, although they maintain a genetic configuration almost identical to that of HIV (Stevenson, 1997; Emerman and Malim, 1998). The role of their regulatory genes has only partially been determined (Fig. 4).

Among the structural genes, *gag* encodes a p55 precursor polyprotein that is processed by a viral protease (which is encoded by the *pol* gene, as is RT) in several small fragments, among which are the peptides forming the viral matrix, the capsid, and the nucleocapsid, which have distinct functions. Expression of Gag proteins is necessary and a sufficient condition for the production of new virions from an infected cell. The inhibition of the maturation of the Gag precursor is the fundamental principle of pharmacological agents that have recently provided a historical step in the therapy against HIV infection: the protease inhibitors (PIs). Cells treated with PIs still produce HIV particles, but as shown by ultramicroscopy these are characterized by an aberrant "donut-shape" morphology instead of the classical characteristics of virions, which have a double lipid bilayer and an electron-dense core with the shape of a cone (Fig. 2).

Among the regulatory genes, two are essential →*tat* and *rev.* These genes, together with *nef,* are considered early genes and are translated from the 2-kb mRNA which has been fully spliced from the original full-length 9-kb RNA molecule (Fig. 5) (Stevenson, 1997; Emerman and Malim, 1998). Tat, the RNA of which also derives, as does Rev, from two exons, represents the most important transactivator of viral transcription and acts predominantly by inhibiting the premature end of the synthesis of HIV RNA that is the result of the RNA polymerase II (RNA pol II). This enzyme, in the absence of Tat, initiates the transcription of the viral RNA, but it falls off after a few hundreds

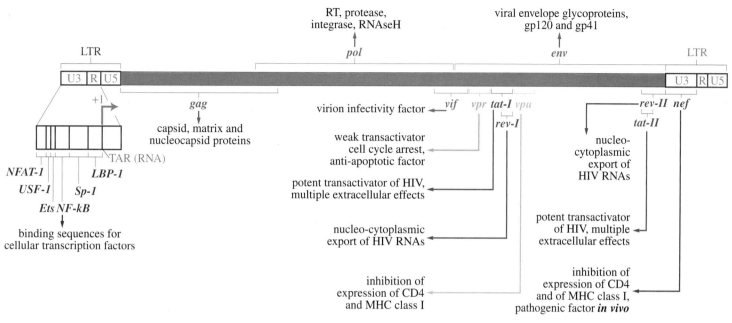

FIGURE 4 Schematic representation of the HIV provirus indicating the structural genes (in green), regulatory genes (in red), accessory genes, and the function of their transcribed products. (Left) The U3 and R regions are shown. The blue arrow indicates the site of initiation of transcription. TAR is the binding site for the TAT protein (see also Fig. 5).

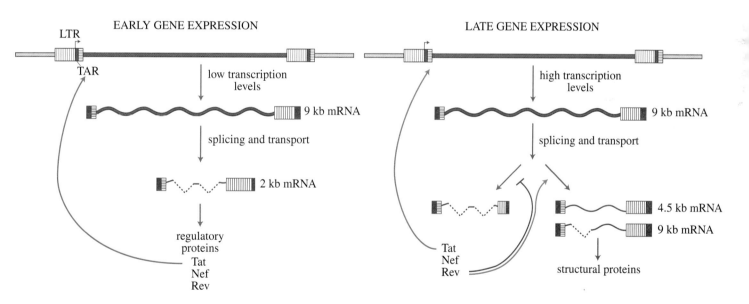

FIGURE 5 Regulation of the expression of early and late HIV genes. Transcription of early viral genes produces long transcripts (9-kb mRNAs); these are initially completely spliced and generate 2-kb mRNA molecules which are translated in the cytoplasm into Tat, Rev, and Nef regulatory proteins. Tat binds to the TAR RNA region (in violet), inhibiting the premature termination of viral mRNA synthesis and thus potentiating HIV transcription. Subsequently, the accumulation of Rev favors the nucleo-cytoplasmic export of the viral mRNA, allowing the expression of transcripts coding for the late gene products, first in the 4.5-kb RNA, which produces Env, VpU, and Vif, and then in the longer 9-kb transcripts, which are translated into the structural proteins Gag and Pol (modified from Cullen and Greene, 1989).

bases, thus generating short and abortive transcripts. The mechanism of action of Tat is dependent on its binding to a RNA sequence, defined as TAR (Tat-binding region), which forms a secondary structure in proximity to the transcription start site (Stevenson, 1997; Emerman and Malim, 1998). The role of Tat in an infected cell represents a classical example of a positive feedback in that, once synthesized, this protein can enter the cell nucleus by virtue of its nuclear localization signal (NLS) and further potentiate HIV transcription (Stevenson, 1997; Emerman and Malim, 1998).

Tat action would be sterile without the contribution of Rev, which is the main posttranscriptional regulator of HIV expression encoded by the virus. Rev, which also possesses a NLS, once within the cell nucleus binds to specific RNA sequences interspersed among the different viral genes named RREs (Rev responsive elements) (Stevenson, 1997; Emerman and Malim, 1998). Binding of Rev to the RRE sequences causes, through a complex mechanism, the export of 4.5 and 9kb HIV RNA from the nucleus to the cytoplasm. Therefore, in sequence, the first RNA transported across the nuclear membrane in the cytoplasm where it is translated into proteins is the 2-kb species, encoding Tat, Rev, and Nef; then, 4.5-kb RNA molecules, encoding Env and other accessory gene products such as Vif and Vpu, appear. Finally, the 9-kb, full-length RNA migrates from the nucleus to the cytoplasm and is translated into Gag and other regulatory proteins. Note that the only viral genes independent from Rev for the export and translation of their RNA are the early genes *tat, nef,* and *rev.*

The function of the other HIV accessory genes remains under investigation and is controversies; therefore, it is almost impossible to provide an appropriate synthesis of their function (Stevenson, 1997), with the exception of the *nef* gene. This gene was initially described as a negative factor (from which its acronym derives) functionally balancing the positive roles of Tat and Rev on viral replication. Subsequently, other studies demonstrated that *nef* was dispensable for virus replication and that the effects of its expression or absence during *in vitro* replication were unclear. Then, it was discovered that the main function attributable to Nef, and also to VpU, *in vitro* is the downregulation from the plasma membrane of important cell surface molecules such as CD4 and major histocompatibility complex (MHC) class I antigens (Ags), which play an important role in the recognition and killing of infected cells from cytotoxic T lymphocytes (CTLs). The role played by Nef *in vivo* has been studied using animal models involving the infection of macaques by the HIV-related simian immunodeficiency virus (SIV), a retrovirus that is unable to cause disease in its natural host, i.e., African monkeys (Fauci and Desrosiers, 1997). Macaques inoculated with SIV that had been deleted of the *nef* coding sequence were infected but did not pre-

sent signs of disease (Kestler *et al.,* 1991). In addition, these animals were resistant to the inoculation with wild-type SIV (Daniel *et al.,* 1992). In other words, these animals appeared vaccinated against pathogenic SIV and had become chronic carriers of an apathogenic infection (Daniel *et al.,* 1992). However, recent evidence both in animals and in humans indicates that attenuation by *nef* deletion or inactivation delays but does not prevent SIV/HIV diseases.

The HIV Life Cycle

The general strategy of HIV replication involves, after infection of CD4$^+$ cells, the reverse transcription of its genomic RNA into a copy of linear DNA with two complete and identical LTRs identically repeated at the 5′ and 3′ ends, its random integration in the chromosomal DNA of the host, and the coordinated expression of viral mRNAs encoding the different HIV proteins. The 9-kb HIV RNA encompasses not only the mRNA for the Gag proteins but also the genomic RNA to be incorporated into the new progeny virions. The newly synthesized viral proteins assemble in correspondence to the internal lipid bilayer of the plasma membrane, which is forced at critical points at which gp120 Env buds in the form of spikes from the plasma membrane (Swanstrom and Wills, 1997). Two RNA molecules are paired and incorporated in the core of the virions in association with the RT enzyme. The budding viral particles are then released from the infected cells after a final sealing process which depends on the small Gag protein (p6), a mutation of which generates the ultrastructural view of infected cells tethered by strings of virions incapable of detaching from one another and from the cell surface (Swanstrom and Wills, 1997). This observation underscores that every step of the HIV life cycle is a target of endogenous regulation by viral proteins as well as exogenous modulation by the host. In this regard, treatment of chronically infected cells by interferon-α (IFN-α) resulted in a block of the release of virions from the cell surface, although, unlike p6 Gag mutants, the particles were perfectly sealed (Poli *et al.,* 1989). The production of new progeny virions thus represents the last step of the virus life cycle and the premise for the infection of new target cells. However, infection of CD4$^+$ cells can also occur, and indeed it does occur with a higher efficiency, as the result of the contact between a productively infected cell and one or more uninfected cells. In this scenario, multiple points of attack will make the infectious process more efficient as a consequence of the physiologic interaction between intercellular adhesion molecules (ICAM), such as ICAM-1, -2, and -3, and their counterreceptor LFA-1. It has been reported that cells obtained from individuals genetically deficient for LFA-1 expression or as a consequence of the functional blockade of LFA-1 or ICAMs by monoclonal antibodies (mAbs) result in the diminished capacity of cell–cell spreading of HIV and, in particular, in the absence of cell fusion induced by the virus and causing the formation of giant cells, also named syncytia (Pantaleo *et al.,* 1991).

The Cellular Receptors
Involved in HIV Recognition

The process of gp120 Env recognition of the CD4 molecule is the crucial event which begins the infection of a cell. In theory, its blockade should protect the healthy cells from HIV and extinguish, or at least strongly contain, the replicative capacity of the virus and, consequently, its pathogenic potential. This hypothesis has been thoroughly investigated to the point of engineering chimeric immunoadhesins made of a soluble form of the constant component of the immunoglobulins and CD4 (sFc–CD4), resulting in increased half-life and stability *in vivo.* These molecules, although very effective *in vitro* in terms of inhibition of the replicative capacity of HIV (Perno *et al.,* 1990), fail in their objective once administered to infected individuals. Subsequently, it was discovered that the efficacy of sCD4 and its derivatives was limited to selected viral strains adapted to growth in laboratory conditions, such as the LAI/IIIB strain, the first isolate of HIV simultaneously identified by the laboratories of L. Montagner at the Pasteur Institute in Paris and of R. C. Gallo at the NIH in 1983 (Barrè-Sinoussi *et al.,* 1983; Gallo *et al.,* 1983). Paradoxically, these molecules favored the replication of primary HIV isolates obtained from infected individuals by increasing their fusogenic capacity. These studies, however, have received new vigor by the recent discovery of the second viral receptor (R's) family for HIV and SIV belonging to the family of chemokine R.

Since the mid-1980s investigators have attempted to identify a surrogate model of HIV infection in small animals that could, either naturally or by genetic manipulation, be infected and possibly develop an AIDS-related disease. After several false announcements, however, it became clear that this was unlikely. With the exception of the chimpanzee, no other animals has been susceptible to HIV infection. In addition, chimpanzees, although infectable, do not develop disease and their use remains limited. Transfection of the human CD4, i.e., the inoculation of a plasmid DNA containing the human gene, in murine cells or in cells of other small animals results in the correct transcription, translation, and membrane expression of the human molecule at densities comparable to those observed in human cells. These cells perfectly bound HIV particles with high affinity. The virus, however, remained attached to the external plasma membrane and it was incapable of proceeding in the infectious process. This observation triggered the search for the missing factor: a coreceptor molecule that, together with CD4, would allow the entry of the virus in the cell. This search remained vane for about a decade until, at the beginning of 1996, a team from the NIH announced the

discovery of the second R: a molecule spanning seven times the plasma membrane encompassing the R for the subclass of cytokines named chemokines because of their common property of eliciting the migration of immune cells. Transfection of this gene, previously cloned and renamed "fusin," conferred the susceptibility to HIV infection to nonhuman cells which became fusogenic if challenged with the viral Env gps in the presence of CD4 (Feng *et al.,* 1996). It was noted, however, that the susceptibility to infection was restricted to the viruses called T lymphotropic (T-tropic), whereas the other HIV subgroup, named macrophage tropic (M-tropic), was unable to infect animal cells coexpressing fusin and CD4. This suggested that the hypothesis that at least one additional coreceptor for this subset of viruses, presumably similar in structure to fusin, was valid. This hypothesis was corroborated by the discovery by another group at the NIH that three chemokines of the CC subfamily (RANTES, MIP-1α, and MIP-1β) were able to inhibit quite selectively the replication of M-tropic HIV (Cocchi *et al.,* 1995). A few months later, the second type of coreceptor was identified in the molecule known as CCR5, which turned out to be the receptor for the three chemokines capable of inhibiting HIV. Finally, in the same year, the ligand of fusin, which had been renamed CXCR4, was discovered in the CXC chemokine SDF-1. At this point, a coherent model of the first critical step of HIV infection of a CD4$^+$ cell could be made (Fig. 6). HIV virions, via gp120 Env, bind to the CD4 molecule, and this event induces a conformational change in gp120 Env which renders this molecule capable of binding the chemokine receptor (CCR5 for M-tropic strains and CXCR4 for T-tropic strains). The interaction among these molecules triggers a drastic conformational change in gp41 Env, which, like a jackknife, inserts itself into the plasma membrane of the target cell, thus allowing cell infection (Dimitrov, 1997). In addition to CCR5 and CXCR4, several other chemokine R's have demonstrated to function as viral coreceptors, although they are infrequently utilized by the virus. The engineering of cells of small animals expressing both CD4 and one or more chemokine R is under way.

Natural History of the Disease
and Therapeutic Strategies

Infection by HIV, contracted horizontally through sexual intercourse, transfusion of blood or derivatives, or vertically from mother to child, is followed for 1 or 2 weeks by using signs and symptoms that have been misinterpreted for years as the consequence of a strong flu of infectious mononucleosis, a disease characterized by painful enlargement of laterocervical lymph nodes, fever, malaise, and night sweats. Indeed, only later it was understood that this acute syn-

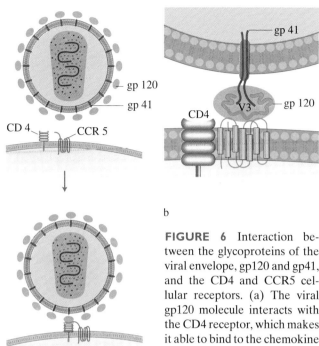

b

FIGURE 6 Interaction between the glycoproteins of the viral envelope, gp120 and gp41, and the CD4 and CCR5 cellular receptors. (a) The viral gp120 molecule interacts with the CD4 receptor, which makes it able to bind to the chemokine receptor CCR5 (as shown in Fig. 6b), inducing a conformational change in the gp41 molecule. This penetrates into the target cell membrane, which then fuses with the viral membrane, and the ribonucleoprotein complex is transferred to the host cell (modified from Cammack, 1998). (b) Interaction of gp120 and the CD4 and CCR5 receptors. The V3 region (in green) of gp120 is responsible for the binding (modified from Vicenzi and Poli, 1998).

a

drome, concomitant with the primary HIV infection (PHI), represents a crucial moment in the disease (Fig. 7) in that it coincides with the systemic dissemination of HIV and its seeding in organs and tissues, particularly in the lymph nodes (LNs), tonsils, and the diffuse lymphoid system such as the gut-associated lymphoid tissue. Ninety-eight percent of lymphocytes reside within these tissues, whereas only 2% of them circulate in the peripheral blood compartment. Studies in the early 1990s demonstrated that a first decrease in circulating CD4$^+$ T cells occurs during PHI, and that this decrease stops or is strongly attenuated during the transition from PHI to the subsequent asymptomatic and apparently silent phase of the disease. PHI ends within 2 or 3 months since the initial moment of infection, whereas the

following asymptomatic phase endures for 7 or 8 years on average (Fig. 8). It is during this time frame, in which an infected individual is in healthy condition and likely does not recall a bad flu that occurred a few months or years before, that HIV diffuses throughout society because the infected individual will continue a normal life and he or she will likely have sexual intercourse with one or more partners who will thus become exposed to the risk of being infected (Hecht *et al.,* 1998). In addition, the infected individual will possibly infect others accidentally as a consequence of blood exchange, either via transfusion in underdeveloped countries or by violent fighting (Brambilla *et al.,* 1997); also, if the infected individual is a young woman, she may become pregnant and deliver children. In this regard, statistical analyses indicate that approximately one of every four or five babies born from infected mothers used to be infected in developed countries, whereas this increases to one of two children in some African regions. It should be noted that treatment of infected women with AZT or Nevizepine, the first antiretroviral agent, during the last trimester of pregnancy or at birth has demonstrated a partial but highly significant clinical efficacy, with a reduction of approximately 50% of the number of infected babies. Finally, at the end of the asymptomatic phase, severe infec-

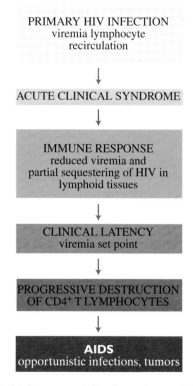

FIGURE 7 Main events following primary HIV infection (PHI). Viremia is measured as the number of molecules of viral RNA present in 1 ml of plasma (modified from Pantaleo *et al.,* 1993a).

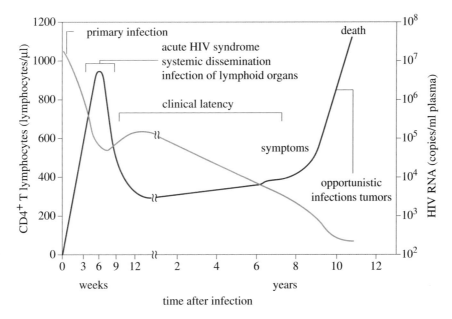

FIGURE 8 Diagram of the evolution of the number of CD4$^+$ T cells (in blue) and viremia (in red). Viremia is measured as the number of molecules of viral RNA present in 1 ml of plasma. The diagram represents a typical evolution occurring in most infected individuals. From the beginning of the infection until its evolution as AIDS, there is a progressive decrease in the number of T cells, whereas viremia levels increase significantly during the first weeks of infection and diminish in concomitance with the phase of clinical latency, to rise again after 8–10 years along with the onset of AIDS symptoms (modified from Pantaleo *et al.*, 1993b).

tious or neoplastic diseases typically associated to AIDS, such as Kaposi's sarcoma and B cell lymphomas, emerge.

The transition from PHI to the asymptomatic phase is associated with, and likely caused by, the onset of the immune response. The ultimate effect of HIV infection of immune cells, namely, the profound immunodeficiency state leading to AIDS, has seeded the implicit belief that the effect of the virus on the immune system is exclusively of a suppressive nature. In contrast, it has been demonstrated that HIV causes a profound activation of the immune system, although the immune response ultimately fails to control the infection and its pathogenic consequences. It is likely that the slow progression of the disease is the result of an equilibrium that is established between the replicative and cytopathic capacity of HIV and the host immune defenses. In individuals called rapid progressors (RPs), representing less than 10% of all infected individuals, the disease has an acute course and leads to death within 2 or 3 years from the moment of infection. The epidemiological counterpart of RP are long-term nonprogressors (LTNPs).

For several years the role played by the virus in the pathogenesis of disease has been highly debated and the discussion is ongoing. On the one hand, some investigators believe that HIV has all the pathogenic determinants to cause AIDS as a consequence of its ability to replicate in

humans. On the other hand, others think that the virus is only a trigger of a disease essentially mediated by autoimmune-like mechanisms. According to this hypothesis, the virus expresses antigenic epitopes shared by human cells, triggering an immune response which ultimately causes the organ disease and the immunological collapse (Susal *et al.*, 1996). Although evidence exists that similar mechanisms are indeed operational and play a role in the pathogenesis of the disease, there is little doubt that HIV plays a direct and central role in it, although this role has not been completely elucidated.

HIV Infection and the Host Immune Response

Once inside the human body, HIV reaches the LN system or the diffuse lymphoid system if the infection occurred through contaminated blood. Even before the formation of HIV-specific Abs, the Env gps are able to activate the complement cascade that can attach to the viral particles. These complexes, as well as the immune complexes formed by Abs and virions that will be formed within a few weeks after infection, are trapped in the network of follicular dendritic cells (FDCs) present in the germinal centers (GCs) of the LN. This cellular net has the function of retaining Ags in their native configuration and, in the case of viruses, as intact virions. Thus, Ags and viral particles are "frozen" in ultrastrucutures known as iccosomes which decor the den-

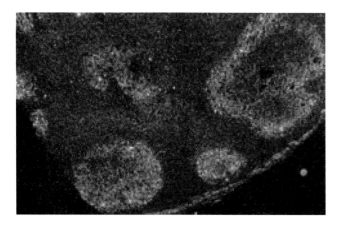

FIGURE 9 Electron micrograph of a section of lymph node, obtained from a patient in an early stage of infection by HIV, showing the presence of the viral RNA (white dots). The picture was obtained by eliminating proteinaceous components (antibodies and complement fragments) forming immunocomplexes which coat the RNA of the viral particles (reproduced with permission from Greene, 1993).

drites of FDCs. It has been demonstrated that virions rescued from FDCs retain their infectious capacity. These events occur during the initial symptomatic PHI and correspond to the clinical phase known as acute retroviral syndrome, with enlargement of the lymph nodes and constitutional symptoms. Suggestive pictures obtained by *in situ* hybridization show the GC decorated with HIV RNA as a result of the deposition and trapping of virions and immune complexes in the FDC network (Pantaleo *et al.,* 1993a) (Fig. 9).

Controversial results have been reported on the possibility that FDCs, which do not express CD4, can be infected by HIV. Their main pathogenic role is the exposure of virions on their cell surface which makes them a very efficient site of infection for $CD4^+$ T lymphocytes that are migrating through the GC. With the progression of the disease, this initial florid phase, with enlargement of GC and clinically resulting in a painful lymphoadenopathy, is replaced by an atrophic phase with destruction of the FDC network, which represents a scaffold for the GC. The mechanism of destruction of GC is unknown, but it is possible that it is the result of a sort of autoimmune attack by $CD8^+$ T lymphocytes restricted in their killing capacity by MHC class I (CTLs). This hypothesis is supported by the demonstration of the infiltration of activated $CD8^+$ T cells in the GC of the LN, a histopathological feature that has not been described in other viral or autoimmune diseases. The destruction of FDCs also results in the loss of a filter function of these cells with a consequent, according to some investigators (Pantaleo *et al.,* 1993a; Fauci, 1996), the direct overflow of HIV particles produced in LN in blood.

Viremia and Pathogenesis

A substantial decrease in the levels of HIV in the blood compartment, which is measured as the number of HIV RNA copies per milliliter of plasma (viremia), occurs during the transition from PHI to the asymptomatic phase in most infected individuals. Because every virion carries two copies of genomic RNA, by dividing the viremia number by 2 it is possible to determine precisely how many virions are present in the blood of an infected individual. The ability to perform quantitative determinations of HIV has resulted in the reproposed direct role of the virus as a critical determinant of the destruction of $CD4^+$ T lymphocytes (the most evident and prognostic marker of HIV disease progression toward AIDS). Studies led by D. D. Ho of the Aaron Diamond AIDS Research Center in New York City and by G. Shaw of the University of Alabama at Birmingham have calculated the viral turnover by using the potent antivirals, including PI, as probes. According to these studies, every infected individual reaches a sort of equilibrium with HIV that is directly revealed by the levels of viremia. This balance between production and destruction of virions, defined as set point, is reached approximately 6 months after PHI and remains relatively constant over time for each individual, at least during the asymptomatic phase (Ho *et al.,* 1995; Wei *et al.,* 1995).

A fundamental study has been conducted retrospectively on samples stored and frozen over time from a prospective cohort of infected individuals that were followed from the moment of seroconversion to the eventual AIDS phase or death. This study established that a single viremia determination performed up to 10 years before the onset of AIDS was capable of predicting with high accuracy who would have developed disease or would have remained symptom free after several years. For those who had levels of viremia higher than 36,000 copies/ml 10 year in the past, the likelihood of developing AIDS and eventually dying in the following years was significantly higher compared to that of individuals that had lower levels (Mellors *et al.,* 1996) (Fig. 10). Based on the concept of the viremia set point (Ho *et al.,* 1995; Wei *et al.,* 1995), Ho and Shaw established that the daily rate of virion production is on the order of 10^{10} and that the origin derives mostly (>90%) from $CD4^+$ T lymphocytes that are acutely infected and rapidly die as a consequence of HIV (Fig. 11).

Among the consequences of these results it has been calculated that an infected individual generates spontaneous viral variants every day because of the error-prone RT enzyme, which lacks the proofreading function possessed by the cellular polymerases. This process may generate drug-resistant variants even in the absence of antiviral agents and, in particular, if suboptimal antiretroviral therapies such as PI or RT inhibitor (RTI) monotherapies are administered.

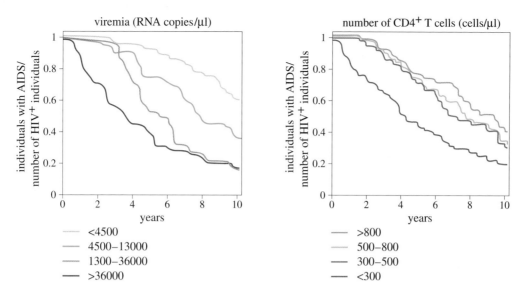

FIGURE 10 Diagrams (Kaplan–Meyer curves) showing the length of the pre-AIDS stage in HIV-infected individuals according to the levels of viremia (a) vs. the number of circulating CD4$^+$ T cells (b) at study entry. Viremia discriminates with high accuracy the likelihood of evolution toward AIDS, whereas only those individuals who had fewer than 300 CD4$^+$ T cells/μl at the beginning of the study demonstrated a higher probability of progression towards AIDS (modified from Mellors *et al.*, 1996).

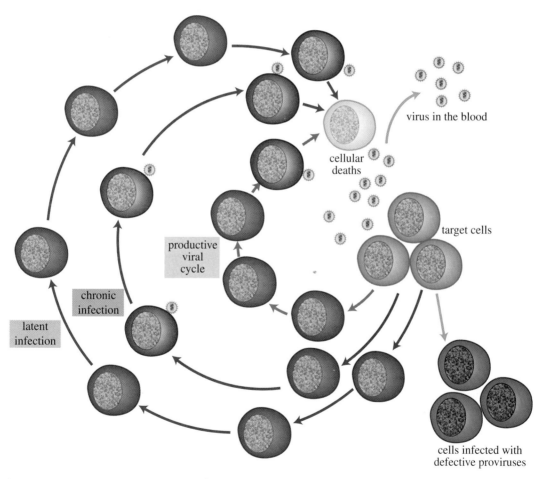

FIGURE 11 Turnover of HIV and CD4$^+$ T lymphocytes in patients infected with HIV. According to Ho and Shaw, most virus production on a daily basis in an infected person, on the order of billions of particles, derives from acutely infected CD4$^+$ T lymphocytes which undergo cell death as a consequence of the infection. The target cells (in light blue) can either undergo a cycle of rapid virus production or be cronically infected producing few virions. Lastly, the target cells can maintain the provirus, both defective (cells in blue) or replicative (cells in violet), in a latent form. (modified from Fauci and Desrosiers, 1997, p. 602).

Therapeutic Strategies

From the previous discussion, the dogma "treat early, treat hard" with potent cocktails of antiretroviral agents, including PI and RTI, has been derived for infected individuals. On the one hand, this strategy suppresses the replicative capacity of the virus, which represents the driving element of disease progression; on the other hand, these agents would diminish the likelihood of emergence of drug-resistant strains that may compromise the efficacy of antiviral agents to be used in the future (Ho *et al.*, 1995; Wei *et al.*, 1995). The residual viral production which is measured after highly aggressive antiretroviral therapy (HAART) has been attributed to different immune cells, including macrophages, latently infected T lymphocytes, and viruses released from FDC (Table 1). Studies based on complex mathematical models extrapolated from results obtained from infected individuals treated with antiretroviral agents (Ho *et al.*, 1995; Wei *et al.*, 1995) have resulted in a significant step forward in our knowledge of the infectious process but have also generated a premature and partially wrong optimism. The fundamental assumption that by completely or near completely blocking the acute replicative capacity of HIV (i.e., that caused by the *ex novo* infection of CD4⁺ T lymphocytes) a progressive elimination of the residual viral production by the effect of immune response or simply by cell senescence would have occurred was unrealistic. Suspension of HAART for up to 24 months of therapy and in the face of a constant suppression of viremia below the detection limit (the threshold of which is today equal to or less than 20 copies/ml of plasma) for the period of treatment has resulted in a discouraging rebound

of viremia to set point levels in a few weeks. In other words, the infection process restarted its natural evolution (Chun *et al.*, 1997; Finzi *et al.*, 1997). In addition, several groups have demonstrated the capacity to isolate infectious HIV from almost all individuals from circulating cells and precisely from latently infected resting memory CD4⁺ T lymphocytes (Chun *et al.*, 1997; Finzi *et al.*, 1997).

Unfortunately, HIV has revealed its retroviral nature, i.e., the capacity of remaining permanently integrated in the host genome ("a retrovirus is forever"). In addition, it is known that memory T lymphocytes can persist for decades, as it has been observed by studying the survivors of the atomic bombs dropped on Hiroshima and Nagasaki during World War II. The pessimistic rebound of these recent results has been further reinforced by the observation that several therapeutic cocktails administered today for obtaining the complete or near-complete suppression of HIV replication involve the self-administration of several pills on a daily basis with major problems of costs and therapeutic compliance. The irregular assumption of these drugs, as mentioned, causes an increase in viremia and, more important, favors the emergence of drug-resistant viruses. The dominance of these latter HIV strains is often the prelude to the total failure of antiretroviral therapy and to the inexorable progression of disease.

Although the possibilities of controlling the infection by pharmacological means have been disappointing (despite the fact that the positive impact on the progression of disease of the new combination therapies represents one of the most important achievements since the beginning of the epidemic) (Fig. 12) some positive and encouraging elements come from the study of the natural history of the disease (Fig. 10) and, in particular, from LTNP and the so-called exposed uninfected (EU) individuals.

LTNP and Resistance to HIV Infection

Epidemiological studies have demonstrated that some infected individuals live in a healthy condition and with a relatively intact immune system after 10 or more years since infection. We know that overall these LTNP, estimated to be between 1 and 5% of all infected individuals, belong to all risk categories (homosexuals, IVDUs, hemophiliacs, and children). The key element of their lower susceptibility not to HIV infection but to its consequences has not been determined. The two fundamental, and partially antagonistic, hypotheses are based on the virus and on the characteristics of the host; these characteristics can be further subdivided into genetic and epigenetic determinants.

Regarding HIV, based on the animal model of SIV-infected macaques discussed previously, most attention has been focused on the role of the *nef* gene. Although the majority of the studies published to date do not indicate the presence of particular deletions or mutations in this or other genes of LTNPs' HIV, combined with the fact that in-

TABLE I

Multiple Forms of HIV in the Infected Individual

HIV	HOST	COMMENTS
Proviral DNA	CD4⁺ T lymphocytes, macrophages	permanent infection, presence of defective viruses
Unintegrated HIV DNA	CD4⁺ T lymphocytes, macrophages	marker of acute infection
Viremia	plasma	marker of ongoing HIV replication
Extracellular pool	FDC (LN), lymphoid tissue	immuno-complexes of HIV, Ab and/or complement
Intracellular transcripts	CD4⁺ T lymphocytes, macrophages	both messenger and genomic RNA, markers of HIV replication

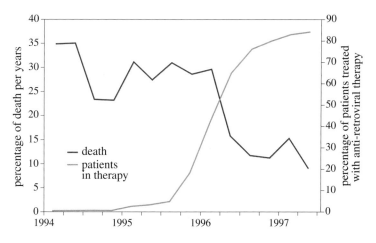

FIGURE 12 Effect of antiretroviral therapy, including an inhibitor of the HIV protease, on the mortality of HIV-positive advanced phase patients (less than 100 CD4$^+$ cells/ml) in the United States from 1994 to 1997 (modified from Palella *et al.*, 1998).

fectious virus could be isolated from most LTNP, some cases of *nef* deletion very similar to that observed in the SIV/macaque model have been reported in Australia (six cases), the United States (one case), and Italy (one case) (Deacon *et al.*, 1995; Salvi *et al.*, 1998). This suggests that the LTNP condition can be the result of infection by defective viruses.

Regarding the host, the most compelling results have been obtained from studies of population genetics. A homozygotic deletion of 32 bp in the gene encoding the CCR5 chemokine R (-32ccr5/-32ccr5) confers a near total resistance to HIV infection (EU condition), whereas the heterzyogotic configuration (-32ccr5/CCR5) is a marker of LTNP (Dean *et al.*, 1996; Huang *et al.*, 1996). The homozygous deletion of 32 bp in CCR5 causes the lack of expression of this receptor on the cell surface; therefore, these studies have further highlighted the importance of CCR5 in HIV infection and encourage the search for specific inhibitors. Simultaneously, these findings have reinforced the observation that the infection is always initiated by a CCR5-dependent HIV, whereas only after several years will viral variants that can use different chemokine R's in addition to CD4 emerge (Huang *et al.*, 1996). Other mutations present in the CCR2 gene (the receptor for MCP-1, -2, -3, and -4) and in the SDF-1 gene (the chemokine which binds fusin/CXCR4) have been related to the LTNP condition or to the lack of progression toward AIDS, respectively (Kostrikis *et al.*, 1998). In the future it is likely that other genetic variants correlated to these or other features of HIV infection will be discovered.

From an epigenetic standpoint, a definitive linkage to the LTNP condition has not been found. Virologically, although LTNPs usually have lower viremia levels compared to progressors, this correlation is not absolute and it is in-

deed not unusual to observe LTNPs characterized by stably high levels of viremia. At the immunological level, LTNPs have been demonstrated to possess a near normal capacity to mount B- or T-dependent responses. In other words, it must be determined if LTNPs are indeed so because of a strong immune response or, on the contrary, their immune system is still functional compared to progressors because these individuals are experiencing a less aggressive infection.

The existence of EU individuals, i.e., people who are resistant to infection in the face of continuous or frequent exposure to the virus, was initially postulated and then demonstrated by epidemiological studies performed on categories of individuals professionally exposed to HIV (mostly nurses), prostitutes, and homosexual men with a high number of partners. As mentioned, the strongest correlate of this lucky condition is represented by individuals homozygous for the 32ccr5 mutation (Dean *et al.*, 1996; Huang *et al.*, 1996). Immunological studies have evidenced that EU individuals possess CTLs directed against HIV, thus confirming that these individuals were actually exposed to the virus and that CTLs may indeed protect from infection. This finding has also been interpreted as an indication that HIV infection is the failure of the immune response, strongly unbalanced towards a nonprotective production of Ab (Clerici and Shearer, 1993), i.e., toward a T_H2 predominance of CD4$^+$ T helper responses (T_H2 lymphocytes mostly help the production of several classes of Abs, whereas T_H1 cells are predominantly involved in the induction of CTLs), a condition that has been linked to the worsening of parasite diseases. A controversial point of this theory is the fact that T_H1 responses are also readily demonstrable in infected individuals and precede the onset of anti-HIV Abs. In addition, the demonstration of CTLs underlies the fact that HIV infected at least some cells because its peptides need to be expressed in conjunction with class I MHC Ags and β_2 microglobulin through the endogenous pathway before being expressed on the cell surface. It is possible that so-called EU individuals (with the exception of those carrying the homozygous 32ccr5 deletion) have either been infected by very low amounts of HIV or these individuals have been exposed to defective viruses; both these hypotheses are quite unlikely in light of the specific risk categories and/or of the sexual habits of several EU individuals. In other words, EU individuals and LTNPs represent fascinating "experiments of nature" indicating that it is possible to achieve virtual complete protection from HIV infection and, in the case of infection, to develop a more benign disease and perhaps even a state of chronic asymptomatic carrier as observed in other viral infections such as that caused by hepatitis B virus. Uncovering the key to these two natural conditions in order to develop new preventive and therapeutic strategies is one of the most important goals of the research in this field.

Conclusions and Perspectives

The study of retroviruses was considered an exoteric science of minimal relevance for humankind (Vogt, 1997). At the beginning of the 1970s the discovery of the RT, an enzyme that subverted the central dogma of biology (i.e., that the information was rigidly passing from DNA to RNA to proteins) demonstrated that genetic information can travel upward from RNA to DNA in order to return in a RNA form. Even the Nobel prize awarded to H. Temin and D. Baltimore for the discovery of RT did not trigger the belief that human pathogenic retroviruses could exist or cause serious problems to mankind, although indications could be derived from a less rigid interpretation of the pathologies of other animal species (Vogt, 1997). The discovery of HTLV-I and HTLV-II, the first human pathogenic retroviruses, resulted from the tenacious belief of a few investigators. These discoveries anticipated the demonstration that AIDS was caused by a human retrovirus. Paraphrasing R. C. Gallo, if the epidemic would have been initiated only a decade before, it would not have been possible to rapidly identify its etiologic agent, to clone its genes, to set up a large-scale screening assay, and to identify antiretroviral agents targeting the steps of the HIV life cycle. The fundamental lesson of HIV has thus an importance which goes beyond the already serious problem of a plague affecting the world on a global scale, synergizing with historical problems of mankind such as poverty, malaria, and tuberculosis. We live in times in which everything must fall under the rigid laws of the market, including biomedical research that is required to provide immediate practical applications. The lesson from HIV is that humankind will possibly have to face new diseases, perhaps caused by retroviruses, if a generous and respectful contribution to the so-called basic research that, almost by definition, may appear as irrelevant for solving today's health problems is discontinued. Let us remember the lesson: A retrovirus is forever.

References Cited

BARRÈ-SINOUSSI, F., CHERMANN, J. C., REY, F., NUGEYRE, M. T., CHAMARET, S., GRUEST, J., DAUGUET, C., AXLER-BLIN, C., VEZINET-BRUN, F., ROUZIOUX, C., ROZENBAUM, W., and MONTAGNIER, L. (1983). Isolation of a T lymphotropic retrovirus from a patient at risk for acquired immune deficiency syndrome (AIDS). *Science* **220**, 868–871.

BRAMBILLA, A., PRISTERA, R., SALVATORI, F., POLI, G., and VICENZI, E. (1997). Transmission of HIV-1 and HBV by head-butting. *Lancet* **350**, 1370.

CAMMACK, N. (1998). Una "sana" competizione. *Ricerca Roche* **20**, 31–32.

CHUN, T. W., STUYVER, L., MIZELL, S. B., EHLER, L. A., MICAN, J. A., BASELER, M., LLOYD, A. L., NOWAK, M. A., and FAUCI, A. S. (1997). Presence of an inducible HIV-1 latent reservoir during highly active antiretroviral therapy. *Proc. Natl. Acad. Sci. USA* **94**, 13193–13197.

CLERICI, M., and SHEARER, G. M. (1993). A TH1–TH2 switch is a critical step in the etiology of HIV infection. *Immunol. Today* **14**, 107–111.

COCCHI, F., DEVICO, A. L., GARZINO DEMO, A., ARYA, S. K., GALLO, R., and LUSSO, P. (1995). Identification of RANTES, MIP-1α and MIP-1β as the major HIV-suppressive factors produced by CD8$^+$ T cells. *Science* **270**, 1811–1815.

CULLEN, B. R., and GREENE, W. C. (1989). Regulatory pathways governing HIV-1 replication. *Cell* **58**, 423.

DANIEL, M. D., KIRCHHOFF, F., CZAJAK, S. C., SEHGAL, P. K., and DESROSIERS, R. C. (1992). Protective effects of a live attenuated SIV vaccine with a deletion in the *nef* gene. *Science* **258**, 1938–1941.

DEACON, N. J., TSYKIN, A., SOLOMON, A., SMITH, K., LUDFORD-MENTING, M., HOOKER, D. J., MCPHEE, D. A., GREENWAY, A. L., ELLETT, A., and CHATFIELD, C. (1995). Genomic structure of an attenuated quasi species of HIV-1 from a blood transfusion donor and recipients. *Science* **270**, 988–991.

DEAN, M., CARRINGTON, M., WINKLER, C., HUTTLEY, G. A., SMITH, M. W., ALLIKMETS, R., GOEDERT, J. J., BUCHBINDER, S. P., VITTINGHOFF, E., GOMPERTS, E., DONFIELD, S., VLAHOV, D., KASLOW, R., SAAH, A., RINALDO, C., DETELS, R., and O'BRIEN, S. J. (1996). Genetic restriction of HIV-1 infection and progression to AIDS by a deletion allele of the CKR5 structural gene. Hemophilia Growth and Development Study, Multicenter AIDS Cohort Study, Multicenter Hemophilia Cohort Study, San Francisco City Cohort, ALIVE Study. *Science* **273**, 1856–1862.

DIMITROV, D. S. (1997). How do viruses enter cells? The HIV coreceptors teach us a lesson of complexity. *Cell* **91**, 721–730.

EMERMAN, M., and MALIM, M. H. (1998). HIV-1 regulatory/accessory genes: Keys to unraveling viral and host cell biology. *Science* **280**, 1880–1884.

FAUCI, A. S. (1996). Host factors and the pathogenesis of HIV-induced disease. *Nature* **384**, 529–534.

FAUCI, A. S., and DESROSIERS, R. C. (1997). Pathogenesis of HIV and SIV. In *Retroviruses* (J. M. Coffin, S. H. Hughes, and H. E. Varmus, Eds.), pp. 587–636. Cold Spring Harbor Laboratory Press, Cold Spring Harbor, NY.

FENG, Y., BRODER, C. C., KENNEDY, P. E., and BERGER, E. A. (1996). HIV-1 entry cofactor: Cloning of a seven-transmembrane G protein-coupled receptor. *Science* **272,** 872–877.

FINZI, D., HERMANKOVA, M., PIERSON, T., CARRUTH, L. M., BUCK, C., CHAISSON, R. E., QUINN, T. C., CHADWICK, K., MARGOLICK, J., BROOKMEYER, R., GALLANT, J., MARKOWITZ, M., HO, D. D., RICHMAN, D. D., and SILICIANO, R. F. (1997). Identification of a reservoir for HIV-1 in patients on highly active antiretrovial therapy. *Science* **278,** 1295–1300.

GALLO, R. C., SARIN, P. S., GELMANN, E. P., ROBERT-GUROFF, M., RICHARDSON, E., KALYANARAMAN, V. S., MANN, D., SIDHU, G. D., STAHL, R. E., ZOLLA-PAZNER, S., LEIBOWITCH, J., and POPOVIC, M. (1983). Isolation of human T-cell leukemia virus in acquired immune deficiency syndrome (AIDS). *Science* **220,** 865–867.

GREENE, W. C. (1993). AIDS and the immune system. *Sci. Am.,* 98–105.

HECHT, F. M., GRANT, R. M., PETROPOULOS, C. J., DILLON, B., CHESNEY, M. A., TIAN, H., HELLMANN, N. S., BANDRAPALLI, N. I., DIGILIO, L., BRANSON, B., and KAHN, J. O. (1998). Sexual transmission of an HIV-1 variant resistant to multiple reverse-transcriptase and protease inhibitors. *N. Engl. J. Med.* **339,** 307–311.

HO, D. D., NEUMANN, A. U., PERELSON, A. S., CHEN, W., LEONARD, J. M., and MARKOWITZ, M. (1995). Rapid turnover of plasma virions and CD4 lymphocytes in HIV-1 infection. *Nature* **373,** 123–126.

HUANG, Y., PAXTON, W. A., WOLINSKY, S. M., NEUMANN, A. U., ZHANG, L., HE, T., KANG, S., CERADINI, D., JIN, Z., YAZDANBAKHSH, K., KUNSTMAN, K., ERICKSON, D., DRAGON, E., LANDAU, N. R., PHAIR, J., HO, D. D., and KOUP, R. A. (1996). The role of a mutant CCR5 allele in HIV-1 transmission and disease progression. *Nature Med.* **2,** 1240–1243.

KESTLER, H. W., RINGLER, D. J., MORI, K., PANICALI, D. L., SEHGAL, P. K., DANIEL, M. D., and DESROSIERS, R. C. (1991). Importance of the *nef* gene for maintenance of high virus loads and for development of AIDS. *Cell* **65,** 651–662.

KOSTRIKIS, L. G., HUANG, Y., MOORE, J. P., WOLINSKY, S. M., ZHANG, L., GUO, Y., DEUTSCH, L., PHAIR, J., NEUMANN, A. U., and HO, D. D. (1998). A chemokine receptor CCR2 allele delays HIV-1 disease progression and is associated with a CCR5 promoter mutation. *Nature Med.* **4,** 350–353.

MELLORS, J. W., RINALDO, C. R., GUPTA, P., WHITE, R. M., TODD, J. A., and KINGSLEY, L. A. (1996). Prognosis in HIV-1 infection predicted by the quantity of virus in the plasma. *Science* **272,** 1167–1170.

PALELLA *et al.* (1998). Declining morbidity and mortality among patients with advanced human immunodeficiency virus infection. HIV Outpatient Study Investigators. *N. Engl. J. Med.* **26,** 338–360.

PANTALEO, G., BUTINI, L., GRAZIOSI, C., POLI, G., SCHNITTMAN, S. M., GALLIN, J. I., and FAUCI, A. S. (1991). LFA-1 is required for HIV-mediated cell fusion, but not rival spreading, in CD4$^+$ T lymphocytes. *J. Exp. Med.* **173,** 511–514.

PANTALEO, G., GRAZIOSI, C., DEMAREST, J. F., BUTINI, L., MONTRONI, M., FOX, C. H., ORENSTEIN, J. M., KOTLER, D. P., and FAUCI, A. S. (1993a). HIV infection is active and progressive in lymphoid tissue during the clinically latent stage of disease. *Nature* (*London*) **362,** 355–358.

PANTALEO, G., *et al.* (1993b). New concepts in the immunopathogenesis of human immunodeficiency virus infection. *N. Engl. J. Med.* **328,** 327–335.

PERNO, C. F., BASELER, M. W., BRODER, S., and YARCHOAN, R. (1990). Infection of monocytes by human immunodeficiency virus type 1 blocked by inhibitors of CD4-gp120 binding, even in the presence of enhancing antibodies. *J. Exp. Med.* **171,** 1043–1056.

POLI, G., ORENSTEIN, J. U., KINTER, A., FOLKS, T. M., and FAUCI, A. S. (1989). Interferon alpha but not AZT suppresses HIV expression in chronically infected cell lines. *Science* **244,** 575–577.

SALVI, R., GARBUGLIA, A. R., DI CARO, A., PULCIANI, S., MONTELL, F., and BENEDETTO, A. (1998). Grossly defective *nef* gene sequences in a human immunodefieciency virus type-1-seropositive long-term nonprogressor. *J. Virol.* **72,** 3646–3657.

STEVENSON, M. (1997). Molecular mechanisms for the regulation of HIV replication, persistence and latency. *AIDS* **11S,** 25–33.

SUSAL, C., DANIEL, V., and OPELZ, G. (1996). Does AIDS emerge from a disequilibrium between two complementary groups of molecules that mimic MHC? *Immunol. Today* **17,** 114–119.

SWANSTROM, R., and WILLS, J. W. (1997). Synthesis, assembly, and processing of viral proteins. In *Retroviruses* (J. M. Coffin, S. H. Hughes, and H. E. Varmus, Eds.), pp. 263–334. Cold Spring Harbor Laboratory Press, Cold Spring Harbor, NY.

UNAIDS/WHO (2001). *Report on the Global HIV/AIDS Epidemic.* UNAIDS/WHO, Geneva.

VICENZI, E., and POLI, G. (1998). HIV: Porte d'ingresso e uscieri buttafuori. *Ricerca Roche* **20,** 9–11.

VOGT, P. K. (1997). Historical introduction to the general properties of retroviruses. In *Retroviruses* (J. M. Coffin, S. H. Hughes, and H. E. Varmus, Eds.), pp. 1–26. Cold Spring Harbor Laboratory Press, Cold Spring Harbor, NY.

WEI, X., GHOSH, S. K., TAYLOR, M. E., JOHNSON, V. A., EMINI, E. A., DEUTSCH, P., LIFSON, J. D., BONHOEFFER, S., NOWAK, M. A., HAHN, B. H., SAAG, M. S., and SHAW, G. M. (1995). Viral dynamics in human immunodeficiency virus type 1 infection. *Nature* (*London*) **373,** 117–122.

General References

BARTLETT, J. G., and MORE, R. D. (1998). Improving HIV therapy. *Sci. Am.* **279,** 84–87.

COFFIN, J. M., HUGHES, S. H., and VARMUS, H. E. (Eds.) (1997). *Retroviruses.* Cold Spring Harbor Laboratory Press, Cold Spring Harbor, NY.

GALLO, R. C. (1986). The first human retrovirus. *Sci. Am.* **255**(6), 88–98.

GALLO, R. C. (1987). The AIDS virus. *Sci. Am.* **256,** 46–96.

VEAZEY, R. S., DEMARIA, M., CHALIFOUX, L. V., SHVETZ, D. E., PAULEY, D. R., KNIGHT, H. L., ROSENZWEIG, M., JOHNSON, R. P., DESROSIERS, R. C., and LACKNER, A. A. (1998). Gastrointestinal tract as a major site of CD4$^+$ T cell depletion and viral replication in SIV infection. *Science* **280,** 427–731.

ANTONELLO COVACCI
RINO RAPPUOLI
Siena, Italy

Vaccination

Vaccination, as described by Herodotus and Thucydides, has its roots in the ancient past and has been obscured for centuries by the symbols with which Medieval and Renaissance thought concealed it. In the past 50 years microbiology and cellular immunology have been reinterpreted due to the introduction of physical and mathematical methods. We can thus describe the molecular basis of immunization, design new molecules, and rationalize their use. In this article, which includes a historical introduction, we describe the foundations of humoral and cellular response, the criteria with which protective antigens are chosen, and their formulation. Moreover, we summarize the methods to identify potential candidates for new-generation vaccines, and we review the results obtained by vaccination experiments performed on animals through the parenteral administration of purified genes instead of their products, the detoxified proteins.

Vaccination in the Classical Tradition

The threshold that separates magic and religion, *teurgia* and *teologia*, is elusive: We have to consider these doctrines as overlapping and interacting. Classical thought assimilated art to technology and religion and natural forces were full of magical values: Every event was divine and immanent. It is not fortuitous that the attempts to modify the biological process or a physical event were carried out in conventicles or sects, and that they were transmitted orally in an initiatic language full of symbolic allusions. The difficulty of understanding has always been represented by the metaphor of an obscure vernacular. An introduction to vaccination must necessarily refer to myth, magic, and religion. In a memorable description of an epidemic, probably typhus, that oc-

curred in Athens in 430 BC, Thucydides narrates that the survivors, who were immune to reinfection, assisted the sick and the dying to attenuate the apparently unarrestable diffusion of the epidemic (Silverstein, 1989). The idea that every survivor was resistant is also present in a writing by Procopius on the pandemia known as the Justinian plague that broke out in 541 AD which had the characteristics of the bubonic plague (Silverstein, 1989). The first information on the practice of vaccination dates back to the Greek, Egyptian, Babylonian, and Indian and Chinese civilizations. In all these cultures it was common practice to control the spreading of smallpox through the introduction, in healthy persons, of organic material extracted from a lesion (Paul, 1993). In China, scabs and exudates were grinded to powder and nasal administration was performed; in India they were introduced subcutaneously; in Persia they were taken orally; finally, in the Arab world, there is mention of the use of fluid to be inoculated by means of a sharp stick. Science historians suggested that this practice had been independently set up in distant geographical areas. However, it is likely that it spread from a single diffusion center that is difficult to define because of the cultural migrations that have linked the Far East to the West, starting from the fifth millennium of the pre-Christian era. Each of these versions was firmly maintained until the eighteenth century. However, Arabic medicine has had immense popularity in Europe due to the contemporary presence of Jews, Christians, and Muslims in Spain. This gave birth to the period in which religion and magic merged with scientific needs (Walker, 1972). The ritual of scrofula healing through the touch of the hands of the Christian kings of France and England was introduced during the Middle Ages. In his most famous book, *Les Rois Taumaturges,* Marc Bloch accurately described the origin and the perpetuation of this false practice. Popular medicine was quickly hidden under the insignia of myth and often linked to magicians and witches. Girolamo Fracastoro described the agents of the disease as corpuscles (*seminaria*) that are capable of reproduction and of diffusion (*Seminario: De Sympathia et Antipathia Rerum Liber Unus,* Venice 1546), displaying an acute intuition on the mechanism of infections. That these beliefs were highly widespread can be deduced a from Athanasius Kircher, the mythomaniacal writer who reported to have seen the bacillus of the plague long before the existence of the instruments that were powerful enough to define the microorganisms. The studies performed by two generations of experts, including F. Yates, E. Garin, D. P. Walker, A. Warburg, P. Zambelli, and P. Rossi, show that the modern scientist put down his roots during the Middle Ages and rose in the Renaissance.

Beginning from the second century AD, when Rome had reached the peak of its wealth, until the ninth century AD, with the outset of the decline of the Muslim Empire, the monotheistic religions were used to justify and reinforce the empires of the world. The tensions between orthodoxy and heresy, inherent to monotheism, broke up the unity of the empires of Byzantium and Baghdad in the pluralistic Christian provinces of the East and of the Islam. Neoplatonists such as Plotinus, Jamblicus, and Proclus provide the reasons for making the ancient Gods compatible with monotheism, reinterpreting them metaphysically as hierarchical emanations of the One. In 1462, Cosimo de' Medici commissioned Marsilio Ficino, a priest and physician but above all a humanist, to translate into Latin Plato's works, starting with the *Corpus Hermeticum:* Ficino regarded Plato as a religious writer. His Renaissance syncretism included all that was common in the ancient world, such as the culture of mystic religions, astrology, and magic — *prisca theologia* (ancient theology). Movements that were apparently harmless grew increasingly stronger and the Christian reli-

gion joined a complex of mystical and esoteric doctrines, the Jewish *Cabala,* and the theory of chemical sympathies and the celestial influences constituted the basis of alchemy and astronomy (Yates, 1964).

The marble intarsia in the Cathedral of Siena, the work of Giovanni di Cosimo (1488) portraying Hermes Trismegistus (Fig. 1), preludes something that is unexpectedly new. The secret knowledge of a mythical character contemporary of Moses justifies and reinforces a religion that is unable to ignore the complexity of the physical world. The heterogeneous *Corpus Hermeticum,* which in the *Poimandres* includes a genesis, constitutes a form of gnostic mysticism, a magical religion dominated by suns, that offers to the esoterists the possibility of being transformed into powerful magicians and to operate on Nature. The idea that man was no longer thoroughly subordinated to God, that man could free himself through knowledge and modify the normal flow of nature, was strongly mediated by the reading of *The Ermetica* [Giovanni Pico della Mirandola: "magia est pars practica scientia naturalis" (Walker, 1972)]. For late medieval society, composing music without abiding to the canons or interfering with the course of a disease were considered outrageous; the passages from *Asclepius* that Ficino translated were the beginning of his divergences with the official religious doctrine. What had been moderate in Ficino and Pico with Agrippa became open rivalry toward Christianity; with Giordano Bruno the magical religion of the ancient Egyptians was relived and Christ became one of the many hermetic magicians. The trial against Galileo was just one of the effects of Bruno's strong philosophy: Even today, his political thought and his anticipation of science are buried by the symbolic and allegoric elements found in his books.

How vaccination was interpreted at the end of the fifteenth century, or how we think it could have been interpreted, is summarized in a famous representation by Lucas Cranach, *The Melancholic Witch* (Fig. 2). The painting is particularly complex and recalls a similar composition by Albrecht Dürer. His symbolism shows various levels of interpretation, some of which are described in the famous *Saturn and Melancholy* by R. Klibansky, E. Panofky, and F. Saxl. The woman sharpening a wooden stick represents an inspired witch. The analogy with the instrument used by the Arabs is clear, but the action can also be interpreted, from a pure alchemistic point of view, as the search for the essence, symbolized by the pith of the wood. Contrary to an inspiration protected by an angelic entity, as in Dürer, in which the senses are symbolized by the quiet lying dog, in Cranach's work the dogs are playing wildly, the senses are active, connected with devilish figures, indicating that the witch is in a sabbat. Other elements suggest that the painting was originally larger and that it was subsequently divided into two parts; the rhetorical elements that emerge express the condemnation of experimental knowledge.

Indeed, medical practice and vaccination were intensely

FIGURE 1 Giovanni di Cosimo: tarsia flooring of the Cathedral of Siena (1488), representing Hermes Trismegistus according to the tradition of Lattanzio. The inscription at his feet defines him as a contemporary of Moses (Moyse). The respectful figure wearing a turban is Moses. The character with a serious attitude represents a high-ranking Egyptian who takes part in the hermetic dialogues — perhaps Asclepius or Tet. Hermes has his left hand on a table ornated by sphinxes on which is inscribed a passage from *Ascelpius* (as quoted by Lattanzio) and from *Poimandres*. The sentence in Moses's book comes from a description of Hermes made by Cicero and reads more or less "dedicate yourselves to the letters and to the law, Egyptians" but probably constitutes an exhortation of the legislator of the Jews to the Egyptians. The tarsia is at the entrance of a Christian building, almost as if to announce its fortune and position in the Renaissance (photo courtesy of Scala).

demonized. Even in the healing of the scrofula, which actually never occurred, the sacred oil was severely prohibited and was substituted by common oil. The practice of the Arabs permeated the agrarian culture and reached Edward Jenner by indirect means (Silverstein, 1989). The philosophy of the chemical sympathies that Paracelsus and van Helmont adopted allowed the consolidation of an acquired immunity through the presence in the blood of balsamic forces able to neutralize infections, thus anticipating the physicochemical vision and repudiating the vitalistic principles. The hermetic ideas heavily influenced the work of great scientists, from Nicolas Copernicus to Galilei, up to Isaac Newton. In the writings of the latter it is explicitly

stated that, together with the understanding of the law of gravity and the system associated with it, Newton was also reevaluating the results obtained by Pythagoras, concealed by Apollo's myth and the seven-string lyre. His effort to establish the plans and the proportions of Solomon's Temple led, in his opinion, directly to God, who in the size of this construction asserted the divine plans of the Universe.

In Europe and the rest of the world, the practice of smallpox "variolation" was common. Born thousands of years before, the use of this practice continued to be ignored by the institutionalized medicine until two Italian doctors, E. Timoni and J. Pilarini, made it public to the Royal Society in London. Their description applied to the use of *alastrim*

FIGURE 2 Lucas Cranach: *The Melancholic Witch.* The painting is part of a series dedicated to the levels of inspiration protected by angelic entities. Note the contrast between the dogs in the foreground that represent sensuality and the dog in the background that represents insensitivity to material things (photo courtesy of the National Gallery of Scotland).

(a disease with a benign course), originally adopted by the Brahmins and practiced for purely lucrative reasons on the beautiful Circassian women destined to the Turkish sultan and to the Persian Sufi (Paul, 1993). In the era of high industrialization the vaccination became a social awareness that affected even the English Royal Family. Following numerous cases that coincided with an epidemic causing some apprehension for the life of the hereditary prince, Sir H. Sloane experimented with vaccination on human subjects. The attempt was defined as "the royal experiment" and the volunteers were granted pardon (Silverstein, 1989). Among the vaccinees, a woman was forced to sleep close to a sick young man and to look after him, which closely recalled the words of Thucydides. On April 1722, the royal

princes were vaccinated. This gave rise to a practice that Jenner brought to perfection in 1798. Jenner consulted many heresy suspects because they were aware that the millers were naturally resistant since they were in close contact with the vaccinic virus coming from infected animals. The extracted pus was inoculated in patients who were subsequently exposed to smallpox.

In 1768 the theory of spontaneous generation was disproved by Lazzaro Spallanzani. Vaccination developed: Louis Pasteur carried out experiments on the protection of poultry with attenuated *Pasteurella aviseptics;* E. von Behring experimented with sublethal doses of *Corynebacterium diphtheriae* filtrates; and P. Ehrlich and A. L.-C. Calmette assayed the immune reaction to the snake poison and to ricine. In 1890, von Behring and S. Kitasato succeeded in inducing a passive immune reaction using hyperimmune sera of tetanus antitoxin. Serotherapy was subsequently applied to other pathogens and the properties of the reactions provoked were described: bacteriolysis for *Vibrio cholerae* in 1894, precipitation for the plague in 1897, and agglutination in 1898. In 1977, after nearly a century, smallpox was definitely eradicated. The period between these two dates witnessed the birth of new vaccines that were able to efficiently combat many of the more dreadful infectious diseases and the development of two principles introduced by Gaston Ramon that are still the basis of modern vaccinology: the detoxification of a toxin into a toxoid following formalin treatment and the introduction of adjuvants, for example, aluminum hydroxide precipitates, slowly releasing the antigen from the inoculation site and inducing primary and secondary antibody responses (Silverstein, 1989).

Viral Vaccines and Bacterial Vaccines

Vaccination and immunization are synonyms: Whereas the first word is used mainly in medical practice, the second is applied in research. In order to be successful, vaccination must have an antigen, which induces a protective response, and an adjuvant that, administered with the antigen, guarantees the bioavailability and favors the immune response in areas otherwise difficult to stimulate. This combination must generate an immunological memory and not induce other effects, such as the appearance of autoimmunity (Kaufmann, 1996). To reach this target, studies are regularly performed on volunteers to evaluate the following: (phase 1) any undesired effects and an antigen response, (phase 2) the level of protection after exposure to the pathogen, and (phase 3) experiments that are carried out on volunteers in an area endemic for the infectious agent. Like any other classification, even this subdivision into phases is somewhat flexible; because of the nature of the microorganism, often too dangerous for direct and extended exposures, sometimes it is necessary to monitor the

TABLE I

Bacterial and Viral Vaccines Currently in Use That Were Obtained through the Use of Whole Organisms, Attenuated or Killed, or of Their Subunit, Purified from Organisms in Toto (Natural) or in Vitro, by Using DNA Recombinant Techniques (Recombinants)

ORGANISMS: BACTERIA AND VIRUSES	WHOLE	SUBUNIT
Adenovirus	Killed	
Bordatella pertussis	Killed	Natural and recombinant
Corynebacterium diphtheriae		Natural
Haemophilus influenzae		Natural
Hepatitis A virus	Killed	
Hepatitis B virus		Recombinant
Influenza virus	Killed	Natural
Measles	Attenuated	
Meningococcus A		Natural
Meningococcus C		Natural
Mumps	Attenuated	
Polio	Attenuated and killed	
Rabies	Killed	
Rubella	Attenuated	
C. tetani		Natural
Salmonella typhi	Attenuated	Natural
Chicken pox	Killed	

appearance and the decrease of protection. The list of available vaccines and those that can be effectively administered (Table 1) does not include all possible pathogens (Woodrow and Levine, 1990; Kaufmann, 1996). For the traditional vaccines not based on the use of genetic manipulation techniques, three formulations are described: attenuated, inactive, and subunit.

Viral Vaccines

Attenuated viral vaccines have shown excellent results, such as the control of poliomyelitis and smallpox; they are most likely to be developed because the knowledge of the biology of viruses has enormously progressed in the past 20 years (Atlas, 1997).

Attenuation is obtained through various strategies: (i) with repeated passages on cell cultures, as in the case of Sabin's polio vaccine which is composed of a mixture of three strains having different antigens (types 1–3), of which we know the mutations (type 1 has the mutations distributed along the length of the genome and practically does not undergo reversion, whereas type 3 more frequently undergoes reversion at a frequency of three doses every million and is responsible for cases of paralysis described in the vaccinees and in some of their families); (ii) administering, as in Jenner's smallpox vaccine, a bovine virus in the human or, as in the case of Marek's chicken disease, the herpes virus of the turkey; (iii) using naturally attenuated strains, as with the polio vaccine type 2; (iv) selecting ts mutants able to grow at lower temperatures and that are inhibited at temperatures higher than 37°C, as in the case of influenza virus type A; and (v) administering the wild-type virus through unnatural routes (Atlas, 1997). Adenoviruses encapsulated in an acid-resistant formulation that is orally administered and is soluble in the intestine have given excellent results (Paul, 1993). Attenuated vaccines confer long-term immunity with low concentration of antigen. The risk associated with their residual biological activity makes them ideal candidates for target manipulations. It is not surprising that most of the veterinary vaccines are attenuated vaccines, and that most viral vaccines currently being developed belong to this group.

Inactivated viral vaccines, such as rabies and polio, that are administered parenterally require antigen doses that are significantly higher and have antibody titers that are proportionally reduced, and, more important, they do not follow completely the natural route (Woodrow and Levine, 1990). Those vaccinated with attenuated polio, however, are prone to intestinal colonization by virulent poliovirus. Often, the high tropism of a virus for a particular tissue complicates the preparation and subsequent large-scale production. In these cases there is a tendency to use a component of a surface structure, as in the case of the first-generation hepatitis B vaccine. The discovery of the HbsAg antigen in the serum of Australian aborigines provided the basis for the preparation of a vaccine containing only viral components produced *in vitro* through the DNA recombinant technique (Lemon and Thomas, 1997).

Bacterial Vaccines

In contrast to the attenuated viral vaccines, bacterial vaccines have not generated promising results. Vaccination against tuberculosis using the Bacillus Calmette–Guérin strain (BCG) seems to have a better protection in Europe than in India; in both cases, however, the rates of success are lower than expected (Ramakrishnan *et al.,* 1997; Kaufmann, 1996) and do not increase even in experiments performed with mutants of *Salmonella enteritidis* serotypes *Typhi, galE,* or *aroA* (Miller *et al.,* 1994). The attenuated vaccines currently used are nevertheless less efficacious than the inactivated bacterial vaccines. Notwithstanding the large amount of research, the cholera or salmonella vaccines, prepared with inactivated cells, induce short-term immunity and are not practical except in subjects exposed for very short periods to the infectious agent.

In contrast, pertussis vaccination has produced positive

results, modifying, where systematically adopted, the incidence of the disease (Rappuoli, 1997). Despite the favorable evidence, many countries have refused the systemic use because of a side effect — an encephalopathy of variable severity displayed sporadically (1 in every 300,000 vaccinees). Historically, pertussis vaccination has been associated, as a trivalent combination, to the diphtheria and tetanus toxoids. The onset of the side effects described and the unfavorable opinion of the mass media, including specialized journals, have prohibited its total acceptance. The consequences are apparent: When it was introduced, vaccination reduced the disease by 85% but, at the moment of its lowest spread in the international health systems, there have been 120,000 cases of the disease recorded in two periods (1978 and 1982), with serious consequences in hundreds of cases and 28 deaths. However, the side effects of the vaccination involved a small number of people, considering that the final result would have been the vaccination of the whole population. In the 1980s, interest of researchers in *Bordetella pertussis* led to the first bacterial recombinant pertussis vaccine. As a result of the identification of the genes coding for the subunit of the pertussis toxin and the calculated introduction of two mutations that nullify toxicity while maintaining its immunogenicity, and with the purification of the virulence factors (such as pertactin and filamentous hemagglutinin), there is currently a vaccine that satisfies all safety criteria and that combines the tetanus and diphtheria toxoids (Pizza *et al.*, 1989).

The successes achieved with a bacterium do not annul the failures that characterize many bacterial vaccines based on inactivated cells. A considerable part of the world population suffers from diseases chronically associated with the infection of the soft tissues, including leprosy (Atlas, 1997). In the past few years, the use of armadillos as experimental animal models has made it possible to propagate, in a living system, an organism that as yet cannot be cultured *in vitro*. The results of much experimentation on the leprosy vaccination, however, are far from definitive.

Many pathogenic bacteria produce toxins that are secreted in the external milieu and exert their action at a distance. Among these, tetanus toxin and diphtheria toxin have a well-defined mechanism of action: that of tetanus toxin is an enzyme able to break a component of the synaptic transmission, the synaptobrevin, and that of diphtheria toxin is a selective inhibitor of the elongation factor (Schiavo *et al.*, 1992). A natural toxoid of diphtheria toxoid (CRM197) is also available.

The toxoids, although they have a structure that is superimposable to the toxins and conserve the capacity of stimulating an immune response, have lost the toxic properties; they are still the basis of the industrial processes for the production of trivalent and even pentavalent vaccines. They can even act as carriers for saccharide haptens (Woodrow and Levine, 1990; Kaufmann, 1996). The haptens per se are

not immunogenic but become so if associated with a macromolecule that acts as carrier. Pasteur's discovery of a capsular structure that surrounds *Streptococcus pneumoniae* and other pathogens, such as *Haemophilus influenzae,* meningococci, *Klebsialla pneumoniae,* and *Bacteroides fragilis,* has drawn attention to vaccines containing purified components of the capsule, such as the polysaccharides (Lipsitch and Moxon, 1997; Virji *et al.,* 1996). When the antibodies bind to the bacterial antigen, they favor phagocytosis through the interaction of the immunoglobulin with a receptor present on the surface of the leukocytes (opsonization). The capsule constitutes a physical barrier able to concentrate the active complement, C3b, in the hidden pockets present on the capsular surface, rarely accessible to the specific receptor and preventing or limiting phagocytosis. The pneumococcus vaccine available on the market contains 23 of the 25 serotypes associated with 95% of the diseases caused by this organism and its administration is recommended in the elderly. The influenza type B vaccine, even though in use, protects only adults, whereas the capsular polysaccharides of *Neisseria meningitidis* and *Escherichia coli* show a scarce immunogenicity in all subjects (Paul, 1993). The current trend is to use the polysaccharides together with a carrier, such as the diphtheria, tetanus, and cholera toxoids, or even with the surface proteins. Obviously, the length of the polysaccharide and the methods used for conjugation are some of the limiting factors (Virji *et al.,* 1996).

In addition to the conventional vaccines, the possibility of inducing therapeutic vaccination has begun to be evaluated. The introduction of techniques of passive immunization, by using humanized monoclonal antibodies, through selection for high affinity is limited to acute infections (Woodrow and Levine, 1990; Kaufmann, 1996).

The Cellular and Molecular Bases of Vaccinology

Infection is often associated with the appearance of specific antibodies that neutralize infectivity, inducing lysis of the pathogen or of the infected cell through the activation of the complement, and redirect the antigen–antibody complex toward specialized cells such as macrophages. It is generally expected that a vaccine will induce a reliable number of neutralizing antibodies. There are substantial differences between the reduction of the infectivity and a specific antibody class: If a virus replicates in a cell that has a receptor for the complement or for the Fc (the constant part of the immunoglobulins), the formation of an antigen–antibody complex can even facilitate infection. In fact, in these cases antibodies or complement act as Trojan horses, favoring the propagation of the virus.

The intricacies of connections that constitute the immune system require antibodies and cells strictly associated with soluble mediators, the cytokines, and cells that have a memory and specificity for the target. For many intracellular pathogens, the extracellular phase is the only one in which the antibody can bind the antigen, activate the complement, or favor phagocytosis (Paul, 1993). In principle, every protein can be transformed into smaller peptides able to bind to the molecules of the major histocompatibility complex (MHC) and also recognized by a T cell receptor (Zinkernagel and Doherty, 1997). Even though there are peptides that bind indifferently to both class I and class II MHC molecules, the majority of proteins exhibit selectivity for one or the other. Peptides that derive from the soluble fraction of the flavovirus selectively stimulate the $CD8^+$ cells, whereas the structural proteins are associated with the class II MHC.

Specific proteins act as superantigens instead of antigens. The staphylococcal enterotoxins are able to bind directly to class II MHC without being transformed into smaller peptides and without being influenced by the MHC polymorphism; they can activate many T cells simultaneously and favor the release of high quantities of cytokines with clinical and pathological effects similar to endotoxic shock. The effector T cells act through direct cytotoxicity or by releasing cytokines in the surrounding environment. Despite the fact that $CD8^+$ cells are generally cytotoxic and $CD4^+$ cells are helpers that participate in the delayed-type hypersensitivity, this distinction is not totally absolute. Furthermore, the isotype of an antibody is closely linked to the T cell response. Following the discovery of the existence of at least three types of helper T cells — Th1, Th2, and Th0 — it was possible to verify that Th1 cells, through Interferon-γ (IFN-γ), increase the production of IgG2a and Th2 via IL-4 increases the IgG1 and IgE and, via IL-5 and IL-6, the antibody responses of the IgA type. Given these circumstances, it is difficult to establish the necessary conditions to selectively activate one class or another.

In conclusion, the peptides that emerge from the fragmentation of the antigen associate with a class I and a class II MHC and are exposed by antigen presenting cells (APCs) to the T cells. The mechanisms according to which the antigens are processed are different: Exogenous proteins are incorporated in the APC through endocytosis and phagocytosis. They are then processed in the endosomes and the lysosomes, associated with the class II glycoproteins that come from the endoplasmatic reticulum and from the Golgi complex, and, finally, exposed on the surface where they are recognized by $CD4^+$ helper T cells by means of the T cell receptor. The endogenous proteins, synthesized *ex novo* in the cytosol, are fragmented in peptides of 8–10 amino acids by the proteosome in the endoplasmic reticulum by specific carrier proteins (TAP1 and TAP2) and associated with class I glycoproteins. The class I MHC complex and the peptides pass across the Golgi complex and reach the cellular surface, where they are recognized by the cytotoxic $CD8^+$ T cells (Zinkernagel and Doherty, 1997).

The mechanism of action of the adjuvants has not been completely clarified: Their putative function is to allow an antigen to produce a consistent and long-lasting immune response. Undoubtedly, they play two important roles: They increase the bioavailability by modifying antigen clearance, and they favor the immune response through the formation of an aggregate in T-dependent areas, able to strengthen the response of the B cells and the macrophages. The ability to direct the antigen toward particular anatomical areas is a critical aspect for every adjuvant. The particulate antigens, for example, are filtered by the M cells of the intestine that are specialized cells capable of phagocytosis of proteins and peptides without degrading them. They are subsequently released in a pocket that usually associates them with macrophages and T and B lymphocytes, and then they are gathered in lymphatic follicles. Soluble antigens, administered orally, are associated invariably with the epithelial cells enriched by class II antigens (Paul, 1993). The B subunit of the cholera toxin, which is the adjuvant of reference, seems not only to bind the ganglioside GM_1 receptor on the M cells but also to inhibit the activation of the TH1 cells with respect to TH2 (Douce *et al.*, 1995). As a result of the work of Freund on the suspension in the mineral oil of killed bacteria, and the subsequent granulomatoid reaction, modern pharmacology has included molecules deriving from the saponine and from squalene that are able to form reticular structures which limit the diffusion of the molecules (i.e., their bioavalibility) (Ott *et al.*, 1995).

In conclusion, every vaccine tends to satisfy the following requirements: the activation of the cells that present the antigen, the association with the antigens of histocompatibility, the activation of the B and T cells, persistence in the lymphatic tissues, the production of a repertoire of TH and TC and associated lymphokines, and an antibody response.

The Future of Vaccines

"Atomic" Vaccines

X-ray crystallography has allowed the three-dimensional description of very complex molecules, such as myoglobin, hemoglobulin, or even viral capsides. The folding of a protein is in part determined by its amino acid sequence: the charges of each residue and the presence of functional groups that are able to react chemically to form strong chemical bonds that dictate the folding. The monomers are flexible components useful for building complex structures in association with other molecules. Parallel computers allow the manipulation of these surfaces and the comparison of them through databases containing biological information organized via relations. The knowledge is the result of a data hierarchy that associated functional and structural motifs to the sequence. Unfortunately, we are unable to mathematically predict the immunological future of a particular region of a molecule. At best, we can predict linear epitopes that are specific for B and T cells, with limitations with regard to complexity and variety. Due to the accumulation of information on and interest in the structure and functions of the biological macromolecules, this approach will undoubtedly be very successful. The mathematical methods become increasingly complex, similar to the processors carrying out the operations that grow increasingly stronger.

An alternative to the computer methods is supplied by an ingenious system of biological filtering that exploits the repertoires of peptides or sequences that codify for small portions of a molecule, expressed on the surface of a bacterial virus and selected for their affinity toward antibody or toward a particular ligand. G. Cesareni and coworkers demonstrated the validity of these principles and the technique developed (Felici et al., 1991). However, it remains to be clarified how the sequences, once identified, will be presented to the immune system and how many times, after having been introduced through a phage, the vector can be reintroduced in the organism without hyperimmunization of the patient. Furthermore, it remains to be determined how many conjugate molecules can coexist in a carrier without causing a decreased instead of an increased response, which is an effect of the interactions between the molecules. Currently, it is believed that systems of possible vaccine candidates, rather than the vaccine, have been identified (Hensel et al., 1995).

Molecular Vaccines

Biologically active molecules can be expressed in bacteria, yeasts, and mammalian cells. The origin of the antigen influences the expression system: For example, molecules that require glycosylation to maintain their antigenic strength cannot be glycosylated in bacteria. Generally, bacterial toxins are expressed in the organism of origin because they need very complex mechanisms for secretion and exportation: The pertussis toxin, for example, requires that a group of genes, known as pertussis toxin liberation genes, be organized as a type IV secretion system (Winans et al., 1996). Viruses or viral empty particles reach the maximum efficacy in cell systems in continuous culture, from which they can be further purified. It does not seem feasible that the trend to purify components expressed in homologous or heterologous systems can be rapidly changed since these systems are undergoing substantial development (Woodrow and Levine, 1990; Kaufmann, 1996). The idea of avoiding purification by directly modifying viruses or bacteria capable of infecting the host is attractive, however. The vaccinic virus or *Salmonella* represent possible candidates because they can multiply and infect various animal species, including man. Unfortunately, we do not know to what extent a vaccinic virus or a *Salmonella*

TABLE 2

Vaccines in the Active Phase of Research (Clinical Phases I–III)[a]

ORGANISM	CLINICAL PHASE	
Blastomyces		
Borrelia	I/II	III
Brugia	—	—
Chlamydia	—	—
Cytomegalovirus	I/II	—
Cholera 0139	I/II	—
Dengue	—	—
Entomeba	—	—
Epstein–Barr	—	—
ETEC	I/II	—
Yellow fever	—	—
HCV	—	—
HDV	—	—
HEV	—	—
Herpes simplex 1, 2	I/II	—
HIV	I/II	—
Papillomavirus	—	—
Legionella	—	—
Leishmania	I/II	III
Malaria	I/II	III
Mycobacterium leprae	I/II	III
Mycobacterium tuberculosis	I/II	III
Mycoplasma	—	—
Salmonella typhi	I/II	III
Treponema pallidum	I/II	—
Varicella zoster	I/II	III

[a]Data are from Appendix A of the Jordan Report (1996) published by the National Institutes of Health.

can be filled with antigens without jeopardizing the expression, presentation, and completeness of the immune response. Even though the final target is the identification of polyvalent vaccines, current research (Table 2) indicates that a recombinant of the vaccinic virus able to vaccinate against both smallpox and hepatitis A must still be produced (Woodrow and Levine, 1990; Paul, 1993).

Transgenic Somatic Vaccination

Purified DNA can be inoculated in animal tissues and transcribed and translated *in situ*, producing quantities of antigens that are measurable and accessible to the immune system (McDonnell and Askari, 1996). This technique is known as transgenic somatic vaccination. The initial experiments were performed by using the reporter genes that usually codify an enzyme that are able to convert a chromogenic substrate. The DNA has been manipulated to develop the expression through the incorporation of inducible and tissue-specific promoters, with appropriate signals for efficient translation. This DNA can be produced in very large quantities and at very high levels of purity. The dimensions of the DNA are not limiting because the bacterial vectors can contain the information for the complex virus or for a whole eukaryotic gene, including the introns that play a part in the regulation of expression in determined biological compartments.

Some viruses, such as HIV, herpes simplex (human papilloma virus), and hepatitis C virus, represent a great challenge for modern vaccinology (Table 2).Promising results have been obtained by immunizing mice with DNA of influenza virus and BCG (Kumar and Sercarz, 1996). After 1 year, these mice showed high antibody titers and protection following reinfection. The permanent integration of the DNA in the host genome should solve the persistence problems because the antigen is produced for a long time. Knowledge of the function of the entire system is vague. The muscular cells preferably express class I antigens and are not suitable for the purpose. It can therefore be concluded that the DNA or the antigen expressed in the myocytes are captured by the macrophages or the dendritric cells, thus stimulating CD4$^+$ and CD8$^+$ cells. With the help of the CD4$^+$ cells, a quantifiable antibody response can be obtained. The antigen–antibody complexes are delivered on follicular dendrit cells and function as promoters of long-term memory. Studies show that TH1 responses can be obtained that favor bacterial immunity. Moreover, recent observations indicate that vaccination with DNA containing nucleoproteins of influenza virus, the least exposed to the antigenic variation, induces protection against lethal doses. Unfortunately, this type of vaccination exposes the cells to the DNA molecules that might be integrated, generating recombination with wild viruses and somatic inactivation of the host genes. These considerations indicate a serious problem because mutation accumulations can lead to very serious secondary pathologies.

A new generation of vectors that can be impaired by enzymatic enucleation of the transgene has appeared in some laboratories. This technique seems to offer great advantages and can be performed in a short period of time. Many genes can be selected simply by analyzing the animals that have shown protection without the researchers having to know beforehand which is the best antigen. This type of information can facilitate the development of vaccines consisting of single purified components.

Antitumour Vaccines and Vaccines for Birth Control

Tumorigenic progression, whatever its cause, implies the loss of homozygocity, the accumulation of punctiform mutations, and the acquisition of dominant characteristics. Cell reprogramming is an event that could take as long as

transformation and cannot be accomplished with current techniques. It is preferable to think that the antigens present on the surface of tumors are located only on the cells to be eliminated. Vaccination with molecules of this kind and the activation of the infiltrate lymphocytes could control and eliminate the malignant cells without a supporting chemotherapy. A few tumoral types are currently controlled with this strategy and the results are conflicting — some are convincing and some are disappointing. This approach cannot replace radiotherapy and chemotherapy, but in the future neoplasias such as melanoma could be cured by immunotherapy.

Some research groups are studying the possibility of accomplishing birth control by means of vaccination against the hormones that are responsible for reproduction or that are able to interfere with the growth of germinal cells. In this case, the use of vaccination implies a modification of a basic function and not an interference with a pathological process; moreover, these modifications can become permanent. While researchers concentrate on the animal world, governments of many countries seem to be interested in extending this type of vaccination to entire human populations and not only to consenting individuals. We have mentioned this extreme use of vaccination to be thorough.

The ancient practice of vaccination offers a rational solution to the circulation of the pathogenic microorganisms — that of eradication. In an era characterized by a marked increase of antibiotic misuse and the consequent massive spread of resistances, the only form of containment seems to be the biological elimination of the pathogenic agent. In these times of reductionism the phrase of the great architect M. van der Rohe, "more is less," seems to dominate even biology.

References Cited

ATLAS, R. M. (1997). *Principles of Microbiology,* 2nd ed. Brown, Dubuque, IA.

DOUCE, G., TURCOTTE, C., CROPLEY, I., ROBERTS, M., PIZZA, M., DOMENGHINI, M., RAPPUOLI, R., and DOUGAN, G. (1995). Mutants of *Escherichia coli* heat-labile toxin lacking ADP-ribosyltransferase activity act as nontoxic, mucosal adjuvants. *Proc. Natl. Acad. Sci. USA* **92,** 1644–1648.

FELICI, F., CASTAGNOLI, L., MUSACCHIO, A., JAPPELLI, R., and CESARENI, G. (1991). Selection of antibody ligands from a large library of oligopeptides expressed on a multivalent exposition vector. *J. Mol. Biol.* **222,** 301–310.

HENSEL, M., SHEA, J. E., GLEESON, C., JONES, M. D., DALTON, E., and HOLDEN, D. W. (1995). Simultaneous identification of bacterial virulence genes by negative selection. *Science* **269,** 400–403.

KAUFMANN, S. H. E. (Ed.) (1996). *Concepts for Vaccine Design.* de Gruyter, New York.

KUMAR, V., and SERCARZ, E. (1996). Genetic vaccination: The advantages of going naked. *Nature Med.* **2,** 857–859.

LEMON, S. M., and THOMAS, D. L. (1997). Vaccines to prevent viral hepatitis. *N. Engl. J. Med.* **336,** 196–204.

LIPSITCH, M., and MOXON, E. R. (1997). Virulence and transmissibility of pathogens: What is the relationship? *Trends Microbiol.* **5,** 31–36.

McDONNELL, W. M., and ASKARI, F. K. (1996). DNA vaccines. *N. Engl. J. Med.* **334,** 42–45.

MILLER, V. L., KAPER, J. B., PORTNOY, D. A., and ISBERG, R. R. (Eds.) (1994). *Molecular Genetics of Bacterial Pathogenesis.* ASM Press, Washington, DC.

OTT, G., BARCHFELD, G. L., and VAN NEST, G. (1995). Enhancement of humoral response against human influenza vaccine with the simple submicron oil/water emulsion adjuvant MF59. *Vaccine* **13,** 1557–1562.

PAUL, W. E. (Ed.) (1993). *Fundamental Immunology,* 3rd ed. Raven Press, New York.

PIZZA, M., COVACCI, A., BARTOLONI, A., PERUGINI, M., NENCIONI, L., DE MAGISTRIS, M. T., VILLA, L., NUCCI, D., MANETTI, R., BUGNOLI, M., and RAPPUOLI, R. (1989). Mutants of pertussis toxin suitable for vaccine development. *Science* **246,** 497–500.

RAMAKRISHNAN, L., VALDIVIA, R. H., McKERROW, J. H., and FALKOW, S. (1997). *Mycobacterium marinum* causes both long-term subclinical infection and acute disease in the leopard frog (*Rana pipiens*). *Infect. Immun.* **65,** 767–773.

RAPPUOLI, R. (1997). Rational design of vaccines. *Nature Med.* **3,** 374–376.

SCHIAVO, G., BENFENATI, F., POULAIN, B., ROSSETTO, O., POLVERINO DE LAURETO, P., DASGUPTA, B. R., and MONTECUCCO, C. (1992). Tetanus and botulinum-B neurotoxins block neurotransmitter release by proteolytic cleavage of synaptobrevin. *Nature* **359,** 832–835.

SILVERSTEIN, A. M. (1989). *A History of Immunology.* Academic Press, San Diego.

VIRJI, M., MAKEPEACE, K., PEAK, I. R., FERGUSON, D. J., and MOXON, E. R. (1996). Pathogenic mechanisms of *Neisseria meningitidis. Ann. N. Y. Acad. Sci.* **797,** 273–276.

WALKER, D. P. (1972). *The Ancient Theology.* Cornell Univ. Press, Ithaca, NY.

WINANS, S. C., BURNS, D. L., and CHRISTIE, P. J. (1996). Adaptation of a conjugal transfer system for the export of pathogenic macromolecules. *Trends Microbiol.* **4,** 64–68.

WOODROW, G. C., and LEVINE, M. M. (Eds.) (1990). *New Generation Vaccines.* Dekker, New York.

YATES, F. A. (1964). *Giordano Bruno and the Hermetic Tradition.* Univ. of Chicago Press, Chicago.

ZINKERNAGEL, R. M., and DOHERTY, P. C. (1997). The discovery of MHC restriction. *Immunol. Today* **18,** 14–17.

MICHAEL J. OWEN

Imperial Cancer Research Fund
Lincoln's Inn Fields
London WC2A 3PX, United Kingdom

The immune system comprises a complex network of interacting pathways in which multiple cell types interact with a myriad of different cytokines and chemokines to generate an immune response that is appropriate for the antigenic insult suffered by the organism. A breakdown of the regulation of this integrated response, or the dysfunction in a particular part of the response, often results in immune deficiency. The ability to manipulate the mouse genome, either by ectopic expression of genes or by the generation of mutants congenitally deficient in gene expression, has enabled immunologists to generate mice that are models of immunodeficiency, some of which recapitulate human pathologies.

Transgenic and Knockout Mice as Models of Immune Deficiency

The Interactive Immune System

The immune system comprises essentially two types of cell. Lymphocytes are responsible for the antigen-specific immune response. B lymphocytes bind intact antigen and mature to produce high-affinity antibody. T lymphocytes recognize cell-bound antigen resulting from the processing of intact antigen and the presentation of peptide fragments bound in the groove of major histocompatibility complex (MHC) class I or class II antigens. Phagocytes engulf and destroy infected cells or microorganisms, processes collectively referred to as the innate immune response. Crucially, however, these arms of the immune system do not act independently. Thus, antigen presenting cells such as macrophages, dendritic cells, and B cells interact with helper T (Th) cells. Cytokines produced by Th cells drive the establishment of the functional program of B cells, cytotoxic T (Tc) cells, and phagocytes. A consequence of this interacting network of immune cells is that disruption of any part is likely to result in immune deficiency, the symptoms of which may be complex and may not give an obvious indication of the primary lesion. The genetic defect resulting in the observed immunodeficiency in humans may have been identified, but the immune dysfunction may be difficult to study because of the problems in obtaining suitable tissues. The ability to manipulate the mouse genome, either by ectopic expression of a gene product or by the generation of targeted mutations, provides

ESTABLISHING A HOMOZYGOUS LINE FOLLOWING BLASTOCYTE INJECTION

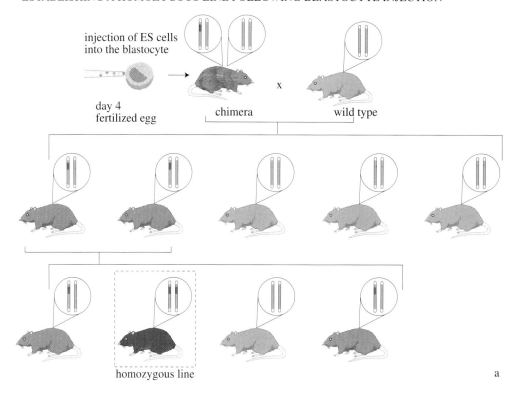

ESTABLISHING A TRANSGENIC FOUNDER LINE

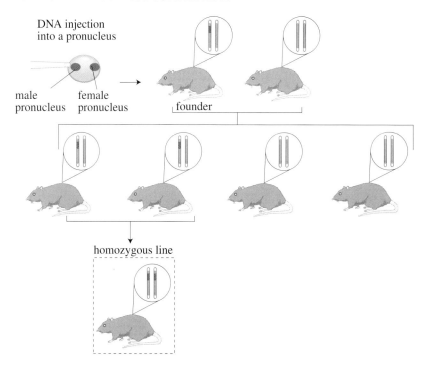

FIGURE 1 The principles of gene targeting and production of transgenic mice. (a) To establish a mouse line harboring a targeted integration event, embryonic stem cell clones (ES cells) containing the targeting event are injected into a blastocyst and the blastocyst is implanted into a pseudopregnant foster mother. Resulting chimeras are bred to wild-type mice to give offspring that are heterozygous for the targeted mutation. Further breeding between heterozygotes produces a homozygous line. Analysis of the various generations of mice is facilitated because the strains from which the ES cell and blastocyst were derived have different skin color genotypes. (b) For random integration of a gene into the mouse genome, a DNA fragment is injected into the male pronucleus of a fertilized egg which is reimplanted into a pseudopregnant foster mother. The resulting offspring are heterozygous for transgene expression. Further breeding produces a homozygous line.

a means of overcoming both of the previously mentioned limitations.

The Principle of Transgenesis and Gene Targeting

The principles of transgenesis and gene targeting are shown in Figure 1. Transgenesis involves the direct injection of a DNA fragment containing the gene of interest with appropriate regulatory elements into the male pronucleus of a fertilized egg. Integration of the input DNA, usually as concatamers, into the genome occurs at random. The expression of the transgene is often influenced by the surrounding chromatin. These position effects can result in the extinguishing or inappropriate expression of the transgene, depending on the site of integration. Thus, it is important to analyze several founder lines to avoid potential artifacts due to position effects.

Gene targeting technology involves the introduction of mutations into the genome of embryonic stem (ES) cells. Targeted ES cells are introduced into blastocysts to generate chimeric mice. If these ES cells contribute to the germ cells, the introduced mutation will be carried through to subsequent generations in which homozygous or heterozygote progeny can be studied.

Immunologists were among the earliest scientists to recognize the power of transgenesis and gene targeting. Early examples were the generation of transgenic mice expressing rearranged T cell receptor (TCR) or Ig genes. These mice have yielded much valuable information on the processes of allelic exclusion, tolerance, and MHC restriction. Gene targeting has proved an even more powerful approach. One of the first targeted mutations of any locus to be transmitted to the germline was in the β_2-microglobulin gene, which encodes one of the components of MHC class I antigens. Examples of genes involved in the immune system that have been targeted are shown in Table 1. These genes encompass multiple functional categories and mice defective in their expression have been invaluable for the dissection of the immune system. It is clearly impossible to discuss in detail the phenotypes of each of these mouse mutants. Rather, the principles and power of this technology for the study of immune dysfunction will be illustrated by reference to some key examples. This article discussed primarily the T lymphocyte lineage to demonstrate the utility of this approach in dissecting the biology of a complex developmental lineage.

Lymphoid Development

T Cell Development

The importance of many genes involved in the regulation of T cell development has been established using gene

TABLE I.
Targeted Mutations in the Immune System

LIGANDS	
Membrane bound	B7 (CD28 ligand)
	CD40L (CD40 ligand)
Cytokines/chemokines	Ifnγ, IL-2, Il-4, IL-6, IL-7, IL-8, IL-10, IL-12, LIF, TNF-β, IL-1β

SURFACE RECEPTORS	
Antigen receptors	IgD, E, M, Igκ, J$_H$-Eμ, J$_H$, TCRα, β, δ, CD3ε, CD3ζ
Adhesion molecules	ICAM-1, LFA-1, α4 integrins
Co-receptors	CD2, CD4, CD8α, CD8β, CD28, CTLA-4, CD40
Cytokine/chemokine receptors	IL-2Rα, β, γ, IL-7R, IL-8R, TNF-R1, TNF-R2
MHC	MHCII, β_2-microglobulin
Others	CD23, CD45, FcRγ

CYTOPLASMIC MOLECULES	
Signal transduction	Abl, p59^{fyn}, p56^{lck}, Vav, ZAP-70, Syk, JAK-1, JAK-3
Antigen processing/ presentation	LMP-7, TAP1, H-2M, Ii
Others	Perforin, iNOS

NUCLEAR FACTORS	
Transcription factors	BSAP/Pax5, IRF-1, IRF-2, NF-κ, Bp50, Oct2, Pu.1, ETS-1, TCF-1, Sox-4, Stat-6, Ikaros
Others	RAG-1, RAG-2, TdT

targeted and transgenic mice (Fig. 2). The generation of mouse mutants deficient in components of the TCR has underscored the importance of this receptor at two key stages of development. The T cell receptor is a multichain complex comprising the clonotypic α and β chains together with the CD3 signal transduction module, CD3γ, δ, ε, and ζ, and is expressed on most mature thymocytes and peripheral T cells. An alternative form of the TCR, expressed on pre-T cells, comprises TCRβ and the CD3 polypeptides together with an additional polypeptide called pre-Tα. Mice lacking the TCRβ gene, the CD3ε or -ζ genes, or the pre-Tα gene show defects in the progression of CD4$^-$8$^-$ [double negative (DN)] to CD4$^+$8$^+$ [double positive (DP)] thymocytes. RAG-deficient mice, which lack expression of RAG-1 or RAG-2, the lineage-specific components of the recombination machinery, also exhibit a block at the DN to DP thymocyte transition. In contrast, thymocytes from

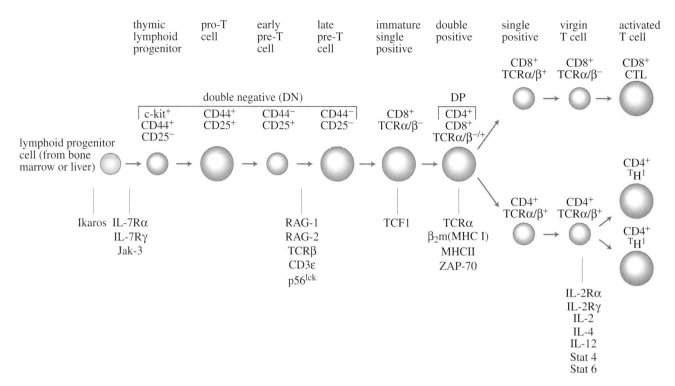

FIGURE 2 Stages in T cell development from the totipotent stem cell. This developmental program can be followed by means of phenotypic markers (indicated by the labels above the cells). Progenitor lymphoid stem cells from the sites of hematopoiesis colonize the thymus and develop to form mature thymocytes. The lymphoid progenitor cells do not show extensive proliferative capacity. Only a proportion of the DP thymocyte subset is proliferative. Mature thymocytes exit the thymus and, upon contact with antigen, undergo further differentiation. Below, the stages are indicated at which mutations rendering specific genes silent cause an arrest in T cell development.

mice deficient in TCRα or MHC class I/class II gene expression arrest at the transition between DP and mature $CD4^+8^-$ or $CD4^-8^+$ [single positive (SP)] thymocytes. The analysis of these mutants elegantly demonstrates that the two isoforms of the TCR control distinct stages of thymocyte development. The pre-TCR drives the DN to DP thymocyte transition and the αβ TCR controls the maturation of DP to SP thymocytes (Fig. 2).

The analysis of transgenic mice expressing a cloned TCR has been invaluable in shaping our knowledge of the DP to SP thymocyte. Many TCR transgenic models have been described. However, the principles and power of this approach can be readily demonstrated by reference to one such model, the H-Y TCR transgenic mouse. The H-Y TCR used in this study recognizes a peptide from the male H-Y antigen complexed to the class I molecule, $H-2K^b$. When expressed in the thymus of male $H-2K^b$ mice, TCR-expressing cells are deleted. This deletional process, called negative selection, is a major mechanism of establishing immunological tolerance. In female mice, TCR-expressing thymocytes are selected by interaction with the $H-2K^b$ molecule presumably complexed to a cross-reacting peptide(s). This

is the process of positive selection which establishes a repertoire of useful, MHC-restricted T cells. A combination of transgenic and knockout experiments have extended these pioneering studies. In elegant experiments using intercrosses of TCR transgenic mice with knockout mice, two groups have shown that positive selection is likely to occur when the TCR exhibits an intermediate affinity for the MHC–peptide ligand (Bevan *et al.*, 1994). The strategy underpinning these experiments utilizes mouse mutants deficient in MHC class I expression. One such mutant lacks a component of the TAP peptide transporter system and the other lacks β_2-microglobulin. In each case, some MHC class I surface expression can be achieved by the addition of peptides that contain class I binding motifs to fetal thymic organ cultures. In these systems, high-affinity peptides that act as antagonists when presented to peripheral T cells expressing the appropriate TCR cause negative selection, whereas lower affinity agonist peptides cause positive selection.

Cytokines and cytokine receptor genes play an important role in the survival, proliferation, and maturation events associated with lymphoid development. Knockout technology has made valuable contributions to our knowledge of cytokine and cytokine receptor function, ruling in or out essential, nonredundant roles for several of these molecules. These studies are illustrated by reference to two cytokine receptors, the IL-2R and the IL-7R, which are members of the same cytokine receptor superfamily. Mice deficient in IL-2Rα or -β gene expression show no gross defects in thymocyte development but do have peripheral T cell abnormalities (Willerford *et al.*, 1995; Suzuki *et al.*, 1995). In contrast, mice lacking expression of the IL-7Rα gene show a profound block in early thymocyte development; thymocyte numbers are reduced by 10- to 100-fold and most thymocytes are early DN cells (Peschon *et al.*, 1994). Mice lacking expression of the IL-2Rγ chain, which is shared between the IL-2R and IL-7R and is indeed diagnostic of this family of receptors, show a reduction in thymocyte numbers but not in the generation of the DP and SP thymocyte subsets (DiSanto *et al.*, 1995). These data clearly demonstrate that the IL-7R plays an essential role in thymocyte development. However, the role of the IL-2R in thymocyte development, if any, is not established in these experiments.

The surface cues that drive any developmental process are relayed to the nucleus by a cascade of cytoplasmic signal transduction components. Signaling molecules have been popular candidates for gene targeting and transgenic studies. These approaches have established the importance of several signaling intermediates in thymocyte development, including the p56^{lck} and ZAP-70 protein tyrosine kinases (PTKs). The expression of a dominant negative p56^{lck} transgene in the thymus results in a block at the DN stage of development (Anderson *et al.*, 1994). This is also the case for p56^{lck} null mice, although the developmental arrest is only partial. In contrast, the expression of a constitutively active p56^{lck} transgene in the thymus results not only in the inhibition of TCRβ rearrangement but also in a developmental arrest at the DP thymocyte checkpoint (Anderson *et al.*, 1994). Together, these experiments underscore the importance of p56^{lck} in the early stages of thymocyte development as a sensor of pre-TCR-mediated signals. Knockout experiments have also provided insight into the key role of the ZAP-70 PTK in progression through the DP thymocyte checkpoint. ZAP-70 null mice are defective in both positive and negative selection (Negishi *et al.*, 1995). Thus, this PTK acts as a sensor of TCR-mediated signals at this stage of development.

The regulation of development of a cell lineage must ultimately be explained in terms of the changes in gene expression patterns that occur. Transgenic technology has been useful in identifying several transcription factors that are important in development. Two examples that illustrate this approach are gene targeting experiments involving Ikaros and TCF-1. Ikaros-deficient mutants lack B and T lymphocytes (Georgopoulos *et al.*, 1994), showing that this helix–loop–helix transcription factor is essential for lymphoid development. Similarly, TCF-1, a member of the HMG family of transcription factors, is essential for the proliferation of late DN thymocytes (Verbeek *et al.*, 1995).

The previous examples illustrate the power of transgenic technology in the study of primary immunodeficiencies resulting from defects in thymocyte development. Some of these mice have also been useful as models of peripheral immunodeficiencies. Examples that illustrate the power of these studies are discussed later.

Mice Defective in Lymphocyte Activation

IL-2R-Deficient Mice

IL-2R is the major growth factor receptor for peripheral T cells. Interaction of the IL-2R with IL-2 is necessary for the orderly progression of an activated T cell into the cell cycle. The high-affinity IL-2R comprises three polypeptide chains, IL-2Rα, -β, and -γ. The IL-2Rγ chain is a common chain that is shared between different members of the cytokine receptor family, including IL-4R and IL-7R. This gene is therefore important for both B and T cell signaling. All three genes encoding the IL-2R have been knocked out and each mutant has yielded an interesting immunodeficiency phenotype.

IL-2Rγ null mice have a markedly reduced number of thymocytes but normal proportions of the various subsets, suggesting that there is a failure of proliferation in these mice (DiSanto *et al.*, 1995). The number of peripheral T cells is reduced and these cells are defective in T cell activation. This mouse mutant partially recapitulates the human IL-2Rγ chain mutation (Noguchi *et al.*, 1993) that results in X-linked severe combined immunodeficiency (XSCID). However, the defect in human XSCID appears more severe, with patients characterized by an almost total absence of both thymocytes and T cells.

In contrast to the situation with the IL-2Rγ gene, null mutations in the IL-2Rα and -β genes have no defects in T cell development, as discussed previously. However, these mutations cause defects in peripheral T cell function resulting in lymphoproliferative disease. Interestingly, these mutant mice have provided insight into the role of IL-2R in inducing apoptosis of activated lymphocytes. Thus, IL-2Rα null mice show polyclonal T and B lymphocyte expansions which, for T cells, are correlated with impaired activation-induced cell death *in vivo* (Willerford *et al.*, 1995). Older IL-2Rα null mice develop autoimmune diseases. A similar phenotype was also observed in IL-2Rβ-deficient mice (Suzuki *et al.*, 1995). Thus, T cells were spontaneously activated and high serum concentrations of autoantibodies causing hemolytic anemia were observed. These results establish the role of IL-2R in maintaining homeostasis and preventing autoimmunity.

CD40- and CD40 Ligand-Deficient Mice

CD40 acts as a costimulatory molecule on B cells that enhances lymphocyte survival and plays an important role in B cell growth and differentiation. It is a member of the tumor necrosis factor (TNF) receptor family and is expressed on all mature B lymphocytes. The ligand for CD40 (CD40L) also belongs to the TNF receptor family and is transiently expressed by activated CD4$^+$ T cells. In accord with its costimulatory role, B cells from CD40-deficient mice fail to proliferate and to undergo isotype switching in response to CD40L in the presence of an appropriate interleukin, and they do not mount an immune response to T cell-dependent antigens (Castigli et al., 1994). They do, however, respond normally to T cell-independent antigens. The lack of functional expression of CD40L on T cells results in a decreased IgM response and a complete abrogation of an IgG1 response to a thymus-dependent antigen, but these mice respond normally to a T-independent antigen (Xu et al., 1994). The phenotype of this mutant recapitulates an X chromosome-linked immunodeficiency in man that is characterized by recurrent bacterial infections, low or absent IgG, -A, and -E, and increased IgM and IgD serum levels. This immunodeficiency has been shown to be caused by point mutations in the CD40L gene that give rise to defective expression of CD40L on T cells (Korthäuer et al., 1993).

ZAP-70-Deficient Mice

ZAP-70 is a cytoplasmic PTK that has been implicated in signaling through the TCR. As discussed previously, this PTK is essential for both positive and negative selection of thymocytes since both of these processes are defective in ZAP-70$^{-/-}$ mice (Negishi et al., 1995). Consequently, ZAP-70 null mice have no peripheral CD4$^+$ or CD8$^+$ T cells. A rare human SCID is associated with genetic mutations in the ZAP-70 gene (Arpaia et al., 1994). These patients, although lacking CD8 peripheral T cells, have normal numbers of nonfunctional CD4 T cells. This difference between ZAP-70-defective mice and humans remains unexplained, but there are at least two possibilities. First, there may be a differential ability of the other known member of this PTK family, Syk, to compensate for the lack of ZAP-70. Alternatively, the human mutations may not result in a null phenotype. Both of these alternatives are potential confounding factors in the interpretation of the phenotype of a knockout mouse and in the assessment of its relevance to a human genetic disease.

Jak-3-Deficient Mice

Jak-3 is a member of the Janus kinase family of tyrosine kinases, the expression of which is restricted to hematopoietic tissues. Jak-3, together with Jak-1, has been implicated in signaling through the cytokine receptor family of molecules, all of which use the IL-2Rγ chain. Mice deficient in Jak3 expression showed a severe block in B cell development at the pre-B cell stage (Thomis et al., 1995). There was also a large decrease in the numbers of thymocytes and T cells, although the more mature thymocyte populations and peripheral T cells were clearly present. Jak-3-deficient peripheral T cells, however, were functionally defective because they were unable to proliferate in response to mitogenic signals.

A group of human autosomal SCID patients that have mutations in the Jak-3 gene have been identified (Russell et al., 1995). These patients have mutations or deletions that lead to markedly reduced levels of Jak-3. Jak-3-deficient individuals exhibit a phenotype that is similar to Jak-3$^{-/-}$ mice and also to mice and humans lacking the common γ chain of this family of receptors. This collection of mutant mice not only allows the mechanisms of the human SCIDs to be analyzed but also provides vehicles for testing gene therapy strategies prior to clinical studies.

Development of Th1 and Th2 Cells

Mature CD4 T cells are classified into two major groups which are defined functionally according to the spectrum of cytokines that they synthesize (Fig. 2). Th1 cells synthesize IL-2, interferon-γ (IFN-γ), and TNF-β, whereas Th2 cells secrete IL-4–IL-6 and IL-10. Functionally, Th1 cells interact preferentially with phagocytic cells, whereas Th2 cells promote B cell proliferation and differentiation. The differentiation of Th1 and Th2 cells from a putative common precursor (a virgin T cell that has not yet responded to antigen) is thought to be controlled, at least in part, by the cytokine milieu. In particular, the relative levels of IL-4 and IL-12 at the time of priming of virgin T cells are critical, with IL-4 essential for routing cells along the Th2 pathway and IL-12 for routing cells along the Th1 pathway.

The importance of IL-4 in Th1/Th2 differentiation is demonstrated by the phenotype of IL-4-deficient mice (Kopf et al., 1993). CD4$^+$ T cells from IL-4$^{-/-}$ mice failed to produce Th2-derived cytokines after stimulation and showed reduced responses to nematode infection, which normally induces a selective activation of Th2 cells. These experiments demonstrate that IL-4 is required for the generation of Th2-derived cytokines and that immune responses dependent on these cytokines are impaired.

The cytokine receptor family, of which IL-4R is a member, transduces signals into lymphocytes via the Jak/Stat signal transduction pathway. Stat proteins are a family of transcription factors that exist in an inactive form in the cytoplasm and are phosphorylated by Jak kinases following cytokine receptor interactions. Once phosphorylated, Stat proteins translocate to the nucleus as dimers, where they activate target genes. Stat6 is specifically activated in response to IL-4. The essential role of Stat6 in IL-4R-mediated signaling is demonstrated by the analysis of Stat6-deficient mice (Takeda *et al.*, 1996). There is a lack of IgE and IgG1 responses and of Th2 cytokine production after *in vivo* stimulation with agents that generate a selective activation of a Th2-type immune response.

This strategy has also established the importance of specific cytokines and signal transduction molecules for the Th1 pathway. Thus, IL-12-deficient mice are impaired both in IFN-γ production and in the ability to mount delayed-type hypersensitivity responses (Magram *et al.*, 1996). However, other Th1 responses, such as cytotoxic T lymphocyte generation and proliferation and the secretion of IL-2 and IL-10 following antigen stimulation, appear normal. Mutation of Stat4, which is activated in response to IL-12, also affects the Th1 pathway. In Stat4 null mice, the Th1 response is abrogated (Thierfelder *et al.*, 1996). These mice failed to show an increase in IFN-γ expression or to develop a Th1 response in response to IL-12 or antigenic stimulus. In contrast, the Th2 response was enhanced.

Transgenic and Knockout Mice as Models of Inflammatory Bowel Disease

As discussed previously, the characterization of TCR- and cytokine-deficient mice has provided major insights into our understanding of lymphocyte development and function. However, an unexpected bonus from the analysis of these mice was the observation that they develop intestinal lesions similar to those of human inflammatory bowel disease (IBD). Chronic intestinal inflammation appears with high penetrance in $TCR\alpha^{-/-}$, $TCR\beta^{-/-}$, and $TCR(\beta \times \delta)^{-/-}$ mutants housed under specific pathogen-free conditions and also in mice with targeted disruptions of cytokine genes, including IL-2, IL-10, and transforming growth factor-β (Powrie, 1995). In contrast, mice lacking RAG gene expression do not exhibit colitis. Analysis of intercrosses of $RAG-2^{-/-}$ with $JH^{-/-}$ mice has demonstrated that T cells, but not B cells, are required for the generation of colitis in IL-2 null mice (Ma *et al.*, 1995). The colitis-like disease in these various mice can be explained by a lack of a regulatory $\alpha \beta$T cell subset which normally prevents the development of immune responses to gut antigens. According to this notion, in TCR- or cytokine-deficient mice, microbial pathogens generate a dysregulated response that leads to a chronic inflammatory condition with symptoms of colitis. The importance of microbial pathogens in this process has been demonstrated by the observation that germ-free colonies of IL-2 or IL-10 null mice fail to develop colitis (Powrie, 1995).

A key issue is the relevance of animal models to human disease. There are several differences between the two

major forms of human IBD, ulcerative colitis and Crohn's disease, and the colitis observed in TCR- and cytokine-deficient mice. In particular, the apparent dependence of the knockout mouse on microbial infection for the observed colitis may make these mice better models for the type of inflammatory intestinal disease commonly seen in globally immunosuppressed patients, such as those suffering from AIDS, rather than models of IBD, believed by many to be largely an autoimmune process. These considerations are obviously of prime importance for the use of mouse models to assess the potential of regimens for the amelioration of the human disease.

Limitations and Problems

The use of transgenic technology has been of immense importance in the study of lymphoid development and function and has provided valuable models of immunodeficiency. However, there are problems and limitations associated with the generation of transgenic mice that must be appreciated in interpreting the observed phenotypes. Some of these pitfalls are listed in Table 2.

Common problems associated with the expression of transgenes in mice are the choice of the regulatory elements used to drive expression of the transgene and the position effects exerted by the surrounding chromatin. The cognate promoter and enhancer elements that regulate the expression of a gene may not be characterized, necessitating the use of heterologous promoters and enhancers. This may in turn result in the inappropriate developmental expression of the gene, giving an artifactual phenotype. Po-

TABLE 2

Problems, Limitations, and Solutions

PROBLEM/LIMITATION	SOLUTION
Lethality	tissue specific KOs
Redundancy	multiple knockouts of family members
	more refined assays
Background effects	multiple backgrounds
Null or dominant negative mutation-different constructs, different phenotypes	analysis of protein expression in mutant mouse
Phenotypes due to effects on neighbouring genes	"hit and run" or cre-lox; re-express transgene
Position effects of transgene	Locus Control Region (LCR); targeted integration using cre/lox

sition effects, in which the surrounding chromatin influences the expression of the transgene, may also result in the inappropriate expression of the transgene, depending on the chromatin context into which the transgene has integrated and, consequently, a variation in phenotype in different founders. Enhancers containing locus control regions (LCRs), operationally defined as conferring position-independent, copy-dependent expression of a transgene, overcome this limitation. Both the CD2 and p56^{lck} enhancers possess LCR activity and are frequently used to express genes in the T cell lineage. Alternatively, it may be possible to target the integration of the transgene in different founder lines to the same chromatin location using the lox/cre system (Fukushige and Sauer, 1992).

There are several problems associated with the generation and analysis of gene-targeted mice. Sometimes, there is no obvious phenotype, as in the case of CD2 null mice. In these cases, it is often argued that the role of the gene is redundant; that is, another gene or genes can substitute for the targeted gene. However, truly redundant genes are free from selective pressure and so their expression would rapidly be lost due to the accumulation of mutations. In the case of the CD2 gene it was calculated that the chance of retaining CD2 expression without selection since primates diverged from rodents is less than 1 in 10^9 (Davis and van der Merwe, 1996). Although a phenotype may never be evident, it is sometimes the case that one appears with age. For example, IL-2-, IL-2Rα, and IL-2Rβ null mice are all grossly normal when young but develop immunopathologies as the mice age.

Alternatively, the disruption of a gene may result in embryonic lethality, a phenotype that is not useful for the study of the immune system. If the lethality is late during embryogenesis it may be possible to use radiation chimeras with fetal liver cells or to analyze thymocyte development using *in vitro* fetal thymic organ culture. An alternative strategy involves the analysis of chimeras generated by the injection of targeted ES cells into blastocysts from RAG-deficient mice. In the resulting chimeric mice, any mature B or T cells will be derived from the targeted ES cell since lymphoid development in RAG-deficient mice is arrested at the pro- to pre-B/T cell stage.

A further complication of the interpretation of a "knock-out" phenotype is the precise nature of the introduced mutation. For example, if the targeting event results in the expression of an altered protein product, the mutation may generate a dominant negative form of the protein. If the gene is part of a family, the dominant negative mutant may inhibit the activity of other family members and thus give a more severe phenotype. Thus, it is essential to characterize thoroughly the products of the targeted gene at the level of RNA and/or protein. The phenotype of a knockout mouse may also vary depending on the genetic background.

Therefore, it is desirable to cross founder mice onto at least two different genetic backgrounds to check for phenotypic differences.

A major consideration in the use of gene-targeted mice to determine the function(s) of a gene is whether the observed phenotype is a consequence of the mutation to the gene that was targeted or whether the targeted mutation has indirectly affected the expression of a neighboring gene. The strategy used for most gene targeting experiments involves the insertion of a drug-resistance gene driven by a suitable promoter and enhancer into an exon. The regulatory elements that control the expression of the exogenous gene may interfere with the expression of the adjacent gene. For example, three different targeting strategies that generated null mutations of the myogenic transcription factor gene, *MRF4,* resulted in different phenotypes. This phenotypic variation resulted, at least in part, from interference with the expression of a neighboring gene, *Myf5,* probably by the regulatory elements of the selection cassette (Olson *et al.,* 1996). Suppression of the genetic mutation by reexpression of a wild-type transgene should demonstrate that the observed phenotype is caused directly by the mutation within the targeted gene. However, the regulatory elements that direct expression of the gene of interest may not be known. An alternative approach is to use gene targeting strategies that result in removal of extraneous DNA sequences. The most sophisticated of these is the "hit and run" technique in which single base pair mutations can be introduced (Hasty *et al.,* 1991). The cre/lox system can also be used to remove the drug-selection gene (Gu *et al.,* 1993). These methods may soon become standard practice in order to ensure that the correct function is ascribed to a gene from gene targeting studies.

Conclusions

Transgenic and gene targeting approaches have been of immense value for the study of immune function and dysfunction. The generation of mouse models of human immunodeficiencies has permitted not only study of the mechanisms of the disease but also a means to analyze the efficiency of corrective gene therapy approaches. These techniques are likely to be of increasing importance as many novel genes involved in the regulation of the immune response are identified by the human and murine genome projects.

To date, most gene targeting strategies have resulted in the generation of a null phenotype. In the future, an allelic series of mutants is likely to prove of greater value for determining the function of a protein. This will most likely require new technological advances to increase the frequency and accuracy of gene targeting. Such advances will surely produce many more models of immunodeficiencies and may identify the genetic origin of human diseases, the etiology of which is currently unknown.

Acknowledgments

Due to limitations on the number of references, it has not been possible to cite all the appropriate publications describing the advances presented here. The author apologizes to all of his colleagues whose work he has been unable to acknowledge.

References Cited

ANDERSON, S. J., LEVIN, S. D., and PERLMUTTER, R. M. (1994). Involvement of the protein tyrosine kinase p56lck in T cell signaling and thymocyte development. *Adv. Immunol.* **56,** 151–178.

ARPAIA, E., SHAHAR, M., DADI, H., COHEN, A., and ROLFMAN, C. M. (1994). Defective T cell receptor signaling and CD8$^+$ thymic selection in humans lacking zap-70 kinase. *Cell* **76,** 947–958.

BEVAN, M. J., HOGQUIST, K. A., and JAMESON, S. C. (1994). Selecting the T cell receptor repertoire. *Science* **264,** 796–979.

CASTIGLI, E., ALT, F. W., DAVIDSON, L., BOTTARO, A., MIZOGUCHI, E., BHAN, A. K., and GEHA, R. S. (1994). CD40-deficient mice generated by recombination-activating gene-2-deficient blastocyst complementation. *Proc. Natl. Acad. Sci. USA* **91,** 12135–12139.

DAVIS, S. J., and VAN DER MERWE, P. A. (1996). The structure and ligand interactions of CD2: Implications for T-cell function. *Immunol. Today* **17,** 177–187.

DISANTO, J. P., MULLER, W., GUY-GRAND, D., FISCHER, A., and RAJWESKY, K. (1995). Lymphoid development in mice with a targeted deletion of the interleukin 2 receptor gamma chain. *Proc. Natl. Acad. Sci. USA* **92,** 377–381.

FUKUSHIGE, S., and SAUER, B. (1995). Genomic targeting with a positive-selection lox integration vector allows highly reproducible gene expression in mammalian cells. *Proc. Natl. Acad. Sci. USA* **89,** 7905–7909.

GEORGOPOULOS, K., BIGBY, M., WANG, J.-H., MOLNAR, A., WU, P., WINANDY, S., and SHARPE, A. (1994). The Ikaros gene is required for the development of all lymphoid lineages. *Cell* **79,** 143–156.

GU, H., ZOU, Y.-R., and RAJEWSKY, K. (1993). Independent control of immunoglobulin switch recombination at individual switch regions evidenced through Cre-loxP-mediated gene targeting. *Cell* **73,** 1155–1164.

HASTY, P., RAMIREZ-SOLIS, R., KRUMLAUF, R., and BRADLEY, A. (1991). Introduction of a subtle mutation into the Hox-2.6 locus in embryonic stem cells. *Nature* **350,** 243–246.

KOPF, M., LE GROS, G., BACHMANN, M., LAMERS, M. C., BLUETHMANN, H., and KOHLER, G. (1993). Disruption of the murine IL-4 gene blocks Th2 cytokine responses. *Nature* **362,** 245–250.

KORTHÄUER, U., GRAF, D., MAGES, H. W., BRIERES, F., PADAYACHEE, M., MALCOLM, S., UGAZIO, A. G., NOTARENGELO, L. D., LEVINSKY, R. J., and KROCZEK, R. A. (1993). Defective expression of T-cell CD40 ligand causes X-linked immunodeficiency with hyper-IgM. *Nature* **361,** 539–543.

MA, B. A., DATTA, M., MARGOSIAN, E., CHEN, J., and HORAK, I. (1995). T cells, but not B cells, are required for bowel inflammation in interleukin 2-deficient mice. *J. Exp. Med.* **182,** 1567–1572.

MAGRAM, J., CONNAUGHTON, S. E., WARRIER, R. R., CARVAJAL, D. M., WU, C.-Y., FERRANTE, J., STEWART, C., SARMIENTO, U., FAHERTY, D. A., and GATELY, M. K. (1996). IL-12-deficient mice are defective in IFN gamma production and type 1 cytokine responses. *Immunity* **4,** 471–481.

NEGISHI, I., MOTOYAMA, N., NAKAYAMA, K.-I., NAKAYAMA, K., SENJU, S., HATAKEYAMA, S., ZHANG, Q., CHAN, A. C., and LOH, D. Y. (1995). Essential role for ZAP-70 in both positive and negative selection of thymocytes. *Nature* **376,** 435–438.

NOGUCHI, M., YI, H., ROSENBLATT, H. M., FILIPOVICH, A. H., ADELSTEIN, S., MODI, W. S., MCBRIDE, O. W., and LEONARD, W. J. (1993). Interleukin-2 receptor gamma chain mutation results in X-linked severe combined immunodeficiency in humans. *Cell* **73,** 147–157.

OLSON, E. N., ARNOLD, H.-H., RIGBY, P. W. J., and WOLD, B. J. (1996). Know your neighbors: Three phenotypes in null mutants of the myogenic bHLH gene MRF4. *Cell* **85,** 1–4.

PESCHON, J. J., MORRISSEY, P. J., GRABSTEIN, K. H., RAMSDELL, F. J., MARASKOVSKY, E., GLINIAK, B. C., PARK, L. S., ZIEGLER, S. F., WILLIAMS, D. E., WARE, C. B., MEYER, J. D., and DAVISON, B. L. (1994). Early lymphocyte expansion is severely impaired in interleukin 7 receptor-deficient mice. *J. Exp. Med.* **180,** 1955–1960.

POWRIE, F. (1995). T cells in inflammatory bowel disease: Protective and pathogenic roles. *Immunity* **3,** 171–174.

RUSSELL, S. M., TAYEBI, N., NAKAJIMA, H., RIEDY, M. C., ROBERTS, J. L., AMAN, J., MIGONE, T.-S., NOGUCHI, M., MARKERT, M. L., BUCKLEY, R. H., O'SHEA, J. J., and LEONARD, W. J. (1995). Mutation of Jak3 in a patient with SCID: Essential role of Jak3 in lymphoid development. *Science* **270,** 797–800.

SUZUKI, H., KUNDIG, T. M., FURLONGER, C., WAKEHAM, A., TIMMS, E., MATSUYAMA, T., SCHMITS, R., SIMARD, J. J. L., OHASHI, P. S., GRIESSER, H., TANIGUCHI, T., PAIGE, C. J., and MAK, T. W. (1995). Deregulated T cell activation and autoimmunity in mice lacking interleukin-2 receptor beta. *Science* **268,** 1472–1476.

TAKEDA, K., TANAKA, T., SHI, W., MATSUMOTO, M., MINAMI, M., KASHIWAMURA, S., NAKANISHI, K., YOSHIDA, N., KISHIMOTO, T., and AKIRA, S. (1996). Essential role of Stat6 in IL-4 signalling. *Nature* **380,** 627–630.

THIERFELDER, W. E., VAN DEURSEN, J. M., YAMAMOTO, K., TRIPP, R. A., SARAWAR, S. R., CARSONS, R. T., SANGSTER, M. Y., VIGNALI, D. A. A., DOHERTY, P. C., GROSVELD, G. C., and IHLE, J. N. (1996). Requirement for Stat4 in interleukin-12-mediated responses of natural killer and T cells. *Nature* **382,** 171–174.

THOMIS, D. C., GURNIAK, C. B., TIVOL, E., SHARPE, A. H., and BERG, L. J. (1995). Defects in B lymphocyte maturation and T lymphocyte activation in mice lacking Jak3. *Science* **270,** 794–802.

VERBEEK, S., IZON, D., HOFHUIS, F., ROBANUS-MAANDAG, E., TE RIELE, H., VAN DE WETERING, M., OOSTERWEGEL, M., WILSON, A., MACDONALD, H. R., and CLEVERS, H. (1995).

An HMG-box-containing T-cell factor required for thymocyte differentiation. *Nature* **374,** 70–74.

WILLERFORD, D. M., CHEN, J., FERRY, J. A., DAVIDSON, L., MA, A., and ALT, F. W. (1995). Interleukin-2 receptor alpha chain regulates the size and content of the peripheral lymphoid compartment. *Immunity* **3,** 521–530.

XU, J., FOY, T. M., LAMAN, J. D., ELLIOTT, E. A., DUNN, J. J., WALDSCHMIDT, T. J., ELSEMORE, J., NOELLE, R. J., and FLAVELL, R. A. (1994). Mice deficient for the CD40 ligand. *Immunity* **1,** 423–431.

General References

BETZ, U. A. K., VOSSHENRICH, C. A. J., RAJEWSKY, K., and MULLER, W. (1996). Bypass of lethality with mosaic mice generated by Cre-loxP-mediated recombination. *Curr. Biol.* **6,** 1307–1316.

MIKLOS, G. L. G., and RUBIN, G. M. (1996). The role of the genome project in determining gene function: Insights from model organisms. *Cell* **86,** 521–529.

SILVIA BIOCCA

Department of Neuroscience
University of Roma Tor Vergata
00133 Rome, Italy

ANTONINO CATTANEO

Biophysics Sector
International School for Advanced Studies
Via Beirut 2-4, 34013 Trieste, Italy

Toward an Artificial Immune System: The Immunotechnology

Immunotechnologies exploit the virtually unlimited repertoire provided by antibodies to engineer modified molecules endowed with properties of specific chemical recognition that find applications in a wide range of different fields. The exploitation of this repertoire has been facilitated by the development of libraries of molecules exposed at the surface of phages, which has led to the construction of artificial immune systems. These systems use the fundamental working principles of the immune system, such as diversity, selection, and amplification. This article describes the developments in the field of in vitro immune systems and in the use of recombinant antibodies as proteins with therapeutical, diagnostic, and catalytic activity or in the field of biosensors. The recombinant antibodies are also utilized as ectopically expressed genes to confer a phenotype of interest in different biological systems from plant biotechnology to functional genomics and gene therapy.

L'universo (che altri chiama la Biblioteca) si compone d'un numero indefinito, e forse infinito, di gallerie esagonali, con vasti pozzi di ventilazione nel mezzo, bordati di basse ringhiere . . . a ciascuna parete di ciascun esagono corrispondono cinque scaffali; ciascuno scaffale contiene trentadue libri di formato uniforme; ciascun libro e' di quattrocentodieci pagine; ciascuna pagina, di quaranta righe; ciascuna riga, di quaranta lettere di color nero.

Jorge Luis Borges
La Biblioteca di Babele

Among the most extraordinary properties of the immune system is the generation of a virtually unlimited repertoire of antibody molecules able to recognize an extremely vast chemical universe made of natural or artificial epitopes. The chemical basis of the biological specificity of the immune system was recognized at the beginning of the twentieth century as a result of the pioneering studies by P. Ehrlich, followed by the fundamental studies of K. Landsteiner in the 1930s. Antibodies share a common basic structure in which the domain responsible for specific recognition (variable regions) is independent and separate from the domains involved in other interactions and in effector functions (constant domains). The diversity of the immune system is the result of combinatorial rearrangements of a restricted number of modules that give rise to the so-called primary repertoire of membrane antibodies expressed at the surface of the corresponding B lymphocytes. The association between antibody and cell is fundamental for the process of selection and somatic affinity maturation, triggered by the binding of the antigen. The affinity of the primary recognition events is not necessarily very high; however, it is sufficiently high to trigger the process of selection of the cells carrying that specificity and the ensuing somatic hypermutation. The process of somatic mutation contributes to an increase in the potential diversity of the antibody repertoire and leads to the so-called affinity maturation of antibodies.

The principal properties of the antibody response are typically Darwinian, such as the generation of a high germinal and somatic diversity, the selection and amplification of appropriate clones mediated by antigen-dependent survival mechanisms, and the hereditability due to the important property that antibodies are selected together with the rearranged genes that encode them.

Immunotechnologies exploit the universal and specific chemical recognition properties of antibody variable regions, coupled to an increasing spectrum of effector functions which is greater than that present in the immune system, and utilize many functioning properties of the immune system. All fields in which some form of specific chemical recognition is necessary or convenient represent areas for potential applications of immunotechnologies.

From Monoclonal Antibodies to Phage Libraries

For many years, antibodies were utilized as complex mixtures of antibody proteins, from which it was difficult to separate those antibodies able to recognize different parts of the same antigen. Only in the case of a pure antigen was it possible to isolate by affinity chromatography that population of antibodies denoted as polyclonal since they resulted from the secretion of antibodies from several lymphocyte clones. Modern immunotechnologies were born with the description of the method of monoclonal antibodies (Koehler and Milstein, 1975), which allows isolation of cell lines that secrete a single antibody species of predetermined specificity, even from complex mixtures of antigens. The hybridoma technology has been an enormous success and has allowed the isolation of a many monoclonal antibodies, and it has been utilized in several different fields, such as diagnostic, therapeutic, and biotechnological. Recombinant DNA technology, together with gene transfer techniques, has made it possible to utilize monoclonal antibodies for further significant development of immunotechnologies. Indeed, hybridoma cell lines can be the sources not only of antibody proteins but also of the genes that encode for these antibodies, allowing the construction of antibody forms, tailored at will, in ways that are only limited by the imagination (Winter and Milstein, 1991).

Antibody Engineering

The structural organization of antibodies, in functionally separate and independent domains, makes these molecules particularly amenable to modification by protein engineering (Fig. 1). In particular, the functional antigen-binding domains, if separated from the domains that mediate the effector functions, can be combined in different ways or fused to heterologous protein domains to build new molecules with recognition properties and new effector properties (Neuberger et al., 1984). Antibody engineering in general provides new adaptor molecules that associate an effector function of interest to a specific recognition event, including new constant regions, enzymatic activities, toxins, and peptides. The opposite is possible as well — that is, to link Fc antibody portions to recognition domains not based on antibody variable regions but, for instance, on ligand-binding domains from membrane receptors (as in the so-called immunoadhesins). In a "classic" antibody, the recognition function is distributed in the hypervariable regions (or complementarity determining regions; CDRs) of the heavy and light chain variable regions (V_H and V_L, respectively). The V_H and V_L variable regions can associate noncovalently in the absence of the constant regions to form the so-called Fv fragments (Fig. 1). Such an association may be stabilized by introducing a linker peptide that connects the V_H and V_L domains into a single polypeptide chain to form the so-called single-chain Fv fragments (scFv) (Bird et al., 1988). The scFv fragments have notable advantages, including increased stability, reduced size, simplified expression, and versatility for the construction of chimeric proteins. For these reasons, scFv fragments represent the basic format for most applications, with particular regard to their expression on the surface of filamentous phage. This format lends itself to the construction of bivalent or bispecific recognition units. If the linker peptide connecting the two variable regions is shorter, the two domains do not associate intramolecularly, whereas they can associate intermolecularly with the corresponding domain of another molecule to form a bivalent (monospecific) domain known as diabody (Holliger et al., 1993). A bispecific dimer recognizing two antigens, A and B, can be obtained from the dimerization of two distinct scFv fragments: $V_HA–V_LB/V_HB–V_LA$ or $V_LA–V_HB/V_LB–V_HA$. Miniantibodies that are bivalent, bispecific, or both can also be engineered by exploiting dimerization domains such as leucine zippers or engineered CH_3 domains. Also, bispecific antibodies directed against adjacent but distinct epitopes on a protein can be obtained by linking two scFv fragments with a suitable linker peptide. In this way, so-called chelating recombinant antibodies are obtained, i.e. antibodies in which the overall binding affinity for the target protein is

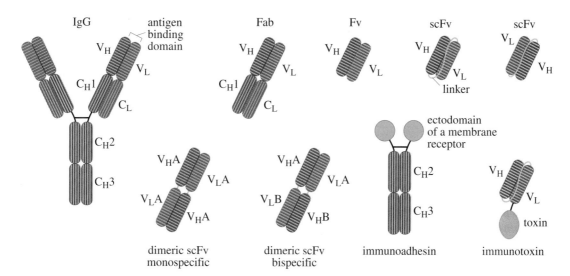

FIGURE 1 Recombinant antibodies. A complete antibody is made of two heavy chains (H, in red) and two light chains (L, in blue) with the two antigen binding regions made up of a heavy (VH) and light (VL) variable region. The VH-CH1/VL-CL molecule is termed a Fab and, in the format shown, is monomeric in its binding ability. Monomeric binding units can also be created by expressing the variable regions, either non-covalently associated (Fv) or joined by a linker peptide (scFv, linker in yellow). This connects the carboxy-terminal portion of the VH to the amino-terminal part of the VL or viceversa. ScFv fragments represent the basic unit exploited in the creation for monospecific monovalent antibodies, bispecific antibodies, immunotoxins and different forms of immunoconjugates. Fc fragments may be linked to binding domains non antibody-based, such as, for example, ectodomains of membrane receptors (immunoadhesins).

higher than that of the individual antibodies due to an avidity effect.

In some cases, single V_H domains maintain antigen recognition properties. However, this is the exception rather than the rule; furthermore, V_H domains expose to the solvent a hydrophobic surface that is normally buried in the interface with the V_L domain, and this is the cause of potentially nonspecific interactions. However, in some animal species, such as camels and lamas, natural antibodies made of heavy chains are expressed and secreted. These natural isolated heavy chains are less hydrophobic, and V_H domains of this nature constitute a potentially versatile source of smaller recognition domains (Davies and Riechmann, 1995). Further possibilities of achieving smaller recognition units are provided by the use of CDR-derived peptides that constitute the starting material for the synthesis of mimetic molecules. The six CDRs of an antibody can be molecularly transplanted from a "donor" to an "acceptor" antibody, carrying with them the recognition specificity of the initial antibody (Jones *et al.*, 1986). This procedure, known as CDR grafting, allows one to graft the antigenic specificity of a monoclonal antibody of murine origin in the context of an acceptor antibody framework more suitable for the particular application envisaged.

Antibody Libraries

Questo pensatore osservo' che tutti i libri, per diversi che fossero, constavano di elementi eguali: lo spazio, il punto, la virgola, le ventidue lettere dell'alfabeto. Stabili', inoltre, un fatto che tutti i viaggiatori hanno confermato: non vi sono, nella vasta Biblioteca, due soli libri identici. Da queste premesse incontrovertibili dedusse che la Biblioteca e' totale, che i suoi scaffali registrano tutte le possibili combinazioni dei venticinque simboli ortografici (numero, anche se vastissimo, non infinito) cioe' tutto cio' ch'e' dato di esprimere, in tutte le lingue. Tutto: la storia minuziosa dell'avvenire, le autobiografie degli arcangeli, il catalogo fedele della Biblioteca, migliaia e migliaia di cataloghi falsi, la dimostrazione della falsita' di questi cataloghi, la dimostrazione della falsita' del catalogo autentico.

Jorge Luis Borges
La Biblioteca di Babele

In classical antibody engineering, the starting material for cloning the antibodies was hybridoma cell lines of known antigenic specificity. With the advent of the polymerase chain reaction (PCR) and the synthesis of universal oligonucleotide primers, it became possible to amplify whole families of variable region genes not only from hybridoma cell lines but also from primary lymphocytes isolated from immunized or nonimmunized mice (Orlandi *et al.,* 1989). The subsequent expression of the cloned variable regions in alternative expression systems to myeloma cells (initially *Escherichia coli* bacterial cells and filamentous bacterio- phages) has allowed the expression of mixtures of poly- clonal antibodies, whose genes are now available, to yield a so-called antibody library. This is an example of an immune system in a test tube (Winter and Milstein, 1991) which contains not only the antibody specificities derived from the original V_H/V_L pairings but also other specificities de- rived from new V_H/V_L combinations. Thus, the library ob- tained is more complex than the starting library since it contains new combinations and hence new specificities. The expression of libraries of variable regions represents a fun- damental milestone in the development of immunotech- nologies. In order to exploit the potential of an antibody li- brary, it is necessary to preserve the physical link between an antibody and the corresponding gene, as occurs in B lym- phocytes. This link between antibody genotype and phe- notype is preserved when antibodies are expressed on the surface of filamentous phage. This technology, initially de- veloped for the expression of libraries of millions of dif- ferent peptides (Smith, 1985), was proven to be extremely powerful for the expression of libraries of antibodies (Mc- Cafferty *et al.,* 1990).

The possibility of expressing antibody fragments on the surface of a virus that infects a bacterial cell (usually the filamentous phage Ff) makes it possible to isolate an anti- body fragment with a single predefined specificity from a complex library containing a vast excess of nonbinding antibody fragments. Filamentous phages have a single- stranded DNA genome and infect *E. coli* through the F pilus. The recognition of the F pilus by the phage protein p3 determines the insertion of the viral DNA in the *E. coli* cell, in which it is replicated through a double-stranded phase. Unlike other phages, which propagate by killing the host cell, the filamentous phage is exported from the cell by a process of active secretion concomitant with the matura- tion of the phage particles, a process which does not kill the cell — only slows its growth. The phage protein most widely used for the expression of antibodies is p3, which is en- coded by the phage gene III. This protein, present in three to five copies per phage particle, has a tripartite structure with three functional domains: Two are important for the penetration in the bacterial cell, and the third domain, which occupies the central portion of the protein, is re- sponsible for the recognition of the F pilus. The antibody fragments are fused to the aminoterminal part of p3. In the resulting fusion protein, the antibody domain is exposed at the surface of the phage without affecting its ability to in-

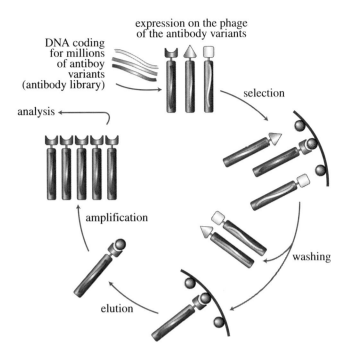

FIGURE 2 The antibody phage display cycle. DNA encoding for millions of antibody variants (antibody library) is cloned into the phage genome as part of one of the phage coat protein p3. From these repertoires phage carrying specific recognition unit can be isolated by a series of recursive cycles of selec- tion on purified antigens immobilized on a solid phase. These cycles involve binding to the antigen, washing the non specific phages, selective elution of the specific phages and their am- plification through infection in E.coli.

fect the bacterial target cell. The principles of the technol- ogy based on the surface expression of antibodies on phage are illustrated in Fig. 2.

The DNA encoding for millions of different antibody domains, i.e., a library of antibody domains, is inserted in the genome of a filamentous bacteriophage as a fusion to gene III. The fusion protein, once expressed, is incorpo- rated in the mature phage particle that is secreted from the bacterial cell, and it is displayed at the phage surface. A phage library is therefore generated in which each phage particle displays a different antibody domain and carries the corresponding genotype. Only phages displaying an antibody domain able to bind the selector antigen are iso- lated and used for a new cycle. In this way, it is possible to select antibody domains directed against antigens of inter- est from libraries with a diversity up to 10^{10}–10^{11}.

The selection of libraries of antibody domains displayed at the phage surface provides the recombinant equivalent of a polyclonal antiserum, with the additional property of being intrinsically linked to the corresponding genes. Since immunotechnologies essentially manipulate the genes that encode for antibodies of interest, it is obvious that the display of antibodies on phage represents a key element in successfully exploiting the potential of these technologies.

The analogy between a library of antibodies displayed on phage and an immune system in the test tube is therefore very appropriate in that a phage displaying at its surface an antibody directed against an antigen X is conceptually analogous to a B lymphocyte expressing the membrane form of the antibody against antigen X. In both cases, the specific antigen–antibody reaction is followed by the amplification and the selection of the corresponding clone.

Antibody libraries can be of various nature and quality, depending on the source of variable regions utilized for their construction. Accordingly, the libraries may be built from variable regions derived from immune or nonimmune donors or from synthetic variable region genes. In the first case, the repertoire derives from immunoglobulin IgG genes cloned from the spleen of animals immunized with the antigen of interest (Clackson *et al.*, 1991), such as for the production of monoclonal antibodies, or from human donors (Burton *et al.*, 1991) in different pathological situations, in which case the sources are bone marrow, tonsils, or circulating lymphocytes. The repertoires derived from immunized lymphocytes are enriched in antibodies specific for the immunizing antigen, even if the combinatorial library contains, in addition to the original V_H/V_L pairings, many new combinations. The construction of libraries from patients is useful for the analysis of the natural immune response, for example, in patients with autoimmune diseases, viral infections, or neoplastic diseases. In any case, these are not universal libraries but rather are built ad hoc.

In analogy with the primary repertoire of the natural immune system, one can construct universal repertoires from which high-affinity antibodies can be isolated and directed against any antigen of interest, independent of the immune history (Marks *et al.*, 1991).

The construction of such universal libraries can be achieved starting from the repertoire of IgM or IgG variable regions isolated from lymphoid tissues of donors not explicitly immunized. As an alternative, universal libraries can be artificially constructed by utilizing a small number of basic variable regions followed by extensive randomization of one or more CDRs (usually CDR3). This procedure generates antibodies with a repertoire of CDRs that is not limited to the CDRs present in the germline variable regions. For these universal libraries, the diversity is a parameter of fundamental importance. Indeed, the probability that the library contains an antibody that recognizes any antigen of interest with an affinity higher than a preassigned value is strictly related to the dimension, or the diversity, of the library. The probability, P, that an epitope is being recognized with an affinity greater than a threshold value, p,

by at least one element of a library of N different antibodies is given by

$$P = 1 - e^{-Np}$$

From this formula it may be deduced that the greater the diversity of the library, the greater the probability of finding a high-affinity antibody against any antigen of interest. Typically, a diversity of 10^{10} different antibodies allows the isolation of antibodies with affinities in the range of 10^{-8}–10^{-10} M. The diversity of the libraries reached by the state-of-the-art technology appears to be satisfactory, and rather than increasing the diversity significantly, it appears to be more important to increase the quality of the libraries (i.e., the proportion of phages in the libraries displaying functional antibodies) or to design molecular scaffolds more suitable for the specific application envisaged and, subsequently, to improve the procedures to select the antibodies. Furthermore, since phage technology allows improvement of the affinity of antibodies isolated from the libraries, diversity does not appear to be the limiting factor to derive high-affinity antibodies with this technology.

The primary repertoire of the immune system has evolved in such a way to avoid antibody specificities potentially damaging for the organism (immune tolerance) or toxic for the cells that express those antibodies. On the contrary, these antibody specificities can be present in the synthetic libraries since they have not undergone the selection process that leads to the phenomenon of immune tolerance. The artificial immune system allows one to obtain a complete repertoire of antibody specificities from which antibodies against any antigen of interest can be isolated, bypassing the use of animals and, hence, the biological immune system. By further reducing the complexity of the system and still preserving the key characteristics (genetic diversity, expression system in which genotype and phenotype are physically linked, amplification, and selection), a new expression system for antibody libraries has been described: the ribosome display system (Fig. 3), a totally *in vitro* system which avoids the use of living cells (Hanes and Plueckthun, 1997; He and Taussig, 1997). The antibody library, in the form of an mRNA library, is *in vitro* translated. The engineered absence of stop codons in the mRNAs has the consequence that the nascent antibody chains remain physically associated to their corresponding mRNA and to the ribosomes. The selection of antibodies directed against an antigen of interest is performed by solid-phase adsorption cycles, similar to those described for phages (Fig. 2), whereas the amplification of the increasingly enriched antibody mRNA is performed by PCR. The ribosome display method is therefore based on cycles of *in vitro* transcription and translation, selection by panning, and amplification. In theory, the diversity obtainable with this method is not limited by the life cycle of the phage–bacteria system or by

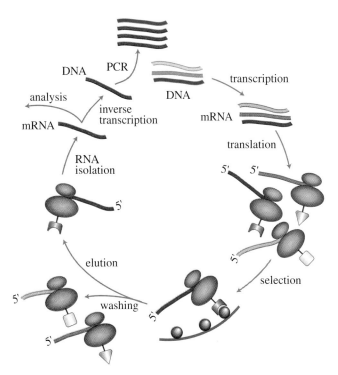

FIGURE 3 The antibody ribosome-display cycle. This approach involves the translation of proteins in vitro and their selection whilst attached to ribosomes. This can occurr if the mRNA lacks a stop codon. In this case, the ribosome is not able to detach and mRNA with the encoding protein remain attached together. The isolation of specific antibodies can be obtained by a series of recursive cycles of selection on purified antigens immobilized on a solid phase. These cycles involve binding, washing, selective elution. After selection, the mRNA from selected ribosomes is converted into cDNA, amplified by PCR, and used for the next transcription, translation and selection round.

technical problems related to the efficiency of transformation. In practice, the potential of this approach remains to be verified.

......................................▼

Methods for Select Antibodies: Finding the Book in the Babel Library

La certezza che un qualche scaffale d'un qualche esagono celava libri preziosi e che questi libri preziosi erano inaccessibili, parve quasi intollerabile.

Jorge Luis Borges
La Biblioteca di Babele

The size of the antibody library displayed on phage is such that the isolation of the rare antibodies with the desired specificity is equivalent to the problem of finding a volume in the Babel library. In fact, the very nature of the expres-

sion system employed provides a solution to the problem, unifying in a sequential way the recognition process with the amplification of the recognized species. All selection methods involve sequential cycles of binding to the antigen, separation from nonspecifically bound phages, elution, and amplification of eluted phages (Fig. 2). Procedures differ in the way antigen is presented. In particular, the antigen can be immobilized on solid phase or it can be soluble, coupled to biotin. Phages bound to the soluble antigen can be recovered by incubation with avidin coupled to magnetic microspheres. The latter method allows one to finely tune the antigen concentration and therefore lends itself to the selection of antibodies with increasing binding affinities. For membrane antigens, the libraries can be selected by panning on living or on fixed cells chemically treated with paraformaldehyde. As was shown for monoclonal antibodies, this allows the discovery of new antigens. In the frequent case in which the antigen is not available in a purified form, but the corresponding gene is available, it can be expressed at the surface of bacterial or mammalian cells and the modified cells can be used as selectors.

In the selection methods described, the recognition of the antigen and the subsequent amplification of phages are separate and distinct events. Two new selection methods have been described in which the recognition event determines in a direct and immediate way the amplification of bound phages (Fig. 4). Krebber *et al.* (1995) determined that the protein p3 is divided in two: The aminoterminal portion is fused to the antigen of interest and expressed in the periplasmic space of *E. coli*, whereas the carboxyterminal portion is fused to an antibody domain (or a library of it) and is expressed on phage (Fig. 4A). The phages that contain the carboxyterminal portion of p3, fused to the antibody, are not infective unless they do not interact in the periplasmic space with the missing part of p3, thereby reconstituting a functional infective p3 protein as a consequence of the antigen–antibody interaction. In the second method (Fig. 4B), the antigen is expressed at the bacterial surface, fused to the F pilus protein traA, which is responsible for the phage recognition and infection (Malmborg *et al.*, 1997). These bacteria can only be infected by those phages that express at their surface an antibody directed against the antigen expressed on the F pilus. In this case, the process of infection is mediated by the antigen–antibody interaction.

The methods of biological selection allow, in principle, one to compare two libraries, an antibody library displayed on phage with an antigen library displayed at the bacterial surface derived, for instance, from the genes expressed in a given tissue. As a result of the selective infection, a set of antigen–antibody pairs will be isolated, which is of much interest for projects of functional genomics.

In perspective, one could envisage selection methods for antibodies that overcome the simple antigen–antibody in-

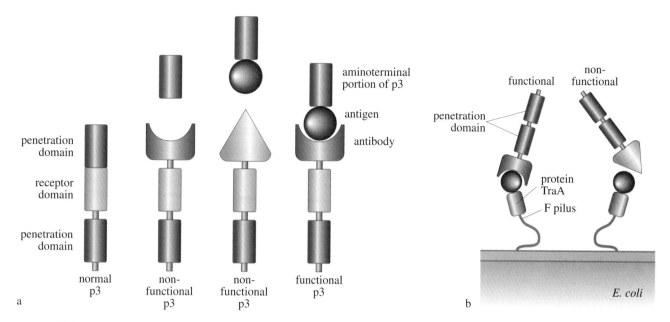

FIGURE 4 Selection strategies of phage display antibody libraries by judicious modification of the phage protein p3. A) The protein p3 is divided in three domains: two penetration domains and one receptor domain. An antigen of interest is fused to the penetration domain and an antibody to the rest of p3 including the receptor domain. Infection is only possible when the two parts are brought close together as a result of an antigen/antibody interaction in the periplasmic space, forming a functional p3 protein. B) The antigen is expressed at the tip of the F pilus, fused to the protein TraA, which is the E.coli receptor interacting with the penetration domain of the phage p3 protein. A large number of antibodies (a library) can be displayed on phages, unable to infect E.coli unless an antigen/antibody interaction is occurred.

teraction *in vitro* to isolate antibodies on the basis of functional properties in different contexts. These methods include selection methods based on enzymatic activities of catalytic antibodies or based on the ectopic expression of intracellular antibodies targeted to different intracellular compartments.

Molecular Evolution of Antibodies

Parlano della Biblioteca febbrile, i cui casuali volumi corrono il rischio incessante di mutarsi in altri.

Jorge Luis Borges
La Biblioteca di Babele

The affinity of antibodies isolated from phage display libraries is generally sufficiently high for most applications but may not be sufficient for some particular applications that require, ideally, picomolar affinities.

Selection schemes with the antigen in solution at decreasing or limiting concentrations as cycles proceed can be exploited to isolate among the pool of antibodies of the desired specificity present in the library those with greater affinity and/or slower dissociation constant. By coupling

these affinity selection procedures with methods to construct variants of a given lead antibody, one can isolate antibodies with greater affinity, mimicking the process of affinity maturation of antibodies in the immune system.

In the immune system, antibody maturation by somatic mutations occurs stepwise with small increments of affinity, which confer a selective advantage to the cells. Affinity maturation involves mutations that modify residues in the antigen combining site, optimizing the interaction, creating new additional contact regions, or eliminating residues with energetically unfavorable interactions. The display of antibodies on phage offers the possibility of improving antibody affinity by many different strategies of molecular evolution. Each method requires a source of diversity, thus creating a secondary library derived from the lead antibody or from a mixture of antibodies, directed against an antigen of interest. The methods with the most success include (i) chain shuffling, (ii) random mutagenesis in mutator strains, (iii) mutagenesis with error-prone PCR, (iv) sexual PCR and DNA shuffling, and (v) site-directed mutagenesis of CDRs (Fig. 5).

In the approach of chain shuffling, one of the variable regions (e.g., V_H) is kept constant and is expressed on phage in association with a library of V_L. In the second method, the diversity is created by propagating the phages express-

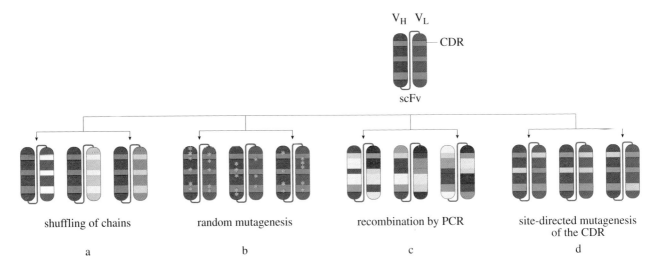

FIGURE 5 Improvement of the affinity of phage antibodies. The scheme shows different methods used to improve the affinity of phage antibodies which rely upon the creation of sub-libraries of mutated antibodies based on the original phage antibody (in this case a scFv fragment): (a) combinatorial shuffling of chains (marked by different intensity of the VL regions, in blue, and the hypervariable domains, in yellow); (b) random mutagenesis, in which the diversity is created by exploiting E.coli mutator strains (random mutations in VH and VL domains are indicated by asterisks); (c) mutagenesis with PCR, by using an error prone reaction of amplification, leading to fragments mutagenized not only in the ipervariable regions (green, yellow and violet bands) but also in the rest of the molecule (different intensity of red and blue); (d) site directed mutagenesis of CDRs, leading to fragments mutagenized only in the hypervariable regions (green, yellow and violet bands) Such libraries are then used as the basis for further cycles of selection for the isolation of improved antibodies.

ing the antibodies of interest in *E. coli* mutator strains. Random mutations create a small sublibrary from selected antibodies, which will be further selected and analyzed for affinity and specificity.

An alternative to the use of mutator strains is the use of error-prone PCR to amplify V regions by using nonproofreading enzymes under conditions of low fidelity. This method has the advantage of targeting the mutations to the V regions only, whereas mutator strains produce mutations in the whole plasmid.

For all these methods the mutations are strictly hierarchical; that is, mutations occurring in separate molecules cannot be combined unless they occur again independently, an event which is extremely unlikely. The methods of sexual PCR or *in vitro* DNA shuffling allow mutations in separate molecules to be combined at random in a single molecule, giving rise to new molecules completely different from the initial population of molecules.

In general, all the previously mentioned methods allow improvement of the affinity of antibodies derived from large phage display libraries up to the nanomolar range. To further improve the affinity of these antibodies, the site-directed mutagenesis of CDRs has been shown to be very efficient, yielding an affinity improvement of 1200-fold over that of the initial antibody. In this case, new libraries are

created in parallel and independently, and in each a single CDR has been mutated. Afterwards, groups of mutagenized CDRs are combined, and new selection schemes are applied to verify them in a recursive process which combines elements of parallel and hierarchical strategies.

The problem associated with improving the affinity of antibodies derived from phage display libraries is not, as in the case of the immune system, the generation of diversity but the selection of better variants against an increasing number of irrelevant mutations. Selection of antibodies with improved affinity is an example of a more general strategy which is now feasible by the use of phage display libraries. In general, schemes of molecular evolution to select antibodies harboring new properties may be conceived, for example, to confer increased stability, resistance to proteolysis, or tolerance to the absence of disulfide bonds.

Expressing Antibodies in Artificial Environments

All the applications of recombinant antibodies involve their expression in environments different from that of the lymphoid cell. Therefore, the design of the various applications is intimately connected with a better scientific understanding of the factors which regulate the biosynthesis of antibodies in artificial conditions. Different applications may require different cellular systems for the expression and purification of the recombinant antibodies. In the case of whole immunoglobulins, the use of mammalian cells is preferable, but for some applications transgenic plants are used as bioreactors. On the other hand, for simpler antibody domains, such as scFv fragments, prokaryotic cells such as *E. coli* are the expression system of choice.

Expression of Antibodies in Lymphoid and Nonlymphoid Mammalian Cells

The biosynthesis of antibodies in lymphoid cells is a highly regulated complex process involving many regulatory events at the trancriptional, translational, and posttranslational levels. In particular, the correct glycosylation is very important for the production of therapeutic or diagnostic antibodies to be used *in vivo* since it affects their half-life (clearance from circulation) and their immunogenicity. Many of these regulatory events are lymphoid cell specific, whereas others are in common with other mammalian cells and can collectively be seen as part of a "quality control system" for antibody synthesis and expression in the secretory pathway. The efficiency of secretion of recombinant antibodies varies from cell to cell; it is very high not only in myeloma but also in nerve cells. In general, antibodies are correctly synthesized, assembled, and secreted in many mammalian cell types.

Expression of Antibodies in Prokaryotes

Antibodies and antibody fragments are increasingly being expressed in bacterial systems not only for their production but also for the selection of new antibody specificities, the engineering new chimeric forms of antibodies, and the construction of antibody phage libraries. Moreover, antibody expression in *E. coli* can provide important information on the process of antibody folding in artificial environments.

For many applications, the periplasmic space of *E. coli* corresponds to the secretory pathway of eukaryotic cells and, for this reason, it is widely used for the expression of antibody domains. In fact, proteins related to mammalian proteins important for folding and assembly (chaperonines) are present in the periplasmic space, such as the enzyme that catalyzes the formation of disulfide bonds. Nevertheless, for most antibody fragments the folding in the periplasm does not proceed quantitatively, and a high, albeit variable, proportion aggregates in an insoluble form. The extent of aggregation depends in a crucial way on the primary sequence of the variable domains. This is more ev-

ident when the antibody domains are expressed in different cellular environments, such as the cytoplasm of mammalian cells.

Protein aggregation during the maturation of newly synthesized polypeptides is a natural phenomenon and represents an off pathway of the normal folding process, greatly limiting the yield of native proteins. Therefore, the study of new strategies aimed at preventing the aggregation of antibodies expressed in different systems and intracellular compartments is particularly important. The results of several studies on the biophysical properties of the folding process of antibody domains in artificial conditions have shown that single point mutations, generally in the framework of the variable regions, can confer solubility to an otherwise aggregating protein or improve its stability and half-life. This allowed the conclusion that some frameworks are intrinsically more suitable than others to be expressed in different artificial conditions and may be used as acceptors of CDRs of interest or of libraries of CDRs (Jung and Plueckthun, 1997).

Catalytic Antibodies

Antibodies can be used as enzymes. The binding specificity of the antibodies for the corresponding antigens may exceed that of the enzymes for their substrates. In fact, as pointed out by Pauling in 1940, the fundamental difference between enzymes and antibodies is that the former selectively bind transition states and the latter bind ground states. In 1969, Jenks suggested that antibodies could be artificially made and used to selectively stabilized the transition state of a chemical reaction. Following the advent of monoclonals, the chemical potential of the immune system as a source of antibodies with catalytical activity was exploited by the research groups of Schultz and Lerner (Lerner *et al.*, 1991). Their approach was based on the design of haptens analogous to the transition states of the reaction of interest for the production of antibodies directed against the transition state. The antibodies selected in this way, so-called catalytic antibodies or abzymes, work as enzymes for the substrate of the reaction. The first experimental demonstrations of this strategy used acyl-transfer reactions and the formation or hydrolysis of carbon–carbon bonds. Since then, many catalytic antibodies have been generated that catalyze a wide array of chemical reactions, with rates 10^3- to 10^6-fold faster than those of uncatalyzed reactions. The catalytic antibodies follow Michaelis–Menten kinetics, display substrate specificity, and bind the transition state analog considerably more tightly than substrate.

The analysis of the thermodynamic values for binding and catalysis confirms the notion that binding interactions can be dramatically different between antibodies and enzymes. In general, the antigen–antibody binding is enthalpically driven, whereas the substrate–enzyme interaction is driven by a favorable entropic variation.

One interpretation of the thermodynamic values for antigen–antibody binding is that the antibody pocket is fairly rigid. From studies on the effect of temperature on catalysis, it appears that abzyme catalysis is also dominated by an enthalpic contribution. Nevertheless, for some catalytic antibodies, entropic effects are also very important.

The great variety of catalytic antibodies isolated to date includes antibodies catalyzing new chemical reactions not documented previously. In all cases, the starting point is the synthesis of a perfect transition state analog in order to elicit an antibody capable of providing a maximum difference in the energy between substrate and hapten binding and converting this binding energy into a catalytic turnover of substrate. Although in general the mechanism of catalytic antibodies involves the stabilization of the transition state, in some cases catalytic antibodies with different modes of action have been obtained, indicating that the diversity of the immune system may introduce catalytic properties not predictable a priori. Experimental schemes for isolating catalytic antibodies on the basis of their chemical reactivity have been described (Smithrud and Benkovic, 1997), such as the so-called active immunization. Whereas in the usual mode of immunization care is taken to ensure that the antigen to be used is as inert as possible so that the resulting antibodies can interact with antigens in their native state, in the reactive immunization the opposite is done. A reactive antigen is designed so that a chemical reaction occurs in the binding pocket of the antibody during its induction on the surface of B lymphocytes. Other strategies have been described for isolating catalytic antibodies with different modes of action. In particular, catalytic antibodies with specific efficient acid-basic catalysis have been isolated. The importance of cofactors in many enzymatic reactions has stimulated the idea to introduce cofactors in the binding sites of the antibodies, further expanding the field of activity of catalytic antibodies.

The use of antibody phage display libraries allows to further extend the possibility of selecting new catalytic antibodies on the basis of their efficiency in catalysis. A method has been recently described in which antibodies are directly selected on the basis of their property of hydrolyzing a chemical bond. A phage library enriched with antibodies directed against the transition state of the reaction of interest undergoes cycles of solid-phase absorption with a multifunctional substrate which, if recognized by a phage–antibody and transformed in the reaction product, will create a covalent bond between that phage and the solid phase. In this way, those phage particles displaying the antibodies able to catalyze the reaction of interest will automatically be trapped, selected, and recovered. Selection schemes to isolate catalytic antibodies from phage libraries will allow

one to obtain antibodies with enhanced catalytic properties by using molecular evolution procedures similar to those described for affinity maturation.

Applications of Antibodies

The immunotechnology developments described previously have provided a new, constantly evolving, field of research for the production of antibodies with many different features to be used in a wide range of applicative fields.

Therapeutical and Diagnostic Antibodies *in Vivo*

After the phase of great hope regarding the therapeutic potential of antibodies, which was triggered by the discovery of monoclonal antibodies, the results of the initial clinical trials have clearly revealed major problems associated with immunotherapies and immunodiagnostics *in vivo*, including the immunogenicity of recombinant molecules, the specificity of the target antigens, and the physical accessibility of the target cells. The formulation and understanding of these problems, and the possibility of engineering new antibody forms and new strategies to overcome these difficulties, have brought a renewed but cautious optimism (Scott and Welt, 1997).

Humanized and Human Antibodies. The clinical utility of rodent monoclonal antibodies is limited by their immunogenicity. The "humanization" of antibodies is the most widely adopted strategy to avoid this problem (Riechmann *et al.,* 1988). Humanized antibodies are commonly created by transplanting the antigen-binding segments (CDRs) from rodent antibodies into human variable regions. In many cases, transplanting the CDR residues is not sufficient to restore completely the antigen binding and it is also necessary to recruit part of the framework region from the rodent antibody. More than 100 antibodies have been humanized by CDR grafting, some of which have been clinically evaluated with some tangible therapeutical successes.

The production of human monoclonal antibodies ab initio would represent the best solution, although it appears not to be feasible because of the impossibility of immunization in humans and because the procedures for *in vitro* immunization of lymphocytes are not generally successful. An alternative approach of interest is the production of human antibody repertoires in transgenic mice (Brueggemann and Neuberger, 1996). These can be obtained by the use of yeast artificial chromosomes containing a large region (larger than 10^3 kb) of the human immunoglobulin loci and a nearly complete set of diversity and joining segments. Transgenic mice carrying this minichromosome are crossed with mice in which the endogenous IgH and Igk loci have been silenced, at the transcriptional level, by homologous recombination. This approach leads to transgenic animals expressing human antibodies and makes it possible to use such strains for the efficient production of human repertoires, essentially without interference by endogenous mouse Ig.

Notwithstanding the successes, these procedures are still very laborious, and the use of libraries of human antibodies displayed on filamentous phages is the most promising technology for the isolation of human antibodies. Human antibodies isolated from phage display libraries are already used for clinical evaluation in a gene therapy and diagnosis perspective.

Human antibodies are generally selected from libraries in scFv or Fabs format, which may be affinity matured and improved in their specificity and expression properties. The use of simple antibody forms provides the additional advantage that effector functions can be easily engineered, tailored for particular therapeutical or imaging applications, thus improving the pharmacokinetic and pharmacodynamic properties (Carter and Mechant, 1997). In particular, effector functions useful in therapy include fusion of antibodies to toxins and to enzymes in order to inactivate drugs or prodrugs and the effector function needed for the action of antibodies in the immune system or bispecificity to recruit cells or molecules. For antibodies used for *in vivo* imaging of cells or tissues, such as tumors, different strategies can be used, such as chelation of scFv fragments with ^{99}mTc (the radionuclide more amenable for *in vivo* imaging) and radiolabeling of antibodies by phosphorylation by specific kinases, or, more generally, by binding to photoactive molecules.

Immunotoxins and Immunoconjugates. Immunotoxins are antibody-based molecules potentially useful in therapy, which bring to mind the original concept of the "magic bullet." They are chimeric proteins made of an antibody domain covalently linked to a toxin or its subunit (Ghetie and Vitetta, 1994). In this case, the specific recognition moiety of the antibodies directed against selected membrane antigens is exploited to link them to a toxin and selectively kill target cells recognized by the antibody. Generally, the subunits from ricin, difteria, or exotoxin *Pseudomonas* are particularly suitable for this purpose. Clinical evaluation of immunotoxins has been beset by problems of immunogenicity and aspecific toxicity. The immunogenicity of the antibody component is minimized by utilizing scFv fragments derived from human antibody libraries. Other molecules, such as human ribonucleases or nonprotein toxins, are being explored as potentially less immunogenic.

A particularly promising application to overcome the lack of specific antitumor antigens is the targeting of immunotoxins to vascular endothelial cells to damage the vasculature that supplies the tumors (neovasculature antigens). This approach has two important advantages: the vascular endothelial cells, unlike the tumor, are directly accessible to administered therapeutic agents and a judicious and well-

controlled vasculature damage is expected to translate into a widespread tumor cell death because each capillary nourishes many tumor cells (Huang *et al.,* 1997).

Immunoliposomes. Liposomes are lipid vesicles potentially useful in therapy for the delivery of drugs, toxins, and DNA. However, translation of liposomes into clinical practice has been hampered by their poor stability and rapid clearance from circulation. Liposomes can be stabilized and targeted to specific tissues by the fusion to an antibody molecule, such as the antitransferrin receptor antibodies which allow selective passage across the blood–brain barrier of specific immunoconjugates.

Bispecific Antibodies. Bispecific antibodies (Fig. 1) are well suited to be utilized for diagnosis and therapy. Thus, the bispecificity allows the expansion of the number of experimental strategies, exploiting one of the two specificities for the recruitment of new effector functions or the creation of bridges between different cells. Bispecific antibodies have been successfully used to direct cytotoxic T cells to kill target tumor cells of different origins by a cell-mediated cytotoxic response. Due to their small size, these molecules are easily diffusable into the tissues and their immunogenicity appears to be limited. Bispecific molecules recognizing a specific receptor or a particular cellular type and a viral antigen (e.g., the capsid protein of the adenovirus) have been successfully utilized to specifically target the viral infection to a cell of interest; thus, there are high expectations for the use of viral vectors in gene therapy.

Intracellular Antibodies

The ectopic expression of antibodies (Cattaneo and Biocca, 1997) in cells, tissues, or different intracellular compartments represents a new approach in which the antibody is delivered to the biological system as a gene and not as a protein, increasing the number of antigens that can be targeted and modulated by antibodies (inter- and intracellular immunization).

Antibodies are normally secreted by lymphoid cells. Following the demonstration that antibody assembly and secretion can occur efficiently in nonlymphoid cells (Cattaneo and Neuberger, 1987), it became clear that the ectopic expression of secreted antibodies could be utilized to perturb or interfere with the function of selected extracellular antigens in other tissues, such as the nervous system (neuroantibodies), or organisms such as plants (plantibodies).

Intercellular Immunization. The neuroantibody approach exploits the efficiency of antibody secretion artificially expressed by different cell types of the nervous system (Cattaneo and Biocca, 1997). This strategy is based on the expression, in the nervous system of transgenic mice, of antibodies directed against extracellular or extracellularly exposed molecules, such as growth factors, neuropeptides, or their receptors. The possibility of achieving a fine control by placing the antibody genes under the transcriptional control of a variety of different promoters makes this approach a useful complement for functional studies in the nervous system as well as in other tissues (Cattaneo and Biocca, 1997)

Intracellular Immunization. The substitution of the N-terminal leader sequence for secretion of the wild-type immunoglobulins with targeting signals for other intracellular compartments allows for an increased number of antigens that can be neutralized, including intracellular gene products (intracellular antibodies) (Biocca *et al.*, 1990, 1994; Cattaneo and Biocca, 1997). This approach is based on the idea that antibody chains, if equipped with suitable targeting signals, can be targeted toward new ectopic intracellular sites. The great wealth of information that has accumulated on the different ways in which normal cellular proteins find their way inside the cell has demonstrated the existence of specific localization signals, many of which are dominant and autonomous. The presence or absence of an N-terminal leader sequence, and the consequent synthesis on membrane-bound or free ribosomes, determines this first choice. Proteins cotranslationally inserted across the membrane of the endoplasmic reticulum (ER) undergo a hierarchically organized sequence of targeting decisions. Other targeting signals, some of which have already been identified, determine the subsequent targeting through the secretory pathway; proteins can be secreted by a constitu-

tive or regulated pathway and transported to the plasma membrane, lysosomes, or various compartments of the Golgi apparatus of the ER, or to vesicles of a different nature. In a similar way, proteins that are synthesized on free ribosomes may diffuse in the cytoplasm or, if equipped with suitable targeting signals, could be targeted to the nucleus, mitochondria, or the inner surface of the plasma membrane. Many of these signals are autonomous and dominant, i.e., they can be grafted to confer a new intracellular location to a reporter protein. The properties of these signals have been exploited to redirect individual antibody chains or antibody domains to different intracellular compartments (Fig. 6) by incorporating them to the amino or carboxy terminal of the antibody molecules. Figure 7 shows scFv fragments targeted to different intracellular compartments of fibroblast cells viewed by immunofluorescence and confocal microscopy. The retargeted antibody domains are able to interact with the corresponding intracellular antigen, thus neutralizing its biological action. This approach has been successfully applied to different biological systems (from yeast to human cells) and for proteins localized in different cellular compartments. In particular, it has been used to block the activity of cytoplasmic proteins involved in signal transduction, cytoplasmic or nuclear proteins involved in the process of replication of pathogenic retroviruses, or membrane receptors.

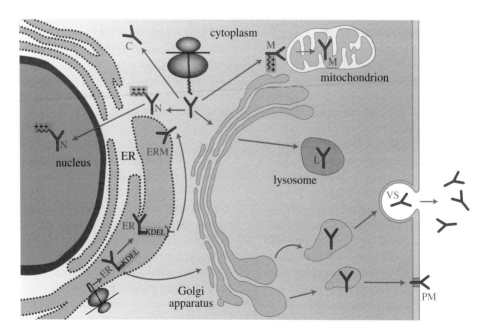

FIGURE 6 Targeting of antibodies to different intracellular compartments. Some of the known targeting signals (in yellow) may be exploited to redirect the intracellular traffic of antibodies in eukariotic cells. The abbreviations near the antibodies indicate the cellular compartments where the antibodies are directed; N: nucleus, C: cytoplasm, ER: endoplasmic reticulum, M: mitochondrion, VS: secretory vescicle, PM: plasma membrane; L: lysosome.

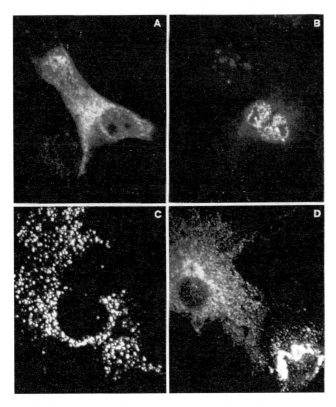

FIGURE 7 Antibody fragments targeted to different intracellular compartments of cells, viewed by indirect immunofluorescence and confocal microscopy. ScFv fragments were equipped with the following targeting signals: (a) cytoplasm, (b) nucleus, (c) mitochondria and (d) secretory pathway.

The intracellular antibodies may act by directly neutralizing the active epitope of the target protein. In a different mode of action, they can act by diverting the intracellular traffic of the antigen. This is the case for membrane receptors, whose appearance at the surface can be inhibited by the interaction with antibody intracellularly retained in the ER by retention signals (intracellular anchors). In perspective, antibodies may be equipped with particular effector functions tailored for particular intracellular immunization applications, including the introduction of signals for proteolytic degradation to target the antigen–antibody complex to the intracellular degradative compartment (suicide antibodies). Antibodies usually act stoichiometrically with respect to the antigen. The use of antibodies with catalytic activity, either intrinsic or introduced by fusing the antibody domain with enzymes, coupled to metabolic selection schemes is an attractive development.

The intracellular antibody strategy has acquired a prominent status in different research and applicative fields, such as the somatic gene therapy against viral or tumor pathologies, plant biotechnology, and basic research.

One of the more promising application in gene therapy for the expression of intracellular antibodies is toward the creation of cellular resistance to viral infection by HIV virus (Marasco, 1997). This has been obtained by the intracellular expression of antibody domains directed against viral proteins in the secretory pathway (gp120) or in the cytoplasm and the nucleus (reverse transcriptase, rev, tat, and integrase). The clinical evaluation of the therapeutical efficacy of this approach is under way. The genes for intracellular antibodies are introduced *ex vivo* in stem cells of the hematopoietic system or in peripheral T lymphocytes, and then the modified cells are reintroduced in patients. The coexpression of more than one antibody domain, in different intracellular compartments, may result in a stronger protection (combined approach).

The success of the technology of intracellular antibodies is crucially dependent, although still unpredictable, on the folding properties of antibody domains under different cellular conditions. For example, either the absence of proteins assisting their folding (chaperonines) or the reducing environment of the cytoplasm would hinder the formation of intrachain disulfide bonds of the variable domains of the heavy and light chains, determining suboptimal conditions for folding. Intrachain disulfide bonds are one of the hallmarks of the antibody domain architecture; in fact, the contributing cysteine residues are perfectly conserved in all known human or mouse germline V regions. From studies on the contribution of the disulfide bridges to the folding stability, it appears that variable domains have a range of folding stabilities and that the overall stability of the fold is contributed to by many critical residues or combinations of residues in the framework regions. It may be surmised that domains that are intrinsically more stable may tolerate the removal or the absence of the disulfide bond and, conversely, that those that do tolerate the removal or the absence of the cysteines are more stable. Identifying those modifications that lead to a more stable antibody with improved folding properties in the absence of disulfide bonds represents an important goal for future applications (Cattaneo and Biocca, 1998). The achievement of this goal will be greatly facilitated by the implementation of selection schemes for intracellular antibodies on the basis of their property to be functionally expressed in the intracellular environment or to induce a selectable phenotype.

Antibody-Mediated Viral Infection. Antibodies directed against membrane receptors have been successfully expressed on virus and retrovirus surface, resulting in specific antigen–antibody-mediated infection of target cells. In gene therapy, the development of these experiments would allow to confer cellular/tissue specificity to the viral or retroviral infection. Also, antibodies could be used as immunoconjugates (as immunoliposomes or antibodies fused to DNA-binding proteins) to obtain the introduction of genes into target cells in an antigen-dependent manner.

Plantibodies

Plants are capable of synthesizing and assembling virtually every kind of antibody molecule, ranging from the smallest antigen-binding domains to full-length and even multimeric antibodies. The principal applications of expression of antibodies in plants relate to the use of transgenic plants as bioreactors of antibodies and to protocols of passive and intracellular immunization.

Plants as Bioreactors. The expression of antibodies in plants presents an important advantage for the large-scale production of antibodies at an extremely competitive cost (Hiatt *et al.*, 1989). In particular, transgenic plants are highly suitable for the expression of multimeric antibody forms (either IgG or IgA), exploiting successive sexual crossings of transgenic plants expressing each individual antibody chain. Therefore, large-scale expression of antibodies in plants indicates that plants can be used as bioreactors on an agricultural scale. For human therapy, the presence of plant-specific glycans might increase the immunogenicity of the recombinant antibody. In this case, it may be necessary to remove the complex glycans either by chemical or by recombinant modifications.

Passive Immunization with Transgenic Plants. Large quantities of specific antibodies are generally required for passive immunization for therapeutical purposes. The efficacy of transgenic plants as a system of expression of recombinant antibodies, together with the increasing number of transformable species of plants including edible plants, has allowed the possibility of utilizing plants as a source of antibodies for topic immunotherapy or for oral administration in a passive immunization protocol (Ma and Hein, 1995). In this case, the secretory IgA (SIgA) is the more suitable isotype since it is the predominant form of antibody in human oral cavity and all mucosal sites and is particularly effective owing to its increased resistance to proteolysis. A SIgA antibody with immunotherapeutical potential has been efficiently expressed in transgenic plants by coexpression of the heavy and light chains together with the J-associated protein and the so-called secretory component. The availability of functional antibodies in palatable plants could reduce or possibly eliminate the need for purification of the plant antibody prior to treatment. Plants can be utilized for local delivery of antibodies, a rapidly developing technology still in the experimental phase for treatment of diseases of the oral cavity and the gastrointestinal tract.

Intracellular Immunization in Plants. The ectopic expression of intracellular antibodies in transgenic plants has been successfully utilized to confer a new phenotype of interest, such as resistance to pathogenic virus (Tavladoraki *et al.*, 1993) or other pathogens. The use of intracellular antibodies for the production of transgenic plants with modified properties, in an applicative or basic research perspective, is a very interesting new field since it represents a combination of immunotechnology and plant biotechnology.

Biosensors

Biosensors exploit the remarkable specificity of biomolecular recognition to provide analytical tools that can measure the presence of a single, very diluted molecular species in a complex mixture. In a biosensor, molecular sensing is provided by a recognition element linked to a molecular signal transduction unit. Detection occurs at two levels: molecular signal transduction, in which the physical properties of a macromolecule change upon binding, and macroscopic signal determination to detect this change. Ideally, a general biosensor should be made of families of molecules that vary in their binding site but retain a constant signal transduction function that is easily measurable by the same detection instrumentation (possibly miniaturizable) (Hellinga and Marvin, 1998).

The repertoire of antibodies provides a readily available source of diversity in binding sites for the construction of biosensors. However, the antigen–antibody recognition unit is not linked to a signal transduction function. In the immune system, transduction is provided by the cellular context. In an antibody-based biosensor, the signal transduction element should be artificially incorporated; one example of an antibody-based biosensor exploits the phenomenon of surface plasmon resonance to measure binding events between molecules. Recently, a very general approach has been developed in which antibodies are coupled to an electrochemical reporter function by macromolecular assembly to construct ion channel switches (Cornell *et al.*, 1997). This is based on the fusion of the antibody to gramicidin, a small peptide that spans half a lipid bilayer (Fig. 8). An ion-conductive transmembrane channel is formed when two peptides respectively located in the inner and outer membrane leaflet of the lipid bilayer align. In this configuration, when single binding molecules of interest are recognized by antibodies linked to gramicidin, a macroscopic variation of current on the order of million of ions for seconds occurs in response to a single molecular interaction event. With this particular biosensor it is possible to measure variations of current occurring in a single channel.

The technology of antibody libraries, by providing a virtually unlimited repertoire of binding domains, is crucial for the engineering of modular biosensors.

Conclusions

Throughout the years, the exploitation of antibodies has moved from complex and poorly controlled collections of immunoglobulins (polyclonal antisera) to individual well-

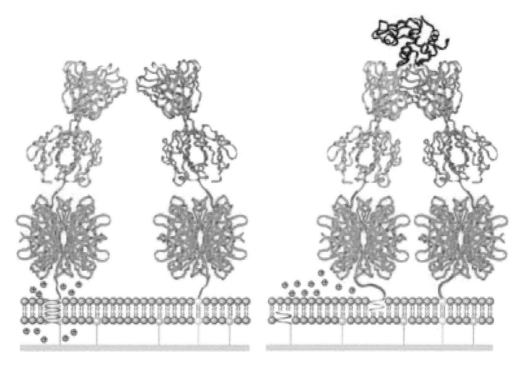

FIGURE 8 Biosensor based on an antibody moiety (blue) and on an ion channel formed by two molecules of gramicidin. The gramicidin channels are formed from two halves, one of which (pink) is tethered to the gold in the inner surface of the membrane and the other (violet) to a mobile antibody. Other antibodies, directed against an antigen which is different from the target antigen, are immobilized via a transmembrane biotinylated lipid tethered to the gold substrate. Antibodies are coupled to their two attachment points using a biotin-streptavidin adaptor (green). In the absence of antigen (a), the mobile gramicidin half-channel is free to form complete channels, but on binding (b) of antigen (red) the antibodies are crosslinked, restricting the formation of gramicidin channels and, consequently, the transmembrane ionic current.

characterized reagents (monoclonal antibodies). The latter have allowed the development of modern immunotechnologies, based on the modification and protein engineering of individual antibodies, in ways and forms only limited by the imagination. Today, the technology of antibody repertoires displayed on phage allows not only the isolation of individual recombinant antibodies but also the handling, in a parallel fashion, of complex collections of antibodies together with the corresponding genes — the recombinant version of polyclonal antisera. In a sense, the technology of phage libraries is at the heart of modern immunotechnologies since it allows the construction and exploitation of artificial immune systems tailored to measure. In perspective, this technology has important implications for the projects of human genome sequencing. Indeed, the speed at which new sequence data are generated is much greater than the speed at which information on the localization and the function of the corresponding protein is generated. The success of the genome sequencing projects will depend on reaching the final goal of understanding the function of the proteins encoded by the genome, the so-called functional genomics. It is clear, therefore, that a system that is intrinsically parallel, such as that of the phage libraries, lends itself to developing selection schemes for antibodies directed against libraries of proteins or peptides (libraries against libraries), rather than against individual proteins or antigens, to study the function of classes of proteins in cells or tissues of interest. This will provide a significant technological contribution to functional genomics.

Quando si proclamo' che la Biblioteca comprendeva tutti i libri, la prima impressione fu di straordinaria felicita'. Tutti gli uomini si sentirono padroni di un tesoro intatto e segreto. Non v'era problema personale o mondiale la cui eloquente soluzione non esistesse.

Jorge Luis Borges
La Biblioteca di Babele

References Cited

BIOCCA, S., NEUBERGER, M., and CATTANEO, A. (1990). Expression and targeting of intracellular antibodies in mammalian cells. *EMBO J.* **9,** 101–108.

BIOCCA, S., PIERANDREI-AMALDI, P., CAMPIONI, N., and CATTANEO, A. (1994). Intracellular immunization with cytosolic recombinant antibodies. *Biotechnology* **12**, 396–399.

BIRD, R. E., *et al.* (1988). Single-chain antigen-binding proteins. *Science* **242**, 423–426.

BRUEGGEMANN, M., and NEUBERGER, M. S. (1996). Strategies for expressing human antibody repertoires in transgenic mice. *Immunol. Today* **17**, 391–397.

BURTON, D. R., BARBAS, C. F., PERSSON, M. A. A., KOENIG, S., CHANOCK, R. M., and LERNER, R. A. (1991). A large array of human monoclonal antibodies to HIV-1 from combinatorial libraries from asymptomatic individuals. *Proc. Natl. Acad. Sci. USA* **88**, 10134–10137.

CARTER, P., and MERCHANT, A. M. (1997). Engineering antibodies for imaging and therapy. *Curr. Opin. Biotechnol.* **8**, 449–454.

CATTANEO, A., and BIOCCA, S. (1997). *Intracellular Antibodies: Development and Applications.* Springer-Verlag, New York.

CATTANEO, A., and BIOCCA, S. (1998). Selection of intracellular antibodies. *Trends Biotechnol.*, in press.

CATTANEO, A., and NEUBERGER, M. S. (1987). Polymeric immunoglobulin M is secreted by transfectables of non-lymphoid cells in the absence of immunoglobulin J chain. *EMBO J.* **6**, 2753–2758.

CLACKSON, T., HOOGEBOOM, H. R., GRIFFITHS, A. D., and WINTER, G. (1991). Making antibody fragments using phage display libraries. *Nature* **352**, 624–628.

CORNELL, B. A., BRAACH-MAKSVYTIS, V. L. B., KING, L. G., OSMAN, P. D. J., RAGUSE, B., WIECZOREK, L., and PACE, R. J. (1997). A biosensor that uses ion-channel switches. *Nature* **387**, 580–583.

DAVIES, J., and RIECHMANN, L. (1995). Antibody VH domains as small recognition units. *Bio/Technology* **13**, 475–479.

GHETIE, M. A., and VITETTA, E. (1994). Recent developments in immunotoxin therapy. *Curr. Opin. Immunol.* **6**, 707–714.

HANES, J., and PLUECKTHUN, A. (1997). *In vitro* selection and evolution of functional proteins using ribosome display. *Proc. Natl. Acad. Sci. USA* **94**, 4937–4942.

HE, M., and TAUSSIG, M. J. (1997). Antibody–ribosome–mRNA (ARM) complexes as efficient selection particles for *in vitro* display and evolution of antibody combining sites. *Nucleic Acid Res.* **25**, 5132–5134.

HELLINGA, H. W., and MARVIN, J. S. (1998). Protein engineering and the development of generic biosensors. *Trends Biotechnol.* **16**, 183–189.

HIATT, A., CAFFERKEY, R., and BOWDISH, K. (1989). Production of antibodies in transgenic plants. *Nature* **342**, 76–78.

HOLLIGER, P., PROSPERO, T., and WINTER, G. (1993). "Diabodies": Small bivalent and bispecific antibody fragments. *Proc. Natl. Acad. Sci. USA* **90**, 6444–6448.

JONES, P. T., DEAR, P. H., FOOTE, J., *et al.* (1986). Replacing the complementarity determining regions in a human antibody with those from a mouse. *Nature* **321**, 522–524.

JUNG, S., and PLUECKTHUN, A. (1997). Improving *in vivo* folding and stability of a scFv antibody fragment by loop grafting. *Prot. Eng.* **10**, 959–966.

KOEHLER, G., and MILSTEIN, C. (1975). Continuous cultures of fused cells secreting antibodies of predefined specificity. *Nature* **348**, 552–554.

KREBBER, C., SPADA, S., DESPLANCQ, D., and PLUECKTHUN, A. (1995). Co-selection of cognate antibody–antigen pairs by selectively-infective phages. *FEBS Lett.* **377**, 227–231.

LERNER, R. A., BENKOVIC, S. J., and SCHULTZ, P. G. (1991). At the crossroads of chemistry and immunology: Catalytic antibodies. *Science* **252**, 659–667.

MA, J. K.-C., and HEIN, M. B. (1995). Immunotherapeutic potential of antibodies produced in plants. *Trends Biotechnol.* **13**, 522–527.

MALMBORG, A. C., SOEDERLIND, E., FROST, L., and BORREBAECK, C. A. K. (1997). Selective phage infection mediated by epitope expression on F pilus. *J. Mol. Biol.* **273**, 544–551.

MARKS, J. D., HOOGENBOOM, H. R., BONNERT, T. P., McCAFFERTY, J., GRIFFITHS, A. D., and WINTER, G. (1991). Bypassing immunization: Human antibodies from V-genes libraries displayed on phage. *J. Mol. Biol.* **222**, 581–597.

MARASCO, W. A. (1997). Intrabodies: Turning the humoral immune system outside in for intracellular immunization. *Gene Ther.* **4**, 11–15.

McCAFFERTY, J., GRIFFITHS, A. D., WINTER, G., and CHISWELL, D. J. (1990). Phage antibodies: Filamentous phage displaying antibody variable domains. *Nature* **348**, 552–554.

NEUBERGER, M. S., WILLIAMS, G. T., and FOX, R. O. (1984). Recombinant antibodies possessing novel effector functions. *Nature* **312**, 604–608.

ORLANDI, R., GUSSOW, D. H., JONES, P. T., and WINTER, G. (1989). Cloning immunoglobulin variable domains for expression by the polymerase chain reaction. *Proc. Natl. Acad. Sci. USA* **86**, 3833–3837.

RIECHMANN, L., CLARK, M., WALDMANN, H., and WINTER, G. (1988). Reshaping human antibodies for therapy. *Nature* **332**, 323–327.

SCOTT, A. M., and WELT, S. (1997). Antibody-based immunological therapies. *Curr. Opin. Immunol.* **9**, 717–722.

SMITH, G. P. (1985). Filamentous fusion phage: novel expression vectors that display cloned antigens on the virion surface. *Science* **228**, 1315–1317.

SMITHRUD, D. B., and BENKOVIC, S. J. (1997). The state of antibody catalysis. *Curr. Opin. Biotechnol.* **8**, 459–466.

TAVLADORAKI, *et al.* (1993) Transgenic plants expressing a functional scFv antibody are specifically protected from virus attack. *Nature* **366**, 469–472.

WINTER, G., and MILSTEIN, C. (1991). Man-made antibodies. *Nature* **349**, 293–299.

General References

MILSTEIN, C. (1990). The Croonian Lecture 1989. Antibodies: A paradigm for the biology of molecular recognition. *Proc. R. Soc. London B* **239**, 1–16.

WINTER, G., GRIFFITHS, A. D., HAWKINS, R. E., and HOOGENBOOM, H. R. (1994). Making antibodies by phage display technology. *Annu. Rev. Immunol.* **12**, 433–455.

Ulrich Behn

Institute for Theoretical Physics
Leipzig University
Leipzig, Germany

Franco Celada

Chair of Immunology
University of Genova
Genova, Italy

Hospital for Joint Diseases
New York, New York 10003
USA

Philip E. Seiden[†]

IBM Thomas J. Watson Research Center, Yorktown
Heights, New York, USA

Computer modeling can be a valuable adjunct in immunological research. It can be used both as a theoretical tool to explore and understand the behavior of immunological processes and in an experimental mode to provide an additional resource — experiments in machina, i.e., experiments in the computer. In this article, which is not a review but a subjective selection of topics, we discuss why, when, and how modeling is done and what immunologists can expect from it.

Computer Modeling in Immunology

....●

A Place for Modeling

Natural evolution, as theorized by Charles Darwin, walks on two opposite and complementary legs — expansion (of diversity) and reduction (by selection). It may not be coincidental that the progress of scientific knowledge also relies on two opposite phases — the generation and the test of ideas. Those of us who are in their sixties were taught in high school that the methodological revolution from which sprung modern science was triggered by Bacon and the English empiricists by substituting deductive with inductive reasoning. We have since realized that science's preferred way of reasoning bypasses the classical antinomy deduction-induction of the Greek masters: Peirce (1931–

1958) defined the novel procedure utilized by scientists (and by private detectives) and called it abduction. Figure 1 illustrates the three ways of reasoning using a minimalistic cartoon devised by Umberto Eco (1986) when looking at a bag that spilled a bean. There is no doubt that abduction followed by experimental test is one central avenue of scientific reasoning. However, Popper's (1935) dogged insistence on the falsification of the reigning theory depicts research as an activity in which the negative critical phase prevails on the acquisition of new facts, new ideas, and novel perspectives — a world in which the hypothesis, with the indication of the experiment apt to refute it, shines until falsified and is substituted by another hypothesis with similar requirements. Even if refutation can be seen as recognition and

[†]Deceased, April 21, 2001.

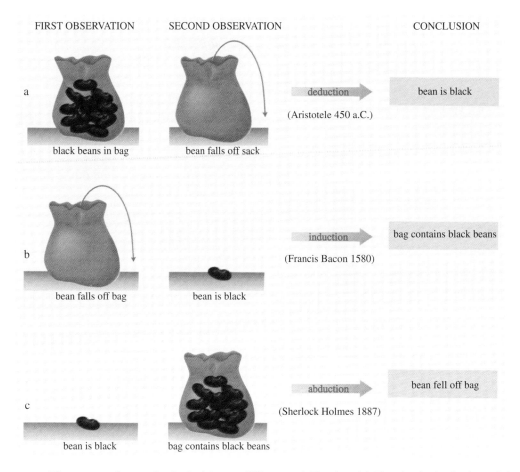

FIRST OBSERVATION   SECOND OBSERVATION   CONCLUSION

a   black beans in bag   bean falls off sack   deduction → bean is black
(Aristotele 450 a.C.)

b   bean falls off bag   bean is black   induction → bag contains black beans
(Francis Bacon 1580)

c   bean is black   bag contains black beans   abduction → bean fell off bag
(Sherlock Holmes 1887)

FIGURE 1   Three ways of reasoning in the history of Western civilization. (a) The classical syllogism. If the first two premises are well formulated, the conclusion is inescapable. However, there is nothing new in the conclusion. This is not the reasoning of scientific progress. (b) The empirical process, from the particular to the general. In our example, the conclusion adds an element of novelty but is not infallible: All that can be said is that some of the beans in the bag were black — at the limit, only one. (c) Abductive (hypothetical) reasoning is imaginative and the conclusion is novel. There is nothing illogical in the hypothesis, but it has no certainty at all; the bean could have fallen from another bag or even directly from the bean plant. Therefore, the conclusion should be tested (corroborated or rejected) by independent information. Sherlock Holmes's procedures (wrongly labeled deduction) were of this nature. The fact that they were always correct (and those of Dr. Watson always wrong) depended on the bias of the author, Sir Arthur Conan Doyle. Abduction is the foremost way of reasoning of modern science. The lack of a similar prejudicial bias is perhaps the reason why numerous hypotheses, even those by the best researchers, are eventually rejected.

correction of an error (Freeman and Skolimowsky, 1974) and Peirce — rather than Popper — marvels at the lively and wild imagination, the game of falsification does not seem to fully describe the heuristic activity of the scientific mind, especially in those disciplines that are in continuous expansion, as immunology has been during the past several decades. The sheer number of investigations occurring simultaneously causes a tumultuous acquisition of data, often not consistent with each other, and the coexistence of several hypotheses in conflict adds to the complication. From the beginning of our theoretical activity, we saw a distinctive need for mathematical modeling arise from this com-

plexity. As it was described then (Celada, 1992), Planet Immunology (Fig. 2) is a structure with a hard core of certainties but a large zone of soft and semisoft material, with a saturnal ring and some navigational problems. We did not know at the time that professional thinkers had used the same image, with a more precise meaning. According to Imre Lakatos (1974), a foremost admirer and critic of Karl Popper, investigators in the field, rather than engaging endless ritual duels between theories and refutation by experiment, "push forward a research program, with a conventionally accepted hard core and with a positive heuristic which defines problems, foresees anomalies and list them

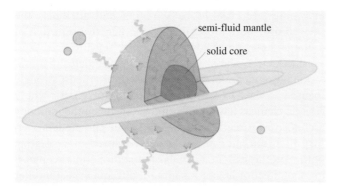

semi-fluid mantle

solid core

FIGURE 2 Planet Immunology is composed of a hard core and a large soft mantle in a semifluid state (the crust is too young to dwell on). The core consists of hard proven facts and dependable figures on which a consensus is almost complete. The mantle contains diverse material assembled without evident order and in constant, irregular movement.

as they appear but, as long as the research program sustains its momentum, survives them." In other words, the hard core of facts and hypotheses withstands assault because it is surrounded with a protective belt of auxiliary notions which may or may not be discarded based on experimental evidence. However, sometimes testing is difficult, expensive, or unpopular, and hypotheses which are logically incompatible coexist for long periods of times. This threatens to slow down the program or bring it to a standstill, which is a sentence of death for such a dynamic endeavor.

It is in this environment that mathematical modeling in all its forms can become a useful tool in the hands of researchers and even an important asset. It does so by creating conditions that stimulate the germination of ideas and by offering preliminary tests of their solidity, thus acting on both phases of the scientific process.

Reproducing Reality and Fostering Expansion

Modeling is possible — as science is possible — because the physical and biological reality is defined by quantities and numerical relations. The first step in the construction of any model is to simulate closely a physical or biological phenomenon using purely mathematical terms. The choice of these terms and of their relations requires an imaginative effort. Obviously, no model will fully reproduce reality, but from the degree of mimicry obtained it will be possible to judge on the completeness of the inputs. Typically, if the degree is low, the model will have to be modified in its starting quantities or in its parameters and variables. Also, this activity is creative: it is part of the expansion phase of the scientific process because introducing a parameter or factor is equivalent to adopting a working hypothesis. Another way by which modeling helps expansion is staged when the model is successful, i.e., it shows a high degree of correspondence with the phenomenon. At this point, the simulation

may project a reconstruction of nature which is accessible in all its facets, easy to observe and measure, with temporal and causal relationships neatly defined and dynamic attractors earmarked. This is the basis for the use of models as didactic tools (e.g., in teaching immunology; Celada and Seiden, 1992). Arguably even more important is their function in the very motor of scientific progress, the generation of hypotheses/abductions.

If science were to use only inductive thinking no model could substitute for the actual observation of natural facts. Instead, abduction is science's highway (Fig. 1c), and an important mission is entrusted to creative imagination. The springboard for imagination can also be clear and quantitative simulation, as the natural phenomenon, which is often concealed or cryptic.

Pleiotropy, Multicasuality, and Minimalism

Certain characteristics of scientific disciplines — and even the attitudes of their members — may be influenced by the evolutionary history of the system they study. For example, the distinctive approaches of microbiologists and/or immunologists can be traced back to the fact that bacteria appeared 3 billion years ago, whereas lymphocytes evolved 2.7 billion years later, making the immune system the youngest cellular organization and the ultimate branch of evolution of the animal kingdom. In Immunology, many possible concurrent and/or alternative causes are usually proposed for each phenomenon as soon as it is demonstrated to exist in cleverly designed experimental environments, and quickly published. Also, for each cause, many effects are recognized, all of which will be published. This produces a wealth of information that may seem to be contradictory at first and cause some discontent. However, some confusion is unavoidable on account of how the young immune system developed abruptly from a nonspecific hit and destroy defense into a full-fledged cognitive system. In Nature's characteristic way (Jacob, 1970), new ways emerged without discarding the old, as apparently obsolete molecular and cellular elements and pathways were reused as support for the new specific responses. This generated many duplications for each function, which is evolutionarily justified as an insurance against system failures.

Although there is no question that *in vivo* there are preferred or master pathways for any function, most potentialities remain dormant. On the other hand, any normally nonactive function may be forced into action under certain conditions. Since it is in the power of the researcher to set the conditions of the experiment, even to such extremes which could never be found in normal life, it is not surprising that every single pathway may be brought to life and demonstrated. This is analytically correct and certainly informative; the drawback is that the hierarchy of pathways is lost and so is their biological perspective and their relative weight in normal life.

Modeling finds an interesting application herein helping to answer the dissecting questions, e.g., what is sufficient to explain a given step, what is essential, and what is commonly used? Modeling, in all its forms, is able to discriminate because its vocation is to simplify, economize, and minimalize — in other words, the opposite of Nature's function. It can be done by introducing one function (of many possible) at a time; running the simulation; repeating the run with function 2, 3, . . . *n;* and comparing the results to evaluate their correspondence to observations in the field in order to establish a pecking order among 1, 2, 3, . . . *n.*

Models as Testing Instruments

The proper test of a hypothesis is the experiment. It consists of reproducing a natural phenomenon under controlled conditions and comparing its dimensions with those predicted by the prevailing theory. To control the experimental conditions means to be able to vary all parameters, one at a time, in order to assess, by state-of-the-art instruments, their effects on the results.

In Immunology as in all biosciences, experiments have become increasingly sophisticated and thus remote from *in vivo* observations. Cell populations, single cells, and subcellular systems are studied *in vitro* or in intermediate systems such as adoptive transfers and transgenic animals. A further extension of this tendency is the performance of experiments *in machina*, that is, using dynamic models of the system to be tested. This kind of experimentation has total flexibility, and the loss of reality due to abstraction is not qualitatively different than the loss suffered between *in vivo* and *in vitro* tests. However, there is a give-and-take between abstraction and demonstrativeness. For this reason, experiments *in machina* cannot substitute for *in vivo* or *in vitro* tests. However, if the parameters introduced are realistic, the results of the simulation can give indications of which directions would be more or less promising to probe with bench experiments. This is very helpful in the dynamics of the gray layer of Planet Immunology, and here experiments *in machina* can save time, money, and scores of laboratory animals.

Because conflicting hypotheses may and do coexist for long periods of time before a definitive experiment can be devised and performed, modeling has been applied as a preliminary clarifying and testing tool. For this purpose, the model must be neutral enough to be able to simulate both alternative postulates. The primary aim of this exercise is to reveal the discriminating parameters on which biological experiments may be fruitfully aimed. An example of this use is the modeling of B cell hypermutation after primary antigenic stimulation. *In vivo* mutations are found with significantly higher frequences on the gene segments coding for the complementarity determining regions (CDRs) of the antibody. One theory assumes an even distribution of mutations over the entire V region and attributes the skewed distribution observed to the negative selection of mutants suffering residue changes in sensitive positions outside CDRs. The alternative theory postulates that mutational events are originally confined, or at least strongly focused to the DNA sequences coding for CDRs, where residue changes have no disruptive effects on the antibody architecture (Celada and Seiden, 1996). The modeling effort discussed later in this article consisted of simulating both hypotheses — in a cellular automaton — and comparing the effect of increasing mutation frequencies (another biological unknown) in the two conditions in terms of affinity maturation of the response.

Continuous and Discrete Methods

The computational methods used for modeling in immunology are categorized into two broad classes. In principle, either technique can be used for any problem in the sense that both methods constitute universal methods. However, it is often much easier, simpler, and more reliable to use one method instead of the other. It is important to exercise care in approaching the immunological problem of interest so that the better approach is taken for the job. Furthermore, this simple bifurcation oversimplifies the problem. Within each technique, there are many subdivisions (and even mixed types) that can be usefully applied to problems of interest. The two approaches, continuous and discrete modeling, are not mutually exclusive. It is easy to imagine a synthesis of both approaches in the future. For example, as models become more complex one could envision an automaton in which some of the dynamical processes occurring in a cell are described by a set of differential equations. The spirit would be to use the best available techniques for the computation at hand.

The continuous techniques involve the use of differential equations to describe the system in question. A typical immunological example would be to describe the time behavior of the populations of cells by rate equations containing terms that specifically describe the various influences on the cellular populations, e.g., birth, death, and transformation into other forms, such as activated and anergic. These are called continuous methods because the variables describing the populations are continuous (real numbers) and not discrete integers characteristic of, e.g., the population of real cells. This generally does not create any problems unless the number of cells is so small that the distinction between plus or minus one cell is important. The advantage of such methods is that they are not hampered by limitation to small sizes so that extrapolation to the infinite limit is easy. Another very important advantage of these techniques is that differential equations have been studied for 300 years and have a very solid mathematical underpinning. An investigator using a differential equation generally has an idea of how the various terms behave and has recourse to extensive mathematical theory in obtaining viable solutions to the equations. Difficulties in using differential equations occur when trying to handle the many special cases and nonlinearities that abound in biological systems. Often, the connection of the various processes embodied are much greater abstractions than for discrete methods, and the approximations necessary to solve the equations need to be mathematical in character and may not have straightforward biological interpretations.

The specific discrete method that has proved useful in immunology is the cellular automaton. With this technique, one describes the entities of interest and their interactions and lets the interactions occur repeatedly over time in a series of time steps, with the results of one time step being the input for the next. One strength of this technique is that the entities and interactions can be described in terms that approximate very closely the biological situation. Furthermore, special cases and nonlinearities pose no problems. Complex ones can result in increased computing time but they usually pose no difficulties for either implementation or solution. On the other hand, the system size is usually quite small compared to real life so that constant attention must be paid to finite size problems, and the extrapolation to the infinite limit is often difficult or impossible. Lastly, cellular automata are relatively new concepts in Mathematics. They originated about 50 years ago and were not used seriously until approximately 20 years ago. Therefore, their mathematical underpinning is quite weak and one does not have the store of previous results and theory to fall back on.

In this article, we give examples of both types. Examples of continuous methods will be described in the next section and examples of discrete methods will be presented in the last section. Readers desiring a more systematic account of the discipline are referred to reviews by Mohler *et al.* (1980), Perelson (1888), and Perelson and Weisbuch (1992, 1996).

Continuous Models

The description of the immune system by differential equations is based on several tacit assumptions. The constituents such as cells (e.g., lymphocytes) and molecules (e.g., secreted antibodies) appear in a macroscopic number. Then it is justified to consider their spatial concentration, i.e., their number per unit volume, as a continuous variable. Furthermore, one neglects the spatial distribution of constituents assuming a well-stirred vessel in which interactions are due to random collisions. This leads to nonlinear differential equations for the mean values of the spatial concentrations of constituents as is typical for mean field theories in statistical physics. In their simplest version the equations are similar to those describing the kinetics of chemical reac-

tions (mass-action law) or to Lotka–Volterra equations describing predator–prey systems, i.e., equations with bilinear interactions.

Typically, the system of nonlinear equations describing a network of interactions has the following structure:

$$\dot{x}(t) = \text{gain} - \text{loss},$$

where $\dot{x}(t)$ denotes the temporal change of the concentrations $x = \{x_i\}$, which is determined by the difference between gain and loss. Gain (e.g., due to secretion of molecules by a cell after stimulation or due to proliferation of cells after stimulation) is described by bilinear or higher nonlinear terms. Gain due to generation of new cells from bone marrow, thymus, etc. is described by time-dependent sources. Loss [e.g., due to complex building (molecule–molecule and molecule–cell) or due to phagocytosis] is described by bilinear or higher nonlinear terms. Finite lifetime leads to linear terms.

What information can be obtained from nonlinear dynamical systems? The state of the system is a point in the state space spanned by the variables x_i. The solution $x(t)$ is represented by a curve in the state space. A family of those curves representing solutions starting from different initial conditions describes a flow in the state space. If gain = loss, x is stationary: $x = x_0$ is called a fixed point. The stability of fixed points is tested by examining the fate of a small deviation from the fixed point: If it increases in time, the fixed point is unstable; if it decreases to zero, the fixed point is stable (attracting). More complicated attractors may exist, e.g., limit cycles (periodic behavior) or strange attractors describing chaotic behavior. In the previously mentioned cases the motion is finite. However, the system may lose the global stability, i.e., it explodes, which may indicate the necessity of a revision of the model. After searching for the attractors of the flow, one investigates the global behavior, which typically requires numerics. Another typical problem is the response of the system to (time-dependent) perturbations which describes, e.g., the immune response to infection or to therapy.

There exist different types of networks relevant for the immune system: the network of idiotypic interactions of B cells, the network of T cell interactions, hormone regulation, etc. They are not independent; for instance, B cell proliferation needs assistance from T cells. In principle, each one may be connected with any other one. However, some of the interactions are strong and others weak, which presents the possibility of defining subsystems. Properties of isolated subsystems are thought to change only slightly if interactions with others are taken into account. This should be contrasted to a possible emergence of holistic properties which cannot be explained by reduction to subsystems or, in physical terms, to collective phenomena. The architecture (or connectivity) of the network is defined by the interaction strength between the constituents. The actual repertoire of constituents is chosen at random from the potential repertoire, which is orders of magnitude larger. To describe the random architecture statistical physics must be employed.

The strength of the interactions involving molecules and cells, called affinity, is thought to be connected with the matching between complementary spatial structures. A mathematical description of matching, however, is extremely complicated, as is the description of the translation of primary structures (genetic code) to secondary structures (chemical structure) and finally into tertiary structures (spatial configuration). A simple model considers binary strings which allow matching with complementary strings as a function of mismatching bits.

The simplest form of terms describing interactions is bilinear in the concentrations of the interacting constituents, i.e., proportional to $x_i x_j$, as in the kinetics of reactions (mass-action law). This is certainly helpful for describing molecule–molecule interactions but it is also used to describe more complex constituents such as those in predator–prey dynamics. Higher nonlinear terms are necessary to model specific properties of cells, such as activation thresholds, saturation, supersaturation, or maturation. Frequently, one uses terms of the form $x_i f(h_i)$, where $f(h)$ is a bell-shaped curve given, for example, by $h^n/(h^n + \vartheta^n)$, where n is the Hill coefficient and h_i is the field stimulating x_i. It should be noted, however, that a derivation of interaction terms in mean field variables describing both cell–cell and cell–molecule interactions starting from first principles is a field of active research.

The validity of mean field equations is subject to a principal limitation: For small concentrations the mean field theory is no longer justified, and fluctuations become important. For example, a limit cycle may lead to very small values of one variable which increase afterwards. However, concentrations of less than one cell per individual are nonsense. In this case, one often sets the concentration to zero by hand. For low concentrations the individual behavior of a single cell may be decisive; a probabilistic description would be more appropriate.

The following are extensions which lead to more realistic models:

Delay: There exists a hierarchy of the levels of description: Intracellular chemistry, cells interacting with cells or molecules, organs, and organisms. Here we are on the second level. Cells are considered as entities which react if properly stimulated with proliferation or secretion of molecules. This needs a certain lapse of time, a delay, which is typically on the order of other characteristic times of the system. The behavior of delayed systems, e.g., the stability of fixed points, may be qualitatively changed.

Metadynamics: The architecture is usually not static: New types of cells are introduced from bone marrow or by hypermutation of proliferating cells. This introduces a stochastic metadynamics.

Partial differential equations: If we omit the assumption of a well-stirred vessel, we allow for a spatial distribution of concentrations leading to diffusion, the appropriate description of which is by use of partial differential equations.

Compartmental models: Bone marrow, thymus, spleen, lymph nodes, etc. are spatially separated and highly specialized organs. To describe these, compartmental models have been developed.

Distributed parameters: Parameters are in general distributed; they may change in time (e.g., due to aging of cells or the organism or to switches between different modes of functioning).

This more or less abstract discussion, along with the plethora of phenomena, makes clear that there is an urgent need for minimalistic models. If we do not understand the minimalistic models we will not understand more sophisticated and more realistic models. Nonlinear dynamics and statistical physics have developed methods to describe phenomena as they typically occur in the immune system. On the other hand, describing these phenomena must be accomplished to further develop this arsenal. The need to estimate the values of various parameters as they appear in modeling (affinities, lifetimes, delay times, etc.) is a serious challenge to experimentalists.

In the following sections, a necessarily subjective selection of topics is presented with the aim of giving a flavor of the range of problems, types of questions which can be addressed, and the simplifications necessary for modeling with differential equations.

Clonal Selection

In a series of papers, Bell (1970) developed a minimal mathematical model of clonal selection and antibody production. According to clonal selection theory (Burnett, 1959) the antigen-reactive cells are selected by contact with antigen for proliferation and antigen production.

The model describes four kinds of cells: target cells (small lymphocytes) which transform when sufficiently stimulated by antigen into proliferating cells, which divide and secrete antibodies as long as they are stimulated. As stimulation by antibodies diminishes they differentiate into plasma cells, which are terminal antibody secreting cells, and memory cells, which are assumed to be similar to target cells, i.e., they produce no antibodies, are capable of being stimulated, and have a long lifetime. Target cells are stimulated if a sufficiently large fraction of receptors (determined by assuming chemical equilibrium) are bound by antigen. The level of free antigen selects certain groups of cells according to their affinity for a particular antigen. The model includes a mechanism for high dose tolerance: Cells which are stimulated too strongly are likely to be killed.

The numerically calculated response of the system to antigen shows a reasonable behavior. In the initial stage of antigen shows a reasonable behavior. In the initial stage of the response there is an exponential increase of antibodies. The system responds more vigorously to a prolonged exposure than to a sudden exposure to the same amount of antigen. The secondary response is much more rapid than the primary response due to a higher number of memory cells which are weighted toward higher affinity due to the selective proliferation. The model offers a possible explanation for low dose tolerance: For small doses of antigen the stimulation leads only to very few or no memory cells so that finally the number of target cells plus memory cells is depleted. The inclusion of multivalent antigens presents the possibility of describing the formation of large aggregates of antigen and antibody molecules which will precipitate.

Additional modifications of the original model (not discussed here) are introduced to obtain a better quantitative agreement with experiment.

Idiotype/Anti-idiotype Interactions and the Internal Image of Antigen

There are several mechanisms for B cell memory (Vitetta *et al.*, 1991): Long-lived memory B cells (as assumed previously), B cells which are stimulated by follicular dendritic cells presenting fragments of antigen, and idiotypic memory B cells stimulated by the internal image of antigen provided by anti-idiotypic B cells. Probably all of these mechanisms exist. The maintenance of memory B cells over very long periods is well established. It is difficult, however, to distinguish by *in vivo* experiments whether the memory B cells are individually long lived or survive as a pool. This is a place for modeling.

In the frame of a model describing the idiotypic network of interactions of B cells, the subsystem of a pair of idiotype/anti-idiotype B cell clones in the presence of antigen was investigated by Behn *et al.* (1993). It was determined that a state of nonzero population of both clones (a memory state) is stable only in the presence of memory cells, thus indicating a possible synergy of different mechanisms: Idiotype/anti-idiotype cycles are stabilized by memory cells which are thought to be stimulated by their anti-idiotypic counterparts. The equations are:

$$\dot{x}_1 = -\gamma x_1 + (1 - \kappa)mx_2x_1 + d_1mx_2 + (d_1 + x_1)\overline{m}y$$
$$\dot{x}_2 = -\gamma x_2 + (1 - \kappa)mx_1x_2 + d_2mx_1$$
$$\dot{y} = \alpha y - \overline{m}x_1y,$$

where x_i and x_2 are the concentrations of antibodies of idiotype $i = 1$ or 2 irrespective of whether they are free or bound as receptors on the cell surface, y denotes the concentration of antigens which multiply with the rate α (virulence) and has the affinity $\overline{m}$ to the idiotype 1. The affinity between idiotypic and anti-idiotypic antibodies is denoted by m, and the inverse lifetime of antibodies is γ. Memory cells are represented by d_i which are supposed to be present after a sufficiently strong stimulation of the corresponding idiotype i; their number is zero in the virgin state. The bilinear terms describe the gain due to stimulation by antigen or anti-idiotypic antibodies and the loss due to repression or complex formation, respectively. The parameter κ allows for an asymmetry between stimulation and inhibition. Here, we chose $\kappa > 1$, i.e., the network is globally inhibitory as advocated by Jerne (1974).

Depending on the parameter setting, the system has three configurations of fixed points describing the virgin state, a healthy immunized state, and a state of chronic infection (Fig. 3). The dynamical behavior shows a secondary response that is faster and stronger than the primary one (Fig. 4). In the state of chronic infection, a repeated exposure to small doses of antigen may drive the system away from the fixed point of chronic infection to a state characterized by a higher concentration of idiotypic antibodies.

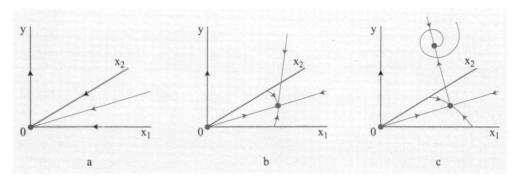

FIGURE 3 Schematic flow diagram for an idiotype/anti-idiotype (x_1–x_2) pair. Antigen y is recognized by x_1. (a) The virgin state: $x = 0$ is stable in the (x_1, x_2) plane. Antibodies will die out without stimulation by antigen. (b) The immunized state: The population of memory cells is large enough so that a nonzero stable fixed point in the (x_1, x_2) plane exists. Antibodies x_2 can be considered as an internal image of the antigen. (c) Chronic infection: If the antigen becomes too virulent, the immunized state loses its stability and a new stable fixed point with a nonzero number of antigen emerges (the spiral indicates the convergence toward a fixed point).

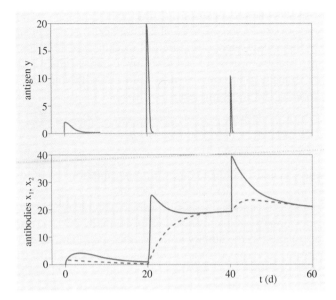

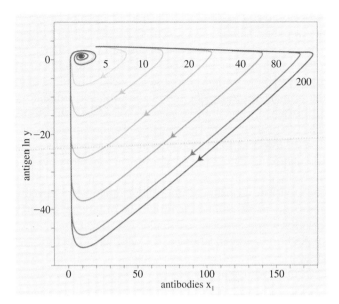

FIGURE 4 The primary and secondary response of an idiotype/anti-idiotype pair (the continuous and dashed lines, respectively) in response to repeated infections (red line). Starting in the virgin state the system is infected at $t = 0$ with a subclinical dose of antigen and relaxes back to the virgin state. At $t = 20$, it is infected with a dose of antigen strong enough to initiate the appearance of memory cells which leads to the formation of a stable idiotype/anti-idiotype pair (primary response). The secondary response to the same dose of antigen at $t = 40$ is stronger and more efficient.

FIGURE 5 Flow diagram for a specific stimulation therapy of chronic infection. The trajectories are shown in the ln y–x_1 plane (x_2 is nearly constant) for increasing total dose of antigen (numbers on the graph) applied by repeated injections of smaller equal doses. After these injections, which drive the system out of the fixed point of chronic infection, the excess of antibodies x_2 reduces the antigen with exponential rate. When the concentration becomes sufficiently small the antigen can be considered extinct.

number of antibodies may be sufficient to cure the chronic infection (Fig. 5). This strategy resembles a therapy from the beginning of this century which tried to cure chronic tuberculosis by provoking it to become acute.

Central Immune System and Control of Autoreactive Clones

As suggested by Coutinho (1989), it is advantageous to distinguish between a central immune system (a highly connected self-referential network) and a peripheral immune system constituted by disconnected clones which produce the response to foreign antigen. The latter was illustrated previously.

Autoreactive antibodies are found in normal healthy individuals despite the negative selection of T lymphocytes in the thymus which abolish autoreactivity mediated by T cells and T-dependent B cells. The central immune system has an internal dynamics not necessarily driven by antigen and might play a role in controlling autoreactive clones as advocated by Jerne (1974) and Coutinho (1989).

To describe this in a mathematical model, a nonlinear dynamics of the constituents of the central immune system based on a simplified architecture was investigated (Sulzer *et al.*, 1994). Idealizing a network architecture proposed by Stewart and Varela (1989), four clones (A–D), distin-

guished by their connectivity, are assumed (Fig. 6). Members of A (the core of the network) have an all-to-all interaction with A–D, B is the mirror group to C, and D interacts with A only. Interactions within A and between B and C are strong, and the others are weak. The dynamical equations are as follows:

$$\dot{X}_i = -\gamma_X X_i + f_p(h_i)X_i - f_d(h_i)X_i$$
$$\dot{x}_i = -\gamma_x x_i + p_x f_d(h_i)X_i - b_A h_i x_i,$$

where X_i and x_i denote the concentrations of large B cells and antibodies of clone i, respectively, and γ_X and γ_x are their inverse lifetimes. The number of B cells increases due to proliferation, $f_p(h_i)X_i$, and decreases by terminal differentiation, $f_d(h_i)X_i$. Both proliferation and differentiation are governed by saturation functions $f_{p/d}(h) = \kappa_{p/d}h/(h + \vartheta_{p/d})$, where the rates $\kappa_{p/d}$ and the thresholds $\vartheta_{p/d}$ are chosen properly so that $f(h) = f_p(h) - f_d(h)$ is bell shaped. Clone i is stimulated by the field $h_i = \sum_{j=1}^{N} m_{ij}x_j$, where the affinities m_{ij} are assumed to be uniformly 1 or $\eta \ll 1$ for strong or weak coupling, respectively.

We first describe stationary properties of the model in the absence of self-antigen. An isolated idiotype/anti-idiotype pair (member of the mirror groups B and C) has three stable steady states. In addition to the virgin state (in which all concentrations are zero), there are two memory states in which the population of one clone is high while those of the other is low.

If we include the coupling to A, idiotype/anti-idiotype pairs qualitatively preserve this property. They provide a stimulation of the core clones A which leads, depending on both the strength of the coupling η and the initial conditions, to four qualitatively different situations which are mutually exclusive. In the independent pair solution each B–C pair attains one of its three possible states and provides a stimulus to A which is either too low or too large to trigger their expansion so that A decreases. In the coupled pair solution the A core and the B–C pairs coexist and mutually contribute to maintain the equilibrium. In the coupled single solution members of B or C clones may be at equilibrium with A and at the same time overstimulate their anti-idiotypic partner. (It is only in this case that D clones may coexist with B or C clones.) Finally, A clones can maintain their equilibrium autonomously forming a polyreactive core.

This scheme does not change qualitatively when some clones recognize self-antigen u_i, which is assumed to have a constant concentration. Self-antigen coupling to i enhances the field to $h_i = \sum_{j=1}^{N} m_{ij}x_j + u_i$. Since the population of A is always low or zero, a coupling of the self-antigen to a member of A causes no autoimmunity. The same is true if the self-antigen couples to a clone of the mirror groups B or C with low population: The corresponding pair is in a toler-

ant state. Autoreactive clones are controlled by their anti-idiotypes. If, however, the self-antigen couples to a clone of the mirror groups with high population the corresponding pair is in an autoimmune state. The high population of auto-antibodies may cause autoimmune reactions. If both the concentrations of auto-antibodies and their anti-antibodies vanish, the pair is in a neutral state: The system is not currently autoimmune but it is also not protected against autoimmunity.

Numerical investigation of the dynamic response against foreign antigen reveals that the system can be thrown out of its tolerant state, depending on the concentration of self-antigen and the initial dose of foreign antigen as well as the removal rate of complexes. It was observed that it is much less likely to induce autoimmunity if the foreign antigen is recognized by the auto-antibodies than if it is recognized by their anti-antibodies. On the other hand, an encounter with an foreign antigen can also induce a transition from an autoimmune state to the tolerant state. In this case, a foreign antigen which couples to the anti-antibodies (thus mimicking the self-antigen) is most effective in provoking a spontaneous cure.

The most important simplification in the previous context was in omitting T lymphocytes as explicit components of the network. This amounts to assuming that the network's B cells always obtain sufficient T cell help. Since we wanted to study idiotypic suppression as a mechanism to control autoreactive B cells, this may be justified. In contrast, the problem studied in the following section needs a detailed and explicit analysis of the network providing T cell regulation.

Hyposensitization Therapy of Allergic Diseases

Hyposensitization is a successful therapy for type I allergic diseases. There are much empirical data on hyposensitization, but this therapy still deserves a theoretical understanding. T cell regulation seems to be one of the crucial issues in allergy. The mathematical model (Behn *et al.*, 2001) discussed here explains hyposensitization as a dynamic phenomenon of the T cell regulation.

There are two subsets of T helper cells, Th1 and Th2, which differ in the secreted cytokine pattern and the immune reactions they support (Mosmann and Sad, 1996; Powrie and Coffmann, 1993). Via their cytokines, Th2 cells encourage the production of antibodies, particularly IgE antibodies, which characterize allergy of type I. The Th1 subset induces delayed-type hypersensivity reactions and, within a certain limit, the production of other antibody types in addition to IgE. The cytokines secreted by T helper cells have autocrine and cross-suppressive effects. Therefore, there is always one T cell subset that dominates. Certainly, more than one reason can exist as to why one antigen causes a Th1-dominated and the other one a Th2-dominated immune response. The presence of cytokines,

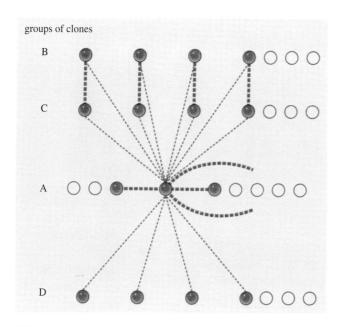

groups of clones

FIGURE 6 Architecture of the central immune system. Shown are the four groups of clones A–D, which are distinguished by their connectivity. Interactions within A and between B and C are strong (thick lines), and the others are weak (thin lines).

the type of antigen presenting cells, the antigen itself, the concentration of the antigen, and probably genetic factors can play important roles.

Allergy of type I is a typical Th2 response. The allergen dose is a major regulator of Th1/Th2 selection. For medium and high antigen concentration the immune response is Th2 dominated, whereas low antigen concentration leads to Th1-dominated response (Romagnani, 1995; Mosmann and Sad, 1996). The therapy of hyposensitization by injecting increasing doses of allergen starts with low allergen levels, thus provoking a Th1-dominated response. Through Th1/Th2 cross-suppression, a Th1 dominance may be supported even for higher allergen levels. Our hypothesis states that hyposensitization leads to a Th1-dominated immune response. After the therapy the number of allergen-specific Th2 cells is reduced because of the dominating Th1 subset. Less Th2 cells means less IgE production, i.e., reduced allergic symptoms.

This mechanism is the base of a mathematical model which consists of six nonlinear differential equations describing the dynamics of the concentrations of naive T cells, Th1 and Th2 cells, the corresponding memory cells, and the allergen (Behn *et al.*, 2001). Allergen contact leads to production of both specific and memory T cells by naive T cells. Autocrine and cross-suppressive effects mediated by the cytokines are included. The cytokine concentration is estimated to be proportional to the number of T cells which secrete it. Cytokines from cells other than T cells and

other immune processes are considered as cytokine background. The allergen elimination is taken as proportional to the product of T cells and allergen. The memory cells are considered as long-lived cells (Bradley *et al.*, 1993), which are not permanently restimulated. The parameters are chosen such that in the beginning, before the therapy, low (high) allergen doses cause a Th1- (Th2-) dominated response. Interestingly, there is a kind of separatrix in the space of states separating Th1- and Th2-dominated immune responses. For a successful therapy it is necessary to cross the separatrix. This leads temporarily to a state which gives a Th1-dominated response to even a high allergen dose, interpreted as a desensitized state (Fig. 7). After some period of time the therapy should be repeated regularly *in vivo* and in the model.

The model is robust against variation of the parameters over some range. It may reproduce surprisingly well qualitative features of both successful and unsuccessful therapies. Model simulations show, for example, that even one missed allergen injection during therapy or one too high allergen dose may endanger the success of hyposensitization. In case of a missed allergen injection, it may be helpful to decrease the dose of the next allergen injection. Th1/Th2 regulation, especially the switch from the dominance of one T cell subset to another, has a special role not just in allergy. Altered profiles of Th1/Th2-type cytokine production are also associated with other diseases in humans, such as candidiasis, AIDS, leishmaniasis, and autoimmune diseases.

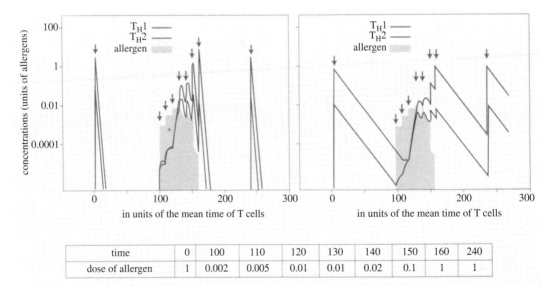

time	0	100	110	120	130	140	150	160	240
dose of allergen	1	0.002	0.005	0.01	0.01	0.02	0.1	1	1

FIGURE 7 Successful hyposensitization. Concentrations, on a logarithmic scale, of T cells (left) and memory T cells (right) as a function of time. The injections of allergen are indicated with arrows. The first peak shows the allergic response dominated by Th2 cells (red curve) in response to a high dose of allergen. After repeated injections of increasing doses of allergen, according to a schedule similar to the therapeutic strategy used for bee venom allergy shown in the table below, the response to a high dose of allergen is switched to Th1 dominated (black curve). The figure highlights the accumulation of allergen (pink area) during the hyposensitization. Injection of higher doses provokes an immediate response which causes the immediate disappearance of the allergen.

Discrete Models

We now discuss discrete dynamics. The advantage of discrete dynamics is that the character of the mathematical nature of the processes being modeled does not have a strong influence on the implementation of the model. For example, nonlinearity will not cause any real difficulties in carrying out the simulation. Second, the occurrence of special cases can be included without difficulty. Lastly, and perhaps most important, the structure of the simulation can bare a very close relationship to the immunological processes being modeled. The parameters and entities can be described in biological terms and the approximations made in setting up the model are generally biological in character rather than mathematical. These advantages are not without their price. First, the simulations will be much smaller in size than those using differential equations. It is generally not possible to approach the infinite limit where finite size effects are unimportant. This is not as serious a problem in immunology as it is in fields such as physics since the real biological systems are often actually quite finite in this mathematical sense. Another disadvantage is that a discrete system can often be a much larger user of computer resources than a system of differential equations (although this is not always the case). Modern powerful computer systems make this much less of a problem than it has been in the past.

Binary Automata

There is a wide range of sizes and complexity for discrete dynamical systems. As mentioned previously, one of the advantages of discrete systems is the ability to handle complexity in a straightforward manner. However, there can be great utility in very simple discrete systems, and in fact we begin our discussion with a very simple but quite useful implementation.

The following example involves the use of very simple automata to follow the dynamics inherent in a set of rules that describe the overall behavior of some process. The use of this technique in immunology was first described by Kaufman *et al.* (1985). Their use of the technique is well described by the title of their article, "Towards a Logical Analysis of the Immune Response." The specific example of the technique that we discuss was described by Chowdhury *et al.* (1990). They described a simple set of rules to account for the response to a foreign antigen. Their model includes antigen, B cells, T helper cells, suppressor cells, and antibodies. The idea is that antibodies clear the antigen, but the generation of antibodies is a multistep process. The process is described by the following set of rules:

Antibodies: Ab ← Ag and B and Th
Suppressors: Ts ← Ts or Th
Helpers: Th ← [Ag and (not Ts)] or Th
B cells: B ← (Ag or B) and Th
Antigen: Ag ← Ag and (not Ab)

The values of the five parameters are binary and the rules are logical rules, i.e., "and," "or," and "not" are logical operators. Therefore, the first rule should be read as saying that the value of Ab will be 1 only if Ag, Ab, and Th are all 1; otherwise Ab is 0. The state of the system at any time can be represented by five binary numbers, each one representing one of the parameters. Table 1 shows the response of a virgin system to antigen. The state is the decimal value of the five-digit binary number representing the five variables. The first line, state 0, is the virgin system.

	TABLE I					
	State of Evolution after Injection of Antigen					
DESCRIPTION	**STATE**	**Ag**	**B**	**Th**	**Ts**	**Ab**
	0	0	0	0	0	0
Inject Ag	16	1	0	0	0	0
	20	1	0	1	0	0
	30	1	1	1	1	0
	31	1	1	1	1	1
	15	0	1	1	1	1
	14	0	1	1	1	0
	14	0	1	1	1	0
Reinject Ag	30	1	1	1	1	0
	31	1	1	1	1	1
	15	0	1	1	1	1
	14	0	1	1	1	0

						CURSOR
TABLE 2						
Fixed Points of the System						
DESCRIPTION	STATE	Ag	B	Th	Ts	Ab	CURSOR STATE
Virgin	0	0	0	0	0	0	4
Low-dose paralysis	2	0	0	0	0	0	6
Vaccinated	6	0	0	1	1	0	6
Memory	14	0	1	1	1	0	14
High-dose paralysis	18	1	0	0	1	0	2

Adding antigen, by setting Ag to 1, results in state 16. The rules are then applied to a given state to get the next state. State 14 is called a fixed point since applying the rules does not change it. If antigen is reinjected into this state as in the bottom of Table 1, we return to the same fixed point, but in this case it took only three steps rather than the five required for the primary injection of antigen. This system has $2^5 = 32$ possible states since each of the five parameters can be 0 or 1. Of these 32 states, 5 are fixed points. They are shown in Table 2, along with names suggesting the corresponding state of the immune system. The last column shows the number of the 32 possible states that evolve to each fixed point. Most of the states evolve to the memory state and the high-dose paralysis state is quite rare: The only other state that evolves to it (besides itself) is 26. However, 26 is a Garden of Eden state, meaning that there is no state that evolves to it. The only states which appear in an evolutionary path, in addition to the fixed points, are 4, 15, 20, 30, and 31. All these eventually evolve to 14 except 4, which evolves to 6. Therefore, the fixed points 0, 2, and 18, are only reached from Garden of Eden states. The vaccinated state is interesting in that it evolves to the memory state after exposure to antigen after four steps.

Now we discuss what happens when the antigen is considered to be self-antigen and is always present, i.e., Ag is 1. The fixed points change to those in Table 3. In this case, there are now only 16 possible states since Ag is always 1. To deal with the self-antigen problem, a killer cell can be added which will eliminate Th cells that respond to the self-antigen. The rule for this cell will be

Killers: $K \leftarrow$ (Ag and Th) and not Ts,

and we modify the helper rule to account for killers as follows:

Helpers: Th ← (not K) and [Ag and (not Ts)] or Th.

The results are given in Table 4. The fixed points are unchanged; however, now only one-eighth of the precursor states lead to disease, whereas without the killer cells three-

						CURSOR
TABLE 3						
Fixed Points for Self-Antigen						
DESCRIPTION	STATE	Ag	B	Th	Ts	Ab	CURSOR STATE
High-dose paralysis	18	1	0	0	1	0	4
Disease	31	1	1	1	1	1	12

							CURSOR
TABLE 4							
Fixed Points for Self-Antigen with Killer Cells							
DESCRIPTION	STATE	K	Ag	B	Th	Ts	Ab	CURSOR STATE
High-dose paralysis	18	0	1	0	0	1	0	12
Disease	31	0	1	1	1	1	1	4

fourths led to disease. In addition, there are 16 new states with a population of killer cells; all of these go to state 18.

These simple models obviously do not give a complete and detailed picture of the dynamics of the system; however, they do give a view of the possible states of the system, their dominance, and how they can be approached. One might consider this technique as giving the existence diagram of what is possible for a particular process. One can also increase the complexity of the system as desired. Although the two states of a binary parameter may not be sufficient, it may be enough to consider just three or four states. The system becomes more complicated but its analysis using a computer is still simple and straightforward.

Cellular Automata

Cellular automata can exist in diverse forms (Wolfram, 1986) but the specific example we discuss here can be envisioned as a checkerboard whose sites can be either empty or occupied by a checker (a binary system). The state of a site can be changed by the influence of the state of its neighboring sites. An interesting illustration of this class of automata is John Conway's Game of Life (Gardner, 1970). The evolution of life is accomplished by the following two rules:

An occupied site will remain occupied on the next time step if and only if it has two or three occupied nearest neighbors.

An empty site will become occupied on the next time step if and only if it has three occupied nearest neighbors.

The neighbors are taken as the eight sites that surround a site on a checkerboard. Note that a cellular automaton is discrete in both space and time and all actions are simultaneous. That is, the rules are applied to all sites at the same time and on the basis of the outcome all sites are updated simultaneously; this is called a time step.

These simple rules produce an extraordinarily rich number of interesting stable forms, a few of which are shown in Fig. 8. One particularly interesting form is the glider, which is a stable oscillator with period 5 and moves across the board.

We now describe a much more complex model that attempts to incorporate a more realistic description of part of the immune system. This model (Seiden and Celada, 1992; Celada and Seiden, 1992a,b), called IMMSIM, is an attempt to model the humoral side of the immune system and to examine questions concerned with clonal selection. It is based on a cellular automaton. The usual rules for cellular automata are the following (Wolfram, 1984):

They consist of a discrete lattice of sites.
They evolve in discrete time steps.
Each site takes on a finite set of possible values.

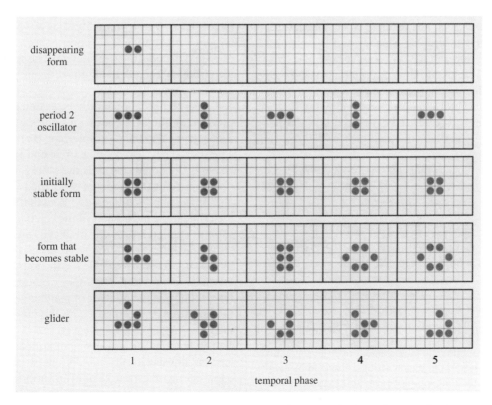

FIGURE 8 Evolution of some simple forms of the Game of Life, in particular a disappearing form, a period 2 oscillator, an initially stable form, a form which evolves to a stable form, and a period 4 glider.

The value of each site evolves according to the same *deterministic* rules.

The rules for the evolution of a site depend only on *a local neighborhood of sites around it.*

The system used for IMMSIM has been called a generalized or hyper cellular automaton. The modifications involve the italicized portions of the previous rules. They are changed as follows:

"Deterministic" is replaced by "probabilistic."

"Local neighborhood of sites around it" is replaced by "the site itself."

We add a sixth rule:

Entities can move from site to site.

The IMMSIM model uses a 2-dimensional triangular grid to represent a small portion of the body. The virtue of this grid is that each site has six identical neighbors rather that the eight neighbors of two different types: sides and corners. Each site, rather than being binary as in the game of life, can be populated with a number of distinct entities of several types. The entities involved consist of B cells, T cells, nonspecific antigen presenting cells (APCs), antigens, antibodies, and antigen–antibody complexes. They are endowed with binary receptors, epitopes, and peptides. Schematics of the three types of cells are shown in Fig. 9 for the case of eight-bit receptors. The segment labeled receptor is the specific clonotypic receptor which T and B cells possess. In addition, B cells and APCs possess class II major histocompatibility complex (MHC) molecules. An APC is nonspecific, so it does not have a specific receptor but it will have the same MHC molecules as possessed by the B cell. In addition, an APC possesses an Fc receptor. It is not spe-

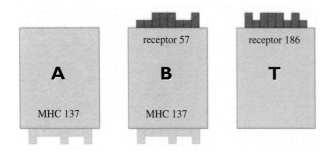

FIGURE 9 A schematic of A, B, and T cells with eight-bit receptors (in blue). The short teeth represent bit = 0, the long teeth bit = 1. The numbers which characterize the segments are the decimal values of the binary strings, obtained by summing the values of the long teeth of each 8-bit segment. From the right the eight positions have values 1, 2, 4, 8, 16, 32, 64, and 128. The different receptor types are shown at the top and the MHC molecules (in yellow) at the bottom.

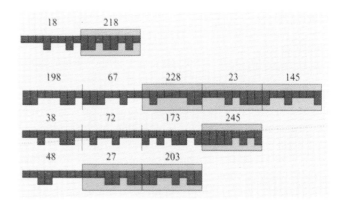

FIGURE 10 Examples of antigens represented by eight-bit segments. The epitopes, which can be recognized by B cell receptors or by antibodies, are indicated as exposed segments and the peptides, which can be recognized by T cell receptors when presented in the context of an MHC molecule, are shown boxed.

cifically denoted on the cell since it is the same for all APCs. However, it is allowed to bind to the Fc region of an antibody when the antibody is complexed to an antigen.

Antigens are constituted by many segments representing epitopes and peptides. The epitopes are the segments to which the B and A cells can bind. A B cell will bind if its receptor matches the antigen in a complementary fashion, i.e., zero binds to one and vice versa. The peptides are the segments of the endocytosed antigen that are presented on the MHC molecule. In reality, they are obtained by breaking down the antigen. However, since no biochemistry is done in the model the peptides must be specified along with the epitopes when defining the antigen. Examples of antigens are shown in Fig. 10.

Under the proper conditions, as described later, B cells produce antibodies. These antibodies will contain an epitope, identical to the receptor of the B cell producing it, and a peptide. They also contain an Fc epitope which need not be explicitly noted since it is the same for all antibodies.

The core of the model is concerned with the capturing and processing of antigen and how that processing affects the population and activity of the cellular components. The steps in antigen processing by B cells are shown in Fig. 11. Antigen capture and processing are performed by B cells that bear a receptor specific for the antigen. First, a B cell attempts to bind to the epitopes of the antigen. This binding depends on the number of matching bits. Typically, the binding strength decreases with the number of mismatching bits.

The binding and processing of antigens can also be carried out nonspecifically by APCs in two ways: directly with a low interaction strength or in an antibody-facilitated fashion (by means of an Fc receptor) with intermediate interaction strength.

Once the antigen is bound by either a B or A cell the pep-

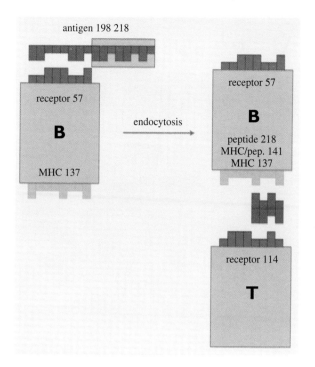

FIGURE 11 Steps in the recognition and processing of antigen by B and T cells. The antigen is recognized by the receptors on the B cells and processed in the cell; the hidden part is presented on the surface by the MHC complex, where it is recognized by the T cell receptors.

tides are processed for presentation by the MHC. The MHC molecule is divided in half. The left-hand four bits represent the bare part of the MHC to which the T cell receptor will bind. The right-hand four bits represent the MHC groove onto which the peptide binds. Binding takes place between the MHC groove and one half of the peptide. The other half of the peptide is presented to the T cell. Therefore, as shown in the right-hand side of Fig. 11, the T cell sees an eight-bit segment consisting of the left-hand four bits of the MHC and the four bits of the peptide which did not bind to the MHC. The T cell receptor is then allowed to bind to this MHC–peptide complex with the same rules applied as those for a B cell receptor binding to an epitope.

When a successful T cell binding takes place, both the T cell and the B cell are allowed to divide to establish clones of their own type (if presentation takes place by an A cell, only the T cell divides). Some of the daughter B cells become plasma cells and produce antibodies.

In the model, self–nonself discrimination is obtained by preselection of one of the two lymphoid populations. The B cells are chosen randomly with complete diversity (i.e., 2^n types, where n is the number of receptor bits). The T cells are also chosen in the same way but then they are passed through the thymus. In the model thymus they are exposed to self-peptides presented on APCs and if they bind they are eliminated proportional to the strength with which they

bind. Since the thymus is a dense organ, we can allow the same T cell to be exposed to all the MHC–peptide combinations many times. If the number of times is great enough the thymus will be very efficient and most of the self-reactive T cells will be eliminated.

A typical experiment is set up as follows. First, the system is populated with the desired number of APCs and B and T cells. Second, a schedule of injections of antigen is chosen. Third, the simulation begins. In each site, all possible interactions are considered. Interactions, which are restricted as to the type of the entities involved, can take place only between entities in the same compartment of the grid. If a single entity is capable of having more than one successful interaction, the one that actually happens is determined stochastically. After all interactions are determined, they are implemented together. Then the birth of new cells occurs, both clonal growth and the birth of new virgin cells. Cells are also allowed to die with a given half-life. Then all entities are given an opportunity to diffuse into the neighboring sites. This constitutes a time step. The process can be repeated for as many time steps as desired.

The previous discussion is only an abbreviated description intended to let the reader understand the nature of the approach. A more detailed description can be found in Seiden and Celada (1992).

Use of the Automaton to Do Immunological Experiments

To show that this system behaves as an immune system, we can examine an immunization experiment. In this experiment, the same antigen is injected twice, once at time step zero and the second at time step 100. The results of the experiment are shown in Fig. 12. At the first injection the response is slow because it takes time for the clones of the responding cells to develop from their small virgin populations. A small amount of antibody is produced but it takes time for the antigen to be removed from the system. At the second injection, however, many of the appropriate B and T memory cells are present so that the response is rapid. Much more antibody is produced and the antigen is eliminated quite rapidly.

Even this simple experiment shows what can be done with the IMMSIM model. Figure 12 only shows the total antibody produced; however, the experiment was done with eight-bit receptors, allowing perfect matches between epitope and receptor, and one-bit mismatches to bind with a relative interaction strength (affinity) of 20 to 1. This means that nine different antibodies are present: one perfect match and eight 1-bit mismatches. The population of each type of antibody is presented in Fig. 13. The solid line in the figure shows the high-affinity antibodies (perfect match) and the other eight curves show the eight low-affinity antibodies (one-bit mismatches). Note that the high-affinity antibodies do not dominate the primary response, and in fact many low-affinity antibodies have a larger population than that

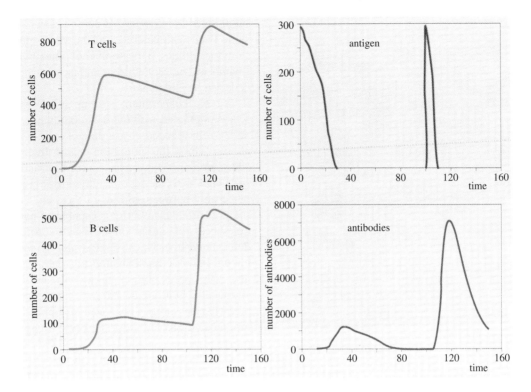

FIGURE 12 Evolution of the populations of T cells, B cells, antigens, and antibodies in a system injected with antigen at conventional time steps 0 and 100.

of the high-affinity antibodies. However, the secondary response is dominated by the high-affinity antibodies, a clear illustration of affinity maturation.

This result is obtained with no hypermutation of B cell receptors upon clonal growth. Because the experiment was done with complete diversity (i.e., all receptor types were present), mutation was not needed. In the actual biological case, the B cell diversity is far from complete; the bone marrow produces only a small fraction of the possible receptor types. Adding mutation to the model allowed the study of how both affinity maturation and hypermutation contribute to an effective immune response. The results of an experiment with mutation are shown in Fig. 14. In this experi-

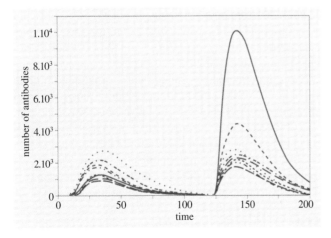

FIGURE 13 Antibody response to an immunization. The high-affinity antibodies (8/8 bit matches) are shown as a continuous line. The other lines represent antibodies of low affinity (7/8 bit matches). Time steps are conventional.

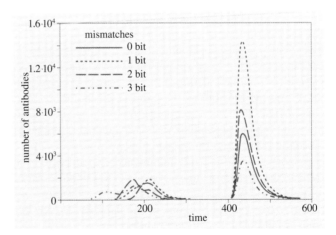

FIGURE 14 Affinity maturation in a system with restricted initial diversity and hypermutation. The production of antibodies with high and intermediate affinity (12/12, 11/12, and 10/12 bit matches) by the B cells is blocked by the bone marrow; therefore, all the clones with these characteristics are the products of mutations. The results are indicated for the four classes of antibody capable of binding antigen. Time steps are conventional.

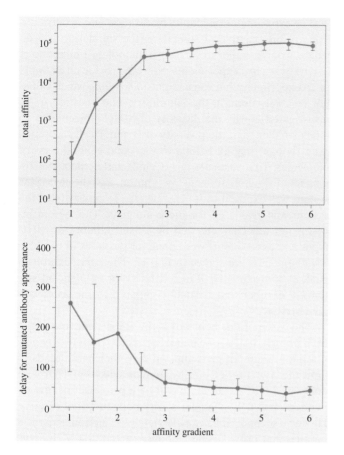

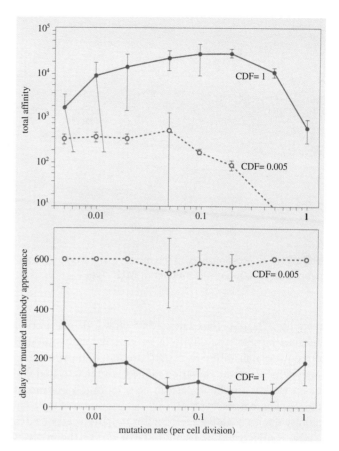

FIGURE 15 Experiments with restricted diversity as a function of enhancement factor. The graphs show the average value of the total affinity and the time of appearance of the first mutated antibody. The vertical bars at each point indicate the standard deviations calculated over 10 separate experiments (with the same parameters but with a different initial random number).

FIGURE 16 Experiments with restricted diversity and maturation as a function of the mutation rate for CDF = 1 and CDF = 0.005. (Top) The average value of the total affinity (calculated by multiplying the number of cells by the average affinity of the cells); (bottom) the time of appearance of the first mutated antibody.

ment, 12-bit receptors were used and up to 3-bit mismatches were allowed. However, the only B cells that arise from the bone marrow are 3-bit mismatches; all the 0-, 1-, and 2-bit mismatches arose from mutation occuring during antigen-stimulated clonal growth.

There are two crucial parameters in the hypermutation/affinity maturation process: mutation rate and enhancement of affinity (the factor by which the affinity is increased per each advantageous mutation). Celada and Seiden (1996) studied these using the IMMSIM model by running many experiments at different mutation rates and affinity enhancements. The results are shown in Figs. 15 and 16. In both figures it is clear that, for some values of the parameters, the immune response, although sometimes ample, can be very unreliable. Both the mutation rate and affinity enhancement have to be large enough to get a sufficient and reliable response.

The effect of mutation on affinity maturation is greatly affected by the degree of focusing of mutations on the complementarity determining regions (CDRs) of the antibody's variable regions. In order to model alternative hypotheses on the subject, we proceeded as follows. Initially, all mutations are effective in changing the receptor and all affect affinity since the entire receptor makes contact with the antigen. This arrangement simulates the hypothetical situation in which hypermutation is limited to the CDRs. To represent the opposite hypothesis, we introduce a second parameter called the complementarity determining fraction (CDF), which ranges from 1 (the case described previously) to 0.005 (where 99.5% of mutations fall outside the CDR in the framework regions). Any mutation of the latter kind is considered disruptive to the antibody and is lethal to the cell. Results obtained for CDF = 1 and 0.005 are shown in Fig. 16. The top of Fig. 16 shows that increasing the mutation rate favors maturation, as indicated by an increasing total affinity which peaks at approximately 0.2 mutations/cell division/paratope for CFD = 1, and then decreases rapidly. The bottom of Fig. 16 shows the time of first appearance of mutated antibodies. The region with a mutation rate between 0.1 and 0.5 gives the most reliable

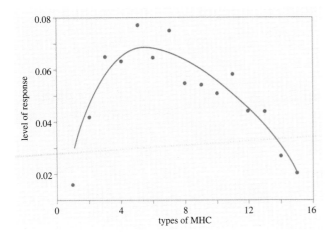

FIGURE 17 Responsivity of the model to a group of 100 antigens as a function of the number of MHC types.

response. The data for CDF = 0.005 show little maturation because most of the affinity is produced by nonmutated three-mismatch antibody. A small amount of mutated antibody appears in the mutation range between 0.05 and 0.2, but it is very unreliable, as can be seen in the bottom of Fig. 16. There is also a progressive deterioration of the total affinity due to negative selection for mutation rates ≥0.05. These results could be useful to favor one or the other of the alternative views described if a reliable measure of mutation rates *in vivo* were available. This is not currently the case but will certainly be so in the near future.

The experiments on affinity maturation and hypermutation are an example of something the simulation can do which is quite similar to what an immunologist might do in *in vivo* or *in vitro* experiments. Now we provide an example of an experiment that can be done *in machina* which would not be easily done in the laboratory. The question to be investigated is why the diversity of MHC molecules is so much smaller than the diversity of B and T cell receptors, even though they all belong to the same general class of molecules. To investigate this (Celada and Seiden, 1992b), a panel of random antigens was chosen, and the IMMSIM system was exposed to it. The average responsivity of the system was evaluated for these antigens. The system investigated was characterized by a fixed number of MHC types and experiments were done for the cases of types 1–15. The results are shown in Fig. 17. The curve exhibits a peak at approximately 4 or 5 MHCs (for an eight-bit system), a number that is small compared to the 256 available MHCs.

The cause of this behavior is the ambivalent role of the MHC which, on the one hand, fosters the response by presenting foreign antigens and, on the other hand, tends to limit the T cell repertoire by thymus selection. That is, if only one MHC exists there will be a whole set of foreign peptides that will engender a very poor response (if any). Having a second MHC can alleviate this since, in general, a peptide that provokes no response on the first MHC type will be able to do so on the second. Therefore, the responsivity increases as we add MHC types. Eventually, however, we reach diminishing returns since each added MHC

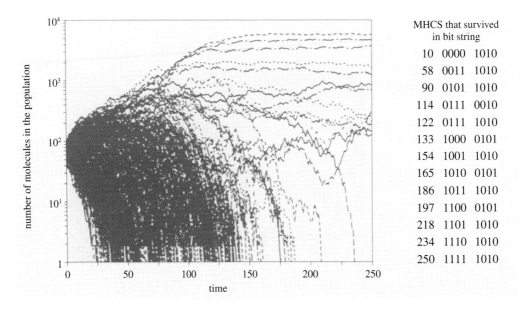

MHCS that survived in bit string		
10	0000	1010
58	0011	1010
90	0101	1010
114	0111	0010
122	0111	1010
133	1000	0101
154	1001	1010
165	1010	0101
186	1011	1010
197	1100	0101
218	1101	1010
234	1110	1010
250	1111	1010

FIGURE 18 Evolution of the population of each MHC specificity. The figure is composed of 256 separate lines, for each MHC present at the conventional time zero. Selection was based on the capacity to respond to a panel of 100 antigens. The MHC types which survived were analyzed for their bit string composition as shown in the table on the right.

type reduces the number of T cell types available due to increased negative selection.

The distribution of MHC types is highly selected, i.e., in addition to the number being small in each individual, the number in the whole population is small compared to the possible diversity. This was examined by observing the evolution of a large population. Ten thousand individuals were created, each having 2 eight-bit MHC types chosen at random from the complete repertoire of 256. The individuals are allowed to mate with each other, and the offspring receive 1 MHC from each parent, chosen at random from the 2 they possess. The mating partners are also chosen at random, but at each generation the number of offspring of each individual is proportional to the responsivity (calculated separately for each combination of MHC types in the previous experiment). The results of such an experiment over 250 generations are shown in Fig. 18.

A Darwinian selection ensues and some MHC types are more successful than others. After 250 generations, only a limited number of MHC types are present in the species. Analysis of the selected MHCs reveals that diversity of the left half of the eight-bit string (the restriction element) is strongly favored. Selective forces prompting this result are the fact that some MHC combinations are better than others and the definite advantage of heterozygosity. The latter is the cause of the diversity of the left half of the MHCs that survive. The left half is where the T cell receptor binds directly to the MHC; therefore, if an individual has 2 MHC types with the same left half, it would appear as almost homozygous. Repeating the experiment with different random numbers (equivalent to a new batch of animals) can change the list of surviving MHCs (generally between 5 and 15), but the diversity of the left-hand side remains.

Acknowledgments

The experiments by Celada and Seiden described in the text have been supported by grants from NIH (1R01AI 42262-02), The Center for Alternatives to Animal Testing (CAAT, #99027), and from the Italian Minister of University and Research (9906114235-001). In writing this review, F. Celada was responsible for the section "A Place for Modeling," U. Behn for "Continuous Models," and P. Seiden for "Discrete Models."

References Cited

BEHN, U., DAMBECK, H., and METZNER, G. (2001). Modelling Th1–Th2 regulation, allergy and hyposensitization. In *Dynamical Modelling in Biotechnology*, 227–243. World Scientific Press, Singapore.

BEHN, U., VAN HEMMEN, J. L., and SULZER, B. (1993). Memory to antigenic challenge of the immune system: Synergy of idiotypic interactions and memory B cells. *J. Theor. Biol.* **165**, 1–25.

BELL, G. I. (1970). Mathematical model of clonal selection and antibody production. *J. Theor. Biol.* **29**, 191–232.

BELL, G. I. (1971a). Mathematical model of clonal selection and antibody production. *J. Theor. Biol.* **33**, 339–379.

BELL, G. I. (1971b). Mathematical model of clonal selection and antibody production. 3. The cellular basis of immunological paralysis. *J. Theor. Biol.* **33**, 379–398.

BRADLEY, L. M., CROFT, M., and SWAIN, S. L. (1993). T cell memory: New perspectives. *Immunol. Today* **14**, 197–199.

BURNETT, F. M. (1959). *The Clonal Selection Theory of Immunity*. Vanderbilt Univ. Press/Cambridge Univ. Press, Nashville/Cambridge, UK.

CELADA. F. (1992). Computer modelling of the immune system: Who are the "Fruitors"? In *Theoretical Immunology* (A. S. Perelson and G. Weisbuch, Eds.), NATO ASI Series, Vol. H66. Springer-Verlag, Berlin.

CELADA, F., and SEIDEN, P. E. (1992a). Teaching immunology: A Montessory approach. In *T Lymphocytes: Structure, Functions, Choices* (F. Celada and B. Pernis, Eds.), NATO ASI Series, Vol. A233, pp. 215–225. Plenum, New York.

CELADA, F., and SEIDEN, P. E. (1992b). A computer model of cellular interactions in the immune system. *Immunol. Today* **13**, 56–62.

CELADA, F., and SEIDEN, P. E. (1996). Affinity maturation and hypermutation in a simulation of the humoral immune response. *Eur. J. Immunol.* **26**, 1350–1358.

CHOWDHURY, D., STAUFFER, D., and CHOUDARY, P. V. (1990). A unified discrete model of immune response. *J. Theor. Biol.* **164**, 207–215.

COUTINHO, A. (1989). Beyond clonal selection and network. *Immunol. Rev.* **110**, 63–67.

ECO, U. (1986). Personal communication.

FREEMAN, E., and SKOLIMOWSKY, H. (1974). The search for objectivity in Pierce and Popper. In *The Philosophy of Karl Popper* (P. A. Schlipp, Ed.). Open Court, La Salle, IL.

GARDNER, M. (1970). Mathematical games. *Sci. Am.* **223** (October), 120–123; **223** (November), 112–117.

JACOB, F. (1970). La nature est un bricoleur. Personal communication.

JERNE, N. K. (1974). Towards a network theory of the immune system. *Ann. Immunol. (Inst. Pasteur)* **125C**, 373–389.

KAUFMAN, M., URBAIN, J., and THOMAS, R. (1985). Towards a logical analysis of the immune response. *J. Theor. Biol.* **114**, 527–561.

LAKATOS, I. (1974). Popper on demarcation and induction. In *The Philosophy of Karl Popper* (P. A. Schlipp, Ed.). Open Court, La Salle, IL.

MOHLER, R. R., BRUNI, C., and GANDOLFI, A. (1980). A system approach to immunology. *Proc. IEEE* **68**, 964–990.

MOSMANN, T. R., and SAD, S. (1996). The expanding universe of T cell subsets: Th1, Th2, and more. *Immunol. Today* **17**, 138–146.

PEIRCE, C. S. (1931–1958). *Collected Papers* (C. Hartshorne and P. Weiss, Eds.). Harvard Univ. Press, Cambridge, MA.

PERELSON, A. L. (Ed.) (1988). *Theoretical Immunology*, Parts I and II, Santa Fe Institute Studies in the Science of Complexity, Vols. II and III. Addison-Wesley, Redwood City, CA.

PERELSON, A. L., and WEISBUCH, G. (Eds.) (1992). *Theoretical and Experimental Insight into Immunology*, NATO ASI Series H, Cell Biology, Vol. 66. Springer-Verlag, Berlin.

PERELSON, A. L., and WEISBUCH, G. (1996). Immunology for physicists. *Rev. Modern Phys.* **69**, 1219–1268.

POPPER K. R. (1935). *Logik der Forschung*. Julius Springer, Vienna.

POWRIE, F., and COFFMANN, R. L. (1993). Cytokine regulation of T cell function: Potential for therapeutic intervention. *Immunol. Today* **14,** 270–274.

ROMAGNANI, S. (1995). Biology of human Th1 and Th2 cells. *J. Clin. Immunol.* **15,** 121–129.

SEIDEN, P. E., and CELADA, F. (1992). A model for simulating cognate recognition and response in the immune system. *J. Theor. Biol.* **158,** 329–357.

STEWART, J., and VARELA, F. J. (1989). Exploring the meaning of connectivity in the immune network. *Immunol. Rev.* **110,** 37–61.

SULZER, B., VAN HEMMEN, .J. L., and BEHN, U. (1994). Central immune system, the self and autoimmunity. *Bull. Math. Biol.* **56,** 1009–1040.

VITETTA, E. S., BERTON, M. T., BURGER, C., KEPRON, M., LEE, W. T., and YIN, X.-M. (1991). Memory B and T cells. *Annu. Rev. Immunol.* **9,** 193–217.

WOLFRAM., S. (1984). Cellular automata. *Physica* **10D,** vii–xii.

WOLFRAM, S. (1986). Theory and *Applications of Cellular Automata: Including Selected Papers 1983–1986.* World Scientific Press, Singapore.

ISBN 0-12-077342-2